MODERN CATALYTIC METHODS FOR ORGANIC SYNTHESIS WITH DIAZO COMPOUNDS

MODERN CATALYTIC METHODS FOR ORGANIC SYNTHESIS WITH DIAZO COMPOUNDS

FROM CYCLOPROPANES TO YLIDES

Michael P. Doyle
University of Arizona
Tucson, Arizona

M. Anthony McKervey
The Queen's University of
Belfast, Northern Ireland

Tao Ye
Nottingham University
United Kingdom

A Wiley-Interscience Publication

JOHN WILEY & SONS, INC.

New York • Chichester • Weinheim • Brisbane • Singapore • Toronto

Library of Congress Cataloging-in-Publication Data
Doyle, Michael P.
   Modern catalytic methods for organic synthesis with diazo
compounds : from cyclopropanes to ylides / Michael P. Doyle and M.
Anthony McKervey, Tao Ye.
   "A Wiley-Interscience publication."
   Includes bibliographical references (p.   –   ) and index.
   ISBN 0-471-13556-9 (cloth : alk. paper)
   1. Organic compounds—Synthesis. 2. Catalysis. 3. Diazo
compounds.   I. McKervey, M. Anthony, 1938–   . II. Ye, Tao, 1963–
. III. Title.
   QD262.D68   1997
   547′ .2—dc21   97–16541

Printed in the United States of America.

10   9   8   7   6   5   4   3   2   1

■ CONTENTS

Our primary purpose in writing this book was to bring together in a single volume for the first time those features of the chemistry of diazocarbonyl compounds that characterize their enduring versatility as intermediates for organic synthesis. We were conscious of several excellent accounts of the general chemistry and properties of diazo compounds, including diazocarbonyl compounds, of which *Diazo Compounds, Properties and Synthesis*, by Regitz and Maas, and *Diazo Chemistry*, Volumes I and II, by Zollinger, are the most comprehensive. In addition, numerous journal reviews and book chapters, several of which are our own, have appeared in recent years, but they have only covered segments of this rapidly expanding topic. The emphasis in our book is firmly on applications in synthesis, particularly through the use of modern catalytic procedures.

Although more than a century has passed since Curtius and Büchner began their systematic study of the reactions of ethyl diazoacetate, modern organic synthesis continues to benefit from the unique versatility of diazocarbonyl compounds in cyclopropanation, unactivated C–H insertion, Wolff rearrangement, ylide formation with its entire family of subsequent transformations, aromatic cycloaddition and substitution, and many other useful reactions. Almost without exception there are intermolecular and intramolecular counterparts and, recently, even macrocyclization. We have passed from an age when reactions of diazocarbonyl compounds, particularly those involving transition metal catalysts, were referred to as "carbenoid," signifying a lack of understanding of reaction intermediates involved in these processes; today we have a relatively precise understanding of the reaction intermediates that control product formation, and this understanding allows us to control reactivity and, especially, selectivity.

There are perhaps four principal reasons for the high level of activity in these areas. First, the enormous number of transformations that can occur with diazocarbonyl compounds makes them extremely versatile reactants. Second, methodology for their synthesis has continued to develop on a broad front, and there are now available well-tested, reliable procedures for the preparation of all the main classes of diazocarbonyl compounds. It is our hope that the information in Chapter 1 will provide a useful starting point for anyone interested in applying diazocarbonyl compounds in synthesis. Third, the introduction of dirhodium(II) catalysts for diazocarbonyl decomposition in the 1970s opened up numerous new opportunities for highly chemoselective transformations that were largely inaccessible with conventional copper catalysts. Chapter 2 focuses on catalysts and provides a comprehensive description of their unique advantages. Fourth, the ever-increasing demands for stereocontrol in the production of molecules of high enantiopurity has led to the introduction of chiral catalysts for asymmetric transformations of diazocarbonyl compounds.

Selectivity, an elusive goal in the 1970s, has been realized in reactions of diazocarbonyl compounds in the 1990s. In addition to high levels of chemoselectivity, regiocontrol can now be achieved without undue difficulty. Furthermore, stereocontrol that encompasses diastereoselectivity and enantioselectivity is a trademark of reactions that utilize diazocarbonyl compounds. The breakthrough required to reach this high level of control has been catalyst design and development, but many new challenges and opportunities remain for the synthetic chemist.

Example procedures are located in all the chapters. They provide access to compounds that may be of direct interest or serve as examples for procedures that might be applied to the construction of related materials. In addition to those for the synthesis of diazo compounds, procedures for the preparation of widely used catalysts and for applications of diazocarbonyl compounds in each of the main categories of their diverse transformations are provided.

We have provided complete reference citations, including titles and inclusive page numbers. This should allow the reader to efficiently judge the appropriateness of the reference for further elaboration of the topic. Furthermore, we recognize that, increasingly, complete reference citations are a requirement for manuscripts and proposals.

We owe a debt of gratitude to many colleagues. We wish to thank Irene Campell and Naomi Nowak for their skill and patience in processing numerous fragments of several chapters. We are also grateful to Simon Davies, Janine Douglas, Conchita Garcia, Michelle Groarke, Kevin Lydon, Hazel Moncrieff, and Nicolas Pierson for their invaluable assistance in preparing the references and to David C. Forbes for reading the draft version. An Alexander von Humboldt Senior Research Award for U.S. Scientists provided Michael P. Doyle with the time and library access at the University of Regensburg, hosted by Professor Dr. Henri Brunner, to begin this project. We wish to thank Helmut Duddeck, Shun-ichi Hashimoto, Shiro Ikegami, Tsutomu Katsuki, Michael Kennedy, Gerhard Maas, Stephen F. Martin, Malachy McCann, Christopher J. Moody, Paul Müller, Peter Meyers, Albert Padwa, Andreas Pfaltz, Hans-Ulrich Reissig, Stanley M. Roberts, and Brian Walker for their encouragement and helpful discussions.

Michael P. Doyle

M. Anthony McKervey

Tao Ye

ACKNOWLEDGMENTS

The representative experimental procedures reproduced in chapters 1–12 for the synthesis and applications of diazocarbonyl compounds are reproduced by kind permission of the publishers.

Chapter 1: page 6 "Excerpted from reference 44, Copyright 1992, Elsevier Science Ltd"; Page 24 "Excerpted from reference 135, Copyright 1980, George Thieme Verlag; Stuttgart, New York"; **Chapter 2:** Page 77 "Excerpted from references 24 and 103, Copyright 1982 and 1994, American Chemical Society"; Page 79 "Excerpted from reference 105, Copyright 1992, Marcel Dekker, Inc."; Page 83 "Excerpted from reference 125, Copyright 1993, American Chemical Society"; Page 88 "Excerpted from reference 130, Copyright 1995, Royal Netherlands Chemical Society"; **Chapter 3:** Page 128 "Excerpted from reference 67, Copyright 1993, American Chemical Society"; Page 140 "Excerpted from reference 107, Copyright 1996, American Chemical Society"; Page 145 "Excerpted from reference 110, Copyright 1994, Elsevier Science Ltd."; **Chapter 4:** Page 172 "Excerpted from reference 82, Copyright 1990, American Chemical Society"; **Chapter 5:** Page 263 "Excerpted from reference 81, Copyright 1996, American Chemical Society"; **Chapter 6:** Page 327 "Excerpted from reference 62, Copyright 1988, American Chemical Society"; Page 339 "Excerpted from reference 73, Copyright 1990, American Chemical Society"; **Chapter 7:** Page 364 "Excerpted from reference 31, Copyright 1984, American Chemical Society"; Page 383 "Excerpted from reference 73, Copyright 1991, American Chemical Society"; Page 403 "Excerpted from reference 122, Copyright 1990, American Chemical Society"; **Chapter 8:** Page 467 "Excerpted from reference 128, Copyright 1990, Royal Society of Chemistry"; Page 472 "Excerpted from reference 138, Copyright 1982, Elsevier Science Ltd."; **Chapter 9:** Page 491 "Excerpted from reference 17, Copyright 1994, Elsevier Science Ltd."; Page 496 "Excerpted from reference 20, Copyright 1995, VCH Verlagsgesellschaft Postfach 101161, D-69451 Weinheim, Germany"; **Chapter 10:** Page 537 "Excerpted from reference 16, Copyright 1992, Elsevier Science Ltd."; Page 543 "Excerpted from reference 26, Copyright 1989, American Chemical Society"; Page 548 "Excerpted from reference 44, Copyright 1995, American Chemical Society"; Page 557 "Excerpted from reference 58, Copyright 1985, American Chemical Society"; **Chapter 12:** Page 604 "Excerpted from reference 10, Copyright 1993, Royal Society of Chemistry"; Page 612 "Excerpted from reference 19, Copyright 1996, American Chemical Society"; Page 621 "Excerpted from reference 43, Copyright 1985, Royal Society of Chemistry"

MODERN CATALYTIC METHODS FOR ORGANIC SYNTHESIS WITH DIAZO COMPOUNDS

CHAPTER 1

Synthesis of α-Diazocarbonyl Compounds

1.1 INTRODUCTION

The first recorded synthesis of an α-diazocarbonyl compound dates back to the work of Curtius[1,2] on diazotization of natural α-amino acids (ethyl diazoacetate was first synthesized in 1883 from glycine) and although Wolff discovered in 1912 the diazocarbonyl rearrangement that now bears his name, simple diazocarbonyl compounds only became readily available in the late 1920s with the discovery by Arndt and Eistert,[3-5] and by Bradley and Robinson,[6] that the key to successful acylation of diazomethane with an acid chloride, a reaction previously believed capable of furnishing chloromethyl ketone only, lies in the use of diazomethane in sufficient excess to sequester the hydrogen chloride liberated and thereby prevent its addition to the diazoketone. Acylation of diazomethane remains the single most important route to acyclic terminal α-diazoketones. The diazo group transfer technique, which is now available for both terminal and nonterminal systems, also occupies an important place in diazocarbonyl methodology.

1.2 ACYLATION OF DIAZOALKANES

The Arndt–Eistert synthesis of diazoketones involves addition of an acyl chloride to ethereal diazomethane[3-5] (at least 1–2 equiv. excess) at or below 0°C (eq. 1); extensive purification of the product is usually unnecessary. Numerous synthetic intermediates containing the diazoketone functional group have been obtained in this way. For example, diazoketones $\mathbf{1}^7$ and $\mathbf{2}^8$ have been prepared by this route (eqs. 2 and 3).

> **Caution:** Diazocarbonyl compounds in general should always be presumed to be toxic and potentially explosive. They should always be handled with care in a well-ventilated fumehood.

1

$$RCOCl + 2CH_2N_2 \xrightarrow[0^{\circ}C]{Et_2O} RCOCHN_2 + CH_3Cl + N_2 \quad (1)$$

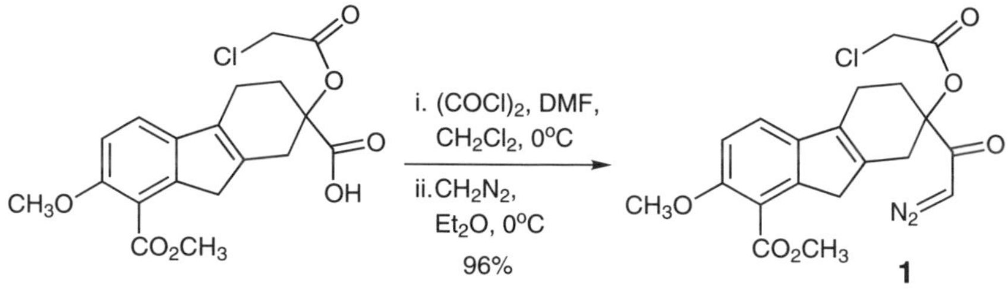

Cl
O
O
O
OH
CH3O
CO2CH3
i. (COCl)2, DMF,
CH2Cl2, 0°C
ii.CH2N2,
Et2O, 0°C
96%
Cl
O
O
O
N2
CH3O
CO2CH3
1

(2)[7]

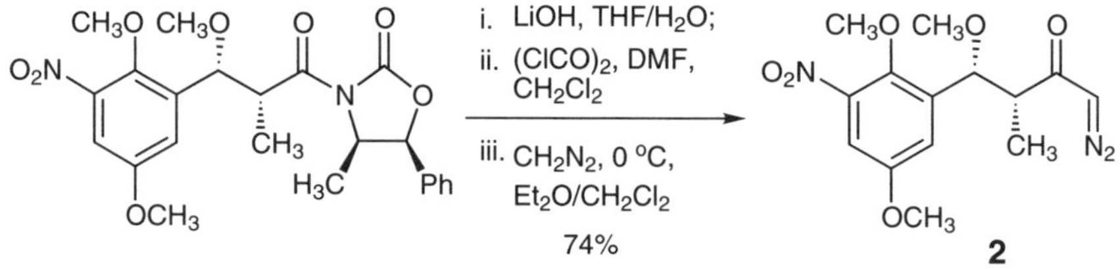

CH3O CH3O
O
O
O2N
N
O
CH3
H3C
Ph
OCH3
i. LiOH, THF/H2O;
ii. (ClCO)2, DMF,
CH2Cl2
iii. CH2N2, 0 °C,
Et2O/CH2Cl2
74%
CH3O CH3O
O
O2N
CH3
N2
OCH3
2

(3)[8]

The use of excessive amounts of diazomethane can be avoided with nonenolizable precursors,[9–11] such as aromatic acyl chlorides, by including one equivalent of triethylamine in the diazomethane solution. Under such conditions, however, enolizable acyl chlorides give only low yields of impure diazoketones, presumably as a consequence of competing ketene formation. This approach is occasionally successful when lower temperatures are employed, as follows:[12]

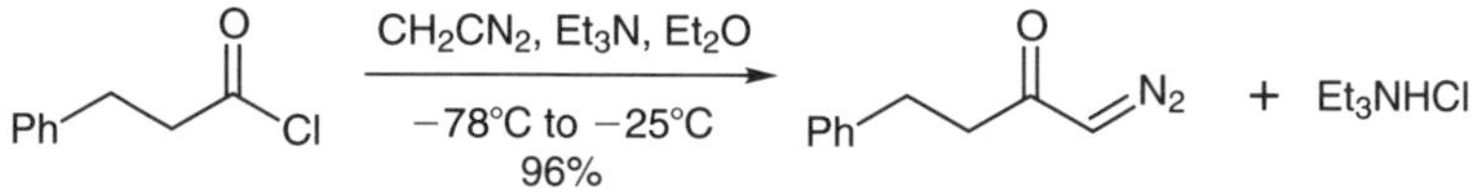

O
Ph
Cl
CH2CN2, Et3N, Et2O
−78°C to −25°C
96%
O
Ph
N2
+ Et3NHCl

(4)

Although quite safe when handled as a dilute solution in an inert solvent, diazomethane presents several safety hazards of which all users of the reagent should be fully informed. It is both extremely toxic and highly irritating. Diazomethane and several of its precursors have been cited as carcinogens. Diazomethane has been known to explode unaccountably. Rough surfaces are proven initiators of detonation. Ground-glass joints and any glassware that have not been fire polished should not be allowed to come in contact with diazomethane. Contact with alkali metals or drying agents such as calcium sulfate can result in explosion. Moisture is best removed from ethereal diazomethane by addition of potassium hydroxide pellets. If the reagent is generated using the proper apparatus and is handled as a dilute solution at low temperature (0°C), the risks cited are minimized. All reactions involving diazomethane should be carried out in an efficient fumehood behind a safety screen.[13]

Statement: Notwithstanding the need to exercise due care in working with diazomethane and diazo compounds in general, their importance as reagents in numerous organic transformation is reflected in the commercial availability of many of the compounds.

The most common and convenient method of generating diazomethane is by base-catalyzed decomposition of *N*-methyl-*N*-nitroso amines of general structure **3**, where *R* represents a sulfonyl, carbonyl, or similar electron-withdrawing substitutent. There are three *Organic Synthesis* procedures for its preparation from three different precursors.[14–16] Currently, the majority of users employ one of two commercially available precursors, *N*-methyl-*N*-nitro-*N*-nitrosoguanidine (MNNG) **4**[17] and *N*-methyl-*N*-nitroso-*p*-toluenesulfonamide (Diazald) **5**, the latter the more popular of the two. MNNG is the recommended precursor when less than one millimole of diazomethane is required. Diazald offers advantages over MNNG when greater than one millimole is needed. Specifically labeled Diazalds for the generation of CD_2N_2, $^{13}CH_2N_2$, and $^{13}CD_2N_2$ are also commercially available.

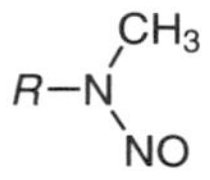

$$R-N\begin{smallmatrix}CH_3\\NO\end{smallmatrix}$$

3

$$R\overset{O}{\underset{NO}{\underset{|}{\overset{||}{C}}}}N{-}CH_3$$

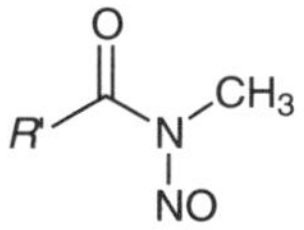

R = alkyl, aryl, H, O-alkyl, O-aryl, NH_2, N(alkyl)$_2$

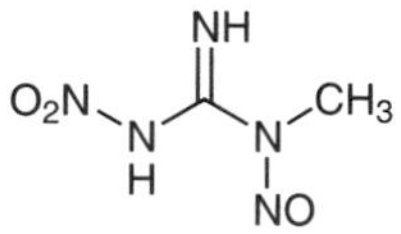

4 MNNG 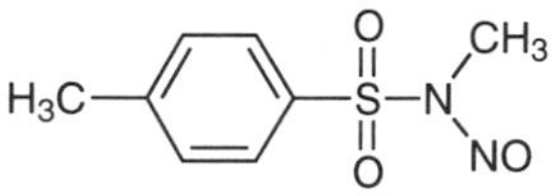   **5 Diazald**

In addition, fairly large amounts of dideuteriodiazomethane can be prepared from diazomethane via a hydrogen–deuterium exchange process.[18] Commercial diazomethane kits with clear glass joints are available for preparing ethereal solutions in ~1 mmol, 1–50 mmol, ~100 mmol and 0.2~0.3 mol quantities.[19] Lombardi[20] has described a technique for the rapid generation of diazomethane and its simultaneous reaction with a substrate. However, this technique contains an element of risk, and at least one report of an explosion has been published.[21] This technique is therefore not recommended for the routine production of diazomethanes. Trimethylsilyldiazomethane,[22–24] also commercially available, is recommended as a safe alternative for diazomethane.

PREPARATION OF DIAZOMETHANE FROM DIAZALD[15]

$$H_3C\text{---}N(NO)\text{---}SO_2\text{---}C_6H_4\text{---}CH_3 \; + \; ROH \; \xrightarrow{KOH} \; CH_2N_2 \; + \; RO\text{---}SO_2\text{---}C_6H_4\text{---}CH_3 \; + \; H_2O$$

A 125-mL distilling flask is fitted with a condenser set for distillation and with a long-stem dropping funnel. The condenser is connected by means of an adapter to a 250-mL Erlenmeyer flask. Through a second hole in the stopper of the Erlenmeyer flask is placed an outlet tube bent so as to pass into and nearly to the bottom of a second Erlenmeyer flask that is not stoppered. Both receivers are cooled in an ice–salt mixture; in the first is placed 10 mL of ether, and in the second 35 mL of ether. The inlet tube passes below the surface of the ether in the second flask.

In the distilling flask are placed a solution of 6 g of potassium hydroxide dissolved in 10 mL of water, 35 mL of 2-(2-ethoxyethoxy)ethanol (carbitol), 10 mL of ether, and the Teflon-coated bar of a magnetic stirrer. The dropping funnel is attached and adjusted so that the stem is just above the surface of the solution in the distilling flask. There is placed in the dropping funnel a solution of 21.4 g (0.1 mole) of p-tolylsulfonylmethylnitrosamide (Diazald) in 125 mL of ether. The distilling flask is heated in a water bath at 65–70°C, the stirrer is started, and the nitrosamide solution is added at a regular rate during 20 min. The rate of addition should about equal the rate of distillation. As soon as all the Diazald solution has been added, additional ether is placeed in the dropping funnel and added at the previous rate until the distillate is colorless. Usually 50–100 mL additional of ether is required. The distillate contains 2.7–2.9 g (64–69%) of diazomethane, as determined by titration.

Pettit and Nelson[25] have designed an apparatus for diazoketone preparation in which the carboxylic acid in ether in one compartment is first treated with oxalyl chloride, triethylamine, and a catalytic amount of dimethylformamide to furnish the acyl chloride. The resulting solution is then filtered into ethereal diazomethane at −78°C. This technique was developed for the synthesis of the anticancer bis-diazoketone, azotomycin **6**.[25] The antibiotic 6-diazo-5-oxo-L-norleucine **7**, commonly known as DON, has been synthesized several times[26,27] from the N-protected acyl chloride and diazomethane, albeit in very low (0.5–1.0%) overall yields. Much of the inefficiency appears to be associated with the formation of

COMMERCIALLY AVAILABLE DIAZOCARBONYL PRECURSORS AND RELATED COMPOUNDS

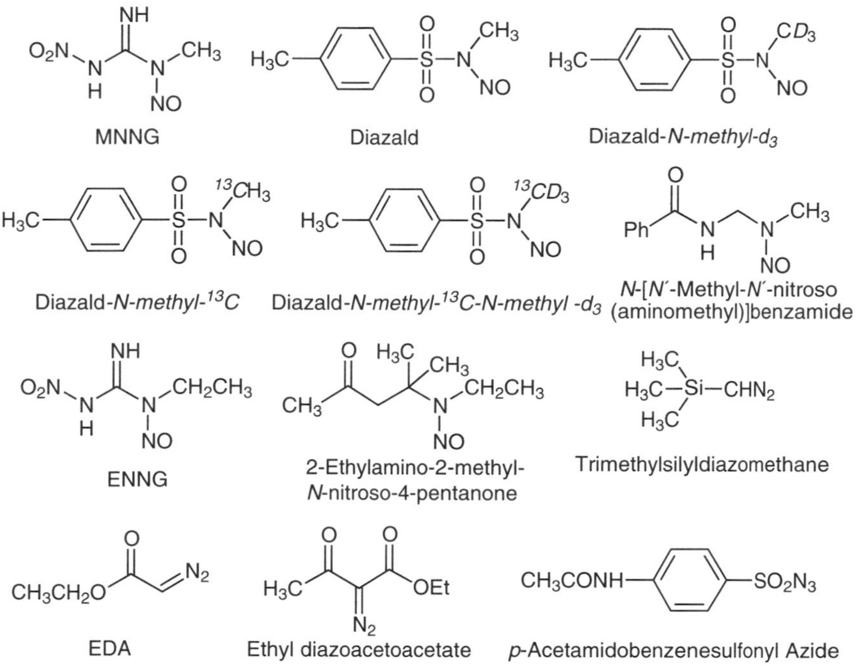

MNNG

Diazald

Diazald-N-methyl-d_3

Diazald-N-methyl-^{13}C

Diazald-N-methyl-^{13}C-N-methyl -d_3

N-[N′-Methyl-N′-nitroso (aminomethyl)]benzamide

ENNG

2-Ethylamino-2-methyl-N-nitroso-4-pentanone

Trimethylsilyldiazomethane

EDA

Ethyl diazoacetoacetate

p-Acetamidobenzenesulfonyl Azide

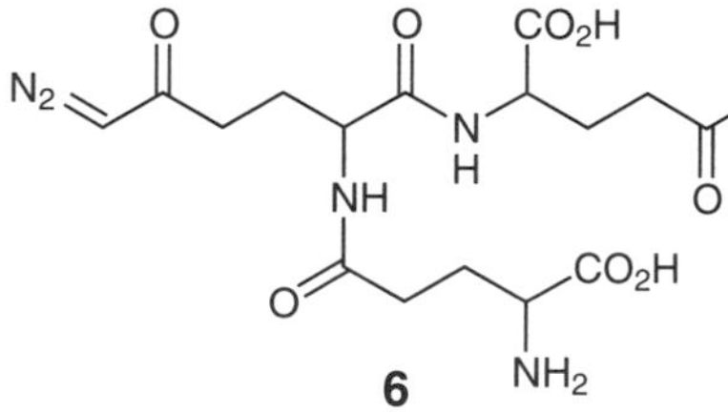

6

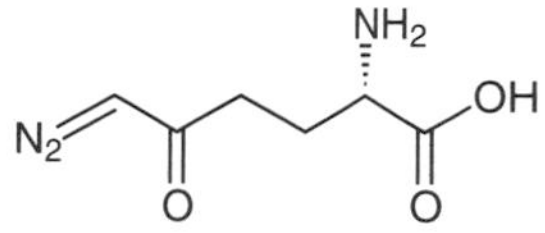

7

the acyl chloride for which the most useful method is the reaction of the dicyclo-hexylammonium salt of N-trifluoroacetyl glutamic acid with oxalyl chloride.[28]

Anhydrides are also suitable acylating agents for diazomethane.[30,31] A convenient procedure involves treatment of the carboxylic acid with dicyclohexylcarbodiimide to form the anhydride, which is then allowed to react with ethereal diazomethane.[32–34] The disadvantage of using a symmetrical anhydride in the formation of a diazoketone is that only half of the carboxy component is converted into the diazoketone. Cyclic anhydrides yield mono-diazoketones.[35] In some cases, formation of diazoketone by the reaction of unsymmetrical anhy-

PREPARATION OF A DIAZOKETONE FROM AN ACID CHLORIDE: SYNTHESIS OF 1-DIAZO-4-PHENYL-2-BUTANONE[29]

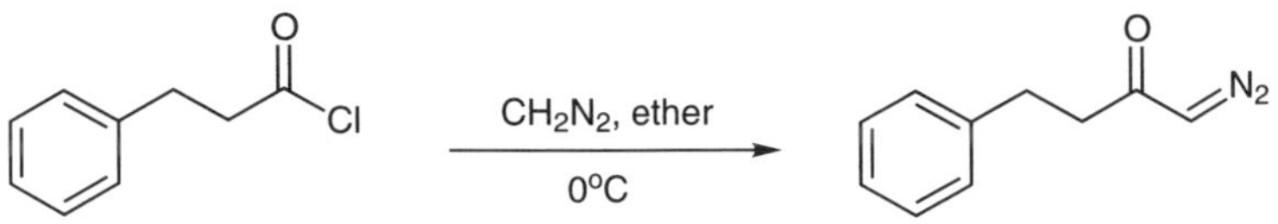

A 1-L Erlenmeyer flask equipped with a 2-in. magnetic stirring bar and a two-hole rubber stopper fitted with a 125 mL Teflon stopcock separatory funnel and a drying tube filled with potasssium hydroxide is charged with a solution of 200 mmol (3.4 equiv.) of diazomethane in 600 mL of dry ether. The solution is cooled to 0°C and stirred at high speed. To this cooled solution, 10.0 g (59 mmol) of 3-phenylpropionyl chloride diluted to 125 mL with anhydrous ether is added dropwise over a 1-h period. The resulting reaction mixture is stirred cold for an additional 0.5 h and then at room temperature for 1 h. After this period of time the reaction is complete, and excess diazomethane is removed by evacuating the Erlenmeyer flask with a water aspirator pump in the hood. The Erlenmeyer flask is evacuated by connecting the aspirator to a one-hole stopper that has been fitted with a plastic or fire-polished glass tube. After the diazomethane has been removed, the remaining ethereal solution is concentrated by rotary evaporation to give 10.5–10.6 g (> 100% crude yield, 90–91% purity) of 1-diazo-4-phenyl-2-butanone as a yellow oil. Pure diazoketone can be obtained by chromatography on silica gel using 15% ethyl acetate/hexane as an eluent, $R_f = 0.37$.

drides with diazomethane is possible. Diazoacetaldehyde[36] has been synthesized via acylation of diazomethane with formic acetic anhydride.

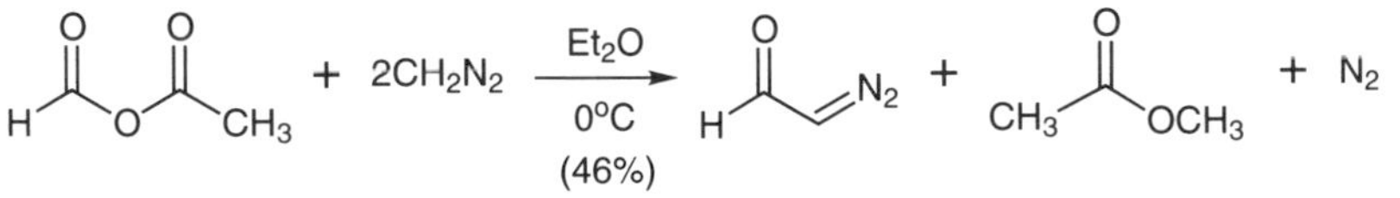

$$\text{(5)}$$

A convenient in situ procedure to form a diazoketone via mixed carbonic anhydride is now available involving carboxylic–carbonic anhydride formation between the carboxylic acid and a chloroformate, followed by treatment with ethereal diazomethane.[33,37–41] Methyl chloroformate and ethyl chloroformate are commonly used, but many workers prefer isobutyl chloroformate. The fact that

---

PREPARATION OF DIAZOKETONES FROM MIXED CARBONIC ANHYDRIDES: SYNTHESIS OF N-*tert*-BUTOXYCARBONYL-L-PHENYLALANYL DIAZOMETHANE[44]

The N-protected amino acid (27.0 mmol) in a solution of dry ether (60 mL) and tetrahydrofuran (60 mL) was stirred at −20°C under a dry nitrogen atmosphere. To this solution triethylamine (3.8 mL, 1 equiv.) followed by isobutyl chloroformate (3.7 mL, 1 equiv.) were added. The solution was stirred for 30 min and then allowed to warm to −10°C. At this temperature ethereal diazomethane (2 equiv.) was added dropwise via a pressure-equalized dropping funnel. The solution was stirred for a further 3 h allowing it to reach room temperature. It was then evaporated to a third of its original volume. The solution was diluted with ether (50 mL) and washed with water (50 mL), saturated aqueous sodium bicarbonate (50 mL), and brine (50 mL). It was then dried and evaporated to give the crude α-diazoketone, which was purified by silica gel chromatography (80% yield).

---

one of the carbonyl groups in the activated intermediate is flanked by two oxygen atoms diminishes its reactivity, so that the diazomethane attack is directed toward the carbonyl group of the original carboxy component. To minimize side reactions, low temperatures and minimal activation times are usually employed. This variation now appears to be the method of choice for the preparation of many different kinds of diazoketones, particularly molecules for which acid chloride formation would be inappropriate or rendered difficult through the reactions of other functional groups in the molecule. Furthermore, compared with the acid chloride route, the mixed carbonic anhydride route requires a much smaller excess of diazomethane. For example, 2-diazoacetyl-2,3-diphenyloxirane has been synthesized by this route (eq. 6).[42,43] This route has also been applied to the pro-

$$\text{(6)}$$

duction of monochiral α-diazoketones from N-protected amino acids and dipeptides.[44,45] Representative examples are shown in **8–11**. Other alternatives for carboxylic acid activation prior to treatment with diazomethane include the for-

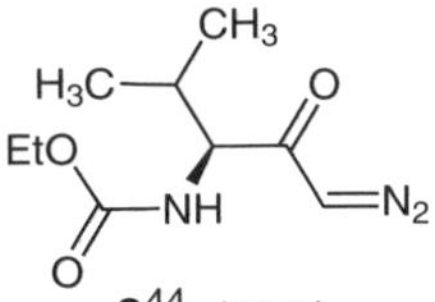

8[44] (78%)

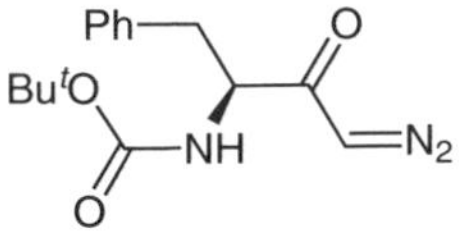

9[44] (80%)

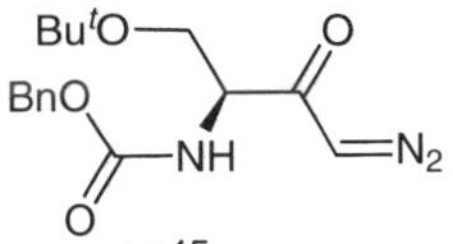

10[45] (89%)

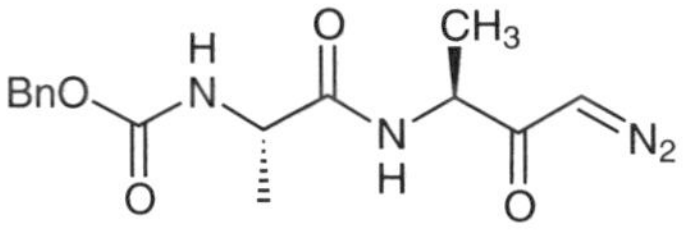

11[45] (41%)

mation of acyl imidazole and various carboxylic acid active esters;[46] none of these alternatives is as commonly used as the mixed carbonic anhydride method.

Acylation of higher diazoalkanes with acyl chlorides and mixed carbonic anhydrides is also possible, though less efficiently than with diazomethane. Diazoethane can be obtained from commercially available precursors, 1-ethyl-3-nitro-1-nitrosoguanidine,[47] 2-ethylamino-2-methyl-N-nitroso-4-pentanone[48] or from readily available N-ethyl-N-nitrosourea.[49] It has been used extensively in acylation to produce diazoethylketones. Numerous synthetic intermediates, for example, diazoethyl ketones **12**,[50] **13**,[51] **14**,[52] and **15**[53] have been prepared by this

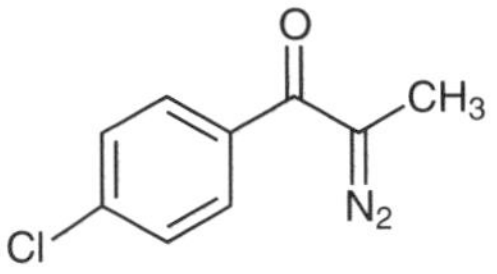

12[50] (79%)

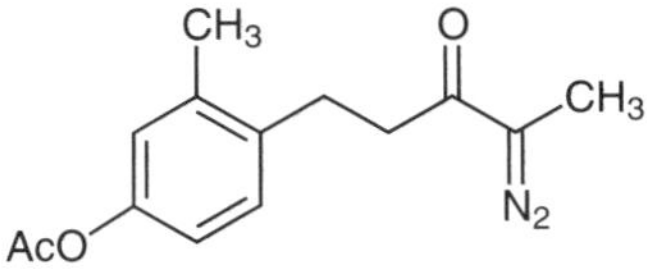

13[51] (75%)

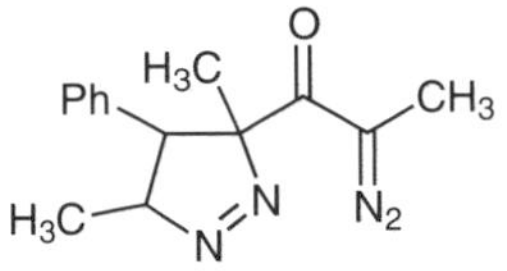

14[52] (54%)

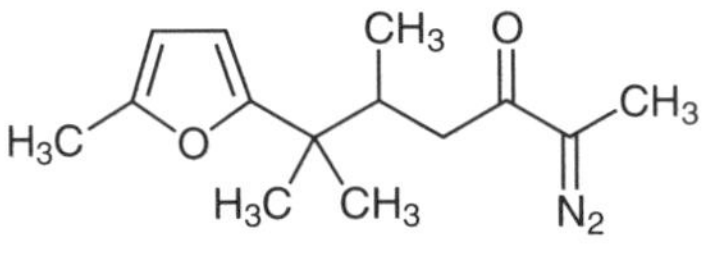

15[53] (50%)

route. Diazoketones prepared from the acylation of higher diazoalkanes other than diazoethane have also been used as key intermediates in synthesis; three examples are shown in eqs. 7,[54] 8,[55] and 9.[56]

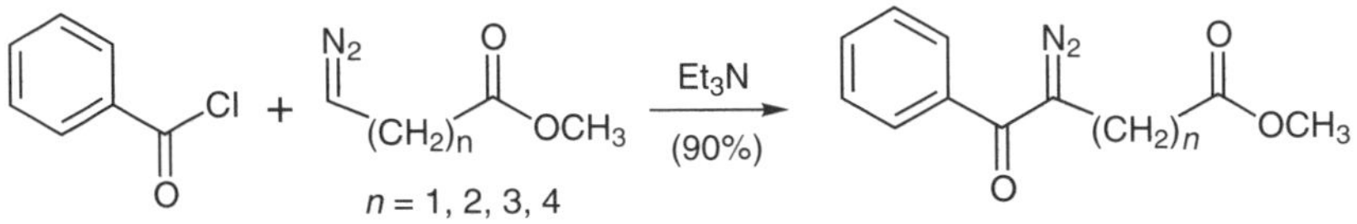

$n = 1, 2, 3, 4$

(7)

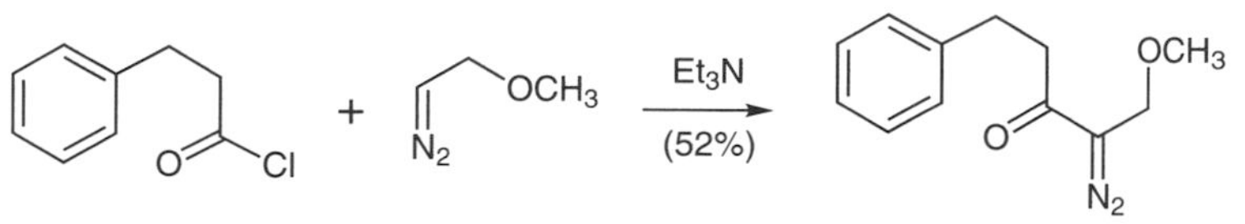

(8)

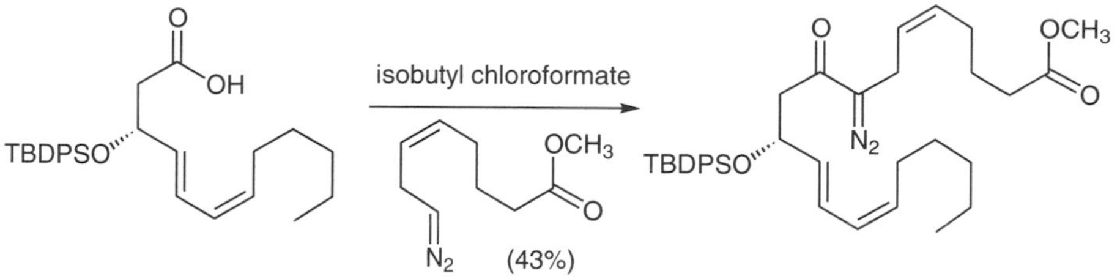

(9)

1.3 DIAZO TRANSFER REACTIONS

An obvious limitation of diazoalkane acylation is that the process is not applicable to cyclic α-diazoketones. Although many routes to cyclic diazoketones have been developed, none competes in usefulness with the diazo transfer technique. The concept of diazo group transfer was first described by Dimroth in 1910[57] and investigated in great detail afterwards by several other groups.[58-61] It is now established as a general method for the preparation of diazo compounds through extensive studies by Regitz and his collaborators.[62-65] Diazo transfer is now the standard route, not only to cyclic α-diazoketones, but to many acyclic systems not accessible by acyl transfer processes. In the broadest sense, diazo transfer refers to the transfer of a complete diazo group from a donor to an acceptor, which for α-diazocarbonyl products must therefore be an acid or ketone derivative. The diazo donor is invariably a sulfonyl azide.[63]

1.3.1 Simple Diazo Transfer Reactions

Diazo transfer to the α-methylene position of a carbonyl compound requires the presence of a base of sufficient strength to deprotonate the substrate. Prior activation of the substrate may however be necessary. Substrates can therefore be divided into two broad categories on the basis of their acidity: those in which the α-methylene position is already sufficiently reactive towards diazo transfer, and those that require prior activation to ensure smooth transfer in the presence of a mild base.[66,67] Of greatest importance in the former group are malonic esters, β-ketoesters, β-ketoamides, and β-diketones, which are readily converted into 2-diazo-1,3-dicarbonyl products by the simple diazo transfer procedure of direct exposure to tosyl azide in dry acetonitrile or ethanol using triethylamine as the

---

SIMPLE DIAZO TRANSFER REACTION:
SYNTHESIS OF *tert*-BUTYL α-DIAZOACETOACETATE[69]

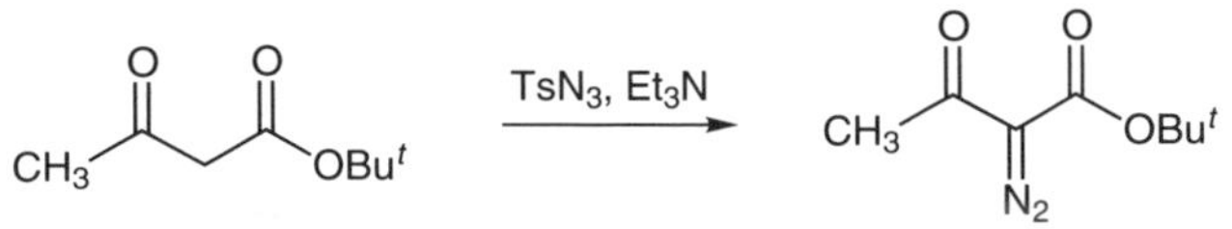

In a 2-L wide-mouthed, Erlenmeyer flask are placed 118.5 g (0.75 mole) of *tert*-butyl acetoacetate, 1 L of anhydrous acetonitrile, and 75.8 g (0.75 mole) of previously distilled triethylamine. The temperature of the mixture is adjusted to 20°C, and 148 g (0.75 mole) of *p*-toluenesulfonyl azide is added dropwise with vigorous stirring over 10–15 min. The addition causes the reaction mixture to warm to 38–40°C and assume a yellow color. After the mixture has been stirred at room temperature for 2.5 h, the solvent is evaporated at 35°C (12 mm). The partially crystalline residue is triturated with 1 L of ether, and the mixture, including the insoluble residue, is placed in a 2-L separatory funnel. The mixture is washed successively with a solution of 45 g of potassium hydroxide in 500 mL of water, a solution of 7.5 g of potassium hydroxide in 250 mL of water, and 250 mL of water. The yellow–orange ethereal phase is dried over anhydrous sodium sulfate, and the solvent is evaporated at 35°C (15 mm) until the residue has attained a constant weight. The yellow–orange diazo ester weighs 130–135 g (94–98%).

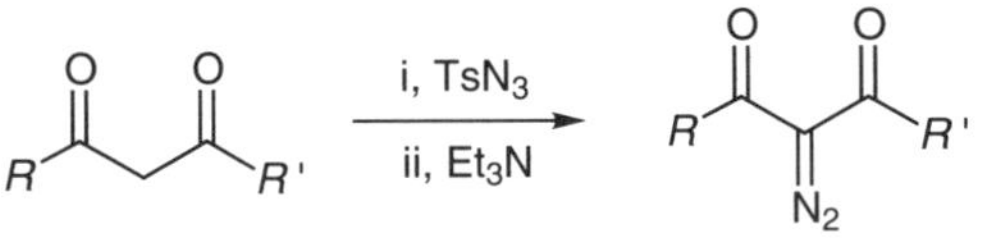

$$(10)$$

base (eq. 10). For example, diazoesters **16**[68] and **17**,[69] diazoamide **18**,[70] diazoketone **19**,[71] diazoketosulfonate **20**,[72] and diazoketophosphonate **21**[73] have all been prepared via this route. Almost all the β-lactam intermediates bearing diazocar-

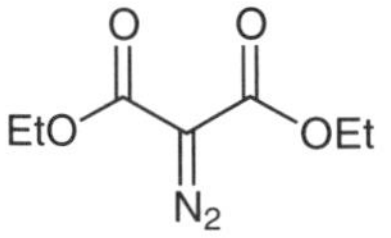

16[68] (95%) 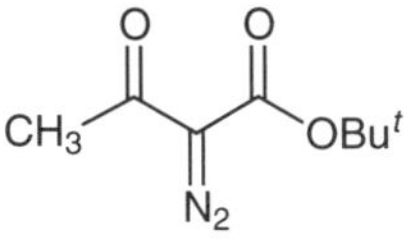   **17**[69] (94%) 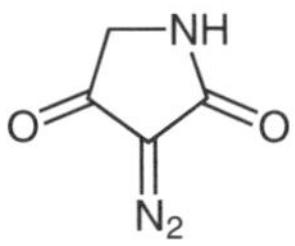   **18**[70] (81%)

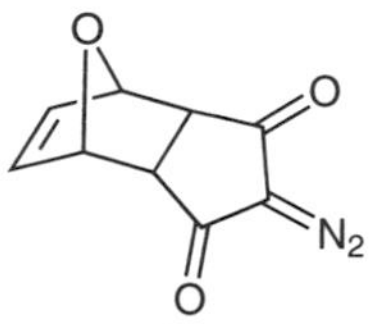

19[71] (81%) 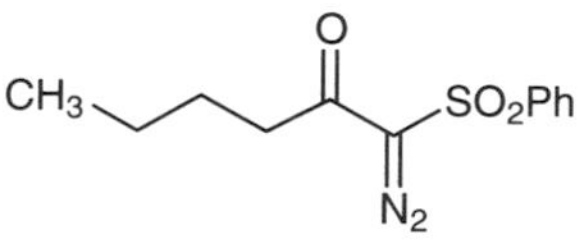   **20**[72] (88%) 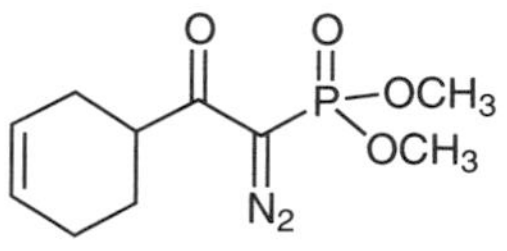   **21**[73] (91%)

bonyl side chains to be discussed in Chapter 8 were obtained by diazo transfer reactions.

1.3.2 Deformylating Diazo Transfer and Related Modifications

While the diazo transfer reaction works extremely well for cases in which the reaction site is activated by two flanking carbonyl functions, the direct Regitz procedure usually fails in cases where the methylene group is only activated by a single carbonyl group. Although satisfactory results can sometimes be achieved with single carbonyl compounds by optimizing the system of base and diazo transfer reagent (vide infra), in general, however, best results are usually obtained by activation in the form of acyl aldehyde formation prior to diazo transfer. This technique, often referred to as Regitz deformylating diazo transfer, has found widespread application since its introduction in 1967.[62-65] It involves Claisen condensation of the ketone with ethyl formate in order to introduce the strongly activating formyl group, which is subsequently released as the sulfonamide in the course of the actual diazo transfer; either the metal salt or the neutral formyl compound can be employed as the activated intermediate (Scheme 1). Variations in the groups R and R' allow most types of acyclic and cyclic α-diazoketones to be synthesized in this way.[62-65,74,75] To demonstrate the broad scope of the deformylating diazo transfer process, representative examples **22–45.** are shown in Table 1.1 with yields in brackets.

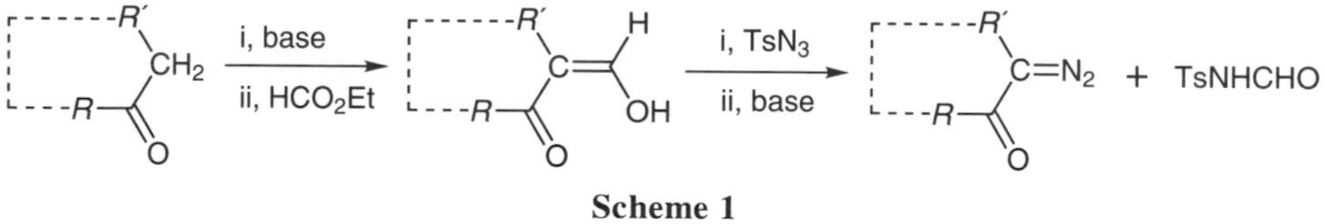

Scheme 1

Isolation of the formylated adduct of the first step of the deformylating diazo transfer process in many cases was not required. In fact, the Regitz deformylating diazo transfer technique can be adapted to a one-pot procedure. Prinzbach and co-workers[98] have obtained bisdiazoketone **47** in 77% yield from the corresponding diketone **46** using a one-pot procedure involving the treatment of the in situ

DEFORMYLATING DIAZO TRANSFER REACTION: SYNTHESIS OF 2-DIAZOCYCLOHEXANONE[74]

2-Hydroxymethylenecyclohexanone.

A mixture of 23 g of sodium metal cut in approximately 1-cm cubes, 2 L of dry ether, 98 g (103 ml, 1 mole) of redistilled cyclohexanone, and 110 g (120 mL, 1 mole) of ethyl formate is placed in a 5-L three-necked flask equipped with a stirrer, which is then placed in a cold water bath. Stirring is continued for 6 h. After standing overnight, 25 mL of ethyl alcohol is added, and the mixture is stirred for an additional hour. After the addition of 200 mL of water, the mixture is shaken in a 3-L separatory funnel. The ether layer is washed with 50 mL of water, and the combined aqueous extracts are washed with 100 mL of ether. The aqueous layer is acidified with 165 mL of 6N hydrochloric acid, and the mixture is extracted twice with 300 mL of ether. The ether solution is washed with 25 mL of saturated sodium chloride solution and then is dried by the addition of approximately 30 g of anhydrous magnesium sulfate powder. The drying agent is removed by suction filtration, and the ether is evaporated on the steam bath. The residue is distilled under reduced pressure using a 6-in. Vigreux column. After a small fore-run there is obtained 88–94 g (70–74%) of 2-hydroxy-methylenecyclohexanone, b.p. 70–72°C/5 mm, n_D^{25} 1.5110.

2-Diazocyclohexanone.

In a 2-L, wide-necked, Erlenmeyer flask are mixed 66.2 g (0.525 mole) of 2-hydroxymethylenecyclohexanone, 400 mL of methylene chloride, and 106 g (1.05 moles) of triethylamine. The flask is cooled in an ice–salt bath at −12 to −15°C, and 98.0 g (0.500 mole) of p-toluenesulfonyl azide is added with vigorous mechanical stirring over a period of approximately 1 h, at such a rate that the temperature of the reaction mixture does not rise above −5°C. Stirring is continued for an additional 2 h as the ice in the cooling bath melts. A solution of 30.8 g (0.55 mole) of potassium hydroxide in 400 mL of water is added, and the mixture is stirred for 15 min at room temperature. The resulting emulsion is placed in a 2-L separatory funnel, the methylene chloride layer is separated after the emulsion has broken, and the aqueous alcoholic layer is washed with two 100-mL portions of methylene chloride. The combined methylene chloride solutions are washed with a solution of 2.8 g of potassium hydroxide in 200 mL of water, and then with 200 mL of water, and dried over anhydrous sodium sulfate. The solvent is removed on a rotary evaporator at 35°C (15 mm) until the weight of the residue is constant. The yield of yellow–orange 2-diazocyclohexanone is 51.5–59.0 g (83–95%).

TABLE 1.1   Diazoketones Produced by the Regitz Procedure

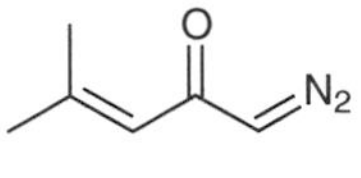

(69%)
22[76]

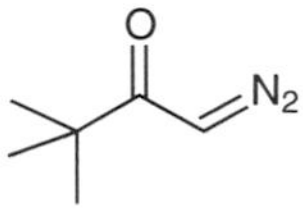

(83%)
23[76]

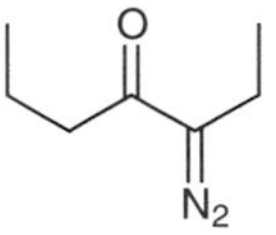

(65%)
24[76]

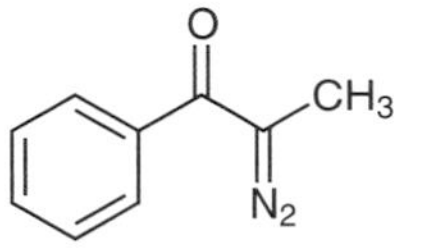

(77%)
25[76]

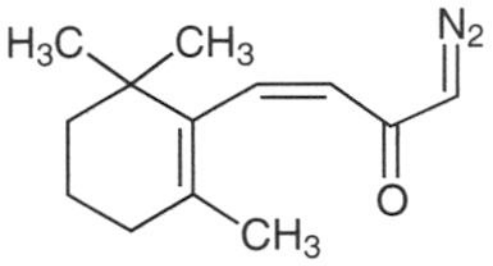

(94%)
26[77]

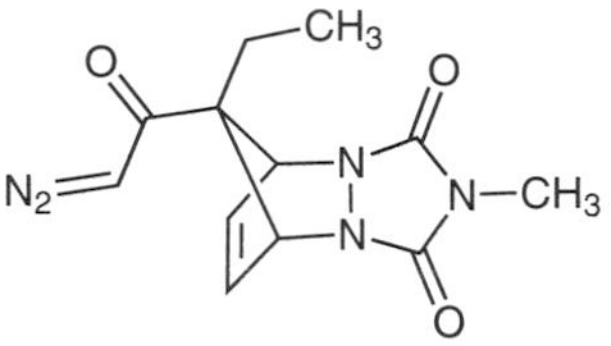

27[78]

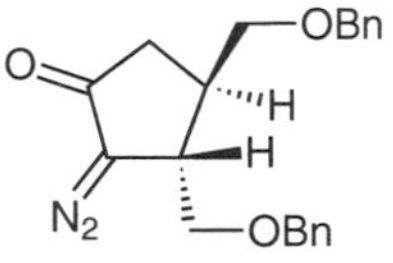

(> 66%)
28[79]

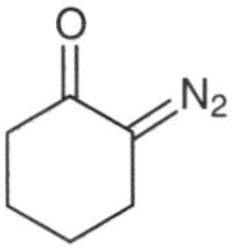

(64%)
29[74]

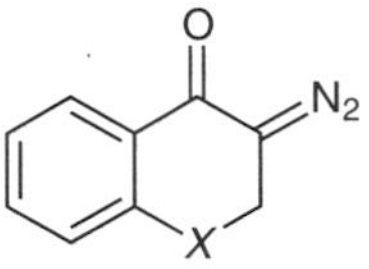

X = S, O, NCH$_3$, CH$_2$ (56–82%)
30[80]

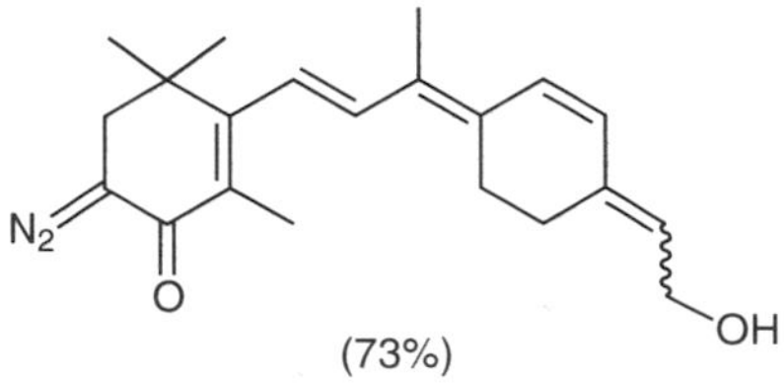

(73%)
31[81]

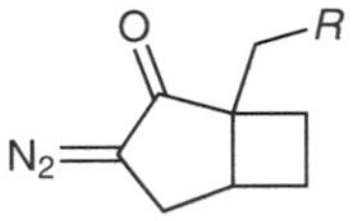

R = Ph (75%);
R = CH$_2$CO$_2$CH$_3$ (> 83%)
32[82,83]

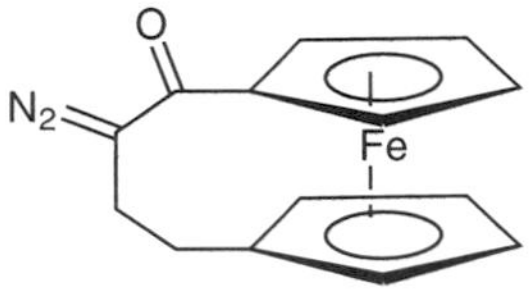

(> 22%)
33[84]

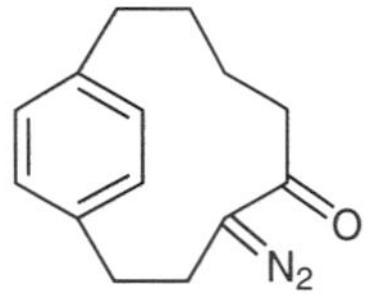

(87%)
34[85,86]

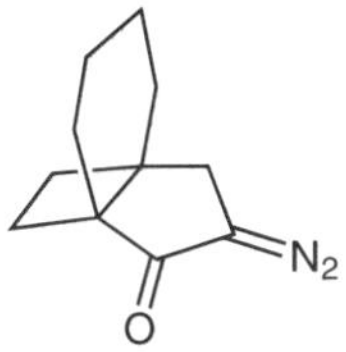

(66%)
35[87]

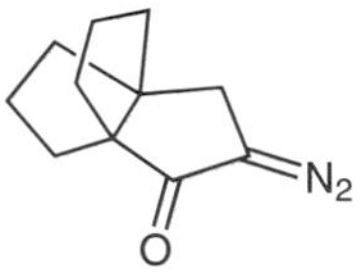

(> 61%)
36[88]

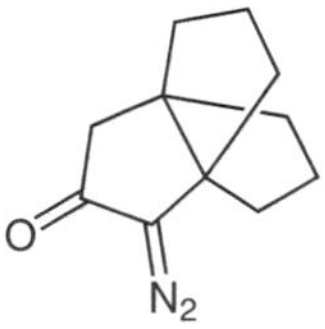

(70%)
37[89]

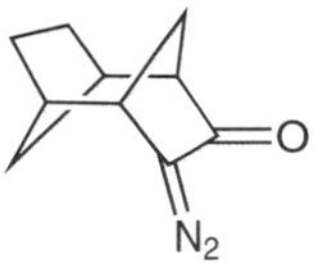

(60%)
38[90]

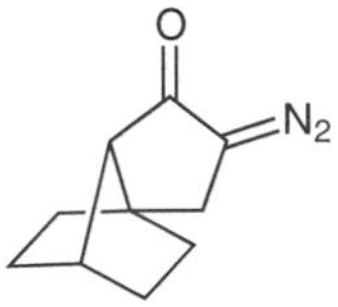

(75%)
39[91]

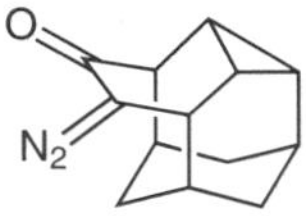

(59%)
40[92]

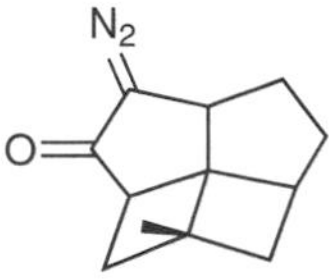

(28%)
41[93]

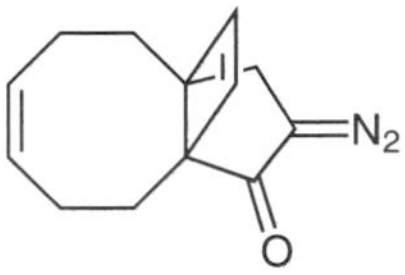

(> 73%)
42[94]

TABLE 1.1 Diazoketones Produced by the Regitz Procedure *(continued)*

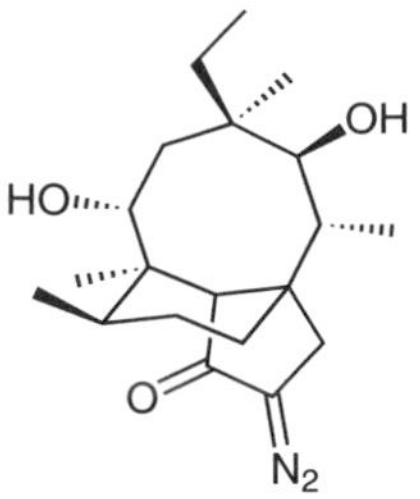
HO
OH
O
N_2

(81%)

43[95]

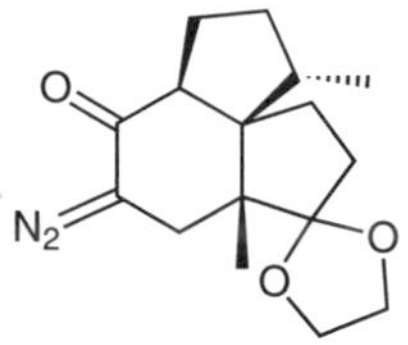
O
N_2
O
O

(88%)

44[96]

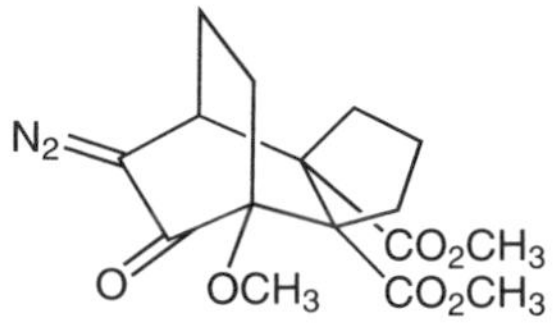
N_2
O
OCH_3
CO_2CH_3
CO_2CH_3

(66%)

45[97]

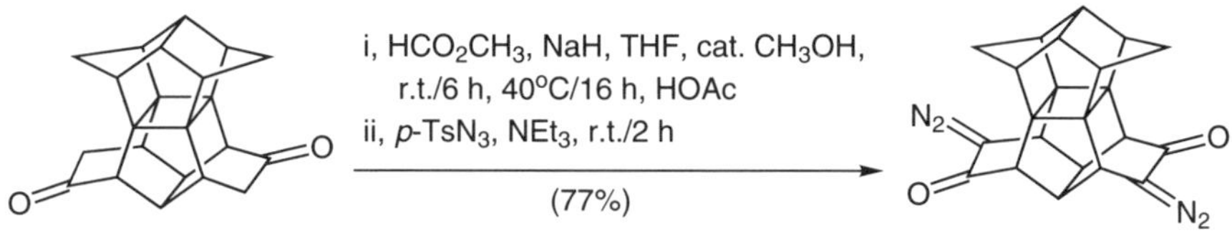
i, HCO_2CH_3, NaH, THF, cat. CH_3OH,
r.t./6 h, 40°C/16 h, HOAc
ii, *p*-TsN_3, NEt_3, r.t./2 h
(77%)
O
O
N_2
O
O
N_2

46 **47**

(11)

formed bisformylated adduct with the diazo transfer reagent in the presence of triethylamine.[98]

Similarly, diazoester compounds can be synthesized via the Regitz deformylating procedure by first treating the ester anion with ethyl formate to produce the doubly activated species. Reaction of this formyl compound with the appropriate diazo transfer reagent accompanied by elimination of the formyl group leads to the diazo ester. For example, compounds **48**,[66,76] **49**[99] and **50**[100] have been prepared by this method.

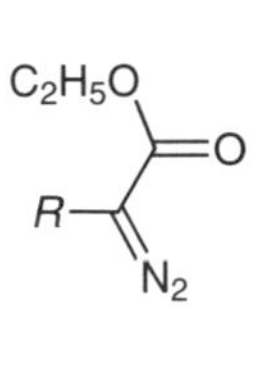
C_2H_5O
O
R
N_2

R = H, (69%);
R = Me, (87%);
R = Et, (43%);
R = *i*-Pr, (51%);
R = *n*-Bu, (48%);
R = *n*-Hex, (70%);
R = Ph, (75%);
R = Bn, (62%)

48 [66,76]

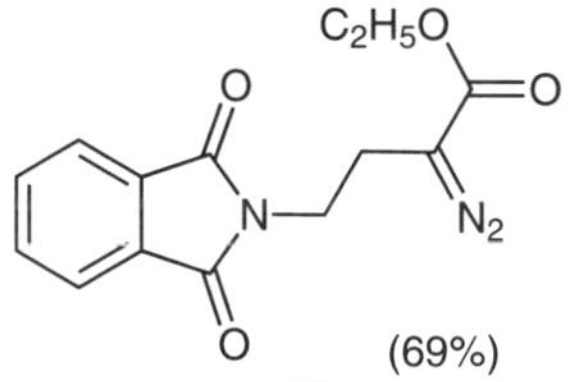
C_2H_5O
O
O
N
N_2
O

(69%)

49 [99]

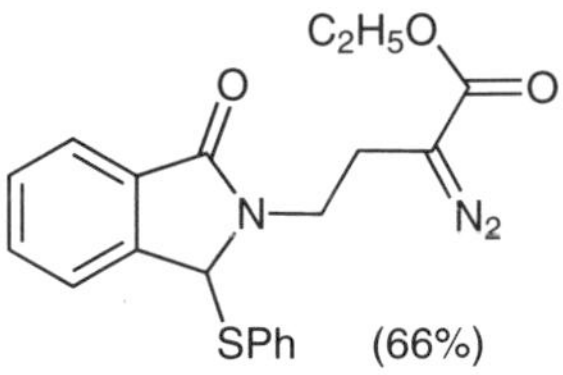
C_2H_5O
O
O
N
N_2
SPh

(66%)

50 [100]

In a useful extension of the Regitz concept, Doyle[101] in 1985 introduced substrate activation via the use of a trifluoroacetyl substitutent to achieve diazo transfer to a base-sensitive acyloxazolidinone derivative (Scheme 2). The trifluoroacetyl group was derived from trifluoroethyl trifluoroacetate.

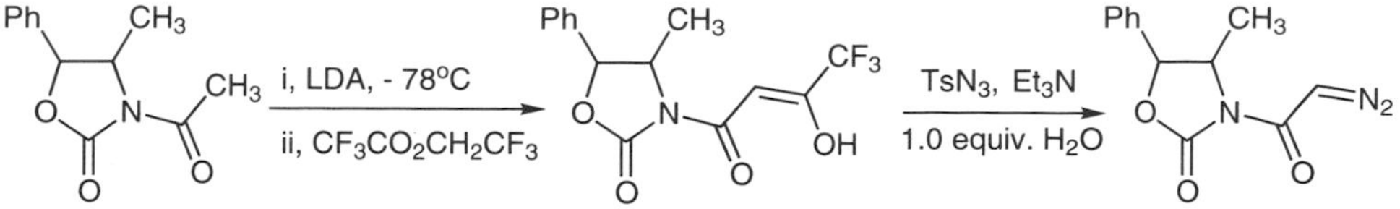

Scheme 2

Danheiser and co-workers[102,103] found that deformylating diazo transfer in a number of crucial cases produced the desired α-diazoketones in relatively low yield. Particularly problematic were reactions involving base-sensitive substrates such as α,β-enones, where the difficulties were attributable in part to the harsh conditions typically required for the Claisen condensation step. By using a very similar strategy to that developed by Doyle, in which the trifluoroacetyl group was employed as an activator (Scheme 3), Danheiser found that the efficiency of

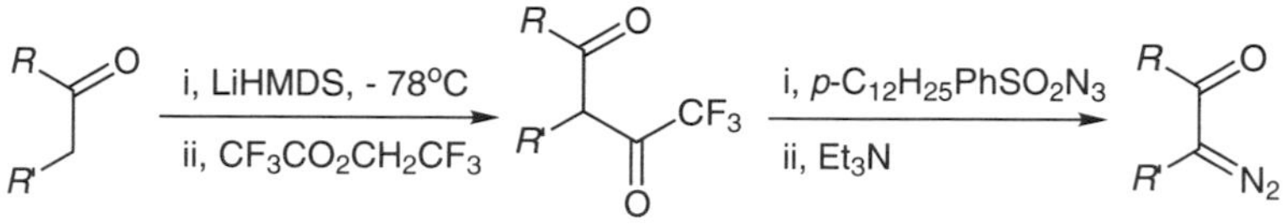

Scheme 3

the diazo transfer reaction for diazoketone formation could be improved, in some cases quite dramatically. The reaction of the ketone enolate with trifluoroethyl trifluoroacetate takes place essentially instantaneously at $-78°C$. This diazo transfer procedure has proved particularly valuable in the preparation of diazo derivatives of α,β-enones and is now described in detail in *Organic Synthesis*.[103] Representative examples of diazoketones **51–63** prepared via this route are shown in Table 1.2. The detrifluoroacetylating diazo transfer procedure has the advantage of providing a regioselective means of effecting diazo transfer to unsymmetrical ketones. As illustrated in eq. 12, diazo transfer to 2-octanone can be achieved with significant regiocontrol when lithium tetramethylpiperidide is employed for the generation of the requisite kinetic enolate.[107] In addition, this procedure is also applicable to the synthesis of α-diazoesters.[107] Four examples involving β-amino acids and peptides are shown in **66–69**.

There are other modifications of the Regitz deformylating procedure that have proved to be synthetically useful for diazoesters and ketones. These routes involve initial activation of the ketone by benzoylation[108,109] and acylation with diethyl oxalate.[110,111] Two examples are shown in Schemes 4[108] and 5.[111] Taber and co-workers[112] have developed a new method for the regioselective construction of unsymmetrical α-diazoketones using benzoylacetone as starting material. This protocol, involving the formation of an α-benzoylated unsymmetrical ketone followed by debenzoylating diazo transfer, is summarized in Scheme 6. Benzoyl-

DETRIFLUOROACETYLATING DIAZO GROUP TRANSFER: SYNTHESIS OF (*E*)-1-DIAZO-4-PHENYL-3-BUTEN-2-ONE[103]

A 500-mL, three-necked, round-bottomed flask is equipped with a mechanical stirrer, nitrogen inlet adapter, and 150-mL pressure-equalizing dropping funnel fitted with a rubber septum. The flask is charged with 70 mL of dry tetrahydrofuran and 15.9 mL (0.075 mol) of 1,1,1,3,3,3-hexamethyldisilazane, and then cooled in a ice-water bath while 28.8 mL (0.072 mol) of a 2.50 M solution of *n*-butyllithium in hexane is added dropwise over 5–10 min. After 10 min, the resulting solution is cooled at −78°C in a dry ice–acetone bath, and a solution of 10.0 g (0.068 mol) of *trans*-4-phenyl-3-buten-2-one in 70 mL of dry tetrahydrofuran is added dropwise over 25 min. The dropping funnel is washed with two 5-mL portions of tetrahydrofuran and then replaced with a rubber septum. The yellow reaction mixture is allowed to stir for 30 min at −78°C, and then 10.1 mL (0.075 mol) of 2,2,2-trifluoroethyl trifluoroacetate (TFEA) is added rapidly in one portion via syringe (over ~5 sec). After 10 min, the reaction mixture is poured into a 1-L separatory funnel containing 100 mL of diethyl ether and 200 mL of 5% aqueous hydrochloric acid. The aqueous layer is separated and extracted with 50 mL of diethyl ether. The combined organic layers are washed with 200 mL of saturated sodium chloride solution, dried over anhydrous sodium sulfate, filtered, and concentrated under reduced pressure using a rotary evaporator to afford 18.61 g of a yellow oil. This yellow oil is immediately dissolved in 70 mL of acetonitrile and transferred to a 500-mL, one-necked flask equipped with a magnetic stirring bar and a 150-mL pressure-equalizing dropping funnel fitted with a nitrogen inlet adapter. Water (1.2 mL, 0.068 mol), triethylamine (14.3 mL, 0.103 mol), and a solution of 4-dodecylbenzenesulfonyl azide (35.74 g, 0.103 mol) in 10 ml of acetonitrile are then sequentially added (each over ~1–2 min) via the dropping funnel. The resulting yellow solution is allowed to stir at room temperature for 6.5 h and then is poured into a 1-L separatory funnel containing 100 mL of diethyl ether and 200 ml of aqueous 5% sodium hydroxide (NaOH). The organic layer is separated, washed successively with three 200-mL portions of 5% aq NaOH, four 200-mL portions of water, 200 mL of saturated sodium chloride, dried over anhydrous sodium sulfate, filtered, and concentrated at reduced pressure using a rotary evaporator to yield 23.17 g of crude reaction product as a light brown oil. The crude reaction product is purified by column chromatography on 230–400 mesh silica gel (30 times by weight, elution with 5–10% diethyl ether–hexane) to furnish 9.54–9.80 g (81–83%) of (*E*)-1-diazo-4-phenyl-3-buten-one (m.p. 68–69°C) as a bright yellow solid.

TABLE 1.2 Diazoketones Prepared by the Detrifluoroacetylating Diazo Transfer Procedure

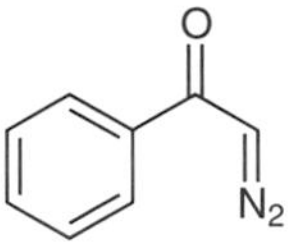

(95%)

51 [103]

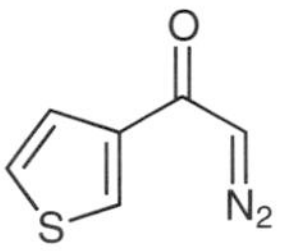

(92%)

52 [103]

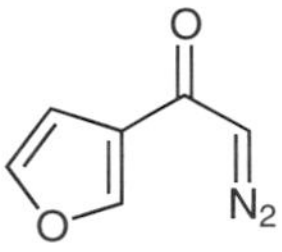

(71%)

53 [102]

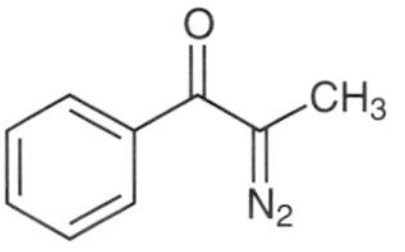

(90%)

54 [103]

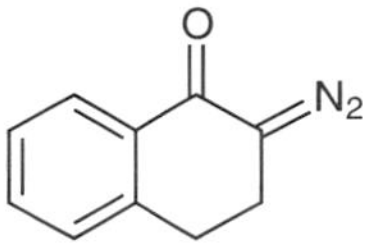

(81%)

55 [103]

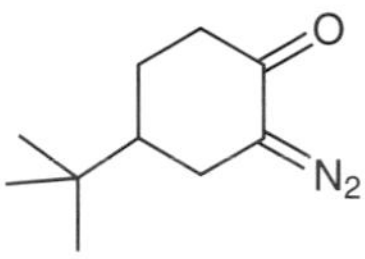

(61%)

56 [103]

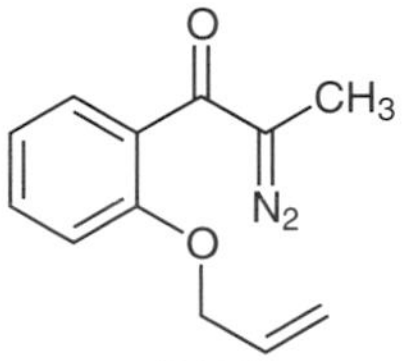

(54%)

57 [104]

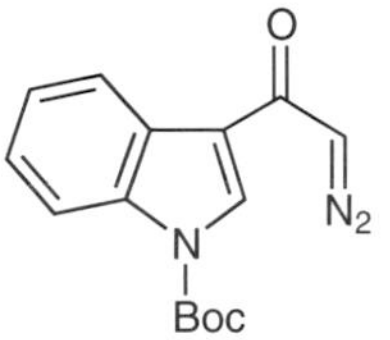

(90%)

58 [105]

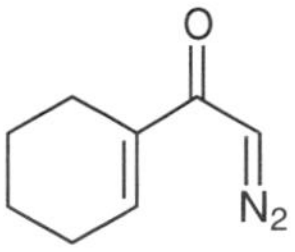

(87%)

59 [103]

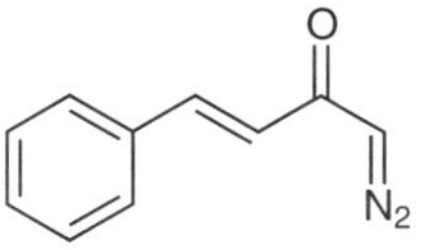

(86%)

60 [103]

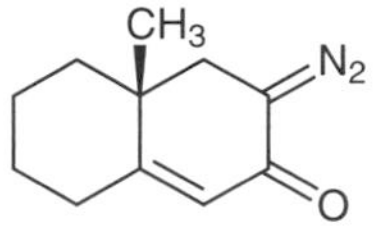

(48%)

61 [102]

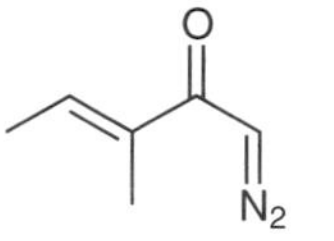

(84%)

62 [103]

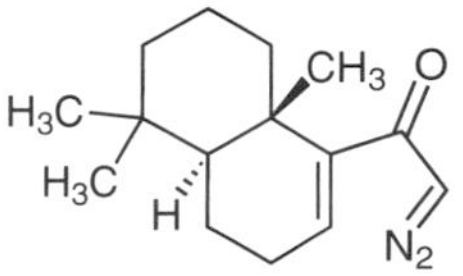

(91%)

63 [106]

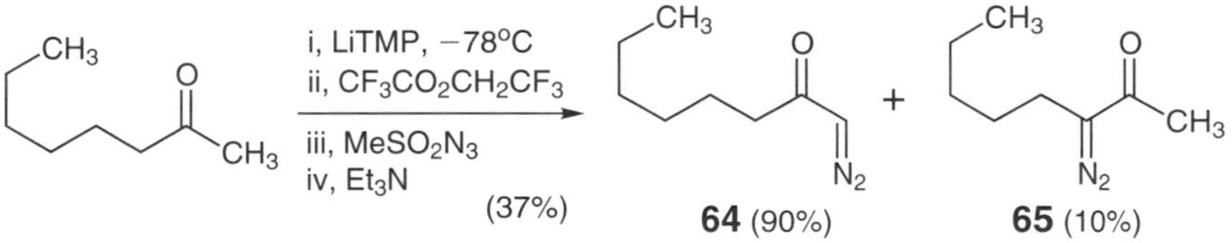

i, LiTMP, −78°C
ii, CF₃CO₂CH₂CF₃
iii, MeSO₂N₃
iv, Et₃N

(37%)

64 (90%)   +   **65** (10%)

(12)

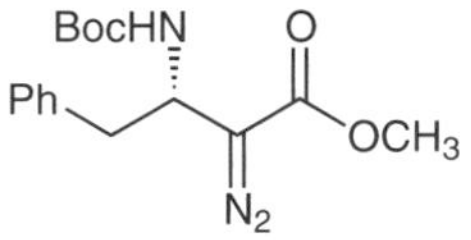

(75%)

66[107]

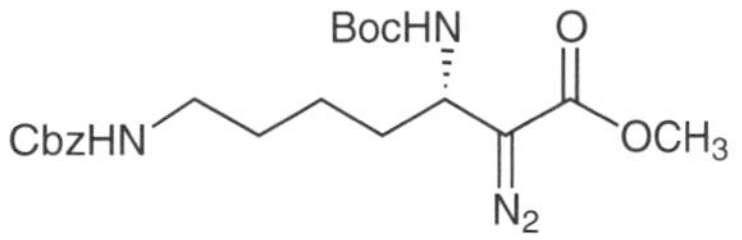

(82%)

67[107]

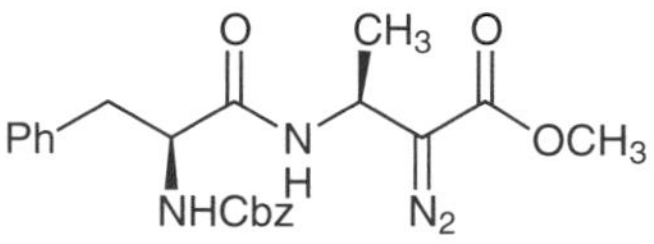

(66%)

68[107]

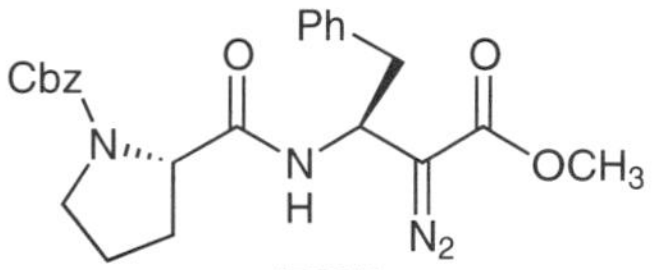

(75%)

69[107]

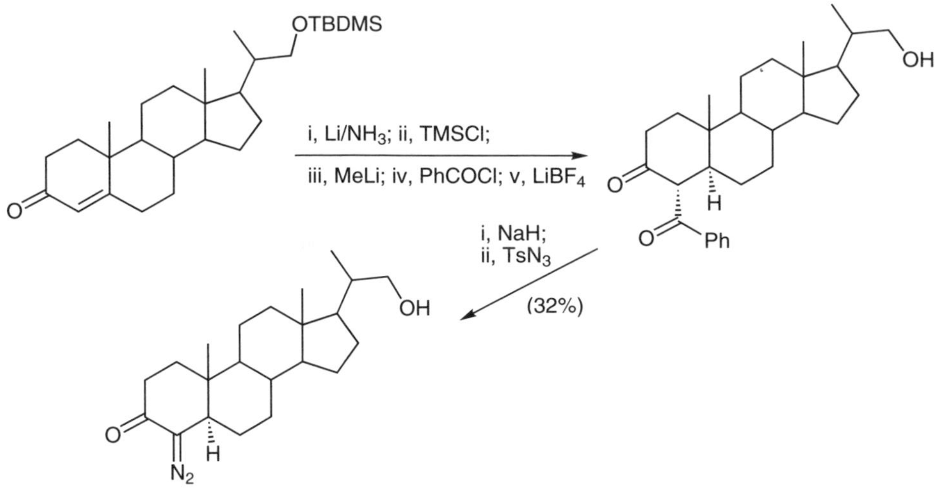
OTBDMS
i, Li/NH3; ii, TMSCl;
iii, MeLi; iv, PhCOCl; v, LiBF4
OH
H
Ph
i, NaH;
ii, TsN3
(32%)
OH
H
N2

Scheme 4

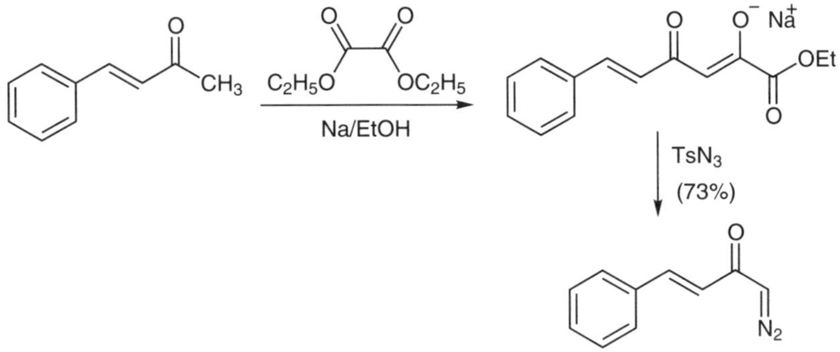
CH3
C2H5O
OC2H5
Na/EtOH
Na+
OEt
TsN3
(73%)
N2

Scheme 5

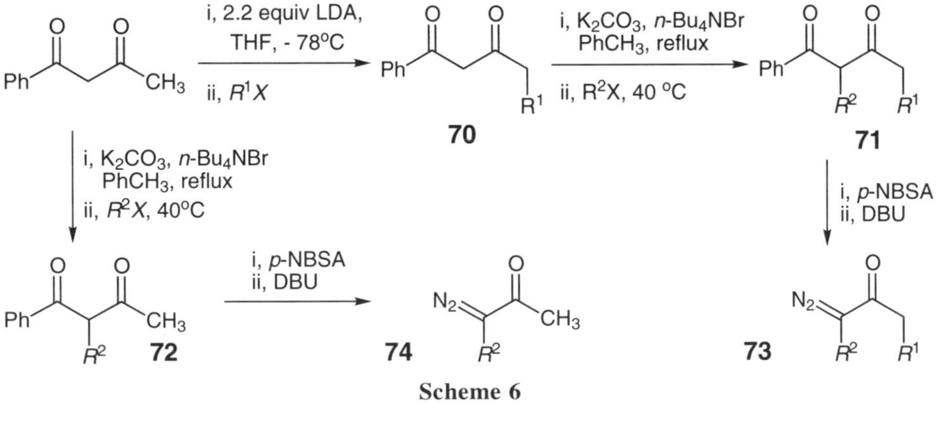
Ph
CH3
i, 2.2 equiv LDA,
THF, - 78°C
ii, R1X
Ph
R1
70
i, K2CO3, n-Bu4NBr
PhCH3, reflux
ii, R2X, 40 °C
Ph
R2
R1
71
i, K2CO3, n-Bu4NBr
PhCH3, reflux
ii, R2X, 40°C
i, p-NBSA
ii, DBU
Ph
CH3
R2
72
i, p-NBSA
ii, DBU
N2
CH3
74
R2
N2
73
R2
R1

Scheme 6

acetone is γ-alkylated as its dianion to provide diketone **70**, which is then α-alkylated to provide a new unsymmetrical diketone **71**. Alternatively, direct α-alkylation of benzolacetone gave the diketone **72**. Diazo group transfer to diketone **71** and **72** with p-nitrobenzenesulfonyl azide (p-NBSA) in the presence of DBU furnished the desired α-diazoketone **73** and **74**, respectively. Diazoketones **75**, **76**, **77**, **78**, and **79** have been synthesized via this route.[112]

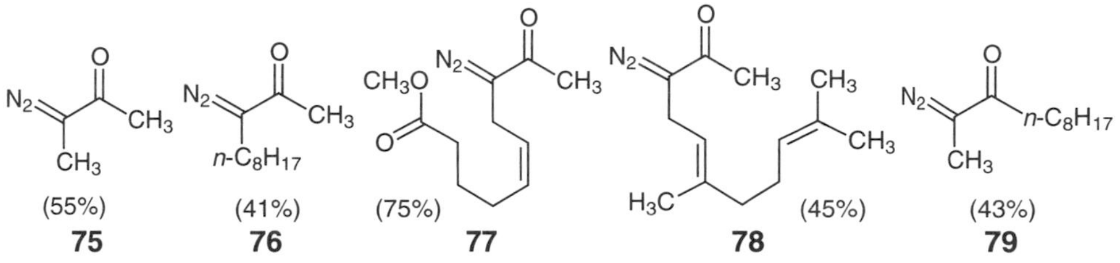

(55%) (41%) (75%) (45%) (43%)
75 **76** **77** **78** **79**

Similarly, α-diazoesters can be synthesized by treating the ester anion with methyl benzoate so as to produce the β-ketoester **80**. The β-ketoester **80** then reacts with the appropriate diazo transfer reagent in the presence of 1,8-diazobicyclo[5.4.0]undec-7-ene (DBU), proceeding to the corresponding diazo ester **81** by a subsequent cleavage of the benzoyl group (Scheme 7).[113] For example, diazoesters **82**, **83**, and **84** were synthesized by this method.[113] α-Diazoesters can also be prepared via an approach analogous to the preparation of α-diazoketone **74** (vide ante), as shown in Scheme 8.[114,115] Thus α-alkylation of an alkyl acetoacetate **85** with an alkyl halide produced an α-substituted β-keto ester **86**, which was then converted into the α-diazoester **87** using an appropriate diazo transfer

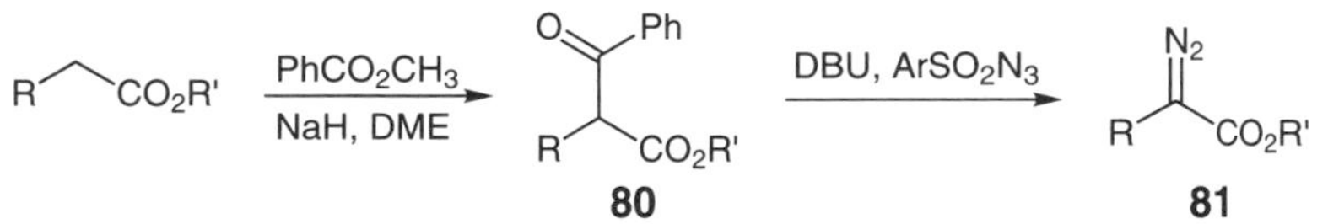

80   **81**

Scheme 7

(88%)   (68%)   (81%)
82   **83**   **84**

85   **86**   **87**

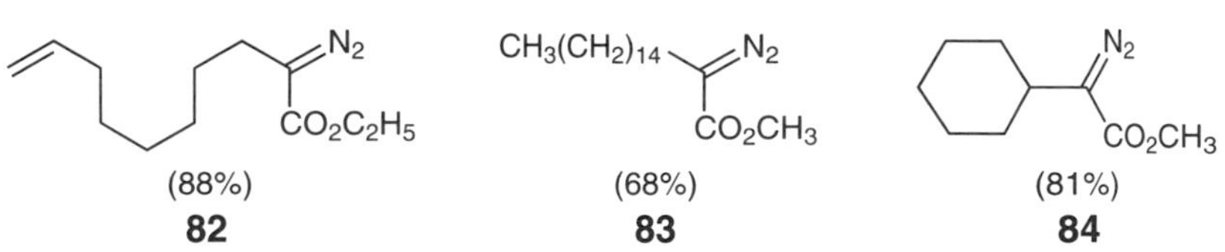

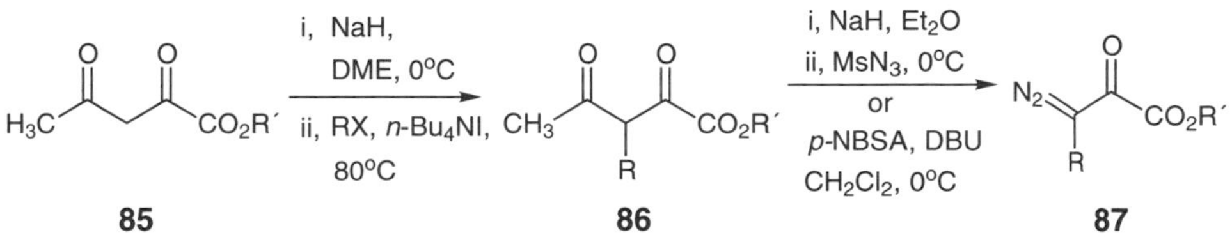

Scheme 8

reagent and base (Scheme 8). For example, diazoesters **88**, **89**, and **90** were synthesized by this method.[114]

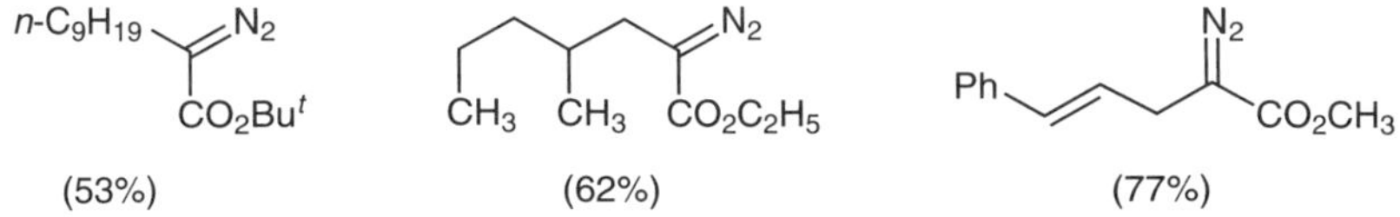

(53%)
88

(62%)
89

(77%)
90

Diazo transfer to enamines provides another option for the synthesis of diazocarbonyl compounds. Treatment of a formyl enamine with tosyl azide gives the corresponding α-diazo aldehyde in good yield.[116,117] Similarly, the bicyclic diazoketone **91** was prepared via the enamine route shown in Scheme 9.[82]

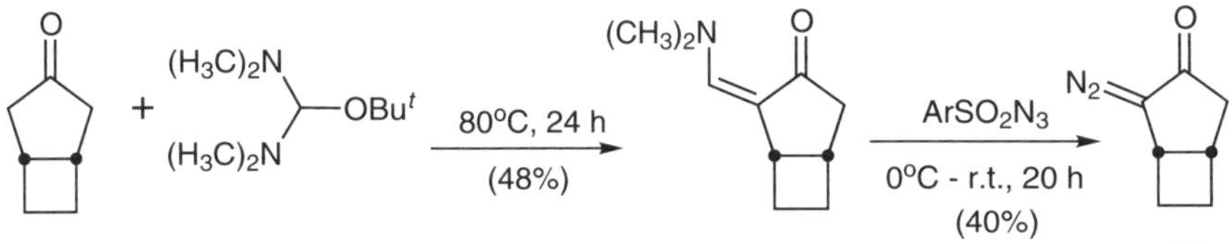

Scheme 9

Norbeck and Kramer[118] have employed diazoketone **94** in their synthesis of (−)-oxetanocin **95** summarized in Scheme 10. Rapid, base-catalyzed epimerization at C-1′ of ketone **92** precluded application of the standard Regitz deformylating diazo transfer protocol to the preparation of diazoketone **94**. However, activation of the 3′ carbon atom was readily achieved by heating the ketone **92** in neat N,N-dimethylformamide dimethyl acetal at 60°C for 15 min to give the enamino ketone **93**. Diazo transfer from triflic azide in 1,2-dichloroethane at 60°C afforded the desired diazoketone **94** in 72% yield (Scheme 10).

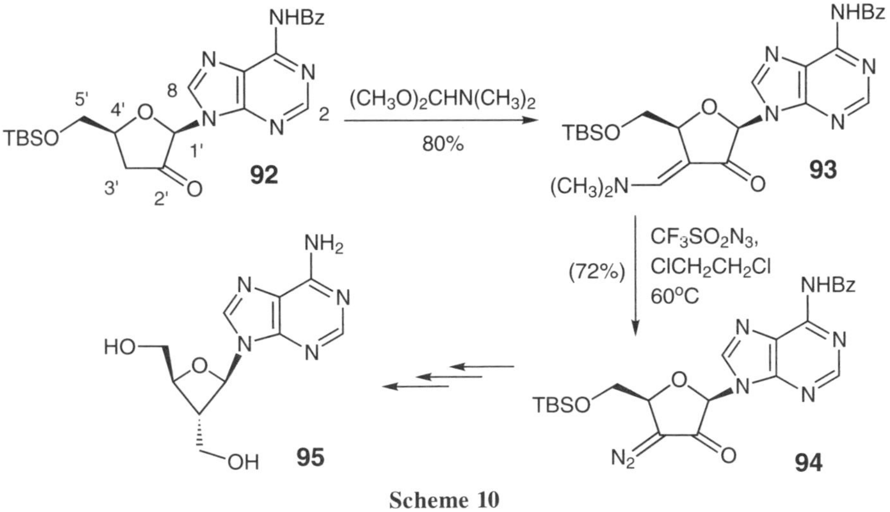

Scheme 10

1.3.3   Effects of Base and Solvent on Diazo Transfer

There have been many studies of the influence of base, solvent, and diazo transfer reagent on both the simple diazo transfer and the Regitz deformylating diazo transfer procedure. Two examples, representative of cyclic and acyclic diazo transfer, in which tosyl azide was the transfer agent and triethylamine the base, were shown in **16**[68] and **29**.[74] In cases where triethylamine is insufficiently basic to cause deprotonation of β-dicarbonyl compounds, Koskinen and Muñoz[119] have advocated the use of potassium carbonate in acetonitrile, under which conditions practically quantitative diazo transfer from tosyl azide to a range of β-keto esters was complete in 1 h at room temperature, the workup simply consisting of filtration of the inorganic salts and the sulfonamide byproduct after addition of diethyl ether. Problems associated with hydrolysis of base-sensitive methyl acetoacetate were not encountered.[120] Similarly, Lee and Yuk[121] have reported that diazo transfer reactions between active methylene compounds and tosyl azide can be efficiently mediated by the presence of cesium carbonate. In this case, the reaction was carried out in tetrahydrofuran and the reported yields, in general, were greater than 90%. Ghosh and Datta[122] claimed that the diazo transfer reaction can be carried out by using potassium carbonate as base and in the solid state. This method involves grinding of the reaction mixture with a pestle and mortar, which contains an element of risk because of the shock sensitivity of sulfonyl azides. Therefore, this method is not recommended for production of diazocarbonyl compounds.

BenAlloum and Villemin[123] have reported a convenient synthesis of diazocarbonyl compounds from tosyl azide and carbonyl compounds with Al_2O_3–KF as a solid base, and Nikolaev and co-workers[124] have used a potassium fluoride–crown ether combination as the base system in the diazo transfer reaction of cyclic and acyclic diketones. The successful use of potassium fluoride in this reaction is apparently due to the facile deprotonation of diketones by the "naked" fluoride anion, even in the case of sterically hindered compounds. Rao and Nagarajan[125] have successfully used 1,8-diazobicyclo[5.4.0]undec-7-ene (DBU) for smooth diazo transfer, notably in the case of hindered compounds where the standard diazo transfer conditions were inefficient, such as in eq. 13. Actually, for the acetoacetamide precursor **96**, diazo transfer with p-carboxybenzenesulfonyl azide (pCBSA) and triethylamine in acetonitrile proved unsuccessful. However, when DBU was used, the reaction was complete in a few hours (eq. 14).[126]

$$\text{(13)}$$

$$\text{(14)}$$

An application of phase-transfer catalysis to the diazo transfer process using an aqueous base has been devised by Ledon.[127] When diazo transfer is complete, the organic layer contains only the diazo compound and traces of the ammonium salt. The latter is eliminated during distillation or by rapid filtration through a silica-gel column. Ledon's procedure for the synthesis of di(*tert*-butyl)diazomalonate **97** (eq. 15) under phase-transfer catalysis is described in *Organic Synthesis*.[120] Thus treatment of di(*tert*-butyl)malonate with tosyl azide and a catalytic amount (2 mol %) of methyl(trioctyl)ammonium chloride in dichloromethane afforded the desired diazoester **97** in 59–63% yield.

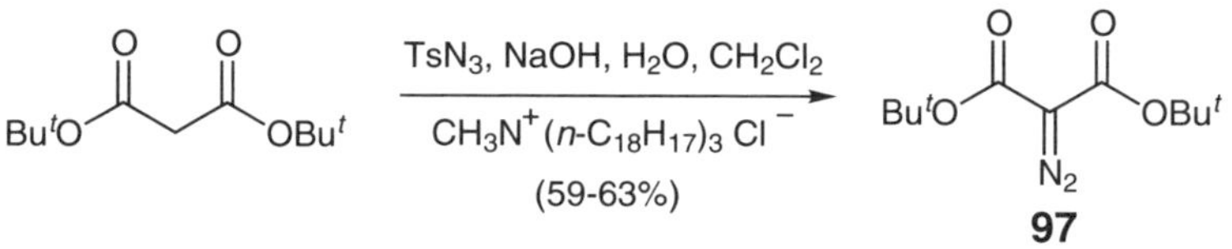

$$\text{(15)}$$

Diazo transfer to an α-methylene group activated by only one carbonyl group requires the presence of a base of sufficient strength to deprotonate the substrate. In most cases, side products are obtained when strong base is employed. However, in certain circumstances, direct diazo transfer can be achieved effectively, and optimized procedures have been extensively used in synthesis. Schöllkopf and co-workers[128] have successfully transferred a diazo group from tosyl azide to lithiated diketopiperazine bislactim ethers. Diazoketones **98** and **99** have been prepared via the direct diazo transfer protocol, when lithium diisopropylamide (eq. 16)[129] or potassium *tert*-butoxide (eq. 17)[61,130,131] was employed as base, and

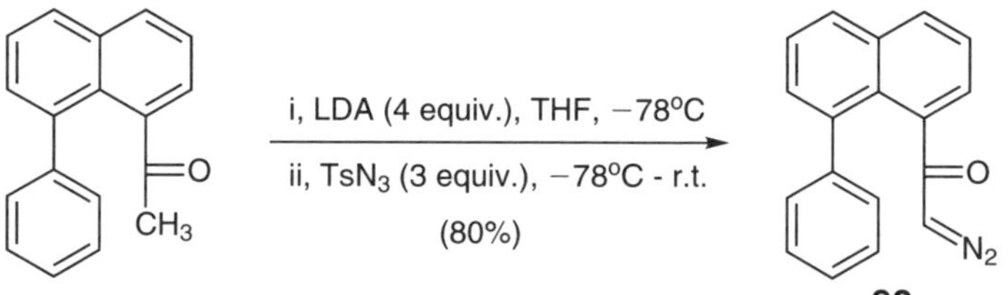

$$\text{(16)}$$

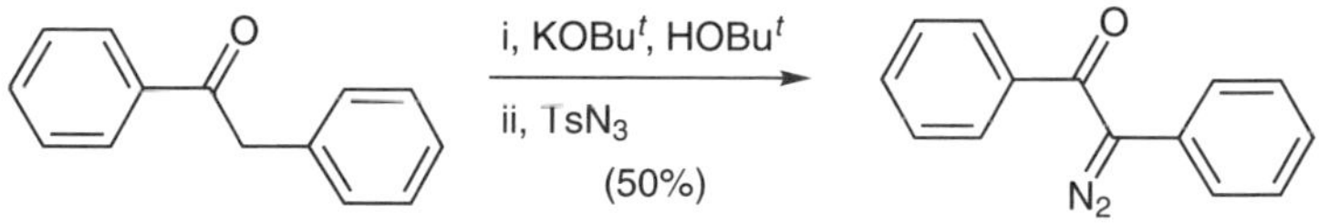

$$\text{(17)}$$

tosyl azide was the diazo donor. In addition, relatively stronger organic bases, such as 1,8-diazobicyclo[5.4.0]undec-7-ene (DBU), are also suitable for the conversion of benzyl ketones to α-diazoketones.[132] Evans and co-workers[133] have reported successful diazo transfer from *p*-nitrobenzenesulfonyl azide (*p*NBSA) to the enolate derivatives of an *N*-acyloxazolidinones and a benzyl ester. Diazocarbonyls **100** and **101** have been obtained via this direct diazo transfer reaction, as shown in eqs. 18 and 19. A further interesting example of the construction of

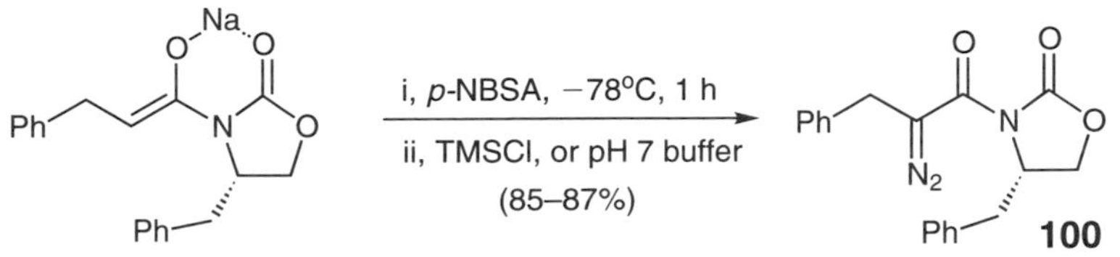

$$(18)$$

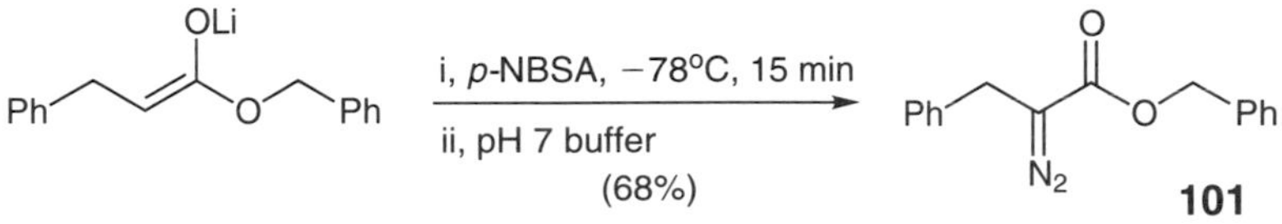

$$(19)$$

diazocarbonyls via the direct diazo transfer protocol is provided by Lim and Sulikowski[134] in their synthetic study of the antitumor antibiotic FR-66979. Thus generation of the sodium enolate derivative of keto **102** was followed by treatment with 2,4-dinitrobenzenesulfonyl azide (DNBSA). Under these conditions, the key intermediate, diazoketone **103**, was obtained in 62% yield (eq. 20).

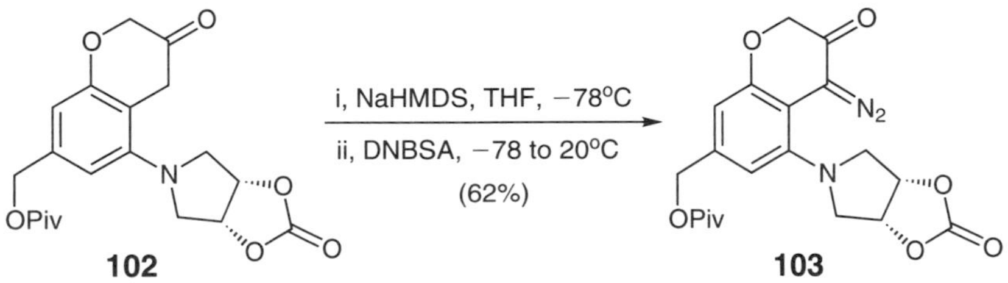

$$(20)$$

Mander and Lombardo[135] have found that when 2,4,6-triisopropylbenzenesulfonyl azide (trisyl azide) is substituted for tosyl azide, some hindered cyclic ketones can be converted into α-diazo derivatives under phase-transfer conditions employing potassium hydroxide, 18-crown-6, and tetrabutylammonium bromide. Some representative examples of cyclic diazoketones prepared via this route are shown in Table 1.3. The utility of the trisyl azide in these conversions stems from the steric hindrance afforded by the ortho isopropyl groups; simpler arylsulfonyl azides are degraded too rapidly under these conditions to be useful. The method appears to be less effective with cyclopentanones,[142,143] and several failures have been reported with such substrates.[93,98,144] The limitation of this method is that it is not suitable for substrates sensitive to aqueous basic hydrolysis, for example, methyl esters. In addition, direct diazo transfer can also be achieved under homogeneous conditions:[143] the slow addition of an equimolar mixture of a ketone and

DIRECT DIAZO TRANSFER USING 2,4,6-TRIISOPROPYLPHENYLSULFONYL AZIDE: SYNTHESIS OF 2-DIAZO-5-METHOXY-1-TETRALONE[135]

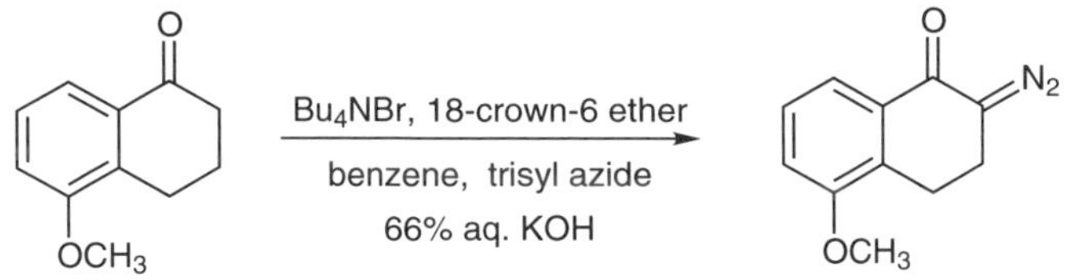

$$\text{(starting material)} \xrightarrow[\substack{\text{benzene, trisyl azide}\\ \text{66\% aq. KOH}}]{\text{Bu}_4\text{NBr, 18-crown-6 ether}} \text{(product)}$$

To a solution of 5-methoxy-1-tetralone (100 mg, 0.57 mmol), tetrabutylammonium bromide (50 mg, 0.16 mmol), 18-crown-6 ether (5 mg), and 2,4,6-triisopropylphenylsulfonyl azide (200 mg, 0.65 mmol) in benzene (10 mL) is added 66% aqueous potassium hydroxide (10 mL). The mixture is stirred vigorously at 35°C for 30 min, further azide (150 mg, 0.49 mmol) added, and stirring continued for 1.5 h, when TLC analysis indicates that the reaction is complete.* Ether (50 mL) is added and the organic layer separated. The aqueous layer is diluted with water (20 mL), extracted with ether (50 mL), then the combined ether layers are washed with water (5 mL) and saturated sodium chloride solution (5 mL) and dried with magnesium sulfate. Removal of solvent and preparative layer chromatography gives a band containing the desired diazoketone (106 mg, 86% purity by ^{1}H-NMR), contaminated with 2,4,6-triisophenylsulfonamide. Recrystallization from ether/pentane gives needles of pure diazoketone; yield: 62 mg (54%); m.p. 86–88°C.

*With less reactive substrates, further sulfonyl azide (50 mg) is added at 2 h intervals until reaction is complete.

2,4,6-triisopropylbenzenesulfonyl azide to a solution of potassium *tert*-butoxide in tetrahydrofuran at −78°C. Diazoketones **111** and **112** were obtained by this method in over 90% yield.

Other examples of α-diazocarbonyl compounds prepared via direct diazo transfer reaction that deserve to be mentioned here are shown in eqs. 21[145] and 22.[146] The presence of a vinyl group or an aryl group apparently enhances the transformation.

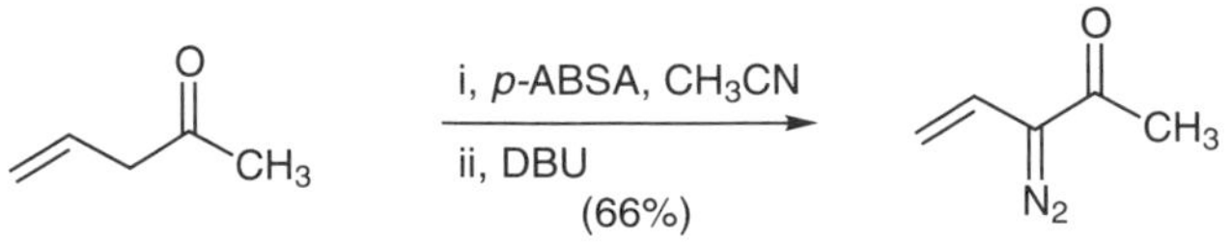

$$\xrightarrow[\substack{\text{ii, DBU}\\ \text{(66\%)}}]{\text{i, }p\text{-ABSA, CH}_3\text{CN}} \qquad (21)$$

TABLE 1.3   Diazoketones Prepared by Mander's Diazo Transfer Procedure

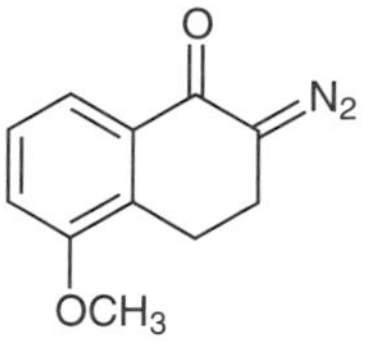

(54%)
104[135]

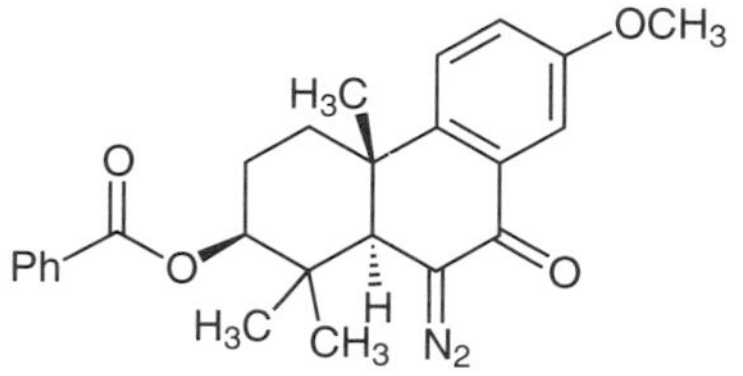

(84%)
105[135]

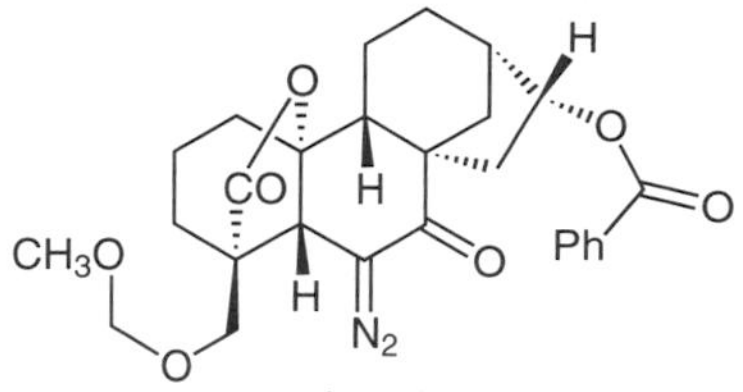

(66%)
106[136]

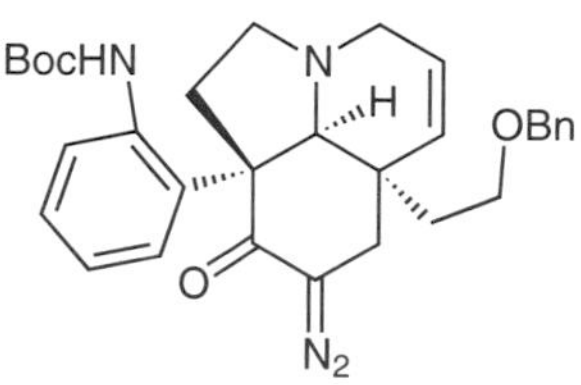

(> 94%)
107[137,138]

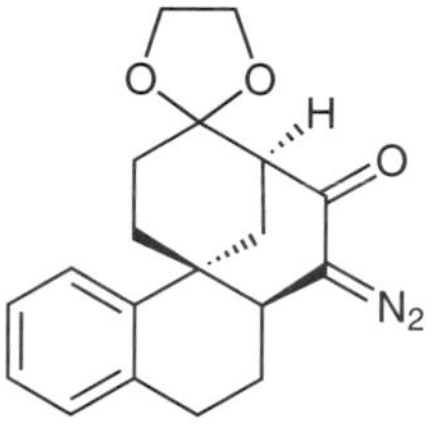

(> 53%)
108[139]

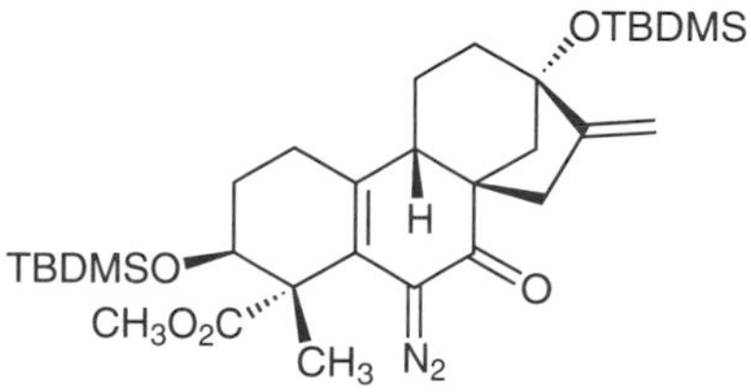

(74%)
109[140]

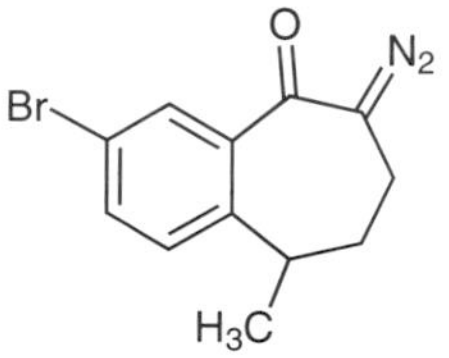

(81%)
110[141]

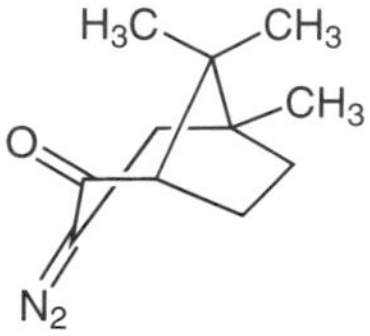

(96%)
111[142]

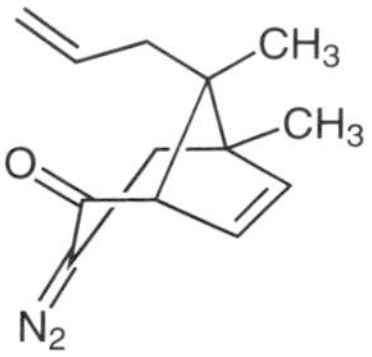

(91%)
112[142]

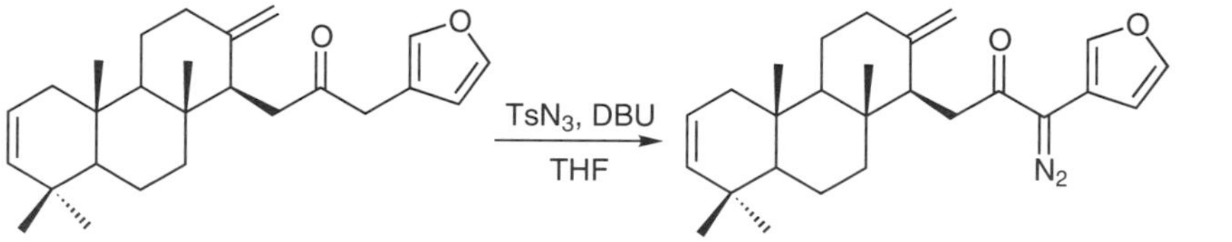

(22)

1.3.4   Effects of Transfer Reagent on Diazo Transfer and Hazards Evaluation

Although tosyl azide is by far the most frequently employed diazo donor and was used in this capacity by Doering and DePuy in 1953 in their synthesis of diazo-cyclopentadiene,[59] workers in the Merck Laboratories[147] have raised doubts re-

SYNTHESIS OF 4-DODECYLBENZENESULFONYL AZIDE[151]

114

4-Dodecylbenzenesulfonyl chloride.

A 250-mL, three-necked, round-bottomed flask, equipped with a mechanical overhead stirrer, a Claisen adapter bearing an immersion thermometer, a pressure-equalizing addition funnel, and reflux condenser, is charged with a solution of 60.00 g (0.184 mol) of 4-dodecylbenzenesulfonic acid in 60 mL of hexane. Stirring is initiated while the mixture is heated to 70°C using a heating mantle, and 22.1 mL (36.24 g, 0.304 mol) of thionyl chloride is added at a rate to maintain controlled reflux. The required addition time is about 1 h. The dark solution is heated an additional 2 h at 70°C and cooled to 40°C. While still warm (~40°C), the mixture is transferred to a 250-mL separatory funnel, and the dark lower layer is separated from the hexane solution. The hexane layer is cooled to 25°C and washed with 60 mL of aqueous 5% sodium bicarbonate solution. The bicarbonate wash is back extracted with 36 mL of hexane and the combined hexane layers are treated with 3 g of carbon and stirred for 2 h at 25°C. The carbon is removed by filtration and the cake is washed with three portions (12 mL each) of hexane. The combined hexane layers plus the hexane washes are used to prepare the azide.

4-Dodecylbenzenesulfonyl azide.

A 500-mL, three-necked, round-bottomed flask fitted with a mechanical overhead stirrer is charged with the hexane solution from the previous step. To this solution is added a solution of 11.6 g (0.178 mol based on the total solids from the sulfonyl chloride above) of sodium azide (NaN_3) in 100 mL of water and 2.0 g of phase-transfer catalyst [Aliquat 336 (tri-n-alkylmethylammonium chloride)]. Stirring is initiated, and the reaction progress is monitored by thin-layer chromatography. Approximately 4 h at 25°C is required to complete the reaction. The two-phase mixture is transferred to a 500-mL separatory funnel and the aqueous layer is removed. The hexane layer is washed with 100 mL of aqueous 5% sodium bicarbonate solution and dried over 28 g of anhydrous sodium sulfate. The concentration and purity of the 4-dodecylbenzenesulfonyl azides are best determined by evaporation of a small sample to an oil of constant weight with visible spectrophotomeric assay for the azide. The hexane solution of dodecylbenzene-sulfonyl azide, when standardized as above, can be used as obtained for most applications. However, if desired, careful concentration of the hexane solution under reduced pressure at toom temperature affords 58.2–61.4 g (90–95%) of the oily mixture of dodecylbenzenesulfonyl azides; corrected for the assay of the azides the yield is usually 95%.

garding safety aspects of this reagent. In the first place, the *Organic Synthesis*[69] preparation of tosyl azide calls for the use of ethanol as the reaction solvent. But as Curphey[148] has pointed out, ethanol is an inappropriate solvent for the reaction of tosyl chloride with sodium azide; replacement with acetone gives a cleaner product not requiring recrystallization. Various sulfonyl azides are now accessible by a similar way using the respective sulfonyl chloride. In general, when the sulfonyl azide formation is complete, the acetone solution may be used directly with appropriate substrates or concentrated at or below room temperature. Well-tested methods for preparing *p*-acetamidobenzenesulfonyl azide **113**[149] and *p*-dodecylbenzenesulfonyl azide **114**[147] are now available in *Organic Synthesis*.[150,151]

The Merck chemists[147,152] have examined the twelve sulfonyl azides **115** to **126** from the standpoint of utility, thermal stability, ease of handling, and safety. In addition to desired performance, the specific heat of decomposition, thermal decomposition temperature, relative rates of decomposition, and impact sensitivity of all twelve compounds were measured (Table 1.4).

Methanesulfonyl (mesyl) azide **115** was the most hazardous of the group, exhibiting the highest specific heat of decomposition and the highest shock sensitivity. In terms of chemical reactivity, mesyl azide is a generally superior reagent to tosyl azide for diazo transfer, its main advantage being the greater ease with

TABLE 1.4   Sulfonyl Azides

Sulfonyl azide	m.p.(°C)	ΔH_D (kcal/mole)	ΔH_D (cal/g)	Approximate initiation temperature (°C)	Relative rate of decomposition	Impact sensiviity (kg cm)	Sulfonylamide m.p. (°C)
115	20	−67.4	−557	125	1.96	50	92
116	19–20	−79.9	−405	120	1.00	50	136–137
117	liq.	−58.9	−168	151	0.36	negative to 150	liq.
118	41–43	−64.5	−277	146	0.96	300	218.5
119	184–186	−53.6	−236	163	2.29	300	295–296
120	39–40	−62.4	−202	136	0.52	negative to 300	119–119.6
121	108–111	−41.3	−172	120		negative to 136.5	215–216
122	39	−78.2	−360	139	0.14	300	143.5
123	54.5–55.5	−74.2	−284	127	0.24	300	166.5
124	Gel	− 7.9	−16.3	240	0.01	negative to 300	
125	178–185	−59	−190	132	0.53	100	no m.p.
126	71–73	−114	−499	118	1.49	50	190.5–191.5

which the sulfonamide byproduct is removed from the reaction mixture by washing with 10% aqueous NaOH solution.[153] Mesyl azide has been prepared by Boyer and co-workers in 1958,[154] and a modified procedure for its preparation has been published by Danheiser and co-workers.[102] An example of the use of mesyl azide in simple diazo transfer is shown in eq. 23.[153]

$$\text{(23)}$$

There have been notices of warning already concerning the dangers associated with the use of mesyl azide, particularly with procedures involving the application of heat. In 1993 two serious explosions were documented in *Chemistry in Britain*.[155] One was a result of an attempt to distill mesyl azide. The circumstances surrounding the second explosion are unclear, but heat was believed to have been a factor. In our own experience of making and using mesyl azide, we have never found it necessary to purify by distillation. The material from reaction of sodium azide and mesyl chloride in acetone has always been of adequate purity for direct use in diazo transfer. Kumar[156] has reported a convenient one-pot procedure for preparing diazocarbonyls using in situ generated mesyl azide under triphase transfer catalysis conditions. This simplified method eliminates the need for direct handling of the mesyl azide.

The Merck study revealed that tosyl azide (**116**) is a close second to mesyl azide in impact sensitivity although more distant in heat of decomposition and explosive power. Some serious accidents with tosyl azide have been reported.[157,158] Despite the greater risk, tosyl azide remains the most popular diazo transfer reagent for activated methylene compounds on a laboratory scale. Several representative examples of its use have already been presented in this chapter (vide ante).

p-Dodecylbenzenesulfonyl azide **117**, which was actually an isomeric mixture within the alkyl side chain, has the lowest specific heat of decomposition, the second highest initiation temperature, and no impact sensitivity at the highest test level. Moreover, the byproduct dodecylbenzenesulfonamides are noncrystalline, facilitating isolation of crystalline diazoketones.[147] *p*-Dodecylbenzenesulfonyl azide **117** has been employed in the formation of α-diazo-β-ketoester **127** in 90% yield (eq. 24). If, on the other hand, the diazo product is a liquid, naphthalene-2-sulfonyl azide **118** can be employed to take advantage of the fact that its byproduct sulfonamide is highly crystalline and of low solubility, making it easily separable from liquid diazocarbonyl products. Hazen et al.[147] successfully employed this diazo transfer reagent in the synthesis of dimethyl diazomalonate **128** (eq. 25).

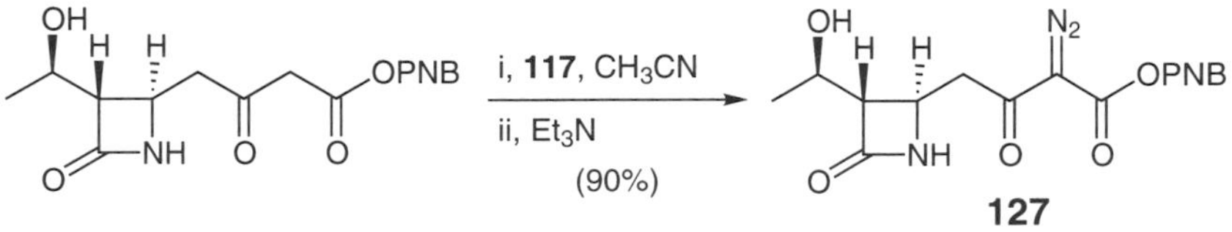

(24)

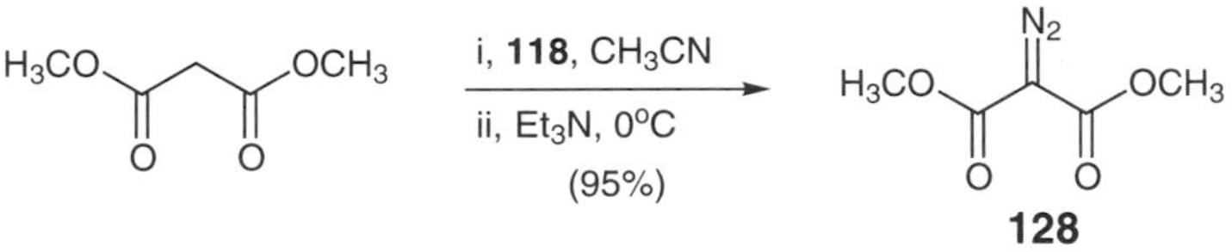

(25)

p-Carboxylbenzenesulfonyl azide (*p*-CBSA) **119** has the highest initiation temperature, has low impact sensitivity, and is water soluble. The lithium and triethylamine salts of this reagent are soluble in tetrahydrofuran and acetonitrile, respectively. The triethylamine salt of its product *p*-carboxylbenzenesulfonamide is essentially insoluble in acetonitrile. Though not suitable for base-sensitive compounds, it does offer a handle for removal from neutral products. Hendrickson and Wolf[75] have synthesized diazo diketone **129**, employing this reagent, in 86% yield:

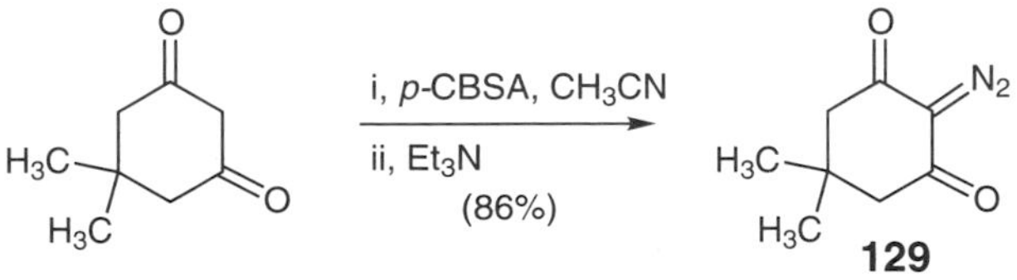

(26)

As mentioned previously, 2,4,6-triisopropylbenzenesulfonyl azide **120** has been used for the direct transfer of the diazo group to ketone enolates under phase-transfer conditions to furnish α-diazoketones[135] This reagent also offers some advantage in safety over tosyl azide.

p-Acetamidobenzenesulfonyl azide (p-ABSA) **121**[150,159] is a practical diazo transfer reagent and offers some advantage in safety and ease of product separation. This reagent is currently commercially available and was preferred over tosyl azide for reasons of safety, yield, and ease of manipulation. The reaction times with **121** are longer than with other azides. An optically active ester derivative of diazoacetoacetic acid **130** is readily available from **121** and triethylamine in good yield (eq. 27).[159] This reagent has also been employed by Davies and co-workers in their synthesis of vinyl diazoesters, the resulting yields are generally good.[149,160,161]

$$\text{(27)}$$

130

p-Chlorobenzenesulfonyl azide **122**[75] was expected to be more reactive than tosyl azide but was generally found to give lower product yields. The chemical reactivity of p-bromobenzenesulfonyl azide **123** would be expected to be similar to that of the p-chloro derivative **122**. Sulfonyl azide **124** containing a substituted naphthalene derivative was so unreactive that it did not act as a diazo transfer reagent.[152]

A polymer-bound sulfonyl azide **125**[162] has also been employed as a diazo transfer agent to a variety of 1,3-dicarbonyl compounds. Although a researcher claimed this reagent is nonexplosive, the Merck scientists found it to be impact sensitive and explosive above 175°C.[152] This reagent is particularly suitable for column operation where the diazocarbonyl compound is a liquid or a highly soluble solid. The sulfonamide byproduct remains attached to the insoluble polymer. However, yields with 1,3-dicarbonyl compounds are slightly lower than with tosyl azide, and yields with monoactivated methylene compounds are much lower still.

o-Nitrobenzenesulfonyl azide **126** is comparable to tosyl azide but more reactive. In the words of Bollinger and Tuma,[152] "As a diazo transfer agent it finds use as one of last resort." p-Nitrobenzenesulfonyl azide (p-NBSA) **131** would be expected to have thermal properties and reaction capabilities similar to the o-nitrobenzenesulfonyl azide **126**. p-NBSA was recommended as a direct diazo transfer reagent by Evans and co-workers,[133] and two examples of its applications have been illustrated earlier in eqs. 18 and 19. The low solubility of both the starting azide (**131**) and the p-nitrobenzenesulfonyl amide produced, and the significantly higher absorption ability of these compounds on silica gel, simplifies the purification of the corresponding diazoketone. Nikolaev and co-workers[124] have synthesized diazodiketone **132** in 91% yield using p-nitrobenzenesulfonyl azide as the diazo donor:

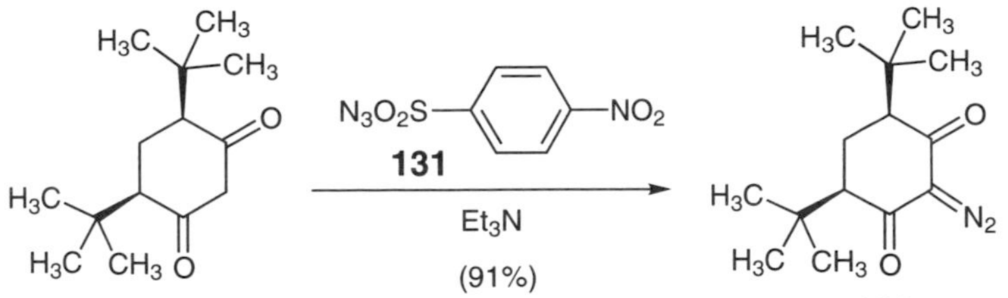

(28)

Finally, it should be mentioned here that trifluoromethanesulfonyl azide (triflic azide)[163] has also been used as a diazo transfer reagent.[163,164] It can be used without base promotion to diazotize active methylene compounds and with base promotion to diazotize less active compounds. An example of the use of this reagent in synthesis was illustrated in Scheme 10.[118] Triflic azide is more hazardous than mesyl azide (**115**).[152]

Several sulfonyl azides offer advantages in safety, facility of product separation, and substrate applicability, yet issues such as reagent stability, toxicity, and economy will confront the organic chemist for choice of the proper diazo transfer reagent.

The preparative scope of the diazo group transfer reaction is considerably widened by using azidinium salts as diazo group donors. Balli and collaborators[165–167] described the use of stable heterocyclic azidinium salts **133** as a diazo transfer reagent for the preparation of base-labile diazocarbonyl compounds. These reagents have the ability to react with weaker nucleophiles such as enols, and have the advantage of reacting with a variety of active methylene compounds in acidic media. Two representative diazocarbonyl compounds prepared using Balli's reagents are shown in **134** and **135**.[166] Monteiro[168] has advocated the use of an azidinium salt **136** generated in situ from relatively inexpensive starting materials for the preparation of diazocarbonyl compounds. For example, α-diazo-β-ketosulfone **137** was obtained in 94% yield using this in situ generated reagent. In addition, azidinium salt **136** has recently been employed by Sezer and Anac[169] in their synthesis of α-diazo-β-oxoaldehydes. Kokel and Viehe[170] showed that the iminium salt **138** was also an effective diazo transfer reagent in acidic medium.

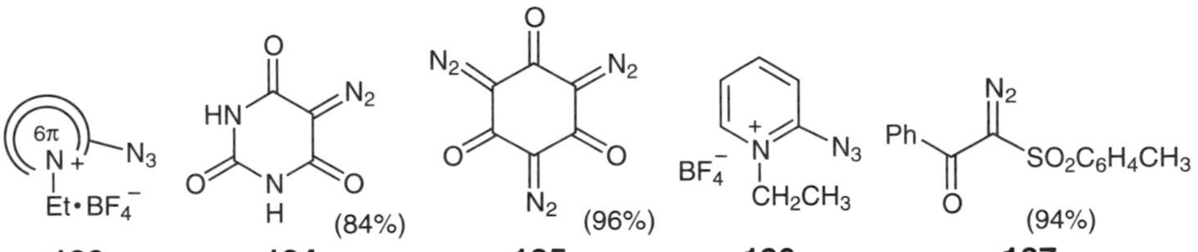

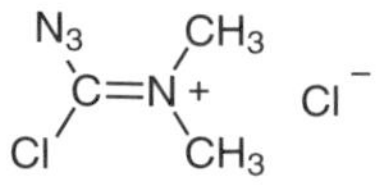

138 **139** (75%) **140** (64%)

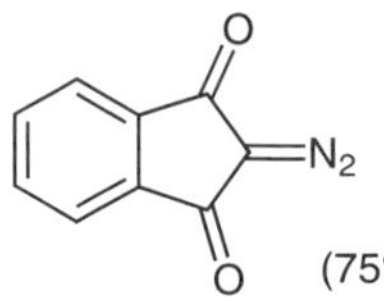

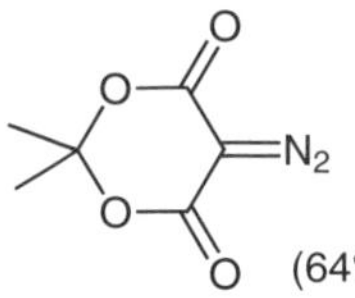

Since the reaction proceeds only in the neutral and acid ranges in which sulfonyl azides are unreactive, it is particularly suitable for diazo compounds that undergo coupling. Two compounds prepared by this method are shown in **139** and **140**.[170] The applicability of diazo group transfer using azidinium salts is limited by the fact that the reaction must be carried out in acidic to neutral medium. This protocol is no longer applicable when the active methylene component does not form an anion under the conditions given, or when the diazo compound is unstable to acid.[62,63]

McGuiness and Shechter[171] have reported that azidotris(diethylamino)phosphonium bromide **141** is an exceptionally safe diazo transfer reagent, which is stable to shock, friction, and rapid heating. The electrophilicity of **141** should lie between that of Balli's reagent and diazo transfer reagents having the azide structure. Acidic methylene compounds are conveniently converted to diazocarbonyl compounds by azidotris(diethylamino)phosphonium bromide in diethyl ether using only a catalytic amount of base. The diazo transfer reaction generates a neutral leaving group, hexaethylphosphortriamide, which is easily removed as its hydrobromide salt. Two examples are illustrated in **142** and **143**.

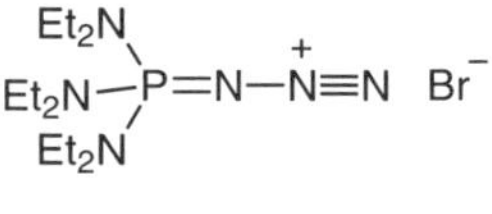

141 **142** (78%) **143** (70%)

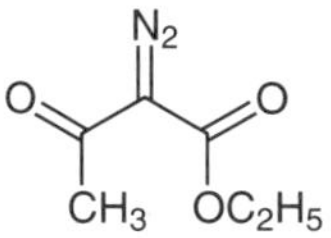

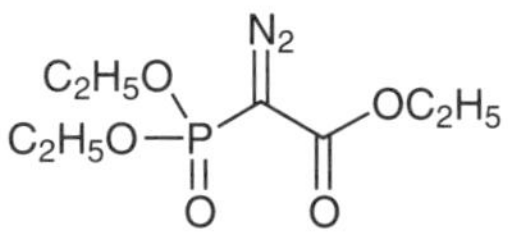

1.4 OTHER ROUTES TO DIAZOCARBONYL COMPOUNDS

Although the other routes to α-diazocarbonyls have diminished somewhat in usefulness since the introduction of the diazo transfer process, some of them remain as complementary to acylation and diazo transfer, and they still have uses in the synthesis of diazocarbonyl compounds. In this section, we will briefly review these classical methods including the Forster reaction, dehydrogenation of hydrazones, the Bamford–Stevens tosylhydrazone decomposition, and diazotization of amines.

The Forster reaction[172] involves oxime formation at the α-methylene position of a ketone followed by reaction with chloramine. This method has been employed extensively in the preparation of α-diazoketones from derivatives of in-

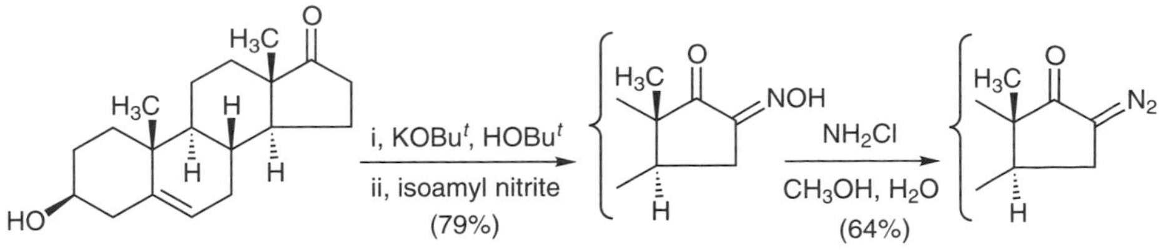

Scheme 11

danone and steroidal ketones. The method is also described for the entry in *Organic Synthesis* shown in Scheme 11.[173] Additional examples of application of the Forster reaction in synthesis are illustrated in **144–148**.[144,174–178]

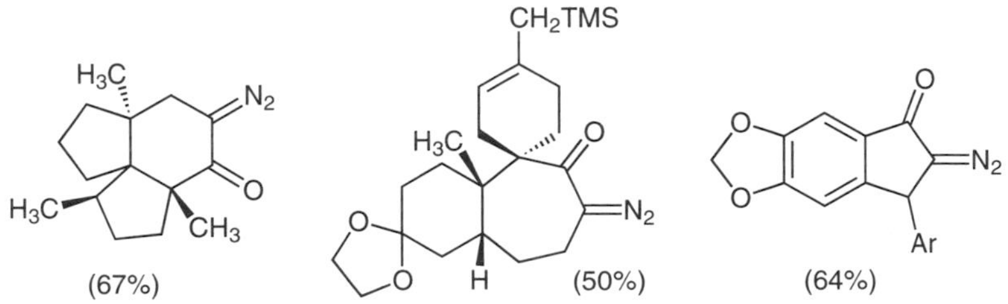

(67%)

144[174,175]

(50%)

145[176]

(64%)

146[177]

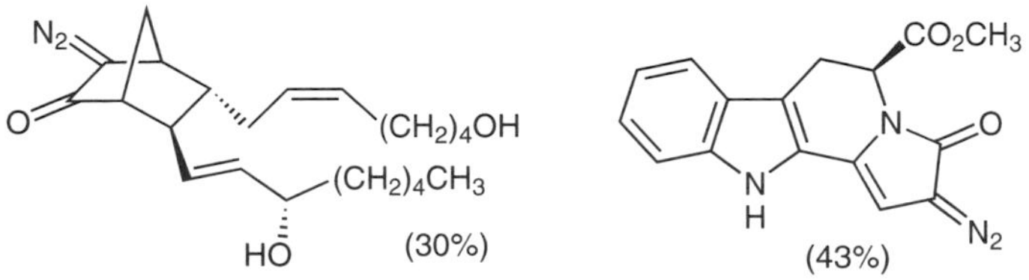

(30%)

147[144]

(43%)

148[178]

Dehydrogenation of hydrazones is one of the oldest methods for the preparation of diazo compounds. Nowdays, direct synthesis of α-diazocarbonyl compounds via this method is not often chosen, mainly due to the limitation of regioselective synthesis of the required hydrazone. Various oxidation agents have been used to convert hydrazones into the corresponding diazo compounds. One representative example is shown in eq. 29.[179] In addition, Holton and Shechter[180]

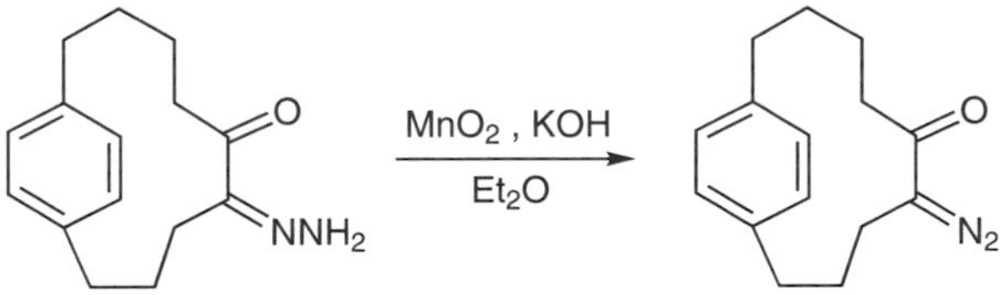

$$\text{(29)}$$

have claimed that lead tetraacetate in a basic environment is an efficient reagent for the preparation of sensitive diazo compounds by oxidation of hydrazones.

The Bamford–Stevens reaction,[181] to convert carbonyl compounds to diazo compounds by cleavage of the derived toluenesulfonyl hydrazones, is closely related to the dehydrogenation of alkylhydrazone discussed previously. Mono hydrazones of dicarbonyl compounds react with base at room temperature, affording the corresponding diazocarbonyl compounds. A typical example is shown in eq. 30.[182] This method still remains as a useful method for producing

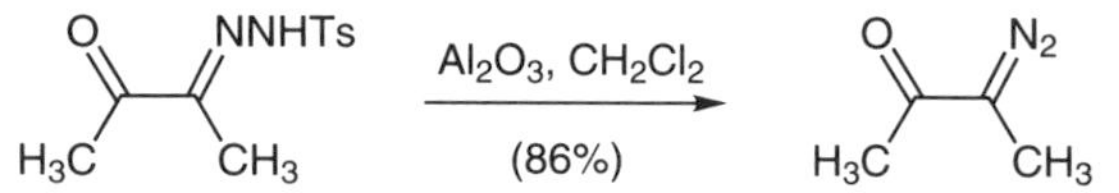

$$\underset{\substack{\text{H}_3\text{C}\quad\text{CH}_3}}{\text{O}\quad\text{NNHTs}} \xrightarrow[\text{(86\%)}]{\text{Al}_2\text{O}_3,\ \text{CH}_2\text{Cl}_2} \underset{\substack{\text{H}_3\text{C}\quad\text{CH}_3}}{\text{O}\quad\text{N}_2} \qquad (30)$$

cyclic diazoketones. Three additional examples of α-diazocarbonyl compounds synthesized via the Bamford–Stevens reaction are illustrated in **149**,[183] **150**,[142] and **151**.[184] Shechter and co-workers[185] have shown that the Bamford–Stevens

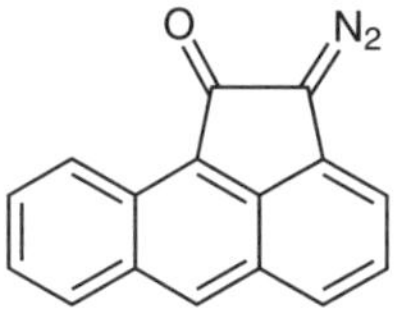

(60%)	(65–76%)	(77%)
149[183]	**150**[142]	**151**[184]

(87%)	(76%)
152[186]	**153**[186]

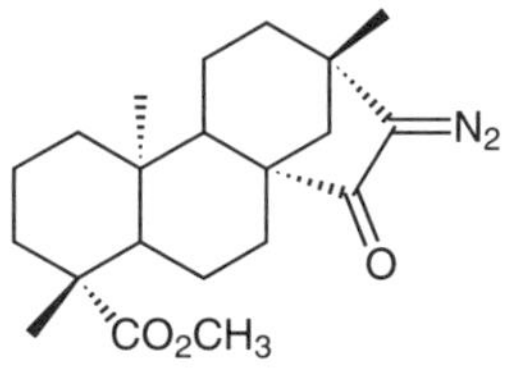

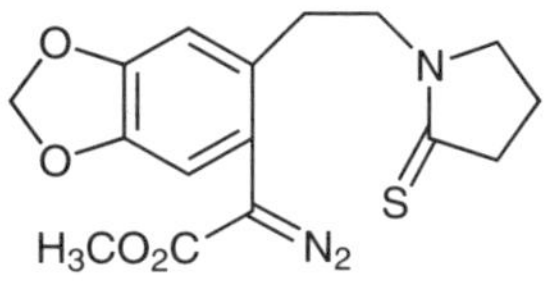

reaction can be carried out via a vacuum pyrolysis method. Two diazoesters obtained via this method were provided by Creary for *Organic Synthesis* (**152** and **153**).[186] Shi and Xu[187] have found a one-pot procedure of the Bamford–Stevens reaction that is suitable for the synthesis of 3-trifluoro-2-diazopropionate (**154**). Thus treatment of a mixture of ethyl trifluoropyruvate and tosylhydrazide in dichloromethane with pyridine and phosphorus oxychloride gave the diazoester **154** in 82% yield (eq. 31).

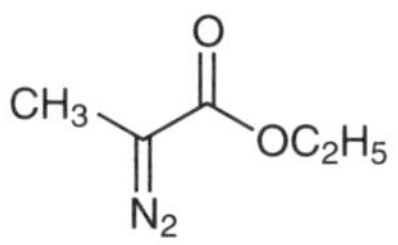

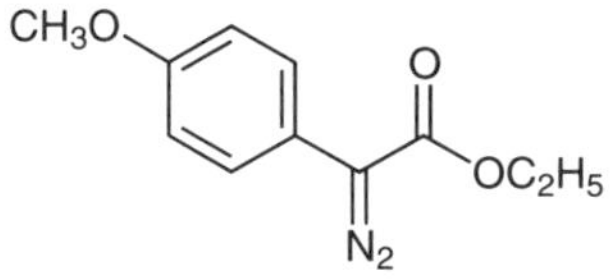

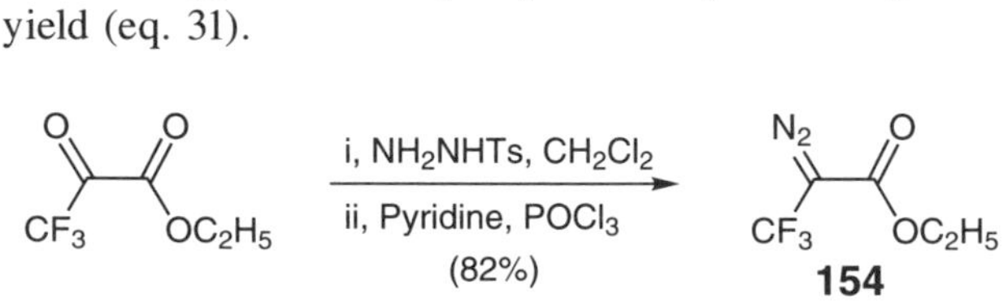

$$\underset{\substack{\text{CF}_3\qquad\text{OC}_2\text{H}_5}}{\text{O}\qquad\text{O}} \xrightarrow[\substack{\text{ii, Pyridine, POCl}_3\\(82\%)}]{\text{i, NH}_2\text{NHTs, CH}_2\text{Cl}_2} \underset{\substack{\text{CF}_3\qquad\text{OC}_2\text{H}_5\\\textbf{154}}}{\text{N}_2\quad\text{O}} \qquad (31)$$

House[188,189] has devised a very useful protocol for the preparation of α-diazo esters not readily accessible by diazotization (Scheme 12). Glyoxylic acid is first

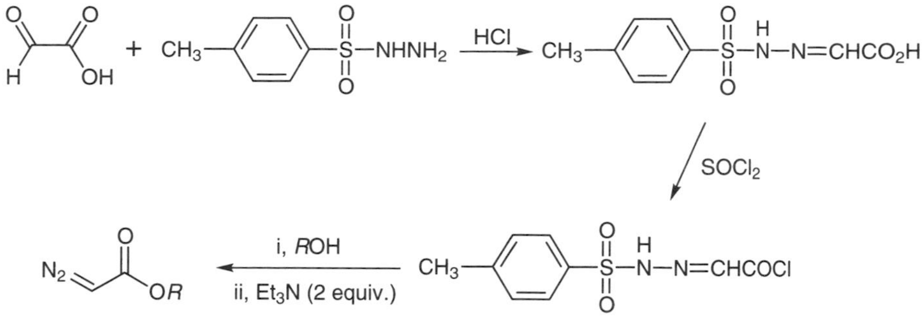

Scheme 12

converted into its tosylhydrazone, which is then treated with thionyl chloride to form the acyl chloride. The acyl chloride is combined with the appropriate alcohol to produce the ester hydrazone, and the process is completed by decomposition of the tosylhydrazone with triethylamine. House's method has been used for the synthesis of diazo esters **155**[190,191] and **156**,[192] which are used for photoaffin-

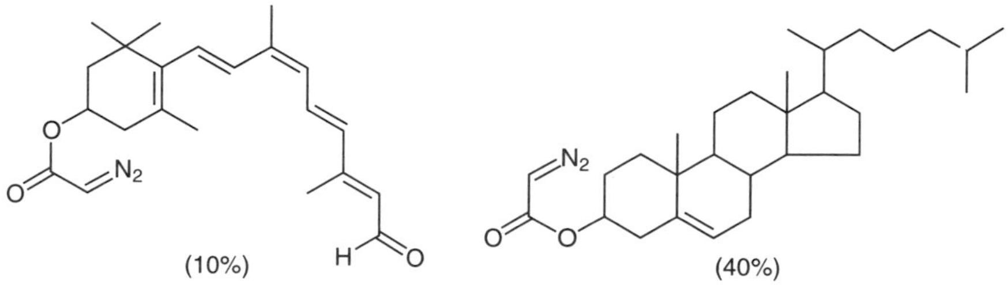

(10%)

155[190,191]

(40%)

156[192]

ity labeling. Typically, two equivalents of triethylamine are used in the final step of House's method. However, sulfinate species have been reported to contaminate the product.[193] Corey and Myers[193] subsequently made a modification in which one equivalent of triethylamine was replaced by the weaker base, *N,N*-dimethylaniline. This modification affords substantial improvement in yields, and the formation of undesired byproduct can be largely circumvented. An example is shown in eq. 32.[194] The Corey modification has been used to produce intermediate **158** in the total synthesis of antheridiogen-An **159** (Scheme 13).[195] Nakanishi and co-workers[196] have developed the glyoxylic acid 2,4,6-triisopropyl-

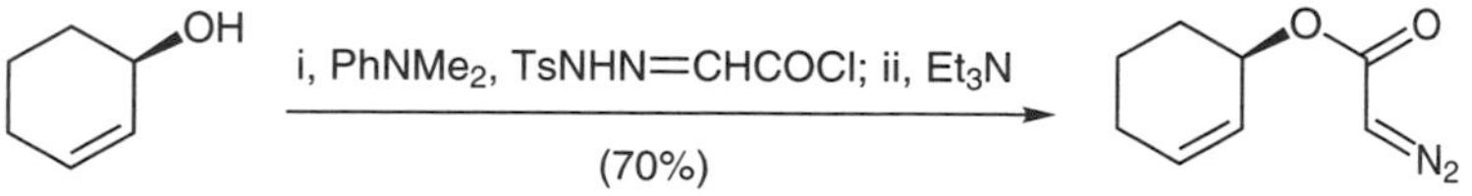

i, PhNMe$_2$, TsNHN=CHCOCl; ii, Et$_3$N

(70%)

(32)

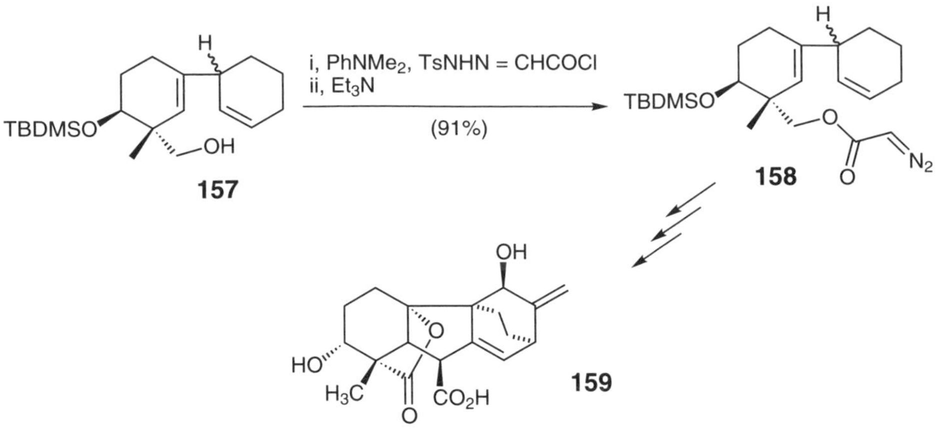

Scheme 13

phenylsulfonylhydrazone to replace the glyoxylic acid tosylhydrazone in House's protocol. This reagent is very stable and can be stored safely at room temperature for years. The tosylhydrazone derived from an α-keto acid is also suitable for the synthesis of α-substituted diazoesters. An example is shown in eq. 33.[197] In addition, a diazoester derived from a phenol can also be prepared via House's protocol, and an example is illustrated in eq. 34.[198]

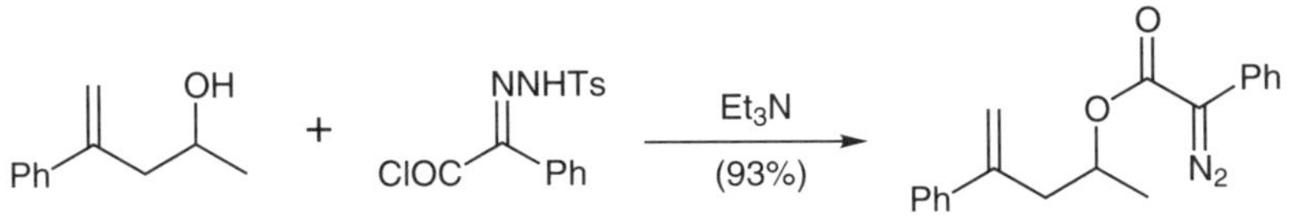

(33)

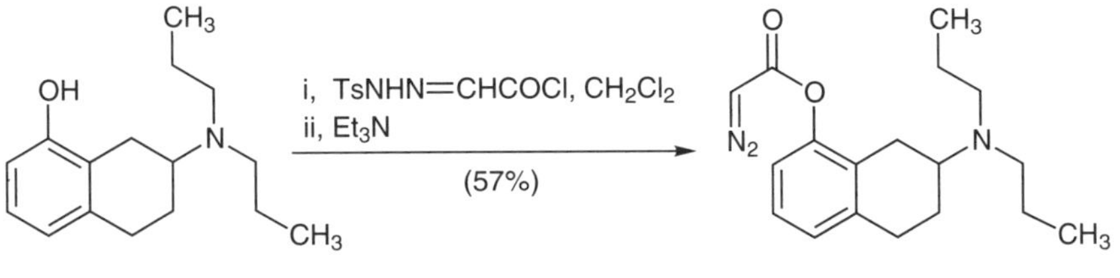

(34)

Diazoamides have also been synthesized via House's method. For example, diazoamide **160** was prepared in 79% yield, as shown in eq. 35.[199] An improved method for the preparation of diazoacetamides includes the cyclohexylcarbodi-

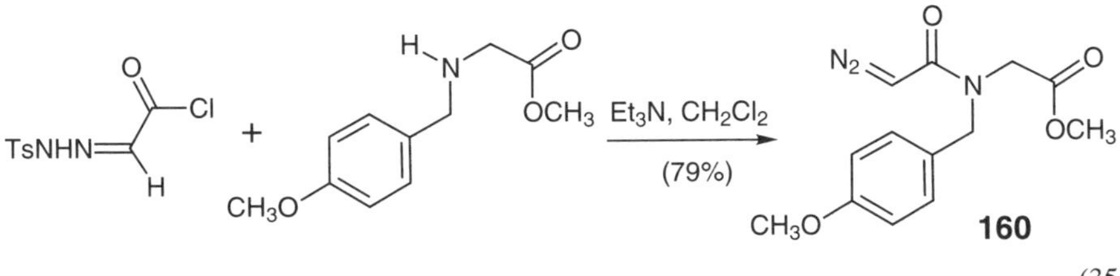

(35)

GLYOXYLIC ACID CHLORIDE *p*-TOLUENESULFONYLHYDRAZONE: PREPARATION OF DIAZOESTER 158[188,189,193,195]
(see Schemes 12 and 13)

Glyoxylic acid *p*-toluenesulfonylhydrazone.

A solution of 46.3 g (0.50 mole) of 80% glyoxylic acid in 500 mL of water is placed in a 1-L Erlenmeyer flask and warmed on a steam bath to approximately 60°C. This solution is then treated with a warm (approximately 60°C) solution of 93.1 g (0.50 mole) of *p*-toluenesulfonylhydrazide in 250 mL (0.63 mole) of aqueous 2.5 M hydrochloric acid. The resulting mixture is heated on a steam bath with continuous stirring until all the hydrazone, which initially separates as an oil, has solidified (about 5 min is required). After the reaction mixture has been allowed to cool to room temperature and then allowed to stand in a refrigerator overnight, the crude *p*-toluenesulfonylhydrazone is collected on a filter, washed with cold water, and allowed to dry for 2 days. The crude product (110–116 g, m.p. 145–149°C dec.) is dissolved in 400 mL of boiling ethyl acetate, filtered to remove any insoluble material, and then diluted with 800 mL of carbon tetrachloride and allowed to cool. After the mixture has been allowed to stand overnight in a refrigerator, the *p*-toluenesulfonylhydrazone is collected and washed with a cold mixture of ethyl acetate and carbon tetrachloride (1:2 by volume). The yield is 92.4–98.5 g (76–81%) of the hydrazone as white crystals, m.p. 148–154°C dec.

Glyoxylic acid chloride *p*-toluenesulfonylhydrazone.

To a suspension of 50.2 g (0.21 mole) of glyoxylic acid *p*-toluenesulfonylhydrazone in 250 mL of benzene is added 30 mL (49 g or 0.42 mole) of thionyl chloride. The reaction mixture is heated under reflux with stirring until vigorous gas evolution has ceased and most of the suspended solid has dissolved (about 1.5–2.5 h is required). The reaction mixture is then cooled immediately and filtered through a Celite mat on a sintered-glass funnel. After the filtrate has been concentrated to dryness under reduced pressure, the residual solid is mixed with 40–50 mL of anhydrous benzene, warmed, and the solid mass is broken up to give a fine suspension. This suspension is cooled and filtered with suction. The crystalline product is washed quickly with two portions of cold benzene to remove most of the residual colored impurities, and then the remaining crude acid chloride is transferred to a flask for recrystallization. The combined benzene filtrates from this initial washing procedure are concentrated under reduced pressure, and the washing procedure with benzene is repeated to give a second crop of the crude acid chloride, which is transferred to a flask for recrystallization (from benzene) (56–68% yield).

Diazoester 158.

Glyoxylic acid chloride *p*-toluenesulfonylhydrazone (15.2 g, 58.3 mmol) was added to an ice-cooled solution of dry alcohol **157** (10.51 g, 31.2 mmol) in 180 mL of dry methylene chloride under an argon atmosphere. Dimethylaniline (7.25 mL, 57.2 mmol) was added and the dark green solution was stirred for 15 min prior to injection of triethylamine (22 mL, 160 mmol). The resulting dark orange suspension was stirred 10 min at 0°C then concentrated in vacuo. Saturated aqueous citric acid (250 mL) and 10% ethyl acetate–hexanes were added and the layers were separated. The organic layer was washed with 250 mL citric acid solution and the combined aqueous layers were extracted with 100 mL 10% ethyl acetate-hexanes. This 100 mL extract was washed with an equal volume of citric acid solution, and the combined organic layers were dried over sodium sulfate. Concentration and flash chromatography (5% ethyl acetate–hexanes) provided **158** as a yellow syrup (11.44 g, 90.5%).

imide-mediated coupling of an amine with glyoxylic acid tosylhydrazone; a representative example is shown in eq. 36.[200]

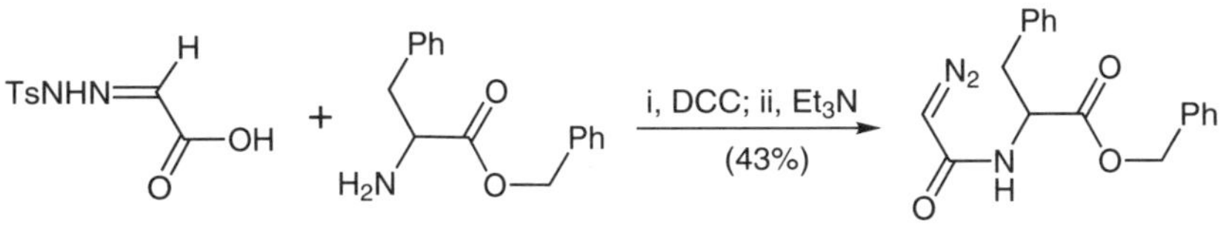

$$\text{(36)}$$

Diazotization remains the method of choice for production of ethyl diazoacetate from glycine ethyl ester (eq. 37).[1,2,201,202] Extension of the diazotization process to other amino acids such as alanine, phenylalanine, methionine, and lysine is also possible (eq. 38)[203] Although diazotization can be effected with sodium ni-

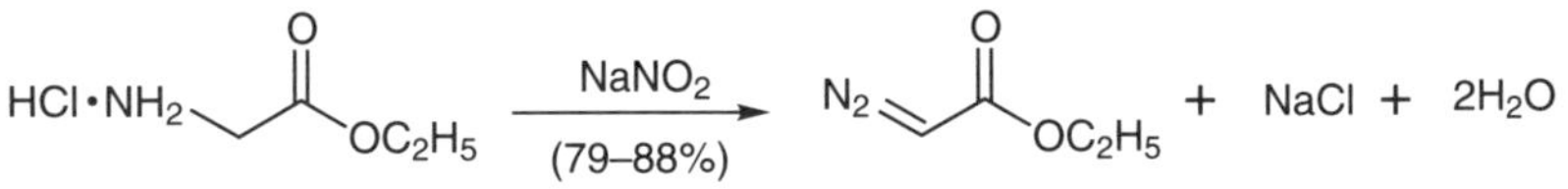

$$\text{(37)}$$

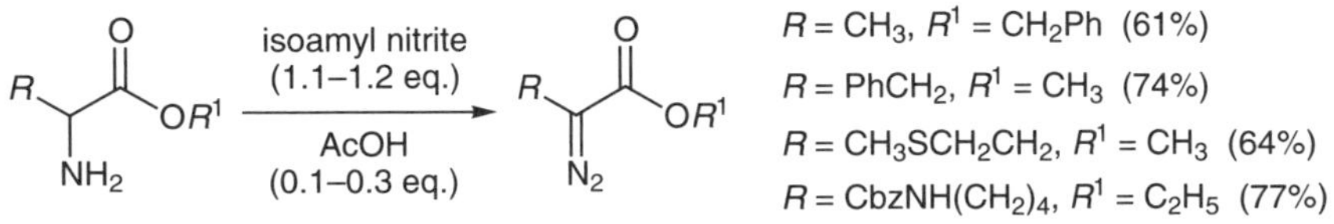

$R = CH_3, R^1 = CH_2Ph$ (61%)
$R = PhCH_2, R^1 = CH_3$ (74%)
$R = CH_3SCH_2CH_2, R^1 = CH_3$ (64%)
$R = CbzNH(CH_2)_4, R^1 = C_2H_5$ (77%)

$$\text{(38)}$$

trite in aqueous acid, isoamyl nitrite is the preferred agent in some cases. Thorsett[204] has applied this method in his synthesis of α-ketoesters by oxidation of the resulting α-diazoester intermediate. The antibiotic azaserine **162**, active against certain tumors, could be prepared by diazotizing the corresponding o-(glycyl)-N-(trifluoroacetyl)-L-serine **161**. Thus treatment of **161** with lithium nitrite in the presence of a catalytic amount of chloroacetic acid and removal of the trifluoroacetyl protecting group with acylase I afforded **162** in 78% overall yield (eq. 39).[205] Similarly, Nishimura and his collaborators[206] synthesized a new anti-

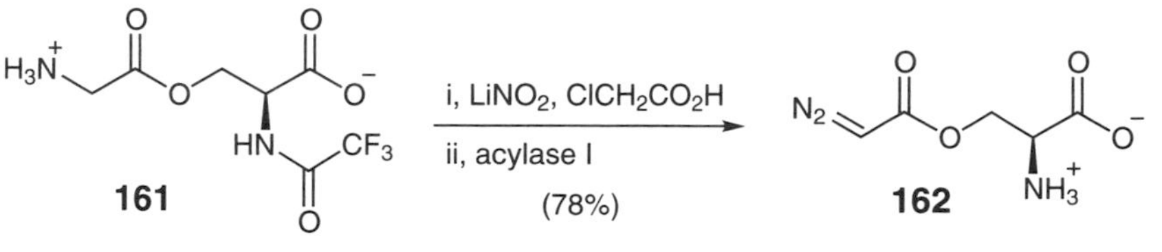

$$(39)$$

tumor antibiotic, *FR900840* (**163**), which was isolated from a culture broth of a strain of streptomyces, by diazotizing the corresponding amino acid (Scheme 14).

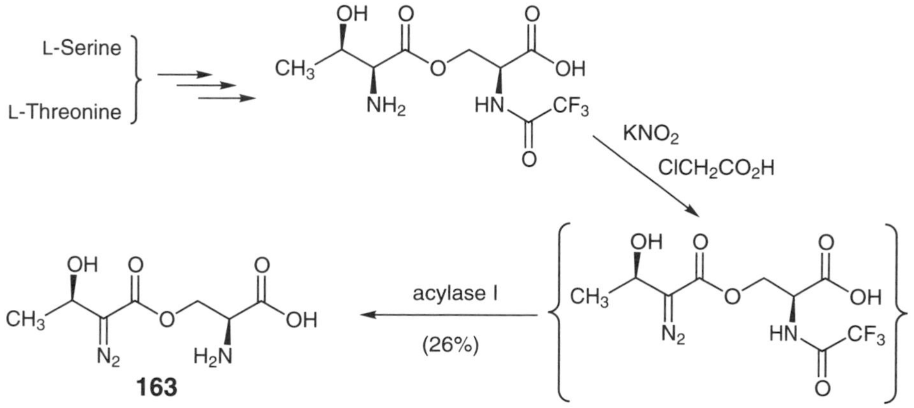

Scheme 14

Challis and Latif[207] have described the synthesis of diazopeptides by aprotic diazotization with dinitrogen tetroxide at $-40°C$. For example, **164** and **165** were synthesized by this method (eqs. 40 and 41):

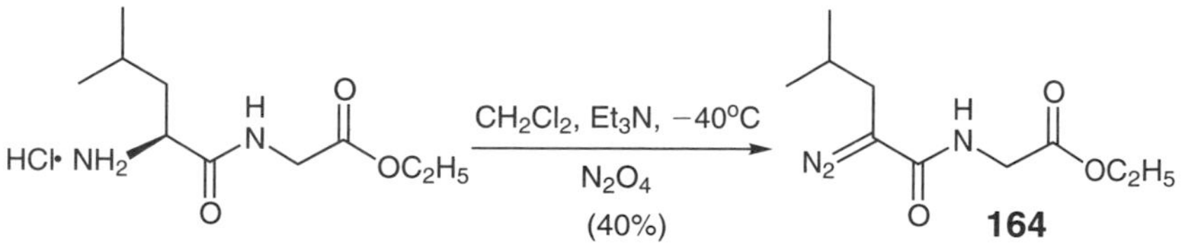

$$(40)$$

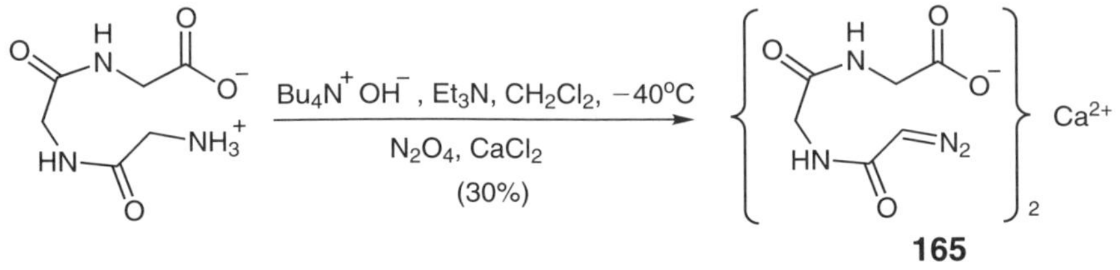

(41)

The synthetically versatile benzyl 6-diazopenicillanate **167** was first prepared from 6-aminopenicillanic acid ester **166** via direct diazotization in nitrous acid but was formed in a low yield (eq. 42).[208] However, a high yield of 6-diazopenicillanate can be obtained via a two-step process developed by Sheehan and co-workers.[209] Thus treatment of β,β,β-trichloroethyl phenylacetamidopenicillanate **168** with dinitrogen tetroxide in dichloromethane gave the corresponding nitroso adduct, which was converted into the 6-diazopenicillanate **169** in 72% yield on treatment with pyridine (eq. 43).

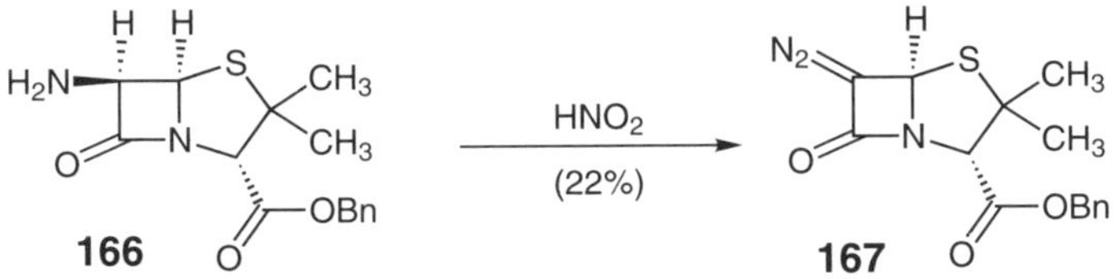

(42)

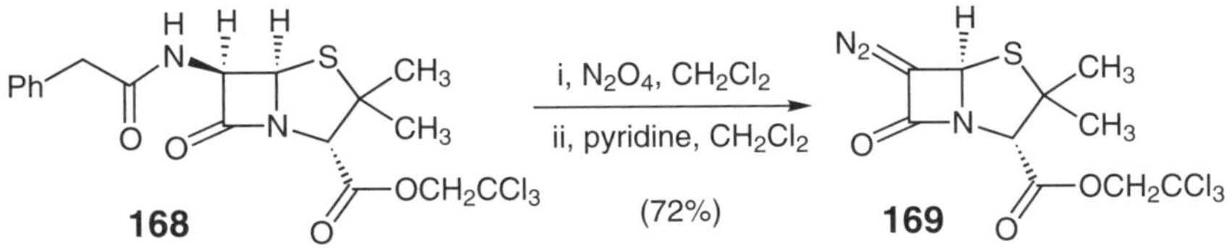

(43)

1.5 CHEMICAL MODIFICATION OF DIAZOCARBONYL COMPOUNDS

Finally, there are a number of examples where a preformed diazocarbonyl function is transferred to another molecule, creating a new diazocarbonyl precursor. Bestmann and Soliman[210] showed that diazoacetyl chloride (**170**), prepared from the reaction of phosgene with diazomethane, is an excellent diazoacetylation reagent. Reaction of this reagent (**170**) with nucleophiles in the presence of triethylamine gave the corresponding diazocarbonyl adducts. Three representative examples are shown in eq. 44. In addition, diazoacetyl chloride **170** is also readily accessible by treatment of diazoacetic acid with tetramethyl-α-chloroenamine.[211]

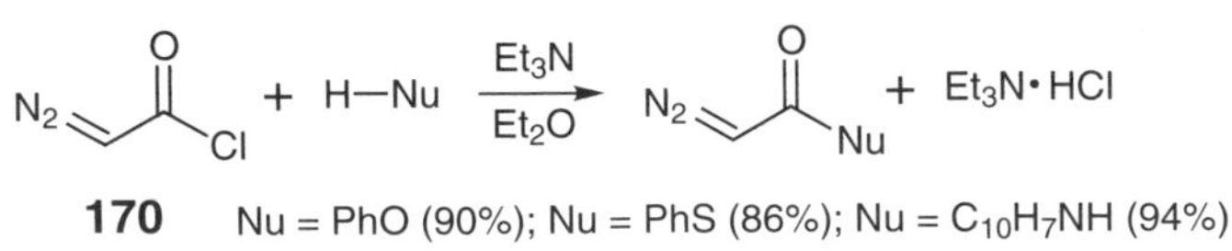

(44)

170   Nu = PhO (90%); Nu = PhS (86%); Nu = $C_{10}H_7NH$ (94%)

Badet and co-workers[212] claimed that the easily obtained and very stable suc-cinimidyl diazoacetate **171** can be used for direct diazoacetylation (see pg. 263) of aromatic or aliphatic amines, phenols, thiophenol and peptides under mild neutral to basic conditions (eq. 45). Therefore, this reagent is particularly suitable for the synthesis of acid-sensitive diazoacetyl compounds. Some diazoacetylated compounds are illustrated by the following examples (**172, 173, 174, 175,** and **176**). An alternative, reproducible procedure for the synthesis of **171** has been de-scribed by Doyle.[213]

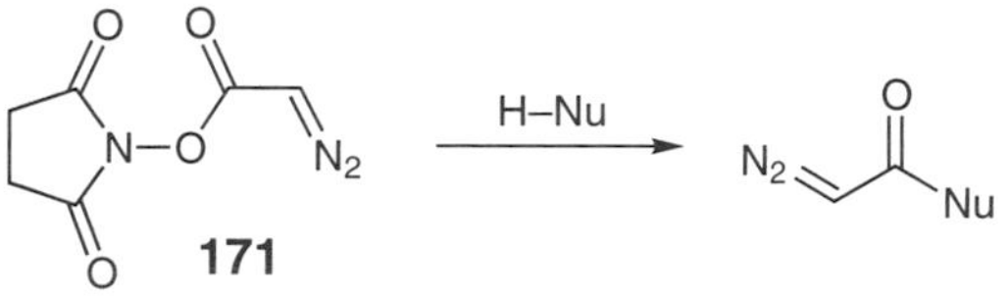

$$\text{(45)}$$

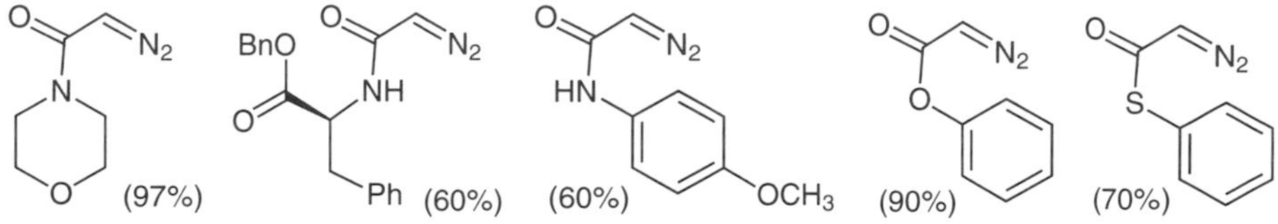

171

172 (97%)   **173** (60%)   **174** (60%)   **175** (90%)   **176** (70%)

Various diazomalonate derivatives can also be synthesized via a diazoacyla-tion protocol. Kido and co-workers[214] have prepared ethyl hydrogen diazoma-lonate (**177**) as a reagent for the synthesis of alkyl ethyl diazomalonates. Treatment of **177** with an appropriate alcohol in dichloromethane and in the pres-ence of 1,3-dicyclohexylcarbodiimide (DCC) and 4-dimethylaminopyridine (DMAP) furnished the corresponding alkyl ethyl diazomalonates (eq. 46). Simi-

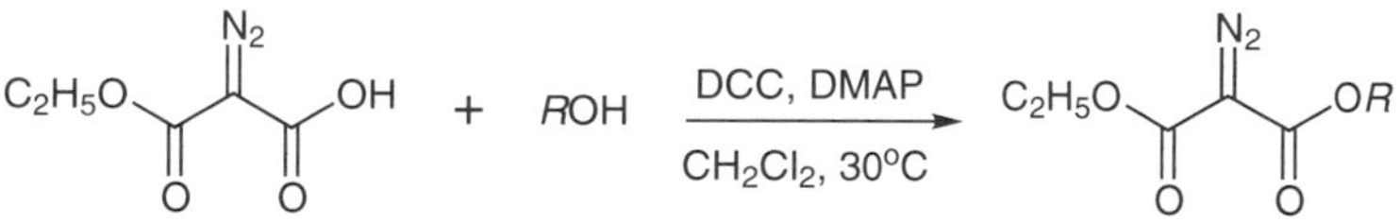

$$C_2H_5O \quad \text{OH} \quad + \quad ROH \xrightarrow[\text{CH}_2\text{Cl}_2, 30°C]{\text{DCC, DMAP}} C_2H_5O \quad \text{OR} \quad \text{(46)}$$

177

larly, Padwa and co-workers[215] have advocated the use of ethyl 2-diazomalonyl chloride (**178**) as an efficient diazoacylating reagent. Ethyl 2-diazomalonyl chlo-ride (**178**) can be easily obtained from ethyl diazoacetate and triphosgene and readily reacts with nucleophilic reagent to give a variety of α-diazocarbonyl compounds (Scheme 15). In addition, *tert*-butyl diazomalonyl chloride (**179**) is readily accessible via a similar method. Representative examples of diazocar-

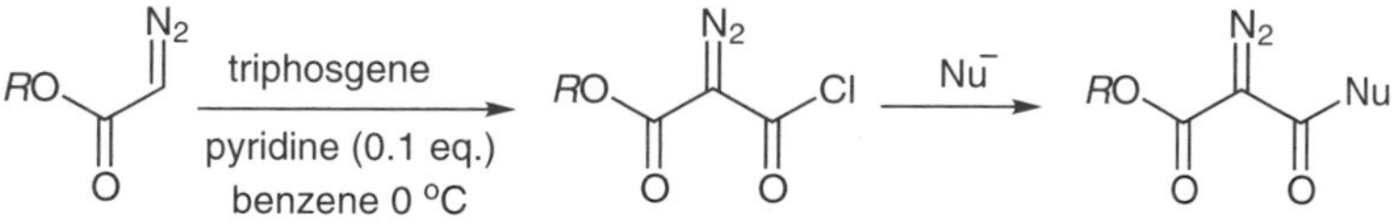

$$RO \xrightarrow[\substack{\text{pyridine (0.1 eq.)} \\ \text{benzene 0 °C}}]{\text{triphosgene}} RO \quad \text{Cl} \xrightarrow{Nu^-} RO \quad Nu$$

178: $R = C_2H_5$ **179** : $R = Bu^t$

Scheme 15

bonyl compounds prepared from reagents **177**, **178**, and **179** are shown in **180**,[214] **181**,[214] **182**,[215] **183**,[215] **184**,[215] **185**,[215] **186**,[215] and **187**.[215]

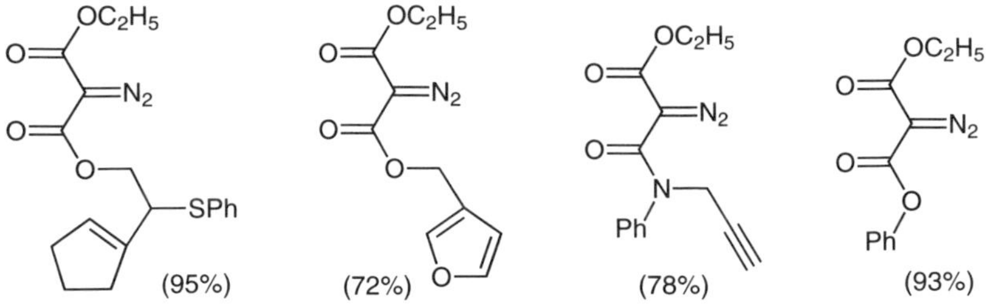

180[214] **181**[214] **182**[215] **183**[215]

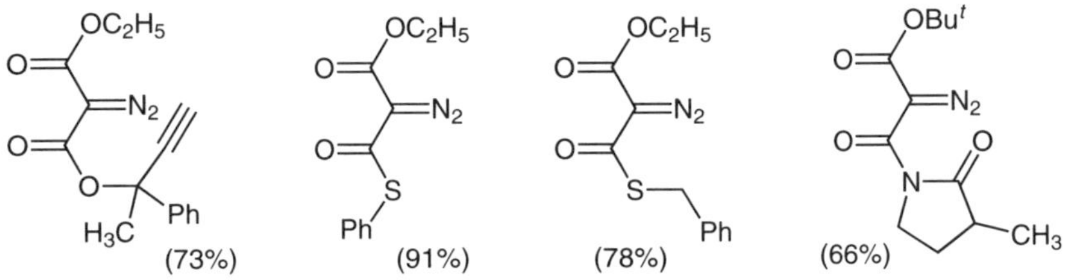

184[215] **185**[215] **186**[215] **187**[215]

Diazoacylation with 2-diazomalonyl chloride has been used by Wydila and Thornton[216] in their synthesis of the galactocerebroside photolabeling reagent **188**. Treatment of diazoester **189** with excess phosgene furnished malonyl chloride **190**. Selective acylation of the amino residue of the psychosine **191** completed the synthesis (Scheme 16).

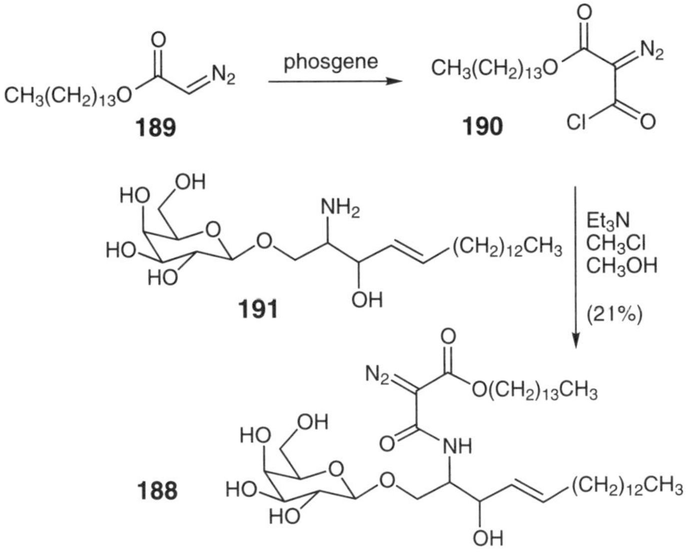

Scheme 16

The silyl enol ethers of α-diazoacetoacetates have been used extensively in the synthesis of carbapenem antibiotics. Karady and co-workers[217] have reported that the silver-mediated coupling reaction between 1-(t-butyldimethylsilyl)-4-chloro-azetidin-2-one **192** and diazo synthon **193** derived from the silylation of benzyl 2-diazoacetoacetate furnished the thienamycin precursor, α-diazo-β-ketoester **194** in 70% yield (eq. 47). The use of 4-acetoxyazetidin-2-one **195** as the alkylat-

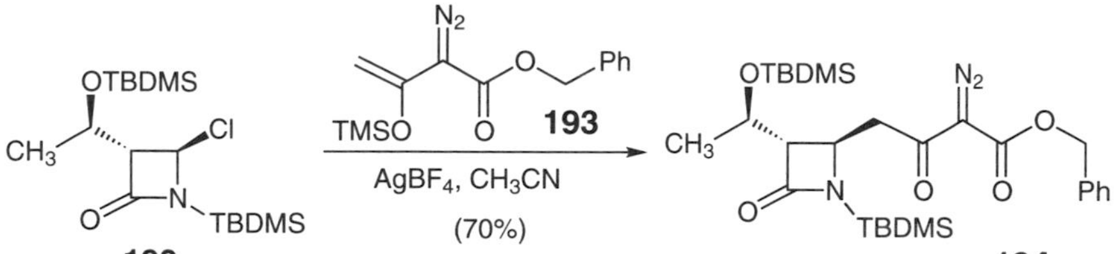

(47)

ing agent has resulted in the synthesis of a variety of carbapenem precursors. In most cases, zinc chloride is the coupling reagent of choice, and a representative example is illustrated in eq. 48.[218-223] In addition, the formation of the diazo syn-

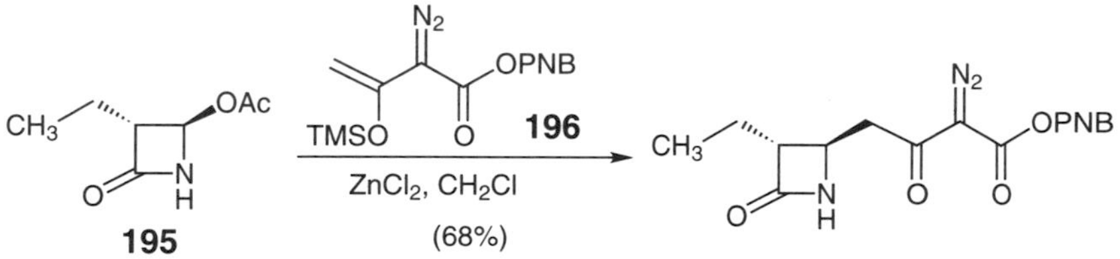

(48)

thon and the coupling reaction can be performed as an efficient one-pot procedure.[224] Thus treatment of a dichloromethane solution of 4-acetoxyazetidin-2-one **197** and 4-nitrobenzyl 2-diazo-3-oxobutanoate **198** with 2.6 equiv. of triethylamine and 2.4 equiv. of trimethylsilyltrifluoromethanesulfonate (TMSOTf) for 15 min at 0°C, rising to room temperature, followed by an additional 0.3 equiv. of TMSOTf smoothly within 15 min provided the α-diazo-β-ketoester **199** in 64% yield (eq. 49).[224]

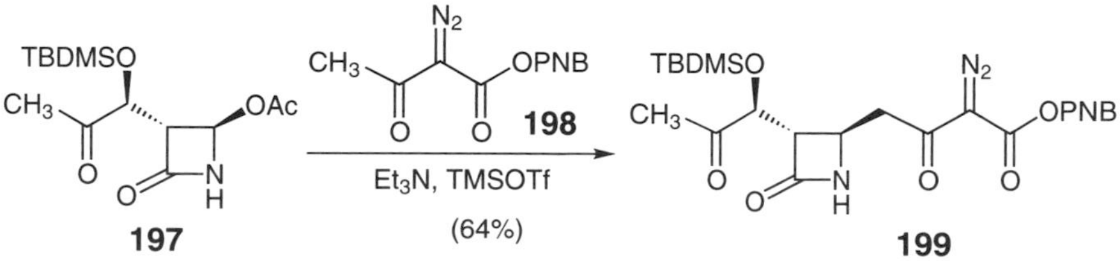

(49)

It has been known that α-diazo-β-dicarbonyl compounds undergo base catalyzed acyl cleavage to produce monocarbonyl diazo compounds.[75,225] In fact, *tert*-

butyl diazoacetate has been prepared in large quantities using this strategy from *tert*-butyl 2-diazo-3-oxobutanoate.[69] One general procedure[226] for the preparation of diazoacetates is now available involving the transformation of the appropriate alcohols or phenols into the corresponding acetoacetic esters by reaction with diketene[227] or the diketene equivalent, 2,2,6-trimethyl-4*H*-1,3-dioxin-4-one.[228] Subsequent diazo transfer followed by base-induced deacylation of the intermediate α-diazoacetoacetic ester then gave the corresponding diazoacetate. Numerous diazoacetates have been synthesized via this route and a representative example is illustrated in eq. 50 (see also pg. 173).[227] Kurth and co-workers[229] claimed that

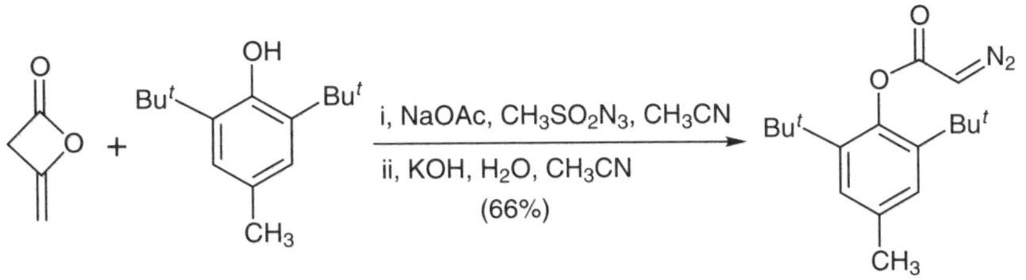

$$(50)$$

the formation of diazoacetate **201** can be accomplished in a one-pot procedure with no isolation of intermediates by sequential addition of 2,2,6-trimethyl-4*H*-1,3-dioxin-4-one, *p*-(*n*-acetylamido)benzenesulfonyl azide/triethylamine/acetonitrile, and pyrrolidine to a xylene solution of alcohol **200** (eq. 51). *N*-Aryl-diazoacetamides have been prepared from secondary amines by the same reaction principle.[230]

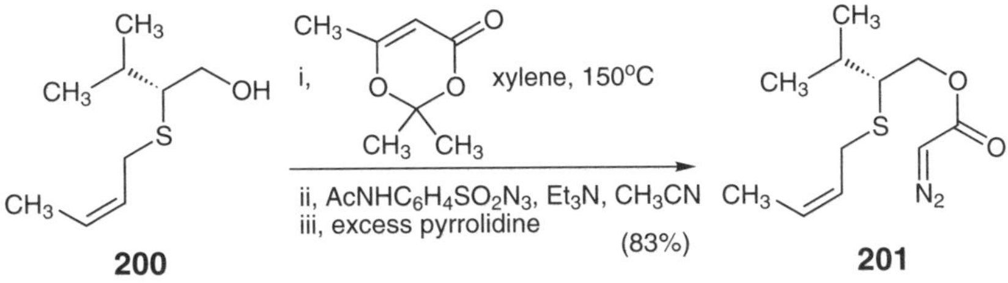

$$(51)$$

α-Diazo-β-dicarbonyl compounds can be obtained from the monosubstituted diazocarbonyl compounds, usually α-diazoesters, by acylation. An example in which the acid chloride **202** is employed as the acylation reagent is illustrated in eq. 52.[231] Acyl isocyanates have also been used as the acylation reagent in this type of transformation.[232,233] An example is shown in eq. 53.[232]

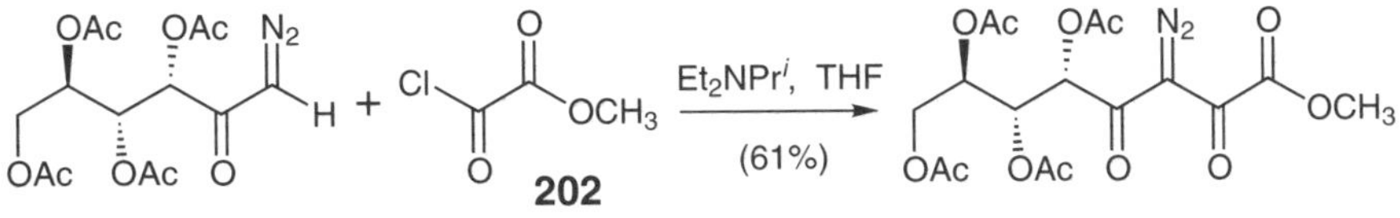

$$(52)$$

$$\text{PhCON=C=O} + \text{N}_2\text{CHCOOC}_2\text{H}_5 \xrightarrow[\text{reflux}]{\text{xylene}} \text{PhCONHCOC(N}_2)\text{COOC}_2\text{H}_5$$

(59%)

(53)

Another aspect of diazocarbonyl synthesis that is receiving increasing attention, mainly through the work of Regitz's group,[65] is the chemical modification of the diazo carbon with retention of the diazo function. Numerous examples of substitution reactions at the diazocarbonyl carbon atom in terminal, acyclic substrates bear witness to the stability of the diazo group, even under quite drastic conditions. In general, the hydrogen atom can be substituted by electrophilic reagents. In addition to halogenation, metallation, nitration, and alkylation processes are possible, leading to new substituted diazocarbonyl compounds. Regitz reviewed these reactions in detail in 1985[234]; just a few representative examples are summarized in Table 1.5. Some uses of the metallated derivatives are discussed in Chapter 10.

TABLE 1.5   Substitution Reaction of α-Diazocarbonyl Compounds

Reaction	Substrate	Conditions	Product	Yield %	Ref.
Metallation	$PhCOCHN_2$	HgO, 20°C	$Hg(CN_2COPh)_2$	97	235
	$N_2CHCOOEt$	n-BuLi, −100°C	$LiCN_2COOEt$	—	236
	$N_2CHCOOEt$	Ag_2O, ≤0°C	$AgCN_2COOEt$	—	237
Halogenation	$Hg(CN_2COOEt)_2$	SO_2Cl_2, −30°C	$ClCN_2CO_2Et$	30	238
	$Hg(CN_2COOEt)_2$	Br_2, ether–THF −100°C	$BrCN_2CO_2Et$	80~90	238
	$Hg(CN_2COOEt)_2$	I_2, 0°C	ICN_2CO_2Et	70~90	238, 239
Nitration	$N_2CHCOOEt$	N_2O_5, CCl_4, −30°C	$O_2NCN_2CO_2Et$	—	240
Alkylation	$AgCN_2CO_2Et$	$H_2C=CH-CH_2-I$ ether, 0°C	$H_2C=CHCH_2C(N_2)CO_2Et$	66	237

1.6   CONCLUSION: SAFETY AND HANDLING OF DIAZOCARBONYL COMPOUNDS

The successful exploitation in organic synthesis of the unique reactivity of α-diazocarbonyl compounds rests crucially on their availability and ease of handling. Since the first synthesis of ethyl diazoactate in 1883, methodology has developed on a broad front and there are now available well-tested procedures for preparing all the main types of diazocarbonyl compounds. It is our hope that the information in Chapter 1 will provide a useful starting point for anyone interested in the use of diazocarbonyl compounds in synthesis, especially those experiencing reluctance due to unfamiliarity in handling these compounds. Although some representative procedures for the synthesis of diazocarbonyl and related compounds have been included in this chapter, when planning experimental work the reader should also consult the original literature for extensive information concerning the detailed experimental techniques including safety, handling, and storage.

Diazo compounds as a class are known to be toxic, and, as with all chemical substances, both known and novel, due care must be taken in preparations. We have repeatedly stressed the precautions required for handling diazomethane, its precursors, and the reagents used in diazo transfer, particularly the sulfonyl azides.

There is a perception that diazocarbonyl compounds in general are thermally unstable and that their properties are comparable to those of diazomethane. In fact, the presence of the carbonyl group in diazocarbonyl compounds has a major *stabilizing* effect on the thermal lability of these compounds. Unfortunately, while there is a very large number of known diazocarbonyl compounds, there are few quantitative data available with which to compare thermal stabilities as measured by the kinetics of thermal decomposition. However, the following data do show a very large difference between diazoalkanes and diazocarbonyl compounds; for example, the half-life for thermal decomposition of 2-diazopropane in diethyl ether at 0°C is approximately 3 h. In contrast, the half-life for the decomposition of ethyl diazoacetate at 100°C in mesitylene is 109 h and under the same conditions and temperature, the half-life for decomposition of 1-diazo-2-cyclohexanone is 32 min. Furthermore, ethyl diazoacetate can be purified by distillation at atmospheric pressure (b.p. 140–141°C).

Diazo compounds are unstable towards strong protonic and Lewis acids, and they can also be decomposed by prolonged exposure to light or by contact with some transition metal salts. Ideally, diazocarbonyl compounds should be stored in the dark at 0°C degrees. Although in the course of numerous experiments involving the controlled decomposition of diazocarbonyl compounds with metal or acid catalysts, we have never experienced any runaway reactions or explosions, nevertheless, we urge that all reactions involving the preparation and use of diazo carbonyl compounds should be routinely carried out behind a safety shield in a well-ventilated fumehood or cupboard.

1.7 REFERENCES

1   Curtius, T., "Ueber die Einwirkung von Salpetriger Säure auf Salzsauren Glycocolläther," *Ber.* **1883**, *16*, 2230–31.

2   Curtius, T., "Ueber Diazoessigsäure und ihre Derivate," *J. Prakt. Chem.* **1888**, *38*, 396–440.

3   Arndt, F.; Eistert, B.; and Partale, W., "Diazo-methan und O-Nitroverbindungen, II.: N-Oxyisatin aus O-Nitro-Benzoylchlorid," *Ber.* **1927**, *60B*, 1364–70.

4   Arndt, F.; and Amende, J., "Synthesen mit Diazo-methan, V.: Über die Reaktion der Säurechloride mit Diazo-methan," *Ber.* **1928**, *61B*, 1122–24.

5   Arndt, F.; Eistert, B.; and Amende, J., "Nachträge zu den 'Synthesen mit Diazo-methan,'" *Ber.* **1928**, *61B*, 1949–53.

6   Bradley, W.; and Robinson, R., "The Interaction of Benzoyl Chloride and Diazomethane Together with a Discussion of the Reactions of the Diazenes," *J. Chem. Soc.* **1928**, 1310-18.

7   Hook, J. M.; Mander, L. N.; and Urech, R., "Studies on Gibberellin Synthesis: The Total Synthesis of Gibberellic Acid from Hydrofluorenone Intermediates," *J. Org. Chem.* **1984**, *49*, 3250–60.

8   Evans, D. A.; Miller, S. J.; and Ennis, M. D., "Asymmetric Synthesis of the Benzoquinoid Ansamycin Antitumor Antibiotics: Total Synthesis of (+)-Macbecin," *J. Org. Chem.* **1993**, *58*, 471–85.

9   Newman, M. S.; and Beal, P., III, "An Improved Method for the Preparation of Aromatic Diazoketones," *J. Am. Chem. Soc.* **1949**, *71*, 1506–07.

10   Berenbom, M.; and Fones, W. S., "An Improved Method for the Preparation of Diazoacetophenone," *J. Am. Chem. Soc.* **1949**, *71*, 1629–29.

11   Bridson, J. N.; and Hooz, J., "Diazoacetophenone," *Org. Synth.* **1973**, *53*, 35–37.

12   Scott, L. T.; and Minton, M. A., "Aliphatic Diazo Ketones. A Modified Synthesis Requiring Minimal Diazomethane," *J. Org. Chem.* **1977**, *42*, 3757–58.

13   For a description of the safety hazards associated with diazomethane, see: Black, T. W., "The Preparation and Reactions of Diazomethane," *Aldrichimica Acta*, **1983**, *16*, 3–10.

14   Arndt, F., "Diazomethane," *Org. Synth. Coll. Vol. 2*, Blatt, A. H., Ed.; John Wiley & Sons: New York, **1943**, 165–67.

15   de Boer, T. J.; and Backer, H. J., "Diazomethane," *Org. Synth. Coll. Vol. 4*, Rabjohn, N., Ed.; John Wiley & Sons: New York, **1963**, 250–53.

16   Moore, J. A.; and Reed, D. E., "Diazomethane," *Org. Synth. Coll. Vol. 5*, Baumgarten, H. E., Ed.; John Wiley & Sons: New York, **1973**, 351–55.

17   McKay, A. F., "A New Method of Preparation of Diazomethane," *J. Am. Chem. Soc.* **1948**, *70*, 1974–75.

18   Gassman, P. G.; and Greenlee, W. J., "Dideuteriodiazomethane," *Org. Synth.* **1973**, *53*, 38–43.

19   Aldrich Chemical Co., Milwaukee, WI: Technical Information Bulletin No. AL-180. Also see: Fales, H. M.; Jaouni, T. M.; and Babashak, J. F., "Simple Device for Preparing Ethereal Diazomethane without Resorting to Codistillation," **1973**, *45*, 2302–3.; Ngan, F.; and Toofan, M., "Modification of Preparation of Diazomethane for Methyl Esterification of Environmental Samples Analysis by Gas Chromatography," **1991**, *29*, 8–10; Hudlicky, M., "An Improved Apparatus for the Laboratory Preparation of Diazomethane," **1980**, *45*, 5377–78.

20   Lombardi, P., "A Rapid, Safe and Convenient Procedure for the Preparation and Use of Diazomethane," *Chem. & Industry* **1990**, 708.

21   Moss, S., "Diazomethane Explosion," *Chem. & Industry* **1994**, 122 & 133.

22   Shioiri, T.; Aoyama, T.; and Mori, S., "Trimethylsilyldiazomethane," *Org. Synth.* **1990**, *68*, 1–7.

23   Aoyama, T.; and Shioiri, T., "New Methods and Reagents in Organic Synthesis. 8. Trimethylsilyldiazomethane. A New, Stable, and Safe Reagent for the Classical Arndt–Eistert Synthesis," *Tetrahedron Lett.* **1980**, *21*, 4461–62.

24   Aoyama, T.; and Shioiri, T., "New Methods and Reagents in Organic Synthesis. 17. Trimethylsilyldiazomethane (TMSCHN$_2$) as a Stable and Safe Substitute for

Hazardous Diazomethane. Its Application to the Arndt–Eistert Synthesis," *Chem. Pharm. Bull.* **1981**, *29*, 3249–55.

25   Pettit, G. R.; and Nelson, P. S., "Synthesis of Azotomycin Synthesis of Antineoplastic Agents. 114.," *J. Org. Chem.* **1986**, *51*, 1282–86.

26   DeWald, H. A.; and Moore, A. M., "6-Diazo-5-oxo-L-norleucine, a New Tumor-Inhibitory Substance. Preparation of L-, D- and DL-Forms," *J. Am. Chem. Soc.* **1958**, *80*, 3941–45.

27   Weygand, F.; Bestmann, H. J.; and Klieger, E., "Synthese des 6-Diazo-5-oxo-L-norleucins und der 7-Diazo-6-oxo-2-L-amino-önanthsäure," *Chem. Ber.* **1958**, *91*, 1037–40.

28   Pettit, G. R.; and Nelson, P. S., "Synthesis of the *Streptomyces-Ambofaciens* Antineoplastic Constituent 6-Diazo-5-oxo-L-norleucine," *J. Org. Chem.* **1983**, *48*, 741–44.

29   Scott, L. T.; and Sumpter, C. A., "Diazo Ketone Cyclization onto a Benzene Ring: 3,4-Dihydro-1(2*H*)-Azulenone," *Org. Synth.* **1990**, *69*, 180–7.

30   Bradley, W.; and Robinson, R., "The Action of Diazomethane on Benzoic and Succinic Anhydrides, and a Reply to Malkin and Nierenstein," *J. Am. Chem. Soc.* **1930**, *52*, 1558–65.

31   Weygand, F.; and Bestmann, H. J., "Neuere Präparative Methoden der Organischen Chemie III; Synthesen unter Verwendung von Diazoketonen," *Angew. Chem.* **1960**, *72*, 535–54.

32   Hodson, D.; Holt, G.; and Wall, D. K., "Diazoketones from the Interaction of Diazoalkanes with Carboxylic Acid–Dicyclohexyl Carbodiimide Mixtures," *J. Chem. Soc. (C)* **1970**, 971–73.

33   Penke, B.; Czombos, J.; Baláspiri, L.; Petres, J.; and Kovács, K., "Synthese von Diazoketonen aus Acylaminosäuren unter Verwendung von gemischten Anhydriden bzw. N,N'-Dicyclohexyl-carbodiimid," *Helv. Chim. Acta* **1970**, *53*, 1057–61.

34   von Horner, L.; and Schwarz, H., "Durch Licht und Radikale Ausgelöste reduktive Eliminierung von Stickstoff aus Benzoyl-Benzaziden und Diazoacetophenonen in Isopropylalkohol," *Liebigs Ann. Chem.* **1971**, *747*, 21–38.

35   Bhati, A., "Reaction of Diazomethane with Some Anhydrides of o-Dicarboxylic Acids," *J. Org. Chem.* **1962**, *27*, 1183–86.

36   Hooz, J.; and Morrison, G. F., "Diazoacetaldehyde," *Org. Prep. Proced. Int.* **1971**, *3*, 227-30.

37   Tarbell, D. S.; and Price, J. A., "Use of Mixed Carboxylic–Carbonic Anhydrides for Acylations on Carbon and Oxygen," *J. Org. Chem.* **1957**, *22*, 245–50.

38   Gordon, E. M.; Godfrey, J. D.; Delaney, N. G.; Asaad, M. M.; von Langen, D.; and Cushman, D. W., "Design of Novel Inhibitors of Aminopeptidases. Synthesis of Peptide-Derived Diamino Thiols and Sulfur Replacement Analogues of Bestatin," *J. Med. Chem.* **1988**, *31*, 2199-211.

39   Ananda, G. D. S.; Steele, J.; and Stoodley, R. J., "Studies Related to Penicillins. Part 24. A Novel Thiazolidine Ring Enlargement of Penam Dioxides," *J. Chem. Soc., Perkin Trans. I* **1988**, 1765–71.

40   Clinch, K.; Marquez, C. J.; Parrott, M.J.; and Ramage, R., "Synthesis of Substituted Tetrahydropyridines and *m*-Hydroxybenzoic Acids," *Tetrahedron* **1989**, *45*, 239–58.

41 Harbeson, S. L.; and Rich, D. H., "Inhibition of Aminopeptidases by Peptides Containing Ketomethylene and Hydroxyethylene Amide Bond Replacements," *J. Med. Chem.* **1989**, *32*, 1378–92.

42 Zwanenburg, B.; and Thijs, L., "Synthesis and Reactions of α,β-Epoxy Diazomethyl Ketones," *Tetrahedron Lett.* **1974**, 2459–62.

43 Thijs, L.; Smeets, F. L. M.; Cillissen, P. J. M.; Harmsen, J.; and Zwanenburg, B., "Synthesis of α,β-Epoxy Diazomethyl Ketones," *Tetrahedron* **1980**, *36*, 2141–43.

44 Ye, T.; and McKervey, M. A., "Synthesis of Chiral *N*-Protected α-Amino-β-Diketones from α-Diazoketones Derived from Natural Amino Acids," *Tetrahedron* **1992**, *48*, 8007–22.

45 Podlech, J.; and Seebach, D., "On the Preparation of β-Amino Acids from α-Amino Acids Using the Arndt–Eistert Reaction: Scope, Limitations and Stereoselectivity. Application to Carbohydrate Peptidation. Stereoselective α-Alkylations of Some β-Amino Acids," *Liebigs Ann.* **1995**, 1217–28.

46 Pettit, G. R.; and Nelson, P. S., "Antineoplastic Agents. 115. Synthesis of Amino Acid Diazoketones," *Can. J. Chem.* **1986**, *64*, 2097–102.

47 Diazoethane can be generated from ENNG by a similar procedure using the MNNG diazomethane-generation apparatus (Aldrich Chemical Co., Milwaukee, WI: Technical Information Bulletin No. AL-180).

48 Adamson, D. W.; and Kenner, J., "Improved Preparations of Aliphatic Diazo-Compounds, and Certain of Their Properties," *J. Chem. Soc.* **1937**, 1551–56.

49 Procedure for the preparation of diazoethane from N-ethyl-N-nitrosourea; see: Marshall, J. A.; and Partridge, J. J., "The Synthesis and Stereochemistry of 5-Substituted 2-Methylcycloheptanones," *J. Org. Chem.* **1968**, *33*, 4090–97. The diazoethane precursor, N-ethyl-N-nitrosourea, can be easily prepared following the similar procedure for the preparation of N-methyl-N-nitrosourea; see: Arndt, F., "Diazomethane," *Org. Synth. Coll. Vol. 2*, Blatt, A. H., Ed.; John Wiley & Sons: New York, **1943**, 165–67.

50 Wilds, A. L.; and Meader, A. L., "The Use of Higher Diazohydrocarbons in the Arndt–Eistert Synthesis," *J. Org. Chem.* **1948**, *13*, 763–79.

51 Kennedy, M.; and McKervey, M. A., "Pseudoguaianolides from Intramolecular Cycloadditions of Aryl Diazoketones: Synthesis of ($\pm$)-Confertin and an Approach to the Synthesis of ($\pm$)-Damsin," *J. Chem. Soc. Perkin Trans 1* **1991**, 2565–74.

52 Veale, C. A.; Rheingold, A. L.; and Moore, J. A., "Heterocyclic Studies. 49. Preparation and Reactions of Dihydro-3,5,7-trimethyl-6-phenyl-1,2-diazepin-4-ones," *J. Org. Chem.* **1985**, *50*, 2141–45.

53 Wenkert, E.; Decorzant, R.; and Näf, F., "84. A Novel Access to Ionone-Type Compounds: (E)-4-Oxo-β-Ionone and (E)-4-Oxo-β-Irone *via* Metal-Catalyzed, Intramolecular Reactions of α-Diazo Ketones with Furans," *Helv. Chim. Acta* **1989**, *72*, 756–66.

54 Von Hauptmann, S.; and Hirschberg, K., "ω-Diazofettsäureester, III: Reaktionen der ω-Diazofettsäureester mit Säuren und Säurechloriden," *J. Prakt. Chem.* **1966**, *[4R], 34*, 262–71.

55 Hudlicky, T.; Olivo, H. F.; Natchus, M. G.; Umpierrez, E. F.; Pandolfi, E.; and Volonterio, C., "Synthesis of β-Methoxy Enones *via* a New Two-Carbon Extension of Carboxylic Acids," *J. Org. Chem.* **1990**, *55*, 4767–70.

56 Taber, D. F.; and Hoerrner, R. S., "Enantioselective Rh-Mediated Synthesis of (−)-PGE₂ Methyl Ester," *J. Org. Chem.* **1992**, *57*, 441–47.

57 Dimroth, O., Über Intramolekulare Umlagerungen," *Ann. Chem.* **1910**, *373*, 336–70.

58 Curtius, T.; and Klavehn, W., Über die Einwirkung von *p*-Tolulsulfonazid auf Malonester und Alkylierte Malonester," *J. Prakt. Chem.* **1926**, *112*, 65–87.

59 von E Doering, W.; and DePuy, C. H., "Diazocyclopentadiene," *J. Am. Chem. Soc.* **1953**, *75*, 5955–57.

60 Fusco, R.; Bianchetti, G.; Pocar, D.; and Ugo, R., "Reaktionen von Arylsulfonylaziden mit Enaminen aus Ketomethylenverbindungen," *Chem. Ber.* **1963**, *96*, 802–12.

61 Rosenberger, M.; Yates, P.; Hendrickson, J. B.; and Wolf, W., "Preparation of α-Diazo Carbonyl Compounds," *Tetrahedron Lett.* **1964**, 2285–89.

62 Regitz, M., "New Methods of Preparative Organic Chemistry, Transfer of Diazo Groups," *Angew. Chem., Int. Ed. Engl.* **1967**, *6*, 733–49.

63 Regitz, M., "Recent Synthetic Methods in Diazo Chemistry," *Synthesis* **1972**, 351–73.

64 Regitz, M., "Transfer of Diazo Groups," in *Newer Methods of Preparative Organic Chemistry; Vol. 6.* Foerst, W., Ed.; Academic Press: New York, **1971**, 81–126.

65 Regitz, M.; and Maas, G., *Diazo Compounds; Properties and Synthesis*, Academic Press, Orlando, **1986**.

66 Regitz, M.; and Menz, F., "Entformylierende Diazogruppen-Übertragung—ein neuer Weg zu α-Diazo-Ketonen—Aldehyden und Carbonsäureestern," *Chem. Ber.* **1968**, *101*, 2622–32.

67 Regitz, M.; and Rüter, J., "Synthese von 2-Oxo-1-diazo-cycloalkanen durch Entformylierende Diazogruppen-Übertragung," *Chem. Ber.* **1968**, *101*, 1263–70.

68 Regitz, M., "Synthese von Diacyl-Diazomethanen durch Diazogruppenübertragung," *Chem. Ber.* **1966**, *99*, 3128–47.

69 Regitz, M.; Hocker, J.; and Liedhegener, A., "*t*-Butyl Diazoacetate," *Org. Synth. Coll. Vol. 5*, Baumgarten, H. E., Ed.; John Wiley & Sons: New York, **1973**, 179–83.

70 Lowe, G.; and Yeung, H. W., "Synthesis of a β-Lactam Related to the Cephalosporins," *J. Chem. Soc., Perkin Trans 1* **1973**, 2907–10.

71 Oda, M.; Kasai, M.; and Kitahara, Y., "Syntheses of 2-Diazo-4-cyclopentene-1,3-dione and 2-Diazo-4,6-cycloheptadiene-1,3-dione. A Marked Difference in the Degree of Resonance Contribution of the Canonical Forms," *Chem. Lett.* **1977**, 307–10.

72 Kennedy, M.; McKervey, M. A.; Maguire, A. R.; and Roos, G. H. P., "Asymmetric Synthesis in Carbon–Carbon Bond Forming Reactions of α-Diazoketones Catalysed by Homochiral Rhodium(II) Carboxylates," *J. Chem. Soc., Chem. Commun.* **1990**, 361–62.

73 Callant, P.; D'Haenens, L.; and Vandewalle, M., "An Efficient Preparation and the Intramolecular Cyclopropanation of α-Diazo-β-ketophosphonates and α-Diazophosphono-acetates," *Synth. Commun.* **1984**, *14*, 155–61. (Sodium hydride was used as base in this diazo transfer reaction.)

74 Regitz, M.; Rüter, J.; and Liedhegener, A., "2-Diazocycloalkanones: 2-Diazocyclohexanone," *Org. Synth.* **1971**, *51*, 86–89. For the preparation of 2-hydrox-

ymethylenecyclohexanone see: Ainsworth, C., "Indazole," *Org. Synth. Coll. Vol. 4*, Rabjohn, N., Ed.; John Wiley & Sons: New York, **1963**, 536–89.

75 Hendrickson, J. B.; and Wolf, W. A., "The Direct Introduction of the Diazo Function in Organic Synthesis," *J. Org. Chem.* **1968**, *33*, 3610–18.

76 Regitz, M.; Menz, F.; and Rüter, J., "Synthese von α-Diazo-carbonylverbindungen durch Entformylierende Diazogruppenübertragung," *Tetrahedron Lett.* **1967**, 739–42.

77 Regitz, M.; Menz, F.; and Liedhegener, A., "Synthese α,β-Ungesättigter Diazoketone durch Entformylierende Diazogruppenübertragung" *Liebigs Ann. Chem.* **1970**, *739*, 174–84.

78 LeBlanc, B. F.; and Sheridan, R. S., "Observation and Substituent Control of Medium-Dependent Hot-Molecule Reactions in Low-Temperature Matrices," *J. Am. Chem. Soc.* **1988**, *110*, 7250–52.

79 Izawa, T.; Ogino, Y.; Nishiyama, S.; Yamamura, S.; Kato, K.; and Takita, T., "A Facile Enantioselective Synthesis of (1*R*,2*R*,3*S*)-1-Amino-2,3-bishydroxymethyl-cyclobutane Derivatives, A Key Synthetic Intermediate of Carbocyclic Oxetanocins," *Tetrahedron* **1992**, *48*, 1573–80.

80 Tamura, Y.; Ikeda, H.; Mukai, C.; Bayomi, S. M. M.; and Ikeda, M., "Photolysis and Thermolysis of 3-Diazothiochroman-4-one and Related Compounds," *Chem. Pharm. Bull.* **1980**, *28*, 3430–33.

81 Zhang, H.; Lerro, K. A.; Yamamoto, T.; Lien, T. H.; Sastry, L.; Gawinowicz, M. A.; and Nakanishi, K., "The Location of the Chromophore in Rhodopsin: A Photoaffinity Study," *J. Am. Chem. Soc.* **1994**, *116*, 10165–73.

82 Wiberg, K. B.; Furtek, B. L.; and Olli, L. K., "Formation of Bicyclo[2.2.0]hexane Derivatives by the Ring Contraction of Bicyclo[3.2.0]heptanones." *J. Am. Chem. Soc.* **1979**, *101*, 7675–79.

83 Wiberg, K. B.; Olli, L. K.; Golembeski, N.; and Adams, R. D. "Tricyclo-[4.2.0.0^{1,4}]octane," *J. Am. Chem. Soc.* **1980**, *102*, 7467–75.

84 Hisatome, M.; Watanabe, J.; Yamashita, R.; Yoshida, S.; and Yamakawa, K., "Bridge Contraction of [4]Ferrocenophanes by Wolff Rearrangement and Synthesis of [3$_4$](1,2,3,4)Ferrocenophane via the Contraction Reaction," *Bull. Chem. Soc. Jpn.* **1994**, *67*, 490–94.

85 Allinger, N. L.; and Walter, T. J., "Synthesis of a [7]Paracyclophane," *J. Am. Chem. Soc.* **1972**, *94*, 9267–68.

86 Allinger, N. L.; Walter, T. J.; and Newton, M. G., "Synthesis, Structure, and Properties of the [7]Paracyclophane Ring System," *J. Am. Chem. Soc.* **1974**, *96*, 4588–97.

87 Tsuji, T.; and Nishida, S., "Photochemical Generation of [4]Paracyclophanes from 1,4-Tetramethylene Dewar Benzenes: Their Electronic Absorption Spectra and Reactions with Alcohols," *J. Am. Chem. Soc.* **1988**, *110*, 2157–64.

88 Kakiuchi, K.; Ue, M.; Takeda, M.; Tadaki, T.; Kato, Y.; Nagashima, T.; Tobe, Y.; Koike, H.; Ida, N.; and Odaira, Y., "Antiproliferating Polyquinanes. V. Di- and Triquinanes Involvling α-Methylene or α-Alkylidene Cyclopentanone, Cyclopentenone, and γ-Lactone Systems," *Chem. Pharm. Bull.* **1987**, *35*, 617–31.

89 Wrobel, J.; Takahashi, K.; Honkan, V.; Lannoye, G.; Cook, J. M.; and Bertz, S. H., "Stereocontrolled Synthesis of (±)-Modhephene *via* the Weiss Reaction," *J. Org. Chem.* **1983**, *48*, 139–41.

90   Otterbach, A.; and Musso, H., "Diasterane (Tricyclo[3.1.1.1^{2,4}]octane)," *Angew. Chem. Int. Ed. Engl.* **1987**, *26*, 554–55.

91   Eaton, P. E.; Jobe, P. G.; and Reingold, I. D., "The 1,7-Cyclobutanonorbornane System," *J. Am. Chem. Soc.* **1984**, *106*, 6437–39.

92   Yamaguchi, R.; Honda, K.; and Kawanisi, M., "3,12-Cycloiceane (Pentacyclo [6.3.1.0^{2,4}.0^{5,10}.0^{7,8}]dodecane)," *J. Chem. Soc., Chem. Commun.* **1987**, 83–84.

93   Rao, V. B.; George, C. F.; Wolff, S.; and Agosta, W. C., "Synthetic and Structural Studies in the [4.4.4.5]Fenestrane Series," *J. Am. Chem. Soc.* **1985**, *107*, 5732–39.

94   Tobe, Y.; Ueda, K.; Kaneda, T.; Kakiuchi, K.; Odaira, Y.; Kai, Y.; and Kasai, N., "Synthesis and Molecular Structure of (Z)-[6]Paracycloph-3-enes," *J. Am. Chem. Soc.* **1987**, *109*, 1136–44.

95   Berner, H.; Schulz, G.; and Fischer, G., "Chemie der Pleuromutiline, 3. Mitt.: Synthese des 14-O-Acetyl-19,20-dihydro-A-nor-mutilins," *Monatsh. Chem.* **1981**, *112*, 1441–50.

96   Ihara, M.; Kawaguchi, A.; Ueda, H.; Chihiro, M.; Fukumoto, K.; and Kametani, T., "Stereoselective Total Synthesis of (±)-3-Oxosilphinene through Intramolecular Diels–Alder Reaction," *J. Chem. Soc., Perkin Trans. 1* **1987**, 1331–37.

97   Saha, G.; and Ghosh, S., "A New Route to the Synthesis of 7-Functionalised Bicyclo[2.2.1]heptane Derivatives," *Synth. Commun.* **1991**, *21*, 2129–36.

98   Fessner, W. D.; Sedelmeier, G.; Spurr, P. R.; Rihs, G.; and Prinzbach, H., "'Pagodane': The Efficient Synthesis of a Novel, Versatile Molecular Framework," *J. Am. Chem. Soc.* **1987**, *109*, 4626–42.

99   Padwa, A.; Kulkarni, Y. S.; and Zhang, Z., "Reaction of Carbonyl Compounds with Ethyl Lithiodiazoacetate. Studies Dealing with the Rhodium(II)-Catalyzed Behavior of the Resulting Adducts," *J. Org. Chem.* **1990**, *55*, 4144–53.

100 Kametani, T.; Yukawa, H.; and Honda, T., "A Novel Synthesis of Pyrrolizidine Alkaloids by Means of an Intramolecular Carbenoid Displacement (ICD) Reaction," *J. Chem Soc., Perkin Trans. 1* **1988**, 833–37.

101 Doyle, M. P.; Dorow, R. L.; Terpstra, J. W.; and Rodenhouse, R. A., "Synthesis and Catalytic Reactions of Chiral *N*-(Diazoacetyl)oxazolidones," *J. Org. Chem.* **1985**, *50*, 1663–66.

102 Danheiser, R. L.; Miller, R. F.; Brisbois, R. G.; and Park, S. Z., "An Improved Method for the Synthesis of α-Diazoketones," *J. Org. Chem.* **1990**, *55*, 1959–64.

103 Danheiser, R. L.; Miller, R. F.; and Brisbois, R. G., "Detrifluoroacetylative Diazo Group Transfer: (*E*)-1-Diazo-4-phenyl-3-buten-2-one," *Org. Synth.* **1996**, *73*, 134–43.

104 Ye, T.; Garcia, C. F.; and McKervey, M. A., "Chemoselectivity and Stereoselectivity of Cyclisation of α-Diazocarbonyls Leading to Oxygen and Sulfur Heterocycles Catalysd by Chiral Rhodium and Copper Catalysts," *J. Chem. Soc., Perkin Trans. 1* **1995**, 1373–79.

105 Danheiser, R. L.; Brisbois, R. G.; Kowalczyk, J. J.; and Miller, R. F., "An Annulation Method for the Synthesis of Highly Substituted Polycyclic Aromatic and Heteroaromatic Compounds," *J. Am. Chem. Soc.* **1990**, *112*, 3093–100.

106 Danheiser, R. L.; Casebier, D. S.; and Firooznia, F., "Aromatic Annulation Strategy for the Synthesis of Angularly-Fused Diterpenoid Quinones. Total Synthesis of (+)-Neocryptotanshinone, (−)-Cryptotanshinone, Tanshinone IIA, and (±)-Royleanone," *J. Org. Chem.* **1995**, *60*, 8341–50.

107 Darkins, P.; McCarthy, N.; McKervey, M. A.; O'Donnell, K.; Ye, T.; and Walker, B., "First Synthesis of Enantiomerically Pure N-Protected β-Amino-α-Keto Esters from α-Amino Acids and Dipeptides," *Tetrahedron: Asymmetry* **1994**, *5*, 195–98.

108 Metcalf, B. W.; Jund, K.; and Burkhart, J. P., "Synthesis of 3-Keto-4-diazo-5-α-dihydrosteroids as Potential Irreversible Inhibitors of Steroid 5-α-Reductase," *Tetrahedron Lett.* **1980**, *21*, 15–18.

109 Abell, A. D.; Brandt, M.; Levy, M. A.; and Holt, D. A., "The Preparation and Evaluation of (±)-*trans*-1-Diazo-8-methoxy-4a-methyl-1,2,3,4,4a,9,10,10a-octahydro-phenanthren-2-one as an Inhibitor of Human Type-1 Steroid 5α-Reductase," *Bioorg. Med. Chem. Lett.* **1996**, *6*, 883–84.

110 Regitz, M.; and Menz, F., "Entformylierende Diazogruppen-Übertragung—ein neuer Weg zu α-Diazo-Ketonen, -Aldehyden und -Carbonsäureestern," *Chem. Ber.* **1968**, *101*, 2622–32.

111 Harmon, R. E.; Sood, V. K.; and Gupta, S. K., "A New Synthesis of α,β-Unsaturated Diazoketones: An Improved Synthesis of 4-Diazo-3-oxo-1-phenyl-1-butene (Diazomethyl Styryl Ketone)," *Synthesis* **1974**, 577–78.

112 Taber, D. F.; Gleave, D. M.; Herr, R. J.; Moody, K.; and Hennessy, M. J., "A New Method for the Construction of α-Diazoketones," *J. Org. Chem.* **1995**, *60*, 2283–85.

113 Taber, D. F.; You, K.; and Song, Y., "A Simple Preparation of α-Diazo Esters," *J. Org. Chem.* **1995**, *60*, 1093–94.

114 Taber, D. F.; Herr, R. J.; Pack, S. K.; and Geremia, J. M., "A Convenient Method for the Preparation of (Z)-α,β-Unsaturated Carbonyl Compounds," *J. Org. Chem.* **1996**, *61*, 2908–10.

115 Taber, D. F.; Hennessy, M. J.; and Louey, J. P., "Rh-Mediated Cyclopentane Construction Can Compete with β-Hydride Elimination: Synthesis of (±)-Tochuinyl Acetate," *J. Org. Chem.* **1992**, *57*, 436–41.

116 Kučera, J.; Janoušek, Z.; and Arnold, Z., "Synthesis of α-Diazoaldehydes," *Collect. Czech. Chem. Commun.* **1970**, *35*, 3618–27.

117 Menicagli, R.; Malanga, C.; Guidi, M.; and Lardicci, L., "Triisobutylaluminum Assisted Reductive Rearrangement of 2-Ethoxy-4-Alkyl-2,3-Dihydrofurans," *Tetrahedron* **1987**, *43*, 171–77.

118 Norbeck, D. W.; and Kramer, J. B., "Synthesis of (−)-Oxetanocin," *J. Am. Chem. Soc.* **1988**, *110*, 7217–18.

119 Koskinen, A. M. P.; and Muñoz, L., "Diazo Transfer-Reactions Under Mildly Basic Conditions," *J. Chem. Soc., Chem. Commun.* **1990**, 652–53.

120 Ledon, H. J., "Diazo Transfer by Means of Phase-Transfer Catalysts: Di-*tert*-Butyl Diazomalonate," *Org. Synth.* **1979**, *59*, 66–71.

121 Lee, J. C.; and Yuk, J. Y., "An Improved and Efficient Method for Diazo-Transfer Reaction of Active Methylene Compounds," *Synth. Commun.* **1995**, *25*, 1511–15.

122 Ghosh, S.; and Datta, I., "Diazo Transfer Reaction in Solid State," *Synth. Commun.* **1991**, *21*, 191–200.

123 BenAlloum, A.; and Villemin, C., "Potassium Fluoride on Alumina: An Easy Preparation of Diazocarbonyl Compounds," *Synth. Commun.* **1989**, *19* , 2567–71.

124 Popic, V. V.; Korneev, S. M.; Nikolaev, V. A.; and Korobitsyna, I. K., "An Improved Synthesis of 2-Diazo-1,3-diketones," *Synthesis* **1991**, 195–98.

125   Rao, Y. K.; and Nagarajan, M., "A Simple Procedure for Preparation of α-Diazocarbonyl Compounds," *Indian J. Chem., Sect. B* **1986**, *25*, 735–37.

126   McClure, D. E.; Lumma, P. K.; Arison, B. H.; Jones, J. H.; and Baldwin, J. J., "1,4-Oxazines *via* Intramolecular Ring Closure of β-Hydroxydiazoacetamides: Phenylalanine to Tetrahydroindeno[1,2-*b*]-1,4-Oxazin-3(2*H*)-ones," *J. Org. Chem.* **1983**, *48*, 2675–79.

127   Ledon, H., "An Improved Preparation of α-Diazocarbonyl Compounds," *Synthesis* **1974**, 347–48.

128   Schöllkopf, U.; Hauptreif, M.; Dippel, J.; Nieger, M.; and Egert, E., "Synthetic Equivalent for Aminocarboxycarbene; Synthesis of 1-Amino-1-cyclopropanecarboxylic Acid Methyl Esters," *Angew. Chem. Int. Ed. Engl.* **1986**, *25*, 192–93.

129   Sugihara, Y.; Yamamoto, H.; Mizoue, K.; and Murata, I., "Cyclohepta[*a*]phenalene: A Highly Electron-Donating Nonalternant Hydrocarbon," *Angew. Chem. Int. Ed. Engl.* **1987**, *26*, 1247–49.

130   Regitz, M., "Eine neue Synthese für α-Diazo-Carbonylverbindungen," *Tetrahedron Lett.* **1964**, 1403–07.

131   Regitz, M., "Reaktionen Aktiver Methylenverbindungen mit Aziden, VI; Eine neue Synthese für α-Diazo-Carbonyl-Verbinungen," *Chem. Ber.* **1965**, *98*, 1210–24.

132   McKervey, M. A.; and Ye, T., "Asymmetric Synthesis of Substituted Chromanones *via* C–H Insertion Reactions of α-Diazoketones Catalysed By Homochiral Rhodium(II) Carboxylates," *J. Chem. Soc., Chem. Commun.* **1992**, 823–24.

133   Evans, D. A.; Britton, T. C.; Ellman, J. A.; and Dorow, R. L., "The Asymmetric Synthesis of α-Amino Acids Electrophilic Azidation of Chiral Imide Enolates, a Practical Approach to the Synthesis of (*R*)-α-Azido and (*S*)-α-Azido Carboxylic Acids," *J. Am. Chem. Soc.* **1990**, *112*, 4011–30.

134   Lim, H.-J.; and Sulikowski, G. A., "Synthesis of the Antitumor Antibiotic FR-66979: Dmitrienko Oxidative Expansion of a Fully Functional Core Structure," *Tetrahedron Lett.* **1996**, *37*, 5243–46.

135   Lombardo, L.; and Mander, L. N., "A One-Step Synthesis of Cyclic α-Diazoketones," *Synthesis* **1980**, 368–69.

136   Mander, L. N.; and Pyne, S. G., "Studies on Gibberellin Synthesis. Assembly of an Ethanophenanthrenoid Lactone and Conversion into a Gibbane Derivative," *Aust. J. Chem.* **1981**, *34*, 1899–911.

137   Overman, L. E.; Robertson, G. M.; and Robichaud, A. J., "Total Synthesis of (±)-Meloscine and (±)-Epimeloscine," *J. Org. Chem.* **1989**, *54*, 1236–38.

138   Overman, L. E.; Robertson, G. M.; and Robichaud, A. J., "Use Of Aza-Cope Rearrangement–Mannich Cyclization Reactions to Achieve a General Entry to *Melodinus* and *Aspidosperma* Alkaloids. Stereocontrolled Total Syntheses of (±)-Deoxyapodine, (±)-Meloscine, and (±)-Epimeloscine and a Formal Synthesis of (±)-1-Acetylaspidoalbidine," *J. Am. Chem. Soc.* **1991**, *113*, 2598–610.

139   Robichaud, A. J.; and Meyers, A. I., "Asymmetric Additions to Chiral Naphthyloxazolines. An Entry into Tetracyclic Terpene Ring Systems Related to Aphidicolin, Scopadulcic Acid, and Kauranes," *J. Org. Chem.* **1991**, *56*, 2607–09.

140   Nuyttens, F.; Hoflack, J.; Appendino, G.; and De Clercq, P. J., "Intramolecular Diels–Alder Reaction with Furan-Diene: Syntheses of Gibberellins (+)-GA$_1$ and (+)-GA$_3$," *Synlett* **1995**, 105–7.

141 Danheiser, R. L.; and Helgason, A. L., "Total Synthesis of the Phenalenone Diterpene Salvilenone," *J. Am. Chem. Soc.* **1994**, *116*, 9471–79.

142 Coates, R. M.; and Kang, H.-Y., "Synthesis and Evaluation of Cyclobutylcarbinyl Derivatives as Potential Intermediates in Diterpene Biosynthesis," *J. Org. Chem.* **1987**, *52*, 2065–74.

143 Uyehara, T.; Takehara, N.; Ueno, M.; and Sato, T., "Rearrangement Approaches to Cyclic Skeletons IX. Stereoselective Total Synthesis of (±)-Camphorenone Based on a Ring-Contraction of Bicyclo[3.2.1]oct-6-en-2-one. Reliable One-Step Diazo Transfer Followed by a Wolff Rearrangement," *Bull. Chem. Soc. Jpn.* **1995**, *68*, 2687–94.

144 Adams, J. L.; and Metcalf, B. W., "The Synthesis of a 3-Diazobicyclo-[2.2.1]heptan-2-one Inhibitor of Thromboxane A_2 Synthetase," *Tetrahedron Lett.* **1984**, *25*, 919–22.

145 Davies, H. M. L.; Saikali, E.; Clark, T. J.; and Chee, E. H., "Anomalous Reactivity of Mono Substituted Rhodium Stabilized Vinylcarbenoids," *Tetrahedron Lett.* **1990**, *31*, 6299–302.

146 Corey, E. J.; Reid, J. G.; Myers, A. G.; and Hahl, R. W., "Simple Synthetic Route to the Limonoid System," *J. Am. Chem. Soc.* **1987**, *109*, 918–19.

147 Hazen, G. G.; Weinstock, L. M.; Connell, R.; and Bollinger, F. W., "A Safer Diazotransfer Reagent," *Synth. Commun.* **1981**, *11(12)*, 947–56.

148 Curphey, T. J., "Preparation of *p*-Toluenesulfonyl Azide—A Cautionary Note," *Org. Prep. Proced. Int.* **1981**, *13*, 112–15.

149 Baum, J. S.; Shook, D. A.; Davies, H. M. L.; and Smith, H. D., "Diazotransfer Reactions with *p*-Acetamidobenzenesulfonyl Azide," *Synth. Commun.* **1987**, *17(14)*, 1709–16.

150 Davies, H. M. L.; Cantrell, W. R.; Romines, K. R.; and Baum, J. S., "Synthesis of Furans *via* Rhodium(II) Acetate-Catalyzed Reaction of Acetylenes with α-Diazocarbonyls: Ethyl 2-Methyl-5-phenyl-3-furancarboxylate," *Org. Synth.* **1992**, *70*, 93–100.

151 Hazen, G. G.; Bollinger, F. W.; Roberts, F. E.; Russ, W. K.; Seman, J. J.; and Staskiewicz, S., "4-Dodecylbenzenesulfonyl Azides," *Org. Synth.* **1996**, *73*, 144–51.

152 Bollinger, F. W.; and Tuma, L. D., "Diazotransfer Reagents," *Synlett* **1996**, 407–13.

153 Taber, D. F.; Ruckle, R. E., Jr.; and Hennessy, M. J., "Mesyl Azide: A Superior Reagent for Diazo Transfer," *J. Org. Chem.* **1986**, *51*, 4077–78.

154 Boyer, J. H.; Mack, C. H.; Goebel, N.; and Morgan, L. R., Jr., "Reactions of Sodium Phenylacetylide and Sodium Alkoxide with Tosyl and Mesyl Azides," *J. Org. Chem.* **1958**, *23*, 1051–53.

155 Brown, R..; Eastwood, F. W.; and Paterson, I., "Words of Warning..." *Chem. Britain* **1994**, *30*, 470.

156 Kumar, S. M., "A Convenient Preparation of Diazo Carbonyl Compounds Under Tri-Phase Phase-Transfer Catalysis Conditions," *Synth. Commun.* **1991**, *21*, 2121–27.

157 Spencer, H., "Explosion of *p*-Toluenesulphonyl Azide," *Chem Brit.* **1981**, *17*, 106.

158 Rewicki, D.; and Tuchscherer, C., "1-Diazoindene and Spiro[indene-1,7′-norcaradiene]," *Angew. Chem., Int. Ed. Engl.* **1972**, *11*, 44–45

159   Hatch, C. E., III; Baum, J. S.; Takashima, T.; and Kondo, K.,"Stereospecific Total Synthesis of the Potent Synthetic Pyrethroid NRDC 182," *J. Org. Chem.* **1980**, *45*, 3281–85.

160   Davies, H. M. L.; McAfee, M. J., and Oldenburg, C. E. M., "Scope and Stereochemistry of the Tandem Intramolecular Cyclopropanation/Cope Rearrangement Sequence," *J. Org. Chem.* **1989**, *54*, 930–36.

161   Davies, H. M. L.; Saikali, E.; Clark, T. J.; and Chee, E. H., "Anomalous Reactivity of Mono Substituted Rhodium Stabilized Vinylcarbenoids," *Tetrahedron Lett.* **1990**, *31*, 6299–302.

162   Roush, W. R.; Feitler, D.; and Rebek, J., "Polymer-Bound Tosyl Azide," *Tetrahedron Lett.* **1974**, *15*, 1391–92.

163   Cavender, C. J.; and Shiner, V. J., Jr., " Trifluoromethanesulfonyl Azide. Its Reaction with Alkyl Amines to Form Alkyl Azides," *J. Org. Chem.* **1972**, *37*, 3567–69.

164   Hakimelahi, G. H.; and Just, G., "Two Simple Methods for the Synthesis of Trialkyl α-Aminophosphono Acetates (3). Trifluoromethanesulfonyl Azide as an Azide Transfer Agent," *Synth. Commun.* **1980**, *10*, 429–35.

165   Balli, H.; Löw, R.; Müller, V.; Rempfler, H.; and Sezen-Gezgin, A., "Einführung der Diazogruppe in Reaktive Methylenverbindungen mit Azidiniumsalzen," *Helv. Chim. Acta* **1978**, *61*, 97–103.

166   Balli, H.; Müller, V.; and Sezen-Gezgin, A., "Einführung der Diazogruppe mit Azidiniumsalzen in Hydroxy-arene und Hydroxy-hetarene," *Helv. Chim. Acta* **1978**, *61*, 104–07.

167   Balli, H.; and Felder, L., "Einführung der Diazogruppe in Amino-arene und Amino-hetarene mit Azidiniumsalzen: Synthese Kondensierter *v*-Triazole," *Helv. Chim. Acta* **1978**, *61*, 108–17.

168   Monteiro, H. J., "Preparation of α-Diazo-β-ketosulfones by Diazo-Transfer Reaction with an *in situ* Generated Azidinium Salt. A Safe and Efficient Procedure for the Diazo-Transfer Reaction in Neutral Medium," *Synth. Commun.* **1987**, *17*, 983–92.

169   Sezer, Ö.; and Anac, O., "Diazoaldehyde Chemistry," *Helv. Chim. Acta* **1994**, *77*, 2323-34.

170   Kokel, B.; and Viehe, H. G., "Iminium-Activated Azides-New Reagents for the Transfer of Diazonium or Diazo Groups," *Angew. Chem., Int. Ed. Engl.* **1980**, *19*, 716–17.

171   McGuiness, M.; and Shechter, H., "Azidotris(diethylamino)phosphonium Bromide—A Self-Catalyzing Diazo Transfer Reagent" *Tetrahedron Lett.* **1990**, *31*, 4987–90.

172   Forster, M. O., "Azotisation by Chloroamine," *J. Chem. Soc.* **1915**, *107*, 260–67.

173   Wheeler, T. N.; and Meinwald, J., "Formation and Photochemical Wolff Rearrangement of Cyclic α-Diazo ketones: D-Norandrost-5-en-3β-ol-16-carboxylic Acids," *Org. Synth.* **1972**, *52*, 53–58.

174   Oppolzer, W.; Bättig, K.; and Hudlicky, T., "The Total Synthesis of (±)-Isocomene by an Intramolecular Ene Reaction," *Helv. Chim. Acta* **1979**, *62*, 1493–96.

175   Oppolzer, W.; Bättig, K.; and Hudlicky, T., "A Total Synthesis of (±)-Isocomene and (±) β-Isocomene by an Intramolecular Ene Reaction," *Tetrahedron* **1981**, *37*, 4359–64.

176 Ireland, R. E.; Dow, W. C.; Godfrey, J. D.; and Thaisrivongs, S., "Total Synthesis of (±)-Aphidicolin and (±)-β-Chamigrene," *J. Org. Chem.* **1984**, *49*, 1001–13.

177 Jung, M. E.; Lam, P. Y.-S.; Mansuri, M. M.; and Speltz, L. M., "Stereoselective Synthesis of an Analogue of Podophyllotoxin by an Intramolecular Diels–Alder Reaction," *J. Org. Chem.* **1985**, *50*, 1087–1105.

178 Palmisano, G.; Danieli, B.; Lesma, G.; and Riva, R., "Bis(indole) Alkaloids. A Nonbiomimetic Approach to the Blue Pigment Trichotomine Dimethyl Ester," *J. Org. Chem.* **1985**, *50*, 3322–25.

179 Allinger, N. L.; and Freiberg, L. A., "The Synthesis of Highly Strained Medium Rings. [8] Paracyclophane-4-carboxylic Acid," *J. Org. Chem.* **1962**, *27*, 1490–91.

180 Holton, T. L.; and Shechter, H., "Advantageous Syntheses of Diazo Compounds by Oxidation of Hydrazones with Lead Tetraacetate in Basic Environments," *J. Org. Chem.* **1995**, *60*, 4725–29.

181 Bamford, W. R.; and Stevens, T. S., "The Decomposition of Toluene-*p*-Sulphonyl-hydrazones by Alkali," *J. Chem. Soc.* **1952**, 4735–40.

182 Muchowski, J. M., "The Use of Alumina in the Synthesis of α-Diazo Carbonyl Compounds," *Tetrahedron Lett.* **1966**, 1773–78.

183 Chang, S.-J.; Ravi Shankar, B. K.; and Shechter, H., "Chemistry of Diazoace-naphthenones and Diazoacenaphthenes," *J. Org. Chem.* **1982**, *47*, 4226–34.

184 Fang, F. G.; Maier, M. E.; and Danishefsky, S. J., "New Routes to Functionalized Benzazepine Substructures: A Novel Transformation of an α-Diketone Thioamide Induced by Trimethyl Phosphite," *J. Org. Chem.* **1990**, *55*, 831–38.

185 Kaufman, G. M.; Smith, J. A.; Vander Stouw, G. G.; and Shechter, H., "Pyrolysis of Salts of *p*-Tosylhydrazones. Simple Methods for Preparing Diazo Compounds and Effecting Their Carbenic Decomposition," *J. Am. Chem. Soc.* **1965**, *87*, 935–37.

186 Creary, X., "Tosylhydrazone Salt Pyrolyses: Phenyldiazomethanes," *Org. Synth.* **1986**, *64*, 207–16.

187 Shi, G.; and Xu, Y., "Ethyl 3-Trifluoro-2-diazo-propionate as a Potentially Useful CF$_3$-Containing Building Block: Preparation and [Rh(OAc)$_2$]$_2$-Catalysed Reaction with Nitriles," *J. Chem. Soc., Chem. Commun.* **1989**, 607–8.

188 House, H. O.; and Blankley, C, J., "Preparation and Decomposition of Unsaturated Esters of Diazoacetic Acid," *J. Org. Chem.* **1968**, *33*, 53–60.

189 Blankely, C. J.; Sauter, F. J.; and House, H. O., "Crotyl Diazoacetate," *Org. Synth.* **1969**, *49*, 22–27.

190 Sen, R.; Carriker, J. D.; Balogh-Nair, V.; and Nakanishi, K., "Synthesis and Binding Studies of a Photoaffinity Label for Bovine Rhodopsin," *J. Am. Chem. Soc.* **1982**, *104*, 3214–16.

191 Sen, R.; Widlanski, T. S.; Balogh-Nair, V.; and Nakanishi, K., "Photoaffinity Labeling of Bacteriorhodopsin with 3-([-1-C-^{14}C]Diazoacetoxy)-*trans*-retinal," *J. Am. Chem. Soc.* **1983**, *105*, 5160–62.

192 Keilbaugh, S. A.; and Thornton, E. R., "Synthesis and Photoreactivity of Cholesteryl Diazoacetate: A Novel Photolabeling Reagent," *J. Am. Chem. Soc.* **1983**, *105*, 3283–86.

193 Corey, E. J.; and Myers, A. G., "Efficient Synthesis and Intramolecular Cyclopropanation of Unsaturated Diazoacetic Esters," *Tetrahedron Lett.* **1984**, *25*, 3559–62.

194   Clive, D. L. J.; and Daigneault, S., "Use of Radical Ring-Opening for Introduction of Alkyl and Substituted Alkyl Groups with Stereochemical Control: A Synthetic Application of Cyclopropylcarbinyl Radicals," *J. Org. Chem.* **1991**, *56*, 3801–14.

195   Corey, E. J.; and Myers, A. G., "Total Synthesis of (±)-Antheridium-Inducing Factor (A_An, 2) of the Fern *Anemia Phyllitidis.* Clarification of Stereochemistry," *J. Am. Chem. Soc.* **1985**, *107*, 5574–76.

196   Ok, H.; Caldwell, C.; Schroeder, D. R.; Singh, A. K.; and Nakanishi, K., "Synthesis of Optically Active 3-Diazoacetylretinals with Triisopropylphenylsulfonylhydrazone," *Tetrahedron Lett.* **1988**, *29*, 2275–78.

197   Zimmerman, H. E.; and Bunce, R. A., "Cyclopropene Photochemistry. Mechanistic and Exploratory Organic Photochemistry," *J. Org. Chem.* **1982**, *47*, 3377–96.

198   Kline, T. B.; Nelson, D. L.; and Namboodiri, K., "Novel [(Diazomethyl)carbonyl]-1,2,3,4-tetrahydronaphthalene Derivatives as Potential Photoaffinity Ligands for the 5-HT$_{1A}$ Receptor," *J. Med. Chem.* **1990**, *33*, 950–55.

199   Iida, T.; Hori, K.; Nomura, K.; and Yoshii, E., "A New Entry to 5-Unsubstituted 3-Acyltetramic Acids from Aldehydes," *Heterocycles* **1994**, *38*, 1839–44.

200   Ouihia, A.; René, L.; and Badet, B., "An Easy Synthesis of Diazoacetamides," *Tetrahedron Lett.* **1992**, *33*, 5509–10.

201   Womack, E. B.; and Nelson, A. B., "Ethyl Diazoacetate," *Org. Synth. Coll. Vol. 3.* Horning, E. C., Ed.; John Wiley & Sons: New York, **1955**, 392–93.

202   Searle, N. E., "Ethyl Diazoacetate," *Org. Synth. Coll. Vol. 4*, Rabjohn, N., Ed.; John Wiley & Sons: New York, **1963**, 424–26.

203   Takamura, N.; Mizoguchi, T.; Koga, K.; and Yamada, S., "Amino Acids and Peptides-XIV. A Simple and Convenient Method for Preparation of α-Substituted α-Diazo Esters," *Tetrahedron* **1975**, *31*, 227–30.

204   Thorsett, E. D., "Conversion of α-Aminoesters to α-Ketoesters," *Tetrahedron Lett.* **1982**, *23*, 1875–76.

205   Curphey, T. J.; and Daniel, D. S., "New Synthesis of Azaserine," *J. Org. Chem.* **1978**, *43*, 4666–68.

206   Nishimura, M.; Nakada, H.; Takase, S.; Katayama, A.; Goto, T.; Tanaka, H.; and Hashimoto, M., "A New Antitumor Antibiotic, FR900840 2. Structural Elucidation of FR900840," *J. Antibiotics* **1989**, *42*, 549–52.

207   Challis, B. C.; and Latif, F., "Synthesis and Characterisation of Some New Diazopeptides," *J. Chem. Soc., Perkin Trans. 1* **1990**, 1005–9.

208   Hauser, D.; and Sigg, H. P., "Desaminierung von 6-Aminopenicillansäure," *Helv. Chim. Acta* **1967**, *50*, 1327–35.

209   Sheehan, J. C.; Lo, Y. S.; Löliger, J.; and Podewell, C. C., "Synthesis of 6-Hydroxy-penicillanates and 7-Hydroxycephalosporanates," *J. Org. Chem.* **1974**, *39*, 1444–45.

210   Bestmann, H. J.; and Soliman, F. M., "Synthesis and Reactions of Diazoacetyl Chloride—Detection of Diazoketene," *Angew. Chem. Int. Ed. Engl.* **1979**, *18*, 947–48.

211   Devos, A.; Remion, J.; Frisque-Hesbain, A.-M.; Colens, A.; and Ghosez, L., "Synthesis of Acyl Halides Under Very Mild Conditions," *J. Chem. Soc., Chem. Commun.* **1979**, 1180-81.

212 Ouihia, A.; René, L.; Guilhem, J.; Pascard, C.; and Badet, B., "A New Diazoacylating Reagent: Preparation, Structure, and Use of Succinimidyl Diazoacetate," *J. Org. Chem.* **1993**, *58*, 1641–42.

213 Doyle, M. P.; and Kalinin, A. V., "Highly Enantioselective Intramolecular Cyclopropanation Reaction of *N*-Allylic-*N*-methyl diazoacetamides Catalyzed by Chiral Dirhodium(II) Carboxamidates," *J. Org. Chem.* **1996**, *61*, 2179–84.

214 Kido, F.; Yamaji, K.; Abiko, T.; and Kato, M., "Ethyl Hydrogen Diazomalonate, a Convenient Reagent for the Synthesis of Alkyl Ethyl Diazomalonates," *J. Chem. Res. (S)* **1993**, 18–19.

215 Marino, J. P., Jr.; Osterhout, M. H.; Price, A. T.; Sheehan, S. M.; and Padwa, A., "Ethyl 2-Diazomalonyl Chloride. An Efficient Diazoacylating Reagent," *Tetrahedron Lett.* **1994**, *35*, 849–52.

216 Wydila, J.; and Thornton, E. R., "Synthesis and Photochemical Studies of a Diazomalonyl-Containing Galactocerebroside Analogue. A Glycolipid Photolabeling Reagent," *J. Org. Chem.* **1984**, *49*, 244–49.

217 Karady, S.; Amato, J. S.; Reamer, R. A.; and Weinstock, L. M., "Stereospecific Conversion of Penicillin to Thienamycin," *J. Am. Chem. Soc.* **1981**, *103*, 6765–67.

218 Reider, P. J.; Rayford, R.; and Grabowski, E. J. J., "Synthetic Approaches to Thienamycin: Carbon–Carbon Bond Formation at C-4 of Azetidin-2-ones," *Tetrahedron Lett.* **1982**, *23*, 379–82.

219 Reider, P. J.; and Grabowski, E. J. J., "Total Synthetic of Thienamycin: A New Approach from Aspartic Acid," *Tetrahedron Lett.* **1982**, *23*, 2293–96.

220 Ueda, Y.; Roberge, G.; and Vinet, V., "A Simple Method of Preparing Trimethylsilyl- and *tert*-Butyldimethylsilyl-Enol Ethers of α-Diazoacetoacetates and Their Use in the Synthesis of a Chiral Precursor to Thienamycin Analogs," *Can. J. Chem.* **1984**, *62*, 2936–40.

221 Hart, D. J.; Lee, C.-S.; Pirkle, W. H.; Hyon, M. H.; and Tsipouras, A., "Asymmetric Synthesis of β-Lactams and the Carbapenem Antibiotic (+)-PS-5," *J. Am. Chem. Soc.* **1986**, *108*, 6054-56.

222 Kim, C. U.; Luh, B.; and Partyka, R. A., "Stereoselective Synthesis of 1-β-Methylcarbapenem," *Tetrahedron Lett.* **1987**, *28*, 507–10.

223 Cainelli, G.; and Panunzio, M., "β-Lactams from Ester Enolates and Silylimines: Enantioselective Synthesis of (+)-PS-5," *J. Am. Chem. Soc.* **1988**, *110*, 6879–80.

224 Häbich, D.; and Hartwig, W., "Synthesis of a New Carbapenem with a 6-Methyl Hydroxyacetate Side Chain," *Tetrahedron Lett.* **1987**, *28*, 781–84.

225 Sundberg, R. J.; and Pearce, B. C., "3-(3-Pyrrolyl)thiopyrrolidones as Precursors of Benzo[1,2-*b*:4,3-*b'*]dipyrroles. Synthesis of Structures Related to the Phosphodiesterase Inhibitors PDE-I and PDE-II," *J. Org. Chem.* **1985**, *50*, 425–32.

226 Doyle, M. P.; Austin, R. E.; Bailey, A. S.; Dwyer, M. P.; Dyatkin, A. B.; Kalinin, A. V.; Kwan, M. M. Y.; Liras, S.; Oalmann, C. J.; Pieters, R. J.; Protopopova, M. N.; Raab, C. E.; Roos, G. H. P.; Zhou, Q.-L.; and Martin, S. F., "Enantioselective Intramolecular Cycloprpanations of Allylic and Homoallylic Diazoacetates and Diazoacetamides Using Chiral Dirhodium(II) Carboxamide Catalysts," *J. Am. Chem. Soc.* **1995**, *117*, 5763–75.

227 Doyle, M. P.; Bagheri, V.; Wandless, T. J.; Harn, N. K.; Brinker, D. A.; Eagle, C. T.; and Loh, K.-L., "Exceptionally High Trans (Anti) Stereoselectivity in Catalytic Cyclopropanation Reactions," *J. Am. Chem. Soc.* **1990**, *112*, 1906–12.

228   Clemens, R. J.; and Hyatt, J. A., "Acetoacetylation with 2,2,6-Trimethyl-4H-1,3-Dioxin-4-one," *J. Org. Chem.* **1985**, *50*, 2431–35.

229   Kurth, M. J.; Tahir, S. H.; and Olmstead, M. M., "A Thioxanone-Based Chiral Template: Asymmetric Induction in the [2,3]-Sigmatropic Rearrangement of Sulfur Ylids. Enantioselective Preparation of Cβ-Chiral Pent-4-enoic Acids," *J. Org. Chem.* **1990**, *55*, 2286–88.

230   Doyle, M. P.; Shanklin, M. S.; Pho, H. Q.; and Mahapatro, S. N., "Rhodium(II) Acetate and Nafion-H Catalyzed Decomposition of *N*-Aryldiazoamides. An Efficient Synthesis of 2(3H)-Indolinones," *J. Org. Chem.* **1988**, *53*, 1017–22.

231   Meyer, B.; Kogelberg, H.; Köll, P.; and Laumann, U., "Synthesis and Properties of Methyl 5,6,7,8-Tetra-O-acetyl-3-deoxy-3-diazo-D-arabino-oct-2,4-diulosonate," *Tetrahedron Lett.* **1989**, *30*, 6641–44.

232   Tsuge, O.; Sakai, K.; and Tashiro, M., "Studies of Acyl and Thioacyl Isocyanates—XII: The Reaction of Benzoyl and Thiobenzoyl Isocyanates with Sulfonium Ylides and with Diazoalkanes," *Tetrahedron* **1973**, *29*, 1983–90.

233   Goerdeler, J.; and Schimpf, R., "Reaktion mit Nitrosobenzol, Diazoalkanen und Nitronen zu 5-gliedrigen Ringen," *Chem. Ber.* **1973**, *106*, 1496–500.

234   Fink, J.; and Regitz, M., "Electrophilic Diazoalkane Substitution," *Synthesis* **1985**, 569–85.

235   Yates, P.; Garneau, F. X.; and Lokensgard, J. P., "Preparation and Spectra of Mercuribis (α-Diazo Ketones)," *Tetrahedron* **1975**, *31*, 1979–83.

236   Schöllkopf, U.; Bánhidai, B.; Frasnelli, H.; Meyer, R.; and Beckhaus, H., "α-Diazo-β-hydroxy-carbonsäureester und -Ketone aus Carbonyl- und Diazolithioverbindungen sowie ihre Umlagerung zu β-Ketocarbonsäureestern und β-Diketonen," *Liebegs Ann. Chem.* **1974**, 1767–83.

237   Schöllkopf, U.; and Rieber, N., "*C*-Alkylierung von Silber-diazoessigsäure-äthylester und Silber-diazoketonen mit S_N1-Aktiven Halogeniden," *Chem. Ber.* **1969**, *102*, 488–93.

238   Schöllkopf, U.; Gerhart, F.; Reetz, M.; Frasnelli, H.; and Schumacher, H., "Halogen-diazoessigsäureäthylester aus Quecksilber-bis-diazo-essigsäureäthylester," *Liebegs Ann. Chem.* **1968**, *716*, 204–6.

239   Gerhart, F.; Schöllkopf, U.; and Schumacher, H., "Ethyl Diazoiodoacetate and its Decomposition to Ethoxycarbonyliodocarene," *Angew. Chem., Int. Ed. Engl.* **1967**, *6*, 74–75.

240   Schöllkopf, U.; Tonne, P.; Schäfer, H.; and Markush, P., "Synthesen von Nitrodiazoessigsäureestern, Nitro-cyan- und Nitro-trifluormethyl-diazomethan" *Liebegs Ann. Chem.* **1969**, *722*, 45–51.

CHAPTER 2

Catalysts for Metal Carbene Transformations

The early development of catalysts for diazo decomposition was accomplished with considerable serendipity, but successes in their applications generated the need to understand the mechanism of their action.[1-4] What were the characteristics of copper bronze and copper(II) sulfate, which were first used with diazo compounds,[1-5] that offered them unique capabilities for dinitrogen loss at temperatures significantly lower than those required for thermal decomposition? What was their interaction with diazo compounds that caused dinitrogen extrusion? Why were selectivities for metal-catalyzed reactions different from those in thermal or photochemical processes? Questions such as these prompted additional experimentation, which led to the construction of new transition metal catalysts that advanced reactions of diazo compounds to a position of prominence in synthetic organic chemistry. In this chapter we will examine the basis for catalytic activity, beginning with the reactivity of diazo compounds with electrophiles, and then elaborate the design of catalysts that have been able to achieve extraordinary levels of selectivity.

2.1 ELECTROPHILIC ADDITION TO DIAZO COMPOUNDS

Diazo compounds are inherently unstable to acid-promoted decomposition, and it is this instability that models their effectiveness for catalytic reactions with transition metal compounds. Diazomethane is the conjugate base of the methanediazonium ion (**1**, eq. 1), whose pK_a value[6] of 10 suggests the driving force for the

$$H_3C-N_2^+ \;\; \rightleftharpoons \;\; H_2C=N_2 + H^+$$

$$\mathbf{1}$$

(1)

high reactivity of diazomethane towards Lewis acids. In contrast, the pK_a values of diazonium ions derived from diazoesters and diazoketones have been estimated to be between -5 and -2, respectively,[7] suggesting the relative stabilities of these diazo compounds to acid-promoted decomposition. The 12–15 order of magnitude difference in diazonium ion pK_a values reflects many of the differences observed in reactions/reactivities between diazocarbonyl compounds and diazomethane or its alkyl/aryl derivatives.

Diazo decomposition results from C-protonation on diazo compounds (eq. 2), which has been recognized to be thermodynamically more favorable than N-

$$H_2C=\overset{+}{N}=N-H \rightleftharpoons H_2C=\overset{+}{N}=\overline{N} + H^+ \rightleftharpoons H_3C-\overset{+}{N}\equiv N \qquad (2)$$

$$\mathbf{2} \qquad\qquad\qquad\qquad\qquad \mathbf{1}$$

protonation.[8,9] Calculations have suggested an energy difference of nearly 40 kcal/mol between C- and N-protonation.[8] N-Protonation (2) is kinetically favored but is not productive. Protonation of diazocarbonyl compounds at low temperatures in superacids ($HF/SbF_5/SO_2$ or $FSO_3H/SbF_5/SO_2$) produces only the O-protonated E- and Z-enoldiazonium ions (eq. 3)[10,11]; neither C-protonation nor

$$\underset{O}{\overset{R}{\diagdown}}C-C\underset{R'}{\overset{N_2}{\diagup}} + H^+ \longrightarrow \underset{HO}{\overset{R}{\diagdown}}C=C\underset{R'}{\overset{N_2^+}{\diagup}} + \underset{HO}{\overset{R}{\diagdown}}C=C\underset{N_2^+}{\overset{R'}{\diagup}} \qquad (3)$$

N-protonation have been observed. However, there is abundant evidence for C-protonation from kinetic and deuterium exchange studies which implicates the C-protonated form as the reactive intermediate in proton-induced diazo decomposition.[12]

2.2 MECHANISM OF CATALYTIC DIAZO DECOMPOSITION. METAL CARBENE GENERATION AND REACTIONS

Transition metal complexes that are effective catalysts for diazo decomposition are Lewis acids.[13] Their catalytic activity depends on coordinative unsaturation at the metal center, which allows them to react as electrophiles with diazo compounds. In the generally accepted mechanism for catalytic decomposition of diazo compounds (Scheme 1),[13-17] electrophilic addition causes the loss of dinitrogen and production of a metal-stabilized carbene (4). Transfer of the electrophilic carbene entity to an electron-rich substrate (S:) regenerates the catalytically active L_nM and completes the catalytic cycle. Spectral evidence for the formation of a diazonium ion adduct (3) has been obtained with iodorhodium(III) tetra-p-tolylporphyrin (Scheme 2), a highly efficient catalyst for cyclopropanation, from which iodoacetate (7) was obtained in the absence of alkene.[18] Whether addition of ethyl diazoacetate (EDA), which results in diazonium ion intermediate 5, or expulsion of dinitrogen,[19] which forms metallocarbene 6, is rate limiting has not yet been firmly established, although saturation kinetics suggests that 3 → 4 is rate limiting with diazo ketones.[20] A stable (porphyrinatorhodium)-diaminocarbene complex, $(TPP)Rh(PhCH_2NC)[:C(NHCH_2Ph)_2]PF_6$, has been

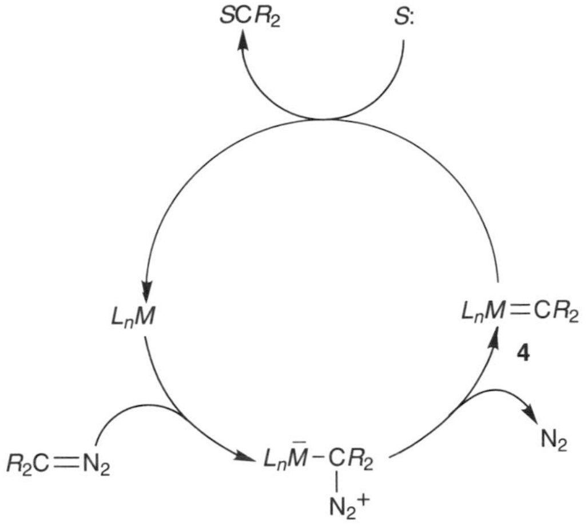

Scheme 1

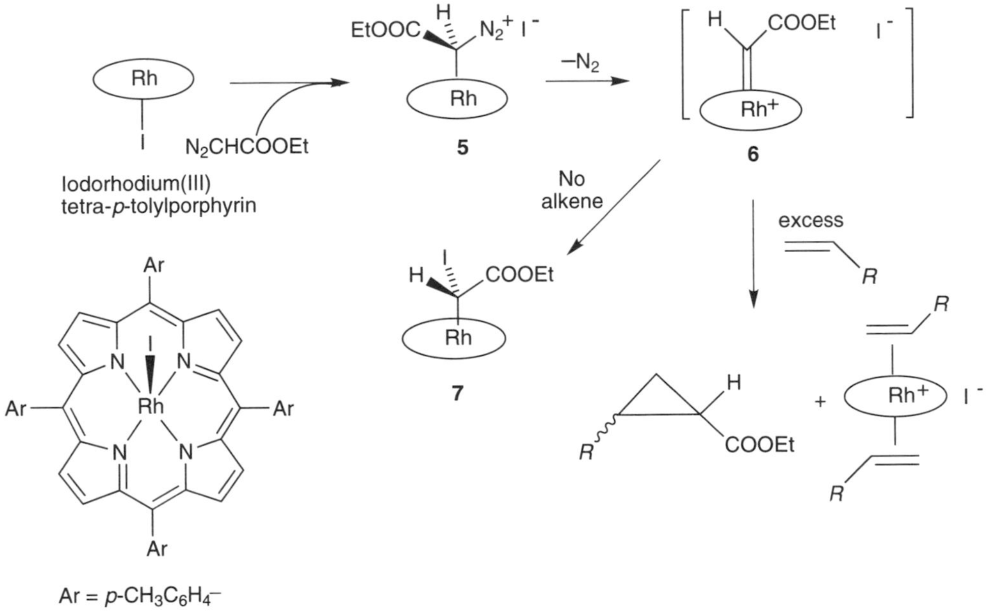

Scheme 2

prepared by nucleophilic addition of an amine to an isocyanide that is coordinated to rhodium, and its crystal structure has been obtained[21]; this characterization, which was the first of a carbene complex in a rhodium macrocycle, further confirms the reliability of the mechanism described in Scheme 1. Subsequently

Werner prepared and characterized a series of square planar and half-sandwich carbene rhodium(I) complexes from the bis(stibane)-rhodium compound *trans*-[RhCl(C$_2$H$_4$)(Sb(*i*-Pr)$_3$)$_2$].[22] Herrmann and co-workers have prepared mono- and dicarbene adducts of 1,3-dimethylimidazoline-2-ylidene (**8**) with late transition metals (Ru^{II}, Rh^{I}, Pd^{II}, Os^{II}, and Ir^{I}) and obtained their X-ray structures (e.g., for **9**)[23]; substituted carbene derivatives can also be obtained. Although these carbene

8   **9**

complexes are chemically very stable and should be regarded as donor adducts with the Lewis-acidic metal fragment, this landmark discovery has significant implications for catalytically generated metal carbenes.

Because of the coordination unsaturation of the active metal catalyst, Lewis bases (*B*:) that can associate with the metal[24] inhibit diazo decomposition (eq. 4).

$$B\text{--}ML_n \quad \underset{+B:}{\overset{-B:}{\rightleftharpoons}} \quad ML_n \quad \underset{-R_2C=N_2}{\overset{+R_2C=N_2}{\rightleftharpoons}} \quad L_n\overset{-}{M}\text{--}\underset{\underset{N_2^+}{|}}{C}R_2 \qquad (4)$$

3

The stability of *BML$_n$* determines the degree to which association with diazo compounds is restricted. Amines, sulfides, and nitriles are generally effective inhibitors for transition metal catalyzed decomposition of diazo compounds, but alkenes or benzene can also play this role with selected transition metal catalysts.[25-27] Halogenated hydrocarbons, typically dichloromethane and 1,2-dichloroethane, are not known to coordinate with catalytically active transition metal complexes, and, therefore, they serve as useful solvents for carbene generation and transfer.

The activities of catalytically active transition metal compounds towards diazo decomposition are dependent on both the electrophilicity of the transition metal compound and on the stability of the diazo compound. Among diazocarbonyl compounds, those with two carbonyl groups that flank the diazomethane carbon are more stable towards transition metal catalyzed decomposition than those with only one carbonyl group.[15] Diazoesters are generally more stable than diazoketones, and diazoamides are more stable than diazoesters. This profile is a useful guide to understanding the reaction conditions required to generate a metal carbene. For example, diazoacetoacetates and diazomalonates require higher temperatures for reactions with transition metal catalysts than do diazoacetates, which can undergo catalytic nitrogen loss at or below room temperature.

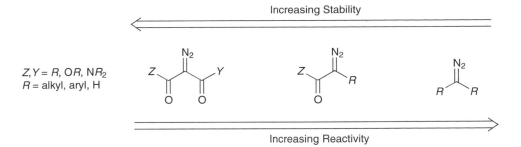

$Z, Y = R, OR, NR_2$
$R = $ alkyl, aryl, H

Diazocarbonyl compounds are the preferred substrates for catalytic diazo decomposition by transition metal complexes.[13] Although they possess three basic sites for potential coordination (Scheme 3), only the one that forms a metal–

$$L_n\bar{M}-O$$
$$RC=CH-N_2^+$$
10

$$L_nM + RCCHN_2$$

$$L_n\bar{M}-N=\overset{+}{N}=CHCR$$
11

$$L_n\bar{M}-CHCOR$$
$$N_2^+$$
3

Scheme 3

carbon bond leading to a metal carbene (**3**) is productive.[18] Although the other two association complexes (**10** and **11**) may be formed and, in fact, serve as inhibitors to metal carbene formation, there is now general agreement that metal carbenes are derived from **3**.[4,13–17]

Dichloromethane is the preferred solvent for diazo decomposition of diazoacetates and diazoketones, but ethyl ether and pentane have also been used.[14] With the less reactive diazomalonates and diazoacetoacetates, higher temperatures are often required, and solvents such as 1,2-dichloroethane, benzene, and toluene are most often employed.[15] Factors that influence the rates of diazo decomposition by transition metal catalysts may include inhibition by the diazo compound (**10** and **11** in Scheme 3) or the inherent stability of the diazo compound.

Dinitrogen extrusion from diazonium ion **3** can be considered to be irreversible since no evidence to the contrary has been described. In the electrophilic process for metal carbene generation (Scheme 1), the metal carbene intermediate is itself electrophilic and primed for reaction with electron-rich substrates (*S*:). Using the formalism of ylide structures, this intermediate can be depicted with two resonance-contributing structures, one a formal metal carbene (**4a**) and the

$$R_2C{=}ML_n \quad \longleftrightarrow \quad R_2\overset{+}{C}{-}\overline{M}L_n$$

4a   **4b**

other a metal-stabilized carbocation (**4b**),[13] the sum of which portray the electrophilic reactivity of the metal carbene intermediate. Ligands on the metal and substituents on the carbene carbon generally influence significantly the electrophilic reactivity of the metal carbene.[13,17] However, iodorhodium(III) porphyrins have been reported to catalyze cyclopropanation of styrene without a detectable substituent effect from the alkene and with a negligible secondary isotope effect,[28] but they appear to be the exceptions. In addition, dissociation of the carbene from the metal (eq. 5) has been proposed to account for the selectivities

$$L_nM{=}CR_2 \quad \xrightarrow{-L_nM} \quad R_2C{:} \quad \xrightarrow{+S{:}} \quad R_2CS \qquad (5)$$

of transition metal catalyzed reactions that are similar to those of thermal processes,[29] and this explanation is based on arguments of limited backbonding from the metal; however, the existence of free carbenes in transition metal-catalyzed reactions is, at best, an exception to general observations that the reactions of diazo compounds in metal-catalyzed reactions are those of metal carbenes.[4,13–17] The relative importance of **4a** and **4b**, the former suggesting limited charge development, considerable backbonding, and minimal substituent effects in the transition state for carbene transfer and the latter suggesting the opposite, is obviously a function of the system under examination.

Transition metal compounds that can effectively catalyze diazo decomposition are coordinatively unsaturated and are capable of stabilizing the metal-bound carbene. Only "late" transition metals in the third and fourth periods, among which are selected compounds of copper, cobalt, iron, palladium, rhodium, and ruthenium, fit these requirements. However, although they are normally thought to catalyze, formally, the generation of metal carbene intermediates, most compounds of cobalt and palladium are sufficiently different from the others in their reactions with diazomethane that a different mechanism for diazo decomposition must be considered.

2.3 MECHANISM OF CATALYTIC DIAZO DECOMPOSITION. ELECTROPHILIC ADDITION BY METAL OLEFIN COMPLEXES

Both copper and rhodium catalysts are effective for diazo decomposition of diazomethane, but polymethylene formation is often the principal outcome. Diazomethane, in particular, and diazoalkanes, in general, do not transfer methylene to reactive substrates (*S*:) as efficiently with these catalysts as with those of palladium.[30] Competition exists between addition of diazomethane to the intermediate

metal carbene and capture by the substrate (eq. 6), and "carbene dimer" formation is often, but not always, the preferred pathway.[31,32]

$$L_nM + R_2CS \xleftarrow{\quad S: \quad} L_nM{=}CR_2 \xrightarrow{\quad R_2C{=}N_2 \quad} L_n\bar{M}{-}CR_2 \xrightarrow{\quad (-N_2) \quad} L_nM + R_2C{=}CR_2$$

$$4 \qquad\qquad \underset{R_2\overset{\displaystyle |}{C}{-}N_2^+}{} \qquad\qquad \text{"carbene dimer"}$$

(6)

Palladium(II) compounds are highly effective catalysts with diazomethane for
the cyclopropanation of alkenes. Since palladium(II) readily coordinates with
alkenes,[33] there is reason to believe that the unique capabilities of PdCl$_2$ and
Pd(OAc)$_2$ to catalyze methylene addition to alkenes[30] is due, not to metal carbene
formation, but to electrophilic addition by the metal olefin complex onto diazomethane.[34] Loss of dinitrogen releases the original olefin as its cyclopropane
derivative (Scheme 4).[35,36] Alternatively, loss of dinitrogen may occur with metal-

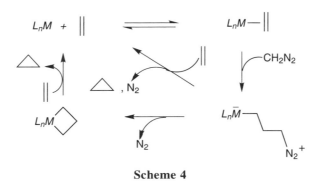

Scheme 4

locyclobutane formation and subsequent cyclopropane formation. Because of
their high reactivities towards electrophilic addition, only diazomethane and
alkyl- or aryl-substituted diazomethanes would be expected to undergo diazo decomposition in this fashion. There are alternative pathways for this process,[37,38]
and although no single mechanism has been firmly established, Scheme 4 does
account for the significant differences in selectivities found with palladium catalysts.[30] Several conditions apply in the evaluation of these mechanistic possibili-
ties: (1) the coordination capability of the transition metal compound with
olefins, (2) the nucleophilicity of the diazo compound, and (3) comparative selectivities in cyclopropanation and other typically metal carbene transformations.
For example, high regioselectivity for the cyclopropanation of dienes and polyenes with diazomethane is one characteristic of palladium catalysts relative to
copper catalysts.

2.4 COPPER CATALYSTS FOR DIAZO DECOMPOSITION

Copper bronze and copper(II) sulfate are the oldest of the copper catalysts employed for diazo decomposition.[2] Copper bronze, which can be freshly prepared by precipitating copper from aqueous copper(II) sulfate solution with zinc,[39] is still one of the most widely used copper catalysts. Both are insoluble in the reaction medium in which decomposition of diazo compounds occurs, and neither is currently regarded to be the active catalyst. Copper(I) chloride, as trialkyl or triaryl phosphite complexes,[40] and copper(II) acetylacetonate[41] were developed in the 1960s as soluble catalyst alternatives to copper bronze and copper(II) sulfate, and both have seen wide uses.[2-4] The introduction of these homogeneous catalysts was motivated, in part, by uncertainties encountered with the use of the insoluble copper compounds that were regarded as heterogeneous catalysts. The synthesis of copper(II) complexes of chiral salicylaldimines by Nozaki and co-workers[42] ushered in the modern era of asymmetric catalysis that, applied to metal carbene transformations, has resulted in an exponential growth in catalyst design and applications.

2.4.1 Oxidation States for Copper

The introduction of copper(I) triflate (CuOTf, OTf $= CF_3SO_3$) by Salomon and Kochi advanced the basic understanding of copper catalysis in metal carbene transformations.[25,43] Copper(I) rather than copper(II) was established as the active catalyst when diazo compounds were found to reduce copper(II) chloride to copper(I) chloride[44] and copper(II) triflate to copper(I) triflate (eq. 7), although

$$CuL_2 + N_2CHR \longrightarrow [CuL_2]^{-} + [N_2CHR]^{+\cdot} \tag{7}$$

extension of this interpretation to other copper catalysts remained controversial for the next several years.[1] Today, however, with the availability of a great variety of copper compounds that are catalytically active for diazo decomposition,[15,35] there is general agreement that the active form of copper is the +1 oxidation state.

There are two methods commonly employed for the reduction of copper(II) complexes to catalytically active copper(I) complexes, especially those with chiral salicylaldimine,[45] semicorrin,[46] or bis-oxazoline ligands.[47-49] Direct addition of small amounts of the diazo compound to the solution containing the copper(II) catalyst (concentrated in CH_2Cl_2 or $CHCl_3$) generally results in a color change, after which time the diazo compound is added to the catalyst solution to effect metal carbene generation and reactions.[50] Phenylhydrazine has also been used (eq. 8) to reduce copper(II),[47,50] but with this reagent near-stoichiometric

$$2CuL_2 + PhNHNH_2 \longrightarrow 2CuL + 2LH + PhH + N_2 \tag{8}$$

amounts should be used in order to avoid accumulation of this catalyst-inhibiting base. However, even DIBAH has been employed for the reduction of copper(II),[51] but its overall suitability has not been established. More recently, copper(I) complexes have been prepared *in situ* by ligand association with either Cu(OTf)[48] or Cu(O^{t}Bu).[49]

Copper(I) compounds possess four coordination sites, and, in the absence of Lewis bases such as chiral nitrogen ligands that strongly coordinate to copper(I), association with alkenes is characteristic. In contrast, copper(II) compounds that include Cu(OTf)$_2$ form only weak association complexes with olefins.[25,43] Kinetic investigations of the influence of olefin coordination on the reactivity of Cu(OTf) towards diazo compounds established olefin inhibition to diazo decomposition,[25] and this evidence is consistent with the interpretation of eq. 4. However, only Cu(OTf) and similar highly electrophilic Cu(BF$_4$) and Cu(PF$_6$) complexes exhibit any tendency to coordinate with olefins in solution. The use of the conveniently handled and easily prepared Cu(CH$_3$CN)$_4$PF$_6$[52] has been reported to hold advantages over Cu(OTf) for catalytic cyclopropanation reactions.[53]

2.4.2   Ligands for Copper

Because of their stabilities and general ease of preparation and handling, copper(II) complexes were initially favored over those of air-sensitive copper(I). Two bidentate ligands such as acetylacetonate are bound to copper(II), but upon reduction to copper(I) one of these bidentate ligands is presumed to dissociate from copper. This argument has been made for copper(II) complexes formed from chiral salicylaldimine (**12**),[45] semicorrin (**13**),[46,54] and bis-oxazoline (**13**)[47–49]

| **12** | **13** | **14** |
| | | (*A* = H or CH$_3$) |

ligands; only one of these ligands remains bound to the active copper(I) catalyst during diazo decomposition. The same would be expected for copper(II) acetylacetonate and its analogs.

The achiral copper compounds that have enjoyed substantial uses as catalysts for diazo decomposition are, in addition to the heterogeneous copper bronze, copper halides, and copper(II) sulfate, homogeneous (trialkyl and triaryl)phosphite copper(I) chloride (**15**),[40,55] bis(acetylacetonato)copper(II) (**16**)[41] and its trifluoro or hexafluoro analogs,[56] copper(I) triflate (**17**),[25] copper(I) hexafluorophosphate,[53] and the *N-tert*-butyl- or *N*-benzylsaliclaldimine complex of copper(II) (**18**).[57]

$(RO)_3PCuCl$

15

$R = i\text{-Pr, Ph}$

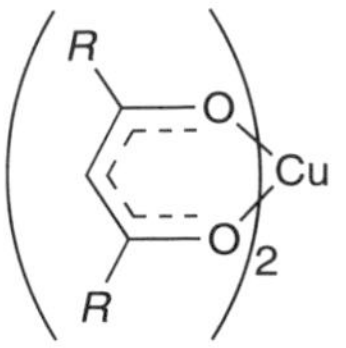

16

$R = CH_3$, Cu(acac)$_2$
CF$_3$, Cu(hfacac)$_2$

$Cu(OSO_2CF_3)$

17

(CuOTf)

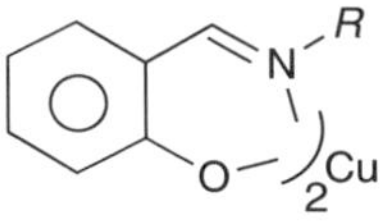

18

$R = Bu^t$, Cu(TBS)$_2$
Bn

Electronic influences from ligands for copper are evident from competitive reactions during diazo decomposition,[4] but there are few reports that provide comparative data. The series of copper(II) acetylacetonates (**16**) is among the most reliable in providing information on the effect of ligand substituents on reactivity and selectivity.[58] Increased fluorine substitution appears to increase reactivity and decrease selectivity in metal carbene transformations, but additional data are required to authenticate the generality of this statement. Copper(II) that has been supported on the perfluorinated ion-exchange polymer Nafion has been shown to be an efficient catalyst for intermolecular cyclopropanation.[59]

Copper catalysts with chiral ligands were developed to effect asymmetric induction in cyclopropanation and other metal carbene transformations. Effective ligands are bi- or tridentate that remain affixed to copper during carbene generation and subsequent transfer to the reacting substrate. Chiral copper complexes that are effective catalysts for enantioselective intermolecular cyclopropanation reactions include those of Aratani's salicyaldamines (**19**),[45] Pfaltz's chiral semicorrins (**20**)[46,50] and 5-aza-semicorrins (**21**),[60] C_2-symmetric bis-oxazolines (**22–25**),[47–49,54] and their analogs (**26, 27**),[61,62] Katsuki's chiral bipyridine (**28**),[63–67] Kanemasa's C_2-symmetric 1,2-diamine (**29**)/Cu(II) complexes,[68] Tolman's chiral

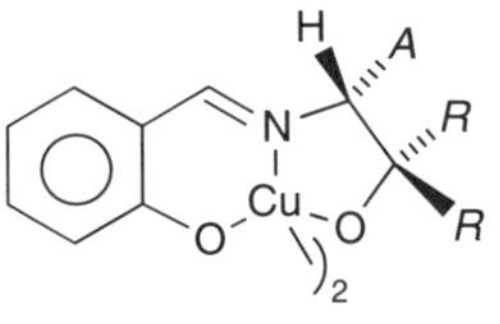

19 $A = CH_3$
      CH$_2$Ph

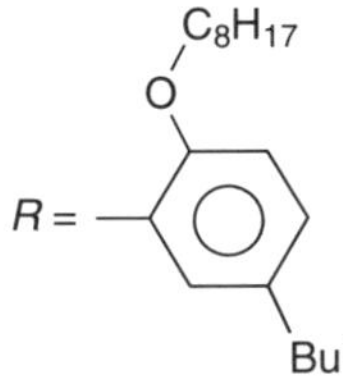

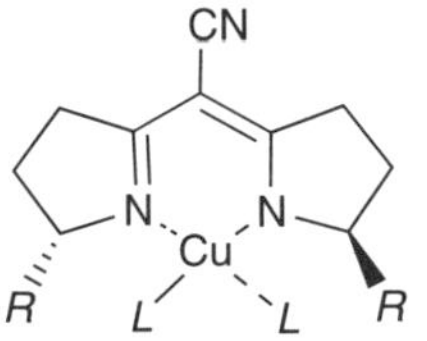

$R = $ [structure] Bu^t, $C_8H_{17}O$

20 $R = CMe_2OH$
      CH$_2$OSiMe$_2$(^{t}Bu)
      COOMe

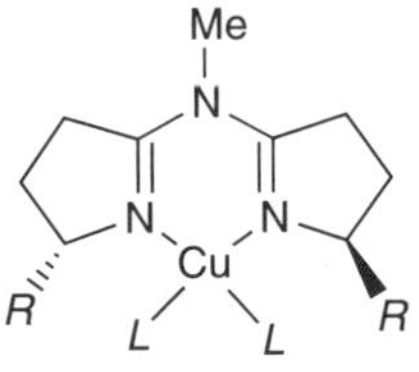

21 $R = CH_2OSiMe_2(^tBu)$
      CMe$_2$OSiMe$_3$

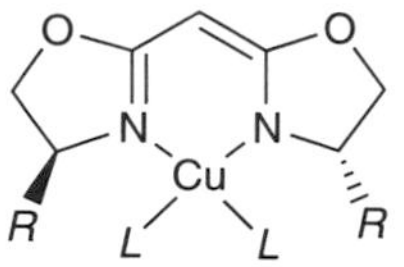

22 $R = $ PhCH$_2$
Me$_2$CH
(*S*)-EtCH(Me)
Ph
Et
Me$_3$C
Me$_2$C(OH)
CH$_2$OH
CH$_2$OSiMe$_2$(^{t}Bu)

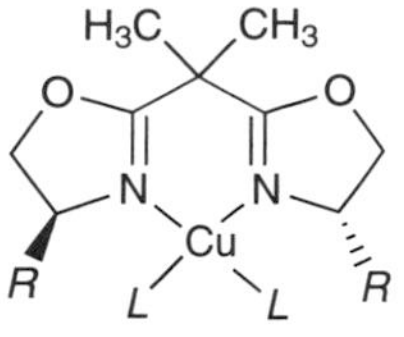

23 $R = $ Me$_2$CH
Me$_3$C

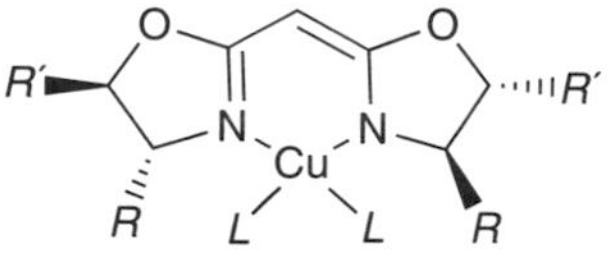

24 $R = $ CH$_3$ $R' = $ Ph
CH$_2$OH   Ph

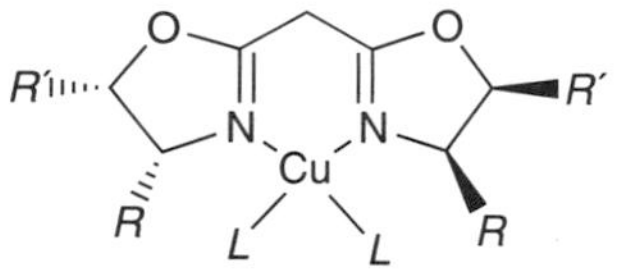

25 $R = R' = $ Ph

$R = R' = $

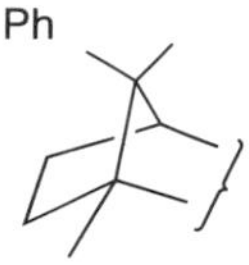

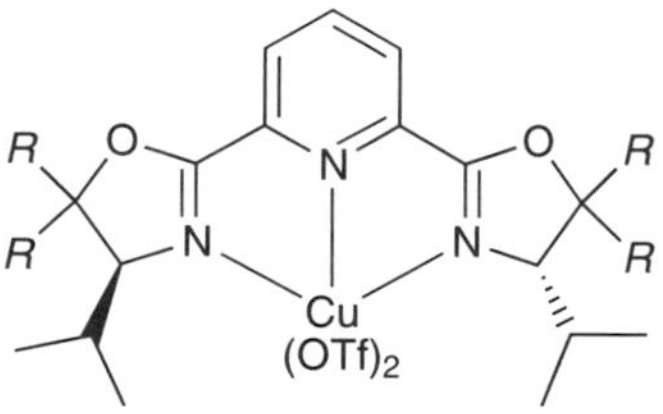

26 $R = $ Ph, Et

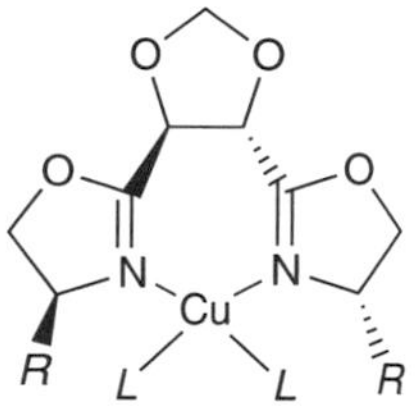

27 $R = $ PhCH$_2$
Me$_2$CH
Me$_3$C

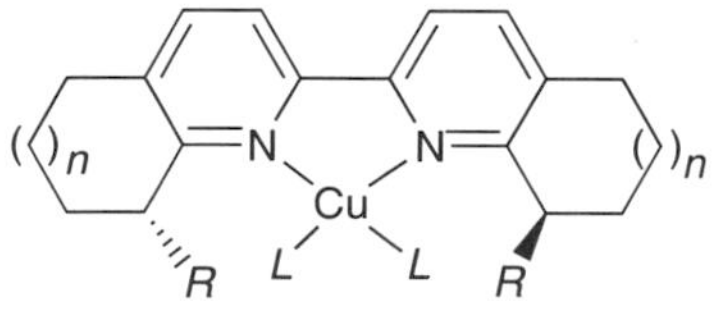

28 $R = $ SiMe$_3$
($n = 0,1$)   SiEt$_3$
Me$_2$COSiMe$_2$(^{t}Bu)
Me$_2$C(OMe)

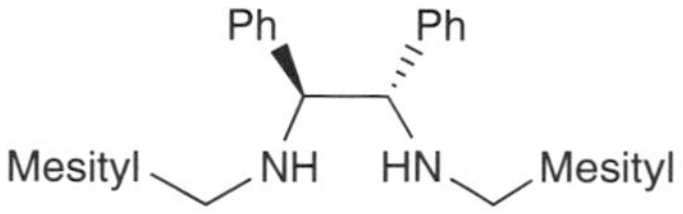

29

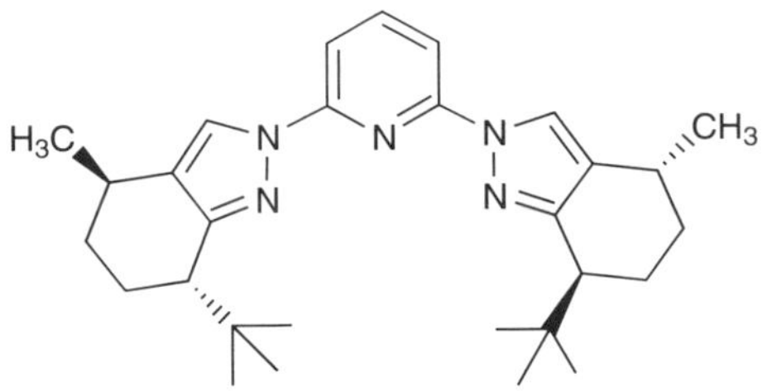

30

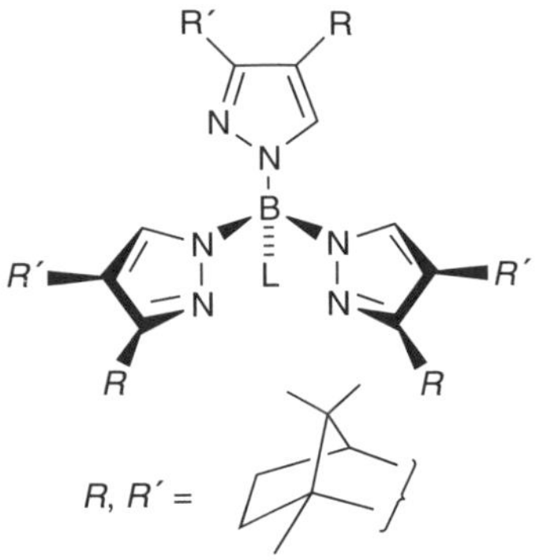

31

bis(pyrazolyl)pyridines (**30**),[69] and the Brunner pyrazolylborate **31**.[70] Thus far, semicorrin (**20**) and bis-oxazoline (**23**) ligated copper complexes have proven to have the highest stereocontrol in intermolecular cyclopropanation reactions. Katsuki's chiral bipyridine catalyst (**28**, $n = 0$) is highly effective for formal intermolecular C–O insertion in an oxonium ylide process. However, few of these ligands have been examined in sufficient detail to draw definite conclusions regarding their overall effectiveness for catalytic metal carbene transformations. What is perhaps surprising is that tridentate and/or C_3-symmetric ligands (**30, 31**) do not appear to offer any advantage in stereocontrol.

The *in situ* method for chiral catalyst generation, widely developed by Brunner[70] and popularized by Evans,[40] has gained wide acceptance as a useful and

---

LIGAND PREPARATION

For **20**: ($R = CH_2OSiMe_2(^tBu)$: Commercially available (Aldrich, Fluka); full synthetic details are in Ref. 72.

For **21**: Full synthetic details are in Ref. 60.

For **23**: ($R = Me_3C$): Prepared from (*S*)-*tert*-leucinol and dimethylmalonyl dichloride from a procedure adapted from that for diethylmalonyl dichloride in Ref. 73:

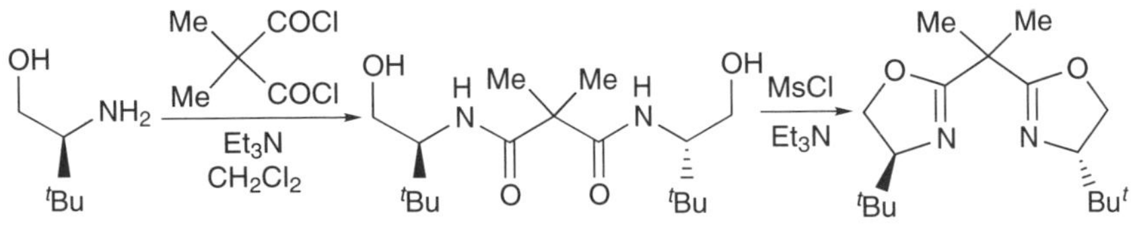

For **28**: ($n = 0$, $R = Me_2COSiMe_2^tBu$): Full synthetic details are in Ref. 66.

convenient methodology. Copper(I) triflate is most often employed, but $Cu(CH_3CN)_4PF_6$ should be preferred.[53] A high-throughput catalyst screening methodology has been described by Burgess for ligand/metal/solvent combinations to optimize selectivity in a C–H insertion reaction,[71] and this protocol may offer advantages for other catalytic metal carbene transformations.

2.5 COBALT CATALYSTS FOR DIAZO DECOMPOSITION

Bidentate bis(dioximato)cobalt(II) complexes, whose chiral bis(α-camphorquinonedioximato)cobalt(II) derivatives (**32**) have been reported by Nakamura

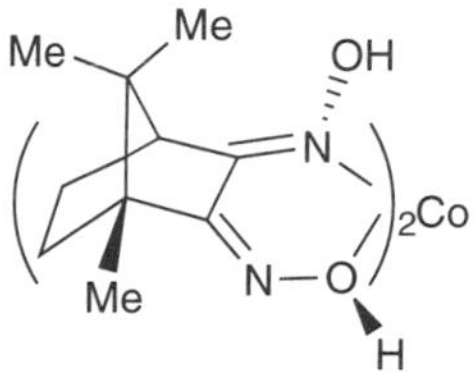

32

and co-workers to induce a high level of enantiocontrol in cyclopropanation reactions with diazo compounds,[74,75] are highly effective catalysts for diazo decomposition. The mechanism of their interaction with diazo compounds, whether or not cyclopropanation involves actual metal carbene formation, has been questioned, and an alternate pathway (Scheme 5, $E = COOR$), analogous to Scheme 4, has been proposed to account for available data.[35,76] Since the 1978 publications by Nakamura, there have been no further published reports of the use of these catalysts for diazo decomposition.

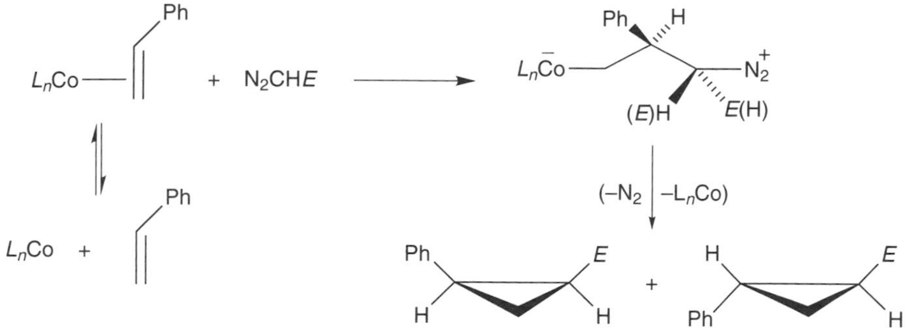

Scheme 5

Neither $Co(OOCPh)_2$ nor $Co_2(CO)_8$ is an effective catalyst for diazodecomposition,[77] but bromocobalt(III) porphyrins decompose 9-diazofluorene to azines.[78] Katsuki recently reported preliminary results with optically active Co(III)-salen

complexes (**33**), which have demonstrated their potential for highly diastereoselective and enantioselective cyclopropanation of styrene and substituted styrenes
using *tert*-butyl diazoacetate.[79] When R^1 = Bu^t the cobalt–salen complex did
not exhibit any catalytic activity, but when R^1 = H, cyclopropanation occurred
smoothly.

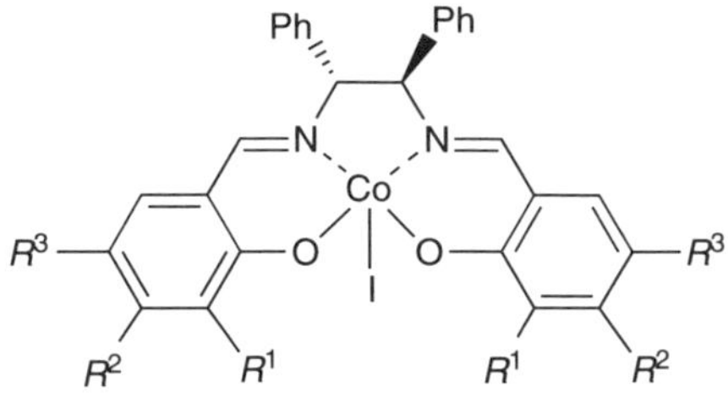

R = Bu^t, H

33

2.6 PALLADIUM CATALYSTS FOR DIAZO DECOMPOSITION

Among transition metal catalysts, palladium(II) chloride and palladium(II) acetate are the most effective for cyclopropanation reactions with diazomethane.[30]
The mechanism of their action is believed to be electrophilic addition by
palladium–olefin complexes to diazomethane,[13,35] rather than to involve initial
formation of a palladium–carbene intermediate, but either mechanism may be
operative in selected cases. Palladium(II) chloride complexes of **34** and **35**[34] have

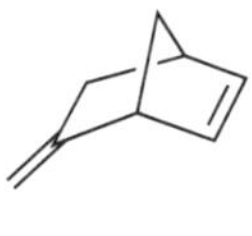

34 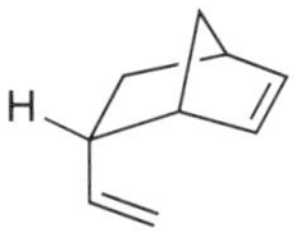   **35**

been shown to be unreactive towards ethyl diazoacetate, causing neither addition
to the olefin complex nor diazo decomposition of ethyl diazoacetate,[80] but the
corresponding experiment with diazomethane has not yet been performed. A
palladium carbene cluster, $Pd_4(\mu\text{-}CPh_2)_4(\mu\text{-}OAc)_4$, has been formed from
$Pd_4(CO)_4(OAc)_4$ and Ph_2CN_2,[81] but its relationship to Pd(II)-catalyzed reactions is
unclear.

Only palladium(II) catalysts are generally effective for cyclopropanation of
α,β-unsaturated esters and nitriles by diazomethane.[30] Dipolar cycloaddition is the
usual outcome of cyclopropanation attempts with other catalysts (Scheme 6),[82] and
there is no definitive evidence that dipolar addition is catalyzed by transition metal
compounds in these cases. Thermal decomposition of the initially formed pyrazolines (**36**) can result in cyclopropane formation, but this methodology for cyclopropane formation is not general.[83] Acids and/or bases catalyze tautomerization

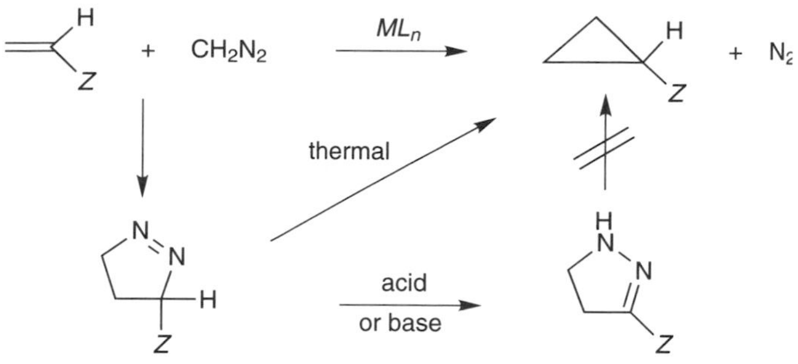

Scheme 6

of pyrazolines to their conjugated counterparts (**37**), and these 2-pyrazolines are not convertible to cyclopropane derivatives.

The palladium catalysts that have been used for diazo decomposition include PdCl$_2$ and its organic solvent–soluble bis-benzonitrile complex, PdCl$_2 \cdot$ 2PhCN,[84] Pd(OAc)$_2$ (which exists as a trimer),[85] Pd(acac)$_2$,[86] and the π-allyl complex (η-C$_3$H$_5$PdCl$_2$).[87] All appear to exhibit similar reactivities and selectivities based on product yields, but reactivity towards diazomethane may differ significantly. Selectivities observed in cyclopropanation reactions are significantly different from those achieved with copper or rhodium catalysts (Section 4.2).

Palladium(II)-catalyzed reactions of diazocarbonyl compounds generally occur in much lower yields and with lower selectivities than those catalyzed by copper or rhodium compounds.[88] The reasons for this difference are unclear, but "carbene dimer" formation (eq 6) is a common outcome, and subsequent reactions of these dimers increase the complexity of the overall reaction. Palladium-catalyzed reactions of diazo esters have no obvious synthetic value.

2.7 RHODIUM CATALYSTS FOR DIAZO DECOMPOSITION

Dirhodium(II) catalysts are the most effective and versatile for diazo decomposition.[4,13–17] The wide variety of their bridging carboxylate or carboxamide ligands provides a degree of control of reactivity and selectivity that is not evident in catalysts derived from copper or palladium. However, although they are generally applicable to diazo decomposition of diazocarbonyl compounds, their effectiveness does not extend to diazomethane. In addition to dirhodium(II) catalysts, other rhodium compounds, including iodorhodium(III) porphyrins[89,90] and hexarhodium hexadecacarbonyl,[91] have proven to be capable of diazo decomposition, often with exceptionally high turnover numbers.

2.7.1 Dirhodium(II) Carboxylates

Prepared and characterized only in the 1960s,[92,93] dirhodium(II) tetraacetate was first introduced as a catalyst for diazo decomposition by Teyssie and co-workers in 1973.[94] Since that time this air-stable compound has become the single most widely used catalyst for metal carbene transformations. Its D_{4h} symmetry with four bridging acetate ligands and one vacant coordination site per metal atom (**38**)[95-97] presents an "octahedral" geometry whose atomic array at each rhodium face resembles that of a circular wall (**39**) whose circumference is electron rich and whose center is electron deficient.

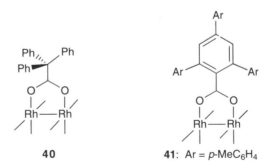

Replacement of the acetate ligand with other carboxylates changes the physical and chemical properties of the dirhodium(II) carboxylate according to the characteristics of its ligands. Dirhodium(II) tetraoctanoate, for example, has high solubility in nonpolar solvents,[98] and rhodium(II) carboxylates of terminally functionalized polyethylenecarboxylic acids have been demonstrated to be effective and reusable catalysts for cyclopropanation and ylide generation.[99] Dirhodium(II) tetra(triphenylacetate) (**40**) has been introduced as a highly efficient catalyst for carbon–hydrogen insertion reactions of α-diazo-β-keto esters,[100] and dirhodium(II) tri- and tetrakis(tri-tolylbenzoate) (**41**) have been investigated for their steric constraints in catalytic cyclopropanation reactions.[101]

Rhodium(II) acetate is conveniently prepared from $RhCl_3 \cdot xH_2O$ by the standard Wilkinson procedure,[102] which involves refluxing rhodium(III) chloride hydrate in acetic acid containing acetic anhydride. The trifluoroacetate, $Rh_2(tfa)_4$,[24] perfluorobutyate, $Rh_2(pfb)_4$,[103] and butyrate, $Rh_2(OOCC_3H_7)_4$,[104] analogs are themselves prepared from $Rh_2(OAc)_4$ by refluxing in a large excess of the replacing acid together with its anhydride. Exchange of the acetate ligand for more complex, less accessible carboxylates has been accomplished in a novel procedure from the tetrasodium salt of dirhodium(II) carbonate, which is prepared from $Rh_2(OAc)_4$ (eq. 9).[105] Using an excess of the carboxylic acid in water, carbon di-

$$Rh_2(OAc)_4 \xrightarrow[\text{H}_2\text{O}]{\text{Na}_2\text{CO}_3} Na_4Rh_2(CO_3)_4 \xrightarrow[\text{H}_2\text{O}]{\text{xs } R\text{COOH}} Rh_2(OOCR)_4 \quad (9)$$

oxide is released, and exchange occurs with observation of a color change from the initial blue of $Na_4Rh_2(CO_3)_4$ to colorless with precipitation of the green aquo adduct of $Rh_2(OOCR)_4$. Product yields are generally greater than 70%.

RHODIUM(II) PERFLUOROBUTYRATE[24,103]

Rhodium(II) acetate (500 mg, 1.13 mmol), dissolved in 10 mL of heptafluorobutanoic acid and 1 mL of heptafluorobutanoic anhydride, was heated to reflux under nitrogen for 1 h. Excess acid and anhydride were distilled and the resulting aqua-green residue was suspended in 5 mL of hexane, cooled to 0°C, and filtered. The isolated solid was dissolved in a minimum amount of boiling toluene and allowed to cool to −20°C overnight. An aqua precipitate was collected by vacuum filtration and washed with ice cold hexane to give a dark green solid, which was dried in a vacuum oven at 90°C to give 530 mg of a light green solid. The mother liquor was evaporated, and the residue was redissolved in a minimum volume of boiling toluene and cooled overnight at −20°C to give, after drying, an additional 444 mg of light green solid, which was found to be identical by HPLC analysis (methanol, μ-Bondapak-CN reverse-phase column) to the first crop of crystals. A total of 974 mg (78% yield) of rhodium(II) perfluorobutyrate was obtained as a light green, hygroscopic solid, which rapidly turns dark green upon exposure to moist air. The light green, nonhydrated material can be regenerated by heating under vacuum at 100°C for 24 h. ^{19}F-NMR (282 MHz, $CDCl_3$): δ 95.86 (t, J = 8.8 Hz, 3 F), 59.28 (q, J = 8.8 Hz, 2 F), 49.64 (m, 2 F). Fw of $Rh_2(OOCC_3F_7)_4$ = 1057.94.

Ligand exchange is also conveniently accomplished from $Rh_2(OAc)_4$ in refluxing chlorobenzene by using an excess of the alternative carboxylic acid (eq. 10).

$$Rh_2(OAc)_4 + 4R\text{COOH} \rightleftharpoons Rh_2(OOCR)_4 + 4HOAc \quad (10)$$

Although codistillation of chlorobenzene–acetic acid can be employed to remove acetic acid and drive the exchange reaction to completion,[106] the alternative use of a Soxhlet extractor whose thimble contains sodium carbonate to trap acetic acid (eq. 11) is preferable.[107] Chlorobenzene is suitable as a solvent because ligand exchange occurs at a reasonable rate, acetic acid is codistilled as it is formed, and

$$HOAc + Na_2CO_3 \longrightarrow NaOAc + NaHCO_3 \tag{11}$$

decomposition of the tetrasubstituted dirhodium(II) does not readily occur under these conditions. The lower-boiling toluene solvent can also be used, but ligand exchange is considerably slower. The rate of ligand exchange is also dependent on concentration, so that limiting the amount of solvent used is of considerable importance. Although reactants and products are not air sensitive, intermediate complexes may be; performing the exchange reactions under nitrogen is advisable.

Changing the electron-withdrawing capabilities of the carboxylate ligand strongly influences the properties of the dirhodium(II) carboxylate. Drago was the first to recognize this effect by comparing association complexes of rhodium(II) butyrate and perfluorobutyrate with Lewis bases such as pyridine and acetonitrile[24,104]; he concluded that the filled, essentially π^*, orbitals of rhodium(II) carboxylates are very effective at π back-donation into π-acceptor ligands. Complexes with olefins have been determined to exist with a variety of dirhodium(II) carboxylates absorbed on gas chromatography columns,[108] but in solution only dirhodium(II) trifluoroacetate[26] and perfluorobutyrate[27] have demonstrated this capability (spectral measurements), and the attachment of only one alkene per dirhodium(II) was observed (eq. 12). However, a bis-olefin complex has been iso-

$$Rh(L)_4Rh \ + \ \overset{R}{\underset{}{\parallel}} \ \rightleftharpoons \ Rh(L)_4Rh\text{---}\overset{R}{\underset{}{\parallel}} \tag{12}$$

lated in the solid state,[109] and applications to chiral recognition of olefins have been reported.[110]

The development of dirhodium(II) catalysts that possess chiral carboxylate ligands was undertaken independently by Brunner, McKervey, and Ikegami with variants of amino acids and α-substituted carboxylic acids. Enantiomerically pure carboxylate ligands $R^1R^2R^3CCOO$ were surveyed by Brunner and co-workers for their effectiveness on dirhodium(II) (**42**) for enantioselective intermolecular

42

43: $Z =$ H, NO$_2$, OMe
Me, ^{t}Bu,
$^nC_{12}H_{25}$

44: $R =$ PhCH$_2$, ^{t}Bu

cyclopropanation reactions of ethyl diazoacetate with styrene; substituents were varied from H, Me, and Ph to OH, NHAc, and CF_3, but with low enantiocontrol (12% ee) in all cases.[111] McKervey and co-workers initiated the preparation of prolinate derivatives of dirhodium(II) (**43**) and were the first to call them "homochiral" because the symmetrically placed ligands, and not the rhodium(II) center, bear the asymmetry.[112] Two of these catalysts (Z = H, ^{t}Bu) have been applied with high enantiocontrol to selected intermolecular[113-115] and intramolecular[116] cyclopropanation and C–H insertion reactions. A broad selection of carboxylate ligands, based mainly on prolinates but also including phenylalanine

---

DIRHODIUM(II) TETRAKIS[4-*tert*-BUTYL-*N*-BENZENESULFONYL-L-PROLINATE]

A slurry of $Na_4[Rh_2(CO_3)_4] \cdot 2.5H_2O$ (5.91 g, 10.2 mmol), prepared from $Rh_2(OAc)_4$ in 92% yield (Wilson, C. K.; Taube, H., *Inorg. Chem.* **1975**, *14*, 405), and 4-*tert*-butyl-*N*-benzenesulfonyl-L-proline (25.24 g, 81.7 mmol) in 150 mL of distilled water was refluxed for 3 h, during which time the original blue color of the carbonate solution changed to green. Approximately half of the water was then removed under reduced pressure, and the solution was slowly cooled to room temperature. The green crystals were isolated by filtration and washed with cold water then dried under vacuum in a desiccator over P_2O_5 for 16 h to yield 14.20 g of the title compound (>95% yield). Recrystallization was performed in a minimal amount of dry methanol, but both recrystallized and unrecrystallized catalyst gave the same % ee values in an intramolecular cyclopropanation reaction.

---

derivatives, were prepared by Roos and McKervey,[105] but they do not have any reported advantage over **43** (Z = H, ^{t}Bu). Davies and co-workers have also constructed homochiral dirhodium(II) carboxylates, including homologs of **43** (Z = ^{t}Bu) with four- and six-membered rings, but only when Z = $^nC_{12}H_{25}$ and ^{t}Bu (**43**) are the highest levels of enantiocontrol achieved.[117] Acyl analogs of the sulfonyl derivatives (**43**) provide lower % ee values in the intermolecular cyclopropanation reaction for which they were examined.[118] Similar catalysts based on phthalimide derivatives of phenylalanine or *tert*-leucine (**44**) were originally developed by Ikegami, Hashimoto, and co-workers[119] and successfully applied to carbon–hydrogen insertion and aromatic substitution reactions.[120-122] The comparative effectiveness of these catalysts for asymmetric induction has not been established, but there is reason to believe that the amide substituents in **43** and **44**, with their carbonyl(sulfonyl) repulsion by the ligated oxygens, are of critical importance for selectivity enhancement. Useful features of chiral dirhodium(II) carboxylates are described in Table 2.1; the X-ray crystal structure of the McKervey prolinate catalyst can be viewed in Figure 2.1.

TABLE 2.1 Features of Selected Dirhodium(II) Tetrakis(carboxylates)

Name (carboxylate)	Structure	Molecular formula	Formula weight (g/mol)	$[\alpha]_D$	X-ray structure
N-benzenesulfonyl-L-prolinate	**43** (Z = H)	$C_{44}H_{48}N_4O_{16}Rh_2S_4$	1222.93	-132.4 (CH$_3$CN, c = 0.50)	Fig. 2.1[a]
4-$tert$-butyl-N-benzenesulfonyl-L-prolinate	**43** (Z = ^{t}Bu)	$C_{60}H_{80}N_4O_{16}Rh_2S_4$	1447.36	-182.4 (CH$_3$CN, c = (0.50))	n.a.[b]
N-phthaloyl-(S)-phenylalaninate	**44** (R = phCH$_2$)	$C_{68}H_{48}N_4O_{16}Rh_2$	1382.11	-107 (CHCl$_3$, c = 0.040)	[c]
(S)-mandelate	**42** (R^1 = Ph, R^2 = OH, R^3 = H)	$C_{32}H_{28}O_{12}Rh_2$	810.38	-194.6 (CH$_3$CN, c = 0.50)	[d]

[a] Ferguson, G.; Lydon, K.; and McKervey, M. A., unpublished results.
[b] Not available.
[c] Hashimoto, S.; Watanabe, N.; Sato, T.; Shiro, M.; and Ikegami, S. *Tetrahedron Lett.* **1993**, *34*, 5109.
[d] Agaskar, P. A.; Cotton, F. A.; Falvello, L. R.; and Ham, S., *J. Am. Chem. Soc.* **1986**, *108*, 1214.

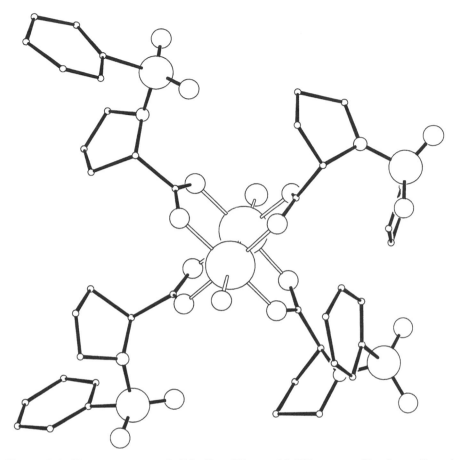

Figure 2.1. X-ray structure of dirhodium(II) tetrakis(*N*-benzenesulfonyl-L-prolinate) [**43** (*Z* = H)] with axial water molecules.

Use of **43** (*Z* = ^{t}Bu and ^{n}C$_{12}$H$_{25}$) in pentane rather than in dichloromethane produces a significant enhancement in % ee estimated to be 0.8 ± 0.2 kcal/mole for intermolecular cyclopropanation reactions of methyl phenyldiazoacetate.[115] Originally discovered by Davies,[113] this solvent effect is specific to **43** and has been interpreted[115] as being due to a change in the alignment of prolinate ligands on dirhodium(II).

2.7.2 Dirhodium(II) Carboxamidates

The first preparations of dirhodium(II) carboxamidates, whose parent structure corresponds to Rh$_2$(acam)$_4$ (**45**), were accomplished by Bear, who in cooperation with Bernal described an unexpected structural selectivity for the bridged structure in which two oxygens and two nitrogens are bound to each rhodium, and the two nitrogens are adjacent to each other (*cis*-2,2).[123] Although only (*cis*-2,2)-

45

(*cis*-2,2)—$Rh_2(acam)_4$

(*cis*—2,2) (*trans*—2,2) (3,1) (4,0)

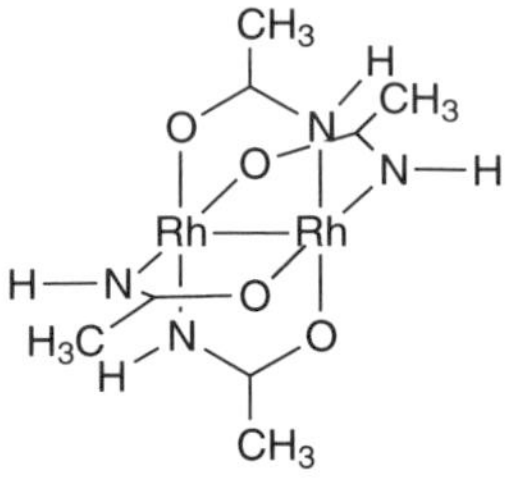

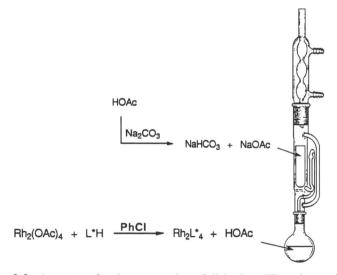

Scheme 7

$Rh_2(acam)_4$ has been isolated and characterized, the (3,1) structure (Scheme 7) for rhodium(II) *N*-phenylacetamidate[124] has also been confirmed. The (*trans*-2,2) isomer is, at best, a very minor constituent, and until recently (vide infra) the (4,0) structure had not been reported.

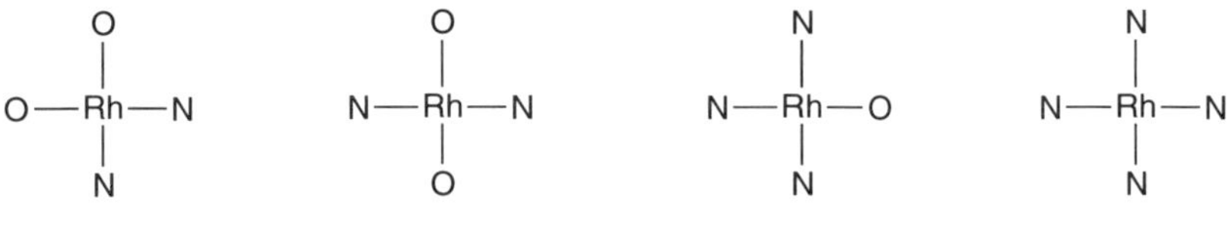

The utility of dirhodium(II) carboxamidates as catalysts for diazo decomposition was made possible by the development of convenient procedures for their synthesis. Ligand replacement on $Rh_2(OAc)_4$ by the carboxamide (two to five-fold molar excess) is accomplished in refluxing chlorobenzene using a Soxhlet extractor containing sodium carbonate to trap the acetic acid (Figure 2.2).[107] Previously, the carboxamidates were prepared in a melt at 140°C, but with acetamide

Figure 2.2. Apparatus for the preparation of dirhodium(II) carboxamidates.

incomplete conversion to dirhodium(II) tetra(acetamidate) occurred, and isolation of the desired product was laborious.[123] The process outlined in Figure 2.2 is semiautomated, and complete conversion to the tetra(carboxamidate) occurs. Dirhodium(II) tetra(acetamidate) has been prepared in this way, but dirhodium(II) tetra(caprolactamate) (**46**),[125] like $Rh_2(oct)_4$ in comparison with $Rh_2(OAc)_4$, is more soluble in organic solvents and is often preferred to $Rh_2(acam)_4$ as a catalyst.

(*cis*-2,2)-$Rh_2(cap)_4$

46

DIRHODIUM(II) CAPROLACTAMATE[125]

In a Soxhlet extraction apparatus, rhodium(II) acetate (0.490 g, 1.11 mmol) and caprolactam (2.560 g, 22.62 mmol) in freshly distilled chlorobenzene were refluxed under a nitrogen atmosphere. The Soxhlet extractor thimble was charged with oven-dried Na_2CO_3 and sand in a 3:1 ratio. The reaction was monitored by HPLC using a μ-Bondapak-CN column with 0.3% CH_3CN in methanol as the eluent, and analysis showed that ligand substitution was >98% complete after 17 h. Chlorobenzene was then removed under reduced pressure, leaving a purple solid. A column was prepared with 17 g of reverse-phase silica (BAKERBOND Cyano 40 mm prep LC packing), and the reaction mixture was loaded on the silica with methanol. A blue band eluted with the methanol. Collection of the blue band, followed by evaporation of methanol, yielded 0.485 g of a blue solid that was pure by HPLC (0.74 mmol, 67% yield) and gave satisfactory elemental analysis for $C_{24}H_{40}O_4Rh_2$; Fw = 654.39.

Chiral dirhodium(II) carboxamidates based on the use of enantiomerically pure α-substituted carboxamides have been developed with considerable diversity by Doyle and co-workers.[126,127] The ligands that have been effectively employed are those based on 2-oxopyrrolidine (**47**),[128,129] 2-oxazolidinone (**48**),[130,131] N-acylimidazolidin-2-one (**49**),[132,133] and 2-azetidinone (**50**),[134] and the most effective for enantiocontrol in metal carbene reactions are those whose attachment (*A*) is a carboxylate ester. For many of those structures described in **47–50**, both enantiomeric forms have been prepared and evaluated. X-ray crystal structures of these dirhodium(II) carboxamidates, as bis-acetonitrile or bis-benzonitrile complexes, reveal that all of them possess the (*cis*-2,2) geometry. Specific features of these catalysts are described in Table 2.2, and X-ray structures of a representative of each structural class (**47–50**) are shown in Figures 2.3–2.6.

47

A = COOMe: Rh$_2$(5S-MEPY)$_4$
A = COOCH$_2$CMe$_3$: Rh$_2$(5S-NEPY)$_4$
A = COO(CH$_2$)$_{17}$CH$_3$: Rh$_2$(5S-ODPY)$_4$
A = CONMe$_2$: Rh$_2$(5S-DMAP)$_4$

48

A = COOMe, R = H: Rh$_2$(4S-MEOX)$_4$
A = COOMe, R = CH$_3$: Rh$_2$(4S-THREOX)$_4$
A = CH$_2$Ph, R = H: Rh$_2$(4R-BNOX)$_4$
A = ^{i}Pr, R = H: Rh$_2$(4R-IPOX)$_4$
A = Ph, R = H: Rh$_2$(4R-PHOX)$_4$

49

A = COOMe, R = CH$_3$: Rh$_2$(4S-MACIM)$_4$
A = COOMe, R = Ph: Rh$_2$(4S-MBOIM)$_4$
A = COOMe, R = PhCH$_2$: Rh$_2$(4S-MPAIM)$_4$
A = COOMe, R = PhCH$_2$CH$_2$: Rh$_2$(4S-MPPIM)$_4$
A = COOMe, R = c-C$_6$H$_{11}$CH$_2$: Rh$_2$(4S-MCHIM)$_4$

50

A = COOCH$_2$Ph: Rh$_2$(4S-BNAZ)$_4$
A = COOCH$_2$CHMe$_2$: Rh$_2$(4S-IBAZ)$_4$

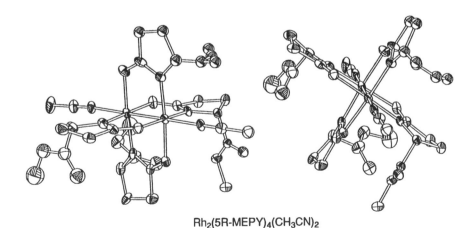

Rh$_2$(5R-MEPY)$_4$(CH$_3$CN)$_2$

Figure 2.3. Two views of the X-ray structure of Rh$_2$(5R-MEPY)$_4$(CH$_3$CN)$_2$ (without ^{i}PrOH).

TABLE 2.2 Features of Selected Dirhodium(II) Tetrakis(carboxamidates)

Name: (carboxamidate)	Abbreviation	Molecular formula	Formula weight	$[\alpha]_D$	X-ray structure
methyl 2-oxopyrrolidine-5(S)-carboxylate	$Rh_2(5S\text{-MEPY})_4$	$C_{31}H_{46}N_6O_{13}Rh_2$ as $(CH_3CN)_2$-(i-PrOH) complex	916.55	−260 (CH_3CN, $c = 0.098$)	R form, Ref. 128
N,N-dimethyl-2-oxopyrrolidine-5(S)-carboxamide	$Rh_2(5S\text{-DMAP})_4$	$C_{32}H_{50}N_{10}O_8Rh_2$ as $(CH_3CN)_2$ complex	908.58	−260 (CH_3CN, $c = 0.068$)	Ref. 129
methyl 2-oxooxazolidin-4(S)-carboxylate	$Rh_2(4S\text{-MEOX})_4$	$C_{24}H_{30}N_6O_{16}Rh_2$ as $(CH_3CN)_2$ complex	864.35	−222 (CH_3CN, $c = 0.120$)	Ref. 130 $(PhCN)_2$ complex
methyl 5(R)-methyl-2-oxooxazolidin-4(S)-carboxylate	$Rh_2(4S\text{-THREOX})_4$	$C_{28}H_{38}N_6O_{16}Rh_2$ as $(CH_3CN)_2$ complex	920.46	−95.5 (CH_3CN, $c = 1.10$)	Ref. 130 $(PhCN)_2$ complex
4(R)-benzyl-2-oxooxazolidine	$Rh_2(4R\text{-BNOX})_4$	$C_{44}H_{46}N_6O_8Rh_2$ as $(CH_3CN)_2$ complex	992.70	+147 (CH_3CN, $c = 0.132$)	Ref. 128
4(R)-phenyl-2-oxooxazolidine	$Rh_2(4R\text{-PHOX})_4$	$C_{40}H_{38}N_6O_8Rh_2$ as $(CH_3CN)_2$ complex	936.62	+323 (CH_3CN, $c = 0.128$)	Ref. 131
methyl 1-acetyl-2-oxoimidazolidine-2-one-4(S)-carboxylate	$Rh_2(4S\text{-MACIM})_4$	$C_{32}H_{42}N_{10}O_{16}Rh_2$ as $(CH_3CN)_2$ complex	1028.58	−325 (CH_3CN, $c = 0.104$)	Ref. 133
methyl 1-benzoyl-2-oxoimidazolidine-4(S)-carboxylate	$Rh_2(4S\text{-MBOIM})_4$	$C_{52}H_{50}N_{10}O_{16}Rh_2$ as $(CH_3CN)_2$ complex	1276.84	−447 (CH_3CN, $c = 0.13$)	Ref. 133
methyl 1-(3-phenylpropanoyl)-2-oxoimidazolidine-4(S)-carboxylate	$Rh_2(4S\text{-MPPIM})_4$	$C_{60}H_{66}N_{10}O_{16}Rh_2$ as $(CH_3CN)_2$ complex	1389.05	−311 (CH_3CN, $c = 0.10$)	Ref. 133
methyl 1-cyclohexylacetyl-2-oxoimidazolidine-4(S)-carboxylate	$Rh_2(4S\text{-MCHIM})_4$	$C_{56}H_{82}N_{10}O_{16}Rh_2$ as $(CH_3CN)_2$ complex	1357.15	−375 (CH_3CN, $c = 0.12$)	Ref. 133

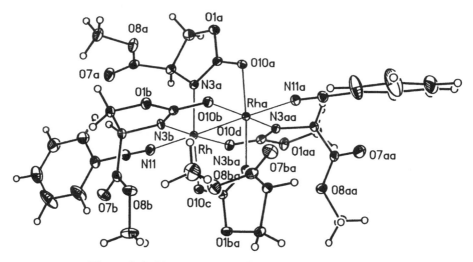

Figure 2.4. X-ray structure of $Rh_2(4S\text{-MEOX})_4(PhCN)_2$.

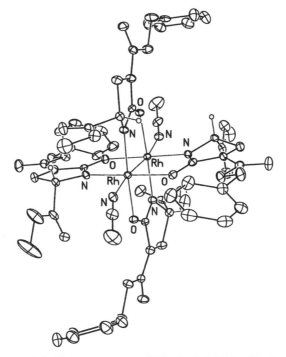

Figure 2.5. X-ray structure of $Rh_2(4S\text{-MPPIM})_4(CH_3CN)_2$.

Although the vast majority of these dirhodium(II) compounds are soluble in CH_2Cl_2, increasing the hydrocarbon content of the ester alkyl group [e.g., $Rh_2(5S\text{-ODPY})_4$] provides solubility in hydrocarbon solvents as well.

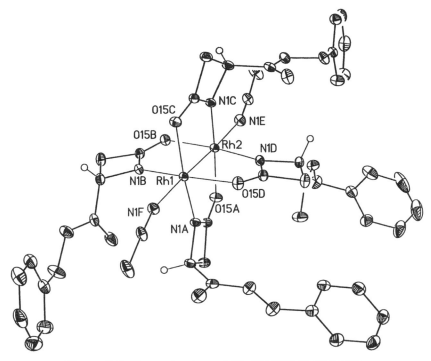

Figure 2.6. X-ray structure of $Rh_2(4S\text{-}BNAZ)_4(CH_3CN)_2$.

The characteristic geometry of each structure is the projection of A from the point of attachment nearly perpendicular from the plane of the ligand, and a useful representation for this configuration is given in Scheme 8, which for

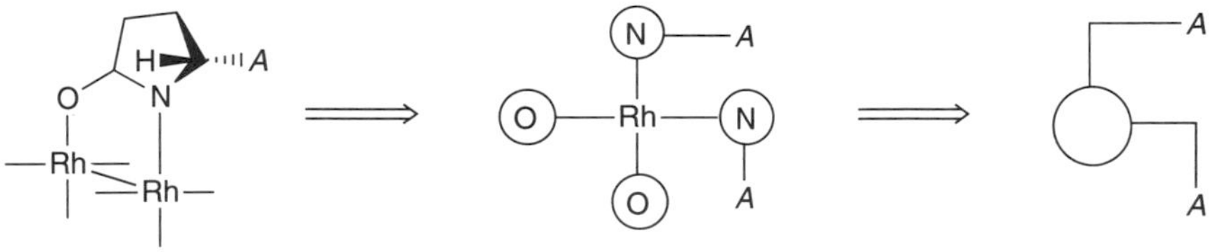

51   **52**   **53**

Scheme 8

$A = COOMe$ would describe the (R) configuration. These structural representations **51–53** portray a rhodium face that has four quadrants, designated by the N,O-ligated atoms: (N,N), (N,O), (O,O), (O,N) clockwise; the attachments A occupy the (N,N) and (N,O) quadrants. In contrast to the C_2-symmetric ligands that are effective for enantiocontrol with copper(I) and whose attachments A are arrayed *trans* (E), the ligand attachments of the dirhodium(II) complexes have a *cis*-(Z)

DIRHODIUM(II) TETRAKIS[METHYL 2-OXOOXAZOLIDINE-4(S)-CARBOXYLATE], Rh$_2$(4S-MEOX)$_4$[130]

A mixture of dirhodium(II) tetraacetate (0.520 g, 1.17 mmol) and methyl (4S)-2-oxazolidinone-4-carboxylate (1.40 g, 9.65 mmole) in 50 mL of freshly distilled chlorobenzene, contained in a round-bottom flask fitted with a Soxhlet extraction apparatus (Figure 2.2) was heated at reflux under nitrogen for 4–6 h. The thimble in the Soxhlet extraction apparatus was charged with an oven-dried mixture (5 g) of two parts sodium carbonate and one part sand. Progress of the ligand replacement was followed by HPLC (μ-Bondapak-CN column, 2% CH$_3$CN in MeOH). As the reaction progressed, the Rh$_2$(OAc)$_4$ band disappeared and was replaced by several bands with longer retention volumes until, after 4–6 h, one principal band in addition to that for the ligand was observed. the reaction rate was critically dependent on the rate of stirring and a rapid reflux rate. The resulting blue solution was cooled, and the solvent was removed under reduced pressure to leave a purple solid/oil residue. Methanol (3–5 mL) was added to the residue; the solid was filtered and washed sequentially with cold methanol (3 × 5–10 mL) and pentane. After drying, 0.90 g of purple solid (97% purity by HPLC) was obtained. Recrystallization from anhydrous acetonitrile (1 mL/100 mg) yielded 0.72 g of a red-orange solid consistent with the formula Rh$_2$(4S-MEOX)$_4$(CH$_3$CN)$_2$ (0.94 mmol, 80% yield): ^{1}H-NMR (CDCl$_3$): δ 4.47–4.33 (m, 4 H), 4.22 (dd, J = 6.7, 4.2 Hz, 1 H), 4.13 (dd, J = 8.8, 3.7 Hz, 1 H), 3.75 (s, 3 H), 3.73 (s, 3 H), 2.20 (CH$_3$CN); ^{13}C-NMR (CDCl$_3$): δ 173.4, 173.3, 168.6, 168.5, 116.0 (CH$_3$CN), 76.3, 76.1, 63.7, 63.2, 52.4, 52.1, 2.4 (CH$_3$CN); $[\alpha]_D^{23}$ −222 (c 0.120, CH$_3$CN).

---

array. A carbene bound to Rh has, by molecular mechanics calculations,[128] two minimum-energy conformations: one in which the carboxylate group of the carbene produced from a diazoester bisects the (O,O)-quadrant (**54**) and the other in which the carboxylate group bisects the (O,N)-quadrant (**55**). Therefore, substrate approach to the carbene center is restricted from one side in each configuration (**54** and **55**), and control of enantioselectivity is determined by the relative rates for substrate attack on **54** and **55**.[128]

54   **55**

The imidazolidinone-ligated dirhodium(II) complexes **49** provide the highest level of complexity that, as will be seen in Section 3.8, greatly influences diastereoselectivity as well as enantioselectivity. The carbonyl groups of the 1-acyl substituents are directed anti from the ligated oxygens of the same ligands (**56**). The carboxylate attachments (*A*) are arrayed perpendicular to each other,

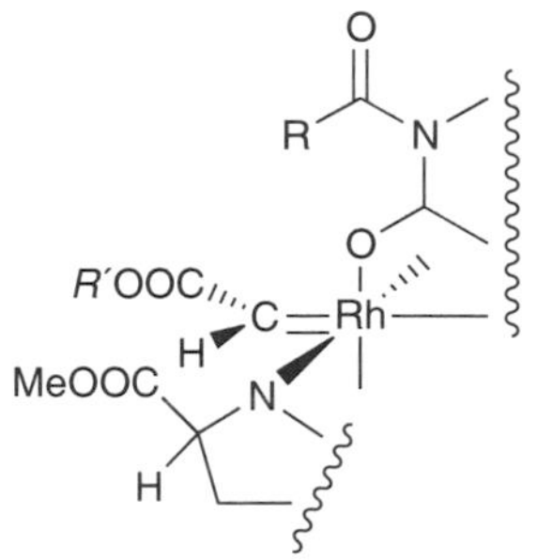

56

and they are closer to the carbene center than are R. The net effect is to enclose the carbene center, inhibiting access from a reacting substrate, and thereby exacting a higher selectivity. This set of dirhodium(II) carboxamidates is the only one from which both (3,1) and (4,0) isomers have been isolated and characterized.[133,135] The catalytic center of the (4,0) isomer is the rhodium bound to four oxygens; so this complex does not provide enantiocontrol in catalytic metal carbene transformations.

Except for azetidinone-ligated catalysts **50** whose Rh–Rh bond distance is 2.53 Å, the Rh–Rh bond lengths for **47–49** are independent of the ligand (2.46 ± 0.01Å), as are their Rh–O and Rh–N bond lengths. However, there are minor differences in internal bond angles expected from the presence of C, N, or O within the ring. The mechanism for the formation of these complexes has been described.[133,135]

Attempts to prepare chiral dirhodium(II) carboxamidates conveniently using acyclic carboxamides have thus far been unsuccessful. Apparently there is a distinct preference for the higher-energy conformation (in acyclic systems), represented by **57**, in the ligand exchange process. In addition to these structural influences, N–H acidity affects the rate of ligand exchange.

57   **58**

The (*R*) and (*S*) forms of Rh$_2$(MEPY)$_4$, Rh$_2$(MEOX)$_4$, and Rh$_2$(MPPIM)$_4$ have been commercialized and are available from Aldrich and (in Europe) Acros. The detailed syntheses of $Rh_2(5R\text{-}MEPY)_4$ has been published in *Organic Synthesis*.[136] All preparations follow the same methodology as described for $Rh_2(cap)_4$, but the chiral dirhodium(II) carboxamidates are generally chromatographed on reverse-phase cyano-linked columns rather than on silica gel, which causes their decomposition, presumably by ester hydrolysis. Although generally unreactive towards water, prolonged contact causes their partial decomposition.

2.7.3   Hexarhodium Hexadecacarbonyl

This black rhodium carbonyl cluster, $Rh_2(CO)_{16}$, has been described as an effective catalyst for diazo decomposition in cyclopropanation and ylide-generation reactions of diazoesters.[77,91] Cyclopropanation selectivities are nearly the same as those achieved with $Rh_2(OAc)_4$.[137] The identity of the catalytically active species, whether it be the hexarhodium cluster or a fragment, is unknown, but quantitative recovery of $Rh_2(CO)_{16}$ following catalytic cyclopropanation occurred when performed under an atmosphere of carbon monoxide.[91]

2.7.4   Dirhodium(II) Phosphates and Orthometallated Phosphines

(S)-$(+)$-$1,1'$-Binaphthyl-$2,2'$-diyl hydrogen phosphate **59** has been reported to be an effective chiral ligand for dirhodium(II) (**60**) in metal carbene reactions.[138,139]

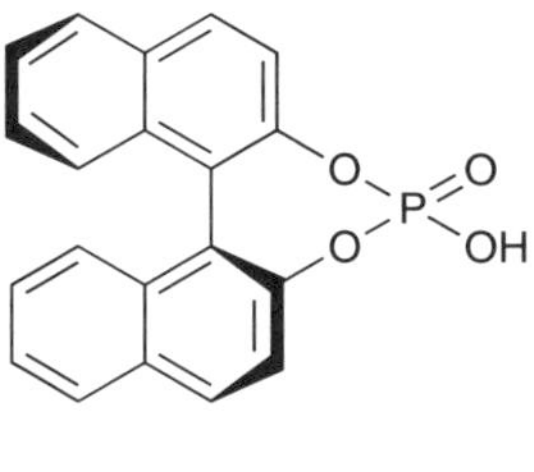

59
(*S*-BNHPH)

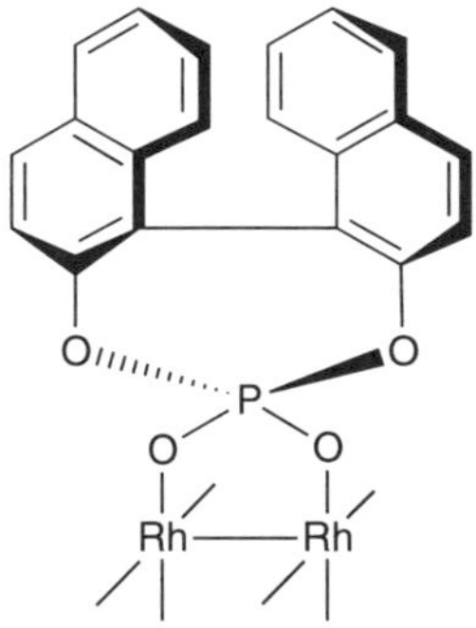

60

The ligand is prepared from $(\pm)$-$1,1'$-bis-2-naphthol through reaction with $POCl_3$, followed by resolution.[140] Using the $Na_4Rh_2(CO_3)_4$ method for ligand exchange (eq. 9), McKervey, McCann, and co-workers formed the disubstituted complex, $Rh_2(S\text{-}BNHP)_2(HCO_3)_2$, which was catalytically active for a variety of metal carbene transformations.[138] Pirrung prepared $Rh_2(R\text{-}BNHP)_4$ by ligand replacement on $Rh_2(OAc)_4$ and found this catalyst to be effective for carbonylcarbene dipolar addition reactions.[139]

A novel set of dirhodium(II) catalysts bearing two carboxylate ligands and two orthometallated phosphine ligands (**61**) has been prepared.[141] The two carboxylate ligands are *cis*, and the two orthometallated phosphines alternate as de-

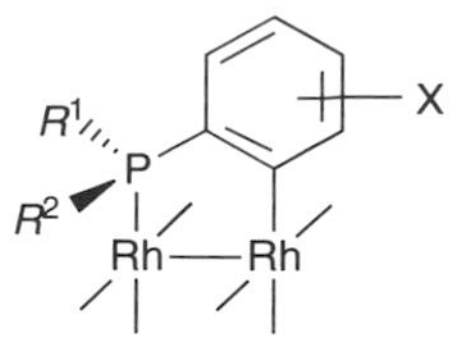

61

X = H, F, CH$_3$CF$_3$

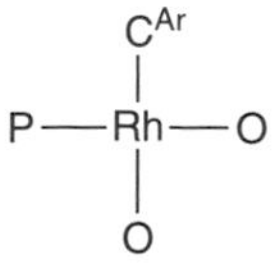

62

picted in **62**. These dirhodium(II) compounds possess C_2 symmetry, and they are especially active towards diazo decomposition of diazoketones in intramolecular cyclopropanation reactions.

2.7.5 Rhodium(III) Porphyrins

The use of metal porphyrins as catalysts for diazo decomposition originated with Callot,[106] who reported the use of iodorhodium(III) mesotetraarylporphyrins (**63**, Ar = phenyl, mesityl) for cyclopropanation. More recently, Kodadek and co-

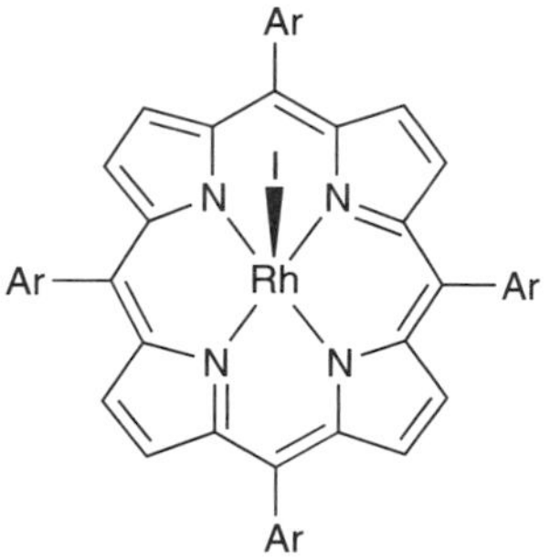

63

workers employed this rhodium(III) porphyrin system (Ar = p-tolyl) to obtain detailed information about the mechanism of metal carbene formation,[18,142] and to evaluate its potential for enantioselective cyclopropanation reactions.[143,144] In the latter case, ligands were substituted with binaphthyl (**64**) or pyrenyl-naphthyl (**65**), and the resulting catalyst has been described as resembling the structure de-

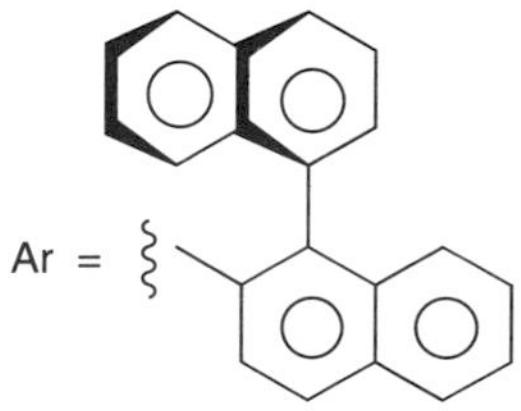

Ar =

64

Ar =

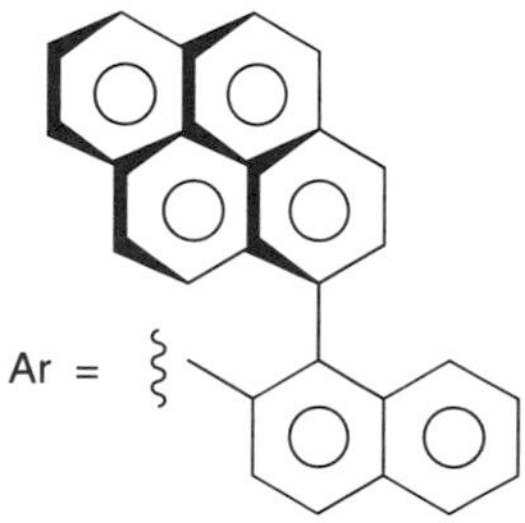

65

picted as **66**. Although enantiocontrol is not high with these chiral catalysts, turnover numbers are exceptional.

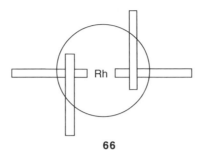

66

2.8   RUTHENIUM CATALYSTS FOR DIAZO DECOMPOSITION

Until recently ruthenium compounds have been overlooked as potentially effective catalysts for diazo decomposition. A polymeric dicarbonylruthenium(I) acetate, $[Ru_2(CO)_4(OAc)_2]_n$, and its bis-acetonitrile complex, $Ru_2(CO)_4(OAc)_2$-$(CH_3CN)_2$, have been reported by Maas to be efficient cyclopropanation catalysts with diazoacetates.[145] Ruthenium carborane complexes that possess triphenylphosphine ligands are also effective for intermolecular cyclopropanation reactions of ethyl diazoacetate,[146] and triruthenium dodecacarbonyl, $Ru_3(CO)_{12}$, has been reported to be effective for both cyclopropanation and ylide generation with ethyl diazoacetate.[77] Recently, $RuCl_2(PPh_3)_3$ has been found to be an effective catalyst for the cyclopropanation of styrene and substituted styrenes with ethyl diazoacetate, but aliphatic alkenes gave low product yields.[147] Ruthenium(II,II) tetracarboxylates have also been examined for their catalytic activity and, although cyclopropanation reactions are observed, diazo decomposition in the presence of olefins also gives metathesis products.[148] Among the diruthenium catalysts, the (I,I) oxidation state, being isoelectronic with dirhodium(II), holds the greatest promise. Thus $Ru_2(CO)_4(OAc)_2$ and its acetate-substituted derivatives may be alternatives to $Rh_2(OAc)_4$, but further demonstrations of reactivity and selectivity are required.

In a recent series of publications, Nishiyama and co-workers have described the development of C_2-symmetric 2,2-bis(2-oxazolin-2-yl)pyridine (Pybox) as a chiral ligand for enantioselective reactions (**67**).[149–152] In applications with dia-

R = Et, ^{i}Pr, ^{s}Bu,
PhCH$_2$, Ph

Pybox
67

zoacetates they found that the combination of **67** ($R = i$-Pr) with ruthenium(II) chloride, but not with CuOTf, $Rh_2(OAc)_4$, or $PdCl_2(PhCN)_2$, exhibited high catalytic activity and excellent enantiocontrol,[152] even when $R = $ Et. The active catalyst is prepared *in situ* by the combination of $[RhCl_2(p$-cymene$)]_2$[153] and **67** in dichloromethane under an atmosphere of ethylene (eq. 13). X-ray crystal struc-

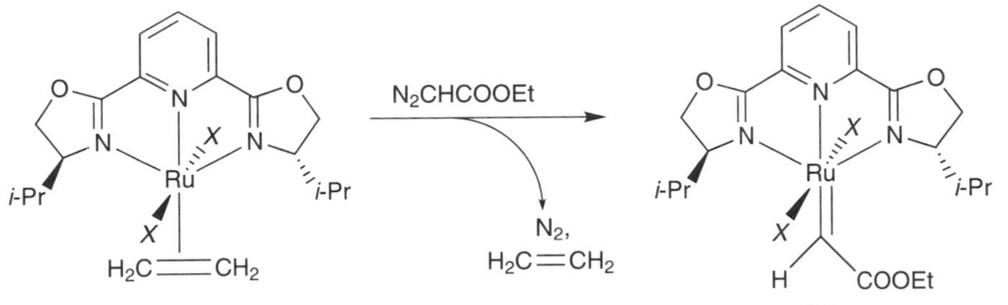

(13)

tures of the ethylene complex of $RuCl_2$(Pybox-*dihydro*)(C_2H_4) and of $Ru(OAc)_2$ (Pybox-^{i}Pr)(CO) confirm the structural representation of the catalyst to be that derived from **68**, where X (Cl, Br, OAc) substituents are *trans*.[152] Changing the substituent X from Cl to Br appears to increase enantiocontrol but decrease diastereocontrol in cyclopropanation reactions.

Treatment of **68** (generated *in situ*) with trimethylsilyl diazomethane formed the corresponding trimethylsilylcarbene complex (**70**) that is thermally and air stable in the solid state but decomposes gradually in solution (Scheme 9).[151] NMR

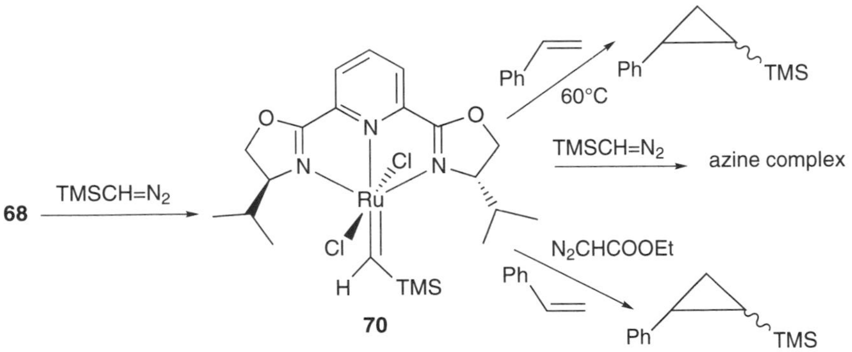

Scheme 9

data confirm the assignment. The carbene complex **70** undergoes cyclopropanation with stryene, forms an azine complex with excess $TMSCH = N_2$, and catalyzes cyclopropanation by ethyl diazoacetate. The stability of **70** suggests that carbene transfer may be the rate-limiting step in many reactions catalyzed by **68**.

Although there is limited information available on these ruthenium catalysts, they appear to have high potential for stereoselective metal carbene transformations, though their reactivities may be too low to expect synthetic versatility.

2.9   OTHER TRANSITION METALS AS CATALYSTS FOR DIAZO DECOMPOSITION

A wide variety of transition metals catalyze diazo decomposition. Some, like osmium porphyrin complexes,[154] provide the expected link between catalytic and stoichiometric metal carbenes. Others that include $[CpFe(CO)_2THF]^+$ are the Lewis acid fragments[155] of "stable" carbenes or their ylide precursors that have been employed for stoichiometric reactions.

2.9.1   Osmium

Complexes developed by Woo and co-workers with *meso*-tetra-*p*-tolylporphyrin (TTP), specifically $[Os(TTP)]_2$, catalyze diazo decomposition and, in the presence of alkenes, form cyclopropane compounds with high trans:cis (anti:syn) ratios.[154] The osmium porphyrin carbene complex, $(TTP)Os = CHCOOEt$, has been isolated and characterized.[156] Although the osmium–porphyrin complexes are isoelectronic with their corresponding rhodium–porphyrin complexes (**63**), they provide markedly different diastereoselectivities in cyclopropanation reactions, and the rhodium(III) porphyrins do not form stable metal carbenes derived from diazoesters as do osmium.

Demonceau and co-workers have discovered that $OsCl_2(PPh_3)_3$[157] and $[OsCl_2(p\text{-cymene})]_2$[158] are catalytically active towards cyclopropanation of styrenes by ethyl diazoacetate performed at 60–80°C. However, linear and cyclic olefins exhibited low reactivity, suggesting that, as is the case with ruthenium catalysts, carbene transfer from the metal carbene intermediates is slow relative to reactions performed with copper and rhodium catalysts.

2.9.2   Iron

Hossian and coworkers have capitalized on the promising selectivities achieved by Brookhart and others[38] using ylide-stabilized iron carbenes by making them catalytic for cyclopropanation reactions. The $Cp(CO)_2Fe^+$ catalyst, employed as its BF_4^- salt in THF,[155,159–162] is formed from $Cp(CO)_2FeCH_3$ by protonation with less than one equivalent of $HBF_4 \cdot OEt_2$ followed by several recrystallizations to ensure the absence of HBF_4.[159] Room temperature or higher is required for catalytic reactions, even with phenyldiazomethane,[160] which suggests that ligand (THF) dissociation is the rate-limiting step.

Mansuy and co-workers have reported the formation of iron–carbene complexes in reactions between diazoalkanes and iron porphyrins,[163,164] but Kodadek and Woo have demonstrated the catalytic activity of Fe(TTP) and its tetrak-is(perfluorophenyl) analog, Fe(PFP), towards cyclopropanation of alkenes with ethyl diazoacetate.[165] Reactions are characterized by high turnover numbers, similar to those found with rhodium–porphyrin complexes,[144] and they exhibit a kinetic isotope effect with styrene. The preparation of these iron–porphyrin systems from Fe(TTP)Cl or Fe(PFP)Cl was carried out *in situ* with the rigorous exclusion of dioxygen.

2.9.3 Platinum and Nickel

Platinum(II) compounds that include Zeise's dimer (**71**) cause diazo decomposition and, in reactions performed on the coordinated olefin (Scheme 4) can result in cyclopropane formation.[166] However, platinum(II) compounds generally form stable platinacyclobutanes (**72**)[167] upon reaction with diazo compounds R_2CN_2,

and this pathway renders such reactions stoichiometric rather than catalytic. Platinacyclobutane compounds react with pyridine and substituted pyridines to form platinum(II) ylides and, as demonstrated by Jennings and co-workers,[168] such ylides are formed directly by the action of platinum(II) halides on diazoalkanes in the presence of pyridine. Platina(IV)cyclobutanes derived from hydrocarbon **73** undergo stereospecific rearrangement to 1,3-divinyl cyclopentane derivatives (eq. 14) upon treatment with diazomethane.[169] Platinum chemistry

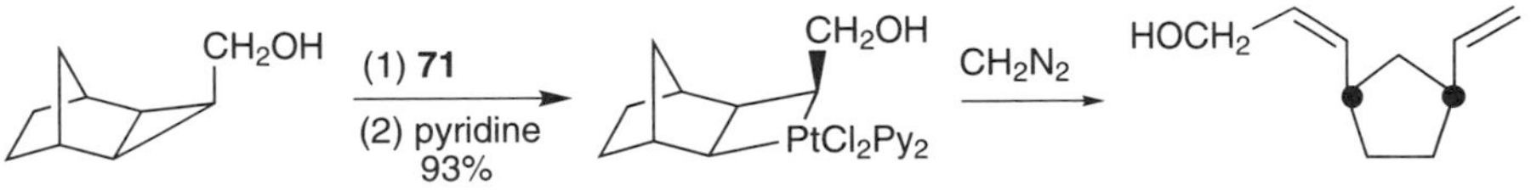

(14)

with diazo compounds is distinctly different from that of palladium(II), and there is currently no obvious advantage to the use of platinum(II) for reactions with diazo compounds.

There are few examples of nickel-promoted diazo decomposition,[4] and most of them are associated with diazomethane reactions.[30] The creation of a suitably electrophilic nickel catalyst that is capable of high turnover numbers and selectivity has yet to be achieved.

2.9.4 The Chromium Triad

Molybdenum hexacarbonyl and dimolybdenum(II) tetraacetate promote cyclopropane formation in reactions that occur between electron-deficient alkenes and
diazocarbonyl compounds, but the principal cause for their success is apparently
prevention of tautomerism of the pyrazoline product formed from dipolar
addition.[82] The pyrazoline (**74**) undergoes loss of dinitrogen to form cyclopropane
and alkenes (eq. 15, Z = COOR', CN, COMe).[83] The pyrazoline mechanism

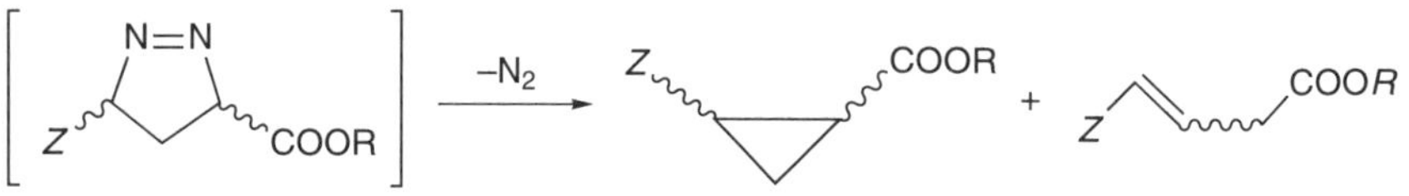

74   **75**   **76**

$$(15)$$

(Scheme 6) for cyclopropanation is indicated by the presence of alkenes (**76**) that
accompany the cyclopropane product. A broad selection of molybdenum compounds and representative compounds of other transition metals (Zr, V, Cr, Re,
Fe, Ru, Os, Ir), generally as carbonyls, was investigated, but without obvious improvement over results obtained with $Mo(CO)_6$ or $Mo_2(OAc)_4$.

Doyle and co-workers were the first to use a stable metal carbene as a *procatalyst* for catalytic metal carbene transformations.[170] The objective was the same as
that subsequently employed by Hossain[155] with $Cp(CO)_2Fe^+$: to use the transition
metal fragment of a stable metal carbene as a catalyst for diazo decomposition.
The development reported by Doyle capitalized on the reactivity of the stable Fischer carbene with diazomethane.[171] Consequently, treatment of $(CO)_5W =$
$C(OMe)Ph$ with ethyl diazoacetate generated the $(CO)_5W$ fragment as the catalytically active species for diazo decomposition (Scheme 10). In principle, any of the
"stable" metal carbenes or their ylide precursors could serve as a procatalyst for
catalytic metal carbene reactions. In practice, however, generation of the highly

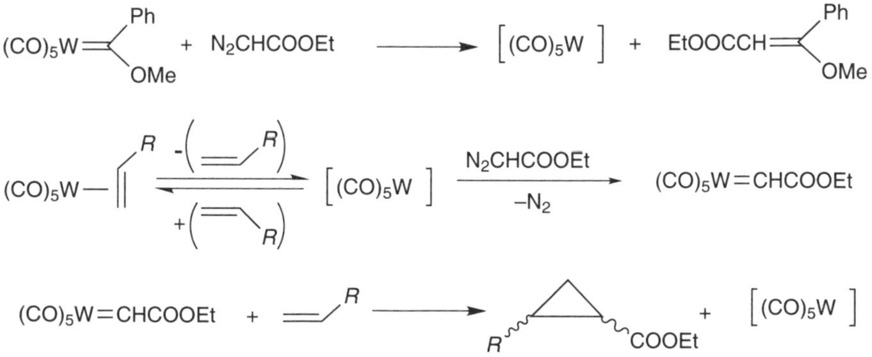

Scheme 10

electrophilic, catalytically active fragment is the rate-limiting step, and the subsequently generated metal carbene may have limited stability under conditions required for its formation.

2.10 METAL CARBENES AS STOICHIOMETRIC REAGENTS

Advances in our understanding of catalytic metal carbene reactions are often aided by insights provided by the parallel development of stoichiometric transformations of stable metal carbenes. That transition metal carbene complexes could serve as carbene transfer reagents had its origins in investigations by Pettit and Jolly of the acid-promoted decomposition of $Cp(CO)_2FeCH_2OMe$,[172] but it was only after subsequent reports by E. O. Fischer and co-workers of cyclopropanation reactions of electron-rich olefins (vinyl ethers) and electron-poor olefins (α,β-unsaturated carbonyl derivatives) by isolable heteroatom-stabilized carbenes in the chromium triad that synthetic advances gained momentum.[38,173,174]

Brookhart has classified transition metal carbene complexes that are employed as stoichiometric reagents by *heteroatom-stabilized* (Fischer-type) and *non-heteroatom-stabilized*.[38] Fischer-type carbenes are generally isolable, and they are stable under ordinary conditions. Non-heteroatom-stabilized metal carbene complexes are considerably less stable and are, consequently, generated *in situ*; however, these metal carbenes are much more reactive than are Fischer-type metal carbenes towards carbene transfer reactions.

Heteroatom-stabilized metal carbenes	Non-heteroatom-stabilized metal carbenes
$(CO)_5M = \overset{Z}{\underset{R}{<}}$	$(CO)_5W = \overset{R}{\underset{R'}{<}}$
M = Cr, Mo, W, Fe	R or R' = H, alkyl, aryl
Z = OR', NR_2, SR'	
R = alkyl, aryl, Z	$\left[Cp(CO)_2Fe = \overset{R}{\underset{R'}{<}} \right]^+$
Fischer-type	R or R' = H, alkyl, aryl

A classification of stability across the spectrum of metal carbenes portrays a continuum for carbene stability versus carbene transfer reactivity (Scheme 11). However, although this classification accounts for the influence of carbene substituents Y, Z, and R, there has not been a similar classification for the transition metal M except to note that the early transition metals are more amenable to the formation of heteroatom-stabilized metal carbenes, whereas the late transition

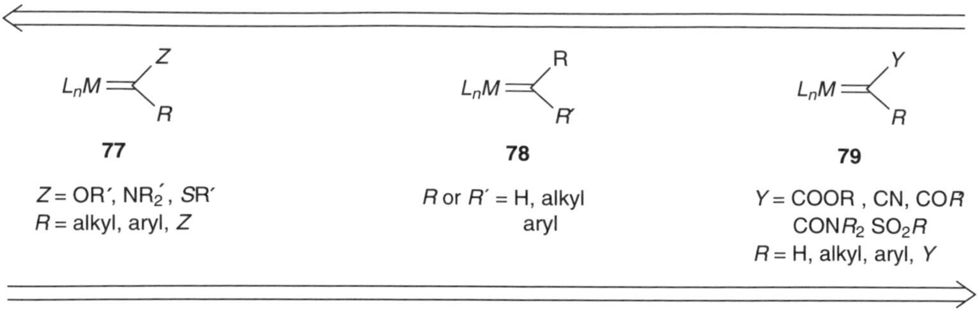

77	78	79
$Z = OR'$, NR_2', SR'	R or $R' = H$, alkyl	$Y = COOR$, CN, COR
$R = $ alkyl, aryl, Z	aryl	$CONR_2$ SO_2R
		$R = H$, alkyl, aryl, Y

Scheme 11

metals are more suitable as catalysts to produce metal carbenes of type **79**. One reason for the relative absence of correlatable information is accessibility: **77** and **78** can be formed by stoichiometric methods while **79** generally cannot,[38] whereas **79** and **78** can be accessed catalytically with diazo compounds, while **77** generally cannot.[12]

Fischer-type carbenes undergo cyclopropanation with alkenes,[175] most often at elevated temperatures under which conditions one molecule of carbon monoxide is extruded, by a reaction mechanism (Scheme 12) that is proposed to involve metallocyclobutane formation. Electron-deficient alkenes ($H_2C = CHA$; α,β-unsaturated esters, nitriles)[176–180] may afford a different regioisomer (**82**) in metallocyclobutane formation than electron-rich alkenes ($H_2C = CHD$; vinyl

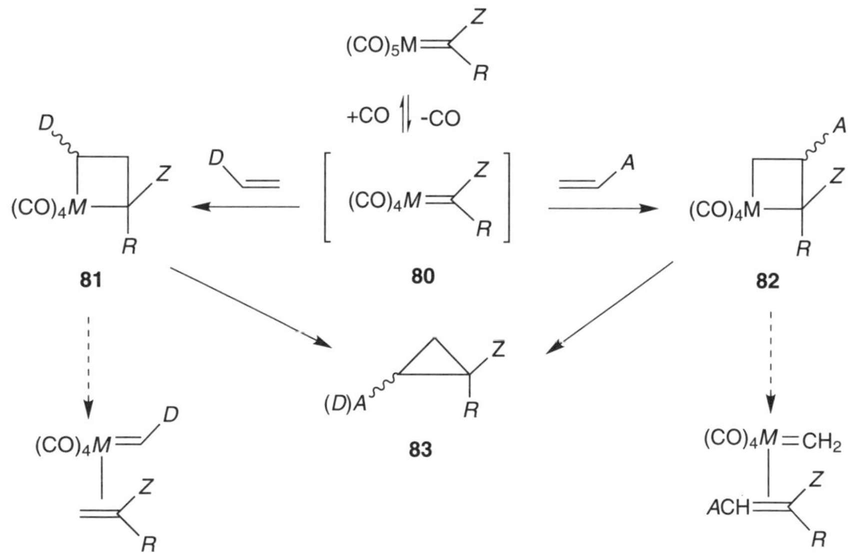

Scheme 12

ethers),[181] but both form cyclopropane product(s) with the same substitution pattern (**83**). In these cases metathesis products from metallocyclobutane intermediates may be observed.[182]

Non-heteroatom-stabilized metal carbenes, with one known exception,[183] undergo metal carbene transformations in the same way as do those that are generated catalytically. Indeed, linear correlations exist with reactivity and selectivity for cyclopropanation between $(CO)_5W=CHPh$ and the carbene generated catalytically from $PhCHN_2$ with $Rh_2(OAc)_4$.[32] Still, because of metallocyclobutane involvement with Fischer-type carbenes and with $(CO)_5W=CPh_2$,[183] there continues to exist a belief that such intermediates are also involved in catalytic reactions. However, no evidence exists to support this contention, and we will avoid metallocyclobutanes as intermediates unless there is convincing evidence for their involvement in metal carbene transformations.

2.11 REFERENCES

1   Wulfman, D. S.; Linstrumelle, G.; and Cooper, C. F., "Synthetic Applications of Diazoalkanes, Diazocyclopentadienes and Diazoazacyclopentadienes" in *The Chemistry of Diazonium and Diazo Groups*; Patai, S., Ed.; Wiley: New York, 1978; Part 2, Chapter 18.

2   Dave, V.; and Warnhoff, E. W., "The Reactions of Diazoacetic Esters with Alkenes, Alkynes, Heterocyclic and Aromatic Compounds," *Org. React. (N.Y.)* **1970**, *18*, 217–401.

3   Marchand, A. P.; and Brockway, N. M., "Carbalkoxycarbenes," *Chem. Rev.* **1974**, *74*, 431–69.

4   Maas, G., "Transition-Metal Catalyzed Decomposition of Aliphatic Diazo Compounds—New Results and Applications in Organic Synthesis," *Top. Curr. Chem.* **1987**, *137*, 76–253.

5   Silberrad, O.; and Roy, C. S., "Gradual Decomposition of Ethyl Diazoacetate," *J. Chem. Soc.* **1906**, *89*, 179–82.

6   McGarrity, J. F.; and Smyth, T., "Hydrolysis of Diazomethane—Kinetics and Mechanism," *J. Am. Chem. Soc.* **1980**, *102*, 7303–8.

7   McGarrity, J. F., "Basicity, Acidity and Hydrogen Bonding," in *The Chemistry of Diazonium and Diazo Groups*; Patai, S., Ed.; Wiley, New York, 1978; Part 1, Chapter 6.

8   Niemeyer, H. M., "Molecular Orbital Studies of the Protonation of Diazomethane," *Helv. Chim. Acta* **1976**, *59*, 1133–39.

9   McGarrity, J. F.; and Cox, D. P., "Protonation of Diazomethane in Superacid Media," *J. Am. Chem. Soc.* **1983**, *105*, 3961–66.

10   Allard, M.; Levisalles, J.; and Sommer, J. M., "Site of Protonation of α-Diazoketones in Super-acids," *J. Chem. Soc., Chem. Commun.* **1969**, 1515.

11   Wentrup, C.; and Dahn, H., "O-Protonation of α-Diazoketones in Super-strong Acids," *Helv. Chim. Acta* **1970**, *53*, 1637–45.

12   Regitz, M.; and Maas, G., *Aliphatic Diazo Compounds—Properties and Synthesis*; Academic Press, New York, 1986.

13 Doyle, M. P., "Catalytic Methods for Metal Carbene Transformations," *Chem. Rev.* **1986**, *86*, 919–39.

14 Padwa, A.; and Krumpe, K. E., "Application of Intramolecular Carbenoid Reactions in Organic Synthesis," *Tetrahedron* **1992**, *48*, 5385–453.

15 Doyle, M. P., "Metal Carbene Complexes in Organic Synthesis: Diazodecomposition—Insertion and Ylide Chemistry," in *Comprehensive Organometallic Chemistry II*; Hegedus, L. S., Ed.; Pergamon Press, New York, 1995; Vol. 12, Chapter 5.2.

16 Davies, H. M. L., "Tandem Cyclopropanation/Cope Rearrangement: A General Method for the Construction of Seven-Membered Rings," *Tetrahedron* **1993**, *49*, 5203–23.

17 Padwa, A.; and Austin, D. J., "Ligand Effects on the Chemoselectivity of Transition Metal Catalyzed Reactions of α-Diazo Carbonyl Compounds," *Angew. Chem., Int. Ed. Engl.* **1994**, *33*, 1797–815.

18 Maxwell, J. L.; Brown, K. C.; Bartley, D. W.; and Kodadek, T., "Mechanism of the Rhodium Porphyrin-Catalyzed Cyclopropanation of Alkenes," *Science* **1992**, *256*, 1544–47.

19 Alonso, M. E.; and Carmen Garcia, M. del, "Kinetics of the Dirhodium Tetraacetate Catalyzed Decomposition of Ethyl Diazoacetate in 1,4-Dioxane. Is Nitrogen Involved in the Transition State?," *Tetrahedron* **1989**, *45*, 69–76.

20 Pirrung, M. C.; and Morehead, A. T., Jr., "Saturation Kinetics in Dirhodium(II) Carboxylate-Catalyzed Decomposition of Diazo Compounds," *J. Am. Chem. Soc.* **1996**, *118*, 8162–63.

21 Boschi, T.; Licoccia, S.; Paolesse, R.; Tagliatesta, P.; Pelizzi, G.; and Vitali, F., "Synthesis and Reactivity towards Nucleophiles of bis(isocyanide)(porphyrinato) rhodium(III) Complexes. Crystal and Molecular Structure of a Novel Carbene Complex: $\{(TPP)Rh(PhCH_2NC)[:C(NHCH_2Ph)_2]PF_6\}$," *Organometal.* **1989**, *8*, 330–36.

22 Werner, H., "Success and Serendipity During Studies Aimed at Preparing Carbenerhodium(I) Complexes," *J. Organometal. Chem.* **1995**, *500*, 331–36.

23 Herrmann, W. A.; Elison, M.; Fischer, J.; Köcher, C.; and Artus, G. R. J., "N-Heterocyclic Carbenes[+]: Generation Under Mild Conditions and Formation of Group 8–10 Transition Metal Complexes Relevant to Catalysis," *Chem. Eur. J.* **1996**, *2*, 772–80.

24 Drago, R. S.; Long, J. R.; and Cosmano, R., "Comparison of the Coordination Chemistry and Inductive Transfer through the Metal–Metal Bond in Adducts of Dirhodium and Dimolybdenum Carboxylates," *Inorg. Chem.* **1982**, *21*, 2196–202.

25 Salomon, R. G.; and Kochi, J. K., "Copper(I) Catalysis in Cyclopropanations with Diazo Compounds. The Role of Olefin Coordination," *J. Am. Chem. Soc.* **1973**, *95*, 3300–10.

26 Doyle, M. P.; Colsman, M. R.; and Chinn, M. S., "Olefin Coordination with Rhodium(II) Trifluoroacetate," *Inorg. Chem.* **1984**, *23*, 3684–85.

27 Doyle, M. P.; Mahapatro, S. N.; Caughey, A. C.; Chinn, M. S.; Colsman, M. R.; Harn, N. K.; and Redwine, A. E., "Olefin Coordination with Rhodium(II) Perfluoroalkanoates in Solution," *Inorg. Chem.* **1987**, *26*, 3070–72.

28 Brown, K. C.; and Kodadek, T., "A Transition-State Model for the Rhodium Porphyrin-Catalyzed Cyclopropanation of Alkenes by Diazo Esters," *J. Am. Chem. Soc.* **1992**, *114*, 8336–38.

29 Pirrung, M. C.; and Morehead, A. T., Jr., "Electronic Effects in Dirhodium(II) Carboxylates. Linear Free Energy Relationships in Catalyzed Decompositions of Diazocompounds and CO and Isonitrile Complexation," *J. Am. Chem. Soc.* **1994**, *116*, 8991–9000.

30 Tomilov, Yu. V.; Dokichev, V. A.; Dzhemilev, U. M.; and Nefedov, O. M., "Catalytic Decomposition of Diazomethane as a General Method for the Methylenation of Chemical Compounds," *Russ. Chem. Rev.* **1993**, *62*, 799–838.

31 Shankar, B. K. R.; and Shechter, H., "Rhodium Ion Catalyzed Decomposition of Aryldiazoalkanes," *Tetrahedron Lett.* **1982**, *23*, 2277–80.

32 Doyle, M. P.; Griffin, J. H.; Bagheri, V.; and Dorow, R. L., "Correlations Between Catalytic Reactions of Diazo Compounds and Stoichiometric Reactions of Transition Metal Carbenes with Alkenes. Mechanism of the Cyclopropanation Reaction," *Organometal.* **1984**, *3*, 53–61.

33 Hartley, F. R., "Thermodynamic Data for Olefin and Acetylene Complexes of Transition Metals," *Chem. Rev.* **1973**, *73*, 163–90.

34 Wipke, W. T.; and Goeke, G. L., "Trans Addition of the Elements Pd-Cl to a Diene," *J. Am. Chem. Soc.* **1974**, *96*, 4244–49.

35 Doyle, M. P., "Asymmetric Cyclopropanation," in *Catalytic Asymmetric Synthesis*; Ojima, I., Ed.; VCH Publishers, New York, 1993, Chapter 3.

36 Hanks, T. W.; and Jennings, P. W., "Platinacyclobutanes on the Route to Cyclopropanation," *J. Am. Chem. Soc.* **1987**, *109*, 5023–25.

37 Lukin, K. A.; Kuznetzova, T. C.; Kozhushkov, S. I.; Piven, V. A.; and Zefirov, N. S., "Oligomethylenation of Bicyclopropylidene by Diazomethane in the Presence of Palladium(II) Acetate," *Zh. Org. Khim.* **1988**, *24*, 1644–48.

38 Brookhart, M.; and Studabaker, W. B., "Cyclopropanes from Reactions of Transition- Metal-Carbene Complexes with Olefins," *Chem. Rev.* **1987**, *87*, 411–32.

39 LaLonde, R. T.; and Tobias, M. A., "Carbon–Carbon Bond Fission in Cyclopropanes. IV. The Acid-Promoted Opening of the Three-Membered Ring in *exo*-7-Methylbicyclo[4.1.0]heptane and *exo*-6-Methylbicyclo[3.1.0]hexane," *J. Am. Chem. Soc.* **1964**, *86*, 4068–73.

40 Moser, W. R., "The Mechanism of the Copper-Catalyzed Addition of Diazoalkanes to Olefins. I. Steric Effects," *J. Am. Chem. Soc.* **1969**, *91*, 1135–40.

41 Nozaki, H.; Moriuti, S.; Yamabe, M.; and Noyori, R., "Reactions of Diphenyldiazomethane in the Presence of Bis(acetylacetonato)copper(II). Modified Diphenylmethylene Reactions," *Tetrahedron Lett.* **1966**, 59–63.

42 Nozaki, H.; Moriuti, S.; Takaya, H.; and Noyori, R., "Asymmetric Induction in Carbenoid Reaction by Means of a Dissymmetric Copper Chelate," *Tetrahedron Lett.* **1966**, 5239–44.

43 Salomon, R. G.; and Kochi, J. K., "Cationic Olefin Complexes of Copper(I) Structure and Bonding in Group Ib Metal–Olefin Complexes," *J. Am. Chem. Soc.* **1973**, *95*, 1889–97.

44 Wittig, G.; and Schwarzenback, K., "Über Methylenierte Metallhalogenide, I," *Justus Liebigs Ann. Chem.* **1961**, *650*, 1–20.

45 Aratani, T., "Catalytic Asymmetric Synthesis of Cyclopropanecarboxylic Acids: An Application of Chiral Copper Carbenoid Reaction," *Pure & Appl. Chem.* **1985**, *57*, 1839–44.

46 Pfaltz, A., "Chiral Semicorrins and Related Nitrogen Heterocycles as Ligands in Asymmetric Catalysis," *Acc. Chem. Res.* **1993**, *26*, 339–45.

47 Lowenthal, R. E.; Abiko, A.; and Masamune, S., "Asymmetric Catalytic Cyclopropanation of Olefins: bis-Oxazoline Copper Complexes," *Tetrahedron Lett.* **1990**, *31*, 6005–8.

48 Evans, D. A.; Woerpel, K. A.; Hinman, M. M.; and Faul, M. M., "bis-Oxazolines as Chiral Ligands in Metal-Catalyzed Asymmetric Reactions. Catalytic, Asymmetric Cyclopropanation of Olefins," *J. Am. Chem. Soc.* **1991**, *113*, 726–28.

49 Müller, D.; Umbricht, G.; Weber, B.; and Pfaltz, A., "C_2-Symmetric 4,4′,5,5′-Tetrahydrobi(oxazoles) and 4,4′,5,5′-Tetrahydro-2,2′-methylenebis[oxazoles] as Chiral Ligands for Enantioselective Catalysis," *Helv. Chim. Acta* **1991**, *74*, 232–40.

50 Fritschi, H.; Leutenegger, U.; and Pfaltz, A., "Semicorrin Metal Complexes as Enantioselective Catalysts. Part 2. Enantioselective Cyclopropane Formation from Olefins with Diazo Compounds Catalyzed by Chiral (Semicorrinato)copper Complexes," *Helv. Chim. Acta* **1988**, *71*, 1553–65.

51 Dauben, W. G.; Hendricks, R. T.; Luzzio, M. J.; and Ng, H. P., "Enantioselectivity Catalyzed Intramolecular Cyclopropanations of Unsaturated Diazo Carbonyl Compounds," *Tetrahedron Lett.* **1990**, *31*, 6969–72.

52 Kubas, G. J., "Tetrakis(acetonitrile)copper(I) Hexafluorophosphate," *Inorg. Synth.* **1979**, *19*, 90–92.

53 Doyle, M. P.; Peterson, C. S.; and Parker, D. L., Jr., "Formation of Macrocyclic Lactones by Enantioselective Intramolecular Cyclopropanation of Diazoacetates Catalyzed by Chiral Cu^I and Rh^{II} Compounds," *Angew. Chem. Int. Ed. Engl.* **1996**, *35*, 1334–36.

54 Lowenthal, R. E.; and Masamune, S., "Asymmetric Copper-Catalyzed Cyclopropanation of Trisubstituted and Unsymmetrical cis-1,2-Disubstituted Olefins: Modified bis-Oxazoline Ligands," *Tetrahedron Lett.* **1991**, *32*, 7373–76.

55 Moser, W. R., "The Mechanism of the Copper-Catalyzed Addition of Diazoalkanes to Olefins. II. Electronic Effects," *J. Am. Chem. Soc.* **1969**, *91*, 1141–46.

56 Bertrand, J.; and Kaplan, R. I. "A Study of Bis(hexafluoroacetylacetonato)copper(II)," *Inorg. Chem.* **1966**, *5*, 489–91.

57 Corey, E. J.; and Myers, A. G., "Efficient Synthesis and Intramolecular Cyclopropanation of Unsaturated Diazoacetic Esters," *Tetrahedron Lett.* **1984**, *25*, 3559–62.

58 Alonso, M. E.; and Fernández, R., "Effect of Catalyst on Zwitterionic Intermediacy in Additions of Dimethyl Diazomalonate to Vinyl Ethers," *Tetrahedron* **1989**, *45*, 3313–20.

59 Nugent, W. A.; and Waller, F. J., "Copper(II) Supported on Nafion Perfluorinated Ion-Exchange Polymer—An Efficient Catalyst for Cyclopropanation Reactions," *Syn. Commun.* **1988**, *18*, 61–68.

60 Leutenegger, U.; Umbricht, G.; Fahrni, C.; von Matt, P.; and Pfaltz, A., "5-Aza-Semicorrins: A New Class of Bidentate Nitrogen Ligands for Enantioselective Catalysis," *Tetrahedron* **1992**, *48*, 2143–56.

61 Gupta, A. D.; Bhuniya, D.; and Singh, V. K., "Synthesis of Homochiral bis(Oxazolinyl)pyridine Type Ligands for Asymmetric Cyclopropanation Reactions," *Tetrahedron* **1994**, *50*, 13725–30.

62 Bedeker, A. V.; and Andersson, P. G., "A New Class of bis-Oxazoline Ligands for the Cu-Catalyzed Asymmetric Cyclopropanation of Olefins," *Tetrahedron Lett.* **1996**, *37*, 4073–76.

63 Ito, K.; and Katsuki, T., "Catalytic Asymmetric Cyclopropanation Using Copper Complex of Optically-Active Bipyridine as a Catalyst," *Tetrahedron Lett.* **1993**, *34*, 2661–64.

64 Ito, K.; and Katsuki, T., "Asymmetric Cyclopropanation of *E*-Olefins Using a Copper Complex of an Optically Active Bipyridine as a Catalyst," *Synlett* **1993**, 638–40.

65 Ito, K.; Yoshitake, M.; and Katsuki, T., "Enantioselective Synthesis of *trans*-Whisky Lactone by Using Newly Developed Asymmetric Ring Expansion Reaction of Oxetane as a Key Step," *Chem. Lett.* **1995**, 1027–28.

66 Ito, K.; Yoshitake, M.; and Katsuki, T., "Enantiospecific Ring Expansion of Oxetanes: Stereoselective Synthesis of Tetrahydrofurans," *Heterocycles* **1996**, *42*, 305–17.

67 Ito, K.; Yoshitake, M.; and Katsuki, T., "Chiral Bipyridine and Biquinoline Ligands: Their Asymmetric Synthesis and Application to the Synthesis of *trans*-Whisky Lactone," *Tetrahedron* **1996**, *52*, 3905–20.

68 Kanemasa, S.; Hamura, S.; Harada, E.; and Yamamoto, H., "C_2-Symmetric 1,2-Diamine/Copper(II) Trifluoromethanesulfonate Complexes as Chiral Catalysts. Asymmetric Cyclopropanations of Styrene with Diazo Esters," *Tetrahedron Lett.* **1994**, *35*, 7985–88.

69 Christenson, D. L.; Tokar, C. J.; and Tolman, W. B., "New Copper and Rhodium Cyclopropanation Catalysts Supported by Chiral bis(Pyrazolyl)pyridines. A Metal-Dependent Enantioselectivity Switch," *Organometal.* **1995**, *14*, 2148–50.

70 Brunner, H.; Singh, U. P.; Boeck, T.; Altman, S.; Scheck, T.; and Wrackmeyer, B., "Asymmetic Catalysis. 80. An Optically-Active Tetrakispyrazolylborate—Synthesis and Use in Cu-Catalyzed Enantioselective Cyclopropanation," *J. Organometal. Chem.* **1993**, *443*, C16-C18.

71 Burgess, K.; Lim, H.-J.; Porte, A. M.; and Sulikowski, G. A., "New Catalysts and Conditions for a C–H Insertion Reaction Identified by High Throughput Catalyst Screening," *Angew. Chem. Int. Ed. Engl.* **1996**, *35*, 220–22.

72 Fritschi, H.; Leutenegger, U.; Siegmann, K.; Pfaltz, A.; Keller, W.; and Kratky, Ch., "Semicorrin Metal Complexes as Enantioselective Catalysts. Part 1. Synthesis of Chiral Semicorrin Ligands and General Concepts," *Helv. Chim. Acta* **1988**, *71*, 1541–52.

73 Denmark, S. E.; Nakajima, N.; Nicaise, O. J.-C.; Faucher, A.-M.; and Edwards, J. P., "Preparation of Chiral Bisoxazolines: Observations on the Effect of Substituents," *J. Org. Chem.* **1995**, *60*, 4884–92.

74 Nakamura, A.; Konishi, A.; Tatsuno, Y.; and Otsuka, S., "A Highly Enantioselective Synthesis of Cyclopropane Derivatives through Chiral Cobalt(II) Complex Catalyzed Carbenoid Reaction. General Scope and Factors Determining the Enantioselectivity," *J. Am. Chem. Soc.* **1978**, *100*, 3443–48, 6544–46.

75 Nakamura, A.; Konishi, A.; Tsujitani, R.; Kudo, M.; and Otsuka, S., "Enantioselective Carbenoid Cyclopropanation Catalyzed by Chiral *vic*-Dioximatocobalt(II) Complexes Prepared from Natural Camphor and β-Pinene. Mechanism and Stereochemistry," *J. Am. Chem. Soc.* **1978**, *100*, 3449–61.

76 Doyle, M. P., "Chiral Catalysts for Enantioselective Carbenoid Cyclopropanation Reactions," *Recl. Trav. Chim. Pays-Bas* **1991**, *110*, 305–16.

77 Tamblyn, W. H.; Hoffman, S. R.; and Doyle, M. P., "Correlation between Catalytic Cyclopropanation and Ylide Generation," *J. Organometal. Chem.* **1981**, *216*, C64-C68.

78 Callot, H. J.; and Schaeffer, E., "Synthèse de Vinylcobalt(III)porphyrines," *Tetrahedron Lett.* **1977**, 239–42.

79 Fukuda, T.; and Katsuki, T., "Co(III)-Salen Catalyzed Asymmetric Cyclopropanation," *Synlett* **1995**, 825–26.

80 Doyle, M. P.; Wang, L. C.; and Loh, K.-L., "Influences of Olefin Coordination on Cyclopropanation Selectivity," *Tetrahedron Lett.* **1984**, *25*, 4087–90.

81 Stromnova, T. A.; Busygina, I. N.; Kochubey, D. I.; and Moiseev, I. I., "Palladium Carbene Cluster—Synthesis, Structure and Reactivity," *J. Organometal. Chem.* **1991**, *417*, 193–204.

82 Doyle, M. P.; Dorow, R. L.; and Tamblyn, W. H., "Cyclopropanation of α,β-Unsaturated Carbonyl Compounds and Nitriles with Diazo Compounds. The Nature of the Involvement of Transition Metal Promoters," *J. Org. Chem.* **1982**, *47*, 4059–68.

83 Engel, P. S., "Mechanism of the Thermal and Photochemical Decomposition of Azoalkanes," *Chem. Rev.* **1980**, *80*, 99–150.

84 Tomilov, Yu. V.; Kostitsyn, A. B.; Shulishov, E. V.; and Nefedov, O. M., "Palladium(II)-Catalyzed Cyclopropanation of Simple Allyloxy and Allylamino Compounds and of 1-Oxy-1,3-butadienes with Diazomethane," *Synthesis* **1990**, 246–48.

85 Mende, U.; Radüchel, B.; Skuballa, W.; and Vorbrüggen, H., "New Simple Conversion of α,β-Unsaturated Carbonyl Compounds into their Corresponding Cyclopropyl Ketones and Esters," *Tetrahedron Lett.* **1975**, *16*, 629–32.

86 Dzhemilev, U. M.; Dokichev, V. A.; Sultanov, S. Z.; Khusnutdinov, R. I.; Tomilov, Yu. V.; Nefedov, O. M.; and Tolstikov, G. A., "Reaction of Diazoalkanes with Unsaturated Compounds. 6. Catalytic Cyclopropanation of Various Unsaturated Hydrocarbons and their Derivatives by Diazomethane," *Izv. Akad. Nauk SSSR, Ser. Khim.* **1989**, 1861–69.

87 Dinulescu, I. G.; Enescu, L. N.; Ghenciulescu, A.; and Avram, M., "The Cyclopropanation of Strained Alkenes with Diazomethane in Presence of π-Alkylpalladium Chloride Complexes," *J. Chem. Res. (S)* **1978**, 456–57.

88 Anciaux, A. J.; Hubert, A. J.; Noels, A. F.; Petiniot, N.; and Teyssie, Ph., "Transition-Metal-Catalyzed Reactions of Diazo Compounds. 1. Cyclopropanation of Double Bonds," *J. Org. Chem.* **1980**, *45*, 695–702.

89 Callot, H. J.; Metz, F.; and Piechocki, C., "Sterically Crowded Cyclopropanation Catalysts. *Syn*-Selectivity Using Rhodium(III) Porphyrins," *Tetrahedron* **1982**, *38*, 2365–69.

90 O'Malley, S.; and Kodadek, T., "Synthesis and Characterization of the "Chiral Wall" Porphyrin: A Chemically Robust Ligand for Metal-Catalyzed Asymmetric Epoxidations," *J. Am. Chem. Soc.* **1989**, *111*, 9116–17.

91 Doyle, M. P.; Tamblyn, W. H.; Buhro, W. E.; and Dorow, R. L., "Exceptionally Effective Catalysis of Cyclopropanation Reactions by the Hexarhodium Carbonyl Cluster," *Tetrahedron Lett.* **1981**, *22*, 1783–86.

92 Nazarova, L. A.; Chernyaev, I. I.; and Morozova, A. S., "Acetate Compounds of Rhodium," *Zh. Neorgan. Khim.* **1965**, *10*, 539–41.

93 Cotton, F. A.; Curtis, N. F.; Harris, C. B.; Johnson, B. F. G.; Lippard, S. J.; Mague, J. T.; Robinson, W. R.; and Wood, J. S., "Mononuclear and Polynuclear Chemistry of Rhenium(III): Its Pronounced Homophilicity," *Science* **1964**, *145*, 1305–7.

94 Paulissenen, R.; Reimlinger, H.; Hayez, E.; Hubert, A. J.; and Teyssie, Ph., "Transition Metal Catalyzed Reactions of Diazocompounds—II. Insertion in the Hydroxylic Bond," *Tetrahedron Lett.* **1973**, 2233–36.

95 Boyar, E. B.; and Robinson, S. D., "Rhodium(II) Carboxylates," *Coord. Chem. Rev.* **1983**, *50*, 109–208.

96 Felthouse, T. R., "The Chemistry, Structure, and Metal–Metal Bonding in Compounds of Rhodium(II)," *Prog. Inorg. Chem.* **1982**, *29*, 73–166.

97 Cotton, F. A.; and Walton, R. A., *Multiple Bonds Between Metal Atoms*; Wiley, New York, 1982.

98 Davies, H. M. L.; Huby, N. J. S.; Cantrell, W. R., Jr.; and Olive, J. L., "α-Hydroxy Esters as Chiral Auxiliaries in Asymmetric Cyclopropanations by Rhodium(II)-Stabilized Vinylcarbenoids," *J. Am. Chem. Soc.* **1993**, *115*, 9468–79.

99 Bergbreiter, D. E.; Morvant, M.; and Chen, B., "Catalytic Cyclopropanation with Transition Metal Salts of Soluble Polyethylene Carboxylates," *Tetrahedron Lett.* **1991**, *32*, 2731–34.

100 Hashimoto, S.; Watanabe, N.; and Ikegami, S., "Dirhodium(II) Tetra(triphenylacetate)—A Highly Efficient Catalyst for the Site-Selective Intramolecular C–H Insertion Reactions of α-Diazo-β-keto-esters," *Tetrahedron Lett.* **1992**, *33*, 2709–12.

101 Callot, H. J.; Albrecht-Gary, A.-M.; Al Joubbeh, M.; Metz, B.; and Metz, F., "Crystallographic Study and Ligand Substitution Reactions of Dirhodium(II) Tris- and Tetrakis(tritolylbenzoate)," *Inorg. Chem.* **1989**, *28*, 3633–40.

102 Rempel, G. A.; Legzdins, P.; Smith, H.; and Wilkinson, G., "Tetrakis(acetato) dirhodium(II) and Similar Carboxylato Compounds," *Inorg. Synth.* **1972**, *13*, 90–91.

103 Doyle, M. P.; and Shanklin, M. S., "Highly Regioselective and Stereoselective Silylformylation of Alkynes under Mild Conditions Promoted by Dirhodium(II) Perfluorobutyrate," *Organometal.* **1994**, *13*, 1081–88.

104 Drago, R. S.; Long, J. R.; and Cosmano, R., "Metal Synergism in the Coordination Chemistry of a Metal-Metal Bonded System: $Rh_2(C_3H_7COO)_4$," *Inorg. Chem.* **1981,** *20*, 2920–27.

105 Roos, G. H. P.; and McKervey, M. A., "A Facile Synthesis of Homochiral Rh(II) Carboxylates," *Syn. Commun.* **1992**, *22*, 1751–56.

106 Callot; H. J.; and Piechocki, C., "Cyclopropanation Using Rhodium(III) Porphyrins: Large *cis* vs. *trans* Selectivity," *Tetrahedron Lett.* **1980**, *21*, 3489–92.

107 Doyle, M. P.; Bagheri, V.; Wandless, T. J.; Harn, N. K.; Brinker, D. A.; Eagle, C. T.; and Loh, K.-L., "Exceptionally High Trans (Anti) Stereoselectivity in Catalytic Cyclopropanation Reactions," *J. Am. Chem. Soc.* **1990**, *112*, 1906–12.

108 Schurig, V., "Relative Stability Constants of Olefin-Rhodium(II) vs. Olefin-Rhodium(I) Coordination As Determined by Complexation Gas Chromatography," *Inorg. Chem.* **1986**, *25*, 945–49.

109 Cotton, F. A.; Falvello, L. R.; Gerards, M.; and Snatzke, G., "Application of Tetrakis(trifluoroacetato)dirhodium(II) to Determination of Chirality: The First Structural Characterization of an Axial Bisolefin Complex of a Dimetal Core," *J. Am. Chem. Soc.* **1990**, *112*, 8979–80.

110 Wypchlo, K.; and Duddeck, H., "Chiral Recognition of Olefins by ^{1}H NMR Spectroscopy in the Presence of a Chiral Dirhodium Complex," *Tetrahedron: Asymmetry* **1994**, *5*, 27–30.

111 Brunner, H.; Kluschanzoff, H.; and Wutz, K., "Enantioselective Catalysis. 47. Rhodium(II)–Carboxylate Complexes and their Use in the Enantioselective Cyclopropanation," *Bull. Chem. Soc. Belg.* **1989**, *98*, 63–72.

112 Kennedy, M.; McKervey, M. A.; Maguire, A. R.; and Roos, G. H. P., "Asymmetric Synthesis in Carbon–Carbon Bond Forming Reactions of α-Diazoketones Catalysed by Homochiral Rhodium(II) Carboxylates," *J. Chem. Soc., Chem. Commun.* **1990**, 361–62.

113 Davies, H. M. L.; and Hutcheson, D. K., "Enantioselective Synthesis of Vinylcyclopropanes by Rhodium(II) Catalyzed Decomposition of Vinyldiazomethanes in the Presence of Alkenes," *Tetrahedron Lett.* **1993**, *34*, 7243–46.

114 Corey, E. J.; and Grant, T. G., "A Catalytic Enantioselective Synthetic Route to the Important Antidepressant Sertraline," *Tetrahedron Lett.* **1994**, *35*, 5373–76.

115 Doyle, M. P.; Zhou, Q.-L.; Charnsangavej, C.; Longoria, M. A.; McKervey, M. A.; and García, C. F., "Chiral Catalysts for Enantioselective Intermolecular Cyclopropanation Reactions with Methyl Phenyldiazoacetate. Origin of the Solvent Effect in Reactions Catalyzed by Homochiral Dirhodium(II) Prolinates," *Tetrahedron Lett.* **1996**, *37*, 4129–32.

116 McKervey, M. A.; and Ye, T., "Asymmetric Synthesis of Substituted Chromanones via C–H Insertion Reactions of α-Diazoketones Catalyzed by Homochiral Rhodium(II) Carboxylates," *J. Chem. Soc., Chem. Commun.* **1992**, 823–24.

117 Davies, H. M. L.; Bruzinski, P. R.; Lake, D. H.; Kong, N.; and Fall, M. J., "Asymmetric Cyclopropanations by Rhodium(II) N-(Arylsulfonyl)prolinate Catalyzed Decomposition of Vinyldiazomethanes in the Presence of Alkenes. Practical Enantioselective Synthesis of the Four Stereoisomers of 2-Phenylcyclopropan-1-amino Acid," *J. Am. Chem. Soc.* **1996**, *118*, 6897–907.

118 Yoshikawa, K.; and Achiwa, K., "Steric and Electronic Effects of Substrates and Rhodium Chiral Catalysts in Asymmetric Cyclopropanation," *Chem. Pharm. Bull.* **1995**, *43*, 2048–53.

119 Hashimoto, S.; Watanabe, N.; and Ikegami, S., "Enantioselective Intramolecular C–H Insertion of α-Diazo β-Keto Esters Catalyzed by Homochiral Rhodium(II) Carboxylates," *Tetrahedron Lett.* **1990**, *31*, 5173–74.

120 Hashimoto, S.; Watanabe, N.; and Ikegami, S., "Enantioselective Intramolecular C–H Insertion Reactions of α-Diazo-β-keto Esters Catalyzed by Dirhodium(II) Tetrakis[*N*-phthaloyl-(S)-phenylalaninate]: The Effect of the Substituent at the Insertion Site on Enantioselectivity," *Synlett* **1994**, 353–55.

121 Watanabe, N.; Ohtake, Y.; Hashimoto, S.; Shiro, M.; and Ikegami, S., "Asymmetric Creation of Quaternary Carbon Centers by Enantiotopically Selective Aromatic C–H Insertion Catalyzed by Chiral Dirhodium(II) Carboxylates," *Tetrahedron Lett.* **1995**, *36*, 1491–94.

122 Watanabe, N.; Ogawa, T.; Ohtake, Y.; Ikegami, S.; and Hashimoto, S., "Dirhodium(II) Tetrakis[*N*-phthaloyl-(S)-*tert*-leucinate]: A Notable Catalyst for Enantiotopically Selective Aromatic Substitution Reactions of α-Diazocarbonyl Compounds," *Synlett* **1996**, 85–86.

123 Ahsan, M. Q.; Bernal, I.; and Bear, J. L., "Reaction of $Rh_2(OOCCH_3)_4$ with Acetamide: Crystal and Molecular Structure of $[Rh(NHOCCH_3)_4 \cdot 2H_2O] \cdot 3H_2O$," *Inorg. Chem.* **1986**, *25*, 260–65.

124 Lifsey, R. S.; Lin, X. Q.; Chavan, M. Y.; Ahsan, M. Q.; Kadish, K. M.; and Bear, J. L., "Reaction of $Rh_2(O_2CCH_3)_4$ with *N*-Phenylacetamide: Substitution Products and Geometric Isomers," *Inorg. Chem.* **1987**, *26*, 830–36.

125 Doyle, M. P.; Westrum, L. J.; Wolthuis, W. N. E.; See, M. M.; Boone, W. P.; Bagheri, V.; and Pearson, M. M., "Electronic and Steric Control in Carbon–Hydrogen Insertion Reactions of Diazoacetoacetates Catalyzed by Dirhodium(II) Carboxylates and Carboxamides," *J. Am. Chem. Soc.* **1993**, *115*, 958–64.

126 Doyle, M. P., "Asymmetric Syntheses with Catalytic Enantioselective Metal Carbene Transformations," *Russ. Chem. Bull.* **1994**, *43*, 1770–82.

127 Doyle, M. P., "Chiral Dirhodium Carboxamidates. Catalysts for Highly Enantioselective Syntheses of Lactones and Lactams," *Aldrichimica Acta* **1996**, *29* (1), 3–11.

128 Doyle, M. P.; Winchester, W. R.; Hoorn, J. A. A.; Lynch, V.; Simonsen, S. H.; and Ghosh, R., "Dirhodium(II) Tetrakis(carboxamidates) with Chiral Ligands. Structure and Selectivity in Catalytic Metal–Carbene Transformations," *J. Am. Chem. Soc.* **1993**, *115*, 9968–78.

129 Doyle, M. P.; Winchester, W. R.; Simonsen, S. H.; and Ghosh, R., "Dirhodium(II) Tetrakis[*N,N*-dimethyl-2-pyrrolidone-(5S)-carboxamide]. Structural Effects on Enantioselection in Metal Carbene Transformations," *Inorg. Chim. Acta* **1994**, *220*, 193–99.

130 Doyle, M. P.; Dyatkin, A. B.; Protopopova, M. N.; Yang, C. I.; Miertschin, C. S.; Winchester, W. R.; Simonsen, S. H.; Lynch, V.; and Ghosh, R., "Enhanced Enantiocontrol in Catalytic Metal Carbene Transformations with Dirhodium(II) Tetrakis[methyl 2-oxoxazolidin-4(S)-carboxylate], $Rh_2(4S\text{-MEOX})_4$," *Recl. Trav. Chim. Pays-Bas* **1995**, *114*, 163–70.

131 Doyle, M. P.; Winchester, W. R.; Protopopova, M. N.; Müller, P.; Bernardinelli, G.; Ene, D.; and Motallebi, S., "Tetrakis[4(S)-4-phenyloxazolidin-2-one]dirhodi-

um(II) and Its Catalytic Applications for Metal Carbene Transformations," *Helv. Chim. Acta* **1993**, *76*, 2227–35.

132   Doyle, M. P.; Austin, R. E.; Bailey, A. S.; Dwyer, M. P.; Dyatkin, A. B.; Kalinin, A. V.; Kwan, M. M. Y.; Liras, S.; Oalmann, C. J.; Pieters, R. J.; Protopopova, M. N.; Raab, C. E.; Roos, G. H. P.; Zhou, Q.-L.; and Martin, S. F., "Enantioselective Intramolecular Cyclopropanations of Allylic and Homoallylic Diazoacetates and Diazoacetamides Using Chiral Dirhodium(II) Carboxamide Catalysts," *J. Am. Chem. Soc.* **1995**, *117*, 5763–75.

133   Doyle, M. P.; Zhou, Q.-L.; Raab, C. E.; Roos, G. H. P.; Simonsen, S. H.; and Lynch, V., "Synthesis and Structures of (2,2-*cis*)-Dirhodium(II) Tetrakis[methyl 1-acyl-2-oxoimidazolidine-4(*S*)-carboxylates]. Chiral Catalysts for Highly Stereoselective Metal Carbene Transformations," *Inorg. Chem.* **1996**, *35*, 6064–73.

134   Doyle, M. P.; Zhou, Q.-L.; Simonsen, S. H.; and Lynch, V., "Dirhodium(II) Tetrakis[alkyl 2-oxoazetidine-4(*S*)-carboxylates]. A New Set of Effective Chiral Catalysts for Asymmetric Intermolecular Cyclopropanation Reactions with Diazoacetates," *Synlett* **1996**, 697–98.

135   Doyle, M. P.; Raab, C. E.; Roos, G. H. P.; Lynch, V.; and Simonsen, S. H., "(4,0)-Dirhodium(II) Tetrakis[methyl 1-acetyl-2-oxoimidazolidine-4(*S*)-carboxylate]. Implication for the Mechanism of Ligand Exchange Reactions," *Inorg. Chim. Acta* **1997**, in press.

136   Doyle, M. P.; Winchester, W. R.; Protopopova, M. N.; Kazala, A. P.; and Westrum, L. J., "(1*R*,5*S*)-(−)-6,6-Dimethyl-3-oxabicyclo[3.1.0]hexan-2-one. Highly Enantioselective Intramolecular Cyclopropanation Catalyzed by Dirhodium(II) Tetrakis[methyl-2-pyrrolidone-5(*R*)-carboylate]," *Org. Syn.* **1996**, *73*, 13–24.

137   Doyle, M. P.; Dorow, R. L.; Buhro, W. E.; Griffin, J. H.; Tamblyn, W. H.; and Trudell, M. L., "Stereoselectivity of Catalytic Cyclopropanation Reactions. Catalyst Dependence in Reactions of Ethyl Diazoacetate with Alkenes," *Organometal.* **1984**, *3*, 44–52.

138   McCarthy, N.; McKervey, M. A.; Ye, T.; McCann, M.; Murphy, E.; and Doyle, M. P., "A New Rhodium(II) Phosphate Catalyst for Diazocarbonyl Reactions Including Asymmetric Synthesis," *Tetrahedron Lett.* **1992**, *33*, 5983–86.

139   Pirrung, M. C.; and Zhang, J., "Asymmetric Dipolar Cycloaddition Reactions of Diazo-Compounds Mediated by a Binaphtholphosphate Rhodium Catalyst," *Tetrahedron Lett.* **1992**, *33*, 5987–90.

140   Kyba, E. P.; Gokel, G. W.; Jong, F. de; Koga, K.; Sousa, L. R.; Siegel, M. G.; Kaplan, L.; Sogah, G. D. Y.; and Cram, D. J., "Host–Guest Complexation. 7. The Binaphthyl Structural Unit in Host Compounds," *J. Org. Chem.* **1977**, *42*, 4173–84.

141   Estevan, F.; Lahuerta, P.; Pérez-Prieto, J.; Stiriba, S.-H.; and Ubeda, M. A., "New Rhodium(II) Catalysts for Selective Carbene Transfer Reactions," *Synlett* **1995**, 1121–22.

142   Bartley, D. W,; and Kodadek, T., "Identification of the Active Catalyst in the Rhodium Porphyrin-Mediated Cyclopropanation of Alkenes," *J. Am. Chem. Soc.* **1993**, *115*, 1656–60.

143   O'Malley, S.; and Kodadek, T., "Asymmetric Cyclopropanation of Alkenes Catalyzed by a 'Chiral Wall' Porphyrin," *Tetrahedron Lett.* **1991**, *32*, 2445–48.

144 Maxwell, J. L.; O'Malley, S.; Brown, K. C.; and Kodadek, T., "Shape-Selective and Asymmetric Cyclopropanation of Alkenes Catalyzed by Rhodium Porphyrins," *Organometal.* **1992**, *11*, 645–52.

145 Maas, G.; Werle, T.; Alt, M.; and Mayer, D., "Polymeric Dicarbonyl Ruthenium(I) Acetate—An Efficient Catalyst for Alkene Cyclopropanation with Diazoacetates," *Tetrahedron* **1993**, *49*, 881–88.

146 Demonceau, A.; Saive, E.; deFroidmont, Y.; Noels, A. F.; Hubert, A. J.; Chizhevsky, I. T.; Lobanova, I. A.; and Bregadze, V. I., "Olefin Cyclopropanation Reactions Catalyzed by Novel Ruthenacarborane Clusters," *Tetrahedron Lett.* **1992**, *33*, 2009–12.

147 Demonceau, A.; Abreu Dias, E.; Lemoine, C. A.; Stumpf, A. W.; Noels, A. F.; Pietraszuk, C.; Gulinski, J.; and Marciniec, B., "Cyclopropanation of Activated Olefins Catalyzed by Ru-Phosphine Complexes," *Tetrahedron Lett.* **1995**, *36*, 3519–22.

148 Noels, A. F.; Demonceau, A.; Carlier, E.; Hubert, A. J.; Márguez-Silva, R.-L.; and Sánchez-Delgado, R. A., "Competitive Cyclopropanation and Cross-Metathesis Reactions of Alkenes Catalyzed by Diruthenium Tetrakis Carboxylates," *J. Chem. Soc., Chem. Commun.* **1988**, 783–84.

149 Brunner, H.; Nishiyama, H.; and Itoh, K., "Asymmetric Hydrosilylation," In *Catalytic Asymmetric Synthesis*; Ojima, I., Ed.; VCH Publishers, New York, **1993**, Chapter 6.

150 Nishiyama, H.; Itoh, Y.; Matsumoto, H; Park, S.-B.; and Itoh, K., "New Chiral Ruthenium bis(oxazolinyl)pyridine Catalyst. Efficient Asymmetric Cyclopropanation of Olefins with Diazoacetates," *J. Am. Chem. Soc.* **1994**, *116*, 2223–24.

151 Park, S.-B.; Nishiyama, H.; Itoh, Y.; and Itoh, K., "Trimethylsilylcarbene and bis(trimethylsilyl) Formaldehyde Azine Complexes of Chiral bis(4-isopropyloxazolinyl)pyridine(dichloro)ruthenium(II)," *J. Chem. Soc., Chem. Commun.* **1994**, 1315–16.

152 Nishiyama, H.; Itoh, Y.; Sugawara, Y.; Matsumoto, H.; Aoki, K.; and Itoh, K., "Chiral Ruthenium(II)-bis(2-oxazolin-2-yl)pyridine Complexes. Asymmetric Catalytic Cyclopropanation of Olefins and Diazoacetates," *Bull Chem. Soc. Jpn.* **1995**, *68*, 1247–62.

153 Bennett, M. A.; and Smith, A. K., "Arene Ruthenium(II) Complexes Formed by Dehydrogenation of Cyclohexadienes with Ruthenium(III) Chloride," *J. Chem. Soc., Dalton Trans.* **1974**, 233–41.

154 Smith, D. A.; Reynolds, D. N.; and Woo, L. K., "Cyclopropanation Catalyzed by Osmium Porphyrin Complexes," *J. Am. Chem. Soc.* **1993**, *115*, 2511–13.

155 Seitz, W. J.; and Hossain, M. M., "Iron Lewis Acid Catalyzed Reactions of Phenyldiazomethane and Olefins—Formation of Cyclopropanes with Very High *cis* Selectivity," *Tetrahedron Lett.* **1994**, *35*, 7561–64.

156 Woo, L. K.; and Smith, D. A., "Synthesis of Osmium *meso*-Tetra-*p*-tolylporphyrin Carbene Complexes (TTP)Os$=$CRR' (R,R' = *p*-Tolyl; R = H, R' = SiMe$_3$, COOEt): Stereoselective, Catalytic Production of Olefins from Substituted Diazomethanes," *Organometal.* **1992**, *11*, 2344–46.

157   Demonceau, A.; Lemoine, C. A.; Noels, A. F.; Chizhevsky, I. T.; and Sorokin, P. V., "Cyclopropanation Catalyzed by RuCl$_2$(PPh$_3$)$_3$ and OsCl$_2$(PPh$_3$)$_3$," *Tetrahedron Lett.* **1995**, *36*, 8419–22.

158   Demonceau, A.; Lemoine, C. A.; and Noels, A. F., "Osmium-Catalyzed Cyclopropanation of Olefins," *Tetrahedron Lett.* **1996**, *37*, 1025–26.

159   Seitz, W. J.; Saha, A. K.; Casper, D.; and Hossain, M. M., "Iron Lewis Acid Catalyzed Reactions of Ethyldiazoacetate with Styrene and α-Methylstyrene: Formation of Cyclopropanes with cis Selectivity," *Tetrahedron Lett.* **1992**, *33*, 7755–58.

160   Seitz, W. J.; and Hossain, M. M., "Iron Lewis Acid Catalyzed Reactions of Phenyldiazomethane and Olefins: Formation of Cyclopropanes with Very High *cis*-Selectivity," *Tetrahedron Lett.* **1994**, *35*, 7561–64.

161   Theys, R. D.; and Hossain, M. M., "Asymmetric Cyclopropanation Reactions via Iron Carbene Complexes Having Chirality at the Carbene Ligand," *Tetrahedron Lett.* **1995**, *36*, 5113–16.

162   Seitz, W. J.; Saha, A. K.; and Hossian, M. M., "Iron Lewis Acid Catalyzed Cyclopropanation Reactions of Ethyl Diazoacetate and Olefins," *Organometallics* **1993**, *12*, 2604–8.

163   Artaud, I.; Gregoire, N.; Battioni, J.-P.; Dupre, D.; and Mansuy, D., "Heme Model Studies Related to Cytochrome P-450 Reactions: Preparation of Iron Porphyrin Complexes with Carbenes Bearing a β-Oxygen Atom and Their Transformation into Iron-N-Alkylporphyrins and Ion-Metallacyclic Complexes," *J. Am. Chem. Soc.* **1988**, *110*, 8714–16.

164   Artaud, I; Gregoire, N.; Leduc, P.; and Mansuy, D., "Formation and Fate of Iron–Carbene Complexes in Reactions Between a Diazoalkane and Iron–Porphyrins: Relevance to the Mechanism of Formation of N-Substituted Hemes in Cytochrome P-450 Dependent Oxidation of Sydnones," *J. Am. Chem. Soc.* **1990**, *112*, 6899–05.

165   Wolf, J. R.; Hamaker, C. G.; Djukic, J.-P.; Kodadek, T.; and Woo, L. K., "Shape and Stereoselective Cyclopropanation of Alkenes Catalyzed by Iron Porphyrins," *J. Am. Chem. Soc.* **1995**, *117*, 9194–99.

166   Hanks, T. W.; and Jennings, P. W., "Platinacyclobutanes on the Route to Cyclopropanation," *J. Am. Chem. Soc.* **1987**, *109*, 5023–25.

167   Puddephatt, R. J., "Platinacyclobutane Chemistry," *Coord. Chem. Rev.* **1980**, *33*, 149–94.

168   Hanks, T. W.; Ekeland, R. A.; Emerson, K.; Larsen, R. D.; and Jennings, P. W., "Reactions of Diazomethane Derivatives with Platinum(II): A Facile Method for Platinum Ylide Preparation," *Organometal.* **1987**, *6*, 28–32.

169   Stewart, F. F.; Neilson, W. D.; Ekeland, R. E.; Larsen, R. D.; and Jennings, P. W., "Synthesis of 1,3-Divinylcyclopentane Derivatives from Platina(IV)cyclobutane Complexes," *Organometal.* **1993**, *12*, 4585–91.

170   Doyle, M. P.; Griffin, J. H.; and Conceicao, J. da, "Procatalysts for Carbenoid Transformations," *J. Chem. Soc., Chem. Commun.* **1985**, 328–29.

171   Casey, C. P.; Bertz, S. H.; and Burkhardt, T. J., "Reaction of Metal–Carbene Complexes with Diazoalkanes. A Versatile Vinyl Ether Synthesis," *Tetrahedron Lett.* **1973**, 1421–24.

172   Jolly, P. W.; and Pettit, R., "Evidence for a Novel Metal–Carbene System," *J. Am. Chem. Soc.* **1966**, *88*, 5044–45.

173 Fischer, E. O.; and Dötz, K. H., "Transition Metal Carbene Complexes. XIX. Synthesis of Cyclopropane Derivatives with Transition Metal Carbonyl Carbene Complexes," *Chem. Ber.* **1970**, *103*, 1273–78.

174 Dötz, K. H., "Carbene Complexes in Organic Synthesis," *Angew. Chem. Int. Ed. Engl.* **1984**, *23*, 587–608.

175 Wulff, W. D., "Transition Metal Carbene Complexes in Organic Synthesis," in *Organometallics in Organic Synthesis*; Liebskind, L. J., Ed.; JAI Press, Greenwich, CT, 1989; Vol. 1, pp. 209–393.

176 Harvey, D. F.; and Brown, M. F., "Molybdenum Carbene Complexes: Cyclopropanation of Electron-Poor Olefins," *Tetrahedron Lett.* **1990**, *31*, 2529–32.

177 Buchert, M.; and Reissig, H.-U., "Highly Functionalized Vinylcyclopropane Derivatives by Regioselective and Stereoselective Reactions of Fischer Carbene Complexes with 1,4-Disubstituted Electron-Deficient 1,3-Dienes," *Chem. Ber.* **1992**, *125*, 2723–29.

178 Herndon, J. W.; and Turner, S. U., "Synthesis of Bicyclopropyl Derivatives from the Reaction of Cyclopropylcarbene–Chromium Complexes with Alkenes," *J. Org. Chem.* **1991**, *56*, 286–94.

179 Buchert, M.; Hoffmann, M.; and Reissig, H.-U., "Regioselectivity and Stereoselectivity of the Carbene Transfer from Fischer Carbene Complexes to Trisubstituted Electron-Deficient 1,3-Dienes," *Chem. Ber.* **1995**, *128*, 605–14.

180 Hoffmann, M.; and Reissig, H.-U., "A Unique Solvent Effect Governing Periselectivity of the Carbene Transfer from Fischer Carbene Complexes: Mechanism, Scope, and Limitation of Their Formal [3 + 2] Cycloaddition to Electron-Deficient Olefins," *Synlett* **1995**, 625–27.

181 Murray, C. K.; Yang, D. C.; and Wulff, W. D., "Cyclopropanation with Acyloxy Chromium Carbene Complexes. A Synthesis of $(+/-)$-Prostaglandin E_2 Methyl Ester," *J. Am. Chem. Soc.* **1990**, *112*, 5660–62.

182 Harvey, D. F.; Lund, K. P.; and Neil, D. A., "Cyclization Reactions of Molybdenum and Chromium Carbene Complexes with 1,6- and 1,7-Enynes—Effect of Tether Length and Composition," *J. Am. Chem. Soc.* **1992**, *114*, 8424–34.

183 Casey, C. P.; and Cesa, M. C., "^{13}CO Exchange Reactions of Metal–Carbene Complexes," *Organometal.* **1982**, *1*, 87–94.

CHAPTER 3

Insertion Reactions

Catalytically generated metal carbenes have proven to be highly versatile for insertion into carbon–hydrogen and heteroatom–hydrogen bonds (eq. 1). Although

$$X-H \quad + \quad L_nM=CR_2 \quad \longrightarrow \quad R_2C \underset{X}{\overset{H}{\diagdown}} \quad + \quad ML_n \qquad (1)$$

this transformation formally extends to O–H, S–H, and N–H insertion reactions, only C–H and Si–H insertion will be included in this chapter; other heteroatom–hydrogen insertion processes are described in Chapter 8. What makes C–H and Si–H bonds unique for insertion is their low bond polarity, and this separates them mechanistically from heteroatom–hydrogen bond insertions that are better described as ylide transformations. Because of their synthetic advantages, C–H insertion reactions will be emphasized in this chapter.

The insertion of a carbene into an unactivated C–H bond is a well-known transformation of "free" carbenes.[1-5] Although generally indiscriminate, this transformation is attractive for carbon–carbon bond formation, and the major challenge for its synthetic development has been control of insertion selectivity. Catalytic diazo decomposition has been the principal methodology for synthetic applications of C–H insertion, and it is the only practical way to bring about selective reaction at unactivated C–H bonds.[6-15] However, its advantages were not evident until the advent of dirhodium(II) catalysts.[9-15]

The mechanism of the transition metal catalyzed carbon–hydrogen insertion reactions of carbenes generated from diazo compounds has been the subject of considerable speculation.[16,17] However, there is general agreement that insertion occurs through an electrophilic metal carbene intermediate,[9-17] and experimental confirmation has been obtained for retention of configuration at the carbon atom of the C–H bond undergoing insertion (eq. 2).[18] The data currently available for C–H insertion reactions catalyzed by dirhodium(II) compounds implicate the mechanism, depicted in Scheme 1, as a suitable rationale for the C–H insertion process,[11] although Taber prefers a transition-state model in which there is transfer of hydrogen to rhodium (2).[19] Overlap of the metal carbene's p orbital with the σ orbital of the reacting C–H bond initiates the process in which, according to Scheme 1, C–C and C–H bond formation with the carbene carbon proceeds as the ligated metal dissociates. According to this model, increased electron with-

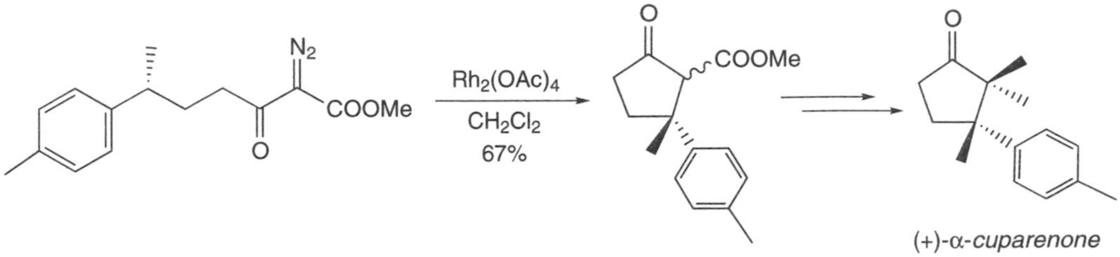

$$(+)\text{-}\alpha\text{-}cuparenone$$

$$(2)$$

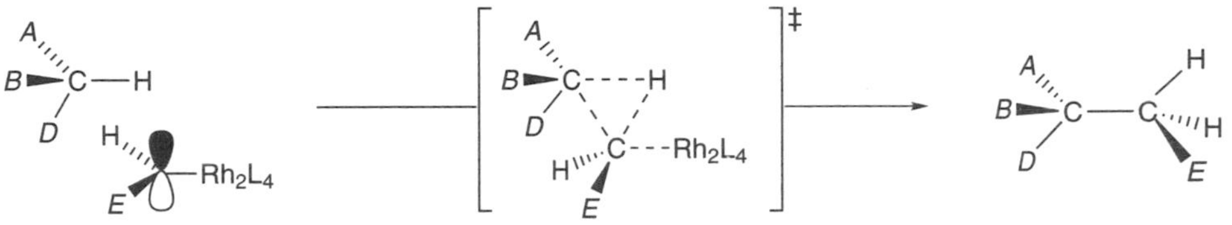

Scheme 1

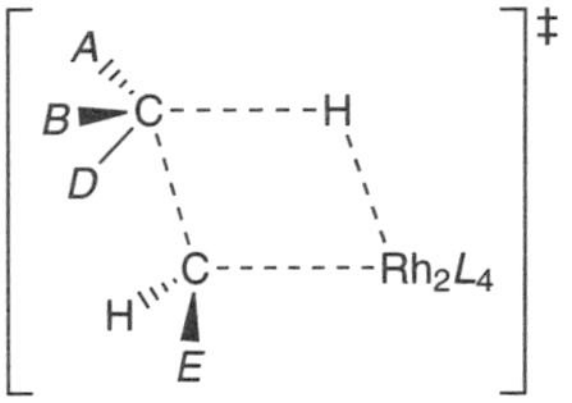

2

drawal by the ligand from the metal increases the electrophilicity of the carbene and causes bond formation to take place at a greater distance from the reacting C–H bond (earlier transition state) with resulting lower selectivity. Decreased electron withdrawal leads to a later transition state and greater reaction selectivity.

Copper catalysts have been employed for C–H insertion reactions, but they have been synthetically useful mainly in geometrically rigid systems (eqs. 3,[20] 4[21]).[8] Product yields are generally low to moderate. These early investigations

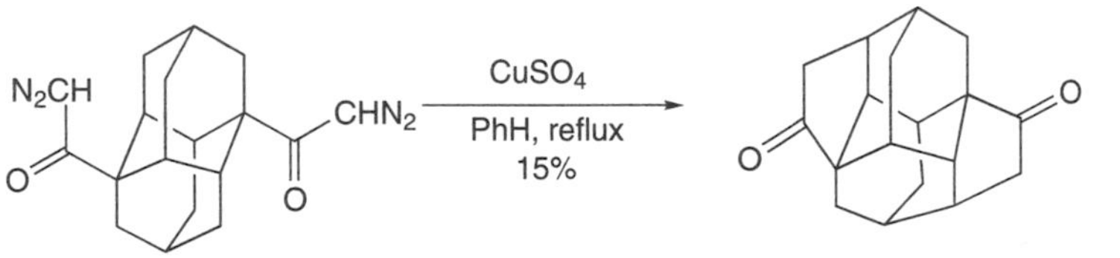

$$(3)$$

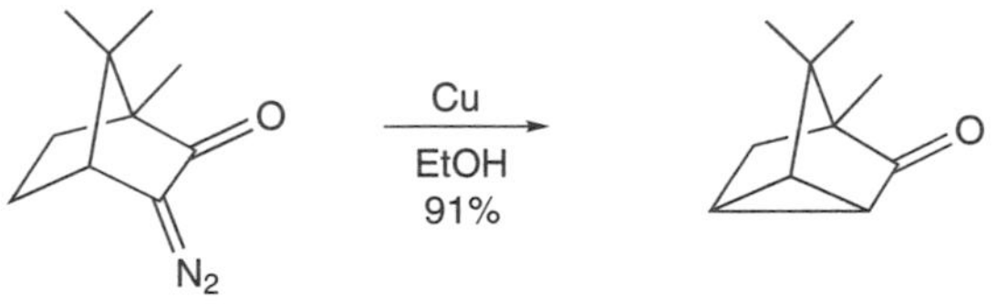

$$(4)$$

established the high preference in intramolecular reactions for five-membered ring formation and, from competition studies, for preferential insertion into tertiary C–H bonds over secondary C–H bonds. However, there have been numerous exceptions to a general pattern of selectivity exhibited by copper catalysts, which the applications of dirhodium(II) catalysts have rectified to a large extent.

3.1 CATALYTIC INTERMOLECULAR CARBON–HYDROGEN INSERTION REACTIONS

Intermolecular C–H insertion into paraffins by diazoacetate esters (eq. 5) has been achieved with dirhodium(II) trifluoroacetate, $Rh_2(tfa)_4$, or dirhodium(II) 9-

$$RH + N_2CHCOOEt \quad \xrightarrow[\substack{or \\ Rh_2(9\text{-}trp)_4}]{Rh_2(tfa)_4} \quad RCH_2COOEt + N_2 \tag{5}$$

triptycenecarboxylate, $Rh_2(9\text{-}trp)_4$, and to a lesser extent with $Rh_2(OAc)_4$, $Rh_2(OOCC_6F_5)_4$, and $Rh_2(OOCCMe_3)_4$ (Scheme 2).[22,23] A large excess of the

$$CH_3-CH_2-CH_2-CH_2-CH_3 \quad \text{(yield)} \qquad\qquad CH_3-\overset{\overset{\displaystyle CH_3}{|}}{CH}-\overset{\overset{\displaystyle CH_3}{|}}{CH}-CH_3$$

$Rh_2(OAc)_4$	33	63	4	(20%)	5	95
$Rh_2(tfa)_4$	31	64	5	(65%)	12	88
$Rh_2(9\text{-}trp)_4$	9	61	30	(86%)	33	67

$$CH_3-\overset{\overset{\displaystyle CH_3}{|}}{CH}-CH_2-CH_3$$

$Rh_2(OAc)_4$	1	8	90	1
$Rh_2(tfa)_4$	5	25	66	4
$Rh_2(5\text{-}trp)_4$	18	18	27	37

Scheme 2

alkane is required to prevent oligomerization of the carbene, and product yields are highly dependent on the catalyst employed, but dirhodium(II) perfluoroalkanoates and $Rh_2(9\text{-}trp)_4$ are clearly most effective. Regioselectivities for these insertion reactions are also catalyst dependent, as seen (Scheme 2) from reactions performed with pentane (at 22°C), 2,3-dimethylbutane (at 60°C), and 2-methylbutane (at 60°C). The preference for insertion into a secondary C–H bond is greater than that for insertion into a primary C–H bond, but insertion into a tertiary C–H bond can be less favorable than secondary C–H bond insertion (see 2-methylbutane). Of all the catalysts examined, $Rh_2(9\text{-}trp)_4$ shows the highest selectivity for insertion into primary C–H bonds, and $Rh_2(OAc)_4$ is most selective for tertiary C–H insertion. Analogous effects have been observed with the use of iodorhodium(III) *meso*-tetra(trimethylphenyl)porphyrin, $Rh(TMPP)I$,[24] which

$$CH_3-CH_2-CH_2-CH_2-CH_3$$

$Rh_2(OOCCMe_3)_4$	33	62	5
Rh(TPP)I	21	71	8
Rh(TMPP)I	14	61	25

also shows a high selectivity for insertion into primary C–H bonds, presumably because of steric influences. Insertion into the methyl groups of toluene, *p*-xylene, and mesitylene also occurs with the use of $Rh(TPP)I$,[25] whereas with these same substrates $Rh_2(tfa)_4$ catalyzes diazodecomposition yielding aromatic cycloaddition products.[26] Intermolecular C–H insertion reactions have mechanistic value,[27] but they are not synthetically useful.

3.2 CATALYTIC INTRAMOLECULAR CARBON–HYDROGEN INSERTION REACTIONS: GENERAL CONSIDERATIONS

Whereas intermolecular C–H insertion reactions usually produce multiple products and generally require highly electrophilic catalysts in order to minimize competitive reactions, intramolecular C–H insertion reactions of diazocarbonyl compounds occur with relative ease, even with moderately electrophilic catalysts, and they are more effective and selective.[11,12,14–17] Both copper and dirhodium(II) catalysts have been used to catalyze diazo decomposition resulting in intramolecular C–H bond insertion, but dirhodium(II) carboxylates and carboxamidates have become the catalysts of choice for these transformations. The development of dirhodium(II) catalysts for intramolecular C–H insertion reactions of diazocarbonyl compounds has been a significant synthetic achievement. β-Keto-α-diazoesters, -phosphonates, and -sulfones have been employed for the construction of cyclopentanone derivatives (eq. 6: $X = Y = CH_2$; $Z = COOEt$, $PO(OR)_2$,

$$\text{(6)}$$

3   4

SO_2Ar), diazoacetates and diazoacetoacetates form γ-lactones (eq. 6: $X = O$, $Y = CH_2$; $Z = H$, $COCH_3$), 3-alkoxy-1-diazoacetates provide access to 2(3H)-dihydrofuranones (eq. 6: $X = CH_2$, $Y = O$, $Z = H$), and diazomalonate esters produce lactones (eq. 6: $X = O$, $Y = CH_2$, $Z = COOR'$).[11] These reactions occur in moderate to high yield and, ordinarily, with an overwhelming preference for the formation of a five-membered ring.[16] Major competing reactions include carbene dimer formation (eq. 7) and water insertion processes (eq. 8), but both can be minimized with experimental modifications.

$$2\ RCHN_2 \quad \xrightarrow[\text{(-N}_2)]{ML_n} \quad RCH{=}CHR \tag{7}$$

$$RCHN_2 + H_2O \quad \xrightarrow[\text{(-N}_2)]{ML_n} \quad RCH_2OH \quad \xrightarrow[ML_n]{RCHN_2} \quad (RCH_2)_2O \tag{8}$$

Regiocontrol in C–H insertion reactions is one of the major advantages of dirhodium(II) catalysts, although the foundations for this selectivity were first observed with copper and, to a lesser extent, with silver catalysts (e.g., eq. 9).[28]

CuSO₄/C₆H₁₂	58%	22%	-
Ag(I)/MeOH	35%	15%	29%

$$\tag{9}$$

For reactions involving five-membered ring formation, there is general agreement that insertion into a tertiary C–H bond is favored over insertion into a secondary C–H bond, and primary C–H insertion, when observed, is least favorable,[11–17] but five-membered ring formation generally takes precedence over these considerations. Heteroatoms such as oxygen can activate an adjacent C–H bond for insertion,[29] and electron-withdrawing functional groups such as COOMe inhibit insertion.[30] There are, however, notable exceptions to these generalizations that limit predictability for the outcome of C–H insertion.

Diastereo- and enantioselectivity set further restrictions on the synthetic utility of C–H insertion reactions. However, both the structure of the diazo compound and, especially, the catalyst employed can provide effective stereocontrol so that a single product can be expected from a transformation that, less than 20 years ago, was considered to be indiscriminate. Chemoselectivity is yet another control feature for which catalyst design has had a major impact.[15]

Diazocarbonyl compounds in which the diazo carbon is bonded to two carbonyl groups are less reactive towards diazo decomposition than their corresponding diazo esters, ketones, and amides in which the diazo carbon is bonded to only one carbonyl group, and, dependent on the catalyst, reaction temperatures of 80°C (refluxing ClCH₂CH₂Cl or PhH) instead of 25°C are usually required. However, product yields from C–H insertion reactions are often higher with

α-diazo-β-ketoesters than with diazo esters and related compounds. In addition, regio- and stereoselectivities are influenced by the structures of the diazo compounds.

3.3 CATALYTIC INTRAMOLECULAR CARBON–HYDROGEN INSERTION REACTIONS OF DIAZO KETONES: REGIOSELECTIVITY FOR CYCLOPENTANONE FORMATION

The advantages of $Rh_2(OAc)_4$ over traditional copper catalysts for diazo decomposition resulting in C–H insertion became evident from early comparative investigations. Wenkert and co-workers were the first to disclose the use of $Rh_2(OAc)_4$ for C–H insertion leading to cyclopentanone formation.[31] The conversion of the isopimaridiene skeleton (**5**) into the 16-keto steroid (**6**) was accomplished (eq. 10) in 60% yield with the use of $Rh_2(OAc)_4$ as the catalyst, but poor

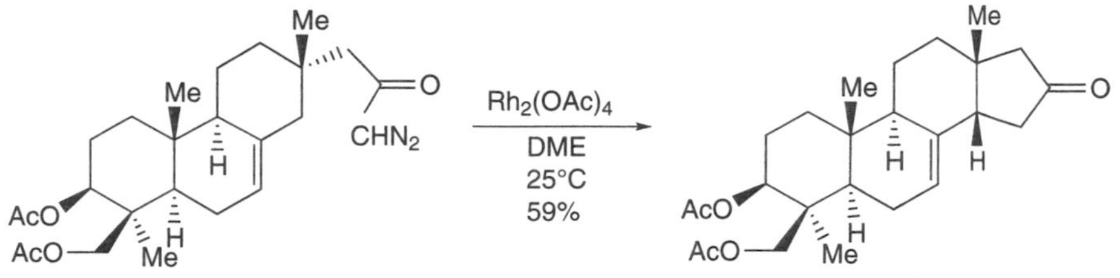

(10)

yields of **6** were obtained when $CuSO_4$ was used. Similar low yields and, also, low selectivity for cyclopentanone formation characterized $CuSO_4$-catalyzed reactions of 1-diazo-2-octanone and 1-diazo-4,4-dimethyl-2-pentanone,[31] and solvent (cyclohexane) insertion was observed.

The same year as Wenkert's initial report of the use of $Rh_2(OAc)_4$ for C–H insertion, Taber and co-workers began publication of a series of communications that more fully elaborated the versatility of this methodology for cyclopentanone formation.[18,32] The formation of five-membered ring ketones occurred to the near exclusion of cyclobutanones or cyclohexanones (eqs. 11 and 12), when such competition was possible.

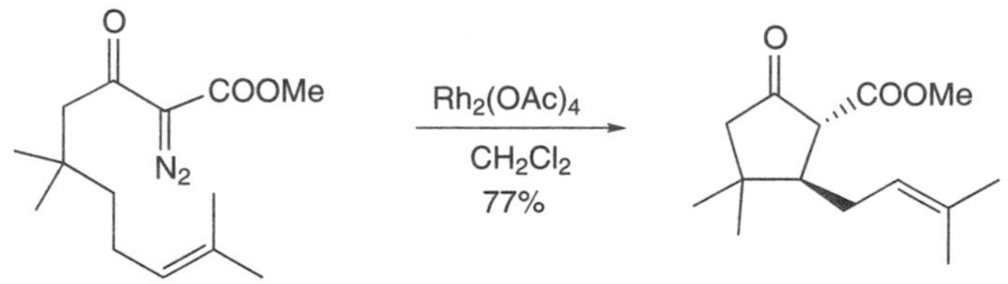

(11)

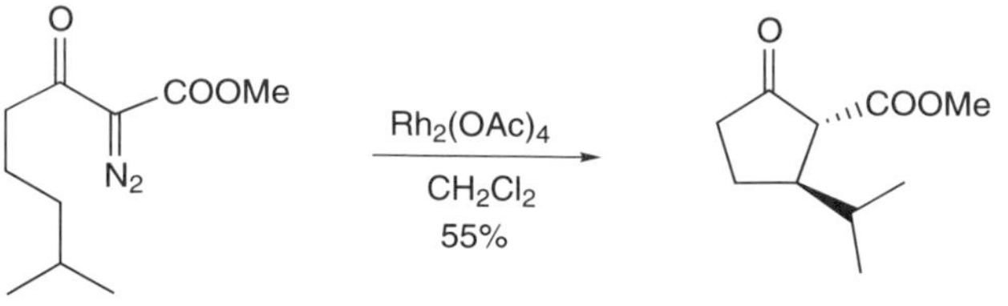

$$(12)$$

Investigations of competitive intramolecular C–H insertion reactions confirmed and extended previously known preferences for 3°>2°>>1° C–H bonds.[33] A strong influence of strongly electron withdrawing groups on regioselectivity was demonstrated by Stork and Nakatani, who found that an ester substituent deactivated both α- and β-methylene groups towards C–H insertion (eq. 13)[30] so

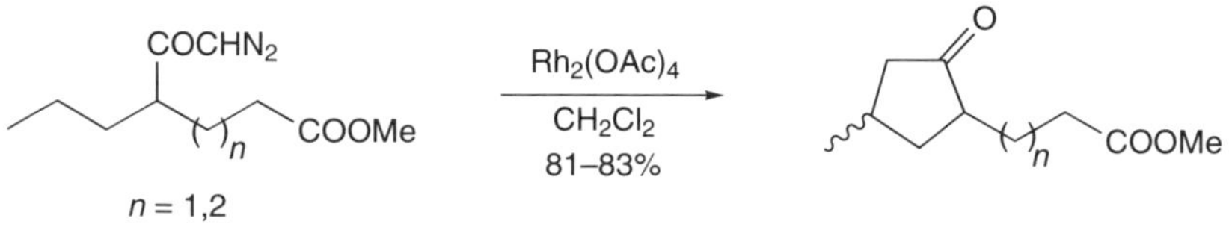

$$(13)$$

that, even when this intramolecular pathway is the only one possible, only carbene dimer formation is realized. The same conclusion could not be made from investigations of β-lactam formation by insertion into C–H bonds α or β to a carboxylate group (see Section 3.6),[34] and there is reason to believe that electronic dipolar repulsion from the face of the catalyst to the carboxylate substituents of diazo compounds may also be involved.

If electron-withdrawing groups can deactivate a C–H bond for insertion, then an electron-donating group should activate a C–H bond for insertion. Adams and co-workers have suggested this as the cause of the extraordinary regiocontrol that they observe in reactions such as that of eq. 14,[29,35] where the net electron-donating

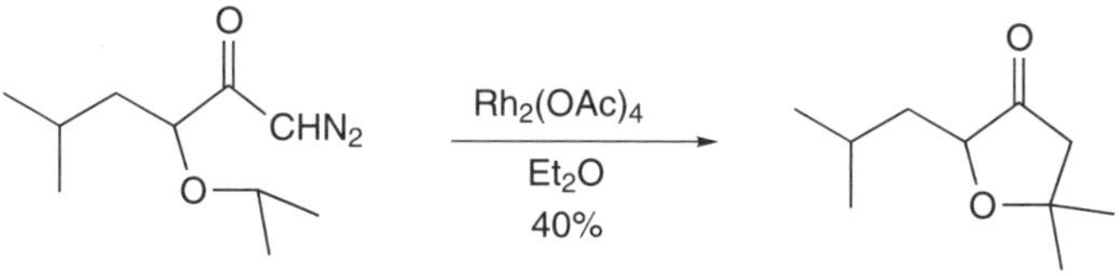

$$(14)$$

influence of the oxygen atom adjacent to the C–H bond undergoing insertion activates that bond for exclusive 3(2H)-furanone production. Ether activation and the regiochemical preference for five-membered ring formation are the controlling principles in the total synthesis of the mushroom metabolite (+)-muscarine (**7**)[36] and the natural insect attractant endo-1,3-dimethyl-2,9-dioxabicyclo-[3.3.1]nonane (**8**)[37] by intramolecular Rh$_2$(OAc)$_4$ catalyzed C–H insertion.

The effect of oxygen substituents on regiocontrol has been described through a series of highly informative investigations involving diazo ketones **9** (eq. 15)[38]

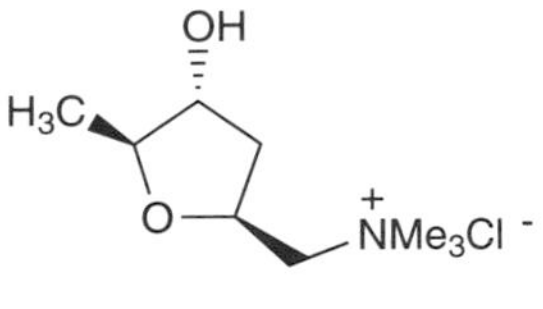

7 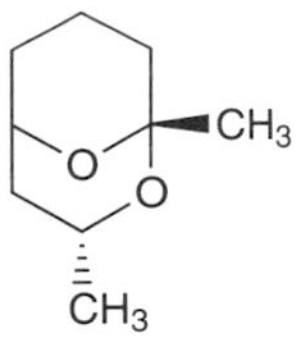   **8**

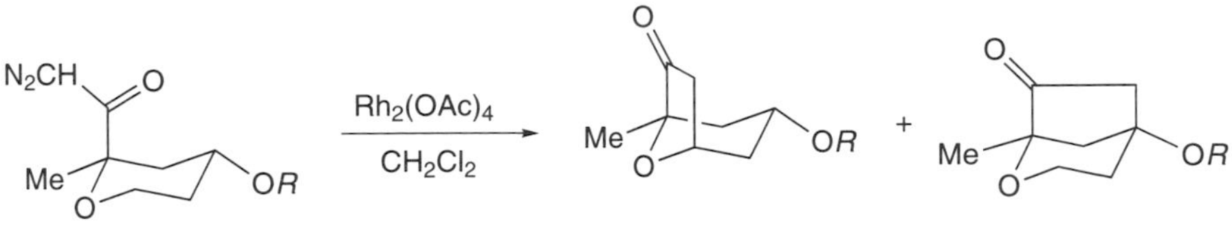

9	R	10, %	11, %
	Me	96	4
${}^{t}BuPh_2Si = TBDPS$	99	1
	Ac	>99	<1
	H	86	14	(15)

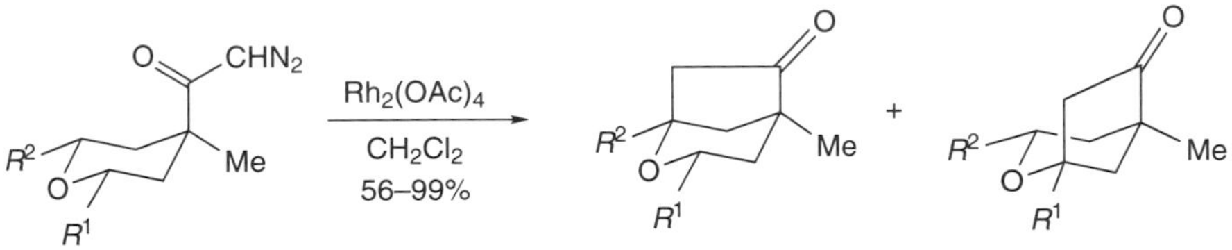

12	R^1	R^2	13, %	14, %
	OAc	OMe	>99	<1
	OH	OMe	42	58
${}^{t}BuMe_2Si = TBDMS$	OMe	69	31
${}^{i}Pr_3Si = TIPS$	OMe	86	14
	N_3	OMe	11	89
	CH_2TMS	H	50	50	(16)

and **12** (eq. 16).[39] The influence of alkyl substitutents by similar investigations using copper catalysts was previously reported (eq. 9).[28] Ether oxygen substituents are highly activating, but acetoxy groups are deactivating. Surprisingly, the azido substituent has a greater directive effect than even methoxy. A small isotope (k_H/k_D) effect was determined for C–H insertion.[38,39] A variety of secondary alcohols have been converted to 3(2H)-furanone products using this methodology (e.g., eq. 17).[40]

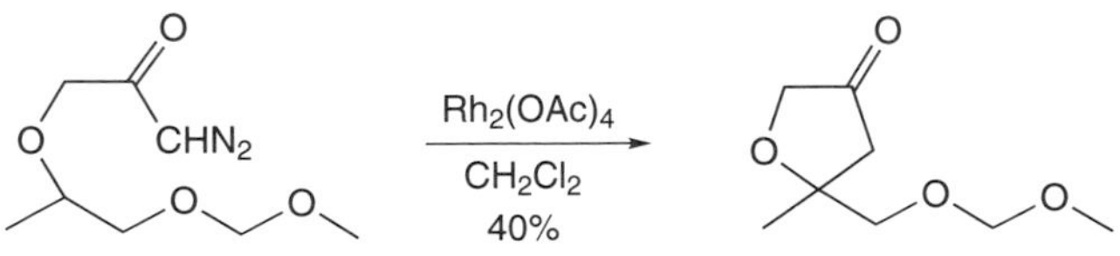

The preference for cyclopentanone ring formation is not always followed in intramolecular C–H insertion reactions of diazoketones. In structurally rigid systems competition with cyclobutanone formation has been observed (eq. 18).[41] Analogs of **5** have produced cyclobutanone (eq. 19) and cyclohexanone products

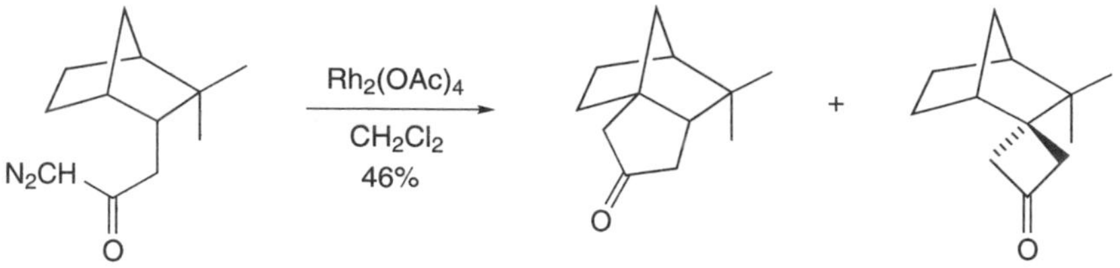

(18)

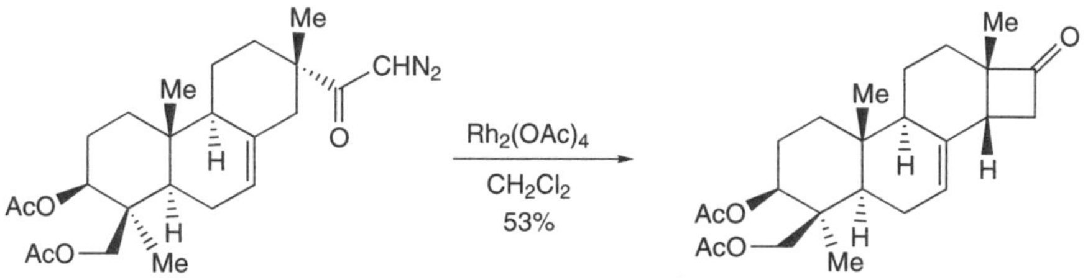

(19)

in addition to the preferred cyclopentanones in dirhodium(II) acetate catalyzed reactions.[42] Subtle changes in the structure of the diazo compound influence regioselectivity (eqs. 20[43] and 21[44]). The carbon–hydrogen insertion reaction yield-

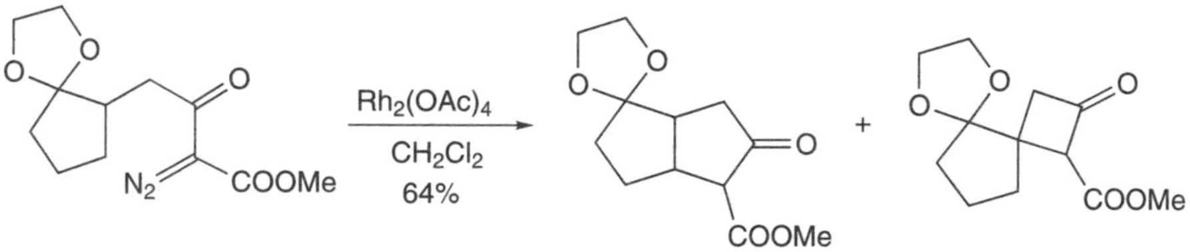

(20)

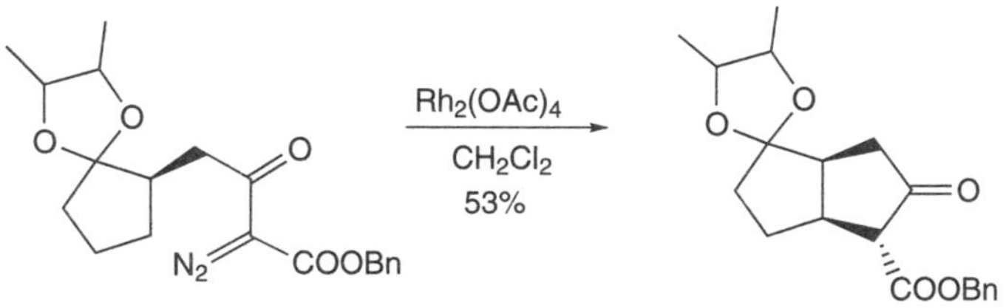

(21)

16

ing **16** was a key step in the synthesis of (+)-isocarbacyclin.[44] Even ethyl 4-cy-clopentyl-2-diazo-3-oxobutyrate, the analog of **15** without the ethylene ketal, produced cyclopentanone and cyclobutanone insertion products (83% yield) in a 3:2 ratio.[45] This preference is the reverse of that found with **15**, but a more substituted derivative (**17**) produced only the cyclobutanone product (eq. 22). Lee has provided a variety of examples that necessitate a review of the factors that influence five-membered ring formation in C–H insertion reactions (eq. 23).[40]

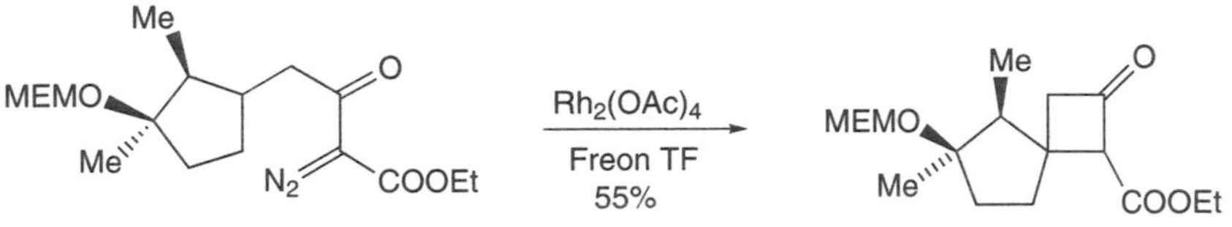

$$\text{(22)}$$

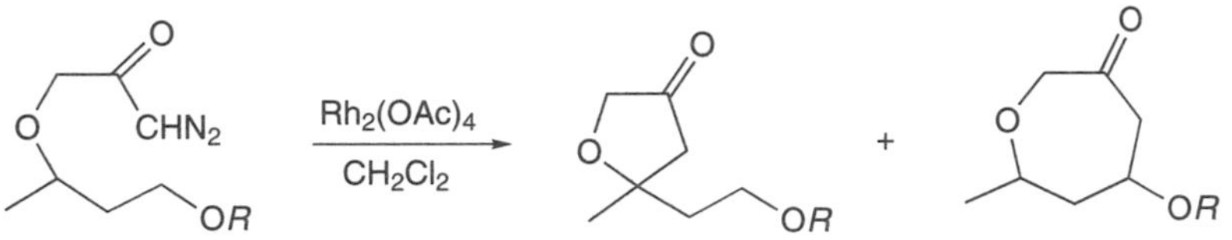

17

R = TBDMS	0%	80% (1:1)
TIPS	0%	82% (4:1)
Bn	58%	0%

$$\text{(23)}$$

Thus, Rh$_2$(OAc)$_4$ catalyzed C–H insertion reactions do not always provide high regioselectivity; so the need for new catalysts to provide higher levels of regiocontrol for these transformations is evident.

3.4 CATALYTIC INTRAMOLECULAR CARBON–HYDROGEN INSERTION REACTIONS OF DIAZOKETONES AND DIAZOESTERS: DIASTEREOSELECTIVITY AND SYNTHETIC APPLICATIONS

Catalytic C–H insertion reactions can provide exceptional diastereocontrol for these carbon–carbon bond-forming reactions as the examples in eqs. 24,[19]

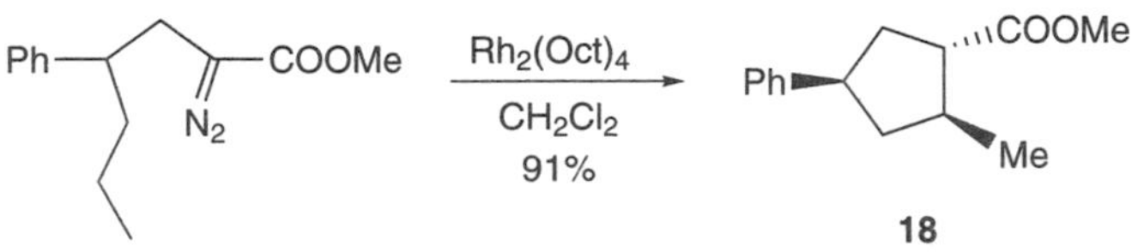

$$\text{(24)}$$

18

25,[19] and 26[46]suggest. Taber has provided a useful conformational evaluation of transition-state structures that allows one to conclude that the product obtained comes from a pseudocyclohexane structure in which the substituents are, prefera-

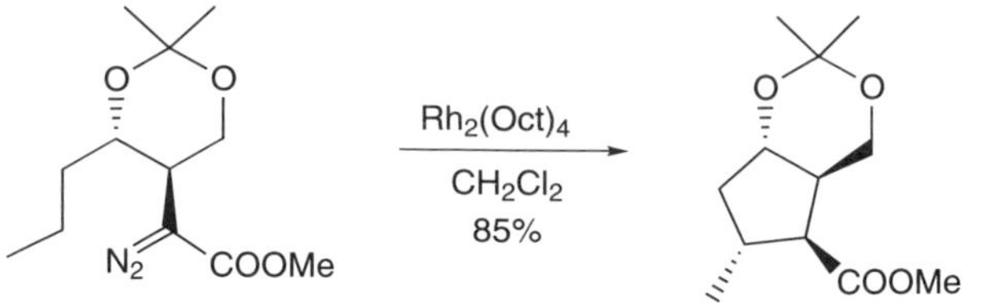

(25)

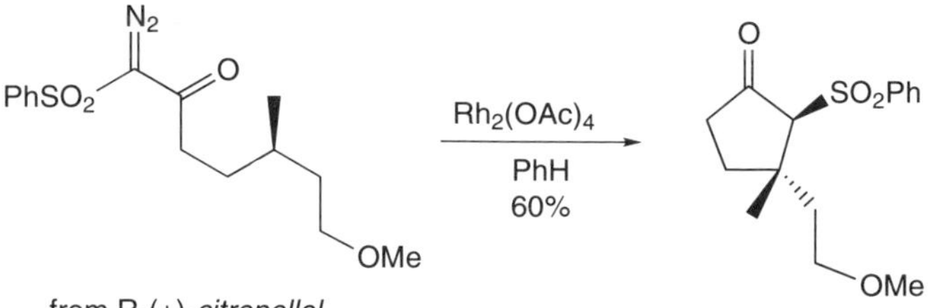

(26)

from R-(+)-*citronellol*

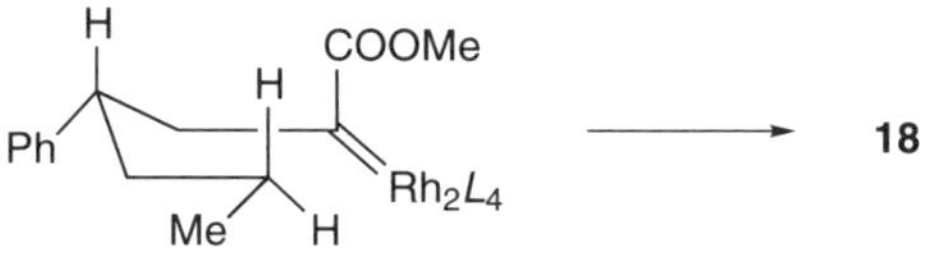

19

bly, in the more stable pseudoequatorial positions (e.g., **19**). This model provides predictability to diastereoselectivity in these transformations that is more general than the few examples that were originally explored.[19]

When there is a C–H bond adjacent to the diazo carbon, β-hydride migration intervenes, and alkene products are formed exclusively or in major competing reactions (eq. 27)[47]; changing the dirhodium(II) ligand from octanoate to trifluoroacetate increases the relative percentage of **21**. The explanation for exclusive *cis*-olefin formation in β-hydride migration is analogous to that provided by

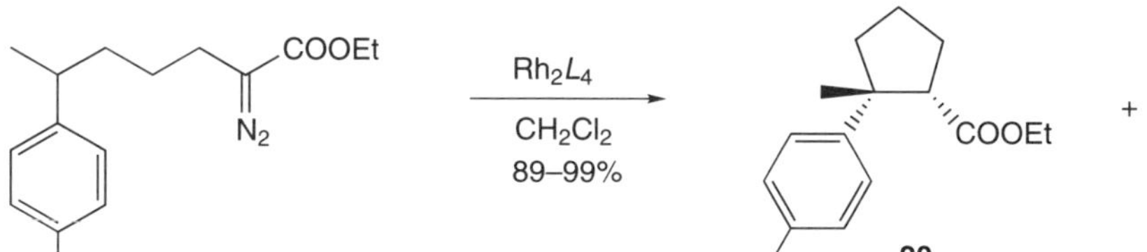

(27)

Catalyst	**20:21**
$Rh_2(OOCCMe_3)_4$	55:45
$Rh_2(Oct)_4$	45:55
$Rh_2(OOCPh)_4$	42:58
$Rh_2(OAc)_4$	37:63
$Rh_2(tfa)_4$	26:74

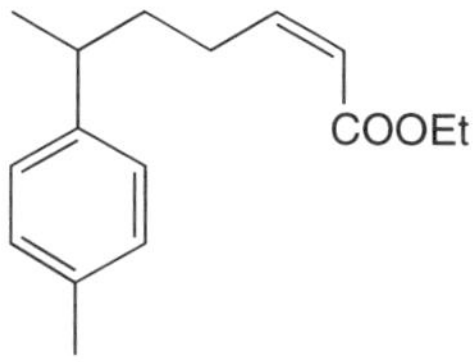

Shechter for preferential *cis*-alkene formation in "carbene dimerization" of diazo compounds catalyzed by $Rh_2(OAc)_4$.[48]

Because of the high diastereocontrol and regiocontrol that could be achieved in dirhodium(II) catalyzed C–H insertion reactions, synthetic applications are growing in number. These include methyl dihydrojasmonate (**22**),[32] (+)-α-cuparenone (eq. 2),[18] (±)-tochuinyl acetate (**23**),[47] (+)-albene (**24**),[41] pentaleno-lactones *E* (**25**)[49] and *F*,[45] estrone methyl ether (**26**),[50] (+)-grandisol (**27**),[46] and the novel fatty acids (±)-dicranenones (**28**).[51] High diastereocontrol is well docu-

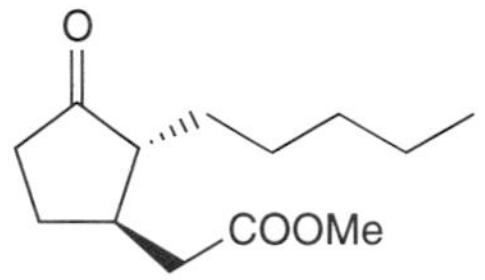

22

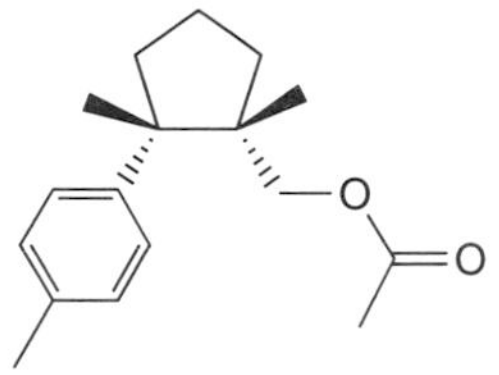

23

24

25

26

27

$A = C \equiv C, CH_2CH_2$

28

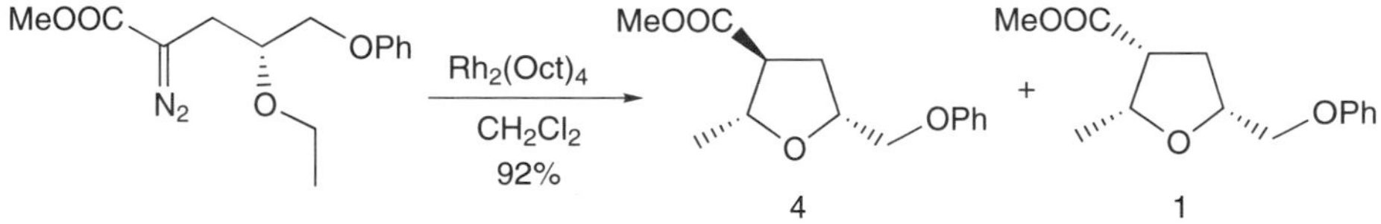

mented for the synthesis of 2,3,5-trisubstituted tetrahydrofurans (eq. 28)[52] and disubstituted 3(2*H*)-furanones (eqs. 29[36] and 30[53]). α-Diazo-β-ketosulfones

4

1

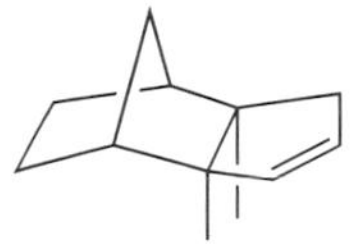

(28)

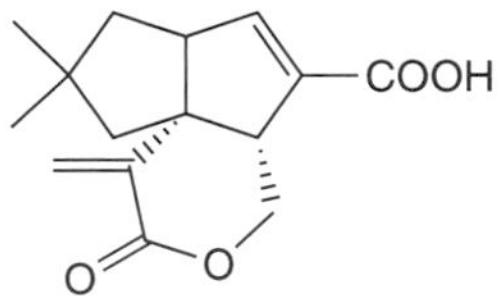

8:1 *cis:trans*

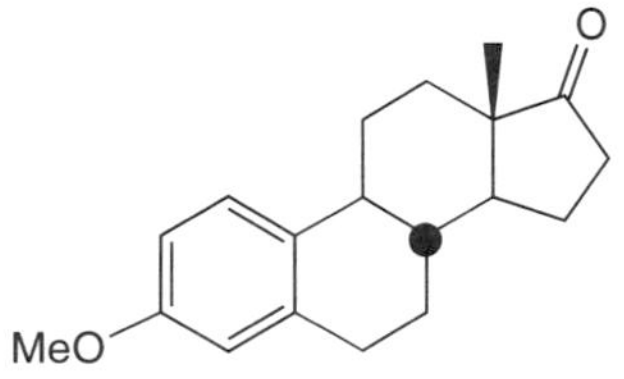

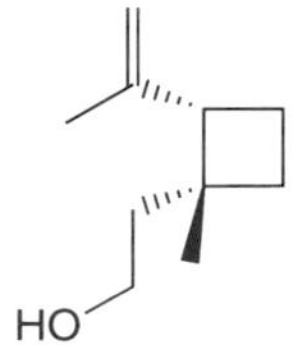

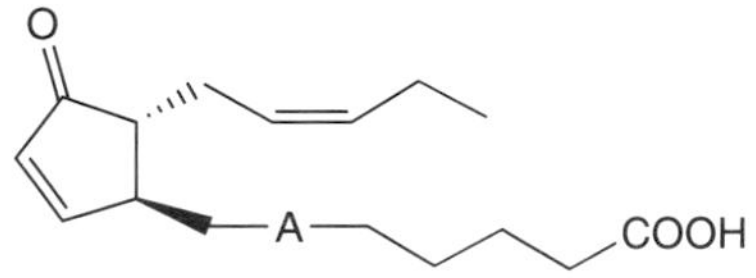

muscarine

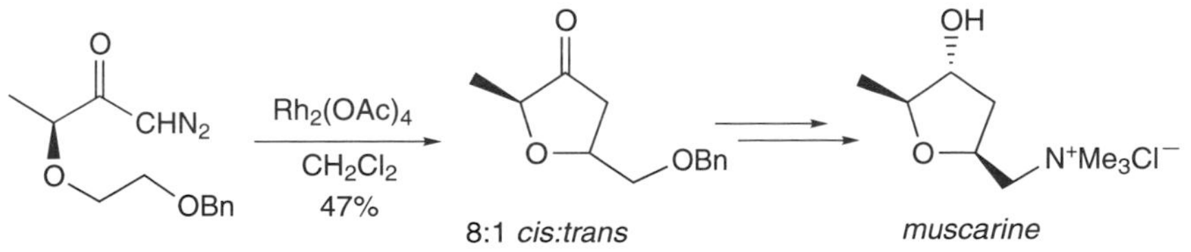

(29)

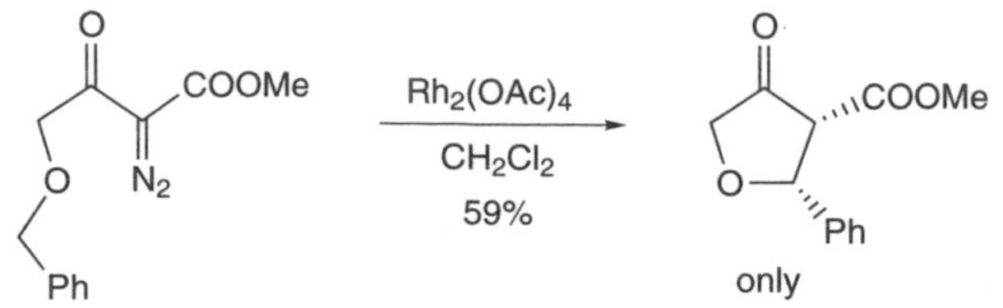

(30)

(eq. 31)[54] and α-diazo-β-ketoalkylphosphonates are similarly effective,[55,56] the latter having been employed for the synthesis (eq. 32) of the antibiotic ($\pm$)-sarkomycin (**29**).

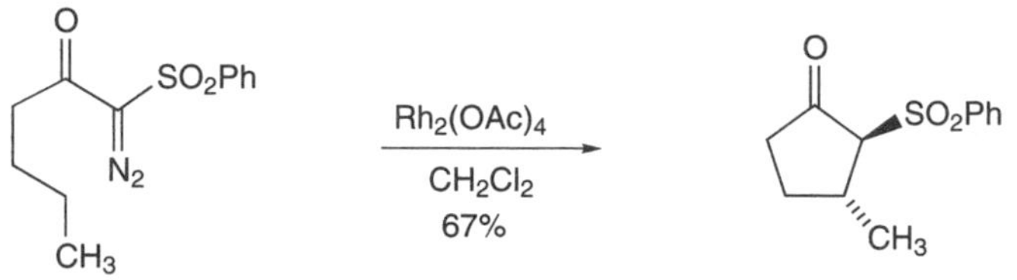

(31)

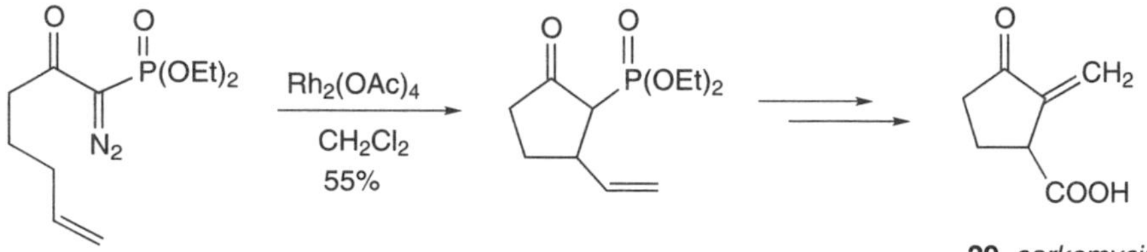

29, *sarkomycin*

(32)

Treatment of diazoester **30** with dirhodium(II) mandelate resulted in the formation of spiro-enone **31**, which was subsequently transformed into a tetra-cyclic C$_{14}$ ginkgolide **32** (eq. 33).[57] In addition, [4,4,4,5]fenestrane and [4,4,4,4]-

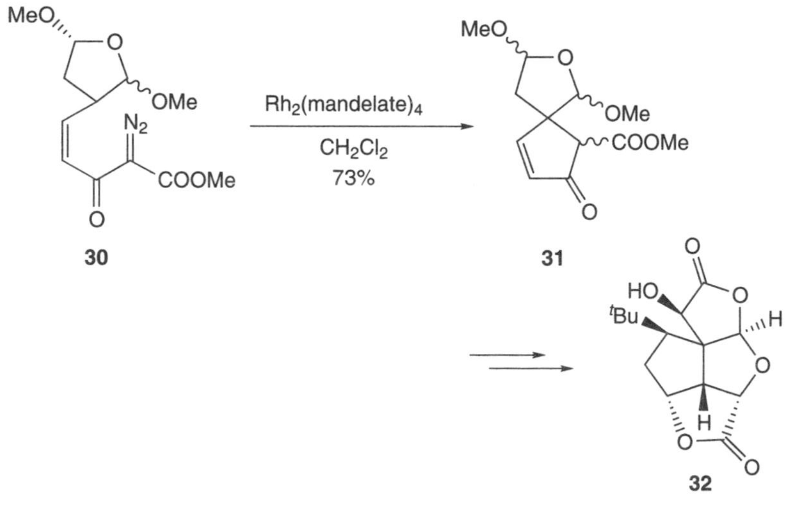

30   **31**   (33)

32

fenestrane derivatives have been synthesized using the C–H insertion process (e.g., eq. 34).[58-60] Other examples[61-65] further demonstrate the versatility of this methodology, especially with dirhodium(II) catalysts, for synthetic applications.

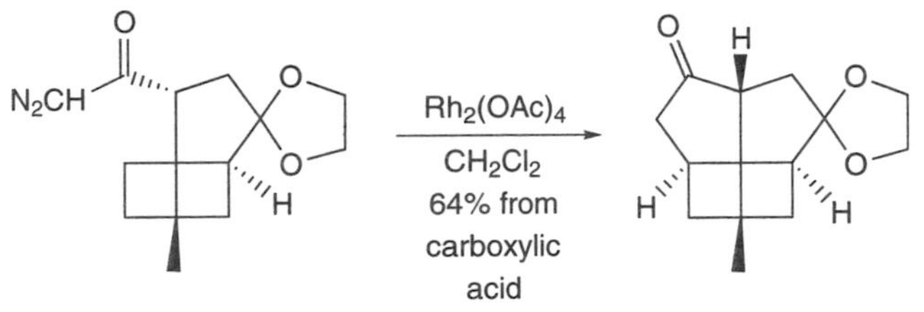

(34)

3.5 CATALYTIC INTRAMOLECULAR CARBON–HYDROGEN REACTIONS OF DIAZOACETATES: LACTONE FORMATION

Highly efficient intramolecular C–H insertion reactions of diazoacetoacetates catalyzed by $Rh_2(OAc)_4$ (eqs. 35 and 36) have been reported.[66,67] Higher reaction

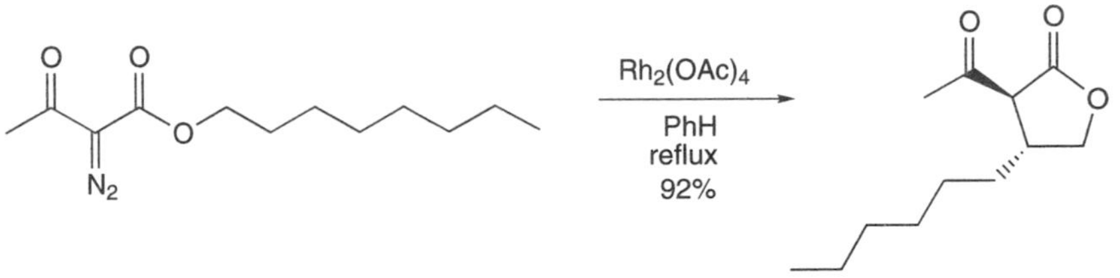

(35)

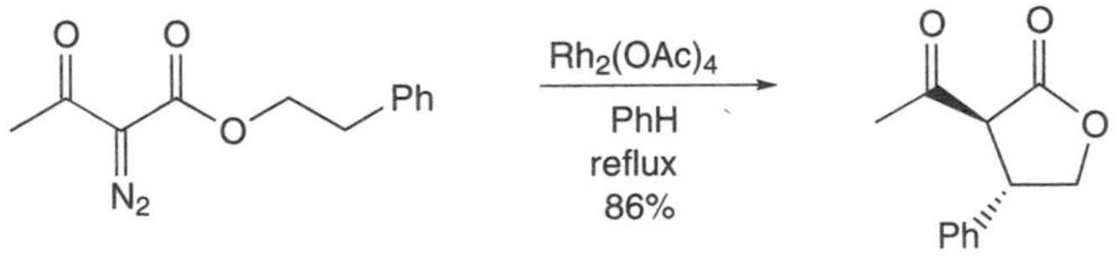

(36)

temperatures are used for their diazo decomposition (PhH or $ClCH_2CH_2Cl$, reflux, rather than CH_2Cl_2) than with diazoacetate compounds, but isolated product yields are generally higher. Electronic factors that control product formation, including the preference for five-membered rings and C–H bond reactivity ($3° > 2° > 1°$), operate with diazoesters in much the same manner as with diazoketones. However, there are exceptions.

Cane and Thomas reported the first example of an intramolecular C–H insertion reaction of a diazoester catalyzed by $Rh_2(OAc)_4$ (eq. 37) but, rather than forming the normally favored γ-lactone, a δ-lactone directed to the synthesis of pentalenolactone E was produced.[45] Lee reported that alkyl methyl diazomalonates form, preferentially in several examples, β-lactones in $Rh_2(OAc)_4$-catalyzed reactions (e.g., eq. 38),[68] and Saba has confirmed this selectivity with a

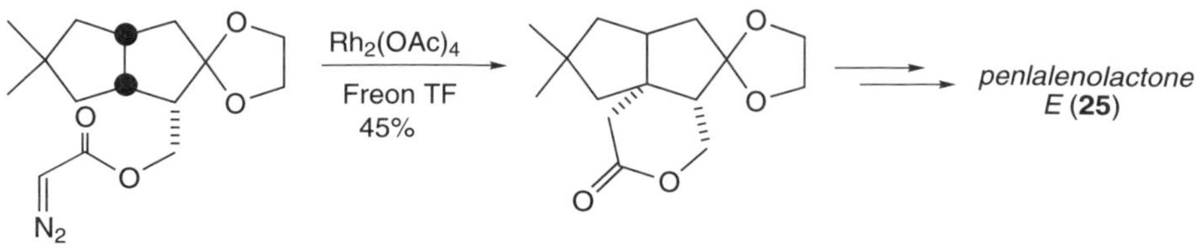

$$(37)$$

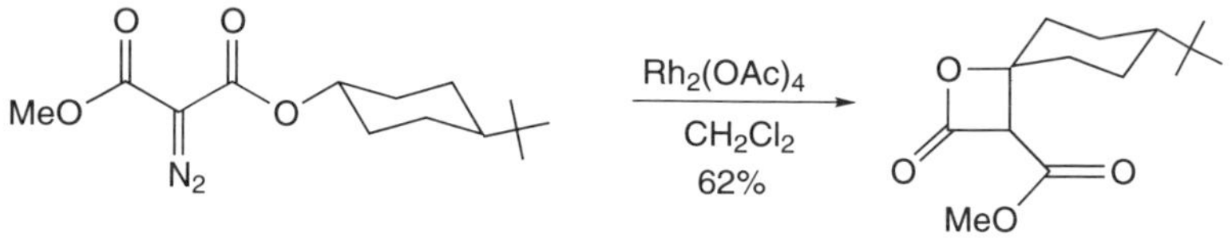

$$(38)$$

series of dicyclohexyl diazomalonic esters.[69] Doyle and Poulter have reported the first example of a macrocyclic C–H insertion reaction, forming an 11-membered ring, catalyzed by $Rh_2(pfb)_4$ (eq. 39);[70] intramolecular cyclopropanation was not competitive.

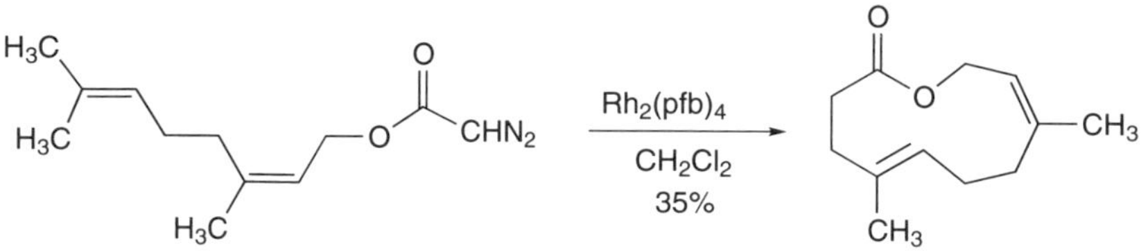

$$(39)$$

Spirolactone formation occurs readily with cycloalkylmethyl diazoacetates in reactions catalyzed by $Rh_2(cap)_4$ (eq. 40),[71] and high selectivities have been achieved for insertion into ether oxygen-activated C–H bonds (eq. 41), even when γ-lactone formation should have been dominant (eq. 42).[72] However, certain cy-

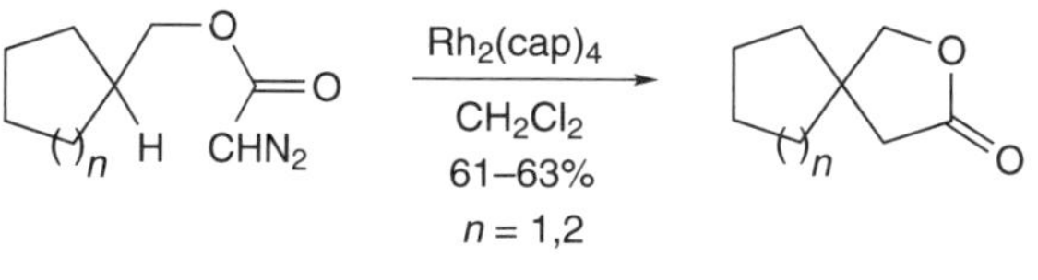

$$(40)$$

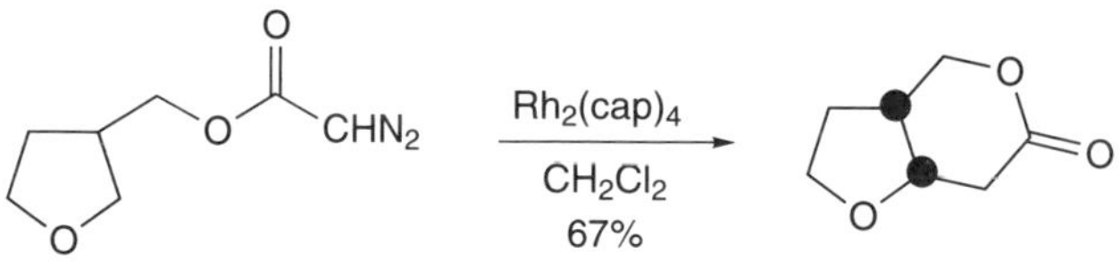

$$(41)$$

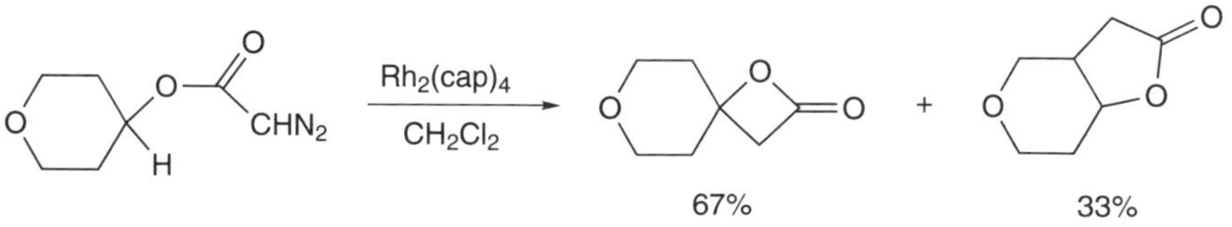

$$(42)$$

clic benzylic and allylic diazoacetates undergo hydride abstraction rather than insertion with formation of ketone and methylenecycloalkane products (eq. 43)[72];

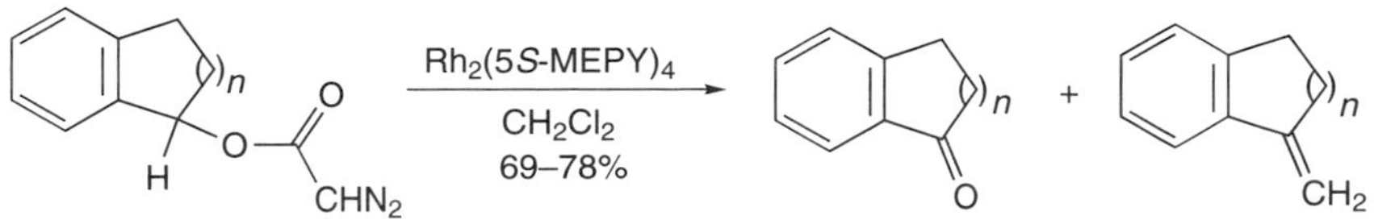

	32	**33**
$n = 1$	83	17
$n = 2$	86	14

(43)

chiral dirhodium(II) carboxamidate catalysts provide higher yields of these products than does $Rh_2(cap)_4$, and with $Rh_2(OAc)_4$ this process is relatively unimportant. Carbon dioxide and ketene are formed along with **32** and **33**, and the mechanism depicted in Scheme 3 accounts for the overall transformation. Saba

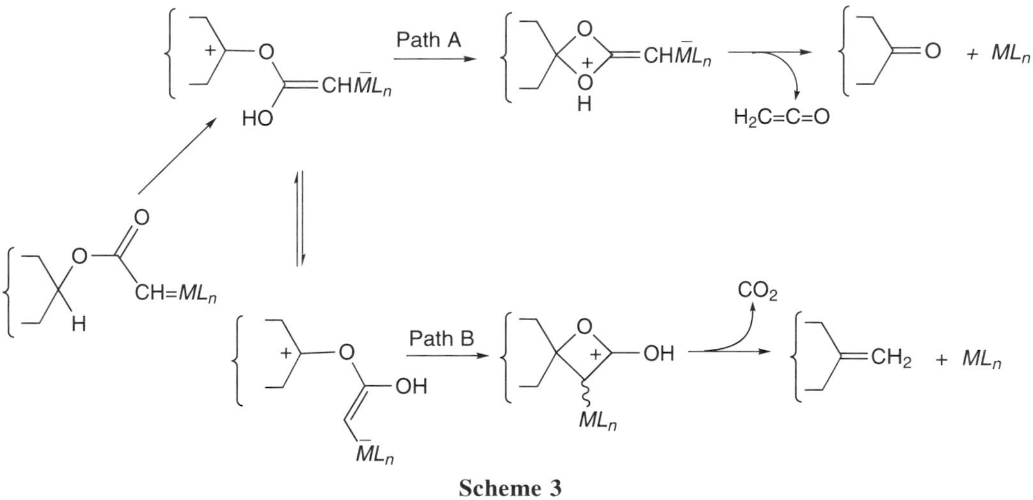

Scheme 3

and co-workers have reported a transannular C–H insertion process that occurs with carbon dioxide extrusion (eq. 44),[73,74] and they described this product's use

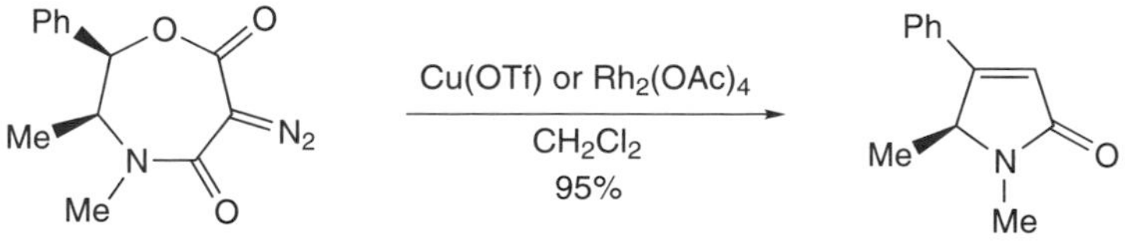

(44)

for the synthesis of 4,5-disubstituted 2-pyrrolidinones, which can be considered as precursors to γ-aminobutyric acid. Diazoketones can undergo similar hydride abstraction reactions. Lee reported that such a process was competitive with C–H insertion (e.g., eq. 45).[40] Highly electrophilic catalysts such as dirhodium trifluo-

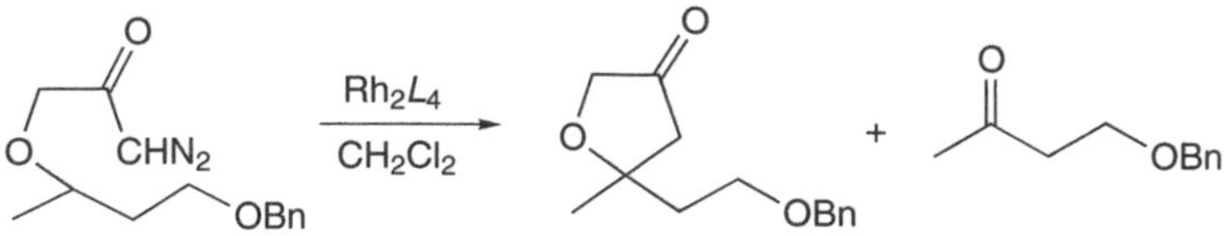

$Rh_2(tfa)_4$	5%	55%
$Rh_2(Oct)_4$	28%	45%
$Rh_2(OAc)_4$	58%	37%
$Rh_2(tpa)_4$	60%	15%
$Rh_2(acam)_4$	35%	0%

(45)

roacetate favor hydride abstraction. This oxidation process formally requires the loss of cyclopropanone, but its mechanism has not yet been established. (46)

Although C–H bond reactivity is normally $3° > 2° > 1°$, *l*-menthyl diazoaceto-acetate and diazomalonate undergo insertion exclusively into the equatorial C–H bond at the secondary position (eq. 46).[66,67,69] This selectivity was originally

CATALYTIC DINITROGEN EXTRUSION FROM *l*-(−)-MENTHYL DIAZOACETOACETATE[67]

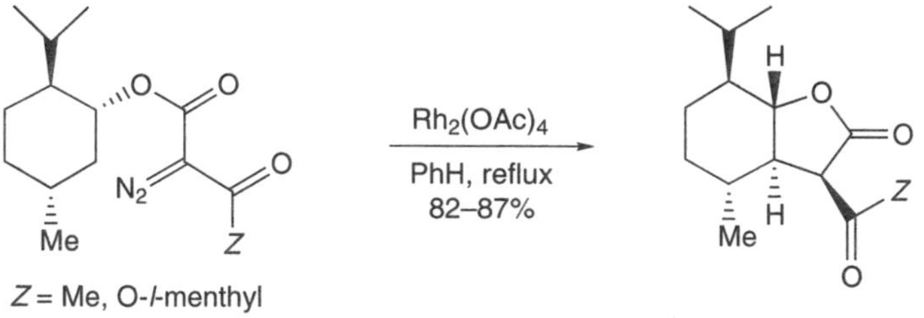

Z = Me, O-*l*-menthyl

To 1.0 mol % of dirhodium(II) acetate (4.4 mg, 1.0×10^{-2} mmol), based on diazo compound, in 10 mL of refluxing anhydrous benzene was added 0.268 g (1.01 mmol) of *l*-(−)-menthyl diazoacetoacetate in 3.0 mL of benzene by syringe pump over a 6-h period. The resulting solution was passed through a short plug of neutral alumina, and the alumina was washed with 15 mL of CH_2Cl_2. The solvent was removed under reduced pressure, and the resulting solid residue was recrystallized from pentane to yield 0.193 g of a white solid (0.81 mmol, 80% yield) identified as (−)-(*1S,4S,5R,6S,9R*)-4-acetyl-6-methyl-9-isopropyl-2-oxabicyclo[4.3.0]nonan-3-one: mp 98–99°C; ^{1}H NMR ($CDCl_3$) δ 3.72 (*t*, *J* = 10.7 Hz, 1 H), 3.44 (*d*, *J* = 12.4 Hz, 1 H), 2.41 (*s*, 3 H), 2.31 (*dt*, *J* = 12.4, 10.7, 10.7 Hz, 1 H), 2.00–1.88 (*m*, 1 H), 1.84–1.62 (*m*, 3 H), 1.58–1.38 (*m*, 1 H), 1.23–1.00 (*m*, 2 H), 0.95 (*d*, *J* = 7.0 Hz, 3 H), 0.89 (*d*, *J* = 7.0 Hz, 3 H), and 0.82 (*d*, *J* = 6.6 Hz, 3 H); $[\alpha]_D^{24} = -33.4$ (*c* = 0.59, $CHCl_3$).

thought to be due to steric effects, with the isopropyl group blocking access to the methine hydrogen. However, as we will see, the observed selectivity is the result of preferential insertion into equatorial C–H bonds (Section 3.9).

3.6 CATALYTIC INTRAMOLECULAR CARBON–HYDROGEN INSERTION REACTIONS OF DIAZOACETAMIDES: LACTAM FORMATION

Ponsford and Southgate were the first to report that $Rh_2(OAc)_4$ was an effective catalyst for intramolecular carbon–hydrogen insertion reactions.[75] Diazoacetoacetamide **34** underwent $Rh_2(OAc)_4$-catalyzed diazo decomposition at room temperature to yield β-lactam **35** in 75% yield (eq. 47); the use of Cu in toluene at

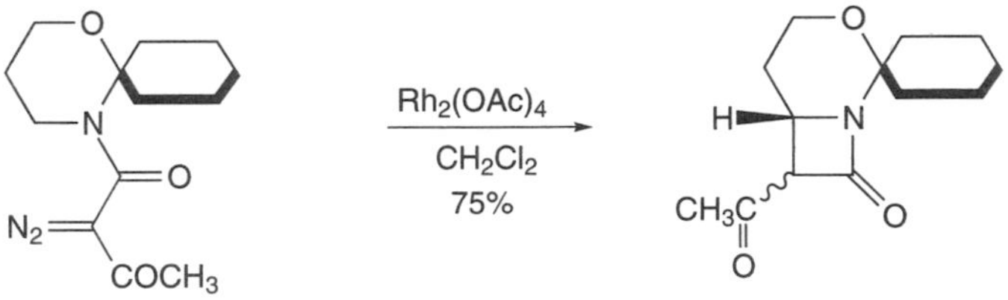

$$(47)$$

90°C gave **35** in only 25% yield. Additional examples within the same structural framework have been reported.[76,77] Previously, photochemical reactions had been employed.[78,79] This heteroatom activation of adjacent C–H bonds has made possible a new methodology for the construction of β-lactam derivatives.[80]

A series of diazoacetoacetamides, prepared by condensation of 2° benzylamines with diketene followed by diazo transfer, has been reported to undergo exclusive β-lactam formation in high yield and with exceptional stereocontrol for the *trans* diastereoisomer (eq. 48).[80] Insertion into C–H bonds that are β or α to

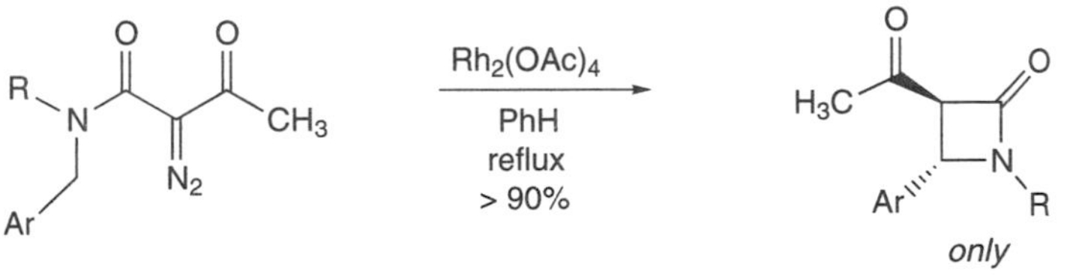

$$(48)$$

$R = {}^t Bu, {}^i Pr, Bn$
$Ar = p\text{-}NO_2C_6H_4, m\text{-}BrC_6H_4, m\text{-}MeOC_6H_4, Ph, 3,4\text{-}(MeO_2C_6H_3)$

strongly electron-withdrawing carboxylate substituents occurs with relative ease in high yield and with high diastereocontrol (eqs. 49 and 50).[34,81] However, unlike the stereochemical course of the reactions described in eqs. 48 and 49, **39** is produced with exclusive *cis* stereochemistry. Insertion into the *tert*-butyl group is not observed.

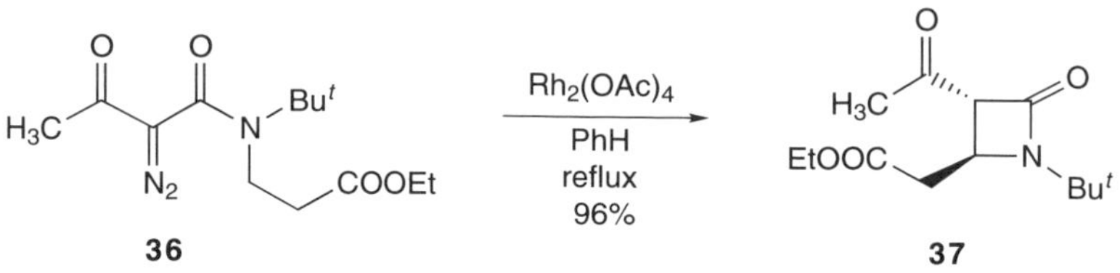

$$(49)$$

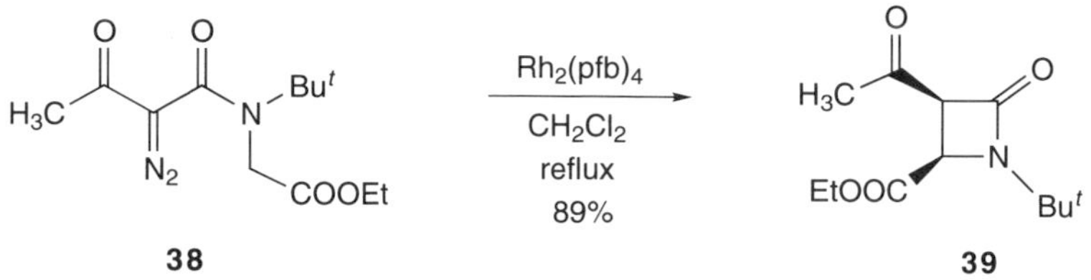

$$(50)$$

The cause of insertion into normally deactivated positions has been attributed to the activating influence of the amide nitrogen in the transition state for C–H insertion. Product diastereoselectivity is dependent on the conformationally dependent orientation of substituents in the intermediate metal carbene.[34] Thus the diazomalonamide analogues of **36** and **38** with the easily removable N-(p-methoxyphenyl) substituent provide strikingly different results (eqs. 51 and 52)

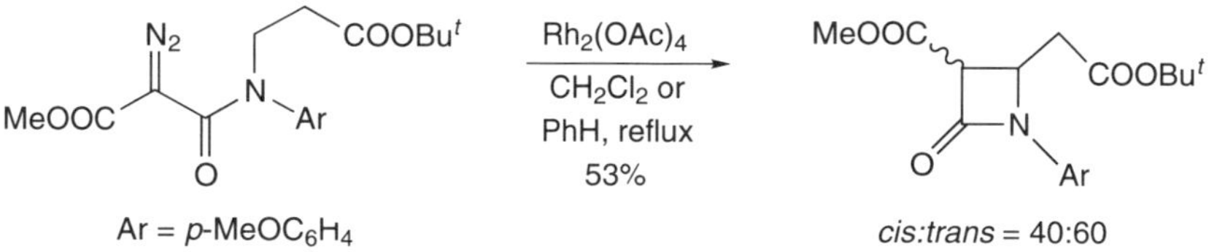

$$(51)$$

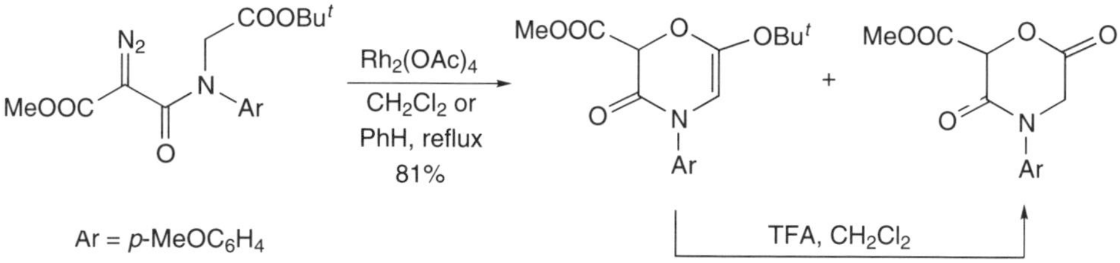

$$(52)$$

from those of eqs. 49 and 50.[82] The carbonyl ylide transformation represented in eq. 52 had been previously reported to be a competing process to insertion in reactions of N-neopentyl and N-(n-butyl) analogues of **36**.[34,81] α-Diazo-β-ketophosphoramidates (e.g., **40**) undergo Rh$_2$(OAc)$_4$-catalyzed production of 2-oxa-1,2-azaphosphetidines (eq. 53), although in low yield.[83]

Diazoamides whose structures allow both β- and γ-lactam formation often form mixtures of these compounds in Rh$_2$(OAc)$_4$-catalyzed reactions (eq. 54).[84]

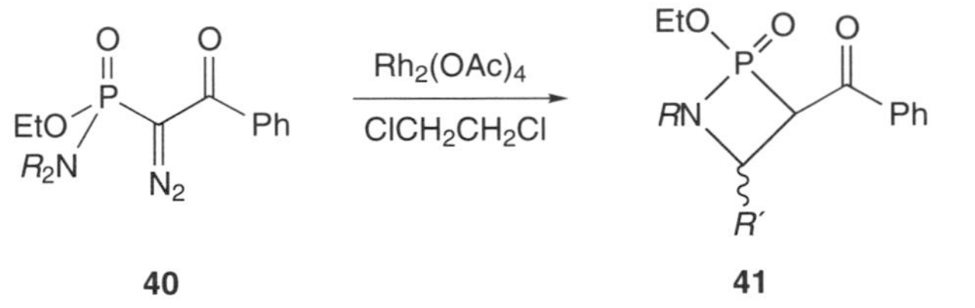

$$(53)$$

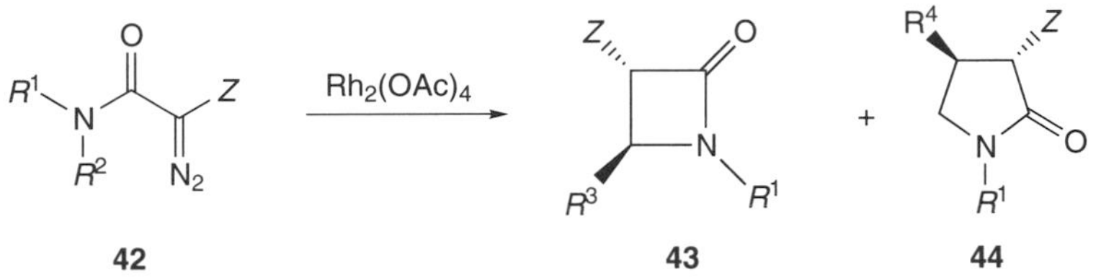

$$(54)$$

TABLE 3.1   Lactam formation from Rh$_2$(OAc)$_4$-catalyzed reactions of 42

R^1	R^2	Z	Yield (%)	43:44	Ref.
^{i}Pr	^{i}Pr	CH$_3$CO	89	>99:<1	80
^{i}Pr	^{i}Pr	H	95	81:19	80
^{t}Bu	^{n}Bu	CH$_3$CO	92	37:63	84
^{t}Bu	^{n}Bu	H	95	<1:>99	84
PMP^a	^{n}Bu	COOMe	76	8:92	82
^{t}Bu	PhCH$_2$CH$_2$	CH$_3$CO	94	49:51	84
^{t}Bu	PhCH$_2$CH$_2$	H	85	<1:32^b	84
PMP^a	PhCH$_2$CH$_2$	COOMe	84	<1:>99	82

^{a}PMP = p-methoxyphenyl.
^{b}68% of product from aromatic cycloaddition (see eq. 59).

Analysis of the data in Table 3.1 allows the following generalizations: (1) β-lactam formation is more pronounced with Z = COCH$_3$ or COOMe than with Z = H and (2) C–H insertion is favored by 3°>2°>>1°. With N-benzyl derivatives the presence or absence of an acyl group uniquely defines the course of the Rh$_2$(OAc)$_4$-catalyzed reaction (eq. 55); the diazoacetoacetamides ($R = {}^t$Bu, S = H, Me, OMe, Br, NO$_2$) give β-lactam products exclusively, whereas diazoac-

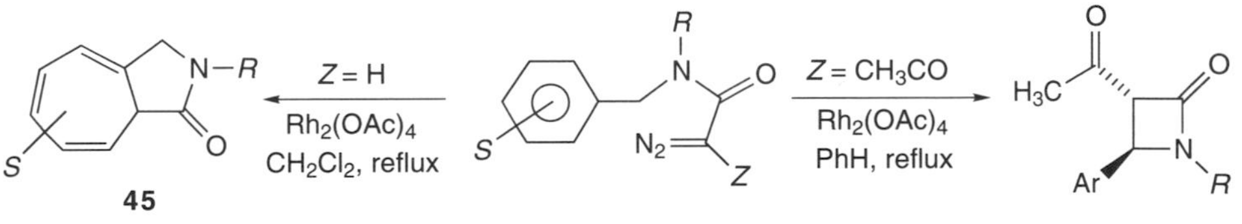

$$(55)$$

etamides (R = ^{t}Bu, S = H, Me, OMe, Br) yield aromatic cycloaddition products (**45**) exclusively.[85]

Diazo decomposition of diazoacetamides derived from L-phenylalanine has revealed that not only does C–H insertion occur, mainly at an N-benzyl group, but that hydride abstraction, similar to that described in eq. 43 and Scheme 3, can be an important competing process (eq. 56).[86] C-Alkylation of the phenylalanine

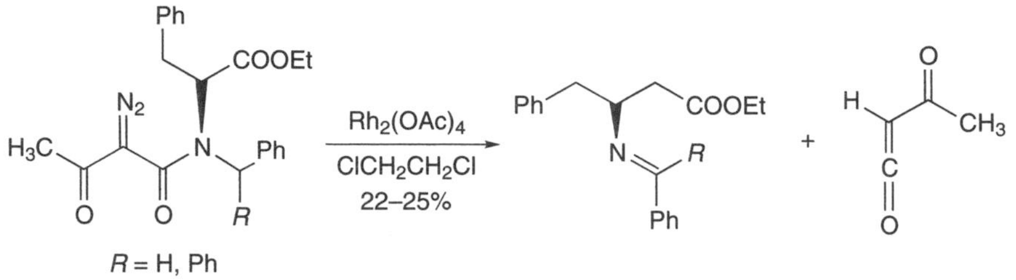

$$R = \text{H, Ph}$$

$$(56)$$

framework was a minor reaction; insertion into the benzylic group and aromatic cycloaddition were dominant.

3.7 CATALYST-DEPENDENT CHEMOSELECTIVITY AND REGIOSELECTIVITY FOR CARBON–HYDROGEN INSERTION

In competitive intramolecular transformations of diazo compounds catalyzed by dirhodium(II) compounds, ligands of the dirhodium(II) catalyst effectively, and often completely, switch reaction preference.[87,88] Relative to $Rh_2(OAc)_4$ in competitive insertion/cyclopropanation (eq. 57), use of dirhodium(II) perfluorobu-

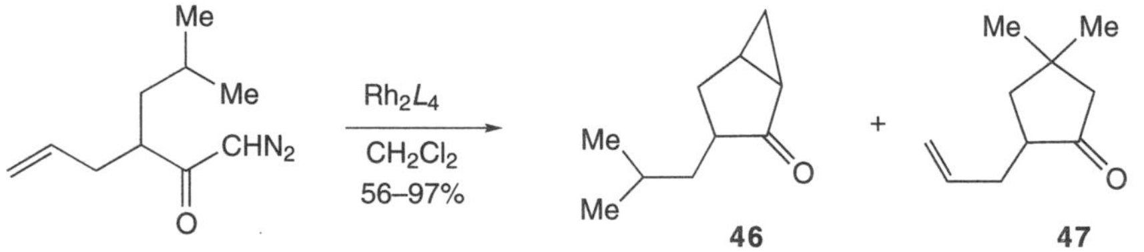

	46	47
$Rh_2(pfb)_4$	0	100
$Rh_2(OAc)_4$	44	55
$Rh_2(cap)_4$	100	0

$$(57)$$

tyrate, $Rh_2(pfb)_4$, produces insertion product **47** exclusively, whereas the cyclopropanation product **46** is the only one from the reaction catalyzed by dirhodium(II) caprolactamate, $Rh_2(cap)_4$. Significant ligand effects on products from competitive insertion and aromatic substitution (eq. 58)[89] and from competi-

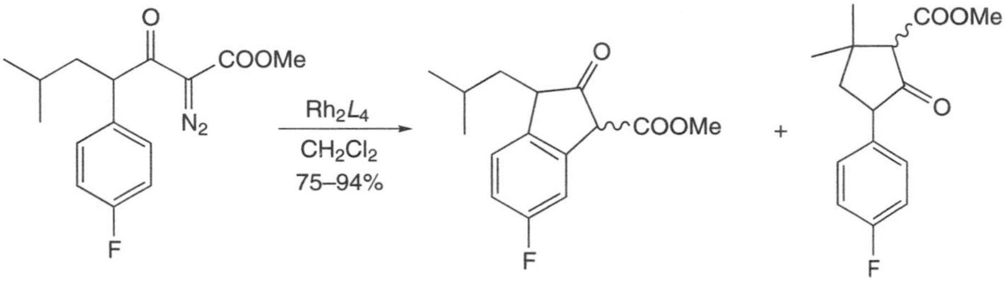

	48	**49**
Rh$_2$(OAc)$_4$	< 1	>99
Rh$_2$(OOCCPh$_3$)$_4$	>99	< 1

$$(58)$$

tive insertion and aromatic cycloaddition (eq. 59)[88] have been reported. Similar preferences for insertion or aromatic substitution and cycloaddition are observed with substituted benzene derivatives.[88,89]

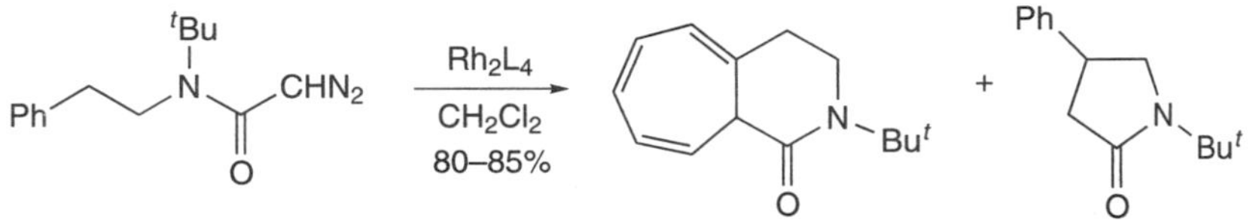

	50	**51**
Rh$_2$(OAc)$_4$	68	32
Rh$_2$(pfb)$_4$	95	5
Rh$_2$(cap)$_4$	3	97
Rh$_2$(acam)$_4$	77	77

$$(59)$$

Doyle and co-workers examined C–H insertion reactions of diazoacetoacetates and diazoacetates and reported exceptional dependence of regioselectivity on the carboxylate or carboxamide ligand of the dirhodium(II) catalyst.[66,67] In competition reactions between insertion into 3° and 1° C–H bonds of diazoacetate **52**, use of Rh$_2$(pfb)$_4$ produced products in relative amounts that were a statistical accounting of the number of 3° and 1° C–H bonds (eq. 60).[66] In contrast,

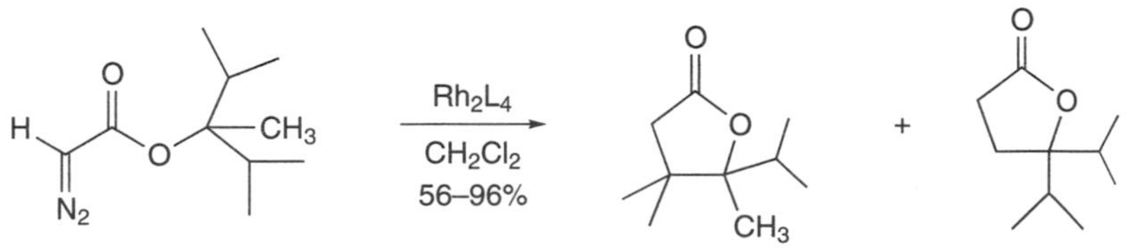

	53	**54**
Rh$_2$(pfb)$_4$	32	68
Rh$_2$(OAc)$_4$	53	47
Rh$_2$(acam)$_4$	>99	<1

$$(60)$$

use of $Rh_2(acam)_4$ only gave the product from insertion into the 3° C–H bond **53**. Similar effects were observed with the corresponding diazoacetoacetamide, the only difference being with $Rh_2(OAc)_4$, where a 90:10 ratio of 3°:1° C–H insertion products was obtained. In the competition between 3° and 2° C–H bond insertion, regioselectivities greater than 35 have been reported using $Rh_2(cap)_4$. These effects are electronic manifestations of the controlling influence of dirhodium(II) ligands on regioselectivity, but they are not always displayed in C–H insertion reactions (eq. 61).[67] In eq. 61 product preference is governed by

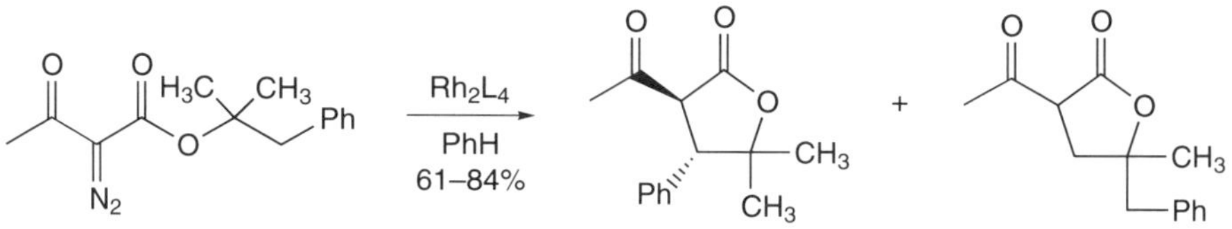

	55	**56**
$Rh_2(pfb)_4$	29	71
$Rh_2(OAc)_4$	29	71
$Rh_2(cap)_4$	30	70

(61)

conformational factors. In general, for systems in which each of the possible sites for insertion can present a C–H bond to the carbene center with equal probability for C–H insertion, electronic influences influence regioselectivity. However, when the possible sites for insertion cannot present their C–H bonds to the carbene center with equal probability for C–H insertion, regioselectivity is governed more by conformational preference than by electronic preference.[67] In practice, when ligand effects from the dirhodium(II) catalyst are negligible, as they are in eq. 61, conformational effects dominate.

Ikegami and co-workers have reported that dirhodium(II) tetra(triphenylacetate), $Rh_2(tpa)_4$, provides a high degree of regiocontrol for C–H insertion reactions (eq. 62).[43] The nature of this regiocontrol appears steric in origin. Whereas

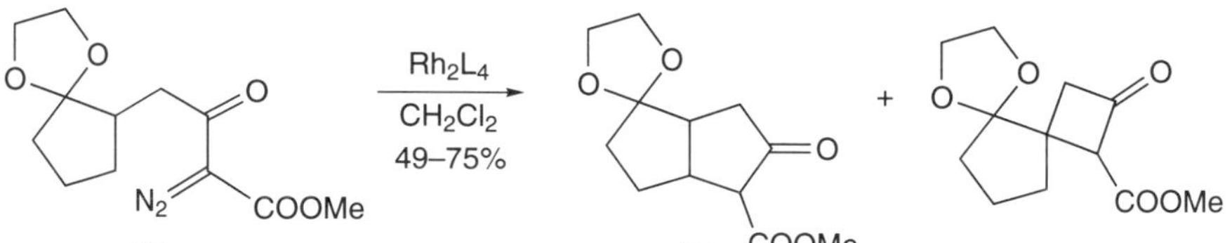

	57	**58**
$Rh_2(tfa)_4$	56	44
$Rh_2(OOCPh)_4$	54	46
$Rh_2(OAc)_4$	37	63
$Rh_2(OOCCMe_3)_4$	37	63
$Rh_2(acam)_4$	14	86
$Rh_2(tpa)_4$	96	4

(62)

the direction of increasing electron withdrawal from the dirhodium(II) ligand takes the **57**:**58** ratio to a value of 56:44, $Rh_2(tpa)_4$, which should be comparable to $Rh_2(OAc)_4$ in its overall electronic influence, increases that ratio to 96:4. Other examples reported by Ikegami reveal similar regioselectivity enhancement.[43]

Whereas **36** generates a single β-lactam product in high yield (eq. 49), its *N*-neopentyl analogue **59** gives a mixture of products (eq. 63) whose amounts are

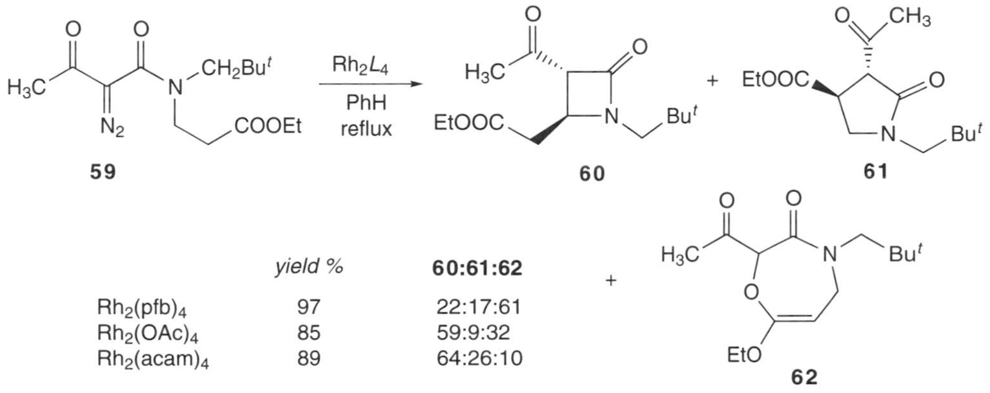

	yield %	60:61:62
$Rh_2(pfb)_4$	97	22:17:61
$Rh_2(OAc)_4$	85	59:9:32
$Rh_2(acam)_4$	89	64:26:10

(63)

significantly influenced by the catalyst employed.[81] Carbonyl ylide formation is the cause of the production of **62**, and this product is favored by the use of $Rh_2(pfb)_4$. However, limited catalyst ligand control is achieved with other analogues of **36** and/or **59**.[81]

Clark has compared three catalysts for their effectiveness in the competition between oxonium ylide formation and C–H insertion (eq. 64),[90] and Cu(hfacac)$_2$

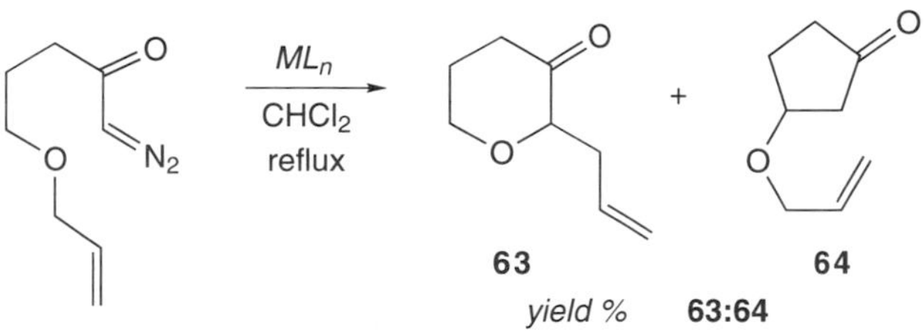

	yield %	63:64
$Rh_2(OAc)_4$	59	69:31
$Cu(acac)_2$	73	84:16
$Cu(hfacac)_2$	80	100:0

(64)

has been found to minimize unwanted C–H insertion and favor ylide generation/[2,3]-sigmatropic rearrangement. Padwa has concluded from his investigations that, in general, copper catalysts show a preference for cyclopropanation and ylide formation over C–H insertion.[15]

3.8 ENANTIOSELECTIVITY IN INTRAMOLECULAR CARBON–HYDROGEN INSERTION REACTIONS

Enantioselective C–H insertion is a recent undertaking, and already significant advances for organic syntheses have been achieved. Since dirhodium(II) catalysts are generally superior to those of copper for C–H insertion, chiral catalyst development has focused on dirhodium(II) compounds. Those from Hashimoto, Ikegami,[91] and McKervey[92] utilize homochiral dirhodium(II) carboxylates that are formed from *N*-protected amino acids (e.g., **65** and **66**), and they exhibit their

65

66

a. Ar = Ph
b. Ar = p-^{t}BuC_6H_4

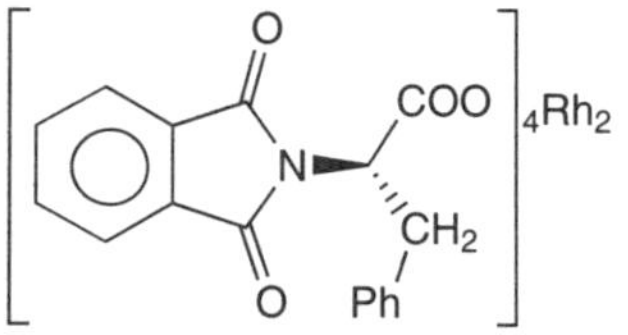

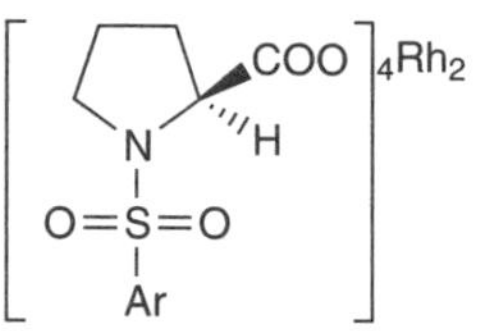

highest levels of enantiocontrol with diazoketones of the structure $RCOCN_2R'$ where $R' \neq$ H. Doyle's chiral dirhodium(II) carboxamidates (**67**) give high enantiocontrol with diazoacetates.[93–95]

67

a. Rh_2(5*S*-MEPY)$_4$: $Z = CH_2$
b. Rh_2(4*S*-MEOX)$_4$: $Z = O$
c. Rh_2(4*S*-MACIM)$_4$: $Z = NCOCH_3$
d. Rh_2(4*S*-MPPIM)$_4$: $Z = NCOCH_2CH_2Ph$
e. Rh_2(4*S*-MCHIM)$_4$: $Z = NCOCH_2(C_6H_{11})^c$

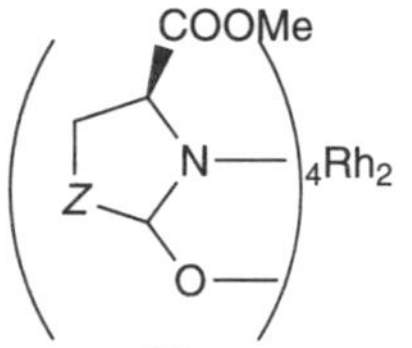

Applications of the homochiral dirhodium(II) carboxylates with α-diazo-β-ketoesters[91] and α-diazo-β-ketosulfones[92] have shown that they are active at or below room temperature, but they generally provide only low-to-moderate enantiomeric excesses for intramolecular C–H insertion reactions. With α-diazo-β-ketosulfone **68** use of the dirhodium(II) prolinate catalyst **66a** gives cyclopentanone **69** in high yield (eq. 65) but with only 12% ee.[92] However, ee values up to

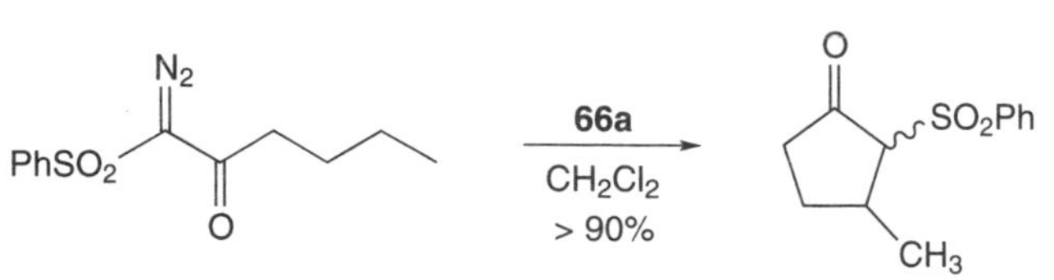

68 **69**, 12% ee (65)

82% and high *cis* selectivity characterize the uses of these catalysts for chromanone formation (eq. 66).[96,97] Catalyst **66a**, compared with a series of prolinate

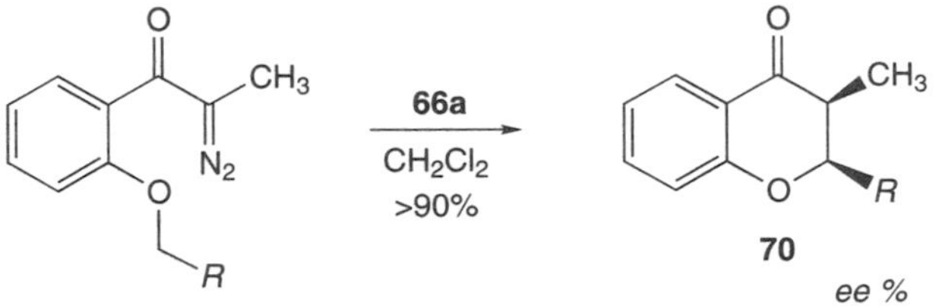

$$(66)$$

	ee %	cis:trans
R = CH$_3$	82	75:25
Ph	62	89:11
CH=CH$_2$	79	93:7

analogues, provides the highest enantiocontrol and diastereocontrol. Noteworthy also is the relative absence of products from oxonium ylide formation/[2,3]-sigmatropic rearrangement (3% with **66a**), but this ylide transformation is the sole outcome of copper-catalyzed reactions.[97] Binaphthyl hydrogen phosphate ligated dirhodium(II) (**71**) also catalyzed formation of chromanone **70** (33% ee,

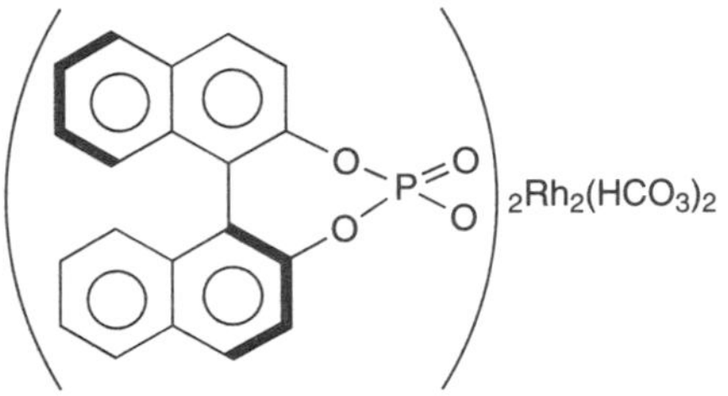

71

94:6 cis:trans), and when this catalyst was used for the conversion of **36** to **37** (eq. 49) the product β-lactam was formed in 93% yield with 26% ee.[98]

   Ikegami and co-workers have investigated asymmetric induction in C–H insertion reactions of α-diazo-β-ketoesters **72**. With the phthalimide-derivatized phenylalaninate dirhodium(II) catalyst **65**, diazo decomposition of the methyl ester gave cyclopentanone derivatives **74** in moderate to good yield (eq. 67), but

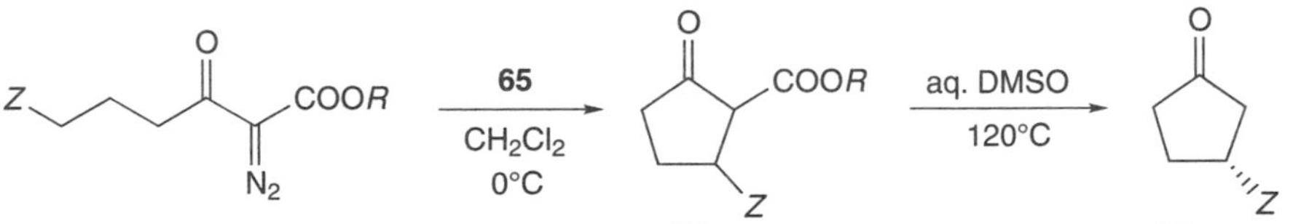

	R = Me		R = C^{i}Pr$_2$Me	
Z =	yield %	ee %	yield %	ee %
Me	76	24	71	32
Pent	43	29	76	35
CH=CH$_2$	44	38	63	53
Ph	96	46	86	76

$$(67)$$

with only modest ee values.[91] However, improvements in enantiocontrol were achieved by increasing the steric bulk of the ester alkyl group from methyl to diisopropylmethylcarbinyl.[99] Use of alanine or phenylglycine derivatives as ligands for dirhodium(II) gave lower % ee values,[91] and the more elaborate 2-alkoxyferrocenecarboxylic acids as chiral ligands for dirhodium(II) offered no improvement.[100] The (S)-phenylalaninate-ligated catalyst produced **74** with the (R)-configuration, and the (R)-phenylalaninate-ligated catalyst formed (S)-**74**. With esters of chiral alcohols including (+)-neomenthol, double diastereoselection occurred, and **74** (Z = Ph) was formed with 80% ee.[101]

Doyle's chiral dirhodium(II) carboxamidate catalysts have proven to be exceptionally versatile for C–H insertion reactions of diazoacetates and certain diazoacetamides. With $Rh_2(5S\text{-MEPY})_4$ and $Rh_2(5R\text{-MEPY})_4$, for example, highly enantioselective intramolecular C–H insertion reactions occur with 2-alkoxyethyl diazoacetates (eq. 68).[102] $Rh_2(5S\text{-MEPY})_4$ gives lactone **75** having

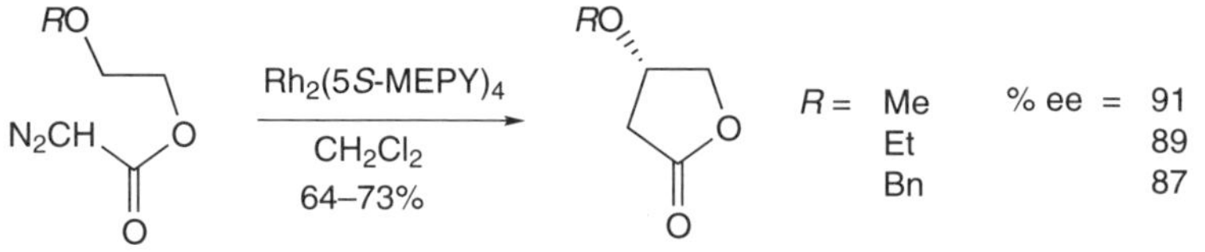

$$(68)$$

the S configuration, and $Rh_2(5R\text{-MEPY})_4$ produces **75** in the R configuration. Polymer-bound $Rh_2(5S\text{-MEPY})_4$ has also been employed for C–H insertion, yielding **75** (R = Me) at 80°C (65–72% ee), and the catalyst was recovered and reused seven times with similar results.[103] Applications with diazoesters that undergo insertion into a C–H bond vicinal to the incipient chiral center (eq. 69)

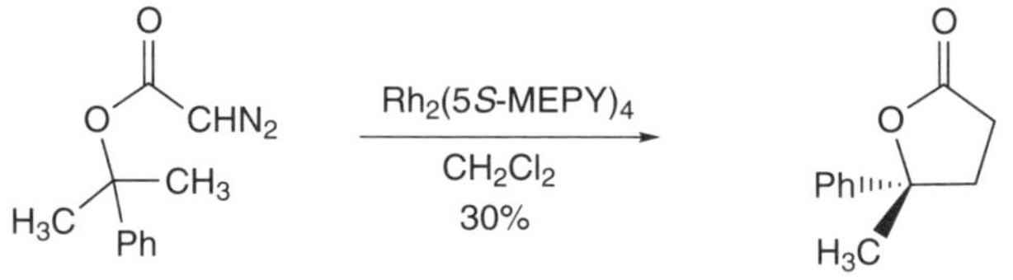

$$(69)$$

demonstrate further advantages of this catalytic methodology for asymmetric synthesis[102]; competition with aromatic cycloaddition reduced the yield of **76**.

The $Rh_2(MEPY)_4$ catalysts also provide moderate enantiocontrol in C–H insertion reactions of N-alkyl-N-(tert-butyl)diazoacetamides.[84,104] Substituents at the 2-position of the N-alkyl group control regioselectivity so that when this substituent is an alkoxy group, only the γ-lactam product is formed (eq. 70). In these

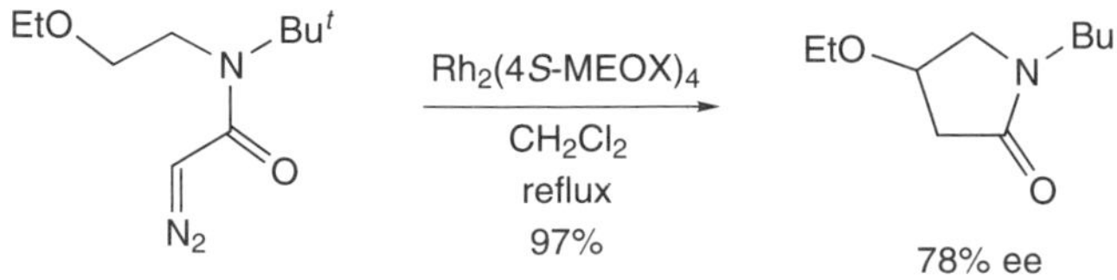

$$(70)$$

systems Rh$_2$(4S-MEOX)$_4$ gave higher % ee's than Rh$_2$(5S-MEPY)$_4$ by as much as 20%. When the N-alkyl substituent at the 2-position is a carboxylate group (dia-zoacetamide analogue of **36**), β-lactam formation occurs (46% ee), but this C–H insertion reaction is in competition with C–H insertion into the *tert*-butyl group.[104] However, when the N-alkyl substituents are tied back to make γ-lactam formation unfavorable, the production of β-lactam products is generally the sole insertion process (eq. 71) and high enantiocontrol can be achieved.[105]

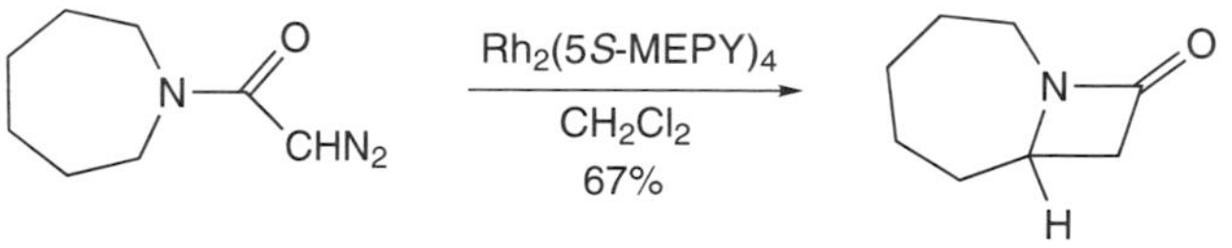

$$\text{(71)}$$

97% ee

Diazoacetates derived from primary alcohols undergo γ-lactone formation (**77**) in moderate to high chemical yield and, from the use of Rh$_2$(4S-MPPIM)$_4$ (**67d**), with exceptional enantiocontrol (eq. 72) and with surprising regiocon-

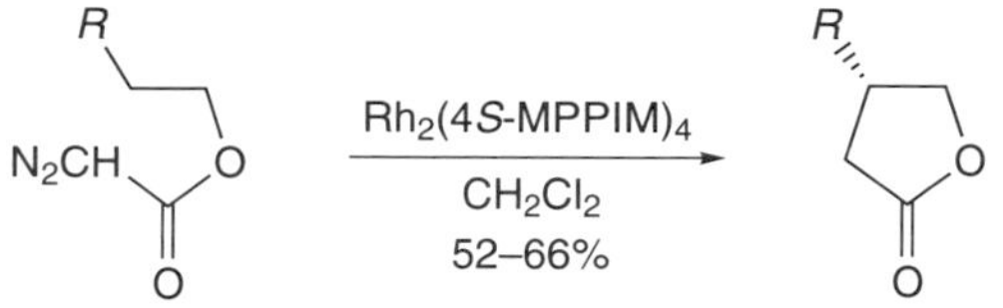

$$\text{(72)}$$

R =	ee %
Et	96
^{i}Bu	95
PhCH$_2$	89
m-MeOC$_6$H$_4$CH$_2$	92
3,4-(MeO)$_2$C$_6$H$_3$CH$_2$	94

trol[106,107]; β-lactone formation was the major competing insertion reaction, but its relative yield was generally less than 5%. This methodology has made possible the synthesis of a series of naturally occurring lignans, among which are (−)-en-terolactone (**78**), (+)-arctigenin (**79**), and (+)-isodeoxypodophyllotoxin (**80**), from cinnamic acid precursors in high enantiomeric purity.[107]

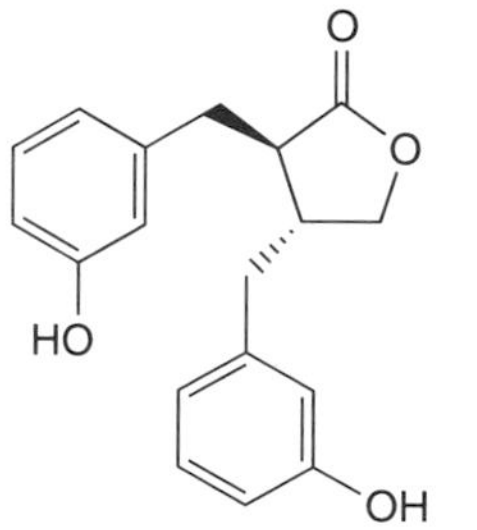

78

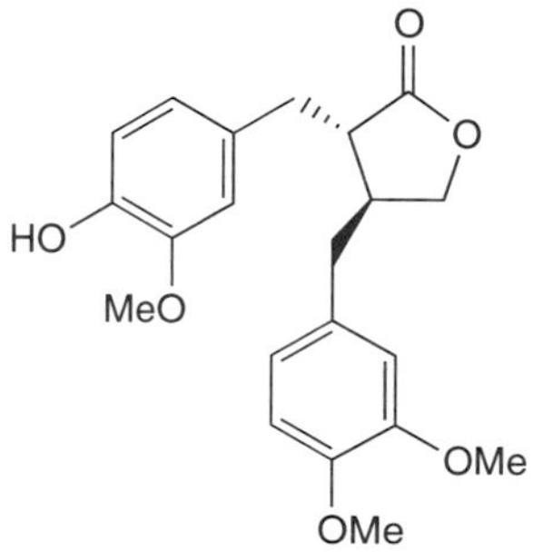

79

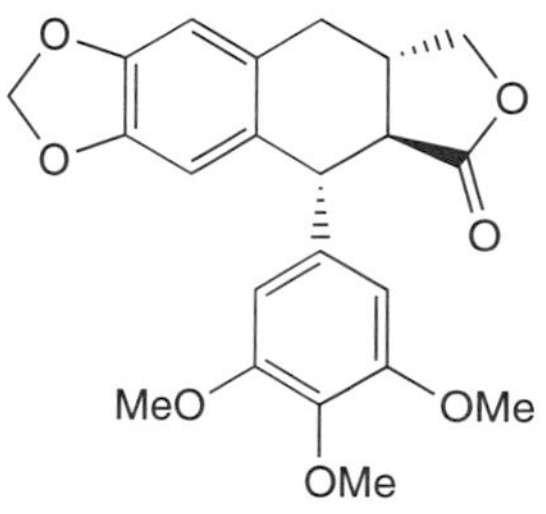

80

SYNTHESIS OF (+)-ARCTIGENIN [107]

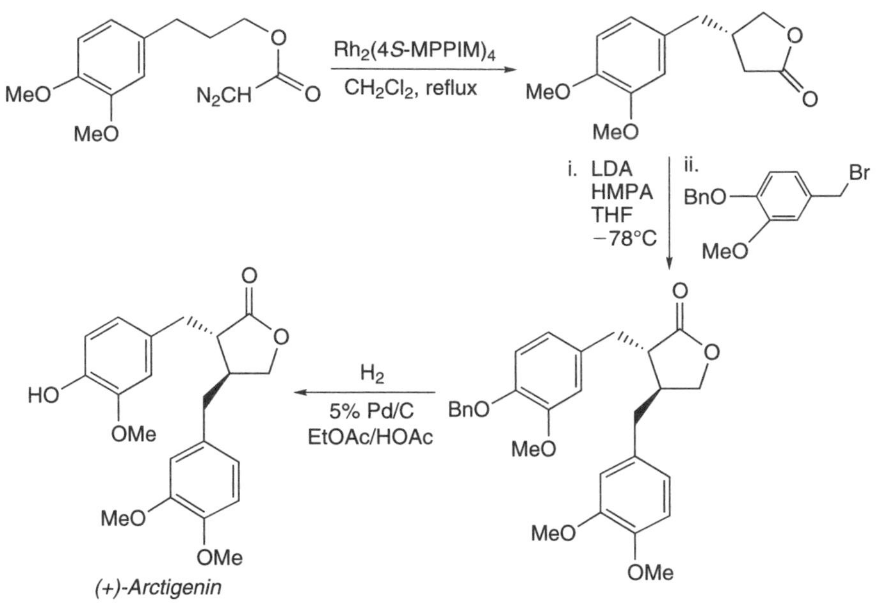

(+)-Arctigenin

4(*S*)-[(3,4-Dimethoxyphenyl)methyl]dihydro-2(3*H*)-furanone.

A solution of 3-(3,4-dimethoxyphenyl)prop-1-yl diazoacetate (0.159 g, 0.60 mmol) in 4 mL of rigorously dried CH_2Cl_2 was added via syringe pump at a rate of 0.4 mL/h to a refluxing solution of Rh_2(4*S*-MPPIM)$_4$ (10.8 mg, 1.3 mol %) in 7 mL of dry CH_2Cl_2. The initial blue/purple color of the reaction solution turned to olive green by the end of the substrate addition. Refluxing was continued for an additional 3 h, the reaction solution was cooled to room temperature, and the catalyst was removed by filtration on a short plug of silica gel (CH_2Cl_2). Removal of the solvent under reduced pressure provided 0.112 g of the title compound (0.47 mmol, 79% yield) as a light yellow oil. Purification by radial chromatography (4:1 hexanes: EtOAc) afforded 89 mg of the lactone (0.37 mmol, 62% yield) as a colorless oil. Enantiomeric excesses werc 94% with baseline separation by GC analysis on a 30-m Chiraldex A-DA column; $[\alpha]_D^{25} = -7.30$ (*c* 1.11, CHCl$_3$) [with 1.3 mol % Rh_2(4*S*-MPPIM)$_4$]; $[\alpha]_D^{24} = +7.50$ (*c* 0.971, CHCl$_3$) [with 2.0 mol % Rh_2(4*R*-MPPIM)$_4$]; ^{1}H-NMR δ 6.81 (*d*, *J* = 8.0 Hz, 2 H), 6.69 (*dd*, *J* = 8.0, 2.0 Hz, 1 H), 6.66 (*d*, *J* = 2.0 Hz, 1 H), 4.34 (*dd*, *J* = 9.1, 6.8 Hz, 1 H), 4.04 (*dd*, *J* = 9.1, 6.0 Hz, 1 H), 3.87 (*s*, 3 H), 3.86 (*s*, 3 H), 2.91–2.76 (*m*, 1 H), 2.74–2.70 (*comp.*, 2 H), 2.61 (*dd*, *J* = 17.4, 7.9 Hz, 1 H), 2.29 (*dd, J* = 17.4, 6.7 Hz, 1 H). Using 2.0 mol % Rh_2(4*R*-MPPIM)$_4$, the (R)-isomer was isolated in 61% yield (94% ee) after purification by radial chromatography. The major byproducts from these catalytic reactions were carbene dimers and the water insertion product. Carbene dimer formation was controlled by adjusting the rate of addition of the diazo compound but varied with the catalyst employed. Water insertion, especially in small-scale reactions, was minimized by using rigorously dried solvents, reagents, and equipment. The solvent CH_2Cl_2 was dried over CaH_2 for 12–20 h prior to use. Diazoacetate and the septa employed were dried in a desiccator over KOH and Drierite for at least 15 h prior to use. All glassware, stirring bars, and needles were oven dried. The weighing of reagents and preparation of solutions took place in a glove bag under N_2.

(3S,4S)-3-[(3-Methoxy-4-hydroxyphenyl)methyl]-4-[3,4-dimethoxyphenyl)methyl]dihydro-2(3H)-furanone. (+)-Arctigenin.

To a rapidly stirred solution of lactone (61 mg, 0.25 mol) in 8 mL of anhydrous THF at −78° C was added 0.35 mL of 1.5 M LDA (in cyclohexane, 0.48 mmol, 1.9 equiv.) and 0.15 g of HMPA (0.76 mmol, 1.6 equiv.). After 0.5 h a solution of 4-benzyloxy-3-methoxybenzyl bromide (0.13 g, 0.41 mmol, 1.6 equiv.) in 1.0 mL of THF was added in one portion, and the resulting mixture was stirred for 12 h at −78°C, warmed to −20°C (2 h) and then to 0°C (2 h). The excess base was quenched at 0°C with 10 mL of saturated aqueous NH_4Cl, and the solution was extracted with ether (10 mL) and EtOAc (2 × 10 mL). The combined organic layer was washed with H_2O (2 × 30 mL) and the solution was extracted with ether (10 mL) and EtOAc (2 × 10 mL) and brine (30 mL), dried over anhydrous $MgSO_4$, and the solvent was removed under reduced pressure. The residue was purified by column chromatography on silica gel (hexanes:EtOAc, 4:1) to afford 92 mg of (3R,4S)-3-[(3-methoxy-4-benzyloxyphenyl)methyl]-4-[3,4-dimethoxyphenyl)methyl]dihydro-2(3H)-furanone (0.20 mmol, 79% yield): ^{1}H NMR δ 7.43–7.26 (comp, 5 H), 6.95–6.47 (comp, 6 H), 5.12 (s, 2 H), 4.14–4.11 (m, 1 H), 3.89–3.86 (m, 1 H), 3.85 (s, 3 H), 3.84 (s, 3 H), 3.80 (s, 3 H), 3.00–2.86 (comp, 2 H), 2.62–2.49 (comp, 4 H).

To a solution of this lactone (0.092 g, 0.20 mmol) in 10 mL of EtOAc and 1 mL of AcOH was added 5% Pd/C (0.05 g, 10 mol %). The resulting mixture was stirred under H_2 (balloon pressure), and the reaction was monitored by TLC (hexanes:EtOAc = 2:1). After 1.5 h, the reaction mixture was combined with 20 mL of EtOAc and 20 mL H_2O. The organic layer was washed sequentially with saturated aqueous $NaHCO_3$ (20 mL) and brine (20 mL), dried over anhydrous $MgSO_4$, and the solvent was removed under reduced pressure to provide 68 mg of (+)-arctigenin (0.18 mmol, 92% yield) as a light brown oil. Further purification by radial chromatography (CH_2Cl_2:MeOH = 99:1) yielded 60 mg (0.16 mmol, 82% yield) of **79** as an amorphous white solid: $[\alpha]_D^{25} = +27.1$ (c 0.56, EtOH, 94% ee); ^{1}H-NMR δ 6.82 (d, J = 7.9 Hz, 1 H), 6.75 (d, J = 8.2 Hz, 1 H), 6.63 (d, J = 1.9 Hz, 1 H), 6.61 (dd, J = 7.9, 1.9 Hz, 1 H), 6.55 (dd, J = 8.2, 1.9 Hz, 1 H), 6.46 (d, J = 1.9 Hz, 1 H), 5.56 (br s), 4.17–4.12 (m, 1 H), 3.91–3.85 (m, 1 H), 3.85 (s, 3 H), 3.82 (s, 6 H), 2.98–2.48 ($comp$, 2 H), 2.67–2.43 ($comp$, 4 H).

According to our view of the mechanism for C–H insertion (see Scheme 1), reaction is initiated by overlap of the metal carbene's carbon p orbital with the σ orbital of the reacting C–H bond. The formation of C–C and C–H bonds is concurrent with dissociation of the dirhodium(II) species (Scheme 4). As hydrogen

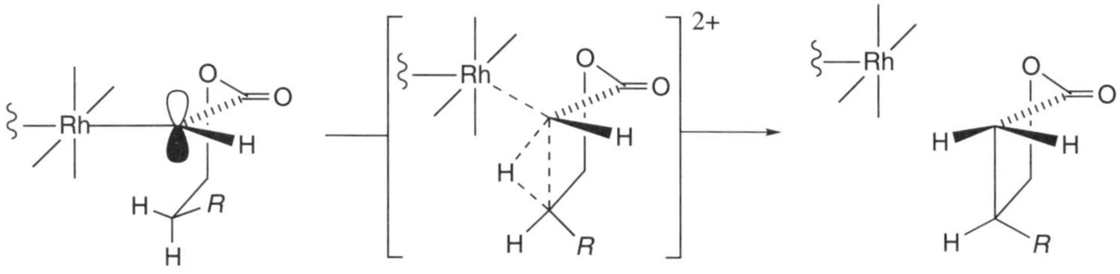

Scheme 4

migrates to the carbene center, the substituents on the carbon where insertion is taking place rotate towards the resting positions that conform to their placement in the product. The absolute configurations of the C–H insertion products formed in the Rh$_2$(MPPIM)$_4$-catalyzed reactions are predictable from the model in Scheme 5, which depicts the S-MPPIM-ligated catalyst with the bound carbene that is positioned to undergo C–H insertion, resulting in the enantiomeric lactones **83** and **84**. The high preference for **83** with Rh$_2$(4S-MPPIM)$_4$, and this catalyst's enhancement of enantiocontrol over Rh$_2$(5S-MEPY)$_4$ or Rh$_2$-(4S-MEOX)$_4$, is consistent with steric repulsion between $anti$-R and the N-3-phenylpropanoyl attachment of the imidazolidinone ligands in **82**. Thus the syn conformer **81** provides the lower-energy transition state for C–H insertion, even when R is as small as ethyl (96% ee).

This methodology has been extended to C–H insertion reactions of secondary cycloalkyl diazoacetates, where diastereoselectivity in the formation of cis- and $trans$-fused bicyclic lactones is a critical control feature.[108] Use of Rh$_2$(5S-MEPY)$_4$ or its enantiomer produced insertion products with a high degree of enantiocontrol, but diastereocontrol was only 3:1 (e.g., eq. 73). However, both high enantiocontrol and nearly complete diastereocontrol were achieved with recently developed dirhodium(II) tetrakis[methyl 1-acetylimidazolidin-2-one-4(S)-

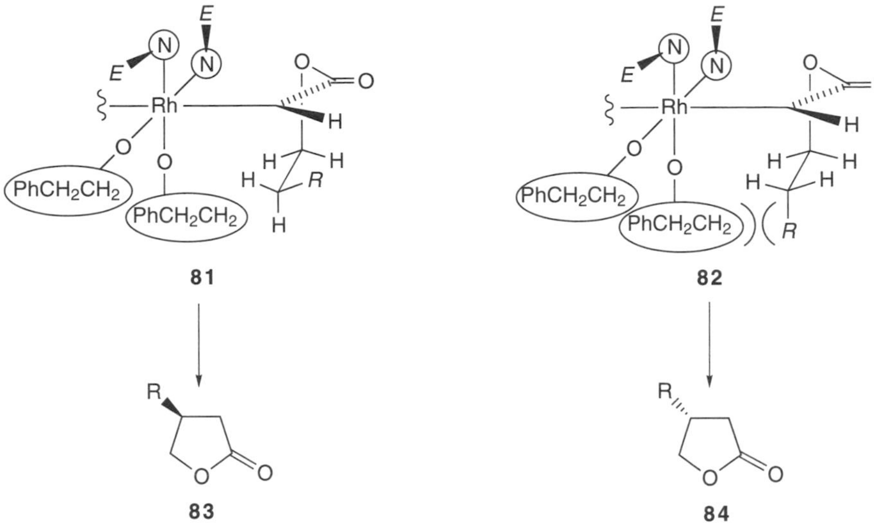

Scheme 5

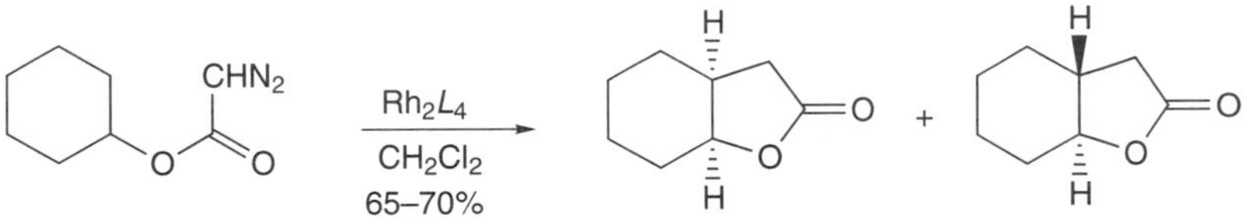

Rh$_2$(OAc)$_4$	40	60
Rh$_2$(4S-MEOX)$_4$	55 (96% ee)	45 (95% ee)
Rh$_2$(5S-MEPY)$_4$	75 (97% ee)	25 (91% ee)
Rh$_2$(5S-MACIM)$_4$	99 (97% ee)	1 (65% ee)

$$(73)$$

carboxylate], Rh$_2$(4S-MACIM)$_4$ (**67c**). The oxazolidinone analog of Rh$_2$(5S-MEPY)$_4$, Rh$_2$(4S-MEOX)$_4$, facilitated high enantiocontrol but significantly lower diastereocontrol. Similarly high enantio- and diastereoselectivities have been achieved with cyclopentyl through cyclooctyl diazoacetates, with *cis-* or *trans*-4-alkylcyclohexyl diazoacetates, where preferential insertion into equatorial C–H bonds (e.g., **85** and **86**) has been demonstrated,[108,109] and with 2-adamantyl diazoacetate (**87**),[94] but in these cases Rh$_2$(MEOX)$_4$ catalysts provided the highest levels of enantiocontrol.

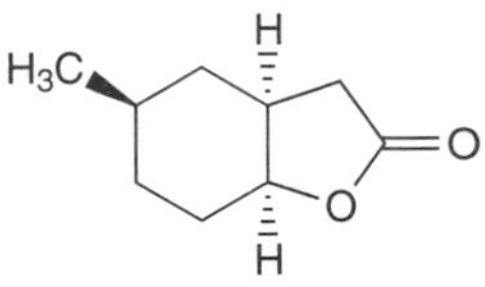

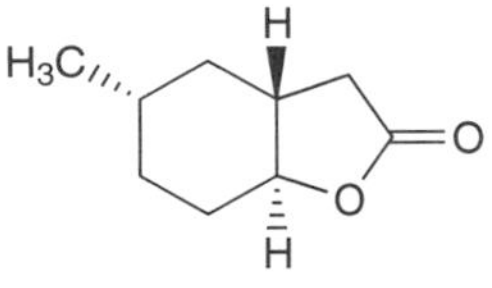

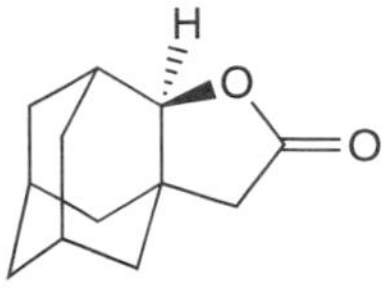

Rh$_2$(4S-MEOX)$_4$:   **85**, 98% ee   **86**, 95% ee   **87**, 98% ee

High enantio- and diastereocontrol in $Rh_2(MEPY)_4$-catalyzed C–H insertion reactions of glycerol-derived diazoacetates have provided a convenient synthesis of pure 2-deoxyxylolactone (**89**, Scheme 6).[110] The reactant diazoester was conve-

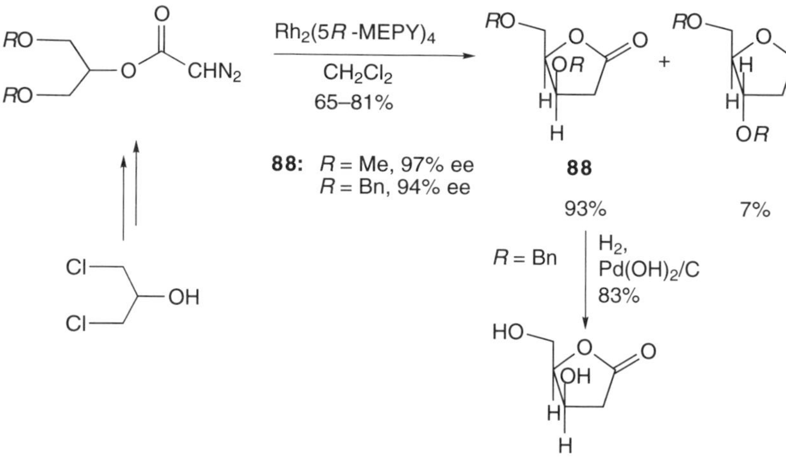

Scheme 6

niently prepared from commercial 1,3-dichloro-2-propanol. As little as 0.1 mol % of catalyst was required to effect complete reaction (1000 turnovers). The success of these C–H insertion reactions is based on ether oxygen activation of adjacent C–H bonds.[35–39] In their absence, enantioselectivity remains high, but diastereocontrol using $Rh_2(MEPY)_4$ catalysis is much lower. However, both $Rh_2(4S$-MPPIM$)_4$ and $Rh_2(4S$-MCHIM$)_4$ (**67e**) provide extraordinary diastereocontrol and impressive enantiocontrol, as exemplified in reactions of 3-pentyl diazoacetate (eq. 74),[111] which suggests that the high level of stereocontrol that

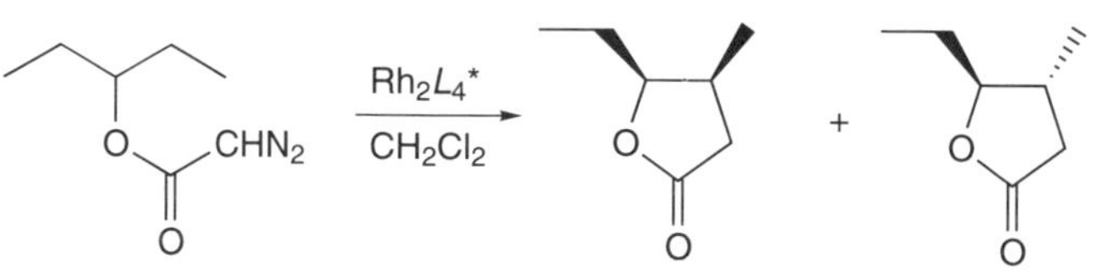

	yield %			(74)
$Rh_2(4S$-MCHIM$)_4$	88	98 (99% ee)	2	
$Rh_2(4S$-MPPIM$)_4$	81	97 (99% ee)	3	
$Rh_2(4S$-MACIM$)_4$	81	94 (86% ee)	6 (36% ee)	
$Rh_2(5S$-MEPY$)_4$	70	78 (98% ee)	22 (71% ee)	
$Rh_2(4S$-MEOX$)_4$	75	69 (98% ee)	31 (92% ee)	

has characterized C–H insertion reactions of cycloalkyl diazoacetates can now be successfully extended to acyclic diazoacetates.

SYNTHESIS OF (4R,5R)-(+)-2-DEOXYXYLOLACTONE
(Scheme 6)[110]

Diazo Decomposition of 1,3-Dialkoxy-2-propyl Diazoacetates.

To the dirhodium(II) catalyst (0.1 mol %) in 50 mL of refluxing dichloromethane was added by syringe pump the diazoester (1.0 mmol) in 10 mL of anhydrous dichloromethane during 10 h (1.0 mL/h). The initial color of the solution containing $Rh_2(5S\text{-MEPY})_4$ or $Rh_2(5R\text{-MEPY})_4$ was blue which generally changed to an olive color by the end of the addition; with $Rh_2(4S\text{-MEOX})_4$ the initial color was light red, which became light yellow at the end of the addition. After addition was complete, CH_2Cl_2 was evaporated under reduced pressure, and the lactone product(s) was isolated by column chromatography on silica (7:3 hexane:ethyl acetate eluent). This product mixture, which was chromatographically pure, was analyzed spectroscopically.

3O,5O-Dibenzyl-2-deoxyxylolactone.

bp 220–230°C (0.2 Torr); $[\alpha]_D^{23} = -5.40$ (c 0.537, MeOH); ^{1}H-NMR δ 7.37–7.22 (m, 10 H), 4.63–4.51 (m, 5 H), 4.44 (d, J = 11.9 Hz, 1 H), 3.86 (d, J = 5.3 Hz, 2 H), 2.72 (dd, J = 17.6, 3.1 Hz, 1 H), 2.62 (dd, J = 17.6, 5.8 Hz, 1 H). ^{13}C-NMR: δ 174.7, 137.8, 137.1, 128.6, 128.5, 128.1, 127.9, 127.8, 127.7, 82.0, 74.5, 73.7, 71.8, 67.7, 35.6; IR (film) 1784 cm^{-1}.

(4R,5R)-(+)-2-Deoxyxylolactone.

The dibenzyl ether (312 mg, 1.00 mmol) from the reaction performed with $Rh_2(5R\text{-MEPY})_4$ dissolved in 50 mL of ethyl acetate was placed in a glass vessel containing 30 mg of 20% $Pd(OH)_2$ on carbon and shaken in a Parr hydrogenator under 30 psi hydrogen for 24 h. The resulting mixture was filtered through Celite, and the solvent was evaporated under reduced pressure. Pure 2-deoxyxylolactone (110 mg, 83% yield) was isolated by column chromatography on silica gel (9:1 ethyl acetate:methanol, R_f of 0.50): $[\alpha]_D^{26} = +56.2$ (c 0.49, MeOH); ^{1}H NMR (CDCl$_3$/acetone-d_6) δ 4.91 (br s, 1 H), 4.65 (ddd, J = 5.9, 4.3, 1.7 Hz, 1 H), 4.49 (td, J = 5.5, 4.3 Hz, 1 H), 3.90 (d, J = 5.5 Hz, 2 H), 3.43 (br s, 1 H), 2.84 (dd, J = 17.6, 5.9 Hz, 1 H), 2.43 (dd, J = 17.6, 1.7 Hz, 1 H).

Tertiary alkyl diazoacetates also undergo facile C–H insertion catalyzed by chiral dirhodium(II) carboxamidates, but here regiocontrol is an important consideration.[112] Enantiocontrol, which is highly dependent on the chiral ligand of the catalyst, is greatly enhanced with the use of $Rh_2(4S\text{-MACIM})_4$ (e.g., eq. 75; **90:91** = 90:10). Regioselectivity, even when competition in C–H insertion is

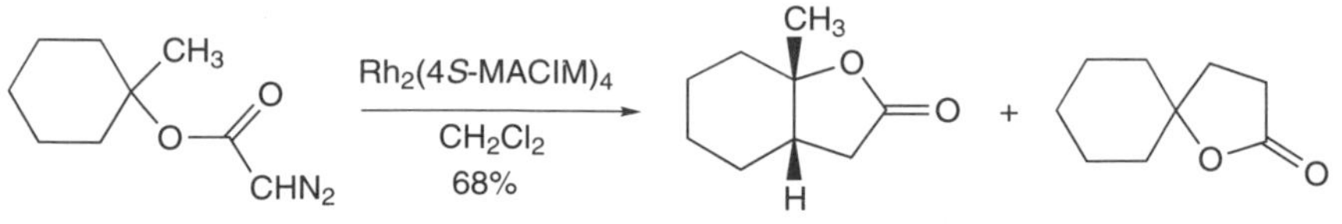

90 (90% ee)   **91**

(75)

with an electronically unfavorable primary C–H bond, varies with the catalyst and application, and conformational restrictions may be responsible for overriding electronic preferences.[67] Even when only one insertion product is formed, catalysts whose ligand configurations are identical can give products whose absolute configurations are mirror images (e.g., eq. 76).

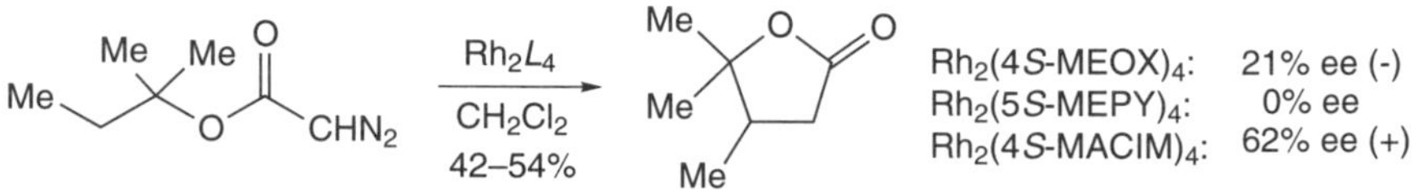

Rh₂(4S-MEOX)₄: 21% ee (-)
Rh₂(5S-MEPY)₄: 0% ee
Rh₂(4S-MACIM)₄: 62% ee (+)

(76)

Sulikowski has reported that Cu(OTf) in combination with chiral bis-oxazolines **92** or **93** provides moderate enantiocontrol in C–H insertion directed to the synthesis of a 1,2-disubstituted mitosene (eq. 77).[113] Diastereoselectivity for **95**

92a   R = ^{i}Pr
92b   R = ^{t}Bu

93

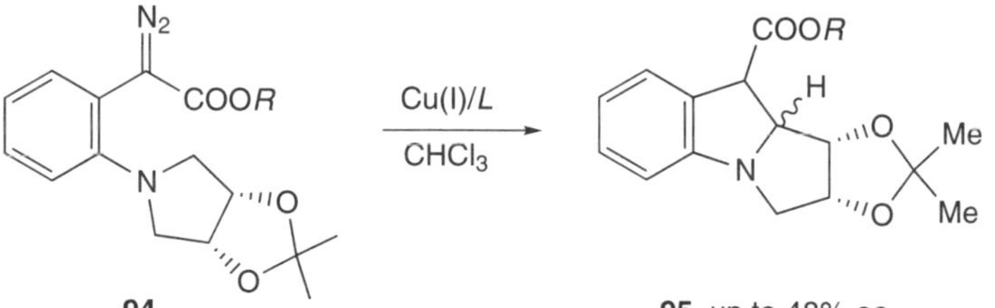

94

95, up to 48% ee

(77)

ranged from 1:2 to 5:1 *anti*:*syn* and from 1:1 to 13:1 *exo*:*endo*; enantioselectivities up to 48% ee (with **92a**) were observed. Both enantio- and diastereoselectivities were solvent dependent, and catalyst screening by Burgess further elaborated solvent and catalyst influences on the diastereomeric ratio for diazo compound **94** with R = *l*-menthyl.[114] Surprisingly, $AgSbF_6$ was found to be nearly as effective as Cu(OTf) for this transformation. Dirhodium(II) catalysts **66b** and **67a** gave low or negligible asymmetric induction.[113]

3.9 DIASTEREOSELECTION AND REGIOSELECTION IN CARBON–HYDROGEN INSERTION REACTIONS PROMOTED BY CHIRAL CATALYSTS

Individual diazoacetate enantiomers matched with the appropriate chiral catalyst can provide exceptional diastereoselectivity and regioselectivity in product formation. Thus, for example, (1*S*,2*R*)-*cis*-2-methylcyclohexyl diazoacetate (**96**) forms the all-*cis* bicyclic lactone **97** when treated with Rh_2(4*R*-MPPIM)$_4$ (eq. 78) but produces **98** when catalyzed by Rh_2(5*S*-MEPY)$_4$, and the mirror image relationship (eq. 79) has been established.[115] Similar examples of chiral catalyst-

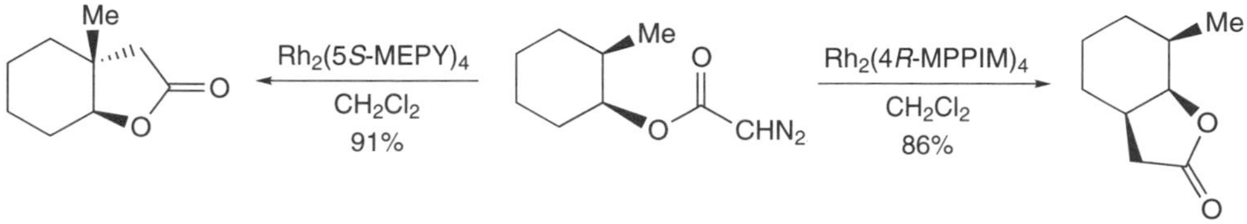

(1*S*)-98 **(1*S*,2*R*)-96** **(1*R*)-97**

(78)

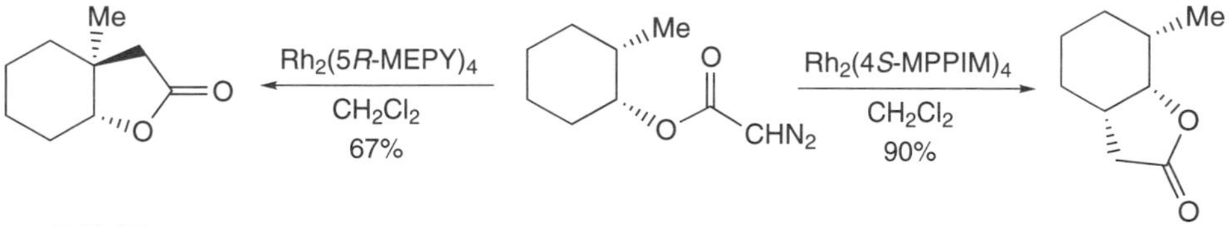

(1*R*)-98 **(1*R*,2*S*)-96** **(1*S*)-97**

(79)

induced product control have been reported for catalytic diazo decomposition of *trans*-2-methylcyclohexyl diazoacetate, *d*-(+)- and *l*-(−)-menthyl diazoacetates, (+)-neomenthyl diazoacetate (eq. 80), and (+)- and (−)-2-octyl diazoacetates.[115]

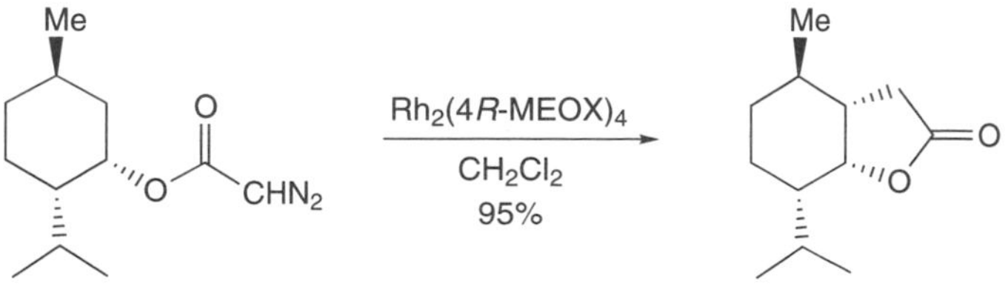

(80)

Dirhodium(II) carboxamidate catalysts (**67**) are generally much more selective than Cu(I)/bis-oxazoline **92b**. The exceedingly high product diastereoselection observed in reactions of cyclohexyl diazoacetates is due to virtually exclusive insertion into equatorial C–H bonds.

The differing influence of chiral ligands of dirhodium(II) on diastereoselectivity and regioselectivity is consistent with catalyst structure. The chiral Rh$_2$(MEPY)$_4$ and Rh$_2$(MEOX)$_4$ catalysts are constructed with two closed (occupied) and two open quadrants on the dirhodium(II) face (e.g., **99**, E = COOMe). The bound carbene takes a resting position so as to minimize interactions with the ligand's ester attachments (e.g., **100**). For its metal carbene reactions,

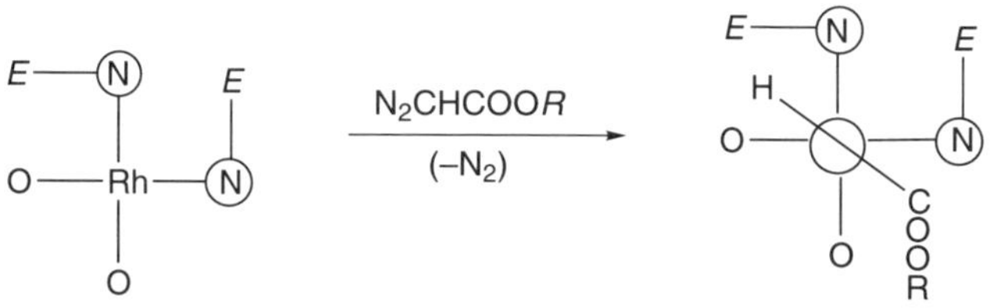

99   **100**

Rh$_2$(MEOX)$_4$ provides a somewhat more open framework for insertion than does Rh$_2$(MEPY)$_4$,[94] and diastereocontrol is often greater with Rh$_2$(MEPY)$_4$ than Rh$_2$(MEOX)$_4$ catalysts, although the reverse is seen for regiocontrol. Selectivity differs considerably from Rh$_2$(MEPY)$_4$- and Rh$_2$(MEOX)$_4$-directed reactions with that from the use of chiral *N*-acylimidazolidinone-ligated dirhodium(II) catalysts. With these structures[95] the open quadrants of **99** are restricted (e.g., **101**) and the bound carbene is subject to steric influences from both the ligand's ester (E = COOMe) and acyl (Ac = NCO*R*) attachments (e.g., **102**).

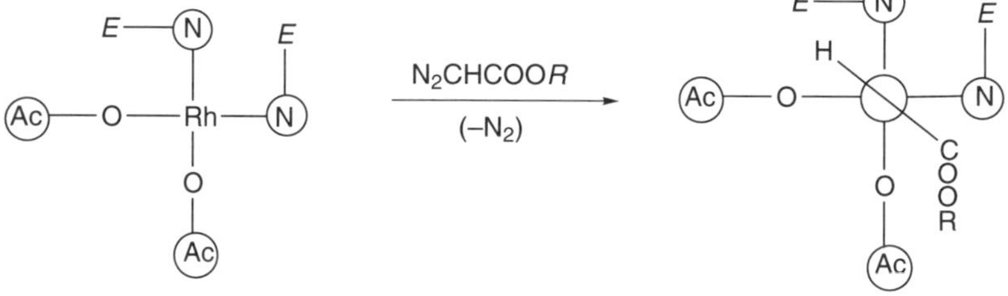

101   **102**

The influence of chiral catalyst on diastereoselection with cycloalkyl diazoacetates is understandable in terms of the conformational descriptions of metal carbene intermediates in Scheme 7. The equatorial–axial conformational equilibrium of the cyclohexyl group provides access of the carbene to equatorial C–H bonds, insertion into which yields the *trans*-fused lactone **105** or the *cis*-fused lactone **106**. (Access to axial C–H bonds is prevented by crowding of the cyclo-

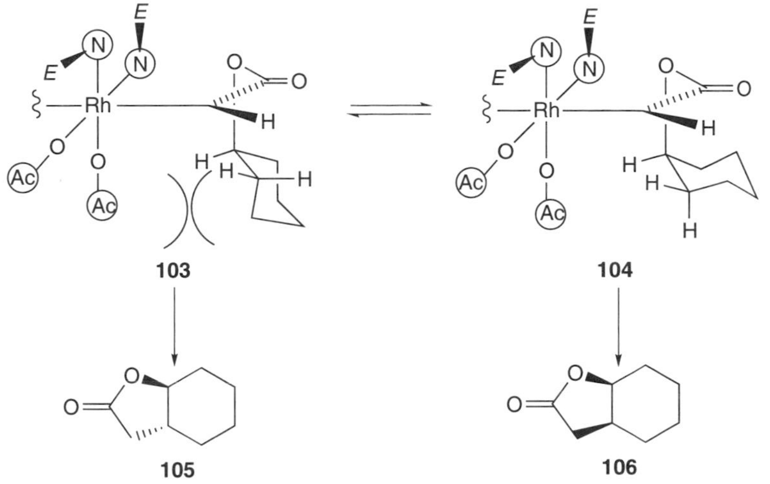

Scheme 7

hexane ring into the catalyst face that would be required for carbene insertion.) In the absence of significant steric influences from the catalyst face adjacent to the cyclohexyl group, both diastereoisomers are produced. However, by the placement of substituents in those quadrants of the catalyst face that destabilize **103** relative to **104** [i.e., with $Rh_2(4S\text{-}MACIM)_4$ and $Rh_2(4S\text{-}MPPIM)_4$], diastereoselection for **106** is significantly enhanced.

Because individual diazoacetate enantiomers undergo diazo decomposition with high diastereo- and regiocontrol, racemic mixtures of diazoacetates can be expected to provide high levels of enantiomer differentiation in reactions catalyzed by chiral dirhodium(II) carboxamidates. This has indeed been found,[116] and, as shown in eq. 81, each product of C–H insertion is highly enantioen-

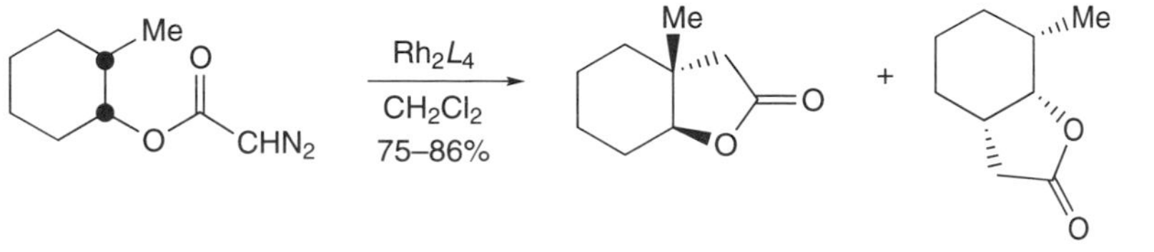

(81)

	(1S)-98	(1S)-97
$Rh_2(5S\text{-}MEPY)_4$	45 (91% ee)	49 (98% ee)
$Rh_2(4S\text{-}MEOX)_4$	40 (99% ee)	47 (99% ee)
$Rh_2(4S\text{-}MACIM)_4$	11 (87% ee)	66 (77% ee)

riched; β-lactone formation accounts for the residue in diastereoselection. Other examples include *rac-trans*-2-methylcyclohexyl diazoacetate and *rac*-2-octyl diazoacetate,[116] and in each case the selectivities obtained are predictable from those obtained with the individual diazoacetate enantiomers.

3.10 DIASTEREOSELECTION IN CARBON–HYDROGEN INSERTION REACTIONS CONTROLLED WITH THE USE OF CHIRAL AUXILIARIES

An alternative to the use of chiral catalysts is the introduction of covalently bound chiral auxiliaries within the diazocarbonyl substrates, usually as an ester of a chiral alcohol. Proximity to the reaction center is a critical control feature that has proven to be more difficult to effect in C–H insertion reactions. However, Taber and co-workers have developed 1-naphthylborneol (**107**) as an effec-

107

tive auxiliary for α-diazo-β-keto-esters, and they have reported moderate levels of diastereoselectivity with its use (e.g., eq. 82); other examples show diastereomer ratios (dr) of 83:17 to 87:13.[50] Compound **108** was further con-

$$(82)$$

108, dr = 92:8

verted to (+)-estrone methyl ether (**26**). Wee and Liu have further established the utility of this chiral auxiliary in C–H insertion reactions of diazomalon-amides.[117] Although when R = alkyl (eq. 83) β-lactam formation competes with

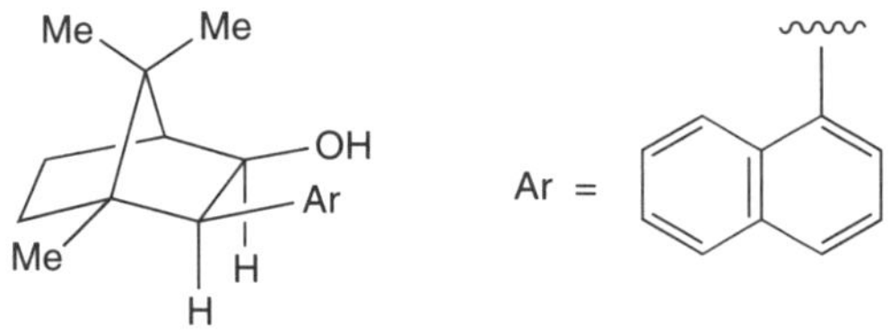

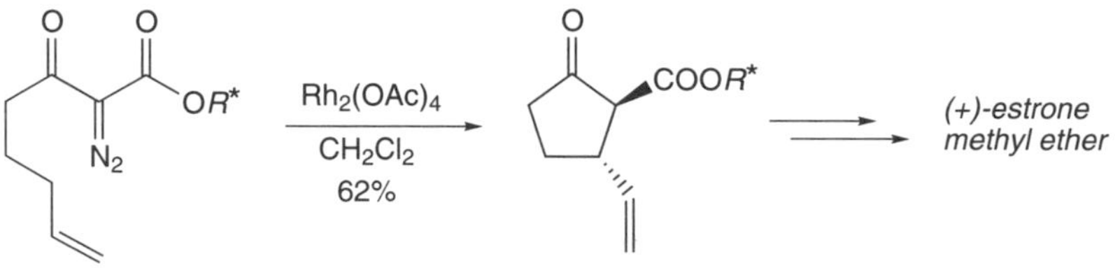

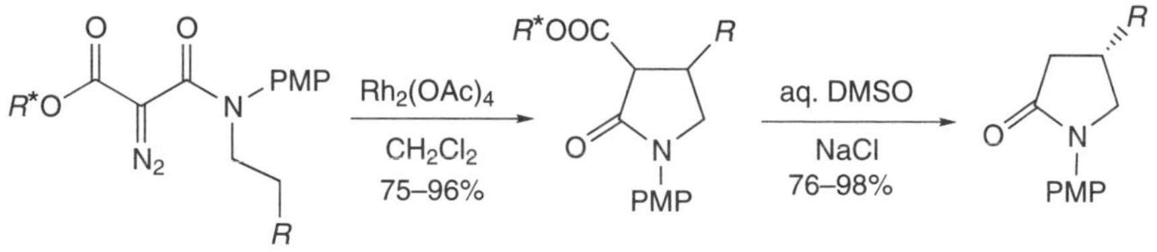

PMP = p-MeOC$_6$H$_4$

R* from **107**

R = ^{n}Hex	45% ee (R)
R = ^{c}Hex	98% ee (S)
R = Ph	79% ee (S)
R = 3,4-(MeO)$_2$C$_6$H$_3$	50% ee (S)
R = 3-NO$_2$C$_6$H$_4$	77% ee (S)

$$(83)$$

the production of γ-lactam products, when R = Ar this competition is negligible. Diastereocontrol with this system was moderate to high, and product configurations are predictable from a simple transition-state model. Ikegami has employed a chiral ketal derived from (R)-methyl 2-oxocyclopentane-1-acetate (96% ee) to ensure regiocontrolled cyclization of **109** using dirhodium(II) tetrakis(triphenylacetate), $Rh_2(tpa)_4$ (eq. 84).[118] This C–H insertion transformation was the key

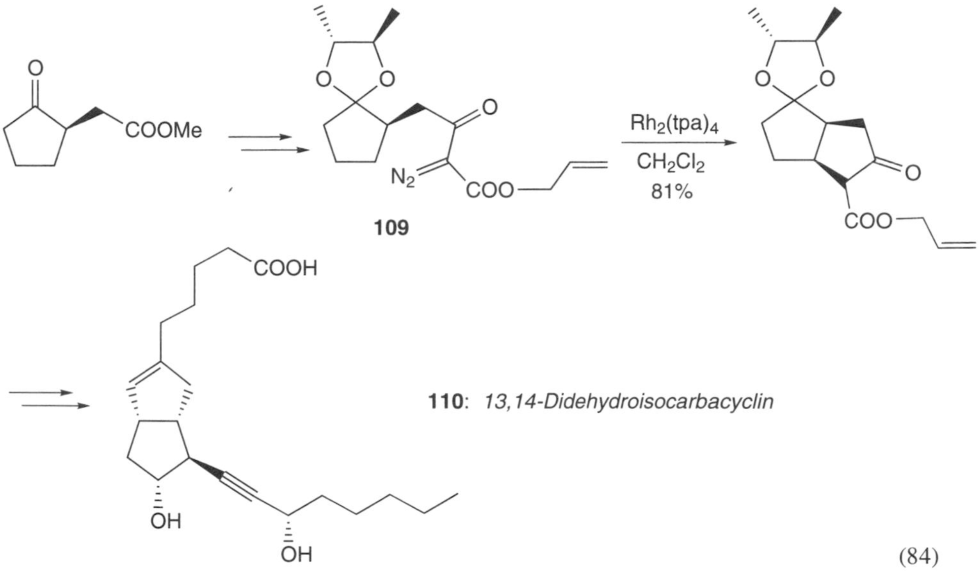

109

110: *13,14-Didehydroisocarbacyclin*

(84)

step in an expeditious synthesis of 13,14-didehydroisocarbacyclin, **110**. Although chiral auxiliaries are viable, chiral catalysts, when applicable, offer economies and efficiencies that diminish the uses of this auxiliary methodology for synthetic applications.

3.11 SILICON–HYDROGEN AND RELATED INSERTION REACTIONS

The first comprehensive report of Si–H insertion in transition metal catalyzed reactions of diazo compounds was presented only recently. Doyle and co-workers described the high-yield reactions of diazocarbonyl compounds with stoichiometric amounts of organosilanes, catalyzed by $Rh_2(OAc)_4$ (eqs. 85 and 86).[119] Silyl

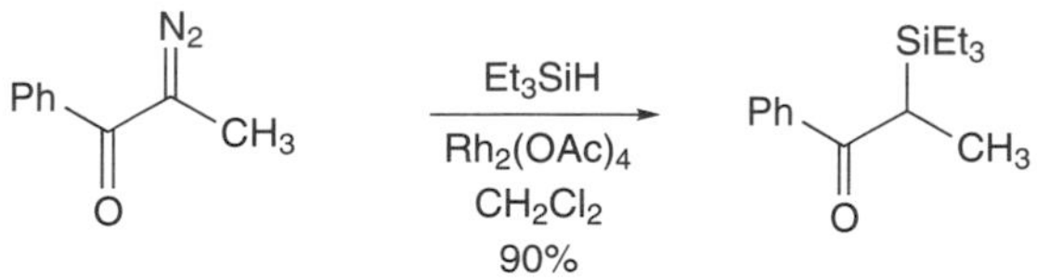

(85)

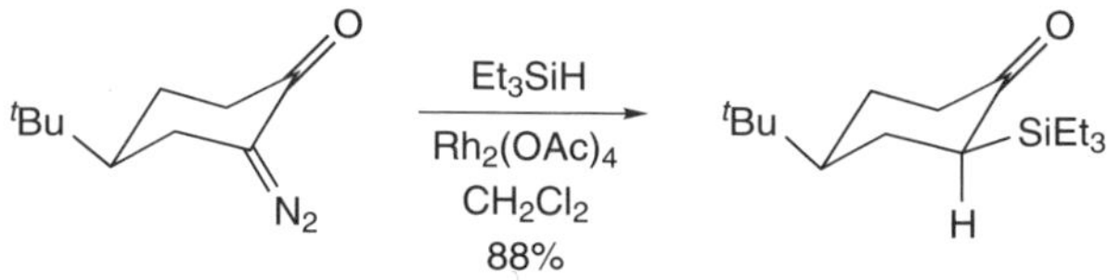

(86)

enol ether formation is avoided using this methodology. As expected, complete retention of configuration was observed [reaction with (*S*)-1-(naphthyl)phenyl-methylsilane]. In many cases Cu(acac)$_2$ is a viable alternative to Rh$_2$(OAc)$_4$ for Si–H insertion.[119] Previous reports had shown that Si–H insertion could compete successfully with cyclopropenation of a carbon–carbon triple bond using CuCl catalysis, and copper-catalyzed carbene insertion into the Ge–H bond of Et$_3$GeH has been described.[13]

Landais has extended this methodology with the primary objective of influencing stereocontrol, generally with the use of chiral auxiliaries (e.g., eqs. 87[120] and 88[121]), including those of *d*- and *l*-menthyl, (*R*)-pantolactone, and camphor

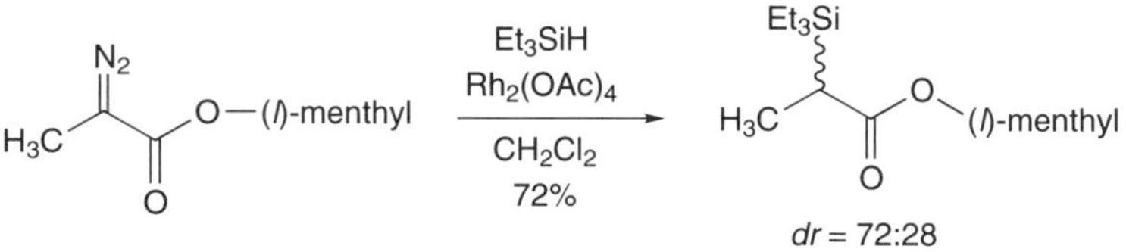

(87)

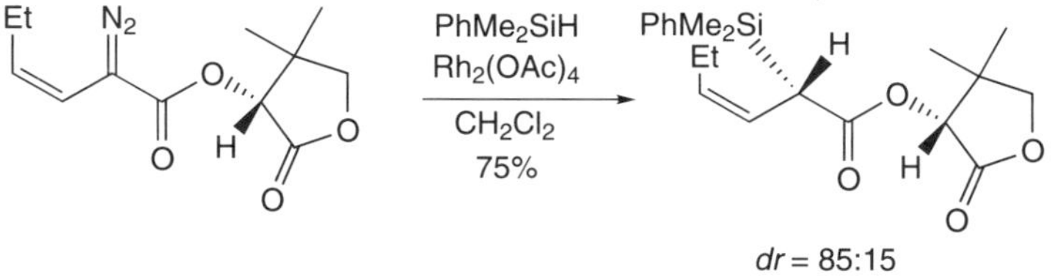

$dr = 72:28$

(88)

$dr = 85:15$

derivatives. The insertion process was suggested to take place by oxidative addition of the organosilane onto the rhodium bound to the carbene.[120] However, there is no reason to expect that the mechanism for Si–H insertion is any different from that for C–H insertion.

The use of dialkyl- or diarylchlorosilanes has made possible the convenient replacement of the silicon functional group by hydroxyl (eq. 89).[122] Di- and trisubstituted tetrahydrofurans (e.g., **111**) have been prepared stereoselectively from

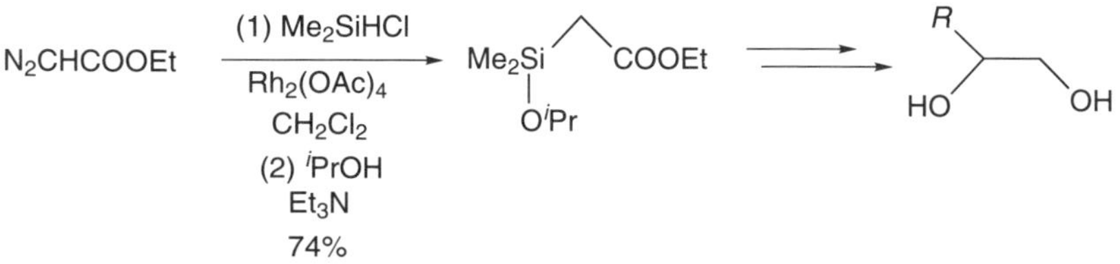

(89)

111   **112**

α-silylacetic esters using electrophile mediated cyclization of β-hydroxyhomoallylsilanes,[123] and diol **112** has been formed from α-styryldiazoacetate by a process employing Si–H insertion in a key step.[124] The Rh$_2$(OAc)$_4$-catalyzed insertion of ethyl diazoacetate into dihydrosilanes provides a convenient route into 3,3-disubstituted 3-silaglutarates, and even insertion resulting in a triester from a trihydrosilane has been achieved.[125] Insertion into the Sn–H bond of Bu$_3$SnH by ethyl diazoacetate has also been reported.[122] Dialkyl phosphonates have been prepared by copper-catalyzed reactions of diazo compounds with dialkyl hydrogen phosphites in moderate to high yield (eq. 90).[126] Dialkyl phospho-

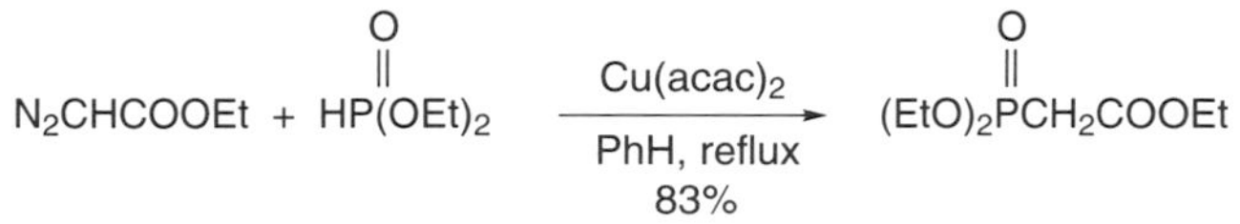

$$N_2CHCOOEt + HP(OEt)_2 \xrightarrow[\substack{PhH,\ reflux \\ 83\%}]{Cu(acac)_2} (EtO)_2PCH_2COOEt \qquad (90)$$

nates have been prepared by copper-catalyzed reactions of diazo compounds with dialkyl hydrogen phosphites in moderate to high yield (eq. 90).[126]

Doyle, Moody and co-workers have recently reported that asymmetric Si–H insertion, catalyzed by a broad range of chiral dirhodium(II) catalysts, occurs in good yield and modest enantioselectivity (eq. 91).[127] Among the catalysts that were

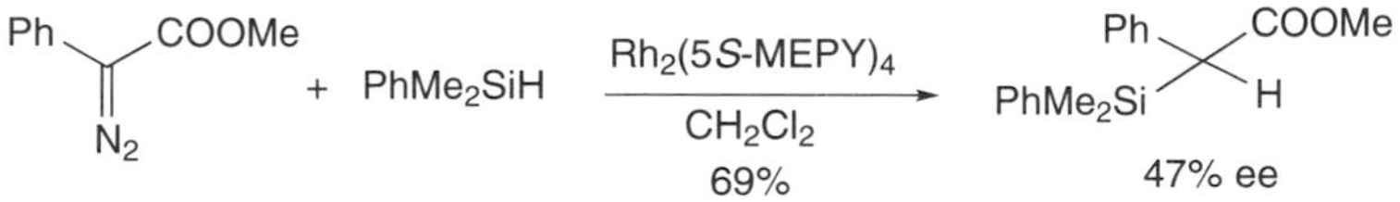

$$(91)$$

employed, Rh$_2$(MEPY)$_4$ gave the highest % ee values. The higher reactivity of the Si–H bond towards insertion makes possible intermolecular reactions that are uncharacteristic of C–H insertion using the same catalysts, but this higher reactivity dictates an earlier transition state that limits catalyst control of enantioselectivity.

3.12 REFERENCES

1   Moss, R. A.; and Jones, M., Jr., "Carbenes," in *Reactive Intermediates*, Vol. 3; Moss, R. A., and Jones, M., Jr., Eds.; Wiley: New York, 1985; Chapter 3.

2   Doyle, M. P., "Selectivity of Carbenes Generated From Diazirines," in *Chemistry of Diazirines*; Liu, M. T. H., Ed.; CRC Press, Inc.; Boca Raton, FL, 1987; pp. 33–74.

3   Doyle, M. P.; Taunton, J.; Oon, S.-M.; Liu, M. T. H.; Soundararajan, N.; Platz, M. S.; and Jackson, J. E., "Reactivity and Selectivity in Intermolecular Reactions of Chlorophenylcarbene," *Tetrahedron Lett.* **1988**, *29*, 5863–66.

4   Moss, R. A.; and Ho, G.-J., "Kinetics of a 1,3-CH Carbene Insertion Reaction: *tert*-Butylchlorocarbene," *J. Am. Chem. Soc.* **1990**, *112*, 5642–44.

5   Bach, R. D.; Su, M. D.; Aldabbagh, E.; Andrés, J. L.; and Schlegel, H. B., "A Theoretical Model for the Orientation of Carbene Insertion into Saturated Hydrocarbons and the Origin of the Activation Barrier," *J. Am. Chem. Soc.* **1993**, *115*, 10237–46.

6   Dave, V.; and Warnhoff, E. W., "The Reactions of Diazoacetic Esters with Alkenes, Alkynes, Heterocyclic and Aromatic Compounds," *Org. React. (N.Y.)* **1970**, *18*, 217–401.

7   Wulfman, D. S.; Linstrumelle, G.; and Cooper, C. F., "Synthetic Applications of Diazoalkanes, Diazocyclopentadienes and Diazoazacyclopentadienes," in *The Chemistry of Diazonium and Diazo Groups*; Patai, S., Ed.; Wiley: New York, 1978; Part 2, Chapter 18.

8   Burke, S. D.; and Grieco, P. A., "Intramolecular Reactions of Diazocarbonyl Compounds," *Org. React. (N.Y.)* **1979**, *26*, 361–475.

9   Maas, G., "Transition Metal Catalyzed Decomposition of Aliphatic Diazo Compounds—New Results and Applications in Organic Synthesis," *Top. Curr. Chem.* **1987**, *137*, 75–253.

10 Doyle, M. P., "Catalytic Methods for Metal Carbene Transformations," *Chem. Rev.* **1986**, *86*, 919–39.

11 Doyle, M. P., "Metal Carbene Complexes in Organic Synthesis: Diazodecomposition—Insertion and Ylide Chemistry," in *Comprehensive Organometallic Chemistry II*; Hegedus, L. S., Ed.; Pergamon Press, New York, 1995; Vol. 12, Chapter 5.2.

12 Ye, T.; and McKervey, M. A., "Organic Synthesis with α-Diazocarbonyl Compounds," *Chem. Rev.* **1994**, *94*, 1091–160.

13 Nefedov, O. M.; Shapiro, E. A.; and Dyatkin, A. B., "Diazoacetic Acids and Derivatives," in *Supplement B: The Chemistry of Acid Derivatives*; Patai, S., Ed.; Wiley: New York, 1992, Chapter 25.

14 Padwa, A.; and Krumpe, K. E., "Application of Intramolecular Carbenoid Reactions in Organic Synthesis," *Tetrahedron* **1992**, *48*, 5385–453.

15 Padwa, A.; and Austin, D. J., "Ligand Effects on the Chemoselectivity of Transition Metal Catalyzed Reactions of α-Diazo Carbonyl Compounds," *Angew. Chem., Int. Ed. Engl.* **1994**, *33*, 1797–815.

16 Taber, D. F., "Carbon–Carbon Bond Formation by C–H Insertion," in *Comprehensive Organic Synthesis: Selectivity, Strategy, and Efficiency in Modern Organic Chemistry*; Trost, B. M.; and Fleming, I., Eds.; Pergamon Press: New York, 1991; Vol. 3, Chapter 4.2.

17 Doyle, M. P., "Electronic and Steric Control in Intramolecular Carbon–Hydrogen Insertion Reactions of Diazo Compounds Catalyzed by Rhodium(II) Carboxylates and Carboxamides," in *Homogeneous Transition Metal Catalysts in Organic Synthesis*; Moser, W. R.; and Slocum, D. W., Eds.; ACS Advanced Chemistry Series 230; American Chemical Society: Washington, D. C., 1992, pp. 443–61.

18 Taber, D. F.; Petty, E. H.; and Ramon, K., "Enantioselective Ring Construction: Synthesis of (+)-α-Cuparenone," *J. Am. Chem. Soc.* **1985**, *107*, 196–99.

19 Taber, D. F., You, K. K.; and Rheingold, A. L., "Predicting the Diastereoselectivity of Rh-Mediated Intramolecular C–H Insertion," *J. Am. Chem. Soc.* **1996**, *118*, 547–56.

20 Burns, W.; McKervey, M. A.; Mitchell, T. R. B.; and Rooney, J. J., "A New Approach to the Construction of Diamondoid Hydrocarbons. Synthesis of *anti*-Tetramantane," *J. Am. Chem. Soc.* **1978**, *100*, 906–11.

21 Yates, P.; and Danishefsky, S., "A Novel Type of Alkyl Shift," *J. Am. Chem. Soc.* **1962**, *84*, 879–80.

22 Demonceau, A.; Noels, A. F.; Hubert, A. J.; and Teyssie, P., "Transition-Metal-Catalyzed Reactions of Diazoesters. Insertion into C–H Bonds of Paraffins by Carbenoids," *J. Chem. Soc., Chem. Commun.* **1981**, 688–89.

23 Demonceau, A.; Noels, A. F.; Hubert, A. J.; and Teyssie, P., "Transition-Metal-Catalyzed Reactions of Diazoesters. Insertion into C–H Bonds of Parrafins Catalyzed by Bulky Rhodium(II) Carboxylates: Enhanced Attack on Primary C–H Bonds," *Bull. Soc. Chim. Belg.* **1984**, *93*, 945–48.

24 Callot, H. J.; and Metz, F., "Homologation of *n*-Alkanes Using Diazoesters and Rhodium(III)porphyrins. Enhanced Attack on Primary C–H Bonds," *Tetrahedron Lett.* **1982**, *23*, 4321–24.

25 Callot, H. J.; and Metz, F., "Selective Two-Carbon Methyl Group Homologation Using Diazo Esters and Rhodium(III) Porphyrins," *Nouv. J. Chim.* **1985**, *9*, 167–71.

26 Anciaux, A. J.; Demonceau, A.; Noels, A. F.; Hubert, A. J.; Warin, R.; and Teyssie, P., "Transition-Metal-Catalyzed Reactions of Diazo Compounds. 2. Addition to Aromatic Molecules: Catalysis of Buchner's Synthesis of Cycloheptatrienes," *J. Org. Chem.* **1981**, *46*, 873–76.

27 Alonso, M. E.; and Carmen Garcia, M. del, "Kinetics of the Dirhodium Tetraacetate Catalyzed Decomposition of Ethyl Diazoacetate in 1,4-Dioxane. Is Nitrogen Involved in the Transition State?," *Tetrahedron* **1989**, *45*, 69–76.

28 Agosta, W. C.; and Wolff, S., "Preparation of Bicyclo[3.2.1]octan-6-ones from Substituted Cyclohexyl Diazo Ketones," *J. Org. Chem.* **1975**, *40*, 1027–30.

29 Adams, J.; and Spero, D. M., "Rhodium(II) Catalyzed Reactions of Diazocarbonyl Compounds," *Tetrahedron* **1991**, *47*, 1765–808.

30 Stork, G.; and Nakatani, K., "Regiocontrol by Electron Withdrawing Groups in the Rh-Catalyzed C–H Insertion of α-Diazoketones," *Tetrahedron Lett.* **1988**, *29*, 2283–86.

31 Wenkert, E.; Davis, L. L.; Mylari, B. L.; Solomon, M. F.; da Silva, R. R.; Shulman, S.; Warnet, R. J.; Ceccherelli, P.; Curini, M.; and Pellicciari, R., "Cyclopentanone Synthesis by Intramolecular Carbon–Hydrogen Insertion of Diazo Ketones. A Diterpene-to-Steroid Skeleton Conversion," *J. Org. Chem.* **1982**, *47*, 3242–47.

32 Taber, D. F.; and Petty, E. H., "General Route to Highly Functionalized Cyclopentane Derivatives by Intramolecular C–H Insertion," *J. Org. Chem.* **1982**, *47*, 4808–9.

33 Taber, D. F.; and Ruckle, R. E., Jr., "Cyclopentane Construction by Rh$_2$(OAc)$_4$-Mediated Intramolecular C–H Insertion: Steric and Electronic Effects," *J. Am. Chem. Soc.* **1986**, *108*, 7686–93.

34 Doyle, M. P.; Taunton, J.; and Pho, H. Q., "Conformational and Electronic Preferences in Rhodium(II) Carboxylate and Rhodium(II) Carboxamide Catalyzed Car-

bon–Hydrogen Insertion Reactions of *N,N*-Disubstituted Diazoacetoacetamides," *Tetrahedron Lett.* **1989**, *30*, 5397–400.

35   Adams, J.; Poupart, M.-A.; Grenier, L.; Schaller, C.; Ouimet, N.; and Frenette, R., "Rhodium Acetate Catalyzes the Addition of Carbenoids α- to Ether Oxygens," *Tetrahedron Lett.* **1989**, *30*, 1749–52.

36   Adams, J.; Poupart, M.-A.; and Grenier, L., "Diastereoselectivity in the Synthesis of 3(2*H*)-Furanones. Total Synthesis of (+)-Muscarine," *Tetrahedron Lett.* **1989**, *30*, 1753–56.

37   Adams, J.; and Frenette, R., "Stereoselective Synthesis of *endo*-1,3-Dimethyl-2,9-dioxabicyclo[3.3.1]nonane," *Tetrahedron Lett.* **1987**, *28*, 4773–74.

38   Spero, D. M.; and Adams, J., "Regiospecific Control of Rh(II) Carbenoids in the C–H Insertion Reaction," *Tetrahedron Lett.* **1992**, *33*, 1143–46.

39   Wang, P.; and Adams, J., "Model Studies of the Stereoelectronic Effect in Rh(II) Mediated Carbenoid C–H Insertion Reactions," *J. Am. Chem. Soc.* **1994**, *116*, 3296–305.

40   Lee, E.; Choi, I.; and Song, S. Y., "Tertiary Alcohol Synthesis from Secondary Alcohols *via* C–H Insertion," *J. Am. Chem. Soc., Chem. Commun.* **1995**, 321–22.

41   Sonawane, H. R.; Bellur, N. S.; Ahuja, J. R.; and Kulkarni, D. G., "Site-Selective Rhodium(II) Acetate Mediated Intramolecular Metal-Carbene Insertions into C–H Bonds of Bicyclo[2.2.1]heptanes: Efficient Syntheses of (+)-Albene and (−) β-Santalene," *J. Org. Chem.* **1991**, *56*, 1434–39.

42   Ceccherelli, P.; Curini, M.; Marcotullio, M. C.; and Rosati, O., "Regiocontrol by the Carbon–Carbon Double Bond in the Rh$_2$(OAc)$_4$ Mediated Carbon–Hydrogen Insertion of α-Diazo-Ketones," *Tetrahedron* **1991**, *47*, 7403–8.

43   Hashimoto, S.; Watanabe, N.; and Ikegami, S., "Dirhodium(II) Tetra(triphenylacetate): A Highly Efficient Catalyst for the Site Selective Intramolecular C–H Insertion Reactions of α-Diazo β-Keto Esters," *Tetrahedron Lett.* **1992**, *33*, 2709–12.

44   Hashimoto, S.; Shinoda, T.; and Ikegami, S., "A Simple Synthesis of (+)-Isocarbacyclin via a Convergent Process," *J. Chem. Soc., Chem. Commun.* **1988**, 1137–39.

45   Cane, D. E.; and Thomas, P. J., "Synthesis of *dl*-Pentalenolactones E and F," *J. Am. Chem. Soc.* **1984**, *106*, 5295–303.

46   Monteiro, H. J.; and Zukerman-Schpector, J., "A New Practical Synthesis of (+)-Grandisol from (+)-Citronellol Using an Intramolecular Carbenoid Cyclization," *Tetrahedron* **1996**, *52*, 3879–88.

47   Taber, D. F.; Hennessy, M. J.; and Louey, J. P., "Rh-Mediated Cyclopentane Construction Can Compete with β-Hydride Elimination: Synthesis of (±)-Tochuinyl Acetate," *J. Org. Chem.* **1992**, *57*, 436–41.

48   Shankar, B. K. R.; and Shechter, H., "Rhodium Ion Catalyzed Decomposition of Aryldiazoalkanes," *Tetrahedron Lett.* **1982**, *23*, 2277–80.

49   Taber, D. F.; and Schuchardt, J. L., "Intramolecular C–H Insertion: Synthesis of (±)-Pentalenolactone E Methyl Ester," *J. Am. Chem. Soc.* **1985**, *107*, 5289–90.

50   Taber, D. F.; Raman, K.; and Gaul, M. D., "Enantioselective Ring Construction: Synthesis of (+)-Estrone Methyl Ether," *J. Org. Chem.* **1987**, *52*, 28–34.

51   Sakai, K.; Fujimoto, T.; Yamashita, M.; and Kondo, K., "Total Synthesis of (±)-Dicranenones, Novel Cyclopentenonyl Fatty Acids," *Tetrahedron Lett.* **1985**, *26*, 2089–92.

52   Taber, D. F.; and Song, Y., "2,3,5-Trisubstituted Tetrahydrofurans by Rh-Mediated Cyclization of an α-Diazo Ester," *Tetrahedron Lett.* **1995**, *36*, 2587–90.

53   Ye, T.; McKervey, M. A.; Brandes, B. D.; and Doyle, M. P., "Stereoselective Synthesis of Disubstituted 3(2*H*)-Furanones via Catalytic Intramolecular C–H Insertion Reactions of α-Diazo-β-keto Esters Including Asymmetric Induction," *Tetrahedron Lett.* **1994**, *35*, 7269–72.

54   Monteiro, H. J., "Synthesis of α-Phenylsulfonyl Cyclopentanones by Intramolecular Carbenoid Cyclization of α-Diazo-β-keto Phenylsulfones," *Tetrahedron Lett.* **1987**, *28*, 3459–62.

55   Corbel, B.; Hernot, D.; Haelters, J.-P.; and Sturtz, G., "Synthesis of α-Phosphorylated Cyclopentanones by Intramolecular Carbenoid Cyclizations of α-Diazo β-Keto Alkylphosphonates and Phosphine Oxides," *Tetrahedron Lett.* **1987**, *28*, 6605–8.

56   Mikolajczyk, M.; Zurawinski, R.; and Kielbasinski, P., "A New Synthesis of (±)-Sarkomycin from a β-Ketophosphonate," *Tetrahedron Lett.* **1989**, *30*, 1143–46.

57   Corey, E. J.; and Kamiyama, K., "A Simple Stereoselective Synthesis of a Tetracyclic C_{14} Ginkgolide," *Tetrahedron Lett.* **1990**, *31*, 3995–98.

58   Rao, V. B.; Wolff, S.; and Agosta, W. C., "Synthesis of [4.4.5.5]Fenestrane," *Tetrahedron* **1986**, *42*, 1549–53.

59   Rao, V. B.; George, C. F.; Wolff, S.; and Agosta, W. C., "Synthetic and Structual Studies in the [4.4.4.5]Fenestrane Series," *J. Am. Chem. Soc.* **1985**, *107*, 5732–39.

60   Rao, V. B.; Wolff, S.; and Agosta, W. C., "Synthesis of Methyl 1-Methyltetracyclo[4.3.1.0^{3,10}.0^{8,10}]decane-7-carboxylate, a Derivative of [4.4.4.5]Fenestrane," *J. Chem. Soc., Chem. Commun.* **1984**, 293–94.

61   Ceccherelli, P.; Curini, M.; Marcotullio, M. C.; Rosati, O.; and Wenkert, E., "A New, General Cyclopentenone Synthesis," *J. Org. Chem.* **1990**, *55*, 311–15.

62   Ceccherelli, P.; Curini, M.; Marcotullio, C.; Rosati, O.; and Wenkert, E., "Regioselectivity of Rhodium(II)-Catalyzed Decomposition of 1-Alkyl-1-(diazoacetyl)alkenes. Synthesis of 2-Alkyl-2-cyclopentenones and 2-Alkylidenecyclopentanones," *J. Org. Chem.* **1991**, *56*, 7065–70.

63   Hashimoto, S.; Watanabe, N.; and Ikegami, S., "Highly Selective Insertion into Aromatic C–H Bonds in Rhodium(II) Triphenylacetate-Catalyzed Decomposition of α-Diazocarbonyl Compounds," *J. Chem. Soc., Chem. Commun.* **1992**, 1508–10.

64   Ceccherelli, P.; Curini, M.; Marcotullio, M. C.; and Rosati, O., "A New Synthetic Route to Methylenomycin B via Rhodium(II)-Mediated Decomposition of α,β-Unsaturated α'-Diazoketones," *Syn. Commun.* **1991**, *21*, 17–23.

65   Engling, G.; Emrick, T.; Hellmann, J.; McElroy, E.; Brandt, M.; and Reingold, I. D., "A Novel Approach to Bicyclo[3.3.0]octane-2,8-dione," *J. Org. Chem.* **1994**, *59*, 1945.

66   Doyle, M. P.; Bagheri, V.; Pearson, M. M.; and Edwards, J. D., "Highly Selective γ-Lactone Synthesis by Intramolecular Carbenoid Carbon–Hydrogen Insertion in Rhodium(II) Carboxylate and Rhodium(II) Carboxamide Catalyzed Reactions of Diazo Esters," *Tetrahedron Lett.* **1989**, *30*, 7001–4.

67   Doyle, M. P.; Westrum, L. J.; Wolthuis, W. N. E.; See, M. M.; Boone, W. P.;
     Bagheri, V.; and Pearson, M. M., "Electronic and Steric Control in Carbon–
     Hydrogen Insertion Reactions of Diazoacetoacetates Catalyzed by Dirhodium(II)
     Carboxylates and Carboxamides," *J. Am. Chem. Soc.* **1993**, *115*, 958–64.

68   Lee, E.; Jung, K. W.; and Kim, Y. S., "Selectivity in the Lactone Formation
     via C–H Insertion Reaction of Diazomalonates," *Tetrahedron Lett.* **1990**, *31*,
     1023–26.

69   Chelucci, G.; and Saba, A., "Intramolecular C–H Insertion or Styrene Cyclopropa-
     nation in Catalytic Decomposition of Dicyclohexyldiazomalonic Esters," *Tetrahe-
     dron Lett.* **1995**, *36*, 4673–76.

70   Doyle, M. P.; Protopopova, M. N.; Poulter, C. D.; and Rogers, D. H., "Macrocyclic
     Lactones from Dirhodium(II)-Catalyzed Intramolecular Cyclopropanation and
     Carbon–Hydrogen Insertion," *J. Am. Chem. Soc.* **1995**, *117*, 7281–82.

71   Doyle, M. P.; and Dyatkin, A. B., "Spirolactones from Dirhodium(II)- Catalyzed
     Diazo Decomposition with Regioselective Carbon–Hydrogen Insertion," *J. Org.
     Chem.* **1995**, *60*, 3035–38.

72   Doyle, M. P.; Dyatkin, A. B.; and Autry, C. L., "A New Catalytic Transformation
     of Diazo Esters: Hydride Abstraction in Dirhodium(II)-Catalyzed Reactions,"
     J. Chem. Soc., Perkin Trans. 1 **1995**, 619–21.

73   Checlucci, G.; and Saba, A., "(*S*)-(+)- and (*R*)-(−)-1,5-Dimethyl-4-phenyl-
     1,5-dihydro-2*H*-pyrrol-2-ones by Carbene Ring Contraction and Decarboxylation
     of (2*R*,3*S*)-(−)- and (2*S*,3*R*)-(+)-6-Diazo-3,4-dimethyl-2-phenyloxazepane-5,7-
     diones," *Angew. Chem. Int. Ed. Engl.* **1995**, *34*, 78–79.

74   Chelucci, G.; Saba, A.; and Valle, G., "Stereospecific Course of a Transannular
     C–H Insertion Process," *Tetrahedron: Asymmetry* **1995**, *6*, 807–10.

75   Ponsford, R. J.; and Southgate, R., "Preparation of 8-Oxo-7-(1-hydroxyethyl)-3-
     oxa-1-azabicyclo[4.2.0]octane Derivatives: Intermediates for Thienamycin Syn-
     thesis," *J. Chem. Soc., Chem. Commun.* **1979**, 846–47.

76   Brown, P.; and Southgate, R., "A Stereocontrolled Route to Optically Active 1-
     Methyl Carbapenems," *Tetrahedron Lett.* **1986**, *27*, 247–50.

77   Smale, T. C., "A Chiral Intermediate for Thienamycin Analogue Synthesis:
     (3*S*,4*R*)-4-(2-Hydroxyethyl)-3-[(*R*)-1-(4-Nitrobenzyloxycarbonyloxy)ethyl]-
     azetidin-2-one," *Tetrahedron Lett.* **1984**, *25*, 2913–14.

78   Brunwin, D. M.; Lowe, G.; and Parker, J., "The Total Synthesis of a Nuclear
     Analogue of the Penicillin-Cephalosporin Antibiotics," *J. Chem. Soc. (C)* **1971**,
     3756–62.

79   Tomioka, H.; Kitagawa, H; and Izawa, Y., "Photolysis of *N,N*-Diethyldiazoac-
     etamide. Participation of a Noncarbenic Process in Intramolecular Carbon–
     Hydrogen Insertion," *J. Org. Chem.* **1979**, *44*, 3072–75.

80   Doyle, M. P.; Shanklin, M. S.; Oon, S.-M.; Pho, H. Q.; van der Heide, F. R.; and
     Veal, W. R., "Construction of β-Lactams by Highly Selective Intramolecular C–H
     Insertion from Rhodium(II) Carboxylate Catalyzed Reactions of Diazoac-
     etamides," *J. Org. Chem.* **1988**, *53*, 3384–86.

81   Doyle, M. P.; Pieters, R. J.; Taunton, J.; Pho, H. Q.; Padwa, A.; Hertzog, D. L.;
     and Precedo, L., "Synthesis of Nitrogen-Containing Polycycles via Rhodium(II)-
     Induced Cyclization-Cycloaddition and Insertion Reactions of *N*-(Diazoace-

toacetyl)amides. Conformational Control of Reaction Selectivity," *J. Org. Chem.* **1991**, *56*, 820–29.

82 Wee, A. G. H.; Liu, B.; and Zhang, L., "Dirhodium Tetraacetate Catalyzed Carbon–Hydrogen Insertion Reaction in *N*-Substituted α-Carbomethoxy-α-diazoacetanilides and Structural Analogues. Substituent and Conformational Effects," *J. Org. Chem.* **1992**, *57*, 4404–14.

83 Afarinkia, K.; Cadogan, J. I. G.; and Rees, C. W., "Synthesis of 1,2-Azaphosphetidines," *J. Chem. Soc., Chem. Commun.* **1992**, 285.

84 Doyle, M. P.; Oon, S.-M.; van der Heide, F. R.; and Brown, C. B., "β-Lactam Formation via Rhodium(II) Catalyzed Carbon–Hydrogen Insertion Reactions of α-Diazo Amides," *Biorg. Med. Chem. Lett.* **1993**, *3*, 2409–14.

85 Doyle, M. P.; Shanklin, M. S.; and Pho, H. Q., "Cycloheptatriene Syntheses through Rhodium(II) Acetate-Catalyzed Intramolecular Addition Reactions of *N*-Benzyldiazoacetamides," *Tetrahedron Lett.* **1988**, *29*, 2639–42.

86 Zaragoza, F.; and Zahn, G., "Rhodium(II) Acetate-Catalyzed Decomposition of 2-Diazo-3-oxobutanamides Derived from L-Phenylalanine," *J. Prakt. Chem./Chem.-Ztg.* **1995**, *337*, 292–98.

87 Padwa, A.; Austin, D. J.; Hornbuckle, S. F.; Semones, M. A.; Doyle, M. P.; and Protopopova, M. N., "Control of Chemoselectivity in Catalytic Carbenoid Reactions. Dirhodium(II) Ligand Effects on Relative Reactivities," *J. Am. Chem. Soc.* **1992**, *114*, 1874–76.

88 Padwa, A.; Austin, D. J.; Price, A. T.; Semones, M. A.; Doyle, M. P.; Protopopova, M. N.; Winchester, W. R.; and Tran, A., "Ligand Effects on Dirhodium(II) Carbene Reactivities. Highly Effective Switching between Competitive Carbenoid Transformations," *J. Am. Chem. Soc.* **1993**, *115*, 8669–80.

89 Hashimoto, S.; Watanabe, N.; and Ikegami, S., "Highly Selective Insertion into Aromatic C–H Bonds in Rhodium(II) Triphenylacetate-Catalysed Decomposition of α-Diazocarbonyl Compounds," *J. Chem. Soc., Chem. Commun.* **1992**, 1508–10.

90 Clark, J. S.; Krowiak, S. A.; and Street, L. J., "Synthesis of Cyclic Ethers from Copper Carbenoids by Formation and Rearrangement of Oxonium Ylides," *Tetrahedron Lett.* **1993**, *34*, 4385–88.

91 Hashimoto, S.; Watanabe, N.; and Ikegami, S., "Enantioselective Intramolecular C–H Insertion of α-Diazo β-Keto Esters Catalyzed by Homochiral Rhodium(II) Carboxylates," *Tetrahedron Lett.* **1990**, *31*, 5173–74.

92 Kennedy, M.; McKervey, M. A.; Maguire, A. R.; and Roos, G. H. P., "Asymmetric Synthesis in Cabon–Carbon Bond Forming Reactions of α-Diazoketones Catalysed by Homochiral Rhodium(II) Carboxylates," *J. Chem. Soc., Chem. Commun.* **1990**, 361–62.

93 Doyle, M. P.; Winchester, W. R.; Hoorn, J. A. A.; Lynch, V.; Simonsen, S. H.; and Ghosh, R., "Dirhodium(II) Tetrakis(carboxamidates) with Chiral Ligands. Structure and Selectivity in Catalytic Metal Carbene Transformations," *J. Am. Chem. Soc.* **1993**, *115*, 9968–78.

94 Doyle, M. P.; Dyatkin, A. B.; Protopopova, M. N.; Yang, C. I.; Miertschin, C. S.; Winchester, W. R.; Simonsen, S. H.; Lynch, V.; and Ghosh, R., "Enhanced Enantiocontrol in Catalytic Metal Carbene Transformations with Dirhodium(II) Tetrakis[methyl 2-oxooxazolidin-4(*S*)-carboxylate], Rh₂(4*S*-MEOX)₄," *Recl. Trav. Chim. Pays-Bas* **1995,** 114, *163–70.*

95   Doyle, M. P.; Zhou, Q.-L.; Raab, C. E.; Roos, G. H. P.; Simonsen, S. H.; and
     Lynch, V., "Synthesis and Structures of (2,2-*cis*)-Dirhodium(II) Tetrakis[methyl
     1-acyl-2-oxoimidazolidine-4(*S*)-carboxylates]. Chiral Catalysts for Highly Stereo-
     selective Metal Carbene Transformations," *Inorg. Chem.* **1996**, *35*, 6064–73.

96   McKervey, M. A.; and Ye, T., "Asymmetric Synthesis of Substituted Chroma-
     nones *via* C–H Insertion Reactions of α-Diazoketones Catalysed by Homochiral
     Rhodium(II) Carboxylates," *J. Chem. Soc., Chem. Commun.* **1992**, 823–24.

97   Ye, T.; Garcia, C. F.; and McKervey, M. A., "Chemoselectivity and Stereoselectiv-
     ity of Cyclisation of α-Diazocarbonyls Leading to Oxygen and Sulfur Heterocy-
     cles Catalysed by Chiral Rhodium and Copper Catalysts," *J. Chem. Soc. Perkin
     Trans. 1* **1995**, 1373–79.

98   McCarthy, N.; McKervey, M. A.; Ye, T.; McCann, M.; Murphy, E.; and Doyle,
     M. P., "A New Rhodium(II) Phosphate Catalyst for Diazocarbonyl Reactions
     Including Asymmetric Synthesis," *Tetrahedron Lett.* **1992**, *33*, 5983–86.

99   Hashimoto, S.; Watanabe, N.; Sato, T.; Shiro, M.; and Ikegami, S., "Enhancement
     of Enantioselectivity in Intramolecular C–H Insertion Reactions of α-Diazo-β-
     keto Esters Catalyzed by Chiral Dirhodium(II) Carboxylates," *Tetrahedron Lett.*
     1993, *34*, 5109–12.

100 Sawamura, M.; Sasaki, H.; Nakata, T.; and Ito, Y., "Synthesis of Optically Active
     Ferrocene Analogues of Salicylic Acid Derivatives and Rhodium(II)-Catalyzed
     Asymmetric Intramolecular C–H Insertion of α-Diazo-β-keto Esters Using New
     Chiral Carboxylato Ligands," *Bull. Chem. Soc. Jpn.* **1993**, *66*, 2725–29.

101 Hashimoto, S.; Watanabe, N.; Kawano, K.; and Ikegami, S., "Double Asymmetric
     Induction in Intramolecular C–H Insertion Reactions of α-Diazo-β-keto Esters,"
     Synth. Commun. **1994**, *24*, 3277–87.

102 Doyle, M. P.; van Oeveren, A.; Westrum, L. J.; Protopopova, M. N.; and Clayton,
     T. W., Jr., "Asymmetric Synthesis of Lactones with High Enantioselectivity by
     Intramolecular Carbon–Hydrogen Insertion Reactions of Alkyl Diazoacetates
     Catalyzed by Chiral Rhodium(II) Carboxamides," *J. Am. Chem. Soc.* **1991**, *113*,
     8982–84.

103 Doyle, M. P.; Eismont, M. Y.; Bergbreiter, D. E.; and Gray, H. N., "Enantioselec-
     tive Metal Carbene Transformations with Polyethylene-Bound Soluble Recover-
     able Dirhodium(II) 2-Pyrrolidone-5(*S*)-carboxylates," *J. Org. Chem.* **1992**, *57*,
     6103–5.

104 Doyle, M. P.; Protopopova, M. N.; Winchester, W. R.; and Daniel, K. L., "Enanti-
     ocontrol and Regiocontrol in Lactam Syntheses by Intramolecular Carbon–Hy-
     drogen Insertion Reactions of Diazoacetamides Catalyzed by Chiral Rhodium(II)
     Carboxamides," *Tetrahedron Lett.* **1992**, *33*, 7819–22.

105 Doyle, M. P.; and Kalinin, A. V., "Highly Enantioselective Route to β-Lactams
     via Intramolecular C–H Insertion Reactions of Diazoacetylazacycloalkanes Cata-
     lyzed by Chiral Dirhodium(II) Carboxamidates," *Synlett* **1995**, 1075–76.

106 Doyle, M. P.; Protopopova, M. N.; Zhou, Q.-L.; Bode, J. W.; Simonsen, S. H.; and
     Lynch, V., "Optimization of Enantiocontrol for Carbon–Hydrogen Insertion with
     Chiral Dirhodium(II) Carboxamidates. Synthesis of Natural Dibenzylbutyrolac-
     tone Lignans from 3-Aryl-1-propyl Diazoacetates in High Optical Purity," *J. Org.
     Chem.* **1995**, *60*, 6654–55.

107 Bode, J. W.; Doyle, M. P.; Protopopova, M. N.; and Zhou, Q.-L., "Intramolecular Regioselective Insertion into Unactivated Prochiral Carbon–Hydrogen Bonds with Diazoacetates of Primary Alcohols Catalyzed by Chiral Dirhodium(II) Carboxamidates. Highly Enantioselective Total Synthesis of Natural Lignan Lactones," *J. Org. Chem.* **1996**, *61*, 9146–55.

108 Doyle, M. P.; Dyatkin, A. B.; Roos, G. H. P.; Cañas, F.; Pierson, D. A.; van Basten, A.; Müller, P.; and Polleux, P., "Diastereocontrol for Highly Enantioselective Carbon–Hydrogen Insertion Reactions of Cycloalkyl Diazoacetates," *J. Am. Chem. Soc.* **1994**, *116*, 4507–8.

109 Müller, P.; and Polleux, P., "Enantioselective Formation of Bicyclic Lactones by Rhodium-Catalyzed Intramolecular C–H Insertion Reactions," *Helv. Chim. Acta* **1994**, *77*, 645–54.

110 Doyle, M. P.; Dyatkin, A. B.; and Tedrow, J. S., "Synthesis of 2-Deoxyxylolactone from Glycerol Derivatives via Highly Enantioselective Carbon–Hydrogen Insertion Reactions," *Tetrahedron Lett.* **1994**, *35*, 3853–56.

111 Doyle, M. P.; Zhou, Q.-L.; Dyatkin, A. B.; and Ruppar, D. A., "Enhancement of Enantiocontrol/Diastereocontrol in Catalytic Intramolecular Cyclopropanation and Carbon–Hydrogen Insertion Reactions of Diazoacetates with Rh_2(4S-MP-PIM)$_4$," *Tetrahedron Lett.* **1995**, *36*, 7579–82.

112 Doyle, M. P.; Zhou, Q.-L.; Raab, C. E.; and Roos, G. H. P., "Improved Enantioselection for Chiral Dirhodium(II) Carboxamide-Catalyzed Carbon–Hydrogen Insertion Reactions of Tertiary Alkyl Diazoacetates," *Tetrahedron Lett.* **1995**, *36*, 4745–48.

113 Lim, H.-J.; and Sulikowski, G. A., "Enantioselective Synthesis of a 1,2-Disubstituted Mitosene by a Copper-Catalyzed Intramolecular Carbon–Hydrogen Insertion Reaction of a Diazo Ester," *J. Org. Chem.* **1995**, *60*, 2326–27.

114 Burgess, K.; Lim, H.-J.; Porte, A. M.; and Sulikowski, G. A., "New Catalysts and Conditions for a C–H Insertion Reaction Identified by High Throughput Catalyst Screening," *Angew. Chem. Int. Ed. Engl.* **1996**, *35*, 220–22.

115 Doyle, M. P.; Kalinin, A. V.; and Ene, D. G., "Chiral Catalyst Controlled Diastereoselection and Regioselection in Intramolecular Carbon–Hydrogen Insertion Reactions of Diazoacetates," *J. Am. Chem. Soc.* **1996**, *118*, 8837–46.

116 Doyle, M. P.; and Kalinin, A. V., "Enantiomer Differentiation in Intramolecular Carbon–Hydrogen Insertion Reactions of Racemic Secondary Alkyl Diazoacetates Catalyzed by Chiral Dirhodium(II) Carboxamidates," *Russ. Chem. Bull.* **1995**, *44*, 1729–34.

117 Wee, A. G. H.; and Liu, B. S., "The $Rh_2(OAc)_4$ Catalysed C–H Insertion in Chiral Ester Diazoanilides," *Tetrahedron Lett.* **1996**, *37*, 145–48.

118 Hashimoto, S.; Miyazaki, Y.; and Ikegami, S., "A Regio- and Stereocontrolled Synthesis of 13,14-Didehydroisocarbacyclin," *Synlett* **1996**, 324–26.

119 Bagheri, V.; Doyle, M. P.; Taunton, J.; and Claxton, E. E., "A New and General Synthesis of α-Silyl Carbonyl Compounds by Si–H Insertion from Transition Metal Catalyzed Reactions of Diazo Esters and Diazo Ketones," *J. Org. Chem.* **1988**, *53*, 6158–60.

120 Landais, Y.; and Planchenault, D., "Asymmetric Metal Carbene Insertion into the Si–H Bond," *Tetrahedron Lett.* **1994**, *35*, 4565–68.

121   Landais, Y.; Planchenault, D.; and Weber, V., "Rhodium(II) Vinylcarbenoid Insertion into the Si–H Bond. A New Stereospecific Synthesis of Allylsilanes," *Tetrahedron Lett.* **1994**, *35*, 9549–52.

122   Andrey, O.; Landais, Y.; Planchenault, D.; and Weber, V., "Synthesis of α-(Alkoxysilyl)acetic Esters. A Route to 1,2-Diols," *Tetrahedron* **1995**, *51*, 12083–96.

123   Andrey, O.; and Landais, Y., "Highly Stereoselective Access to 2,4- and 2,4,5-Substituted Tetrahydrofurans from α-Silylacetic Esters. A Study of Homoallylic Stereocontrol" *Tetrahedron Lett.* **1993**, *34*, 8435–38.

124   Angelaud, R.; Landais, Y.; and Maignan, C., "The Dimethyl (1-Phenylthio)-cyclopropylsilyl Group as a Masked Hydroxyl Group," *Tetrahedron Lett.* **1995**, *36*, 3861–64.

125   Barnier, J. P.; and Blanco, L., "Reaction of Dihydrosilanes with Ethyl Diazoacetate—Synthesis of 3,3-Disubstituted 3-Silaglutarates," *J. Organometal. Chem.* **1996**, *514*, 67–71.

126   Polozov, A. M.; Polezhaeva, N. A.; Mustaphin, A. H.; Khotinen, A. V.; and Arbuzov, B. A., "A New One-Pot Synthesis of Dialkyl Phosphonates from Diazo Compounds and Dialkyl Hydrogen Phosphites," *Synthesis* **1990**, 515–17.

127   Buck, R. T.; Doyle, M. P.; Drysdale, M. J.; Feris, L.; Forbes, D. C.; Haigh, D.; Moody, C. J.; Pearson, N. D.; and Zhou, Q.-L., "Asymmetric Rhodium Carbenoid Insertion into the Si–H bond," *Tetrahedron Lett.* **1996**, *37*, 7631–34.

CHAPTER 4

Intermolecular Cyclopropanation and Related Addition Reactions

An increasing realization of the importance of three-membered ring compounds in diverse areas of organic chemistry and biology has drawn considerable attention to methods for their synthesis.[1-5] Cyclopropanes occur as structural subunits in

Cyclopropane **Cyclopropene** **Aziridine**

biologically active natural and unnatural products,[1,6-9] they are frequently used as mechanistic probes to define specific details of reaction pathways,[10-14] and they are increasingly valuable as synthetic intermediates.[15-18] Biomolecules that possess a cyclopropene ring, including fatty acids and sterols, have been isolated and characterized,[19-21] and there is expanding interest in cyclopropenes as synthetic intermediates.[3-5] Aziridines have long been known for their synthetic usefulness,[22] but only recently have catalytic methods for their construction become available. This chapter will describe catalytic methods that are effective for the synthesis of these three-membered ring compounds by intermolecular transformations.

Although the origins of transition metal catalyzed cyclopropanation extend back to 1906,[22] this methodology evolved slowly until the late 1960s, when catalyst development began in earnest. Today the attractiveness of catalytic methods for diazo decomposition resulting in addition to alkenes and alkynes is due not only to the economy of chemical catalysis but also to the high levels of selectivity that can be achieved in these addition reactions.[23-29] Control of selectivity in cyclopropanation, unlike in epoxidation, encompasses diastereoselection as well as regioselection and enantioselection. This chapter will focus on how the highest levels of selectivity can be achieved in catalytic transformations of diazo compounds.

4.1 CYCLOPROPANATION WITH DIAZOMETHANE

Palladium(II) compounds are the most effective catalysts for the cyclopropanation of alkenes with diazomethane.[30] Their utility for addition to α,β-unsaturated ketones, aldehydes, amides, and esters (e.g., eq. 1,[31] eq. 2,[30] and eq. 3[37]) is evident in publications of Vorbrüggen,[31-33] Tomilov,[30,34] and others.[35-39] Palladium(II)

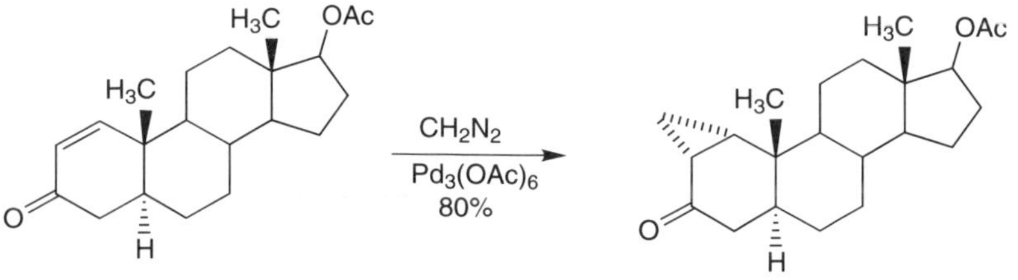

(1)

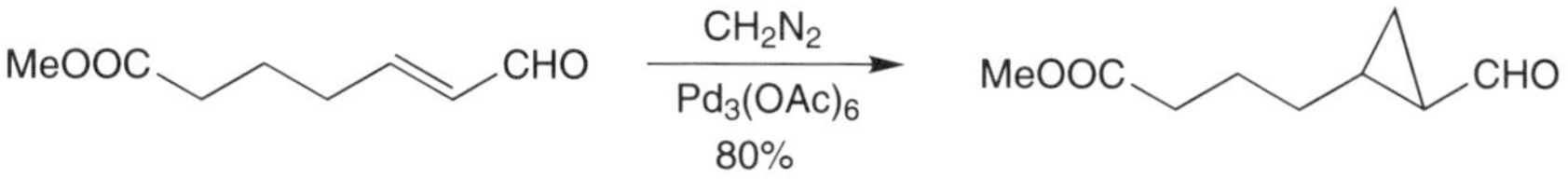

(2)

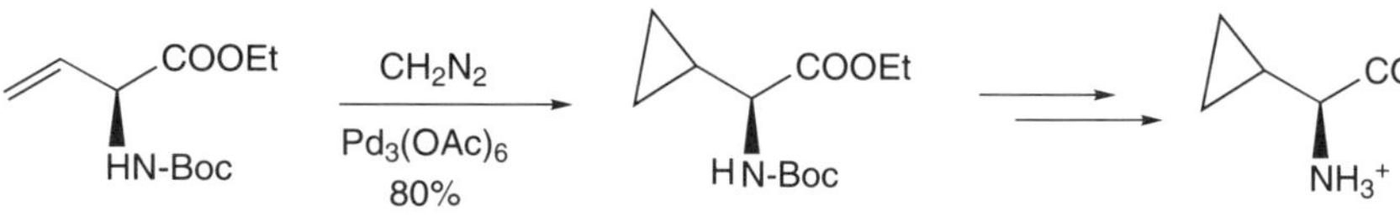

(3)

acetate, ordinarily considered to be in its trimeric form, is the preferred catalyst, and the amount of diazomethane usually employed is less than 3 molar equivalents based on the alkene. Electron-poor alkenes are often more reactive than electron-rich alkenes; methyl methacrylate has been reported to be as reactive as styrene towards cyclopropanation with diazomethane catalyzed by palladium(II) and approximately 200 times more reactive than cyclohexene.[40]

Ring strain enhances reactivity in palladium(II)-catalyzed cyclopropanation reactions with diazomethane.[30] High product yields are characteristic of reactions with medium ring alkenes, norbornenes and related strained bicyclic and polycyclic alkenes, and allenes.[41-43] Low product yields are characteristic of reactions with cyclohexenes, cyclopropenes, and many simple alkenes. *exo*-Addition occurs virtually exclusively in Pd(II)-catalyzed reactions of norbornenes with diazomethane (>98%); in contrast, mixtures of *endo* and *exo* products are formed in reactions catalyzed by Cu(I) or Rh(II).[44] The enhanced reactivity that occurs with ring strain (**1**) leads to some remarkable examples of regioselection (eq. 4[34]), and, as suggested by eq. 5,[31,34,35] increasing substitution of the carbon–carbon

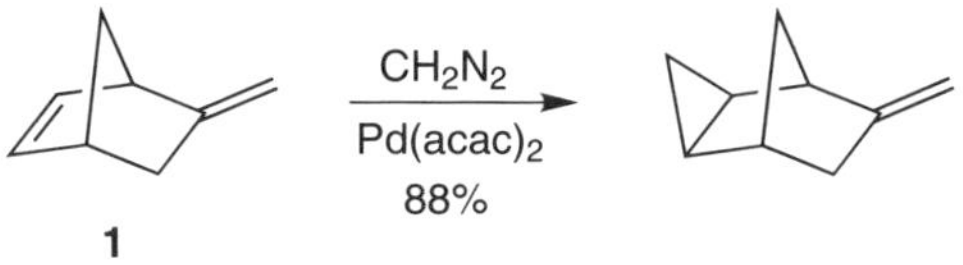

(4)

1

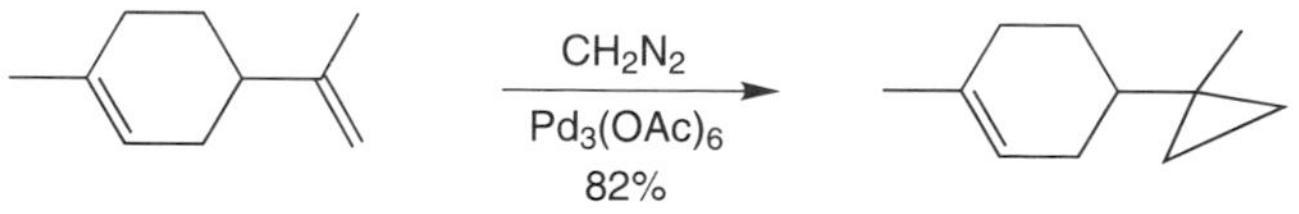

$$(5)$$

double bond (**2**) decreases its reactivity towards cyclopropanation. In contrast, copper(I) chloride catalyzed reactions show limited regioselection with the same or similar olefinic substrates,[30] but Cu(OTf)$_2$ has been reported to catalyze cyclopropanation of diene **3** with a regiocontrol that is suggestive of Pd(II)-like reactivity (eq. 6).[45] Examples that suggest the synthetic potential of Cu(OTf)$_2$ or

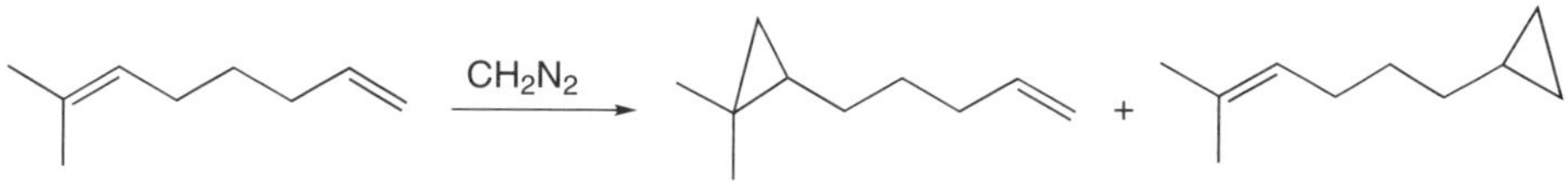

	34	66
Cu(OTf)$_2$	34	66
Cu(acac)$_2$	86	14

$$(6)$$

Cu(OTf) are rare,[30] but this catalytic system holds promise as an alternative to palladium(II) compounds for cyclopropanation reactions with diazomethane.

Allylamines, allyl ethers, and allyl acetates (**4**) undergo exclusive methylene cyclopropanation, catalyzed by PdCl$_2$(PhCN)$_2$, without evidence of ylide generation (eq. 7), and neither O–H nor N–H insertion is observed with the alcohol or

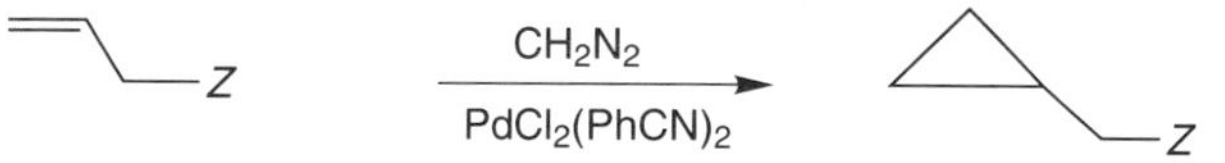

Z = OH, OMe, OAr, OAc   70–78%
Z = NH$_2$, NMe$_2$, NHPh   62–68%

$$(7)$$

amines.[46] This selectivity contrasts with that of dirhodium(II) compounds whose catalysis provides competitive or exclusive ylide formation/[2,3]-sigmatropic rearrangement with some of these same compounds.[24,47] The facility with which the palladium-catalyzed reactions occur suggests that heteroatom coordination enhances Pd(II) activation of the carbon–carbon double bond (**5**) for diazomethane addition (Scheme 1). Details of this mechanism for cyclopropanation are discussed in Section 2.3.

Chiral auxiliaries, including Oppolzer's bornane[10,2]sultam (**6**)[48,49] and L-ephedrine (**8**),[50] have been employed to achieve diastereocontrol in palladium(II) acetate catalyzed cyclopropanation reactions with diazomethane (eqs. 8 and 9). With sultam derivatives diastereomeric ratios of cyclopropane products (**7**) increase significantly when the reaction temperature is raised from −30 to 0°C and

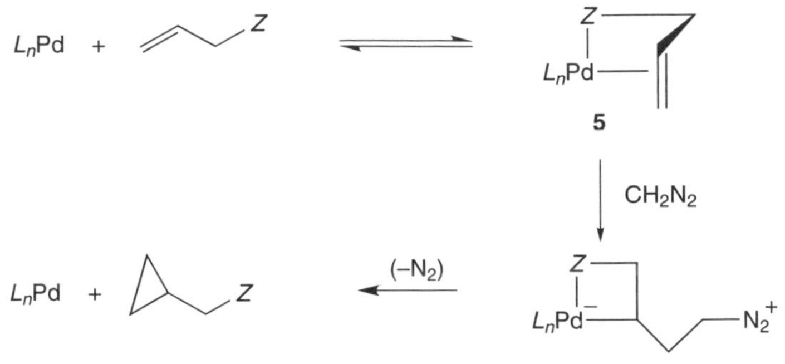

Scheme 1

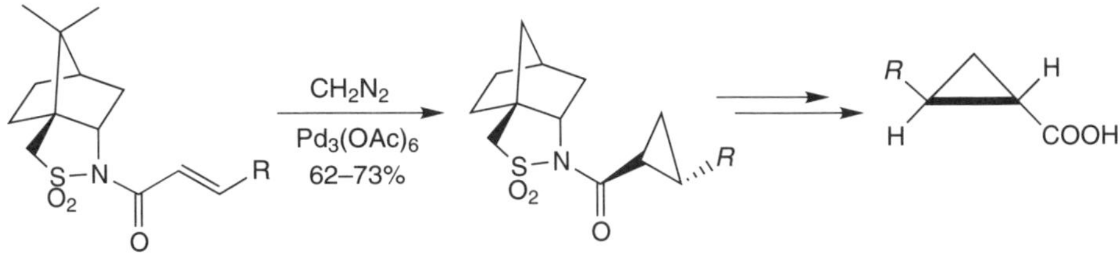

(8)

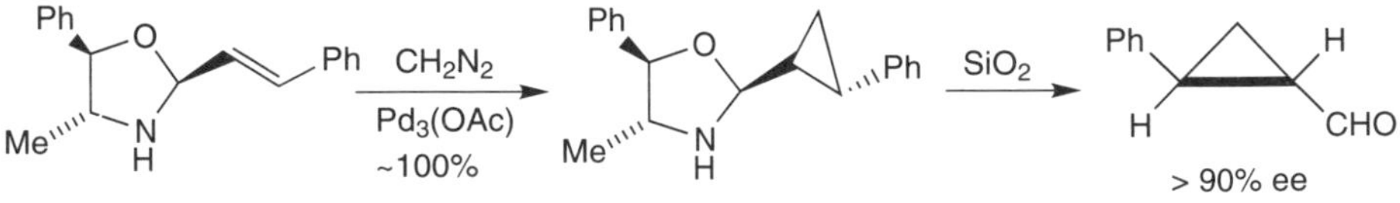

(9)

remains nearly the same when increased further to 30°C (R = Ph, from 76:24 to 93:7 dr). Selected cyclopropyl-L-amino acids (**10**) have also been prepared by Pd(II)-catalyzed cyclopropanation of vinyl-substituted L-amino acid derivatives with diazomethane (eq. 10).[51] An alternative methodology using the Simmons–

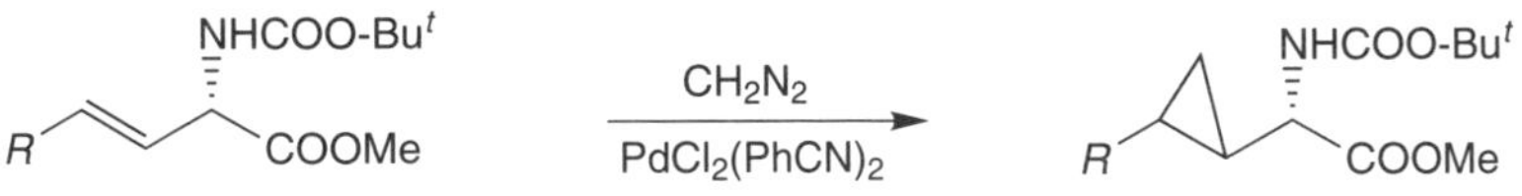

R = H (98%), R = CH$_2$OAc (68%)   **10**   (10)

Smith reaction and carbohydrates as chiral auxiliaries also provides high levels of diastereocontrol.[52,53] Asymmetric cyclopropanation of 1,2-*trans*-disubstituted olefins with diazomethane using chiral semicorrin copper catalysts (70–80% ee) has been announced,[54] but full details have yet to be provided.

One of the major problems associated with catalytic methylenation is the production and handling of diazomethane. However, Tomilov has developed a methodology whereby CH_2N_2 is generated *in situ*.[55] According to this procedure N-methyl-N-nitrosourea is added gradually to a mixture of the alkene, the palladium catalyst, organic solvent, and 40–50% KOH solution. Less than 2 molar equivalents of N-methyl-N-nitrosourea is generally required, and high product yields have been achieved. This methodology provides the opportunity to employ catalytic methylenation on a preparative scale with appropriate reactants.

4.2 CYCLOPROPANATION REACTIONS WITH DIAZOCARBONYL COMPOUNDS. GENERAL

Diazocarbonyl compounds, the preferred substrates for catalytic diazo decomposition, can be prepared by numerous methods with relative ease (Chapter 1).[56] Their uses for the synthesis of a vast array of cyclopropane compounds that are natural products (**11–14**), pyrethroid fragments (**13** and **15**),[61] natural and synthetic amino acids (**16–18**)[62–64] and carbocyclic nucleosides (**19**)[65] are now well

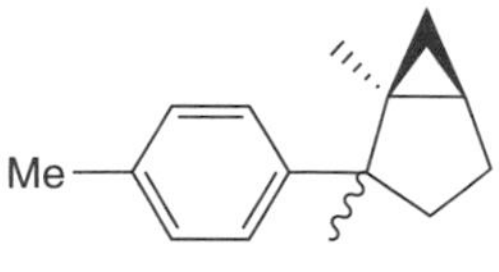

11
(+)-cyclolaurene[57]

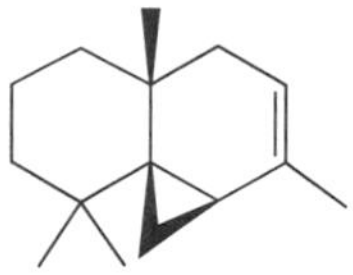

12
(+)-thujopsene[6]

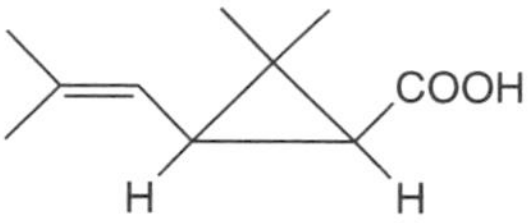

13
cis-chrysanthemic
acid[58]

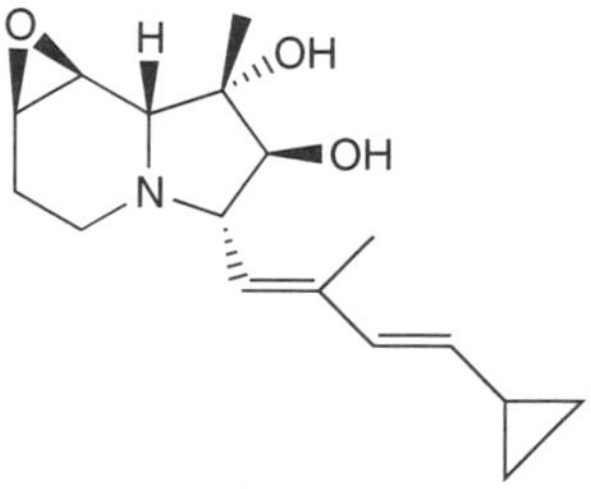

14
cyclizidine[59]
(indolizidine antibiotic)

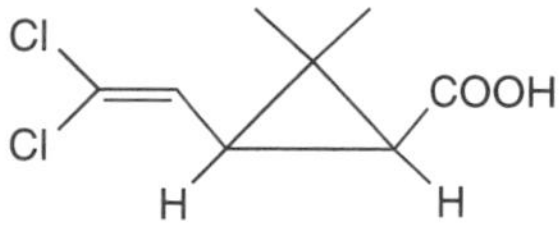

15
permethrinic acid[60]

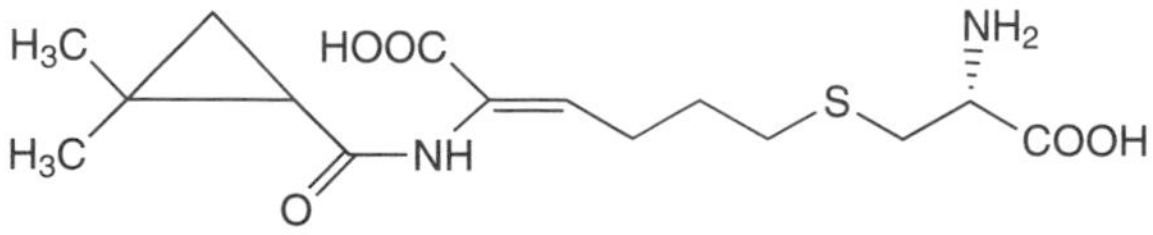

16
cilastatin[62]

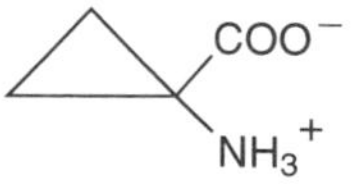

17[63]

18[64]
conformationally constrained
analog of L-glutamate

19[65]
a cyclopropyl
carbocyclic L-nucleoside

documented, and the construction of cyclopropane compounds as synthetic intermediates is becoming increasingly common.[1,2,27] Amino acid and amido derivatives of C_{60}, among others, are prepared by intermolecular cyclopropanation.[66,67]
With synthetic demands for increasing complexity in molecular design, the development of new catalysts and methods for highly selective cyclopropanation has
taken on a new priority, and significant advances have been made.[23-29]

Most early work in catalytic cyclopropanation reactions with diazocarbonyl
compounds focused on intermolecular transformations and, with a limited selection of catalysts and a limited understanding of the catalytic process,[68-70] it is perhaps not surprising that these reactions were thought to provide limited synthetic
opportunities. Interest in catalytic cyclopropanation reactions increased after the
first report of an intramolecular process by Stork in 1961,[6,71] but major advances
were realized only after the development of homogeneous catalysts.[23,24] The organic solvent-soluble catalysts allowed reactions to take place at lower temperatures than were feasible with their heterogeneous counterparts and, by ligand
modification, provided the opportunity to control reaction selectivity.[29]

The rate-limiting step in cyclopropanation reactions of diazocarbonyl compounds is diazo decomposition,[23-27] and both the diazo compound and catalyst
ligands influence the rate of reaction. Increasing carbonyl substitution at the
diazo carbon decreases the nucleophilic reactivity of the diazocarbonyl compound (Scheme 2),[72] and decreasing reactivity is found in the series: amide <

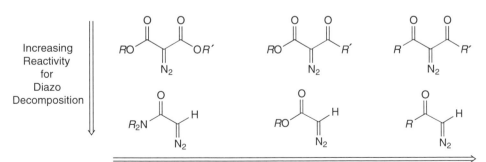

Increasing Reactivity for Diazo Decomposition

Scheme 2

ester < ketone. Diazophosphonates and diazosulfones are less reactive than dia-zoesters. Higher temperatures are generally required for diazo decomposition of an α-diazo-β-ketoester than a diazoacetate. Catalyst ligands are important deter-minants of the electrophilic reactivity of the transition metal compound; repre-sentative trends are given in Scheme 3 for copper and rhodium catalysts without

$$\text{Cu(acac)}_2 \ < \ \text{CuCl}\bullet\text{P(O}R)_3, \text{Cu(salicylate)}_2 \ < \ \text{Cu(OTf)}, \text{CuPF}_6$$

Increasing Reactivity for Diazo Decomposition

$$\text{Rh}_2(\text{acam})_4, \text{Rh}_2(\text{cap})_4 \ < \ \text{Rh}_2(\text{OAc})_4, \text{Rh}_2(\text{oct})_4 \ < \ \text{Rh}_2(\text{pfb})_4, \text{Rh}_2(\text{tfa})_4$$

Scheme 3

chiral ligands. Copper(I) hexafluorophosphate, as its acetonitrile complex,[73] is superior to Cu(OTf) in air stability and handling ease. Consideration of both the diazo compound and the catalyst allows an estimate of the reaction conditions that are required for cyclopropanation.

The use of iodonium ylides [e.g., PhI=C(COOMe)$_2$] as possible alternatives to diazo compounds has been investigated by Müller and Fernandez,[74] who found product compositions in dirhodium(II) catalyzed reactions to be identical to those from the use of the corresponding diazo compounds in a variety of inter- and intramolecular metal carbene transformations, and product yields are com-parable. Further work is warranted to establish if there is any advantage to the use of iodonium ylides.

4.3 INTERMOLECULAR CYCLOPROPANATION REACTIONS OF DIAZOCARBONYL COMPOUNDS. DIASTEREOSELECTIVITY

Intermolecular cyclopropanation of alkenes by the catalytic addition of a carbene is remarkably insensitive to structural influences.[24] This insensitivity is associ-ated with the inherently high electrophilic reactivity of the intermediate metal carbene, which commences bond formation with the carbon–carbon double bond at distances sufficiently removed from the carbene center (e.g., **20**) to allow lim-ited steric interaction between the carbene substituent (COZ) and the alkene sub-stituent (R). In the absence of overriding effects, the double bond approaches the

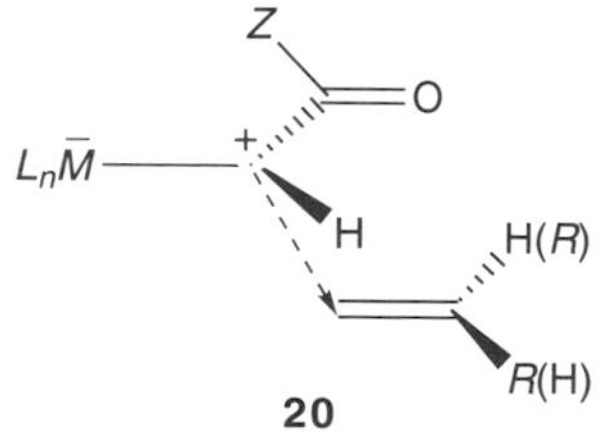

20

carbene center so that electrophilic addition can take place in a Markovnikov fashion, thus allowing the olefin substituent R to stabilize the developing electropositive center. With trisubstituted olefins the orientation of the double bond with respect to the catalyst, consistent with Markovnikov addition, places the olefinic carbon that elicits diastereoisomer formation toward the catalyst face (**21**)

21

so that differentiation actually relies on both the carbene substituent (COZ) and the structure of the catalyst. In either case, however, steric influences from olefin substituents are predicted to be small, and they are (Table 4.1).[75]

TABLE 4.1   Diastereoselectivies in Catalytic Intermolecular Cyclopropanation Reactions of Representative Alkenes with Ethyl Diazoacetate[75]

Alkene	Rh$_2$(OAc)$_4$		Rh$_6$(CO)$_{16}$		CuCl·P(O–i-Pr)$_3$		PdCl$_2$(PhCN)$_2$	
	Yield (%)	t/c^a	Yield (%)	t/c	Yield (%)	t/c	Yield (%)	t/c
PhCH=CH$_2$	93	1.6	86	1.7	88	2.8	52	1.6
EtOCH=CH$_2$	88	1.7	62	1.7	61	1.9	43	1.5
t-BuCH=CH$_2$	87	4.2	42	4.5	23	7.3	34	2.5
cyclohexene	90	(3.8)	88	(3.9)	28	(6.8)	31	(2.2)
dihydropyran	91	(6.5)	82	(6.8)	18	(6.3)	41	(3.8)
(Me$_2$C=CH)$_2$	81	1.8	87	1.9	55	2.7	20	2.3

^{a}*trans:cis* or (*anti:syn*) ratios.

The small variation in diastereoselectivities for olefin cyclopropanation is reflected by the similarly small variation in relative reactivities from reactions with ethyl diazoacetate.[76,77] As expected from an electrophilic addition reaction, vinyl ethers are more reactive than styrene which, in turn, is more reactive than 1-hexene [range in relative reactivity = 8.3 for Rh$_2$(OAc)$_4$].[76] The difference in reactivity between monosubstituted, disubstituted, and trisubstituted ethylenes is small and does not necessarily follow an expected order.

Greater variations in diastereoselectivities result from changes in the carbene substituents (eq. 11). Electronic influences are particularly noticeable in reactions performed with styrene (Table 4.2).[75,76,78–80] The more polar substituent (COPh > SO$_2$Tol > NO$_2$ > COOEt > HC=CHR > Ph) determines the predominant stereochemistry, (**22**) > (**23**); these influences suggest a role for the polar carbene substituent in directing the approach of the alkene to the carbene

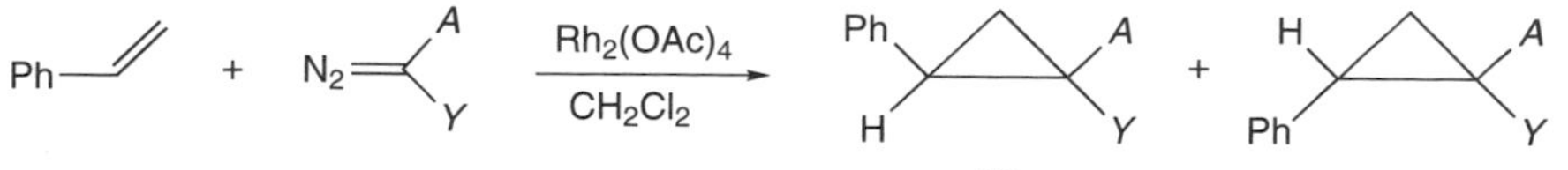

(11)

TABLE 4.2 Dependence of Diastereoselectivity on Carbene Substituents in Rh$_2$(OAc)$_4$-Catalyzed Cyclopropanation Reactions with Styrene

A	Y	Yield (%)	22:23	Ref.
H	COOEt	93	62:38	75
H	CONMe$_2$	74	69:31	78
Ph	COOMe	94	95:5	79
HC=CHCOOEt	COOEt	96	89:11	79
HC=CHPh	COOEt	94	>95:5	79
COOEt	NO$_2$	75	89:11	80
H	NO$_2$	54	71:29	80
CN	NO$_2$	55	75:25	80
COO–t-Bu	NO$_2$	83	67:33	80
COPh	NO$_2$	75	14:86	80
SO$_2$	NO$_2$	73	30:70	80
H	Ph	38	23:77	76

center. As initially proposed by Doyle,[76] the nucleophilic oxygen of the more polar carbene substituent can stabilize the developing electropositive center of the reacting alkene (**24**). This "secondary" effect, not unlike that which controls

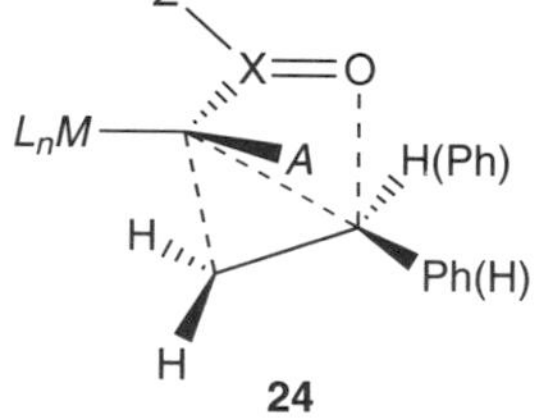

endo selectivity in Diels–Alder reactions, accounts for the observed diastereocontrol in these cyclopropanation reactions. Few reports have described uses of RCOCHN$_2$, N$_2$CHCN, and N$_2$CHSO$_2$Tol in cyclopropanation reactions,[81] and their diastereoselectivities are not known.

The diazoacetate ester substituent, if sufficiently large, can have a profound effect on diastereoselectivity. A variety of diazoacetates have been investigated over the years, but 2,6-di-*tert*-butyl-4-methylphenyl (BHT) is the most effective in enhancing *trans* (*anti*) selectivity in cyclopropanation reactions.[82] The dia-

CYCLOPROPANATION REACTIONS CATALYZED BY RH₂(OAc)₄

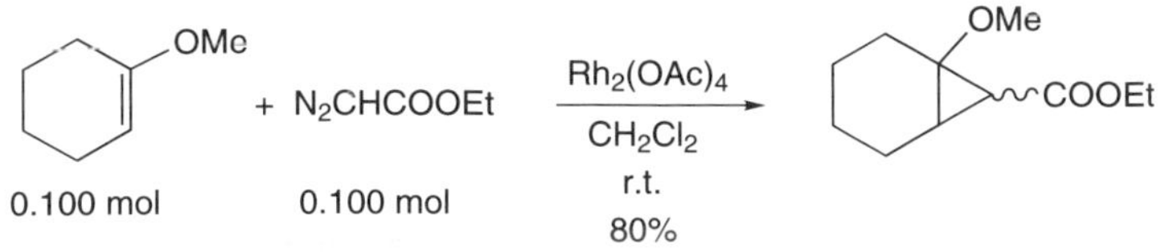

0.100 mol	0.100 mol	80%

A solution of ethyl diazoacetate (10.41 g, 0.100 mL) in 30 mL of anhydrous CH₂Cl₂ was added via syringe pump (2.0 mL/h) to a stirred blue-colored solution of Rh₂(OAc)₄ (22 mg, 0.050 mmol, 0.05 mol %) and 1-methoxycyclohexene (11.00 g, 0.100 mmol). The needle of the addition syringe was placed just below the surface of the reaction solution, and nitrogen evolution commenced with each fractional addition of EDA. After addition of the first 15 mL of EDA solution, the rate of addition was decreased to one-half of the initial rate. [Significantly faster rates of addition, up to 10 mL/h, could be used when Rh₂(OAc)₄ was employed at 0.5 mol %.] When addition was complete, solvent was removed under reduced pressure, and the residue was distilled (bp 68–78°C at 0.5 Torr) to produce 22.88 of clear colorless ethyl 1-methoxybicyclo[4.1.0]heptane-7-carboxylate (0.080 mol, 80% yield) having a 71:29 *anti:syn* isomer ratio (Doyle, M. P.; van Leusen, D.; and Tamblyn, W. H., *Synthesis* **1981**, 787). The only detectable impurities were the carbene dimers diethyl maleate and diethyl fumarate (<5%).

zoacetate, referred to as BDA, is a yellow crystalline solid with a melting point of 153°C, whose synthesis is accomplished in two steps from BHT (eq. 12) in acetonitrile. Use of Rh₂(acam)₄ provided the highest selectivities (Scheme 4), which,

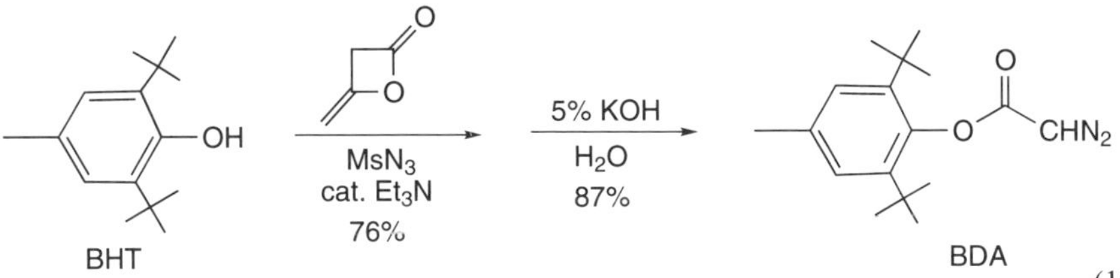

(12)

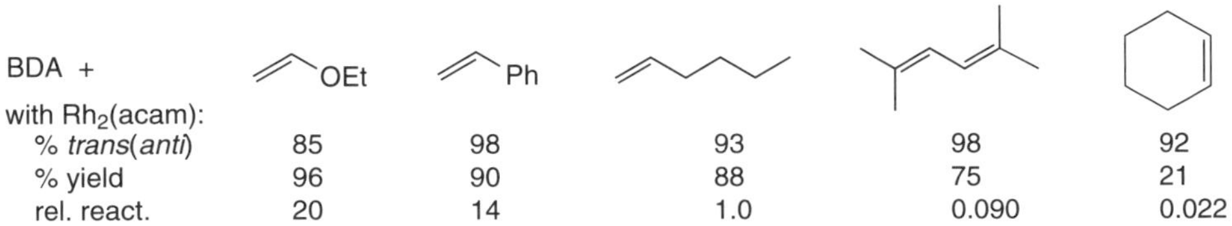

BDA + with Rh₂(acam):	OEt	Ph			
% *trans(anti)*	85	98	93	98	92
% yield	96	90	88	75	21
rel. react.	20	14	1.0	0.090	0.022

Scheme 4

with the exception of ethyl vinyl ether, were >90% *trans(anti)*; $Rh_2(cap)_4$, which is more soluble in organic solvents, can be expected to provide similarly high selectivities with BDA. Relative reactivities for cyclopropanation reflect the steric nature of this diastereocontrol. The disadvantage of the use of BDA is that product cyclopropanecarboxylates are not hydrolyzable, although $LiAlH_4$ reduction is effective at removing BHT.[82] To circumvent this problem Masamune and Lowenthal have used dicyclohexylcarbinyl esters,[83] which are readily hydrolyzed.

Metal ligands can have a significant influence on diastereoselectivity in cyclopropanation reactions. Among carboxylate and carboxamidate ligands for rhodium, the greater the acid strength of the ligand's conjugate acid, the higher is the reactivity of the catalyst and the lower is its selectivity in cyclopropanation reactions.[82] Thus the order of catalyst effectiveness for diastereocontrol among dirhodium(II) catalysts is $Rh_2(cap)_4$, $Rh_2(acam)_4 > Rh_2(OAc)_4$, $Rh_2(oct)_4$, $Rh_2(NHCOCF_3)_4 > Rh_2(tfa)_4$, $Rh_2(pfb)_4$. A similar trend for ligands of copper is expected, and the limited stereochemical data available support this influence. With $[Cu(CH_3CN)_4][BF_4]$, bulky phosphine ligands increase *trans* selectivity for cyclopropanation of styrene using EDA.[84] In general, diastereoselectivities for catalytic cyclopropanation reactions with ethyl diazoacetate catalyzed by copper, palladium, and rhodium are linearly correlatable,[75,85] but synthetically significant differences are not generally found with the use of EDA.

**2,6-DI-*TERT*-BUTYL-4-METHYLPHENYL
DIAZOACETATE (BDA)[82]**

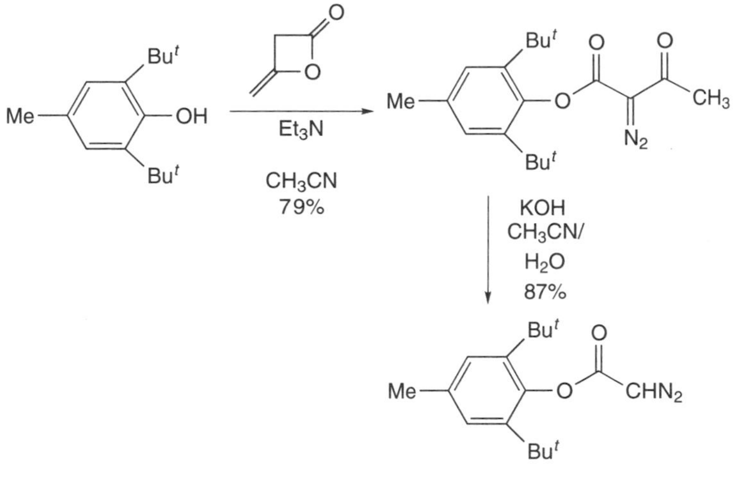

Diketene (7.63 g, 90.8 mmol) in 10.0 mL of anhydrous acetonitrile was added dropwise over 30 min to a rapidly stirred solution of 2,6-di-*tert*-butyl-4-methylphenol (10.0 g, 45.4 mmol), triethylamine (0.441 g, 4.4 mmol), and methanesulfonyl azide (7.14 g, 59.0 mmol) in 25 mL of refluxing acetonitrile contained in a 250-mL three-neck flask fitted with a reflux condenser and an addition funnel. After addition was complete, the resulting brown solution was cooled to room temperature and stirring was continued overnight. The diazoacetate was isolated by addition of water and extraction with ether, washing the ether extract with a 15% aqueous potassium hydroxide solution and then drying the extract over anhydrous magnesium sulfate. Evaporation of the solvent under reduced pressure afforded 11.8 g (35.8 mmol, 79% yield) of 2,6-di-*tert*-butyl-4-methylphenyl diazoacetoacetate, which was recrystallized to a light yellow solid: mp 129–131°C; ^{1}H-NMR (CDCl$_3$) δ 7.13 (*s*, 2 H), 2.51 (*s*, 3 H), 2.33 (*s*, 3 H), and 1.34 (*s*, 18 H).

Acetyl cleavage was performed with the addition of 50 mL of 5% aqueous potassium hydroxide to 4.00 g of 2,6-di-*tert*-butyl-4-methylphenyl diazoacetoacetate (12.1 mmol) in 50 mL of acetonitrile. The solution was stirred for 2 h at room temperature, during which time a yellow precipitate formed in the reaction flask; then 75-mL of ether was added, and the aqueous solution was separated and washed twice with 75-mL portions of ether. The combined ether solution was washed with saturated aqueous sodium chloride, dried over anhydrous magnesium sulfate, and the ether was removed under reduced pressure to reveal the yellow BDA (3.05 g, 10.5 mmol, 87% yield). Recrystallization from ether afforded pure BDA, mp 151–153°C (dec); ^{1}H-NMR (CDCl$_3$) δ 7.11 (*s*, 2 H), 5.01 (*s*, 1 H), 2.31 (*s*, 3 H), and 1.35 (*s*, 18 H); ^{13}C-NMR (CDCl$_3$) δ 142.5, 134.8, 127.0, 35.3, 31.5, and 21.5. IR (KBr): 2116 and 1708 cm^{-1}.

Although *trans* (*anti*) diastereoselectivity could be achieved by predictable steric and electronic influences on the transition state for cyclopropanation, *cis* (*syn*) selectivity should be elusive (although it is the direct result in intramolecular reactions). In spite of this, several efforts have been undertaken to achieve *cis* selectivity, and, although synthetically useful results are rarely obtained, these investigations have been useful for understanding the mechanism of cyclopropanation. Callot was the first to report an abnormal preference for the *cis* (*syn*) isomer in reactions with EDA catalyzed by iodorhodium(III) mesotetraarylporphyrins (**25**: *M* = RhI; Ar = Ph, mesityl).[86] Although only modest *cis* (*syn*) preference was found, subsequent efforts by Kodadek and co-workers using **25** with much larger aryl substituents revealed increases in *cis* (*syn*) diastereoselectivities that with **25d** (*M* = RhI) reached 71% *cis* in reactions of styrene with EDA.[87–89] Even more dramatic results were reported by Hossain for [Cp(CO)$_2$Fe(THF)]$^+$ BF$_4^-$.[90] Table 4.3 describes the variations in diastereoselectivities that are found with representative cyclopropanation catalysts upon reaction of styrene with EDA. Noteworthy is the relatively narrow range of selectivities that is found with such a diversity of catalysts; exceptions are those

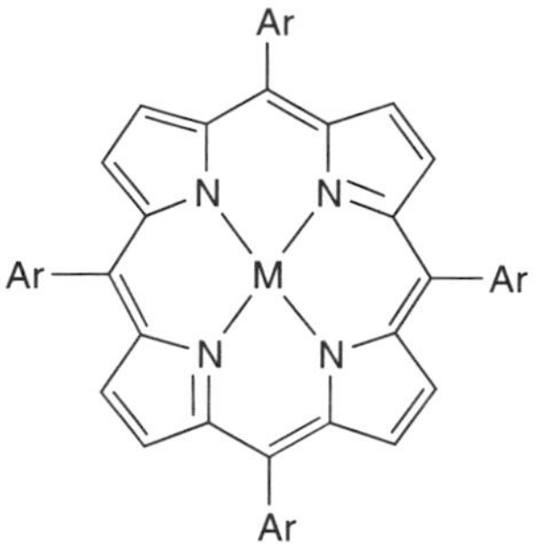

Ar = (a) Ph- (TPP)
(b) p-CH$_3$C$_6$H$_4$- (TTP)
(c) 2,4,6-(CH$_3$)$_3$C$_6$H$_2$- (TMP)
(d) 1,1′-binaphth-2-yl
(e) 1′-pyrenyl-1-naphth-2-yl

25

TABLE 4.3. Catalyst Influence on Diastereoselectivity in the Cyclopropanation of Styrene by Ethyl Diazoacetate (EDA)

Catalyst	Ref.	Yield (%)	*trans*:*cis*
25a (M = RhI)	86		47:53
25d (M = RhI)	87	(2000)[a]	30:70
25e (M = RhI)	88	(1600)[a]	29:71
Rh$_2$(pfb)$_4$	82	88	52:48
Rh$_2$(OAc)$_4$	75	93	62:38
Rh$_2$(acam)$_4$	82	86	68:32
Rh$_2$(MEPY)$_4$	91	59	56:44
Rh$_2$(PHOX)$_4$	92	41	34:66
Rh$_2$(IBAZ)$_4$	93	62	36:64
Ruthenacarborane cluster	94	93	58:42
RuCl$_2$(PPh$_3$)$_3$	95	93	56:44
RuCl$_2$(Pybox-ip)	96	73	91:9
OsCl$_2$(p-cymene)$_2$	98	78	62:38
25a (M = Os)	97	79	91:9
Cu(semicorrin)	99	65	73:27
Cu(bis-oxazoline)	100	77	73:27
Cu(acac)$_2$	75	71	72:28
CuCl · P(OPh)$_3$	75	84	72:28
Cu(OTf)$_2$	75	97	65:35
Co(α-cqd)$_2$ · H$_2$O	101	92	46:54
Co(salen)I	102	76	98:2
"(CO)$_5$W"	103	41	62:38
[Cp(CO)$_2$Fe(THF)]$^+$	90	40	16:84
PdCl$_2$(PhCN)$_2$	75	52	62:38
25b (M = Fe)	90	(1300)[a]	90:10

[a]Number of turnovers; % yield not reported.

few that fall outside the range of 50:50 to 75:25 in the *trans*:*cis* ratio, and these will be discussed in Section 4.6.

4.4 INTERMOLECULAR CYCLOPROPANATION REACTIONS OF DIAZOCARBONYL COMPOUNDS. REGIOSELECTIVITY AND RELATIVE REACTIVITIES

Regiocontrol in catalytic cyclopropanation reactions is dependent on those same electronic and steric factors that were discussed with regard to diastereoselectivity, but in this facet of cyclopropanation selectivity, the nature of the double bond and its substituents can have a significant influence on the preferred site for cyclopropanation. With nonconjugated systems, relative reactivities for cyclopropanation of monoenes indicate the selectivity that can be achieved with dienes or polyenes. There have, however, been few determinations of relative reactivities, and those that have been reported generally describe a narrow range with ethyl diazoacetate (Tables 4.4 and 4.5). Few generalizations can be made: (1) among monosubstituted ethylenes, vinyl ethers are more reactive than styrene, which is more reactive than 1-alkenes; and (2) in the absence of overriding steric effects, 1,1-disubstituted ethylenes are more reactive than monosubstituted ethylenes.

TABLE 4.4 Relative Reactivities for Olefin Cyclopropanation Reactions of EDA Catalyzed by Dirhodium(II) Compounds[76,82]

	Rel. reactivity with	
Alkene	$Rh_2(OAc)_4$	$Rh_2(acam)_4$
n-butyl vinyl ether	8.6	15
styrene	3.5	10
cyclohexene	2.5	1.0
2,5-dimethyl-2,4-hexadiene	2.1	2.0
2-methyl-2-butene	1.5	
vinyl acetate	1.1	
1-hexene	1.0	1.0
3,3-dimethyl-1-butene		
2,5-dimethyl-2,4-hexadiene	0.67	

TABLE 4.5 Relative Reactivities for Olefin Cyclopropanation Reactions of EDA Catalyzed by RhTTPI and RhTMPI[87]

	Rel. reactivity with	
Alkene	RhTTPI	RhTMPI
styrene	3.5^a	2.5^a
1-hexene	1.0	1.0
2-methyl-2-pentene	1.0	0.36
$trans$-2-heptene	0.83	0.28
cis-2-heptene	0.62	0.33
2,3-dimethyl-2-butene	0.19	0.02

^{a}Determined relative to 1-decene.

Electronic Effects: $ROCH{=}CH_2$ > $PhCH{=}CH_2$ > $RCH{=}CH_2$

$RCH{=}CH_2$ < $R_2C{=}CH_2$

Both of these generalizations reflect the importance of electronic influences that are related to carbocation stabilization, and the degree of relative reactivity differences is a function of the amount of charge development in the transition state for cyclopropanation. Steric effects often control differences in reactivity of *cis*- and *trans*-alkenes, as well as of di-, tri-, and tetrasubstituted ethylenes. In general, however, tetrasubstituted ethylenes are less reactive than are their less substituted counterparts.[87]

To date the most significant differences in relative reactivities have resulted from the use of iron *meso*-tetrakis(pentafluorophenyl)porphyrin chloride, Fe(PFP)Cl, that with EDA gave a 74-fold enhancement in styrene cyclopropanation over that with 1-decene.[104] Indene was 26 times less reactive than styrene towards cyclopropanation. The strong electronic effect was further indicated by Hammett plot correlation from reactions of EDA with substituted styrenes ($\rho = -0.7$). However, differences in relative reactivities, although directly expressed in regioselectivities, are not necessarily reflected in diastereoselectivities (Table 4.6), and both regiocontrol and diastereocontrol are required for synthetic viability.

TABLE 4.6 Relative Reactivities and Diastereocontrol

Catalyst	Diazo compound	Rel. reactivity: styrene/1-hexene	Cyclopropane *trans:cis*[a]
Fe(PFP)Cl	EDA	74[b]	90:10
Rh$_2$(acam)$_4$	BDA	20	98:02
Rh$_2$(acam)$_4$	EDA	15	72:28
Rh$_2$(OAc)$_4$	EDA	8.3	62:38
RhTTPI	EDA	3.5	49:51[c]

[a]Cyclopropanation of styrene.
[b]Styrene/1-decene.
[c]Ref. 105.

Regiocontrol has been examined with dirhodium(II) catalysts for cyclopropanation reactions of EDA with a variety of dienes,[24] and, although ligand effects were shown to influence regioselectivity, only the report of Rh$_2$(acam)$_4$-catalyzed reactions of BDA[82] has provided a synthetically relevant example (eq. 13).

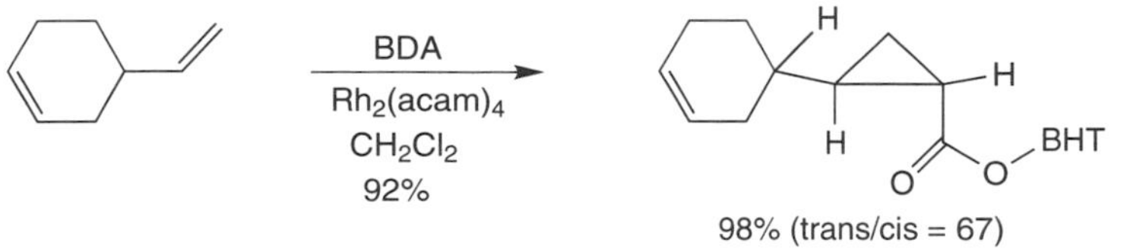

98% (trans/cis = 67)

TABLE 4.7 Regioselectivities for Cyclopropanation of Monosubstituted 1,3-Butadienes by Ethyl Diazoacetate

Rh$_2$(OAc)$_4$	61	39	12	88	69	31	2	98
Rh$_6$(CO)$_{16}$	62	38	11	89	71	28	2	98
CuCl · P(O−i-Pr)$_3$	66	34	9	91	76	34	2	98
PdCl$_2$ · 2PhCN	35	65	23	77	47	53	4	96
Ruthenacarborane cluster[94]	86	14	—	—	—	—	—	—

Regioselectivity in intermolecular cyclopropanation reactions of 1,3-butadienes displays frontier molecular-orbital control, as exemplified by the results described in Table 4.7.[76] A distinct correlation exists between regioselection from reactions catalyzed by Rh$_2$(OAc)$_4$, Rh$_2$(CO)$_{16}$, CuCl · P(O−i-Pr)$_3$, and PdCl$_2$ · 2PhCN, which suggests that the same factors, enhanced or diminished, control catalyst promoted cyclopropanation.[106] 2-Methoxy-1,3-butadiene undergoes exclusive cyclopropanation at the more substituted double bond, and 4-methyl-1,3-pentadiene reacts exclusively at the less substituted double bond.[107,108] Selectivity enhancements with the dienes described in Table 4.7 can be expected with Rh$_2$(acam)$_4$ or Rh$_2$(cap)$_4$ and especially with those, like Fe(PFP)Cl, that increase charge development in the transition state for cyclopropanation. A relatively high selectivity for the cyclopropanation of isoprene has been reported for a ruthenacarborane cluster.[94]

4.5 MECHANISM OF CYCLOPROPANATION. STEREOCHEMISTRY OF CYCLOPROPANE FORMATION

A basic understanding of the mechanism of catalytic cyclopropanation has evolved from proposals of free carbene addition and the involvement of metallo-cyclobutane intermediates[94,95] to that described by Doyle[24,72,76,85] and, more recently, by Kodadek,[104,105] in which the function of the metal is to provide the template upon which carbene transfer occurs. In each of the proposals the orientation of the olefin with respect to the carbene controls the relative stereochemistry of the cyclopropanation reaction. Carbene substituents and the catalyst "face," which constitutes a "wall" of ligands in which the metal is embedded (Scheme 5), control this orientation. In **26** the metal carbene is depicted with a p orbital on the carbene carbon, and the ligated metal is shown to have a uniform surface. Increasing the size of the ester alkyl/aryl R of the carbene carboxylate group, for example, places steric demands on the approaching alkene so that for monosubstituted ethylenes (R^3 = H) the preferred orientation will be that one in which the larger carbon–carbon double bond attachment will be at R^1 rather than

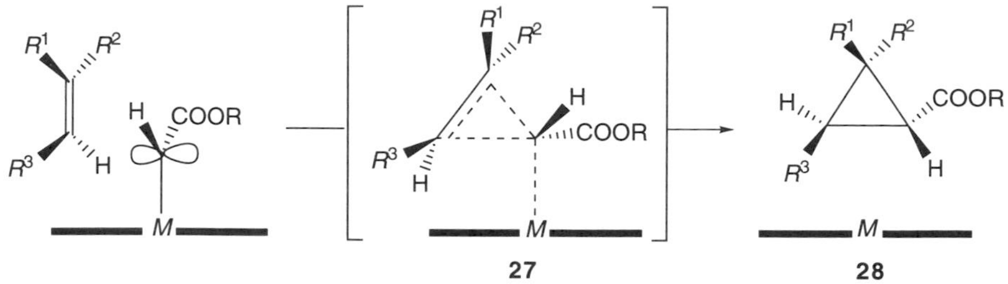

Scheme 5

R^2, and the *trans* (*anti*)-substituted cyclopropane product will prevail. Maximum overlap occurs through the alignment described for **27** in Scheme 5, but even at this stage the olefin is moving away from the catalyst face as its proximity to the carbene carbon increases. In the absence of overriding steric effects, the optimum transition-state orientation can be represented by **29**. Steric interactions be-

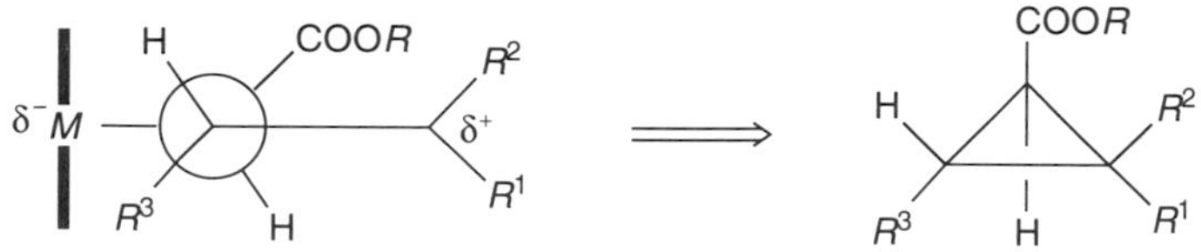

tween R^3 and the catalyst "face" (**31** and **32**) or between COOR and either R^1 or R^2 (Scheme 6) will obviously distort the optimum transition-state orientation. This

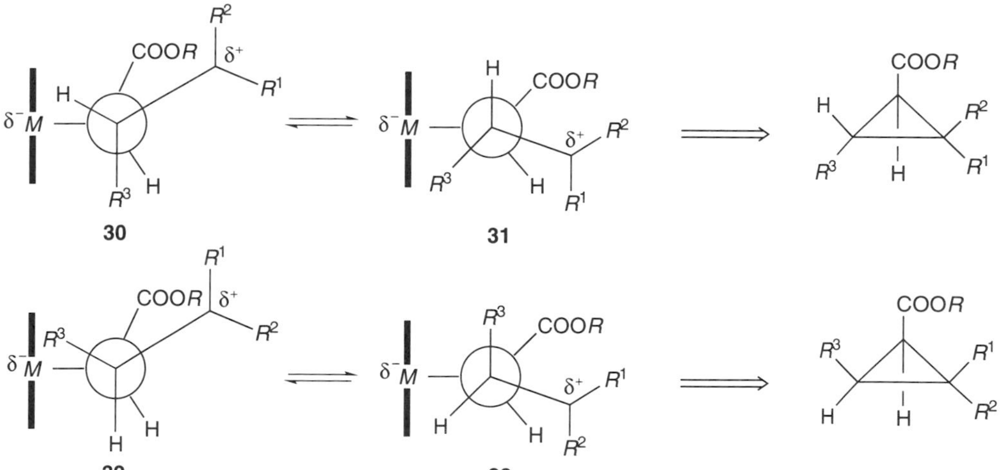

Scheme 6

distortion creates still other steric interactions so that a complete analysis must take into account relative contributions. Noteworthy, however, is the expectation from **30–33** that alkene substituents R^1 and R^2 are sufficiently far removed from the carbene substituent COOR as to exact only limited diastereocontrol, except when R is very large (e.g., from BDA).[82]

This model, which had its origins in two accounts of stereocontrol in cyclopropanation reactions,[24,85] has been used to explain the preference for the *cis*-cyclopropane geometry in reactions of $(CO)_5W$=CHPh with alkenes[109,110] and, by inference, of $Rh_2(OAc)_4$-catalyzed cyclopropanation of alkenes using PhCHN$_2$, whose reactivity–stereoselectivity correlations with those from $(CO)_4WCHPh$ reinforced the notion of metal carbene intermediates in catalytic reactions.[76] Selectivities observed by Dailey for nitrocarbenecarboxylates[80] and by Davies for vinylcarbenecarboxylates (Table 4.2)[79] are consistent with stabilization of the developing electrophilic carbon of the reacting olefin (**34**, where RX=O is

34

NO$_2$ > COOR) in the transition-state, which was originally proposed by Doyle to explain the predominant *trans* (*anti*) stereoselectivities of catalytic reactions with diazoesters. Attachments to catalyst ligands [such as the aryl groups of rhodium(III) porphyrins (**25**)] that extend out into the region where carbene addition is taking place influence olefin orientation,[92] and by favoring (**31/33**) over (**30/32**) removes the electronic stabilization (**34**) that accounts for *trans* (*anti*) preference in cyclopropanation reactions.

Kodadek has proposed a model for rhodium porphyrin catalyzed cyclopropanation of alkenes in which the substrate is envisioned as approaching the metallocarbene in a perpendicular orientation relative to the metal–carbon bond (**35**) axis and then rotating either clockwise or counterclockwise (Scheme 7), eventually reaching the arrangement of atoms in *syn* and *anti* cyclopropane products.[104,105] In this presentation R_L is the larger substituent, and R_S is the smaller substituent. Consistent with the absence of isotope effects or substituent effects on the rate for cyclopropanation of substituted styrenes with EDA catalyzed by RhTTPI,[105] the transition state is early, there is virtually no charge development on the alkene carbons in the transition state, and, presumably, bond formation from the carbene carbon to each of the olefinic carbons has occurred to approximately the same degree. For iron porphyrin systems that effect cyclopropanation from a later transition state,[104] a model is invoked that, like **29**, accounts for stereocontrol on the basis of steric effects without invoking stereoelectronic influences similar to that proposed by Doyle with **34**.

With reactive metal carbenes such as [Cp(CO)$_2$Fe$=$CHR]$^+$ the mechanism for cyclopropanation has been proposed to follow a pathway similar to that origi-

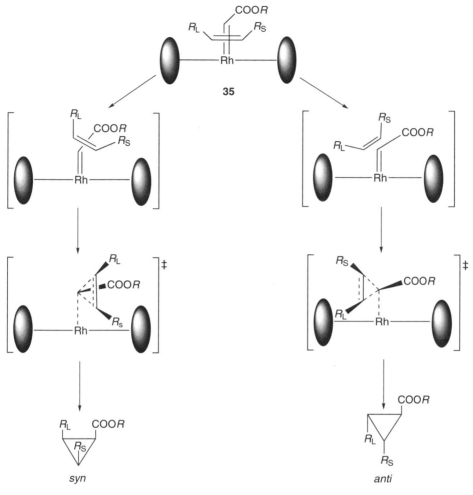

Scheme 7

nally proposed by Doyle.[24,85] Confirmation that cyclopropanation involves backside displacement by the developing electrophile (Scheme 8) was recently provided by Brookhart[111] and Casey,[112] who generated the reaction intermediate **37** from $[Cp(CO)_2FeCHDCHDCH_2LG]^+$, where LG is a leaving group (e.g., Me_2S). Evaluation of the stereochemistry of the cyclopropane product provided the conclusion that cyclopropane formation occurred by backside electrophilic displacement of the ligated metal with net inversion of stereochemistry at C_α. The cause for the high *cis* (*syn*) stereocontrol observed in carbene transfer from $(CO)_5W{=}CHR$ or $[Cp(CO)_2Fe{=}CHR]^+$, where R = Ph or alkyl, has been explained by steric effects on limiting conformations of **36** resulting in cyclopropane formation (Scheme 9).[112] According to this interpretation, which is remarkably similar to that in Scheme 6, the key determinant of stereocontrol is

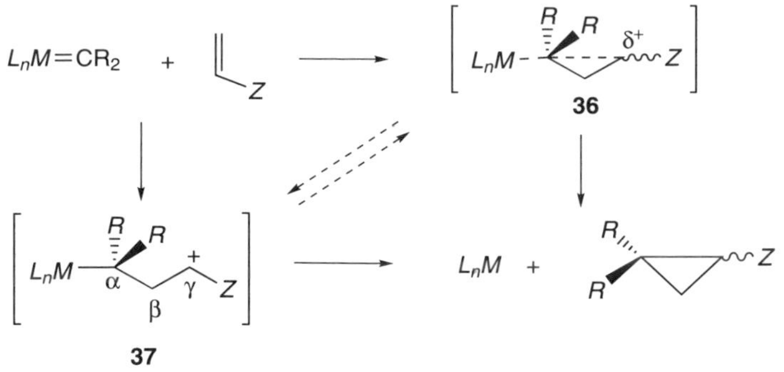

Scheme 8

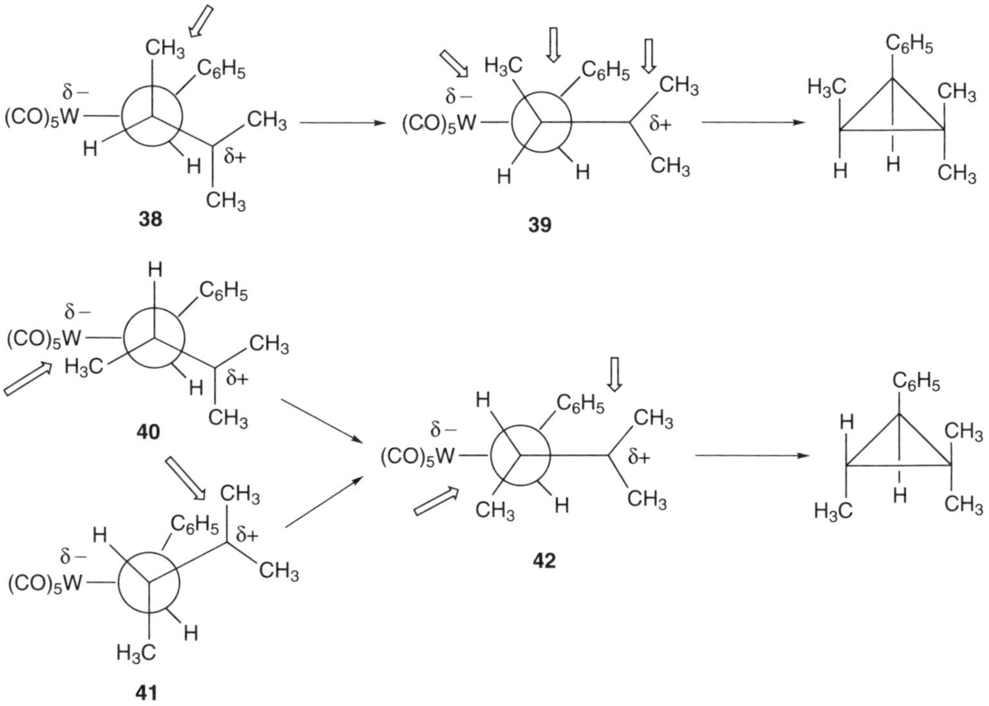

Scheme 9

the steric interactions of the carbene–alkene complex (**38**, **40**, and **41**) rather than those of **39** or **42**, which represent the transition state and maximum orbital overlap. Steric interactions in **38** are much less severe than those in **40** or **41**, accounting for the 94:1 *cis*:*trans* stereochemical preference that has been observed.

This mechanistic treatment also explains the extensive loss of stereochemistry reported by Brookhart for the reactions of $[Cp(CO)_2Fe{=}CHCH_3]^+$ with electron-rich *cis*-HDC$=$CHAr (Ar $= p$-C_6H_4OMe).[113] In this case, positive charge development on C_γ is sufficiently complete that **37** in Scheme 8 is able to undergo isomerization prior to cyclopropane formation. From this analysis, vinyl ethers cannot be expected to be synthetically useful substrates for cyclopropanation with cationic iron carbene complexes.

4.6 INTERMOLECULAR CYCLOPROPANATION REACTIONS OF DIAZOCARBONYL COMPOUNDS. ENANTIOSELECTIVITY

The first report that a homogeneous chiral catalyst could be used for asymmetric induction was published in 1966 by Nozaki and co-workers.[114] Their target was the cyclopropanation of styrene, and although initial results with their copper-ligated chiral salicylaldimines showed low enantiomeric excesses, this and subsequent papers in the series [114–116] promoted extensive efforts in chiral ligand design for copper(II) catalysts that culminated in the development of chiral salicyl-aldimine–copper catalysts by Aratani. Further development was slow, caused in large part by the complexity of the cyclopropanation transformation and by uncertainties about the factors that control selectivity. However, recent advances have raised cyclopropanation technology to a high level of achievement for selected applications, but numerous challenges remain.

4.6.1 Salicylaldimine–Copper Catalysts

Salicylaldimine ligands for copper(II) derived from chiral amino alcohols (**43**) were found to have a remarkable influence on enantioselectivity in cyclopropanation reactions directed to the syntheses of pyrethroids and, especially, to the construction of cilastatin (**44**), which is produced commercially in a joint venture between Merck and Sumitomo (eq. 14). Reported over a 10-year period by

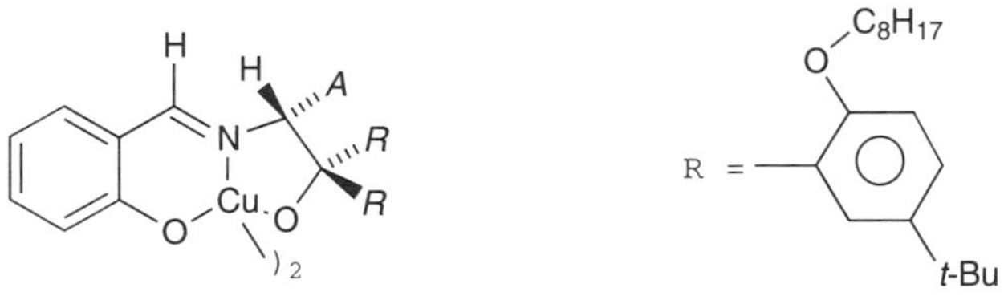

43

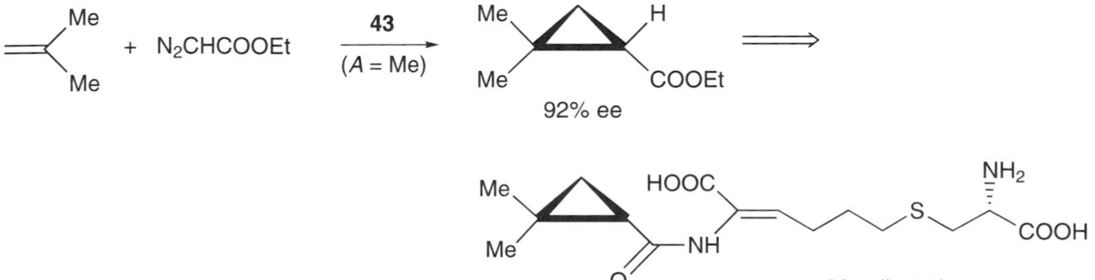

$$\text{(14)}$$

44: *cilastatin*

Aratani and co-workers,[117-120] the applications of these catalysts became the standard against which other chiral cyclopropanation catalysts were measured.

Large R groups on the ligand are required for high enantiocontrol, but the size of the alkyl substituent at the chiral carbon of the amino alcohol fragment does not influence enantioselectivity in an order anticipated by steric control (A = CH_3 > CH_2Ph > i-Pr > i-Bu).[120] An explanation for these influences is suggested for the reaction of styrene with the intermediate metal carbene, depicted as a metal-stabilized carbocation, in Scheme 10.[28] Notice that the bulky R groups

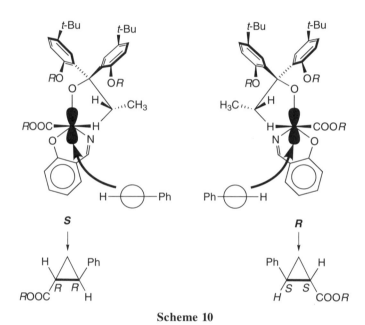

Scheme 10

of **43** prevent attack by styrene from the top side and that the catalyst alkyl substituent (A = CH_3) plays no vital role in orienting either the carbene carboxylate group or styrene for cyclopropane formation. Instead, the configuration of the original amino alcohol, with the hydrogen bonded to the asymmetric carbon pointing out towards the carbene carbon, determines the preferred conformation of the carbene. The (S) configuration of **43** leads preferentially to *trans*-2-phenylcyclopropanecarboxylate having the (1R,2R) configuration (81% de when R = l-menthyl), and the (R) configuration of **43** leads preferentially to *trans*-2-phenylcyclopropanecarboxylate having the (1S,2S) configuration (69% de when R = l-menthyl).[120]

Increasing the steric bulk of the ester alkyl substituent of diazoacetate esters has a strong influence on enantioselectivity and diastereoselectivity for cyclopropanation reactions with 2,5-dimethyl-2,4-hexadiene, leading to chrysanthemic acid derivatives (eq. 15). A striking influence from halide substituents

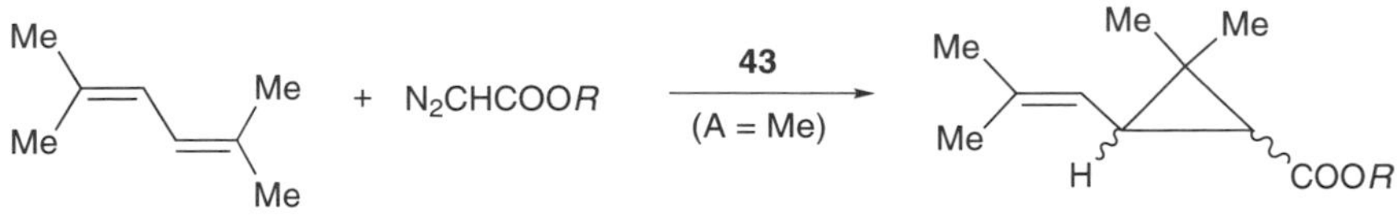

Me
Me
Me
Me
+ N$_2$CHCOOR → (43, A = Me) Me, Me, Me, Me, H, COOR

	trans:cis	% ee (trans)	% ee (cis)
R = Et	51:49	68	62
t-Bu	75:25	75	46
l-adamantyl	84:16	85	46
l-menthyl (de)	93:7	94	46

(15)

remote from the carbon–carbon double bond on stereocontrol in these cyclopropanation reactions has been noted (eq. 16), where the normal *trans* preference

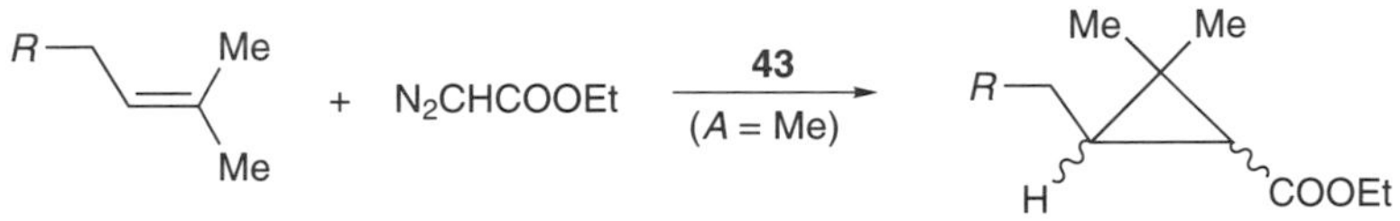

R, Me, Me
+ N$_2$CHCOOEt → (43, A = Me) Me, Me, R, H, COOEt

	trans:cis	% ee (cis)	% ee (trans)
R = CH$_2$Cl	16:84	90	51
CHCl$_2$	12:88	85	31
CCl$_3$	15:85	91	11

(16)

is inverted,[119] and this abnormal *cis* selectivity has been confirmed[121,122] but not adequately explained. Enantioselectivities for cyclopropanation of monosubstituted olefins such as styrene are lower than those for trisubstituted olefins, and, characteristic of these catalysts, the *trans*-substituted cyclopropane isomers generally have higher % ee values than do the corresponding *cis* isomers.

Using modified Aratani catalysts (R = Ph and A = CH$_2$Ph), Reissig and Kunz achieved enantiomeric excesses up to 48% in cyclopropanation reactions of silyl enol ethers with chiral menthyl diazoacetates.[123] Other chiral Schiff bases have been examined for enantioselection using the *in situ* method for catalyst preparation that was pioneered by Brünner, but enantioselectivities were low.[124] The active form of copper in these catalytic reactions is copper(I).

4.6.2 Semicorrin–Copper Catalysts

The next major advance in the development of chiral copper catalysts for highly enantioselective intermolecular cyclopropanation reactions came from reports by Pfaltz and co-workers, beginning in 1986, of the synthesis and applications of chiral semicorrin copper(II) complexes (**45**).[54,99,125–128] The semicorrin ligands are effective in coordinating metal ions, their planar π system with two rigid five-

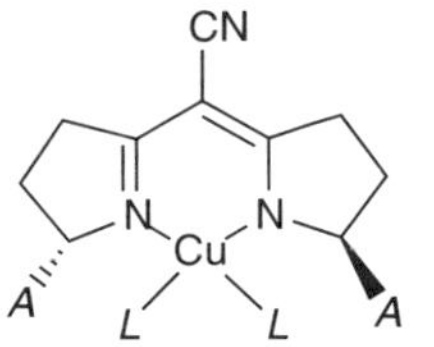

$$A = C(CH_3)_2OH$$
$$CH_2OSiMe_2Bu^t$$
$$COOMe$$

45

membered rings confine the conformational flexibility of the ligand framework, and their C_2 symmetry is designed to influence stereoselectivity.[54] The potential of the Pfaltz semicorrin copper catalysts is suggested by the high enantioselectivities achieved in reactions of styrene with diazoacetates (eq. 17).[99] Increasing the size of A increases enantiocontrol, and $A = CMe_2OH$ was found to be optimal.

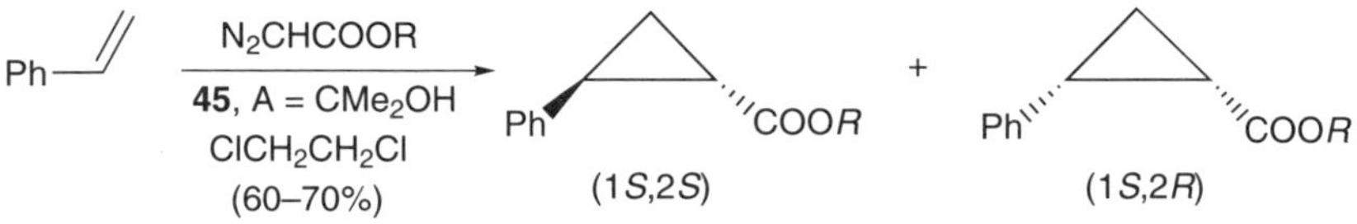

	% ee (trans)	trans:cis	% ee (cis)
$R =$ Et	92	73:27	80
t-Bu	93	81:19	92
d-menthyl	97	82:18	95
l-menthyl	91	85:15	90

(17)

As with the Aratani catalysts, enantioselectivities are higher for the *trans* isomers than for *cis* isomers, although the differences are significantly less with semicorrins (**45**) than with salicylaldimines (**43**). Other monosubstituted olefins, including dienes, and a limited number of disubstituted olefins undergo diazoacetate cyclopropanation with high enantioselectivities.[54,126] The cyclopropanation

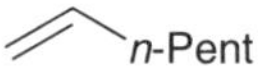

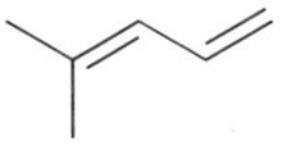

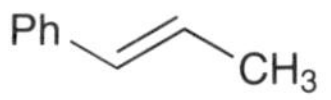

R = d-menthyl	R = d-menthyl	R = Et
trans:cis = 82/18	trans:cis = 63:37	trans:cis (to CH₃) = 80:20
92% ee (trans)	97% ee (trans)	97% ee (trans)
92% ee (cis)	97% ee (cis)	7% ee (cis)

of 1,2-*trans*-disubstituted olefins with diazomethane has been reported to occur with 70–80% ee. Aza-semicorrin copper complexes, especially (**46**), have extended the utility of this class of C_2-symmetric ligands and provided even higher enantioselectivities (up to 99% ee in cyclopropanation of styrene by d-menthyl diazoacetate, 89% yield).[128]

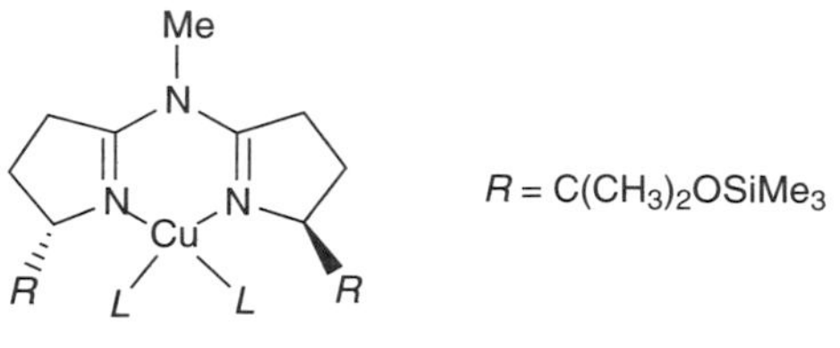

$R = C(CH_3)_2OSiMe_3$

46

The active catalyst is presumed to be a mono(semicorrinato)copper(I) complex that is formed *in situ* (a) by heating **45** or **46** in the presence of the diazo compound or by reduction with phenylhydrazine at room temperature,[99] or, as in the case of **46**, prepared *in situ* from copper(I) triflate and a slight excess of the chiral ligand.[128] These catalysts do not strongly influence *trans:cis* selectivity, although increasing the size of the diazoester alkyl group does increase *trans* selectivity.

4.6.3 bis(Oxazoline)–Copper Catalysts

bis-Oxazolines are C_2-symmetric ligand alternatives to the semicorrins. They have been synthesized in the laboratories of Evans (e.g., **47**),[100,129] Masamune

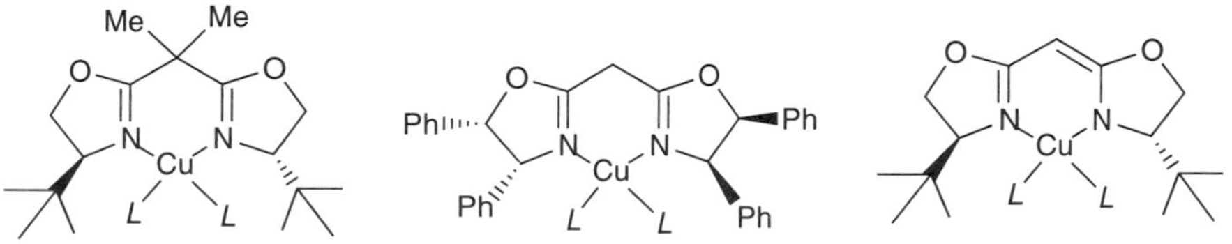

47   **48**   **49**

(e.g., **48**),[83,130] and Pfaltz (e.g., **49**),[131] and in the same cyclopropanation reactions their comparable complexes with copper(I) have similar or greater enantiocontrol. Using **47** the cyclopropanation of styrene (eq. 17) by EDA gave the *trans* isomer in 99% ee and the *cis* isomer in 97% ee, but the *trans:cis* ratio was only 73:27. However, with 2,6-di-*tert*-butyl-4-methylphenyl diazoacetate (BDA)[82] the same enantiocontrol was achieved, but the *trans:cis* ratio was now 94:6.[100] The cyclopropanation of isobutylene with EDA (eq. 14) catalyzed by only 0.1 mol % of **47** produced the corresponding cyclopropane product in greater than 99% ee (91% yield),[100] which further expresses the improvement in enantiocontrol over that achieved with the Aratani catalyst.

Evans developed an *in situ* method for the generation of the bis(oxazoline)–copper(I) catalyst to circumvent the redox chemistry previously employed for the conversion of the copper(II) complex to the catalytically active copper(I) complex. Accordingly, to CuOTf suspended in CHCl$_3$ was added a slight excess of the bis(oxazoline) ligand, and the resulting mixture was subsequently filtered.

The alkene and then, dropwise, the solution of diazoacetate were added to the bis(oxazoline)–copper(I) catalyst solution.

The structure of the ligand is critical to its application. For example, the synthesis of chrysanthemic acid esters by cyclopropanation of 2,5-dimethyl-2,4-hexadiene (eq. 15) gives low % ee values with the use of **47** or **49**. However, high enantioselectivities have been reported when **48** is employed (94% ee).[83] Further advantages of **48** are evident in the results with other di- and trisubstituted olefins (eq. 18). The use of dicyclohexylmethyl diazoacetate makes possible hy-

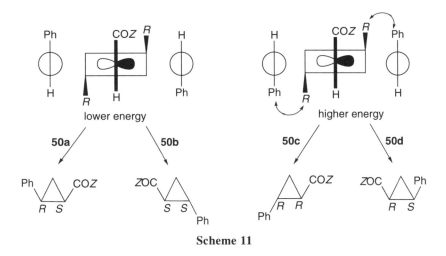

$R = (^cC_6H_{11})_2CH$

trans:cis = 99:1
% ee(trans) = 92

$$(18)$$

drolytic removal of the ester alkyl group, which is not possible with BDA. Factors that influence enantiocontrol for cyclopropanation of *cis*-disubstituted or trisubstituted alkenes appear, in general, to require different catalyst ligand designs than for cyclopropanation of monosubstituted alkenes.

Enantiocontrol with bis(oxazoline)– and semicorrin–copper(I)-catalyzed cyclopropanation reactions can be viewed (Scheme 11) in a manner that is similar

Scheme 11

to that employed to describe enantioselection with the Aratani catalyst (Scheme 10).[132] Accordingly, the metal carbene is viewed as a metal-stabilized carbocation, and the carbene ligands are oriented to minimize steric interaction with substituents *R* of the ligand and to maximize interactions with the approaching styrene. Four limiting conformations can be envisioned in this front view of the reacting system (**50a–d**), of which the two (**50c,d**) that have the olefin substituent Ph on the same side as the ligand's R group are less stable. Because of the C_2 symmetry of the semicorrin ring, the same enantiomers are formed from these models when the COZ substituent is positioned up (Scheme 11) or down. These models show that the observed high enantioselectivities arise from interactions of the ligand's R substituent with olefin substituents (e.g., Ph), and they suggest that monosubstituted and 1,1-disubstituted alkenes should be the preferred substrates. Increasing the size of carbene substituent COZ probably changes the angle of approach by the alkene so that the influence of the ligand substituent R is magnified. Furthermore, the depictions in Scheme 11 suggest that the major determinant of diastereocontrol should be the carbene COZ substituent, and this is experimentally observed.

4.6.4 Pyridine-Ligated Copper Catalysts

The success of semicorrin- and bis(oxazoline)-ligated copper catalysts have generated other designs for chiral C_2-symmetric copper ligands. Katsuki and co-workers have synthesized optically active bipyridines (**51**)[133] and, by *in situ*

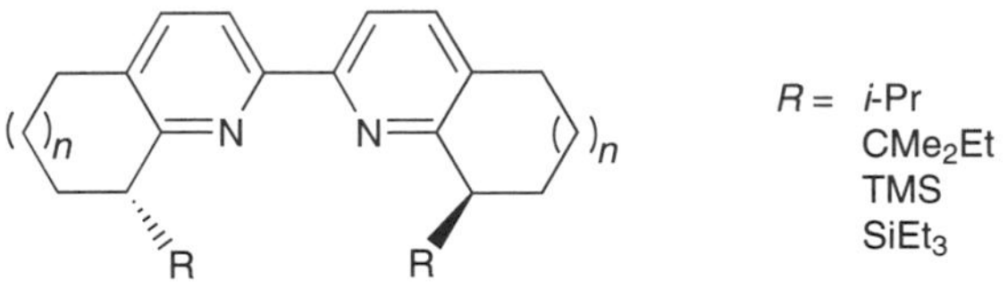

R = *i*-Pr
CMe_2Et
TMS
SiEt_3

51 (*n* = 0,1)

catalyst preparation with CuOTf and using *tert*-butyl diazoacetate, have found exceptional enantiocontrol for cyclopropanation of *E*-olefins (eq. 19).[134,135] With

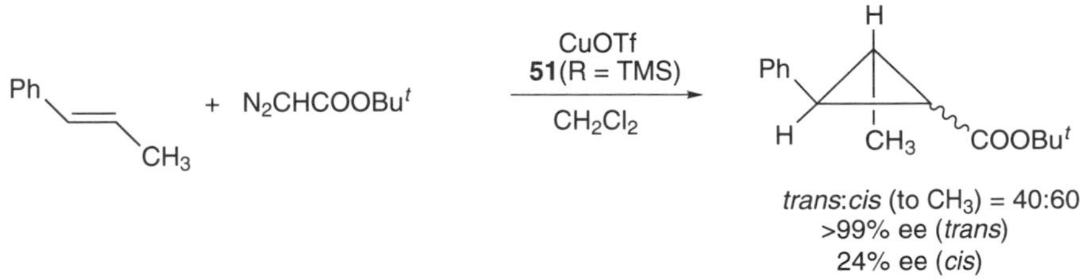

Ph + N_2CHCOOBu*t* → Ph, H, CH_3, COOBu*t*

CuOTf
51(R = TMS)
CH_2Cl_2

trans:cis (to CH_3) = 40:60
>99% ee (*trans*)
24% ee (*cis*)

(19)

monosubstituted olefins % ee values, even with the use of *tert*-butyl diazoacetate, are somewhat lower than are found with the semicorrin- or bis(oxazoline)-ligated copper catalysts.

Singh reported[136] that the chiral 2,6-bis[4(S)-isopropyloxazolin-2-yl)pyridine (Pybox-*ip*) ligand (**52**), originally developed by Brunner and Nishiyama,[137] was

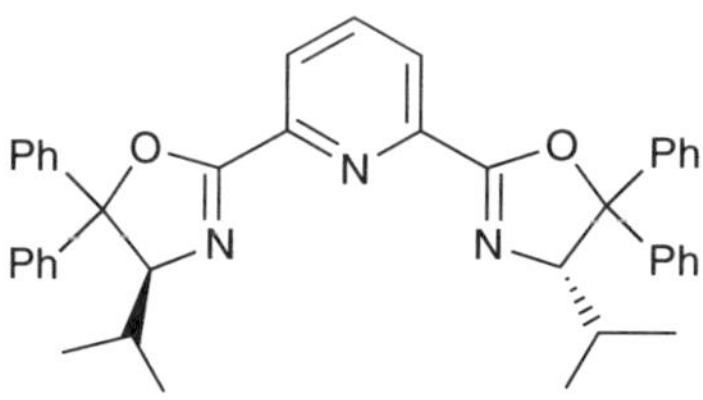

52

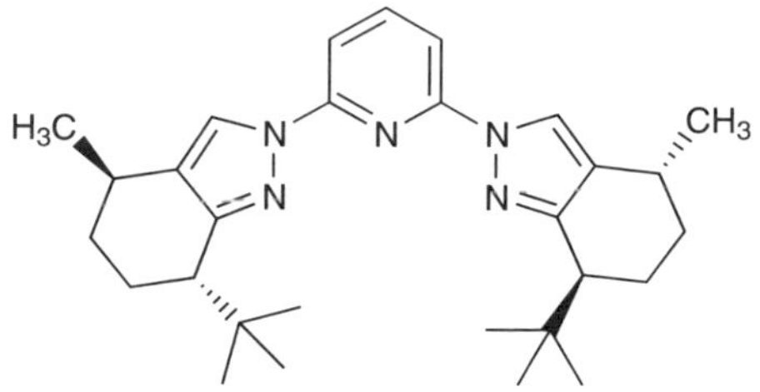

53

relatively ineffective for enantioselective cyclopropanation reactions using copper(II) as the coordinating metal. The extension of this design to chiral bis-(pyrazolyl)pyridine ligands (53) with Cu(OTf), developed by Tolman and co-workers,[138] appears to be more promising with enantiomeric excesses for cyclopropanation of styrene using EDA approaching 70%. With either ligand, however, diastereocontrol is low (2:1 *trans*:*cis* with styrene), and the synthetic advantages of this approach are not yet evident.

4.6.5   Other Chiral Copper Catalysts

The report by Kanemasa and co-workers[139] that C_2-symmetric 1,2-diamine/copper(II) trifluoromethanesulfonate complexes (from 54) are effective chiral catalysts for asymmetric cyclopropanation of styrene and di- or trisubstituted olefins offers considerable promise for future developments. With a molar ratio of 54/Cu(OTf)$_2$ of 2–3 (catalyst activation with phenylhydrazine) cyclopropanation occurred with high enantiocontrol and *trans*:*cis* cyclopropane ratios (eq. 20)

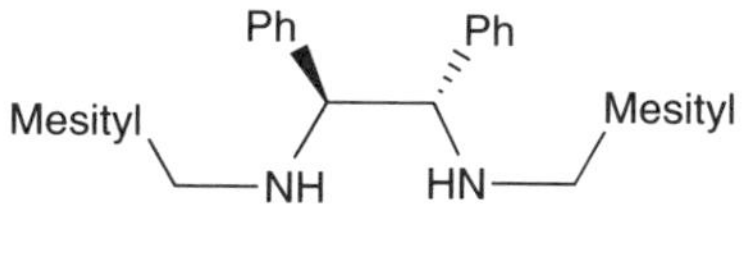

54

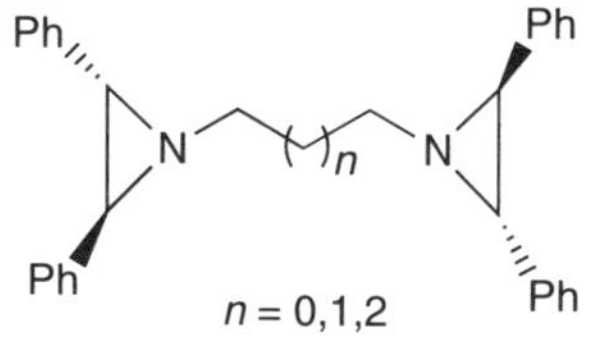

55

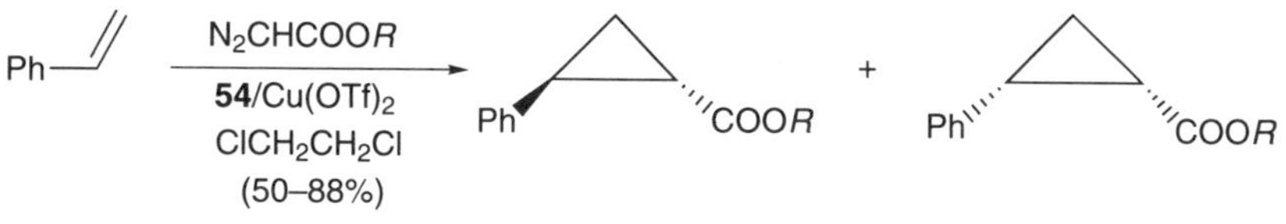

$$Ph—\!\!=\!\!= \xrightarrow[\substack{\textbf{54}/\text{Cu(OTf)}_2 \\ \text{ClCH}_2\text{CH}_2\text{Cl} \\ (50\text{–}88\%)}]{\text{N}_2\text{CHCOO}R} \quad \underset{\text{Ph} \quad\quad \text{COO}R}{\triangle} \quad + \quad \underset{\text{Ph} \quad\quad \text{COO}R}{\triangle} \quad\quad (20)$$

	% ee (trans)	trans:cis	% ee (cis)
R = Et	86	74:26	58
l-menthyl	94	91:9	–
d-menthyl	96	93:7	66

comparable or superior to those from salicylaldimine–copper catalysts.[120] Tanner and co-workers found the C_2-symmetric bis(aziridines) 55 to be moderately effective for the same reaction.[140]

Brunner has reported that combination of an optically active tetrakispyra-zolylborate (**56**) with Cu(OTf) provided modest enantiocontrol (up to 62% ee)

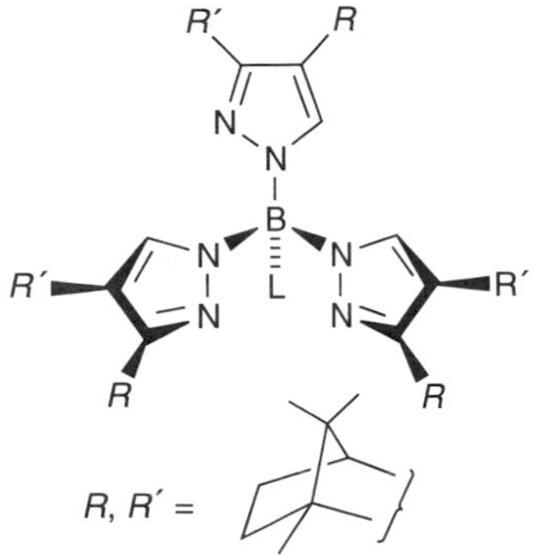

56

and diastereoselectivity in cyclopropanation of styrene with EDA.[141] With this catalyst, however, the predominant stereoisomer is *cis* (*cis*:*trans* = 76:24), which, as we have seen, is highly unusual for cyclopropanation catalysts. An analogous achiral copper(I) hydrotris(3,5-dimethyl-1-pyrazolyl)borate complex, also effective as a cyclopropanation catalyst, exhibits high stereocontrol only for cyclopropanation of *cis*-cycloctene (1:99 *syn*:*anti* with EDA).[142]

4.6.6 Chiral Dirhodium(II) Carboxylates

The prolinate catalyst **57**, developed by McKervey,[143] exhibits limited enantio-control in intermolecular cyclopropanation reactions of alkenes with diazoac-etates. However, Davies has reported that the *p-tert*-butylphenylsulfonamide derivative **57a** is especially effective for highly enantioselective cyclopropanation reactions with vinyl diazomethane **58** (eq. 21).[144,145] Enantiomeric excesses of

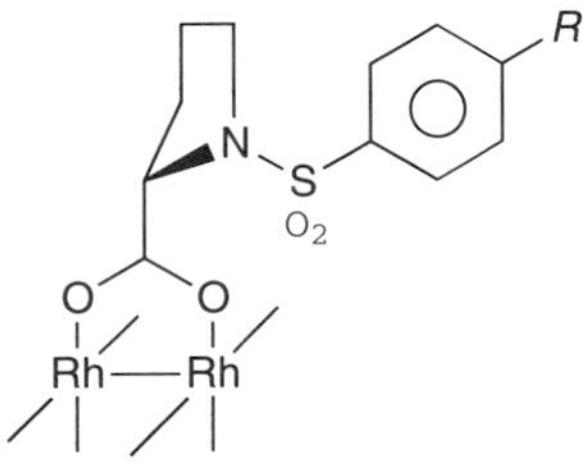

57

a: $R = {}^t\mathrm{Bu}$
b: $R = {}^n\mathrm{C}_{12}\mathrm{H}_{25}$

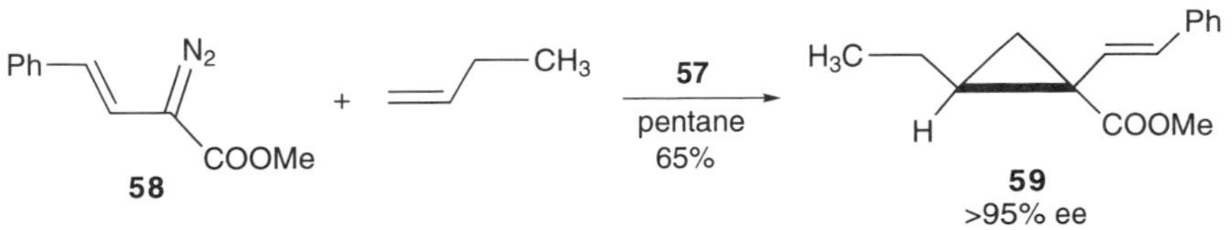

58

59
>95% ee

(21)

59% (EtOCH=CH$_2$) and higher, but generally at or greater than 90%, were observed for reactions with monosubstituted alkenes performed at room temperature, and only the depicted diastereoisomer (**59**) was observed. Higher enantiomeric excesses were reported for reactions in pentane than for those in methylene chloride or benzene. Reactions performed with monosubstituted alkenes in pentane at −78°C using the more soluble catalyst **57b** uniformly give enantiomeric excesses >90%, even the highly reactive ethyl vinyl ether (93% ee).[145] Corey has employed this methodology for the synthesis of the antidepressant sertraline (Scheme 12).[146] In contrast with ethyl diazoacetate using the same

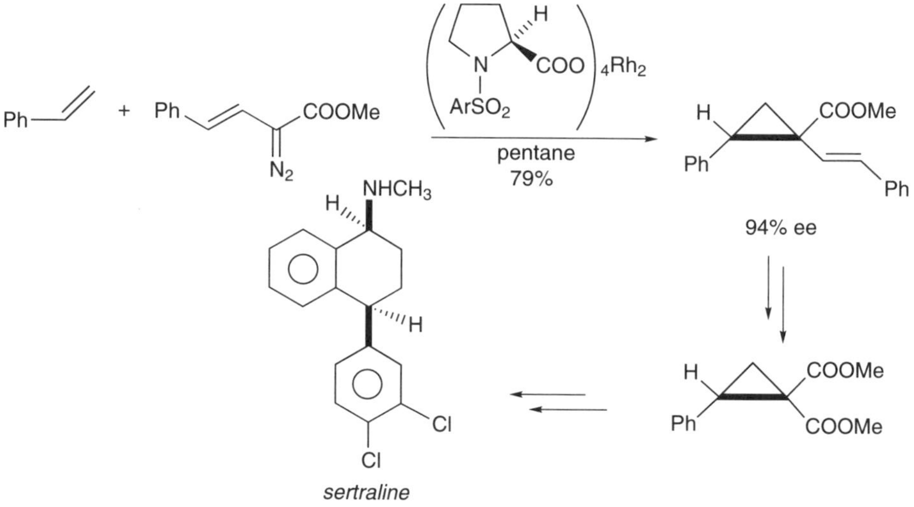

sertraline

Scheme 12

catalyst, low enantiomeric excesses were observed for the products from reactions with styrene (6% ee, *trans*; 30% ee, *cis*).[144]

Comparison of chiral copper(I) and dirhodium(II) catalysts for the cyclopropanation of styrene by methyl phenyldiazoacetate showed that **57a** in pentane provided the highest enantiomeric excess (85%),[147] even compared to **47** and to chiral dirhodium(II) carboxamidates. 1,1-Diphenylethene produced the corresponding cyclopropane in 97% ee using this catalytic methodology (eq. 22). With a selection of monosubstituted alkenes, methyl phenyldiazoacetate formed cyclopropane products with high diastereocontrol and enantiomeric excesses ranging from 64 to 87%[148]; vinyl or phenyl substituents on alkyl diazoacetates are essential to high enantiocontrol, but increasing the size of the ester alkyl group decreases % ee.[145] The influence of the solvent on enantiocontrol is reported to be the result of alignment of prolinate ligands on dirhodium(II) to increase enantiocontrol; the more rigid structural arrangement of chiral dirhodium(II) carboxamidates prevents a similar solvent effect.[147]

4.6.7 Chiral Dirhodium(II) Carboxamidates

Enantiocontrol in intermolecular cyclopropanation reactions of diazoacetate esters is generally lower for carboxamidate-ligated dirhodium(II) catalysts derived from chiral 2-oxopyrrolidines (**60**) and 2-oxooxazolidines (**61**), 1-acyl-2-oxoimidazolidines (**62**), and 2-oxoazetidines (**63**) than for reactions using the optimal chiral salicylaldimine (**43**), semicorrin (**45**), or bis(oxazoline) (**47-49**) ligated copper catalysts for the same alkene/diazoester combination.[28,91–93,149,150] However, unlike these copper catalysts, for which the *trans* cyclopropane isomer from intermolecular reactions with menthyl diazoacetates generally has a higher enan-

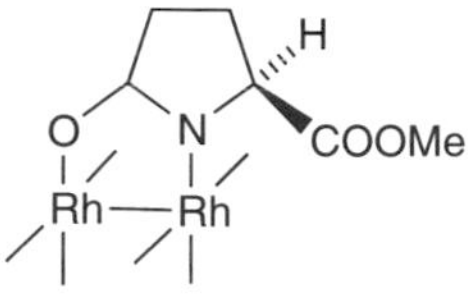

Rh₂(5*S*-MEPY)₄
60

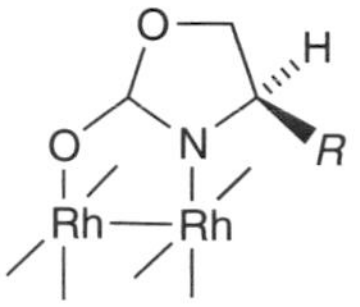

61

R = COOMe: Rh₂(4*S*-MEOX)₄
R = CH₂Ph: Rh₂(4*R*-BNOX)₄
R = CHMe₂: Rh₂(4*R*-IPOX)₄
R = Ph: Rh₂(4*S*-PHOX)₄

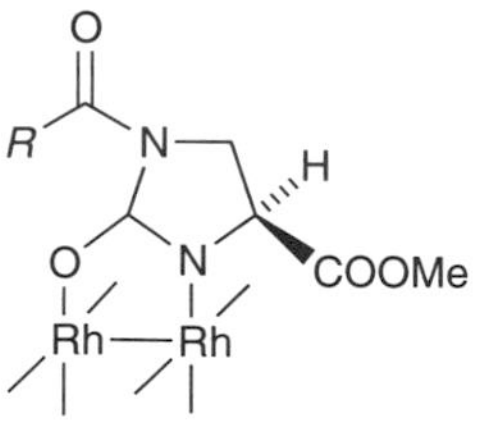

62

R = CH₃: Rh₂(4*S*-MACIM)₄
R = PhCH₂CH₂: Rh₂(4*S*-MPPIM)₄

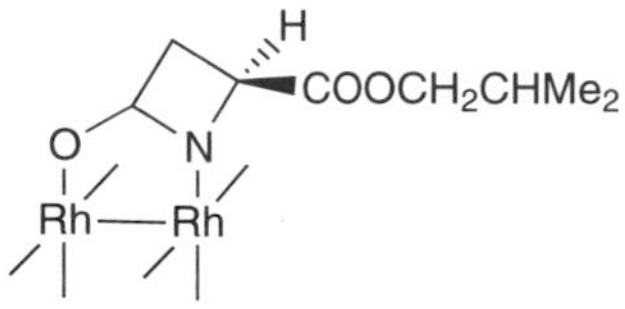

Rh₂(4*S*-IBAZ)₄
63

SYNTHESIS OF METHYL 1,2,2-TRIPHENYLCYCLOPROPAN-ECARBOXYLATE[147]

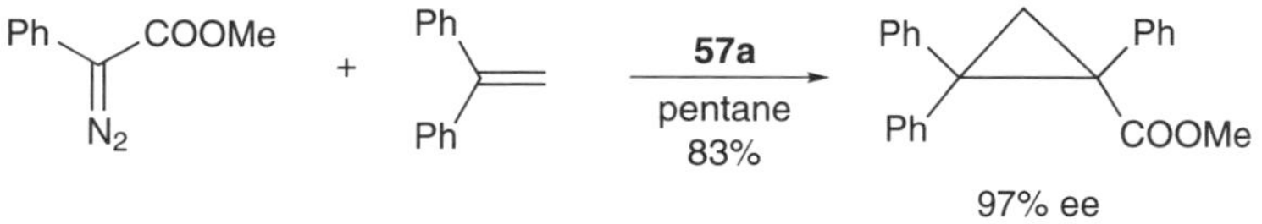

(22)

97% ee

To the *tert*-butylbenzenesulfonylprolinate catalyst **57a** (7.3 mg, 5.0 μmol) and 1,1-diphenylethene (901 mg, 5.0 mmol) in 15 mL of refluxing pentane was added methyl phenyldiazoacetate (88 mg, 0.50 mmol) in 10 mL of pentane over 6 h. Chromatographic isolation on silica gel (10:1 hexanes:EtOAc) allowed recovery of unreacted alkene and isolation of the title compound (136 mg, 83% yield): mp 94.5°C; $[\alpha]^{25}_D$ +264 (c 1.25, CHCl₃). An enantiomeric excess of 97 ± 3% was determined with the use of Eu(hfc)₃.

tiomeric excess than the *cis* cyclopropane isomer, with chiral dirhodium(II) carboxamidates this preference is reversed.[132] Carboxylate substituents on the ligated oxazolidinone rings provide far greater enantiocontrol in these reactions than do benzyl or isopropyl substituents.[92] However, the PHOX ligand (**61**) affords enantiocontrol that nearly matches that from carboxylate-substituted ligands and, in addition, provides a surprisingly high preference for the *cis* stereoisomer in intermolecular cyclopropanation reactions of styrene (eq. 23).[92,150] Noteworthy

$R =$ *d*-menthyl	$Rh_2(5S\text{-MEPY})_4$	57 (31% de)	43 (88% de)
Et	$Rh_2(5S\text{-MEPY})_4$	56 (58% ee)	44 (33% ee)
l-menthyl	$Rh_2(4S\text{-PHOX})_4$	27 (40% de)	73 (72% de)
Et	$Rh_2(4S\text{-PHOX})_4$	34 (24% ee)	66 (57% ee)
d-menthyl	$Rh_2(4R\text{-BNOX})_4$	67 (34% de)	33 (62% de)
Et	$Rh_2(4R\text{-BNOX})_4$	46 (8% ee)	54 (13% ee)
$(^cC_6H_{11})_2CH$	$Rh_2(4S\text{-IBAZ})_4$	34 (77% ee)	66 (95% ee)
Et	$Rh_2(4S\text{-IBAZ})_4$	36 (47% ee)	64 (73% ee)
Et	$Rh_2(4S\text{-MACIM})_4$	43 (30% ee)	57 (37% ee)

(23)

are results from the use of $Rh_2(4S\text{-IBAZ})_4$ for which the *cis* cyclopropane isomer is dominant, and, using dicyclohexyl diazoacetate, the enantiomeric excess of the *cis* isomer is 95%.[93] Similar results are obtained for cyclopropanation of 4-methyl-1,3-pentadiene (eq. 24).

trans:cis = 46:54

66% ee (*trans*)
91% ee (*cis*)

(24)

Molecular modeling has been employed to evaluate the minimum-energy conformations for the carbene bound to rhodium in these catalysts.[151] There are two limiting conformations, depicted in Scheme 13 (A = attachment), for the rhodium-bound carbene from reactions with diazoacetates: one (**64**) whose attachment is in the (O,O) quadrant and another (**65**) whose attachment is in the (O,N) quadrant. For $Rh_2(5R\text{-MEPY})_4$ metal carbene conformation **64** (A = COOMe) is favored over **65** by about 3 kcal/mol, but for $Rh_2(4S\text{-BNOX})_4$ (A = CH_2Ph) the difference in energy is negligible. However, modeling a transition state that includes styrene as the olefinic substrate showed that attack on the

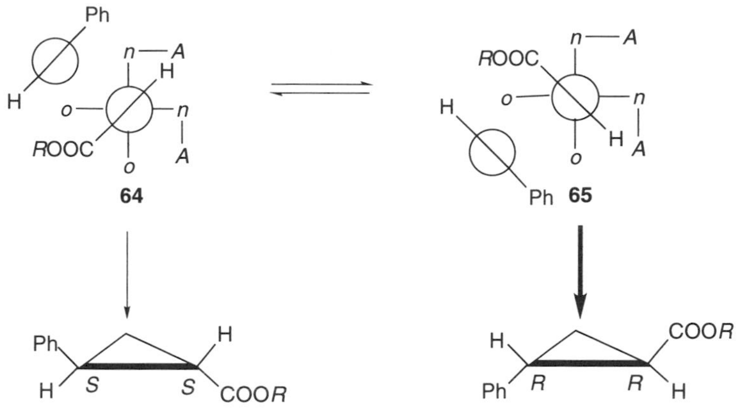

Scheme 13

less conformationally stable **65** was preferred over **64**, and this is consistent with experimental observations. Note that, in contrast to chiral semicorrin/bis-oxazoline ligated copper(I) catalysts (Scheme 11), enantiocontrol with chiral dirhodium(II) carboxamidates is not derived from direct interaction of the ligand's chiral attachment with the alkene substituent. Rather, enantiocontrol is due to the influence of A on metal carbene conformational populations in the transition state for addition.

Analogous in reaction design to results reported by Doyle, McKervey, and Davies with **57a**, methyl phenyldiazoacetate undergoes cyclopropanation of styrene (eq. 25) to give **66** as the dominant cyclopropanation product (>95:5 d.r.)

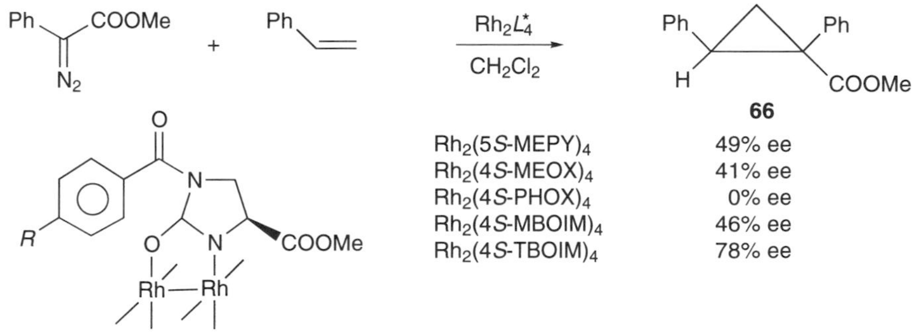

Rh$_2$(5S-MEPY)$_4$	49% ee
Rh$_2$(4S-MEOX)$_4$	41% ee
Rh$_2$(4S-PHOX)$_4$	0% ee
Rh$_2$(4S-MBOIM)$_4$	46% ee
Rh$_2$(4S-TBOIM)$_4$	78% ee

67
Rh$_2$(4S-MBOIM)$_4$, R = H
Rh$_2$(4S-TBOIM)$_4$, R = t-Bu

(25)

but with only moderate ee's from reactions catalyzed by Rh$_2$(5S-MEPY)$_4$, Rh$_2$(4S-MEOX)$_4$, and Rh$_2$(4S-MBOIM)$_4$ (**67**).[147] Increasing the steric bulk of **67** with R = t-Bu, Rh$_2$(4S-TBOIM)$_4$, increased enantiocontrol, but Rh$_2$(4S-PHOX)$_4$, in which phenyl has replaced the carboxylate group, produced **66** with 0% ee. An explanation for this divergence in selectivity is given in Scheme 14. Carboxylate

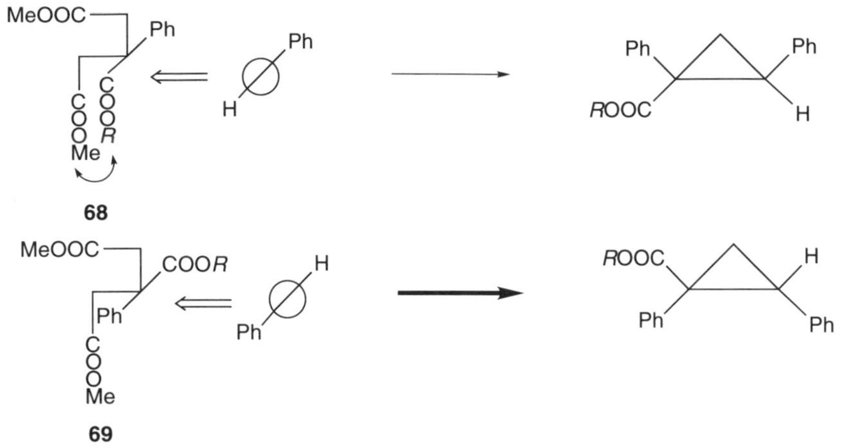

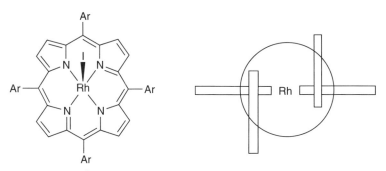

Scheme 14

repulsions render **68** of sufficiently higher energy than **69** so that enantiocontrol is primarily determined by **69**. Without this dipolar repulsion, as in the case of Rh$_2$(4S-PHOX)$_4$-catalyzed reactions, both reacting conformers are of nearly equal energy, and little or no differentiation is possible.

4.6.8 Other Chiral Rhodium Catalysts

Kodadek's "Chiral Wall" (**70a**), and "Chiral Fortress" (**70b**) porphyrins[87,88,152] have provided unique designs (Scheme 15) for enantiocontrol in catalytic cyclo-

70a: Ar = 1,1′-binaphth-2-yl
70b: Ar = 1′-pyrenyl-1-naphth-2-yl

Scheme 15

propanation reactions, but enantiocontrol in reactions with ethyl diazoacetate was moderate to low (<60% ee). However, turnover numbers (>1800) and cis:trans (syn:anti) diastereoselectivities (>70:30 syn:anti) are relatively high. Although informative, results obtained with this system do not suggest that further improvements in stereocontrol can be achieved. The $Rh_2(4S\text{-}IBAZ)_4$ catalyst (**63**) provides comparable diastereocontrol and significantly improved enantioselectivity.

4.6.9 Pybox–Ruthenium Catalysts

Originally developed as a chiral ligand of rhodium for enantioselective hydrosilylation of ketones,[137] the C_2-symmetric 2,6-bis(2-oxazolin-2-yl)pyridine (Pybox) complex of ruthenium has been successfully applied to asymmetric cyclopropanation by Itoh, Nishiyama, and co-workers.[96] *In situ* combination of **71** with $RuCl_2(p\text{-}cymene)$ produces the catalytically active species, which, under an atmosphere of ethylene, forms the stable complex **72** (eq. 26), whose X-ray struc-

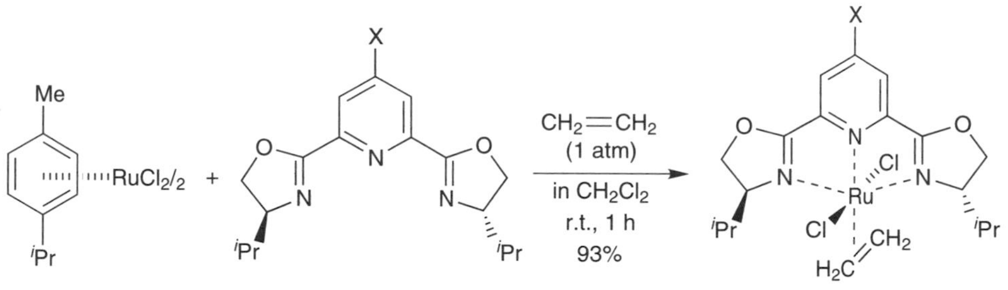

(26)

ture has been reported.[153] The advantage of this catalyst appears to lie in the high diastereoselection for the *trans*-cyclopropane isomer that accompanies high enantiocontrol (eq. 27). Even the nonchiral analog of **72**, however, provides high

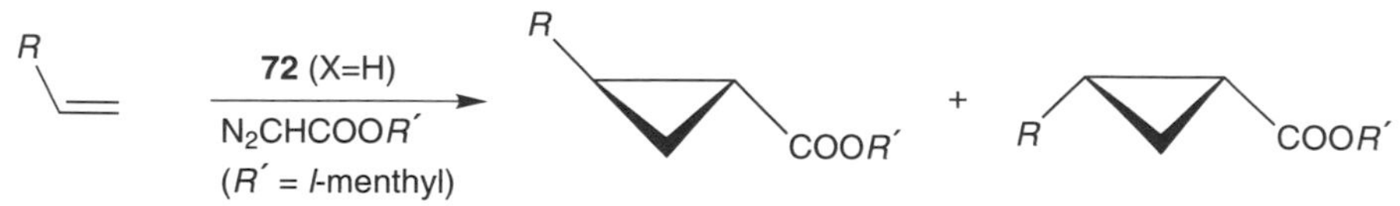

R =			
Ph	87%	95 (95% ee)	5 (76% ee)
CH_2Ph	45%	93 (97% ee)	7
n-Pent	54%	92 (98% ee)	8 (94% ee)
$Me_2C=CH$	86%	79 (98% ee)	21 (79% ee)

(27)

diastereocontrol, suggesting that the core catalyst has intrinsic potential for high *trans*-diastereoselectivity. According to Nishiyama, both *cis* and *trans* cyclopropane products result from alkene attack at the *re* face of the intermediate metal carbene (Scheme 16).[153] The actual involvement of a metal carbene in these

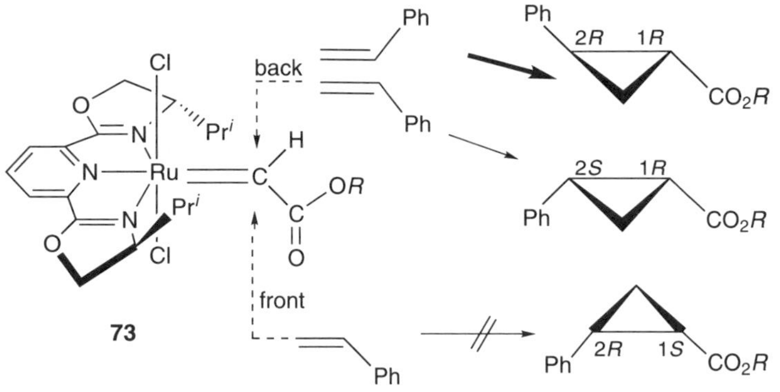

Scheme 16

reactions is certified by the isolation of a stable trimethylsilylcarbene complex of RuCl$_2$(Pybox-*ip*) which could undergo silylcarbene transfer to styrene,[154] and by stable complexes formed from BDA (eq. 12) or similar aryl diazoacetates with bulky 2,6-substituents.[155] The transfer of the carbene to styrene took place in high yield with complete diastereocontrol for the *trans* isomer and, using R = 2,6-diisopropylphenyl in **73**, with 97% ee.

Electronic influences from 4-substituents of pyridine in **72** affect relative reactivities and enantioselectivities but not diastereoselectivities.[156] Electron-withdrawing groups increase catalytic activity and % ee values from reactions of diazoesters with styrene. Correlation of the log of enantiomer ratios with Hammett σ_p values shows ρ values of 0.36 to 0.52, and $\rho = +1.53$ for correlation with relative reaction rates. The influence of remote substituents here is opposite to that expected, and this system is deserving of further investigations.

4.6.10 Cobalt Catalysts

Even before Aratani communicated his first report of a highly enantioselective copper catalyst for cyclopropanation reactions, Nakamura, Otsuka, and co-workers reported their discovery that bis[(−)-camphorquinone-α-dioximato] cobalt(II), Co(α-cqd)$_2$·H$_2$O, was a highly effective catalyst for enantioselective cyclopropanation,[157] and subsequent publications elaborated their results.[101,158] Optical yields as high as 88% ee were achieved for the cyclopropanation of styrene by neopentyl diazoacetate,[101] and this catalyst was amenable to cyclopropane formation from dienes and even α,β-unsaturated nitriles or esters (eq. 28). Reactions were generally performed neat in the alkene, but solvents were found to have a significant influence on stereoselectivity, increasing the *trans*:*cis* ratio and reaction rates with increasing solvent polarity. The addition of a donor base

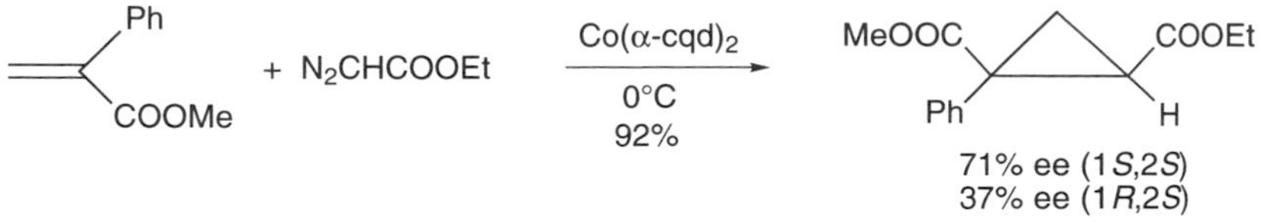

71% ee (1S,2S)
37% ee (1R,2S)

(28)

to Co(α-cqd)$_2 \cdot$ H$_2$O to replace water decreased the enantioselectivity of the catalyst. However, since these publications there have been no further reports on applications of these catalysts.

The rate of diazo decomposition is directly proportional to the concentration of the alkene, but at low alkene concentrations the rate of reaction falls towards zero.[158] In the absence of olefins Co(α-cqd)$_2$ decomposes diazoacetate very slowly at 20°C. The use of deuterated styrene showed isomerization and demonstrated that bond rotation occurs during the course of cyclopropane formation. The combined data were interpreted in terms of metal carbene formation and addition to alkenes. However, Doyle has reinterpreted these results with the suggestion that the mechanism of reaction involves addition of the diazo compound to a metal–olefin complex to form a diazonium ion intermediate, which then collapses to the cyclopropane.[28,132]

The recent report of chiral cobalt(III)–salen complexes that are effective for asymmetric cyclopropanation provides new opportunities for catalyst development. The simplest derivative, **74**, catalyzes the cyclopropanation of styrene by

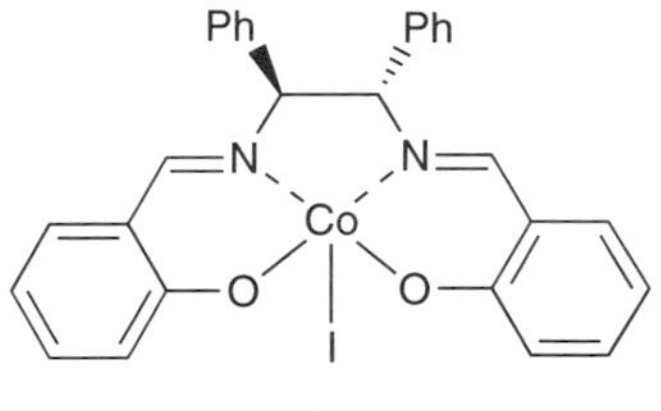

74

tert-butyl diazoacetate with exceptional *trans* diastereoselectivity (*trans*:*cis* = 96:4) and moderate enantioselectivity (74% ee).[159] The mechanism of cyclopropanation is suggested to involve a metal carbene intermediate.

4.7 DONOR–ACCEPTOR CYCLOPROPANES IN ORGANIC SYNTHESIS

The placement of a heteroatom on a carbon atom vicinal to a carboxylate group activates the cyclopropane for ring-opening reactions. These same heteroatoms (N,O,S) increase the reactivity of the carbon–carbon double bond to which they are attached and, in diene systems, regiocontrol is governed by FMO considerations. Doyle and van Leusen prepared a wide range of 2-alkoxycyclopropane-carboxylates by $Rh_2(OAc)_4$-catalyzed reactions of vinyl ethers with ethyl diazoacetate.[81] Using $PtCl_2 \cdot 2PhCN$, $[Rh(CO)_2Cl]_2$, or $[Ru(CO)_3Cl_2]_2$ as the most active catalysts, they effected ring opening of these donor–acceptor cyclopropanes without solvent under mild conditions and in high isolated yields (eq. 29).

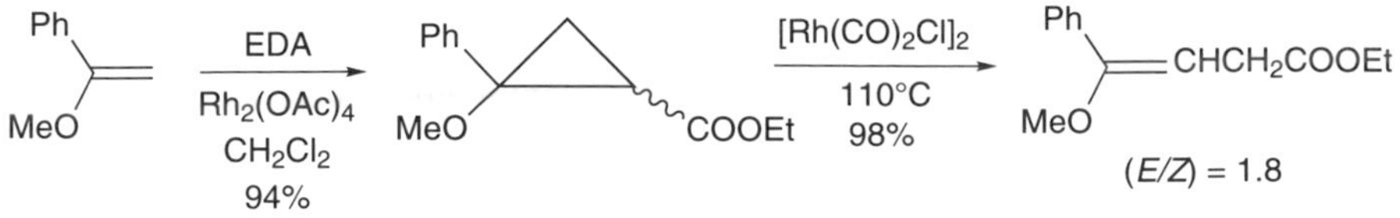

$$(29)$$

Reissig and co-workers were the first to prepare 2-siloxy-substituted cyclopropanecarboxylates, initially with $Cu(acac)_2$[160] and subsequently with $Rh_2(OAc)_4$. Unique applications of these reaction intermediates for the synthesis of 4-oxobutanoic acid esters (eq. 30)[161,162] and γ-keto-δ-amino acids (eq. 31)[163]

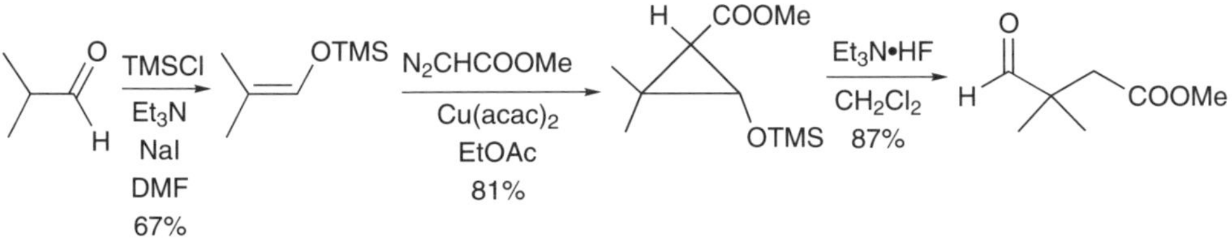

$$(30)$$

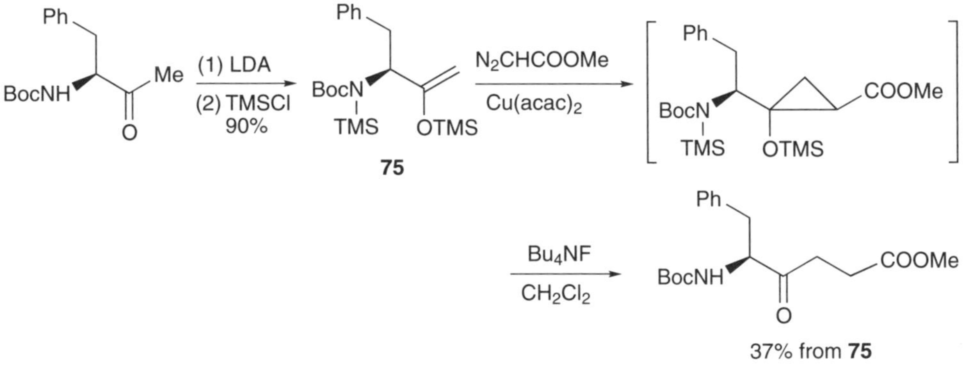

$$(31)$$

have been reported. Investigations of enantiocontrolled cyclopropanation of **76** (eq. 32) using variants of the Aratani catalyst showed that the highest % ee's could be achieved with **43** ($R = o\text{-}MeOC_6H_4$ and $A = PhCH_2$).[164] Increasing the size of the diazo ester alkyl group gave lower yields and did not improve enantioselectivity, although diastereocontrol for the *cis* cyclopropane isomer increased

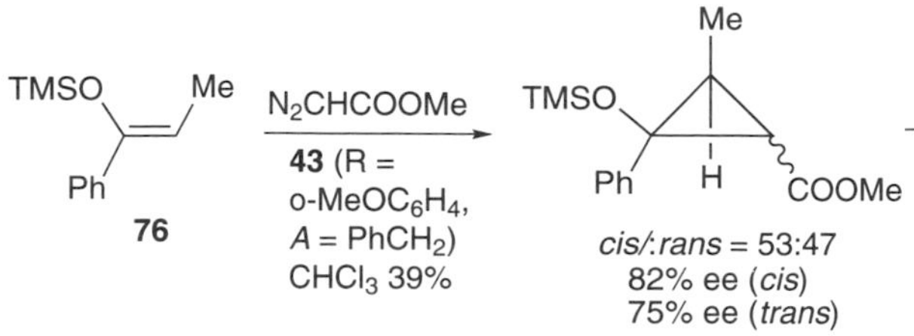

$$(32)$$

to 85:15 with $N_2CHCOOBu^t$; no reaction occurred with BDA, all of which is consistent with Scheme 10. Variations in the alkene, including the use of tetra-substituted alkenes and those with bulky *tert*-butyl substituents, gave moderate levels of enantiocontrol but low diastereoselectivity.[165]

Siloxycyclopropanecarboxylates have been employed as protected enone equivalents for intramolecular Diels–Alder reactions (Scheme 17),[166,167] for

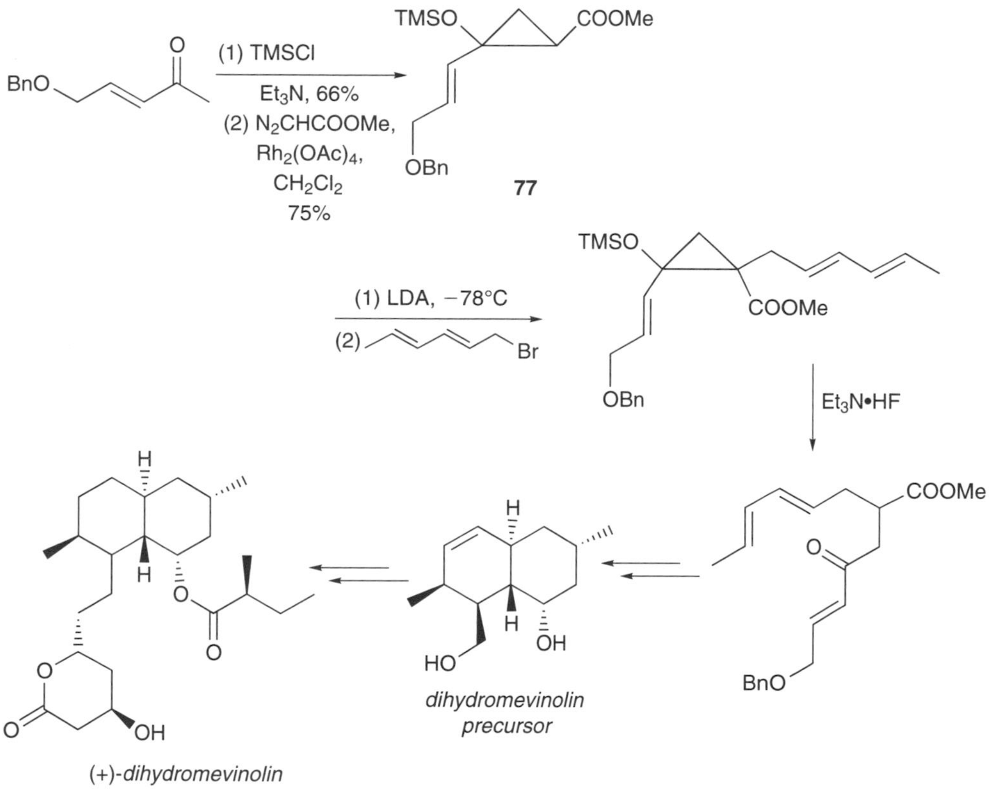

Scheme 17[166]

[4 + 1] cycloadditions (eq. 33),[168] in efficient one-pot conversions to γ-butyrolac-tones (eq. 34),[169] and for the synthesis of functionalized cyclodecenone derivatives

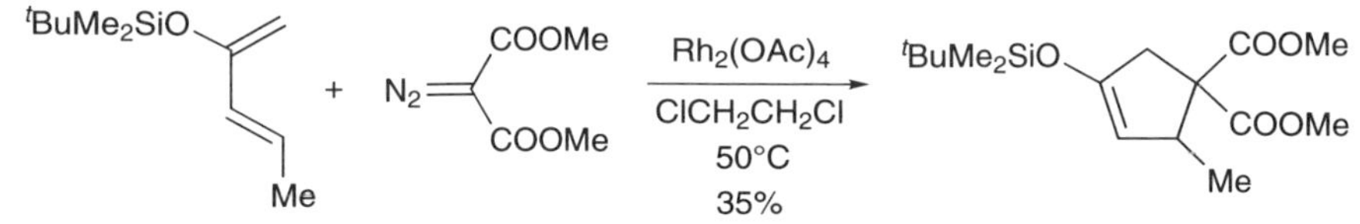

(33)

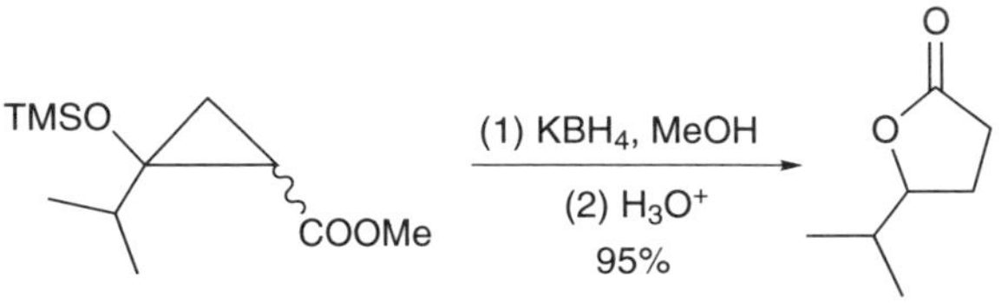

(34)

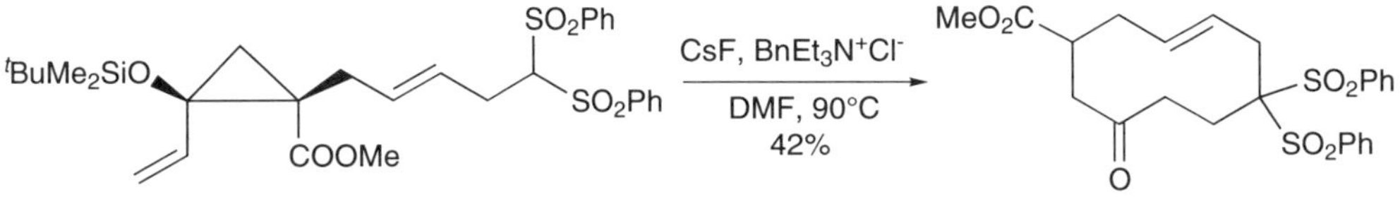

(35)

(eq. 35).[170] Estrone derivatives have been prepared from these donor–acceptor cyclopropanes.[171] Alkylation of cyclopropanes such as **77** shows high diastereocontrol for placement of the alkyl group *syn* to the siloxy group; reduction of the carboxylate ester by LiAlH$_4$ forms stable carbinols without affecting the siloxy group.[172] Swern oxidation of the carbinols formed by LiAlH$_4$ reduction results in the formation of 2,5-dihydrooxepines, which readily undergo ring contraction to 2-vinyl-2,3-dihydrofurans (eq. 36).[173]

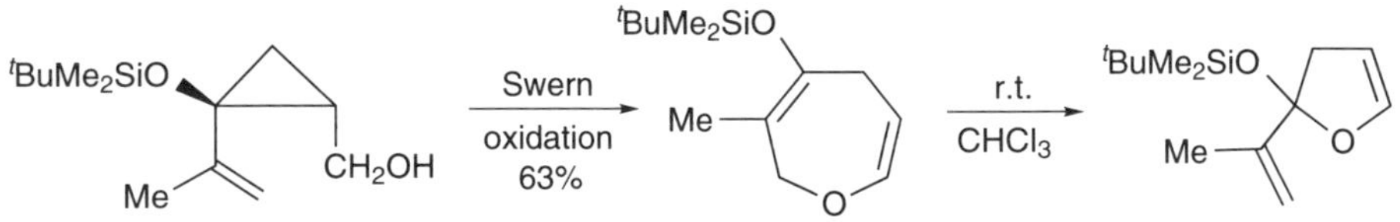

(36)

Treatment of unsaturated sugars with ethyl diazoacetate catalyzed by Rh$_2$(OAc)$_4$ afforded cyclopropane products with high diastereocontrol (eq. 37).[174]

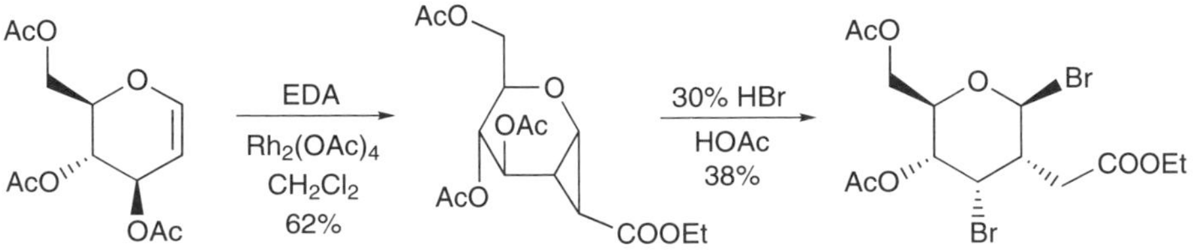

(37)

Ring opening occurred by cleavage of the cyclopropane at the expected position. Swenton has reported unusual rearrangement reactions of donor–acceptor cyclopropanes that occur photochemically and thermally, whose outcomes are dependent on the geometry of the cyclopropane derivatives.[175]

The nitrogen analogs of oxygen in donor–acceptor cyclopropanes have been sparsely studied, but those that have been reported suggest a rich chemistry, especially as directed to the synthesis of amino acid analogs. Pellicciari and coworkers have prepared a proline-γ-acetic acid by such a synthetic scheme (eq. 38).[176] Using *N*-silylated enamine **78**, cyclopropanation with EDA occurred in good yield (eq. 39), but similar reactions with other enamines were unsuccessful.[177] The basicity of amines limits reactivities because of catalyst inhibition and, potentially, ylide formation.

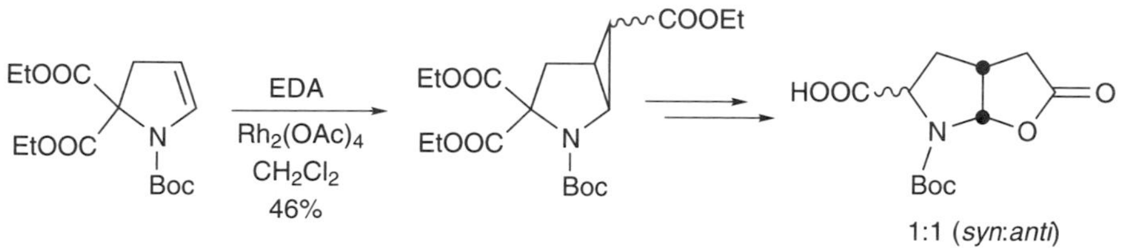

$$(38)$$

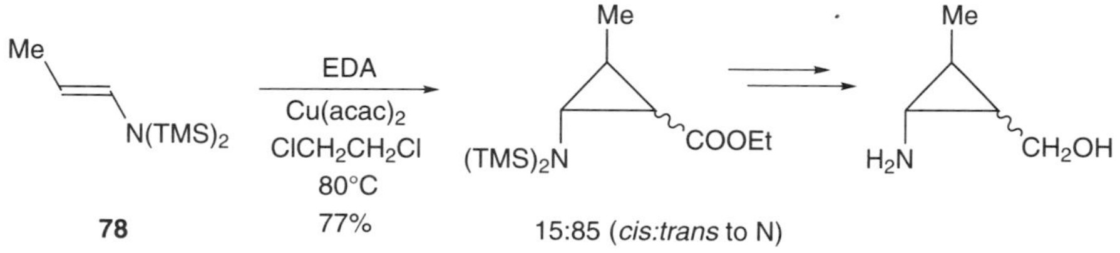

$$(39)$$

4.8 VINYLDIAZOACETATES—INTERMOLECULAR ANNULATION REACTIONS

4.8.1 Formal [3+4]-Cycloaddition

The capability of vinyldiazoacetates to undergo cyclopropanation of dienes with predominant *cis*-1,2-divinyl diastereoselection (Table 4.2) makes possible subsequent facile [3,3]-sigmatropic rearrangement and entry into 1,4-cycloheptadienes (eq. 40)[178] or bicyclic dienes (eq. 41),[179] dependent on the diene that is employed. These transformations have been investigated in detail by Davies, who has also defined optimum catalysts and solvent conditions.[180] Reactions performed in pentane often lead to superior results.[181,182] Vinyldiazoacetates **79** with electron-withdrawing groups for *R* are indefinitely stable at 0°C, but those with vinyl substituents of H, alkyl, or alkoxy must be used immediately after preparation to avoid their rearrangement to 3*H*-pyrazoles.[183]

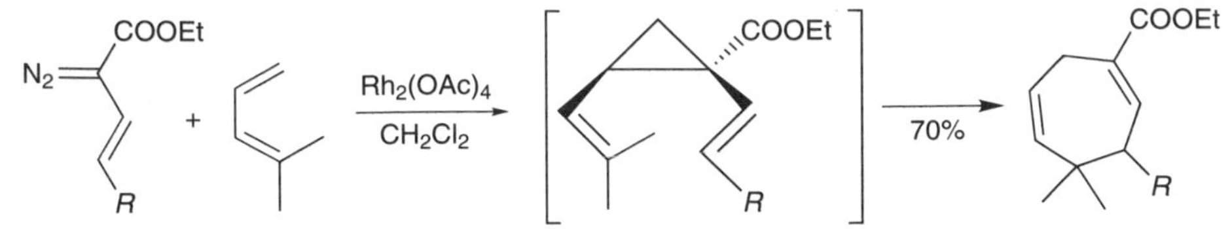

79

R = CH =CHPh

(40)

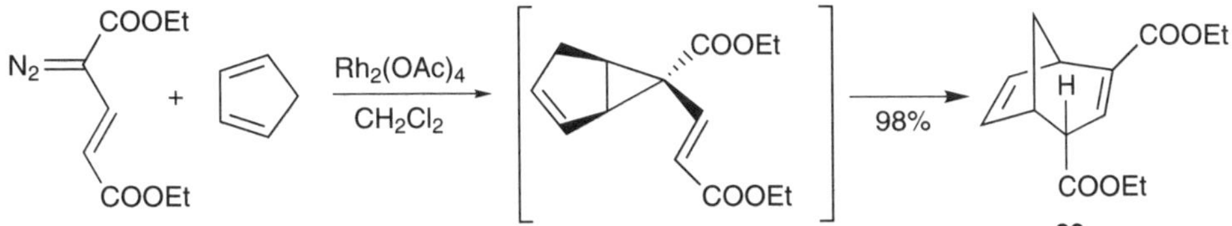

80

(41)

This tandem cyclopropanation/Cope rearrangement methodology has been employed as key steps in the synthesis of tropones (**81**),[182] tropolones (**82**),[184] and hydroazulenes (**83**),[185] as well as tropanes (**84**), from reactions with pyrroles,[181,186] and their hydrocarbon (**79**)[179] and ether (**85**) analogs.[187] Enantioselective cyclo-

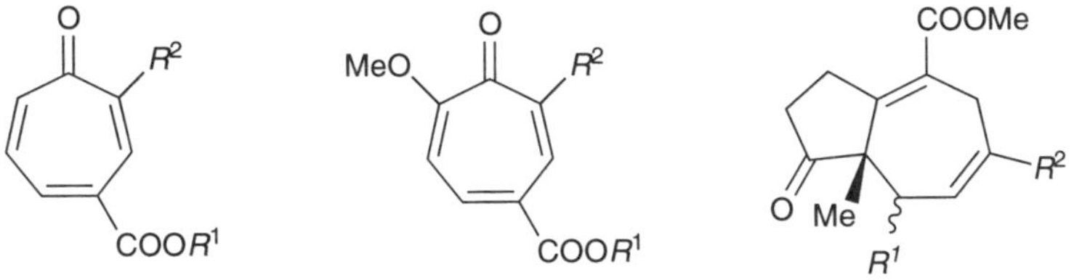

81 **82** **83**

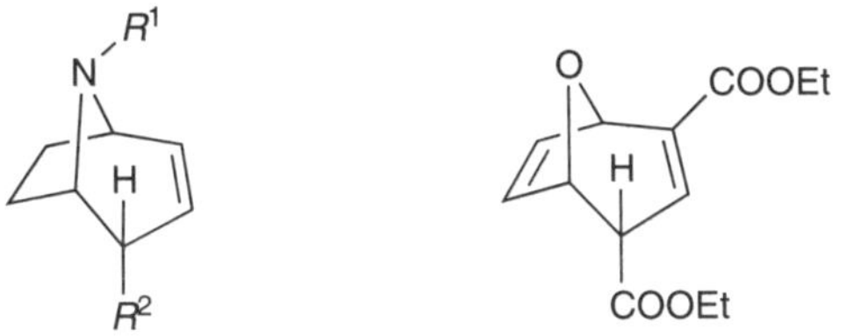

84 **85**

propanation catalyzed by **57** followed by Cope rearrangement has led to bicyclic products analogous to **80** in enantiomeric excesses of typically 42–91% (eq. 42);[188] 1,4-cycloheptadienes were formed with similarly high enantioselectivity (eq. 43). Detailed analyses of the factors that influence enantiocontrol in reactions of vinyldiazoacetates with alkenes catalyzed by **57** have been reported.[145]

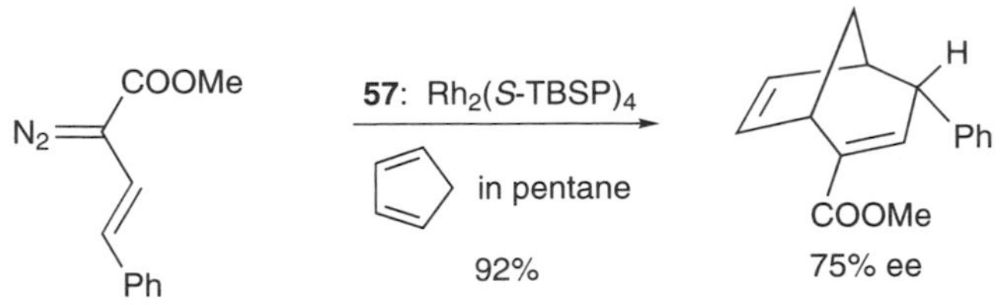

(42)

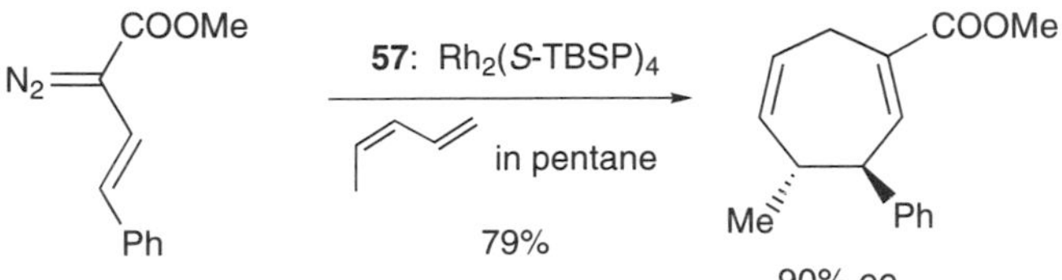

(43)

Alternative chiral dirhodium(II) carboxylate (*N*-acylprolinate) catalysts for diazo decomposition of **79** in the presence of styrene, ethyl vinyl ether, and 1-chloro-1-fluoroethene with pentane as the solvent did not show any advantage over **57** ($R = {}^tBu$).[189]

4.8.2 Formal [3+2]-Cycloaddition

Dirhodium(II)-catalyzed cyclopropanation of vinyl ethers by vinylcarbenes derived from vinyldiazoacetates, followed by direct or Lewis acid catalyzed rearrangement, provides a synthetically useful, stereoselective entry to highly substituted cyclopentene compounds (eqs. 44 and 45).[190-192] Competition between the

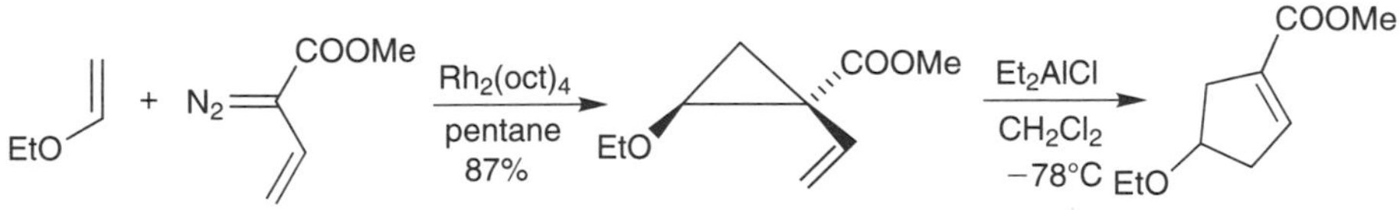

(44)

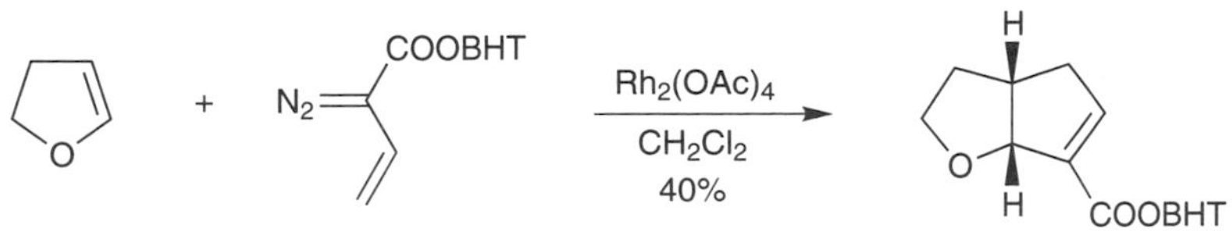

(45)

pathways leading to vinylcyclopropane or directly to the cyclopentene can be controlled by the selection of catalyst and solvent, as is evident in the examples. The use of Et$_2$AlCl to promote the vinylcyclopropane–cyclopentene rearrangement[193] replaces the harsh thermal conditions that were previously employed.[7]

With the *tert*-butyl ester of the vinyldiazoacetate, thermal or Lewis acid (BBr$_3$) promoted rearrangement of vinylcyclopropane compounds generated from vinyl ethers and vinyldiazomethane derivatives directs the formation of α-alkylidenebutyrolactones (eq. 46).[194,195] Syntheses of the ether analogues of the furanone antibiotic ($\pm$)-acetomycin have been accomplished using this methodology.[195]

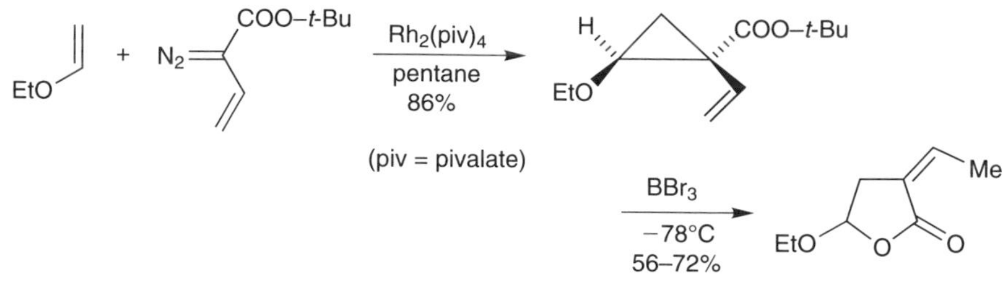

(piv = pivalate)

(46)

4.9 1,3-DIPOLAR KETOCARBENE ADDITION

Certain diazoketones, but not diazoesters, undergo formal 1,3-dipolar ketocarbene addition to electron-rich alkenes, particularly vinyl ethers.[24,196,197] Developed by Wenkert and co-workers as an advantageous methodology for the synthesis of dihydrofurans, this transformation (e.g., eq. 47)[198] occurs with a

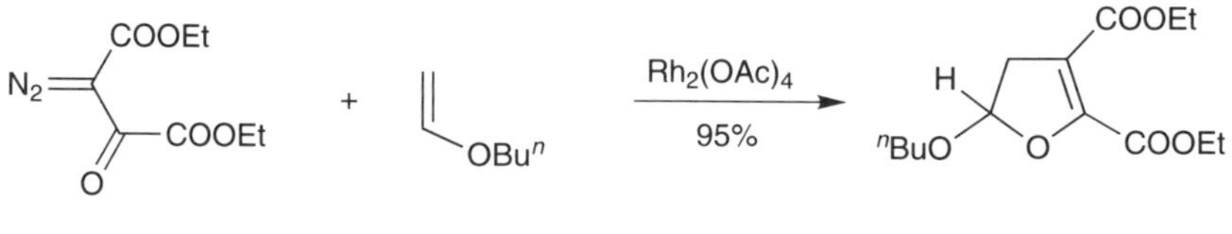

(47)

broad representation of diazoaldehydes and diazoketones, including CH$_3$COCN$_2$COOEt, N$_2$CHCOCOOEt, PhCOCHN$_2$, and EtOOCCN$_2$CHO.[76,198,199] The mechanism of reaction has been represented as initial cyclopropanation followed by catalytic rearrangement,[199,200] but another interpretation of the accumulated results is that outlined in Scheme 18. Charge development in metal carbene addition to electron-rich alkenes affords, in the limit, a reaction intermediate (**86**) capable of undergoing cyclopropane formation (**87**) or of reacting with the nucleophilic carbonyl oxygen to form dihydrofurans (**88**).[24,72] A similar mechanism accounts for the "apparent allylic or vinyl C–H insertion" reaction (**89**) that is a characteristic transformation of diazomalonates with vinyl ethers (e.g., eq. 48);[196,197,200] in this case, however, hydrogen transfer is competitive with cyclopropanation.

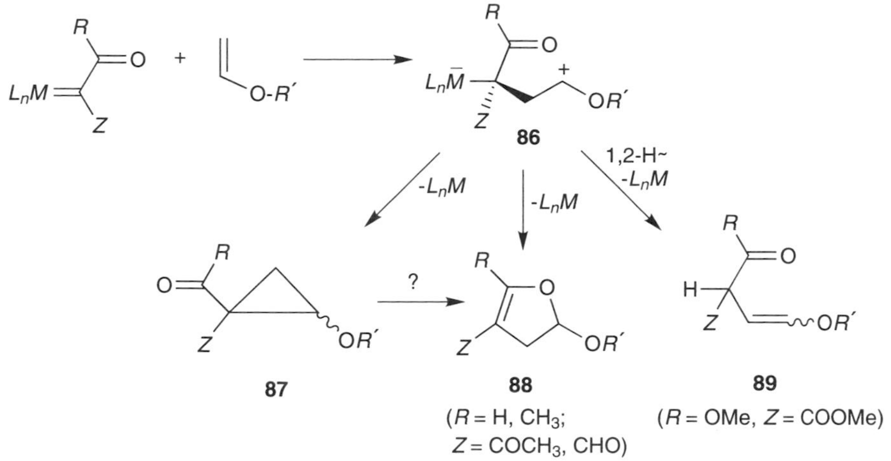

Scheme 18

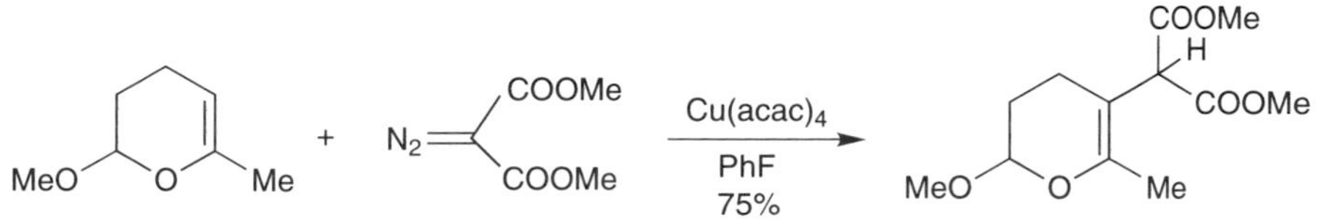

Compound **86**

1,2-H~
-L$_n$M

Compound **87**

? → Compound **88**

($R = H$, CH$_3$;
$Z =$ COCH$_3$, CHO)

Compound **89**

($R =$ OMe, $Z =$ COOMe)

$$(48)$$

Pirrung has developed the 1,3-dipolar addition reaction with cyclic diazodike-tones as a useful synthetic methodology for the construction of furan-fused natu-ral products. Dirhodium tetraactate is the preferred catalyst, and reactions with dihydrofurans,[201] vinyl acetates (eq. 49)[202] and vinyl ethers,[203] aromatic heterocy-cles,[204] and acetylenes[205] have been reported. This methodology has been used for the synthesis of pseudosemiglabrin (**90**, eq. 50), a platelet aggregation antagonist,

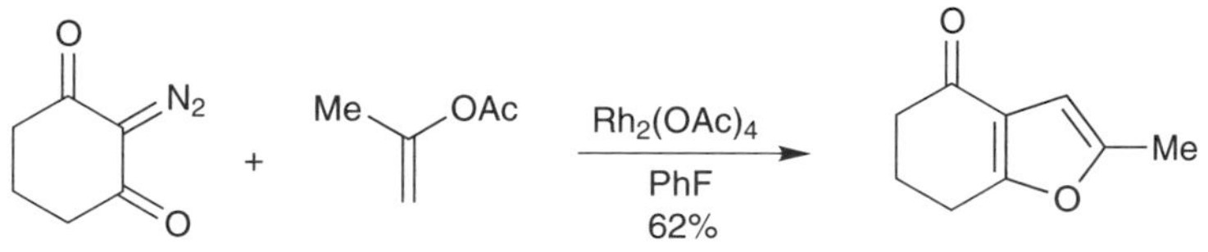

$$(49)$$

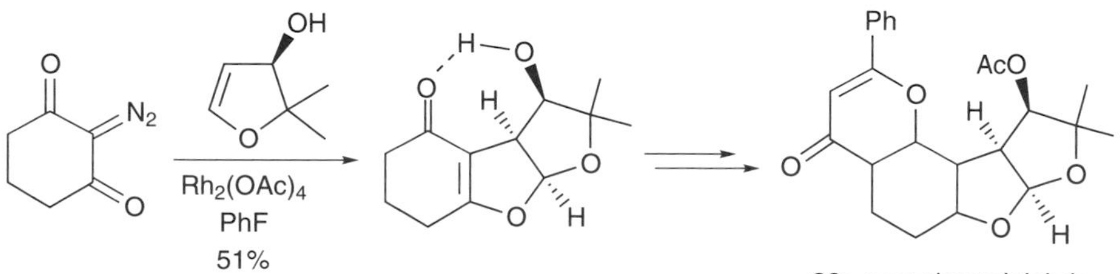

90: *pseudosemiglabrin*

$$(50)$$

and its diastereoisomer semiglabrin,[206] and for the synthesis of furanocou-marins,[207] furanoflavone, and furanochalcone natural products.[208] A formal total synthesis of (±)-aflatoxin B$_2$ (**91**) has been achieved with the key step being Rh$_2$(OAc)$_4$-catalyzed dipolar cycloaddition (eq. 51).[209] The mechanism of these

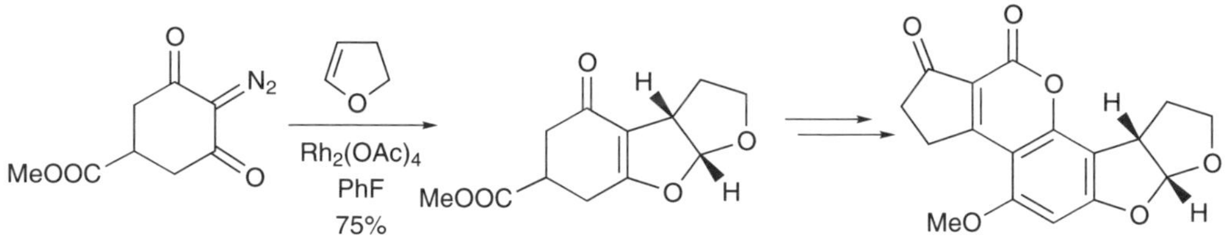

91: aflatoxin B$_2$

(51)

reactions is suggested as occurring by initial cyclopropanation (**92**) followed by ring opening to a zwitterionic intermediate (**93**), which collapses to the observed products (Scheme 19).[203] The distribution to products through paths a and b is de-

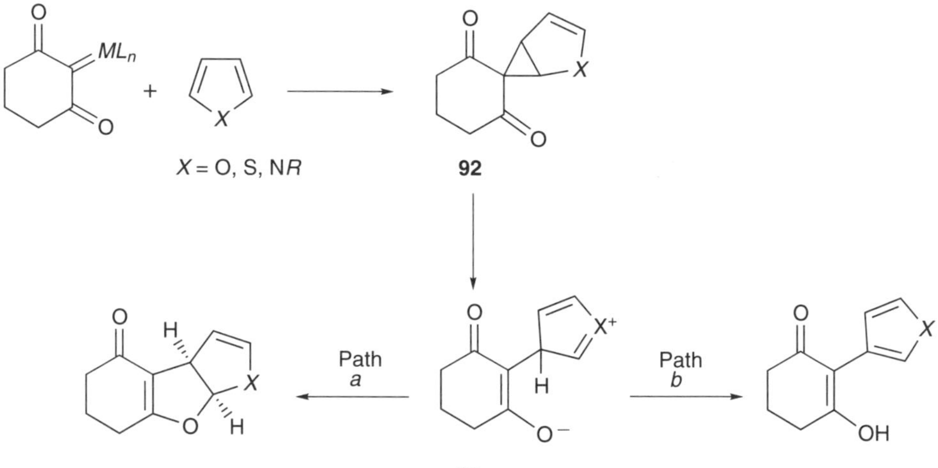

Scheme 19

pendent on the reactant; path a is favored when $X = O$. The results of Shechter and co-workers,[210] who observed cyclopropane products in the Rh$_2$(OAc)$_4$-cata-lyzed reactions of 2-diazo-1,3-indandione with alkenes and their subsequent con-version to "apparent allylic C–H insertion" products (Scheme 20) suggest the validity of the reaction pathway of Scheme 19. These reactions are reported to be remarkably free of byproducts and exhibit the high regiocontrol expected from electrophilic addition.

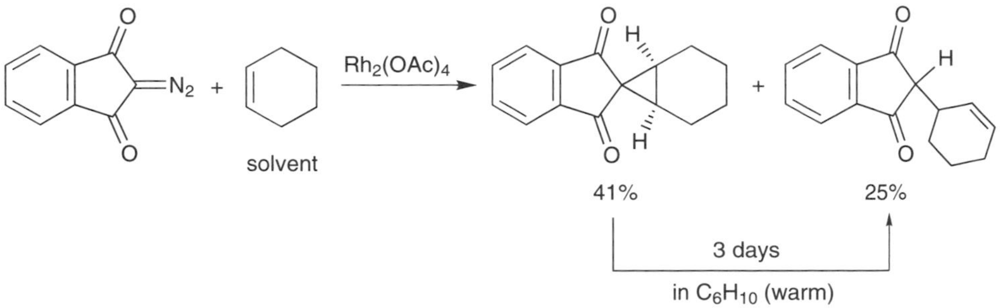

Scheme 20

Asymmetric dipolar addition to dihydrofuran has been achieved with the use of the binaptholphosphate rhodium catalyst developed by Pirrung and Zhang (eq. 52).[211] Modest enantioselectivities were achieved, but chiral copper catalysts did

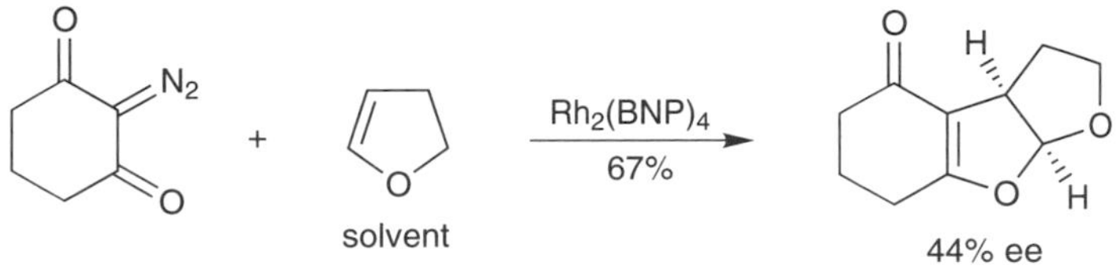

$$(52)$$

not cause diazo decomposition, and use of the $Rh_2(MEPY)_4$ catalysts was reported to give inferior ee's. Further efforts to achieve enantiocontrol in these reactions have not been reported.

4.10 INTERMOLECULAR CYCLOPROPANATION AND SUBSEQUENT REACTIONS IN SYNTHESIS

The versatility of cyclopropanes as intermediates in organic synthesis has relied, in large part, on a diversity of transformations that are specific for cyclopropane derivatives, among which are the cyclopropylcarbinyl–homoallylic rearrangement, the vinylcyclopropane–cyclopentene rearrangement (eqs. 44 and 45), and the Cope (eqs. 40–43) and other sigmatropic rearrangements. Reviews that describe some of the constituent advantages of cyclopropane compounds in organic synthesis have been published recently, but none of them covers the topic with sufficient breadth to be called thorough.[1,2,7,9,15] Pyrethroids derived from chrysanthemic acid (**94**) were the first commercially important synthetic targets for catalytic cyclopropanation, and, together with permethrenic acid (**95**), from which an even more active pyrethroid insecticide is produced, they have received considerable attention.[61,212] Cilastatin (**44**), which is produced from nonracemic 2,2-dimethylcyclopropanecarboxylic (**96**), is another example of a cyclopropane compound having considerable commercial advantage.[120]

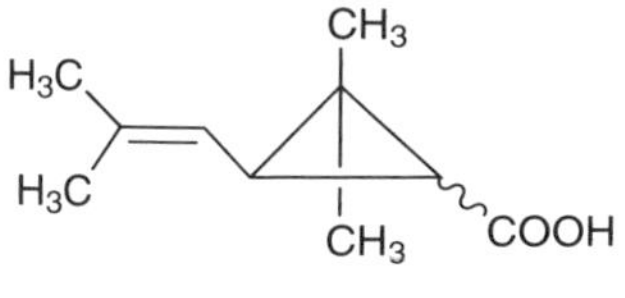

94
chrysanthemic acid

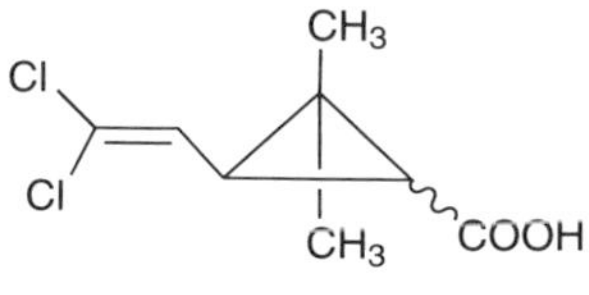

95
permethrenic acid

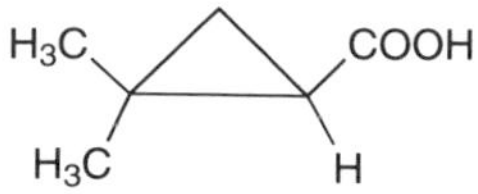

96
precursor to *cilastatin*

Applications of intermolecular cyclopropanation reactions in total synthesis include the construction of pentalenolactone E methyl ester (**97**, Scheme 21),[213]

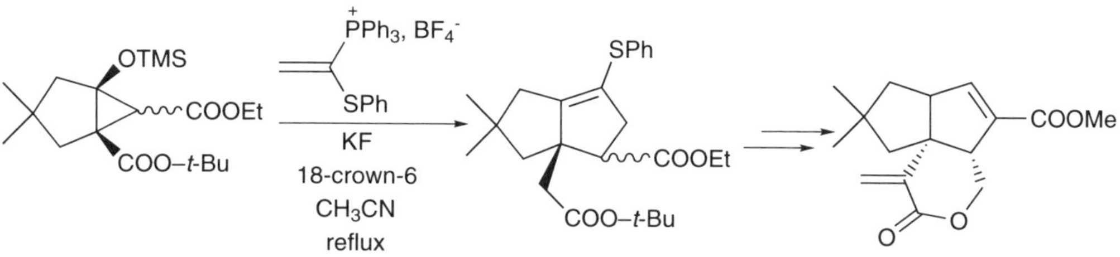

97: *pentalenolactone E methyl ester*

Scheme 21

the formation of a bicyclic derivative along a synthetic pathway to the pentacyclic alkaloid (+)-eburnamonine (**98**, Scheme 22),[214] the formal synthesis of (+)-dihy-

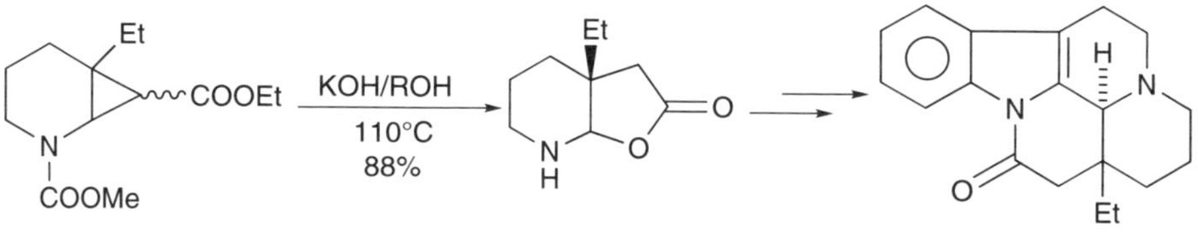

98: (+)-*eburnamonine*

Scheme 22

dromevinolin (Scheme 17),[215] and the preparation of estrone analogs (**99**, Scheme 23).[171] Intermolecular cyclopropanation of vinyl ethers with diazoacetates provides a direct route to donor–acceptor–substituted cyclopropane compounds (Section 4.7)[160–177,216,217] that are also valuable reactants for the synthesis of 1,4-dicarbonyl and β,γ-unsaturated carbonyl compounds.[81] Thermal and photochemical rearrangements are dependent on the stereochemical relationship of the donor–acceptor cyclopropane substituents.[175]

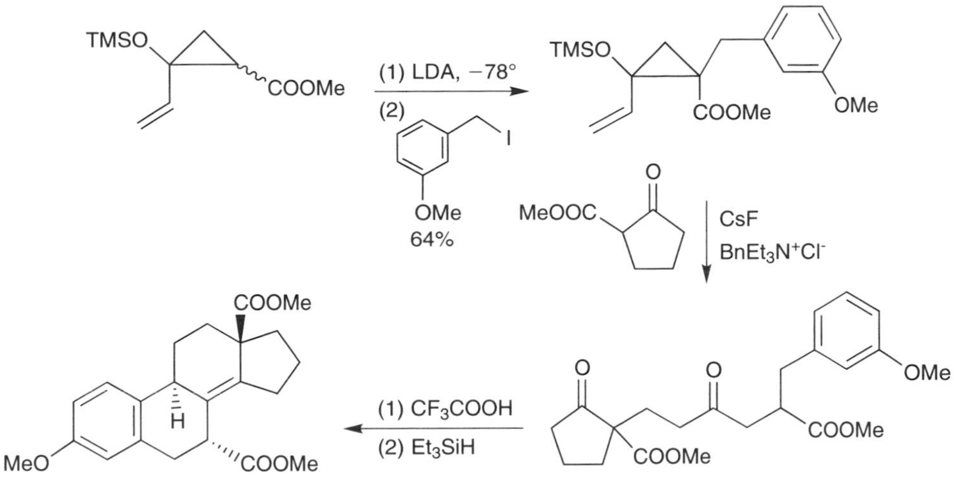

Scheme 23

Precursors to butanolides,[217] bicyclo[3.3.0]octane derivatives,[218] dienes,[219] and cyclopropane-substituted amino acids (eqs. 53[220] and 54[223])[9,63,64,220–224] have also been reported. Methylenecyclopropanes prepared from allenes are versatile reagents for [3+2]-methylenecyclopentane annulations (eq. 55).[225] The diastereo-

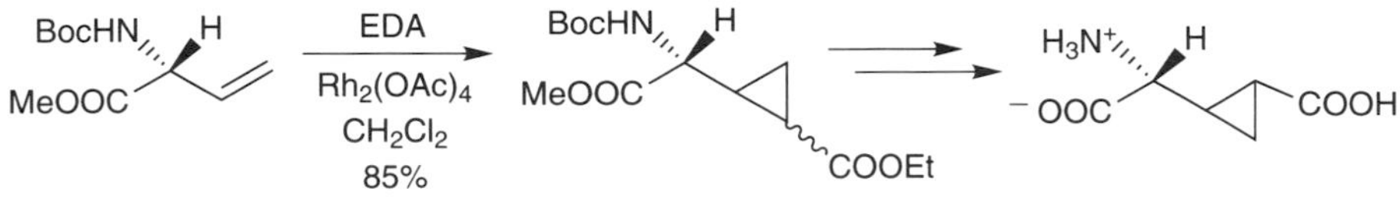

(53)

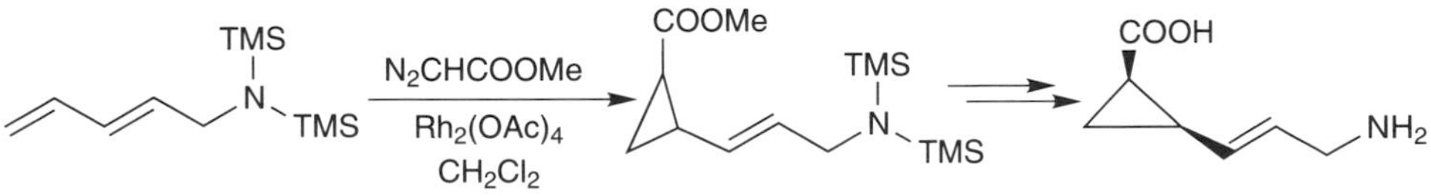

(54)

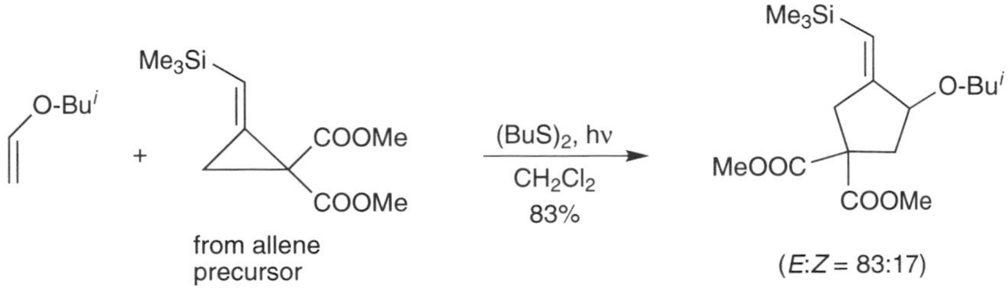

(55)

selective cyclopropanation of glycals[174,226] provides a convenient methodology for the synthesis of a diverse array sugar derivatives. Vinylcyclopropane-cyclopentene, iminocyclopropane and carbonyl-cyclopropane, cyclopropylcarbinyl, transition metal catalyzed, and free-radical rearrangements are available for use in organic synthesis.[15] Even with this diversity of applications, however, intermolecular cyclopropanation reactions have proven to be less versatile than their intramolecular counterparts.

Nitrocyclopropanes and their α-carboxylate-substituted derivatives have been prepared from nitrodiazomethane and nitrodiazoacetate (**100**), respectively, in moderate to good yields (eq. 56).[80,227] The 1-nitrocyclopropanecarboxylates are

$$
\begin{array}{ccc}
\underset{\textbf{100}}{O_2N\diagup\!\!\!\overset{N_2}{\diagdown}COOBu^t} + \underset{}{=\!\!\!\diagup Ph} & \xrightarrow[\substack{CH_2Cl_2 \\ 83\%}]{Rh_2(OAc)_4} & \underset{(E\!:\!Z=2\!:\!1)}{Ph\diagup\!\!\bigtriangleup\!\!\diagdown\substack{NO_2 \\ COOBu^t}}
\end{array}
\tag{56}
$$

precursors to cyclopropaneamino acids. Rhodium(II) acetate is the preferred catalyst since use of copper catalysts gives markedly lower yields. Substituents other than carboxylate for nitrodiazomethane, including PhCO, TolSO$_2$, NC, and F$_3$C have been prepared,[80] and they undergo catalytic cyclopropanation reactions with alkenes.

Diazomalonates,[196,197] (*p*-tolylsulfonyl)diazomethane,[81] diazoacetonitrile,[81] trifluoromethyl-α-diazopropionate,[228] chlorovinyldiazomethanes,[229] and phenyldiazomethane[76] have all been effectively employed for catalytic intermolecular cyclopropanation reactions. Ordinarily, these diazo compounds will undergo diazo decomposition at or above ambient temperature, and cyclopropanation occurs with electron-rich olefins ranging from vinyl ethers to 1-alkenes. With phenyldiazomethane, and, presumably, chlorovinyldiazomethane, however, good yields of cyclopropane products are only obtained with vinyl ethers.[76,229]

4.11   CYCLOPROPENATION OF ALKYNES

With their high strain energy cyclopropenes are often regarded as exotic and highly reactive organic compounds. However, since the isolation and characterization of the cyclopropene fatty acid sterculic acid (**101**),[230] numerous biomolecules

101
sterculic acid

that contain a cyclopropene ring, particularly in sterols and fatty acids, have been identified (e.g., **102**).[19-21] Furthermore, their high reactivity in addition and cy-

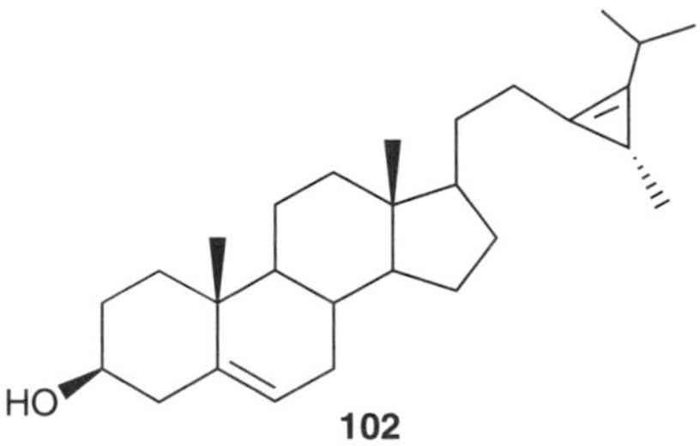

102

calysterol[231]
a sponge sterol

cloaddition reactions makes cyclopropene compounds potentially valuable synthetic intermediates for the construction of carbocyclic and heterocyclic compounds.[3–5]

Rhodium(II) acetate catalyzed reactions of diazoesters with alkynes have recently made available an array of 3-cyclopropenecarboxylate esters that are stable at or, with most systems, even above room temperature (eq. 57).[232,233] Moderate

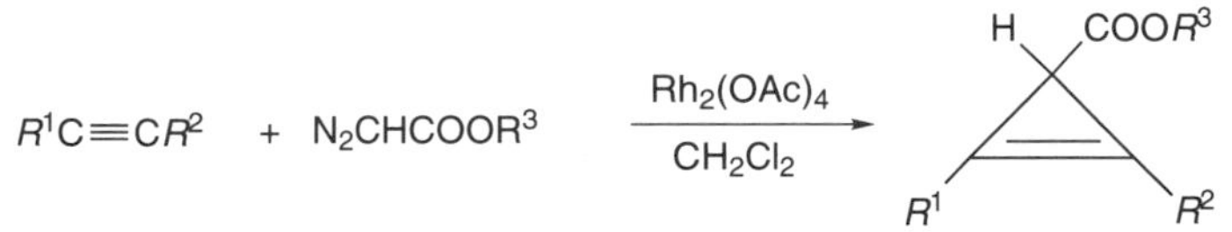

$$R^1C \equiv CR^2 \quad + \quad N_2CHCOOR^3 \quad \xrightarrow[\text{CH}_2\text{Cl}_2]{\text{Rh}_2(\text{OAc})_4} \quad \text{(57)}$$

to high yields of cyclopropene compounds have been achieved with both terminal and internal alkynes. Previously, copper catalysts provided low to moderate yields of cyclopropenes in reactions of diazo esters with disubstituted acetylenes,[5] but the higher temperatures required with these catalysts often led to thermal or catalytic ring opening and products derived from vinylcarbene intermediates.[4,233] 1,3-Enynes undergo cyclopropenation preferentially, but the vinylcyclopropene products derived from these catalytic metal carbene reactions are unstable and form [2+2] cycloaddition products (eq. 58).[234,235] However, with isolated C≡C

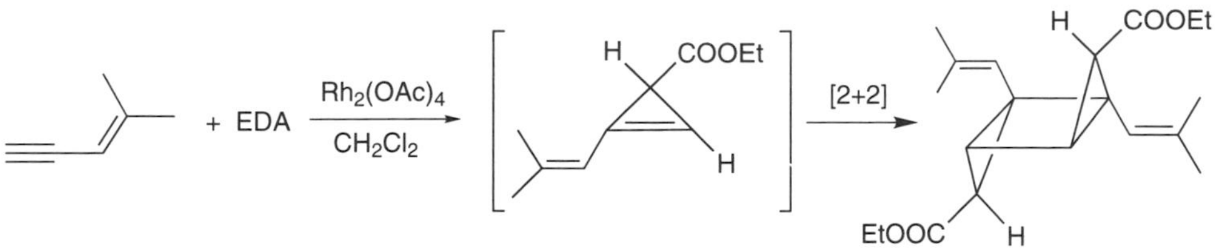

$$+ \text{ EDA} \quad \xrightarrow[\text{CH}_2\text{Cl}_2]{\text{Rh}_2(\text{OAc})_4} \quad \xrightarrow{[2+2]} \quad \text{(58)}$$

and C=C bonds, addition to the carbon–carbon double bond can be the preferred reaction, espcially for silylated 1-alkynes; $\text{Rh}_2(\text{OAc})_4$ provides higher selectivity for addition to the carbon–carbon double bond than do copper catalysts.[236] Reactions of diazoesters with phenylacetylene, catalyzed by either copper or rhodium(II) catalysts, do not yield stable cyclopropene products.

Chiral dirhodium(II) carboxamidates are exceptional catalysts for highly enantioselective intermolecular cyclopropenation reactions.[237] With alkyl diazoacetates and a series of 1-alkynes, use of $Rh_2(MEPY)_4$ catalysts gives 1-substituted alkyl cyclopropene-3-carboxylates (**103**, eq. 59) with enantiomeric excesses

$$\equiv\!\!-R \quad + \quad N_2CHCOOR' \quad \xrightarrow[\quad CH_2Cl_2 \quad]{Rh_2(5S\text{-}MEPY)_4} \quad \underset{R}{\triangle}\!\!\overset{H_{,,}\quad COOR'}{}$$

103

R =	R' =	yield, %	% ee	
$CH(OEt)_2$	Me	42	>98	(59)
CH_2OMe	Bu^t	52	78	
CH_2OMe	Et	73	69	
Bu^n	Et	70	54	
Bu^t	Et	85	57	

ranging from 48 to ≥98% in good yields. Catalysis by $Rh_2(5S\text{-}MEPY)_4$ results in the predominant production of the (S)-cyclopropene enantiomers, and catalysis by $Rh_2(5R\text{-}MEPY)_4$ forms the (R)-cyclopropene enantiomers. With menthyl diazoacetates (MDA) and the same series of alkynes, diastereodifferentiation is ≥97:3 dr when R = CH_2OMe and is 93:7 dr when R = Bu^n. No additional enhancement of selectivity has been observed with diazoacetates of other chiral auxiliaries.[238] With MDA double diastereoselectivity was pronounced; whereas the combination of l-MDA with propargyl methyl ether catalyzed by $Rh_2(5S\text{-}MEPY)_4$ gave the corresponding cyclopropene with 72:28 dr, use of d-MDA gave the cyclopropene having ≥97:3 dr (also with propargyl acetate). Similar results were obtained from reactions of MDA with 1-hexyne and 3,3-dimethyl-1-propyne. That stereocontrol for cyclopropenation is greater with propargyl methyl ether than with 1-hexyne or 3,3-dimethyl-1-butyne suggests that polar interactions of the alkyne with the catalyst ligands may be operative. Alternative dirhodium(II) carboxamidate catalysts, including $Rh_2(4S\text{-}BNOX)_4$, $Rh_2(4S\text{-}IPOX)_4$, $Rh_2(4S\text{-}MEOX)_4$, and $Rh_2(4S\text{-}PHOX)_4$, provided only a fraction of the enantioselection obtained with $Rh_2(MEPY)_4$ catalysts. Mechanistic details have been discussed.[237]

The use of disubstituted acetylenes gave lower product yields and low enantiocontrol (≤20% ee). Also, a chiral(semicorrinato)-copper(II) catalyst (**45**, R = $CH_2OSiMe_2Bu^t$) was not effective for enantioselective cyclopropenation; both % ee values and product yields were low.[239] Diimide reduction[237] or catalytic hydrogenation of the substituted cyclopropenes (**103**) from enantioselective cyclopropenation produces cis-disubstituted cyclopropane products exclusively and in high yield (eq. 60) with enantiomeric excesses that are the same as those of the cyclopropene reactant.

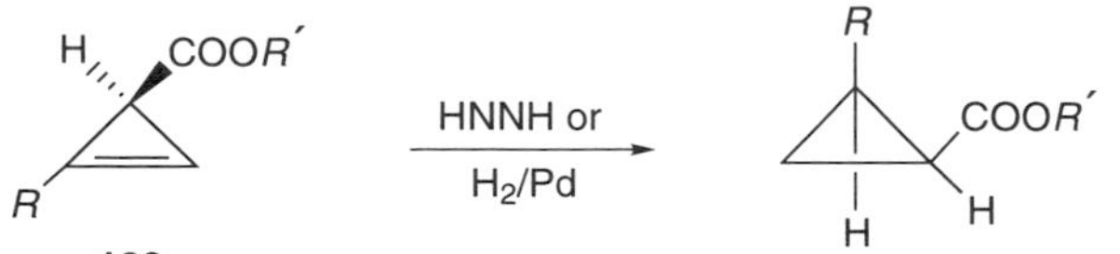

(60)

103

Cyclopropenes are particularly reactive substrates for dipolar cycloaddition and Diels–Alder reactions, and these and other cyclopropane transformations have been reviewed.[3-5] Various transition metal compounds, but particularly those of Cu and Rh(II), catalyze ring-opening reactions of cyclopropenes that occur through intermediate metal carbenes (e.g., eq. 61).[233] Often, copper(I) and rhodium(II) catalysis result in different products (e.g., eqs. 62 and 63)[240,241] that,

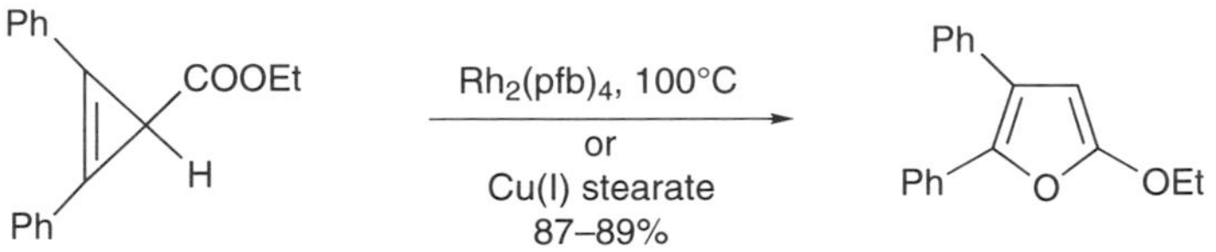

(61)

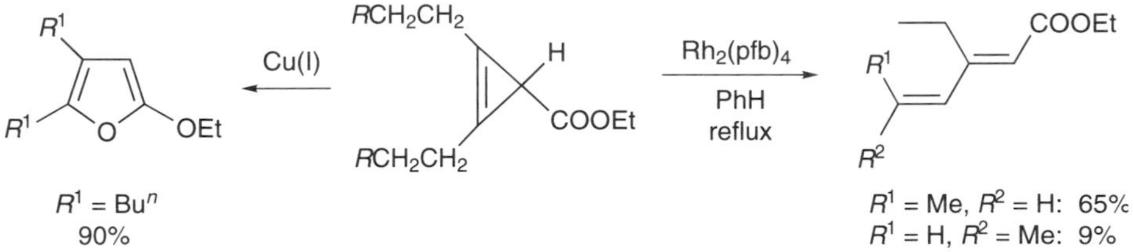

(62)

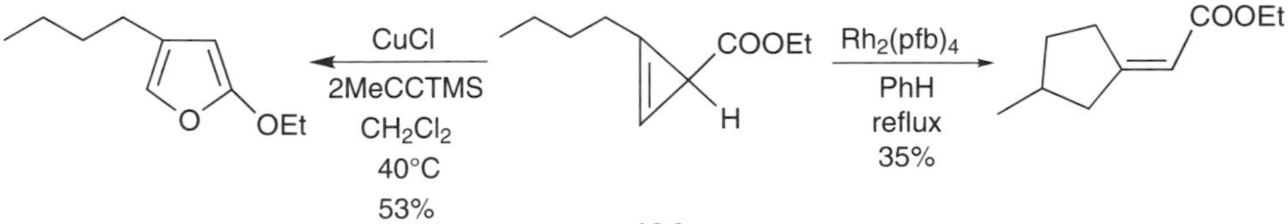

(63)

according to Müller and co-workers,[242] can be attributed to different geometrical isomers of intermediate metal-complexed vinyl carbenes (e.g., **107** and **108** in Scheme 24). Vinylcarbene complexes of titanocene[243] have been generated from cyclopropenes,[244] and intriguing uses of the transition metal vinyl carbene complexes[245,246] have made cyclopropenes even more versatile reactants in organic synthesis. A stable cobaltacyclobutene has been formed "directly" by the addition of EDA to an alkyne complex of CpCo(PPh₃),[247] but the involvement of a cyclopropene intermediate cannot be disregarded.

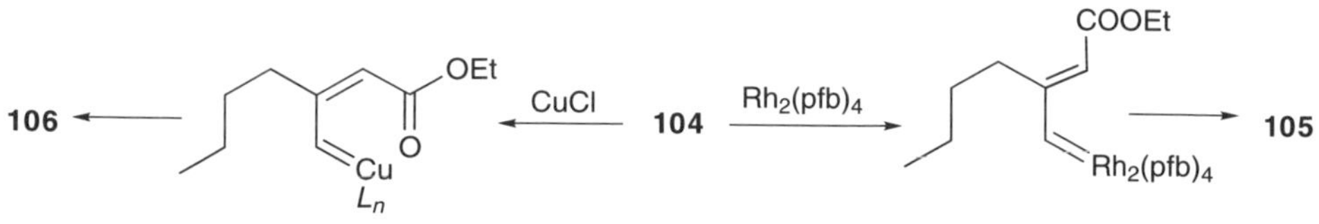

Scheme 24

Reissig has prepared novel GABA analogs and dipeptides incorporating the cyclopropene ring by $Rh_2(OAc)_4$-catalyzed cyclopropanation followed by selective transformations of the cyclopropene product (Scheme 25).[248,249] These reactions complement those from olefin cyclopropanation (eq. 54) but add synthetic versatility derived from the reactive cyclopropane.

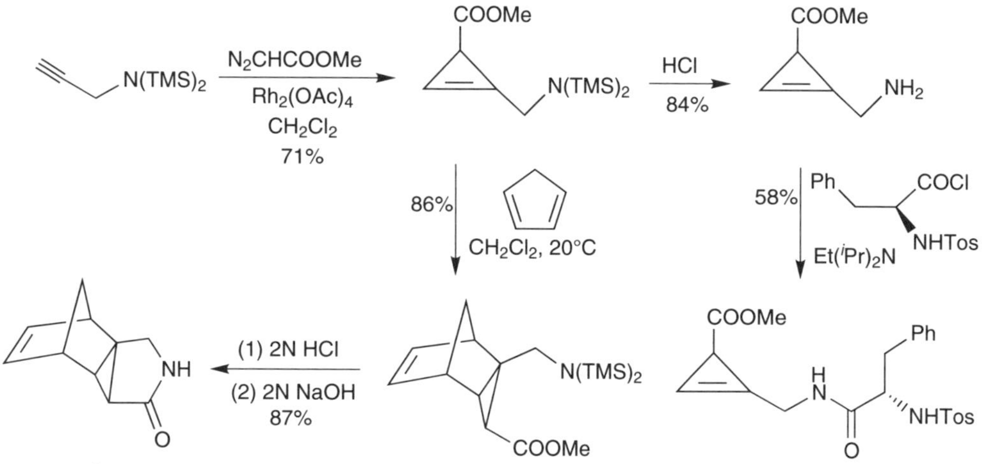

Scheme 25

4.12 AZIRIDINATION

Aziridines are valued synthetic reactants, and methods for their synthesis are of considerable interest.[250,251] Two approaches by addition to a double bond (Scheme 26) are possible—carbene addition to imines and nitrene addition to alkenes—and both have been achieved catalytically.

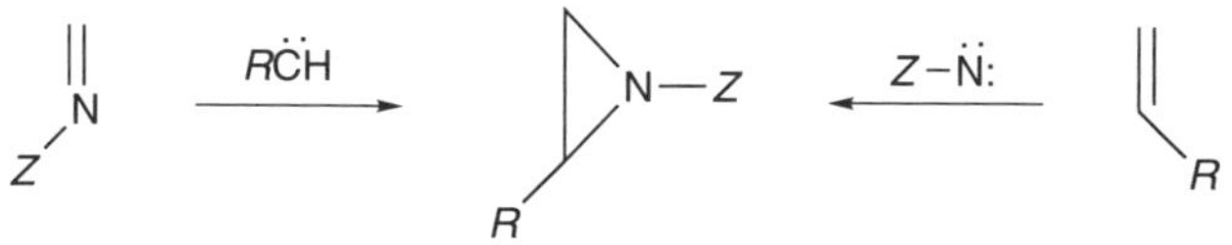

Scheme 26

4.12.1 Addition to Imines

Although the initial publication reported that alkyl diazoacetates underwent $Cu(OTf)_2$- or $Rh_2(OAc)_4$-catalyzed carbene addition to carbodiimides to form iminoaziridines,[252] a subsequent publication by the same group reinterpreted the data, together with X-ray analysis, to document the production of iminooxazolines (eq. 64).[253] Whether or not an iminoaziridine was a reaction intermediate was

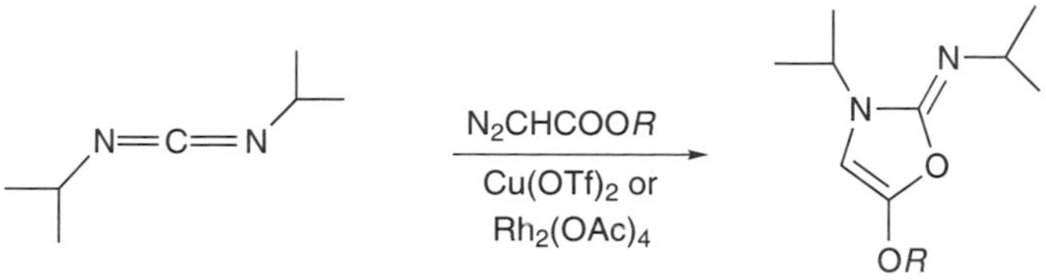

$$\text{(64)}$$

not established. Recently, however, there have been reports of successful catalytic metal carbene additions to imines that suggest opportunities for further development. Jorgensen has communicated that $Cu(OTf)_2$ is effective when used in catalytic amounts for reactions between imines and EDA (eq. 65).[254] Yields are

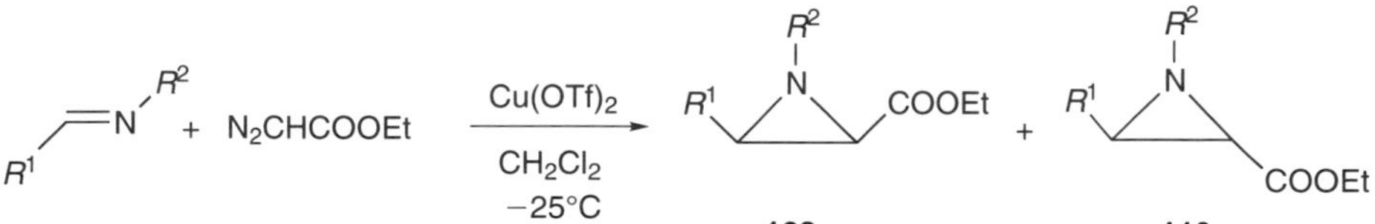

R^1	R^2	yield (%)	109:110
Ph	Ph	80	63:37
Bu^t	Ph	90	38:62
Ph	TMS	35	>95:5

$$\text{(65)}$$

generally high, and, in the absence of overriding steric effects, there is a moderate diastereomeric preference for the *cis* isomer. The use of *l*-MDA, instead of EDA, and $H_2C=NPh$ provided the corresponding azine with a 62:38 dr. Attempts to use chiral bis-oxazolines such as **47** as ligands for copper gave moderate product yields but low ee's.

Jacobsen and co-workers have described moderate ee's for similar reactions of N-benzylideneanilines with EDA catalyzed by $Cu(MeCN)_4PF_6$ in the presence of chiral bis-oxazolines **111** (eq. 66).[255] Substituents on the aromatic rings have a

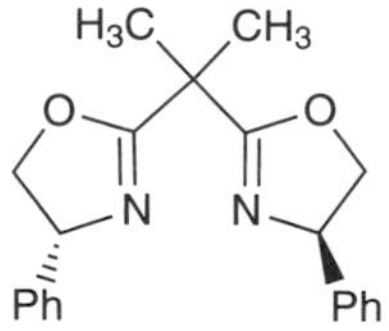

111

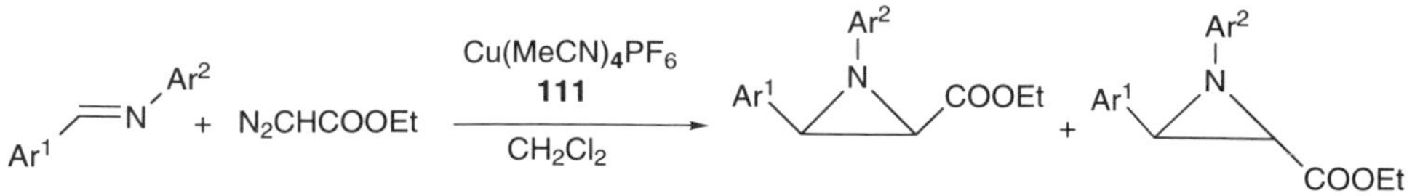

Ar^1	Ar^2	yield (%)	ee (%)	112:113	ee (%)	
Ph	p-MeOC$_6$H$_4$	23	67	9	32	
Ph	p-ClC$_6$H$_4$	34	49	4	22	
Ph	Ph	37	44	4	35	(66)

significant effect on enantioselection. Low turnover numbers and low yields are characteristic, however.

The mechanism of this reaction is suggested to proceed via a metal associated azomethine ylide (Scheme 27).[255] The associated metal with its chiral ligands

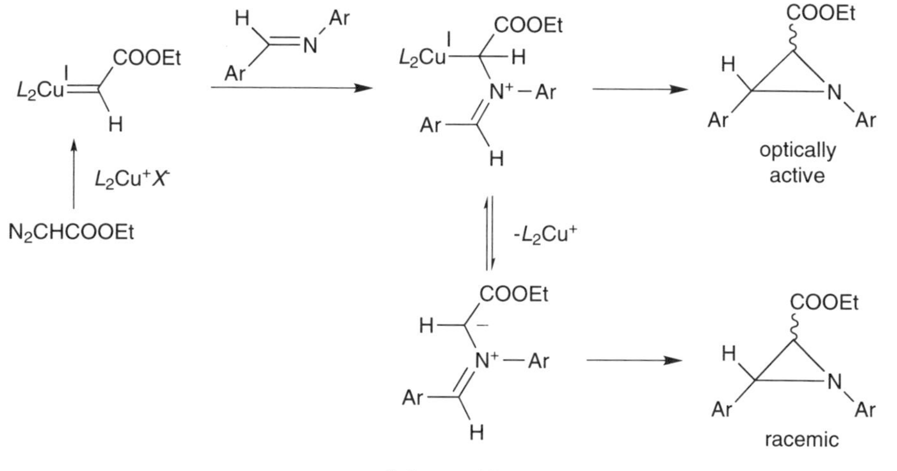

Scheme 27

mediates enantiocontrolled formation of the aziridine. Dissociation of the metal leads to racemic product. Müller has reported that aziridines formed by catalysis with Rh$_2$(5S-MEPY)$_4$ or with the chiral Ikegami phthalimidophenylalaninate gave racemic product.[256]

4.12.2   Nitrene Addition to Alkenes

The alternative to carbene addition to imines for aziridination is the addition of nitrenes to alkenes. This latter approach has the advantage that the stereochemistry of the alkene is generally retained in the addition transformation. Although organic azides were originally considered to be potential sources for metal nitrenes in catalytic reactions,[257] it was not until (*N*-arylsulfonylimino)phenyliodin-anes, PhI = NSO$_2$Ar, were used as nitrene sources that significant progress could be made in aziridination. When *p*-TsN = IPh and alkenes were combined in the presence of Cu(acac)$_2$, Cu(OTf)$_2$, or Cu(CH$_3$CN)$_4$ClO$_4$, the latter generally giving the higher yields, aziridination products were formed in yields ranging from 55 to 95% (eq. 67).[258,259] Olefins ranging from cyclohexene to methyl acrylate (con-

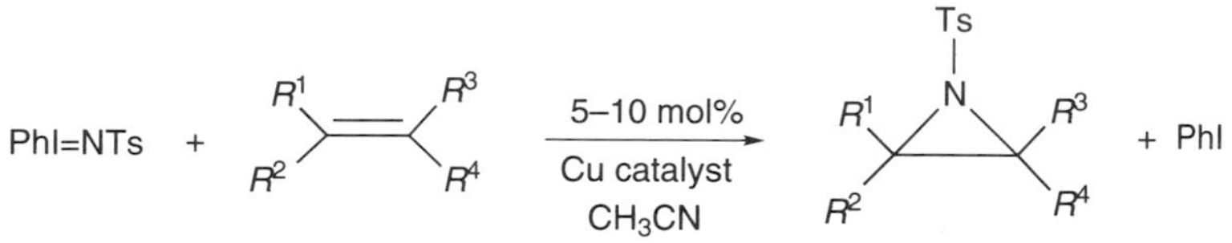

$$\text{PhI=NTs} \quad + \quad \overset{R^1}{\underset{R^2}{\Big\rangle}} = \overset{R^3}{\underset{R^4}{\Big\langle}} \quad \xrightarrow[\substack{\text{Cu catalyst} \\ \text{CH}_3\text{CN}}]{\text{5–10 mol\%}} \quad \overset{R^1}{\underset{R^2}{\Big\rangle}}\overset{\overset{\text{Ts}}{|}}{\underset{}{N}}\overset{R^3}{\underset{R^4}{\Big\langle}} \quad + \text{ PhI} \quad (67)$$

trast to metal carbene reactions) underwent addition; silyl enol ethers formed α-aminoketone derivatives directly (eq. 68). With *cis*-stilbene, reaction at room

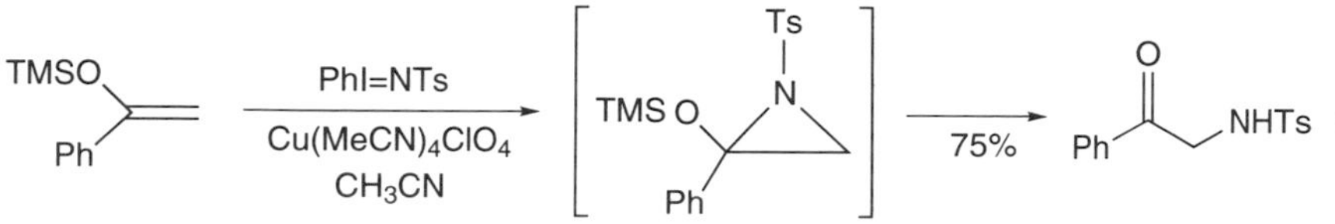

$$(68)$$

temperature formed a 2:3 ratio of *cis*- and *trans*-1,2-diphenylaziridines, but a 9:1 ratio of these two products was obtained at −20°C.

Catalysts that promote cyclopropanation are also effective for aziridination of alkenes. A copper complex of hydrotris(3,5-dimethyl-1-pyrazolyl)borate has been employed,[142] and those known to catalyze reactions of diazo compounds can be expected to be effective with (arylsulfonylimino)phenyliodinanes. Although initially discounted as a relatively ineffective catalyst,[258] Rh$_2$(OAc)$_4$ has been shown to offer advantages for aziridination when [*N*-(*p*-nitrobenzenesulfonyl)imino]-phenyliodinane, PhI = NNs, was used; with this catalyst in 2 mol %, rather than the 5 mol % required with copper catalysts, aziridination proceeded stereospecifically and in moderate to good yield.[260]

Both bis-oxazoline **111** and salen **114** as chiral ligands for copper(I) offer moderate to high levels of enantiocontrol in catalytic aziridination reactions (e.g.,

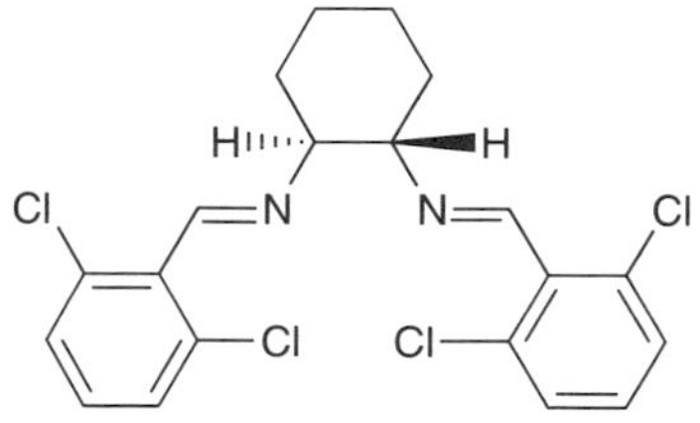

114

eq. 69[261], and eq. 70[262]), and their utility for the formation of amino acid derivatives has been examined.[262] In these cases catalyst turnover remains a limitation

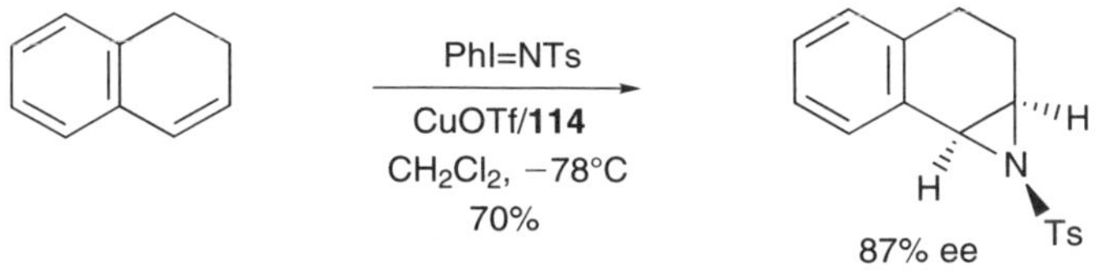

$$
\begin{array}{c}
\xrightarrow[\substack{\text{CuOTf/114} \\ \text{CH}_2\text{Cl}_2, \, -78^\circ\text{C} \\ 70\%}]{\text{PhI=NTs}}
\end{array}
\tag{69}
$$

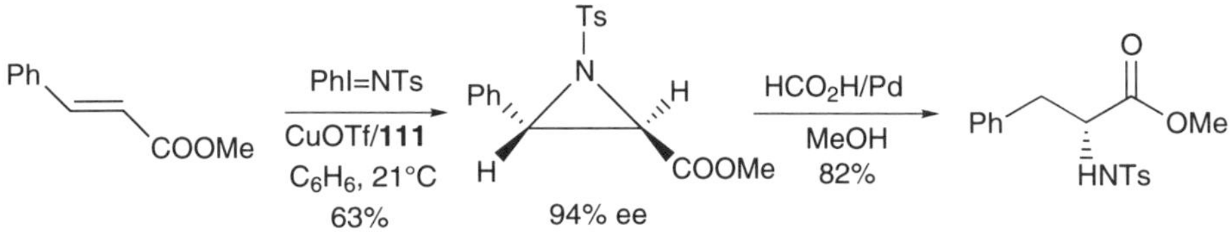

87% ee

$$
\xrightarrow[\substack{\text{CuOTf/111} \\ \text{C}_6\text{H}_6, \, 21^\circ\text{C} \\ 63\%}]{\text{PhI=NTs}}
\qquad
94\% \text{ ee}
\qquad
\xrightarrow[\substack{\text{MeOH} \\ 82\%}]{\text{HCO}_2\text{H/Pd}}
\tag{70}
$$

(5–10 mol % of catalyst required) and, until this problem is resolved, practical uses of this promising methodology are currently restricted. By comparison, (salen)manganese(III) catalysts are ineffective (low yields and low % ee).[263] However, use of the Rh$_2$(5S-MEPY)$_4$ catalyst has been reported to be promising; with PhI=NNs and *cis*-β-methylstyrene, the aziridine product was formed in 80% yield with 73% ee.[260]

4.13   REFERENCES

1   *The Chemistry of the Cyclopropyl Group*; Rappoport, Z., Ed.; Wiley: New York, 1987; Parts 1 and 2.

2   Salaün, J., "Optically Active Cyclopropanes," *Chem. Rev.* **1989**, *89*, 1247–70.

3   Baird, M. S., "Functionalized Cyclopropenes as Synthetic Intermediates," *Top. Curr. Chem.* **1988**, *144*, 137–209.

4   Binger, P.; and Büch, H. M., "Cyclopropenes and Methylenecyclopropanes as Multifunctional Reagents in Transition Metal Catalyzed Reactions," *Top. Curr. Chem.* **1987**, *135*, 77–151.

5   Protopopova, M. N.; and Shapiro, E. A., "Carbene Synthesis and Chemical Reactions of Cyclopropene-3-carboxylate Esters—Promising Intermediates for Organic Synthesis," *Russ. Chem. Rev.* **1989**, *58*, 667–81.

6   Burke, S. D.; and Grieco, P. A., "Intramolecular Reactions of Diazocarbonyl Compounds," *Org. React.* **1979**, *26*, 361–475.

7   Hudlicky, T.; Kutchan, T. M.; and Naqvi, S. M., "The Vinylcyclopropane–Cyclopentene Rearrangement," *Org. React.* **1985**, *33*, 247–335.

8   Ho, T.-L., *Carbocycle Construction in Terpene Synthesis*; VCH: New York, 1988, Chapter 10, pp. 516–31.

9   Burgess, K.; Ho, K.-K.; and Moye-Sherman, D., "Asymmetric Syntheses of 2,3-Methanoamino Acids," *Synlett* **1994**, 575–83.

10 Suckling, C. J., "The Cyclopropyl Group in Studies of Enzyme Mechanism and Inhibition," *Angew. Chem., Int. Ed. Engl.* **1988**, 27, 537–52.

11 Newcomb, M.; and Chestney, D. L., "A Hypertensive Mechanistic Probe for Distinguishing between Radical and Carbocation Intermediates," *J. Am. Chem. Soc.* **1994**, *116*, 9753–54.

12 Silverman, R. B.; Ding, C. Z.; Borrillo, J. L.; and Chang, J. T., "Mechanism-Based Enzyme Inactivation via a Diactivated Cyclopropane Intermediate," *J. Am. Chem. Soc.* **1993**, *115*, 2982–83.

13 Caldwell, R. A.; and Zhou, L., "Are Perpendicular Alkene Triplets Just 1,2-Biradicals? Studies with the Cyclopropylcarbinyl Clock," *J. Am. Chem. Soc.* **1994**, *116*, 2271–75.

14 Husbands, S.; Suckling, C. A.; and Suckling, C. J., "Latent Inhibitors. Part 10. The Inhibition of Carboxypeptidase A by Tetrapeptide Analogs Based on 1-Aminocyclopropane Carboxylic Acid," *Tetrahedron* **1994**, *50*, 9729–42.

15 Wong, H. N. C.; Hon, M.-Y.; Tse, C.-W.; Yip, Y.-C.; Tanko, J.; and Hudlicky, T., "Use of Cyclopropanes and Their Derivatives in Organic Synthesis," *Chem. Rev.* **1989**, *89*, 165–98.

16 Davies, H. M. L., "Addition of Ketocarbenes to Alkenes, Alkynes and Aromatic Systems," in *Comprehensive Organic Synthesis*; Trost, B. M., Ed.; Pergamon Press: New York, 1991, Chapter 4.8, 1031–67.

17 Reissig, H.-U., "Formation of C–C Bonds by [2+1] Cycloaddition," in *Stereoselective Synthesis* of *Houben-Weyl Methods of Organic Chemistry*, Vol. E 21c; Helmchen, G.; Hoffmann, R. W.; Mulzer, J.; and Schaumann, E., Eds.; Georg Thieme Verlag: New York, 1995.

18 "Carbene," Regitz, M., Ed.; Band E19b in *Houben–Weyl Methoden der Organischen Chemie*; Georg Thieme Verlag: New York, 1995.

19 Baird, M. S.; Dale, C. M.; Lytollis, W.; and Simpson, M. J., "A New Approach to Cyclopropene Fatty Acids," *Tetrahedron Lett.* **1992**, *33*, 1521–22.

20 Baird, M. S.; and Grehan, B., "A New Approach to Cyclopropene Fatty Acids Involving 1,2-Deiodination," *J. Chem. Soc., Perkin Trans. 1* **1993**, 1547–48.

21 Doss, G. A.; and Djerassi, C., "Sterols in Marine Invertebrates. 60. Isolation and Structure Elucidation of Four New Steroidal Cyclopropenes from the Sponge *Calyx podatypa*," *J. Am. Chem. Soc.* **1988**, *110*, 8124–28.

22 Silberrad, O.; and Roy, C. S., "Gradual Decomposition of Ethyl Diazoacetate," *J. Chem. Soc.* **1906**, *89*, 179–82.

23 Maas, G., "Transition-Metal Catalyzed Decomposition of Aliphatic Diazo Compounds—New Results and Applications in Organic Synthesis," *Top. Curr. Chem.* **1987**, *137*, 76–253.

24 Doyle, M. P., "Catalytic Methods for Metal Carbene Transformations," *Chem. Rev.* **1986**, *86*, 919–39.

25 Padwa, A.; and Krumpe, K. E., "Application of Intramolecular Carbenoid Reactions in Organic Synthesis," *Tetrahedron* **1992**, *48*, 5385–453.

26 Nefedov, O. M.; Shapiro, E. A.; and Dyatkin, A. B., "Diazoacetic Acids and Derivatives," in *Supplement B: The Chemistry of Acid Derivatives*; Patai, S., Ed.; Wiley: New York, 1992, Chapter 25.

27 Ye, T.; and McKervey, M.A., "Organic Synthesis with α-Diazocarbonyl Compounds," *Chem. Rev.* **1994**, *94*, 1091–160.

28 Doyle, M. P., "Asymmetric Cyclopropanation," in *Catalytic Asymmetric Synthesis*; Ojima, I., Ed.; VCH Publishers: New York, **1993**, Chapter 3.

29 Padwa, A.; and Austin, D. J., "Ligand Effects on the Chemoselectivity of Transition Metal Catalyzed Reactions of α-Diazo Carbonyl Compounds," *Angew. Chem., Int. Ed. Engl.* **1994**, *33*, 1797–815.

30 Tomilov, Yu. V.; Dokichev, V. A.; Dzhemilev, U. M.; and Nefedov, O. M., "Catalytic Decomposition of Diazomethane as a General Method for the Methylenation of Chemical Compounds," *Russ. Chem. Rev.* **1993**, *62*, 799–838.

31 Mende, U.; Raduchel, B.; Skuballa, W.; and Vorbrüggen, H., "New Simple Conversion of α,β-Unsaturated Carbonyl Compounds into Their Corresponding Cyclopropyl Ketones and Esters," *Tetrahedron Lett.* **1975**, 629–32.

32 Raduechel, B.; Mende, U.; Cleve, G.; Hoyer, G.-A., and Vorbrüeggen, H., "Prostaglandin Analogs. I. Synthesis of 13,14-Dihydro-13,14-methylene-PGF$_{2\alpha}$ and PGE$_2$," *Tetrahedron Lett.* **1975**, 633–36.

33 Kottwitz, J.; and Vorbrüggen, H., "Formation of Cyclopropanes from Strained Alkenes with Diazomethane/Palladium(II) Acetate," *Synthesis* **1975**, 636–37.

34 Dzhemilev, U. M.; Dokichev, V. A.; Sultanov, S. Z.; Khursan, S. L.; Nefedov, O. M.; Tomilov, Yu. V.; and Kostitsyn, A. B., "Interaction of Diazoalkanes and Unsaturated Compounds. 11. Relative Reactivity of Olefins in Cyclopropanation by Diazomethane with Palladium Catalysts," *Izv. Akad. Nauk., Ser. Khim.* **1992**, 2353–61.

35 Suda, M., "Cyclopropanation of Terminal Olefins Using Diazomethane/Palladium(II) Acetate," *Synthesis* **1981**, 714.

36 Majchrzak, M. W.; Kotelko, A.; and Lambert, J. B., "Palladium(II) Acetate, An Efficient Catalyst for Cyclopropanation Reactions with Ethyl Diazoacetate," *Synthesis* **1983**, 469–70.

37 Hallinan, K. O.; Crout, D. H. G.; and Errington, W., "Simple Synthesis of L- and D-Vinylglycine (2-Aminobut-3-enoic Acid) and Related Amino Acids," *J. Chem. Soc., Perkin Trans. 1* **1994**, 3537–43.

38 Shimamoto, K.; and Ohfune, Y., "New Routes to the Syntheses of *cis*-α-(Carboxycyclopropyl)glycines from L-Glutamic Acid. Conformationally Restricted Analogs of the Excitatory Neurotransmitter L-Glutamic Acid," *Tetrahedron Lett.* **1989**, *30*, 3803–4.

39 Shimamoto, K.; Ishida, M.; Shinozaki, H.; and Ohfune, Y., "Synthesis of Four Diastereomeric L-2-(Carboxycyclopropyl)glycines. Conformationally Constrained L-Glutamate Analogs," *J. Org. Chem.* **1991**, *56*, 4167–76.

40 Dzhemilev, U. M.; Dokichev, V. A.; Sultanov, S. Z.; Khursan, S. L.; Nefedov, O. M.; Tomilov, Yu. V.; and Kostitsyn, A. B., "Interaction of Diazoalkanes and Unsaturated Compounds. 11. Relative Reactivity of Olefins in Cyclopropanation by Diazomethane with Palladium Catalysts," *Bull. Russ. Acad. Sci.* **1992**, *41*, 1846–52.

41 Tomilov, Yu. V.; Bordakov. V. G.; Dolgii, I. E.; and Nefedov, O. M., "Reaction of Diazoalkanes with Unsaturated Compounds. 2. Cyclopropanation of Olefins by Diazomethane in the Presence of Palladium Compounds," *Izv. Akad. Nauk., Ser. Khim.* **1984**, *33*, 533–38.

42 Zefirov, N. S.; Kozhushkov, S. I.; Kuznetsova, T. S.; Kokoreva, O. V.; Lukin, K. A.; Ugrak, B. I.; and Tratch, S. S., "Triangulanes: Stereoisomerism and General Method of Synthesis," *J. Am. Chem. Soc.* **1990**, *112*, 7702–7.

43 Zefirov, N. S.; Kozhushkov, S. I.; Kuznetsova, T. S.; Lukin, K. A.; and Kazimirchik, I. V., "Vinylspiropentane," *Zh. Org. Khim.* **1988**, *24*, 673–78.

44 Dzhemilev, U. M.; Dokichev, V. A.; Maidanova, I. O.; Nefedov, O. M.; and Tomilov, Yu. V., "Interaction of Diazoalkanes with Unsaturated Compounds. 12. Stereochemistry of Cyclopropanation of Norbornenes with Diazomethane in the Presence of Transition Metal Complexes," *Russ. Chem. Bull.* **1993**, *42*, 697–700.

45 Solomon, R. G.; and Kochi, J. K., "Copper(I) Catalysis in Cyclopropanations with Diazo Compounds. The Role of Olefin Coordination," *J. Am. Chem. Soc.* **1973**, *95*, 3300–10.

46 Tomilov, Yu. V.; Kostitsyn, A. B.; Shulishov, E. V.; and Nefedov, O. M., "Palladium(II)-Catalyzed Cyclopropanation of Simple Allyloxy and Allylamino Compounds and of 1-Oxy-1,3-butadienes with Diazomethane," *Synthesis* **1990**, 246–48.

47 Doyle, M. P.; Tamblyn, W. H.; and Bagheri, V., "Highly Effective Catalytic Methods for Ylide Generation from Diazo Compounds. Mechanism of the Rhodium and Copper Catalyzed Reactions with Allylic Compounds," *J. Org. Chem.* **1981**, *46*, 5094–102.

48 Vallgaarda, J.; and Hacksell, U., "Stereoselective Palladium-Catalyzed Cyclopropanation of α,β-Unsaturated Carboxylic Acids Derivatized with Oppolzer's Sultam," *Tetrahedron Lett.* **1991**, *32*, 5625–28.

49 Vallgaarda, J.; Appelberg, U.; Csöregh, I.; and Hacksell, U., "Stereoselectivity and Generality of the Palladium-Catalyzed Cyclopropanation of α,β-Unsaturated Carboxylic Acids Derivatized with Oppolzer's Sultam," *J. Chem. Soc. Perkin Trans. 1* **1994**, 461–70.

50 Kurokawa, N.; and Ohfune, Y., "The Palladium(II)-Assisted Synthesis of (±)-α-(Methylenecyclopropyl)glycine and (±)-*trans*-α-(Carboxycyclopropyl)glycine, Two Bioactive Amino Acids," *Tetrahedron Lett.* **1985**, *26*, 83–84.

51 Abdallah, H.; Gree, R.; and Carrie, R., "Synthesis Asymetriques a l'aide d'Oxazolidines Chirales Derives de l'Ephedrine. Preparation de Formyl Cyclopropanes Chiraux," *Tetrahedron Lett.* **1982**, *23*, 503–6.

52 Charette, A. B.; and Côté, B., "Asymmetric Cyclopropanation of Allylic Ethers: Cleavage and Regeneration of the Chiral Auxiliary," *J. Org. Chem.* **1993**, *58*, 933–36.

53 Charette, A. B.; Côté, B.; and Marcoux, J.-F., "Carbohydrates as Chiral Auxiliaries: Asymmetric Cyclopropanation Reaction of Acyclic Olefins," *J. Am. Chem. Soc.* **1991**, *113*, 8166–67.

54 Pfaltz, A., "Chiral Semicorrins and Related Nitrogen Heterocycles as Ligands in Asymmetric Catalysis," *Acc. Chem. Res.* **1993**, *26*, 339–45.

55 Nefedov, O. M.; Tomilov, Yu. V.; Kostitsyn, A. B.; Dzhemilev, U. M.; and Dokitchev, V. A., "Cyclopropanation of Unsaturated Compounds with Diazomethane Generated *in situ*. A New Efficient and Practical Route to Cyclopropane Derivatives," *Mendeleev Commun.* **1992**, 13–15.

56 Regitz, M.; and Maas, G., *Aliphatic Diazo Compounds—Properties and Synthesis*; Academic Press: New York, 1986.

57 Srikrishna, A.; and Krishnan, K., "Total Synthesis of (±)-Cyclolaurene, (±)-Epicyclolaurene, and (±)-β-Cuparenones," *Tetrahedron* **1992**, *48*, 3429–36.

58 Yadav, J. S.; Mysorekar, S. V.; and Rao, A. V. R., "Synthesis of (1*R*)-(+)-*cis*-Chrysanthemic Acid," *Tetrahedron* **1989**, *45*, 7353–60.

59 Leeper, F. J.; Padmanabhan, P.; Kirby, G. W.; and Sheldrake, G. N., "Biosynthesis of the Indolizidine Alkaloid, Cyclizidine," *J. Chem. Soc., Chem. Commun.* **1987**, 505–6.

60 Artt, D.; Jautelat, M.; and Lantzsch, R., "Syntheses of Pyrethroid Acids," *Angew. Chem., Int. Ed. Engl.* **1981**, *20*, 703–22.

61 Bowers, W. S.; Ebing, W.; Martin, D.; and Wegler, R., Eds., *Chemistry of Plant Protection*, "Synthetic Pyrethroid Insecticides: Structures and Properties"; Springer: Berlin, 1990; Vol. 4.

62 Noyori, R., "Chiral Metal Complexes as Discriminating Molecular Catalysts," *Science* **1990**, *248*, 1194–99.

63 Stammer, C. H., "Cyclopropane Amino Acids," *Tetrahedron* **1990**, *46*, 2231–54.

64 Schimamoto, K.; and Ohfune, Y., "Synthesis of 3′-Substituted-2-(Carboxycyclopropyl)glycines via Intramolecular Cyclopropanation. The Folded Form of L-Glutamate Activates the Non-NMDA Receptor Subtype," *Tetrahedron Lett.* **1990**, *31*, 4049–52.

65 Lee, M.; Lee, D.; Zhao, Y.; Newton, M. G.; Chun, M. W.; and Chu, C. K., "Enantioselective Synthesis of Cyclopropyl Carbocyclic L-Nucleosides," *Tetrahedron Lett.* **1995**, *36*, 3499–502.

66 Skiebe, A.; and Hirsch, A., "A Facile Method for the Synthesis of Amino Acid and Amido Derivatives of C_{60}," *J. Chem. Soc., Chem. Commun.* **1994**, 335–36.

67 Hummelen, J. C.; Knight, B. W.; LePeq, F.; and Wudl, F., "Preparation and Characterization of Fulleroid and Methanofullerene Derivatives," *J. Org. Chem.* **1995**, *60*, 532–38.

68 Wulfman, D. S.; Linstrumelle, G.; and Cooper, C. F., "Synthetic Applications of Diazoalkanes, Diazoazacyclopentadienes, and Diazoacyclopentadienes," in *The Chemistry of Diazonium and Diazo Groups*; Patai, S., Ed.; Wiley, New York, 1978; Part 2, Chapter 18.

69 Dave, V.; and Warnhoff, E., "The Reactions of Diazoacetic Esters with Alkenes, Alkynes, Heterocyclic and Aromatic Compounds," *Org. React. (N.Y.)* **1970**, *18*, 217–401.

70 Marchand, A. P.; and Brockway, N. M., "Carbalkoxycarbenes," *Chem. Rev.* **1974**, *74*, 431–69.

71 Stork, G.; and Ficini, J., "Intramolecular Cyclization of Unsaturated Diazoketones," *J. Am. Chem. Soc.* **1961**, *83*, 4678.

72 Doyle, M. P., "Metal Carbene Complexes in Organic Synthesis: Cyclopropanation," in *Comprehensive Organometallic Chemistry II*, Vol. 12; Hegedus, L. S., Ed.; Pergamon Press: New York, 1995, Chapter 5.1.

73 Kubas, G. J., "Tetrakis(acetonitrile)copper(I) Hexafluorophosphate," *Inorg. Synth.* **1979**, *19*, 90–92.

74 Müller, P.; and Fernandez, D., "Carbenoid Reactions in Rh(II)-Catalyzed Decomposition of Iodonium Ylides," *Helv. Chim. Acta* **1995**, *78*, 947–58.

75 Doyle, M. P.; Dorow, R. L.; Buhro, W. E.; Griffin, J. H.; Tamblyn, W. H.; and Trudell, M. L., "Stereoselectivity of Catalytic Cyclopropanation Reactions. Cata-

lyst Dependence in Reactions of Ethyl Diazoacetate with Alkenes," *Organometal.* **1984**, *3*, 44–52.

76 Doyle, M. P.; Griffin, J. H.; Bagheri, V.; and Dorow, R. L., "Correlations between Catalytic Reactions of Diazo Compounds and Stoichiometric Reactions of Transition-Metal Carbenes with Alkenes. Mechanism of the Cyclopropanation Reaction," *Organometal.* **1984**, *3*, 53–61.

77 Anciaux, A. J.; Hubert, A. J.; Noels, A. F.; Petiniot, N.; and Teyssie, P., "Transition-Metal-Catalyzed Reactions of Diazo Compounds. 1. Cyclopropanation of Double Bonds," *J. Org. Chem.* **1980**, *45*, 695–702.

78 Doyle, M. P.; Loh, K.-L.; DeVries, K. M.; and Chinn, M. S., "Enhancement of Stereoselectivity in Catalytic Cyclopropanation Reactions," *Tetrahedron Lett.* **1987**, *28*, 833–36.

79 Davies, H. M. L.; Clark, T. J.; and Church, L. A., "Stereoselective Cyclopropanations with Vinylcarbenoids," *Tetrahedron Lett.* **1989**, *30*, 5057–60.

80 O'Bannon, P. E.; and Dailey, W. P., "Nitrocyclopropanes from Nitrodiazomethanes. Preparation and Reactivity," *Tetrahedron* **1990**, *46*, 7341–58.

81 Doyle, M. P.; and van Leusen, D., "Rearrangements of Oxocyclopropanecarboxylate Esters to Vinyl Ethers. Disparate Behavior of Transition-Metal Catalysts," *J. Org. Chem.* **1982**, *47*, 5326–39.

82 Doyle, M. P.; Bagheri, V.; Wandless, T. J.; Harn, N. K.; Brinker, D. A.; Eagle, C. T.; and Loh, K.-L., "Exceptionally High Trans (Anti) Stereoselectivity in Catalytic Cyclopropanation Reactions," *J. Am. Chem. Soc.* **1990**, *112*, 1906–12.

83 Lowenthal, R. E.; and Masamune, S., "Asymmetric Copper-Catalyzed Cyclopropanation of Trisubstituted and Unsymmetrical cis-1,2-Disubstituted Olefins: Modified bis-Oxazoline Ligands," *Tetrahedron Lett.* **1991**, *32*, 7373–76.

84 Green, J.; Sinn, E.; Woodward, S.; and Butcher, R., "Mechanistic Insights into Catalytic Cyclopropanation by Copper(I) Phosphine Complexes. X-ray Crystal Structures of $[Cu(F-BF_3)(PCy_3)_2]$ (Cy = cyclo-C_6H_{11}) and $Cu(MeCN)_2\{1,2-C_6H_4CH_2NMe_2(PPh_2)\}]BF_4$," *Polyhedron* **1993**, *12*, 991–1001.

85 Doyle, M. P., "Electrophilic Metal Carbenes as Reaction Intermediates in Catalytic Reactions," *Acc. Chem. Res.* **1986**, *19*, 348–56.

86 Callot, H. J.; and Piechocki, C., "Cyclopropanation Using Rhodium(III) Porphyrins: Large *cis* vs *trans* Selectivity," *Tetrahedron Lett.* **1980**, *21*, 3489–92.

87 Maxwell, J. L.; O'Malley, S.; Brown, K. C.; and Kodadek, T., "Shape-Selective and Asymmetric Cyclopropanation of Alkenes Catalyzed by Rhodium Porphyrins," *Organometal.* **1992**, *11*, 645–52.

88 O'Malley, S.; and Kodadek, T., "Asymmetric Cyclopropanation of Alkenes Catalyzed by a Rhodium 'Chiral Fortress' Porphyrin," *Organometal.* **1992**, *11*, 2299–302.

89 Bartley, D. W.; and Kodadek, T., "Identification of the Active Catalyst in the Rhodium Porphyrin-Mediated Cyclopropanation of Alkenes," *J. Am. Chem. Soc.* **1993**, *115*, 1656–60.

90 Seitz, W. J.; Saha, A. K.; and Hossain, M. M., "Iron Lewis Acid Catalyzed Cyclopropanation Reactions of Ethyl Diazoacetate and Olefins," *Organometal.* **1993**, *12*, 2604–8.

91 Doyle, M. P.; Brandes, B. D.; Kazala, A. P.; Pieters, R. J.; Jarstfer, M. B.; Watkins, L. M.; and Eagle, C. T., "Chiral Rhodium(II) Carboxamides. A New

Class of Catalysts for Enantioselective Cyclopropanation Reactions," *Tetrahedron Lett.* **1990**, *31*, 6613–16.

92 Doyle, M. P.; Winchester, W. R.; Protopopova, M. N.; Müller, P.; Bernardinelli, G.; Ene, D.; and Motallebi, S., "Tetrakis[4(*S*)-4-phenyloxazolidin-2-one]dirhodium(II) and Its Catalytic Applications for Metal Carbene Transformations," *Helv. Chim. Acta* **1993**, *76*, 2227–35.

93 Doyle, M. P.; Zhou, Q.-L.; Simonsen, S. H.; and Lynch, V., "Dirhodium(II) Tetrakis[alkyl 2-oxoazetidine-4(*S*)-carboxylates]. A New Set of Effective Chiral Catalysts for Asymmetric Intermolecular Cyclopropanation Reactions with Diazoacetates," *Synlett* **1996**, 697–98.

94 Demonceau, A.; Saive, E.; de Froidmont, Y.; Noels, A. F.; Hubert, A. J.; Chizhevsky, I. T.; Lobanova, I. A.; and Bregadze, V. I., "Olefin Cyclopropanation Reactions Catalysed by Novel Ruthenacarborane Clusters," *Tetrahedron Lett.* **1992**, *33*, 2009–12.

95 Demonceau, A.; Abrcu Dias, E.; Lemoine, C. A.; Stumpf, A. W.; Noels, A. F.; Pietraszuk, C.; Gulinski, J.; and Marciniec, B., "Cyclopropanation of Activated Olefins Catalyzed by Ru–Phosphine Complexes," *Tetrahedron Lett.* **1995**, *36*, 3519–22.

96 Nishiyama, H.; Itoh, Y.; Matsumoto, H.; Park, S.-B.; and Itoh, K., "New Chiral Ruthenium bis(Oxazolinyl)pyridine Catalyst. Efficient Asymmetric Cyclopropanation of Olefins with Diazoacetates," *J. Am. Chem. Soc.* **1994**, *116*, 2223–24.

97 Smith, D. A.; Reynolds, D. N.; and Woo, L. K., "Cyclopropanation Catalyzed by Osmium Porphyrin Complexes," *J. Am. Chem. Soc.* **1993**, *115*, 2511–13.

98 Demonceau, A.; Lemoine, C. A.; and Noels, A. F., "Osmium Catalyzed Cyclopropanation of Olefins," *Tetrahedron Lett.* **1996**, *37*, 1025–26.

99 Fritschi, H.; Leutenegger, U.; and Pfaltz, A., "Semicorrin Metal Complexes as Enantioselective Catalysts. Part 2. Enantioselective Cyclopropane Formation from Olefins with Diazo Compounds Catalyzed by Chiral (Semicorrinato)copper Complexes," *Helv. Chim. Acta* **1988**, *71*, 1553–65.

100 Evans, D. A.; Woerpel, K. A.; Hinman, M. M.; and Faul, M. M., "bis-(Oxazolines) as Chiral Ligands in Metal-Catalyzed Asymmetric Reactions. Catalytic, Asymmetric Cyclopropanation of Olefins," *J. Am. Chem. Soc.* **1991**, *113*, 726–28.

101 Nakamura, A.; Konishi, A.; Tatsuno, Y.; and Otsuka, S., "A Highly Enantioselective Synthesis of Cyclopropane Derivatives through Chiral Cobalt(II) Complex Catalyzed Carbenoid Reaction. General Scope and Factors Determining the Enantioselecitvity," *J. Am. Chem. Soc.* **1978**, *100*, 3443–48, 6544–46.

102 Fukuda, T.; and Katsuki, T., "Co(III)-Salen Catalyzed Asymmetric Cyclopropanation," *Synlett* **1995**, 825–26.

103 Doyle, M. P.; Griffin, J. H.; and Conceicão, J. da, "Procatalysts for Carbenoid Transformations," *J. Chem. Soc., Chem. Commun.* **1985**, 328–29.

104 Wolf, J. R.; Hamaker, C. G.; Djukic, J.-P.; Kodadek, T.; and Woo, L. K., "Shape and Stereoselective Cyclopropanation of Alkenes Catalyzed by Iron Porphyrins," *J. Am. Chem. Soc.* **1995**, *117*, 9194–99.

105 Brown, K. C.; and Kodadek, T., "A Transition-State Model for the Rhodium Porphyrin-Catalyzed Cyclopropanation of Alkenes by Diazo Esters," *J. Am. Chem. Soc.* **1992**, *114*, 8336–38.

106 Doyle, M. P.; Dorow, R. L.; Tamblyn, W. H.; and Buhro, W. E., "Regioselectivity in Catalytic Cyclopropanation Reactions," *Tetrahedron Lett.* **1982**, *23*, 2261–64.

107 Pfaltz, A., "Chiral Semicorrins and Related Nitrogen Heterocycles as Ligands in Asymmetric Catalysis," *Acc. Chem. Res.* **1993**, *26*, 339–45.

108 Nishiyama, H.; Itoh, Y.; Sugaware, Y.; Matsumoto, H.; Aoki, K.; and Itoh, K., "Chiral Ruthenium(II)-bis(2-oxazolin-2-yl)pyridine Complexes. Asymmetric Catalytic Cyclopropanation of Olefins and Diazoacetates," *Bull. Chem. Soc. Jpn.* **1995**, *68*, 1247–62.

109 Casey, C. P.; Polichnowski, S. W.; Shusterman, A. J.; and Jones, C. R., "Reactions of $(CO)_5WCHC_6H_5$ with Alkenes," *J. Am. Chem. Soc.* **1979**, *101*, 7282–92.

110 Brookhart, M.; and Studabaker, W. B., "Cyclopropanes from Reactions of Transition-Metal–Carbene Complexes with Olefins," *Chem. Rev.* **1987**, *87*, 411–32.

111 Brookhart, M.; and Liu, Y., "Investigation of the Stereochemistry of Fe–Cα Bond Cleavage When Phenylcyclopropane Is Generated by γ-Ionization of Stereospecifically Deuterated $C_5H_5(CO)_2FeCHDCHDCH(OCH_3)C_6H_5$ Complexes. A Transition-State Model for Transfer of the Carbene Ligand from $C_5H_5(CO)_2Fe{=}CHR^+$ to Alkenes," *J. Am. Chem. Soc.* **1991**, *113*, 939–44.

112 Casey, C. P.; and Vosejpka, L. J. S., "Stereochemistry and Mechanism of Cyclopropane Formation from Ionization of $C_5H_5(CO_2)Fe(CH_2)_3X$," *Organometal.* **1992**, *11*, 738–44.

113 Brookhart, M.; Kegley, S. E.; and Husk, G. R., "Transfer of Ethylidene from η-$C_5H_5(CO_2)Fe{=}CHCH_3^+$ to Para-Substituted Styrenes. Loss of Stereochemistry about the C_α–C_β Double Bond of *cis*-β-Deuterio-*p*-methoxystyrene," *Organometal.* **1984**, *3*, 650–52.

114 Nozaki, H.; Moriuti, S.; Takaya, H.; and Noyori, R., "Asymmetric Induction in Carbenoid Reaction by Means of a Dissymmetric Copper Chelate," *Tetrahedron Lett.* **1966**, 5239–44.

115 Nozaki, H.; Takaya, H.; Moriuti, S.; and Noyori, R., "Homogeneous Catalysis in the Decomposition of Diazo Compounds by Copper Chelates," *Tetrahedron* **1968**, *24*, 3655–69.

116 Noyori, R.; Takaya, H.; Nakanisi, Y.; and Nozaki, H., "Partial Asymmetric Synthesis of Methylenecyclopropanes and Spiropentanes," *Can. J. Chem.* **1969**, *47*, 1242–45.

117 Aratani, T.; Yoneyoshi, Y.; and Nagase, T., "Asymmetric Synthesis of Chrysanthemic Acid. An Application of Copper Carbenoid Reaction," *Tetrahedron Lett.* **1975**, 1707–10.

118 Aratani, T.; Yoneyoshi, Y.; and Nagase, T., "Asymmetric Synthesis of Chrysanthemic Acid. An Application of Copper Carbenoid Reaction," *Tetrahedron Lett.* **1977**, 2599–602.

119 Aratani, T.; Yoneyoshi, Y.; and Nagase, T., "Asymmetric Synthesis of Permethric Acid. Stereochemistry of Chiral Copper Carbenoid Reaction," *Tetrahedron Lett.* **1982**, *23*, 685–88.

120 Aratani, T., "Catalytic Asymmetric Synthesis of Cyclopropane Carboxylic Acids: An Application of Chiral Copper Carbenoid Reaction," *Pure & Appl. Chem.* **1985**, *57*, 1839–44.

121 Becalski, A.; Cullen, W. R.; Fryzuk, M. D; Herb, G.; James, B. R.; Kutney, J. P.; Piotrowska, K.; and Tapiolas, D., "The Chemistry of Thujone. XII. The Synthesis of Pyrethroid Analogues via Chiral Cyclopropanation," *Can. J. Chem.* **1988**, *66*, 3108–15.

122 Laidler, D. A.; and Milner, D. J., "Asymmetric Synthesis of Cyclopropane Carboxylates. Catalysis of Diazoacetate Reactions by Copper(II) Schiff Base Complexes Derived from α-Amino Acids," *J. Organometal. Chem.* **1984**, *270*, 121–29.

123 Kunz, T.; and Reissig, H.-U., "Enantioselective Synthesis of Siloxycyclopropanes and of γ-Oxocarboxylates by Asymmetric Catalysis," *Tetrahedron Lett.* **1989**, *30*, 2079–82.

124 Brunner, H.; and Miehling, W., "Asymmetric Catalysis. 21. Enantioselective Cyclopropanation of 1,1-Diphenylethylene and Diazoacetic Acid Ester with Copper Catalysts," *Monatsh. Chem.* **1984**, *115*, 1237–54.

125 Fritschi, H.; Leutenegger, U.; and Pfaltz, A., "Chiral Copper–Semicorrin Complexes as Enantioselective Catalysts for the Cyclopropanation of Olefins by Diazo Compounds," *Angew. Chem. Int. Ed. Engl.* **1986**, *25*, 1005–6.

126 Pfaltz, A., "Enantioselective Catalysis with Chiral Cobalt and Copper Complexes," in *Modern Synthetic Methods 1989*; Scheffold, R., Ed.; Springer: Berlin-Heidelberg, 1989, pp. 199–248.

127 Fritschi, H.; Leutenegger, U.; Siegmann, K.; Pfaltz, A.; Keller, W.; and Kratky, Ch., "Semicorrin Metal Complexes as Enantioselective Catalysts. Part 1. Synthesis of Chiral Semicorrin Ligands and General Concepts," *Helv. Chim. Acta* **1988**, *71*, 1541–52.

128 Leutenegger, U.; Umbricht, G.; Fahrni, C.; von Matt, P.; and Pfaltz, A., "5-Aza-Semicorrins: A New Class of Bidentate Nitrogen Ligands for Enantioselective Catalysis," *Tetrahedron* **1992**, *48*, 2143–56.

129 Evans, D. A.; Woerpel, K. A.; and Scott, M. J., "'bis(Oxazolines)' as Ligands for Self-Assembling Chiral Coordination of Polymers—Structure of a Copper(I) Catalyst for the Enantioselective Cyclopropanation of Olefins," *Angew. Chem. Int. Ed. Engl.* **1992**, *31*, 430–32.

130 Lowenthal, R. E.; Abiko, A.; and Masamune, S., "Asymmetric Catalytic Cyclopropanation of Olefins: bis-Oxazoline Copper Complexes," *Tetrahedron Lett.* **1990**, *31*, 6005–8.

131 Müller, D.; Umbricht, G.; Weber, B.; and Pfaltz, A., "C_2-Symmetric 4,4',5,5'-Tetrahydrobi(oxazoles) and 4,4',5,5'-Tetrahydro-2,2'-methylenebis[oxazoles] as Chiral Ligands for Enantioselective Catalysis," *Helv. Chim. Acta* **1991**, *74*, 232–40.

132 Doyle, M. P., "Chiral Catalysts for Enantioselective Carbenoid Cyclopropanation Reactions," *Recl. Trav. Chim. Pays-Bas* **1991**, *110*, 305–16.

133 Ito, K.; Tabuchi, S.; and Katsuki, T., "Synthesis of New Chiral Bipyridine Ligands and Their Application to Asymmetric Cyclopropanation," *Synlett* **1992**, 575–76.

134 Ito, K.; and Katsuki, T., "Asymmetric Cyclopropanation of E-Olefins Using a Copper Complex of an Optically Active Bipyridine as a Catalyst," *Synlett* **1993**, 638–40.

135 Ito, K.; and Katsuki, T., "Catalytic Asymmetric Cyclopropanation Using Copper Complex of Optically Active Bipyridine as a Catalyst," *Tetrahedron Lett.* **1993**, *34*, 2661–64.

136 Gupta, A. D.; Bhuniya, D.; and Singh, V. K., "Synthesis of Homochiral Bis-oxazolinyl) Pyridine Type Ligands for Asymmetric Cyclopropanation Reactions," *Tetrahedron* **1994**, *50*, 13725–30.

137 Brunner, H.; Nishiyama, H.; and Itoh, K., in *Catalytic Asymmetric Synthesis*; Ojima, I., Ed.; VCH Publishers: New York, 1993, Ch. 6.

138 Christenson, D. L.; Tokar, C. J.; and Tolman, W. B., "New Copper and Rhodium Cyclopropanation Catalysts Supported by Chiral Bis(pyrazolyl)pyridines. A Metal-Dependent Enantioselectivity Switch," *Organometal.* **1995**, *14*, 2148–50.

139 Kanemasa, S.; Hamura, S.; Harada, E.; and Yamamoto, H., "C_2-Symmetric 1,2-Diamine/copper(II) Trifluoromethanesulfonate Complexes as Chiral Catalysts. Asymmetric Cyclopropanations of Styrene with Diazo Esters," *Tetrahedron Lett.* **1994**, *35*, 7985–88.

140 Tanner, D.; Andersson, P. G.; Harden, A.; and Somfai, P., "C_2-Symmetric Bis(Aziridines): A New Class of Chiral Ligands for Transition Metal-Mediated Asymmetric Synthesis," *Tetrahedron Lett.* **1994**, *35*, 4631–34.

141 Brunner, H.; Singh, U. P.; Boeck, T.; Altmann, S.; Scheck, T.; and Wrackmeyer, B., "Asymmetric Catalysis. 80. An Optically-Active Tetrakispyrazolylborate: Synthesis and Use in Cu-Catalyzed Enantioselective Cyclopropanation," *J. Organometal. Chem.* **1993**, *443*, C16–C18.

142 Pérez, P. J.; Brookhart, M.; and Templeton, J. L., "A Copper(I) Catalyst for Carbene and Nitrene Transfer to Form Cyclopropanes, Cyclopropenes, and Aziridines," *Organometal.* **1993**, *12*, 261–62.

143 Kennedy, M.; McKervey, M. A.; Maguire, A. R.; and Roos, G. H. P., "Asymmetric Synthesis in Carbon–Carbon Bond Forming Reactions of α-Diazoketones Catalysed by Homochiral Rhodium(II) Carboxylates," *J. Chem. Soc., Chem. Commun.* **1990**, 361–62.

144 Davies, H. M. L.; and Hutcheson, D. K., "Enantioselective Synthesis of Vinylcyclopropanes by Rhodium(II) Catalyzed Decomposition of Vinyldiazomethanes in the Presence of Alkenes," *Tetrahedron Lett.* **1993**, *34*, 7243–46.

145 Davies, H. M. L.; Bruzinski, P. R.; Lake, D. H.; Kong, N.; and Fall, M. J., "Asymmetric Cyclopropanations by Rhodium(II) N-(Arylsulfonyl)prolinate Catalyzed Decomposition of Vinyldiazomethanes in the Presence of Alkenes. Practical Enantioselective Synthesis of the Four Stereoisomers of 2-Phenylcyclopropan-1-amino Acid," *J. Am. Chem. Soc.* **1996**, *118*, 6897–907.

146 Corey, E. J.; and Grant, T. G., "A Catalytic Enantioselective Synthetic Route to the Important Antidepressant Sertraline," *Tetrahedron Lett.* **1994**, *35*, 5373–76.

147 Doyle, M. P.; Zhou, Q.-L.; Charnsangavej, C.; Longoria, M. A.; McKervey, M. A.; and Garcia, C. F., "Chiral Catalysts for Enantioselective Intermolecular Cyclopropanation Reactions with Methyl Phenyldiazoacetate. Origin of the Solvent Effect in Reations Catalyzed by Homochiral Dirhodium(II) Prolinates," *Tetrahedron Lett.* **1996**, *37*, 4129–32.

148 Davies, H. M. L.; Bruzinski, P. R.; and Fall, M. J., "Effect of Diazoalkane Structure on the Stereoselectivity of Rhodium(II) (S)-N-(Arylsulfonyl)prolinate Catalyzed Cyclopropanations," *Tetrahedron Lett.* **1996**, *37*, 4133–36.

149 Doyle, M. P.; Raab, C. E.; and Zhou, Q.-L., unpublished results.

150 Müller, P.; Baud, C.; Ené, D.; Motallebi, S.; Doyle, M. P.; Brandes, B. D.; Dyatkin, A. B.; and See, M. M., "Enantioselectivity and *cis/trans*-Selectivity in Dirhodi-

um(II)-Catalyzed Addition of Diazoacetates to Olefins," *Helv. Chim. Acta* **1995**, *78*, 459–70.

151 Doyle, M. P.; Winchester, W. R.; Hoorn, J. A. A.; Lynch, V.; Simonsen, S. H.; and Ghosh, R., "Dirhodium(II) Tetrakis(carboxamidates) with Chiral Ligands. Structure and Selectivity in Catalytic Metal–Carbene Transformations," *J. Am. Chem. Soc.* **1993**, *115*, 9968–78.

152 O'Malley, S.; and Kodadek, T., "Asymmetric Cyclopropanation of Alkenes Catalyzed by a 'Chiral Wall' Porphyrin," *Tetrahedron Lett.* **1991**, *32*, 2445–48.

153 Nishiyama, H.; Itoh, Y.; Sugawara, Y.; Matsumoto, H.; Aoki, K.; and Itoh, K., "Chiral Ruthenium(II)-Bis(2-oxazolin-2-yl)pyridine Complexes. Asymmetric Catalytic Cyclopropanation of Olefins and Diazoacetates," *Bull. Chem. Soc. Jpn.* **1995**, *68*, 1247–62.

154 Park, S.-B.; Nishiyama, H.; Itoh, Y.; and Itoh, K., "Trimethylsilylcarbene and Bis(trimethylsilyl) Formaldehyde Azine Complexes of Chiral Bis(4-isopropyloxazolinyl)pyridine(dichloro)ruthenium(II)," *J. Chem. Soc., Chem. Commun.* **1994**, 1315–16.

155 Park S.-B.; Sakata, N.; and Nishiyama, H., "Aryloxycarbonylcarbene Complexes of Bis(oxazolinyl)pyridineruthenium as Active Intermediates in Asymmetric Catalytic Cyclopropanations," *Chem. Eur. J.* **1996**, *2*, 303–6.

156 Park, S.-B.; Murata, K.; Matsumoto, H.; and Nishiyama, H., "Remote Electronic Control in Asymmetric Cyclopropanation with Chiral Ru–Pybox Catalysts," *Tetrahedron: Asymmetry* **1995**, *6*, 2487–94.

157 Tatsuno, Y.; Konishi, A.; Nakamura, A.; and Otsuka, S., "Enantioselective Synthesis of 2-Phenylcyclopropanecarboxylates through Chiral Cobalt Chelate Complex-Catalyzed Carbenoid Reactions," *J. Chem. Soc., Chem. Commun.* **1974**, 588–89.

158 Nakamura, A.; Konishi, A.; Tsujitani, R.; Kudo, M.; and Otsuka, S., "Enantioselective Carbenoid Cyclopropanation Catalyzed by Chiral *vic*-Dioximatocobalt(II) Complexes Prepared from Natural Camphor and β-Pinene. Mechanism and Stereochemistry," *J. Am. Chem. Soc.* **1978**, *100*, 3449–61.

159 Fukuda, T.; and Katsuki, T., "Co(III)-Salen Catalyzed Asymmetric Cyclopropanation," *Synlett* **1995**, 825–26.

160 Kunkel, E.; Reichelt, I.; and Reissig, H.-U., "Synthesis of 2-Siloxysubstituted Methyl Cyclopropanecarboxylates," *Liebigs Ann. Chem.* **1984**, 512–30.

161 Reissig, H.-U.; Reichelt, I.; and Kunz, T., "Methoxycarbonylmethylation of Aldehydes via Siloxycyclopropanes: Methyl 3,3-Dimethyl-4-oxobutanoate," *Org. Syn.* **1992**, *71*, 189–99.

162 Kunz, T.; Janowitz, A.; and Reissig, H.-U., "A Flexible Synthesis of Methyl 4-Oxobutanoates and Their Derivatives," *Synthesis* **1990**, 43–47.

163 Radunz, H.-E.; Reissig, H.-U.; Schneider, G.; and Rietmüller, A., "Synthesis of Enantiomerically Pure γ-Keto-δ-amino Acids, Central Building Blocks of Enzyme Inhibitors," *Liebigs Ann. Chem.* **1990**, 705–7.

164 Dammast, F.; and Reissig, H.-U., "Synthesis of Optically Active Siloxycyclopropanes by Asymmetric Catalysis. I. Influence of the Catalyst on the Cyclopropanation of (Z)-1-Phenyl-1-(trimethylsiloxy)prop-1-ene," *Chem. Ber.* **1993**, *126*, 2449–56.

165 Dammast, F.; and Reissig, H.-U., "Synthesis of Optically Active Siloxycyclopropanes by Asymmetric Catalysis. II. Influence of the Silyl Enol Ether Structure on the Enantioselective Cyclopropanation with Methyl Diazoacetate," *Chem. Ber.* **1993**, *126*, 2727–32.

166 Schnaubelt, J.; and Reissig, H.-U., "Chelate Controlled Intramolecular Diels–Alder Reaction as Key for Synthesis of a Dihydromevinolin Precursor," *Synlett* **1995**, 452–54.

167 Zschiesche, R.; Frey, B.; Grimm, E.; and Reissig, H.-U., "Stereoselectivity of Intramolecular Diels–Alder Reactions of 1,7,9-Decatrien-3-ones to Octalone Derivates," *Chem. Ber.* **1990**, *123*, 363–74.

168 Schnaubelt, J.; Marks, E.; and Reissig, H.-U., "[4+1]Cycloadditions of the Rhodium Di(methoxycarbonyl) Carbenoid to 2-Siloxy-1,3-dienes," *Chem. Ber.* **1996**, *129*, 73–75.

169 Grimm, E. L.; and Reissig, H.-U., "2-Siloxy-Substituted Methyl Cyclopropanecarboxylates as Building Blocks in Synthesis: Efficient One-Pot Conversion to γ-Butyrolactones," *J. Org. Chem.* **1985**, *50*, 242–44.

170 Schnaubelt, J.; Ullmann, A.; and Reissig, H.-U., "An Efficient One-Pot Synthesis of Functionalized Cyclodecenone Derivatives from Siloxycyclopropanes," *Synlett* **1995**, 1223–25.

171 Schnaubelt, J.; Zschiesche, R.; Reissig, H.-U.; Linder, H. J.; and Richter, J., "Syntheses of Estrone Derivatives from Siloxycylopropanecarboxylic Acid Esters," *Liebigs Ann. Chem.* **1993**, 61–70.

172 Hofmann, B.; and Reissig, H.-U., "Synthesis of New Alkenyl-Substituted 2-(*tert*-Butyldimethylsiloxy)cyclopropanecarboxylates and Their Diastereoselective Conversion into (Hydroxymethyl)cyclopropanes," *Chem. Ber.* **1994**, *127*, 2315–25.

173 Hofmann, B.; and Reissig, H.-U., "Swern Oxidation of Alkenyl-Substituted 2-(*tert*-Butyldimethylsiloxy)-1-(hydroxymethyl)cyclopropanes: A Novel and Flexible Route to Functionalized 2,5-Dihydrooxepines," *Chem. Ber.* **1994**, *127*, 2327–35.

174 Hoberg, J. O.; and Claffey, D. J., "Cyclopropanation of Unsaturated Sugars with Ethyl Diazoacetate," *Tetrahedron Lett.* **1996**, *37*, 2533–36.

175 Biggs, T. N.; and Swenton, J. S., "Thermal and Photochemical Rearrangements of Cyclopropyl Ethers of *p*-Quinols. Competing Reaction Pathways Leading to Five- and Six-Membered Spirocyclic Ketones," *J. Org. Chem.* **1992**, *57*, 5568–73.

176 Arenare, L.; De Caprariis, P.; Marinozzi, M.; Natalini, B.; and Pellicciari, R., "Synthesis of 2-Azabicyclo[3.1.0]hexane Tricarboxylate and its Transformation into a New Proline-γ-acetic Acid Equivalent," *Tetrahedron Lett.* **1994**, *35*, 1425–26.

177 Paulini, K.; and Reissig, H.-U., "An Efficient Route to GABA-Analogous Amino Acids: Cyclopropanation of *N*-Silylated Allylamines and Enamines," *Liebigs Ann. Chem.* **1991**, 455–61.

178 Davies, H. M. L.; Clark, T. J.; and Smith, H. D., "Stereoselective Synthesis of Seven-Membered Carbocycles by a Tandem Cyclopropanation/Cope Rearrangement between Rhodium(II)-Stabilized Vinylcarbenoids and Dienes," *J. Org. Chem.* **1991**, *56*, 3817–24.

179 Davies, H. M. L.; Smith, H. D.; and Korkor, O., "Tandem Cyclopropanation/Cope Rearrangement Sequence. Stereospecific [3+4] Cycloaddition Reaction of Vinylcarbenoids with Cyclopentadiene," *Tetrahedron Lett.* **1987**, *28*, 1853–56.

180 Davies, H. M. L., "Tandem Cyclopropanation/Cope Rearrangement: A General Method for the Construction of Seven-Membered Rings," *Tetrahedron* **1993**, *49*, 5203–23.

181 Davies, H. M. L.; Saikali, E.; and Young, W. B., "Synthesis of (±)-Ferruginine and (±)-Anhydroecgonine Methyl Ester by a Tandem Cyclopropanation/Cope Rearrangement," *J. Org. Chem.* **1991**, *56*, 5696–700.

182 Davies, H. M. L.; Clark, T. J.; and Kimmer, G. F., "Versatile Synthesis of Tropones by Reaction of Rhodium(II)-Stabilized Vinylcarbenoids with 1-Methoxy-1-[(trimethylsilyl)oxy]buta-1,3-diene," *J. Org. Chem.* **1991**, *56*, 6440–47.

183 Davies, H. M. L.; Hougland, P. W.; and Cantrell, W. R., Jr., "Convenient Synthesis of Vinyldiazomethanes from α-Diazo-β-keto Esters and Related Systems," *Synth. Commun.* **1992**, *22*, 971–78.

184 Davies, H. M. L.; and Clark, T. J., "Synthesis of Highly Functionalized Tropolones by Rhodium(II)-Catalyzed Reactions of Vinyldiazomethanes with Oxygenated Dienes," *Tetrahedron* **1994**, *50*, 9883–92.

185 Cantrell, W. R., Jr.; and Davies, H. M. L., "Stereoselective Convergent Synthesis of Hydroazulenes via an Intermolecular Cyclopropanation/Cope Rearrangement," *J. Org. Chem.* **1991**, *56*, 723–27.

186 Davies, H. M. L.; Saikali, E.; Huby, N. J. S.; Gilliatt, V. J.; Matasi, J. J.; Sexton, T.; and Childers, S. R., "Synthesis of 2β-Acyl-3β-aryl-8-azabicyclo[3.2.1]octanes and Their Binding Affinities at Dopamine and Serotonin Transport Sites in Rat Striatum and Frontal Cortex," *J. Med. Chem.* **1994**, *37*, 1262–68.

187 Davies, H. M. L.; Clark, D. M.; Alligood, D. B.; and Eiband, G. R., "Mechanistic Aspects of Formal [3+4] Cycloadditions between Vinylcarbenoids and Furans," *Tetrahedron* **1987**, *43*, 4265–70.

188 Davies, H. M. L.; Peng, Z.-Q.; and Houser, J. H., "Asymmetric Synthesis of 1,4-Cycloheptadienes and Bicyclo[3.2.1]octa-2,6-dienes by Rhodium(II) N-[p-(tert-Butyl)phenylsulfonyl]prolinate Catalyzed Reactions Between Vinyldiazomethanes and Dienes," *Tetrahedron Lett.* **1994**, *35*, 8939–42.

189 Yoshikawa, K.; and Achiwa, K., "Steric and Electronic Effects of Substrates and Rhodium Chiral Catalysts in Asymmetric Cyclopropanation," *Chem. Pharm. Bull.* **1996**, *43*, 2048–53.

190 Davies, H. M. L.; and Hu, B., "Regioselective [3+2] Annulations with Rhodium(II) Stabilized Vinylcarbenoids," *Tetrahedron Lett.* **1992**, *33*, 453–56.

191 Davies, H. M. L.; and Hu, B., "Highly Stereoselective 3+2 Annulations by Cyclopropanation of Vinyl Ethers with Rhodium(II)-Stabilized Vinylcarbenoids Followed by a Formally Forbidden 1,3-Sigmatropic Rearrangement," *J. Org. Chem.* **1992**, *57*, 3186–90.

192 Davies, H. M. L.; Hu, B.; Saikali, E.; and Bruzinski, P. R., "Carbenoid versus Vinylogous Reactivity in Rhodium(II)-Stabilized Vinylcarbenoids," *J. Org. Chem.* **1994**, *59*, 4535–41.

193 Corey, E. J.; and Kigoshi, H., "A Route for the Enantioselective Total Synthesis of Antheridic Acid, the Antheridium-Inducing Factor from *Anemia phyllitidis*," *Tetrahedron Lett.* **1991**, *32*, 5025–28.

194   Davies, H. M. L.; and Hu, B., "Ring Expansion of *tert*-Butyl 1-Vinylcyclopropane-1-carboxylates to α-Ethylidenebutyrolactones," *J. Org. Chem.* **1992**, *57*, 4309–12.

195   Davies, H. M. L.; and Hu, B. H., "Synthesis of Ether Analogs of (±)-Acetomycin," *Heterocycles* **1993**, *35*, 385–93.

196   Wenkert, E., "Oxocyclopropanes in Organochemical Synthesis," *Acc. Chem. Res.* **1980**, *13*, 27–31.

197   Wenkert, E., "Furan Synthesis," *Heterocycles* **1980**, *14*, 1703–8.

198   Wenkert, E.; Ananthanarayan, T. P.; Ferreira, V. F.; Hoffmann, M. G.; and Kim, H. S., "Simple Syntheses of β-Furoic Esters and γ-Pyrone," *J. Org. Chem.* **1990**, *55*, 4975–76.

199   Alonso, M. E.; Jano, P.; Hernandez, M. I.; Greenberg, R. S.; and Wenkert, E., "A Dihydro-α-furoic Ester Synthesis by the Catalyzed Reactions of Ethyl Diazopyruvate with Enol Ethers," *J. Org. Chem.* **1983**, *48*, 3047–50.

200   Alonso, M. E.; Morales, A.; and Chitty, A. W., "Studies on the Origin of Dihydrofurans from α-Diazocarbonyl Compounds. Concerted 1,3-Dipolar Cycloaddition vs. Nonsynchronous Coupling in the Copper Chelate Catalyzed Reactions of α-Diazocarbonyl Compounds with Electron-Rich Olefins," *J. Org. Chem.* **1982**, *47*, 3747–54.

201   Pirrung, M. C.; and Lee, Y. R., "Hydroxy Direction of the Rhodium-Mediated Dipolar Cycloaddition of Cyclic Carbenoids with Vinyl Ethers," *J. Chem. Soc., Chem. Commun.* **1995**, 673–74.

202   Pirrung, M. C.; and Lee, Y. R., "Dipolar Cycloaddition of Rhodium Carbenoids with Vinyl Esters. Total Synthesis of Pongamol and Lanceolatin B," *Tetrahedron Lett.* **1994**, *35*, 6231–34.

203   Pirrung, M. C.; Zhang, J.; Lackey, K.; Sternbach, D. D.; and Brown, F., "Reaction of a Cyclic Rhodium Carbenoid with Aromatic Compounds and Vinyl Ethers," *J. Org. Chem.* **1995**, *60*, 2112–24.

204   Pirrung, M. C.; Zhang, J.; and McPhail, A. T., "Dipolar Cycloaddition of Cyclic Rhodium Carbenoids to Aromatic Heterocycles," *J. Org. Chem.* **1991**, *56*, 6269–71.

205   Pirrung, M. C.; Zhang, J.; and Morehead, A. T., Jr., "Dipolar Cycloaddition of Cyclic Rhodium Carbenoids to Diagonal Carbon. Synthesis of Isoeuparin," *Tetrahedron Lett.* **1994**, *35*, 6229–30.

206   Pirrung, M. C.; and Lee, Y. R., "Total Synthesis and Absolute Configuration of Pseudosemiglabrin, a Platelet Aggregation Antagonist, and its Diastereomer Semiglabrin," *J. Am. Chem. Soc.* **1995**, *117*, 4814–21.

207   Lee, Y. R., "A Concise New Synthesis of Angular Furanocoumarins: Angelicin, Oroselone & Oroselol," *Tetrahedron* **1995**, *51*, 3087–94.

208   Lee, Y. R.; and Morehead, A. T., Jr., "A New Route for the Synthesis of Furanoflavone and Furanochalcone Natural Products," *Tetrahedron* **1995**, *51*, 4909–22.

209   Pirrung, M. C.; and Lee, Y. R., "Formal Total Synthesis of (±)-Aflatoxin B_2 Utilizing the Rhodium Carbenoid Dipolar Cycloaddition," *Tetrahedron Lett.* **1996**, *37*, 2391–94.

210   Rosenfeld, M. J.; Shankar, B. K. R.; and Shechter, H., "Rhodium(II) Acetate Catalyzed Reactions of 2-Diazo-1,3-indandione and 2-Diazo-1-indanone with Various Substrates," *J. Org. Chem.* **1988**, *53*, 2699–705.

211   Pirrung, M. C.; and Zhang, J., "Asymmetric Dipolar Cycloaddition Reactions of Diazocompounds Mediated by a Binaphtholphosphate Rhodium Catalyst," *Tetrahedron Lett.* **1992**, *33*, 5987–90.

212   Elliott, M.; and Janes, N. F., "Synthetic Pyrethroids —A New Class of Insecticide," *Chem. Soc. Rev.* **1978**, *7*, 473–505.

213   Marino, J. P.; Silveira, C.; Comasseto, J.; and Petragnani, N., "A Total Synthesis of Pentalenolactone E Methyl Ester via a [3+2] Annulation Strategy," *J. Org. Chem.* **1987**, *52*, 4139–40.

214   Wenkert, E.; and Hudlicky, T., "Synthesis of Eburnamonine and Dehydroaspidospermidine," *J. Org. Chem.* **1988**, *53*, 1953–57.

215   Schnaubelt, J.; and Reissig, H.-U., "Chelate Controlled Intramolecular Diels–Alder Reaction as a Key for Synthesis of a Dihydromevinolin Precursor," *Synlett* **1995**, 452–53.

216   Zschiesche, R.; and Reissig, H.-U., "Efficient Syntheses of Polyfunctional Nitro Compounds Using 2-Alkenyl-Substituted Methyl 2-Siloxycyclopropanecarboxylates as Key Building Blocks," *Justus Liebigs Ann. Chem.* **1988**, 1165–68.

217   Reissig, H.-U., "Donor–Acceptor–Substituted Cyclopropanes: Versatile Building Blocks in Organic Synthesis," *Top. Curr. Chem.* **1988**, *144*, 73–135.

218   Kirmse, W.; Hellwig, G.; and van Chiem, P., "Addition of Diazocyclopropanes to Carbonyl Compounds," *Chem. Ber.* **1986**, *119*, 1511–24.

219   Wilson, S. R.; and Zucker, P. A., "Silicon-Mediated Skipped Diene Synthesis. Application to the Melon Fly Pheromone," *J. Org. Chem.* **1988**, *53*, 4682–93.

220   Pellicciari, R.; Natalini, B.; Marinozzi, M.; Monahan, J. B.; and Snyder, J. P., "D-3,4-Cyclopropylglutamate Isomers as NMDA Receptor Ligands—Synthesis and Enantioselective Activity," *Tetrahedron Lett.* **1990**, *31*, 139–42.

221   Pellicciari, R.; Natalini, B.; Marinozzi, M.; Sadeghpour, B. M.; Cordi, A. A.; Lanthorn, T. H.; Hood, W. F.; and Monahan, J. B., "Synthesis, Absolute-Configuration and Activity at *N*-Methyl-D-aspartic Acid (NMDA) Receptor of the Four D-2-Amino-4,5-methanoadipate Diastereoisomers," *Farmaco* **1991**, *46*, 1243–64.

222   Cativiela, C.; Diaz-de-Villegas, M. D.; and Jiménez, A. I., "A Simple Synthesis of (−)-(1*S*,2*R*)-Allocoronamic Acid in its Enantiomerically Pure Form," *Tetrahedron: Asymmetry* **1995**, *6*, 177–82.

223   Paulini, K.; and Reissig, H.-U., "Preparation of Novel Lipophilic GABA Analogs Containing Cyclopropane Rings via Cyclopropanation of *N*-Silylated Unsaturated Amines," *J. Prakt. Chem.* **1995**, *337*, 55–59.

224   Marinozzi, M.; Natalini, B.; Costantino, G.; Pellicciari, R.; Bruno, V.; and Nicoletti, F., "Synthesis of 6,6-Dicarboxy-3,4-Methano-L-Proline, A New Constrained Glutamate Analog Endowed with Neuroprotective Properties," *Farmaco* **1996**, *51*, 121–24.

225   Huval, C. C.; and Singleton, D. A., "Versatile [3+2] Methylene-cyclopentane Annulations of Unactivated and Electron-Rich Olefins with [(Trimethylsilyl)-methylene]cyclopropanedicarboxylates," *J. Org. Chem.* **1994**, *59*, 2020–24.

226   Timmers, C. M.; Leeuwenburgh, M. A.; Verheijen, J. C.; van der Marel, G. A.; and van Boom, J. H., "Rhodium(II) Catalyzed Asymmetric Cyclopropanation of Glycals with Ethyl Diazoacetate," *Tetrahedron: Asymmetry* **1996**, *7*, 49–52.

227 O'Bannon, P. E.; and Dailey, W. P., "Catalytic Cyclopropanation of Alkenes with Ethyl Nitrodiazoacetate. A Facile Synthesis of Ethyl 1-Nitrocyclopropanecarboxylates," *J. Org. Chem.* **1989**, *54*, 3096–101.

228 Shi, G. Q.; and Xu, Y. Y., "Trifluoromethyl-Substituted Carbethoxy Carbene as a Novel CF_3-Containing a^2 Synthon Equivalent for the Preparation of 2-(Trifluoromethyl)-4-oxocarboxylic Ester Derivatives: Highly Functionalized Synthetic Building Blocks Bearing a CF_3 Group," *J. Org. Chem.* **1990**, *55*, 3383–86.

229 DeMeijere, A.; Schulz, T. J.; Kostikov, R. R.; Graupner, F.; Murr, T.; and Bielfeldt, T., "New Cyclopropyl Building-Blocks for Organic Synthesis. 4. Dirhodium(II) Tetraacetate Catalyzed (Chlorovinyl)cyclopropanation of Enol Ethers and Dienol Ethers—A Route to Donor-Substituted Vinylcyclopropanes, Ethynylcyclopropanes and Cycloheptadienes," *Synthesis* **1991**, 547–60.

230 Nunn, J. R., "The Structure of Sterculic Acid," *J. Chem. Soc.* **1952**, 313–18.

231 Fattorusso, E.; Magno, S.; Mayol, L.; Santacroce, C.; and Sica, D., "Calysterol: A C_{29} Cyclopropene-Containing Marine Sterol from the Sponge *Calyx Nicaensis*," *Tetrahedron* **1975**, *31*, 1715–16.

232 Petiniot, N.; Anciaux, A. J.; Noels, A. F.; Hubert, A. J.; and Teyssié, Ph., "Rhodium Catalysed Cyclopropenation of Acetylenes," *Tetrahedron Lett.* **1978**, *14*, 1239–42.

233 Müller, P.; Pautex, N.; Doyle, M. P.; and Bagheri, V., "Rh(II) Catalyzed Isomerizations of Cyclopropenes. Evidence for Rh(II)-Complexed Vinylcarbene Intermediates," *Helv. Chim. Acta* **1990**, *73*, 1233–41.

234 Shapiro, E. A.; Kalinin, A. V.; and Nefedov, O. M., "Selective Catalytic Methoxycarbonylmethylenation of the Triple Bond of Vinylacetylene with Methyl Diazoacetate," *Mendeleev Commun.* **1992**, 116–17.

235 Shapiro, E. A.; Kalinin, A. V.; Ugrak, B. I.; and Nefedov, O. M., "Regioselective $Rh_2(OAc)_4$-Promoted Reactions of Methyl Diazoacetate with Terminal Triple Bond Enynes," *J. Chem. Soc., Perkin Trans. 2* **1994**, 709–13.

236 Shapiro, E. A.; Kalinin, A. V.; Platonov, D. N.; and Nefedov, O. M., "Catalytic Reaction of Methyl Diazoacetate with Silylated Enynes," *Russ. Chem. Bull.* **1993**, *42*, 1191–95.

237 Doyle, M. P.; Protopopova, M.; Müller, P.; Ene, D.; and Shapiro, E. A., "Effective Uses of Dirhodium(II) Tetrakis[methyl 2-oxopyrrolidine-5(R or S)-carboxylate] for Highly Enantioselective Intermolecular Cyclopropenation Reactions," *J. Am. Chem. Soc.* **1994**, *116*, 8492–98.

238 Doyle, M. P.; Protopopova, M. N.; Brandes, B. D.; Davies, H. M. L.; Huby, N. J. S.; and Whitesell, J. K., "Diastereoselectivity Enhancement in Cyclopropanation and Cyclopropenation Reactions of Chiral Diazoacetate Esters Catalyzed by Chiral Dirhodium(II) Carboxamides," *Synlett* **1993**, 151–53.

239 Protopopova, M. N.; Doyle, M. P.; Müller, P.; and Ene, D., "High Enantioselectivity for Intermolecular Cyclopropenation of Alkynes by Diazo Esters Catalyzed by Chiral Dirhodium(II) Carboxamides," *J. Am. Chem. Soc.* **1992**, *114*, 2755–57.

240 Müller, P.; and Gränicher, C., "Structural Effects on the Rh(II)-Catalyzed Rearrangement of Cyclopropenes," *Helv. Chim. Acta* **1993**, *76*, 521–34.

241 Müller, P.; and Gränicher, C., "Selectivity in Rhodium(II)-Catalyzed Rearrangements of Cycloprop-2-ene-1-carboxylates," *Helv. Chim. Acta* **1995**, *78*, 129–44.

242 Müller, P.; Gränicher, C.; Klärner, F.-G.; and Breitkopf, V., "Metallocarbene Intermediates in Rh(II)-catalyzed Rearrangements of Ethyl 2-Butylcycloprop-2-ene-1-carboxylate," *Gazz. Chim. Ital.* **1995**, *125*, 459–63.

243 Binger, P.; Müller, P.; Benn, R.; and Mynott, R., "Vinylcarbene Complexes of Titanocene," *Angew. Chem. Int. Ed. Engl.* **1989**, *28*, 610–11.

244 Binger, P.; Müller, P.; Herrmann, A. T.; Philipps, P.; Gabor, B.; Langhauser, F.; and Krüger, C., "Metallocyclobutenes via η^2-Cyclopropene Complexes of Titanocene and Zirconocene," *Chem. Ber.* **1991**, *124*, 2165–70.

245 Johnson, L. K.; Grubbs, R. H.; and Ziller, J. W., "Synthesis of Tungsten Vinyl Alkylidene Complexes via the Reactions of $WCl_2(NAr)(PX_3)_3$ ($X = R$, OMe) Precursors with 3,3-Disubstituted Cyclopropenes," *J. Am. Chem. Soc.* **1993**, *115*, 8130–45.

246 Nguyen, S. T.; Johnson, L. K.; and Grubbs, R. H., "Ring-Opening Metathesis Polymerization (ROMP) of Norbornene by a Group VIII Carbene Complex in Protic Media," *J. Am. Chem. Soc.* **1992**, *114*, 3974–75.

247 O'Connor, J. M.; Ji, H.; Iranpour, M.; and Rheingold, A. L., "Formation of a Stable Metallacyclobutene Complex from α-Diazocarbonyl and Alkyne Substrates," *J. Am. Chem. Soc.* **1993**, *115*, 1586–88.

248 Paulini, K.; and Reissig, H.-U., "Efficient Synthesis of a Novel GABA Analogue Incorporating a Cyclopropene Ring," *Synlett* **1992**, 505–6.

249 Paulini, K.; and Reissig, H.-U., "Synthesis of Dipeptides Containing Novel Cyclopropyl- and Cyclopropenyl-Substituted β- and γ-Amino Acids," *Liebigs Ann. Chem.* **1994**, 549–54.

250 Padwa, A.; and Woolhouse, A. D., "Aziridines, Azines and Fused Ring Derivatives," in *Comprehensive Heterocyclic Chemistry*; Lwowski, W., Ed.; Pergamon Press: Oxford, 1984, Vol. 7.

251 Tanner, D., "Chiral Aziridines—Their Synthesis and Use in Stereoselective Transformations," *Angew. Chem. Int. Ed. Engl.* **1994**, *33*, 599–619.

252 Hubert, A. J.; Feron, A.; Warin, R.; and Teyssie, P., "Synthesis of Iminoaziridines from Carbodiimides and Diazoesters," *Tetrahedron Lett.* **1976**, 1317–18.

253 Drapier, J.; Feron, A.; Warin, R.; Hubert, A. J.; and Teyssie, P., "Novel Iminooxazolines from Reactions of Diazoacetates with Carbodiimides (A Revision)," *Tetrahedron Lett.* **1979**, 559–60.

254 Rasmussen, K. G.; and Jorgensen, K. A., "Catalytic Formation of Aziridines from Imines and Diazoacetate," *J. Chem. Soc., Chem. Commun.* **1995**, 1401–2.

255 Hansen, K. B.; Finney, N. S.; and Jacobsen E. N., "Carbenoid Transfer to Imines: A New Asymmetric Catalytic Synthesis of Aziridines," *Angew. Chem. Int. Ed. Engl.* **1995**, *34*, 676–78.

256 Moran, M.; Bernardinelli, G.; and Müller, P., "Reactions of Diazo Compounds with Imines," *Helv. Chim. Acta* **1995**, *78*, 2048–52.

257 Kwart, H.; and Kahn, A. A., "Copper-Catalyzed Decomposition of Benzenesulfonyl Azide in Cyclohexene Solution," *J. Am. Chem. Soc.* **1967**, *89*, 1951–53.

258 Evans, D. A.; Faul, M. M.; and Bilodeau, M. T., "Copper-Catalyzed Aziridination of Olefins by [N-(p-Toluenesulfonyl)imino]phenyliodinane," *J. Org. Chem.* **1991**, *56*, 6744–46.

259   Evans, D. A.; Faul, M. M.; and Bilodeau, M. T., "Development of the Copper-Catalyzed Olefin Aziridination Reaction," *J. Am. Chem. Soc.* **1994**, *116*, 2742–53.

260   Müller, P.; Baud, C.; and Jacquier, Y., "A Method for Rhodium(II)-Catalyzed Aziridination of Olefins," *Tetrahedron* **1996**, *52*, 1543–48.

261   Li, Z.; Conser, K. R.; and Jacobsen, E. N., "Asymmetric Alkenes Aziridination with Readily Available Chiral Diimine-Based Catalysts," *J. Am. Chem. Soc.* **1993**, *115*, 5326–27.

262   Evans, D. A.; Faul, M. M.; Bilodeau, M. T.; Anderson, B. A.; and Barnes, D. M., "Bis(oxazoline)–Copper Complexes as Chiral Catalysts for the Enantioselective Aziridination of Olefins," *J. Am. Chem. Soc.* **1993**, *115*, 5328–29.

263   Noda, K.; Hosoya, N.; Irie, R.; Ito, Y.; and Katsuki, T., "Asymmetric Aziridination by Using Optically Active (Salen)mangenese(III) Complexes," *Synlett* **1993**, 469–71.

CHAPTER 5

Intramolecular Cyclopropanation and Related Addition Reactions

Since the first report of catalytic intramolecular cyclopropanation by Stork and Ficini in 1961,[1] a veritable wealth of synthetic applications with diazo compounds has been described.[2-8] The range of applicable diazo compounds extends from vinyl substitution to carbonyl substitution, with the vast majority being diazocarbonyl reactants (**1**). Catalysts that are effective are mainly those of copper(I) and dirhodium(II), although others may be useful but are untested. Intramolecular cyclization appears to prefer five-membered ring formation (**2**, $n = 1$),[7] although ring sizes up to 20 have been achieved (eq. 1).[9] Unlike intermolecular cyclopropa-

$$Z = H, SO_2R, COOR, COCH_3$$
$$Y = O, NR, CR_2$$
$$n = 0, 1, 2 \ldots.$$

1	**2**	(1)

nation reactions for which control of diastereoselectivity is a major consideration, with intramolecular reactions metal carbene addition gives only one stereoisomer ($n = 1,2,3$). The versatility of the methodology is due in large part to the accessibility of the diazocarbonyl reactants and to the stereocontrol inherent in the cyclization. The adaptability of the bicyclic product (**2**) to further structural modification provides this methodology with considerable synthetic advantage.

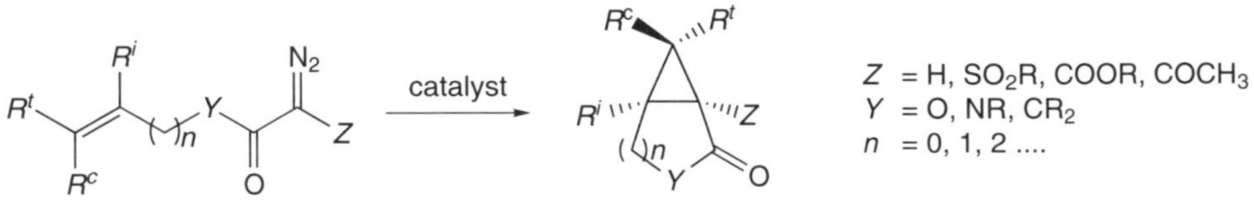

With catalytic intramolecular cyclopropenation, additional ring strain imparted by the bicyclic intermediate (**4**) facilitates metal-catalyzed ring opening (eq. 2).[5] Reaction products are often derived from a vinylogous metal carbene (**5**),

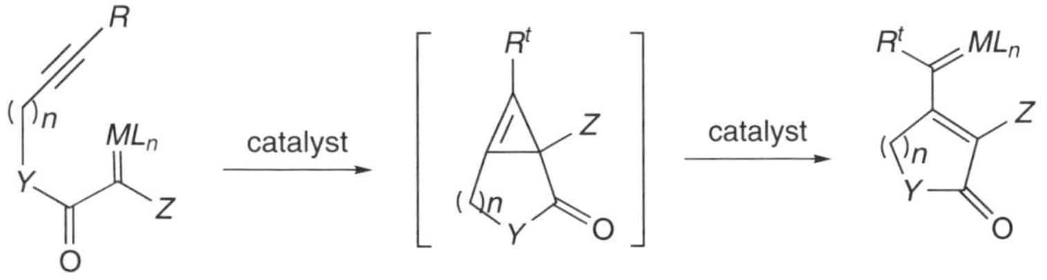

3	**4**	**5** (2)

which may undergo subsequent intra- or intermolecular reaction. The outcome of these reactions can thus result from a cascade of metal carbene transformations.

5.1 INTRAMOLECULAR CYCLOPROPANATION OF DIAZOKETONES

The first example of intramolecular cyclopropanation was established with the diazoketone derived from 5-hexenoic acid (eq. 3),[1] and this methodology became

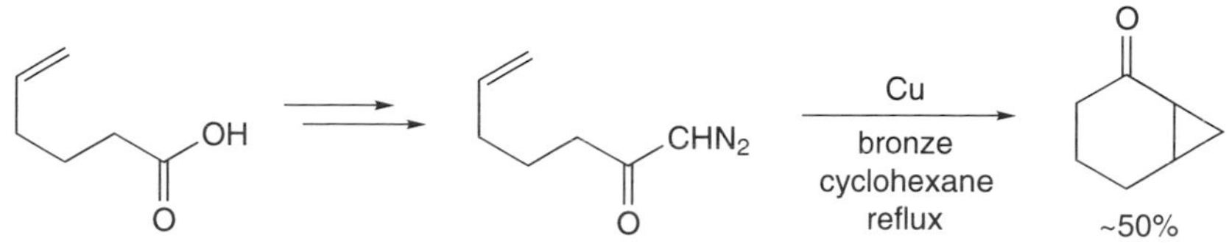

(3)

the model for the synthesis of diverse unnatural and natural products,[2] including those of theoretical interest such as bullvalene[10] and twistane.[11] As exemplified by the syntheses of thujopsene (**6**)[12] and sirenin (**7**)[13] (eqs. 4 and 5), intramolecu-

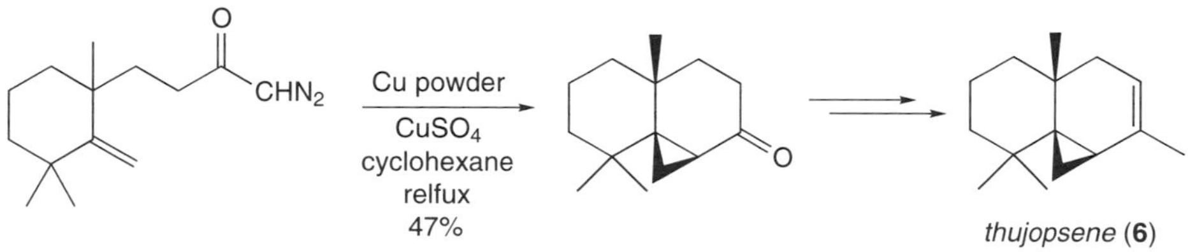

thujopsene (**6**)

(4)

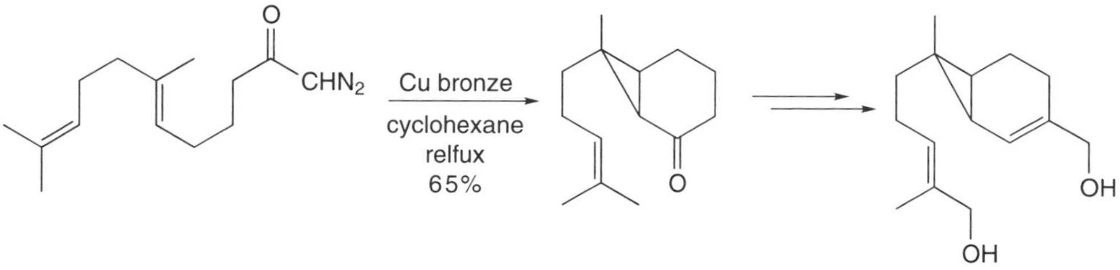

(+)-*sirenin* (**7**)

(5)

lar cyclization provides a defined stereochemistry to the cyclopropane carbon skeleton. Subsequent reactions used to modify the ketone functionality may keep intact the adjacent cyclopropane ring or, as seen in eqs. 6–8, effect ring-opening and keep intact the ketone functionality through the vinylcyclopropane–

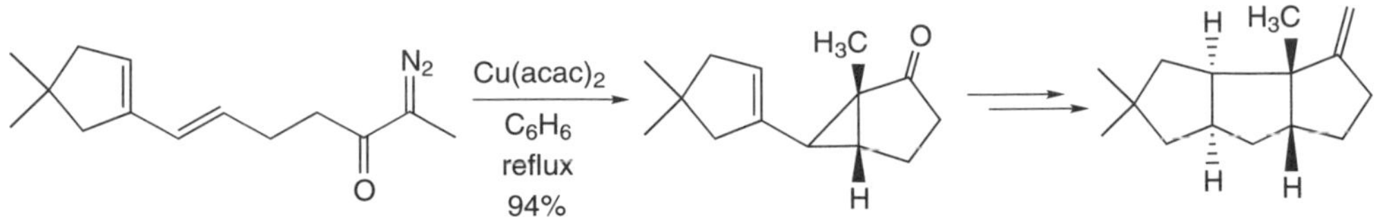

(+)-hirsutene (**8**)

(6)

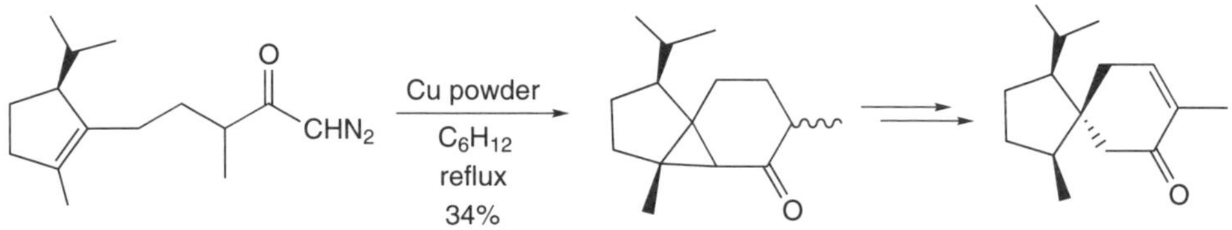

(+)-acorenone B (**9**)

(7)

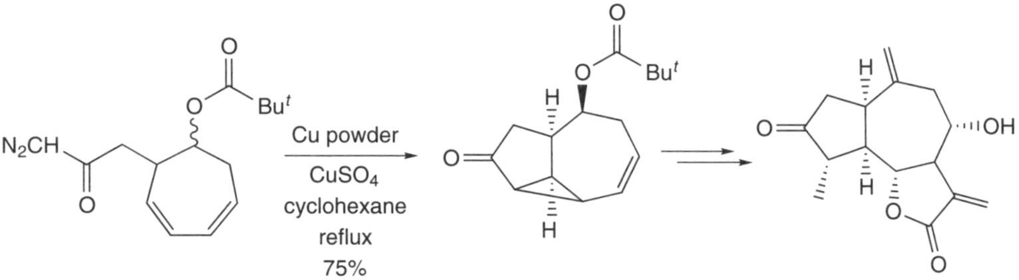

(+)-grosshemin (**10**)

(8)

cyclopentene rearrangement (**8**),[14] catalytic hydrogenation (**9**),[15] or electrophilic ring opening with $Ac_2O/BF_3 \cdot OEt_2$ (**10**).[16] In cyclopropane ring opening reactions, α-ketocyclopropanes are generally more versatile and significantly more reactive than are the corresponding esters.[17] Other examples of intramolecular cyclopropanation reactions of diazoketones leading to specific target molecules, reviewed by Burke and Grieco,[2] include those for the synthesis of aristolone (**11**), longicyclene (**12**), sesquicarene (**13**), the spiro-sesquiterpene epihinesol (**14**), and the tetracyclic diterpene kaurene (**15**).

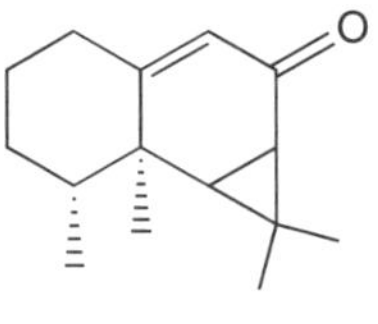

aristolone (**11**) 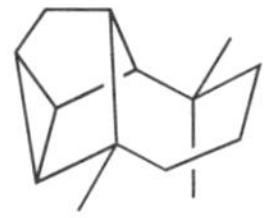   longicyclene (**12**) 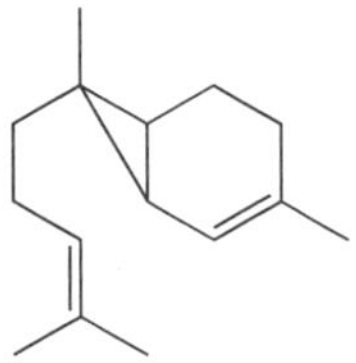   sesquicarene (**13**)

epihinesol (**14**) 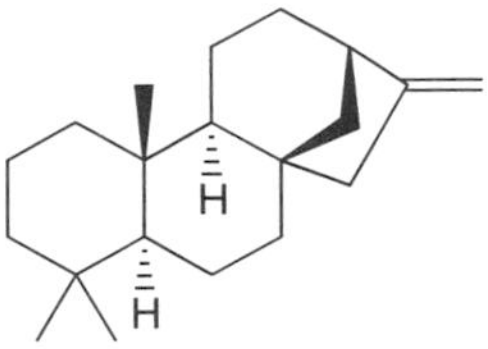   *kaurene* (**15**)

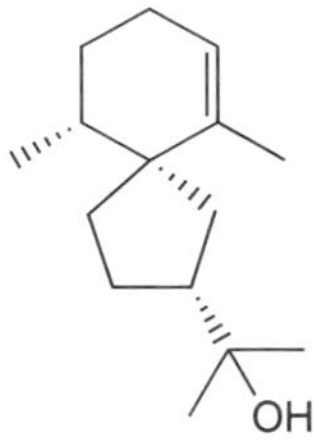

The catalysts employed for intramolecular cyclopropanation of diazoketones were initially heterogeneous copper powder, copper bronze, or cupric sulfate,[2] and, despite advances achieved during the past three decades with homogeneous copper and dirhodium(II) carboxylate catalysts, the heterogeneous copper catalysts are still employed today, often with good results. Few catalyst comparisons have been made. However, $Rh_2(OAc)_4$ is being used with increasing frequency and, often, with exceptional yields and selectivities. Adams and Belley reported the synthesis of eucalyptol (**16**) using $Rh_2(OAc)_4$ as the catalyst for diazo decomposition (eq. 9),[18] and they stated that the traditional copper catalysts provided

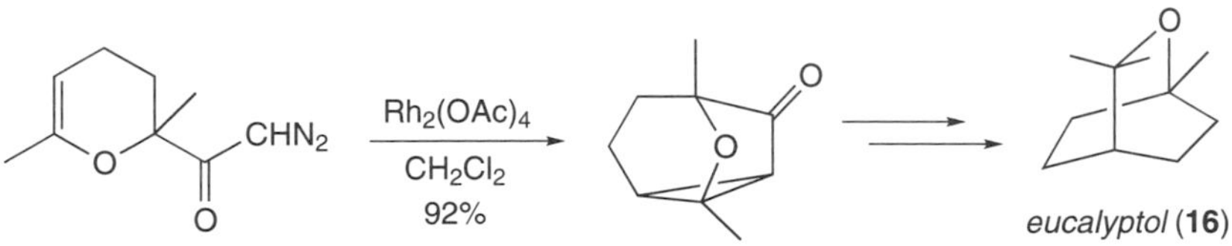

only decomposition products. The same methodology, but with acid-catalyzed ring opening of the cyclopropane intermediate, was employed by Adams for the synthesis of β-chamigrene.[19] The synthesis of $(-)$-8-*epi*-PGE_2 methyl ester and prostaglandin E_2 methyl ester (**17**) by Taber and Hoerrner was accomplished, with the key step being that of $Rh_2(OAc)_4$-catalyzed intramolecular cyclopropanation (Scheme 1).[20] As had been previously observed by Hudlicky (e.g., eq. 6) with diazoketones that had a C–H bond alpha to the diazo functionality,[14] hydrogen migration that results in α,β-unsaturated ketones (e.g., **18**) is a minor competing reaction; this process appears to be important only when "electrophilic" catalysts such as $Rh_2(OOCCF_3)_4$ are employed.[19] Mander achieved the synthesis of gibberellins A_{19} (**19**), A_{36}, and A_{37} by intramolecular cyclopropanation using copper catalysis (eq. 10[21]).[21,22] Doyle and Trudell have reported that $Cu(OTf)_2$ in nitro-

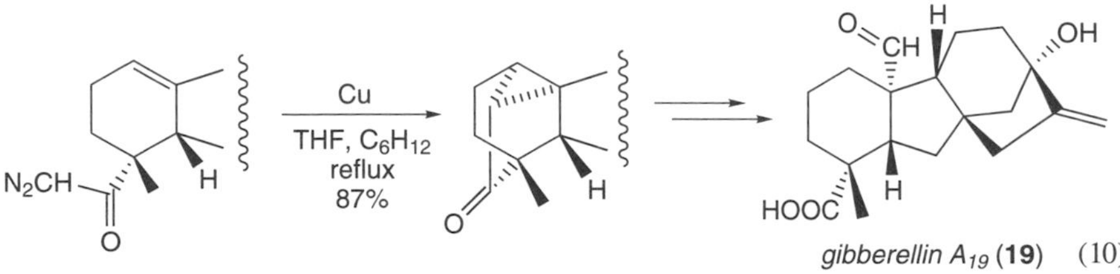

gibberellin A_{19} (**19**)   (10)

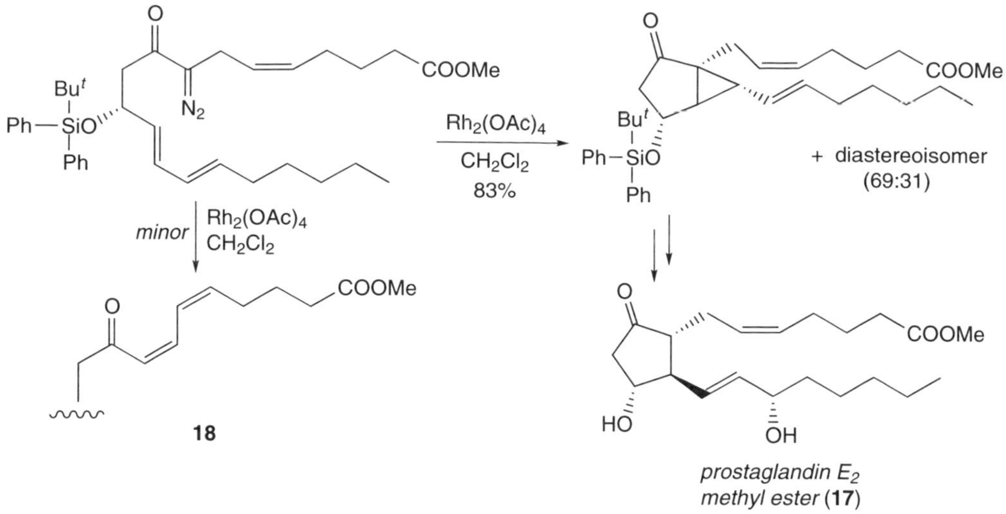

Scheme 1

methane was significantly more effective for intramolecular cyclopropanation of **20** (eq. 11) than was $Rh_2(OAc)_4$ in the same solvent or in ether.[23]

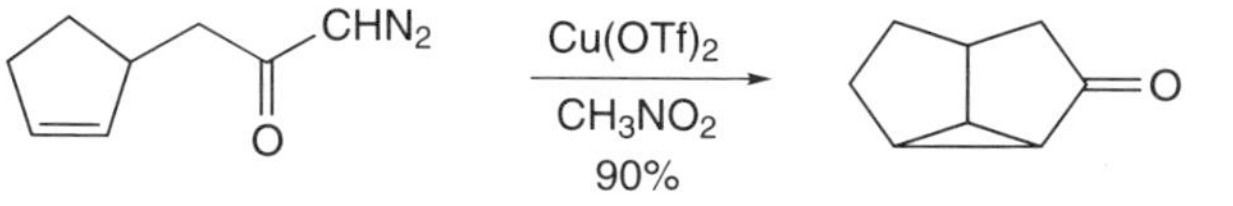

$$(11)$$

Although catalytic diazo decomposition of γ,δ-unsaturated diazoketones such as **20** readily forms products from intramolecular cyclopropanation, a β,γ-unsaturated diazoketone (**21**) yielded hydrinden-2-one **22** under $Cu(OTf)_2$ catalysis (eq. 12).[23] Product yield (50%) was similar with $BF_3 \cdot OEt_2$, suggesting

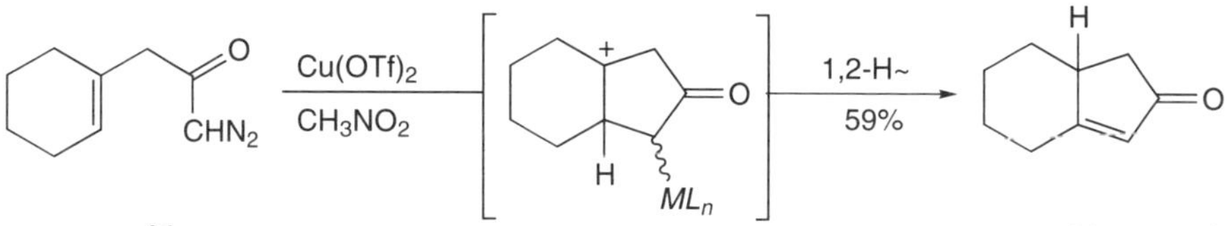

$$(12)$$

electrophilic addition, but a lower yield (26%) was realized with the use of $Rh_2(OAc)_4$. Honda and co-workers have capitalized on this electrophilic addition reaction for the synthesis of natural products 4-*epi*-isovalerenol (**23**, eq. 13)[24] and (−)-neonepetalactone (**24**, eq. 14);[25] $Rh_2(OAc)_4$ was superior to $Cu(OTf)_2$ for these reactions.

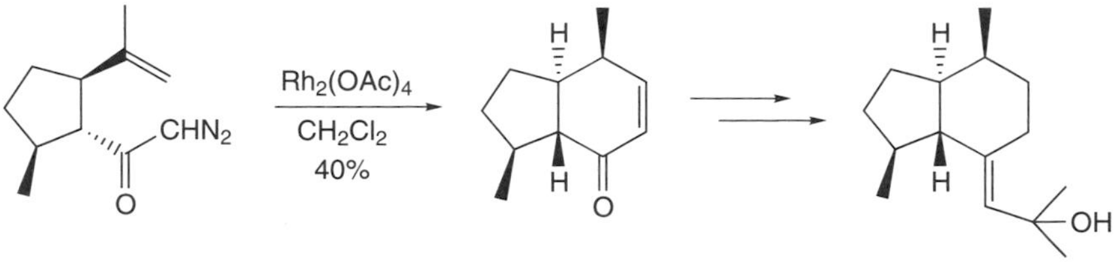

$$4\text{-}epi\text{-}isovalerenenol \ (\mathbf{23})$$

(13)

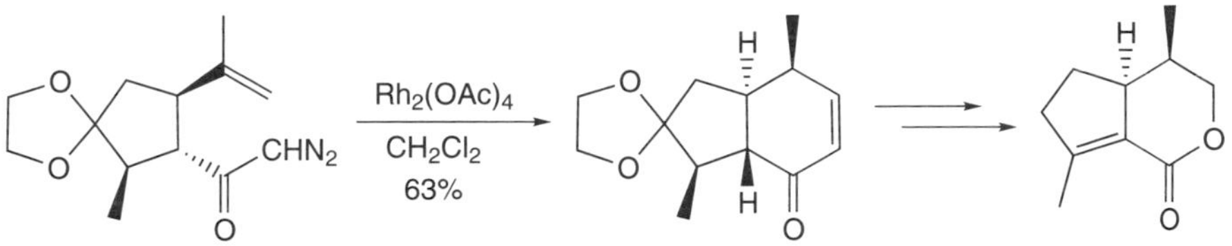

$$(\text{-})\text{-}neonepetalactone \ (\mathbf{24})$$

(14)

The influence of ring size in cyclopropanation reactions catalyzed by $Rh_2(OAc)_4$ has been addressed by Adams and co-workers using $C–H$ insertion as the competing transformation (eq. 15).[26] By replacing R = H with R = Me in **25**,

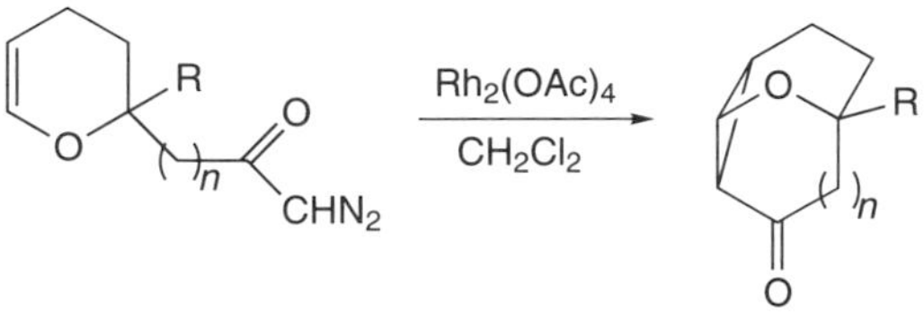

$n = 0, R = $ H: 58%
$n = 1, R = $ H: 61%
$n = 2, R = $ H: 0%
$n = 2, R = $ Me: 95%
$n = 3, R = $ Me: 95%   (15)
$n = 4, R = $ H: 0%

25   **26**

even the eight-membered ring product (**26**, n = 3) could be obtained in high yield. However, the formation of five-membered ring ketones and lactones is generally preferred. Diazo compounds that are designed to undergo intramolecular formation of bicyclo[2.1.0]pentan-2-ones generally proceed through a vinylogous Wolff rearrangement (see Section 9.4) to relatively unstable ketene intermediates that react with alcohols to form 4-pentenoate esters (eq. 16);[27–29] however, there

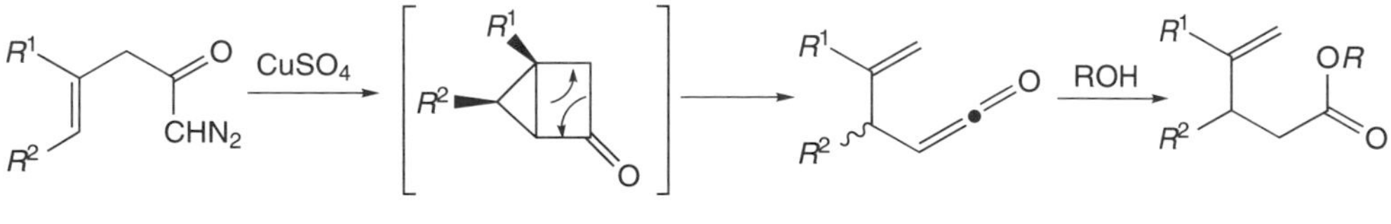

(16)

are limited examples of the use of the intramolecular cyclopropanation strategy with β,γ-unsaturated diazoketones for the synthesis of highly strained products such as **28**[30] and **29**.[31]

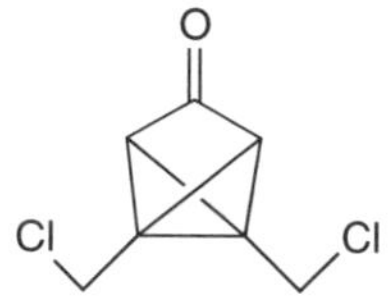

28

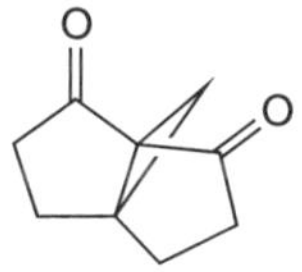

29

Intramolecular cyclopropanation of 3-furyl diazoketone **30** could be achieved with Cu(TBS)$_2$ (eq. 17), but the use of Rh$_2$(OAc)$_4$ yielded only the carbene

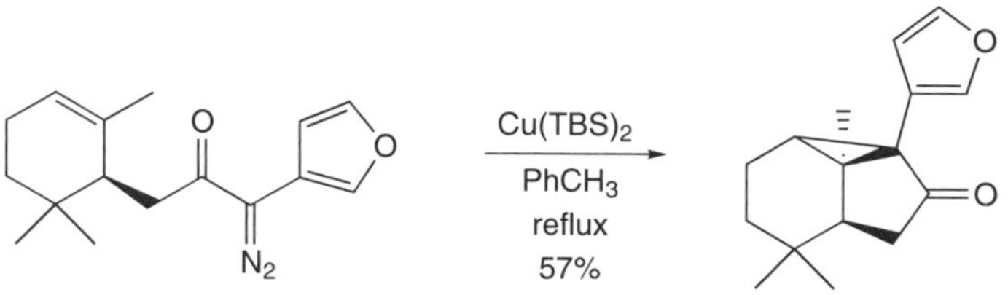

$$\text{Cu(TBS)}_2 \quad \text{PhCH}_3 \quad \text{reflux} \quad 57\%$$

(17)

30

dimer.[32] With the diazoketone in which the 3-furyl group is replaced by hydrogen and the double bond does not have methyl substitution, Rh$_2$(OAc)$_4$-catalyzed intramolecular cyclopropanation occurred without significant dimer formation. Corey had previously employed a 3-furyl diazoketone (**31**) in one route for the stereocontrolled synthesis of the liminoid skeleton (**32**), but in this case Rh$_2$(OAc)$_4$ was the catalyst of choice (eq. 18).[33] Lithium-in-ammonia reduction of

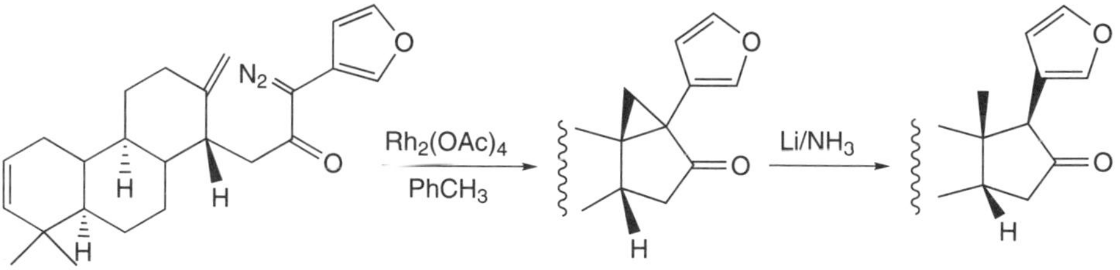

$$\text{Rh}_2\text{(OAc)}_4 \quad \text{PhCH}_3$$

$$\text{Li/NH}_3$$

31

32

(18)

α-ketocyclopropanes is a commonly used methodology for cyclopropane ring opening.

5.2 VINYLCYCLOPROPYLCARBONYL COMPOUNDS IN ORGANIC SYNTHESIS

Intramolecular cyclopropanation reactions have been employed as key steps for the construction of fused and bridged polycyclic systems.[6] Hudlicky and co-workers pioneered the combination of intramolecular cyclopropanation of dienes with the vinylcyclopropane–cyclopentene rearrangement[17] for a net [4+1]-cycloaddition strategy in the synthesis of (±)-hirsutene (eq. 6),[14] (±)-isocomenic acid (**33**),[34] (±)-epiisocomenic acid (**34**),[34] (±)-retigeranic acid (**35**),[35] (±)-epiisocomene,[36] pentalenic acid,[37,38] and pentalenene.[37,38] Noteworthy is the intramolecular cyclopropanation onto an α,β-unsaturated ester (Scheme 2), which is normally unreactive towards metal carbene addition; in these cases the combination of Cu(acac)$_2$ and excess CuSO$_4$ was uniquely effective.

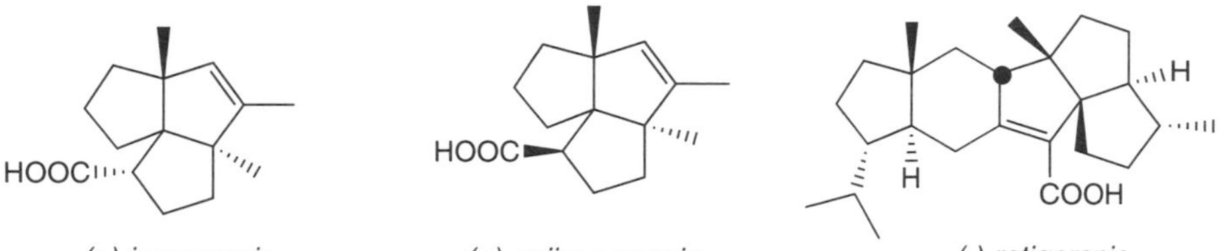

(+)-isocomenic acid (**33**)

(+)-epiisocomenic acid (**34**)

(-)-retigeranic acid (**35**)

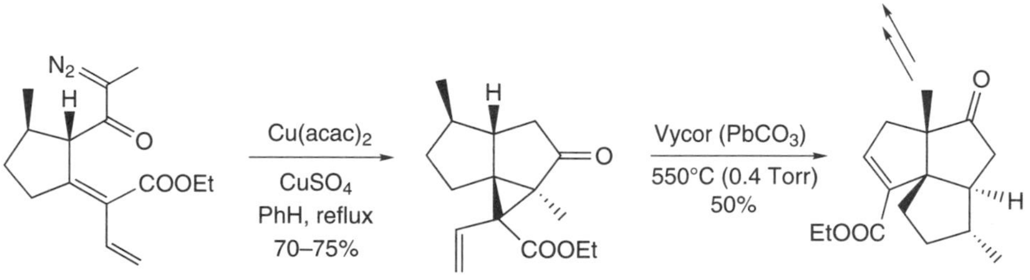

Scheme 2

Tandem cyclopropanation/Cope rearrangement provides entry to seven-membered rings.[39] Davies and co-workers have established the effectiveness of this methodology in the formation of cyclohepta-1,4-dienes (e.g., eq. 19) for a net

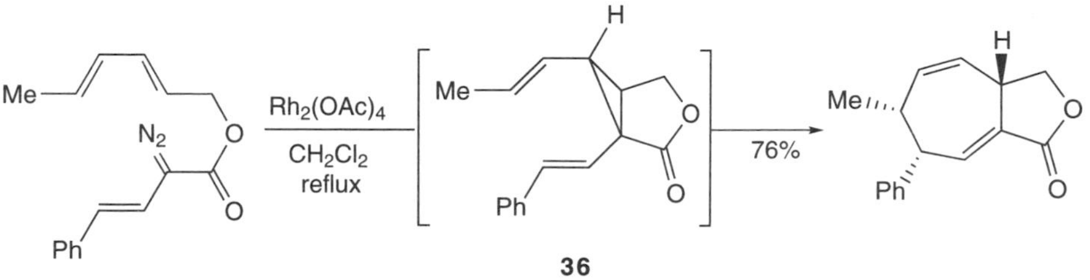

(19)

[3+4]-cycloaddition with vinyldiazoesters (see also Section 4.8).[40] The intermediate cyclopropane product (**36**) undergoes [3,3]-cycloaddition under the conditions employed for catalytic intramolecular cyclopropanation. Product stereochemistry is controlled by the initial diene geometry. Analogous intramolecular cyclopropanation resulting in the formation of a *trans*-divinylcyclopropane requires heating of the product to 150°C in order to undergo the Cope rearrangement (presumably via rearrangement to the *cis*-divinylcyclopropane).

Fused 7-azabicyclo[4.2.0]octadienes (**38**) have been formed by dirhodium(II)-catalyzed diazo decomposition of **37** (eq. 20).[41,42] Here, instead of undergoing a

$$\text{(20)}$$

Cope rearrangement, the intermediate effects cleavage of the pyrrole ring and, through a series of electrocyclic reactions, forms the observed products in good yield. There is a high probability that a cyclopropane intermediate is not formed in this case and, like the transformation described in eqs. 12–14, a carbocation intermediate is produced instead.

Corey and co-workers have provided elegant examples of the uses of vinylcyclopropylcarbonyl compounds in the synthesis of (±)-cafestol (**40**)[43] (Scheme 3) and (±)-atractyligenin (**41**).[44] Copper-catalyzed intramolecular cyclopropanation followed by a cyclopropylcarbinyl rearrangement coupled with electrophilic cyclization gave entry to pentacycle **39**, which was functionalized to produce (±)-cafestol (**40**). This synthetic strategy was further employed for the synthesis of (±)-atractyligenin (**41**).

5.3 INTRAMOLECULAR CYCLOPROPANATION OF DIAZOESTERS AND DIAZOAMIDES

A major advantage of intramolecular cyclopropanation over intermolecular cyclopropanation is that diastereocontrol is built into the transformation. The construction of permethrenic acid (**44**) by intramolecular cyclopropanation of **42**, for example, provides lactone **43**, which is directly converted to *cis*-**44** and then to the potent synthetic pyrethroid NRDC 182 (**45**, Scheme 4).[45] In contrast, intermolecular cyclopropanation reactions give mixtures of *cis*- and *trans*-**44**. An analogous strategy was employed for the synthesis of (1R)-*cis*-chrysanthemic acid (**46**), the chiral framework for which was derived from (R,R)-

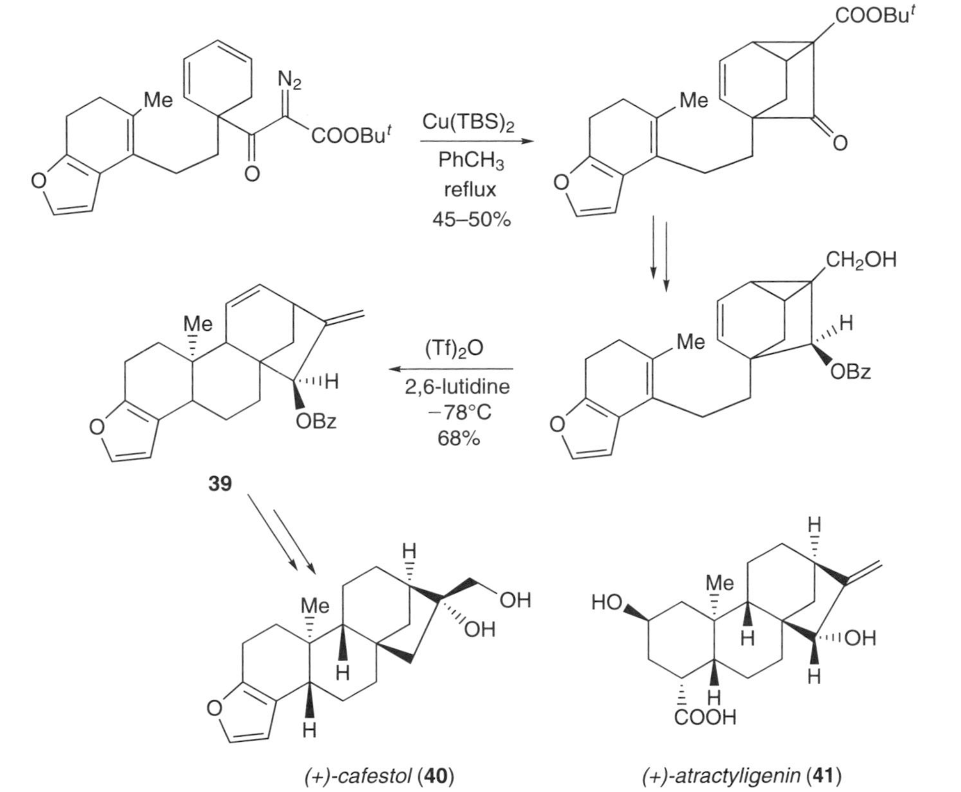

(+)-cafestol (**40**)

(+)-atractyligenin (**41**)

Scheme 3

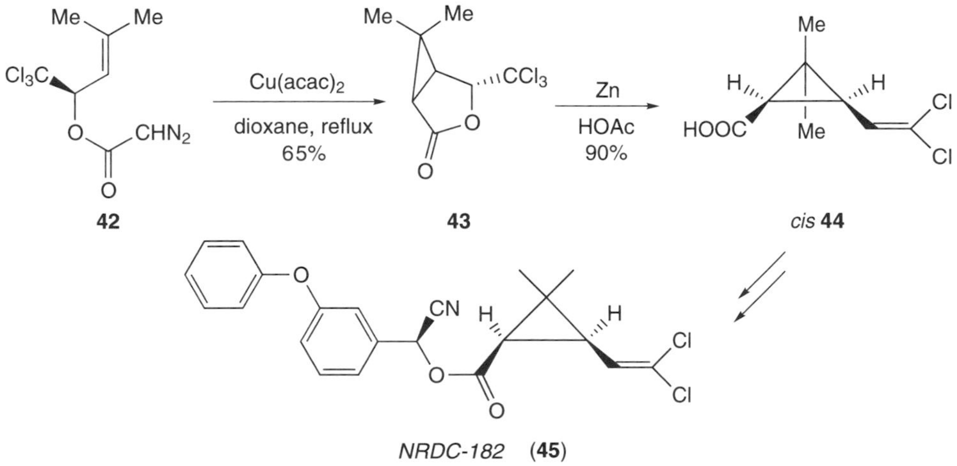

42

43

cis **44**

NRDC-182 (**45**)

Scheme 4

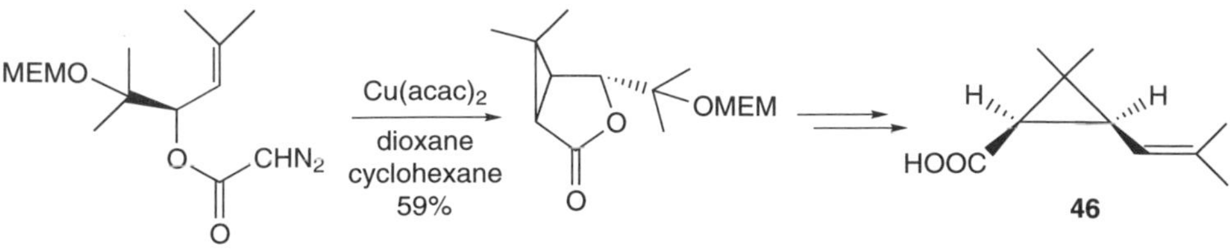

(21)

tartaric acid (eq. 21).[46] Similarly, carbacerulenin (**47**), an inhibitor of the biosynthesis of lipids in fungal cells, was prepared exclusively as the cis-stereoisomer by intramolecular cyclopropanation of an allylic diazoacetate followed by subsequent amidation and oxidation (eq. 22);[47] addition to the more remote carbon–carbon double bonds was not observed. Lactones **48** have provided access to α-aminocyclopropanecarboxylic acids (**49**) of defined stereochemistry (eq. 23).[48]

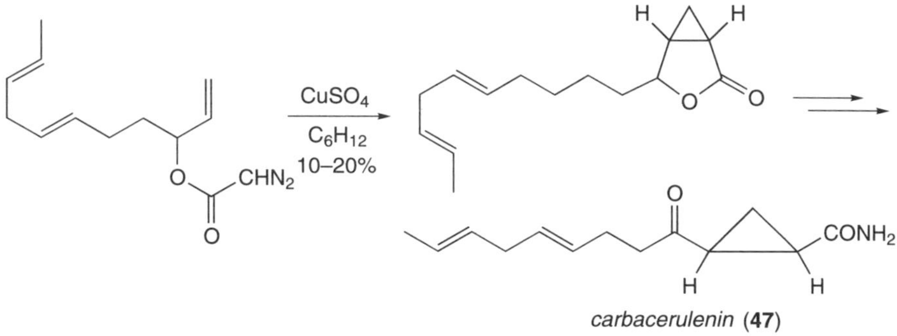

carbacerulenin (**47**)

(22)

$R^c = R^t = H$, Me
$R^c = H$, $R^t = Ph$, Me

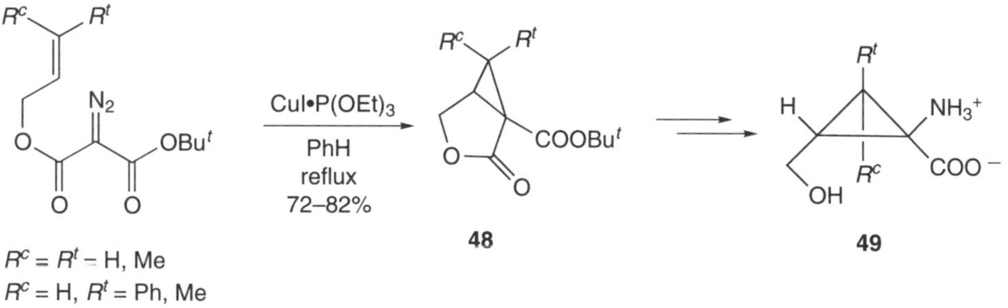

(23)

α-(Carboxycyclopropyl)glycine (**51**) has been prepared via **50** by intramolecular cyclopropanation (eq. 24),[49] and cis-caronaldehyde (**52**) has been formed from prenyl diazoacetate (eq. 25).[50] Without the carbonyl group within the ring (cf.

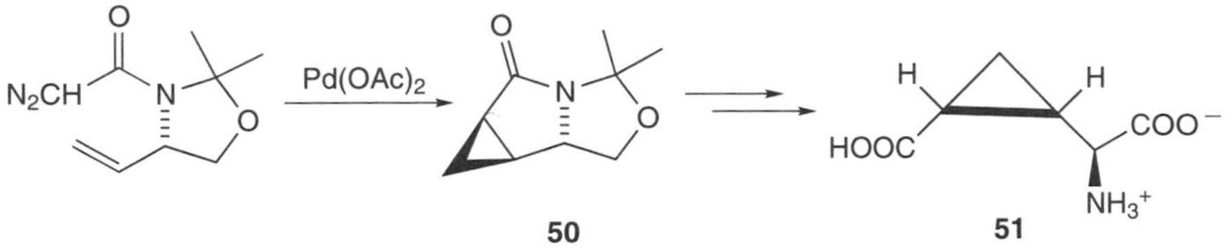

50 51

(24)

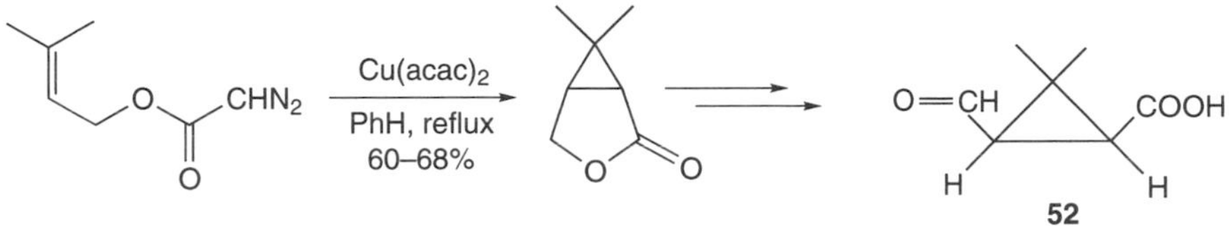

52

(25)

eq. 16), highly strained bicyclo[1.1.0]butane **53** has been prepared in high yield by intramolecular cyclopropanation (eq. 26).[51]

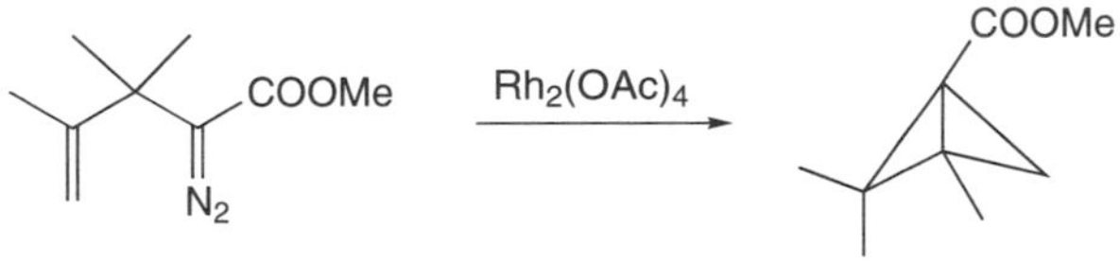

53

(26)

Dauben and co-workers have reported a diastereoselective intramolecular cyclopropanation of chiral α-diazo-β-ketoester **54** using $Rh_2(OAc)_4$ catalysis (eq. 27)[52] directed to the synthesis of vitamin D_3. The ratio of **55a** to **55b** was

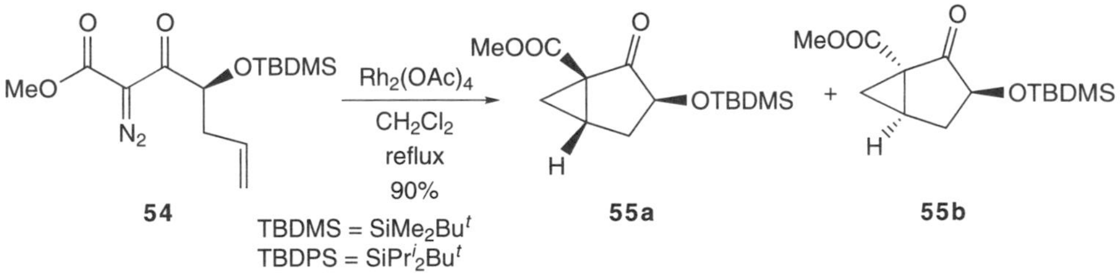

54 55a 55b

(27)

3:1, and the alkoxy substituent was said to have little influence on diastereoselectivity (TBDPS gave 3.5:1 for **55a**:**55b**). Wilson and co-workers had previously described various attempts to generate de-alkoxy analogs of **55** using ester chiral auxiliaries and Rh(TPP)Cl catalysis as a new methodology for the synthesis of *A*-ring precursors to vitamin D$_3$ derivatives (**56**);[53] high diastereocontrol (91:9) was realized when the catalyst was used in 5 mol %.

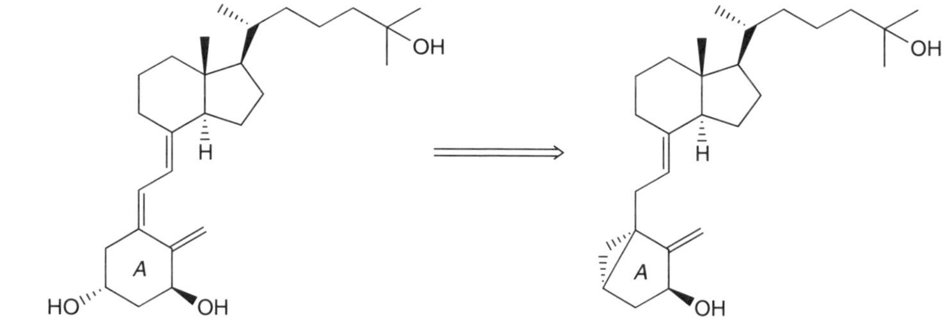

1α,25-(OH)$_2$ vitamin D$_3$ (**56**)

Intramolecular cyclopropanation reactions of α-diazo-β-ketoesters have been employed for the synthesis of (±)-juvabione (**58**), a sesquiterpene with juvenile hormone activity (eq. 28),[54] and for the stereoselective synthesis of (±)-clavukerin A (**60**) and its isomeric (±)-isoclavukerin A (eq. 29) from Okinawan soft coral.[55]

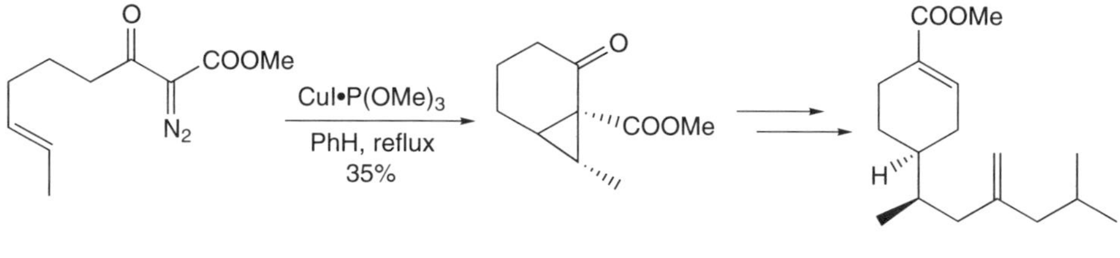

57

(+)-juvabione (**58**)

$$(28)$$

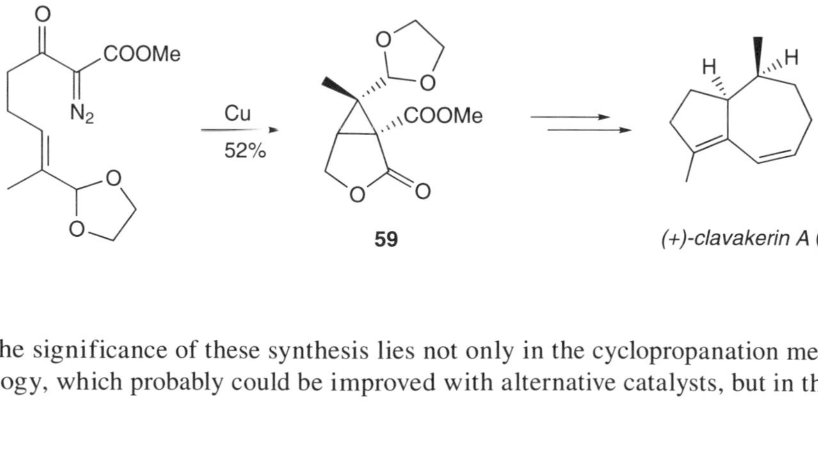

59

(+)-clavakerin A (**60**)

$$(29)$$

The significance of these synthesis lies not only in the cyclopropanation methodology, which probably could be improved with alternative catalysts, but in the use of the constituent β-ketoesters of **57** and **59** for homoconjugate organocuprate addition (eq. 30) and palladium-induced reductive cleavage (eq. 31). The potential

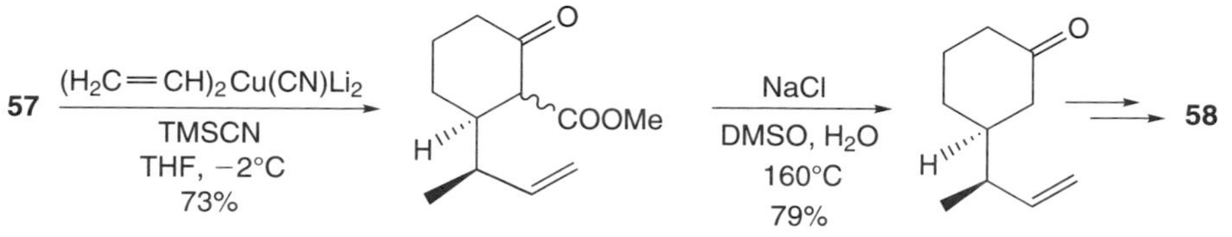

(30)

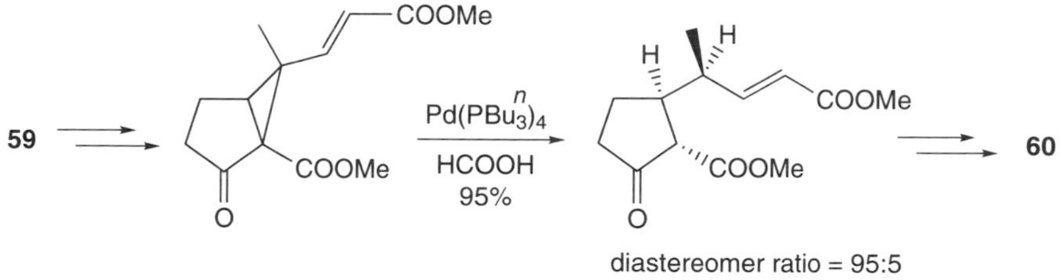

diastereomer ratio = 95:5

(31)

generality of this latter methodology for selective hydrogenolysis had been previously established by Shimizu.[56] Homoconjugate organocuprate addition had been used by Taber[57] and also by Heathcock[58] for the construction of 3-substituted cyclopentanones and cyclohexanones.

Danishefsky and co-workers utilized intramolecular cyclopropanation of diazomalonate **61** in their methodology for the synthesis of pyrrolizidine bases *dl*-hastanecine (**62**) and *dl*-dihydroxyheliotridane (**63**) with complete diastereocontrol (Scheme 5).[59] Diazomalonates were also the reactants for intramolecular

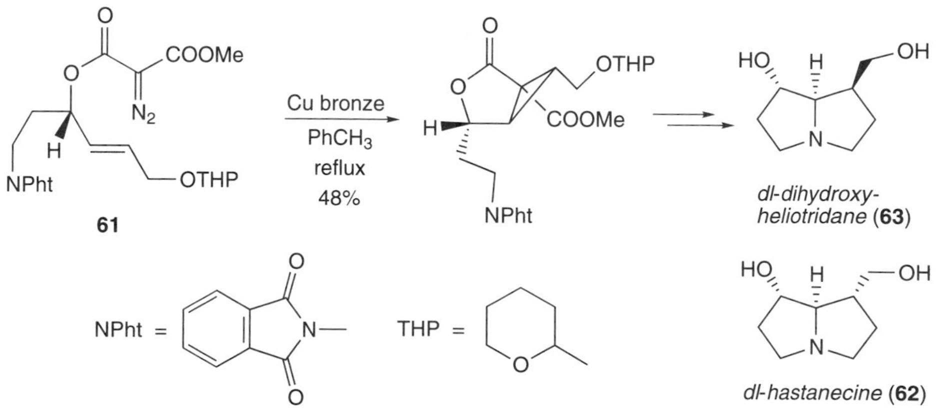

Scheme 5

cyclopropanation reactions directed to the synthesis of substituted 1-aminocyclo-propane-1-carboxylic acids (eq. 23),[48] including protected derivatives of carnosadine and its stereoisomers from **48** ($R^c = R^t =$ H).[60,61] Fukuyama employed this methodology for the stereocontrolled synthesis of (−)-hapalindole G (**63**) from (−)-carvone using a novel decarbomethoxylation strategy to introduce chlorine at the hindered C-13 position (Scheme 6); Cu(TBS)$_2$ was reported to be the only catalyst to give a satisfactory yield for cyclopropanation.[62]

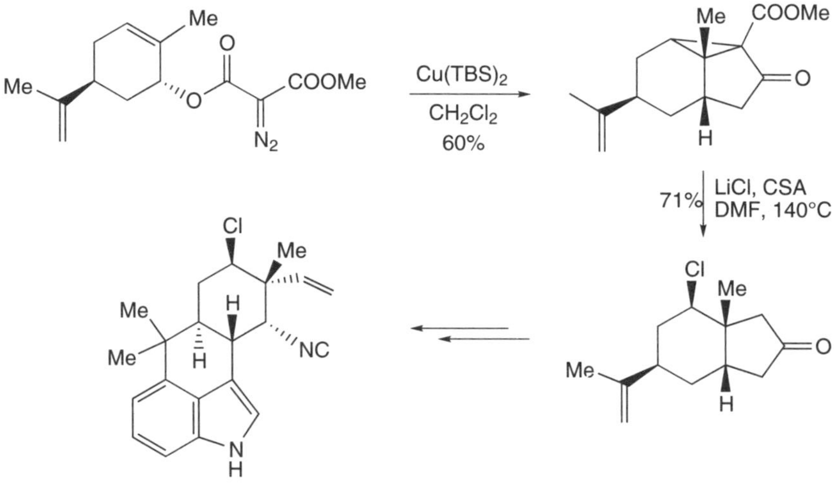

(-)-hapalindole G (**63**)

Scheme 6

5.4 REGIOSELECTIVITY AND CHEMOSELECTIVITY IN INTRAMOLECULAR CYCLOPROPANATION REACTIONS

Formation of cyclopropane-fused five- and six-membered rings is the recognized preference for intramolecular cyclopropanation, as the examples of eqs. 32,[63] eq. 33,[64] and eq. 34[65] demonstrate. Equation 34 suggests that five-membered ring formation is favored over six-membered ring formation, but limited data prohibit generalization. The formation of eight-membered rings has been reported (eq. 15) but is not common. Other examples (eqs. 19 and 22; Schemes 1 and 3)

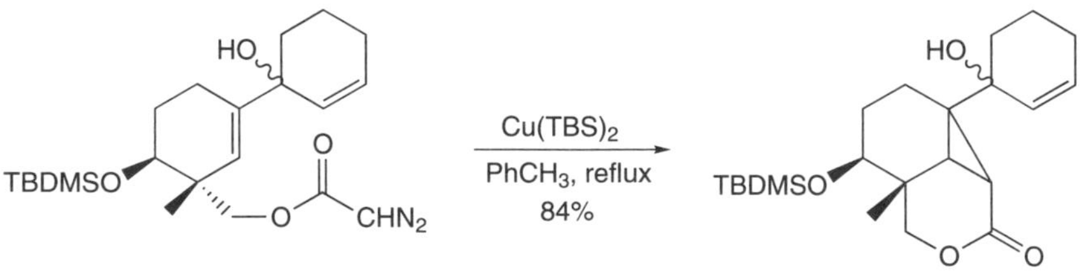

(32)

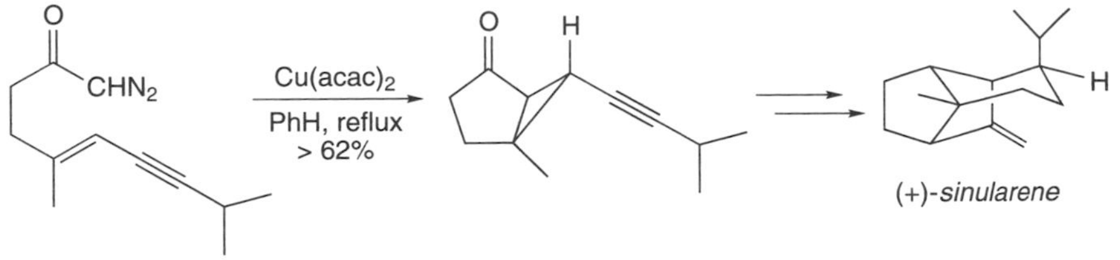

(+)-sinularene

(33)

(+)-quadrone

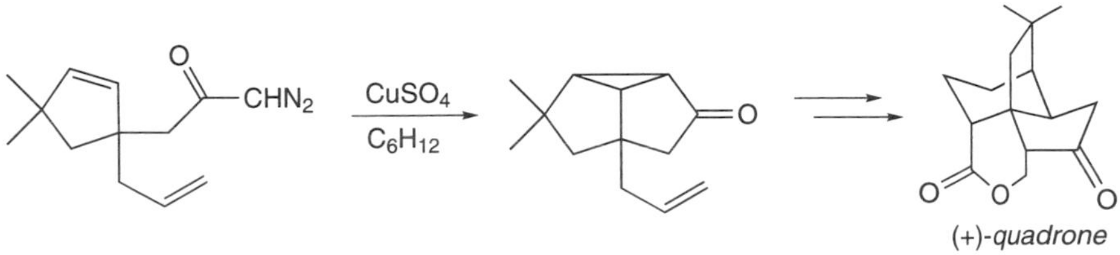

(34)

are consistent with overall five- or six-membered ring preferential regioselectivity. However, as we shall see (Section 5.6), regioselectivity in intramolecular cyclopropanation reactions is by no means firmly established or predictable.

The preference for five- or six-membered ring formation in the broad selection of catalytic metal carbene reactions sets limits on the feasibility of certain transformations. When two different functional groups occupy sites that are nearly equidistant from the reaction center, at which functional group will the metal carbene intermediate undergo preferential addition/association/insertion? Answers to this question have been forthcoming from competitive experiments such as those described in eq. 35,[66] eq. 36,[66] eq. 37,[67,68] and eq. 38.[68] Both the di-

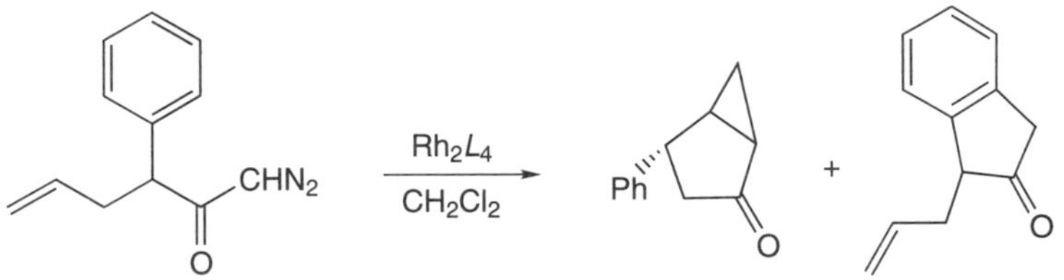

(35)

Rh_2L_4	yield (%)	**64**	**65**
$Rh_2(pfb)_4$	95	0	100
$Rh_2(OAc)_4$	99	67	33
$Rh_2(cap)_4$	72	100	0
$Rh_2(tpa)_4$[69]	83	0	100
$Rh_2(OAc)_2(PC)_2$[70]	90	100	0

tpa = Ph_3CCOO

PC =

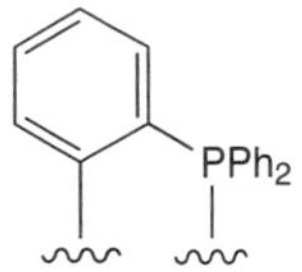

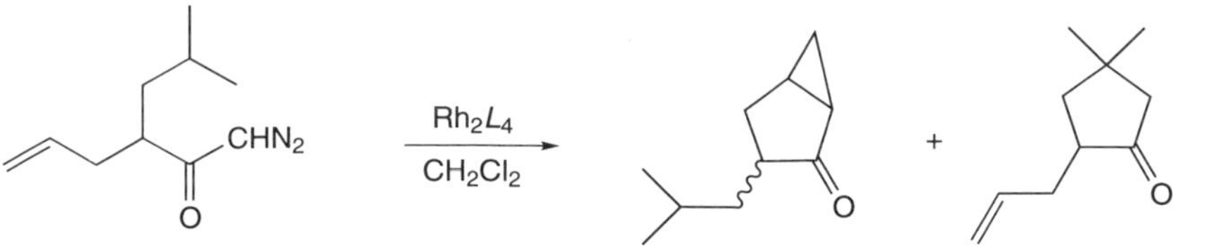

Rh_2L_4	yield (%)	**66**	**67**
$Rh_2(pfb)_4$	56	0	100
$Rh_2(OAc)_4$	97	44	56
$Rh_2(cap)_4$	76	100	0
$Rh_2(OAc)_2(PC)_2$[70]	85	100	0

(36)

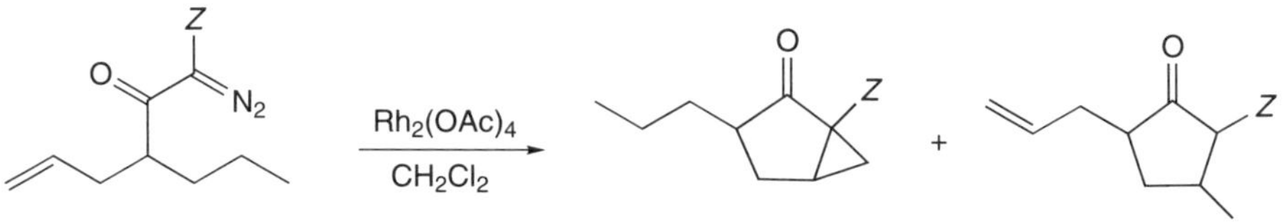

Z	yield (%)	**68**	**69**
COOEt[67]	78	72	28
H[68]	61	92	8

(37)

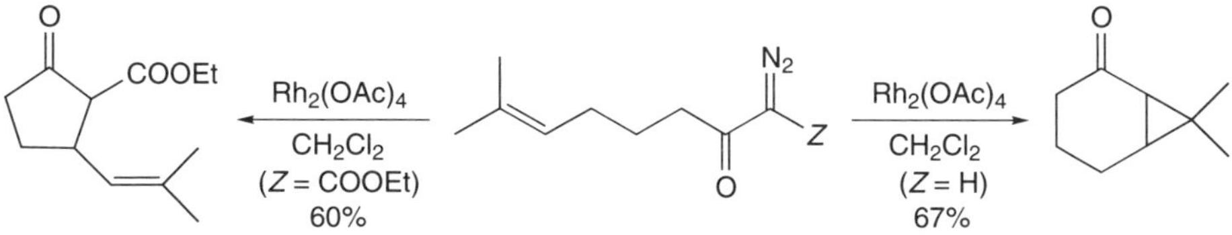

(38)

azo compound and the catalyst[71] have a profound influence on chemoselectivity. The order of reactivity for metal carbenes generated from $Rh_2(pfb)_4$ is aromatic substitution > tertiary C–H insertion > cyclopropanation, whereas that from $Rh_2(cap)_4$ is cyclopropanation > tertiary C–H insertion, and aromatic substitution was not observed as a competing process.[66] The large difference in chemical behavior between $Rh_2(pfb)_4$ and $Rh_2(cap)_4$ has been attributed to the large differences in charge and frontier orbital properties.[66] Qualitatively, these ligand-induced differences in selectivity reflect the relative importance of resonance-contributing structures **70a** and **70b** for the intermediate metal carbene. Increasing the electron-withdrawing capability of metal ligand L increases the contribution from **70b**; electron-donation from L increases the contribution from **70a**. The greater the contribution from **70a**, the lower the reactivity of the metal carbene and the higher is its selectivity. Competitive experiments for cyclopropanation versus carbonyl ylide formation have also been reported,[72,73]

$$L_nM = CR^1R^2 \quad \longleftrightarrow \quad L_n\overset{-}{M} - \overset{+}{C}R^1R^2$$

70a   **70b**

but the influence of dirhodium(II) ligands on chemoselectivity was small. The explanation for the absence of ligand influence in these cases may be that ylide formation involves reversible formation of a dipolar intermediate, whereas cyclopropanation is concerted and irreversible.

Catalyst ligand influences on chemoselectivity has been limited thus far to diazoketones and α-diazo-β-ketoesters. That similar effects can be realized with diazoacetates and diazomalonates has not been tested. There may be similar influences with ligands for copper(I), but, except for an investigation of diazo decomposition of diazoketone **71** in which cyclopropanation (**72**), C–H insertion (**73**), aromatic substitution (**74**), and Wolff rearrangement were competitive,[74] examples here are virtually nonexistent.

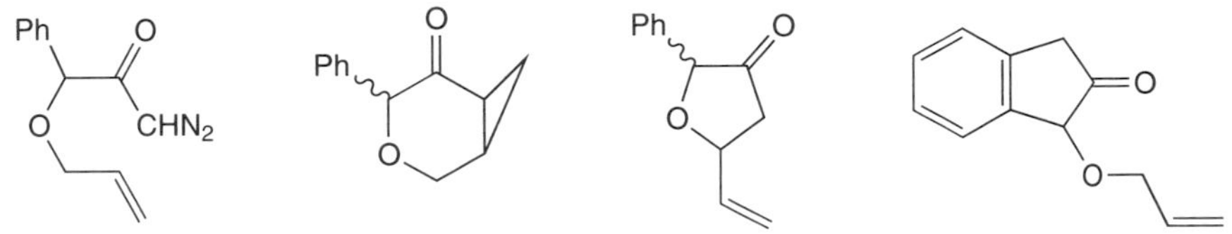

71   **72**   **73**   **74**

5.5 ENANTIOSELECTIVE INTRAMOLECULAR CYCLOPROPANATION REACTIONS

The development of chiral dirhodium(II) carboxamidates has had its greatest impact on intramolecular cyclopropanation reactions of diazoacetates (eq. 39) and diazoacetamides (eq. 40), where these catalysts provide the highest levels of

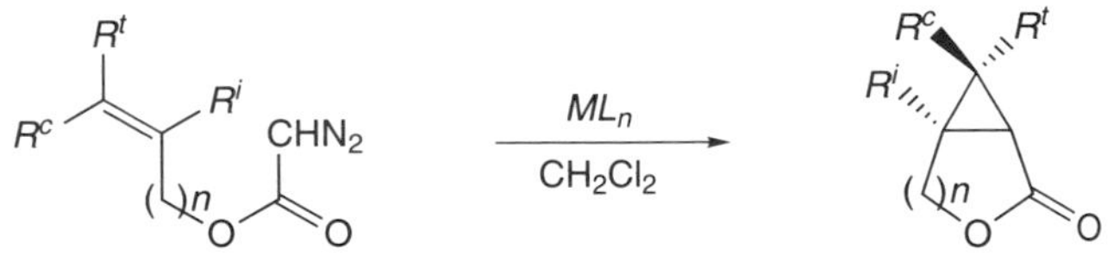

$$\begin{array}{ccc} \textbf{75} & \xrightarrow[\text{CH}_2\text{Cl}_2]{ML_n} & \textbf{76} \end{array} \qquad (39)$$

$$\begin{array}{ccc} \textbf{77} & \xrightarrow[\text{CH}_2\text{Cl}_2]{ML_n} & \textbf{78} \end{array} \qquad (40)$$

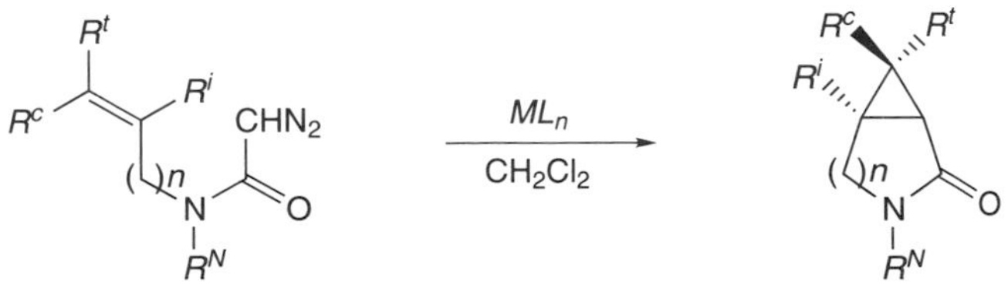

enantiocontrol.[75-81] In the simplest case, allyl diazoacetate, the use of $Rh_2(5S$-MEPY$)_4$ and $Rh_2(5R$-MEPY$)_4$ in catalytic amounts as low as 0.1 mol % causes the formation of the enantiomeric 3-oxabicyclo[3.1.0]hexan-2-ones (**79**) with 95% ee in good yields following distillation (eq. 41).

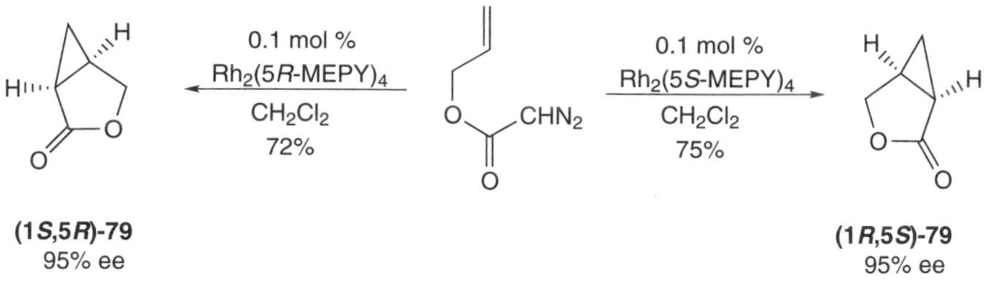

| (1S,5R)-79 | | (1R,5S)-79 |
| 95% ee | | 95% ee |

$$(41)$$

With the $Rh_2(MEPY)_4$ catalysts consistently high levels of enantiocontrol ($\geq$93% ee) are achieved with *cis*-disubstituted allylic diazoacetates and with trisubstituted systems (Table 5.1), including those prepared from nerol (93% ee) and geraniol (95% ee), but with *trans*-disubstituted allylic diazoacetates use of the $Rh_2(MEPY)_4$ catalysts provides lower ee values. However, the steric bias provided by the N-acylimidazolidinone-ligated catalysts enhances enantiocontrol in these cases, so that with $Rh_2(4S$-MPPIM$)_4$ enantioselectivities are extended to $\geq$95%.[79] Similarly, the exceptionally low enantioselectivities observed in $Rh_2(5S$-

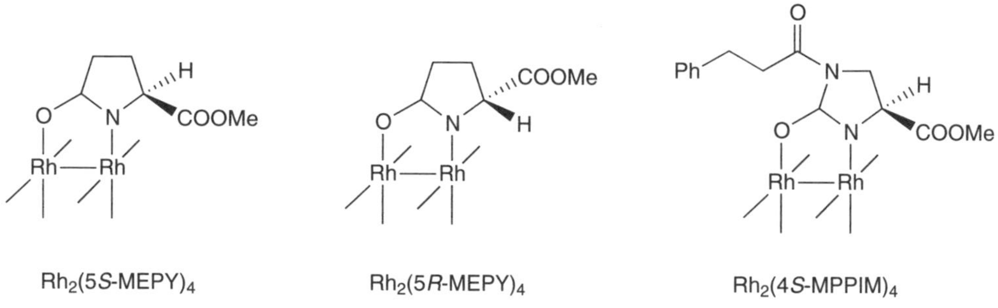

| $Rh_2(5S$-MEPY$)_4$ | $Rh_2(5R$-MEPY$)_4$ | $Rh_2(4S$-MPPIM$)_4$ |

MEPY$)_4$ catalyzed reactions of methallyl and (n-butyl)allyl diazoacetate ($R^i =$ Me, ^nBu) can be circumvented with the use of $Rh_2(4S$-MPPIM$)_4$ (Table 5.1).

Among the pharmacologically relevant compounds whose syntheses have been reported using this methodology, with either or both $Rh_2(5S$-MEPY$)_4$ and $Rh_2(5R$-MEPY$)_4$, are 1,2,3-trisubstituted cyclopropanes as conformationally restricted peptide isosteres (Scheme 7) for renin (**80**)[83] and collagenese[84] inhibitors, presqualene alcohol (**81**) from farnesyl diazoacetate,[85] and the GABA analog 3-azabicyclo[3.1.0]hexan-2-one (98% ee) formed by intramolecular cyclopropanation from N-allyldiazoacetamide (eq. 40, $R =$ H).[78] In addition, the products of these cyclopropanation reactions are synthetic precursors to *cis*-chrysanthemic acid[86] and to the pheromone R-($-$)-dictyopterene C.[87]

TABLE 5.1 Enantioselective Intramolecular Cyclopropanations of Allylic Diazoacetates (**75**, *n* = 1)

Catalyst	R^c	R^t	R^i	76 (*n* = 1) Yield (%)	ee (%)	Configuration	Ref.
Rh$_2$(5S-MEPY)$_4$	CH$_3$	CH$_3$	H	89	98	(1S,5R)	78
Rh$_2$(5R-MEPY)$_4$	CH$_3$	CH$_3$	H	88	93	(1R,5S)	80
Rh$_2$(5S-MEPY)$_4$	Ph	H	H	70	≥94	(1R,5S)	78
Rh$_2$(5S-MEPY)$_4$	CH$_3$CH$_2$	H	H	88	≥94	(1R,5S)	78
Rh$_2$(5S-MEPY)$_4$	PhCH$_2$	H	H	80	≥94	(1R,5S)	78
Rh$_2$(5S-MEPY)$_4$	(CH$_3$)$_2$CHCH$_2$	H	H	73	≥94	(1R,5S)	78
Rh$_2$(5S-MEPY)$_4$	(CH$_3$)$_2$CH	H	H	85	≥94	(1R,5S)	78
Rh$_2$(5S-MEPY)$_4$	(^{n}Bu)$_3$Sn	H	H	79	≥94	(1R,5S)	78
Rh$_2$(5S-MEPY)$_4$	I	H	H	78	≥94	(1R,5S)	78
Rh$_2$(5S-MEPY)$_4$	H	Ph	H	78	68	(1R,5S)	79
Rh$_2$(4S-MPPIM)$_4$	H	Ph	H	61	96	(1R,5S)	78
Rh$_2$(5S-MEPY)$_4$	H	CH$_3$CH$_2$CH$_2$	H	93	85	(1R,5S)	78
Rh$_2$(4S-MPPIM)$_4$	H	CH$_3$CH$_2$CH$_2$	H	83	95	(1R,5S)	79
Rh$_2$(5S-MEPY)$_4$	H	I	H	70	67	(1R,5S)	78
Rh$_2$(5S-MEPY)$_4$	CH$_3$	Me$_2$C=CH(CH$_2$)$_2$	H	88	95	(1S,5R)	78
Rh$_2$(5S-MEPY)$_4$	Me$_2$C=CH(CH$_2$)$_2$	CH$_3$	H	79	93	(1S,5R)	78
Rh$_2$(5S-MEPY)$_4$	H	H	CH$_3$	72	7	(1R,5S)^a	78
Rh$_2$(4S-MPPIM)$_4$	H	H	CH$_3$	75	89	(1S,5R)^a	79
Rh$_2$(5S-MEPY)$_4$	H	H	^{n}Bu	72	35	(1S,5R)	82
Rh$_2$(4S-MPPIM)$_4$	H	H	^{n}Bu	82	93	(1S,5R)	82

^{a}Configurational assignment by L. A. Overman from X-ray structure of 1-(1-naphthyl)ethylamide derivative (private communication).

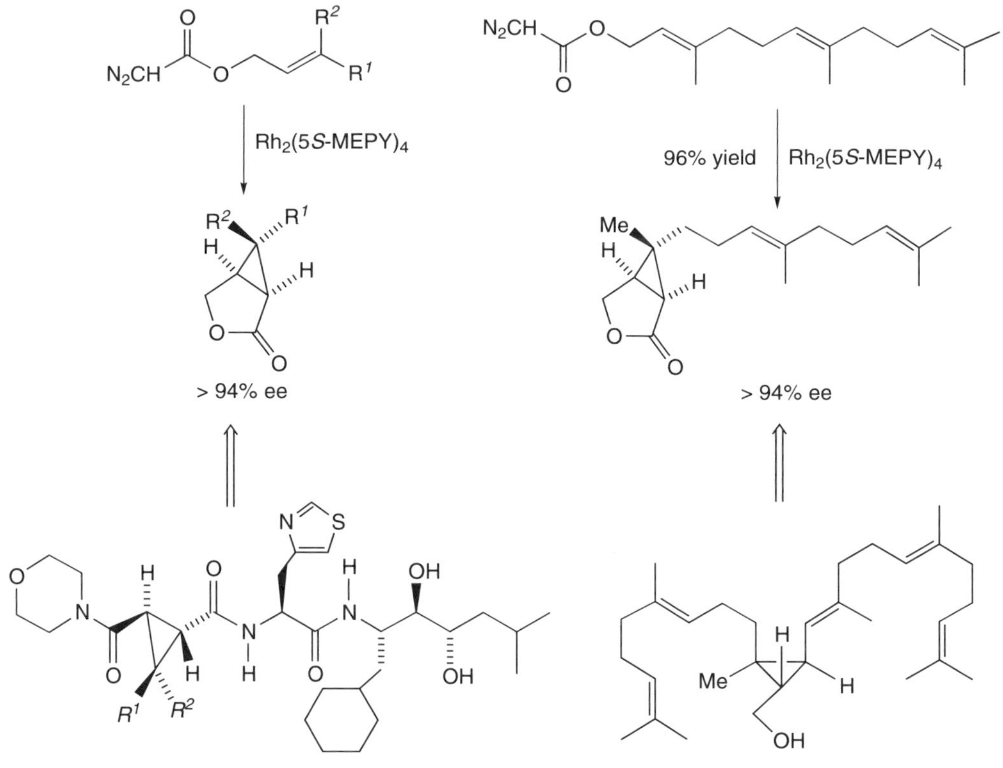

80, *renin inhibitor*

81, *presqualene alcohol*

Scheme 7

TABLE 5.2   Enantioselective Intramolecular Cyclopropanation of Homoallylic Diazoaetates (75, $n = 2$) Catalyzed by Rh$_2$(5S-MEPY)$_4$[78]

			76 ($n = 2$)		
R^c	R^t	R^i	Yield (%)	ee (%)	Configuration
H	H	H	80	71	(1R,6S)
CH$_3$	CH$_3$	H	74	77	(1S,6R)
Ph	H	H	73	88	(1S,6R)
CH$_3$CH$_2$	H	H	80	90	(1S,6R)
c-C$_6$H$_{11}$CH$_2$	H	H	77	80	(1S,6R)
PhCH$_2$	H	H	68	80	(1S,6R)
Me$_3$Si	H	H	65	86	(1S,6R)
H	Ph	H	55	73	(1S,6R)
H	CH$_3$CH$_2$	H	65	82	(1S,6R)
H	H	CH$_3$	76	83	(1R,6S)

SYNTHESIS OF (1R,5S)-(−)-6,6-DIMETHYL-3-OXABICYCLO-[3.1.0]HEXAN-2-ONE[80]

To a three-necked, 1-liter round-bottom flask fitted with an addition funnel and a reflux condenser, to which is attached a drying tube, is added Rh$_2$(5R-MEPY)$_4$(CH$_3$CN)$_2$(i-PrOH) (0.203 g, 0.221 mmol) dissolved in 150 mL of freshly distilled, anhydrous dichloromethane, and this solution is heated at reflux under nitrogen. 3-Methyl-2-buten-1-yl diazoacetate (14.9 g, 96.7 mmol) in 450 mL of anhydrous dichloromethane is added dropwise over a 30-h period to the refluxing blue catalyst solution. The flow of external nitrogen is terminated after 1 h, and the flow of nitrogen evolved from the reaction solution is used to monitor the progress of the reaction. Refluxing is continued for an additional hour following complete addition, and, after cooling to room temperature, the solvent is removed from the green solution under reduced pressure. Kugelrohr distillation of the reside (80°C, 0.15 mm) affords 12.1 g of a lightly colored liquid that, following chromatographic purification on silica–alumina (100 g) with hexanes:ethyl acetate (80:20 to 70:30), yields 10.2 g (81.0 mmol, 84% yield) of >99% pure (1R,5S)-(−)-6,6-dimethyl-2-oxabicyclo[3.1.0]hexan-2-one, bp 70°C (0.15 Torr), $[\alpha]_D^{26}$-86.2 (CHCl$_3$, c 2.5), whose optical purity was determined on a Chiraldex B-PH capillary column to be 96.5% ee. The solid residue remaining after Kugelrohr distillation is dissolved in a minimal volume of methanol and purified by chromatography. Isolation of the red band afforded 98 mg of the recovered catalyst.

With homoallylic diazoacetates there is a moderate reduction in enantioselectivity (Table 5.2) from that achieved with their allylic counterparts.[76,78] However, the % ee values achieved with use of Rh$_2$(MEPY)$_4$ catalysts were much less subject to the substitution pattern on the carbon–carbon double bond. Use of Rh$_2$(4S-MEOX)$_4$ did not increase the % ee value for cyclopropanation of 3-methyl-3-buten-1-yl diazoacetate,[88] but reaction with Rh$_2$(4S-MPPIM)$_4$ gave decreased enantiocontrol.[79] Similarly high enantiocontrol has been achieved in intramolecular cyclopropanation reactions of allylic α-diazopropionates (eq. 42),[89]

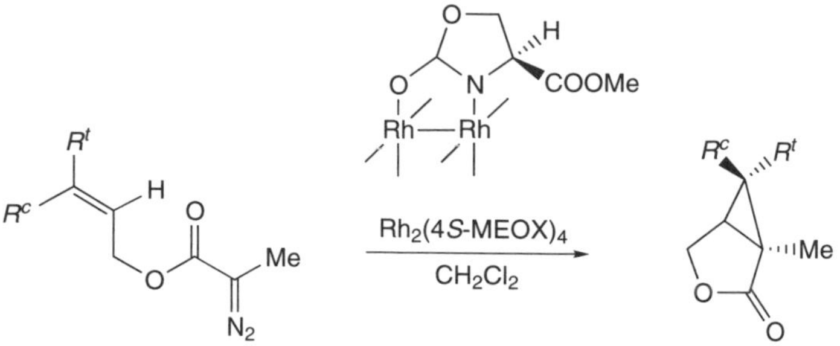

(42)

R^c	R^t	yield (%)	ee (%)
Me	Me	81	71
^nPr	H	62	85
Ph	H	65	78
H	^nPr	46	52
H	Ph	70	43

but with this system $Rh_2(4S\text{-MEOX})_4$ was superior to other dirhodium(II) carboxamidate catalysts. 1,2-Hydrogen migration (see also Scheme 1) was in most cases a minor (<20%) competing process, but its occurrence sets limitations on the applicability of this methodology. With the even less reactive α-diazophenylacetates, dirhodium(II) carboxamidates were ineffective; however, chiral dirhodium(II) carboxylates are promising.[90]

Enantiocontrol in intramolecular cyclopropanation reactions of diazoacetamides (eq. 40) is similar to that found with diazoacetates (Table 5.3), but with diazoacetamides primary consideration must be given to the conformational rigidity of the diazoacetamide (eq. 43). If equilibration between **82a** and **82b** or

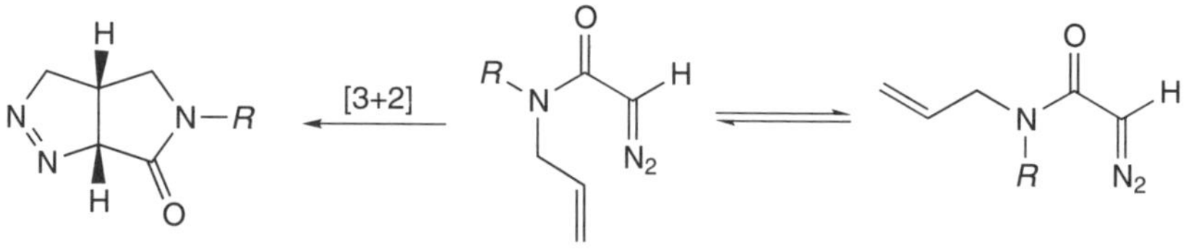

(43)

82a   **82b**

their respective metal carbenes is slow on the reaction time scale, only **82a** is capable of undergoing intramolecular cyclopropanation. Furthermore, **82a** undergoes intramolecular dipolar addition under conditions similar to those used for catalytic diazo decomposition.[77] When $R = {}^tBu$, **82b** is favored, and when $R = H$, **82a** is the lower-energy form. As seen by the data in Table 5.3, $R = Me$ represents a compromise between these two extremes that leads to improved yields and % ee values that are nearly identical with those found for their diazoacetate counterparts.[81]

Chiral dirhodium(II) carboxamidates show amazing selectivity for enantiomer differentiation in intramolecular cyclopropanation reactions of racemic

TABLE 5.3 Enantioselective Intramolecular Cyclopropanation of Allylic and Homoallylic Diazoacetamides (77, n = 1,2)

Catalyst	n	R^c	R^t	R^i	R^N	Yield (%)	ee (%)	Configuration[a]	Ref.
Rh$_2$(4S-MEOX)$_4$	1	H	H	H	H	40	98	(1R,5S)	78
Rh$_2$(5S-MEPY)$_4$	1	H	H	H	CH$_2$CH=CH$_2$	50	72	(1R,5S)	77
Rh$_2$(5S-MEPY)$_4$	1	H	H	H	Me	62	93	(1R,5S)	81
Rh$_2$(4S-MEOX)$_4$	1	H	H	H	Me	45	86	(1R,5S)	81
Rh$_2$(4S-MEOX)$_4$	1	Me	Me	H	Me	91	94	(1S,5R)	81
Rh$_2$(4S-MPPIM)$_4$	1	Me	Me	H	Me	88	94	(1S,5R)	81
Rh$_2$(4S-MPPIM)$_4$	1	^{n}Pr	H	H	Me	88	95	(1R,5S)	81
Rh$_2$(4S-MPPIM)$_4$	1	H	^{n}Pr	H	Me	93	92	(1R,5S)	81
Rh$_2$(4S-MPPIM)$_4$	1	Me	Me$_2$C=CH(CH$_2$)$_2$	H	Me	95	93	(1S,5R)	81
Rh$_2$(4S-MPPIM)$_4$	1	H	H	Me	Et	85	44	(1S,5R)	81
Rh$_2$(5S-MEPY)$_4$	2	H	H	H	^{t}Bu	60	60	(1R,6S)	77
Rh$_2$(5S-MEPY)$_4$	2	Me	Me	H	^{t}Bu	75	75	(1S,6R)	77
Rh$_2$(5S-MEPY)$_4$	2	Et	H	H	^{t}Bu	94	90	(1S,6R)	77
Rh$_2$(5S-MEPY)$_4$	2	H	Et	H	^{t}Bu	62	67	(1S,6R)	77
Rh$_2$(5S-MEPY)$_4$	2	H	H	Me	^{t}Bu	87	78	(1R,6S)	77

[a]Assignments made for $n = 2$ are by analogy with those of Table 5.2.

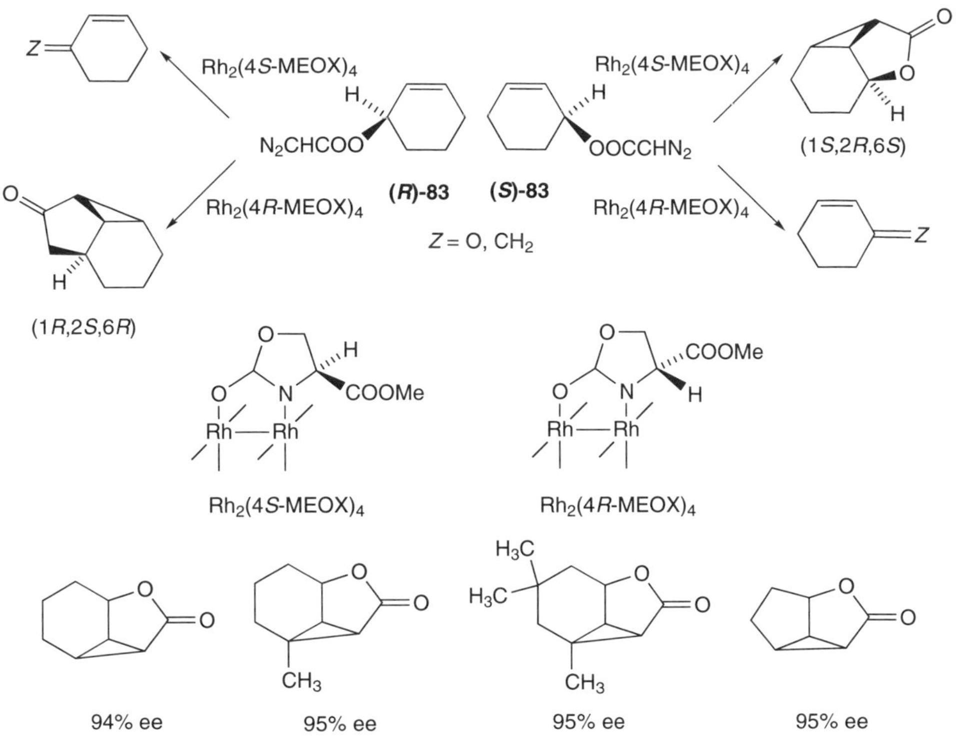

Scheme 8

secondary allylic diazoacetates.[91] In reactions catalyzed by $Rh_2(4S\text{-MEOX})_4$, for example (Scheme 8), (1S)-cycloalk-2-en-1-yl diazoacetates [e.g., (S)-**83**] undergo cyclopropanation (up to 40% isolated yields), whereas (1R)-cycloalk-2-en-1-yl diazoacetates form 2-cycloalkenones and 1-methylene-2-cycloalkenes, these latter being products from intramolecular hydride abstraction from the allylic position alpha to oxygen.[92] With $Rh_2(4R\text{-MEOX})_4$ the (1R)-enantiomer (R)-**83** undergoes intramolecular cyclopropanation ("match"), and the (1S)-enantiomer forms products resulting from hydride abstraction ("mismatch"). With acyclic racemic secondary allylic diazoacetates, enantiomer differentiation occurs through the formation of exo- and endo-diastereoisomers whose opposite preferential configurations, (4R)-endo and (4S)-exo from reactions with $Rh_2(4S\text{-}MEOX)_4$, demonstrate enantiomer differentiation. Martin and co-workers had previously reported exceptional diastereocontrol in intramolecular cyclopropanation reactions of chiral secondary diazoacetates with the use of $Rh_2(MEPY)_4$ catalysts (eq. 44),[93] and data obtained from the single enantiomers were consistent with the product distributions obtained from reactions with the racemic secondary diazoacetates.

SYNTHESIS OF (1*S*,5*R*)-(+)-3,6,6-TRIMETHYL-3-AZABICYCLO[3.1.0]HEXAN-2-ONE[81]

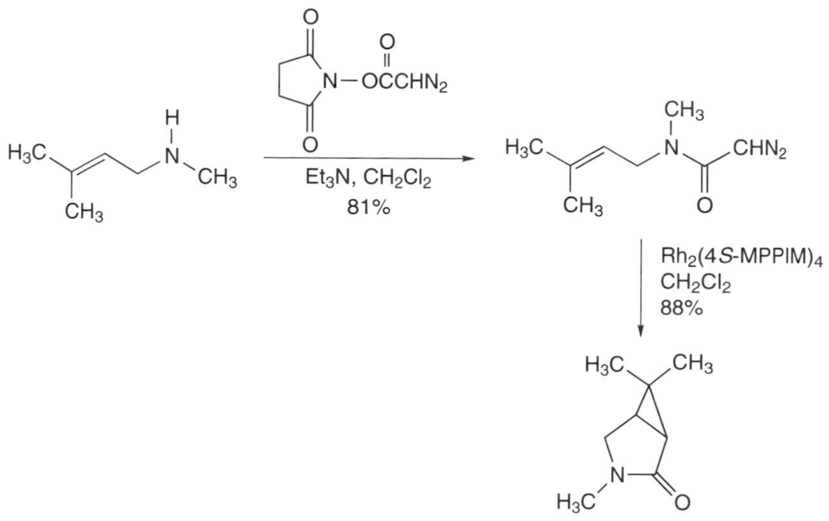

Preparation of *N*-Methyl-*N*-(3-methyl-2-buten-1-yl)diazoacetamide.

To an ice-bath-cooled solution of the *N*-(3-methyl-2-buten-1-yl)-*N*-methyl-amine (0.950 g, 9.6 mmol) and triethylamine (1.21 g, 12.0 mmol) in 20 mL of anhydrous CH$_2$Cl$_2$ was added over 20 min a solution of succinimidyl di-azoacetate (1.46 g, 7.98 mmol) in 30 mL of the same solvent. The mixture was stirred for 30 min at 0°C and then for 1 h at room temperature. After concentration of the solution under vacuum, the residue was purified by flash chromatography on silica gel (1:1 hexanes:EtOAc) to provide the yel-low diazoacetamide (81% yield): IR (film) 2102 (C=N$_2$), 1618 (C=O) cm^{-1}. All operations were conducted at or below room temperature, and the diazoacetamide was stored in a refrigerator. NMR spectra revealed signifi-cant broadening of absorptions for atoms near nitrogen, suggesting re-stricted rotation.

Catalytic Cyclopropanation of *N*-Allylic-*N*-methyldiazoacetamides. General Procedure.

A solution of the diazoacetamide (1.00 mmol) in 10 mL of anhydrous CH$_2$Cl$_2$ was added via syringe pump over 10 h (1.0 mL/h) to a solution of the catalyst (0.1–1.0 mol %) in 10 mL of CH$_2$Cl$_2$. After addition was com-plete, the reaction solution was filtered through a silica gel plug to remove the dirhodium(II) catalysts, and the solvent was then evaporated under reduced pressure. Distillation of the residue provided analytically pure **(1S,5R)-3,6,6-trimethyl-3-azabicyclo[3.1.0]hexan-2-one**: colorless oil, bp 75–80°C (0.4 Torr); chromatographic purification with 1:4 hexanes:EtOAc; enantiomer separation on a 30-m Chiraldex G-TA column operated at 130°C (94% ee); $[\alpha]_D^{21} = +100.4$ (c 2.78, CHCl$_3$) for 94% ee; ^{1}H-NMR δ 3.52 (dd, J = 10.9, 6.6 Hz, 1 H), 3.10 (d, J = 10.9 Hz, 1 H), 2.73 (s, 3 H), 1.79 (dd, J = 6.6, 1.8 Hz, 1 H), 1.60 (t, J = 6.6 Hz, 1 H), 1.10 (s, 3 H), 0.99 (s, 3 H); ^{13}C-NMR δ 172.7, 48.2, 33.1, 28.5, 25.6, 23.9, 21.5, 13.7; IR (film) 1667 (C=O) cm^{-1}.

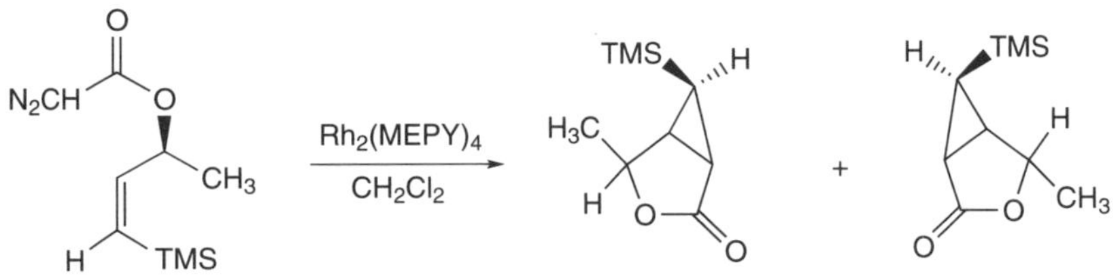

	endo-85	exo-85
Rh$_2$(5S-MEPY)$_4$	>95	<5
Rh$_2$(5R-MEPY)$_4$	37	63

(44)

As expected from the composite results, diazoacetates of prochiral secondary divinyl carbinols (**86**) undergo intramolecular cyclopropanation with exceptional enantiocontrol and, in certain cases, with high diastereocontrol (eq. 45).[93] The use

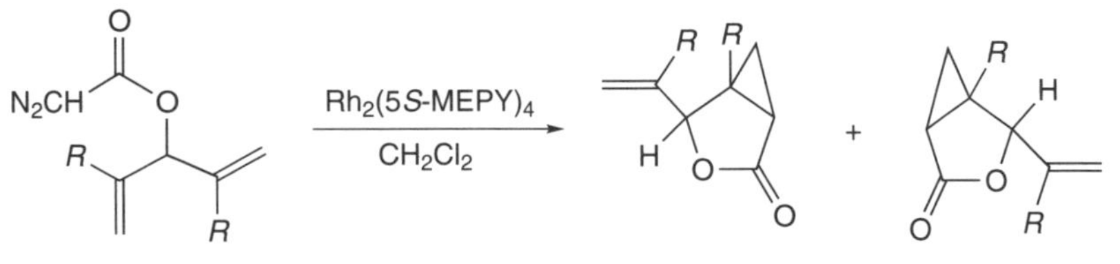

86		**endo-87**	**exo-87**
R = H	75% yield	>95 (>94% ee)	<5
R = Me	73% yield	45 (92% ee)	55 (91% ee)

(45)

of alternative chiral dirhodium carboxamidate catalysts such as Rh$_2$(MEOX)$_4$ is expected to enhance diastereocontrol further while retaining high enantiocontrol.

Nishiyama has reported a high level of enantiocontrol for select intramolecular cyclopropanation reactions of allylic diazoacetates with his Ru(pybox)-Cl$_2$(ethene) catalyst system (**88**),[94,95] and, together with results from CuPF$_6$/(**89**)[82] (Table 5.4), they provide an informative array of comparative information on

88   **89**

TABLE 5.4   Comparative Enantioselectivities for Rh$_2$(5S-MEPY)$_4$,[78] Ru(pybox)Cl$_2$(ethene) (81),[82,84] and CuPF$_6$/(89)[82]

75 ($n = 1$)			Rh$_2$(5S-MEPY)$_4$		88		CuPF$_6$/89	
R^c	R^t	R^i	Yield (%)	ee (%)^a	Yield (%)	ee (%)^a	Yield (%)	ee (%)^a
H	H	H	75	95 (R,S)	—	—	61	20 (R,S)
Me	Me	H	89	98 (S,R)	91	76 (R,S)	80	13 (R,S)
Pr^n	H	H	88	≥94 (R,S)	54	21 (R,S)	82	37 (S,R)
Ph	H	H	70	≥94 (R,S)	79	24 (R,S)	—	—
H	Pr^n	H	93	85 (R,S)	68	78 (R,S)	74	29 (S,R)
H	Ph	H	78	68 (R,S)	93	86 (S,R)	77	4 (R,S)
H	H	Me	72	7 (R,S)	NR^b	—	58	87 (S,R)
H	H	Bu^n	72	35 (S,R)	NR^b	—	73	82 (S,R)

^{a}Absolute configuration (1-position, 5-position).
^{b}No reaction under multiple conditions. Carbene dimer formed.

enantiocontrol as well as about the mechanism for intramolecular cyclopropanation. A key comparison with methallyl and (n-butyl)allyl diazoacetates shows that catalysis with **88** does not produce any cyclopropane product, use of Rh$_2$(5S-MEPY)$_4$ gives low % ee values, but CuPF$_6$/(**89**) provides relatively high enantiocontrol. With Rh$_2$(4S-MPPIM)$_4$, however, high enantiocontrol is restored to each of these systems (Table 5.1), and the chiral dirhodium(II) carboxamidate catalysts can be seen to provide the highest levels of enantiocontrol for all the allylic diazoacetate intramolecular cyclopropanation reactions.

The cause of the selectivity seen in these intramolecular cyclopropanation reactions can be evaluated with reference to the transition-state geometries depicted in Scheme 9. Four limiting structures **90a–d** lead to two enantiomers **76a,b**. The role of the catalyst is twofold: (1) effect approach of the olefin to the carbene from side A (**90a,90b**) or side B (**90c,90d**) and (2) orient the olefin with the respect to the carbene to minimize interactions between catalyst ligands and olefin substituents (**90a/90d** versus **90b/90c**). With **90a/90d** higher enantiocontrol is expected for allylic diazoacetates with R^t and R^i substituents; with **90b/90c**

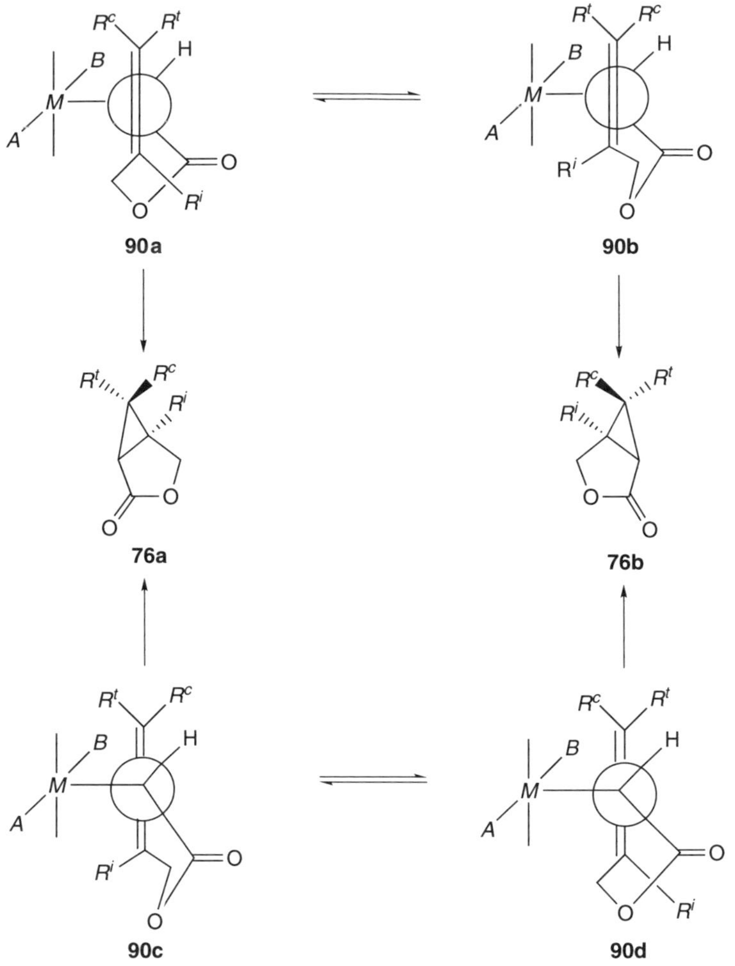

Scheme 9

higher enantiocontrol would be expected with R^c substituents. Enhanced interactions with R^t (**90e**) or R^i (**90f**) magnify the influence of these olefin substituents.

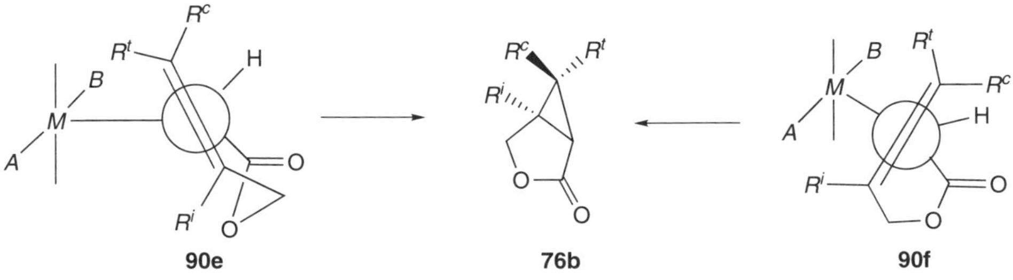

The orientation and degree of substituent interaction can be determined from data such as are presented in Table 5.4. Thus with $Rh_2(5S\text{-MEPY})_4$, **90b** is the dominant transition-state geometry, and **90a** is the dominant geometry for $CuPF_6/(89)$. The Nishiyama catalyst operates according to **90a**, and the absence of intramolecular cyclopropanation when R^i = Me, Bu^n suggests that other steric factors are also involved.

Although diazoacetates and diazoacetamides generally undergo intramolecular cyclopropanation with high enantiocontrol, the same is not true for diazoketones. Early efforts with the Aratani catalyst by Hirai[96] and by Dauben[97] suggested that high enantiocontrol would be difficult to achieve, and only with results obtained by Pfaltz and co-workers with semicorrin–copper complex **84** had anyone achieved results comparable to those from diazoacetates.[98] In the simplest cases (eq. 46) bicyclic cyclopropane products were isolated in modest

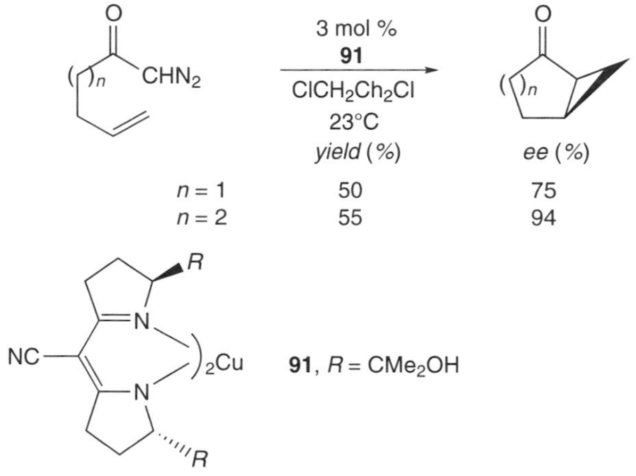

	yield (%)	ee (%)
n = 1	50	75
n = 2	55	94

91, R = CMe$_2$OH

(46)

yield but, in contrast to dirhodium(II) catalysts (Tables 5.1 and 5.2) or to the Aratani catalyst,[97] with higher enantiocontrol when $n = 2$ than when $n = 1$. With methyl-substituted diazoketones, however, very different results are found (eq. 47), and no generalizations can be made regarding factors that influence se-

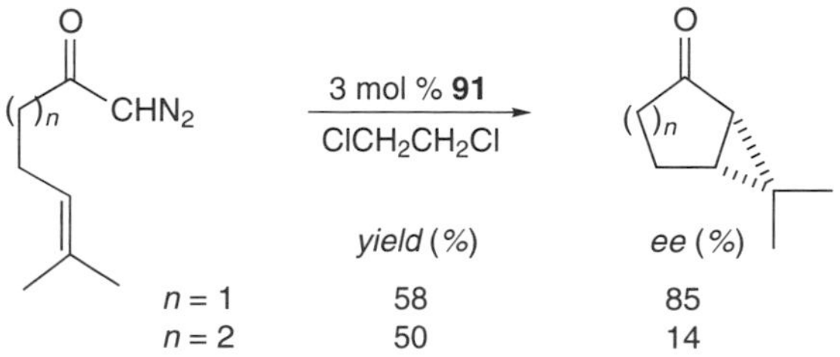

	yield (%)	ee (%)
n = 1	58	85
n = 2	50	14

(47)

lectivity. In efforts designed to effect enantioselective synthesis of cyclopropane–amino acids (e.g., eq. 23), Koskinen and Hassila found that with **91** (R = Bn) *tert*-butyl allyl diazomalonate **92** underwent intramolecular cyclopropanation in 72% yield and 32% ee,[99] and Pfaltz reported that **93** underwent cyclopropanation

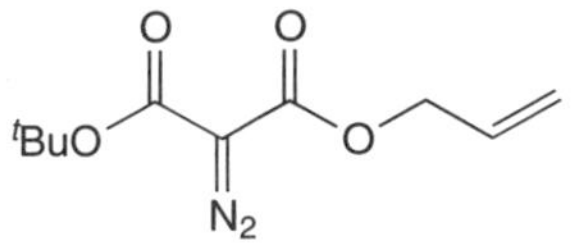

92

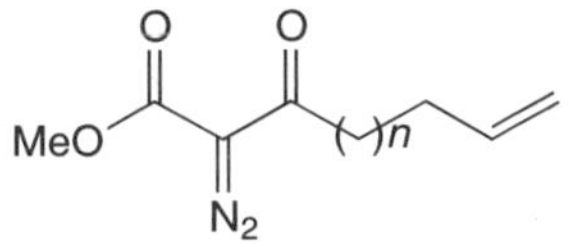

93 (n = 1,2)

in 50% yield with 35–40% ee using **91** ($R = CMe_2OH$).[98] Attempts to effect highly enantioselective intramolecular cyclopropanation reactions with diazoketones, including those of eqs. 46 and 47, using chiral dirhodium(II) carboxamidates have thus far been unsuccessful (<40% ee), even though product yields are high.[100]

A comprehensive study of Cu(I)/bis-oxazoline catalysts for enantioselective intramolecular cyclopropanation of diazoketone **94** (eq. 48) has been reported[101]

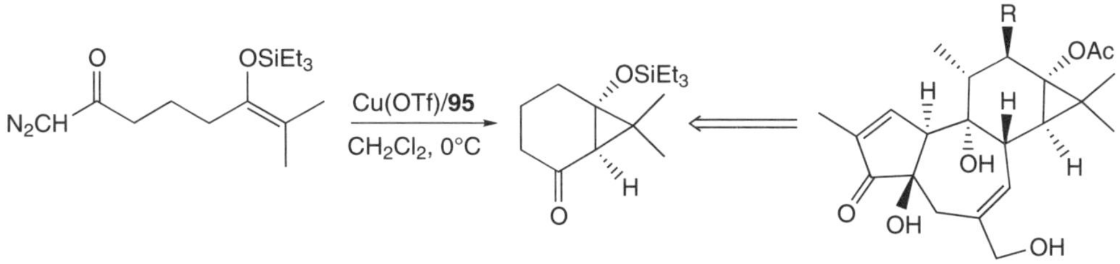

94

96

97

prostratin (R = H)

(48)

as part of the strategy for the construction of the CD-ring of phorbol derivatives **97**, including anti-HIV active prostratin ($R = H$). Enantioselectivity increased with the increasing size of R^2 (**95**), and the data suggest that R^1, as either H or

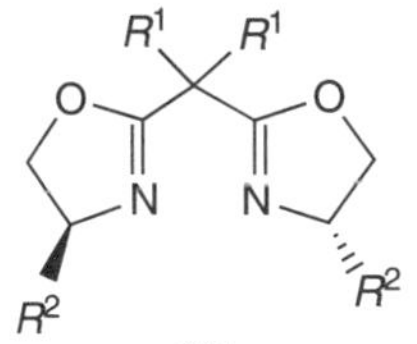

95

R^1	R^2	yield (%)	ee (%) (**96**)
H	^tBu	85	77
Me	Ph	38	13
Me	^nBu	55	36
Me	^tBu	82	78
Me	CMe_2OTMS	70	92

Me, has little influence on enantioselectivity. In this example, 5 mol % of Cu(OTf) and 15 mol % of **95** were employed.

Catalyst refinements often lead to improvements in enantiocontrol, and those of Corey and co-workers illustrate how innovative adaptions can be successful for enhancement of optical purity.[102] Their target was sirenin (**98**), a potent sperm attractant of the water mold *Alomyces*. Diazo decomposition of the vinyldiazomethane **99** produced the intramolecular cyclopropanation product **100** (eq. 49). Enantioselectivity was determined with a variety of chiral catalysts, but only the bis-oxazoline on a biphenyl backbone (**101**) as a ligand for Cu(I) provided ee values that were synthetically meaningful. Ligand **101** and its enantiomer repre-

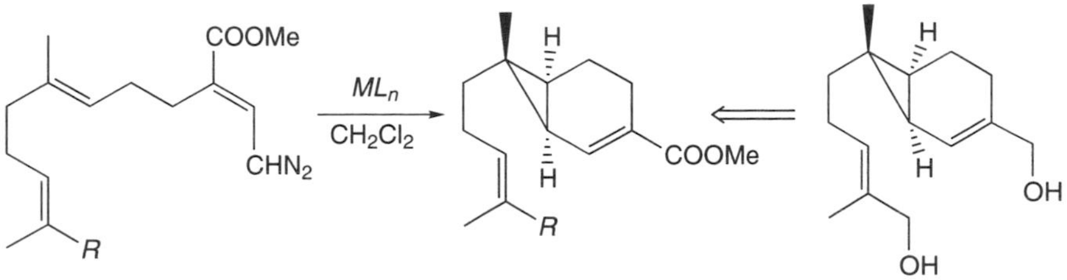

(49)

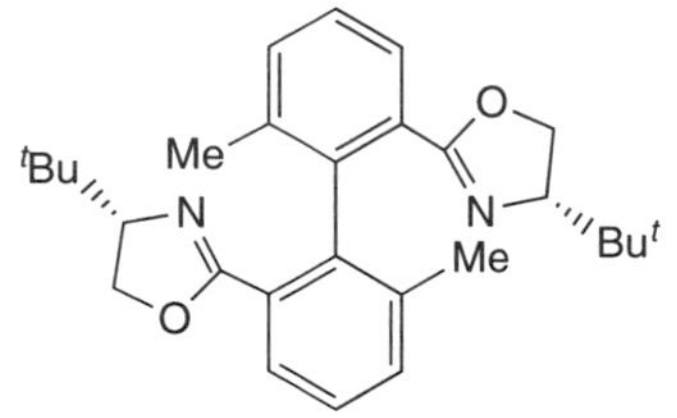

101 (*S,S,S*)

ML_n	% ee (**100**)
95 (R^1 = Me, R^2 = ^tBu)/Cu(OTf)	60
95 (R^1 = Me, R^2 = Ph)/Cu(OTf)	68
Rh$_2$(5S-MEPY)$_4$	30
101/Cu(OTf)	90

sent a new design that can be expected to have additional advantages for catalytic asymmetric syntheses.

5.6 DIASTEREOCONTROL IN INTRAMOLECULAR CYCLOPROPANATION REACTIONS

As stated earlier, Martin and co-workers demonstrated that Rh$_2$(5S-MEPY)$_4$ and Rh$_2$(5R-MEPY)$_4$ exhibited unique diastereocontrol in intramolecular cyclopropanation reactions of secondary allylic diazoacetates (eq. 44).[93] Using enantiomerically pure **102** having the S configuration, catalysis by Rh$_2$(5S-MEPY)$_4$ gave high *endo*:*exo* selectivities, whereas with Rh$_2$(5R-MEPY)$_4$ low diastereoselectivity and low product yields were observed (eq. 50). Preference for the *endo* stereoisomer exists even when R^c is relatively large; however, changing the methyl substituent in **102** to a larger group decreases selectivity for the more crowded *endo* isomer. Use of Cu(TBS)$_2$ gave low diastereocontrol (*endo*>*endo*) with **102**.

Doyle, Martin, and co-workers examined racemic secondary allylic diazoacetates, and they found moderate diastereoselectivities and enantioselectivities.[91]

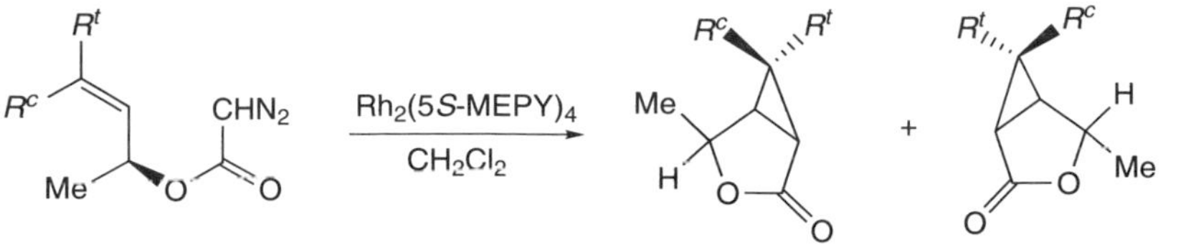

102 *endo*-**103** *exo*-**103**

R^t	R^c	*endo:exo*-103	yield, %
H	^{n}Bu	>95:5	80
^{n}Bu	H	86:14	77
H	TMS	>95:5	74
TMS	H	95:5	76

(50)

Here, the optimum catalyst for high *endo* isomer % ee's was $Rh_2(4S\text{-MEOX})_4$, but $Rh_2(5S\text{-MEPY})_4$ produced higher % ee's for the *exo* isomer (e.g., eq. 51). The

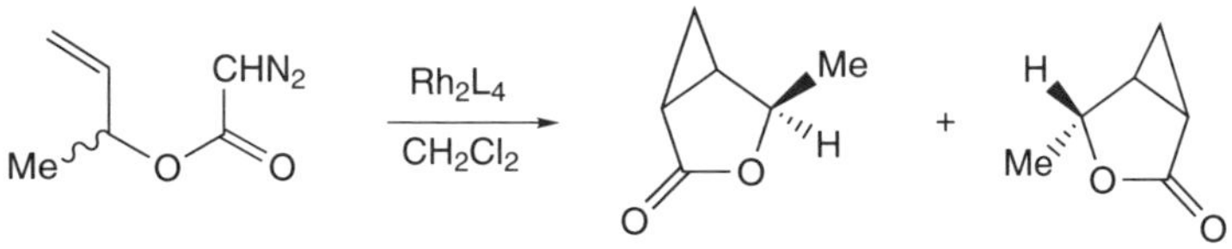

rac-**104** *endo*-**105** *exo*-**105** (51)

	yield (%)	*endo:exo*-105	*endo*-105	*exo*-105
$Rh_2(5S\text{-MEPY})_4$	80	83:17	31% ee	84% ee
$Rh_2(4S\text{-MEOX})_4$	63	70:30	70% ee	30% ee

predominant enantiomer of the diastereomeric pair have opposite configurations, (4*R*)-*endo*-**105** and (4*S*)-*exo*-**105**, demonstrating enantiomer differentiation in which one enantiomer of *rac*-**104** forms *endo*-**105** predominantly, whereas the other enantiomer forms mainly *exo*-**105**.

Significant diastereocontrol has been observed in the formation of **64** (eq. 35) with both $Rh_2(OAc)_4$ and $Rh_2(cap)_4$.[66] Lahuerta's orthometallated phosphine catalysts, $Rh_2(PC)_2(OAc)_2$, where PC = orthometallated phosphine,[70] however, form **64** exclusively but with *exo*:*endo* selectivities of 50:50 to 90:10, and an additional example (eq. 36) confirms this selectivity. Ohfune has prepared the four 3′-substituted 2-(carboxycyclopropyl)glycines by diastereoselective intramolecular cyclopropanation using $Pd(OAc)_2$ with a Boc-protected 1,3-oxazolidine to direct selectivity (eqs. 52 and 53).[103] Only the *exo* diastereoisomer was formed in eq. 53, but this compound was subsequently isomerized to a derivative of the *endo* isomer.

Although there are yet few examples of diastereocontrol in intramolecular cyclopropanation reactions, those presented here suggest that such control can be realized. Factors that influence diastereoselection are intimately associated with

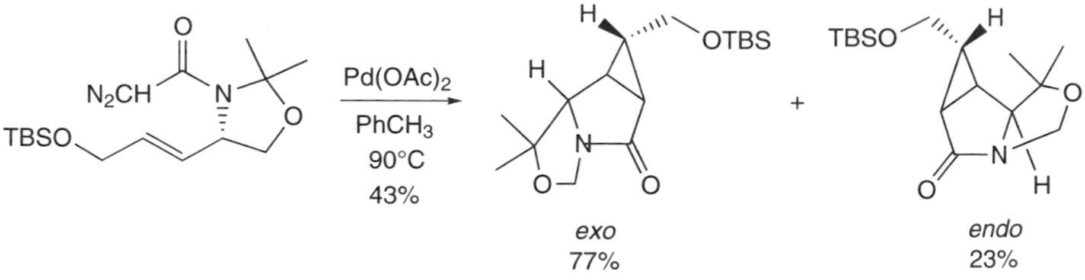

(52)

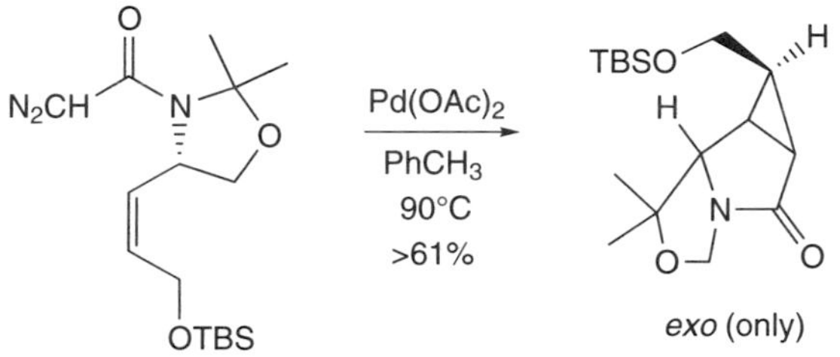

(53)

the catalyst, but specific electronic and steric influences have not been fully defined.

5.7 MACROCYCLIC CYCLOPROPANATION

Since the first report of catalytic intramolecular cyclopropanation in 1961,[1] there has been a general understanding that only five-, six-, or (rarely) seven-membered ring cyclic compounds are accessible via metal carbene intermediates. A brief report in 1982 of the synthesis of (d,l)-casbene involving intramolecular cyclopropanation by the vinyldiazomethane (**106**) derived from all-trans-geranylgeraniol (eq. 54)[104] was the only report prior to 1995 that macrocyclization

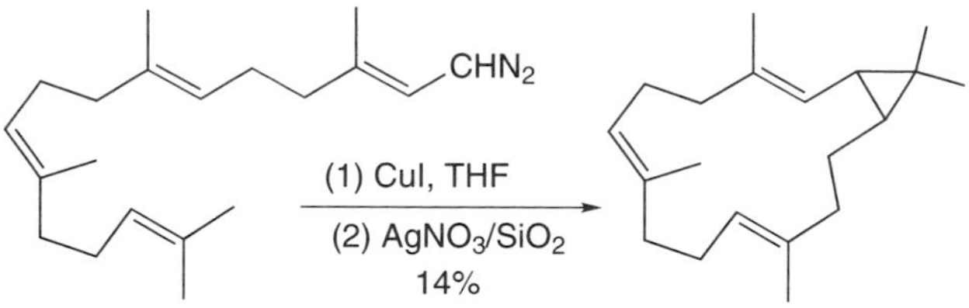

(54)

was a possibility, and the low product yield together with the absence of additional examples relegated this methodology to obscurity. In 1995 Doyle, Poulter, and co-workers reported that macrocyclic lactones are produced in good yield using moderately electrophilic dirhodium(II) catalysts even when γ-lactone formation is competitive.[105] trans,trans-Farnesyl diazoacetate, for example, undergoes exclusive allylic cyclopropanation in reactions catalyzed by dirhodium(II) carboxamidates, except $Rh_2(NHCOCF_3)_4$ (Scheme 10), but formation of a 13-mem-

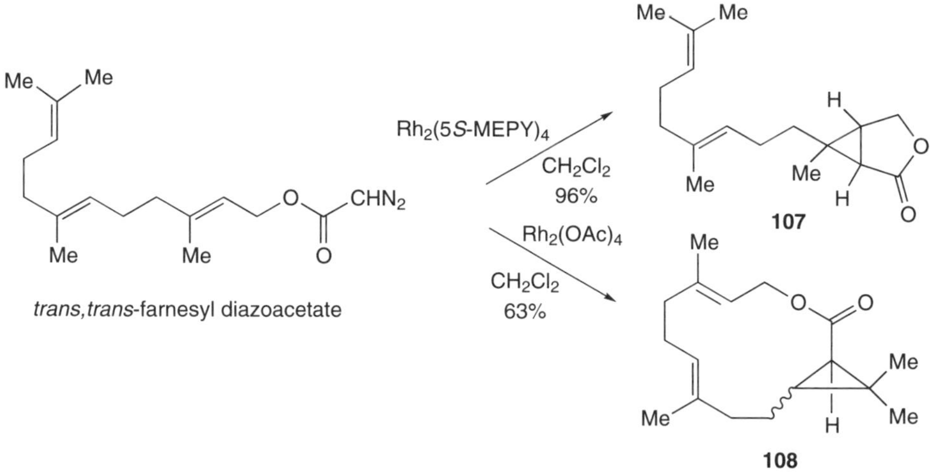

	107:108 (t/c)
Rh$_2$(5S-MEPY)$_4$[85]	100:0
Rh$_2$(cap)$_4$	100:0
Rh$_2$(OAc)$_4$	0:100 (86/14)
Rh$_2$(NHCOCF$_3$)$_4$	0:100 (75/25)
Rh$_2$(pfb)$_4$	0:100 (51/49)

Scheme 10

bered ring lactone is the sole outcome of rhodium(II) carboxylate–catalyzed reactions that occur without high dilution; the internal, nonallylic, double bond of farnesyl diazoacetate was not essential for macrocyclization since its removal did not alter the results, except in product yield. Other examples, including intramolecular cyclopropanation of the diazoacetate of (±)-cis-nerolidol (Scheme 11), suggest the generality of the transformation.

Macrocyclic cyclopropanation can occur with high levels of enantiocontrol, diastereocontrol, and regiocontrol, as the example in eq. 55 amply demonstrates.[9]

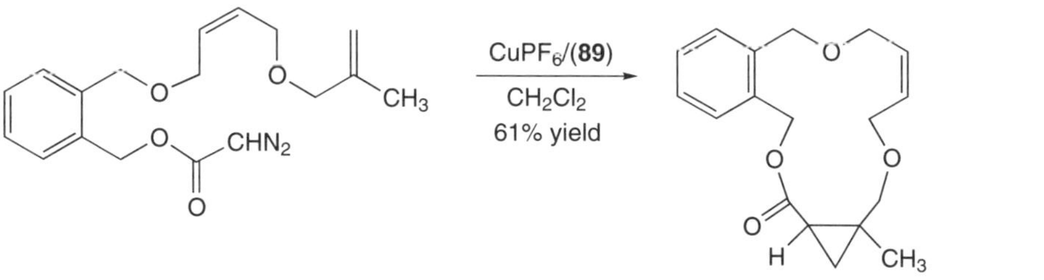

94: 90% ee
>50:1 regioselectivity
>99:1 dr

(55)

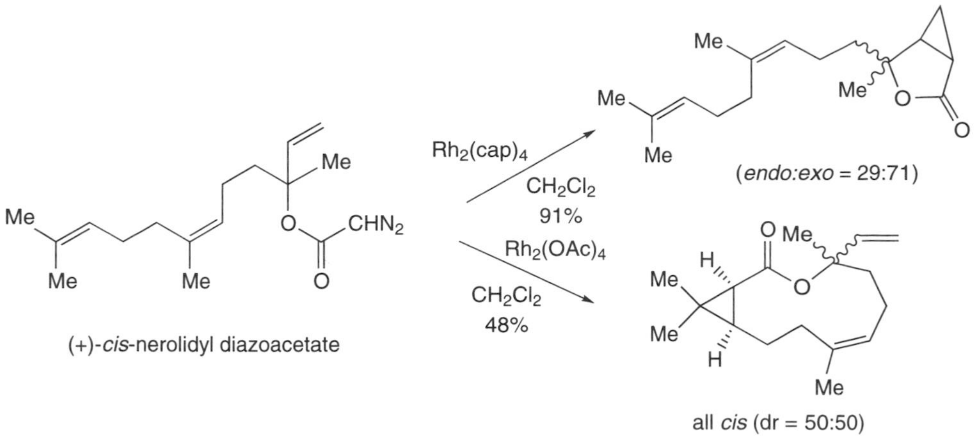

Scheme 11

Only the *cis*-cyclopropane product is formed, and regioselectivity for formation of the 15-membered ring occurs with at least a 50:1 preference over formation of the 10-membered ring. Use of $CuPF_6/($**89**$)$ provides access to macrocycles since its reactivity is similar to dirhodium(II) carboxylates, rather than dirhodium(II) carboxamidates which prefer γ-lactone formation (e.g., eq. 56). The enantiocon-

	110	**111**
$CuPF_6/($**89**$)$	43% (87% ee)	19% (41% ee)
Rh_2(5*S*-MEPY)$_4$	0%	84% (96% ee)
(**88**)	0%	45% (17% ee)

$$(56)$$

trol that is observed is similar to that found in intramolecular allylic cyclopropanation reactions (Table 5.4). Saturation of the *cis*-disubstituted double bond of **109** did not inhibit macrocyclization: With $CuPF_6/($**89**$)$ the corresponding saturated macrocycle is formed in 43% yield with 91% ee.[9]

Macrocyclization is consistent with the formation of an intermediate π complex (**112**) between the carbon double bond and the carbene center (Scheme 12), an explanation initially advanced by Doyle to explain stereoselectivity in inter-

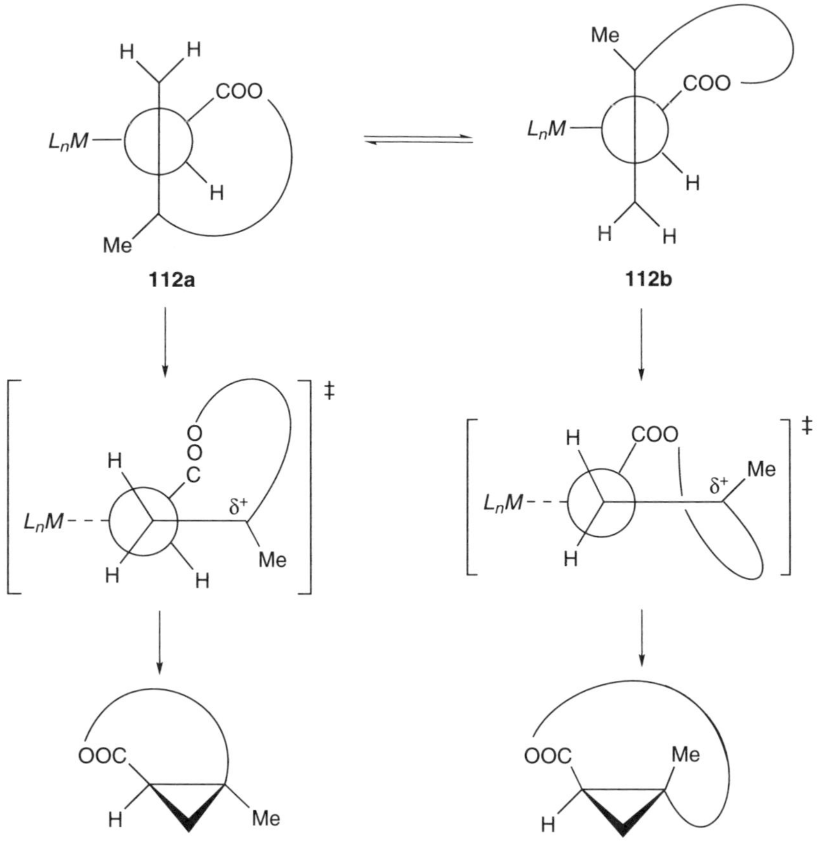

Scheme 12

molecular cyclopropanation reactions.[106,107] Rotation of the alkene on the carbene center so as to maximize charge development at the more substituted carbon of the carbon–carbon double bond in the transition state provides the orientation conducive to cyclopropane bond formation. Allylic double bonds are sterically prevented from such complexation with the net effect that macrocyclization is favored over allylic cyclopropanation for those catalytic systems in which metal carbene contributing structure **70b** plays a prominent directive role in bond formation. In this respect, π-complex formation effectively lowers the activation energy for cyclopropanation and allows macrocyclization to compete with allylic cyclopropanation, which is favored entropically.

Macrocyclic cyclopropanation is an open area that is ripe for development. Some of the structural frameworks that have been prepared thus far are described as **113–118**.[108] The experimental procedures employed for their formation are the same as those for allylic cyclopropanation, and, consequently, the general utility of catalytic cyclopropanation is greatly expanded.[109]

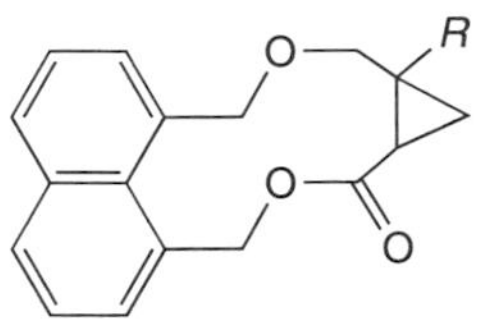

113 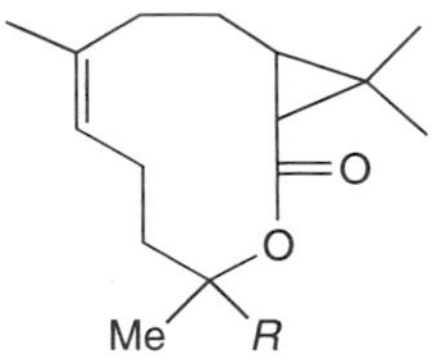   114 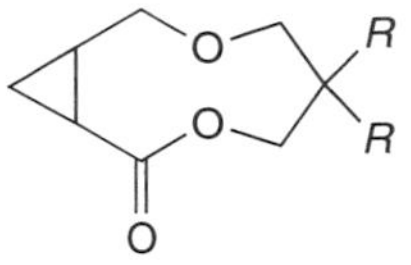   115

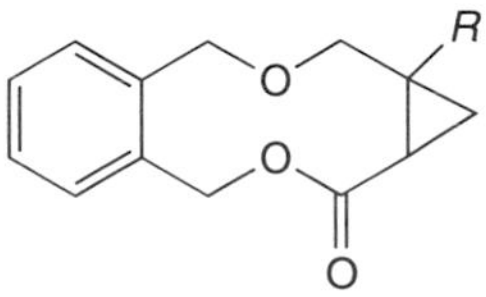

116

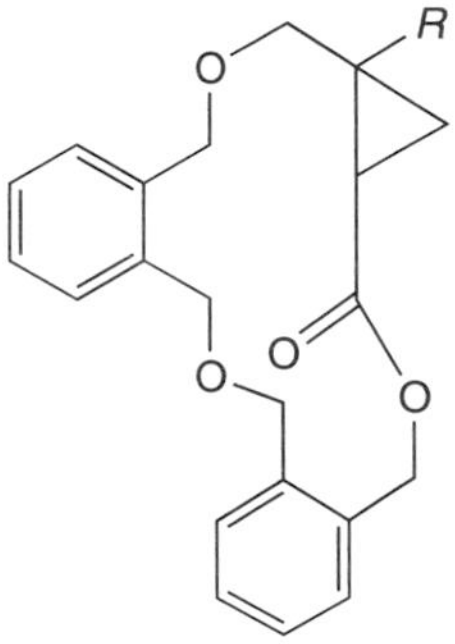

117 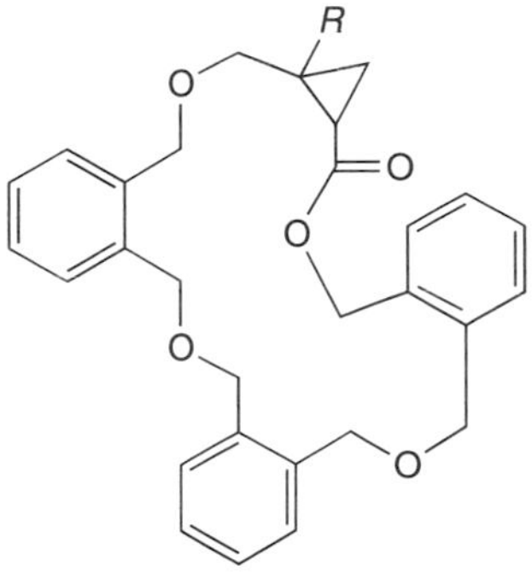   118

5.8 INTRAMOLECULAR CYCLOPROPENATION REACTIONS: TANDEM/CASCADE PROCESSES

Whereas intermolecular reactions form relatively stable cyclopropene products, those that occur intramolecularly do not result in isolable cyclopropenes but instead rearrange by as yet uncertain pathways to vinyl carbenes that undergo subsequent reactions[110] (Scheme 13). For example, propargyl diazoketoacetate and acetamide systems undergo cyclization to **119** (eq. 57) in good yields by in-

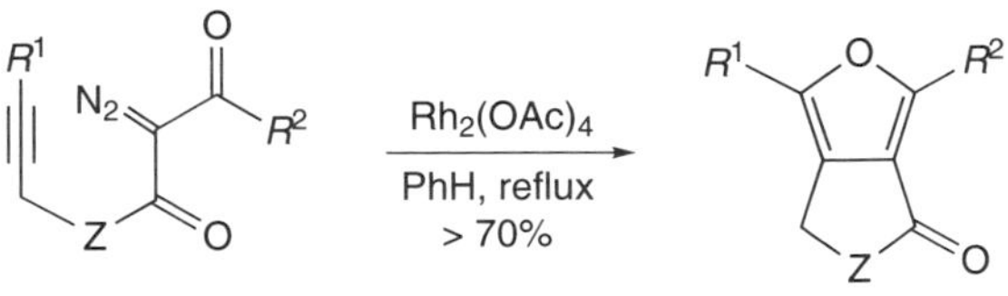

R^1 = H, Me, Ph, SiMe$_3$, C$_5$H$_{11}$
R^2 = Me, Ph
Z = O, NMe

(57)

tramolecular addition, rearrangement to a vinyl carbene intermediate, and then dipolar cyclization with the keto functionality.[111] Tandem reactions of vinyl carbene intermediates occur via intramolecular cyclopropanation–Cope rearrangement (eq. 58),[112] carbon–hydrogen insertion (eq. 59),[113,114] aromatic substitution (eq. 60),[115] cyclopropenation/dimer formation or dipolar addition,[116,117] and inter-

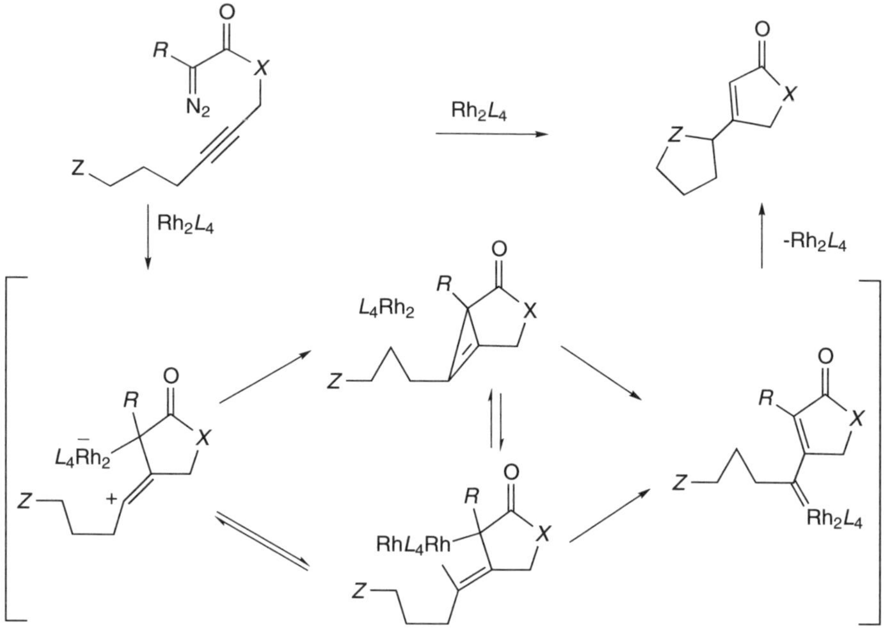

Scheme 13

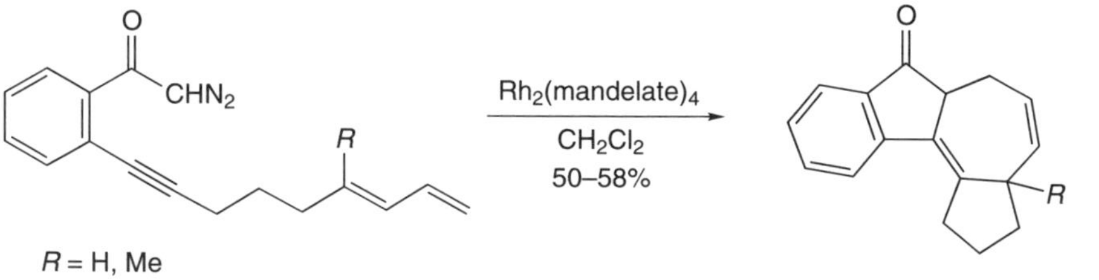

$R = $ H, Me

(58)

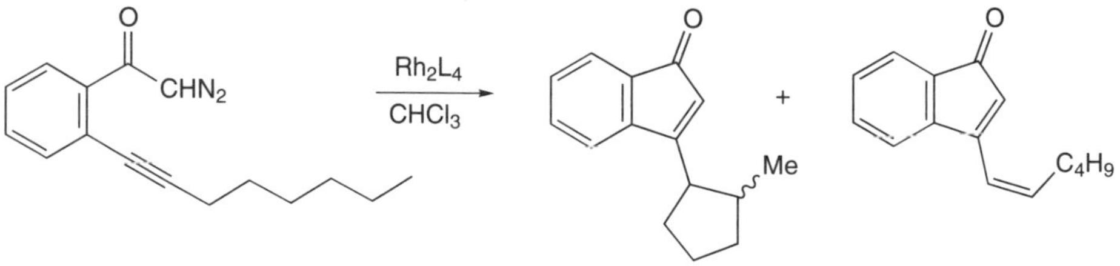

	120	**121**
Rh$_2$(pfb)$_4$	10	90
Rh$_2$(OAc)$_4$	66	33
Rh$_2$(Oct)$_4$	80	20

(59)

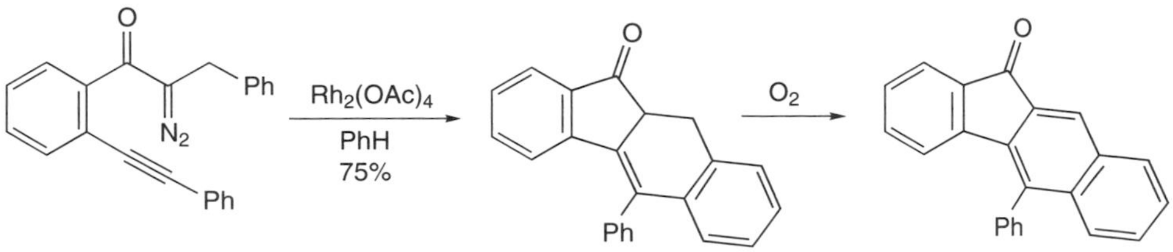

(60)

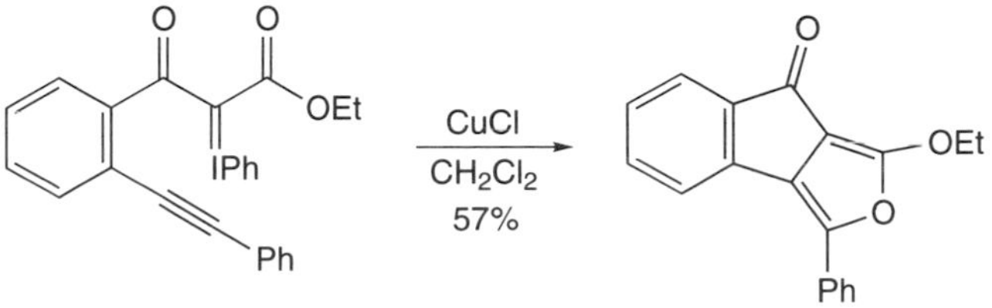

(61)

molecular cyclopropanation.[116] Iodonium ylides undergo similar reactions (eq. 61) in good yields that are similar to those from dirhodium(II)-catalyzed reactions of identically substituted diazo compounds.[118] However, N-propargylindole systems did not undergo addition to the triple bond, and the cause for this has not been established.[119]

Hoye and co-workers provided comparisons with selected tandem reactions of carbene generation from different sources[120] and from diazoketones with different transition metal catalysts (eq. 62).[121] Catalysts amounts were generally

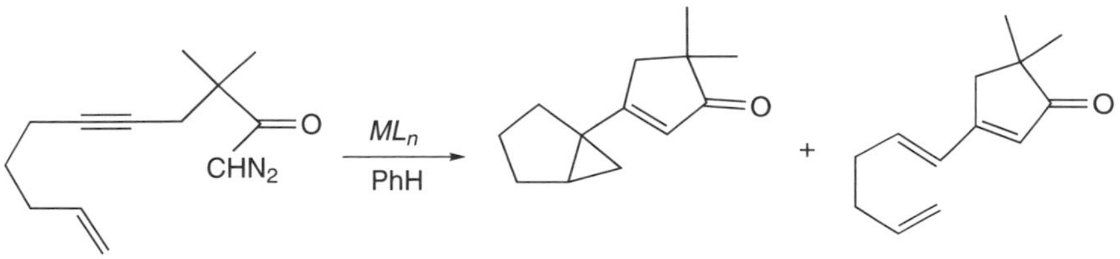

	122	123
Rh$_2$(OAc)$_4$	18	28
Pd(acac)$_2$	87	3
Cu(acac)$_2$	29	7
Rh$_6$(CO)$_{16}$	59	–

(62)

greater than 10 mol %, and, from the products that were obtained, conclusions were drawn that suggested a mechanism similar to that of Scheme 13. Both intramolecular sulfonium ylide[122] and intermolecular addition reactions (eq. 63)[123] were used to trap the intermediate vinylcarbene. Higher yields and, for specific transformations, greater versatility were suggested for Fischer carbenes (eq. 64).[124] A variety of methods have been developed for intramolecular addition that

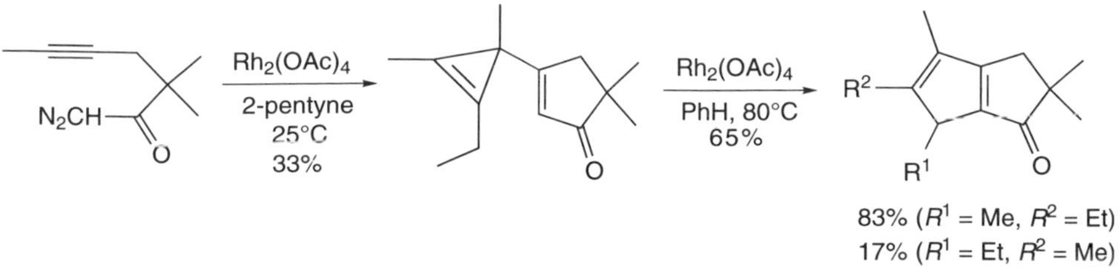

$$83\% \ (R^1 = Me, R^2 = Et)$$
$$17\% \ (R^1 = Et, R^2 = Me)$$

(63)

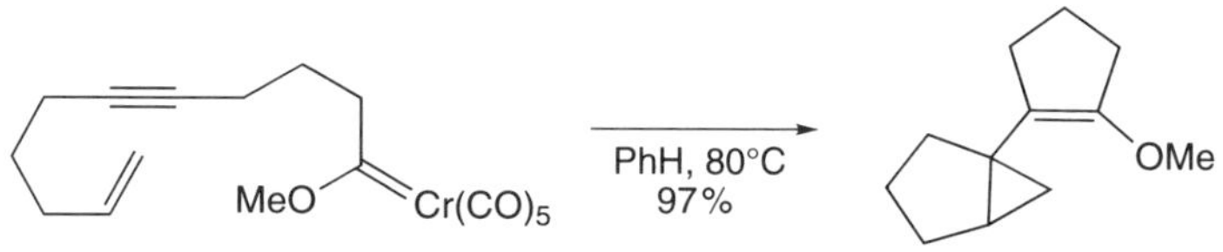

(64)

results in the formation of bicyclic and tricyclic products through Fischer carbene intermediates,[125-127] and the involvement of metallocyclobutane intermediates has been suggested,[128] but the relationship of the mechanism for these reactions to those performed catalytically with diazo compounds has not been firmly established. A recent descriptive summary of this methodology reports its synthetic applicability.[129]

Cyclopropene ring openings catalyzed by dirhodium(II) have been shown to produce vinylcarbene intermediates,[130] and the overall process has been explained by electrophilic addition rather than by metallocyclobutene intermediates (Scheme 14).[130,131] The most effective catalyst is the highly electrophilic

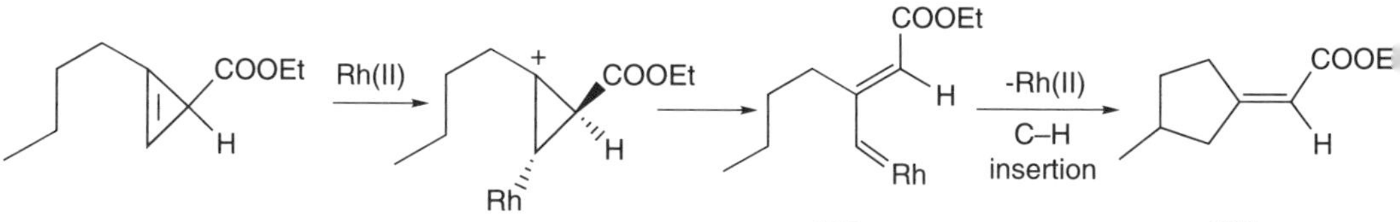

Scheme 14

Rh$_2$(pfb)$_4$. Electrophilic addition *trans* to the ester functionality produces the more stable carbocation; disrotatory ring opening provides vinylcarbene **124**, which undergoes intramolecular C–H insertion. The absence of the (Z) isomer of **125** is consistent with this mechanism. Ring opening reactions catalyzed by Rh(I) (e.g., eq. 65),[132] however, may involve metallocyclobutene intermediates.[132,133]

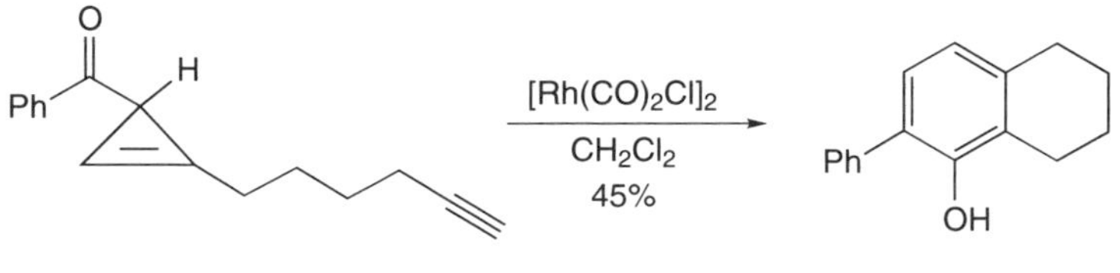

(65)

The utility of the catalytic methodology for tandem cyclization continues to expand as additional applications are reported.[134-136] Although some mechanistic details remain unresolved, the involvement of a vinyl metallocarbene intermediate is certain. This tandem process suggests other possible arrays and defines vinylcarbenes as versatile reaction intermediates for intramolecular transformations.

5.9 REFERENCES

1   Stork, G.; and Ficini, J., "Intramolecular Cyclization of Unsaturated Diazoketones," *J. Am. Chem. Soc.* **1961**, *83*, 4678.

2   Burke, S. D.; and Grieco, P. A., "Intramolecular Reactions of Diazocarbonyl Compounds," *Org. React. (N.Y.)* **1979**, *26*, 361–475.

3   Maas, G., "Transition-Metal Catalyzed Decomposition of Aliphatic Diazo Compounds—New Results and Applications in Organic Synthesis," *Top. Curr. Chem.* **1987**, *137*, 75–253.

4   Doyle, M. P., "Catalytic Methods for Metal Carbene Transformations," *Chem. Rev.* **1986**, *86*, 919–39.

5   Padwa, A.; and Krumpe, K. E., "Application of Intramolecular Carbenoid Reactions in Organic Synthesis," *Tetrahedron* **1992**, *48*, 5385–453.

6   Ye, T.; and McKervey, M. A., "Organic Synthesis with α-Diazocarbonyl Compounds," *Chem. Rev.* **1994**, *94*, 1091–160.

7   Doyle, M. P., "Metal Carbene Complexes in Organic Synthesis: Cyclopropanation," in *Comprehensive Organometallic Chemistry II*, Vol. 12; Hegedus, L. S., Ed.; Pergamon Press: New York, 1995; Chapter 5.1.

8   Reissig, H.-U., "Formation of C–C Bonds by [2+1] Cycloaddition," in *Stereoselective Synthesis* of *Houben–Weyl Methods of Organic Chemistry*, Vol. E 21c; Helmchen, G.; Hoffmann, R. W.; Mulzer, J.; and Schaumann, E., Eds.; Georg Thieme Verlag: New York, 1995; Section 1.6.1.5.

9   Doyle, M. P.; Peterson, C. S.; and Parker, D. L., Jr., "Formation of Macrocyclic Lactones by Enantioselective Intramolecular Cyclopropanation of Diazoacetates Catalyzed by Chiral Copper(I) and Rhodium(II) Compounds," *Angew. Chem. Int. Ed. Engl.* **1996**, *35*, 1334–36.

10   Doering, W. von E.; Ferrier, B. M.; Fossel, E. T.; Hartenstein, J. H.; Jones, M., Jr.; Klumpp, G.; Rubin, R. M.; and Saunders, M., "A Rational Synthesis of Bullvalene, Barbaralone and Derivatives; Bullvalone," *Tetrahedron* **1967**, *23*, 3943–63.

11   Tichy, M., "On the Absolute Configuration of Tricyclo(4,4,0,0$^{3.8}$)decane (Twistane)," *Tetrahedron Lett.* **1972**, 2001–4.

12   Mori, K.; Ohki, M.; Kobayashi, A.; and Matsui, M., "Synthesis of Mono- and Sesquiterpenoids-III. ($\pm$)-Thujopsene," *Tetrahedron* **1970**, *26*, 2815–19.

13   Bhalerao, U. T.; Plattner, J. J.; and Rapoport, H., "Synthesis of *dl*-Sirenin and *dl*-Isosirenin," *J. Am. Chem. Soc.* **1970**, *92*, 3429–33.

14   Hudlicky, T.; Kutchan, T. M.; Wilson, S. R.; and Mao, D. T., "Tota! Synthesis of ($\pm$)-Hirsutene," *J. Am. Chem. Soc.* **1980**, *102*, 6351–53.

15   White, J. D.; Ruppert, J. F.; Avery, M. A.; Torii, S.; and Nokami, J., "Spiroannelation via Intramolecular Ketocarbenoid Addition. Stereocontrolled Synthesis of

(−)-Acorenone B and (±)-α-Chamigrene," *J. Am. Chem. Soc.* **1981**, *103*, 1813–21.

16   Rigby, J. H.; and Senanayake, C., "Total Synthesis of (±)-Grosshemin," *J. Am. Chem. Soc.* **1987**, *109*, 3147–49.

17   Wong, H. N. C.; Hon, M.-Y.; Tse, C.-W.; Yip, Y.-C.; Tanko, J.; and Hudlicky, T., "Use of Cyclopropanes and their Derivatives in Organic Synthesis," *Chem. Rev.* **1989**, *89*, 165–98.

18   Adams, J.; and Belley, M., "Formation of Reactive Tricyclic Intermediates via the Intramolecular Cyclopropanation of Dihydropyrans. Synthesis of Eucalyptol," *Tetrahedron Lett.* **1986**, *27*, 2075–78.

19   Adams, J.; Lepine-Frenette, C.; and Spero, D. M., "Intramolecular Cyclopropanation-Ring Fragmentation Leading to Spirocyclic Ring Construction: A Stereoselective Synthesis of β-Chamigrene," *J. Org. Chem.* **1991**, *56*, 4494–98.

20   Taber, D. F.; and Hoerrner, R. S., "Enantioselective Rh-Mediated Synthesis of (−)-PGE$_2$ Methyl Ester," *J. Org. Chem.* **1992**, *57*, 441–47.

21   Dawe, R. D.; Mander, L. N.; and Turner, J. V., "Stereocontrolled Synthesis of Gibberellin A$_{19}$ from Gibberellic Acid," *Tetrahedron Lett.* **1985**, *26*, 363–66.

22   Dawe, R. D.; Mander, L. N.; Turner, J. V.; and Xinfu, P., "Synthesis of C$_{20}$ Gibberellin A$_{36}$ and A$_{37}$ Methyl Esters from Gibberellic Acid," *Tetrahedron Lett.* **1985**, *26*, 5725–28.

23   Doyle, M. P.; and Trudell, M. L., "Catalytic Role of Copper Triflate in Lewis Acid Promoted Reactions of Diazo Compounds," *J. Org. Chem.* **1984**, *49*, 1196–99.

24   Honda, T.; Ishige, H.; Tsubuki, M.; Naito, K.; and Suzuki, Y., "Novel Carbon–Carbon Bond Formation by Means of a Rhodium Acetate-Catalysed Reaction of γ,δ-Unsaturated Diazoketone and its Application to the Synthesis of 4-*epi*-Isovalerenenol," *J. Chem. Soc., Perkin Trans. 1* **1991**, 954–55.

25   Honda, T.; Ishige, H.; Tsubuki, M.; Naito, K.; and Suzuki, Y., "An Enantioselective Synthesis of (−)-Neonepetalactone," *Chem. Pharm. Bull.* **1991**, *39*, 1641–43.

26   Adams, J.; Frenette, R.; Belley, M.; Chibante, F.; and Springer, J. P., "Intramolecular Cyclopropanation of Enol Ethers: Synthetic Approach to Medium-Sized Carbocycles," *J. Am. Chem. Soc.* **1987**, *109*, 5432–37.

27   Smith, A. B., III; Toder, B. H.; and Branca, S. J., "Vinylogous Wolff Rearrangement. 4. General Reaction of β,γ-Unsaturated α′-Diazo Ketones" *J. Am. Chem. Soc.* **1984**, *106*, 3995–4001.

28   Smith, A. B., III; Toder, B. H.; Richmond, R. E.; and Branca, S. J., "Vinylogous Wolff Rearrangement. 5. Mechanistic Studies," *J. Am. Chem. Soc.* **1984**, *106*, 4001–9.

29   Saha, B.; Bhattacharjee, G.; and Ghatak, U. R., "Vinylogous Wolff Rearrangement of Cyclic β,γ-Unsaturated Diazomethyl Ketones: A New Synthetic Method for Angularly Functionalized Polycyclic Systems," *J. Chem. Soc., Perkin Trans. 1* **1988**, 939–44.

30   Dowd, P.; Schappert, R.; Garner, P.; and Go, C. L., "Halogenation and Rearrangement Reactions of Substituted Tricyclo[2.1.0.0^{2,5}]pentan-3-ones," *J. Org. Chem.* **1985**, *50*, 44–47.

31   Reingold, I. D.; and Drake, J., "[3.3.1]-Propellane-2,8-dione," *Tetrahedron Lett.* **1989**, *30*, 1921–22.

32 Fernandez Mateos, A.; and Lopez Barba, A. M., "Limonoid Model Insect Anti-feedants. A Stereoselective Synthesis of Azadiradione C, D, and E Fragments through Intramolecular Diazo Ketone Cyclization," *J. Org. Chem.* **1995**, *60*, 3580–85.

33 Corey, E. J.; Reid, J. G.; Myers, A. G.; and Hahl, R. W., "Simple Synthetic Route to the Limonoid System," *J. Am. Chem. Soc.* **1987**, *109*, 918–19.

34 Short, R. P.; Revol, J.-M.; Ranu, B. C.; and Hudlicky, T., "General Method of Synthesis of Cyclopentanoid Terpenic Acids. Stereocontrolled Total Synthesis of (±)-Isocomenic Acid and (±)-Epiisocomenic Acid," *J. Org. Chem.* **1983**, *48*, 4453–61.

35 Hudlicky, T.; Fleming, A.; and Radesca, L., "[2+3] and [3+4] Annulation of Enones. Enantiocontrolled Total Synthesis of (−)-Retigeranic Acid," *J. Am. Chem. Soc.* **1989**, *111*, 6691–707.

36 Hudlicky, T.; Kwart, L. D.; Tiedje, M. H.; Ranu, B. C.; Short, R. P.; Frazier, J. O.; and Rigby, H. L., "General Methodology for the Topologically Selective Preparation of Linear and Nonlinear Tricyclopentanoids of Hirsutane and Isocomene Type via Claisen Rearrangement/Cyclopentene Annulation Sequence. Total Synthesis of (±)-Epiisocomene," *Synthesis* **1986**, 716–27.

37 Hudlicky, T.; Natchus, M. G.; and Sinai-Zingde, G., "Stereocontrolled Total Synthesis of Pentalenenes via [2+3] and [4+1] Cyclopentene Annulation Methodologies," *J. Org. Chem.* **1987**, *52*, 4641–44.

38 Hudlicky, T.; Sinai-Zingde, G.; Natchus, M. G.; Ranu, B. C.; and Papadopolous, P., "System Oriented Design of Triquinanes: Stereocontrolled Synthesis of Pentalenic Acid and Pentalenene," *Tetrahedron* **1987**, *43*, 5685–721.

39 Davies, H. M. L., "Tandem Cyclopropanation/Cope Rearrangement: A General Method for the Construction of Seven-Membered Rings," *Tetrahedron* **1993**, *49*, 5203–23.

40 Davies, H. M. L.; McAfee, M. J.; and Oldenburg, C. E. M., "Scope and Stereochemistry of the Tandem Intramolecular Cyclopropanation/ Cope Rearrangement Sequence," *J. Org. Chem.* **1989**, *54*, 930–36.

41 Davies, H. M. L.; and Matasi, J. J., "Rhodium(II) Catalyzed Intramolecular Reactions between Vinyldiazomethanes and Pyrroles. Novel Synthesis of Fused 7-Azabicyclo[4.2.0]octadienes," *Tetrahedron Lett.* **1994**, *35*, 5209–12.

42 Davies, H. M. L.; Matasi, J. J.; and Ahmed, G., "Divergent Pathways in the Intramolecular Reactions between Rhodium-Stabilized Vinylcarbenoids and Pyrroles: Construction of Fused Tropanes and 7-Azabicyclo[4.2.0]octadienes," *J. Org. Chem.* **1996**, *61*, 2305–13.

43 Corey, E. J.; Wess, G.; Xiang, Y. B.; and Singh, A. K., "Stereospecific Total Synthesis of (±)-Cafestol," *J. Am. Chem. Soc.* **1987**, *109*, 4717–18.

44 Singh, A. K.; Bakshi, R. K.; and Corey, E. J., "Total Synthesis of (±)-Atractyligenin," *J. Am. Chem. Soc.* **1987**, *109*, 6187–89.

45 Hatch, C. E., III; Baum, J. S.; Takashima, T.; and Kondo, K., "Stereospecific Total Synthesis of the Potent Synthetic Pyrethroid NRDC 182," *J. Org. Chem.* **1980**, *45*, 3281–85.

46 Yadav, J. S.; Mysorekar, S. V.; and Rao, A. V. R., "Synthesis of (*1R*)-(+)-*cis*-Chrysanthemic Acid," *Tetrahedron* **1989**, *45*, 7353–60.

47 Ramaiah, M.; and Nagabhushan, T. L., "Synthesis of Carbacerulenin. A General Method for Syntheses of *E,E*-1,4-Dienols," *Synth. Commun.* **1986**, *16*, 1049–57.

48 Koskinen, A. M. P.; and Muñoz, L., "Intramolecular Cyclopropanation: Stereospecific Synthesis of (*E*)- and (*Z*)-1-Aminocyclopropane-1-carboxylic Acids," *J. Org. Chem.* **1993**, *58*, 879–86.

49 Yamanoi, K.; and Ohfune, Y., "Syntheses of *trans*- and *cis*-α-"(Carboxycyclopropyl)glycines. Novel Neuroinhibitory Amino Acids as L-Glutamate Analogue," *Tetrahedron Lett.* **1988**, *29*, 1181–84.

50 Shapiro, E. A.; Dyatkin, A. B.; and Nefedov, O. M., "Synthesis of Prenyl Diazoacetate and its Intramolecular Catalytic Deazotization to 6,6-Dimethyl-3-oxabicyclo[3.1.0]hexan-2-one," *Izv. Akad. Nauk SSSR, Ser. Khim.* **1990**, *39*, 1999.

51 Baird, M. S.; and Hussain, H. H., "The Preparation and Decomposition of Alkyl 2-Diazopent-4-enoates and 1-Trimethylsilyl-1-diazobut-3-enes," *Tetrahedron* **1987**, *43*, 215–24.

52 Dauben, W. G.; Hendricks, R. T.; Pandy, B.; Wu, S. C.; Zhang, X.; and Luzzio, M. J., "Stereoselective Intramolecular Cyclopropanations: Enantioselective Syntheses of 1α,25-Dihydroxyvitamin D$_3$ A-Ring Precursors," *Tetrahedron Lett.* **1995**, *36*, 2385–88.

53 Wilson, S. R.; Venkatesan, A. M.; Augelli-Szafvan, C. E.; and Yasmin, A., "A New Synthetic Route to 1,25-Dihydroxy-vitamin D$_3$," *Tetrahedron Lett.* **1991**, *32*, 2339–42.

54 He, M. Q.; Tanimori, S.; and Nakayama, M., "Homoconjugate Addition of Organocuprate to Activated Cyclopropane and Its Application to the Synthesis of (±)-Juvabione," *Biosci. Biotech. Biochem.* **1995**, *59*, 900–2.

55 Shimizu, I.; and Ishikawa, T., "Stereoselective Synthes: cf (±)-Clavukerin A and (±)-Isoclavukerin A Based on Palladium-Catalyzed Reductive Cleavage of Alkenylcyclopropanes with Formic Acid," *Tetrahedron Lett.* **1994**, *35*, 1905–8.

56 Shimizu, I.; and Aida, F., "Palladium-Catalyzed Selective Hydrogenolysis of Alkenylcyclopropanes Having Two Electron Withdrawing Groups Using Ammonium Formate," *Chem. Lett.* **1988**, 601–4.

57 Taber, D. F.; Saleh, S. A.; and Korsmeyer, R. W., "Preparation of Cyclohexanones and Cyclopentanones of High Optical Purity," *J. Org. Chem.* **1980**, *45*, 4699–702.

58 Clark, R. D.; and Heathcock, C. H., "Synthesis of Spiro Lactones and Ketones via Conjugate Addition to Cyclopropyl Malonates and β-Keto Esters," *Tetrahedron Lett.* **1975**, 529–32.

59 Danishefsky, S.; McKee, R.; and Singh, R. K., "Stereospecific Total Synthesis of *dl*-Hastanecine and *dl*-Dihydroxyheliotridane," *J. Am. Chem. Soc.* **1977**, *99*, 7711–13.

60 Burgess, K.; Lim, D.; Ho, K.-K.; and Ke, C.-Y., "Asymmetric Syntheses of Protected Derivatives of Carnosadine and Its Stereoisomers as Conformationally Constrained Surrogates for Arginine," *J. Org. Chem.* **1994**, *59*, 2179–85.

61 Burgess, K.; and Lim, D. Y., "Asymmetric Synthesis of the Stereoisomers of Protected 2,3-Methanoglutamine," *Tetrahedron Lett.* **1995**, *36*, 7815–18.

62 Fukuyama, T.; and Chen, X., "Stereocontrolled Synthesis of (−)-Hapalindole G," *J. Am. Chem. Soc.* **1994**, *116*, 3125–26.

63 Corey, E. J.; and Myers, A. G., "Efficient Synthesis and Intramolecular Cyclopropanation of Unsaturated Diazoacetic Esters," *Tetrahedron Lett.* **1984**, *25*, 3559–62.

64 Piers, E.; and Jung, G. L., "Thermal Rearrangement of Functionalized 6-*exo*-(1-Alkenyl)bicyclo[3.1.0]hex-2-enes. A Total Synthesis of (±)-Sinularene," *Can. J. Chem.* **1985**, *63*, 996–98.

65 Imanishi, T.; Matsui, M.; Yamashita, M.; and Iwata, C., "Anomalous Transformation of Tricyclo[3.3.0.0^{2,8}]octan-3-ones into Bicyclo[3.2.1]octan-3-ones, a Novel Route to (±)-Quadrone," *J. Chem. Soc., Chem. Commun.* **1987**, 1802–4.

66 Padwa, A.; Austin, D. J.; Price, A. T.; Semones, M. A.; Doyle, M. P.; Protopopova, M. N.; Winchester, W. R.; and Tran, A., "Ligand Effects on Dirhodium(II) Carbene Reactivities. Highly Effective Switching between Competitive Carbenoid Transformations," *J. Am. Chem. Soc.* **1993**, *115*, 8669–80.

67 Taber, D. F.; and Ruckle, R. E., Jr., "Cyclopentane Construction by Rh$_2$(OAc)$_4$-Mediated Intramolecular C–H Insertion: Steric and Electronic Effects," *J. Am. Chem. Soc.* **1986**, *108*, 7686–93.

68 Ceccherelli, P.; Curini, M.; Marcotullio, M. C.; and Rosati, O., "Rh$_2$(OAc)$_4$-Mediated Decomposition of Diazocarbonyl Compounds: A Comparison of α-Diazo Ketones and α-Diazo-β-keto Esters Reactivity," *Tetrahedron* **1992**, *48*, 9767–74.

69 Hashimoto, S.; Watanabe, N.; and Ikegami, S., "Highly Selective Insertion into Aromatic C–H Bonds in Rhodium(II) Triphenylacetate–Catalyzed Decomposition of α-Diazocarbonyl Compounds," *J. Chem. Soc., Chem. Commun.* **1992**, 1508–10.

70 Estevan, F.; Lahuerta, P.; Pérez-Prieto, J.; Stiriba, S.-E.; and Ubeda, M. A., "New Rhodium(II) Catalysts for Selective Carbene Transfer Reactions," *Synlett* **1995**, 1121–22.

71 Padwa, A.; and Austin, D. J., "Ligand Effects on the Chemoselectivity of Transition Metal Catalyzed Reactions of α-Diazo Carbonyl Compounds," *Angew. Chem. Int. Ed. Engl.* **1994**, *33*, 1797–815.

72 Padwa, A.; Austin, D. J.; Hornbuckle, S. F.; and Price, A. T., "Rhodium(II) Catalyzed Intramolecular Dipolar Cycloaddition Reactions of Carbonyl Ylides—Computational and Empirical Studies of the Regio- and Chemoselective Effect of Catalyst Ligand," *Tetrahedron Lett.* **1992**, *33*, 6427–30.

73 Cox, G. G.; Moody, C. J.; Austin, D. J.; and Padwa, A., "Chemoselectivity of Rhodium Carbenoids. A Comparison of the Selectivity for O–H Insertion Reactions or Carbonyl Ylide Formation versus Aliphatic and Aromatic C–H Insertion and Cyclopropanation," *Tetrahedron* **1993**, *49*, 5109–26.

74 Hon, Y.-S.; and Chang, R.-C., "The Studies of Carbenoid Reactions of α-*O*-Allyl- or α-*N*-Allyl-α'-diazopropanone Derivatives Catalyzed by Various Metal Ions," *Heterocycles* **1991**, *32*, 1089–99.

75 Doyle, M. P.; Pieters, R. J.; Martin, S. F.; Austin, R. E.; Oalmann, C. J.; and Müller, P., "High Enantioselectivity in the Intramolecular Cyclopropanation of Allyl Diazoacetates Using a Novel Rhodium(II) Catalyst," *J. Am. Chem. Soc.* **1991**, *113*, 1423–24.

76 Martin, S. F.; Oalmann, C. J.; and Liras, S., "Enantioselective, Rhodium Catalyzed Intramolecular Cycloproponations of Homoallylic Diazoacetates," *Tetrahedron Lett.* **1992**, *33*, 6727–30.

77 Doyle, M. P.; Eismont, M. Y.; Protopopova, M. N.; and Kwan, M. M. Y., "Enantioselective Intramolecular Cyclopropanation of N-Allylic and N-Homallylic Diazoacetamides Catalyzed by Chiral Dirhodium(II) Catalysts," *Tetrahedron* **1994**, *50*, 4519–28.

78 Doyle, M. P.; Austin, R. E.; Bailey, A. S.; Dwyer, M. P.; Dyatkin, A. B.; Kalinin, A. V.; Kwan, M. M. Y.; Liras, S.; Oalmann, C. J.; Pieters, R. J.; Protopopova, M. N.; Raab, C. E.; Roos, G. H. P.; Zhou, Q.-L.; and Martin, S. F., "Enantioselective Intramolecular Cyclopropanations of Allylic and Homoallylic Diazoacetates and Diazoacetamides Using Chiral Dirhodium(II) Carboxamide Catalysts," *J. Am. Chem. Soc.* **1995**, *117*, 5763–75.

79 Doyle, M. P.; Zhou, Q.-L.; Dyatkin, A. B.; and Ruppar, D. A., "Enhancement of Enantiocontrol/Diastereocontrol in Catalytic Intramolecular Cyclopropanation and Carbon–Hydrogen Insertion Reactions of Diazoacetates with $Rh_2(4S$-MP-PIM)$_4$," *Tetrahedron Lett.* **1995**, *36*, 7579–82.

80 Doyle, M. P.; Winchester, W. R.; Protopopova, M. N.; Kazala, A. P.; and Westrum, L. J., "(1R,5S)-(−)6,6-Dimethyl-3-oxabicyclo[3.1.0]hexan-2-one. Highly Enantioselective Intramolecular Cyclopropanation Catalyzed by Dirhodium(II) Tetrakis[methyl 2-pyrrolidone-5(R)-carboxylate]," *Org. Syn.* **1996**, *73*, 13–24.

81 Doyle, M. P.; and Kalinin, A. V., "Highly Enantioselective Intramolecular Cyclopropanation Reactions of N-Allylic-N-methyldiazoacetamides Catalyzed by Chiral Dirhodium(II) Carboxamidates," *J. Org. Chem.* **1996**, *61*, 2179–84.

82 Doyle, M. P.; Peterson, C. S.; Zhou, Q.-L.; and Nishiyama, H., "Comparative Evaluation of Enantiocontrol for Intramolecular Cyclopropanation of Diazoacetates with Chiral Cu(I), Rh(II) and Ru(II) Catalysts," *J. Chem. Soc., Chem. Commun.* **1997**, 211–12.

83 Martin, S. F.; Austin, R. E.; Oalmann, C. J.; Baker, W. R.; Condon, S. L.; deLara, E.; Rosenberg, S. H.; Spina, K. P.; Stein, H. H.; Cohen, J.; and Kleinert, H. D., "1,2,3-Trisubstituted Cyclopropanes as Conformationally Restricted Peptide Isosteres: Application to the Design and Synthesis of Novel Renin Inhibitors," *J. Med. Chem.* **1992**, *35*, 1710–21.

84 Martin, S. F.; Oalmann, C. J.; and Liras, S., "Cyclopropanes as Conformationally Restricted Peptide Isosteres. Design and Synthesis of Novel Collagenase Inhibitors," *Tetrahedron* **1993**, *49*, 3521–32.

85 Rogers, D. H.; Yi, E. C.; and Poulter, C. D., "Enantioselective Synthesis of (+)-Presqualene Diphosphate," *J. Org. Chem.* **1995**, *60*, 941–45.

86 Mukaiyama, T.; Yamashita, H.; and Asami, M., "An Asymmetric Synthesis of Bicyclic Lactones and its Application to the Asymmetric Synthesis of (1R,3S)-*cis*-Chrysanthemic Acid," *Chem. Lett.* **1983**, 385–88.

87 Schotten, T.; Boland, W.; and Jaenicke, L., "Enantioselective Synthesis of Dictyopterene C, 6R-(−)-Butyl-2,5-cycloheptadiene. The Pheromone of Several Dictyotales (Phaeophyceae)," *Tetrahedron Lett.* **1986**, *27*, 2349–52.

88 Doyle, M. P.; Dyatkin, A. B.; Protopopova, M. N.; Yang, C. I,.; Miertschin, C. S.; Winchester, W. R.; Simonsen, S. H.; Lynch, V.; and Ghosh, R., "Enhanced Enantiocontrol in Catalytic Metal Carbene Transformations with Dirhodium(II) Tetrakis-[methyl 2-oxooxazolidin-4(S)-carboxylate], $Rh_2(4S$-MEOX)$_4$," *Recl. Trav. Chim. Pays-Bas* **1995**, *114*, 163–70.

89   Doyle, M. P.; and Zhou, Q.-L., "Enantioselective Catalytic Intramolecular Cyclopropanation of Allylic α-Diazopropionates Optimized with Dirhodium(II) Tetrakis[methyl 2-oxazolidinone-4(S or R)-carboxylate]," *Tetrahedron: Asymmetry* **1995**, *6*, 2157–60.

90   Private communication with H. M. L. Davies.

91   Doyle, M. P.; Dyatkin, A. B.; Kalinin, A. V.; Ruppar, D. A.; Martin, S. F.; Spaller, M. R.; and Liras, S., "Highly Selective Enantiomer Differentiation in Intramolecular Cyclopropanation Reactions of Racemic Secondary Allylic Diazoacetates," *J. Am. Chem. Soc.* **1995**, *117*, 11021–22.

92   Doyle, M. P.; Dyatkin, A. B.; and Autry, C. L., "A New Catalytic Transformation of Diazo Esters: Hydride Abstraction in Dirhodium(II)-Catalyzed Reactions," *J. Chem. Soc., Perkin Trans. 1* **1995**, 619–21.

93   Martin, S. F.; Spaller, M. R.; Liras, S.; and Hartmann, B., "Enantio- and Diastereoselectivity in the Intramolecular Cyclopropanation of Secondary Allylic Diazocetates," *J. Am. Chem. Soc.* **1994**, *116*, 4493–94.

94   Nishiyama, H.; Itoh, Y.; Sugawara, Y.; Matsumoto, H.; Aoki, K.; and Itoh, K., "Chiral Ruthenium(II)-bis(2-oxazolin-2-yl)-pyridine Complexes. Asymmetric Catalytic Cyclopropanation of Olefins and Diazoacetates," *Bull. Chem. Soc. Jpn.* **1995**, *68*, 1247–62.

95   Park, S.-B.; Murata, K.; Matsumoto, H.; and Nishiyama, H., "Remote Electronic Control in Asymmetric Cyclopropanation with Chiral Ru-Pybox Catalysts," *Tetrahedron: Asymmetry* **1995**, *6*, 2487–94.

96   Hirai, H.; and Matsui, M., "Synthesis of Optically Active Dihydrochrysanthemolactone by Asymmetric Decomposition of Unsaturated Diazoacetic Ester," *Agr. Biol. Chem.* **1976**, *40*, 169–74.

97   Dauben, W. G.; Hendricks, R. T.; Luzzio, M. J.; and Ng, H. P., "Enantioselectively Catalyzed Intramolecular Cyclopropanations of Unsaturated Diazo Carbonyl Compounds," *Tetrahedron Lett.* **1990**, *31*, 6969–72.

98   Piqué, C.; Fähndrich, B.; and Pfaltz, A., "Enantioselective Intramolecular Cyclopropanation Catalyzed by Semicorrin–Copper Complexes," *Synlett* **1995**, 491–92.

99   Koskinen, A. M. P.; and Hassila, H., "Asymmetric Catalysis in Intramolecular Cyclopropanation," *J. Org. Chem.* **1993**, *58*, 4479–80.

100   Doyle, M. P.; Eismont, M. Y.; and Zhou, Q.-L., "Enantiocontrol in Intramolecular Cyclopropanation of Diazo Ketones. Conformational Control of Metal Carbene Alignment," *Russ. Chem. Bull.*, in press.

101   Tokunoh, R.; Tomiyama, H.; Sodeoka, M.; and Shibasaki, M., "Catalytic Asymmetric Intramolecular Cyclopropanation of Enol Silyl Ether. Synthesis of the Phorbol CD-Ring Skeleton," *Tetrahedron Lett.* **1996**, *37*, 2449–52.

102   Gant, T. G.; Noe, M. C.; and Corey, E. J., "The First Enantioselective Synthesis of the Chemotactic Factor Sirenin by an Intramolecular [2+1] Cyclization Using a New Chiral Catalyst," *Tetrahedron Lett.* **1995**, *36*, 8745–48.

103   Shimamoto, K.; and Ohfune, Y., "Syntheses of 3'-Substituted-2-(carboxycylopropyl)glycines via Intramolecular Cyclopropanation. The Folded Form of L-Glutamate Activates the Non-NMDA Receptor Subtype," *Tetrahedron Lett.* **1990**, *31*, 4049–52.

104   Toma, K.; Miyazaki, E.; Murae, T.; and Takahashi, T., "Biomimetic Short-Step Synthesis of (±)-Casbene from Geranylgeraniol," *Chem. Lett.* **1982**, 863–64.

105   Doyle, M. P.; Protopopova, M. N.; Poulter, C. D.; and Rogers, D. H., "Macrocyclic Lactones from Dirhodium(II)-Catalyzed Intramolecular Cyclopropanation and Carbon–Hydrogen Insertion," *J. Am. Chem. Soc.* **1995**, *117*, 7281–82.

106   Doyle, M. P.; Griffin, J. H.; Bagheri, V.; and Dorow, R. L., "Correlations between Catalytic Reactions of Diazo Compounds and Stoichiometric Reactions of Transition Metal Carbenes with Alkenes. Mechanism of the Cyclopropanation Reaction," *Organometal.* **1984**, *3*, 53–61.

107   Doyle, M. P., "Electrophilic Metal Carbenes as Reaction Intermediates in Catalytic Reactions," *Acc. Chem. Res.* **1986**, *19*, 348–56.

108   Doyle, M. P.; Peterson, C. S.; Marnett, A. B.; Ene, D. G.; Parker, D. L., Jr.; and Stanley, S. A., unpublished results.

109   Reissig, H.-U., "Recent Developments in the Enantioselective Syntheses of Cyclopropanes," *Angew. Chem., Int. Ed. Engl.* **1996**, *35*, 971–73.

110   Padwa, A.; and Weingarten, M. D., "Cascade Processes of Metallo Carbenoids," *Chem. Rev.* **1996**, *96*, 223–69.

111   Padwa, A.; and Kinder, F. R., "Rhodium(II)-Catalyzed Cyclization of 2-Alkynyl 2-Diazo-3-oxobutanoates as a Method for Synthesizing Substituted Furans," *J. Org. Chem.* **1993**, *58*, 21–28.

112   Padwa, A.; Krumpe, K. E.; Gareau, Y.; and Chiacchio, U., "Rhodium(II)-Catalyzed Cyclization Reaction of Alkynyl-Substituted α-Diazo Ketones," *J. Org. Chem.* **1991**, *56*, 2523–30.

113   Padwa, A.; Chiacchio, U.; Garreau, Y.; Kassir, J. M.; Krumpe, K. E.; and Schoffstall, A. M., "Generation of Vinylcarbenes by the Intramolecular Addition of α-Diazo Ketones to Acetylenes," *J. Org. Chem.* **1990**, *55*, 414–16.

114   Padwa, A.; Krumpe, K. E.; and Kassir, J. M., "Rhodium Carbenoid Mediated Cyclization of *o*-Alkynyl Substituted α-Diazoacetophenones," *J. Org. Chem.* **1992**, *57*, 4940–48.

115   Mueller, P. H.; Kassir, J. M.; Semones, M. A.; Weingarten, M. D.; and Padwa, A., "Rhodium Carbenoid Mediated Cyclizations. Intramolecular Cyclopropanation and C–H Insertion Reactions Derived from Type II *o*-Alkynyl Substituted α-Diazoacetophenones," *Tetrahedron Lett.* **1993**, *34*, 4285–88.

116   Padwa, A.; Austin, D. J; Gareau, Y.; Kassir, J. M.; and Xu, S. L., "Rearrangement of Alkynyl and Vinyl Carbenoids via the Rhodium(II)-Catalyzed Cyclization Reaction of α-Diazo Ketones," *J. Am. Chem. Soc.* **1993**, *115*, 2637–47.

117   Padwa, A.; Austin, D. J.; and Xu, S. L., "Rhodium(II) Catalyzed Cyclizations of α-Diazo Substituted Alkynes. A New Mode of Reaction," *Tetrahedron Lett.* **1991**, *32*, 4103–6.

118   Fairfax, D. J.; Austin, D. J.; Xu, S. L.; and Padwa, A., "Alternatives to α-Diazo Ketones for Tandem Cyclization–Cycloaddition and Carbenoid–Alkyne Metathesis Strategies. Novel Cyclic Enol–Ether Formation via Carbonyl Ylide Rearrangement Reactions," *J. Chem. Soc. Perkin Trans. 1* **1992**, 2837–3844.

119   Jones, G. B.; Moody, C. J.; Padwa, A.; and Kassir, J. M., "Inter- and Intramolecular Reactions of Indol-2-yl Carbenes and Related Species. Preparation of 1,1a,2,8b-Tetrahydroazirino[2′,3′:3,4]pyrrolo[1,2-a]indoles," *J. Chem. Soc. Perkin Trans. 1* **1991**, 1721–27.

120 Hoye, T. R.; and Dinsmore, C. J., "Rhodium(II) Acetate Catalyzed Alkyne Insertion Reactions of α-Diazo Ketones: Mechanistic Inferences," *J. Am. Chem. Soc.* **1991**, *113*, 4343–45.

121 Hoye, T. R.; Dinsmore, C. J.; Johnson, D. S.; and Korkowski, P. F., "Alkyne Insertion Reactions of Metal Carbenes Derived from Enynyl α-Diazo Ketones [R'CN$_2$COCR$_2$CH$_2$C≡C(CH$_2$)$_{n-2}$CH═CH$_2$]," *J. Org. Chem.* **1990**, *55*, 4518–20.

122 Hoye, T. R.; and Dinsmore, C. J., "Tandem Alkyne Insertion and Allyl Sulfonium Ylide Rearrangement of γ,δ-Alkynyl-α'-Diazoketones," *Tetrahedron Lett.* **1992**, *33*, 169–72.

123 Hoye, T. R.; and Dinsmore, C. J., "Double (Internal/External) Alkyne Insertion Reactions of α-Diazoketones," *Tetrahedron Lett.* **1991**, *32*, 3755–58.

124 Korkowski, P. F.; Hoye, T. R.; and Rydberg, D. B., "Fischer Carbene Mediated Conversions of Enynes to Bi- and Tricyclic Cyclopropane-Containing Carbon Skeletons," *J. Am. Chem. Soc.* **1988**, *110*, 2676–78.

125 Hoye, T. R.; and Rehberg, G. M., "Reactions of (CO)$_5$Cr═C(Me)N(CH$_2$CH$_2$)$_2$ with Enynes: Mechanistic Insight and Syntethic Value of Changing a Carbene Donor Group from Alkoxy to Dialkylamino," *Organometal.* **1989**, *8*, 2070–71.

126 Hoye, T. R.; and Rehberg, G. M., "(1-Oxidoalkylidene)pentacarbonylchromium Anion [R(O$^-$)C═Cr(CO)$_5$] ↔ Acylpentacarbonylchromate [Acyl-Cr$^-$(CO)$_5$] Chemistry: In Situ Preparation and Reactions with Alkynes and Enynes," *J. Am. Chem. Soc.* **1990**, *112*, 2841–42.

127 Hoye, T. R.; and Rehberg, G. M., "Manganese Fischer Carbene Chemistry: Reactions of Cp'(CO)$_2$Mn═C(OMe/OLi)R with Enynes, 1-Hexyne, and Acrylates," *Organometal.* **1990**, *9*, 3014–15.

128 Hoye, T. R.; and Suriano, J. A., "Reactions of Pentacarbonyl(1-methoxyethylidene)molbydenum and -tungsten with α,ω-Enynes: Comparison with the Chromium Analogue and Resulting Mechanistic Ramifications', *Organometal.* 1992, **11**, 2044–50.

129 Hoye, T. R.; and Vyvyan, J. R., "Polycyclic Cylopropanes from Reactions of Alkene-Containing Fischer Carbene Complexes and Alkynes: A Formal Synthesis of (±)-Carabrone," *J. Org. Chem.* **1995**, *60*, 4184–95.

130 Müller, P.; Pautex, N.; Doyle, M. P.; and Bagheri, V., "Rh(II) Catalyzed Isomerizations of Cyclopropanes. Evidence for Rh(II)-Complexed Vinylcarbene Intermediates," *Helv. Chim. Acta* **1990**, *73*, 1233–41.

131 Müller, P.; and Gränicher, C., "Structural Effects on the Rh^{II}-Catalyzed Rearrangement of Cyclopropenes," *Helv. Chim. Acta* **1993**, *76*, 521–34.

132 Padwa, A.; and Xu, S. L., "A New Phenol Synthesis from the Rhodium(I)-Catalyzed Reaction of Cyclopropenes and Alkynes," *J. Am. Chem. Soc.* **1992**, *114*, 5881–82.

133 Padwa, A.; Kassir, J. M.; and Xu, S. L., "Rhodium-Catalyzed Ring-Opening Reaction of Cyclopropenes—Control of Regioselectivity by the Oxidation State of the Metal," *J. Org. Chem.* **1991**, *56*, 6971–72.

134 Padwa, A.; Kassir, J. M.; Semones, M. A.; and Weingarten, M. D., "A Tandem Cyclization—Onium Ylide Rearrangement—Cycloaddition Sequence for the Synthesis of Benzo-Substituted Cyclopentenones," *J. Org. Chem.* **1995**, *60*, 53–62.

135   Padwa, A.; Krumpe, K. E.; and Weingarten, M. D., "An Unusual Example of a 6-*Endo-Dig* Addition to an Unactivated Carbon–Carbon Triple Bond," *J. Org. Chem.* **1995**, *60*, 5595–603.

136   Nakatani, K.; Maekawa, S.; Tanabe, K.; and Saito, I., "α-Diazo Ketones as Photochemical DNA Cleavers: A Mimic for the Radical Generating System of Neocarzinostatin Chromophore," *J. Am. Chem. Soc.* **1995**, *117*, 10635–44.

Cycloaddition and Substitution Reactions with Aromatic and Heteroaromatic Compounds

Diazocarbonyl compounds react with a variety of aromatic compounds to give products that have found useful applications in synthesis. These reactions can be metal catalyzed or acid catalyzed. Purely thermal or photochemical processes are also known but are of limited use due to lack of selectivity. In this chapter attention is focused on the use of metal catalysts for aromatic cycloaddition and aromatic substitution. Substrates include benzenoid and heteroaromatic compounds. Reactions involving acid catalysis that lead to electrophilic aromatic substitution *via* carbocation intermediates are discussed, together with acid-catalyzed cyclization of unsaturated diazocarbonyl compounds, in Chapter 11.

6.1 INTERMOLECULAR REACTIONS WITH BENZENE AND ITS DERIVATIVES

6.1.1 Diazoesters

Shortly after Curtius published his synthesis of ethyl diazoacetate (EDA) in 1883,[1] Büchner[2] commenced an investigation of its reactions with unsaturated hydrocarbons. Initially, Büchner believed that thermal decomposition of EDA in benzene furnished a single ester product, though he was later to discover that alkaline hydrolysis yielded a mixture of several isomeric carboxylic acids to which he tentatively assigned norcaradiene structures.[3] Although the subject of debate,[4] the norcaradiene formulation persisted until the 1950s, when Doering and coworkers reexamined the reaction.[5] We now know that the Büchner reaction in its original form produced four cycloheptatrienyl esters, **1–4** (Scheme 1). Photochemically the reaction behaves in much the same way. The contemporary interpretation of both processes is that carboethoxycarbene addition to benzene proceeds *via* an unstable norcaradiene intermediate **5**, which is in mobile equilibrium with its more stable cycloheptatriene tautomer **1**; the remaining products **2–4** are isomers of **1** formed by thermally or photochemically induced sigma-

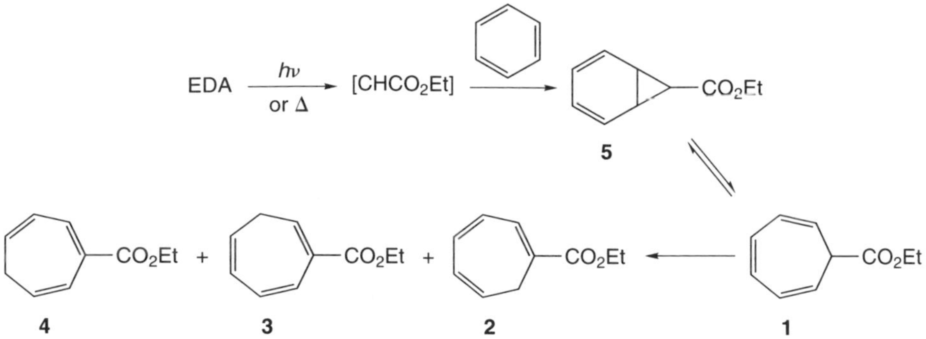

Scheme 1

tropic rearrangement. The pyrolysis route affords not more than 30% of **1**. Yields are better in the photolysis route, especially when pyrex-filtered light is used.

Much of the early work on thermal or photochemical Büchner reactions was characterized by the isolation of complex mixtures of cycloheptatrienyl esters, which usually were neither separated nor individually identified. Toluene and EDA produced three ring-expanded products,[6] and a recent reinvestigation of the reaction with anisole revealed the presence of seven products.[7] With increasing alkyl substitution of the benzene nucleus (xylene → prechnitene), insertion of the carboethoxycarbene into the carbon–hydrogen bonds of the methyl groups gradually increases at the expense of ring expansion.

Low yields and separation problems notwithstanding, the original Büchner re-action represented a singularly effective route to a vast range of seven-membered carbocycles many of which, as mixtures of isomers, were suitable for elaboration into natural and unnatural products.[8] The relationship between the anisole–EDA reaction and tropone derivatives was first recognized by Johnson and co-workers,[9] who used it as a model for their synthesis of stipitatic acid from 1,2,4-trimethoxy-benzene (Scheme 2); for clarity only the cycloheptatriene ester appropriate to the target molecule is shown. Application of the Büchner reaction to bi- and tricyclic aromatics opened up new routes to condensed cycloheptatrienyl systems, which

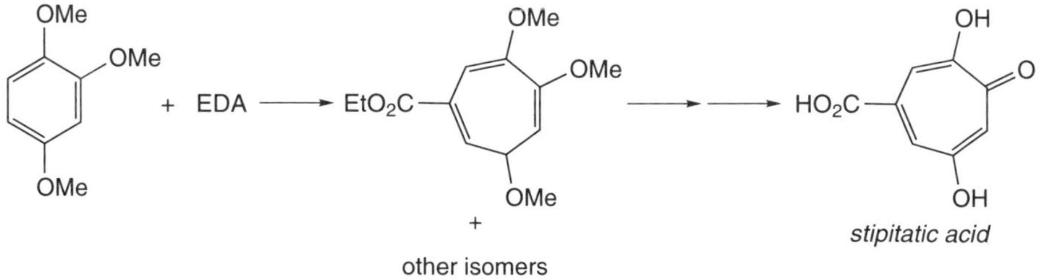

Scheme 2

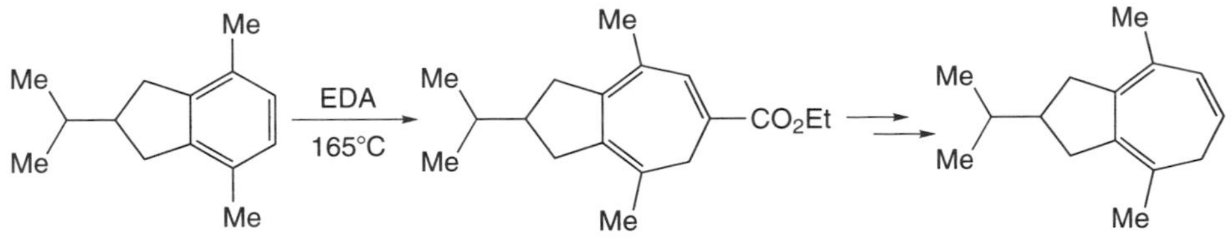

vetivazulene

Scheme 3

were often transformed through decarboxylation–dehydrogenation sequences into azulenes. A typical sequence leading to vetivazulene is shown in Scheme 3.[10]

The problems endemic to the photochemical and thermal reactions were solved comprehensively in 1980, when the Belgian group extended their study of rhodium(II) catalysis of diazocarbonyl reactions to include the Büchner process.[11,12] A measure of the improvement in both selectivity and efficiency can be quickly quantified by comparing the thermal reaction of EDA with anisole (7 products, 35% yield) with its rhodium(II) trifluoroacetate-catalyzed counterpart (2 products, 73% yield). And although the methoxy substituent of anisole clearly exerts a directive effect in favor of the 4-isomer (*cf.* Table 6.1), both

TABLE 6.1 Products of Reaction of Benzene Derivatives with Methyl Diazoacetate Catalyzed by $Rh_2(OCOCF_3)_4$[a]

Substrate	Products (distribution %)	Yield (%)
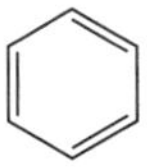	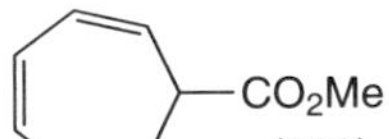 (100)	100
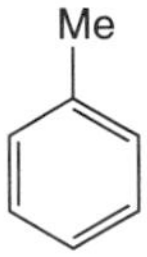	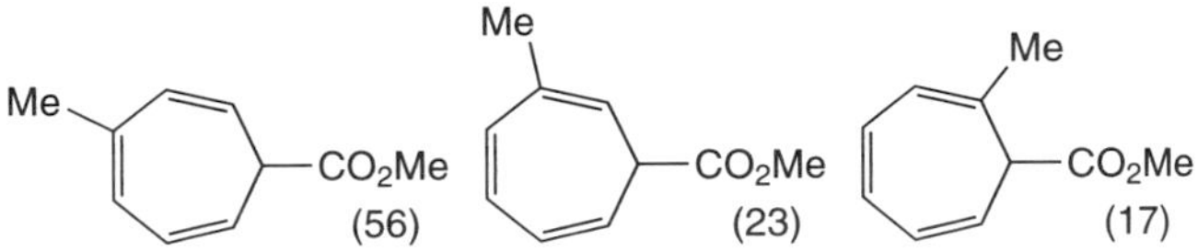 (56), (23), (17)	95
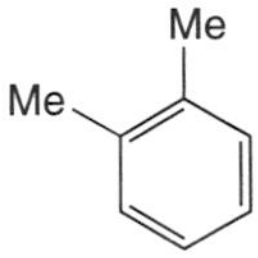	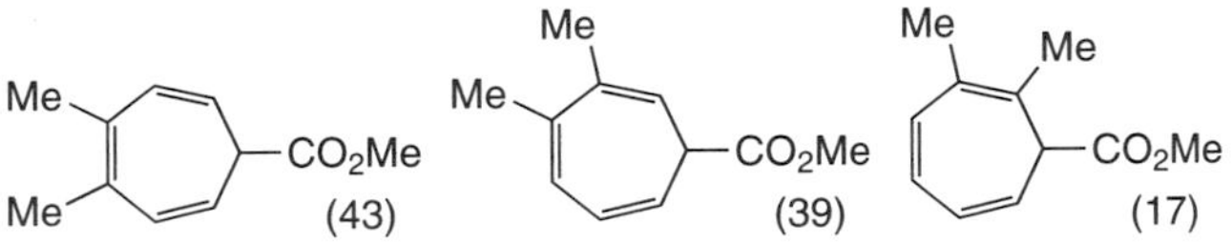 (43), (39), (17)	80
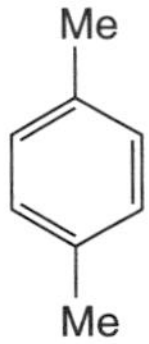	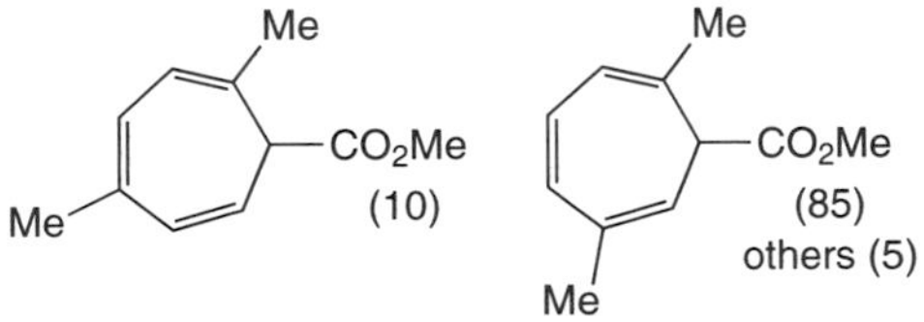 (10), (85), others (5)	90

TABLE 6.1 (*Continued*)

Substrate	Products (distribution %)	Yield (%)
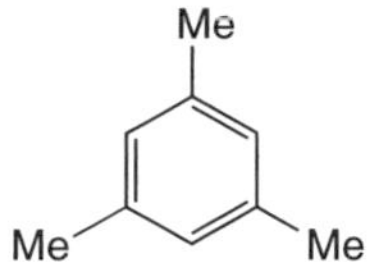 Me (1,3,5-trimethylbenzene)	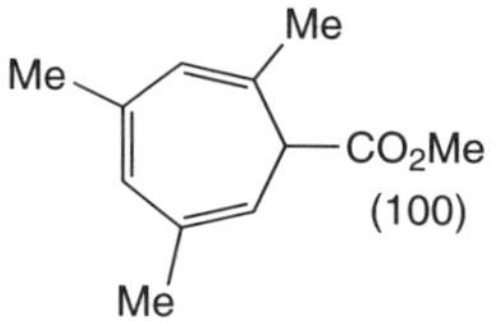 product (100)	60
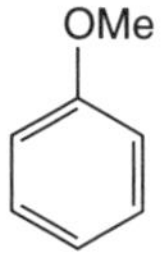 OMe (anisole)	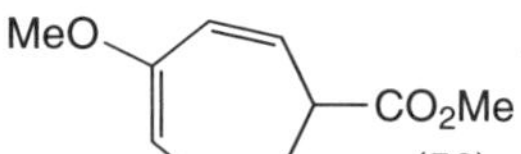 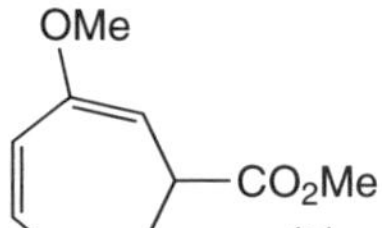 MeO-cycloheptatriene-CO₂Me (56), OMe-cycloheptatriene-CO₂Me (8)	73
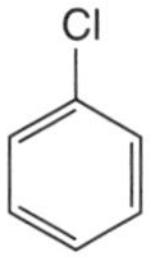 Cl (chlorobenzene)	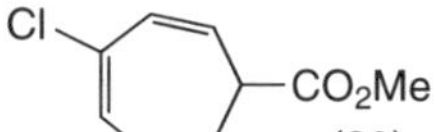 Cl-cycloheptatriene-CO₂Me (80), 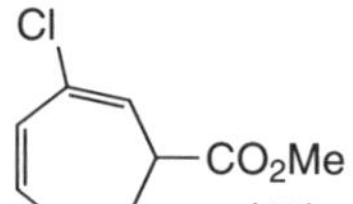 (15), 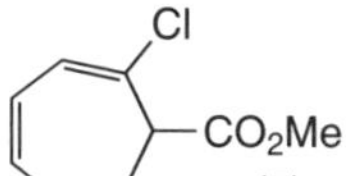 (5)	72
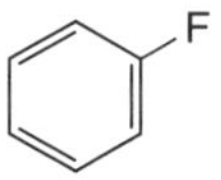 F (fluorobenzene)	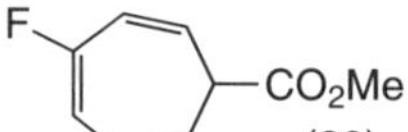 F-cycloheptatriene-CO₂Me (80), 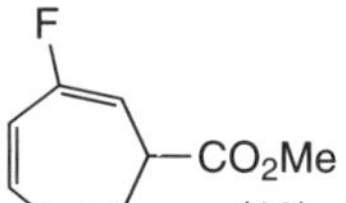 (12), 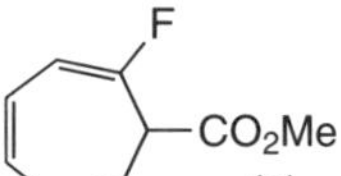 (8)	46
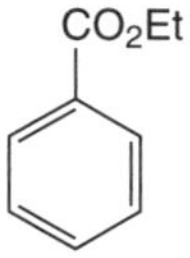 CO₂Et (ethyl benzoate)	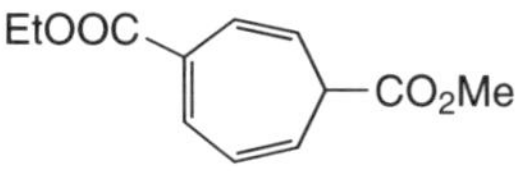 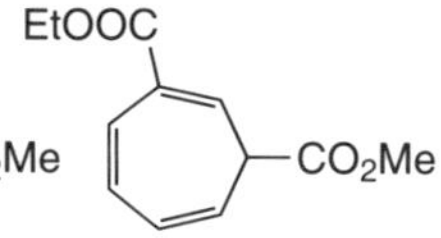 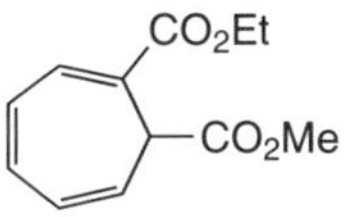 EtOOC-cycloheptatriene-CO₂Me, (EtOOC) CO₂Me, CO₂Et CO₂Me	10
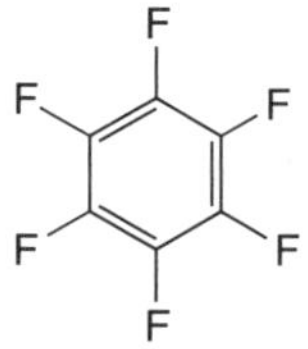 F₆ (hexafluorobenzene)	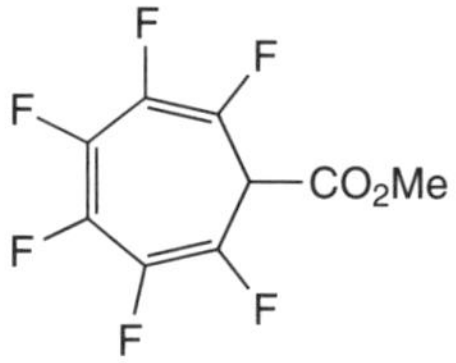 fluorinated cycloheptatriene-CO₂Me	5

*Reaction temperature 22°C [substrate]:[MDA]:[catalyst] = 5000:250:1.[12]

products of the catalyzed (room-temperature) reaction are kinetically controlled unconjugated esters, neither of which was even detected in the thermal reaction. The results of a study of the reactivity of methyl diazoacetate towards several simple benzene derivatives with $Rh_2(OCOCF_3)_4$ as catalyst are summarized in Table 6.1. In parallel studies it was shown that $Rh_2(OCOCF_3)_4$ was the optimum catalyst; lower yields were obtained with $Rh_2(OCOC_6F_5)_4$, $Rh_2(OCOCH_2OCH_3)_4$,

$Rh_2(OCOCH_3)_4$ and $Rh_2(OCOCMe_3)_4$.[12] The products were all unrearranged cycloheptatrienyl esters, with benzene itself furnishing ester **1** as the sole product in quantitative yield. The retarding effect of electron-withdrawing substitutents, revealed in Table 6.1 by lower yields, support the assumption of an electrophilic metal carbene intermediate. However, product distributions, and the additional fact that reactions with EDA and *tert*-butyl diazoacetate gave inferior yields to those with methyl diazoacetate, suggest that steric effects may also be a factor, especially with the methylbenzenes. The catalyzed reaction of EDA with polystyrene (eq. 1) represents an interesting extension of the Büchner process to the formation of a cycloheptatriene-bound polymer.[12]

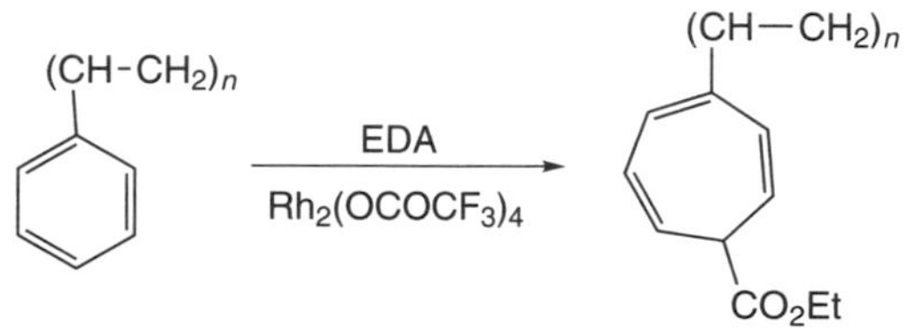

$$(1)$$

Among other substituted benzenes which have been subjected to intermolecular cycloaddition with EDA are 1,2-methylenedioxybenzene (eq. 2) and 1,2-dimethoxybenzene (eq. 3)[13]. Rhodium(II) acetate was the preferred catalyst. In

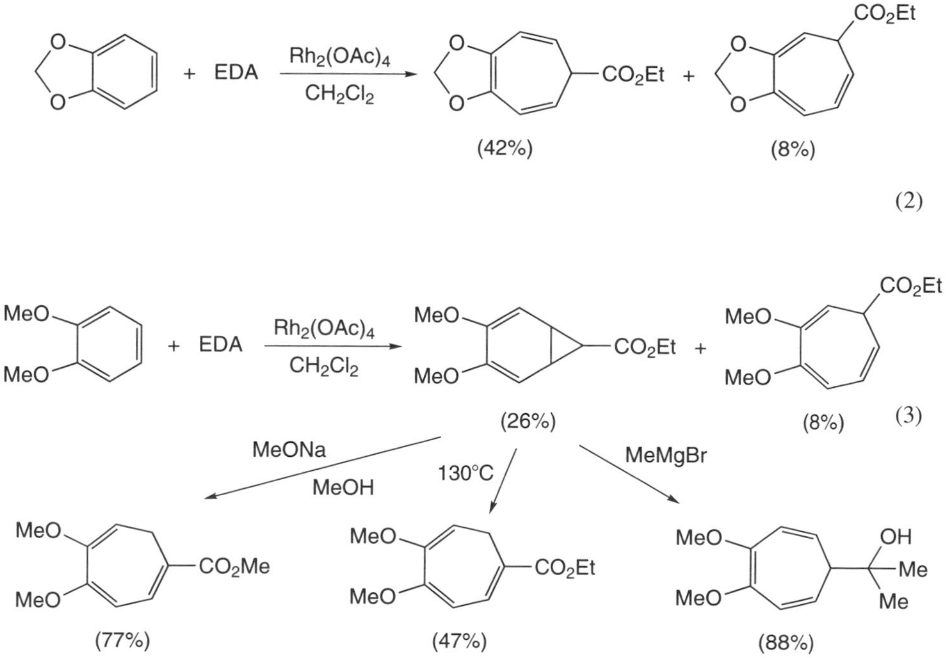

$$(2)$$

$$(3)$$

Scheme 4

the former, the product consisted of a mixture of two cycloheptatrienyl esters, resulting from attack on the 3,4- and 4,5-carbon atoms of the benzene ring. The latter reaction produced a bicyclic adduct and a cycloheptatrienyl ester and is mechanistically significant in that it provides direct evidence for a norcaradiene intermediate in the catalyzed intermolecular Büchner reaction. Earlier work by Saba and co-workers[14,15] had demonstrated that norcaradiene intermediates could be detected at low temperature by NMR spectroscopy (*vide infra*). Interestingly, in the example shown in eq. 3, there was no evidence of the existence of an equilibrium between the two forms at temperatures up to 80°C, though heating the norcaradiene ester at 130°C did cleave the 1–6 bond to form the isomeric cycloheptatriene ester shown in 47% yield. Replacement of the ester moiety of the norcaradiene in eq. 3 by a tertiary alcohol through reaction with methylmagnesium bromide caused spontaneous ring opening of the cyclopropane, as did transesterification using sodium methoxide in methanol (Scheme 4). These two observations suggest that methoxy substituents alone are insufficient to maintain the norcaradiene structure, but that the ethoxycarbonyl groups and the methoxy groups together are required. Although in general the norcaradiene structure is less stable than the cycloheptatriene structure, examples are known where the former is preferred.[16] There is good evidence to suggest that an effective electronic way of stabilizing the norcaradiene form can be achieved by placing π-acceptors such as carbonyl or cyano at the 7-position.[17–19] Interaction between the π-bond at the 7-position and the Walsh orbital of the cyclopropane ring weakens the 1–6 antibounding orbital and consequently strengthens the 1–6 bond. π-Donors at the 3- and 4-positions of the 6-membered ring are also expected to stabilize the norcaradiene structure by strengthening the 1–6 bonds. The significant difference between the nature of the products of the two reactions in eqs. 2 and 3 may reflect the different degrees of conjugation between the lone pairs of electrons on the oxygen atoms and the π-system of the norcaradiene. For the methylenedioxy (eq. 2) case the oxygen atoms may not provide sufficient overlap to give preference to the norcaradiene.

6.1.2 Diazoketones

Intermolecular cycloaddition of diazoketones with benzene derivatives is also known. Some representative examples involving benzene as substrate and $Rh_2(OCOCF_3)_4$ as catalyst are shown in Table 6.2.[20] The cycloaddition products formed from diazokctones are susceptible to acid-catalyzed rearrangement with rearomatization. Comparable rearrangement with the cycloaddition products formed from diazoesters does not occur. In general, the reaction represents an efficient route to cycloheptatrienyl ketones with an additional attractive feature not shared with the cycloheptatrienyl esters in Table 6.1. In the presence of an acid catalyst, such as trifluoroacetic acid, these cycloheptatrienyl ketones undergo a smooth rearomatization to form benzyl ketones.[20] In fact, the overall process can be run as a one-pot, two-stage method for converting benzene into a benzyl ketone. Table 6.2 identifies several of the benzyl ketones that have been produced

TABLE 6.2   Products of Rh$_2$(OCOCF$_3$)$_4$-Catalyzed Reaction of Benzene with Diazoketones with Subsequent Rearrangement to Benzyl Ketones

Entry	Diazoketones	Cycloheptatriene (yield %)	Benzyl Ketone (yield %)
1	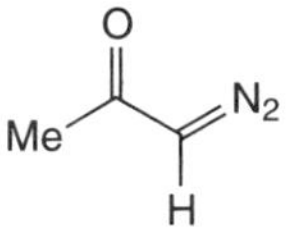	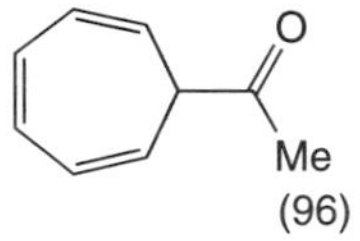 (96)	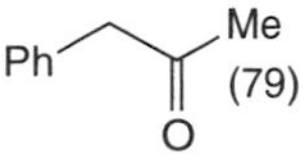 (79)
2	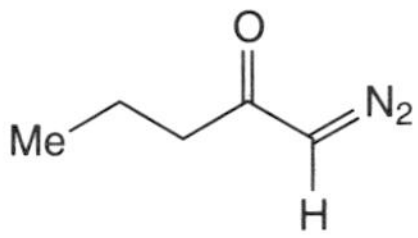	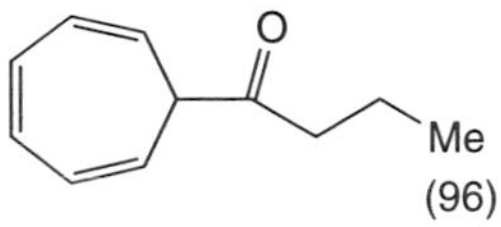 (96)	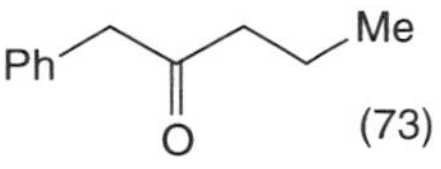 (73)
3	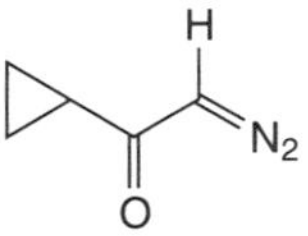	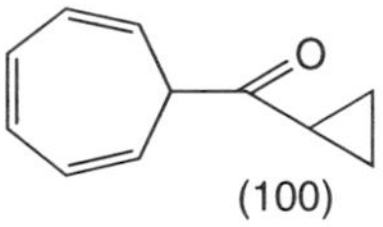 (100)	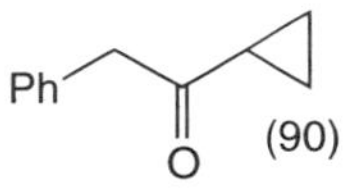 (90)
4	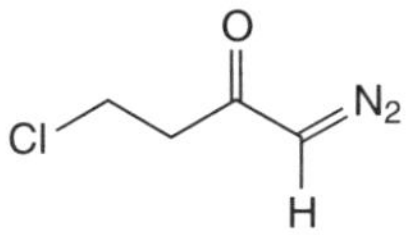	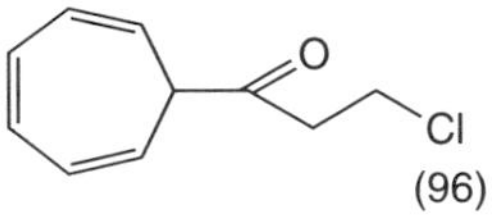 (96)	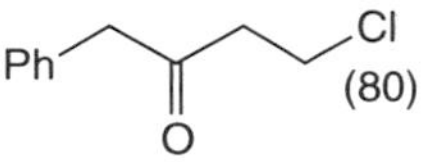 (80)
5	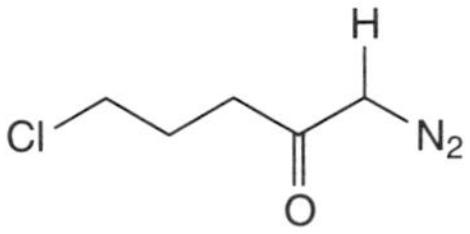	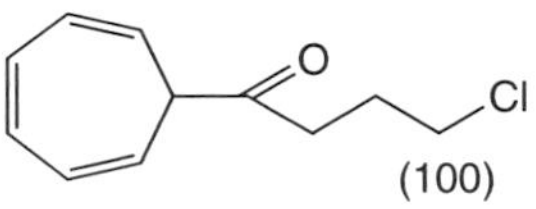 (100)	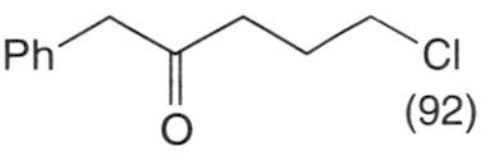 (92)
6	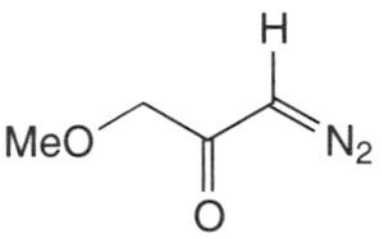	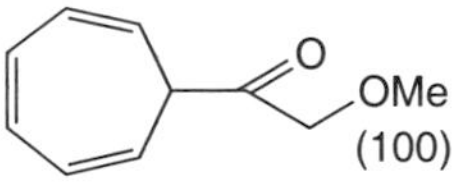 (100)	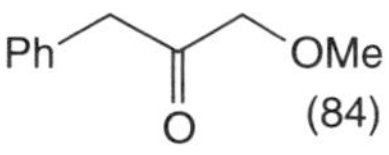 (84)
7	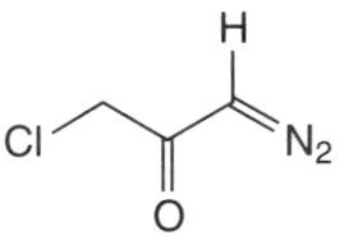	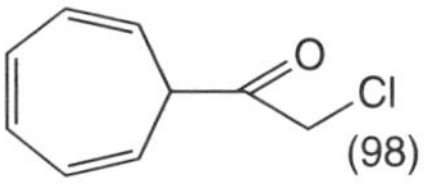 (98)	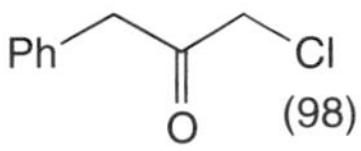 (98)
8	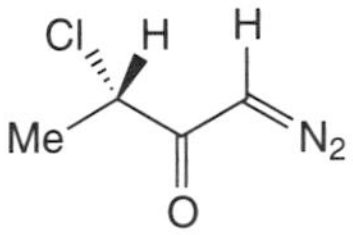	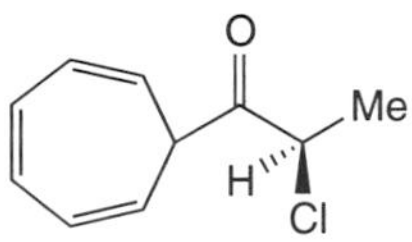	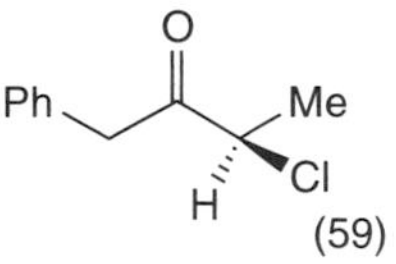 (59)
9	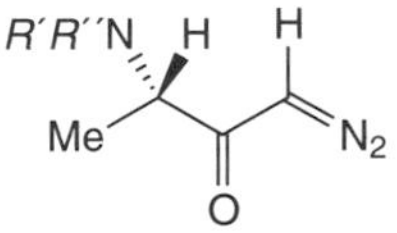 R',R'' = phthaloyl	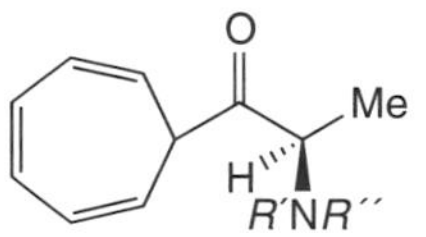	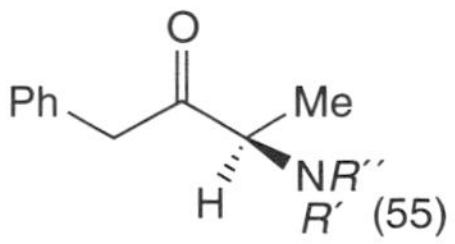 (55)

TABLE 6.2 (*Continued*)

Entry	Diazoketones	Cycloheptatriene (yield %)	Benzyl Ketone (yield %)
10	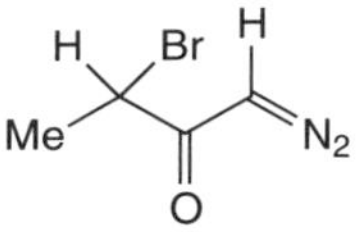	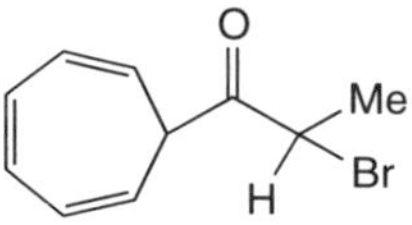	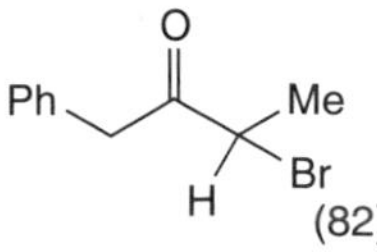 (82)
11	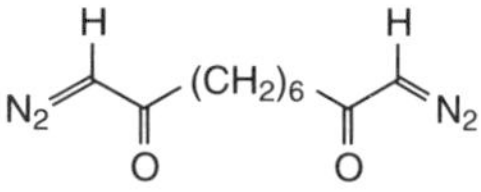	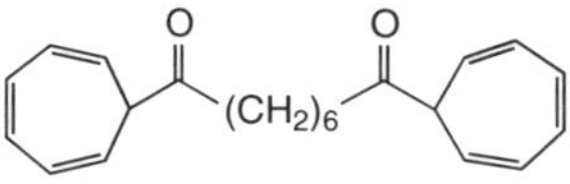	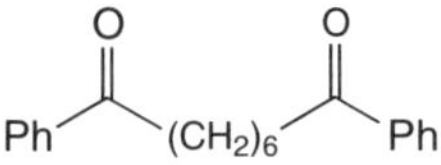 (70)
12	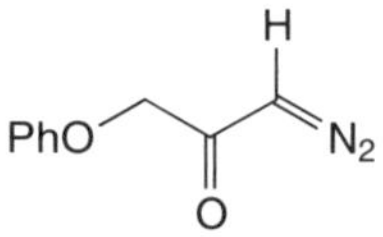	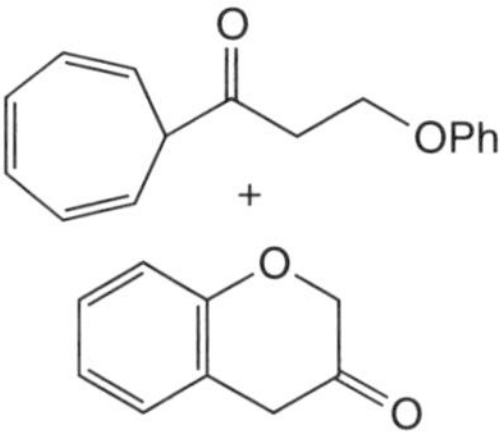 +	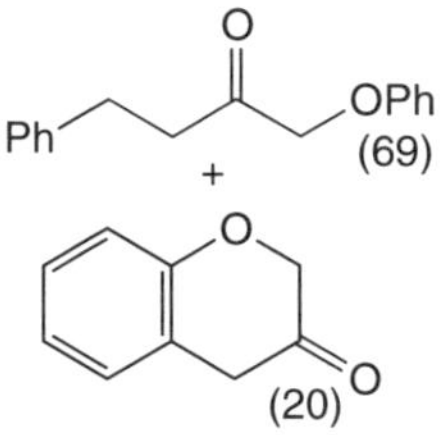 (69) + (20)
13	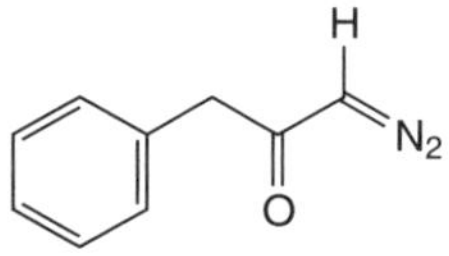	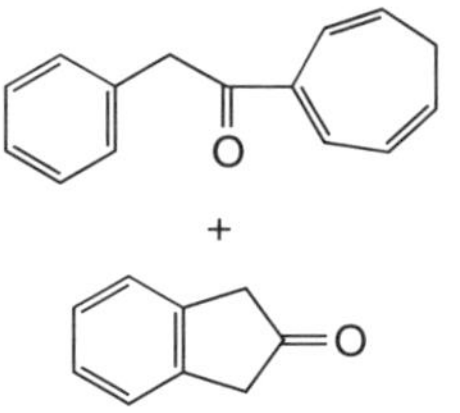 +	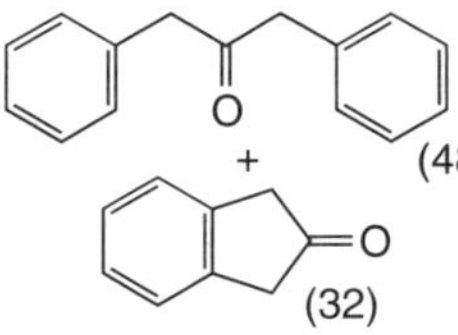 (48) + (32)
14	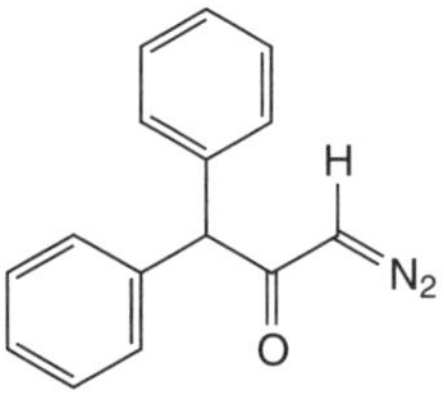	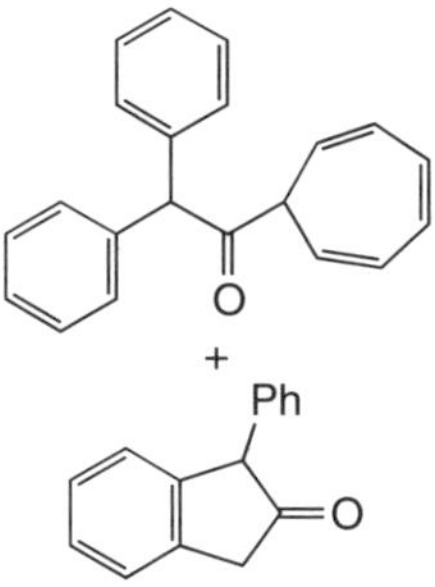 +	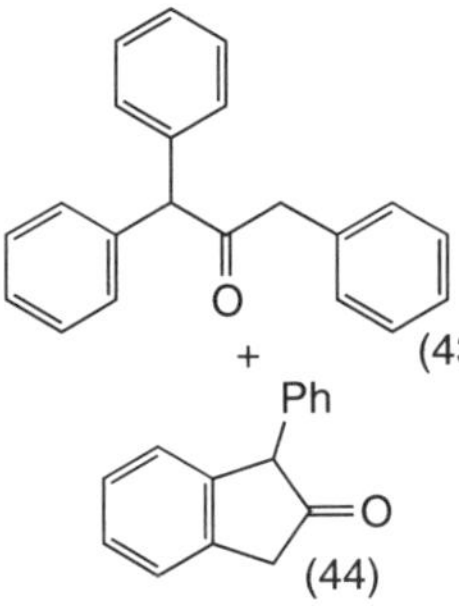 (43) + (44)

in this way. To some extent the ease of the aromatization step is substituent dependent. This is particularly so with the chloromethyl, chloroethyl, and bromoethyl derivatives in entries 4, 7, 8, and 10, respectively, where the instability of the cycloheptatriene adducts relative to benzyl ketones was already evident at the end of the cycloaddition stage. Evidently, adventitious acid is sufficient to cause rearrangement. Mechanistically, the aromatization step bears testimony to the equilibrium presumed to exist between norcaradiene and cycloheptatriene

(Scheme 5) since protonation of the carbonyl group in the former provides a pathway for cyclopropane ring opening, which is completed by aromatization and keto–enol tautomerization.

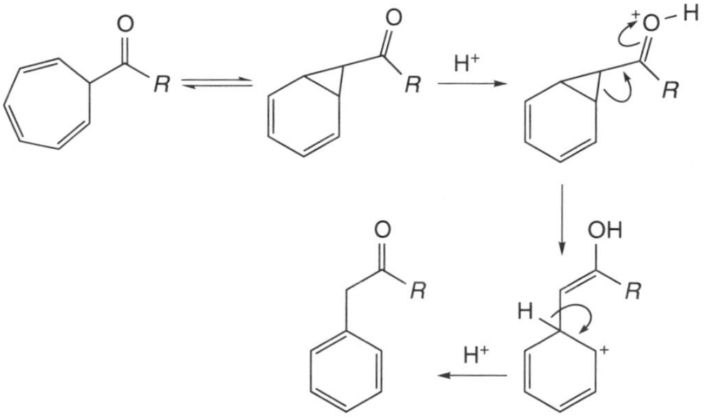
H+
H+

Scheme 5

Of the other entries in Table 6.2, 8 and 9 illustrate the use of enantiopure diazoketones for the production of enantiopure α-substituted benzyl ketones. Entry 11 exemplifies the elaboration of an α,ω-bis-diazoketone, and in entries 11–14, intramolecular cycloaddition (*vide infra*) competes with the intermolecular reaction. Intermolecular cycloaddition of diazoketones to substituted benzenes is also possible. Two representative examples involving toluene, each showing a preference for *para* substitution, are shown in eqs. 4a and 4b.[21] There is one example of intermolecular cycloaddition of a diazoamide to a benzene derivative reported by Chan and Matlin (eq. 5).[22]

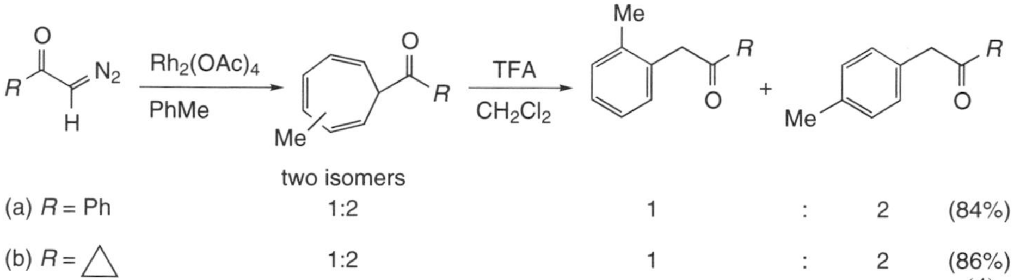
Rh2(OAc)4
PhMe
TFA
CH2Cl2
two isomers
(a) R = Ph 1:2 1 : 2 (84%)
(b) R = △ 1:2 1 : 2 (86%)

(4)

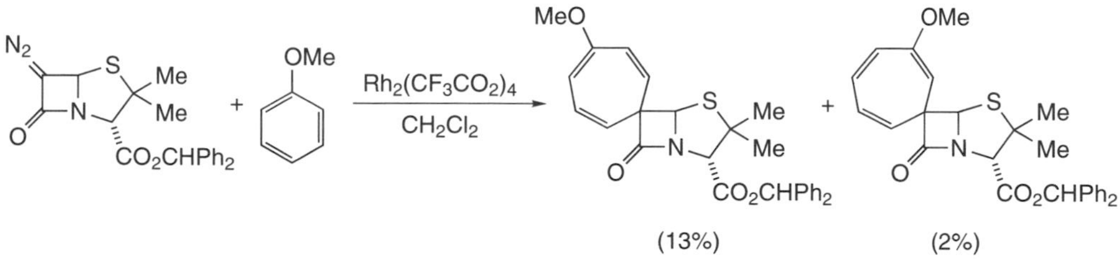
Rh2(CF3CO2)4
CH2Cl2
(13%)
(2%)

(5)

6.2 INTRAMOLECULAR CYCLOADDITION REACTIONS WITH BENZENE DERIVATIVES

6.2.1 Diazoketones

Only a few examples of intramolecular Büchner reactions were known prior to the introduction of rhodium catalysts. All employed copper catalysts in one form or another and although yields of ring-expanded bicyclic products were low, the process was recognized as a potentially useful route to a variety of novel bicyclic structures. Cyclization of 4-phenyl-1-diazobutan-2-one (eq. 6) with a copper catalyst in hot decalin furnished a mixture of products from which azulenone **6** was isolated in 13% yield.[23] The p-methoxybenzyl diazomalonate in eq. 7, on the

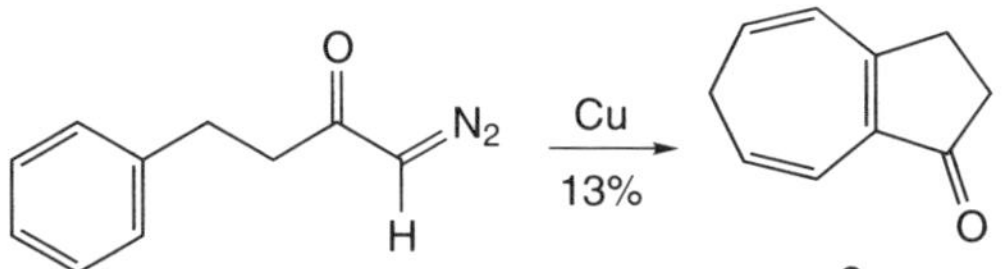

(6)

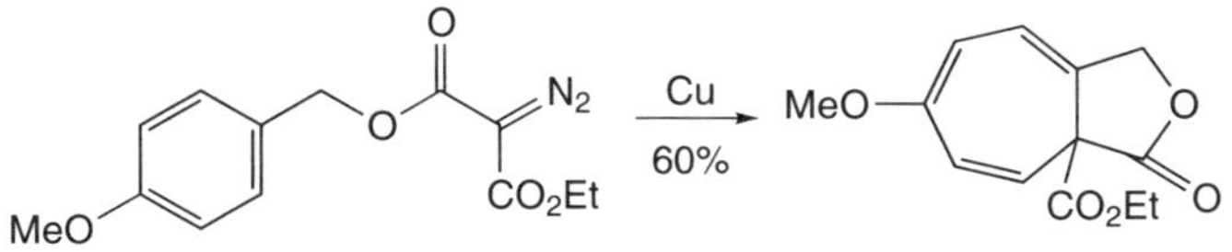

(7)

other hand, produced a bicyclic lactone with an arrangement of double bonds different from that in **6**.[24] Here movement of the double bonds is restricted by the presence of a bridgehead substituent. Vogel and Reel[25] employed copper powder in refluxing benzene to form the cycloaddition product in Scheme 6 en route to a bridged [14]annulene. In this example the product of cycloaddition was a norcaradiene; removal of the carbonyl group caused ring expansion to the annulene.

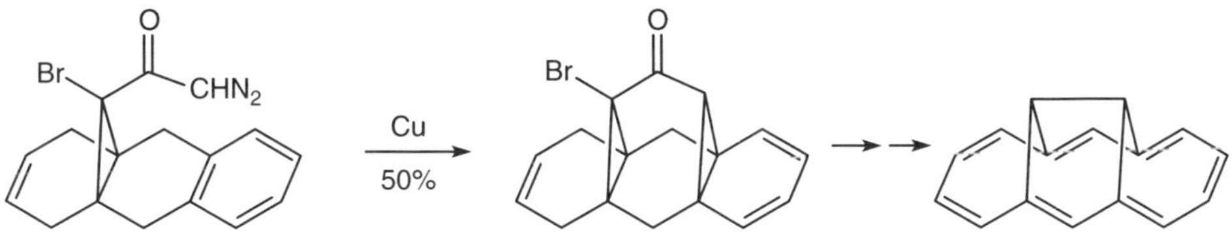

Scheme 6

We now return to the parent system in eq. 6. Clearly, triene **6**, isolated from high-temperature cyclization by Julia and co-workers,[23] is not the kinetic product of cyclization (cf. Scheme 1), but is the result of sigmatropic shifts within the cycloheptatrienyl system. Scott[26,27] reinvestigated the reaction and detected by

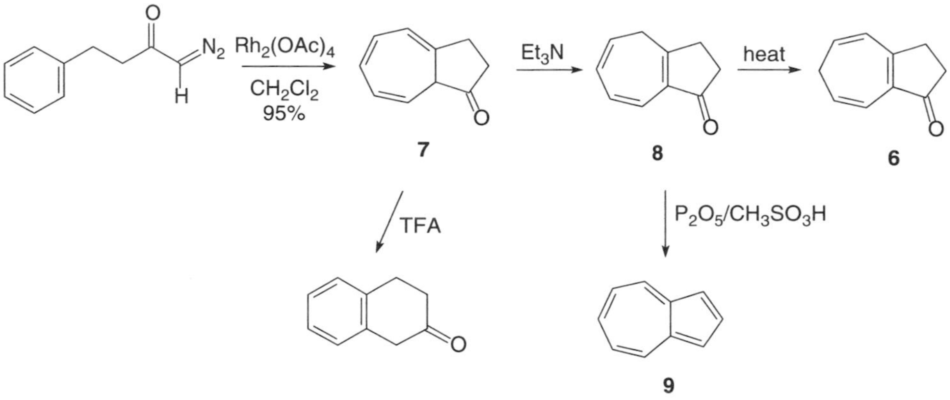

Scheme 7

^{1}H-NMR spectroscopy at low temperatures trienone **7**, the kinetic product (Scheme 7), but during the isolation procedure, which involved exposure to chromatographic alumina, this isomer was transformed into the more conjugated isomer **8**. This isomer differs from isomer **6**, the product of high-temperature cyclization. Scott and Minton also observed that dehydration of isomer **8** with phosphorus pentoxide in methanesulfonic acid produced azulene **9**.[27] The kinetic product of cyclization was finally uncovered when 1-diazo-4-phenylbutan-2-one was cyclized with rhodium(II) acetate in dichloromethane at room temperature.[28,29] Trienone **7** was produced as the sole isomer in 95% yield (Scheme 7). Under triethylamine catalysis isomer **7** isomerized to isomer **8**; while in contact with silica gel, or more conveniently, with trifluoroacetic acid, trienone **7** rearranged quantitatively to 2-tetralone (Scheme 8). This aromatization process probably proceeds via the tricyclic norcaradiene valence tautomer presumed to be in equilibrium with the trienone. All these transformations are summarized in Schemes 7 and 8.

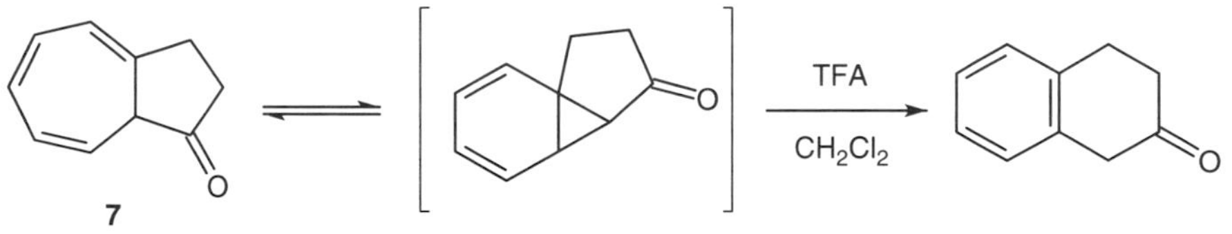

Scheme 8

Trienone **8** has found uses in synthesis, in particular in the construction of analogues of the *A,B* ring carbocyclic framework of taxol.[31] Cyclopropanation of **8** using the Corey dimethylsulfoxide ylide in DMSO furnished a tricyclic derivative that on exposure to lead tetraacetate underwent ring opening to the diacetoxy bicyclo[5.3.1]undecadiene shown in Scheme 9.

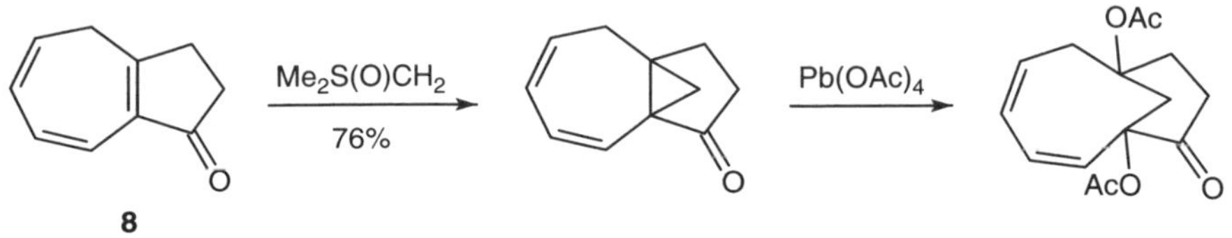

Scheme 9

DIAZOKETONE CYCLIZATION ONTO A BENZENE RING:
3,4-DIHYDRO-1(2H)-AZULENONE 8[30]
[1(2H)-AZULENONE, 3,4-DIHYDRO-]

3,4-Dihydro-1(2H)-azulenone 8.

A 250-mL, one-necked round-bottomed flask is equipped with an egg-shaped magnetic stirring bar and a high dilution trident. The high-dilution trident is further equipped with a 100-mL pressure-equalizing addition funnel attached to a nitrogen inlet and an efficient reflux condenser attached to a nitrogen outlet. The round-bottomed flask is charged with 100 mL of dry freshly distilled methylene chloride and 12 mg of dirhodium(II) acetate. This heterogeneous mixture is stirred at high speed and heated to a rapid reflux without bumping. The addition funnel is charged with a solution of 8.7 g (50 mmol) of 1-diazo-4-phenyl-2-butanone diluted to 50 mL with methylene chloride. As soon as the high dilution trident reservoir (20 mL) fills up and begins to overflow back into the round-bottomed flask, dropwise addition of the diazo ketone solution is initiated (1:20, one drop of diazo ketone solution to every 20 drops of solvent entering the trident reservoir from the condenser). After the addition is complete (2.5–3 h), the reaction mixture is allowed to reflux for an additional 1 h. The reaction mixture is then cooled, and the yellow-green solution of the initially formed unstable trienone is suction filtered through 110 g of neutral alumina in a 250-mL fitted glass funnel to isomerize the β,γ-double bond into conjugation with the carbonyl group and to remove the rhodium diacetate dimer. The alumina is then washed with 100 mL of ethyl acetate, and the combined organic filtrates are concentrated by rotary evaporation to give a yellow oil. Vacuum distillation of this oil through a short-path distillation head gives 5.5–5.7 g (75–78% yield) of a colorless to slightly green oil that solidifies at 0°C, bp 73–75°C/0.2 mm. This material is sufficiently pure for most purposes. Recrystallization from hexane (80 mL per gram of trienone) yields white needles, mp 28.5–29.0°C.

A study of substituent effects on rhodium(II)-catalyzed Büchner reactions of phenyl diazoalkanones of the type shown in Scheme 7 has established a number of points that are very pertinent to their use in synthesis.[29] First, a methyl group

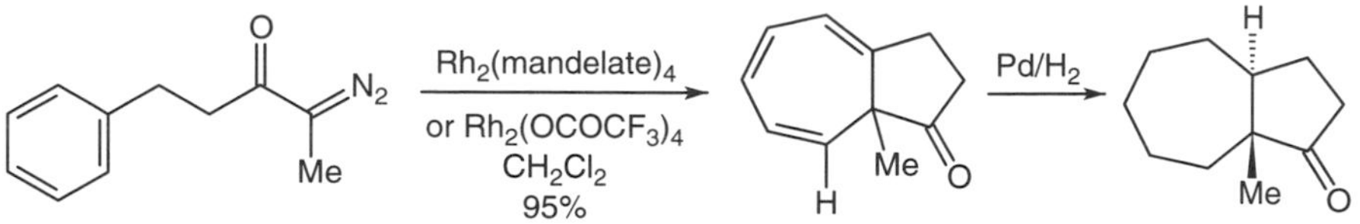

Scheme 10

can be tolerated on the diazocarbon atom as in Scheme 10. Rhodium(II)-cata-lyzed cyclization of this diazoketone yielded a methylazulenone that on hydroge-nation opened up a short stereospecific route to the *trans*-perhydroazulenone system (60% yield from the diazoketone). In this cyclization rhodium(II) trifluo-roacetate and rhodium(II) mandelate were much more effective as catalysts than rhodium(II) acetate. A feature of interest in the trienone in Scheme 10 is the ex-tent to which it is adequately represented as a bicyclic structure. While the spec-troscopic data for this trienone do possess the general features consistent with the bicyclic cycloheptatrienyl form, the presence of two IR carbonyl absorptions at 1745 and 1715 cm^{-1} and the apparent ^{1}H-NMR chemical shift (δ 4.31) for the doublet representing the vinylic proton adjacent to the bridgehead methyl group must be considered rather atypical. However, Hannemann's analysis of the ^{1}H-NMR spectra of several cycloheptatrienes in equilibrium with their norcaradi-ene counterparts does clarify the situation greatly.[32] It seems that the bicyclic cy-cloheptatriene structure is indeed an inadequate representation. The chemical shift of the proton in question above is more consistent with its location in a tri-cyclic norcaradiene tautomer, there existing a rapid equilibration between the two forms with the tricyclic form the dominant tautomer at room temperature (Scheme 11). In fact, Saba has shown that the tricyclic norcaradiene form can be detected at low temperatures by ^{1}H-NMR spectroscopy and trapped in a Diels–Alder reaction with 4-phenyl-1,2,4-triazoline-3,5-dione (PTAD) (Scheme 11).[15]

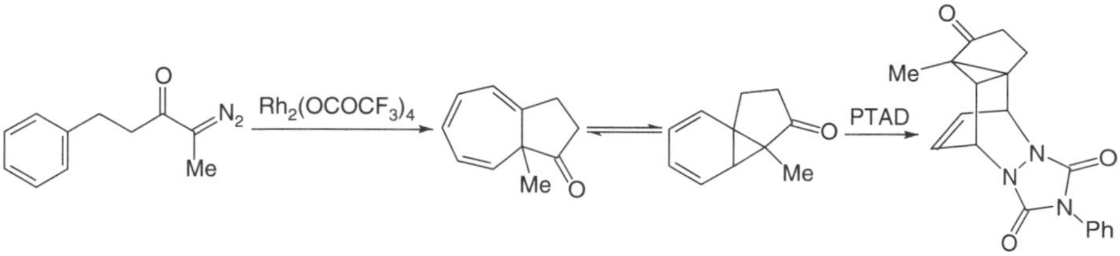

Scheme 11

A second important aspect of the catalyzed Büchner reaction is the effect of substituents on the aromatic ring of the precursor. A comprehensive survey by McKervey and co-workers[29] produced the observations summarized in Schemes 12–16. Thus three *para*-substituted phenyl diazobutanones cyclize smoothly with rhodium(II) acetate to furnish bicyclic trienones in excellent yield. Furthermore, all three diazoketones gave excellent yields of 2-tetralones, *viz.* 7-methyl-2-tetralone, 7-methoxy-2-tetralone, and 7-acetoxy-2-tetralone, when

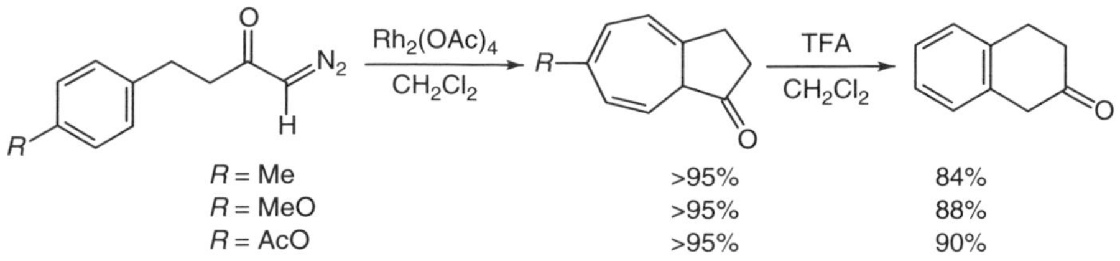

R = Me	>95%	84%
R = MeO	>95%	88%
R = AcO	>95%	90%

Scheme 12

subjected to the consecutive action of rhodium(II) acetate and TFA. The presence of *ortho* substitution introduces the possibility of a directive effect on cyclization. The situation with an *ortho*-methyl group is illustrated in Schemes 13 and 14. The product of rhodium(II)-catalyzed cyclization of the terminal diazoketone (Scheme 13) is exclusively that of cyclization in the direction away from the *ortho*-methyl group, and, accordingly, 5-methyl-2-tetralone was the result of subsequent TFA treatment. A somewhat similar situation was encountered with the decomposition of the *o*-tolyl-disubstituted diazoketone in Scheme 14, although

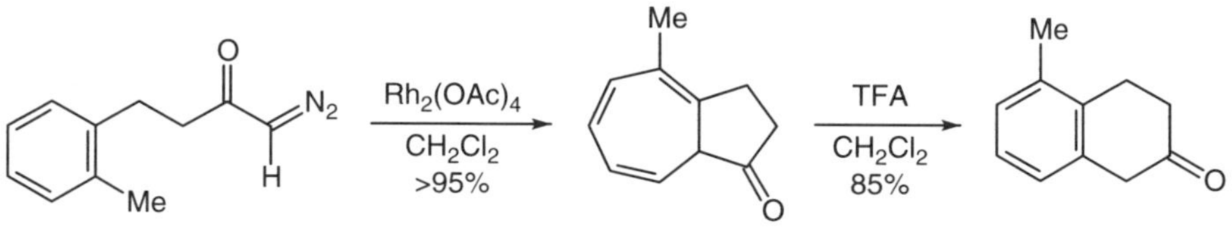

Scheme 13

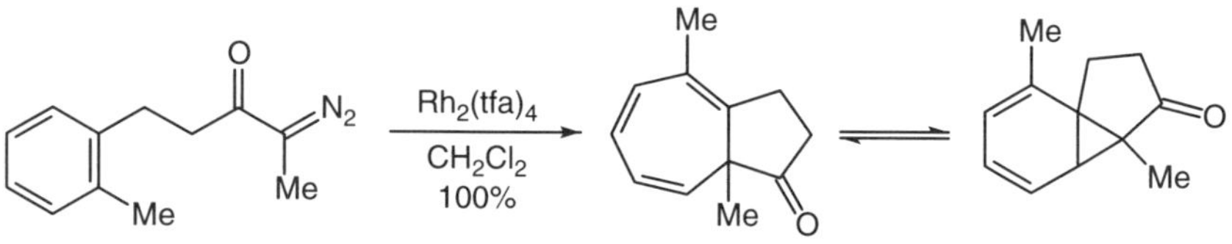

Scheme 14

here cyclization of the disubstituted metal carbene required rhodium(II) trifluoroacetate. The product was regiochemically homogeneous and not only was it that of cyclization in the same direction as that of its monosubstituted counterpart in Scheme 13, its ^{1}H-NMR and IR features indicated that at room temperature it existed predominantly in the tricyclic norcaradiene form (Scheme 14). The behavior of the diazoketones in Schemes 13 and 14 in intramolecular cyclization is thus consistent with the behavior of toluene in intermolecular cyclization with EDA, where little attack occurs adjacent to the methyl substituent. Kennedy and McKervey[33,34] were able to take advantage of this directive effect in their use of the Büchner reaction to synthesize the *pseudo*guaianolide (±)-confertin

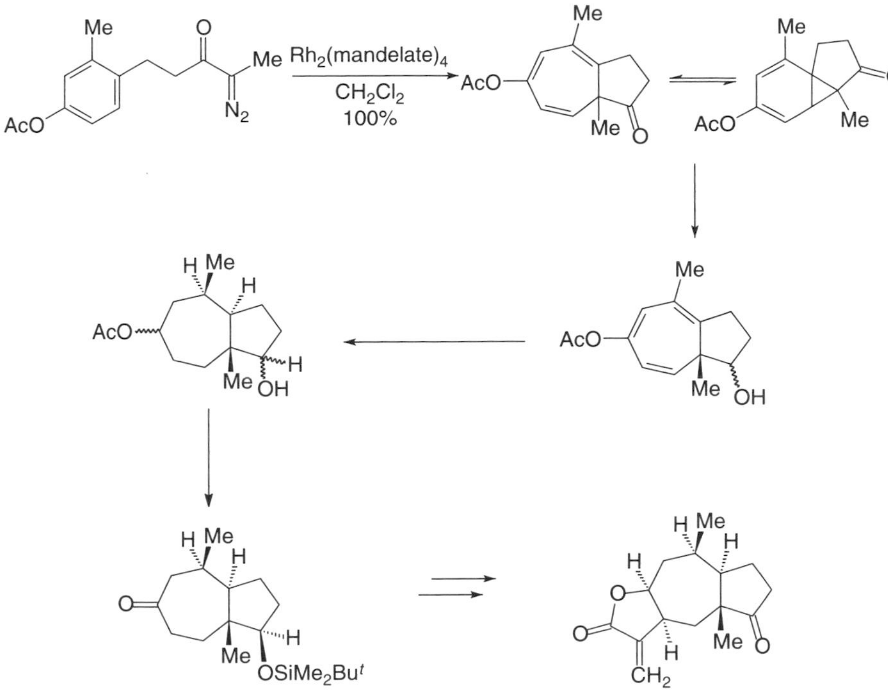

Scheme 15

(Scheme 15). For the purpose of further elaboration of the Büchner product, the phenyl diazopentan-3-one precursor carried a *p*-acetoxy substituent in addition to an *ortho*-methyl substituent. Cyclization of this precursor was accomplished quantitatively using rhodium(II) mandelate to furnish a ring-expanded trienone, which again was in equilibrium with the norcaradiene form. Reduction of the keto function with lithium tri-*t*-butoxyaluminohydride caused the norcaradiene form to disappear, and the epimeric alcohols thus produced were then elaborated into an advanced confertin intermediate that had previously been converted into (±)-confertin by building on the γ-lactone ring (Scheme 15). The feasibility of using this methodology to construct polyfuctional hydroazulenes with γ-lactone rings already built on to the benzenoid precursor has also been demonstrated. Two examples are summarized in Scheme 16.[35]

There have been conflicting reports on the directive effect of an *ortho*-methoxy group on the cyclization of phenyldiazobutanone derivatives. Whereas Kennedy *et al.*[29] proposed that the rhodium(II) acetate–catalyzed reaction occurred with the metal carbene attacking the benzene ring adjacent to the substituent, French workers have claimed that the product is in fact 5-methoxy-2-tetralone,[36] resulting from metal carbene attack away from the substituent (Scheme 17).

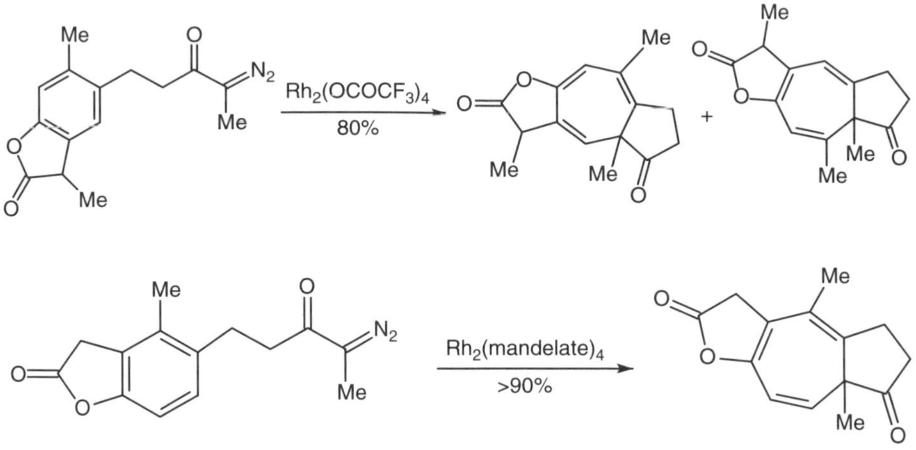

Scheme 16

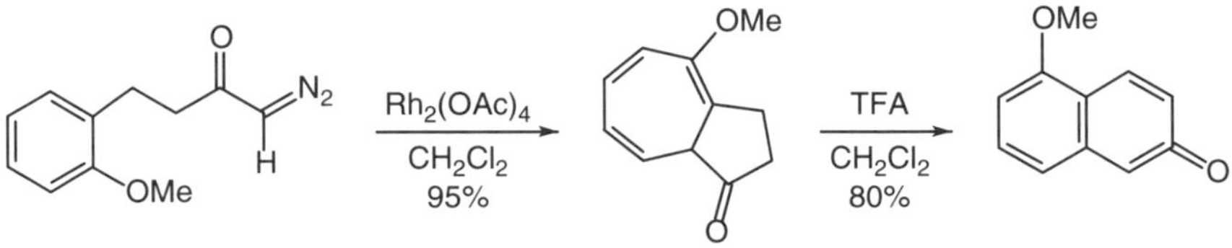

Scheme 17

The behavior of *m*-substituted benzenoid precursors illustrates yet another feature of the intramolecular Büchner reaction.[29] When the substituent is methoxy (as in Scheme 18A) the product of rhodium(II) acetate–catalyzed reaction is 6-methoxy-2-tetralone with no trace of the putative trienone. The *m*-acetoxy diazoketone (Scheme 18B), on the other hand, does furnish an isolable trienone that readily gives tetralone on TFA treatment. The behavior of the *m*-methyl diazoketone (Scheme 18C) falls between that of its methoxy and acetoxy counterparts. Here the product consists of two trienones and two tetralones, which after TFA treatment is simplified to 6-methyl-2-tetralone (70%) and 8-methyl-2-tetralone (30%). This pattern of behavior is also observed with diazoketones in which there is at least one alkoxy substituent *meta* to the reacting side chain. Thus the 3,4-dimethoxy derivative in Scheme 19A cyclizes to an 80:20 mixture of two tetralones, the 3,4-methylenedioxy derivative (Scheme 19B) similarly produces tetralones, and the 3,4,5-trimethoxy derivative (Scheme 19C) furnishes 6,7,8-trimethoxy-2-tetralone.

The existence of a mobile equilibrium between cycloheptatrienyl and norcaradienyl forms helps explain the substituent effects on tetralone formation: In the absence of an electron-donating group at the *meta* position, electrocyclic ring opening of the 1,6-bond of the norcaradiene produces stable cycloheptatrienones;

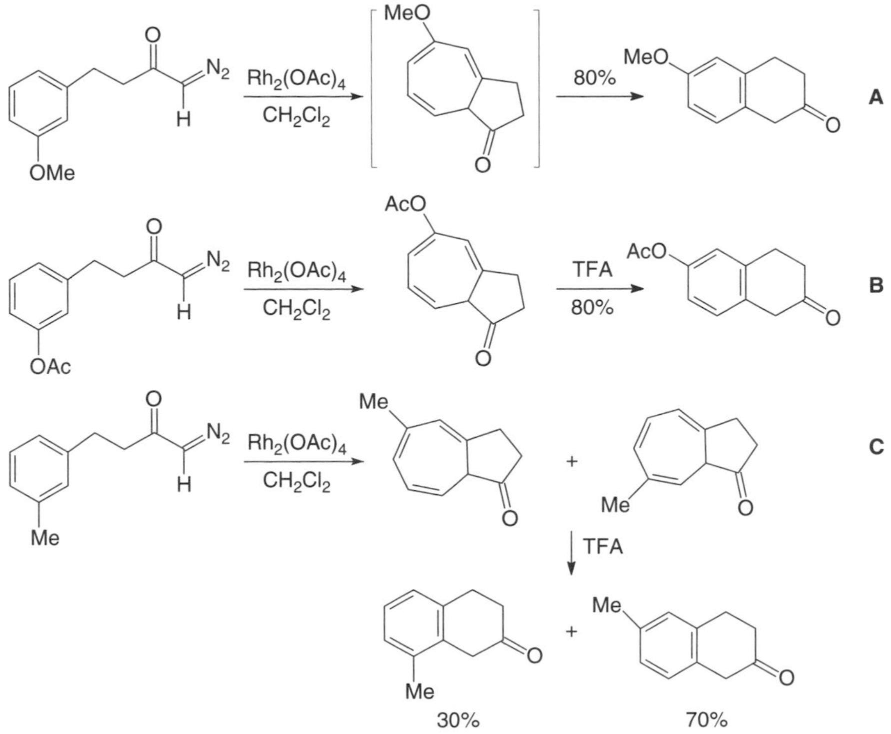

Scheme 18 A, B, C

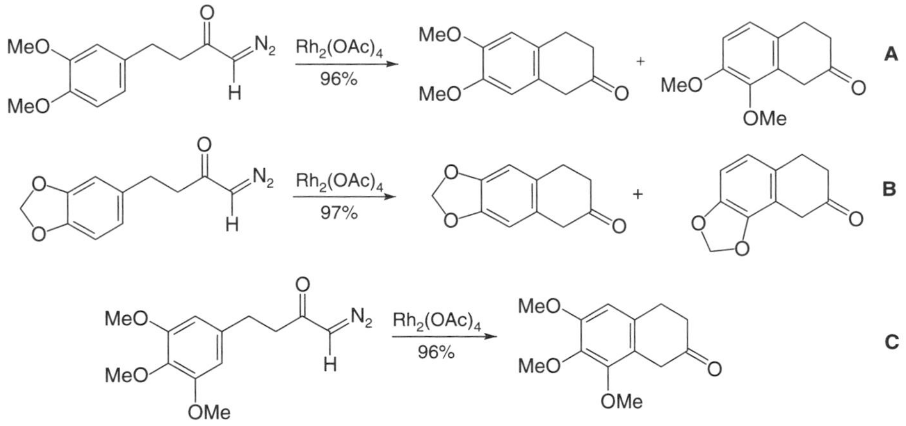

Scheme 19 A, B, C

a *meta*-methoxy group, or to a lesser degree, a *meta*-methyl group, through electron donation, promotes the alternative bond-breaking process terminating in tetralones; the latter may be assisted by adventitious acid. With *para*-phenolic groups in the diazoketone precursor, yet another cyclisation pathway may be followed. For example, the diazoketone in eq. 8 on exposure to copper(I) chloride

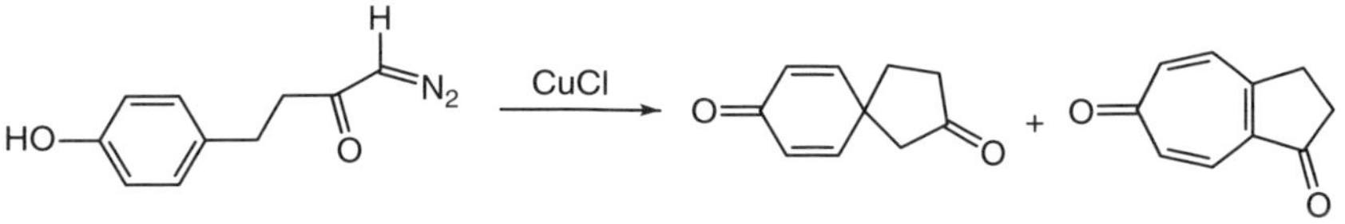

(8)

affords both cycloheptatriene and spirodienone[37,38]; depending on reaction conditions, the spirodienone may be the sole product (eqs. 9 and 10). Other active cata-

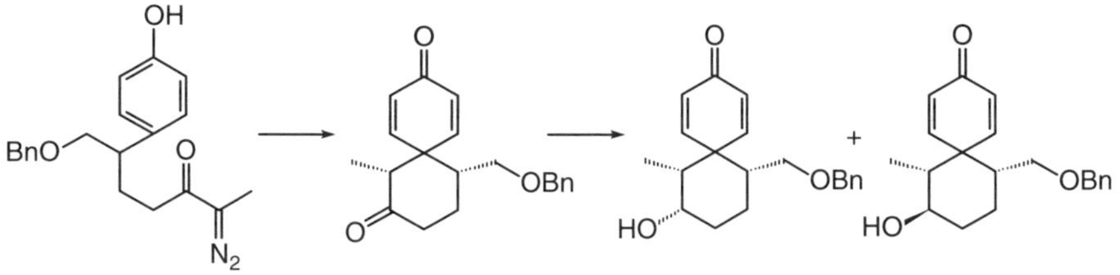

(9)

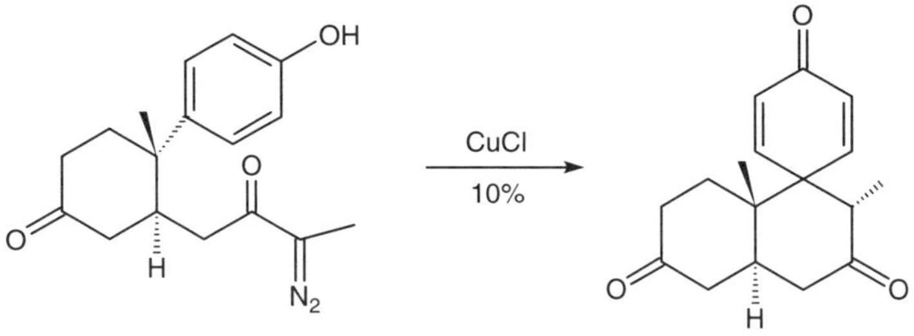

(10)

lysts include rhodium and palladium carboxylates.[37,38] This route to spirodienones forms the basis of total synthesis of the solavetivone,[39] aphidicolin,[40] and stemodin terpenes[41] (Scheme 20).

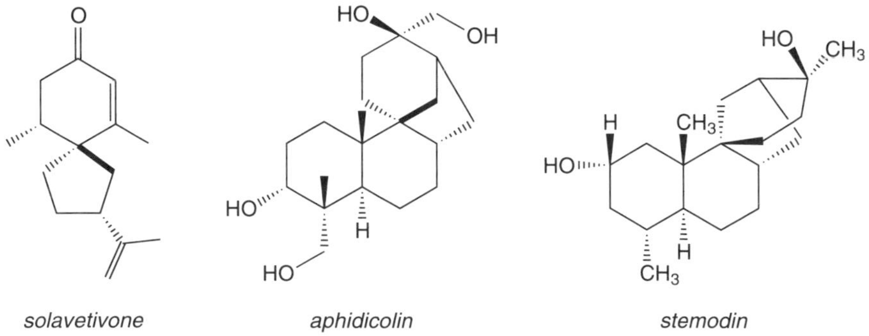

solavetivone　　　*aphidicolin*　　　*stemodin*

Scheme 20

As we have just seen in the cyclization of 4-aryl-1-diazobutan-2-one and of several of its derivatives (Schemes 7, 10–19), the mild conditions associated with rhodium(II) catalysis, as compared with copper catalysts, do provide more favorable conditions for observing unstable primary reaction products. Another case in point is the formation of phenanthrol from the biphenyl diazomethyl ketone in Scheme 21. Under copper catalysis, phenanthrol is the sole product, superficially

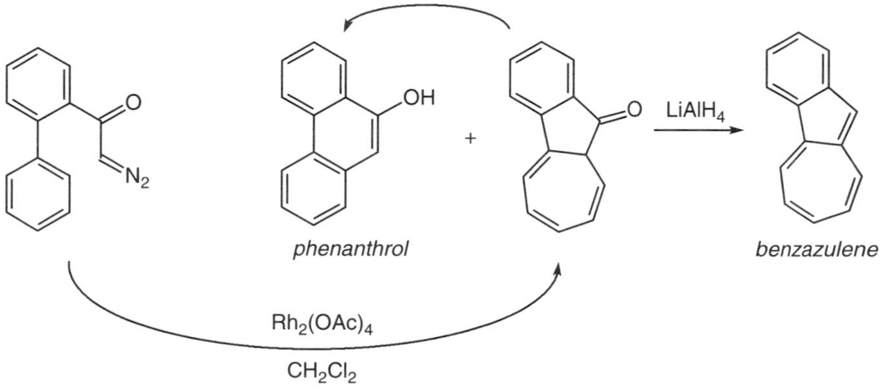

phenanthrol   *benzazulene*

Rh$_2$(OAc)$_4$

CH$_2$Cl$_2$

Scheme 21

supporting the view that the reaction is one of aromatic C–H insertion or electrophilic substitution.[42] When rhodium(II) acetate is employed as catalyst, phenanthrol is still the major reaction product, but it is accompanied by about 15% of an unstable substance that slowly rearranges to phenanthrol on standing at room temperature.[43] Furthermore, the observation that lithium aluminum hydride reduction of the product mixture yields benzazulene (5%) strongly suggests that the minor component of cyclization is a ring-expanded trienone and that the process is indeed of the Büchner type. Interestingly, a major change of behavior occurs in the biphenyl series when a methyl group is added to the diazocarbon atom, as in the precursor in Scheme 22.[43] Decomposition of this diazoketone em-

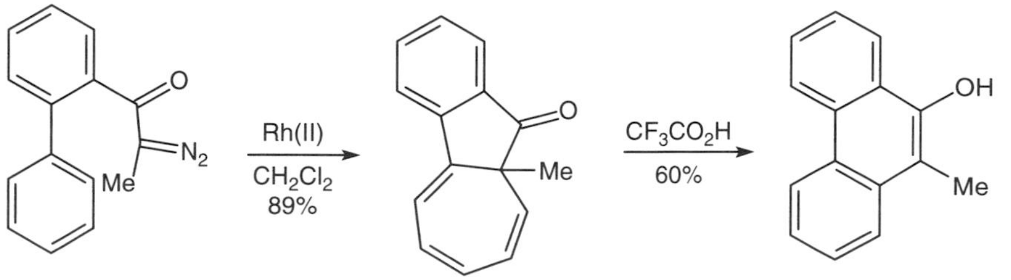

Rh(II)
CH$_2$Cl$_2$
89%

CF$_3$CO$_2$H
60%

Scheme 22

ploying rhodium(II) mandelate furnished as the sole product a stable tricyclic benzoazulenone. Furthermore, this product did not rearomatize easily, requiring treatment with hot TFA to bring about conversion to 10-methyl-9-phenanthrol. The diazocarbonyl precursor in Scheme 23, a simple homologue of the precursor in Scheme 21, also undergoes rhodium(II)-catalyzed decomposition to give a ring-expanded tricyclic product.[43]

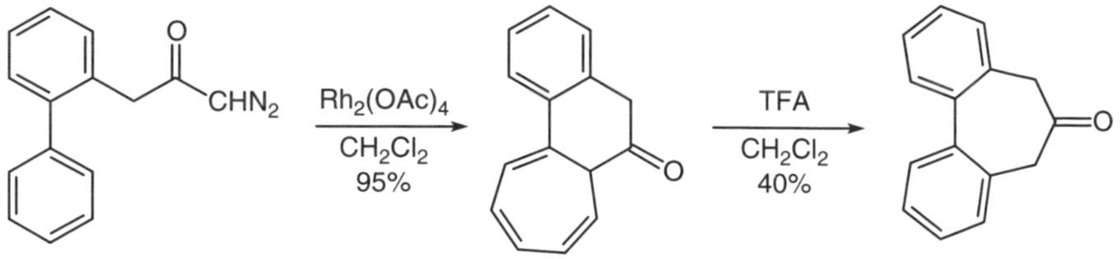

Scheme 23

The next higher homologue in the biphenyl series, the diazoketone in Scheme 24, did not, however, continue the trend and produce a colchicine-like

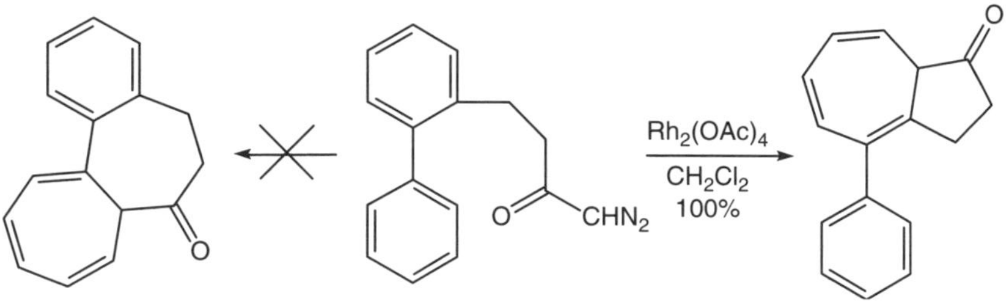

Scheme 24

benzodicycloheptyl product.[43] Rather, the product was a bicyclic phenylazulenone, indicating that the distal benzene ring in the precursor acts as a substituent and not as a site for metal carbene attack. There is also a limit to the ring sizes accessible in the Büchner reactions, although in fact the interest up until now has been predominantly in the formation of bicyclo[5.3.0]decane systems. Lengthening the chain between benzene ring and metal carbene by one methylene unit gives the diazoketone precursor $R = H$ shown in Scheme 25. On exposure to rhodium(II) mandelate this substrate afforded a ring-expanded [5.4.0]undecane derivative, which could be easily transformed by exposure to TFA into 2-benzosuberone (89%) or into its more conjugated isomer (76%) by exposure to triethylamine.[29] However, with the nonterminal diazoketone ($R = $ Me) in Scheme 25, again with rhodium(II) mandelate, the product is *trans*-2-methyl-3-phenylcyclopentanone.[29] In other words, the reaction pathway switches from one of cycloaddition to one of C—H insertion. C—H insertion is also the pathway with the next higher homologue in eq. 11; 3-benzylcyclopentanone is the sole product of the reaction. The Büchner route to benzosuberone derivatives has been

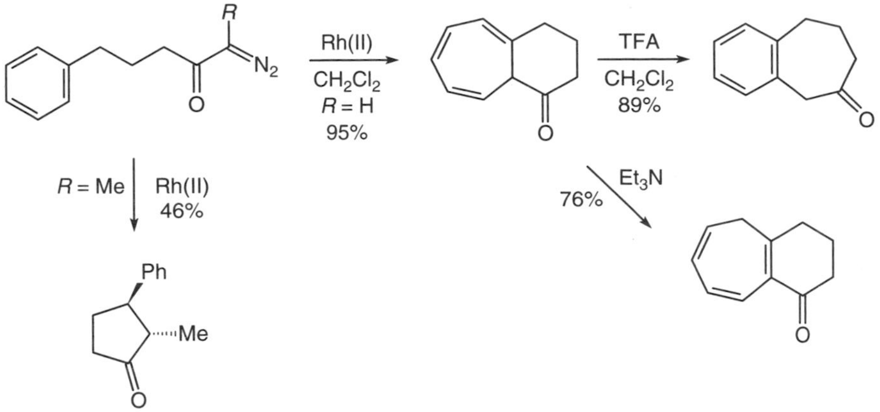

Scheme 25

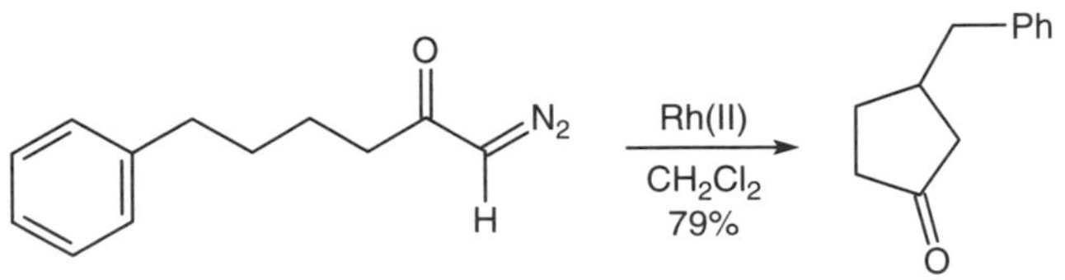

$$(11)$$

exploited by Sonawane *et al.*[44] in their synthesis of (+)-*ar*-himachalene (Scheme 26). Rhodium(II) acetate catalysis afforded a bicyclo[5.4.0]undecane,

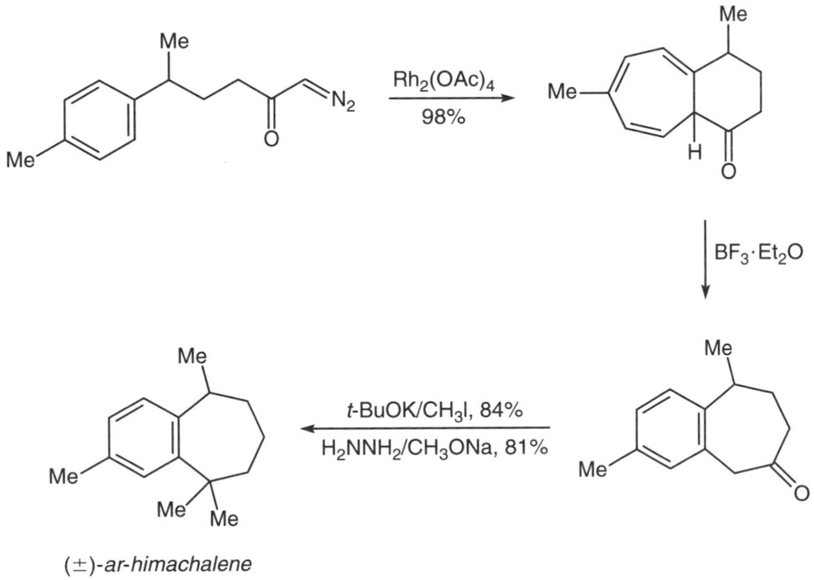

(±)-*ar*-himachalene

Scheme 26

which on exposure to $BF_3 \cdot Et_2O$ rearomatized to a dimethylbenzosuberone. Dimethylation at the benzylic position with methyl iodide and potassium t-butoxide followed by Wolff–Kishner reduction completed the synthesis of the sesquiterpene. Among other studies with binuclear or fused aromatic precursors, that of Manitto and co-workers[45] found further evidence for the existence of a norcaradiene as the first intermediate in the catalytic intramolecular Büchner reaction. Arguing that the isolation of a norcaradiene-like tautomer should be possible if the ring-opening reaction to cycloheptatriene is made thermodynamically and/or kinetically disfavored (e.g., by concomitant dearomatization), these workers subjected 1-diazo-4-(2-naphthyl)butan-2-one (Scheme 27) to rhodium(II) ace-

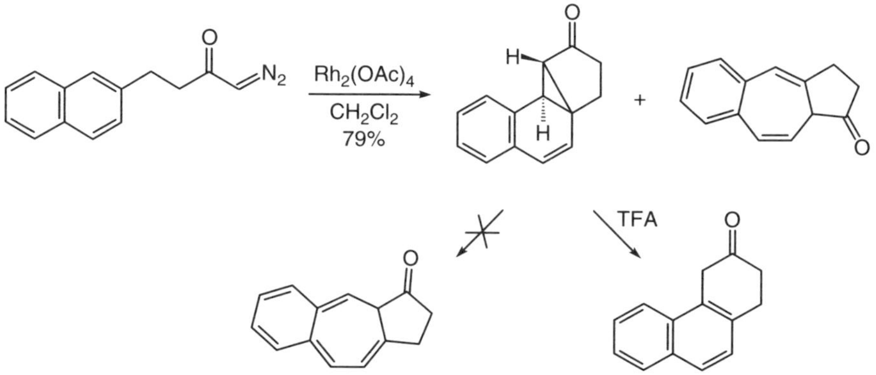

Scheme 27

tate catalysis in dichloromethane. The principal product (71%) of rhodium(II) acetate–catalyzed cyclization was indeed a stable tetracyclic ketone with the cyclopropane unit intact. When exposed to TFA the product immediately rearranged to a dihydrophenanthrenone. A second minor product (8%) was identified as that of attack of the metal carbene on the alternative (2,3) side of the naphthalene nucleus followed by double-bond migration in the tricyclic nonconjugated product. Thus, whereas resistance to dearomatization secures the stability of the major norcaradiene product in Scheme 27, aromatization is clearly the driving force for opening the norcaradine, leading to the minor reaction product. A contrasting picture emerges from the rhodium(II)-catalyzed decomposition of the diazoketone in Scheme 28, which leads directly to a naphthol derivative.[46] It is

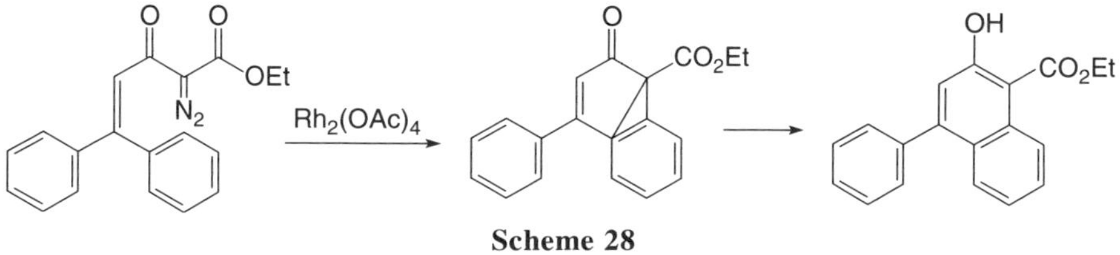

Scheme 28

possible that this product is formed *via* a norcaradiene intermediate, which should be much more prone to aromatization than to ring expansion. In contrast, Wenkert and Liu[47] found that rhodium(II) acetate and the diazoketone in eq. 12

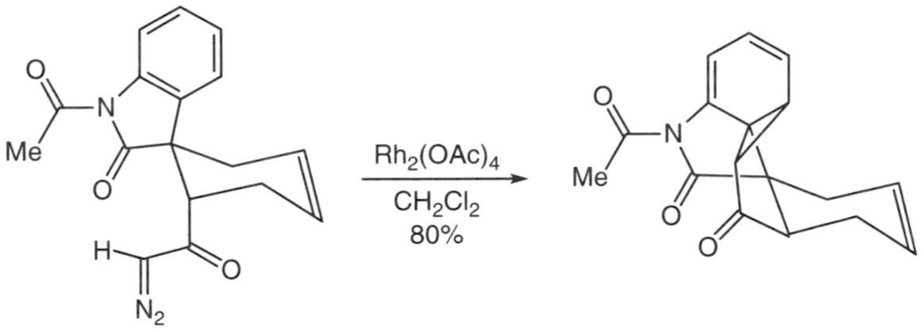

$$Rh_2(OAc)_4 \quad CH_2Cl_2 \quad 80\%$$

(12)

furnished a stable norcaradiene product. Among other examples of the use of intramolecular Büchner reactions in multicyclic systems is the construction by Murata and co-workers[48] of cyclohepta[*a*]phenylene, a highly electron-donating nonalternate hydrocarbon (Scheme 29). Rhodium(II) acetate–catalyzed decom-

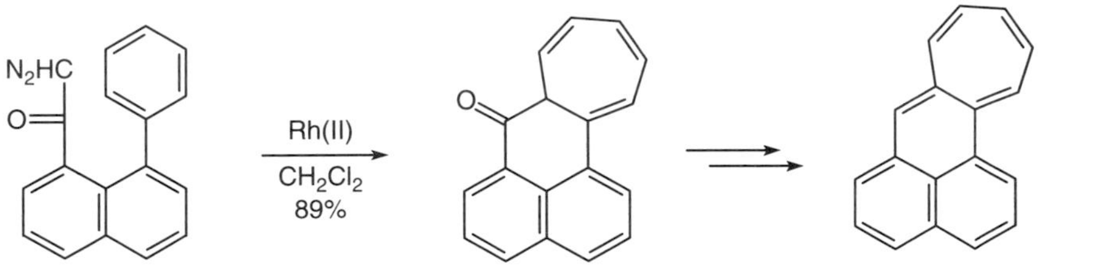

$$Rh(II) \quad CH_2Cl_2 \quad 89\%$$

cyclohepta[a]phenylene

Scheme 29

position of the diazoketone precursor furnished the necessary tetracyclic framework in 89% yield, and reductive removal of the oxygen function followed by dehydrogenation completed the synthesis.

Mander and his group[49] have looked in some detail at the potential of catalyzed Büchner reactions for the construction of complex natural polycyclic diterpenes, one particular target being the lactone harringtonolide (Scheme 30), where

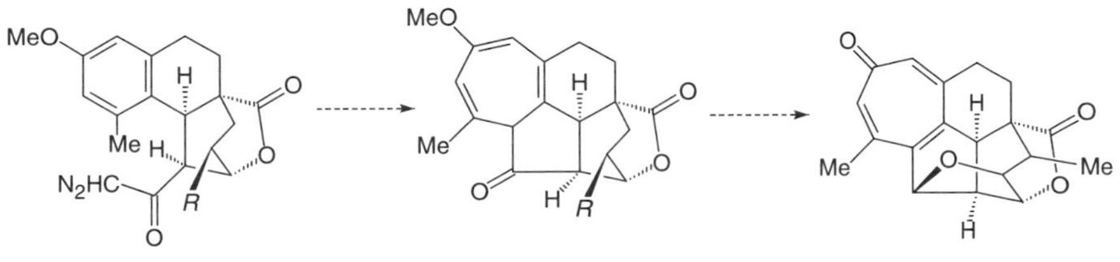

(±)-*harringtonolide*

Scheme 30

ring expansion of a tetralin-derived diazoketone was envisaged as the pivotal step in constructing the fused troponoid framework. Preliminary studies with model substrates and rhodium(II) catalysts were undertaken to assess factors that might be detrimental to chemoselectivity such as the steric crowding around the aromatic ring or the possibility of competing C–H insertion reactions. One of the features of the catalyzed decomposition of the model diazoketone in Scheme 31

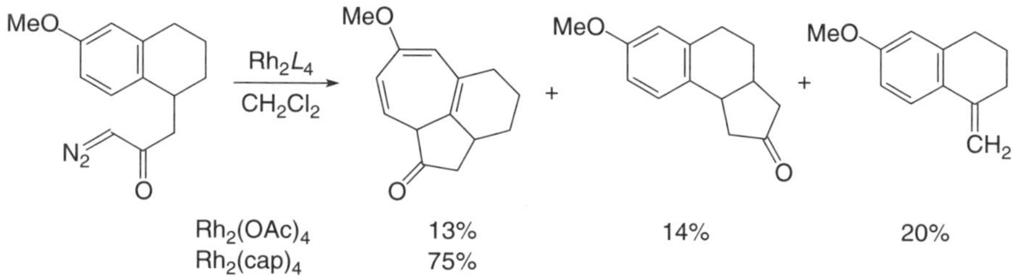

| Rh$_2$(OAc)$_4$ | 13% | 14% | 20% |
| Rh$_2$(cap)$_4$ | 75% | | |

Scheme 31

was the manner in which product distribution depended on the rhodium(II) catalyst used. Whereas rhodium(II) acetate afforded only 13% of the desired ring-expanded cycloheptatrienyl derivative, with larger amounts of product of C–H insertion (14%) and fragmentation (20%), the rhodium(II) caprolactam complex

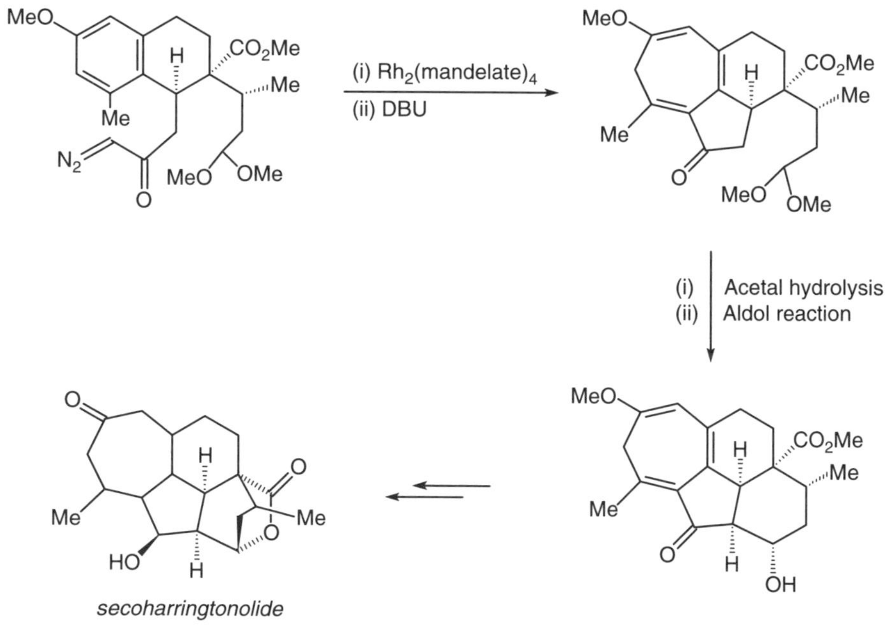

secoharringtonolide

Scheme 32

[$Rh_2(cap)_4$] furnished the cycloheptatriene in 75% yield. The fragmentation reaction leading to the alkene (with the release of ketene) was unexpected and was attributed to the intervention of a benzylic carbocation formed within the metal carbene by a hydride transfer reaction. With the more advanced diazocarbonyl precursor in Scheme 32 and rhodium(II) mandelate, an excellent yield of ring-expanded product was obtained and was subsequently stabilized by conversion into its more conjugated isomer by exposure to DBU. After selective hydrolysis of the acetal function this product smoothly underwent the aldol reaction to afford a ketol that was successfully transformed into secoharringtonolide. In a further study of the use of benzenoid synthons for the construction of complex polycyclic molecules, Mander and co-workers[49] examined the use of tetralin-2-diazoketones in catalyzed Bücher reactions, the ultimate target here being the antheridogens (e.g., antheridic acid). With 2-tetralin diazocarbonyl side chains (eq. 13) it was anticipated that geometric constraints would render norcaradiene products more stable than their cycloheptetriene tautomers, since the latter would necessarily have bridgehead double bonds. In that event, exposure of the 6-methoxytetralin-2-diazoketone (entry 1 in Table 6.3) to rhodium(II) acetate furnished the nor-

TABLE 6.3 Competition between Cyclopropanation and C–H Insertion with Tetralin Diazocarbonyl Substrates

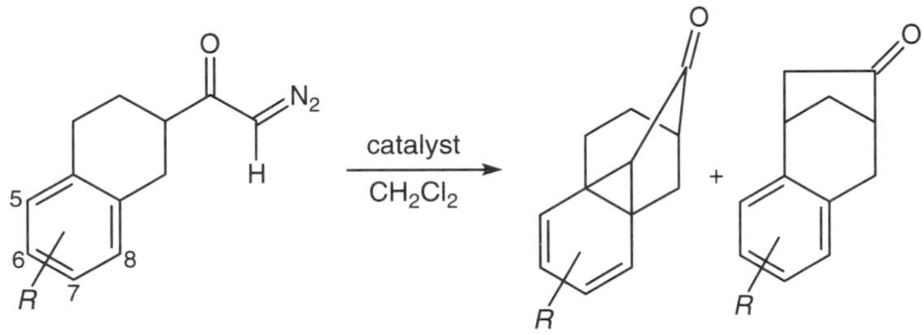

Entry		Catalyst	Yield (%)	Distribution (%)	
1	R=6-OMe	$Rh_2(OAc)_4$	92	83	17
2		$Rh_2(cap)_4$	84	65	35
3		$Rh_2(pfb)_4$	68	25	75
4		$Rh_2(TPA)_4$	97	54	46
5		$Cu(acac)_2$	82	79	21
6	R=5-OMe	$Rh_2(OAc)_4$	86	48	52
7	R=7-OMe	$Rh_2(OAc)_4$	100	51	49
8		$Rh_2(cap)_4$	97	22	78
9		$Rh_2(TPA)_4$	99	10	90
10		$Cu(acac)_4$	68	96	4
11	R=8-OMe	$Rh_2(OAc)_4$	80	30	70
12		$Rh_2(TPA)_4$	75	0	100
13	R=H	$Rh_2(OAc)_4$	87	47	53

caradiene product predominantly. Minor products included a cyclopentanone, the result of C–H insertion. Trace amounts of a cyclobutanone, also the result of C–H insertion, and a dihydronaphthalene, the result of fragmentation, were also formed. The 5-, 6-, 7-, and 8-methoxytetralin substrates and a range of rhodium(II) and copper (II) catalysts were then used to establish the scope and limitation of this type of Büchner reaction. The results, summarized in Table 6.3, proved to be extremely variable, with the rhodium complexes producing the desired norcaradiene with selectivities ranging from 0 to 83%, depending on the substitution pattern of the substrate and the choice of catalyst. In contrast to the results obtained with rhodium catalysts, Cu(acac)$_2$ showed the highest preference for the norcaradiene products. Yields of norcaradienes generally appear to be enhanced by the higher electron densities associated with methoxy substitution, except in the case of the 8-methoxy derivative, where steric repulsion involving the *peri* substituent and one of the neighboring benzylic hydrogen atoms in the transition state for cycloaddition may have an adverse effect. Having established these substituent and ligand effects on the chemoselectivity of intramolecular cyclopropanation of two-tetralin diazocarbonyl compounds, Mander and co-workers then proceeded to assemble a substrate possessing a vinyl group at position 5 and a methoxy group at position 8 of the aromatic ring. Cyclization of this using Cu(II)(acac)$_2$ furnished the norcardiene intermediate shown in Scheme 33. This intermediate was then incorporated into a total synthesis of (±)-antheridic acid.

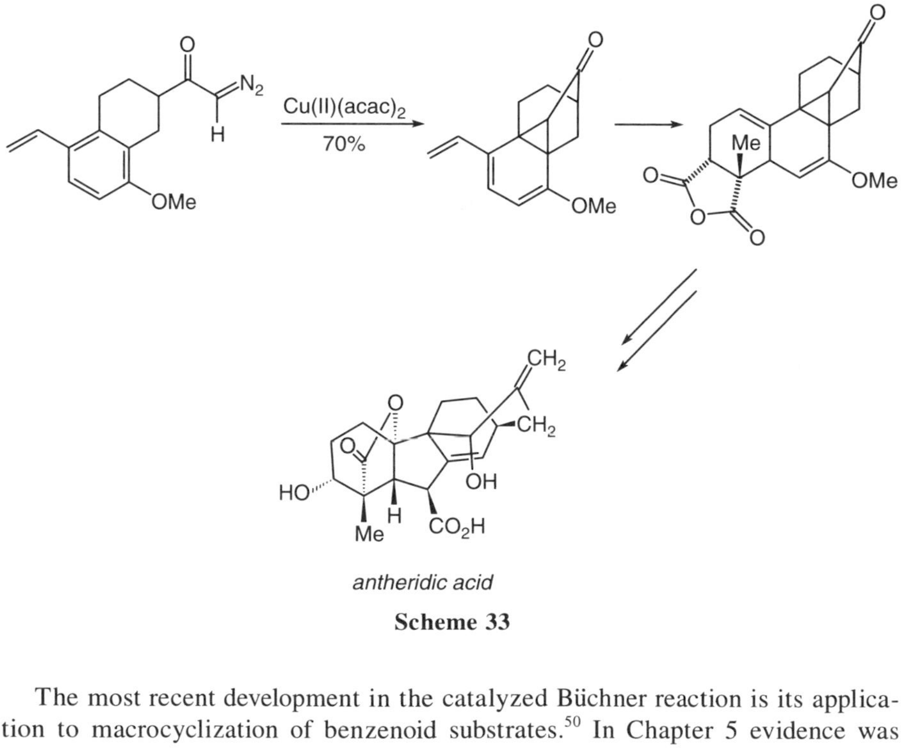

antheridic acid

Scheme 33

The most recent development in the catalyzed Büchner reaction is its application to macrocyclization of benzenoid substrates.[50] In Chapter 5 evidence was presented for the feasibilty of macrocyclization in intramolecular cyclopropana-tion of double bonds sufficiently remote from the metal carbene center to form cyclopropanes fused to 11- to 13-membered rings. Macrocyclization by in-tramolecular cycloaddition on remote aromatic rings is also feasible and has been found to be a general transformation of diazoesters and diazoketones, even in competition with proximal cyclopropanation or C–H insertion. Doyle, McKervey, and co-workers[50] have shown that diazoacetates derived from 1,2-benzenemethanol (Schemes 34 and 35) and *cis*-2-buten-1,4-diol (Scheme 36)

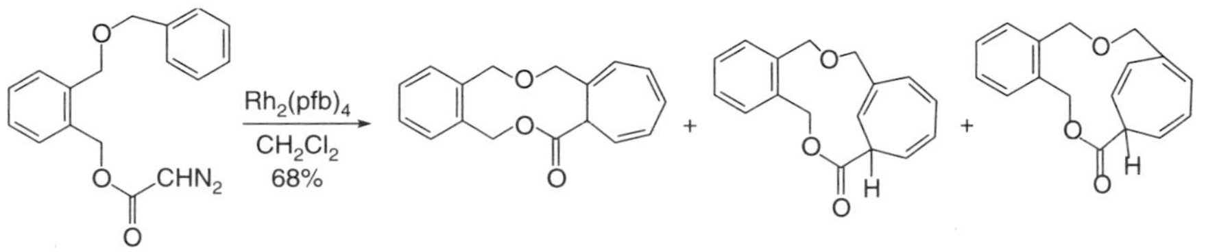

Scheme 34

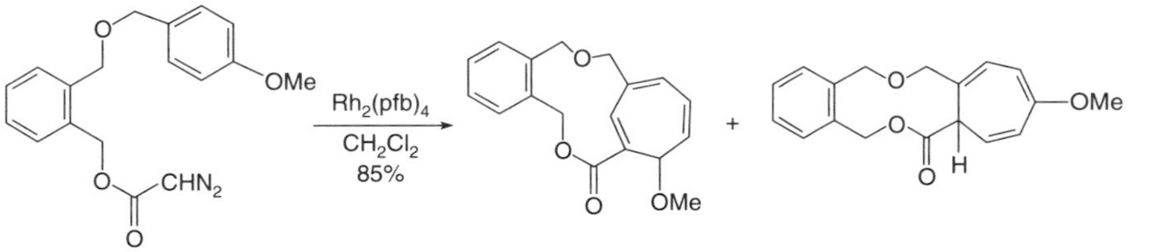

Scheme 35

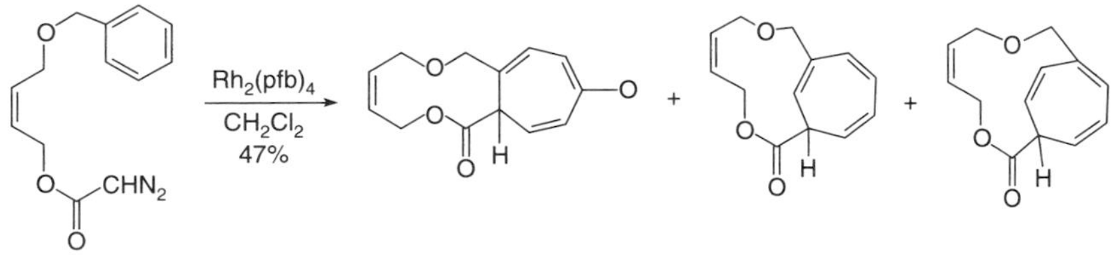

Scheme 36

followed by O-benzylation undergo rhodium(II) perfluorobutyrate–catalyzed cy-cloaddition to form macrocyclic lactones in moderate to high, isolated yields. With diazoacetates addition to the remote benzyl group occurred mainly at the 1,2 position to form disubstituted cycloheptatrienes in the bicyclo[8.5.0] frame-work, but with *p*-methoxybenzyl derivatives (Scheme 35) addition was mainly at the 3,4 position to form a bicyclo[8.3.2] framework. This regiochemical preference is consistent with that of intermolecular aromatic cycloaddition of *p*-disubstituted benzene derivatives[12] and suggests its suitability as a synthetic methodology for remote functionalization with the analogous *cis*-2-buten-1,4-diyl derivative in Scheme 36. Here, $Rh_2(pfb)_4$-catalyzed diazo decomposition produces modest amounts (47% yield) of three cycloheptatriene products, one each from addition to the 1,2-, 2,3-, and 3,4-positions of the benzene ring.

The product ratio of 31:21:48 is remarkably close to that obtained for intermolecular aromatic cycloaddition of EDA to toluene catalyzed by dirhodium(II) trifluoroacetate (18:24:58) and suggests a conformational freedom for carbene addition in the intramolecular reaction that parallels that for the intermolecular reaction. Surprisingly, cyclopropanation of the allylic double bond did not occur in reactions catalyzed by $Rh_2(pfb)_4$, but this was the sole pathway for the dirhodium(II) caprolactamate, $Rh_2(cap)_4$, catalyzed reaction. With $Rh_2(OAc)_4$ neither macrocyclization nor cyclopropanation was observed.

With the *p*-methoxybenzyl analogue of the *cis*-2-buten-1,4-diyl system in Scheme 36, both the product of macrocyclization at the 3,4 position and that of macrocyclization of the 1,2 position were formed (32% yield) in the $Rh_2(pfb)_4$-catalyzed reaction, with the former the dominant isomer (87:13). This macrocyclization methodology is also applicable to suitably structured diazoketones. For example, the diazoketone in Scheme 37 with $Rh_2(pfb)_4$ yields products of cycloaddition at the 1,2 position and at the 3,4 position in 30 and 9% yields, respectively.

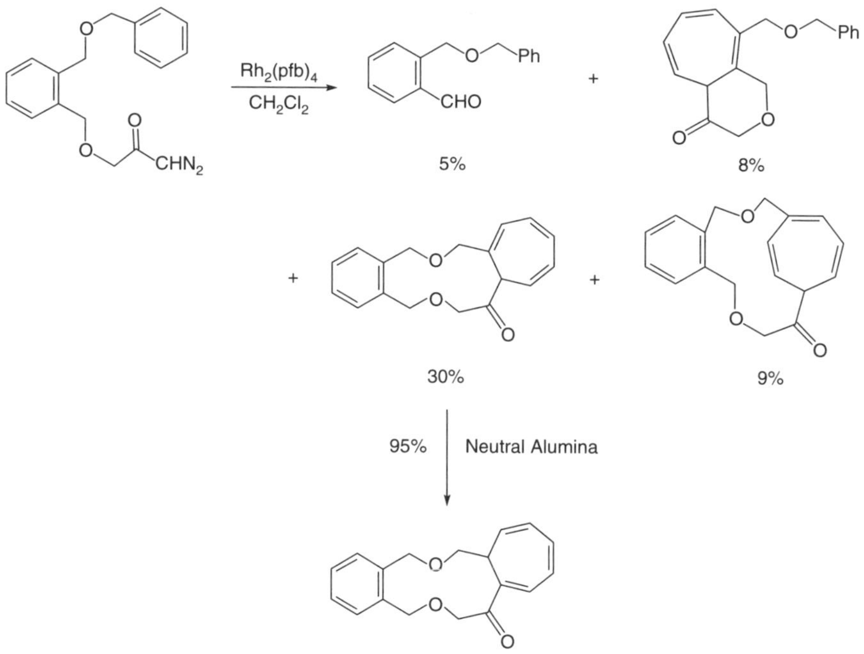

Scheme 37

The former product isomerized to its more conjugated isomer in contact with alumina. Minor products of the reaction (Scheme 37) included those of fragmentation to a benzaldehyde derivative and of aromatic cycloaddition onto the proximal disubstituted benzene ring.

6.2.2   Diazoesters and Diazoamides

In the foregoing discussion of aromatic cycloaddition of benzenoid diazocarbonyl substrates, the emphasis has been on the formation of carbocyclic systems. There are also examples of catalyzed cycloaddition of the type shown in eq. 14 leading

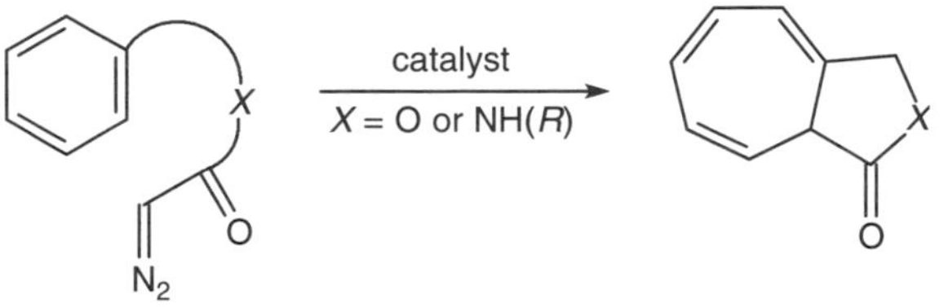

$$X = O \text{ or } NH(R) \tag{14}$$

to heterocyclic systems, where X is an oxygen or nitrogen atom. Saba and his co-workers[51] found that diazoketones with α-phenoxy substituents undergo copper(II)-catalyzed cyclization to furnish cycloheptatrienyl furanones and chromanones in proportions that varied with the position and nature of substituents in the precursors. Several examples are summarized in Scheme 38. Since the fura-

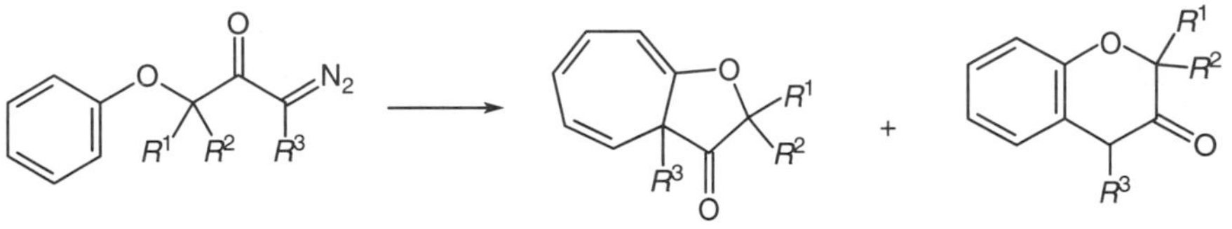

$R^1 = R^2 = H, R^3 = Me$	65%	2 : 3
$R^1 = R^2 = R^3 = Me$	95%	9 : 1
$R^1 = Me, R^2 = H, R^3 = Me$	95%	none
$R^1 = Ph, R^2 = H, R^3 = Me$	88%	none

catalyst = Cu(hexafluoroacetonate)$_2$

Scheme 38

nones were observed to rearomatize easily to chromanones, such as in contact with silica gel, it is very likely that they were the kinetic products of reaction and that the isolated product ratios reflect the relative ease of rearrangement to chromanones. The situation is thus reminiscent of that described previously for the formation of azulenones and β-tetralones. Doyle and co-workers[52] have examined in some detail similar routes to nitrogen heterocycles *via* rhodium(II)-catalyzed decomposition of N-benzyldiazoacetamides and have determined the controlling influence of amide substitution on the chemoselectivity of the process. The diazoacetamides were prepared from the appropriate secondary amine by diketene addition, followed by diazo transfer and deacylation. The t-butyl derivative (eq. 15 and entry 1 in the associated table) cyclized quantitatively with rhodium(II) acetate to afford 2-t-butyl-3,8a-dihydro-2H-cyclohepta[c]pyrrol-1-one. Similarly, the benzyl derivative (entry 2 in the table) cyclized very efficiently with rhodium(II) acetate, but the methyl derivative furnished only 37% of

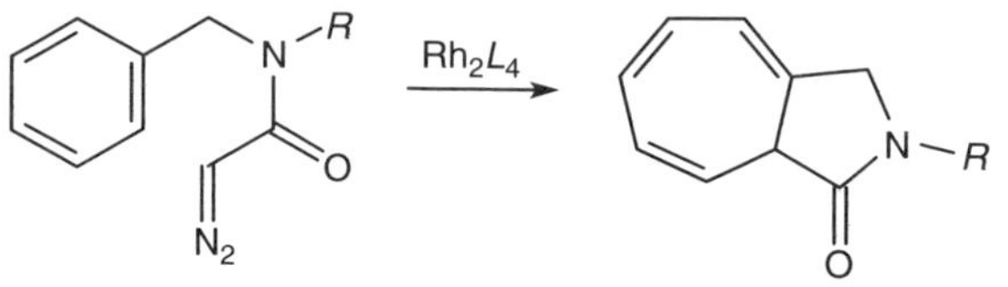

(15)

Entry	R	Ligand L	Yield (%)
1	*t*-butyl	acetate	100
2	phenyl	acetate	93
3	methyl	acetate	37[a]
4	methyl	perfluorobutyrate	54[a]
5	methyl	acetamide	21[a]

[a]At least 2 other major products were formed, both of which were dimeric but were not fumarate or maleate derivatives.

the expected product (entry 3); use of rhodium(II) perfluorobutyrate (entry 4) as catalyst improved the yield to 54%, while rhodium(II) acetamide (entry 5) gave only 21%. Although *para* substituents in the benzyl groups do not influence the selectivity of carbene addition to the aromatic nucleus, *meta* substituents do. Thus rhodium(II) acetate–catalyzed decomposition of the *meta*-methoxy–substituted diazoacetamide in Scheme 39 produced two isomeric azabicyclic products in 93%

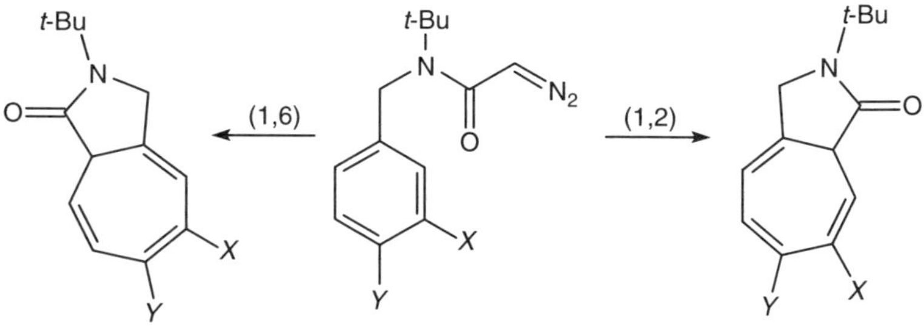

1	94%	X = OMe, Y = H	6%
2	66%	X = OMe, Y = OMe	34%
3	70%	X = Br, Y = H	30%

Scheme 39

yield (entry 1), with a strong preference (94:6) for the regioisomer resulting from addition to the 1,6 position of the aromatic nucleus. With the 3,4-dimethoxy substrate (entry 2) the 1,6 and 1,2 products (94% yield) were isolated in a 2:1 ratio. Thus the methoxy group at the 4 position diminishes the selectivity caused by the electron-donating influence of the *m*-methoxy substituent. Addition at the 1,6 position was also favored (70:30) for cyclization of the *m*-bromo precursor in entry 3. Unlike the azulenones and cycloheptatrienylfuranones, discussed previously, these azabicyclo[5.3.0]decane derivatives showed no tendency to rearomatize (to dihydroisoquinolin-3-ones) on treatment with either trifluoroacetic acid or boron trifluoride.

If in these diazoacetamide cyclizations the relative reactivity is a function of the electron-donating ability of substituents, then preferential addition might be expected to occur on the more activated aromatic nucleus of an unsymmetrically substituted N,N-dibenzyldiazoacetamide. In fact, Doyle's study[52] of the N-benzyl-N-(3,4-dimethoxybenzyl) substrate in Scheme 40 revealed that for three different

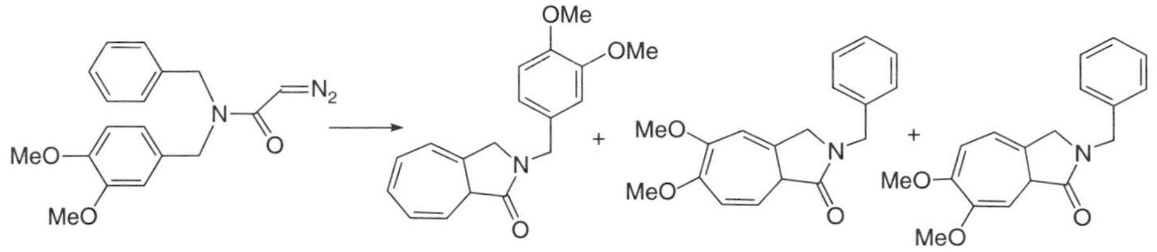

Rh$_2$(OAc)$_4$	61%	26%	13%
Rh$_2$(NHCOCH$_3$)$_4$	57%	30%	13%
Rh$_2$(pfb)$_4$	51%	27%	22%

Scheme 40

rhodium catalysts just the opposite preference was shown with addition to the unsubstituted benzene ring occurring in the higher yield [61:39 with rhodium(II) acetate; the perfluorobutyrate and acetamide catalysts behaved similarly]. These observations suggest that the preferred orientation of the intermediate metal carbene is that shown in **A** rather than in **B** and that addition to the aromatic nucleus

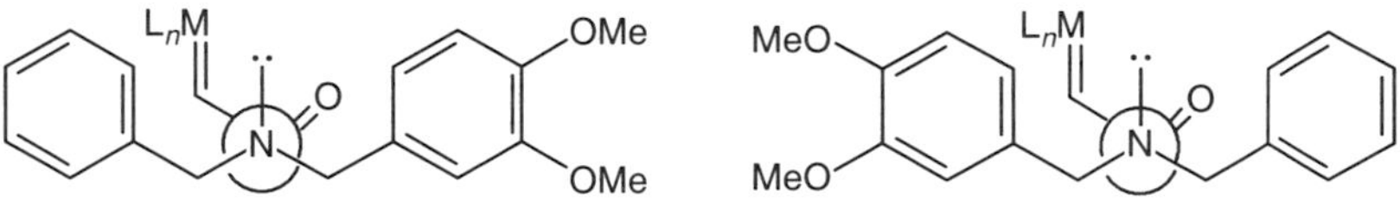

from either conformation occurs on a time scale that is more rapid than the rate of rotation around the carbonyl–nitrogen bond of the amide moiety. This interpretation also accounts for the fact that the yield of cyclization with the N-t-butyl diazoacetamide is much better than that of its N-methyl counterpart, since the orientation favored with the former will be that in which the benzyl group is closer to the rhodium carbene center. In all these cyclizations to azabicyclic products the diazocarbonyl precursor has the diazo carbon at the terminal position. When otherwise substituted, especially with an acetyl group as in diazoacetoamides, aromatic cycloaddition is no longer the pathway followed. Rather, a βC–H insertion into the methylene group leads very efficiently to β-lactam formation (Chapter 3).

6.2.3 Chemoselectivity in Aromatic Cycloaddition

As we have already emphasized in earlier chapters, there are numerous examples of transformations involving metal carbene intermediates where more than one reaction pathway is accessible and mixtures of products result. Choosing a catalyst for a specific transformation, including asymmetric versions, in a complex substrate where several types of catalyzed reactions are possible is of paramount importance. One might, for example, wish to bring about an aromatic cycloaddition in a diazocarbonyl precursor where cyclopropanation, C–H insertion, or ylide formation are all possible. The process in Scheme 36 provides an illustrative example in which the objective was to have the rhodium carbene attack the distal benzene ring in a substrate where there was the possibility of intramolecular cyclopropanation and/or intramolecular C–H insertion. In that event, rhodium(II) perfluorobutyrate catalyzed the aromatic cycloaddition pathway, as intended, to form macrocyles. There was no evidence of intramolecular cyclopropanation, though this was the only reaction pathway with rhodium(II) caprolactamate. Thus ligand effects may enable selective switching between competitive metal carbene transformations.

Aromatic cycloaddition reactions are expected to be similar to cyclopropanation reactions of olefinic bonds, and competition between the two should reflect the difference in reactivity of a metal carbene intermediate towards an aromatic ring and a C=C bond. The effect of ligand switching on the metal is nicely illustrated by the reaction in eq. 16, where the competition between aromatic cycload-

Catalyst	Yield (%)		
$Rh_2(pfb)_4$	100	35	65
$Rh_2(OAc)_4$	99	53	47
$Rh_2(cap)_4$	100	100	0

(16)

dition and cyclopropanation is evenly balanced when rhodium(II) acetate is the catalyst.[53] With rhodium(II) caprolactamate as catalyst, however, the product was that of cyclopropanation exclusively, while rhodium(II) perfluorobutyrate favored the aromatic cycloaddition product by a factor of *ca.* 2:1. Clearly there are different requirements for cycloaddition to an olefinic bond and cycloaddition to an aromatic ring, and these differences are subject to the catalyst employed with selectivity consistent with frontier orbital control.

Doyle and Padwa and their respective collaborators[53] have also probed competition between aromatic cycloaddition and carbon–hydrogen insertion using the diazoketoamide in eq. 17, in which one of the potential reaction sites is offset

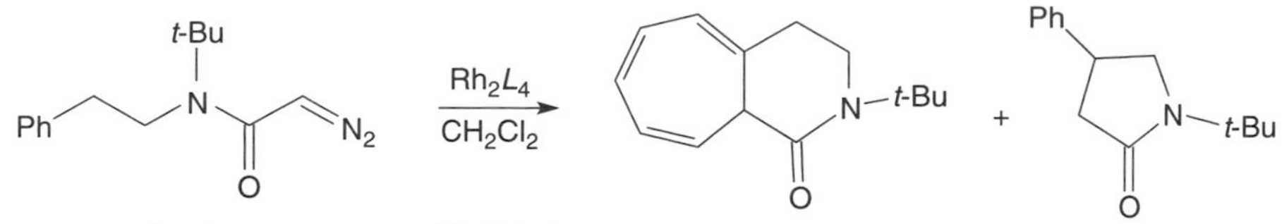

Catalyst	Yield (%)		
$Rh_2(pfb)_4$	80	95	5
$Rh_2(OAc)_4$	85	65	32
$Rh_2(cap)_4$	82	3	97
$Rh_2(acam)_4$	80	23	77

(17)

from the optimal (six- versus five-membered) ring formation. Thus formation of the aromatic cycloaddition product requires cyclization to a six-membered ring fused norcaradiene that rearranges to a bicyclo[5.4.0]cycloheptatriene derivative, whereas carbon–hydrogen insertion produces the preferred five-membered ring γ-lactam. Consistent with the earlier results in eq. 16, $Rh_2(cap)_4$ catalyzed C–H insertion to the virtual exclusion of aromatic cycloaddition, while rhodium(II) acetamidate and rhodium(II) acetate gave mixtures of both products. With rhodium(II) perfluorobutyrate the product was that of aromatic cycloaddition almost exclusively. Substitution at the para position by a methoxy group in the diazoamide in equation 17 gives essentially the same results as those observed with the para hydrogen series, indicating that the effect of a para CH_3O group parallels that of p-hydrogen on aromatic cycloaddition and C–H insertion. In contrast, a para nitro group (eq. 18) inhibits aromatic cycloaddition relative to C–H insertion

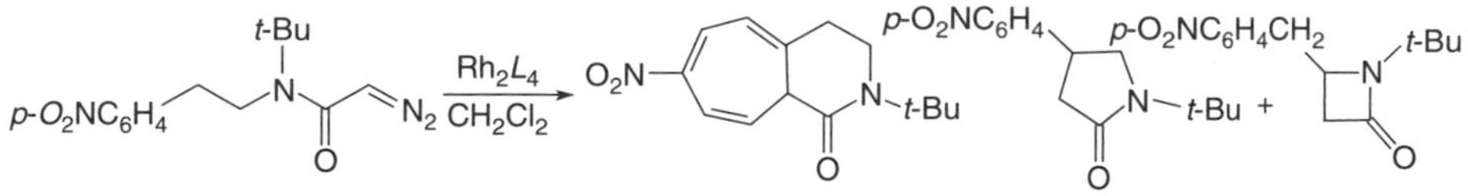

Catalyst	Yield (%)		
$Rh_2(OAc)_4$	80	8	77	15
$Rh_2(pfb)_4$	92	27	60	15
$Rh_2(acam)_4$	92	0	97	3

(18)

and additionally promotes, though to a very limited extent, formation of the previously absent strained β-lactam, the result of C–H insertion at the non-benzylic system.

It has already been mentioned that N-benzydiazoacetamides undergo aromatic cycloaddition virtually exclusively (Schemes 39 and 40) when catalyzed by $Rh_2(OAc)_4$, $Rh_2(acam)_4$, or $Rh_2(pfb)_4$. When these reactions were repeated in dichloromethane at 25°C with N-benzyl-N-t-butyldiazoacetamide (eq. 19), essentially complete aromatic cycloaddition was observed with $Rh_2(OAc)_4$, $Rh_2(acam)_4$, and $Rh_2(cap)_4$. However, when the $Rh_2(cap)_4$-catalyzed reaction was

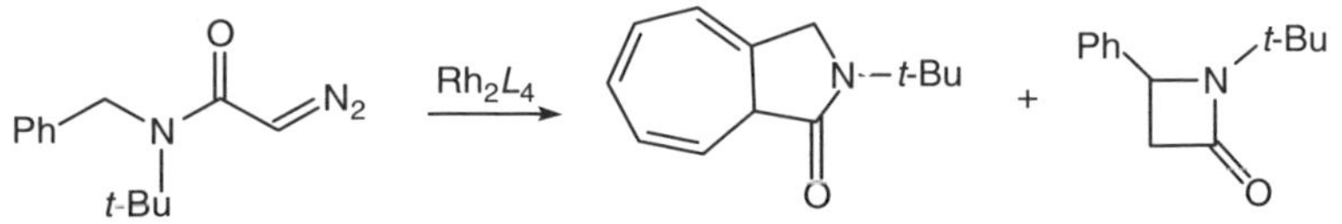

Catalyst	Yield (%)		
Rh$_2$(OAc)$_4$ (25°C)	98	>99	<1
Rh$_2$(cap)$_4$ (25°C)	97	98	2
Rh$_2$(cap)$_4$ (40°C)	99	50	50
Rh$_2$(5S-MEPY)$_4$	96	30	70

(19)

repeated in dichloromethane under reflux (40°C), the products of aromatic cycloaddition and benzylic C–H insertion were now formed in equal amounts. This unprecedented influence of a 15°C temperature change can be attributed to restricted rotation of the phenyl substituent around the benzylic carbon–nitrogen bond.[53] With Rh$_2$(OAc)$_4$ at 40°C aromatic cycloaddition was the sole product. Use of Rh$_2$(5S-MEPY)$_4$ increased the proportion of C–H insertion product, suggesting that chemoselectivities for C–H insertion beyond those obtained with Rh$_2$(cap)$_4$ may be attainable by further fine tuning of the dirhodium(II) environment. Among other examples of competition involving aromatic cycloaddition is that involving the N,N-dibenzyl diazoacetoacetate in Scheme 41. As ex-

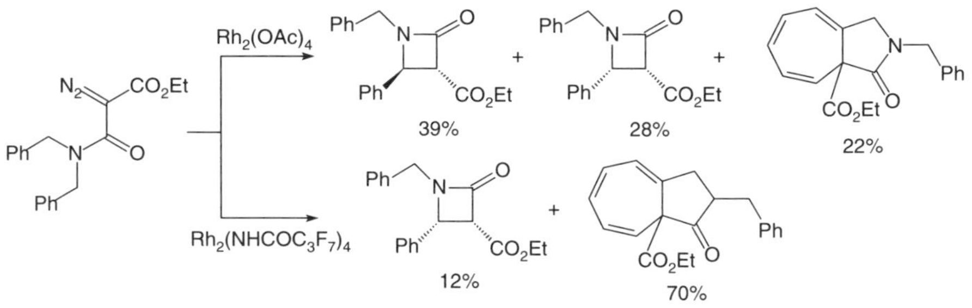

Scheme 41

pected, use of the perfluorobutyramide ligand favored aromatic cycloaddition over C–H insertion, in contrast to Rh$_2$(OAc)$_4$, which promoted the latter process predominantly.

The influence on the carbene carbon atom of substituents other than those on the metal center may also contribute to chemoselectivity. This is well demonstrated by the two sets of reactions in Scheme 42, in which one substrate is a terminal diazoamide (i.e., the diazo carbon carries a hydrogen atom), whereas the second has an acetyl group at that position. With the former and Rh$_2$(OAc)$_4$ only aromatic cycloaddition was observed in the three cases studied. With the same catalyst, the acetyl-substituted substrate furnished only C–H insertion. This contrasting substituent-dependent chemoselectivity has been attributed to conformational influence of the acetyl group, which inhibits approach of the carbenic

Z = H, Br, OMe

Scheme 42

centre to the aromatic ring and thereby shuts down the cycloaddition option.[52] Alternatively, Wee et al.[54] contend that the substituent effect is electronic rather than steric, a creditable argument, which, however, is recognition of the fact that although COMe and CO_2Me substituents on a carbenic carbon atom have very similar electronic influence, they do lead to different products in certain cases. A final example of aromatic cycloaddition in competition with C–H insertion is illustrated by the behavior of the phenylalanine-derived diazoacetoacetamide in Scheme 43, in which the nature of the substituent in the nitrogen atom influences the product distribution.[55]

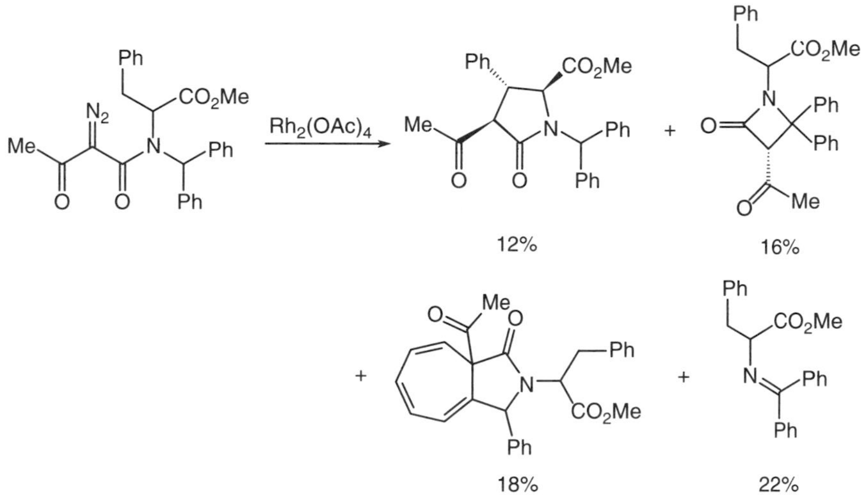

12% 16%

18% 22%

Scheme 43

In summary, these studies suggest that the point has now virtually been reached where it is possible to conduct a specific catalyzed reaction of a diazocarbonyl compound by selecting the dirhodium(II) catalyst whose ligand constitution favors that particular transformation. Either charge or HOMO/LUMO control or both may be operating in these transformations. With certain substrates conformational and/or steric effects of substituents at or near the metal-bound carbon atom undoubtedly play a role in determining the chemoselectivity of product formation.[53]

6.2.4   Asymmetric Synthesis in Aromatic Cycloaddition

It has already been emphasized in Chapters 3, 4 and 5 that there are numerous opportunities for asymmetric synthesis employing diazocarbonyl substrates, and although notable success has been achieved in enantioselective cyclopropanation and intramolecular C–H insertion through the use of chiral catalysts, applications of asymmetric catalysis to other diazocarbonyl reactions are still very much at the exploratory stage. Nevertheless, there are already encouraging indications in aromatic cycloaddition and substitution that catalyst development will lead to synthetically useful levels of stereocontrol in these areas.

The first example of asymmetric synthesis in aromatic cycloaddition was observed in 1990 by McKervey and his co-workers[56] in the decomposition of 2-diazo-5-phenylpentan-2-one to the dihydroazulenone shown in Scheme 44.

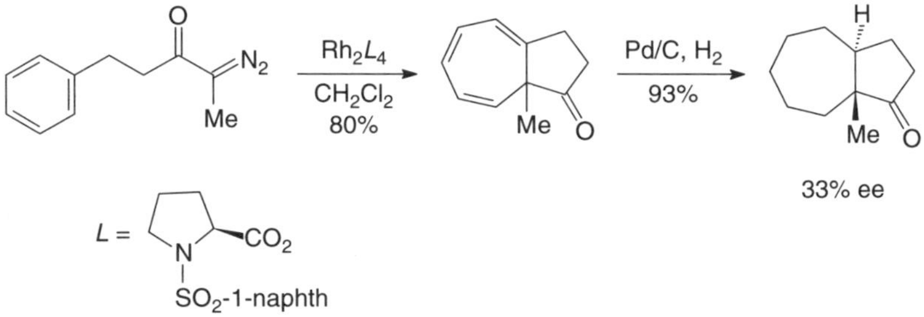

Scheme 44

Several rhodium(II) carboxylates were catalytically active in this reaction with the N-1-naphthalene-sulfonyl-L-prolinate providing the highest enantiomeric excess (33% ee). Hydrogenation of the azulenone over palladium on carbon afforded the fully saturated *trans*-bicyclo[5.3.0]decanone in 93% yield, thus completing a short partial asymmetric synthesis of this compound from a readily accessible diazoketone. Asymmetric intramolecular cycloaddition has also been observed with the biphenyl-derived diazoketone in eq. 20.[57] High-yielding conversion to a tricyclic ketone occurred with several chiral rhodium(II) catalysts,

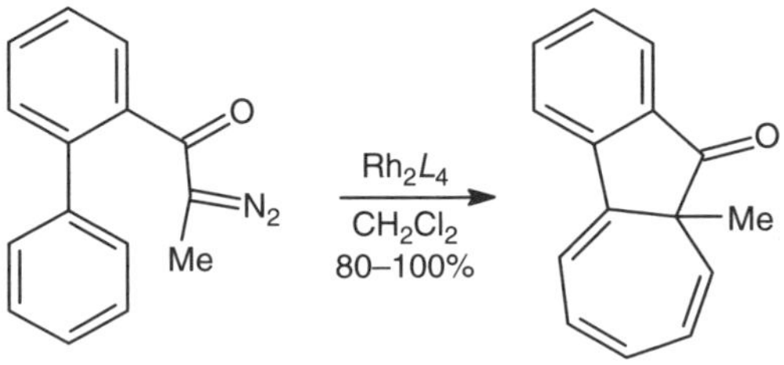

(20)

Catalyst	Temp. (°C)	ee (%)
$Rh_2(5S\text{-MEPY})_4$	0	2
$Rh_2(S\text{-mandelate})_4$	0	8
$Rh_2(L\text{-prolinate})_4$	0	54
$Rh_2(L\text{-prolinate})_4$	−30	79
$Rh_2((+)\text{-phos})_4$	−30	60
$Rh_2(ABO)_4$	−30	71

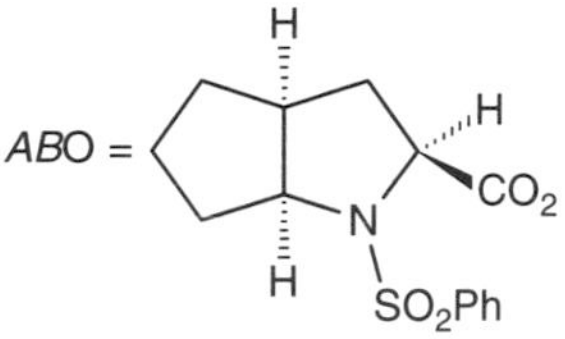

$ABO =$

with enantioselection ranging from very low ee values with $Rh_2(S\text{-MEPY})_4$ and $Rh_2(S\text{-mandelate})_4$ to 79% with $Rh_2(N\text{-benzenesulfonyl-L-prolinate})_4$.

6.3 SUBSTITUTION REACTIONS WITH AROMATIC COMPOUNDS

Regardless of the relative stabilities of cycloheptatriene or norcaradiene intermediates in individual cases, the processes described in the preceding sections can be collectively classified as aromatic cycloaddition of metal carbenes. There is a related intramolecular reaction of aromatic diazocarbonyl compounds, which leads to annulated bicyclic products apparently without the intervention of norcaradiene or cycloheptatriene intermediates. This process has sometimes been described as aromatic C–H insertion, although mechanistically, electrophilic aromatic substitution is probably a more accurate characterization of the reaction pathway. The norcaradiene pathway cannot be completely excluded; it would, however, require the imposition of considerable ring strain in reactions leading to five–six bicyclic systems, the most common form of product. It is probably also significant that there are examples of this reaction for which rhodium(II) carboxylates and protonic acids are catalytically active, suggesting that electrophilic aromatic substitution is a plausible mechanistic alternative to either "aromatic C–H insertion" or "aromatic cycloaddition." Acid-catalyzed cyclization of unsaturated and aromatic diazocarbonyl compounds is discussed in Chapter 11.

The majority of reactions classified here as intramolecular aromatic substitution are of the type illustrated in eq. 21, in which a benzene nucleus is annulated

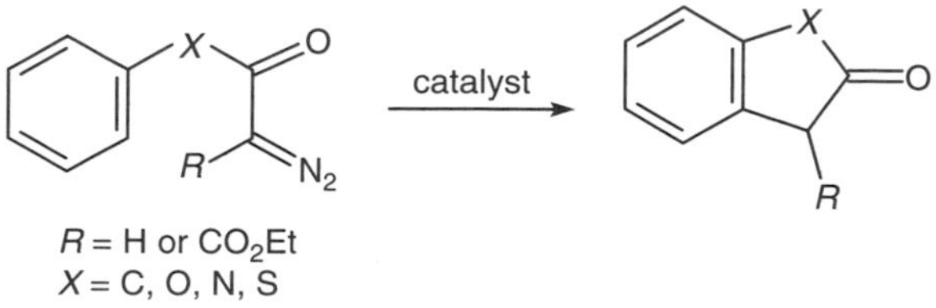

$R = H$ or CO_2Et
$X = C, O, N, S$

(21)

with a five-membered carbocyclic or heterocyclic ring. The two six-membered ring annulations (see Schemes 28 and 38) described by Taylor and Davies[46] and by Saba,[51] respectively, may also belong to this category, although the available evidence does not exclude them from the cycloaddition section.

Durst and his co-workers have studied these annulations extensively using aryl diazoacetoacetates (eq. 22),[58] aryl diazosulfonyl esters (eq. 23)[59,60] and N-aryl diazoacetoacetamides (eq. 24)[61] as precursors with rhodium(II) acetate as the

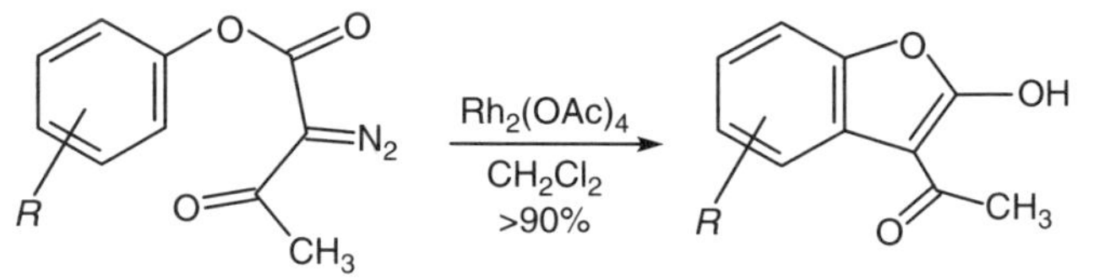

$$\text{(22)}$$

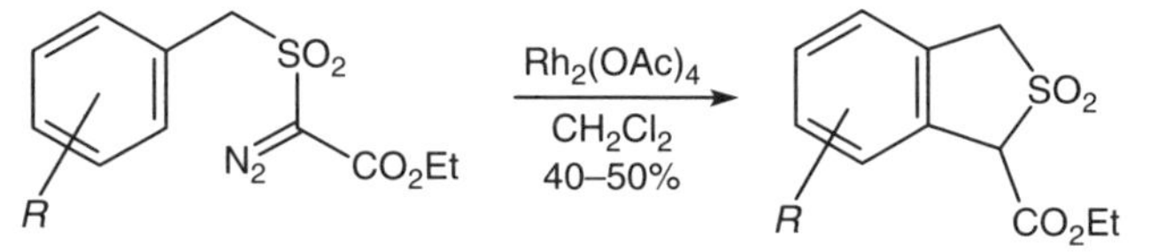

$$\text{(23)}$$

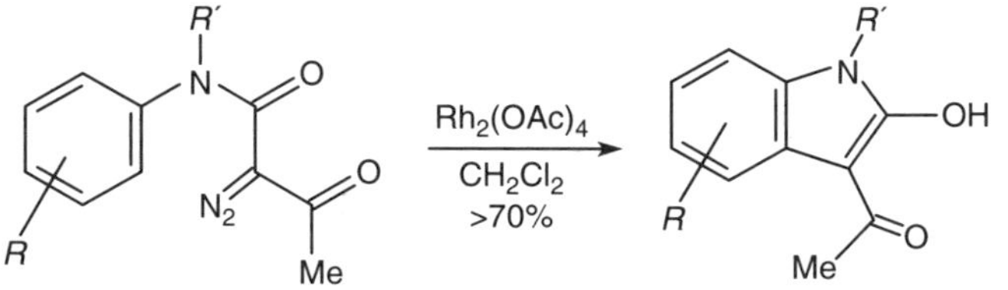

$$\text{(24)}$$

catalyst in dichloromethane. Substituted benzofurans, benzothiophene dioxides and indole derivatives were produced in low to moderate yield. Independently, Doyle and his co-workers[62] studied the cyclisation of N-aryldiazoacetamides and N-aryl diazoacetoacetamides (eq. 25), varying the catalyst, solvent, and the pat-

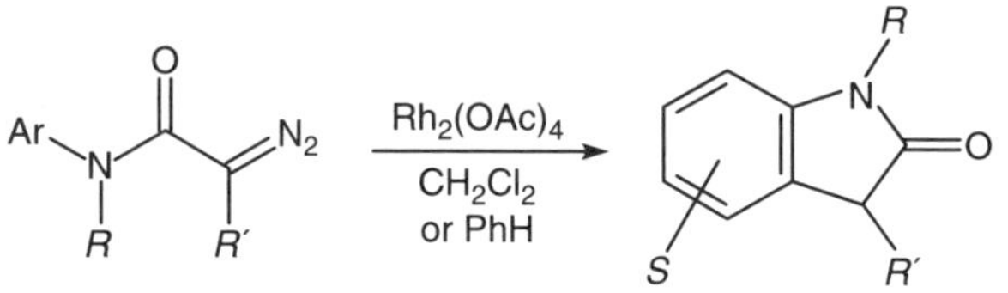

$$\text{(25)}$$

Ar in precursor	R	R'	Yield of indolinones (%)
C_6H_5	Me	H	86
C_6H_5	$PhCH_2$	H	87
$o\text{-}CH_3C_6H_4$	Et	H	86
α-naphthyl	Et	H	98
C_6H_5	Me	$COCH_3$	86
C_6H_5	$PhCH_2$	$COCH_3$	80
$m\text{-}CH_3C_6H_4$	Et	$COCH_3$	84
α-naphthyl	Et	$COCH_3$	70

RHODIUM(II) ACETATE–CATALYZED DECOMPOSITION OF 2-DIAZO-3-OXOBUTANAMIDES[62]

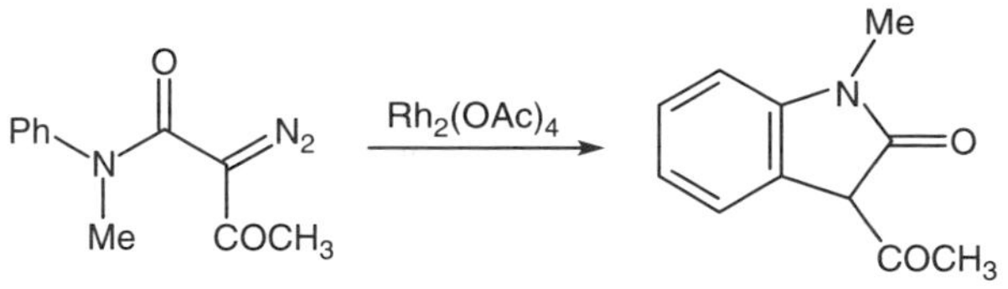

A solution containing 1.0 mmol of the 2-diazo-3-oxobutanamide in 5 mL of dry benzene was added dropwise over 30 min to a mixture of 4 mg of rhodium(II) acetate (1.0 mol percent) in 10 mL of refluxing benzene. The resulting solution was heated at reflux under nitrogen for 3 h. After cooling the solution was filtered through neutral alumina, and benzene was removed under reduced pressure. The crude product was purified by column chromatography on neutral alumina with 5:1 hexane/ether eluent (yield 92%).

^{1}H-NMR (CDCl$_3$, 100 MHz): δ 7.37 (*d*, *J* = 7.6 Hz, 1H), 7.23 (*d* of *t*, *J* = 7.6, 1.2 Hz, 1H), 7.11 (*d* of *t*, *J* = 7.6, 1.1 Hz, 1H) 6.95 (*d*, *J* = 7.6 Hz, 1H), 3.35 (*s*, CH$_3$N) and 2.46 (CH$_3$CO). ^{13}C-NMR (CDCl$_3$, 300 MHz): δ 172.8, 171.0, 138.9, 125.2, 122.2, 122.0, 119.6, 108.3, 101.7, 25.6, and 20.2. IR (neat): 3436 (O–H) and 1661 (C=O) cm^{-1}.

---

tern of substituents in the substrate in an effort to optimize the conditions for efficient cyclization. Whereas the *N*-aryl diazoacetamides were rapidly decomposed by rhodium(II) acetate in dichloromethane at room temperature, the diazoacetoacetamides were stable under these conditions but did decompose readily in refluxing benzene. A selection of the results of these studies are summarized in eq. 25. Clearly, this is an exceptionally efficient route to a wide variety of indolinones. *N*-Methyl, *N*-ethyl, and *N*-benzyl amides were compatible with these transformations. Neither the reactants nor products inhibited the catalytic activity of rhodium(II) acetate, nor were there any byproducts resulting from C–H insertion or from intermolecular cycloaddition to the solvent when the reaction was conducted in benzene.

A study of the effect of substituents on the aromatic nucleus on the regioselectivity of these reactions revealed that: (1) a *meta*-methyl group exerts virtually no effect in the diazoacetamide series (eq. 26) with a modest preference for its *para*

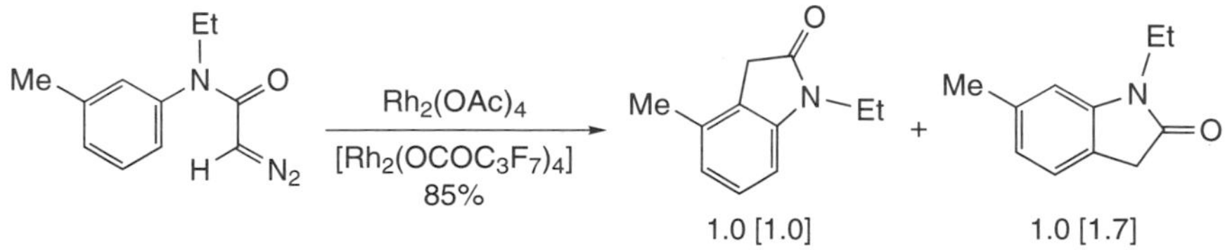

1.0 [1.0]   1.0 [1.7]

(26)

position in the diazoacetoacetamide series; (2) a 1-naphthyl group substitutes exclusively at the β position (eq. 27); a *meta*-methoxy substituent directs substitution at its *para* position exclusively (eq. 28); (3) a 3,4-methylenedioxy group in the diazoacetamide series shows exceptional selectivity for the 6 position (eq. 29)

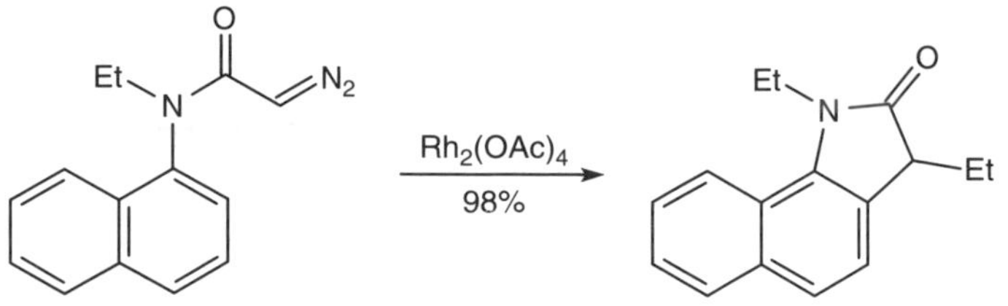

$$\text{Rh}_2(\text{OAc})_4 \quad 98\% \tag{27}$$

$$\text{Rh}_2(\text{OAc})_4 \quad 100\% \tag{28}$$

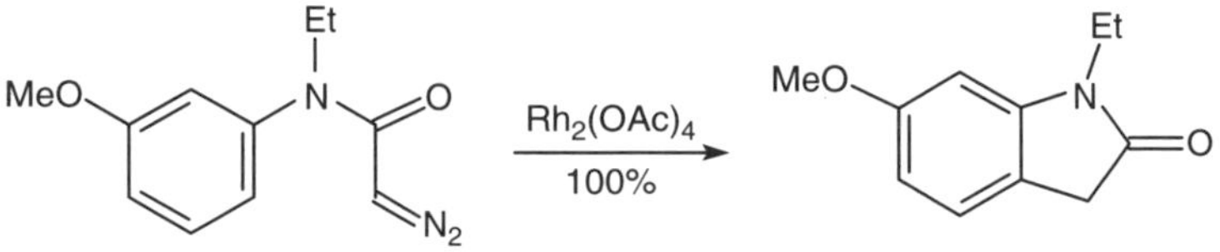

$$\text{Rh}_2(\text{OAc})_4 \quad 90\% \tag{29}$$

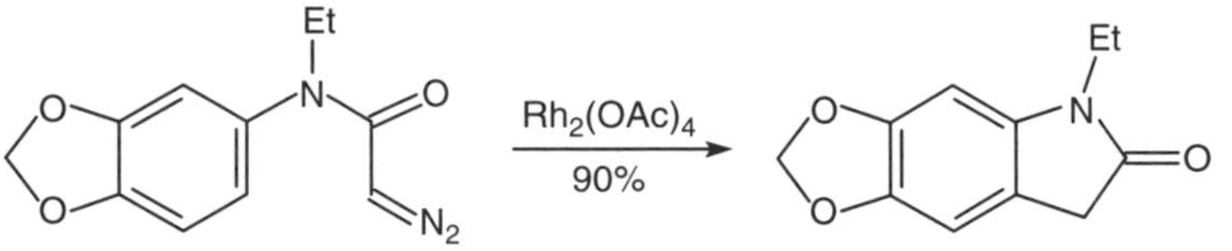

and in the diazoacetoacetamide series for this position also but not exclusively; so clearly the electronic effects of strongly donating substituents are important factors in product selectivity. Furthermore, changing the catalyst from $\text{Rh}_2(\text{OAc})_4$ to $\text{Rh}_2(\text{OCOC}_3\text{F}_7)_4$ or $\text{Rh}_2(\text{NHCOCH}_3)_4$ produces a change in the product ratio in the *meta*-methyl series (eq. 26), which is consistent with a mechanism involving a rhodium carbene intermediate in the electrophilic substitution step (Scheme 45).

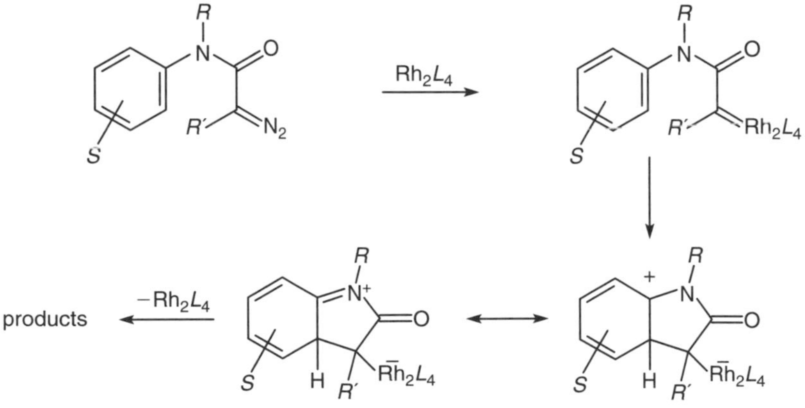

Scheme 45

To explore further the synthetic potential of this method for intramolecular aromatic substitution of diazoamides, Doyle's group compared the catalytic power of rhodium(II) acetate with that of boron trifluoride etherate and Nafion-H, the DuPont patented perfluorinated ion-exchange resin. Boron trifluoride proved to be a much inferior substitute for rhodium(II) acetate in molar equivalents required, reaction selectivity and product yields. In contrast, Nafion-H was an exceptionally suitable alternative for aromatic substitution in the diazoacetamide series. Although the reactions were slow at room temperature, they proceeded smoothly in refluxing chloroform containing catalytic quantities of the resin. Product yields were comparable or superior to those obtained using rhodium(II) acetate, and product isolation was simpler. Diazoacetoacetamides, on the other hand, were generally stable to Nafion-H in refluxing chloroform but did undergo aromatic substitution in toluene at 90°C. However, uncatalyzed decomposition occurred at this temperature and at a comparable rate, indicating that the influence of Nafion-H on the rate of diazoacetoacetamides is marginal. This may be a steric phenomenon whereby the diazoacetoacetamide moiety is denied access to the protonic sites on the resin backbone, whereas the sterically less demanding diazoacetamides are not. However, the relative reactivity of the two in electrophilic substitution may also be a factor. The general conclusion of this study is that whereas Nafion-H is the catalyst of choice for electrophilic aromatic substitution reactions of *N*-aryldiazoacetamides, rhodium(II) acetate offers greater regioselectivity in reactions where two indolinone products are formed. In a related, more recent, investigation of the use of Nafion-H in diazocarbonyl reactions, Wee and Liu[63] investigated the cyclization of α-carbomethoxy-diazoacetamides. In terms of relative reactivity, these substrates should be comparable to the diazoacetoacetamides rather than the diazoacetamides discussed previously, and this was in fact confirmed by the results. Cyclization was slow at 92°C in toluene and the product selectivity was poor. For example, thermal decomposition of the *p*-methoxy diazo substrate in Scheme 46 at 92°C furnished the indolinone (49%),

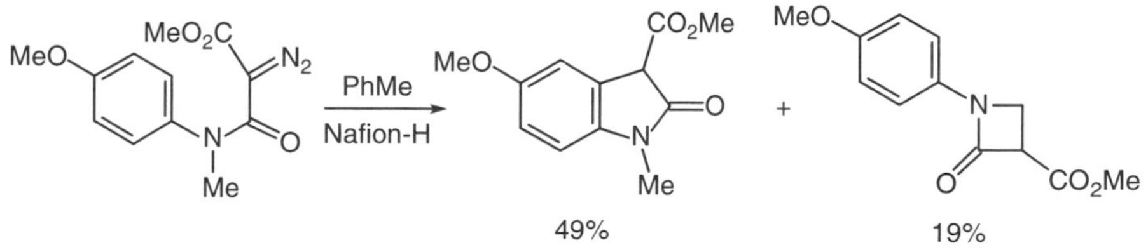

49%   19%

Scheme 46

the 2-azetidinone (19%), and starting material (25%). Repetition of the reaction at 110°C, again in the absence of Nafion-H, led to a decrease in the amount of aromatic substitution and an increase in the amount of C–H insertion leading to the azetidinone. In the presence of Nafion-H there was little improvement in yields or selectivity. Yet another protonic acid that has been employed to catalyze aromatic substitution in diazoacetamides is trifluoroacetic acid. Rishton and

Schwartz[64] found that this acid afforded good yields of 1,4-dihydroisoquinolines (eq. 30) and 1,4,5-trihydro-3-benzazepin-2-ones (eq. 31) from *N*-benzyl- and *N*-phenylethyldiazoacetamides, respectively.

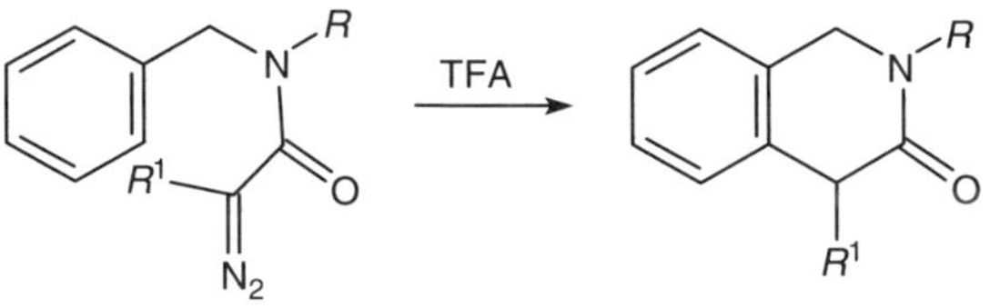

$$(30)$$

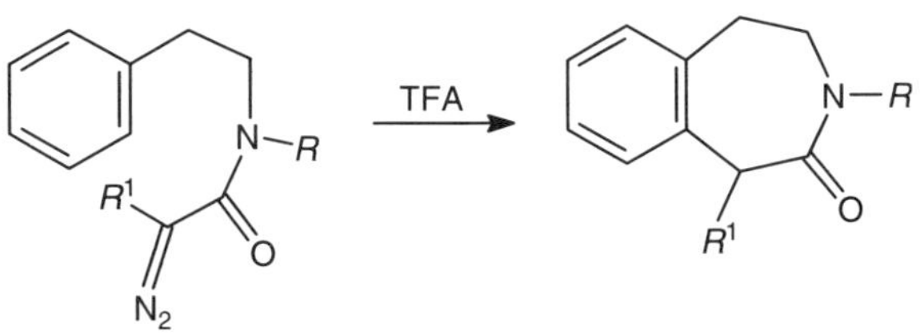

$$(31)$$

Intramolecular aromatic substitution of diazocarbonyl compounds has also been used to generate carbocyclic systems, though much less extensively than the heterocyclic systems described previously. Two examples from the work of Taber and Ruckle[65] and Nakatani[66] that illustrate the formation of indanone derivatives are shown in eqs. 32 and 33, respectively. The catalyst was rhodium(II) acetate in both reactions.

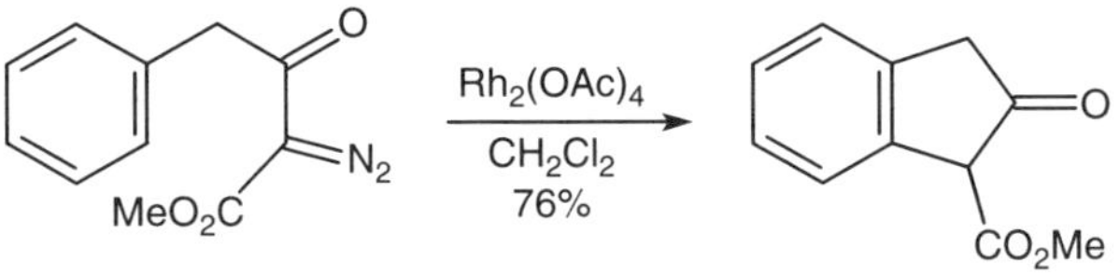

$$(32)$$

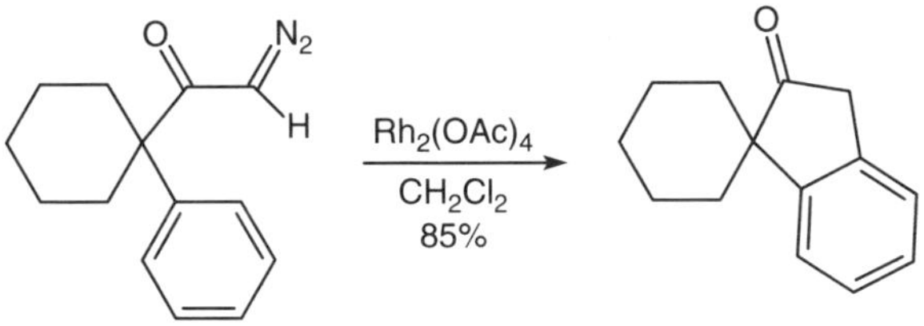

$$(33)$$

6.3.1 Chemoselectivity in Aromatic Substitution

In the preceding section on aromatic substitution in diazoamides the substrates were such that chemoselectivity was not an issue inasmuch as almost all cyclizations leading to indolinones were high-yielding reactions with little or no byproducts. In the carbocyclic series, however, there are examples where control of chemoselectivity is an important factor if the objective is to have synthetically

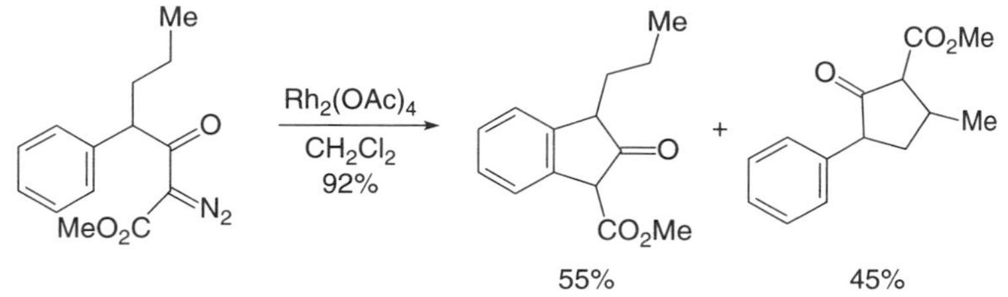

$$(34)$$

55%   45%

useful processes. The two cyclizations in eqs. 32 and 34 from the work of Taber and Ruckle[65] provide useful illustrative examples. In the former, rhodium(II) acetate was an effective catalyst for the conversion of an acyclic diazoketo ester into methyl 2-indanone-1-carboxylate. Here, chemoselectivity for aromatic substitution is complete; the possible alternatives of aromatic cycloaddition or Wolff rearrangement were unlikely to be competitive. However, in eq. 34 there is another possible reaction site, an *n*-propyl side chain within reach of the rhodium carbene, resulting in the formation of significant amounts of a C–H insertion product with a concomitant reduction in chemoselectivity for aromatic substitution. In contrast, the reaction in eq. 33 reported by Nakatani,[66] again with rhodium(II) acetate catalysis, led exclusively to the product of aromatic substitution with no competition from C–H insertion into the neighboring cyclohexane ring.

Several research groups have selected a series of rhodium(II) carboxylates with which to decompose individual diazocarbonyl substrates for the purpose of probing chemoselectivity in aromatic substitution in competition with other reaction pathways, specifically C–H insertion or cyclopropanation. Ikegami and co-workers[67] selected nine rhodium(II) catalysts and screened them for chemoselectivity in the cyclisation of the diazoketo ester in eq. 35. In all nine re-

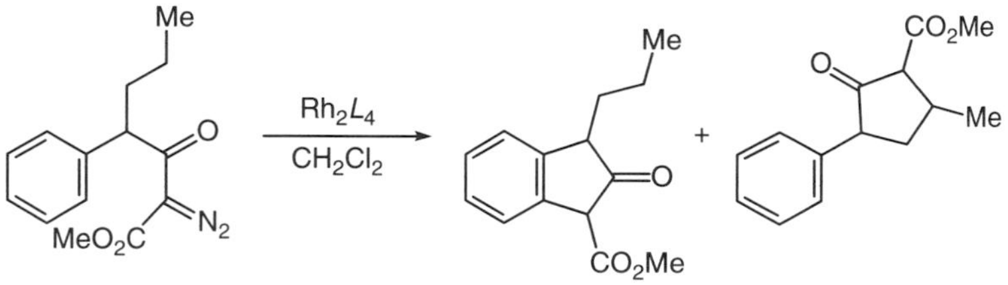

$$(35)$$

Entry	Ligand	Yield (%)		
1	OAc	92	54	46
2	OCOCF$_3$	74	79	21
3	OCOPh	84	76	24
4	OCOCMe$_3$	81	56	44
5	OCO-1-ad^a	86	54	46
6	NHCOMe	71	71	29
7	OCOCHPh$_2$	90	85	15
8	OCOCMePh$_2$	84	96	4
9	OCOCPh$_3$	92	96	4

^{a}1-ad = 1-adamantyl.

actions there was a preference for the product of aromatic substitution over that of C–H insertion. The product ratios also clearly demonstrate that the chemoselectivity is strongly influenced by choice of catalyst with rhodium(II) diphenylpropionate and the corresponding triphenylacetate essentially "switching off" the C–H insertion pathway while maintaining excellent yields of indanone from the aromatic substitution pathway (entries 8 and 9). The fact that rhodium(II) diphenylacetate (entry 7), presumed to have similar electronic properties, was less selective than either the diphenylpropionate or the triphenylacetate was interpreted as evidence that the steric bulk of the ligand on rhodium might retard the methylene C–H insertion in favor of aromatic substitution. This effect was more pronounced when a methine C–H bond was put in competition with aromatic substitution by changing the diazoketo ester to that in eq. 36. Cyclization cata-

(36)

R	L	Yield (%)	Product distribution	
H	tpa	78	>99	<1
F	OAc	94	<1	>99
F	tpa	75	>99	<1

lyzed by rhodium(II) triphenylacetate gave >99% of aromatic substitution. Since methine insertion is regarded as electronically favored but sterically disfavored when compared with methylene insertion, this behavior can be taken as an indication that steric shielding by bulky ligands on the metal contributes more to chemoselectivity than electronic effects, with C–H insertion more prone to steric retardation than aromatic substitution. A particularly effective demonstration of "on–off switching" is revealed by the p-fluoro-diazoketo ester substrate in eq. 36. With rhodium(II) acetate as catalyst, >99% of the reaction pathway is C–H insertion; with rhodium(II) triphenylacetate the outcome is >99% aromatic substitution. Similar effects can be found with simple diazoketones. For example, the cyclizations in eq. 37 can be brought about by rhodium(II) acetate with the outcome for $R = H$ in favor of indanone formation (65 vs. 35%).

With rhodium(II) triphenylacetate as catalyst and $R = H$ or Me, the selectivity for aromatic substitution is virtually complete. Doyle and Padwa and their respective collaborators[53] have also probed aromatic substitution versus C–H insertion using the substrate with $R = $ Me in eq. 37 but with different catalysts. Rhodium(II) perfluorobutyrate catalyzed indanone formation exclusively in 96% yield. This result, combined with other competitive studies, identified the order of reactivity of $Rh_2(OCOC_3F_7)_4$-generated metal carbenes as aromatic substitu-

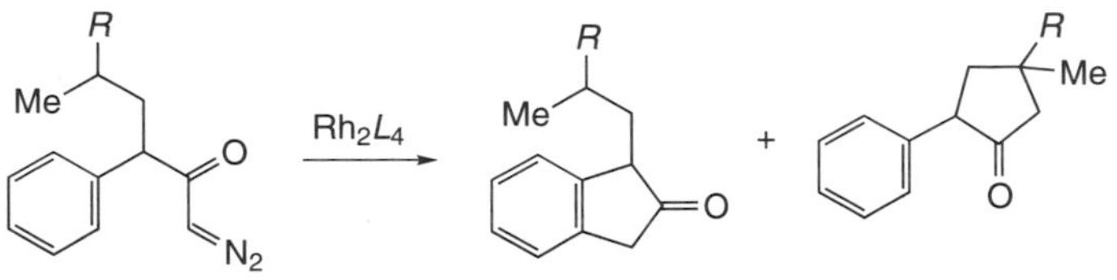

$$\tag{37}$$

R	L	Yield (%)	Product distribution	
H	OAc	78, 96	65	35
H	OCOCPh$_3$	80	>99	<1
Me	OCOCPh$_3$	84	>99	<1
Me	OCOC$_3$F$_7$	96	100	0
Me	cap	64	59	41

tion > tertiary C–H insertion > cyclopropanation. Rhodium(II) caprolacta-mate, on the other hand, revealed little choice between aromatic substitution and C–H insertion (59 vs. 41). The fact that rhodium(II) triphenylacetate and rhodium(II) perfluorobutyrate are about equally efficient in catalyzing the aro-matic substitution in eq. 37 suggest that the steric argument advanced in favor of the former catalyst is not the only determinant in chemoselectivity control.

The question of chemoselectivity in aromatic substitution versus cyclopropa-nation is equally important in the development of synthetically useful diazocar-bonyl transformations.[53] Here also there are strong indications that changing the ligands on the rhodium(II) core is a very effective way of switching from one pathway to the other. This is well illustrated by the behavior of the diazoketones in eq. 38, in which there are two potentially reactive centers available.

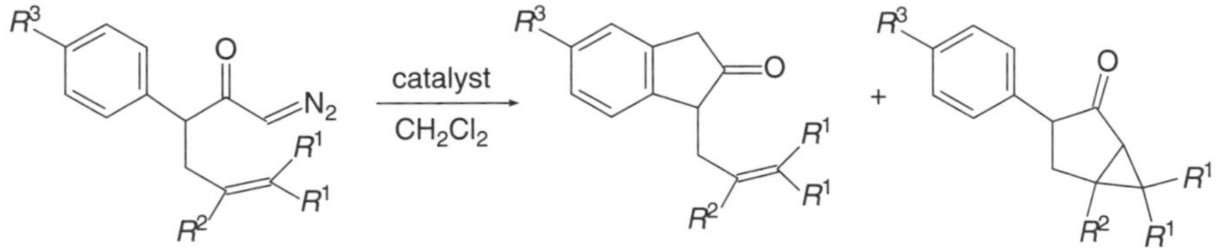

$$\tag{38}$$

	Catalyst	Yield (%)	Indanone	Cyclopropane
$R^1 = R^2 = R^3 = H$	Rh$_2$(OAc)$_4$	99	52, 63	46, 26
	Rh$_2$(pfb)$_4$	95	100	0
	Rh$_2$(cap)$_4$	72	0	100
	Rh$_2$(tpa)$_4$	83	100	0
$R^1 = R^3 = H; R^2 = Me$	Rh$_2$(cap)$_4$	75	0	100
$R^1 = Me; R^2 = R^3 = H$	Rh$_2$(pfb)$_4$	82	100	0
$R^1 = H; R^2 = Me; R^3 = OMe$	Rh$_2$(cap)$_4$	80	0	100

With rhodium(II) acetate as catalyst there is a 2:1 preference for aromatic substitution over cyclopropanation. In contrast, rhodium(II) perfluorobutyrate and rhodium(II) triphenylacetate catalyzed aromatic substitution exclusively, whereas with rhodium(II) caprolactamate there was a complete switch in favor of cyclopropanation. This extraordinary selectivity control is virtually independent of the substituents on the double bond or on the aromatic ring. Although there are minor variations in product distributions from the rhodium(II) acetate–catalyzed reactions, for the four substrates examined, rhodium(II) perfluorobutyrate produced the aromatic substitution products exclusively and rhodium(II) caprolactamate produced cyclopropanes exclusively. While the chemoselectivity associated with rhodium(II) triphenylacetate could be attributed to the steric bulk of the ligand, the preference for aromatic substitution is more likely to be electronic in origin. The large difference in chemoselectivity between the perfluorobutyrate and caprolactamate catalysts can be attributed to the large difference in charge and frontier orbital properties, while the intermediate behavior of rhodium(II) acetate may result from its contrasting charge or frontier orbital properties, where experimental results serve as an indicator of the relative charge and frontier orbital control. Among other studies relevant to the development of new catalysts for high levels of chemoselectivity control are those of Lahuerta and co-workers,[68] who screened a series of nine rhodium(II) complexes $Rh_2(PC)_2(OCOR)_2$ (see Chapter 5), where PC = a metallated phosphine and $R = CH_3$ or C_3F_7, for their efficacy as catalysts for decomposition of the diazoketone in eq. 38 ($R^1 = R^2 = R^3$). Without exception, cyclopropanation was the favored pathway to the exclusion of aromatic substitution with yields in the range 93–99%.

6.3.2 Asymmetric Synthesis in Aromatic Substitution

Intramolecular aromatic substitution is yet another reaction of diazocarbonyl compounds where there is great potential for asymmetric synthesis through the use of chiral catalysts, and although examples are still not numerous, there is evidence already of some singularly successful applications. Hashimoto and co-workers[69] used enantiopure N-protected amino acids as chiral auxiliaries in rhodium(II)-catalyzed aromatic substitutions of the type shown in eq. 39. These particular reactions are in effect examples of enantioselection between two enantiotopic benzene rings undergoing substitution by a metal carbene. Rhodium(II) N-phthaloyl-L-phenylalanate [$Rh_2(S$-PTPA$)_4$] proved to be particularly efficacious, affording indanones with ee values up to 95% depending on the substitution pattern of the diazocarbonyl precursor. The best combination was one of two unsubstituted benzene rings and an ethyl side chain on the carbon atom alpha to the carbonyl group. With an n-propyl side chain the ee value fell to 30%. Experiments with the corresponding diazoacetoacetates (entries 6 and 7) revealed that the nature of the alkyl residue in the ester also had a profound effect on the degree of enantioselection. The authors used steric arguments augmented by the X-ray diffraction structure of the catalyst and the intervention of a rhodium car-

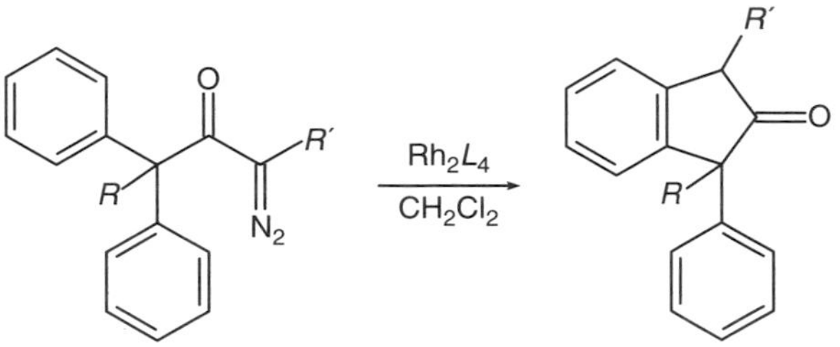

(39)

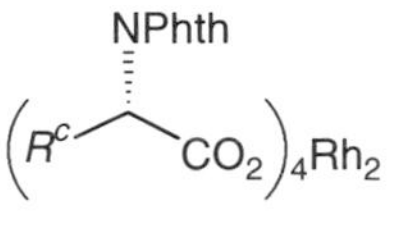

R^c = PhCH$_2$: Rh$_2$[(S)-PTPA]$_4$
R^c = Me : Rh$_2$[(S)-PTA]$_4$
R^c = i-Pr : Rh$_2$[(S)-PTV]$_4$
R^c = t-Bu : Rh$_2$[(S)-PTTL]$_4$

Entry	R	R'	L	Temp. (°C)	Yield (%)	Indanone % ee
1	Me	H	(S)-PTPA	−20	64	77
2	Me	H	(S)-PTPA	0	67	65
3	Et	H	(S)-PTPA	−20	86	95
4	n-Pr	H	(S)-PTPA	−10	57	33
5	allyl	H	(S)-PTPA	−10	41	55
6	Me	CO$_2$CH[CH(CH$_3$)$_2$]	(S)-PTPA	26	85	6
7	Me	CO$_2$Me	(S)-PTPA	0	97	74

R	R'	Rh$_2$[(S)-PTA]$_4$			Rh$_2$[(S)-PTV]$_4$			Rh$_2$[(S)-PTTL]$_4$		
		Temp.	Yield %	ee %	Temp.	Yield %	ee %	Temp.	Yield %	ee %
Et	H	−10	84	69	−20	77	65	−20	84	90
n-Pr	H	−20	85	92	−20	91	74	−20	74	98
Me	H	−10	75	53	10	72	41	10	75	88
allyl	H	10	81	60	0	65	48	0	70	88

bene partially surrounded by the *N*-phthaloyl walls of the ligand to predict the formation of 1-alkyl-1-phenyl-2-indanones with the *S* configuration.

Recognizing that while these experiments did provide a highly efficient route to a chiral center at a quaternary carbon atom, they were very limited in scope, Hashimoto and co-workers[70] extended the catalyst range to include other amino acid auxiliaries such as L-alanine, L-valine, and L-*tert*-leucine, all with N-phthaloyl protection. Four terminal diazoketones with methyl, ethyl, *n*-propyl, or allyl side chains were screened for enantioselective aromatic substitution with four amino acid–based rhodium(II) catalysts. All four catalysts were active, producing enantioselectivities in the range 33–98%. The consistently best performance was that of rhodium(II) *N*-phthaloyl-(*S*)-*tert*-leucinate, which produced ee values of 88–98% (eq. 39 and the table). With a highly enantioselective catalytic route to (*S*)-1-methyl-1-phenyl-2-indanone in hand, Hashimoto was to exploit this chemistry in the synthesis of (*S*)-1-methyl-1-phenyl-1,2,3,4-tetrahydroiso-

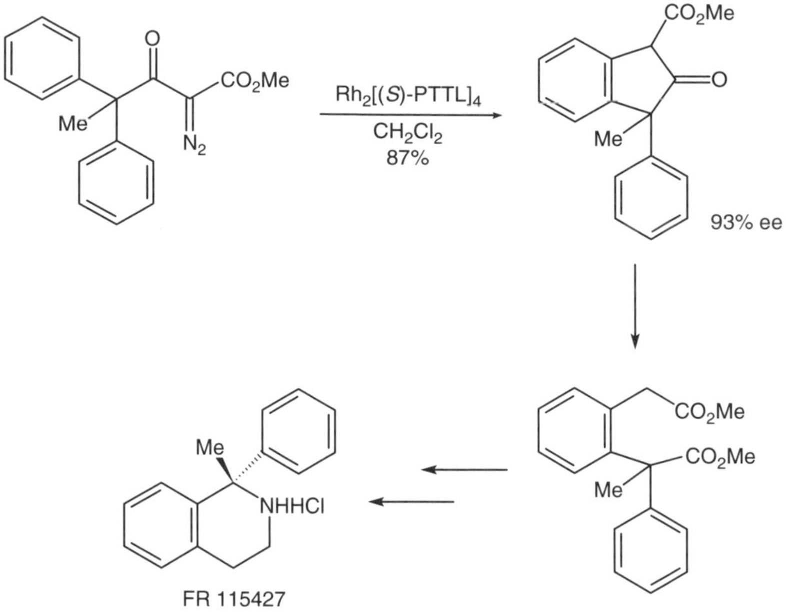

Scheme 47

quinoline hydrochloride, the aspartate receptor antagonist FR 115427[70] (Scheme 47).

6.4 CYCLOADDITION AND SUBSTITUTION REACTIONS WITH HETEROCYCLIC AROMATIC COMPOUNDS

Among other aromatics whose cycloaddition and substitution reactions with diazocarbonyl compounds have found applications in synthesis are the heterocycles furan, pyrrole, and thiophene. These substrates can also act as acceptors for carbenes acting as 1,3-dipoles, an aspect of their chemistry that is addressed in Chapter 12. Here, as with the benzenoid substrates discussed in the preceding section, the emphasis is on metal-catalyzed cycloaddition and metal-catalyzed electrophilic substitution; processes in which cyclopropanes are implicated as intermediates, *en route* to more stable, ring-opened products, are also included.

6.4.1 Furans

Novác and Šorm[71] were among the first to attempt the cyclopropanation of furan by EDA with copper catalysis. The principal product was not the cyclopropane but the *Z,E*-diene presumed to have been formed from it *via* an oxygen-assisted unravelling process of the type shown in Scheme 48; adventitious acid or indeed the copper catalyst may have facilitated ring opening. The furan–EDA cycload-

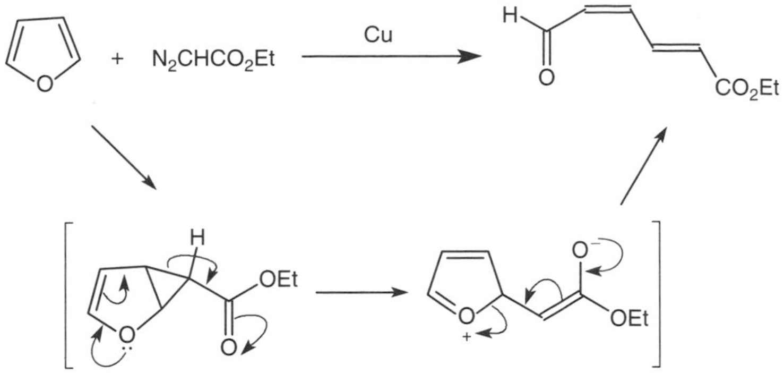

Scheme 48

duct has been generated photochemically and isolated, and shown to undergo acid-catalyzed isomerization to the *Z,E*-diene.[72] The fact that this diene can be further isomerized *in situ* to the *E,E*-isomer enhances the usefulness of the process. Most recent applications involve rhodium(II) rather than copper catalysts, under which conditions the initial cyclopropane adducts are often isolable. Wenkert *et al.*[73] used various combinations of diazoester/diazoketone with furan and substituted furans in dichloromethane or in neat furan to establish the product distribution patterns shown in Scheme 49. Rhodium(II) acetate was the catalyst in all reactions. The furan–ethyl diazoacetate combination furnished a mixture of the cyclopropane adduct, two isomers of the ring opened diene, and a single isomer of a methylene dihydrofuran of undetermined geometry. A mecha-

Products of Rhodium(II) Acetate–Catalyzed Reaction of Furans with EDA

Furan	Intermediate products (%)	Final products (%)

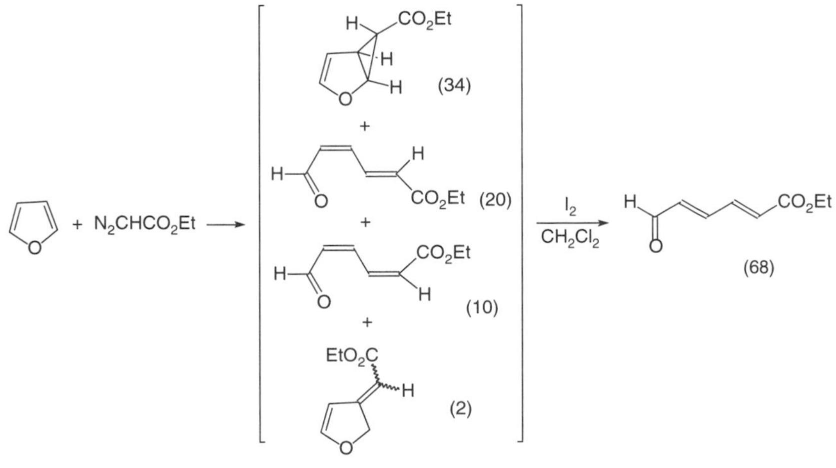

Scheme 49

Products of Rhodium(II) Acetate–Catalyzed Reaction of Furans with EDA

Furan	Intermediate products (%)	Final products (%)

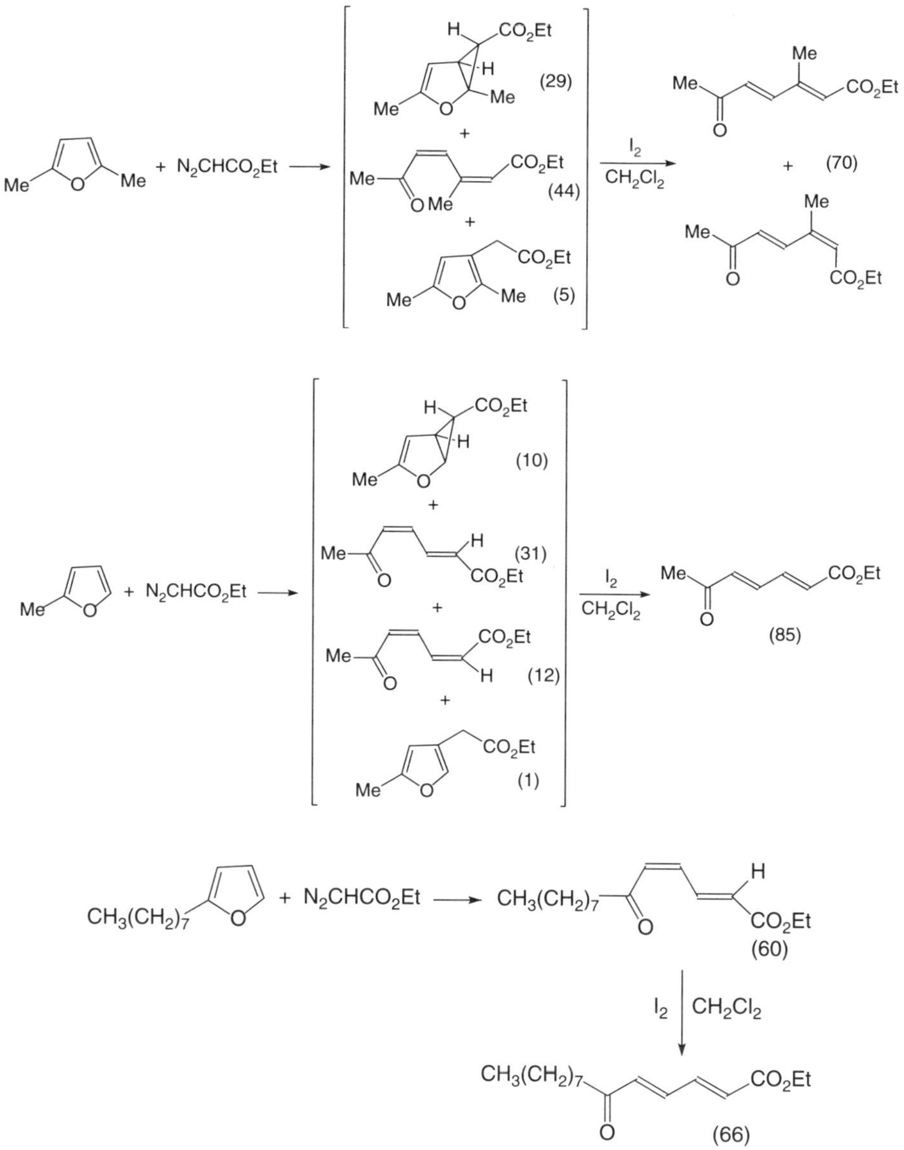

Scheme 49 (*continued*)

nistic interpretation of the origin of these products is outlined in Scheme 50, the key feature being addition of the strongly polarized olefinic bond of furan to the rhodium-complexed carbene. Of the two nonequivalent orientations for this cyclo-

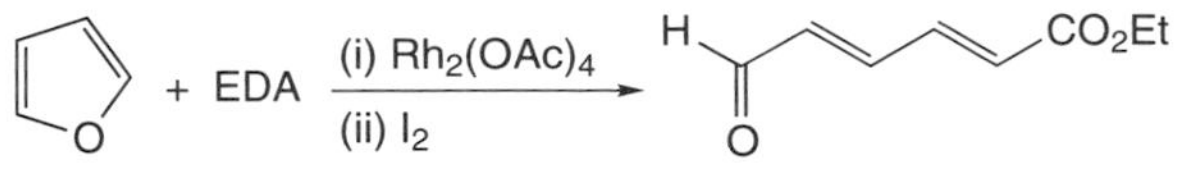

REACTION OF EDA WITH FURANS[73]

A solution of 20 mmol of EDA in 2 mL of the furan was added dropwise over a 18-h period to a stirring suspension of 0.01 mmol of dirhodium tetraacetate in 8 mL of furan. The mixture was filtered through a Florisil pad and the filtrate was evaporated under vacuum.

A solution of the crude product mixture of the furan–diazoester reaction and two iodine crystals in 15 mL of dry methylene chloride was kept at room temperature for 6 h and then evaporated. An ether solution of the residue was washed with 10% sodium thiosulfate solution and with brine, dried, and evaporated. MPLC of the residue derived from furan and elution with 20:1 hexane–ethyl acetate afforded a pale yellow liquid ethyl 6-oxo-2(E),4(E)-hexadienoate: 68% IR $C=O$ 1715 (s), 1685 (s), $C=C$ 1635 (w), 1600 (m) cm^{-1}; ^{1}H-NMR δ 1.31 (t, $J = 7$ Hz, Me), 4.19 (q, $J = 7$ Hz, OCH_2), 6.23 (d, $J = 15$ Hz, H-2), 6.38 (dd, $J = 15$, 8 Hz, H-5), 7.14 (dd, $J = 15$, 12 Hz, H-3), 7.40 (dd, $J = 15$, 12 Hz, H-4), 9.62 (d, $J = 8$ Hz, H-6; ^{13}C-NMR δ 14.0, 60.9, 129.8, 136.8, 140.1, 147.1, 165.3, 192.8; exact mass m/z 154.0619 (calcd. for $C_8H_{10}O_3$ 154.0630).

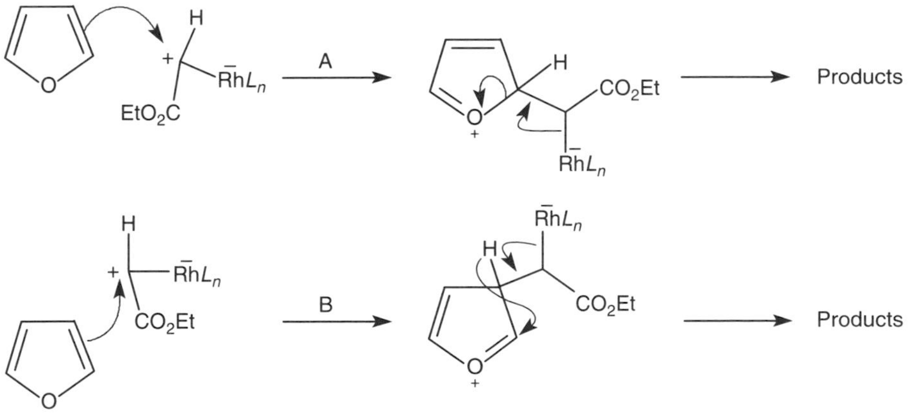

Scheme 50

addition, that in *A* is the favored for electronic reasons, leading to cyclopropane and dienes that are the preponderant reaction products. The behavior of 2,5-dimethylfuran (Scheme 49) was very similar to that of furan itself. 2-Methylfuran, with unequal reaction sites, displayed a 19:1 selectivity with ethyl diazoacetate in favor of the less substituted olefinic bond; otherwise, the product distribution was also very similar to that obtained with furan. Notwithstanding the obvious complexity of these reactions, when coupled with the simplifying acid-catalyzed isomerization/rearrangement described previously, they do constitute a highly efficient synthesis of functionalized (*E,E*)-butadienes from furans. In practice, exposure of the crude product mixtures of the reactions of furan, 2,5-dimethylfuran, 2-methylfuran and 2-*n*-octylfuran in dichloromethane solution to iodine led to the *E,E*-butadienes directly (Scheme 49).

Electron-withdrawing substituents on the furan do not inhibit the cycloaddition reaction, but there is a diminished tendency for the cyclopropane adduct to unravel. Thus treatment of methyl furoate and methyl β-(α-furyl)acrylate with ethyl diazoacetate and rhodium(II) acetate in dichloromethane furnished the cyclopropanes in Schemes 51 and 52, respectively. Whereas the latter cycloadduct

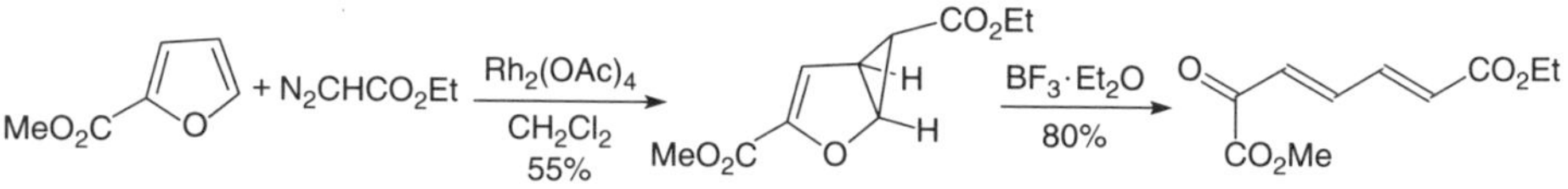

Scheme 51

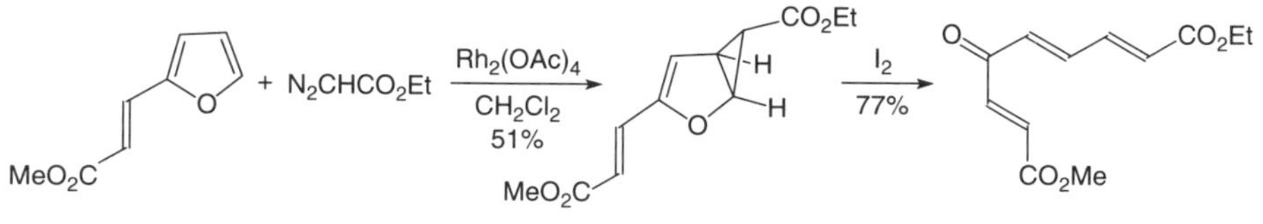

Scheme 52

could be easily unraveled with iodine to afford the triene in 77% yield, the former was inert to iodine but did respond to boron trifluoride to afford the *E,E*-diene.

Several groups have recognized the potential of the furan-diazoester unravelling process in the synthesis of polyunsaturated natural products. Rokach *et al.*[74] provided an early example with their synthesis of a leukotriene (Scheme 53) from

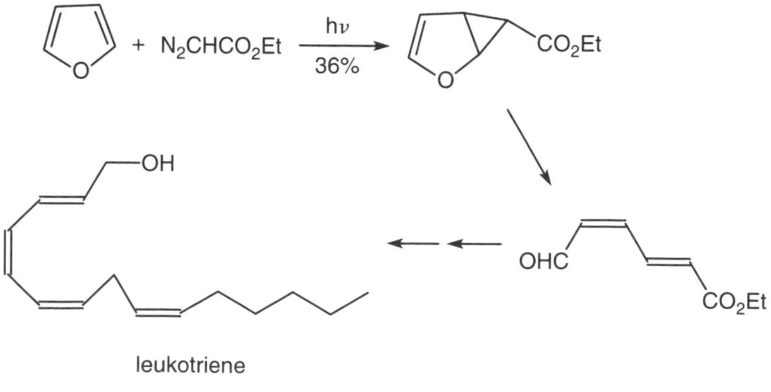

leukotriene

Scheme 53

the *E,E*-diene. In a similar manner, Wenkert *et al.*[73] recognized the connection between the adducts of furan and ethyl α-diazopropionate and fragments suitable for the construction of β-carotene (Scheme 54) and retinol (Scheme 55). In prac-

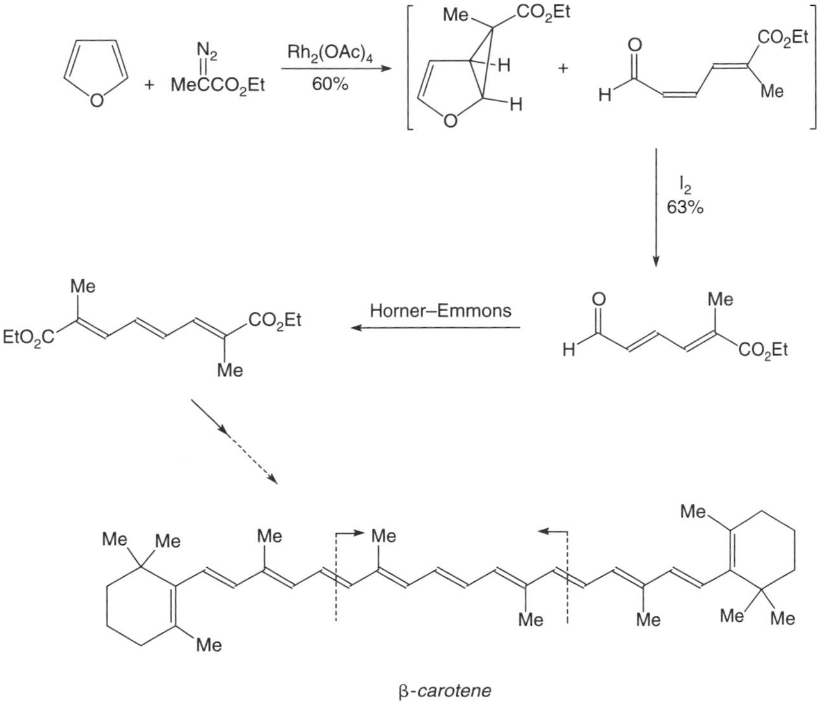

β-*carotene*

Scheme 54

retinol

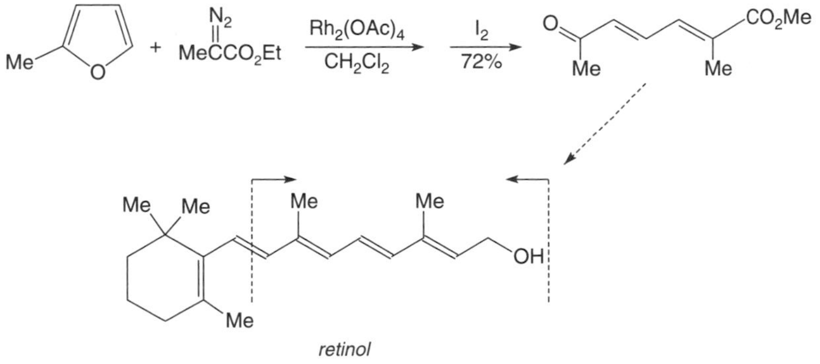

Scheme 55

tice, sequential treatment of furan with the diazopropionate in the presence of rhodium(II) acetate and iodine furnished an aldehydodiene with the correct geometry, and further Horner–Emmons elaboration with the sodium salt of triethyl

α-phosphonopropionate yielded a triene diester representative of the central section of β-carotene (Scheme 54). In a similar manner, the cycloadduct of 2-methylfuran and ethyl α-diazopropionate was used to construct a ten-carbon fragment of retinol (Scheme 55).

Yet another application of this methodology is found in the synthesis of the polyolefinic dicarboxylic acid corticrocin. Ten of the carbon atoms were first assembled in the form of difurylethane (Scheme 56), with the remaining four pro-

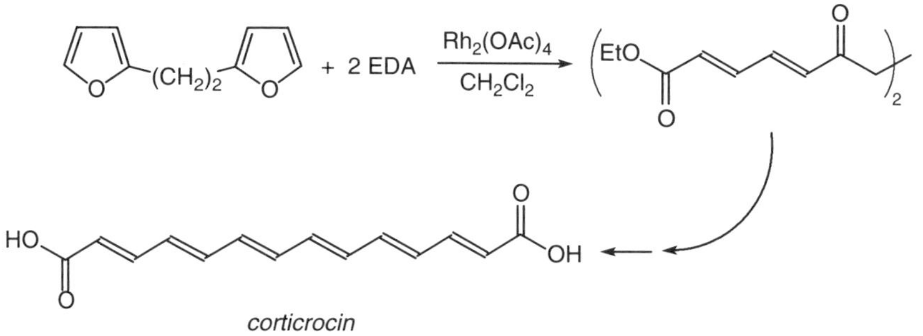

corticrocin

Scheme 56

vided by two molecules of EDA. The double cycloaddition was catalyzed by rhodium(II) acetate to give an acyclic product that on exposure to boron trifluoride etherate isomerized to an all-*trans*-tetraene. The two internal keto functions were then used to introduce the remaining two double bonds of corticrocin.

Even before Wenkert's study of the furan–diazoester cycloaddition, Adams and his colleagues at Merck Frost[75-77] had applied the furan–diazoketone combination to the construction of intermediates suitable for elaboration into leukotrienes. Diazoketones containing appropriate saturated or unsaturated side chains were synthesized and added to furan under rhodium(II) acetate catalysis. The acyclic dienols thus released were then elaborated into leukotrienes. Scheme 57 contains a summary of some of these transformations. In a similar

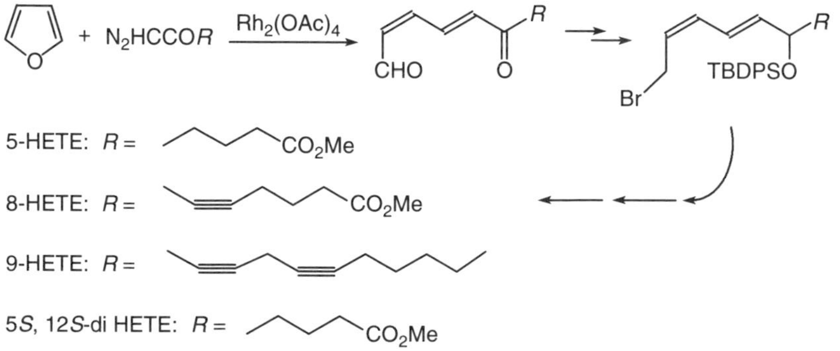

5-HETE: $R = $ /\/\$CO_2Me$

8-HETE: $R = $ \===/\/\$CO_2Me$

9-HETE: $R = $ \===/\===/\/\

5S, 12S-di HETE: $R = $ /\/\$CO_2Me$

Scheme 57

way, the cyclopropanation of 2-*n*-hexylfuran with an α,ω-diazoketoester was used by Sheu and co-workers[78] to construct the cytotoxic fatty acid, ostopanic acid (Scheme 58). Exposure of the reaction mixture to a catalytic amount of io-

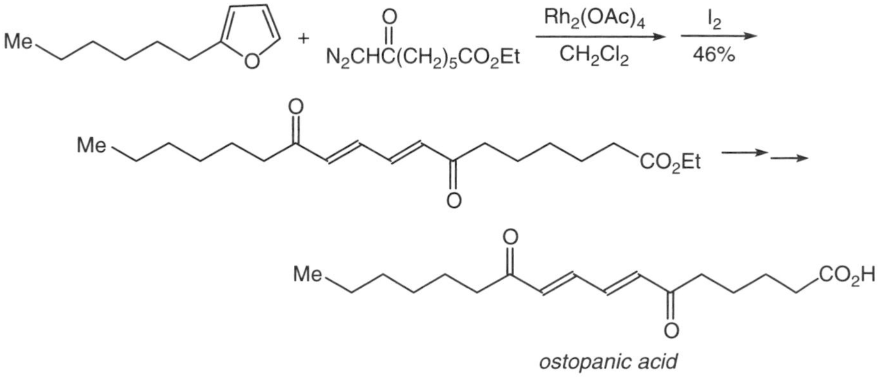

ostopanic acid

Scheme 58

dine in dichloromethane afforded pure ethyl ostopanate, from which ostopanic acid was obtained. A final example of the intermolecular process is that in Scheme 59.[79] Here, an enantiopure diazoketone derived from N-ethoxycarbonyl-

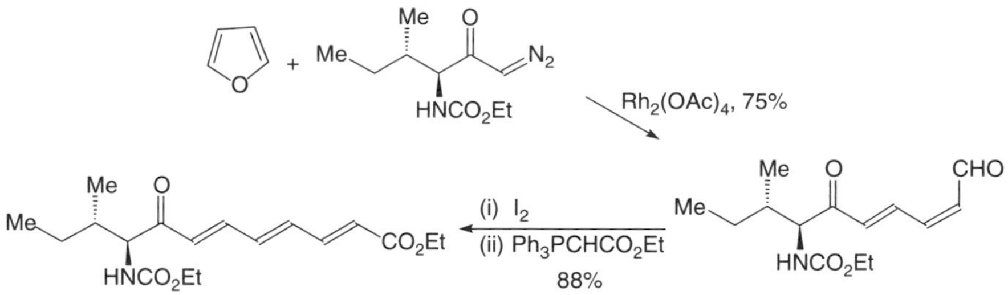

Scheme 59

L-isoleucine was used to form a ring-opened adduct with furan under $Rh_2(OAc)_4$ catalysis. The adduct was further elaborated in a Wittig reaction to form a poly-functional, enantiopure product.

The intramolecular version[76] of the furan–diazocarbonyl reaction is an attractive route to carbocycles. Furans with side chains at C(2) of various lengths terminating in diazomethyl keto functions are known to undergo rhodium-catalyzed unraveling with the production of 2-cyclopentenones, 2-cyclohexenones, and 2-cycloheptenones to each of whose olefinic β-carbon atoms is attached an acrylaldehyde unit. Some representative examples from the work of Wenkert[80,81] and Padwa[82,83] and their respective groups are shown in Schemes 60–63.

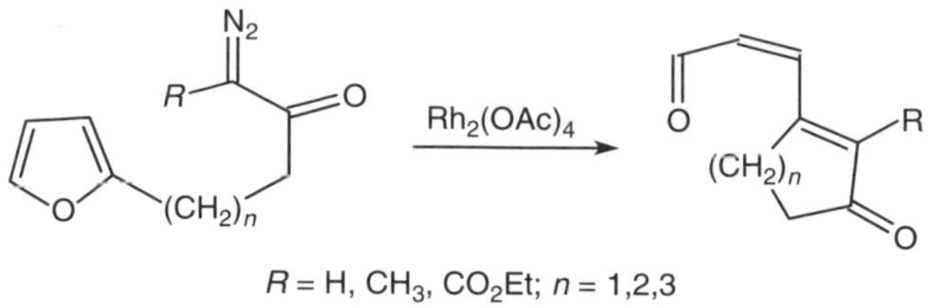

R = H, CH$_3$, CO$_2$Et; n = 1,2,3

Scheme 60

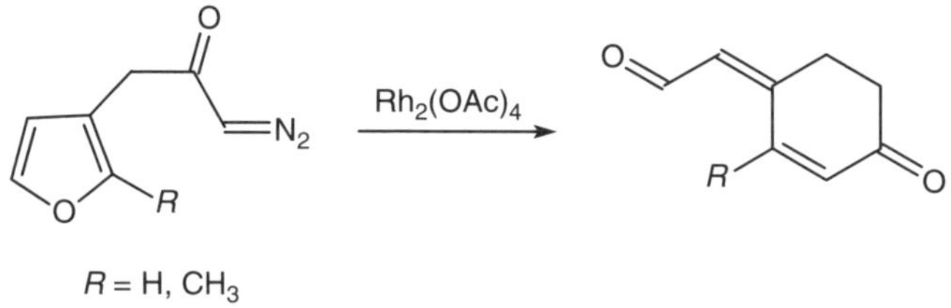

R = H, CH$_3$

Scheme 61

R^1 = R^2 = H, 62%
R^1 = H, R^2 = CH$_3$, 66%
R^1 = R^2 = CH$_3$, 45%

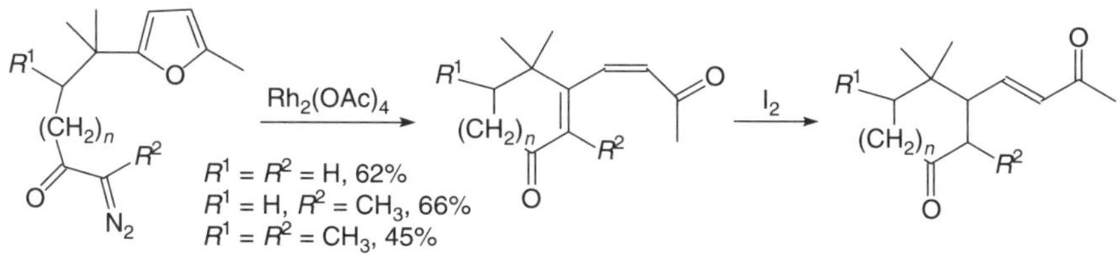

Scheme 62

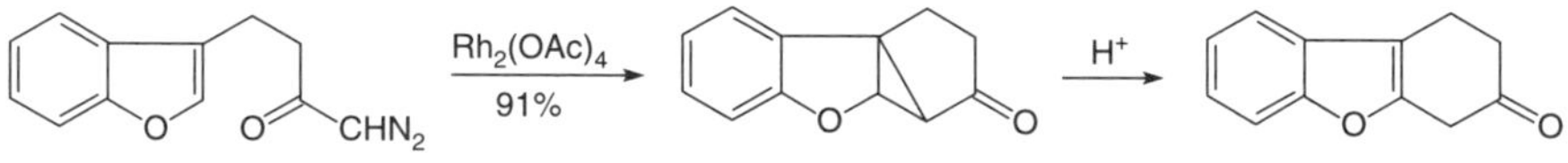

Scheme 63

Yet another pathway for furan–diazocarbonyl cycloadditions is the [3+4] combination explored in detail by Davies and his group.[84,85] They argued that while vinyl carbenes have a strong propensity for intramolecular reactions, their reactivity might be quite different if they were generated as metal carbene complexes by metal-catalyzed decomposition of vinyl diazo precursors. Such complexes should be more stable than free vinyl carbenes. Undesirable intramolecular side reactions should therefore be minimized. Davies tested this idea by preparing the vinyl diazoester in eq. 40 and exposing it to rhodium(II) acetate in the presence of furan at room temperature. The major reaction product was the bicyclic ether. Subsequent studies revealed that formation of adducts of this type is the result of a tandem cyclopropanation–Cope rearrangement process.

$$(40)$$

6.4.2 Pyrroles

Pyrrole and its derivatives exhibit a range of reactivity in catalyzed reactions with diazocarbonyl compounds. This heterocycle possesses the additional feature that the electron supply can be modulated by N-substitution. Unlike furan, pyrrole, and its N-alkyl derivatives, favor ring alkylation rather than cyclopropanation. In fact, the copper-catalyzed reaction of ethyl diazoacetate with pyrrole or pyrrole derivatives has long been recognized as one of the most reliable routes to pyrrole-2-acetic ester.[86-88] Substitution at the 2 position is preferred, although not exclusively so. For example, N-methylpyrrole and EDA give products of both 2 and 3 substitution in a 10:1 ratio. As part of the research directed towards improved synthetic processes for tolmetin and zomepirac, Maryanoff[89-91] has carried out a detailed study of the reaction of EDA, dimethyl diazomalonate, and ethyl 2-diazoacetoacetate with N-methylpyrrole. All reactions showed high regiochemical preference for the formation of 2-substituted adducts. Rhodium(II) acetate gave products in high-yield from the dimethyl diazomalonate and ethyl 2-diazoacetoacetate (eq. 41). For the reaction of EDA, copper(II) compounds are

$$(41)$$

$R = R' = OCH_3$ (83%)
$R = CH_3;\ R' = OC_2H_5$ (72%)

the catalysts of choice. Substitution reactions with pyrrole and alkylpyrroles are presumably favored because the electron-releasing ability of the nitrogen atom stabilizes a zwitterionic intermediate, and consequently cyclopropanation does not occur. Alternatively, cyclopropanation may occur but is rapidly followed by ring opening. The preference for ring alkylation of pyrrole and N-methylpyrrole is also observed with the vinyl metal carbene precursors of Davies.[92] For a more extensive discussion of Davies' work on the use of vinyl metal carbene intermediates in cyclopropanation of five-membered ring aromatics and their use in tandem with Cope rearrangement to produce bridged ring products including several alkaloids, see Section 4.8.

Finally, the outcome of intramolecular aromatic cycloaddition not only depends upon the nature of the diazocarbonyl precursor, but also depends upon the catalyst employed in the reaction. Decomposition of the indole-derived diazoke-

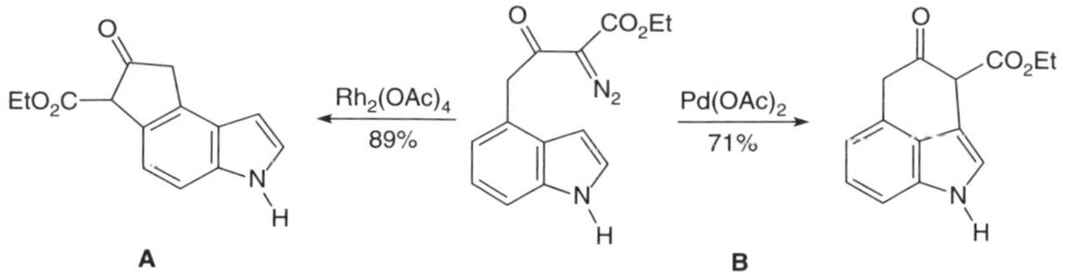

Scheme 64

toester in Scheme 64 with rhodium(II) acetate led to the regioisomer A of the product, whereas palladium(II) acetate furnished regioisomer B.[93]

The intramolecular version of the pyrrole–diazocarbonyl alkylation reaction has been explored by Jefford and Johncock in a very efficient route to another important group of alkaloids, the indolizidines.[94] These workers were among the first to show that simple N-alkylpyrroles having the diazocarbonyl moiety in the side chain, of the type shown in eq. 42, undergo intramolecular alkylation to form

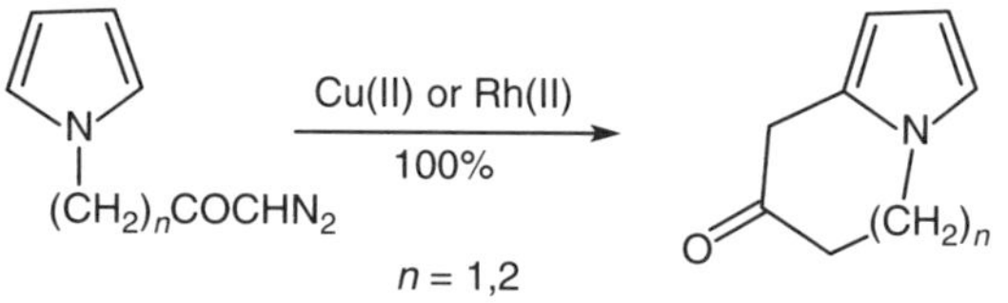

$$\text{(42)}$$

pyrrolizinones with either copper or rhodium catalysis.[95] Applications of this route to the synthesis of ($\pm$)-ipalbidine (Scheme 65), (+)-monomorine (Scheme 66), ($-$)-indolizidine 167B, and ($-$)-indolizidine 209D (Scheme 67) are

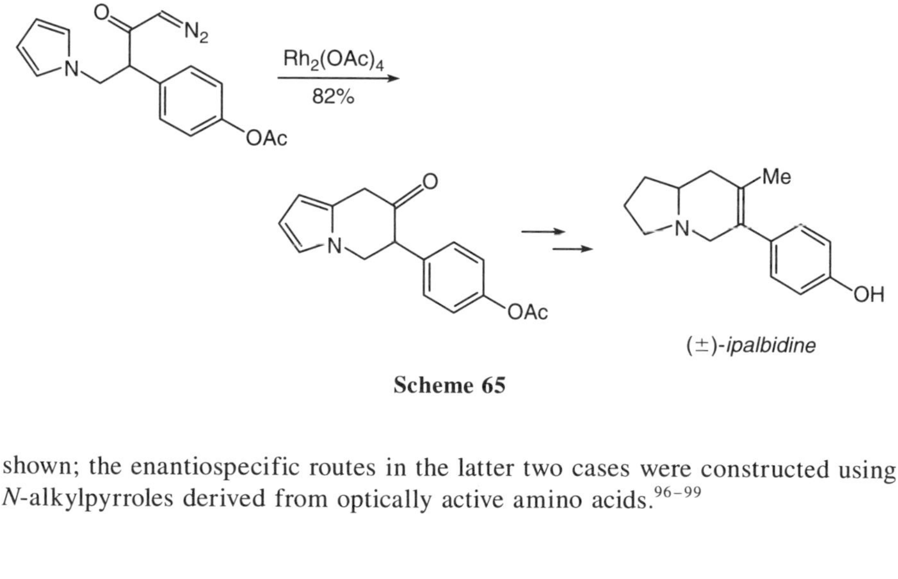

($\pm$)-*ipalbidine*

Scheme 65

shown; the enantiospecific routes in the latter two cases were constructed using N-alkylpyrroles derived from optically active amino acids.[96-99]

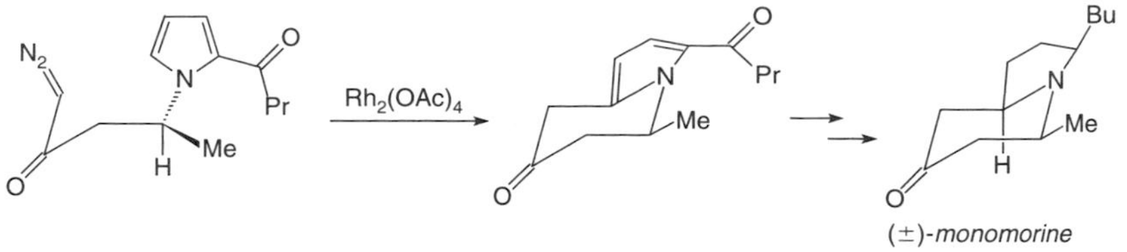

Scheme 66

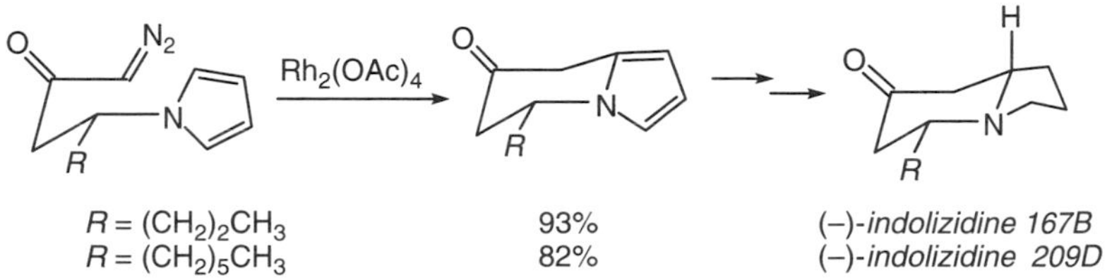

$R = (CH_2)_2CH_3$ 93%   (−)-*indolizidine 167B*
$R = (CH_2)_5CH_3$ 82%   (−)-*indolizidine 209D*

Scheme 67

6.4.3 Thiophenes

Catalytic reaction between thiophenes and diazocarbonyls has been known for several years. Rhodium(II) salts are the catalysts of choice for this reaction type.[100] In general, three possible processes, ylide formation, cyclopropanation, and C–H insertion, may be involved in the reaction. The outcome of the reaction depends upon the diazocarbonyl precursors and reaction conditions. In addition, the intramolecular version of this reaction is also possible. Some representative examples are shown in Scheme 68.[82,101,102]

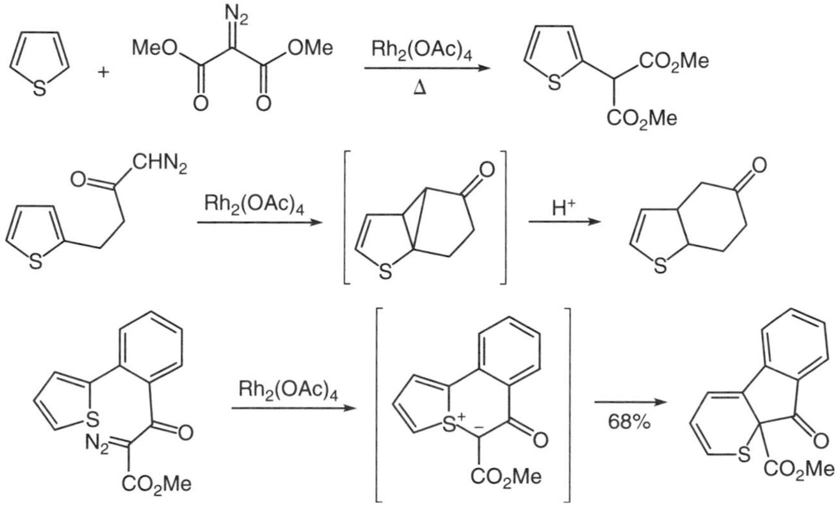

Scheme 68

6.5 REFERENCES

1 Curtius, T., "Ueber die Einwirkung von salpetriger Säure auf salzsauren Glyco-colläther," *Ber. Dtsch. Chem. Ges.* **1883**, *16*, 2230–31.

2 Büchner, E.; and Curtius, T., "Synthese von Ketonsäureäthern aus Aldehyden und Diazoessigäther," *Ber. Dtsch. Chem. Ges.* **1885**, *18*, 2371–77.

3 Büchner, E., "Pseudophenylacetic Acid," *Leibigs Ann. Chem.* **1896**, *29*, 106–9.

4 Grundmann, Ch.; and Ottmann, G., "Über Darstellung und Konstitution der Iso-meren Cycloheptatrien-carbonsäuren," *Ann.* **1953**, *582*, 163–77.

5 von E. Doering, W.; Laber, G.; Vonderwahl, R.; Chamberlain, N.; and Williams, R.B., "The Structure of the Büchner Acids," *J. Am. Chem. Soc.* **1956**, *78*, 5448–48.

6 Alder, K.; Muders, R.; Krane, W.; and Wirtz, P., "Über die Konstitution Photo-chemisch Dargestellter Norcaradien-carbonsäureester," *Ann.* **1959**, *627*, 59–78.

7 Garst, M. E.; and Roberts, V. A., "Synthesis of 3- and 4-(Carboethoxy)-2,4,6-Cycloheptatrien-1-one," *J. Org. Chem.* **1982**, *47*, 2188–90.

8 Pommer, H., "Über den Stand der Forschung auf dem Gebiet der Azulene," *Angew. Chem.* **1950**, *62*, 281–89.

9 Bartels-Keith, J. R.; Johnson, A. W.; and Taylor, W. I., "A Synthesis of Stipitatic Acid," *J. Chem. Soc.* **1951**, 2352–56.

10 Pfau, A. St.; and Plattner, Pl. A., "Volatile Plant Constituents. VIII. Synthesis of Vetivazulene," *Helv. Chim. Acta.* **1939**, *22*, 202–8.

11 Anciaux, A. J.; Demonceau, A.; Hubert, A. J.; Noels, A. F.; Petinoit, N.; and Teyssié, Ph., "Catalytic Control of Reactions of Dipoles and Carbenes; an Easy and Efficient Synthesis of Cycloheptatrienes from Aromatic Compounds by an Extension of Büchner's Reaction," *J. Chem. Soc., Chem. Commun.* **1980**, 765–66.

12 Anciaux, A. J.; Demonceau, A.; Noels, A. F.; Hubert, A. J.; Warin, R.; and Teyssié, Ph., "Transition-Metal-Catalyzed Reactions of Diazo Compounds. 2.[1] Addition to Aromatic Molecules: Catalysis of Büchner's Synthesis of Cyclohepta-trienes," *J. Org. Chem.* **1981**, *46*, 873–76.

13 Matsumoto, M.; Shiono, T.; Mutoh, H.; Amano, M.; and Arimitsu, S., "$Rh_2(OAc)_4$-Catalysed Cycloaddition of Ethyl Diazoacetate to 1,2-Dialkoxyben-zenes: a New Type of Stable Norcaradiene," *J. Chem. Soc., Chem. Commun.* **1995**, 101–2.

14 Saba, A.; and De Lucchi, O., "Cycloheptatriene Furanones and Bicyclo[5.3.0]-decatrienone—Unusual Cycloaddition with 4-Phenyl-1,2,4-triazoline-3,5-dione (PTAD)," *Heterocycles*, **1987**, *26*, 2339–42.

15 Saba, A., "7-Methyltricyclo[5.3.0.0[1,6]]deca-2,4-dien-8-ones: The First Examples of Tricyclonorcaradienones in Tautomeric Equilibrium with the 7-Methyl-bicyclo[5.3.0]1,3,5-decatrien-8-ones," *Tetrahedron Lett.* **1990**, *31*, 4657–60.

16 Okamura, W. H.; and Delera, A. R., in *Comprehensive Organic Synthesis*; Trost, B. M.; and Fleming, I., Eds.; Pergamon: Oxford, 1991; Vol. 5, 699–750.

17 Reich, H. J.; Ciganek, E.; and Roberts, J. D., "Nuclear Magnetic Resonance Spec-troscopy. Kinetics of a 7,7-Dicyanonorcaradiene Valence Tautomerism[1]," *J. Am. Chem. Soc.* **1970**, *92*, 5166–69.

18 Wehner, R.; and Gunther, H., "Direct Observation of *"Büchner's Acid"* Using ^{13}C and [1]H Nuclear Magnetic Resonance Spectroscopy," *J. Am. Chem. Soc.* **1975**, *97*, 923–24.

19 Takeuchi, K.; Fujimoto, H.; and Okamoto, K., "Primary Dependence of 7-Aryl-2,5-di-*tert*-butyl-1,3,5-cycloheptatriene Valence Tautomerism on the π-Acceptor Strength of the 7-Aryl Substituent," *Tetrahedron Lett.* **1981**, *22*, 4981–84.

20 McKervey, M. A.; Russell, D. N.; and Twohig, M. F., "Alkylation of Benzene with α-Diazoketones *via* Cycloheptatrienyl Intermediates," *J. Chem. Soc., Chem. Commun.* **1985**, 491–92.

21 McKervey, M. A.; and Twohig, M. F., unpublished results.

22 Chan. L.; and Matlin, S. A., "New Spiro Derivatives of Penicillin", *Tetrahedron Lett.* **1981**, *22*, 4025–28.

23 Costantino, A.; Linstrumelle, G.; and Julia, S., "Etude des Produits Formés en Traitant la Diazo-1 Phényl-4 Butanone-2 par la Poudre de Cuivre. Synthèse de la Tricyclo[5.3.0.0^{1,6}]décanone-8," *Bull. Soc. Chim. Fr.* **1970**, 907–12.

24 Ledon, H.; Cannic, G.; Linstrumelle, G.; and Julia, S. "Cyclisation des Carbenoides Issus de Diazomalonates Mixtes de Methyle et de Benzyles Substitues," *Tetrahedron Lett.* **1970**, 3971–74.

25 Vogel, E.; and Reel, H., "Bent [14]Annulenes. Synthesis of 1,6 : 8,13-Ethanediylidene[14]annulene," *J. Am. Chem. Soc.* **1972**, *94*, 4388–89.

26 Scott, L. T., "Azulenes: A Synthesis Based on Intramolecular Carbene Addition," *J. Chem. Soc., Chem. Commun.* **1973**, 882–83.

27 Scott, L. T.; Minton, M. A.; and Kirms, M. A., "A Short New Azulene Synthesis," *J. Am. Chem. Soc.* **1980**, *102*, 6311–14.

28 McKervey, M. A.; Tuladhar, S. M.; and Twohig, M. F., "Efficient Synthesis of Bicyclo[5.3.0]decatrienones and of 2-Tetralones *via* Rhodium(II) Acetate–Catalysed Cyclisation of α-Diazoketones Derived from 3-Arylpropionic Acids," *J. Chem. Soc., Chem. Commun.* **1984**, 129–30.

29 Kennedy, M.; McKervey, M. A.; Maguire, A. R.; Tuladhar, S. M.; and Twohig, M. F., "The Intramolecular Büchner Reaction of Aryl Diazoketone. Substituted Effects and Scope in Synthesis," *J. Chem. Soc., Perkin Trans. 1* **1990**, 1047–54.

30 Scott, L. T.; and Sumpter, C. A., "Diazo Ketone Cyclization onto a Benzene Ring: 3,4-Dihydro-1(2*H*)-azulenone," *Org. Synth.* **1990**, *68*, 180–87.

31 Kumar, P.; Rao, A. T.; Saravanan, K.; and Pandey, B., "Selective Bond Cleavage of [5.3.1]Propellanes by Lead Tetraacetate: A Facile Entry into the Carbocyclic Frame [A, B ring] of Taxol," *Tetrahedron Lett.* **1995**, *36*, 3397–400.

32 Hannemann, K., "Formation of Cycloheptatriene/Norcaradiene Systems in the Decomposition of Diaryldiazomethanes in Benzene," *Angew. Chem. Int. Ed. Engl.* **1988**, *27*, 284–85.

33 Kennedy, M.; and McKervey, M. A., "A New Route to Functionalised Hydroazulenes. Synthesis of ($\pm$)-Confertin," *J. Chem. Soc., Chem. Commun.* **1988**, 1028–30.

34 Kennedy, M.; and McKervey, M. A., "Pseudoguaianolides from Intramolecular Cycloadditions of Aryl Diazoketones: Synthesis of ($\pm$)-Confertin and an Approach to the Synthesis of ($\pm$)-Damsin," *J. Chem. Soc., Perkin Trans. 1.* **1991**, 2565–74.

35 Duddeck, H.; Ferguson, G.; Kaitner, B.; Kennedy, M.; McKervey, M. A.; and Maguire, A. R., "The Intramolecular Büchner Reaction of Aryl Diazoketones. Synthesis and X-Ray Crystal Structures of Some Polyfunctional Hydroazulene Lactones," *J. Chem. Soc., Perkin Trans. 1* **1990**, 1055–63.

36 Cordi, A. A.; Lacoste, J.-M.; and Hennig, P., "A Reinvestigation of the Intramolecular Büchner Reaction of 1-Diazo-4-phenylbutan-2-ones Leading to 2-Tetralones," *J. Chem. Soc., Perkin Trans. 1* **1993**, 3–4.

37 Iwata, C.; Yamada, M.; Shinoo, Y.; Kobayashi, K.; and Okada, H., "Studies on the Syntheses of Spirodienone Compounds. VII. Novel Synthesis of the Spiro[4.5]decane Carbon Framework," *Chem. Pharm. Bull. Tokyo* **1980**, *28*, 1932–34.

38 Iwata, C.; Miyashita, K.; Imao, T.; Masuda, K.; Kondo, N.; and Uchida, S., "A Modified Procedure for the Synthesis of Spirodienones from Phenolic α-Diazoketones," *Chem. Pharm. Bull. Tokyo* **1985**, *33*, 853–55.

39 Iwata, C.; Fusaka, T.; Fujiwara, T.; Tomita, K.; and Yamada, M., "Total Synthesis of (±)-Solavetivone; X-Ray Crystal Structure of 2-Hydroxy-6,10-dimethyl-spiro[4.5]dec-6-en-8-one," *J. Chem. Soc., Chem. Commun.* **1981**, 463–65.

40 Iwata, C.; Morie, T.; Maezaki, N.; Shimamura, H.; Tanaka, T.; and Imanishi, T., "Stereoselective Synthesis of the B/C/D Ring Systems of Aphidicolane and Stemodane Diterpenes," *J. Chem. Soc., Chem. Commun.* **1984**, 930–32.

41 Iwata, C.; Morie, T.; and Tanaka, T., "Synthetic Studies on Aphidicolane and Stemodane Diterpenes. 1. Synthesis of (2′R*, 4a′S*, 8a′R*)-2′, 8a′-Dimethyl-4a′,5′,8′,8a′-tetrahydrospiro[2,5-cyclohexadiene-1, 1′(2H)-naphthalene]-3′(4′H), 4,6′(7′H)-trione," *Chem. Pharm. Bull. Tokyo* **1985**, *33*, 944–49.

42 Chattergee, J. N.; Sinha, A. K.; and Bhakta, C., "Cyclization to Phenanthrol," *Indian J. Chem., Sect. B* **1979**, *17*, 329.

43 Duddeck, H.; Kennedy, M.; McKervey, M. A.; and Twohig, M. F., "Rhodium(II)-Catalysed Cyclisation of Diazoketones derived from Biphenyl; A New Route to Benz[a]azulenes and Related Systems," *J. Chem. Soc., Chem. Commun.* **1988**, 1586–88.

44 Sonawane, H. R.; Bellur, S. N.; and Sudrick, S. G., "Efficient Synthesis of (±)-*ar*-Himachalene: Applications of Chemoselective Rhodium-Carbenoid Reaction with Aromatic Nucleus," *Ind. J. Chem., Sect. B* **1992**, *31*, 606–07.

45 Manitto, P.; Monti, D.; and Speranza, G., "Rhodium-Catalysed Decomposition of 1-Diazo-4-(2-naphthyl)butan-2-one. Direct Chemical Evidence for the Formation of the Norcaradiene System in the Intramolecular Buchner Reaction," *J. Org. Chem.* **1995**, *60*, 484–85.

46 Taylor, E. C.; Davies, H. M. L., "Rhodium(II) Acetate–Catalyzed Reaction of Ethyl 2-Diazo-3-oxopent-4-enoates: Simple Routes to 4-Aryl-2-hydroxy-1-naphthoates and β,γ-Unsaturated Esters. The Dianion of Ethyl 4-(Diethylphosphono)acetoacetate as a Propionate Homoenolate Equivalent," *Tetrahedron Lett.* **1983**, *24*, 5453–56.

47 Wenkert, E.; and Liu, S., "Intramolecular Reactions of Oxindolyl Diazo Ketones," *Synthesis* **1992**, 323–27.

48 Sugihara, Y.; Yamamoto, H.; Mizoue, K.; and Murata, I., "Cyclohepta[a]phenalene: A Highly Electron-Donating Nonalternant Hydrocarbon," *Angew. Chem., Int. Ed. Engl.* **1987**, *26*, 1247–49.

49 Rogers, D. H.; Morris, J. C.; Roden, F. S.; Frey, B.; King, G. R.; Russkamp, F.-W.; Bell, R. A.; and Mander, L. N., "The Development of More Efficient Syntheses of Polcyclic Diterpenes through Intramolecular Cyclopropanation of Aryl Rings in Diazomethylketones," *Pure Appl. Chem.* **1996**, *68*, 515–22.

50 Doyle, M. P.; Protopopova, M. N.; Peterson, C. S.; Vitale, J. P.; McKervey, M. A.; and Garcia, C. F., "Formation of Macrocycles by Catalytic Intramolecular Aromatic Cycloaddition of Metal Carbenes to Remote Arenes," *J. Am. Chem. Soc.* **1996**, *118*, 7865–66.

51 Pusino, A.; Saba, A.; and Rosnati, V., "Catalytic Decomposition of Branched α-Phenoxy-α-diazo Ketones Affording 2,8*H*-Cyclohepta[b]furan-3-one and 2*H*-Cyclohepta[b]furan-3a(3a*H*)methyl-3-one," *Tetrahedron* **1986**, *42*, 4319–24.

52 Doyle, M. P.; Shanklin, M. S.; and Pho, H. Q., "Cycloheptatriene Syntheses through Rhodium(II) Acetate Catalysed Intramolecular Addition Reactions of *N*-Benzyldiazoacetamides," *Tetrahedron Lett.* **1988**, *29*, 2639–42.

53 Padwa, A.; Austin, D. J.; Price, A. T.; Semones, M. A.; Doyle, M. P.; Protopopova, M. N.; Winchester, W. R.; and Tran, A., "Ligand Effects on Dirhodium(II) Carbene Reactivities. Highly Effective Switching between Competitive Carbenoid Transformations," *J. Am. Chem Soc.* **1993**, *115*, 8669–80.

54 Wee, A. G. H.; Liu, B.-S.; and Zhang, L., "Dirhodium Tetraacetate Catalysed Carbon–Hydrogen Insertion Reaction in *N*-Substituted α-Carbomethoxy-α-diazoacetanilides and Structural Analogues. Substituent and Conformational Effects," *J. Org. Chem.* **1992**, *57*, 4404–14.

55 Zaragoza, F., "Remarkable Substituent Effects on the Chemoselectivity of Rhodium(II) Carbenoids Derived from *N*-(2-diazo-3-oxobutyryl)-L-phenylalanine Esters," *Tetrahedron* **1995**, *51*, 8829–34.

56 Kennedy, M.; McKervey, M. A.; Maguire, A. R.; and Roos, G. H. P., "Asymmetric Synthesis in Carbon–Carbon Bond Forming Reactions of α-Diazoketones Catalysed by Homochiral Rhodium(II) Carboxylates," *J. Chem. Soc., Chem. Commun.* **1990**, 361–62.

57 McKervey, M. A.; McCarthy, N., unpublished results.

58 Hrytsak, M.; and Durst, T., "Preparation of 3-Acetylbenzofuran-2(3*H*)-ones and 3-Acetylnaphthofuran-2(3*H*)-ones via Intramolecular Rhodium Carbenoid Insertion," *J. Chem. Soc., Chem. Commun.* **1987**, 1150–51.

59 Hrytsak, M.; Etkin, N.; and Durst, T., "Intramolecular Rhodium Carbenoid Insertions into Aromatic C–H Bonds. Preparation of 1-Carboalkoxy-1,3-dihydrobenzo[c]thiophene-2,2-Dioxides," *Tetrahedron Lett.* **1986**, *27*, 5679–82.

60 Babu, S. D.; Hrytsak, M. D.; and Durst, T., "Intramolecular Rhodium Carbenoid Insertions into Aromatic C–H Bonds. Preparation of 1,3-Dihydrothiophene 2,2-dioxides Fused onto Aromatic Rings," *Can. J. Chem.* **1989**, *67*, 1071–76.

61 Etkin, N.; Babu, S. D.; Fooks, C. J.; and Durst, T., "Preparation of 3-Acetyl-2-Hydroxyindoles via Rhodium Carbenoid Aromatic C–H Insertion," *J. Org. Chem.* **1990**, *55*, 1093–96.

62 Doyle, M. P.; Shanklin, M. S.; Pho, H. Q.; and Mahapatro, S. N., "Rhodium(II) Acetate and Nafion-H Catalysed Decomposition of N-Aryldiazoamides. An Efficient Synthesis of 2(3*H*)-Indolinones," *J. Org. Chem.* **1988**, *53*, 1017–22.

63 Wee, A. G.; and Liu, B., "The Nafion-H Catalysed Cyclization of α-Carbo-methoxy-α-Diazoacetanilides. Synthesis of 3-Unsubstituted-2-Indolinones," *Tetrahedron* **1994**, *50*, 609–26.

64 Rishton, G. M.; and Schwartz, M. A., "Acid-Catalyzed Cyclizations of Aromatic Diazoacetamides—Synthesis of Spirodienone Lactams, Isoquinolinones, and Benzazepinones," *Tetrahedron Lett.* **1988**, *29*, 2643–46.

65 Taber, D. F.; and Ruckle, R. E., "Cyclopentane Construction by $Rh_2(OAc)_4$-Mediated Intramolecular C–H Insertion: Steric and Electronic Effects," *J. Am. Chem. Soc.* **1986**, *108*, 7686–93.

66 Nakatani, K., "Synthesis of 2-Indanones by Intramolecular Insertion of α-Diazoketones," *Tetrahedron Lett.* **1987**, *28*, 165–66.

67 Hashimoto, S.-I.; Watanabe, N.; and Ikegami, S., "Highly Selective Insertion into Aromatic C–H Bonds in Rhodium(II) Triphenylacetate-Catalysed Decomposition of α-Diazocarbonyl Compounds," *J. Chem. Soc., Chem. Commun.* **1992**, 1508–10.

68 Estevan, F.; Lahuerta, P.; Pérez-Prieto, J.; Stiriba, S.-E.; and Ubeda, M. A., "New Rhodium(II) Catalysts for Selective Carbene Transfer Reaction," *Synlett* **1995**, 1121–22.

69 Watanabe, N.; Ohtake, Y.; Hashimoto, S.-I.; Shiro, M; and Ikegami, S., "Asymmetric Creation of Quaternary Carbon Centres by Enantiotopically Selective Aromatic C–H Insertion Catalyzed by Chiral Dirhodium(II) Carboxylates," *Tetrahedron Lett.* **1995**, *36*, 1491–94.

70 Watanabe, N.; Ogawa, T.; Ohtake, Y.; Ikegami, S.; and Hashimoto, S.-I., "Dirhodium(II) Tatrakis[*N*-phthaloyl-(*S*)-*tert*-leucinate]: A Notable Catalyst for Enantiotopically Selective Aromatic Substitution Reactions of α-Diazocarbonyl Compounds," *Synlett* **1996**, 85–86.

71 Novác, J.; and Šorm, F., "Reactions of Diazoketones. III. Reaction of Diazoacetone with Furan and its Homologs. A Simple Synthesis of 3,5-Octadiene-2,7-dione," *Collect. Czech. Chem. Commun.* **1958**, *23*, 1126–31.

72 Schenck, G. O.; and Steinmetz, R., "Photochemische Bildungsweisen und Umlagerungen von Thiopheno- und Furano-cyclopropan-carbonsäureestern," *Liebigs Ann. Chem.* **1963**, *668*, 19–25.

73 Wenkert, E.; Guo, M.; Lavilla, R.; Porter, B.; Ramachandran, K.; and Sheu, J.-H., "Polyene Synthesis. Ready Construction of Retinol Carotene Fragments, (±)-6(*E*)-LTB₃ Leukotrienes, and Corticrocin," *J. Org. Chem,* **1990**, *55*, 6203–14.

74 Rokach, J.; Girard, Y.; Guidon, Y.; Atkinson, J. G.; Larue, M.; Young, R. X.; Maseon, P.; and Holme, G., "The Synthesis of a Leukotriene with SRS—Like Activity," *Tetrahedron Lett.* **1980**, 1485–88.

75 Rokach, J.; Adams, J.; and Perry, R., "A New General Method for the Synthesis of Lipoxygenase Products: Preparation of (±)-5-HETE," *Tetrahedron Lett.* **1983**, *24*, 5185–88.

76 Adams, J.; and Rokach, J., "Synthesis of (±)- 8- and 9-HETES," *Tetrahedron Lett.* **1984**, *25*, 35–38.

77 Adams, J.; Leblanc, Y.; and Rokach, J., "Synthesis of 5*S*,12*S*-diHETE(LTB$_x$)," *Tetrahedron Lett.* **1984**, *25*, 1227–30.

78 Sheu, J.-H.; Yen, C.-F.; Huang, H.-C.; and Hong, Y.-L.V., "Total Synthesis of Ostopanic Acid, A Plant Cytotoxin, via Cyclopropanation of 2-*n*-Hexylfuran," *J. Org. Chem.* **1989**, *54*, 5126–28.

79 McKervey, M. A.; and O'Sullivan, M. B., unpublished results.

80 Wenkert, E.; Guo, M.; Pizzo, F.; and Ramachandran, K., "Synthesis of 2-Cycloalkenones (Parts of 1,4-Diacyl-1,3-Butadiene Systems) and of a Heterocyclic Analogue by Metal-Catalyzed, Decomposition of 2-Diazoacylfurans," *Helv. Chim. Acta* **1987**, *70*, 1429–38.

81 Wenkert, E.; Decorzant, R.; and Näf, F., "A Novel Access to Ionone-Type Compounds: (*E*)-4-oxo-β-ionone and (*E*)-4-oxo-β-Irone via Metal-Catalyzed, Intramolecular Reactions of α-Diazo Ketones with Furans," *Helv. Chim. Acta* **1989**, *72*, 756–66.

82 Padwa, A.; Wisnieff, T. J.; and Walsh, E. J., "Synthesis of Cycloalkenones via the Intramolecular Cyclopropanation of Furanyl Diazo Ketones," *J. Org. Chem.* **1986**, *51*, 5036–38.

83 Padwa, A.; Wisnieff, T. J.; and Walsh, E. J., "Intramolecular Cyclopropanation Reaction of Furanyl Diazo Ketones," *J. Org. Chem.* **1989**, *54*, 299–308.

84 Davies, H. M. L.; Clark, D. M.; and Smith, T. K., "[3+4] Cycloaddition Reactions of Vinyl Carbenoids with Furans," *Tetrahedron Lett.* **1985**, *26*, 5659–62.

85 Davies, H. M. L.; Smith, H. D.; and Korkor, O., "Tandem Cyclopropanation/Cope Rearrangement Sequence. Stereospecific [3+4] Cycloaddition Reaction of Vinylcarbenoids with Cyclopentadiene," *Tetrahedron Lett.* **1987**, *28*, 1853–56.

86 Nenitzescu, C. D.; and Solomonica, E., "Action of Aliphatic Diazo Compounds on Pyrrole and its Homologues," *Ber. Dtsch. Chem. Ges.* **1931**, *64*, 1924–31.

87 Gossauer, A., *Die Chemie der Pyrrole (The Chemistry of Pyrrole)*; Springer-Verlag: Berlin, 1974; pp. 126–27.

88 Jones, R. A.; and Bean G. P., *The Chemistry of Pyrroles*; Academic Press: London, 1977; pp. 269–70.

89 Maryanoff, B. E., "β-Substitution of Simple Pyrroles in Metal Assisted Reactions with Ethyl Diazoacetate," *J. Heterocycl. Chem.* **1977**, *14*, 177–78.

90 Maryanoff, B. E., "Carbenoid Chemistry. Reaction of Pyrrole Derivatives with Ethyl Diazoacetate," *J. Org. Chem.* **1979**, *44*, 4410–19.

91 Maryanoff, B. E., "Reaction of Dimethyl Diazomalonate and Ethyl 2-Diazoacetoacetate with *N*-Methylpyrrole," *J. Org. Chem.* **1982**, *47*, 3000–2.

92 Davies, H. M. L.; Young, W. B.; and Smith, H. D., "Novel Entry to the Tropane System by Reaction of Rhodium(II) Acetate Stabilized Vinylcarbenoids with Pyrroles," *Tetrahedron Lett.* **1989**, *30*, 4653–56.

93 Matsumoto, M.; Watanabe, N.; and Kobayashi, H., "Metal Catalysed Intramolecular Cyclisation of 2-Diazo-4-(4-indolyl)-3-oxobutanoic Acid-Esters," *Heterocycles* **1987**, *26*, 1479–82.

94 Jefford, C. W.; and Johncock, W., "Intramolecular Carbenoid Reactions of Pyrrole Derivatives, Efficient Syntheses of Pyrrolizinone and Dihydroindolizinone," *Helv. Chim. Acta* **1983**, *66*, 2666–71.

95 Jefford, C. W.; and Zaslona, A., "Competitive Intramolecular Carbenoid Reactions of Pyrrole Derivatives," *Tetrahedron Lett.* **1985**, *26*, 6035–38.

96 Jefford, C. W.; Kubota, T.; and Zaslona, A., "Intramolecular Carbenoid Reactions of Pyrrole Derivatives. A Total Synthesis of (±)-Ipalbidine," *Helv. Chim. Acta* **1986**, *69*, 2048–61.

97 Jefford, C. W.; Tang, Q.; and Zaslona, A., "A Short, Simple Synthesis of (±)-Monomorine," *Helv. Chim. Acta* **1989**, *72*, 1749–52.

 98   Jefford, C. W.; Tang, Q.; and Zaslona, A., "Short, Enantiogenic Syntheses of (-)-Indolizidine 167B and (+)-Monomorine," *J. Am. Chem. Soc.* **1991**, *113*, 3513–18.

 99   Jefford, C. W.; and Wang, J. B., "Enantiospecific Syntheses of Indolizidines 167B and 209D," *Tetahedron Lett.* **1993**, *34*, 3119–22.

100   Gillespie, R. J.; and Porter, A. E. A., "The Reaction of Diazoalkanes with Thiophen," *J. Chem. Soc., Perkin Trans. 1* **1979**, 2624–28.

101   Storflor, H.; Skramstad, J.; and Nordenson, S., "Stevens Rearrangement of a Potentially Aromatic Thiophenium Ylide: Formation of 10-Methoxycarbonyl-10*H*-Benzo[3,4]cyclopenta[1,2-*b*]thiopyran-9-one," *J. Chem. Soc., Chem. Commun.* **1984**, 208–9.

102   Matsumoto, M.; Kobayashi, H.; and Watanabe, N., "A Short Step Synthesis of Clavicipitic Acids from 4-Cyanomethylindole," *Heterocycles* **1987**, *26*, 1197–202.

CHAPTER 7

Generation and Reactions of Ylides from Diazocarbonyl Compounds

7.1 INTRODUCTION

Metal carbenes derived from α-diazocarbonyl compounds are highly electrophilic. They readily react with an available Lewis base (B:) to effect ylide formation (Scheme 1). Catalytically generated electrophilic metal carbenes add to

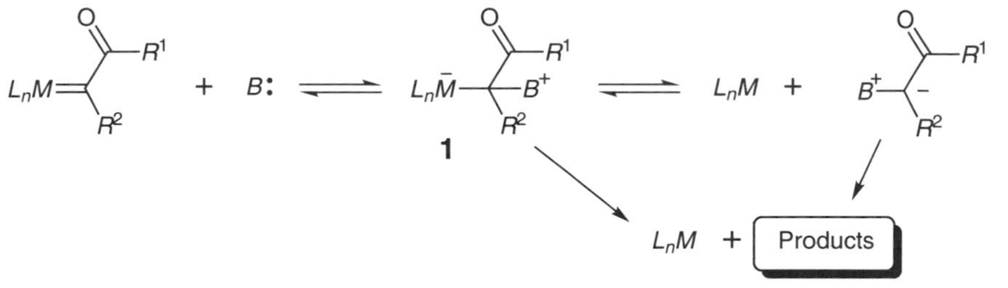

Scheme 1

Lewis bases, characteristically heteroatom-substituted organic compounds, to form adducts **1**, which can either dissociate to form rearranged products (or form a "free" ylide) and the catalyst, or revert to the metal carbene and Lewis base. With catalytically generated metal carbenes, especially those of copper or rhodium, the metal carbon bond of the intermediate metal-stabilized ylide is generally weaker than the $R_2C–B$ bond, and the preferred cleavage is that of the metal–carbon bond. The ease with which catalytically generated metal carbenes transfer the carbene entity to a heteroatom of an organic base is the basis for the synthetic utility to this diazo decomposition methodology.

The most common reactions of catalytically generated ylides are: [2,3]-sigmatropic rearrangement of allyl-substituted ylide intermediates; [1,2] insertion or Stevens rearrangement; β-hydride elimination; and dipolar cycloaddition.[1] These processes, which can proceed intermolecularly or intramolecularly, have shown great versatility in the synthesis of natural products. To demonstrate the broad scope of catalytic ylide formation and subsequent reactions, representative examples have been highlighted in the following sections.

7.2   FORMATION OF SULFUR YLIDES

The generation of sulfur ylides and their reaction chemistry have received considerable attention because of their applications in synthesis and their probable involvement in biochemical processes.[2,3] Both direct and indirect routes for the formation of sulfur ylide **3** from diazocarbonyl compounds have been developed. The direct approach involves the addition of a carbene **2** derived from a diazocarbonyl compound to a sulfide (eq. 1).[1] Thermal, photochemical,[4] and catalytic

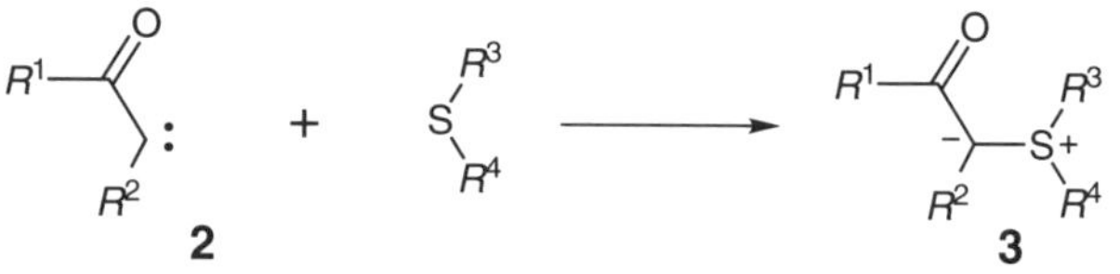

$$(1)$$

methods have been used, but catalytic methodologies offer the greatest advantages. A less direct approach is a base-promoted methodology, which consists of a two-step sequence. The diazocarbonyl compound **4** serves as an alkylation agent in the first step to form a sulfonium salt **5**, which is then deprotonated to the ylide by reaction with a base (Scheme 2).[5,6]

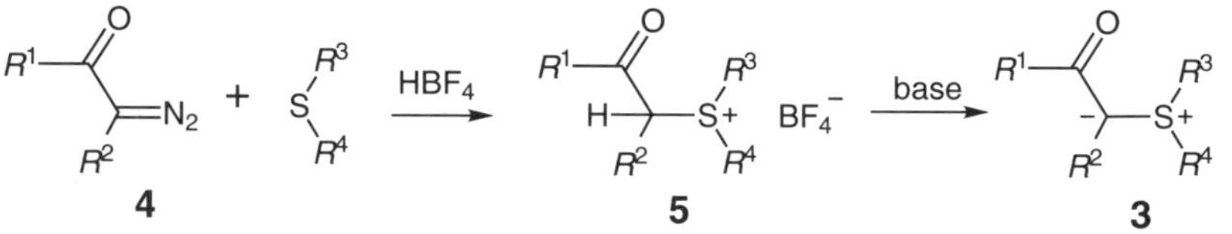

Scheme 2

In competitive experiments with substrates containing both olefinic and thio ether functions, copper-[7,8] and rhodium[9]-stablized carbenes react preferentially with the sulfur atom. Just as sulfides are good nucleophiles in substitution reactions, so too are they good ligands for catalytically active transition metal compounds. By associating with the catalyst (Scheme 3), sulfides inhibit reaction

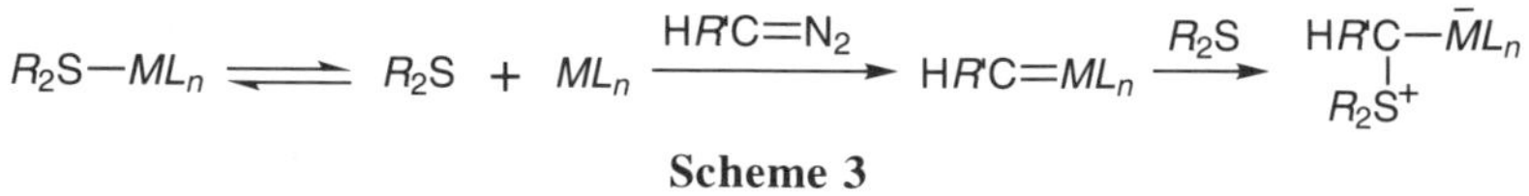

$$R_2S-ML_n \rightleftharpoons R_2S + ML_n \xrightarrow{HR'C=N_2} HR'C=ML_n \xrightarrow{R_2S} \begin{array}{c} HR'C-\bar{M}L_n \\ | \\ R_2\overset{+}{S} \end{array}$$

Scheme 3

with the diazocarbonyl compound, and higher temperatures than those used with ethers, ketones, or halides are required for effective catalysis.[9]

7.2.1 Formation of Stable Sulfonium and Sulfoxonium Ylides from Diazocarbonyl Precursors

Sulfonium ylides can be isolated as stable reaction products if two electron-withdrawing groups stabilize the ylide carbon and their structures do not facilitate subsequent rapid reactions as, for example, the [2,3]-sigmatropic rearrangement. Porter and co-workers have prepared sulfonium ylides **6** from dirhodium(II) acetate–catalyzed decomposition of dimethyl diazomalonate in the presence of thiophene or 2,5-dichlorothiophene (eq. 2)[10-14] These ylides are obtained in high

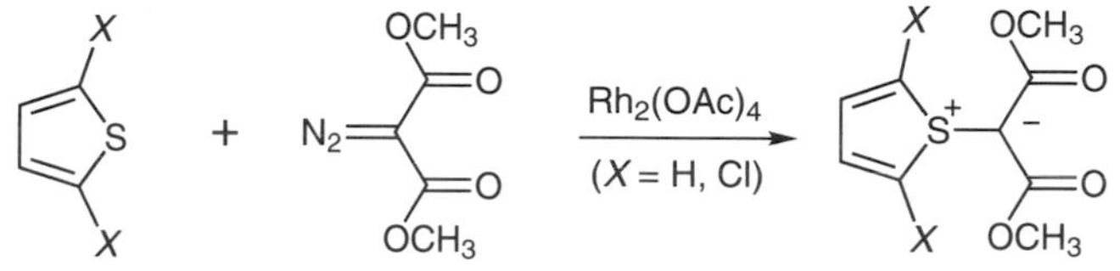

$$ (2) $$

yield, and **6** ($X = $ H) can be isolated as a stable crystalline solid.[10] Certain heterocycles and arenes activated by one or more strongly electron-donating substituents undergo a copper-catalyzed reaction with **6** ($X = $ Cl) to give moderate to excellent yields of aryl malonates.[14] A similar transformation with thioxanthene **7** using $CuSO_4$ (eq. 3) afforded ylide **8**, whose X-ray crystal structure revealed the *trans* stereochemistry.[15]

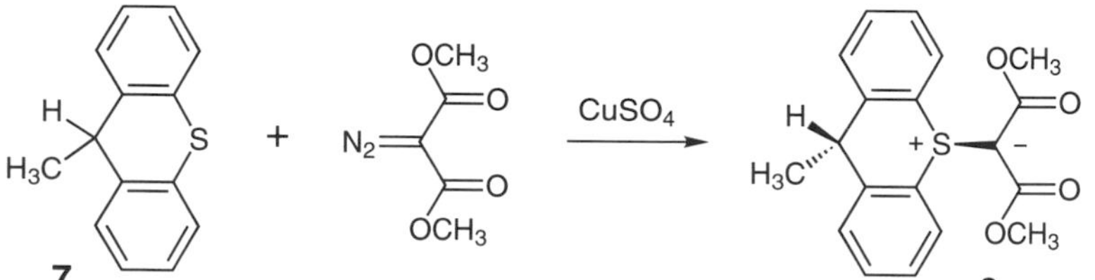

$$ (3) $$

Phenylsulfonium ylides, formed by dirhodium(II) acetate–catalyzed intramolecular cyclization of β-keto-α-diazoesters in good yield (eq. 4), are stable

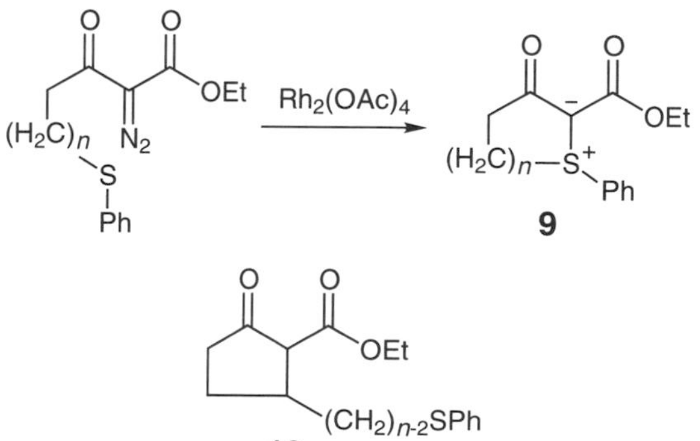

$$ (4) $$

at room temperature.[16] Ring size compatibility with these stable ylides extends from five- to seven-membered rings ($n = 1$–3), although C–H insertion is competitive with the formation of the seven-membered ring ylide structure (**9**, $n = 3$); longer-chain reactants furnish the cyclopentanone derivative **10** exclusively *via* C–H insertion. Upon heating these ylides to 160°C, none forms the products expected from the Stevens rearrangement. However, Moody and co-workers[17] have shown that the corresponding benzylsulfonium ylide **11**, formed from $Rh_2(OAc)_4$-catalyzed diazodecomposition of an β-keto-α-diazoester, does undergo thermal Stevens rearrangement (Scheme 4).

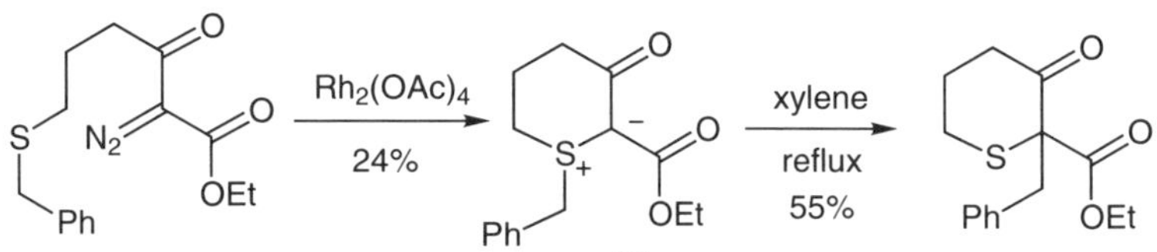

Scheme 4

Transition-metal-catalyzed decomposition of diazocarbonyl compounds in the presence of sulfoxides also produces stable ylides.[18–20] Representative examples of sulfoxonium ylides, obtained from both intermolecular and intramolecular reactions, are shown in eqs. 5–7. The X-ray structure of **12c** shows the sulfoxonium ylide in a pyramidal configuration with an axial sulfoxide oxygen.[19]

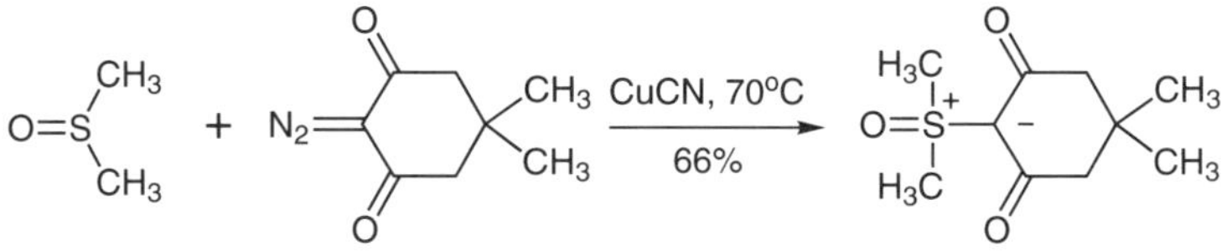

(5)

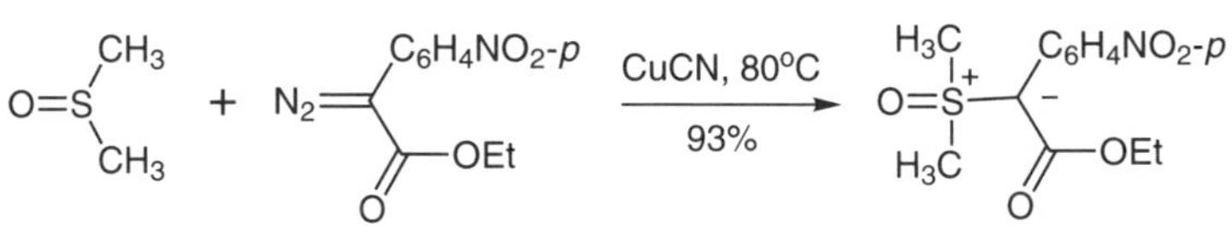

(6)

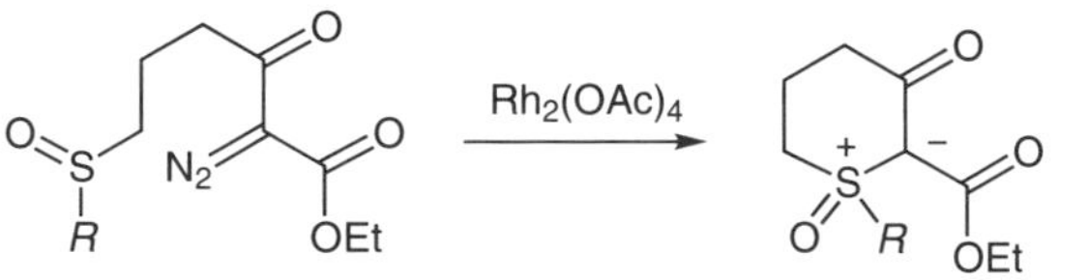

(7)

a: R = Et (78%); **b**: R = CH_2Ph (76%);
c: R = $CH_2CH=CH_2$ (84%); **d**: R = $CH_2CH=CHPh$ (54%)

7.2.2 Intermolecular Formation of Sulfur Ylides and Subsequent Reactions

An exceedingly versatile and mild approach to the formation of sulfur ylides involves the intermolecular reaction of a sulfide with a diazocarbonyl compound under metal catalysis. These catalytic methods are at least as numerous as those formed through the base-promoted methodologies.[2,21] The advantages of catalytic methodologies for sulfonium ylide generation include the simplicity of the approach and the neutral conditions, especially for compounds with more than one acidic center.

7.2.2.1 *[2,3]-Sigmatropic Rearrangement* The [2,3]-sigmatropic rearrangement of allyl-substituted ylides is one of the most versatile bond reorganization processes in organic synthesis, and sulfonium ylides have played a central role in their development.[2] The process proceeds with complete allylic inversion as demanded by orbital symmetry control. Copper catalysts have been employed for the generation of allylsulfonium ylides and their subsequent [2,3]-sigmatropic rearrangement in a variety of synthetic applications. Some representative examples are illustrated in eq. 8,[22] eq. 9,[23] eq. 10,[25] eq. 11,[26,27] eq. 12,[29] eq. 13,[30] and eq. 14.[9] [2,3]-Sigmatropic rearrangement of allylic sulfur ylides derived from copper-stabilized carbenes has been used in the synthesis of trisubstituted olefins (eq. 8). There is a high level of stereoselectivity in this transformation, with a 9:1 preference for the (*E*) over the (*Z*) alkene in this example (eq. 8).[22] The inter-

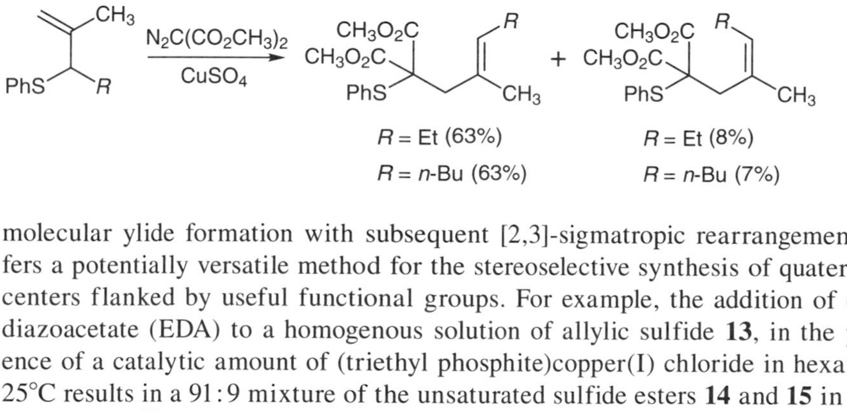

$$R = \text{Et } (63\%) \qquad R = \text{Et } (8\%)$$
$$R = n\text{-Bu } (63\%) \qquad R = n\text{-Bu } (7\%)$$

(8)

molecular ylide formation with subsequent [2,3]-sigmatropic rearrangement offers a potentially versatile method for the stereoselective synthesis of quaternary centers flanked by useful functional groups. For example, the addition of ethyl diazoacetate (EDA) to a homogenous solution of allylic sulfide **13**, in the presence of a catalytic amount of (triethyl phosphite)copper(I) chloride in hexane at 25°C results in a 91:9 mixture of the unsaturated sulfide esters **14** and **15** in 59% yield (eq. 9).[23]

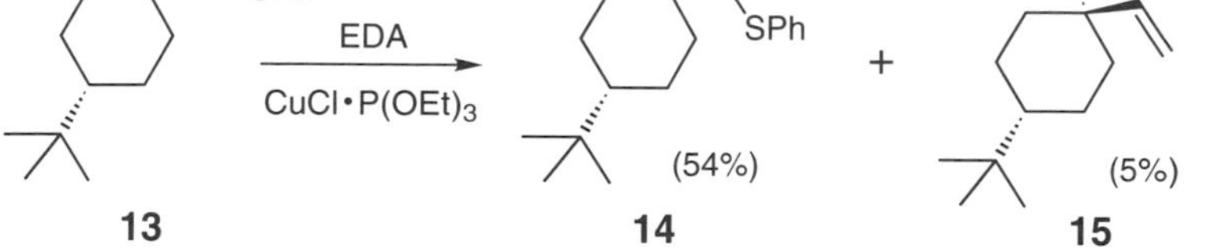

(9)

Vedejs and co-workers[24] have developed a way to enlarge, by three carbons, a sulfur-containing ring. Typically, their methodology converts a vinyl cyclic sulfide to a transient sulfur ylide that isomerizes to a ring-expansion product through the [2,3]-sigmatropic rearrangement. For example, ring expansion of the allylic sulfide substrate **16**, by reaction of diazomalonate under the copper-catalyzed conditions gave the corresponding eleven-membered adduct **17** (eq. 10).[25] Similarly, bridged alkenes, as shown in eq. 11, can also be prepared *via* the [2,3]-

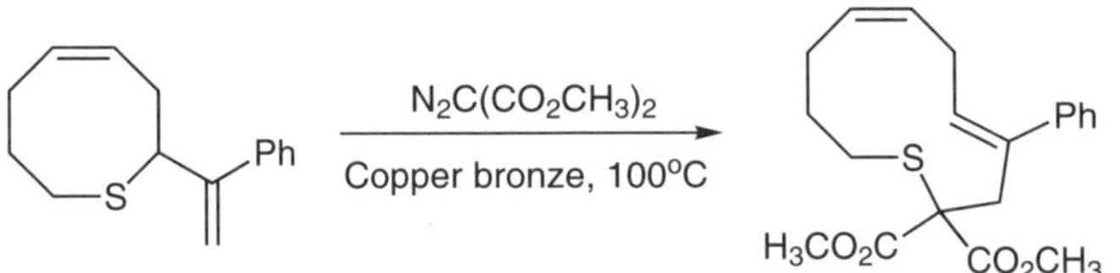

16 **17** (10)

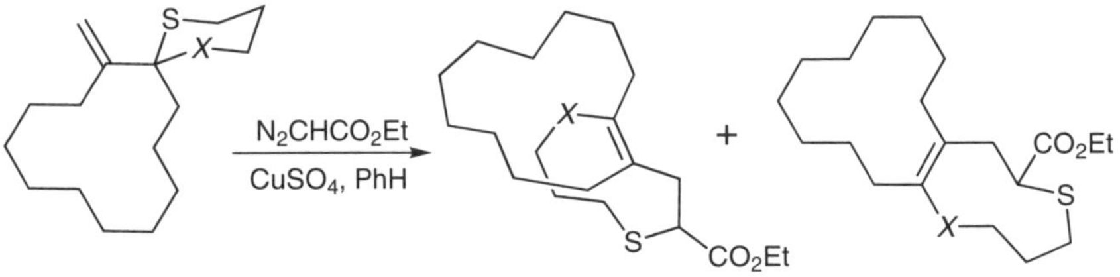

$X = CH_2$ (ref. 25)	$X = CH_2$ (29%)	$X = CH_2$ (29%)
X =S (ref. 26)	X = S (46%)	X = S (12%)

(11)

sigmatropic rearrangement of ylides derived from copper carbenes. In this case, the presence of an additional sulfur atom enhances the stereoselectivity.[26, 27]

Functionalization at C-5 of pyrone **18** has been achieved through the intermolecular ylide formation and subsequent [2,3]-sigmatropic rearrangement. Thus treatment of the pyrone with an excess of EDA in the presence of copper(II) acetylacetonate furnished the 5-substituted 2-pyrone **20** in moderate yield. The sigmatropic rearrangement afforded the intermediate **19**, which was then transformed into the 5-substituted 2-pyrone **20** by treatment with silica gel (Scheme 5).[28] In one of the first examples of sulfonium ylide generation and rear-

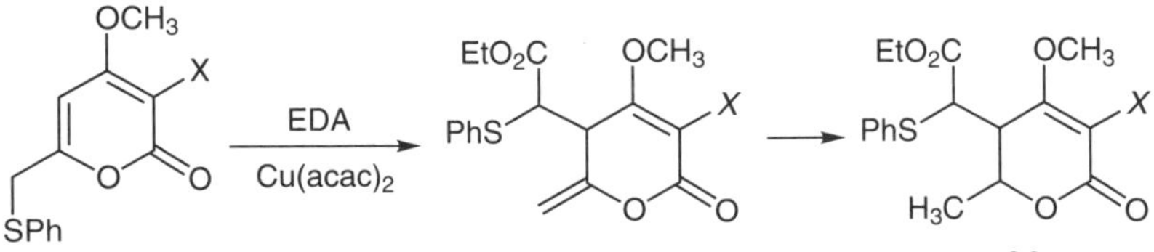

18 **19** **20**

X = H (51%); X = Br (40%)

Scheme 5

rangement with β-lactams, the [2,3]-sigmatropic rearrangement was used to convert a cephalosporin (**21**) into a penicillin (**22**), although product yields with copper were low, improvements could be anticipated through the use of dirhodium catalysts (eq. 12).[29] Suprafacial [2,3]-sigmatropic rearrangement of the intermediate sulfur ylide accounts for the observed product stereochemistry. Additional examples of reaction with β-lactams involving the intermolecular ylide formation and [2,3]-sigmatropic rearrangement are shown in eq. 13.[30]

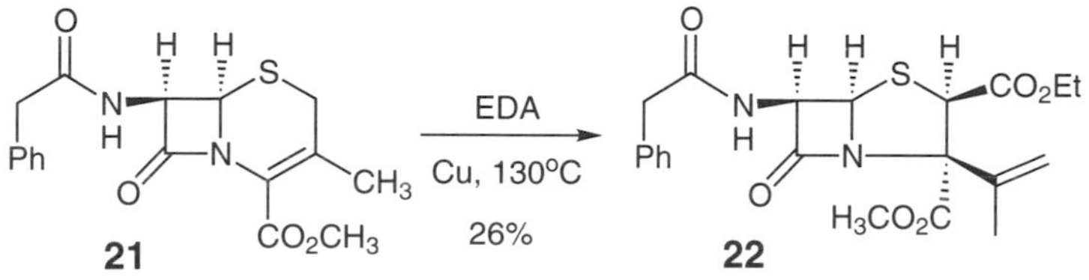

$$(12)$$

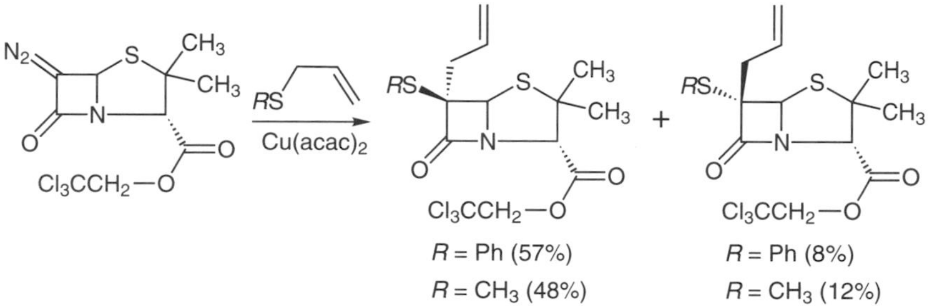

R = Ph (57%) R = Ph (8%)
R = CH$_3$ (48%) R = CH$_3$ (12%)

$$(13)$$

However, as Vedejs has pointed out,[21] copper catalysts are not particularly successful for these transformations, possibly because of the high temperatures that are often required. Rhodium(II) acetate has shown greater potential for these transformations; its use has permitted diazo decomposition to take place under milder conditions and has resulted in higher product yields for some processes. Doyle and co-workers have shown that the rhodium(II) acetate–catalyzed reaction of ethyl diazoacetate with allyl methyl sulfide affords the corresponding [2,3]-sigmatropic rearranged product in excellent yield (eq. 14).[9] 2-Ethenyl-

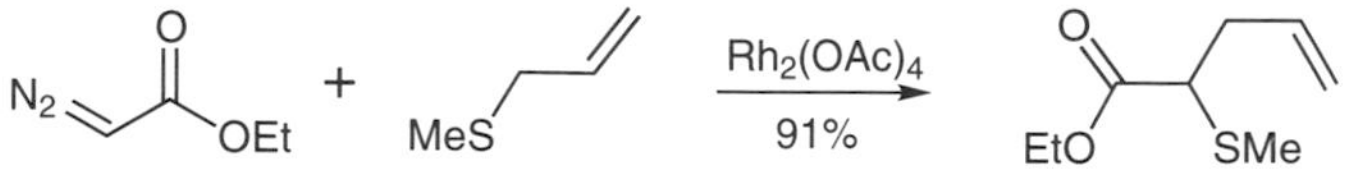

$$(14)$$

2-methyl-1,3-dithiane **23** undergoes ylide generation in the rhodium(II) acetate–catalyzed reaction with EDA with subsequent production of ring-expanded products (as well as some elimination) by the [2,3]-sigmatropic rearrangement (eq. 15).[31] The thiacyclononenes produced from **23** were a mixture of E–Z isomers.[31] Xu and co-workers[32] have employed the rhodium acetate–mediated carbene reaction in their preparation of functionalized CF$_3$-containing γ,δ-unsaturated carboxylic esters. The novel ethyl 3,3,3-trifluoro-2-diazopropionate undergoes decomposition with rhodium acetate in the presence of allyl, propargyl,

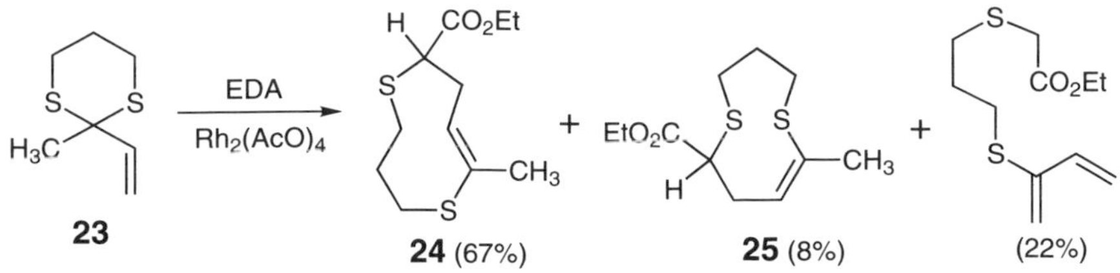

(15)

or allenic sulfides to form the products of [2,3]-sigmatropic rearrangement in high yield (eqs. 16–20).[32]

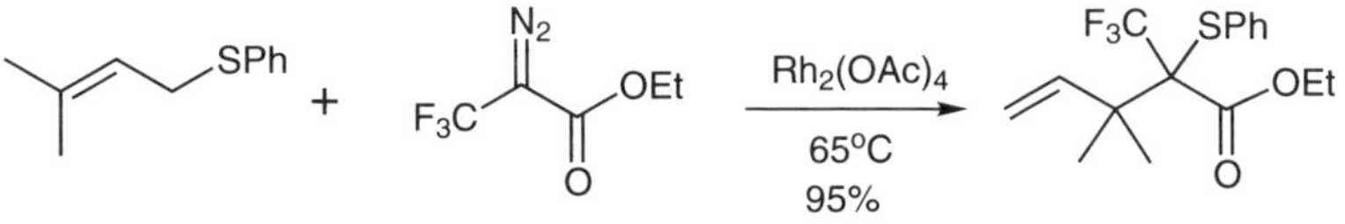

(16)

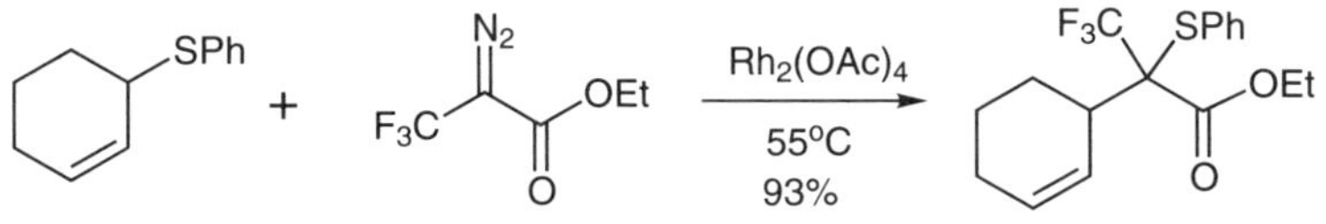

(17)

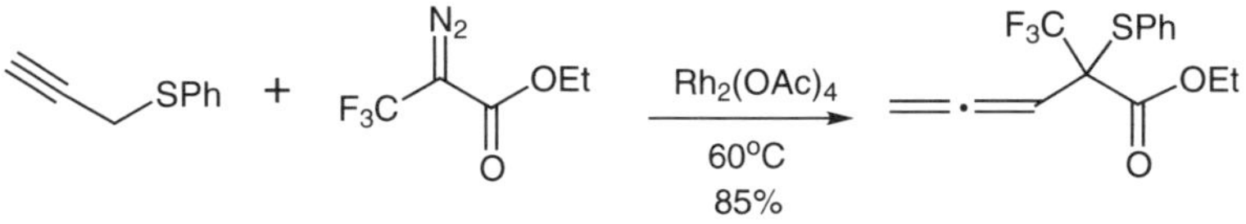

(18)

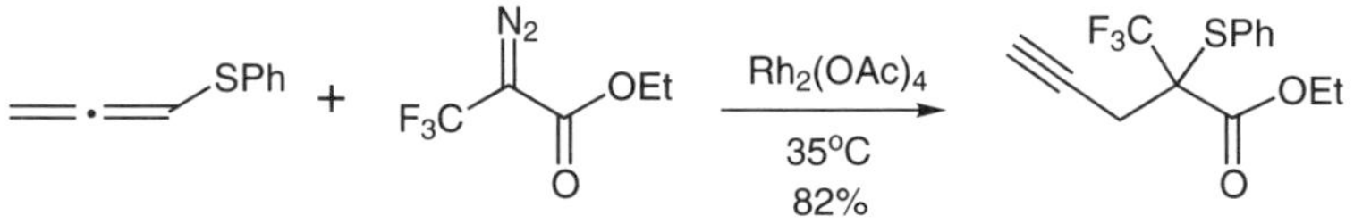

(19)

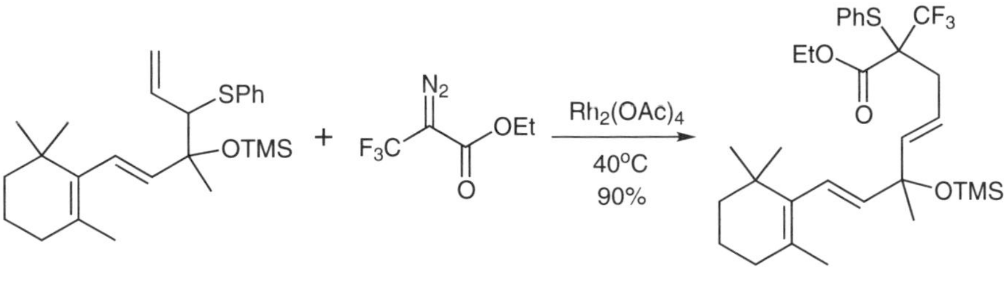

(20)

As mentioned before, diazocarbonyl compounds can act as alkylation agents for the formation of sulfonium salts, which may be deprotonated to sulfonium ylides.[6] For example, α-vinyl cyclic sulfide **26** is converted into a carbonyl-stabilized ylide **27**, as shown in Scheme 6. The resulting ylide undergoes [2,3]-

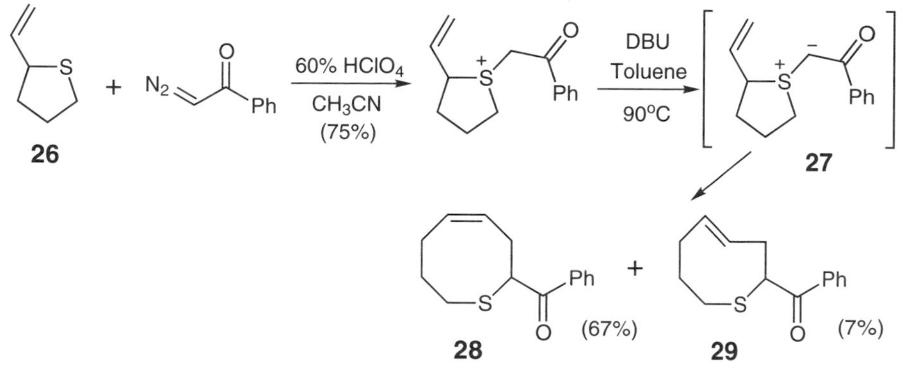

Scheme 6

rearrangement to give a mixture of ring-expansion products **28** and **29**.[25] Warren and co-workers[6] have shown that the allyl 4-methoxyphenyl sulfide **30** can be converted into a sulfonium ylide which undergoes [2,3]-sigmatropic rearrangement in high yield (eq. 21).[6] Similarly, allyl sulfides **31** and **32** are converted into the corresponding homoallylic sulfides **33** and **34** in 56 and 70% yield, respectively, by the same sequence (eq. 22).[6]

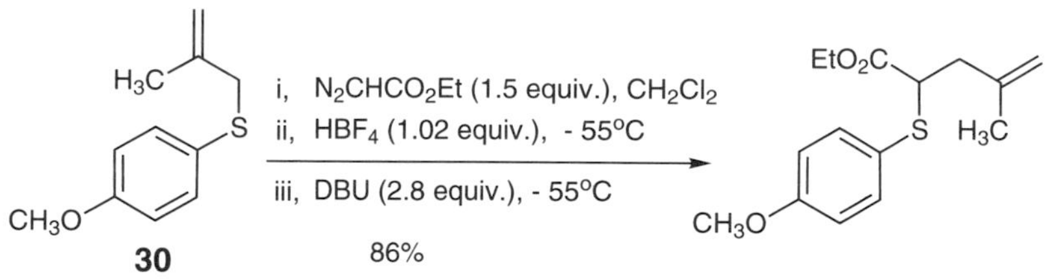

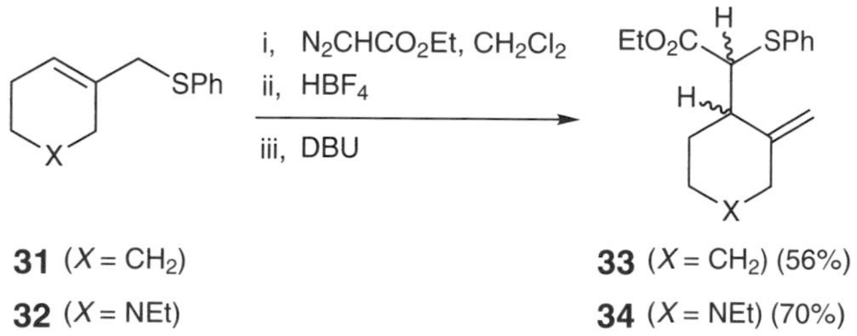

7.2.2.2 [1,2]-Insertion (Stevens Rearrangement) and Related Reactions
Although concerted [1,2] insertions are forbidden processes according to the Woodward–Hoffmann rules, there are many examples of apparent [1,2]-

insertion of ylides derived from reaction of metal carbenes and heteroatoms. Such rearrangements may occur *via* a homolysis–recombination mechanism. The [1,2] insertion is a useful synthetic methodology for carbene insertion into a C–S bond. In these reactions the leaving group ability of the intermediate sulfonium ion is important. 2-Phenyl-1,3-dithiane **35** undergoes ylide generation with EDA in the presence of rhodium(II) acetate to give the ring-expanded [1,2]-insertion product **36** with a diastereomeric ratio of 1.1 (eq. 23).[31]

REACTION OF 2-PHENYL-1,3-DITHIANE WITH EDA[31]

(23)

35   **36**

To a purple solution of phenyl-1,3-dithiane (785.3 mg, 4.00 mmol) and $Rh_2(OAc)_4$ (4.4 mg, 0.01 mmol) in 10 mL of $ClCH_2CH_2Cl$ that was maintained at 60°C was added ethyl diazoacetate (127 mg, 1.00 mmol) in 5 mL of $ClCH_2CH_2Cl$ over 1 h. After addition was complete, the reaction solution was refluxed for 15 min then cooled to room temperature. Filtration of the solution through a short plug of silica to remove the catalyst followed by chromatography on silica gel (5 : 1 hexane : ethyl acetate) to separate reactant dithiane afforded 206 mg (73% yield) of a 1.00 : 1.07 mixture of diastereomers. Major diastereoisomer: ^{1}H-NMR ($CDCl_3$) δ 7.55–7.10 (*m*, Ph), 4.28 (*d, J* = 10.7 Hz, $CHCO_2Et$), 3.92 (*q, J* = 7.1 Hz, CH_2O), 3.84 (*d, J* = 10.7 Hz, $CHPh$), 3.60–2.85 (*m, CH_2S*, 2.11 (quin, *J* = 6.0 Hz, $SCH_2CH_2CH_2S$), 0.96 (*t, J* = 7.1 Hz, CH_3CH_2O). Minor diastereoisomer: ^{1}H-NMR ($CDCl_3$) δ 7.55–7.10 (*m*, Ph), 4.51 (*d, J* = 4.4 Hz, $CHCO_2Et$), 3.94 (*q, J* = 7.1 Hz, CH_2O), 3.92 (*d, J* = 4.4 Hz, $CHPh$), 3.60–2.85 (*m*, CH_2S, 2.10 (quin, *J* = 6.0 Hz, $SCH_2CH_2CH_2S$), 0.99 (*t, J* = 7.1 Hz, CH_3CH_2O).

The carbenes derived from rhodium(II) acetate–catalyzed decomposition of dimethyl diazomalonate reacts with isothiazol-3(2*H*)-one **37** to form the sulfur ylide, which then undergoes ring expansion *via* a [1,2]-shift process (eq. 24).[33]

Kametani and co-workers[34] developed a stereoselective *C*-glycosilation reaction based on ylide formation and subsequent rearrangement. For example, treatment of 1-β-(phenylthio)triacetylglycoside **38** with dimethyl diazomalonate in the presence of a catalytic amount rhodium(II) acetate gave the *C*-glycoside **39** in 71% yield. The *C*-glycoside **39** served as the key intermediate in a total synthesis

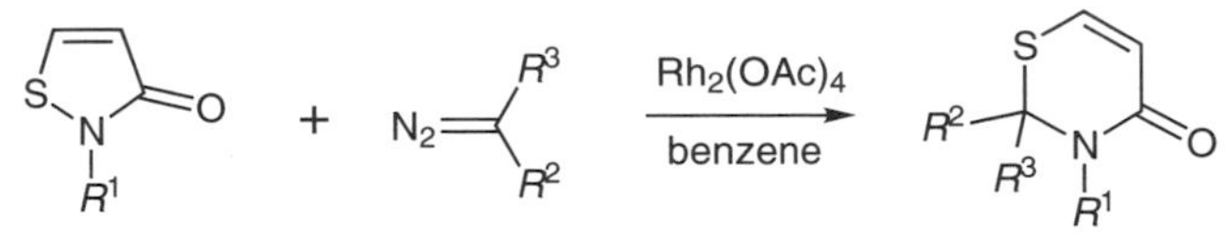

$$R^1 = \text{Et}; R^2 = R^3 = CO_2CH_3 \ (70\%)$$
$$R^1 = \text{Et}; R^2 = \text{COMe}; R^3 = CO_2\text{Et} \ (74\%)$$
$$R^1 = CO_2\text{Et}; R^2, R^3 = -CO_2CMe_2O_2C- \ (95\%)$$

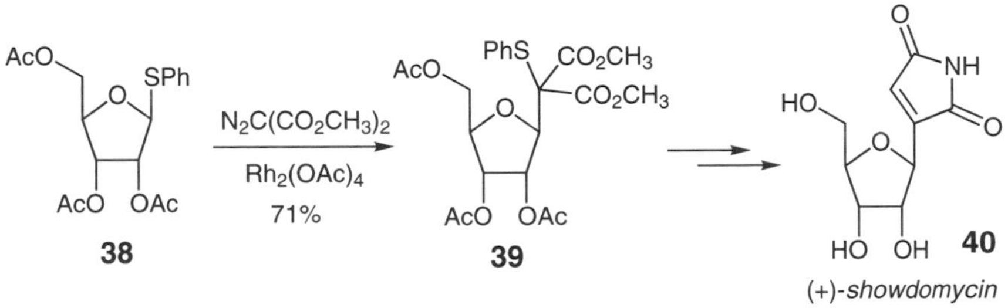

(24)

Scheme 7

of (+)-showdomycin **40** (Scheme 7).[34] Some additional examples of a C-glycosilation reaction are shown in eqs. 25–27.[34] Both α- and β-anomeric phenylthio-

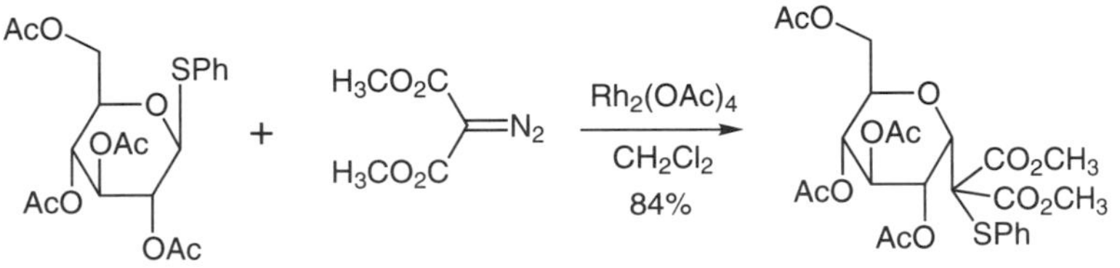

(25)

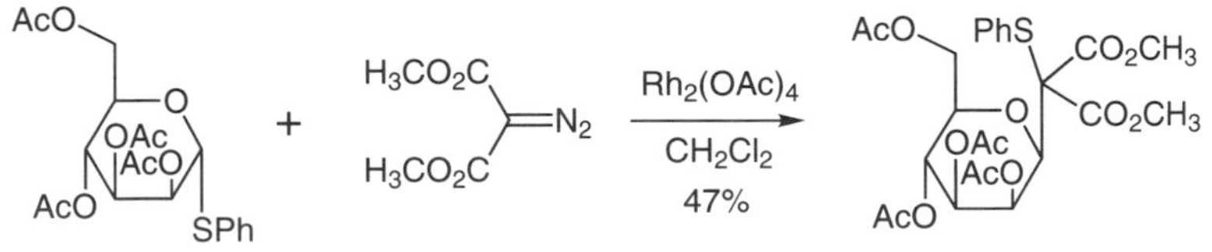

(26)

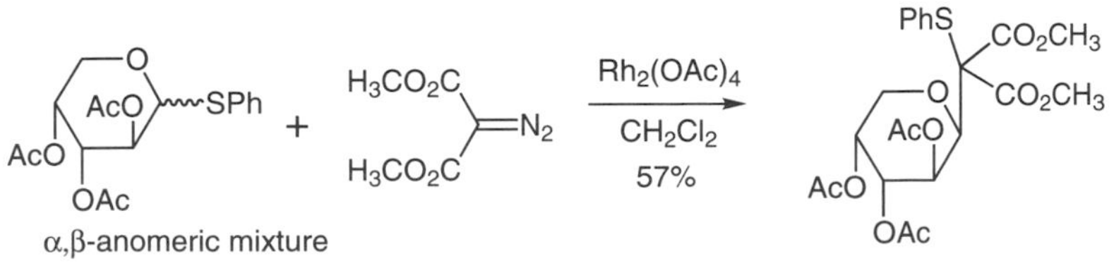

(27)

glycosides undergo sulfonium ylide formation with rearrangement. The stereochemical outcome at the anomeric center was retained in the reaction of the furanoside (Scheme 7), However, the stereochemistry at the anomeric position in the products derived from the reaction of six-membered counterparts was inverted (eqs. 25–27), and this stereochemical control seemed to be affected by an anomeric effect. The observed stereoselectivity giving only one anomer of the product can be rationalized by the possible intramolecular participation as shown in Scheme 8.

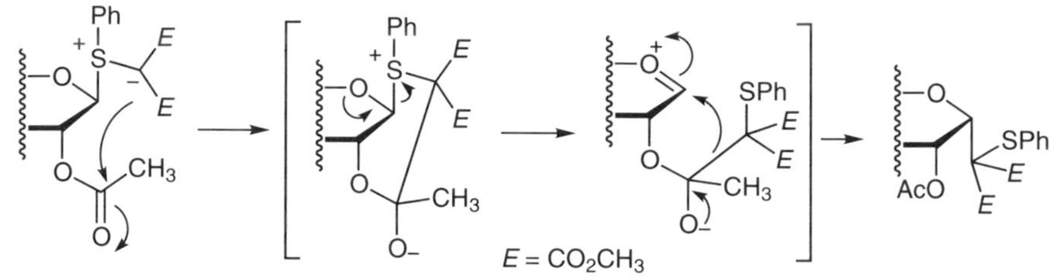

$E = CO_2CH_3$

Scheme 8

Two pyrrolizidine alkaloids, (+)-heliotridine **45** and (+)-retronecine **46**, have recently been synthesized by Kametani's group (Scheme 9).[35,36] The crucial step

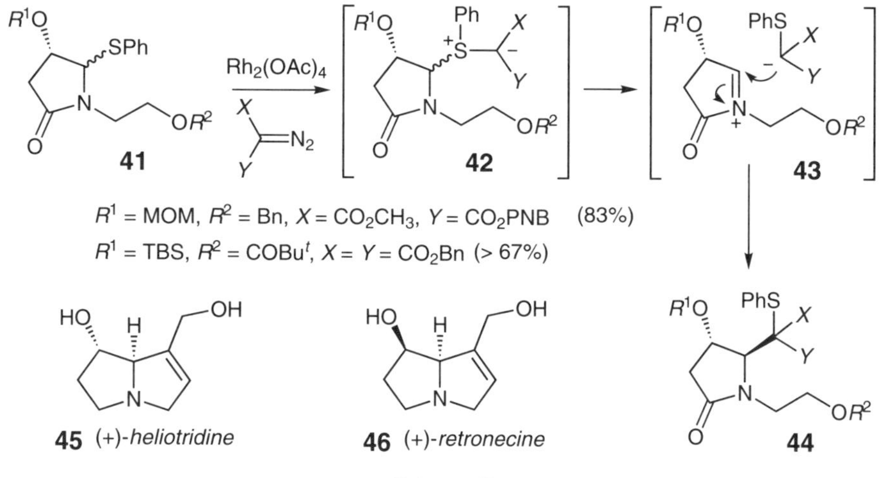

R^1 = MOM, R^2 = Bn, X = CO_2CH_3, Y = CO_2PNB (83%)

R^1 = TBS, R^2 = COBu^t, $X = Y = CO_2Bn$ (> 67%)

45 (+)-heliotridine   **46** (+)-retronecine   **44**

Scheme 9

of both synthetic sequences was the stereoselective formation of the sulfur ylide **42** *via* rhodium acetate catalysed decomposition of diazomalonate in the presence of the optically active sulfide **41**. The ylide **42** then rearranged through the acyliminium salt (**43**) to afford the key intermediate **44**. Further functional group manipulations furnished both target alkaloids.

7.2.3 Intramolecular Formation of Sulfur Ylides and Subsequent Reactions

The intramolecular formation of ylides and subsequent reactions have received considerable attention and have shown great versatility in the synthesis of natural products. In this section, we will briefly review the two major metal-stabilized ylide reactions: [2,3] rearrangement and [1,2] insertion (Stevens rearrangement).

7.2.3.1 [2,3]-Sigmatropic Rearrangement Cyclic allylsulfonium ylides are formed from the catalytic decomposition of diazocarbonyl compounds containing allylic sulfide substituents. Rhodium(II) carboxylates have been used as catalysts for intramolecular S-allyl sulfonium ylide generation with subsequent [2,3]-sigmatropic rearrangement. Three representative examples are illustrated in eqs. 28 and 29[17] and 30.[37]

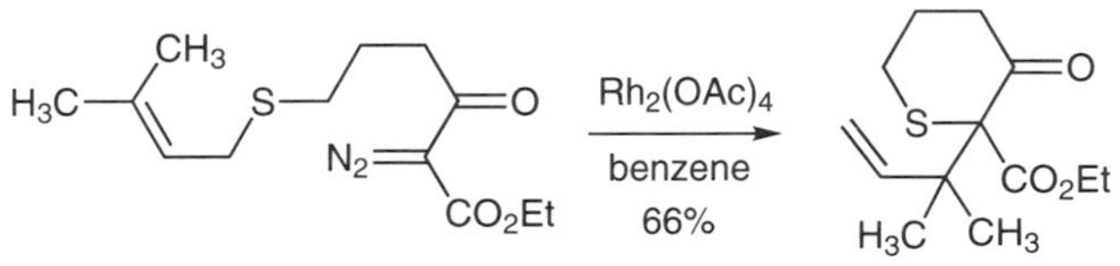

$$(28)$$

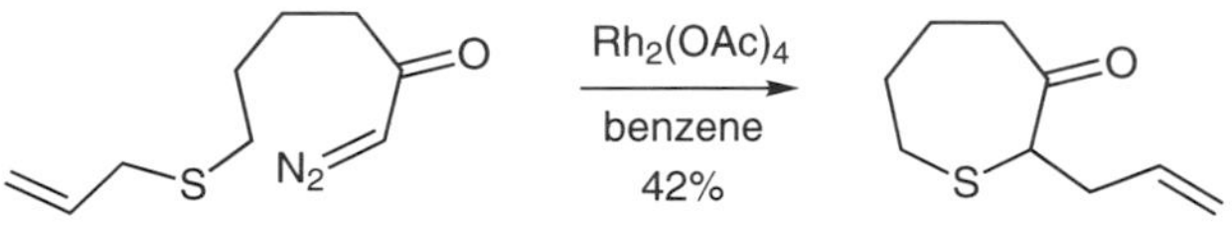

$$(29)$$

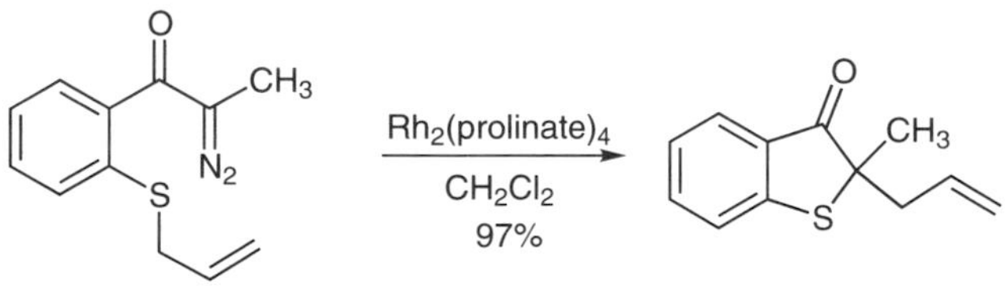

$$(30)$$

An attractive method for construction of five- and six-membered lactones, based on intramolecular allylic sulfur ylide formation and subsequent [2,3]-sigmatropic rearrangement, has been reported by Kido and Yoshikoshi's group (Schemes 10–12).[38–40] The synthesis of five- and six-membered lactones **48** involves a transannular reaction with the initial formation of an eight- or nine-membered cyclic allyl sulfonium ylide followed by a [2,3]-sigmatropic rearrangement. The yields are generally good, and the stereosectivities are modest (1:3–4). In addition, a stereochemically fixed substituent group R in **47** was found to be an important factor in controlling the stereoselectivity in the rearrangement process. As shown in Scheme 11, treatment of diazomalonate **49** with Rh₂(OAc)₄ in benzene furnished the valerolactone **50** as the sole rearranged product in 92% yield.[40] The formation of intermediate ylides from diazocarbonyl

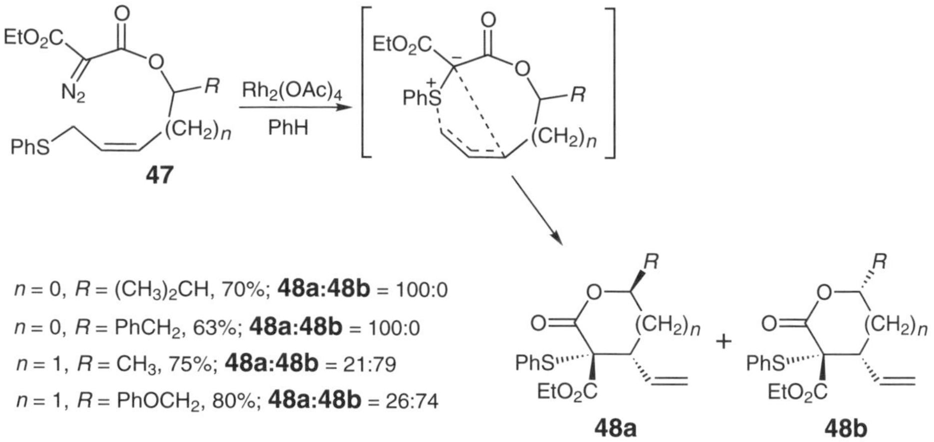

EtO2C
N2
PhS
R
(CH2)n
47
Rh2(OAc)4
PhH
PhS
R
(CH2)n
n = 0, R = (CH3)2CH, 70%; 48a:48b = 100:0
n = 0, R = PhCH2, 63%; 48a:48b = 100:0
n = 1, R = CH3, 75%; 48a:48b = 21:79
n = 1, R = PhOCH2, 80%; 48a:48b = 26:74
48a
48b

Scheme 10

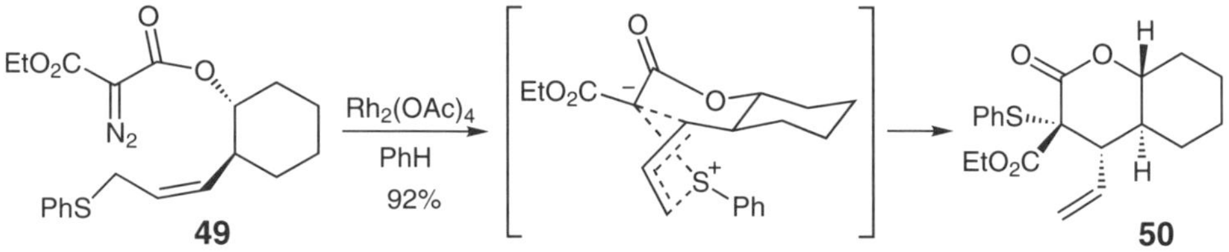

EtO2C
N2
PhS
49
Rh2(OAc)4
PhH
92%
Ph
PhS
EtO2C
H
50

Scheme 11

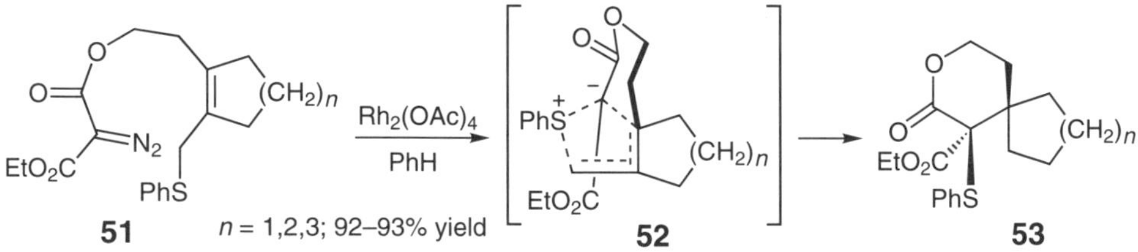

O
(CH2)n
N2
EtO2C
PhS
51
Rh2(OAc)4
PhH
n = 1,2,3; 92–93% yield
PhS
EtO2C
52
53

Scheme 12

precursors **47** and **49** depends on the presence of *Z* and *E* double bond. (See Section 7.2.5.)

Spiro-fused lactones **53** have also been synthesised *via* a similar strategy (Scheme 12).[41] Thus treatment of diazomalonate **51** with a catalytic amount of rhodium acetate in refluxing benzene resulted in [2,3]-sigmatropic rearrangement via the nine-membered cyclic transition state structure **52** to produce **53** in excellent yield (92–93%). This spiroannulation reaction has been extended to the synthesis of a spiro carbocyclic compound by employing β-ketoester **54** as starting material, and has also been used in the stereoselective synthesis of the sesquiter-

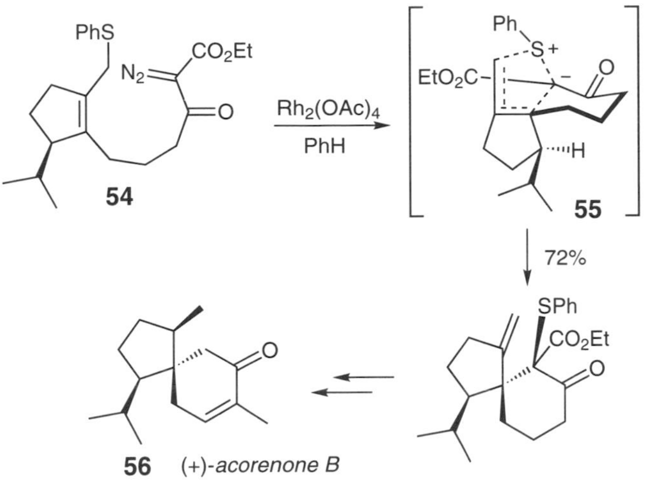

56 (+)-*acorenone B*

Scheme 13

pene (+)-acorenone B **56** (Scheme 13).[42] The remarkable stereoselectivity in the rearrangement process may result from the metal carbene approaching the reaction site from the less hindered side, away from the isopropyl group, as seen in the transition state **55** (Scheme 13).

More recently, the highly substituted cyclohexanone **59** has been obtained by [2,3]-sigmatropic rearrangement *via* the stereocontrolled cyclic transition state **58** derived from the rhodium(II)-catalyzed cyclization of optically active acyclic α-diazo-β-ketoester **57** (Scheme 14).[43,44] Cyclohexanone **59** is a versatile intermediate for enantioselective synthesis of elemanoid sesquiterpenes. Further elaboration

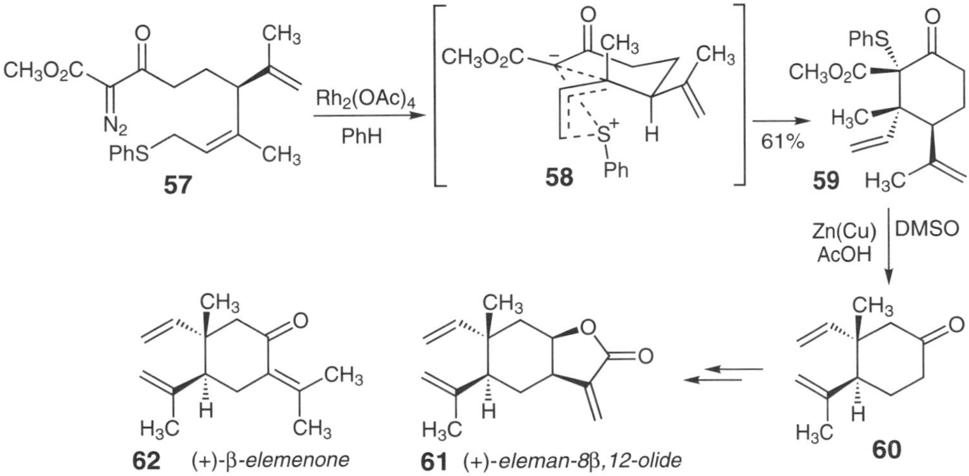

62 (+)-β-*elemenone* 61 (+)-*eleman-8β,12-olide*

Scheme 14

of **59** into the cyclohexanone **60** accomplished a formal asymmetric syntheses of (+)-elemenone **61** and (+)-eleman-8β,12-olide **62** (Scheme 14).[43,44]

Seven-membered lactones **64a** and **64b** have been synthesized *via* an approach involving the intramolecular formation of six-membered cyclic sulfonium ylides followed by [2,3]-sigmatropic rearrangement (Scheme 15).[45,46] This approach is

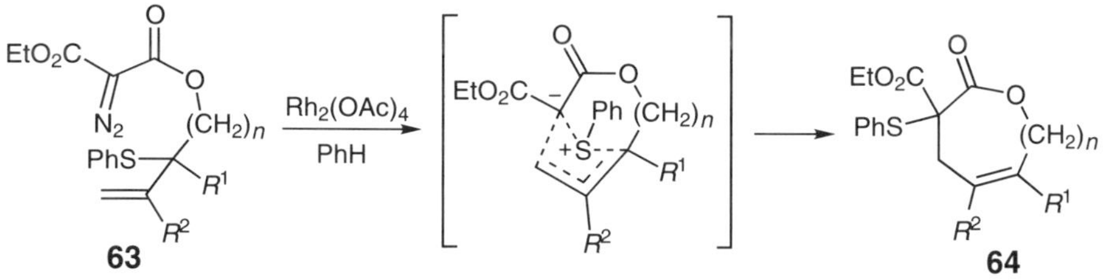

64 a : n = 1, R^1 = H, R^2 = CH$_3$, 53%   **64 b** : n = 1, R^1 = PhCH$_2$, R^2 = H, 49%

64 c : n = 2, R^1 = R^2 = H, 19%

Scheme 15

analogous to the "ring-growing sequence" described in eqs. 10 and 11. The yields in the formation of the seven-membered lactones **64a** and **64b** were good, whereas a similar reaction that afforded the corresponding eight-membered lactone **64c** proceeded in poor yield. When the olefinic group was part of a cyclic system, as illustrated in eq. 31,[46] ylide formation and rearrangement afforded the bicyclic adduct **65** in high yield.[45,46]

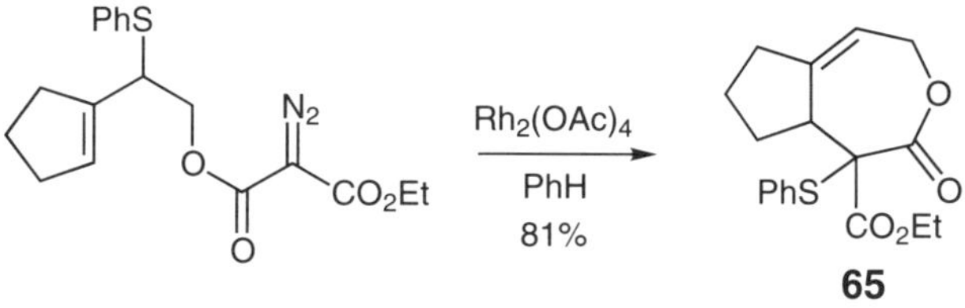

(31)

This synthetic methodology has been extended to the efficient synthesis of the bridged δ-lactone **68** (Scheme 16).[47] In this case, the bridged δ-lactone **68** arose

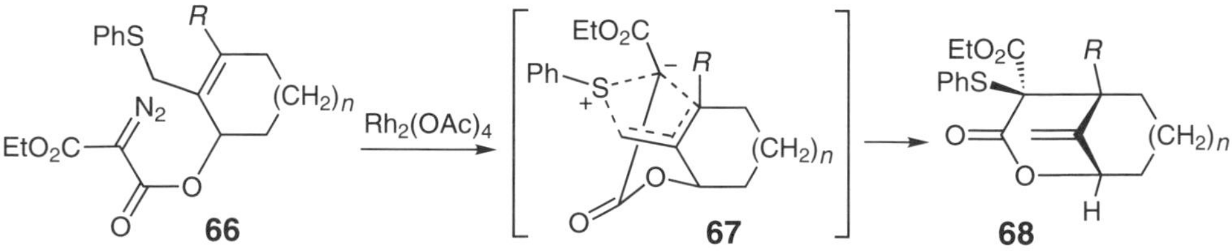

68 a : n = 0, R = H, 34%; **68 b** : n = 1, R = Me, 55%

Scheme 16

from the [2,3]-sigmatropic rearrangement of a stereocontrolled seven-membered cyclic transition state **67**. Because the [2,3]-sigmatropic rearrangement usually proceeds through a highly ordered cyclic transition state, comformational analysis has been applied to predict the stereochemical outcome.

7.2.3.2 [1,2]-Insertion (Stevens Rearrangement) and Related Reactions

The [1,2]-shift process with cyclic sulfonium ylides has recently received considerable attention. As mentioned earlier, benzylsulfonium ylide **11**, derived from $Rh_2(OAc)_4$-catalyzed diazo decomposition of β-keto-α-diazoester, undergoes thermal Stevens rearrangement (Scheme 4). Copper catalysts have also been successfully applied in this type transformation, and a representative example is shown in eq. 32.[48] In these reactions the presence of groups with good migratory

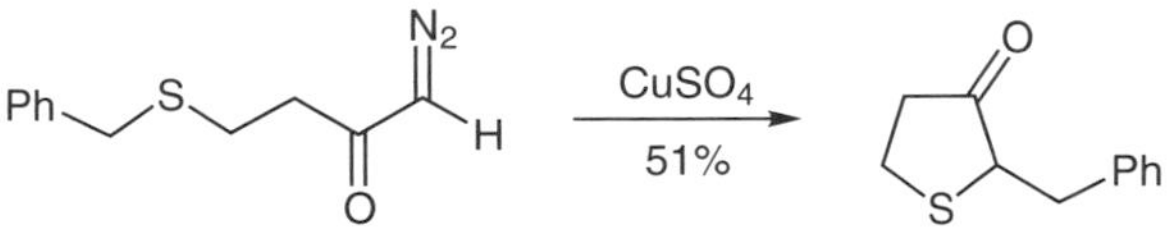

$$(32)$$

aptitude, such as the benzyl group attached to the intermediate sulfonium ion, is important.

Two aromatic sesquiterpenes, ($\pm$)-laurene **71** and ($\pm$)-cuparene **72**, were synthesized by a reaction sequence in which the key step was the formation of a new carbon–carbon bond at a benzylic position from the sulfonium ylide **69** followed by the [1,2]-shift process to give **70** (Scheme 17).[49,50]

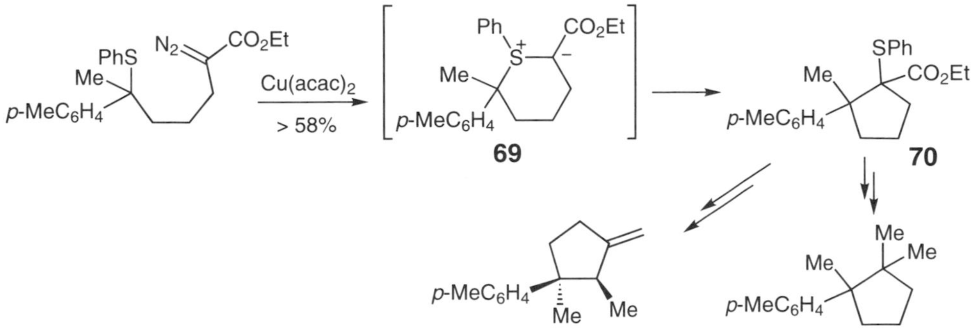

Scheme 17

Sulfonium ylide **74** derived from catalytic decomposition of diazoester **73** undergoes a [1,2]-shift process through the acyliminium salt (**75**) to form a new five-membered ring adduct **76**.[51,52] The bicyclic compound **76** was the key intermediate in the construction of three pyrrolizidine alkaloids, ($\pm$)-trachelanthamidine **77**, ($\pm$)-isoretronecanol **78**, and ($\pm$)-supinidine **79** (Scheme 18).[51,52] In

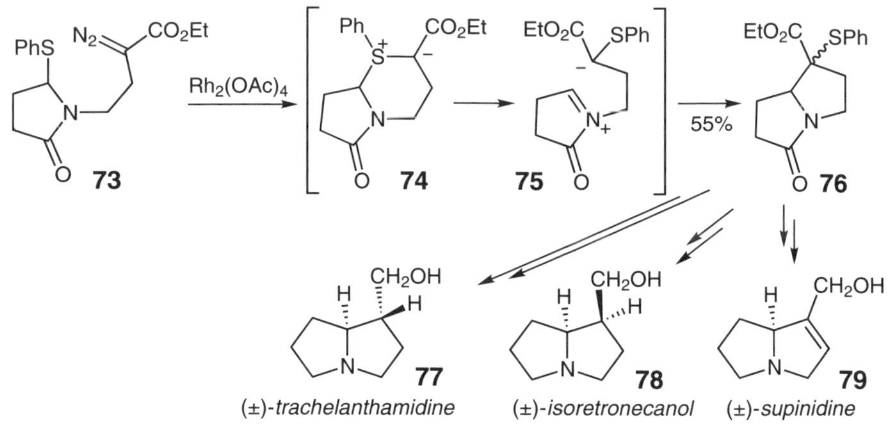

Scheme 18

addition, this strategy has also been used to structurally modify a β-lactam.[53] Copper-catalyzed decomposition of diazoketone **80** facilitates Stevens rearrangement, *via* the iminium ion intermediate **81**, to form the tricyclic product **82** in good yield (Scheme 19).[54,55] The involvement of the iminium ion **81** is consistent with the change in stereochemistry at position 4.

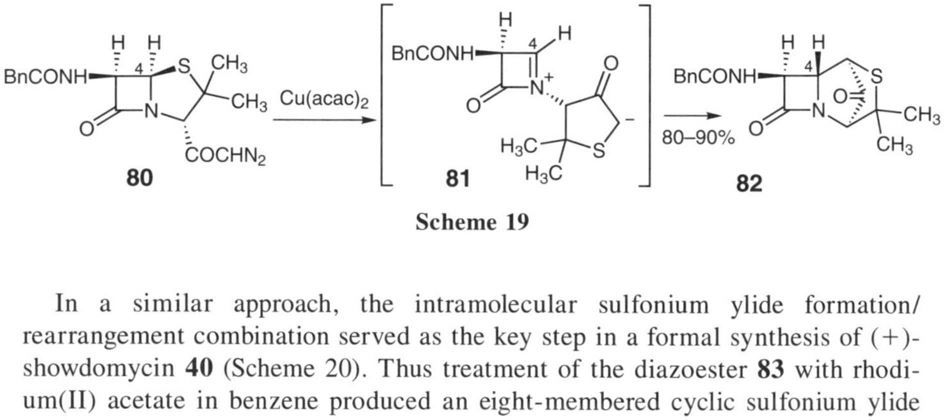

Scheme 19

In a similar approach, the intramolecular sulfonium ylide formation/ rearrangement combination served as the key step in a formal synthesis of (+)-showdomycin **40** (Scheme 20). Thus treatment of the diazoester **83** with rhodium(II) acetate in benzene produced an eight-membered cyclic sulfonium ylide **84**, which rearranged to an oxonium intermediate **85** and then furnished the lactone **86** in 48% yield *via* a stereoselective intramolecular β-C-glycosilation.[56]

7.2.4 β-Elimination and Related [1,4]-Rearrangement

In addition to [2,3] and [1,2] rearrangement, the other major reaction pathway for decomposition of sulfonium ylides is β-elimination. Some sulfonium ylides undergo thermal intramolecular eliminations, especially those carrying large alkyl groups on the sulfur atom. The literature on the β-elimination of stabilized sul-

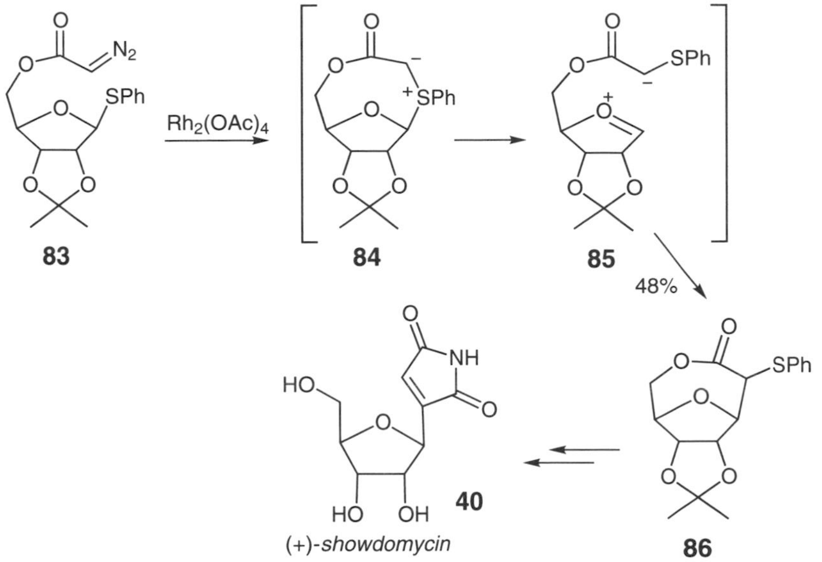

Scheme 20

fonium ylides up to 1977 has been reviewed by Ando.[4] Representative examples of a β-elimination reaction of metal-stabilized sulfoxonium ylides, obtained from both intermolecular and intramolecular reactions, are shown in eq. 33[31] and Schemes 21 and 22. Thus decomposition of S-ethyl diazo sulfide **87** in boiling benzene in the presence of rhodium(II) acetate gave the ylide **88**, which on further heating in xylene eliminated ethylene to give ethyl 3-oxothiane-2-carboxylate **89** in good yield (Scheme 21).[17] Doyle and co-workers have reported that the elimination reaction of the metal-stabilized ylide, in some cases, can be competitive with [2,3]-sigmatropic rearrangement. As shown in eq. 33,[31] rhodium

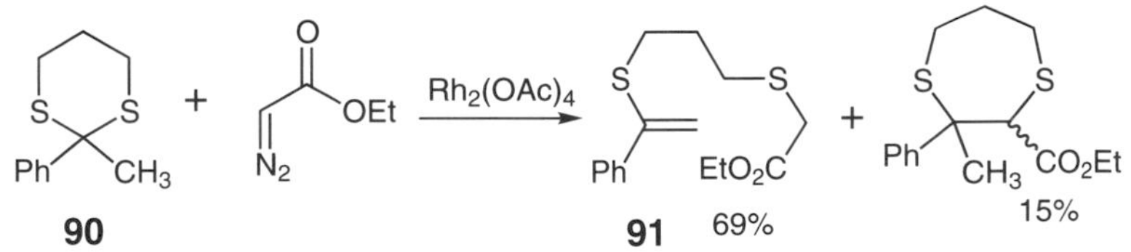

(33)

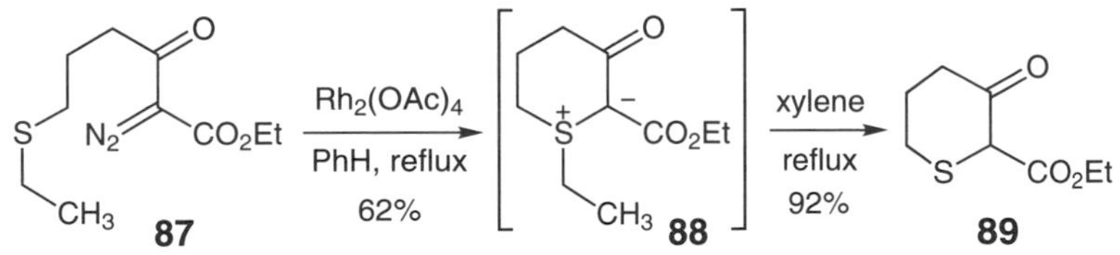

Scheme 21

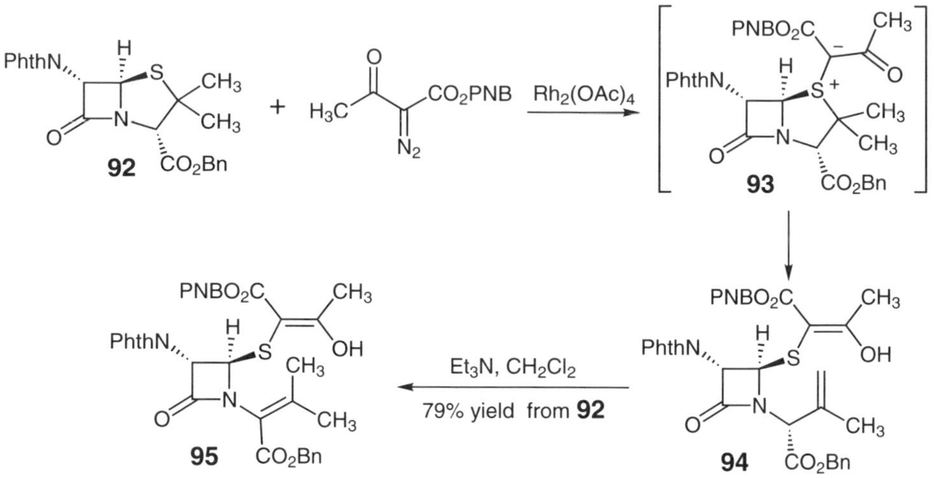

Scheme 22

acetate–catalyzed reaction of ethyl diazoacetate with 1,3-dithiane **90** resulted in the formation of **91** as the major product along with a minor product derived from a [1,2]-shift.[31] An interesting example where the β-elimination reaction competes with the Stevens rearrangement is shown in Scheme 22. In this case, the sulfonium ylide **93** undergoes elimination to afford **94**, which was isomerized to give the α,β-unsaturated ester **95** in 79% yield from **92** (Scheme 22).[57] In contrast, intramolecular reaction with a similar derivative **80** occurs without elimination and, instead, undergoes a [1,2]-shift pathway (Scheme 19).

When the amino functional group in **92** (Scheme 22) was protected as an amide (**96**), the rearrangement of ylide **97** occurred without elimination of the β-hydrogen. Instead, the bicyclic oxa-derivative **98** was formed (Scheme 23).[58]

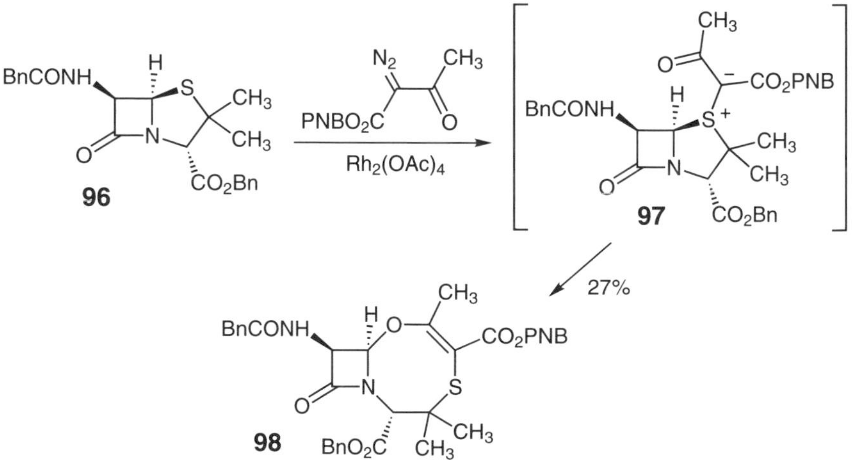

Scheme 23

Compound **98** results from a [1,4] rearrangement that is an extension of the Stevens rearrangement. Kametani and co-workers[59] have also shown that the reaction of heterocyclic sulfide derivatives with α-diazoesters in the presence of rhodium(II) catalyst give [1,4]-rearranged products **99** (Scheme 24).

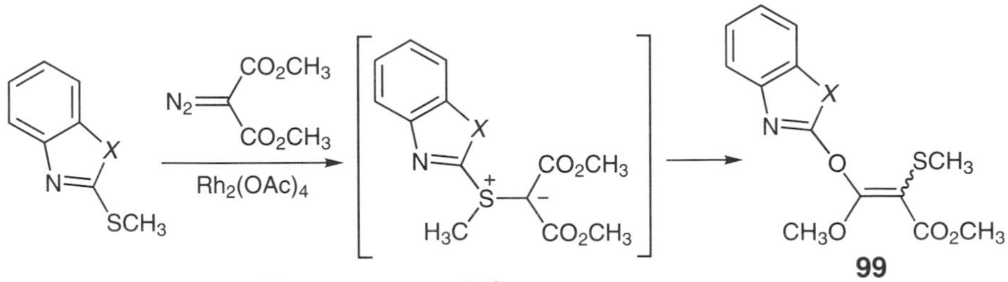

99a: $X = O$, 76%; **99b**: $X = NCH_3$, 93%

Scheme 24

7.2.5 Chemo- and Stereoselectivity in Sulfur Ylide Formation and Subsequent Rearrangement

Metal carbenes derived from the metal-catalyzed decomposition of diazocarbonyl compounds undergo various chemical transformations. Control of chemoselectivity by choice of the appropriate catalyst has significantly increased the synthetic viability of catalytic sulfonium ylide generation. As mentioned earlier, in the intermolecular reactions of both copper[7,8] and rhodium[9] carbenes with substrates containing both olefinic and thioether functions, formation of the sulfonium ylide is the favored process. However, the chemoselectivity of the intramolecular reaction is strongly dependent upon the nature of the diazocarbonyl compound and the choice of metal catalyst. Examples are known where intramolecular reactions of metal carbenes with the allylsulfide functional group result in competition between formation of cyclic sulfonium ylide and cyclopropanation. Rhodium acetate–catalyzed decomposition of α-diazo-β-ketoester **100** bearing an E-olefinic bond afforded the cyclopropane derivative **101** leaving the allylic sulfide group intact (Scheme 25),[40] whereas the corresponding Z-olefinic bond-containing α-diazo-β-ketoester **57** yielded the cyclohexanone **59** *via* a cyclic sulfonium ylide intermediate (Scheme 14).

More recently, McMills and co-workers[60] have reported the synthesis of substituted azabicyclic ring systems based on the [2,3]-sigmatropic rearrangement of catalytically-generated cyclic sulfonium ylides (eqs. 34 and 35).[60] Their studies also demonstrated that both the catalyst and solvent, as well as the diazocarbonyl substrate, are critical in determining the pathway, followed by the metal-stabilized carbene intermediate. As shown in eq. 34,[60] decomposition of *cis*-diazoamide **102** with catalytic $Rh_2(OAc)_4$, $Cu(hfacac)_2$, and $Pd(OAc)_2$ in benzene resulted mainly in the formation of the pyrrolizidine–cyclopropane **103**, whereas rhodium caprolactamate in fluorobenzene as solvent resulted in a complete switch of the product ratio to the ylide pathway to produce mainly **104**. The olefin

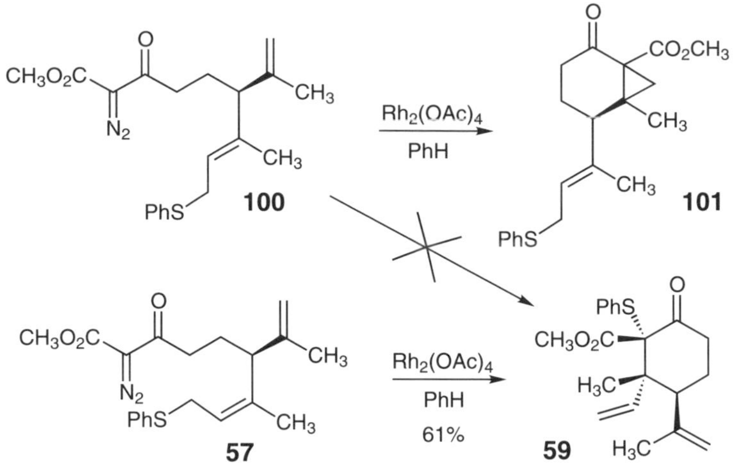

Scheme 25

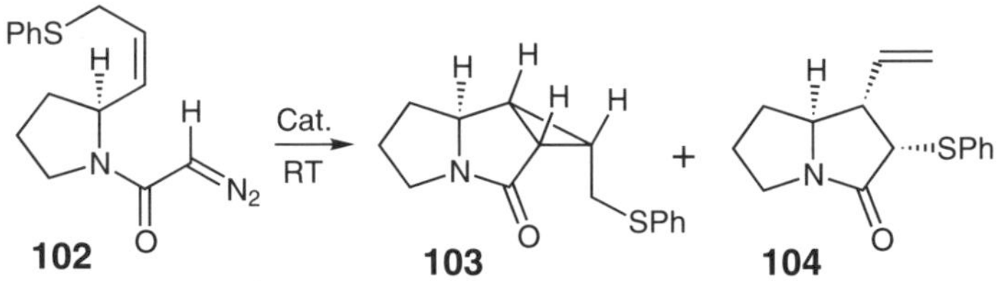

(34)

Catalyst	Solvent	Yield	**103:104**
Rh₂(OAc)₄	PhH	62%	66:34
Cu(hfacac)₂	PhH	61%	92:8
Pd(OAc)₂	PhH	63%	98:2
Rh₂(cap)₄	PhF	60%	6:94

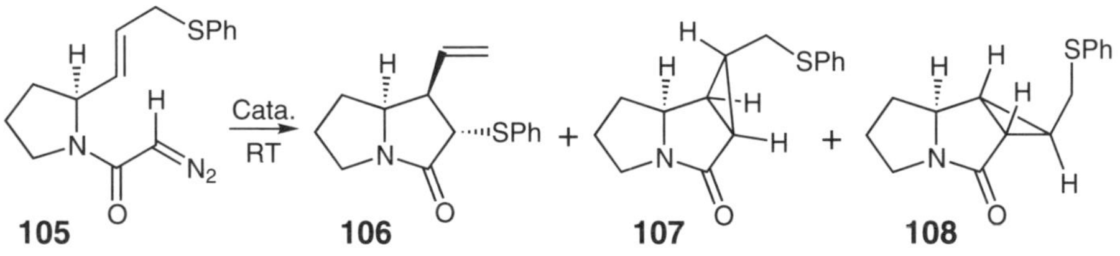

(35)

Catalyst	Solvent	Yield	**106:107:108**
Rh₂(OAc)₄	PhH	70%	0:69:31
Cu(hfacac)₂	PhH	68%	0:68:32
Pd(OAc)₂	PhH	78%	0:73:27
Rh₂(cap)₄	PhH	71%	90:0:10

geometry also plays a role in determining the reaction outcome. Decomposition of the *trans*-diazoamide **105** with $Rh_2(OAc)_4$, $Cu(hfacac)_2$, and $Pd(OAc)_2$ produced only cyclopropanation product **107** and **108**. However, rhodium caprolactamate furnished the *anti*-azabicyclooctane **106** as the dominant product (eq. 35).[60]

In some cases, C–H insertion of the rhodium carbene can be competitive with cyclic sulfonium ylide formation and [2,3]-sigmatropic rearrangement. For example, treatment of diazoketone **109** with rhodium(II) acetate resulted in the formation of both **110** and **111** (eq. 36).[61] In this case, the ratio of ylide formation to C–H insertion (**110**:**111**) was 9:1.

(36)

109   **110** 89%   **111** 10%

The stereochemistry observed in product formation reflects comformational preferences in the transition state for the sigmatropic rearrangement. The [2,3]-sigmatropic rearrangement of cyclic sulfonium ylides generally proceed stereospecifically with respect to the geometry of the olefin group. Examples illustrating this are included in Section 7.2.3.1. Where the reaction is presumed to involve fragmentation of the initially formed ylide into an ion pair, the direction of attack of the nucleophilic partner determines the stereochemistry of the product as shown in eqs. 25–27, and in Schemes 7, 9, 19, and 20.

Asymmetric induction *via* sulfonium ylide formation and tandem [2,3]-sigmatropic rearrangement should be possible through the use of chiral auxiliaries incorporated into diazocarbonyl precursors. Kurth and co-workers[5] have reported the utilization of a thioxanone-based [2,3]-sigmatropic rearrangement strategy to construct chiral-3-methylpent-4-enoic acid **117**. Thus diazocarbonyl **113** derived from the chiral auxiliary **112** was treated with a Lewis acid to give the corresponding thioxonium salt **114**, which was then transformed to ylide **115** by treatment with base. The ylide underwent the [2,3]-sigmatropic rearrangement to give **116** in 66% yield with 94:6 dr. The thioxonone **116** was finally disassembled to give the chiral 3-methyl-4-pentenoic acid **117** (Scheme 26).

7.3 FORMATION OF OXONIUM YLIDES

Early investigations of catalytic reactions between diazocarbonyl compounds and ethers often portrayed them as relatively unreactive towards oxonium ylide generation and unsuitable for applications in organic synthesis.[62–64] Unlike the sulfonium ylides described in Section 7.2.1, stable oxonium ylides have not yet been reported, and their relative instability is due to their equilibrium position favoring the metal carbene (Scheme 27).[31] Ethyl ether has been successfully employed

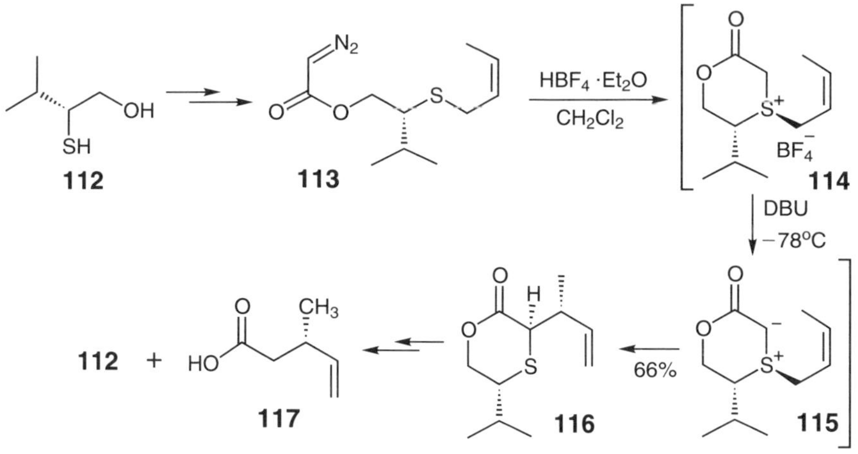

Scheme 26

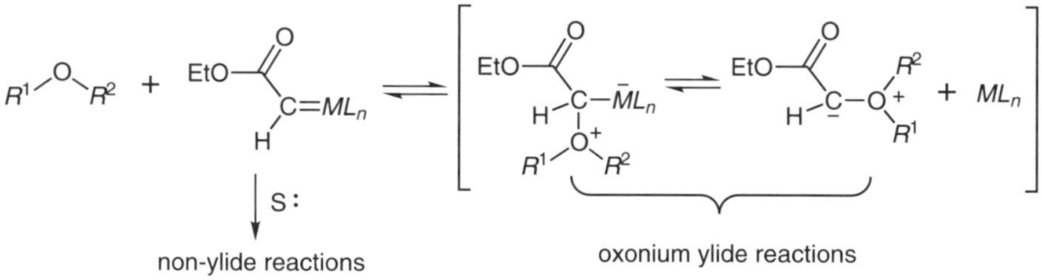

non-ylide reactions   oxonium ylide reactions

Scheme 27

as a solvent for rhodium acetate–catalyzed cyclopropanation reactions,[65] demonstrating that even olefins can overwhelm oxonium ylide formation. However, during the past 10 years, an increasing number of investigations have provided evidence that oxonium ylides are viable and that they have significant synthetic utility.

Rhodium carboxylates and homogenous copper complexes have emerged as highly efficient catalysts for the generation of metal-stabilized oxonium ylides. The use of rhodium carboxylates as catalysts for diazodecomposition made possible the generation of metal carbenes at more moderate temperatures than were previously possible with most copper catalysts.[7]

Useful reactions of oxonium ylides fall into two classes: [2,3]-sigmatropic rearrangement, and [1,2]-insertion Stevens rearrangement. In addition, oxygen transfer processes, which are generally associated with epoxides, and β-elimination reactions also occur with oxonium ylides.

7.3.1 Intermolecular Formation of Oxonium Ylides and Subsequent Reactions

One of the earliest examples of the formation of oxonium ylides was reported by Nozaki and co-workers.[66,67] Tetrahydrofuran **119** was obtained *via* the [1,2]-Stevens rearrangement of the oxonium ylide **118**, as shown in Scheme 28. Later,

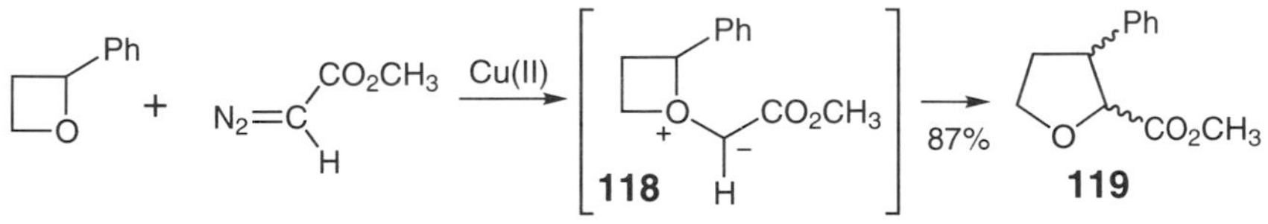

Scheme 28

Ando and co-workers[7] reported that copper sulfate–catalyzed decomposition of dimethyl diazomalonate in allyl ethers gave mainly products derived from the [2,3]-sigmatropic rearrangement of an oxonium ylide intermediate. In this case, the reaction was carried out at high temperature and the yields were modest. Doyle and co-workers[68] have found that rhodium(II) acetate is a more effective catalyst for the intermolecular generation of allylic oxonium ylides from diazocarbonyl compounds and allyl ethers (Scheme 29). With substituted allyl ethers

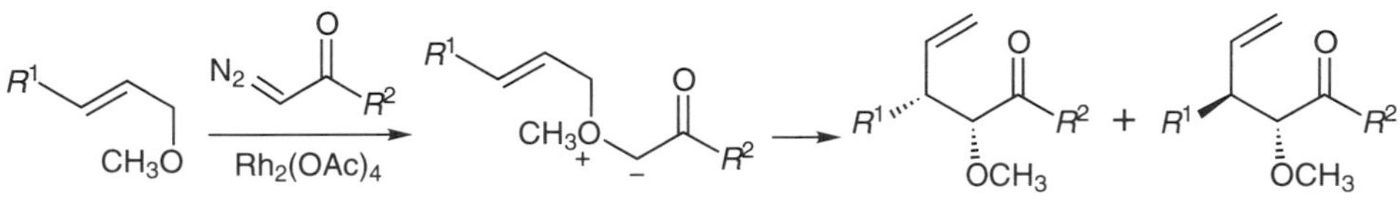

R^1 = Ph, R^2 = Ph	86% yield	**121:122** = 91:9
R^1 = Ph, R^2 = OEt	95% yield	**121:122** = 83:17
R^1 = Ph, R^2 = p-MeOC$_6$H$_4$	85% yield	**121:122** = 91:9
R^1 = Me$_3$Si, R^2 = OEt	68% yield	**121:122** = 97:3

Scheme 29

and diazoketones, oxonium ylide (**120**) generation occurs almost exclusively, and the subsequent [2,3]-sigmatropic rearrangement products **121** and **122** exhibit high degrees of diastereocontrol, favoring the *erythro* isomer (**121**) (Scheme 29) with *trans*-olefins, and favoring the *threo* isomer (**122**) with *cis*-olefins to nearly the same degree. These results can be explained by steric and/or electronic influences in the transition states for the [2,3]-sigmatropic rearrangement. In addition, allyl acetals undergoes ylide generation in rhodium(II) acetate–catalyzed reactions with diazoesters, with subsequent production of 2,5-dialkoxy-4-alkenoates by the [2,3]-sigmatropic rearrangement in moderate to good yields.[31] One representative example is shown in eq. 37.[31] Thus treatment of the dimethyl acetal of

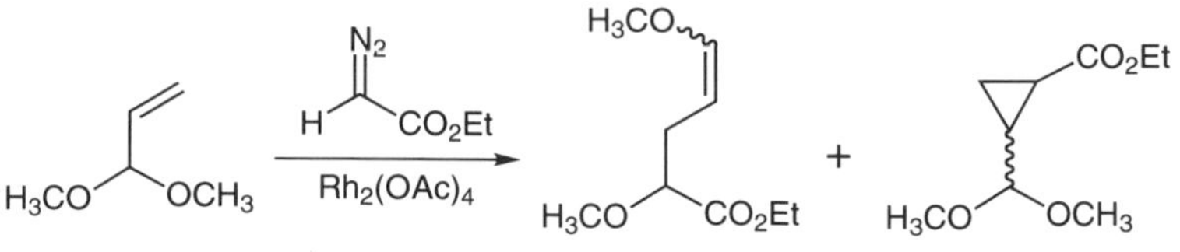

$$
\begin{array}{ccc}
& & \textbf{123 } 58\% \qquad\qquad \textbf{124 } 17\%
\end{array}
\tag{37}
$$

acrolein with EDA at 25°C in the presence of catalytic amounts of rhodium(II) acetate resulted in the formation of **123** as the major product along with a minor amount of the cyclopropane derivative **124**. Products such as **123** have considerable synthetic versatility, but the relatively low selectivity for their formation limits their use.

7.3.2 Intramolecular Formation of Oxonium Ylides and Subsequent Reactions

Cyclic oxonium ylides are readily generated by intramolecular catalytic reactions of diazocarbonyl compounds containing suitably positioned ethereal oxygen atoms. The [2,3]-sigmatropic rearrangement or [1,2]-insertion of cyclic oxonium ylides offers a novel approach to the construction of substituted carbocycles and substituted cyclic ethers. The first examples of intramolecular oxonium ylide generation *via* rhodium(II) acetate–catalyzed diazodecomposition established this methodology as ripe for exploitation. Pirrung and Werner[69] reported the effective use of rhodium acetate for the generation of cyclic allylic oxonium ylides and their subsequent [2,3]-sigmatropic rearrangements (eqs. 38–40),[69] and

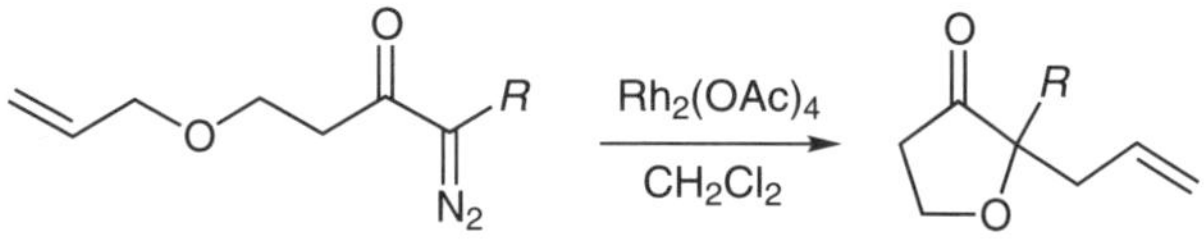

$$
R = \text{H}, 70\%; \ R = \text{CO}_2\text{CH}_3, 91\%
\tag{38}
$$

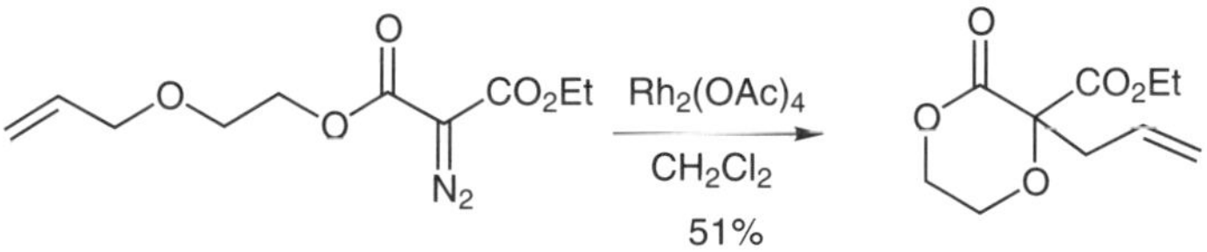

$$
51\%
\tag{39}
$$

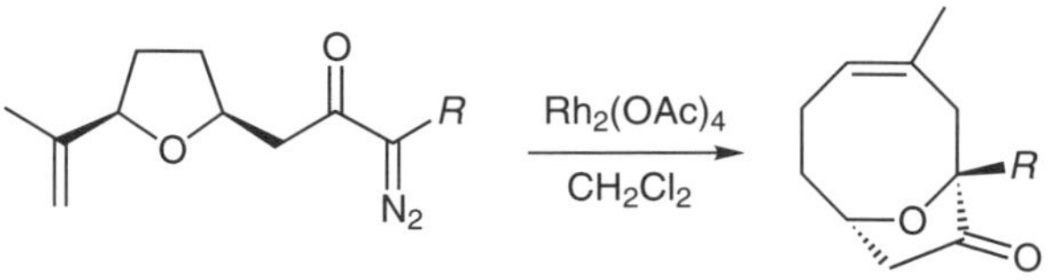

$$
R = \text{H}, 81\%; \ R = \text{CO}_2\text{CH}_3, 67\%
\tag{40}
$$

Roskamp and Johnson[70] independently provided a set of additional examples as shown in equations 41–43.[70,72] Both five- and six-membered ring oxonium ylides

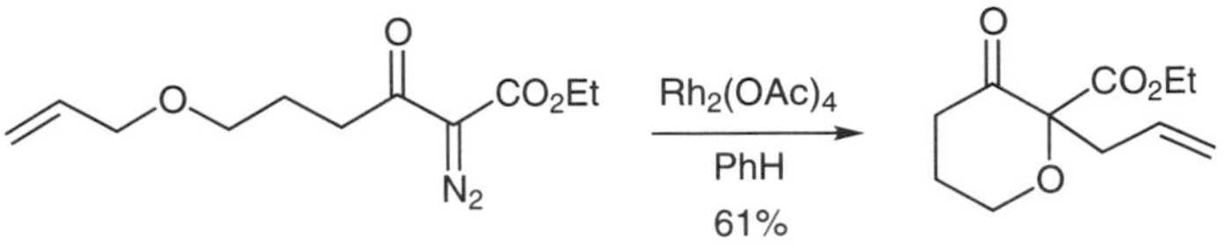

$$\text{(41)}$$

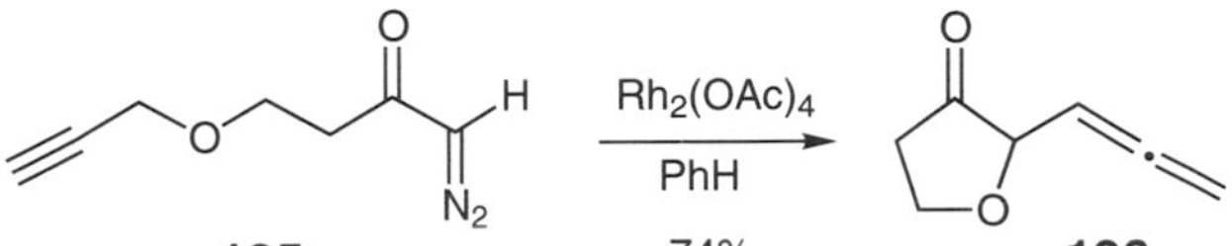

125   74%   **126**   (42)

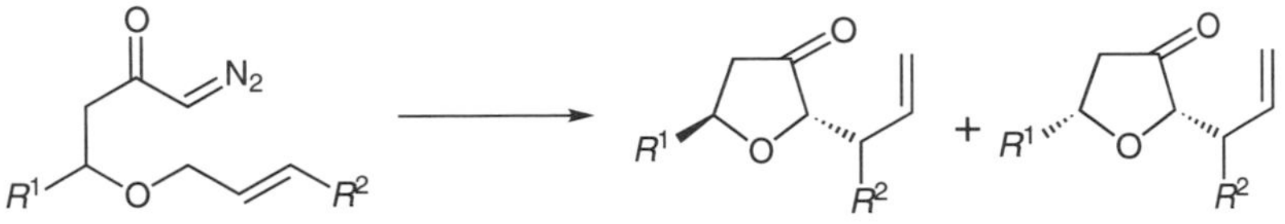

127[70]	$R^1 = CH_3, R^2 = H$	$Rh_2(OAc)_4$	**128** 60%		**129** 5%
130[70]	$R^1 = H, R^2 = CH_3$	$Rh_2(OAc)_4$	**131** 63%		
132[72]	$R^1 = (CH_3)_2CH, R^2 = H$	$Cu(acac)_2$	**133** 82%		**134** 2.5%
135[72]	$R^1 = CH_3(CH_2)_2, R^2 = H$	$Cu(acac)_2$	**136** 82%		**137** 2.5%

are accessible. The preference for [2,3]-sigmatropic rearrangement over [1,2]-insertion is demonstrated by the results of reactions in eqs. 42[70] and 43.[70,72] An additional example is to be found in the studies of West and co-workers,[71] where the rhodium acetate–catalyzed decomposition of diazoketone **138** afforded the cyclooctanone **139** in 54% yield through [2,3]-sigmatropic rearrangement (eq. 44).[71]

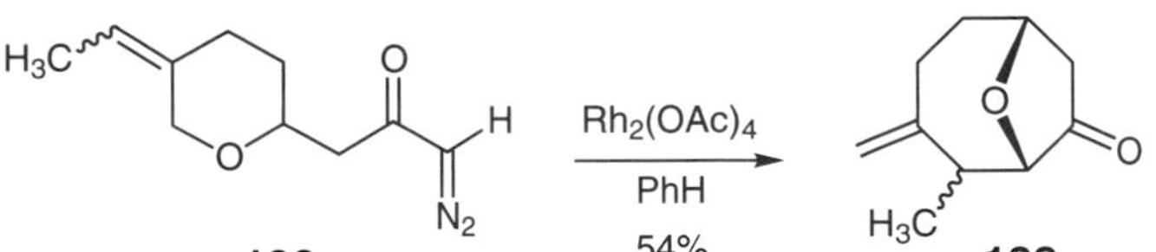

138   54%   **139**   (44)

Futhermore, rhodium acetate–catalyzed decomposition of diazoketone **127** afforded furanones **128** and **129** with high levels of diastereoselectivity (eq. 43). In a similar instance, Clark[72] claimed that $Cu(acac)_2$ is the catalyst of choice for the decomposition of diazoketones **132** and **135**; high yields and excellent levels of diastereocontrol were obtained in both cases, as shown in eq. 43.

The intramolecular generation and [2,3]-sigmatropic rearrangement of oxonium ylides were used by Pirrung and co-workers[73] in the preparation of (+)-griseofulvin **143**. Decomposition of diazoester **140** using rhodium(II) pivalate as the catalyst in refluxing benzene provided the rearranged product **142**, which served as a key step in the total synthesis. The stereochemistry of this process can be understood in terms of a transition-state model **141** that resembles an oxabicyclo[3.3.0]octane ring system with the key stereochemistry-defining methyl group located on the convex face (Scheme 30). The [2,3]-sigmatropic rearrange-

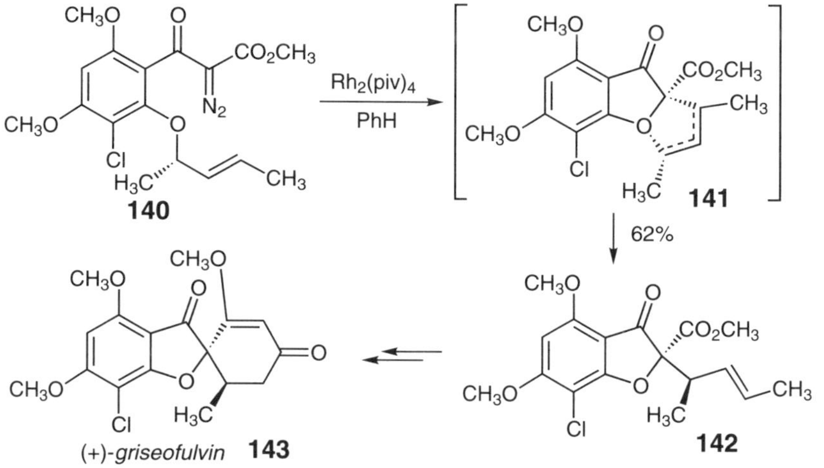

Scheme 30

ment of oxonium ylides has also been used by Clark and co-workers in their synthesis of (±)-decarestrictine L **147**[74] and the bicyclic core **148**[75] of the sesquiterpene neoliacinic acid. Thus reaction of diazoketone **144** with a catalytic amount of copper(II) hexafluoroacetylacetonate in dichloromethane at reflux afforded the tetrahydropyran-3-one **145** as the major diastereoisomer. The conversion of **145** into (±)-decarestrictine L **147** was completed in four steps (Scheme 31). In the second example, the [2,3]-sigmatropic rearrangement of an oxonium ylide has been employed in the synthesis, as illustrated in Scheme 32.

It would be misleading, however, to give the impression that cyclic allylic oxonium ylides always undergo [2,3]-sigmatropic rearrangement. Roskamp and Johnson[70] have explored an example where a cyclic allylic oxonium ylide **149** resulted in the dominant product **150** via [1,2] insertion (Scheme 33). Recently,

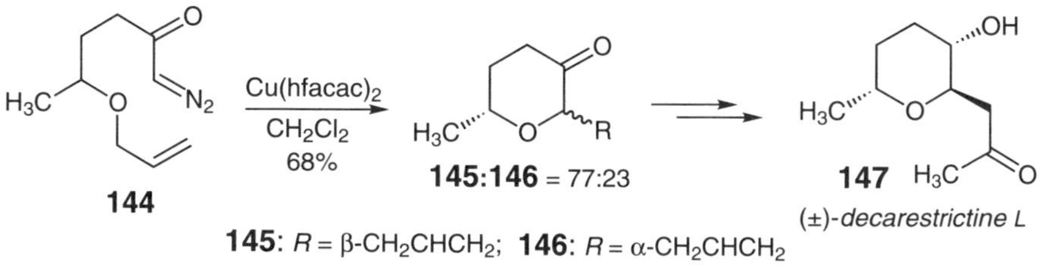

145: $R = \beta\text{-}CH_2CHCH_2$; **146:** $R = \alpha\text{-}CH_2CHCH_2$

Scheme 31

[2,3]-SIGMATROPIC REARRANGEMENT OF A CYCLIC OXONIUM YLIDE: SYNTHESIS OF 8-CHLORO-5,7-DIMETHOXY-2-CARBOMETHOXY-2-[1-(S)-METHYL-2-BUTENYL]BENZO-FURAN-3-ONE[73]

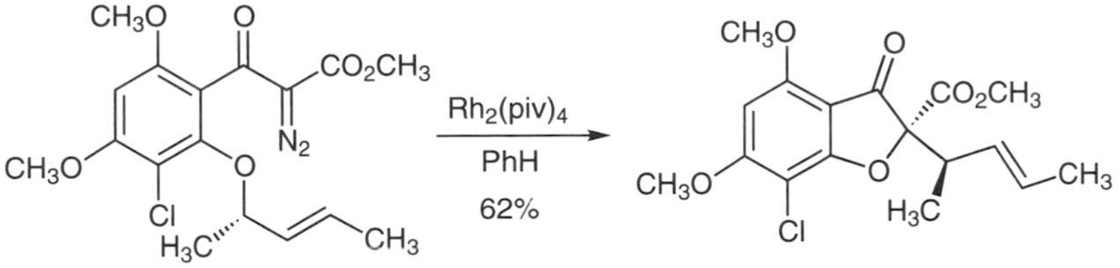

A solution of the diazoester **140** (100 mg, 0.26 mmol), Rh$_2$(piv)$_4$ (7 mg, 0.013 mmol) in dry benzene (30 mL) was refluxed for 45 min–1 h under nitrogen atmosphere. After the disappearence of the starting material the reaction was cooled to room temperature and the solvent was removed under reduced pressure. The residue was chromatographed on a short silica column using hexame:ethyl acetate (7:3) to give **142**. Yield: 58 mg (62%). ^{1}H-NMR (300 MHz, CDCl$_3$) δ 1.16 (d, J = 6.9 Hz, 3 H), 1.44 (dd, J = 5.72, 0.84 Hz, 3 H), 3.30 (quin, J = 7.2 Hz, 1 H), 3.72 (s, 3 H), 3.90 (s, 3 H), 3.98 (s, 3 H), 5.06 (dd, J = 15.3 Hz, 1,2 Hz, 1 H), 5.54 (m, 1 H), 6.09 (s, 1 H; ^{13}C-NMR (75 MHz, CDCl$_3$) δ 15.5, 17.6, 41.4, 52.9, 56.1, 56.3, 89.4, 95.4, 97.0, 104.1, 126.9, 129.1, 157.5, 164.1, 165.6, 168.6, 189.8; IR (CCl$_4$) 1718.0, 1755.2 cm^{-1}; $[\alpha]_D^{20}$ −59.9 (c = 11.5, CHCl$_3$).

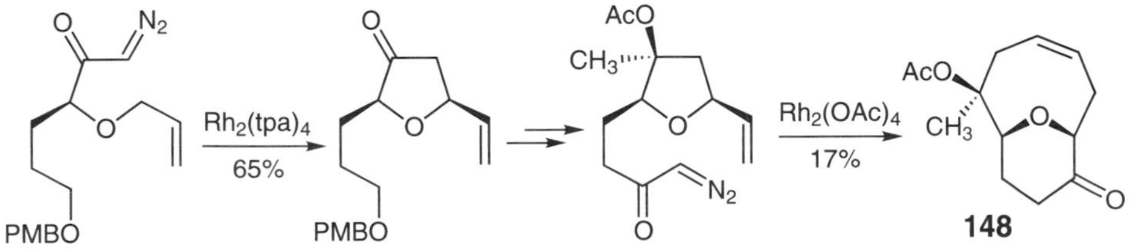

Scheme 32

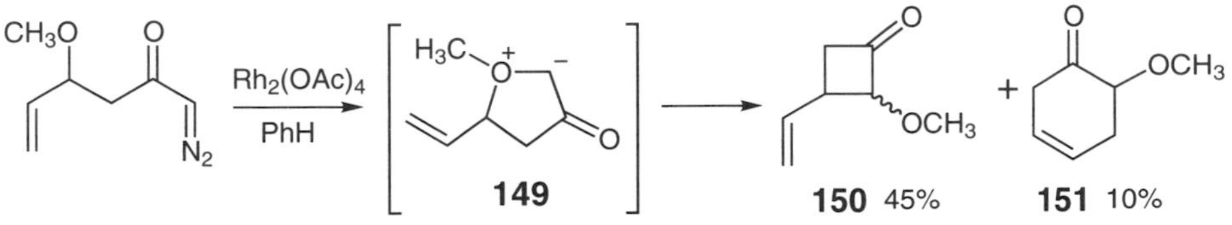

Scheme 33

Brogan and co-workers[76] demonstrated that a six-membered ring ylide **153** derived from the decomposition of diazoester **152** in the presence of either a rhodium or a copper catalyst undergoes [1,2]-insertion to give compound **154** exclusively. In this case, the oxonium ylide **153** was formed by interaction of metal carbene with the least sterically hindered oxygen, and the [1,2]-insertion had occurred at the exocyclic, allylic position, as illustrated in Scheme 34.

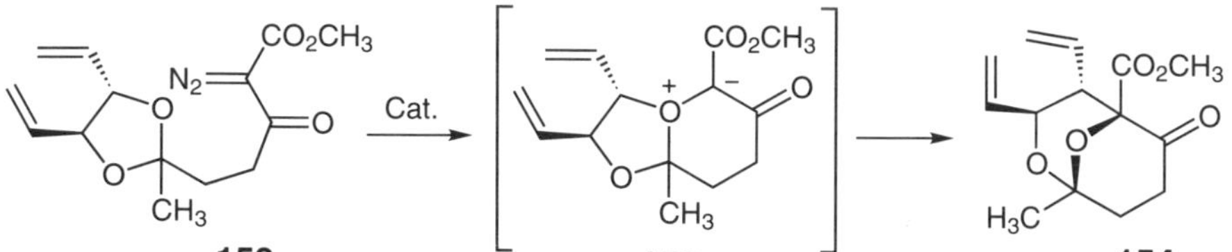

Cat.: Rh$_2$(OAc)$_4$, 25°C, 34% yield   Cat.: Cu(hfacac)$_2$, 80°C, 64% yield

Scheme 34

Synthetic utilization of the [1,2] Stevens rearrangement of oxonium ylides has been advanced through examples provided by Roskamp and Johnson[70] (eqs. 45[70] and 46[70]). Rhodium acetate–catalyzed decomposition of diazoketone **155** yielded

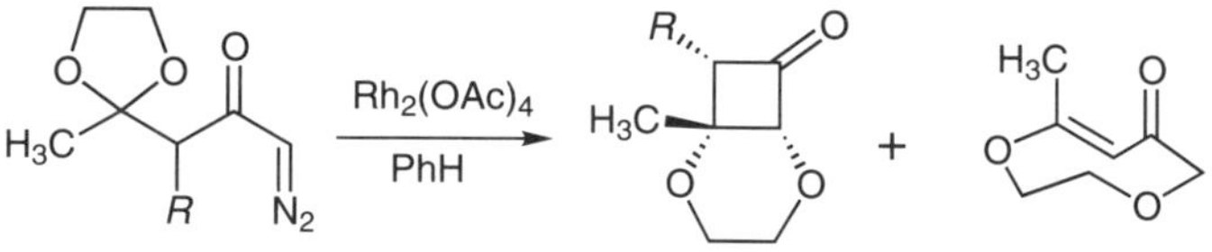

| **155** $R = H$ | **156** 68% | **157** 16% | (45) |
| **158** $R = CH_3$ | **159** 54% | | |

| **160** | **161** | **162** | (46) |

Rh$_2$(OAc)$_4$	74% yield	**161:162** = 77:23
Cu(acac)$_2$	—	**161:162** = 14:86
RhCl(Ph$_3$P)$_3$	—	**161:162** = 9:91

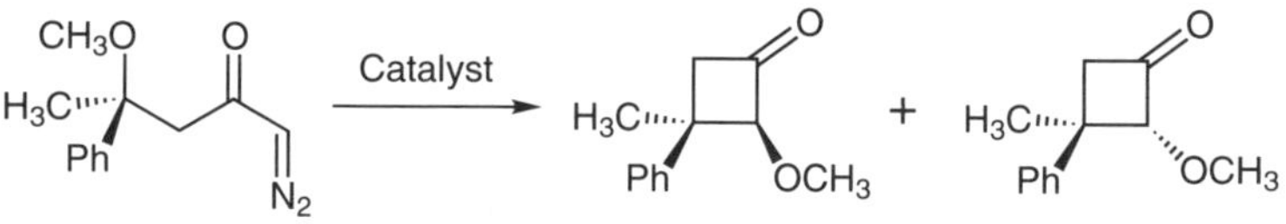

a major product **156** (68%) and a minor product **157** (16%), whereas the α-substituted diazoketone **158** afforded product **159**, with excellent diastereoselectivity (eq. 45).[70] Catalytic decomposition of optically active diazoketone **160** afforded cyclobutanones **161** and **162**. The ratios of **161** to **162** varied from 1:3 to 10:1 depending on the catalyst employed (eq. 46).[70] Diazoester **163**, with a dioxolane functional group similar to diazoketone **155**, produced the ring-fused product **164** in 51% yield, on treatment with anhydrous copper(II) hexafluoroacetylaceto-nate (eq. 47).[76]

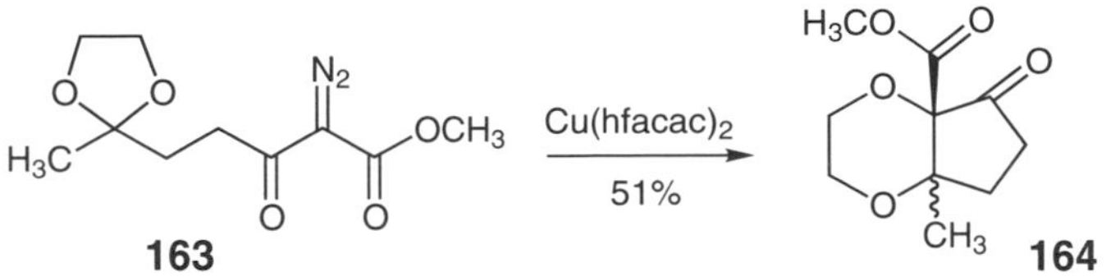

$$ \text{(47)} $$

163   Cu(hfacac)$_2$ 51%   **164**

West and co-workers[77] have successfully utilized the tandem cyclic oxonium ylide generation/Stevens [1,2]-shift protocol in the synthesis of functionalized te-trahydrofuranones. Catalytic decomposition of alkoxy-α-diazo ketone **165** forms the cyclic oxonium ylide **166**, with subsequent generation of cyclic ether **167** (Scheme 35). O-Bridged medium-sized carbocyclic rings have been synthesized

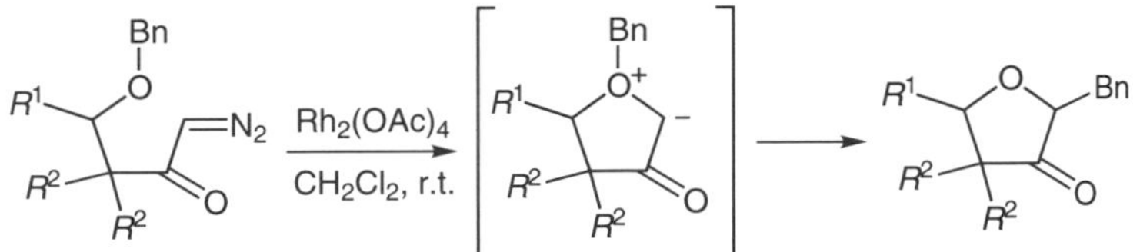

165   Rh$_2$(OAc)$_4$ CH$_2$Cl$_2$, r.t.   **166**   **167**

$R^1 = R^2 = H, 64\%; R^1 = H, R^2 = CH_3, 65\%; R^1 = CH_3, R^2 = H, 52\%$

Scheme 35

via the same methodology. Fused bicyclic oxonium ylide **169**, generated with cat-alytic amounts of rhodium acetate from diazoketone **168**, undergoes a [1,2] shift to give O-bridged seven-membered carbocycles **170** and **171** with a high degree of stereoselectivity (**170**:**171** = 95:5) (Scheme 36).[71]

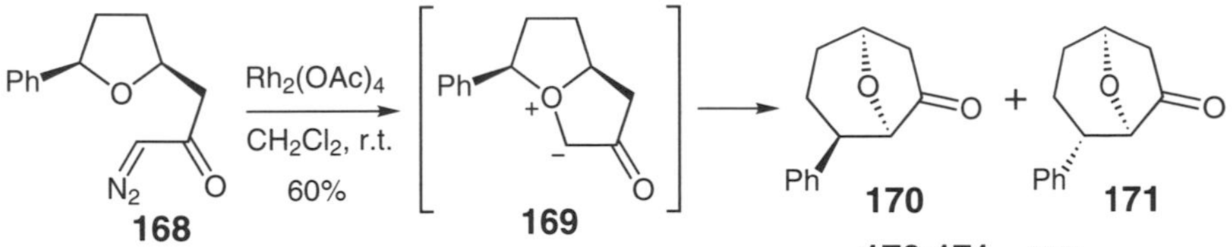

168   Rh$_2$(OAc)$_4$ CH$_2$Cl$_2$, r.t.   60%   **169**   **170** + **171**

170:**171** = 95:5

Scheme 36

Padwa and co-workers[78] have reported that the rhodium mandelate–catalyzed decomposition of diazoketone **172** afforded dibenzocyclononenone **174** in 73% yield. The reaction proceeded *via* an initially formed six-membered cyclic oxonium ylide **173**, which rearranges further by means of the [1,2]-Stevens rearrangement (Scheme 37).

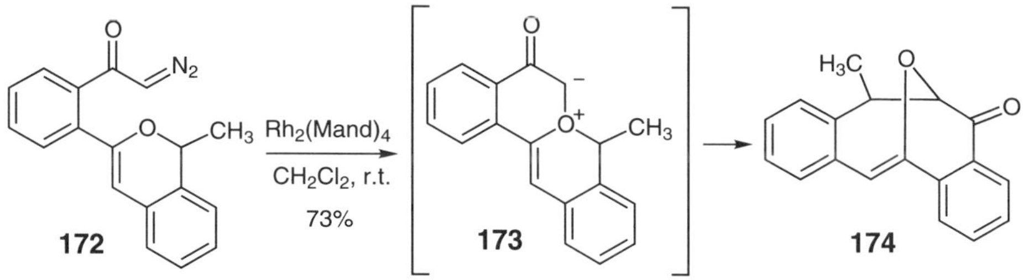

Scheme 37

The β-elimination reaction is also possible with cyclic oxonium ylides, and one representative example is shown in eq. 48.[37,70,80] An interesting application of the intramolecular oxonium ylide formation and subsequent β-elimination reaction is found in the synthesis of oxacephams **177** and **178** (Scheme 38). Thus treatment of diazoester **175** with a catalytic amount of rhodium acetate furnished oxacephams **177** and **178** in 44–63% yield as a 3:1 mixture of diastereoisomers (Scheme 39).[79]

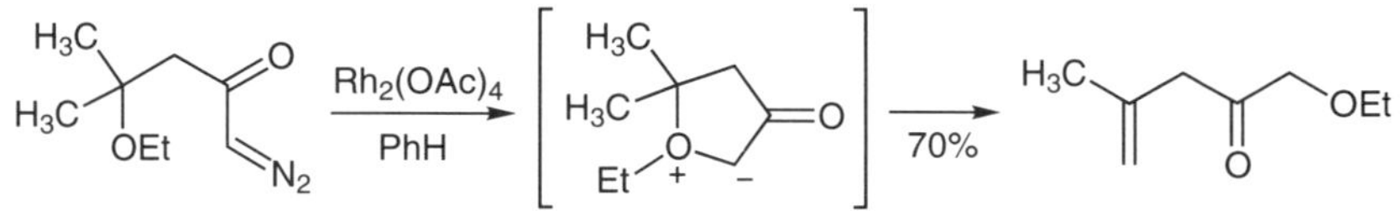

Scheme 38

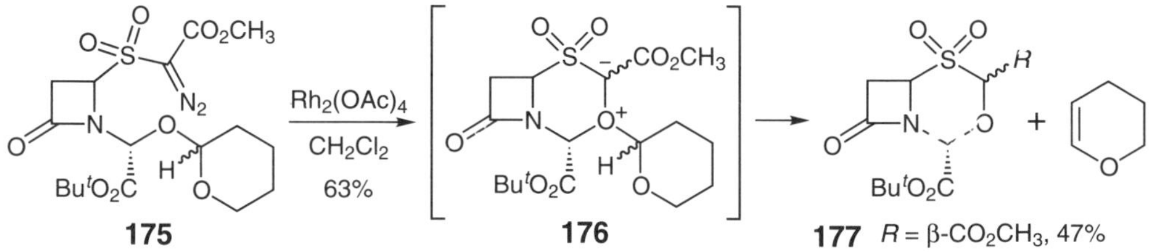

177 $R = \beta\text{-}CO_2CH_3$, 47%
178 $R = \alpha\text{-}CO_2CH_3$, 16%

Scheme 39

7.3.3 Chemo-, Diastereo-, and Enantioselectivity in Oxonium Ylide Formation

C–H insertion is the major competitive reaction in intramolecular cyclic oxonium ylide formation. The nature of the catalyst allows switching between oxonium ylide generation and C–H insertion.[37,80] Thus decomposition of diazoketone **179** with dirhodium catalysts furnished predominantly *cis*-disubstituted chromanone **180**, the product of C–H insertion, with minor amounts of benzofuranone **181**, the product of oxonium ylide [2,3]-sigmatropic rearrangement. When copper complexes were employed as catalysts, the corresponding reaction yielded benzofuranone **181** exclusively (eq. 48).[37,80] Clark and co-workers[81] have reported that

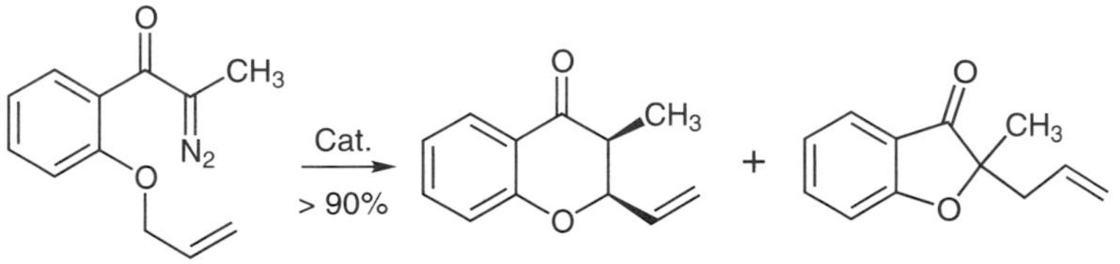

(48)

179	Catalyst	**180**	**181**
	Rh(II) carboxylates	82–97%	3–18%
	Cu(II) Complexes	0%	100%

Cu(hfacac)$_2$ is able to minimize the C–H insertion reaction and favor oxonium ylide generation relative to rhodium(II) acetate. Thus medium-sized cyclic ethers **183** were obtained from Cu(hfacac)$_2$-catalyzed decomposition of α-diazoketones **182** by intramolecular oxonium ylide generation and subsequent [2,3]-sigmatropic rearrangement (eq. 49).[81] In competitive intramolecular reactions between oxo-

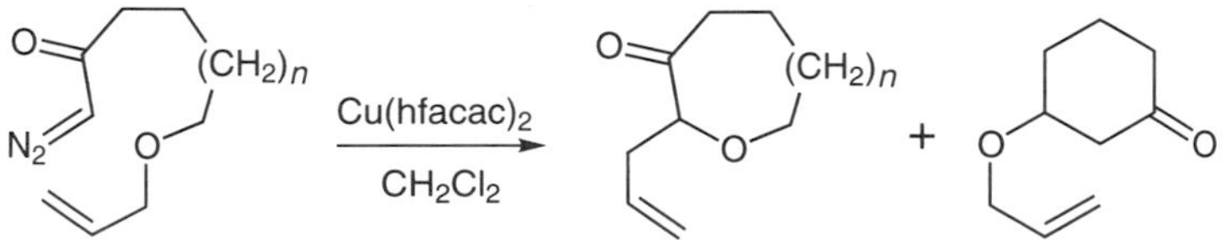

(49)

182		**183**	**184**
	$n = 1$	76%	3%
	$n = 2$	40%	—

nium ylide formation and C–H insertion, the transition metal effectively, and in some cases, completely, switches the reaction preference. An extreme example of catalyst-dependent reactivity is seen with the decomposition of α-diazoketone **185** (eq. 50).[82] Thus, while rhodium acetate catalysis led to only C–H insertion

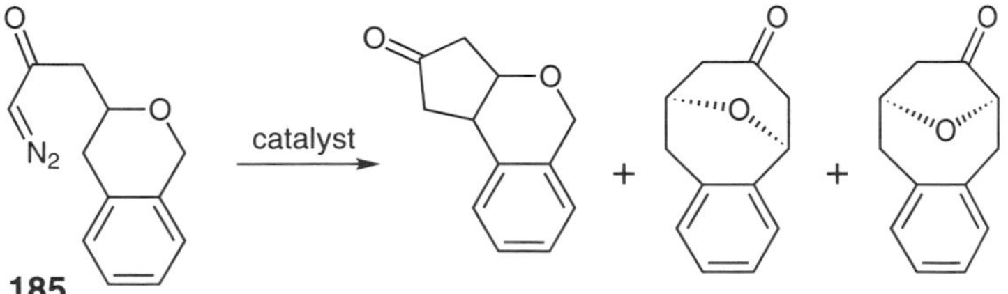

(50)

185		**186**	**187**	**188**
	Rh$_2$(OAc)$_4$	40%	10%	0%
	Cu(hfacac)$_2$	0%	0%	95%

products **186** and **187**, Cu(hfacac)$_2$ furnished only **188**, the product of the oxonium ylide [1,2] shift, in 95% yield.

These examples (eqs. 48–50) appear to reinforce our early conclusion[83] that the copper catalysts are in general less efficient for metal carbene transformations when compared to their rhodium conterparts; they usually favor the ylide formation process.

Cyclopropanation often occurs in competition with ylide generation in intermolecular transformations of allylic substrates. Detailed investigations have been carried out by Doyle and co-workers (eq. 37).[31] Catalyst-dependent chemoselectivity for oxonium ylide generation over cyclopropenation has been described for reactions of propargyl ethers with diazocarbonyl compounds.[84] In the simplest case (eq. 51),[84] the use of dirhodium(II) perfluorobutyrate, Rh$_2$(pfb)$_4$, leads to the

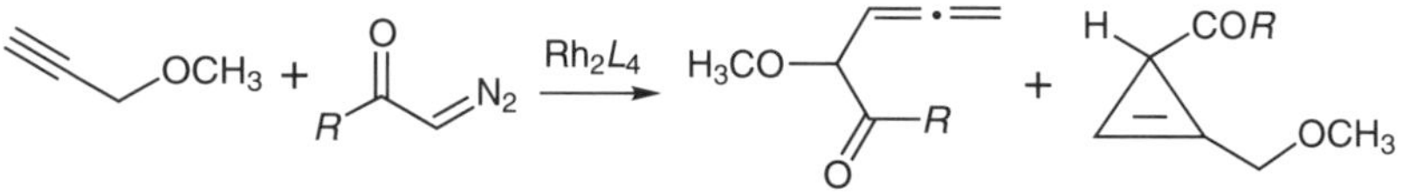

(51)

R	Catalyst	Yield	**189**	**190**
Ph	Rh$_2$(pfb)$_4$	80%	82	18
Ph	Rh$_2$(OAc)$_4$	32%	5	95
OEt	Rh$_2$(pfb)$_4$	67%	61	39
OEt	Rh$_2$(OAc)$_4$	65%	1	99

product from ylide generation and [2,3]-sigmatropic rearrangement with significant chemoselectivity, which is higher for α-diazoacetophenone than for EDA. In contrast, the use of rhodium acetate results in virtually exclusive cyclopropenation. Substituted propargyl methyl ethers give higher yields of allene products, even with rhodium acetate catalysis.

McKervey and co-workers[85] discovered catalyst-dependent asymmetric induction in the cyclic oxonium ylide formation and subsequent [2,3]-sigmatropic rearrangement. Intramolecular oxonium ylide generation from **191** in dichloromethane resulted in the production of **192** in high yield with a 30% enantiomeric excess (eq. 52),[85] and for an aliphatic analogue 9% enantiomeric excess of a [2,3]-

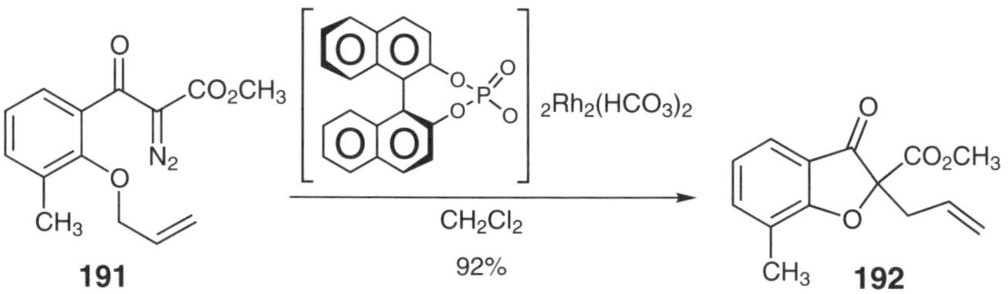

(52)

sigmatropic rearrangement production was measured. These results are the first examples of chiral catalyst–induced asymmetric synthesis *via* the [2,3]-sigmatropic rearrangement, and they suggest involvement of the dirhodium(II) catalyst in the product-forming step. Presumably the catalyst is not fully dissociated from the ylide prior to bond formation in the sigmatropic rearrangement step.

As mentioned earlier, tetrahydrofuran **119** was obtained by Nozaki and co-workers[66,67] upon treatment of (±)-2-phenyloxetane with methyl diazoacetate in the presence of copper catalyst (Scheme 27). When a chiral salicylaldimine copper catalyst was employed in this transformation, enantiomer differentiation was observed but the optical yields were not determined at that time. Almost 30 years later, this reaction was reexamined by Katsuki and co-workers[86,87] through the use of *tert*-butyl diazoacetate as the carbene precursor and copper-based chiral bipyridine **195** as the catalyst (eq. 53).[87] Reaction of *dl*-2-phenyloxetane with 0.5

$$(53)$$

	Yield % (ee%) recovered **193**	Yield % (**196a**/**196b**)	ee % (**196a**)	ee % (**196b**)
dl-**193**	30 (<5)	36 (59/41)	75(2*S*,3*R*)	81(2*S*,3*S*)
R-**193** (89% ee)	30 (87)	35 (89/11)	92(2*S*,3*R*)	16(2*S*,3*S*)
S-**193** (85% ee)	36 (87)	30 (25/75)	11(2*S*,3*R*)	93(2*S*,3*S*)

equiv. of *tert*-butyl diazoacetate in the presence of **195** gave *trans*- and *cis*-tetrahydrofuran derivatives in almost equimolar amounts; the optical purity of these *trans*- and *cis*-tetrahydrofurans was as high as 75 and 81% ee, respectively. The optical purity of the unreacted oxetane was low. Further investigation revealed that the reaction of (*R*)-2-phenyloxetane of 89% ee and 0.5 equiv. of *tert*-butyl diazoacetate in the presence of **195** provided (2*S*,3*R*)-*tert*-butyl 3-phenyltetrahydrofuran-2-carboxylate as a major product in 92% ee, while that of (*S*)-2-phenyloxetane of 85% ee provided (2*S*,3*S*)-*tert*-butyl 3-phenyltetrahydrofuran-2-carboxylate in 93% ee. These results supported the proposal that the reaction proceeded through the oxonium ylide metal complex instead of free oxonium ylide. Katsuki and co-workers[88] also applied this asymmetric ring-expansion reaction of oxetane **197** in their enantioselective synthesis of *trans*-Whisky lactone **198**, as shown in Scheme 40.

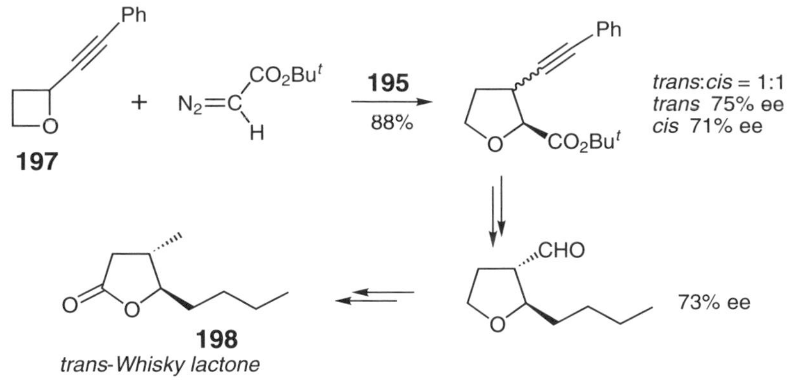

Scheme 40

7.4 FORMATION OF NITROGEN YLIDES FROM DIAZOCARBONYL COMPOUNDS

The generation and reactions of nitrogen ylides are similar to those for sulfur ylides. The catalysis of diazo decomposition for nitrogen ylide generation is a useful alternative to the widely employed base-promoted methodology.[89] Because of their basicity, amines and imines are good ligands for catalytically active transition metal compounds, and reaction temperatures required for diazo decomposition in the presence of amines or imines are, like those with sulfides, higher than for reactions in the presence of substrates that possess other heteroatoms. Because of its varied oxidation states, nitrogen offers a wealth of substrates for ylide generation and reactions.

7.4.1 Intermolecular Formation of Nitrogen Ylides and Subsequent Reactions

Doyle and co-workers[9] reported several examples of intermolecular ammonium ylide generation from diazoacetates, catalyzed by $Rh_2(OAc)_4$ or $Rh_6(CO)_{16}$. These intermediates underwent [2,3]-sigmatropic rearrangement to produce **199** in good yield, with a moderate degree of selectivity (eq. 54).[9] Propargyldimeth-

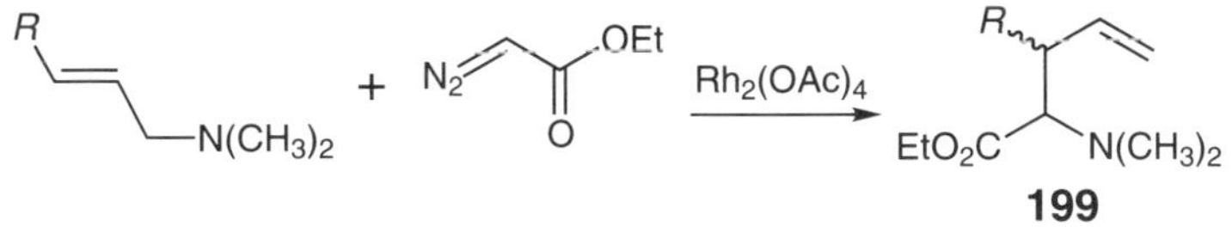

R	Yield	**199**
CH_3	79%	dr: 75:25
Ph	59%	dr: 75:25

ylamine underwent exclusive $Rh_2(OAc)_4$-catalyzed ylide generation and [2,3]-sigmatropic rearrangement to form allene **200** in high yield (eq. 55).[84] More

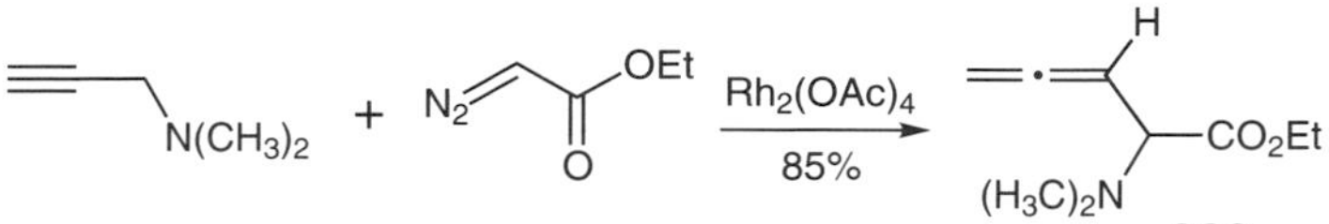

recently, Burger and co-workers[89] have employed methyl 3,3,3-trifluoro-2-diazopropanoate as the carbene precursor in ammonium ylide generation; subsequent [2,3]-sigmatropic rearrangement of the ylide produced an α-amino acid derivative **201** in 72% yield (eq. 56).[89]

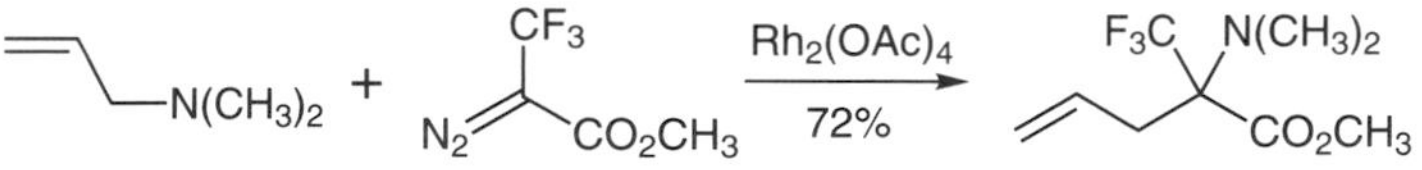

Diazocarbonyl compounds react with tertiary amines in the presence of transition metal catalysts to form ammonium ylides. These ylides undergo facile [1,2] shift of the best migrating group from N to C. Hata and Watanabe[90] reported an early example as shown in eq. 57.[90] Thus treatment of 1-benzylaze-

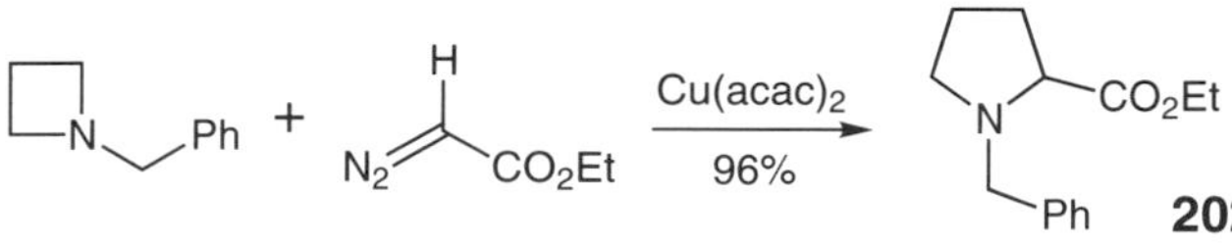

tidine with EDA in the presence of $Cu(acac)_2$ produced pyrrolidine **202** *via* the Stevens rearrangement (eq. 57).[90] In contrast, Burger and co-workers[89] claimed that the reaction of 1-benzylazetidine with methyl 3,3,3-trifluoro-2-diazopropanoate did not undergo ring enlargement; instead, a [1,2]-benzyl migration occurred, and the α-trifluorophenylalanine derivative **203** was obtained in 65% yield (eq. 58).[89]

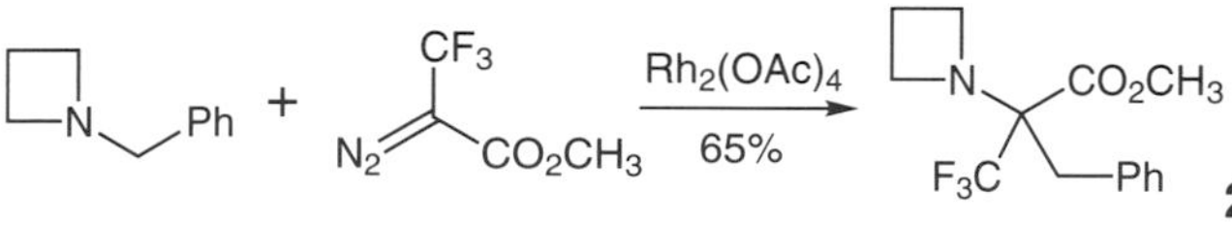

West and co-workers[91] carried out an extensive investigation on intermolecular ammonium ylide generation followed by rearrangement and concluded that α-(N,N-dialkylamino)esters and ketones can be synthesized in one step from simple reactants. Some examples are illustrated in eqs. 59–61.[91] The potential draw-

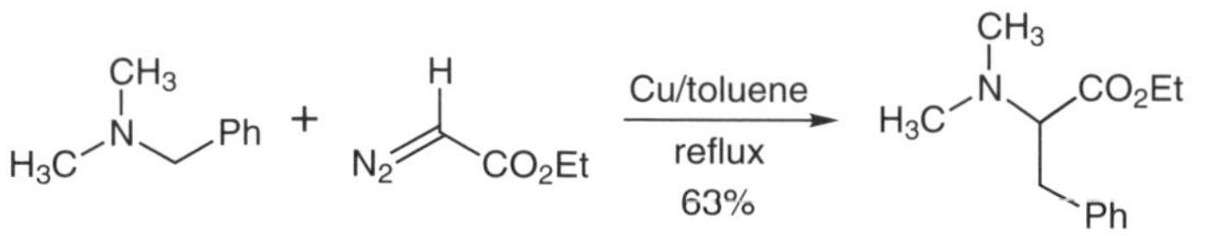

$$(59)$$

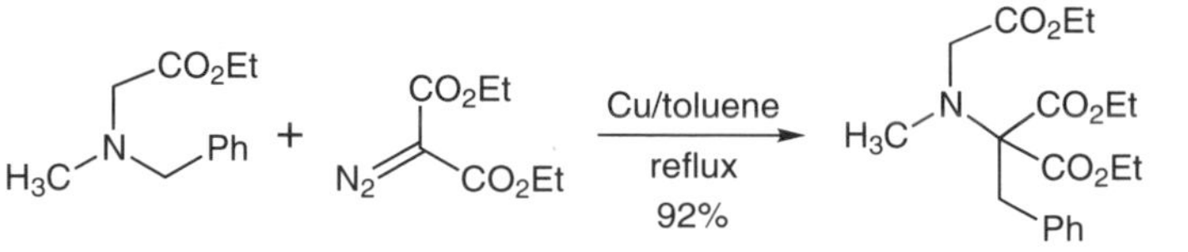

$$(60)$$

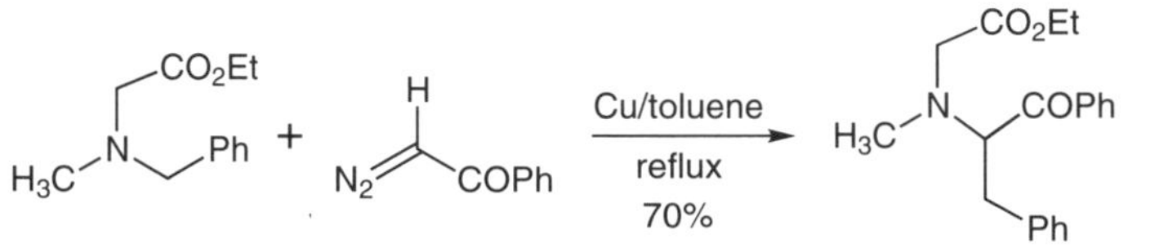

$$(61)$$

backs of this approach include the requirement for a large excess of amine and, in some case, the unsuitability of the resulting tertiary amines for subsequent transformations.

Azomethine ylides can be generated from the reaction of imines with diazocarbonyl compounds in the presence of a transition metal catalyst. These azomethine ylides undergo facile 1,3-dipolar cycloadditions to π-bonds to give pyrrolidines, which are useful in the synthesis of alkaloids. For example, thiazolazetidinone **204** reacted with EDA in the presence of copper(II) acetylacetonate and dimethyl fumarate to form first an azomethine ylide that subsequently underwent 1,3-dipolar cycloaddition to give adduct **205** stereoselectively (eq. 62).[92]

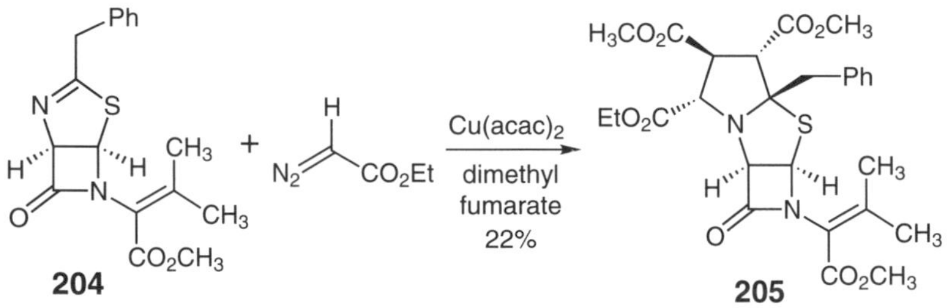

$$(62)$$

Additional examples involving intermolecular azomethine ylide formation coupled to dipolar cycloaddition by dimethyl acetylenedicarboxylate or N-phenylmaleimide were reported by Padwa and co-workers.[93]

7.4.2 Intramolecular Formation of Nitrogen Ylides and Subsequent Reactions

The intramolecular version of nitrogen ylide generation and subsequent rearrangement has been explored by several groups[94–99] to provide a general method for the preparation of cyclic amines. West and Naidu[94] reported that rhodium(II) acetate–catalyzed decomposition of diazoketones **206** yielded the corresponding ammonium ylide **207**, which undergoes the Stevens [1,2] shift to give the substituted piperidines **208** in excellent yield (Scheme 41). Exposure of α-diazo-β-

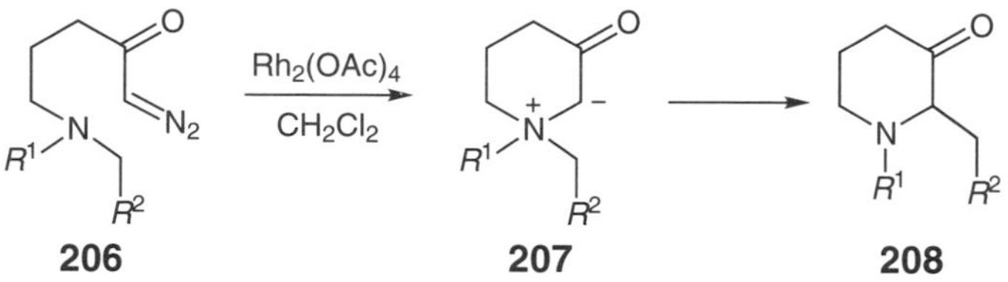

R^1 = CH$_3$, R^2 =Ph; 99% yield; R^1 = CH$_3$, R^2 =CO$_2$Et; 94% yield

R^1 = Et, R^2 =Ph; 91% yield; R^1 = CH$_3$, R^2 =p-CH$_3$OC$_6$H$_4$; 71% yield

Scheme 41

ketoester **209a** to catalytic rhodium(II) acetate in dichloromethane at room temperature failed to produce the desired morpholinone product **210a**, though slow addition to rhodium(II) acetate in refluxing toluene did lead to a modest yield (34%) of this product. On the other hand, the conversion of **209** into **210** occurred smoothly in toluene at reflux when 15 mol% of copper powder or Cu(acac)$_2$ was used as catalyst.[95] Related examples are shown in Scheme 42

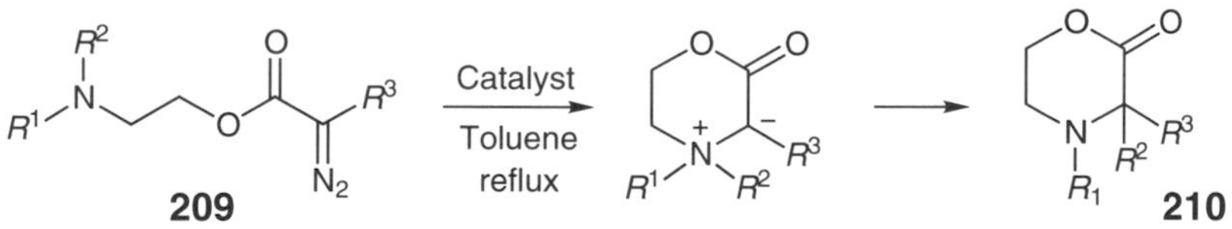

a : R^1 = CH$_3$, R^2 = PhCH$_2$, R^3 = Ac; Catalyst: 15 mol % Cu(0); 74% yield

b : R^1 = CH$_3$, R^2 = 4-CH$_3$OC$_6$H$_4$CH$_2$, R^3 = Ac; Catalyst: 50 mol % Cu(0); 76% yield

c : R^1 = CH$_3$, R^2 = PhCH$_2$, R^3 = H; Catalyst: 5 mol % Cu(acac)$_2$; 64% yield

d : R^1 = R^2 = PhCH$_2$, R^3 = H; Catalyst: 5 mol % Cu(acac)$_2$; 64% yield

Scheme 42

(b–d). The cyclic ammonium ylide [1,2]-shift methodology was employed by West and Naidu as the key step in an enantioselective synthesis of (−)-epilupinine **213**.[96] Thus addition of a toluene solution of diazoketone **211** to a refluxing solution of Cu(acac)$_2$ (5 mol%) in toluene gave the quinolizidine skeleton **212** as a 95:5 ratio of diastereomers in 84% yield. The major diastereoisomer **212a**, formed in 75% enantiomeric excess, was converted subsequently into the (−)-epilupinine **213** via thioketalization, reduction, and desulfurization (37% overall yield from diazoketone **211**) (Scheme 43).

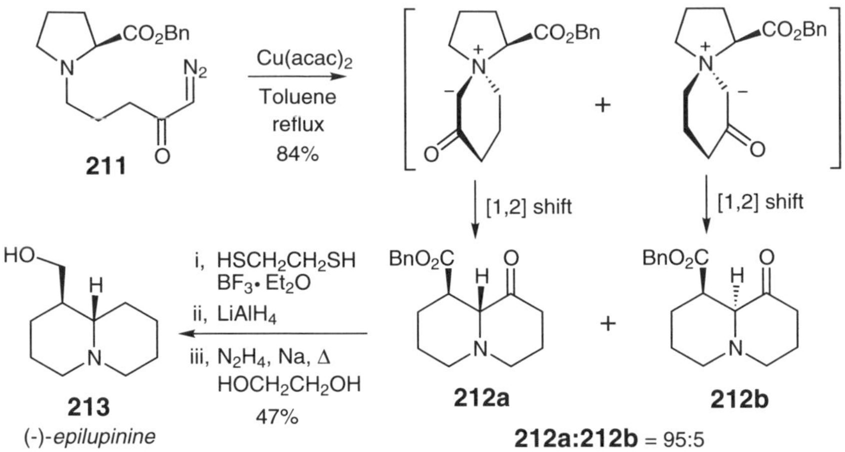

Scheme 43

Tandem intramolecular formation and [2,3]-sigmatropic rearrangement of ammonium ylides from copper-catalyzed decomposition of diazoketones have been studied by Clark and Hodgson.[97] Representative examples are shown in Scheme 44. Copper(II) acetylacetonate was the catalyst of choice for these transformations and the yields were generally good; when rhodium(II) acetate was employed as catalyst, lower yields of the cyclic amines were obtained. Application of this methodology resulted in an efficient enantioselective synthesis of the azabicyclo[6.3.0]undecane **216**, corresponding to the CE ring system found in the alkaloid manzamine A (Scheme 45).[98]

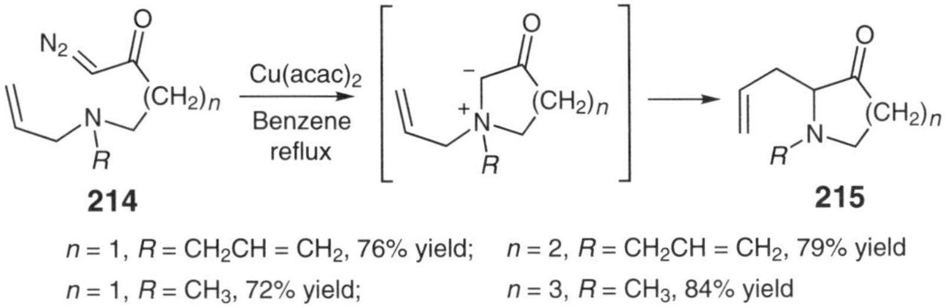

$n = 1$, $R = CH_2CH = CH_2$, 76% yield; $\quad$ $n = 2$, $R = CH_2CH = CH_2$, 79% yield
$n = 1$, $R = CH_3$, 72% yield; $\quad\quad\quad$ $n = 3$, $R = CH_3$, 84% yield

Scheme 44

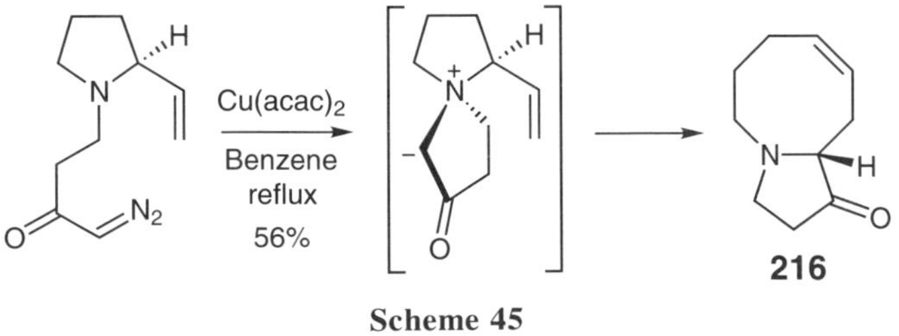

Scheme 45

More recently, McMills[99] and co-workers have reported that the azacyclooctene **219** and azacyclononene **221**–containing compounds can be constructed *via* tandem [2,3]-sigmatropic rearrangement of spirocyclic ammonium ylides derived from decomposition of the corresponding diazoesters (**217** and **218**). In this case, copper(II)-catalyzed decomposition of diazoester **217** yielded the azacyclooctene **219** along with a substantial amount of a product **220** presumably arising from a Stevens [1,2]-rearrangement of the intermediate ylide; the combined yields were good, and the ratio of **219** to **220** was 2.5–4 to 1, as shown in eq. 63.[99] In contrast to the behavior of **217**, the decomposition of diazoester

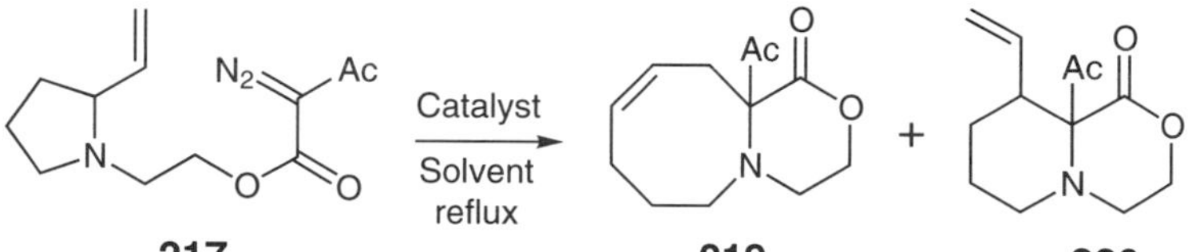

$$\text{(63)}$$

Catalyst: Cu(acac)$_2$, Solvent: PhH, 33% yield; **219:220** = 4:1
Catalyst: Cu(acac)$_2$, Solvent: PhCH$_3$, 62% yield; **219:220** = 3:1
Catalyst: Cu(hfacac)$_2$, Solvent: PhCH$_3$, 70% yield; **219:220** = 2.5:1

218 under the same conditions afforded [2,3]-sigmatropic rearrangement products **221** and **222** exclusively. However, with the larger piperidine ylide both the *cis-* and *trans*-azacyclononenes (**221** and **222**) were produced in varying ratios (eq. 64).[99]

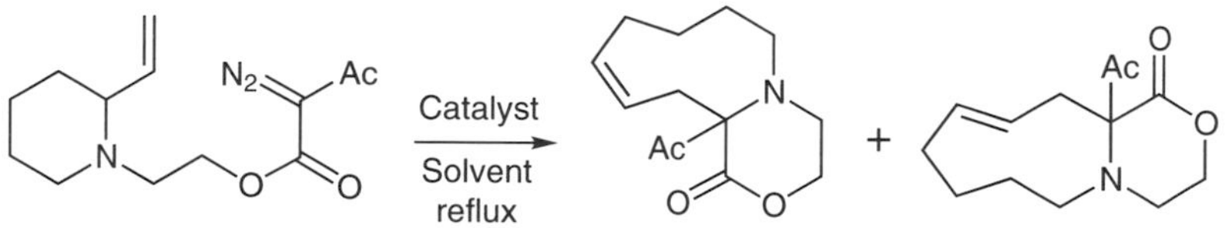

$$\text{(64)}$$

Catalyst: Cu(acac)$_2$, Solvent: PhH, 39% yield; **221:222** = 3:2
Catalyst: Cu(acac)$_2$, Solvent: PhCH$_3$, 62% yield; **221:222** = 5:1
Catalyst: Cu(hfacac)$_2$, Solvent: PhCH$_3$, 65% yield; **221:222** = 5:1

The intramolecular reaction of metal carbenes derived from α-diazo-β-ketoesters attached to an *N*-alkoxy β-lactam provides a novel route to bicyclic structures. This cyclization protocol involves interaction of an initially generated metal carbene with the nitrogen atom of the β-lactam to form the ylide. Intramolecular abstraction of a proton from the alkoxy group on nitrogen, followed by N–O bond cleavage with release of benzaldehyde, afforded the cyclized product. Miller and co-workers[100,101] have applied this methodology to the asymmetric synthesis of **225**, a key intermediate for the preparation of the β-lactam antibiotic

PS-5 **226**. Thus rhodium-catalyzed cyclization of optically active **223** afforded carbapenam **225**, which was then converted into **226**. The formation of the intermediate ylide **224** was the key step of this cyclization (Scheme 46).

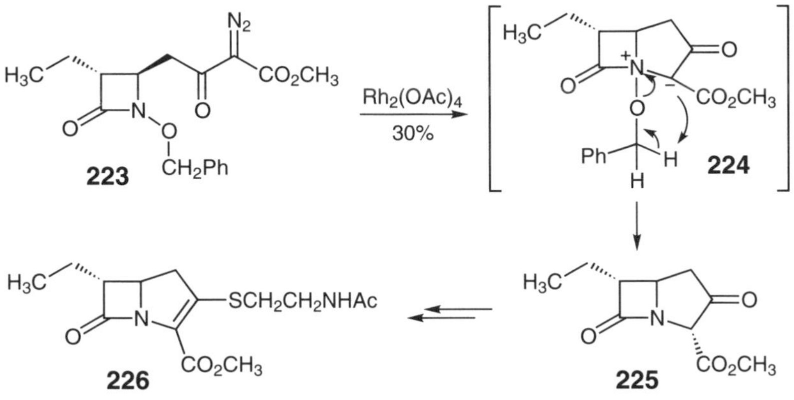

Scheme 46

Rhodium(II)-catalyzed decomposition of the diazocarbonyl β-lactam **227** in Scheme 47 resulted in two cyclized products, **229** and **230**.[102] Both products appear to result from formation of ylide **228**, in which the metal carbene bonds to the more nucleophilic of the two nitrogen atoms (starred in **228**). Product **229** is that from sigmatropic rearrangement of the ylide, whereas product **230** may have resulted from fragmentation of the ylide followed by dimerization. Interestingly, there was no evidence of product arising from N–H insertion into the amide group, a common reaction pathway with related β-lactam substrates (Scheme 47).

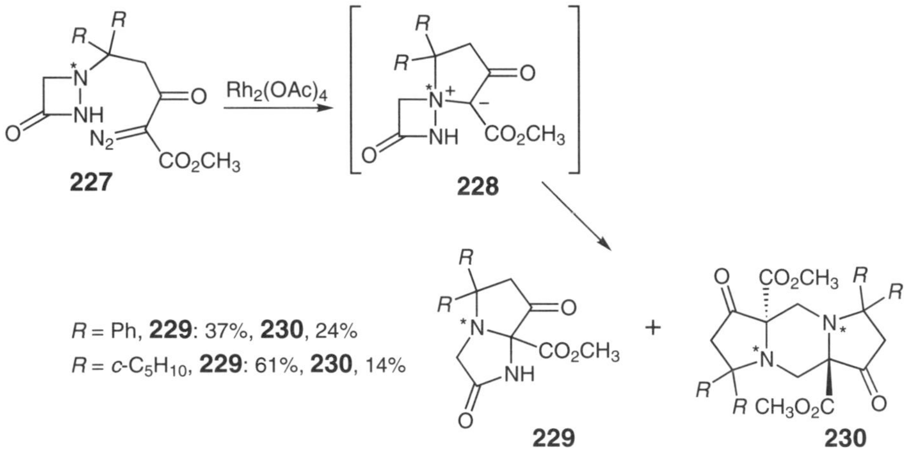

R = Ph, **229**: 37%, **230**, 24%
R = c-C₅H₁₀, **229**: 61%, **230**, 14%

Scheme 47

7.5 FORMATION OF CARBONYL YLIDE FROM DIAZOCARBONYL COMPOUNDS

Carbonyl ylides (**231**) are dipolar reaction intermediates that possess significant versatility in their chemical reactions. Ab initio calculations on the simplest car-

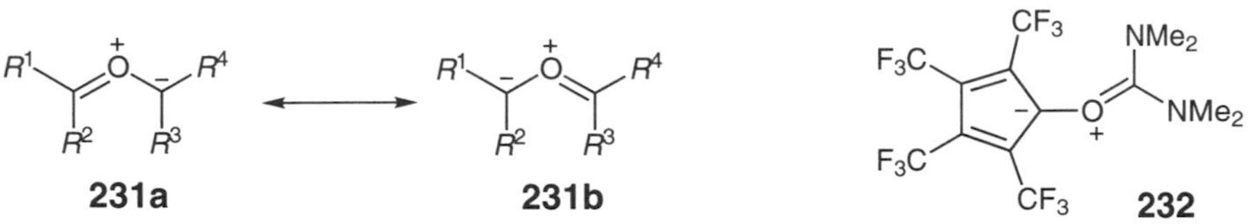

231a **231b** **232**

bonyl ylide, formaldehyde O-methylide, portray this compound as having equal C–O bond lengths and allyl-type resonance.[103] The heat of formation of the carbonyl ylide derived from methylene and acetone has been determined to be 4.5 kcal/mol, and its lifetime in acetone has been measured and found to be 53 ± 5 ns.[104] The activation energy associated with fragmentation of this ylide into methylene and acetone is estimated to be 45 kcal/mol.[105] Few stable carbonyl ylides have been reported; carbonyl ylide **232**, which is a crystalline solid and has unequal C–O bond lengths, is uncharacteristic.[106]

The earliest examples of carbonyl ylide formation in catalytic reactions of diazo compounds performed in the absence of dipolarophiles show that intramolecular proton transfer is a characteristic route to stable products (Scheme 48),[107] and this transformation has been employed in the preparation of $3(2H)$-furanones in good yield (Scheme 49).[108] However, copper salts were employed in these early

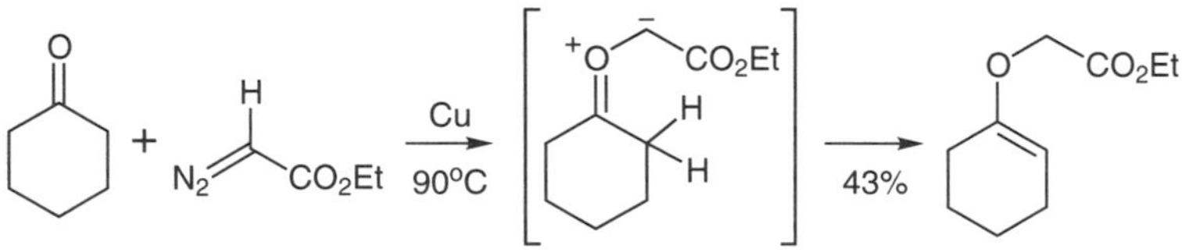

Scheme 48

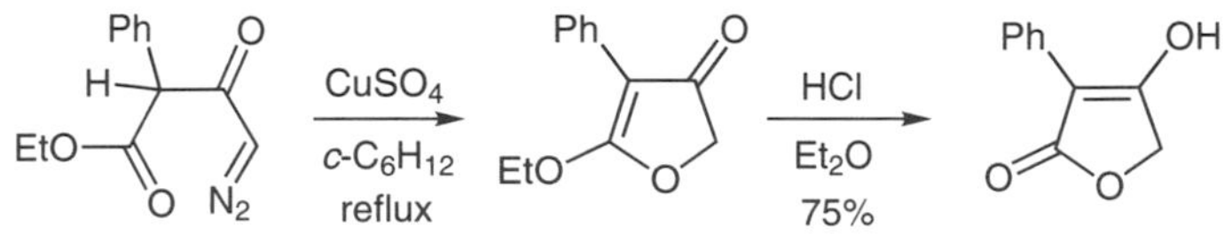

Scheme 49

examples, and relatively high temperatures were required for diazo decomposition. Use of dirhodium(II) catalysts, which cause diazo decomposition to occur under much milder conditions, has allowed the expansion of this methodology beyond proton transfer transformations to more general synthetic utilization.[1]

The catalytic route to carbonyl ylides from α-diazocarbonyl compounds is a facile process that can be either intermolecular or intramolecular. Their subsequent reactions can likewise be intermolecular or intramolecular. Carbonyl ylides are efficiently trapped by dipolarophiles (*vide infra*). However, in the absence of trapping agents, the fate of carbonyl ylides has not been thoroughly studied.

7.5.1   Intermolecular Carbonyl Ylide Formation and Subsequent Reactions

Fewer examples of intermolecular carbonyl ylide formation exist than of their intramolecular counterparts (*vide infra*), but those that do mainly involve trapping by a carbonyl compound, either in an intermolecular process (eq. 65)[109] to produce dioxolanes[109] or in intramolecular 1,3-dipolar cycloaddition (eq. 66)[110–112] to

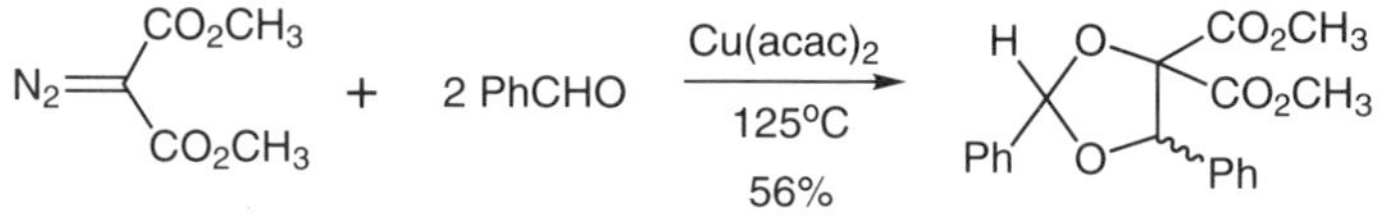

$$\text{(65)}$$

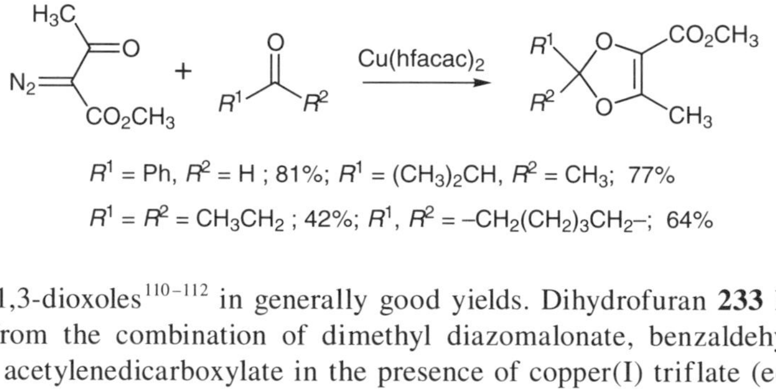

$$\text{(66)}$$

R^1 = Ph, R^2 = H ; 81%; R^1 = $(CH_3)_2CH$, R^2 = CH_3; 77%

R^1 = R^2 = CH_3CH_2 ; 42%; R^1, R^2 = $-CH_2(CH_2)_3CH_2-$; 64%

produce 1,3-dioxoles[110–112] in generally good yields. Dihydrofuran **233** has been formed from the combination of dimethyl diazomalonate, benzaldehyde, and dimethyl acetylenedicarboxylate in the presence of copper(I) triflate (eq. 67).[113]

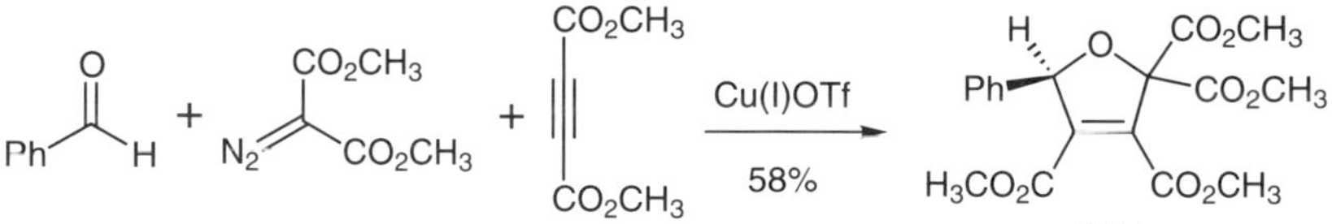

$$\text{(67)}$$

233

Rhodium perfluorobutyrate-catalyzed decomposition of diazoester **234** in the presence of benzaldehyde generated carbonyl ylide **235**, which can be trapped with dimethyl maleate to give tetrahydrofuran **236** in 49% yield (Scheme 50).[114] In contrast, the unsaturated α-diazo-α-(trimethylsilyl)acetate **237** with two

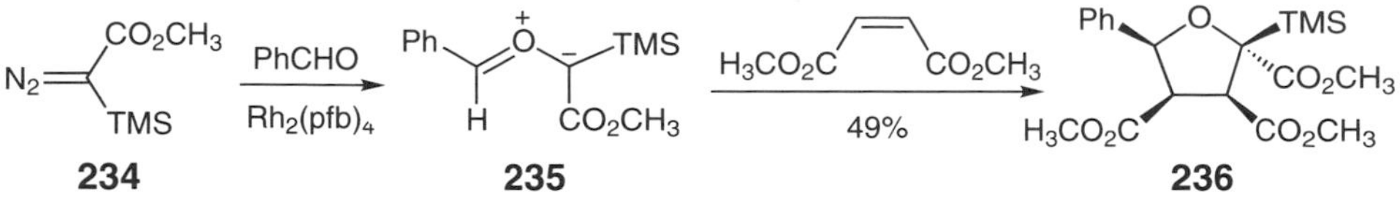

Scheme 50

equivalents of acetaldehyde under the catalytic action of rhodium perfluorobutyrate produced the 1,3-dioxolane **238** in good yield (eq. 68).[115]

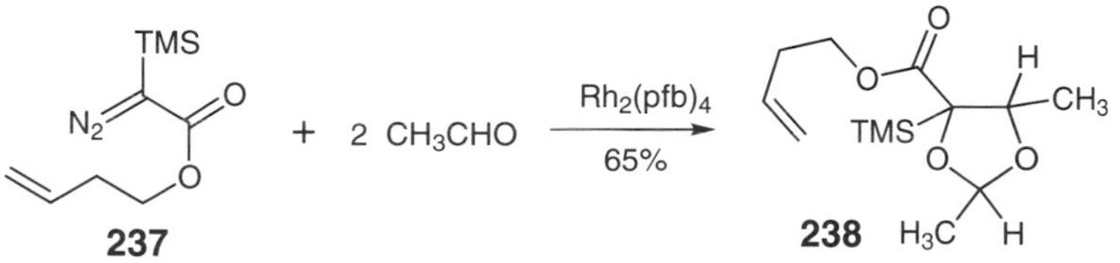

Cyclization of carbonyl ylides containing an α,β-unsaturated carbon–carbon bond provides a facile route for the construction of the furan ring. This method has served as a key step in the total synthesis of the tetracyclic furanoid diterpene methyl vinhaticoate **241**.[116] Reaction of the ketone **239** with the metal carbene derived from ethyl diazoacetate furnished the key intermediate **240**, which was then selectively hydrolyzed and decarboxylated to give **241** (Scheme 51).

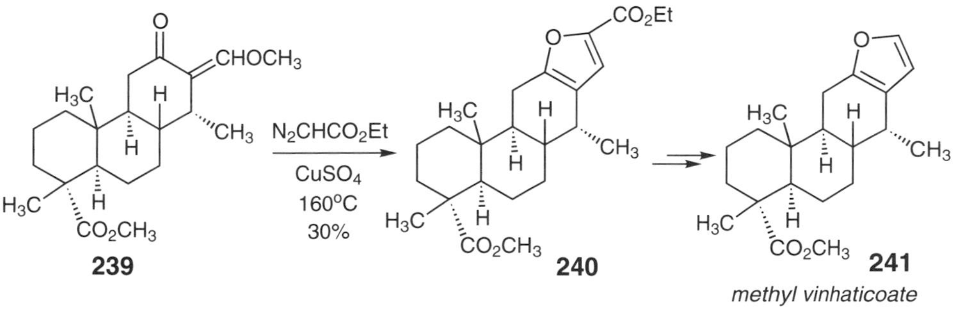

Scheme 51

7.5.2 Intramolecular Carbonyl Ylide Formation and Subsequent Reactions

Intramolecular carbonyl ylide formation has proved to be a very versatile methodology for the construction of a diverse set of highly functionalized organic compounds. Initially described by Ibata and co-workers,[117,118] this methodology has been advanced and extended by Padwa and co-workers[119] with dirhodium(II) catalysts using both intermolecular dipolar cycloaddition and tandem intramolecular processes. In this section an outline of the intramolecular generation of carbonyl ylide and their subsequent reactions in organic synthesis is presented.

7.5.2.1 Carbonyl Ylide Formation from Intramolecular Reactions of Metal Carbene and Ketones

Carbonyl ylides derived from metal-catalyzed decomposition of diazocarbonyl compounds can be readily trapped by dipolarophiles in 1,3-cycloaddition reactions leading to oxygen heterocycles. Although for intramolecular carbonyl ylide formation five- and six-membered ring intermediates are favored (eqs. 69[120] and 70[120]), seven-membered ring ylides are also

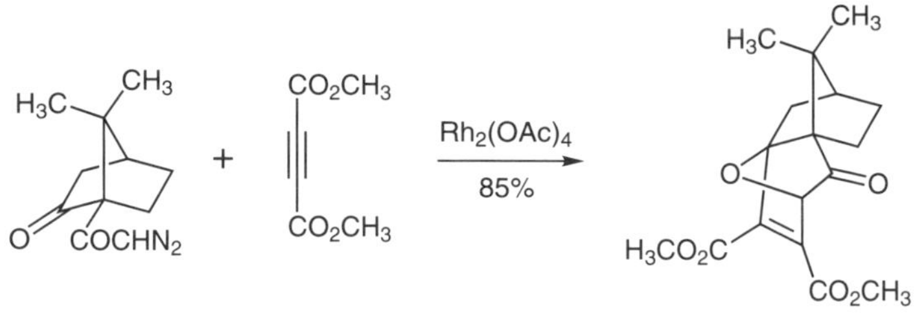

(69)

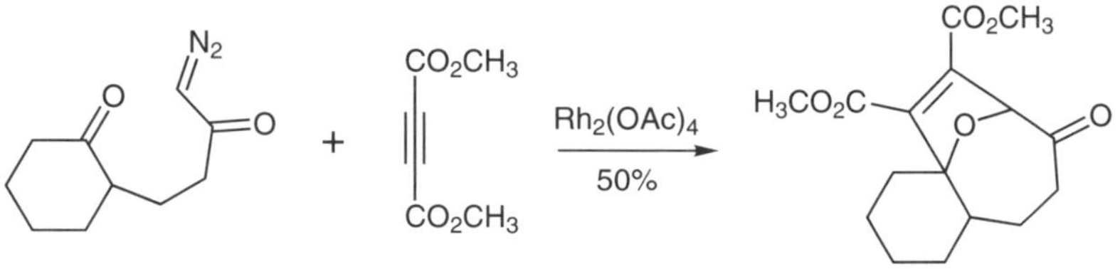

(70)

known.[120-122] In general the carbonyl group of a ketone is much more reactive towards ylide formation than is the carbonyl group of an ester. The scope and mechanistic details of the intramolecular carbonyl ylide formation from diazoketone and subsequent 1,3-dipolar cycloaddition have been extensively studied.[119-122] The ease of trapping depends on substrate structure and the absence of competition from alternative intramolecular reaction pathways. For example, the five-membered ring carbonyl ylides (**243**), generated by rhodium(II)-catalyzed decomposition of 1-diazobutanediones **242** can be trapped by dipolarophile $A = B$ to form cycloadduct **245** when both R^2 and R^3 are alkyl substitients.[120] If R^2 is a hydrogen atom, proton transfer with elimination to form furanone **244** occurs faster than intermolecular 1,3-cycloaddition (Scheme 52).[120] In terms of

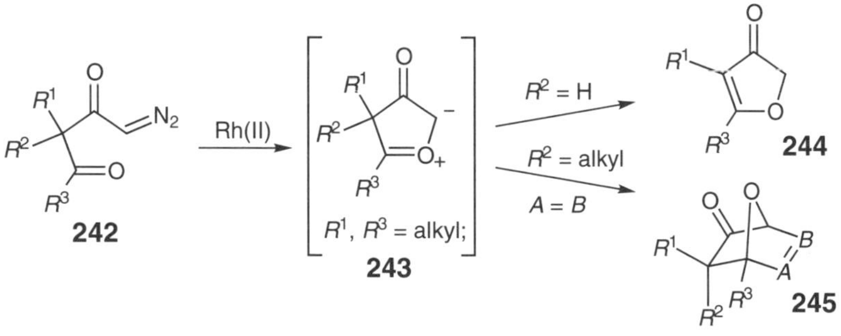

Scheme 52

synthetic applications, the main attraction of this bimolecular cycloaddition lies in the ease with which it can be applied to the rapid assembly of multifunctional cycloadducts. The combination of carbonyl ylide formation and 1,3-dipolar cycloaddition has been used as the key step in the synthesis of several sesquiterpenes.[123–126] For example, treatment of diazoketone **246** with a catalytic amount of rhodium(II) acetate in the presence of 4-bromo-5,5-dimethyl-2-cyclopentenone **247** gave cycloadduct **248** as a single diastereomer in 49% yield; further functional group manipulations furnished the sesquiterpene (±)-illudin M **249** in 2.5% overall yield from **246** (Scheme 53).[123] The dipolar-cycloaddition strategy has also been successfully applied by Padwa and co-workers[124] to the syntheses of pterosin H, I, and Z, as illustrated in Scheme 54.[125,126] The rhodium(II)-

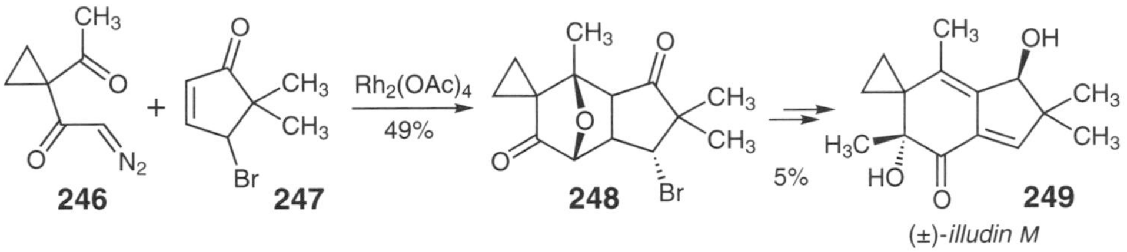

Scheme 53

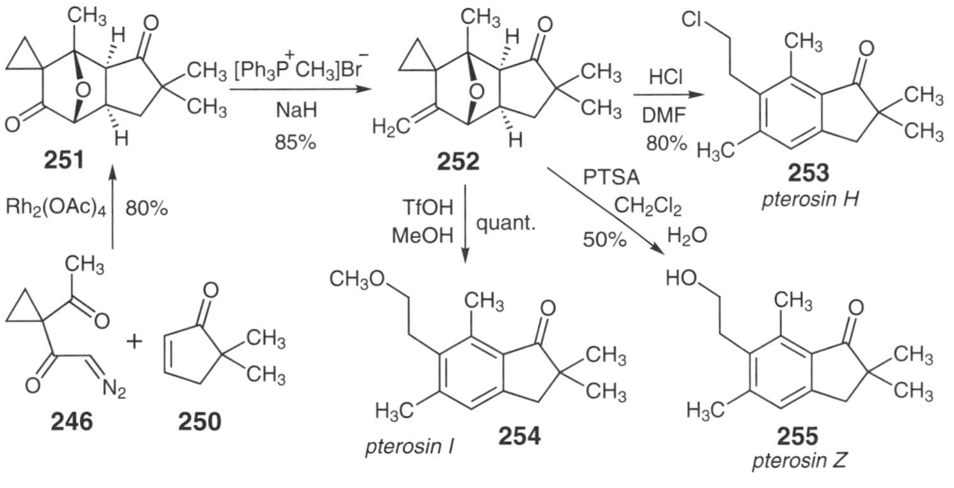

Scheme 54

catalyzed reaction of diazoketone **246** with 5,5-dimethyl-2-cyclopenten-1-one **250** gave exclusively the *exo* cycloadduct **251** in 80% isolated yield. Wittig olefination of **251** afforded the advanced intermediate **252** in 85% yield. The pterosins (**253–255**) were then obtained by treatment of **252** with an appropriate acid–solvent combination, which caused acid-catalyzed ring opening of the bridged ether followed by dehydration, reprotonation, and a subsequent cyclo-propyl carbinyl cation rearrangement. The overall transformation represents a highly effective means by which simple starting materials can be converted into the core skeleton of the pterosins in relatively few steps (Scheme 54).[126]

Intermolecular reaction of cyclic carbonyl ylides with dipolarophiles is an at-tractive protocol for the synthesis of tetrahydrofurans. This methodology has been used in the synthesis of brevicomin,[121,122] a key component of the aggrega-tion pheromone of the female Western pine beetle. Thus treatment of diazo dione **256** with rhodium(II) acetate, in the presence of propionaldehyde, afforded bi-cyclic compounds as a chromatographically separable mixture of *exo* (**257**) and *endo* isomers (**258**). Reduction of these isomers leads to *exo*-(**259**) and *endo*-brevicomin (**260**). Similarly, 6,8-dioxabicyclo[3.2.1]octane **261**, the precursor for the synthesis of solenopsin A **262**, was prepared by the same methodology (Scheme 55).

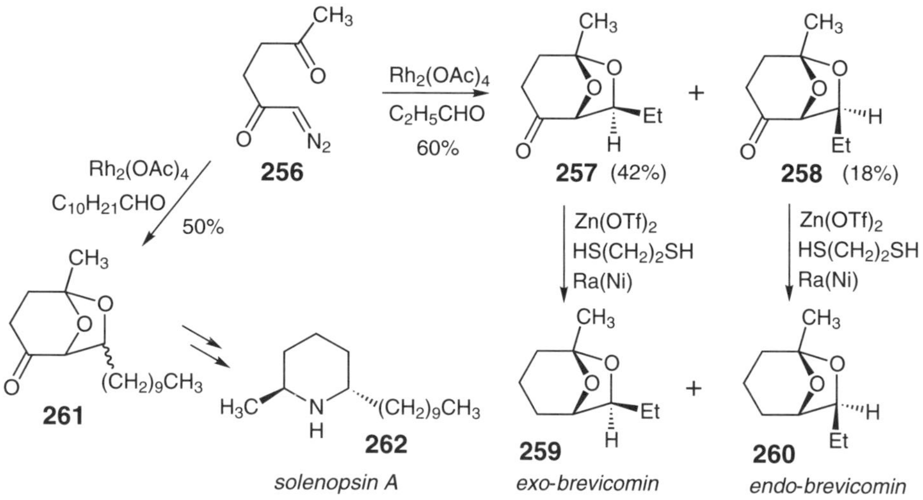

Scheme 55

The six-membered cyclic carbonyl ylide **264**, derived from rhodium(II) acetate–catalyzed decomposition of diazoester **263**, undergoes an intermolecular 1,3-dipolar cycloaddition with an enol ether dipolarophile to give the carbocyclic analog of zaragozic acid core **265** in good yield, (Scheme 56). In contrast, when ester **266** was employed as the precursor, the product **268**, a bicyclic core of zaragozic acid, was obtained in much lower yield, in keeping with the lower reac-tivity of ester carbonyls compared with ketones (Scheme 56).[127] Interestingly, di-azoketoester **263a** reacted with methyl glyoxalate in the presence of catalytic amounts of rhodium acetate to give a single cycloadduct **269a** in 60% yield;[128] the diazoketone **263b** with aromatic aldehydes gave only poor results (eq. 71).[127,128]

INTRAMOLECULAR CARBONYL YLIDE FORMATION AND SUBSEQUENT 1,3-DIPOLAR CYCLOADDITION REACTION: SYNTHESIS OF 6,8-DIOXABICYCLO[3.2.1]OCTANE RING SYSTEMS 257 AND 258[21]

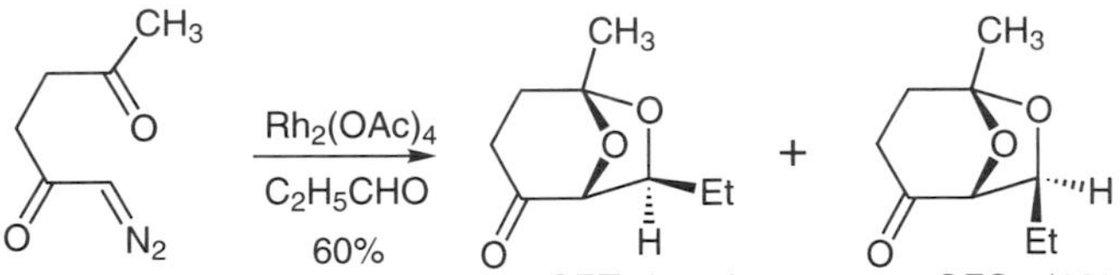

256 **257** (42%) **258** (18%)

A 0.05 M benzene solution containing 1-diazohexane-2,5-dione (**256**) and 1.2 equiv. of propionaldehyde was degassed under a nitrogen atmosphere. To this solution was added a catalytic amount of rhodium(II) acetate dimer. The yellow solution was allowed to stir under a nitrogen atmosphere at room temperature for 1 h until no more nitrogen was evolved. The solvent was removed under reduced pressure, and the residue was subjected to silica gel chromatography by using an ethyl acetate–hexane mixture as the eluent to give the 7-*exo*-ethyl-5-methyl-6,8-dioxabicyclo[3.2.1]octan-2-one (**257**) as the major product in 42% isolated yield along with the minor *endo* isomer (**258**) in 18% yield. **257**: IR (neat) 1740, 1395, 1270 cm^{-1}; ^{1}H-NMR (300 MHz, $CDCl_3$) δ 0.95 (*t*, 3 H, J = 7.4 Hz), 1.59 (qd, 2 H, J = 7.4 and 6.6 Hz), 1.60 (*s*, 3 H), 2.10–2.20 (*m*, 2 H), 2.40–2.60 (*m*, 2 H), 3.98 (*t*, 1 H, J = 6.6 Hz), and 4.14 (*s*, 1 H); ^{13}C-NMR (75 MHz, $CDCl_3$) δ 9.4, 24.2, 27.6, 32.4, 35.0, 80.1, 83.6, 107.7, and 205.8. **258**: IR (neat) 1740, 1395, 1240 cm^{-1}; ^{1}H-NMR (300 MHz, $CDCl_3$) δ 0.98 (*t*, 3 H, J = 7.4 Hz), 1.44 (dpd, 1 H, J = 14.4, 7.4, and 7.2 Hz), 1.59 (dqd, 1 H, J = 14.4, 7.4, and 7.2 Hz), 1.60 (*s*, 3 H), 2.05–2.18 (*m*, 2 H), 2.37 (ddd, 1 H, J = 18.5, 8.5, and 8.3 Hz), 2.50 (ddd, 1 H, J = 18.5, 8.0, and 5.5 Hz), 4.04 (td, 1 H, J = 7.2 and 4.7 Hz), and 4.31 (*d*, 1 H, J = 4.7 Hz); ^{13}C-NMR (75 MHz, $CDCl_3$) δ 10.3, 22.4, 24.1, 33.2, 34.1, 80.0, 83.2, 107.3, and 205.4 ppm.

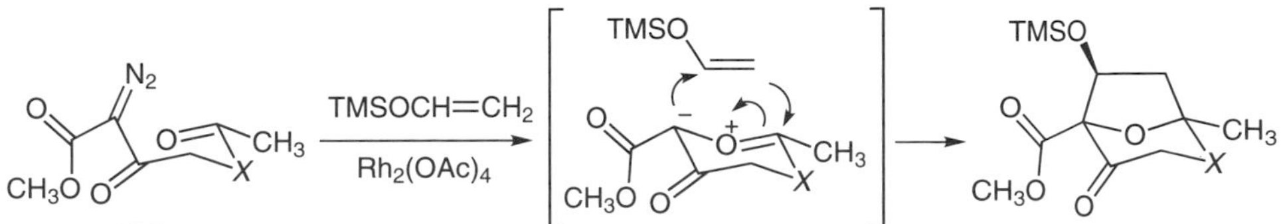

263: $X = CH_2$ **264:** $X = CH_2$ **265:** $X = CH_2$; 66%

266: $X = O$ **267:** $X = O$ **268:** $X = O$; 16%

Scheme 56

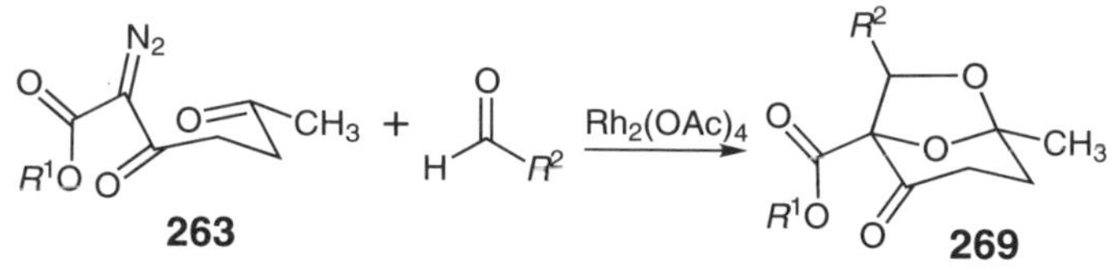

$$(71)$$

a: $R^1 = CH_3CH_2$, $R^2 = CO_2CH_3$; 60% yield
b: $R^1 = CH_3$, $R^2 = Ph$; 7% yield
c: $R^1 = CH_3$, $R^2 = p\text{-}CH_3OC_6H_4$; 16% yield

The intramolecular trapping of carbonyl ylide dipoles with an alkene or alkyne represents an effective method for the synthesis of complex polycyclic heterocycles. Varying the length of the group that separates the olefin or the alkyne from the carbonyl ylide dipoles allows for the synthesis of a variety of interesting oxopolycyclic ring systems.[119] In fact, the intramolecular cycloaddition exhibits enhanced reactivity and stereoselectivity over its intermolecular counterparts. A representative example in which a five-membered carbonyl ylide **271** underwent an intramolecular cycloaddition to give the polycyclic adduct **272** is shown in Scheme 57.[126]

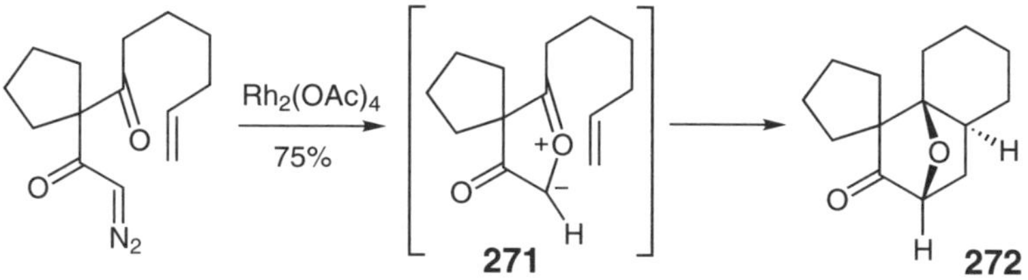

Scheme 57

Dauben and co-workers[129] have used this intramolecular trapping strategy as the central step in their synthesis of the tigliane ring system. Thus carbonyl ylide **274**, generated from the diazocarbonyl **273** in the presence of a catalytic amount of rhodium(II) acetate, underwent an intramolecular addition with the olefin to form the C_6,C_9-oxido-bridged tigliane ring system **275** (Scheme 58). The two new

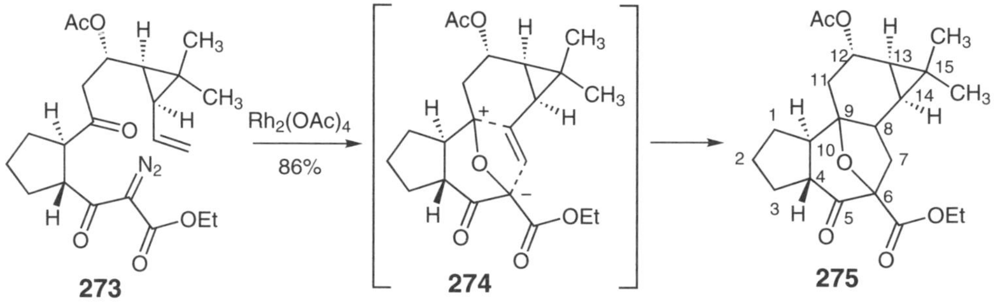

Scheme 58

stereocenters at C-8 and C-9 were formed with configurations relative to C-14 and C-15 that are present in the natural tigliane compounds. The high stereoselectivity in the ring-closure reaction could be related to steric interactions and/or the introduction of conformational strain in the tether, which does not favor a transition state leading to the cyclopropane ring and the oxido-bridge to be on the same side of the molecule. A further example is found in a synthesis of a phorbol analog, where the tandem cyclization–cycloaddition process was again the key transformation.[130]

7.5.2.2 Carbonyl Ylide Formation from Intramolecular Reactions of Metal Carbene and Esters

As mentioned earlier, the carbonyl group of an ester is much less reactive towards ylide formation than is the carbonyl group of a ketone. Scheme 59 illustrates a case in which ester participation would lead to

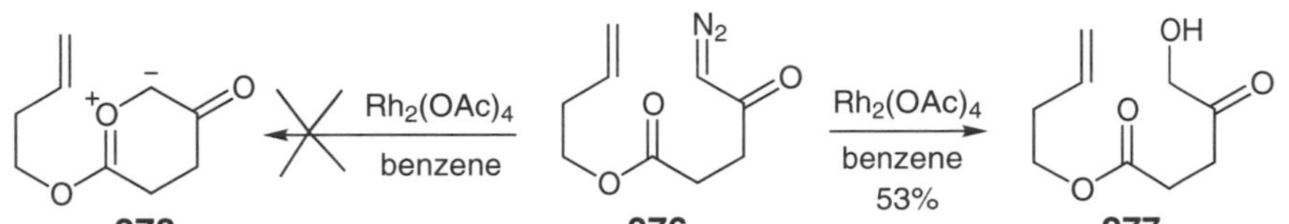

Scheme 59

products derived from ylide **278** formed from diazoketone **276**. In fact this was not observed. The principal product was hydroxyketone **277** presumed to result from O–H insertion of adventitious water.[61] However, under certain circumstances, the intramolecular reaction of the catalytically generated metal carbene and an ester function do produce cyclic carbonyl ylides, an early example of which was shown in Scheme 49. Ibata and co-workers[117,118,131] demonstrated that the cyclic carbonyl ylide **280** derived from copper-catalyzed decomposition of O-(alkoxycarbonyl)-α-diazoacetophenone **279** can be trapped by various dipolarophiles; an example is shown in Scheme 60.

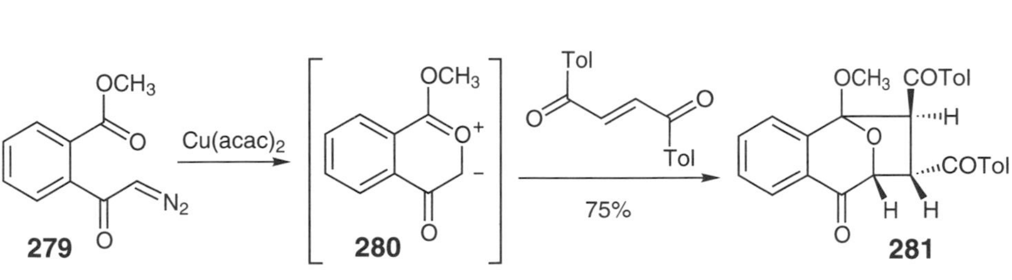

Scheme 60

Intramolecular trapping of the ester-derived cyclic carbonyl ylide provided polycyclic adducts, and two examples are shown in eq. 72[132] and Scheme 61.[133] Thus rhodium(II) acetate–catalyzed decomposition of diazocarbonyl precursors

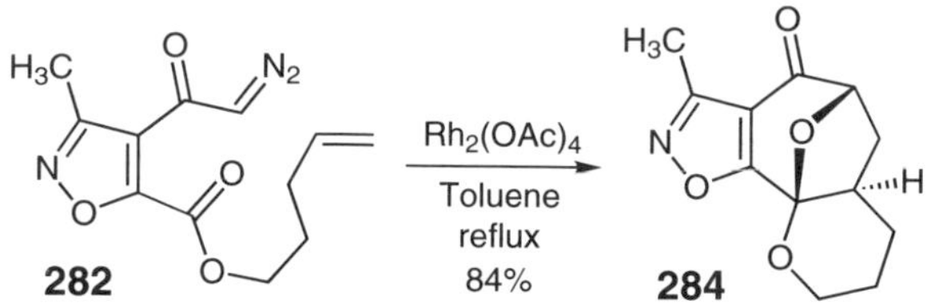

(72)

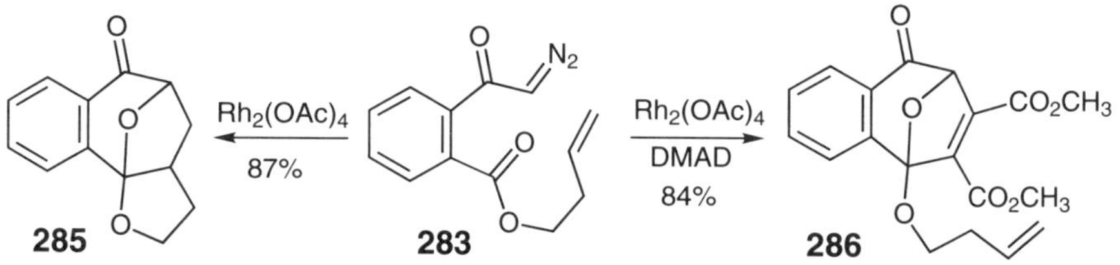

Scheme 61

282 and **283** produced cycloadducts **284** and **285** in good yield. The cyclic carbonyl ylide can be intercepted by intermolecular cycloaddition if the dipolarophile is stronger than that tethered to the substrate. For example, in the presence of dimethyl acetylenedicarboxylate (DMAD), the intramolecular cycloaddition reaction of **283** was completely suppressed, and only the dipolar cycloadduct **286** derived from the bimolecular pathway was isolated in 84% yield (Scheme 61).[133] Finally, the 6-functionalized-11-oxasteroid **289** has been synthesized through the intramolecular rearrangement of the cyclic carbonyl ylide **288** derived from copper-catalyzed decomposition of diazoester **287** (Scheme 62).[134]

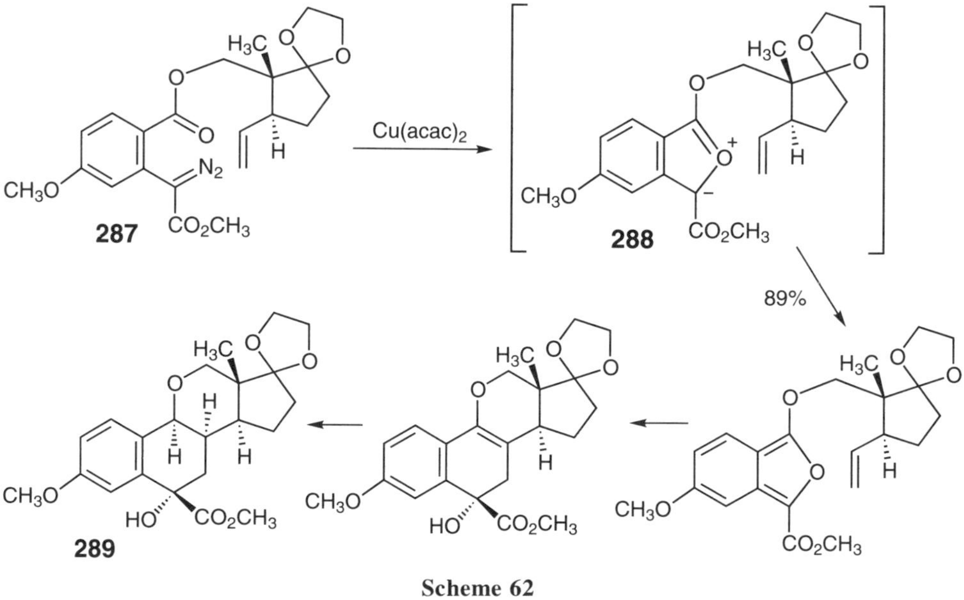

Scheme 62

7.5.2.3 Carbonyl Ylide Formation from Intramolecular Reactions of Metal Carbene and Amides, Imides, or Carbamates

Intramolecular addition of catalytically generated metal carbenes to the carbonyl group of amides can also lead to cyclic carbonyl ylides. These ylides can undergo subsequent dipolar cycloaddition (Scheme 63), intramolecular proton transfer (Scheme 64), or

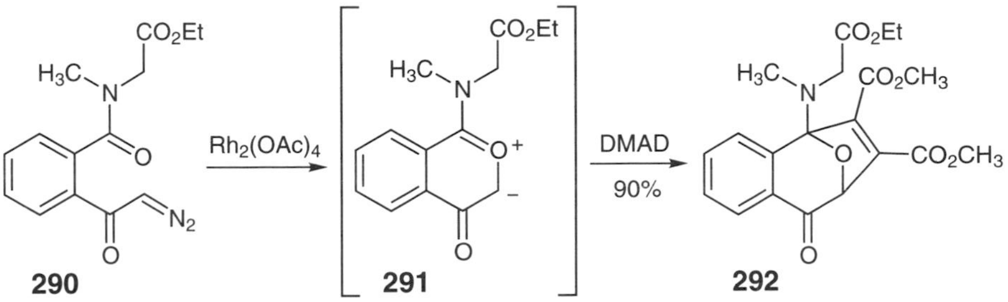

Scheme 63

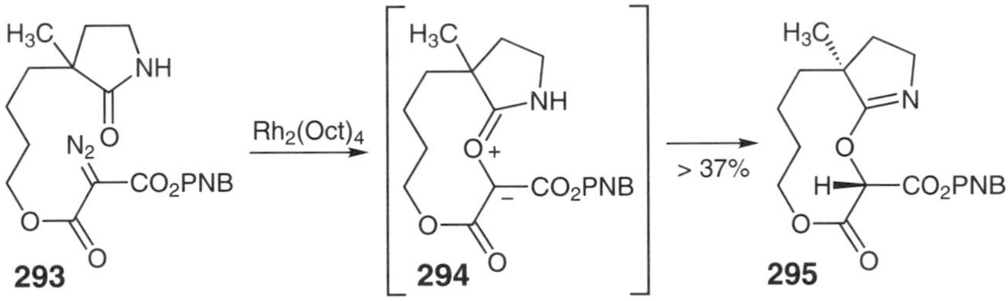

Scheme 64

dipole cascade reaction (Scheme 65), depending upon the nature of the substrates. In the case of α-diazo keto amide **290**, the cyclic carbonyl ylide **291** is sufficiently stabilized *via* resonance to be trapped by dimethyl acetylenedicarboxylate (DMAD) to afford cycloadduct **292** in 90% yield (Scheme 63).[135] Intramolecular proton transfer is also possible with cyclic carbonyl ylides, and an example is shown in Scheme 64. Thus rhodium(II) octanoate–catalyzed decomposition of diazocarbonyl compound **293** afforded the carbonyl ylide **294**, which underwent an intramolecular proton-transfer process to give the fused bicyclic imino ether **295** in modest yield (Scheme 64).[136]

The carbonyl ylide dipole, generated by rhodium(II)-catalyzed diazoketone cyclization onto a neighboring amide carbonyl group, can undergo a proton shift to form the thermodynamically more stable azomethine ylide. The 1,3-dipolar cycloaddition of the azomethine ylide and subsequent rearrangement provides a potentially useful method for the synthesis of nitrogen-based heterocycles.[137–139] Catalytic decomposition of diazoketone **296** produced the carbonyl ylide **297**, which isomerized to the azomethine ylide **298**. 1,3-Dipolar cycloaddition with dimethyl acetylenedicarboxylate produced **299**, and a subsequent 1,3-alkoxy shift generated the tricyclic dihydropyrrolizine **300**, which fragmented to produce pyrrole **301** (Scheme 65).[137,138] Furthermore, rhodium acetate–catalyzed decom-

296: R = H
296a: R = alkyl
297: R = H
297a: R = alkyl
298

87% DMAD

301
300
299

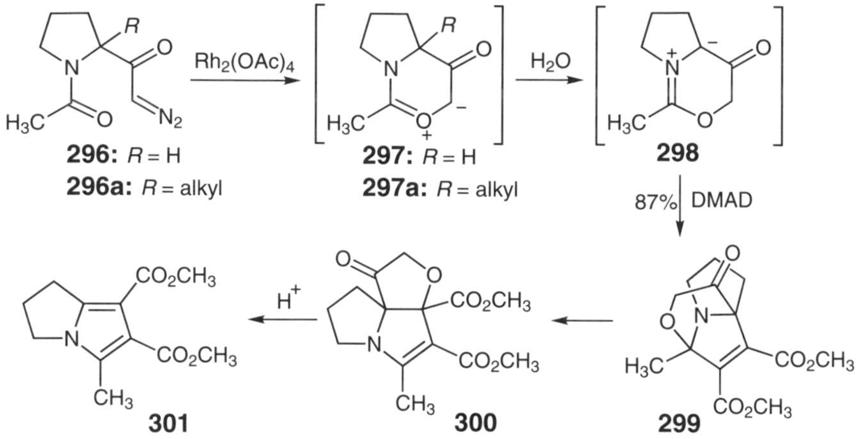

Scheme 65

position of the similar diazoketo amide **302** did not produce any cycloadducts (e.g., **303**) that arise from an intramolecular dipolar cycloaddition process involving distal alkene. When diazodecomposition was carried out in the presence of dimethyl acetylenedicarboxylate (DMAD), the cycloadduct **304** resulting from intermolecular reaction was isolated in 85% yield. In this case, the dipole cascade occurred at a faster rate than intramolecular cycloaddition onto the alkenyl π-bond (Scheme 66). Further interesting examples of the dipole cascade process

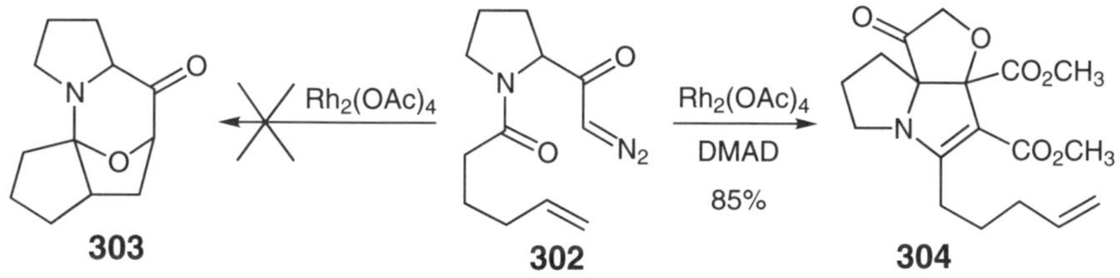

303
302
304

Scheme 66

have been provided by Rodgers and co-workers,[139] as shown in Scheme 67. Thus rhodium acetate–catalyzed decomposition of diazocarbonyl compound **305** afforded the stabilized cyclic carbonyl ylide **306**. Collapse of **306** to the epoxide **307** followed by ring opening gave the zwitterion **308**. Attack of the oxygen on the more electrophilic carbonyl and carbon bond migration then gave the product **309**.

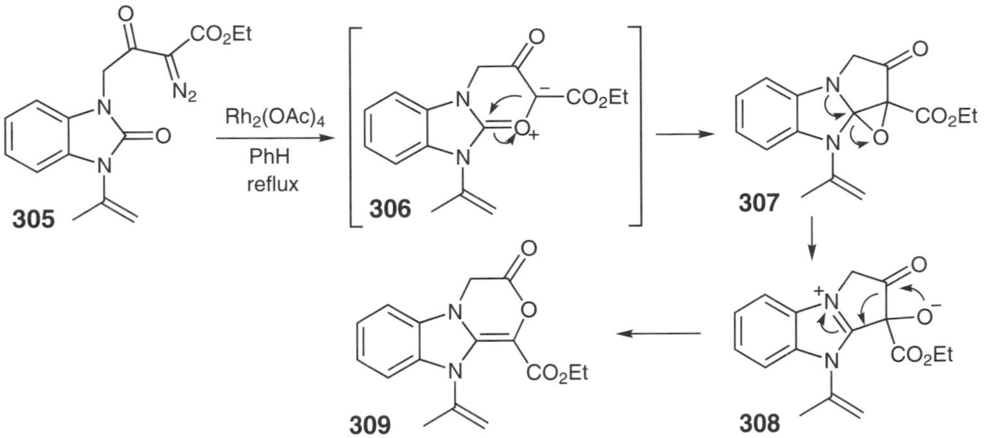

Scheme 67

It should be noted that a proton must be removed from the α-carbon atom in order to promote the dipole cascade reaction. Cycloaddition would be expected to take place exclusively from the carbonyl ylide dipole (e.g., **297a** in Scheme 65) if the α-carbon was attached to an alkyl group (e.g., **296a** in Scheme 65).[138]

Metal-catalyzed decomposition of suitable diazo imides provides an effective route to the 1,3-oxazolium 4-oxides (isomünchnone) **311** (Scheme 68). This type of mesoionic oxazolium ylide is the cyclic equivalent of a carbonyl ylide and undergoes 1,3-dipolar cycloaddition. The first example of a catalytically generated isomünchnone ylide and its subsequent 1,3-dipolar cyclization, described by Ibate and Hamaguchi, is shown in Scheme 69.[140,141]

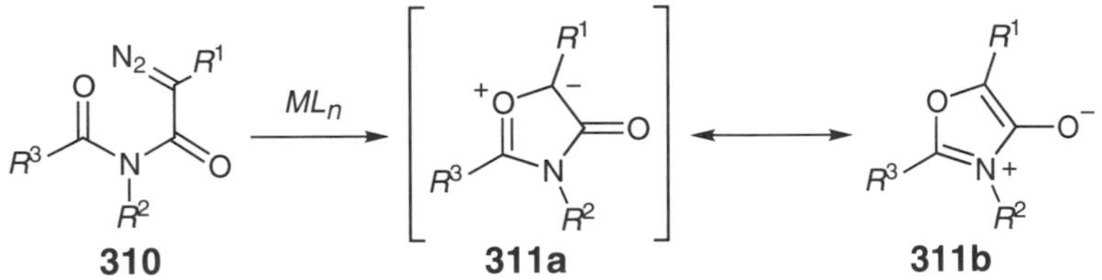

R^1, R^2 = alkyl or acyl group, R^3 = alkyl, alkoxy, or amino group

Scheme 68

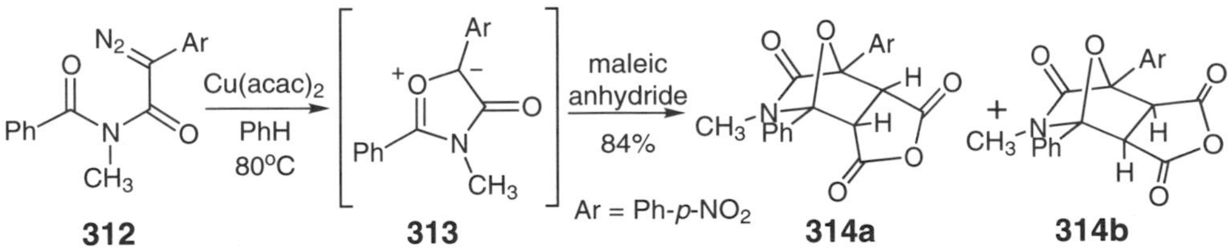

Scheme 69

Later, Maier[142,143] and Padwa[144-154] and their respective co-workers used rhodium(II) carboxylates as catalysts for the generation of isomünchnones from diazo imides. Subsequent cycloaddition reactions of isomünchnones can be intermolecular or intramolecular. The isomünchnone **316**, derived from 2-diazoacetyl-2-pyrrolidinone **315**, reacts readily with both electron-deficient (*N*-phenyl maleimide) and electron-rich (diethyl ketene acetal) dipolarophiles (Scheme 70).[144]

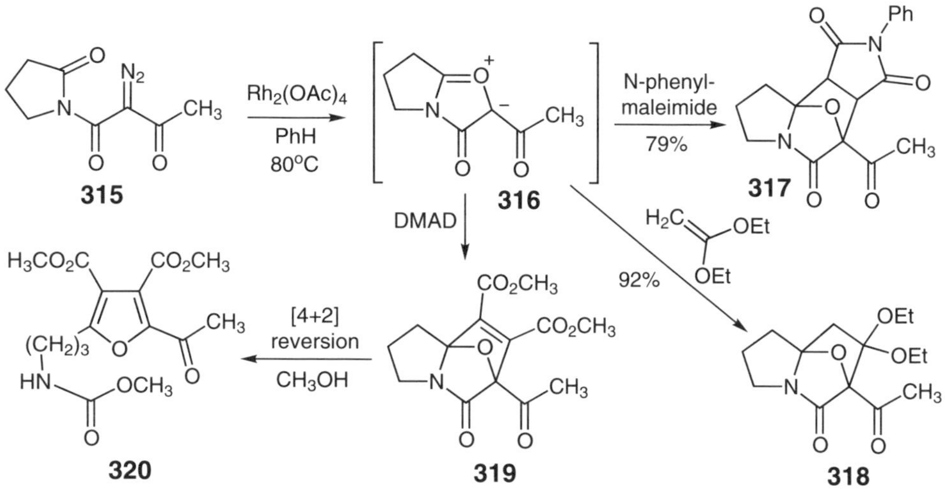

Scheme 70

When an acetylenic dipolarophile, such as DMAD, was used as the trapping agent, the resulting dipolar cycloadduct **319** was not isolated, but underwent a [4+2] cycloreversion instead to furnish furanoisocyanate, which then reacted with methanol to give the furan derivative **320**.[145] Furthermore, intermolecular trapping the catalytically generated isomünchnone with buckminsterfullerene C_{60} afforded a dipolar cycloadduct in modest yield.[146]

Intramolecular cycloaddition of isomünchnones containing π-bonds suitably placed within the molecule has emerged as an attractive sequence to polycyclic structures and has been examined in some detail.[142,143,147-154] Carbon-carbon double[142,147-152] and triple bonds,[143,148] and heteroaromatic π-systems[147,153,154] act as dipolarophiles in these intramolecular cycloaddition reactions with isomünchnones. As shown in Scheme 71, treatment of the alkenyl-substituted diazoimide **321** with a catalytic amount of rhodium(II) acetate in benzene (80°C) gave cycloadduct **323** in 86% yield.[148] Padwa and co-workers[152] have employed a similar intramolecular cycloaddition reaction as the key step in their synthesis of the quinoline skeleton present in the ergot alkaloid family.

Intramolecular reactions of acetylenic isomünchnones afford annulated furans and a representative example is illustrated in Scheme 72. Thus catalytic decomposition of diazoimide **324** with rhodium(II) acetate in refluxing toluene fur-

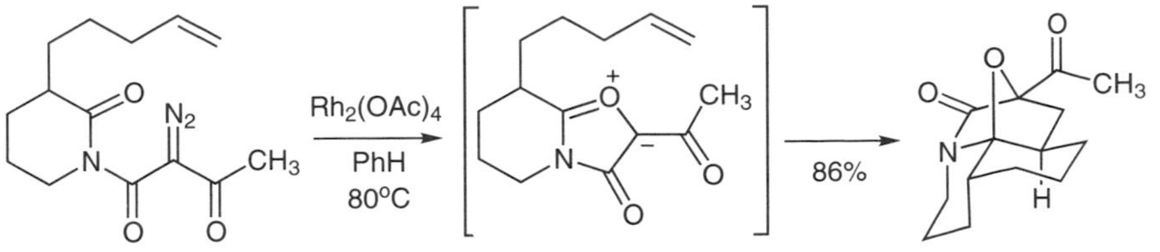

321   **322**   **323**

Scheme 71

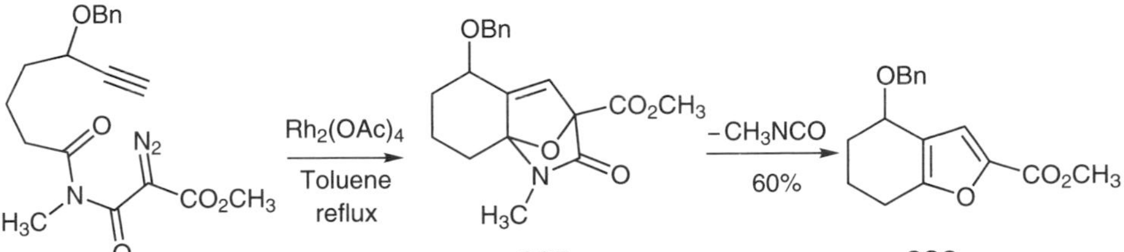

324   **325**   **326**

Scheme 72

nished cycloadduct **325**, which then underwent a subsequent cycloreversion reaction to give furan **326** (Scheme 72).[143] Although the reactivity of heteroaromatic dipolarophiles is decreased because of the loss of aromaticity in the cycloaddition process, the intramolecular 1,3-dipolar cycloaddition of the isomünchnones across a tethered furan, thiophene, and indole ring provides a route to a variety of novel polyheterocyclic compounds in good yield and in a stereocontrolled fashion.[153,154] The scope and limitations of the intramolecular cycloaddition of isomünchnones with thiophene has been studied in detail, and the facility of the cycloaddition is critically dependent on conformational factors in the transition state. As outlined in Scheme 73, the oxabicyclic intermediate **329**, related to the

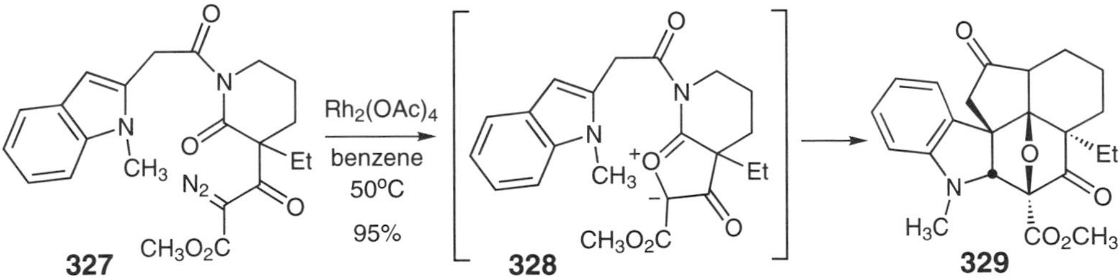

327   **328**   **329**

Scheme 73

pentacyclic skeleton of the aspidosperma alkaloids, was prepared *via* the tandem cyclization–cycloaddition process. Thus treatment of diazoimide **327** with a catalytic quantity of rhodium(II) acetate in benzene at 50°C gave **329**, *via* isomünchnone **328**, in 95% yield as a single diastereomer.[154]

By incorporating an internal nuclophile into the tether, annulation of the original dipolar cycloadduct allows for the construction of a nitrogen-based polycyclic systems.[150] This annulation sequence was employed by Padwa and co-workers,[151] as the key transformation in a formal synthesis of (±)-lycopodine (335), as shown in Scheme 74. The reaction of α-diazoimide 330 with a catalytic quantity

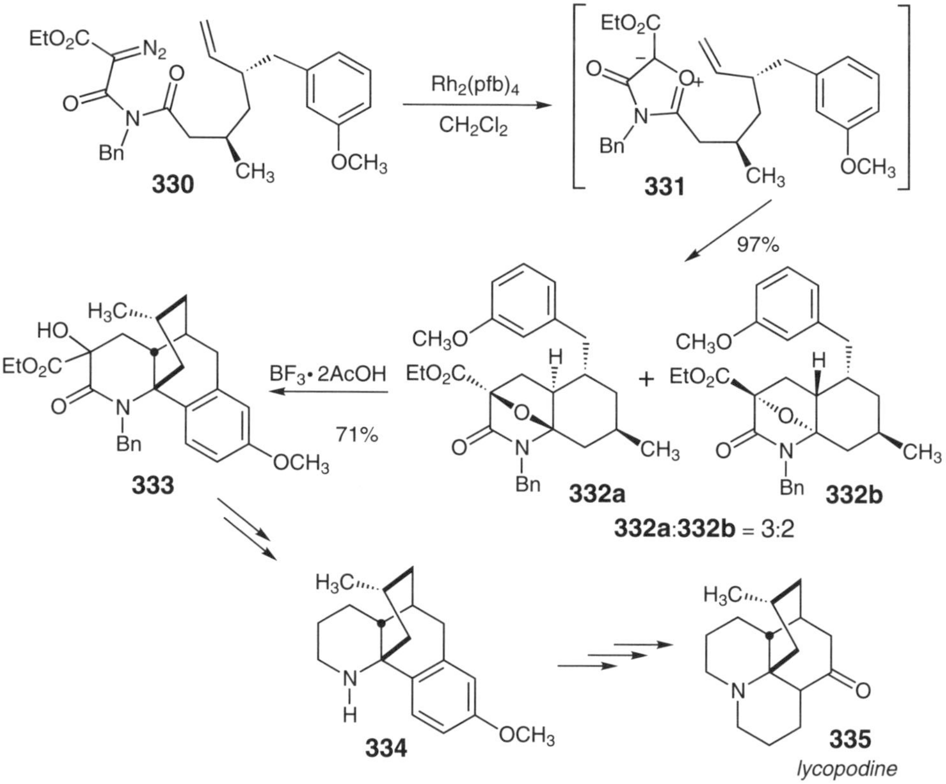

Scheme 74

of rhodium(II) perfluorobutyrate in dichloromethane at 25°C provided cycloadduct as a 3:2 mixture of *endo*-diastereomers (332a and 332b) in 97% yield. Exposure of the diastereomeric mixture to BF$_3$·2AcOH generated the tetracyclic amide 333 as a 4:1 mixture of diastereoisomers (71%) about the tertiary hydroxyl center. Further synthetic modification of lactam 333 led to the key intermediate 334, which had been converted into (±)-lycopodine (335) by Stork and co-workers.[155]

7.5.3 Chemoselectivity in Carbonyl Ylide Formation

For diazocarbonyl compounds possessing several functional groups, predictable control of chemoselectivity constitutes an important prerequisite for synthetic applications. Extensive studies in this area clearly revealed that both the molecular structure and the nature of the catalyst are the crucial factors in controlling the different reaction pathway. Cyclopropanation, C–H insertion, and aromatic cycloaddition are frequently the major competing reactions in intramolecular cyclic carbonyl ylide formation. One of the earlier examples involving competition between the formation of cyclopropane and carbonyl ylide is shown in eq. 73.[156]

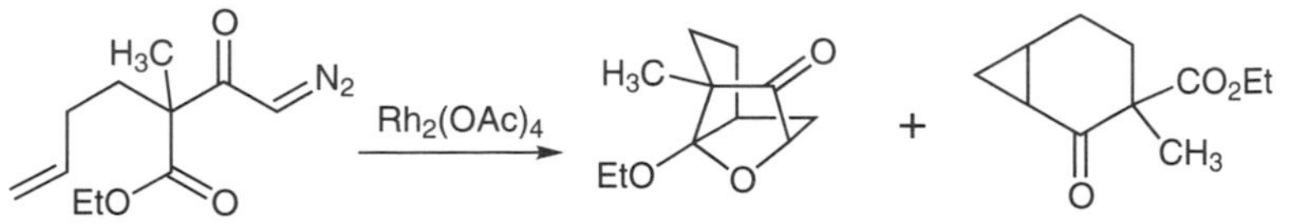

336   **337** (8%)   **338** (26%)   (73)

Thus rhodium(II) acetate–catalyzed decomposition of α-diazoketone **336** resulted in the formation of tricyclic adduct **337** derived from the cycloaddition of a carbonyl ylide intermediate in poor yield. The major product **338** arose from an intramolecular cyclopropanation reaction. Similarly, treatment of α-diazoketone **339** with rhodium(II) acetate led to a 1:1 mixture of the internal dipolar cycloadduct **340** as well as the cyclopropanated product **341** (eq. 74).[157,158] The choice of

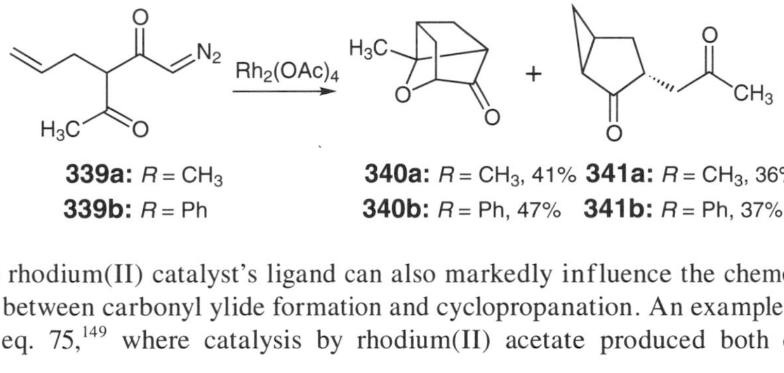

339a: $R = CH_3$ **340a:** $R = CH_3$, 41% **341a:** $R = CH_3$, 36%
339b: $R = Ph$ **340b:** $R = Ph$, 47% **341b:** $R = Ph$, 37%

the rhodium(II) catalyst's ligand can also markedly influence the chemoselectivity between carbonyl ylide formation and cyclopropanation. An example is shown in eq. 75,[149] where catalysis by rhodium(II) acetate produced both cyclopro-

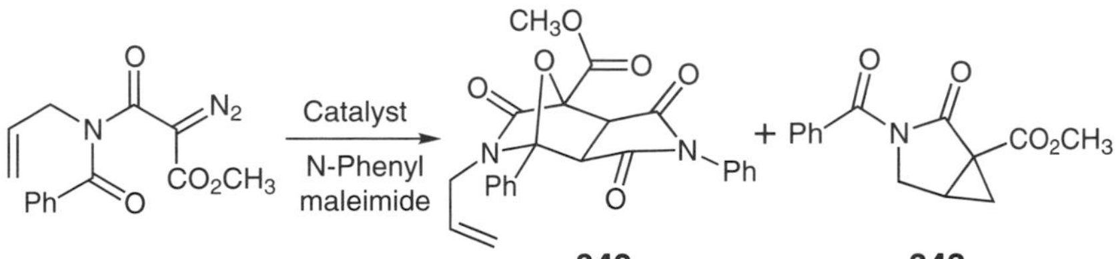

342   **343**

	342	**343**
$Rh_2(OAc)_4$	40%	60%
$Rh_2(tfam)_4$	0%	100%

panated product **342** and bimolecular cycloadduct **343** in comparable amounts. However, only the bimolecular cycloadduct **343** was formed with rhodium(II) trifluoroacetamide (eq. 75).[149]

Examples are known where intramolecular metal carbene reactions result in competition between carbonyl ylide formation and aromatic substitution. For example, treatment of diazoketone **344** with rhodium(II) acetate and DMAD (1.2 equiv.) gave the dipolar cycloadduct **345** (60%) as well as benzocyclopentanone **346** (20%). However, with rhodium(II) caprolactamate, the tandem carbonyl ylide formation–dipolar cycloaddition was the only process that occurred to give **345** in 90% yield. In contrast, with rhodium(II) perfluorobutyrate, ketone **346** was formed (85%) to the virtual exclusion of cycloadduct **345** (eq. 76).[157] Metal

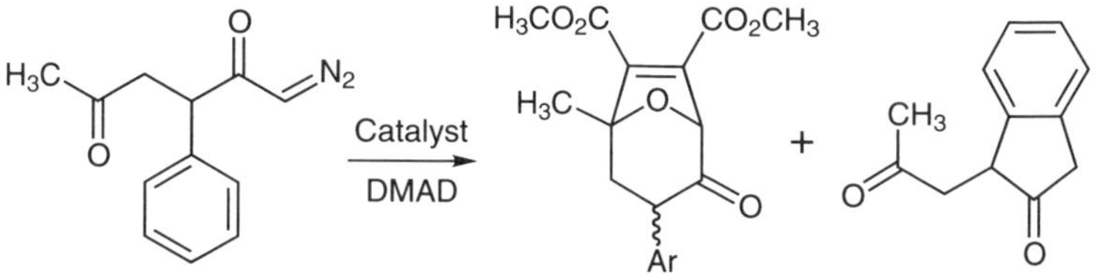

	345	**346**
Rh$_2$(OAc)$_4$	60%	20%
Rh$_2$(pfb)$_4$	0%	85%
Rh$_2$(cap)$_4$	90%	0%

$$(76)$$

carbene transformations involving competition between intramolecular carbonyl ylide formation and C–H insertion have also been investigated.[145,159] Doyle and co-workers[145] have reported that rhodium-catalyzed decomposition of diazoacetoacetamide **347** afforded products derived from both a carbonyl ylide intermediate and intramolecular C–H insertion (eq. 77).[145] As shown in eq. 77, the relative

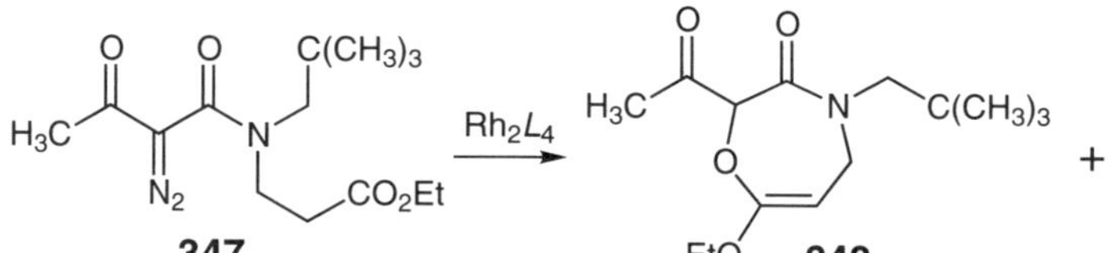

	yield **348**	**349**	**350**
Rh$_2$(pfb)$_4$	97%	61%	22%	17%
Rh$_2$(OAc)$_4$	85%	32%	59%	9%
Rh$_2$(acam)$_4$	89%	10%	64%	26%

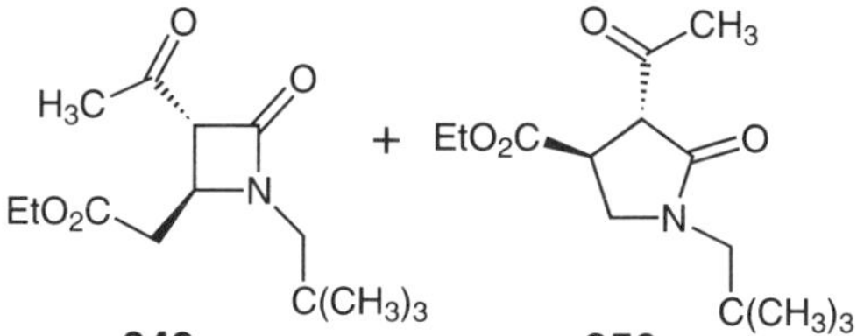

$$(77)$$

yield of the carbonyl ylide product **348** is dependent on electronic influences of the ligands of rhodium(II). In contrast to the above result, in which carbonyl ylide formation is favored over C–H insertion, rhodium(II) acetate–catalyzed decomposition of diazocarbonyl compound **351** afforded C–H insertion product **352** exclusively. When either rhodium(II) perfluorobutyrate or rhodium(II) caprolactamate was employed as catalyst, the reaction simply gave slightly reduced yields of **352**, and no product resulting from an intermediate carbonyl ylide was observed (Scheme 75).[159]

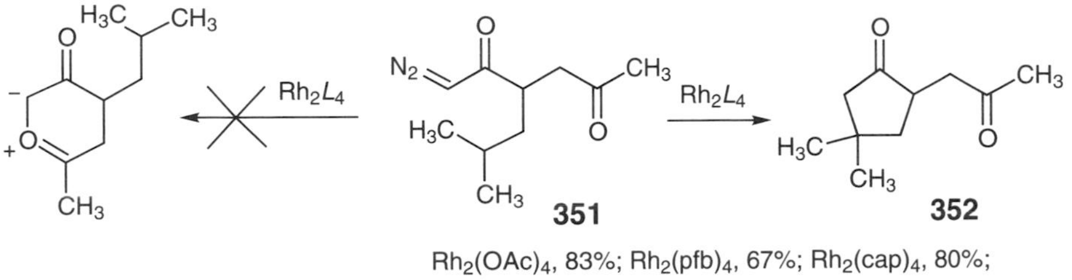

Rh$_2$(OAc)$_4$, 83%; Rh$_2$(pfb)$_4$, 67%; Rh$_2$(cap)$_4$, 80%;

Scheme 75

In cases where more than one tethered carbonyl group is present in the molecule, the possibility of selective addition of the metal carbene to one of them arises. The type of carbonyl group, the ring size of the resulting carbonyl ylide, and the nature of the catalyst employed are the key to control of this type of reaction. A particularly interesting example is shown in Scheme 76, where carbonyl

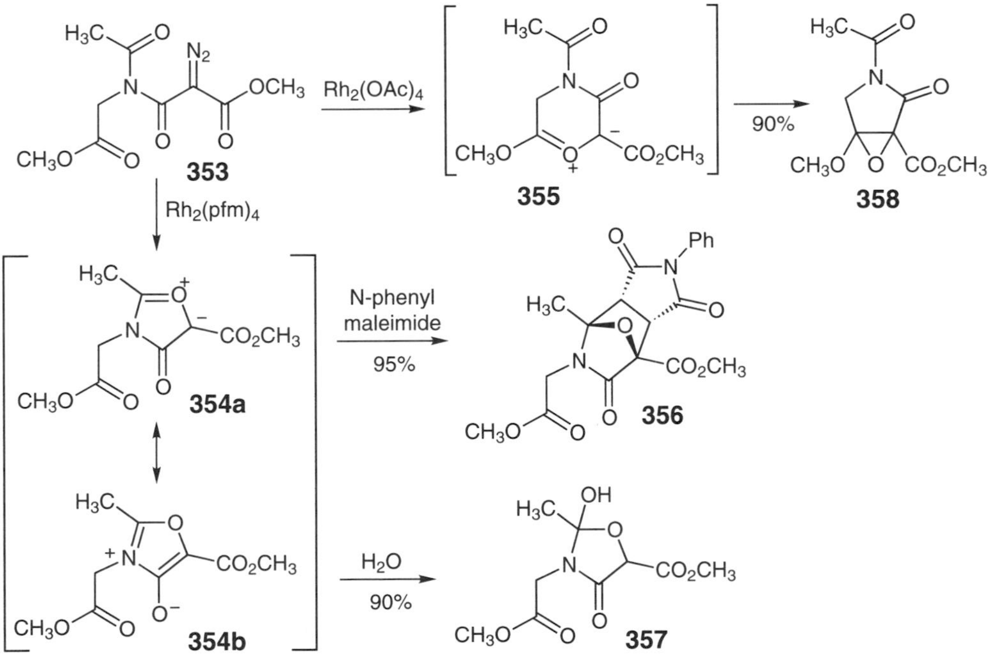

Scheme 76

ylide formation can be selectively controlled by the proper choice of catalyst. Rhodium(II) perfluorobutyramidate–catalyzed decomposition of α-diazoimide **353** produced the intermediate isomünchnone **354**, which was trapped by N-phenylmaleimide to give cycloadduct **356** in excellent yield. In the absence of dipolarophile, isomünchnone **354** reacted with water to give an equilibrium mixture of the diastereomers of hemiketal **357**. When rhodium(II) acetate was employed as the catalyst, the major product was epoxide **358** produced by attack of the rhodium carbene on the ester carbonyl group, as shown in Scheme 76.[160]

7.6 FORMATION OF THIOCARBONYL YLIDE FROM DIAZOCARBONYL COMPOUNDS

Thiocarbonyl ylides have been the subject of much interest in recent years due to their potential role as intermediates in a variety of reactions, including the formation of episulfides and novel heterocyclic ring systems. Only recently have catalytic methods been applied to the generation of thiocarbonyl ylides, a delay in the exploration of their chemistry that may have been due to the presumption that the coordinating ability of sulfur compounds with transition metal catalysts would inhibit their ability to cause diazo decomposition. However, dirhodium(II) carboxylate catalysts have proven to be remarkably effective.[161–164] Takano and co-workers[161] have reported the intermolecular formation of thiocarbonyl ylides *via* diazocarbonyl intermediates. Reaction of (R)-5-(benzyloxymethyl)tetrahydro-2-furanthione **359** with diazocarbonyl compounds **360** in the presence of rhodium(II) acetate afforded the sulfur-free 2-(acymethylene)tetrahydrofuran derivatives **365** in good yields. The desulfurization process has been postulated as proceeding via the mechanism shown in Scheme 77. Thus thiocarbonyl ylide **361** collapses to give the episulfide **362**, which reacts with excess metal carbene to produce sulfonium ylide **363**; rearrangement of the sulfonium ylide **363** accounts for the formation of **365**. This methodology has been used as key step in the synthesis of (+)-nonactic acid **368**, as shown in Scheme 78.[165]

Under certain circumstances, episulfides may be isolated from the rearrangement of catalytically generated thiocarbonyl ylides.[166–168] For example, reaction of thioketone **369** with dimethyl diazomalonate in the presence of rhodium(II) acetate afforded thiirane **372**, malonate **373**, and spirocyclic 1,3-dithiolanetetracarboxylate **376**.[168] Compound **376** results from a consecutive 1,3-dipolar cycloaddition reaction of intermediate **371** or **375**, as shown in Scheme 78. Thiocarbonyl ylides derived from intermolecular reaction of metal carbenes with di-*tert*-butylthioketene undergo both 1,5- and 1,3-cyclization reaction.[166,167] As shown in Scheme 79, the rhodium(II) acetate–catalyzed reaction of diazoester **377** with di-*tert*-butylthioketene gave oxathiolanone **381** and allene episulfide **382** in 69% and 21% yield, respectively. The oxathiolanone **381** was formed through the 1,5-cyclization of the thiocarbonyl ylide **379** to give oxathiole **380** followed by Claisen rearrangement (Scheme 80).

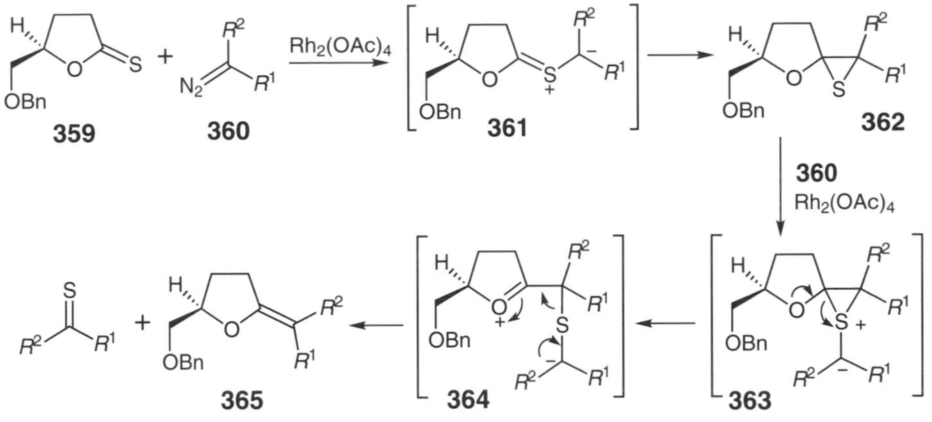

Scheme 77

365a: $R^1 = R^2 = CO_2Et$, 74% yield;
365b: $R^1 = CO_2Et$, $R^2 = P(O)(OEt)_2$, 74% yield

365c: $R^1 = R^2 =$ 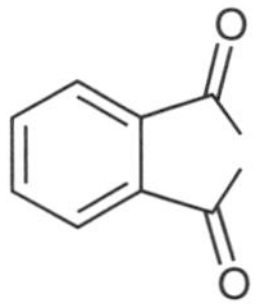 , 96% yield

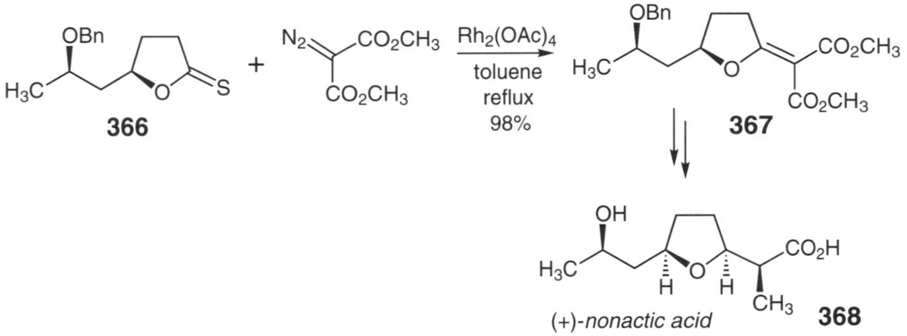

Scheme 78

Cyclic thiocarbonyl ylides derived from metal-catalyzed decomposition of diazocarbonyl compounds cyclize to give episufides, which readily extrude a sulfur atom to produce olefins. Some representative examples are shown in eqs. 78–

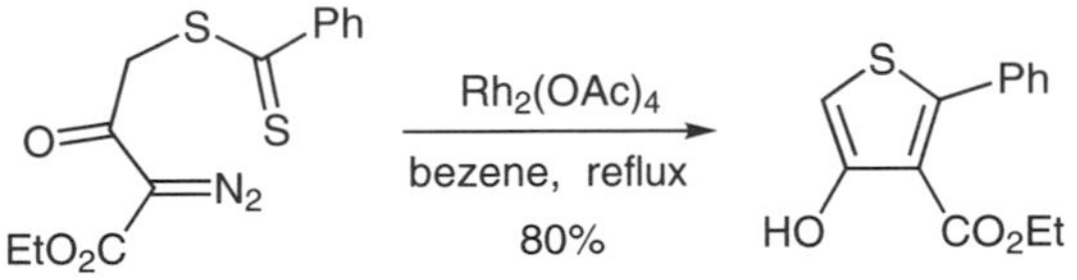

(78)

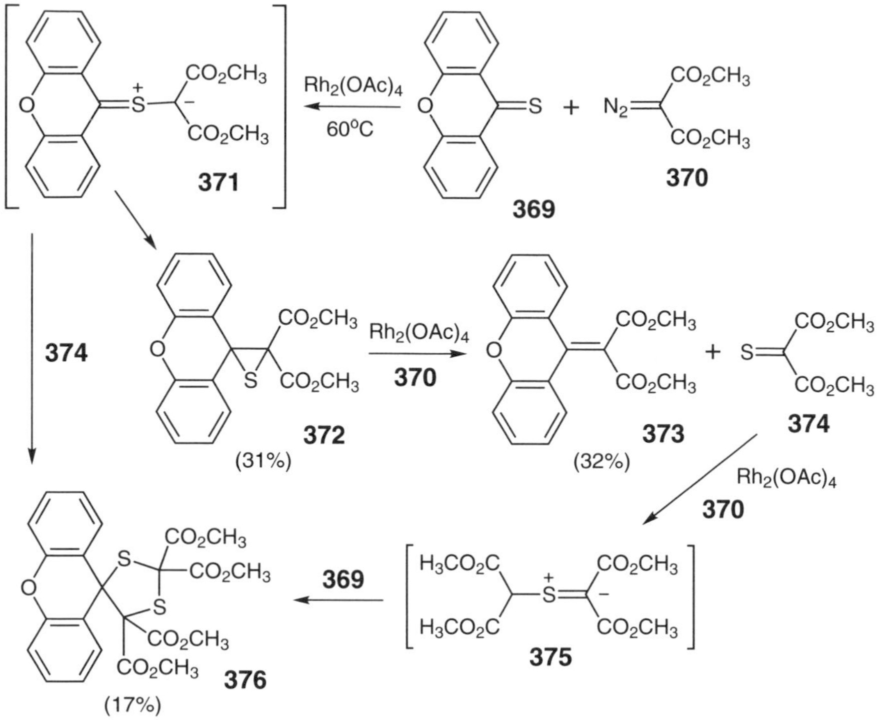

Scheme 79

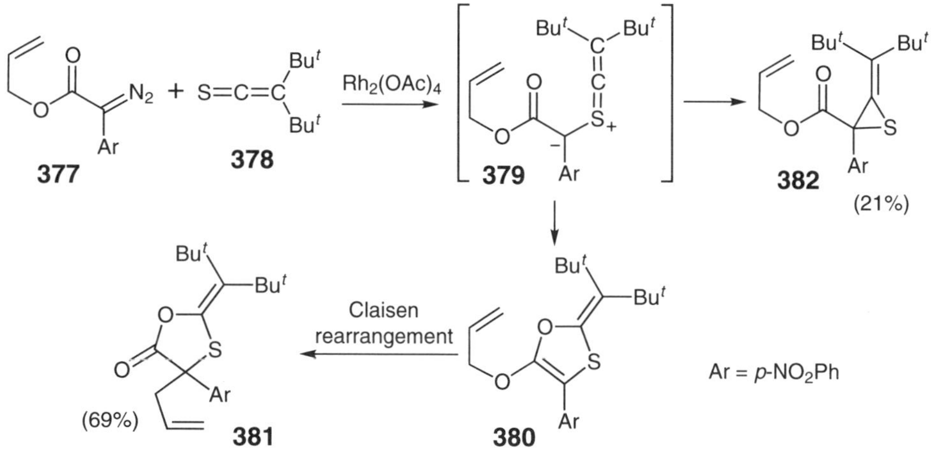

Scheme 80

80.[163,164,169] Equation 80 illustrated the use of this strategy in the synthesis of car-bomethoxy-substituted benzazepines. In a similar approach, the intramolecular metal carbene–thiocarbonyl coupling reaction served as the key step in the syn-

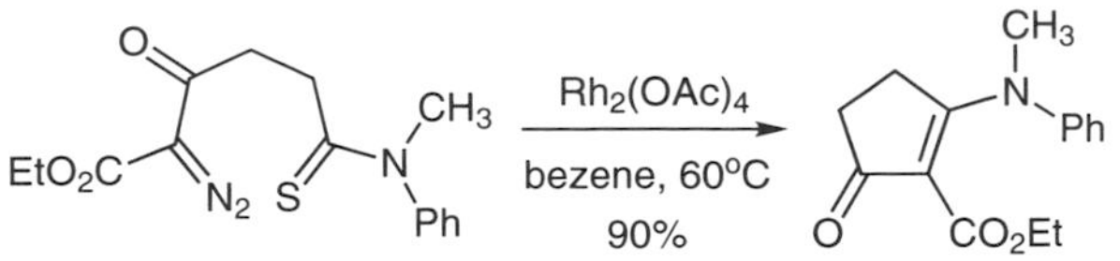

$$\text{(79)}$$

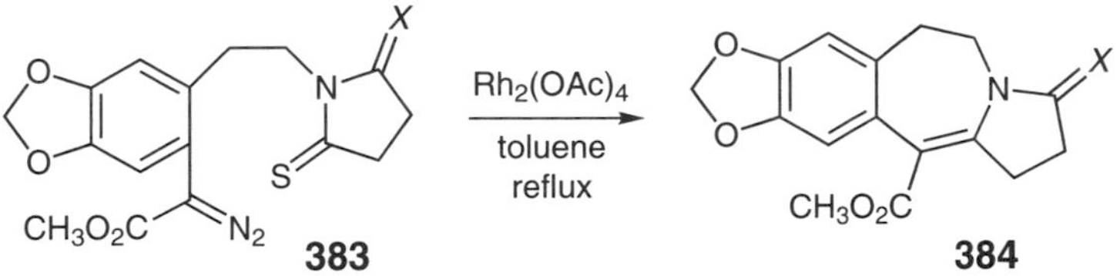

$$\text{(80)}$$

383   **384**

X = O, 77% yield; X = H₂, 79% yield

thesis of (±)-supinidine **387**.[170] Thus treatment of the diazoester **385** with rhodium(II) acetate in refluxing toluene produced pyrrolizidine skeleton **386** in 73% yield. Further chemical manipulation completed the synthesis (Scheme 81).

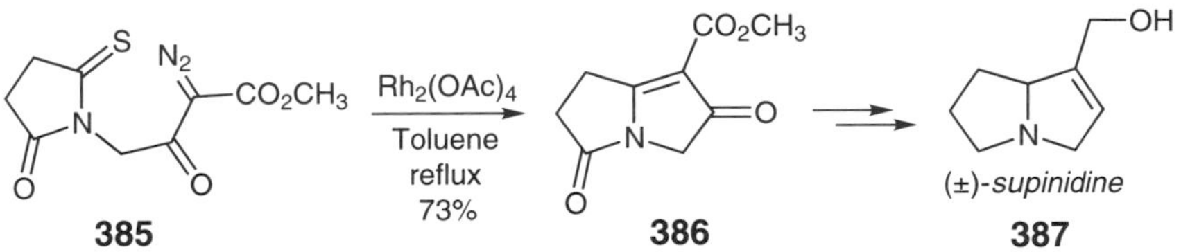

385   **386**   **387**

(±)-*supinidine*

Scheme 81

In contrast to the examples discussed above, the mesoionic thioisomünchnones **389**, derived from rhodium(II) acetate–catalyzed cyclization of diazothioamide **388**, undergo 1,3-dipolar cycloaddition with *N*-phenylmaleimide (Scheme 82).[164]

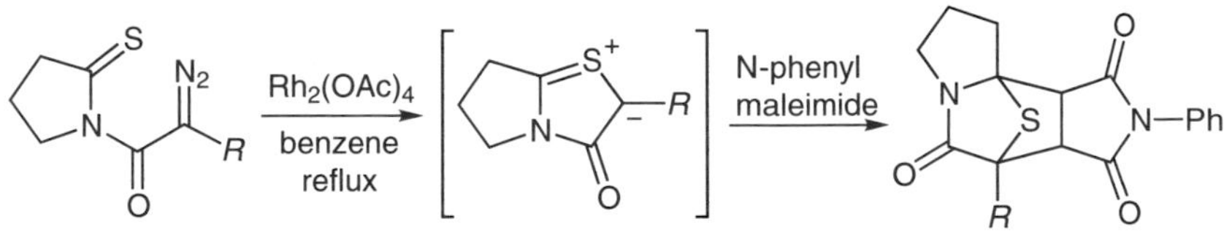

388   **389**   **390**

R = CO₂CH₃, 64% yield, R = COCH₃, 75% yield

Scheme 82

The facility with which the cycloaddition occurs is undoubtedly related to the stability of the aromatic mesoionic dipole **389**. In this particular case, ring closure to an episulfide would not only destroy the aromatic character of the dipole, but would lead to the formation of a highly strained ring.

Danishefsky and co-workers[162] have reported the construction of nitrogen heterocyclic compounds *via* the thiocarbonyl ylide formation process. Intramolecular reaction of the metal carbene derived from diazoketone **391** gave the cyclic thiocarbonyl ylide **392**. Cyclization of **392** led to episufide **393**, which was isomerized to product **394**; the latter was then transformed to dihydropyridone **395** by treatment with partially deactivated W-2 Raney nickel (Scheme 83). This

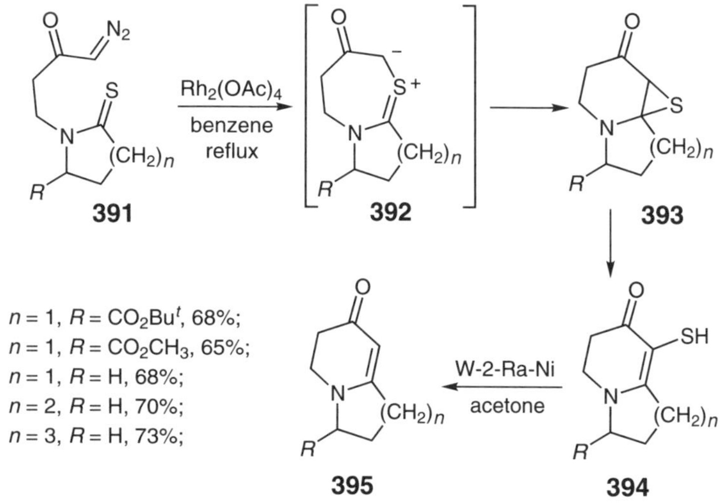

$n = 1, R = CO_2Bu^t, 68\%;$
$n = 1, R = CO_2CH_3, 65\%;$
$n = 1, R = H, 68\%;$
$n = 2, R = H, 70\%;$
$n = 3, R = H, 73\%;$

Scheme 83

strategy has been employed as the key step in Danishefsky's total synthesis of indolizomycin **398**.[171,172] Thus treatment of α-diazoketone **396** with rhodium(II) acetate in refluxing benzene followed by exposure to the W-2 Raney nickel in acetone afforded vinylogous amide **397** in 66% yield (Scheme 84). A similar

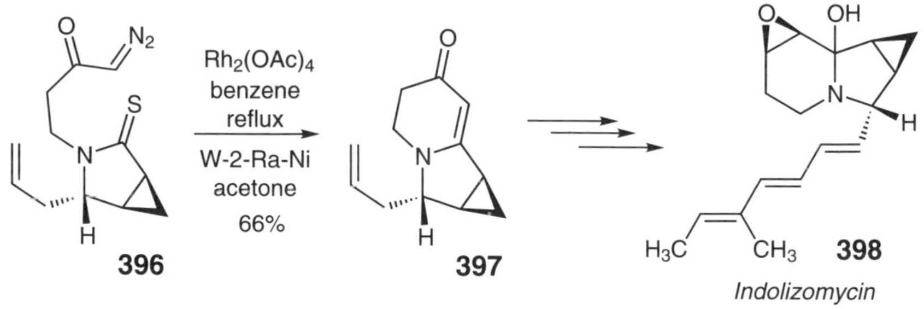

Scheme 84

strategy, involving a thioamide-diazoketone cyclocondensation as the key step, has been applied in a total synthesis of $(-)$-(1S,2R,7S,8aR)-1,2,7-trihydroxyindolizidine **399** (Scheme 85).[173]

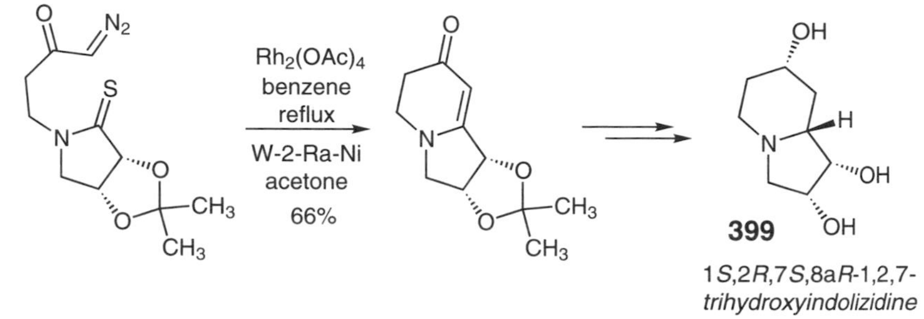

Scheme 85

As depicted in Scheme 83, the episufide **393** undergoes subsequent isomerization to produce vinylthiol **394**. In contrast to this isomerization process, the episufide **401**, derived from the reaction of thiolactam **400** with EDA, undergoes a reductive elimination to give vinylogous urethane **402** (Scheme 86).[172]

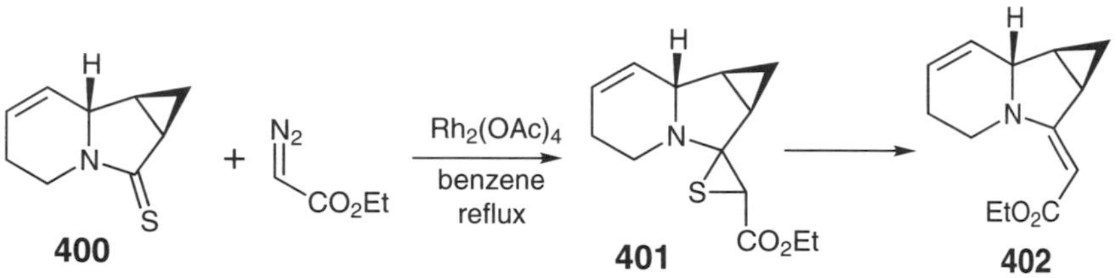

Scheme 86

7.7 REFERENCES

1 Padwa, A.; and Hornbuckle, S. F., "Ylide Formation from the Reaction of Carbenes and Carbenoids with Heteroatom Lone Pairs," *Chem. Rev.* **1991**, *91*, 263–309.

2 Trost, B. M.; and Melvin, L. S., *Sulfur Ylides: Emerging Synthetic Intermediates*; Academic Press: New York, **1975**.

3 Huxtable, R. J., *Biochemistry of Sulfur*; Plenum Press: New York, **1986**.

4 Ando, W., "Ylide Formation and Rearrangement in the Reaction of Carbene with Divalent Sulfur Compounds," *Acc. Chem. Res.* **1977**, *10*, 179–85.

5 Kurth, M. J.; Tahir, S. H.; and Olmstead, M. M., "A Thioxanone Based Chiral Template: Asymmetric Induction in the [2,3]-Sigmatropic Rearrangement of Sulfur Ylides. Enantioselective Preparation of Cβ-Chiral Pent-4-enoic Acids," *J. Org. Chem.* **1990**, *55*, 2286–88.

6 Hartley, R. C.; Warren, S.; and Richards, I. C., "A New and Effective Method for the Low Temperature Generation of Sulfonium Ylides from Allyl Sulfides," *J. Chem. Soc., Perkin Trans. 1* **1994**, 507–13.

7 Ando, W.; Kondo, S.; Nakayama, K.; Ichibori, K.; Kohoda, H.; Yamato, H.; Imai, I.; Nakaido, S.; and Migita, T., "Decomposition of Dimethyl Diazomalonate

in Allyl Compounds Containing Heteroatoms," *J. Am. Chem. Soc.* **1972**, *94*, 3870–76.

8   Ando, W.; Higuchi, H.; and Migita, T., "Reactions of Unsaturated Sulfides with Carbenes. 22. Reactivities of Sulfur and Double Bond, and Formation of Unsaturated Sulfonium Ylides," *J. Org. Chem.* **1977**, *42*, 3365–72.

9   Doyle, M. P.; Tamblyn, W. H.; and Bagheri, V., "Highly Effective Catalytic Methods for Ylide Generation from Diazo Compounds—Mechanism of the Rhodium-Catalyzed and Copper-Catalyzed Reactions with Allylic Compounds," *J. Org. Chem.* **1981**, *46*, 5094–102.

10 Gillespie, R. J.; Murray-Rust, J.; Murray-Rust, P.; and Porter, A. E. A., "Rhodium(II)-Catalysed Addition of Dimethyl Diazomalonate to Thiophen: A Simple Synthesis of Thiophenium Bismethoxycarbonylmethylides and Crystal and Molecular Structure of the Unsubstituted Methylide," *J. Chem. Soc., Chem. Commun.* **1978**, 83–84.

11 Gillespie, R. J.; and Porter, A. E. A., "The Reaction of Diazoalkanes with Thiophen," *J. Chem. Soc., Perkin Trans. 1* **1979**, 2624–28.

12 Murray-Rust, P.; McManus, J.; Lennon, S. P.; Porter, A. E. A.; and Rechka, J. A., "An Expeditious Synthesis of 4-Alkoxycarbonyl-5-hydroxy-1,2,3-triazoles: The Crystal and Molecular Structure of the 2-Thienylammonium Salt of 5-Hydroxy-4-methoxycarbonyl-1-(2-thienyl)-1,2,3-triazole," *J. Chem. Soc., Perkin Trans. 1* **1984**, 713–16.

13 Cuffe, J.; Gillespie, R. J.; and Porter, A. E. A., "2,5-Dichlorothiophenium Bismethoxycarbonylmethylide: a Bismethoxycarbonylcarbene Equivalent," *J. Chem. Soc., Chem. Commun.* **1978**, 641–42.

14 Gillespie, R. J.; and Porter, A. E. A., "The Reaction of Actived Arenes with 2,5-Dichlorothiophenium," *J. Chem. Soc., Chem. Commun.* **1979**, 50–51.

15 Tamura, Y.; Mukai, C.; Nakajima, N.; Ikeda, M.; and Kido, M., "Synthesis, Stereochemistry, and Rearrangement of Thioxanthen-10-io(bismethoxycarbonyl)-methanides and Their 9-Alkyl Derivatives," *J. Chem. Soc., Perkin Trans. 1* **1981**, 212–17.

16 Davies, H. M. L.; and Crisco, L. V. T., "Synthesis and Pyrolysis of Cyclic Sulfonium Ylides," *Tetrahedron Lett.* **1987**, *28*, 371–74.

17 Moody, C. J.; and Taylor, R. J.,"Rhodium Carbenoid Mediated Cyclisations. Synthesis and Rearrangement of Cyclic Sulphonium Ylides," *Tetrahedron Lett.* **1988**, *29*, 6005–8.

18 Dost, F.; and Gosselck, J., "Zur Reaktion von α-Carbonyldiazoverbindungen mit Dimethylsulfoxid," *Tetrahedron Lett.* **1970**, 5091–93.

19 Moody, C. J.; Slawin, A. M. Z.; Taylor, R. J.; and Williams, D. J., "Rhodium Carbenoid Mediated Cyclisations. Synthesis and X-Ray Structures of Cyclic Sulphonium Ylides," *Tetrahedron Lett.* **1988**, *29*, 6009–12.

20 Moody, C. J.; and Taylor, R. J., "Rhodium Carbenoid Mediated Cyclisations, Part 5. Synthesis and Rearrangement of Cyclic Sulphonium Ylides; Preparation of 6- and 7-Membered Sulfur Heterocycles," *Tetrahedron* **1990**, *46*, 6525–44.

21 Vedejs, E., "Sulfur-Mediated Ring Expansions in Total Synthesis," *Acc. Chem. Res.* **1984**, *17*, 358–64.

22 Grieco, P. A.; Boxler, D.; and Hiroi, K., "A Stereoselective *trans*-Trisubstituted Olefin Synthesis *via* Rearrangement of Allylic Sulfonium Ylides," *J. Org. Chem.* **1973**, *38*, 2572-73.

23 Andrews, G.; and Evans, D. A., "The Stereochemistry of the Rearrangement of Allylic Sulphonium Ylides: A New Method for the Stereoselective Formation of Asymmetry at Quaternary Carbon," *Tetrahedron Lett.* **1972**, 5121–24.

24 Vedejs, E.; and Krafft, G. A., "Cyclic Sulfides in Organic Synthesis," *Tetrahedron* **1982**, *38*, 2857–81.

25 Vedejs, E.; and Hagen, J. P., "Macrocycle Synthesis by Repeatable 2,3-Sigmatropic Shifts. Ring-Growing Reactions," *J. Am. Chem. Soc.* **1975**, *97*, 6878–80.

26 Ceré, V.; Paolucci, C.; Pollicino, S.; Sandri, E.; and Fava, A., "Doubly Bridged S-Heterocyclic Ethylenes *via* Stereospecific [2,3]-Sigmatropic Rearrangement. Avenue to Short-Bridged Betweenanenes," *J. Org. Chem.* **1981**, *46*, 486–90.

27 Nickon, A.; Rodriguez, A. D.; Ganguly, R.; and Shirhatti, V., "Betweenanenes with Vinylic Heteroatoms. Route to Sulfur Analogues *via* [2,3]-Sigmatropic Rearrangement," *J. Org. Chem.* **1985**, *50*, 2767–77.

28 De March, P.; Moreno-Mañas, M.; and Ripoll, I., "Functionalization of 4-Hydroxy-6-Methyl-2-Pyrone at C-5 through [2,3]-Sigmatropic Rearrangements of Allylic Sulphonium Ylides," *Synth. Commun.* **1984**, *14*, 521–31.

29 Yoshimoto, M.; Ishihara, S.; Nakayama, E.; and Soma, N., "Studies on β-Lactam Antibiotics I. Skeletal Conversion of Cephalosporin to Penicillin," *Tetrahedron Lett.* **1972**, 2923–26.

30 Giddings, P. J.; John, D. I.; Thomas, E. J.; and Williams, D. J., "Preparation of 6-α-Monosubstituted and 6,6-Disubstituted Penicillanates from 6-Diazopenicillanates—Reactions of 6- Diazopenicillanates with Alcohols, Thiols, Phenylseleninyl Compounds, and Allylic Sulfides, and Their Analogs," *J. Chem. Soc., Perkin Trans. 1* **1982**, 2757–66.

31 Doyle, M. P.; Griffin, J. H.; Chinn, M. S.; and van Leusen, D., "Rearrangements of Ylides Generated from Reactions of Diazo Compounds with Allyl Acetals and Thioketals by Catalytic Methods—Heteroatom Acceleration of the [2,3]-Sigmatropic Rearrangement," *J. Org. Chem.* **1984**, *49*, 1917–25.

32 Shi, G.; Xu, Y.; and Xu, M., "Carbenoid Entry into Trifluoromethylated Molecules: Preparation of Functionalized CF_3-Containing γ,δ-Unsaturated Carboxylic Esters by Rhodium-Catalyzed Reaction of Ethyl 3,3,3-Trifluoro-2-Diazo propionate with Allylic Sulfides and Their Further Facile Conversion to Trifluoromethylated Conjugated Dienoic Esters," *Tetrahedron* **1991**, *47*, 1629–48.

33 Crow, W. D.; Gosney, I.; and Ormiston, R. A., "Ring Transformation of 2-Substituted Isothiazol-3(2*H*)-Ones to 3,4-Dihydro-1,3-thiazin-4(2*H*)-ones by a Novel Carbene Addition-Ring Expansion Sequence," *J. Chem. Soc., Chem. Commun.* **1983**, 643–44.

34 Kametani, T.; Kawamura, K.; and Honda, T., "New Entry to the C-Glycosilation by Means of Carbenoid Displacement Reaction. Its Application to the Synthesis of Showdomycin," *J. Am. Chem. Soc.* **1987**, *109*, 3010–17.

35 Kametani, T.; Yukawa, H.; and Honda, T., "Stereoselective Synthesis of Pyrrolizidine Alkaloids, (+)-Heliotridine and (+)-Retronecine, by Means of an Intermolecular Carbenoid Displacement Reaction," *J. Chem. Soc., Chem. Commun.* **1988**, 685–87.

36 Kametani, T.; Yukawa, H.; and Honda, T., "Stereocontrolled Synthesis of Necine Bases, (+)-Heliotridine and (+)-Retronecine," *J. Chem. Soc., Perkin Trans. 1* **1990**, 571–77.

37   Ye, T.; García, C. F.; and McKervey, M. A., "Chemoselectivity and Stereoselectivity of Cyclisation of α-Diazocarbonyls Leading to Oxygen and Sulfur Heterocycles Catalysed by Chiral Rhodium and Copper Catalysts," *J. Chem. Soc., Perkin Trans. 1* **1995**, 1373–79.

38   Kido, F.; Sinha, S. C.; Abiko, T.; and Yoshikoshi, A., "Stereoselective Synthesis of Contiguously Substituted Butyrolactones Based on the Cyclic Allylsulfonium Ylide Rearrangement," *Tetrahedron Lett.* **1989**, *30*, 1575–78.

39   Kido, F.; Sinha, S. C.; Abiko, T.; Watanabe, M.; and Yoshikoshi, A., "New Entry to the Perhydrofuro[2,3-b]furan Ring-System," *J. Chem. Soc., Chem. Commun.* **1990**, 418–20.

40   Kido, F.; Sinha, S. C.; Abiko, T.; Watanabe, M.; and Yoshikoshi, A., "Stereoselectivity in the Sigmatropic Rearrangement of 8-Membered and 9-Membered Cyclic Allylsulfonium Ylides — Synthesis of Vinyl-Substituted Butyrolactones and Valerolactones," *Tetrahedron* **1990**, *46*, 4887–906.

41   Kido, F.; Abiko, T.; Kazi, A. B.; Kato, M.; and Yoshikoshi, A., "An Efficient and Convenient Route to Spiro-Fused γ-Butyro and δ-Valerolactones," *Heterocycles* **1991**, *32*, 1487–90.

42   Kido, F.; Abiko, T.; and Kato, M., "Spiroannulation by the [2,3]-Sigmatropic Rearrangement *via* the Cyclic Allylsulfonium Ylide — A Stereoselective Synthesis of (+)-Acorenone-B," *J. Chem. Soc., Perkin Trans. 1* **1992**, 229–33.

43   Kido, F.; Yamaji, K.; Sinha, S. C.; Yoshikoshi, A.; and Kato, M., "Steric Control Based on Allyl Substituents in the [2,3]-Sigmatropic Rearrangement of Nine-Membered Allylsulfonium Ylides. A New Entry to the Stereoselective Synthesis of Elemane-Type Sesquiterpenoids," *J. Chem. Soc., Chem. Commun.* **1994**, 789–90.

44   Kido, F.; Yamaji, K.; Sinha, S. C.; Abiko, T.; and Kato, M., "Carbocyclic Construction by the [2,3]-Sigmatropic Rearrangement of Cyclic Sulfonium Ylides. A New Entry for the Stereoselective Synthesis of Substituted Cyclohexanones," *Tetrahedron* **1995**, *51*, 7697–714.

45   Kido, F.; Kazi, A. B.; and Yoshikoshi, A., "New Entry to γ,δ-Unsaturated 7-Membered Lactones," *Chem. Lett.* **1990**, 613–16.

46   Kido, F.; Kazi, A. B.; Yamaji, K.; Kato, M.; and Yoshikoshi, A., "An Efficient Synthetic Route to γ,δ-Unsaturated Seven-Membered Lactones," *Heterocycles* **1992**, *33*, 607–18.

47   Kido, F.; Kawada, Y.; Kato, M.; and Yoshikoshi, A., "An Efficient and Convenient Synthesis of Bridged δ-Lactones," *Tetrahedron Lett.* **1991**, *32*, 6159–162.

48   Kondo, K.; and Ojima, I., "Reactions of Carbene Bearing Sulphide Linkages at the δ-Position," *J. Chem. Soc., Chem. Commun.* **1972**, 860–61.

49   Kametani, T.; Kawamura, K.; Tsubuki, M.; and Honda, T., "Synthesis of an Aromatic Sesquiterpene, (±)-Cuparene, Via Construction of a Quaternary Carbon Center by an Intramolecular Carbenoid Displacement Reaction," *J. Chem. Soc., Chem. Commun.* **1985**, 1324–25.

50   Kametani, T.; Kawamura, K.; Tsubuki, M.; and Honda, T., "Synthesis of Aromatic Sesquiterpenes, (±)-Cuparene and (±)-Laurene by Means of an Intramolecular Carbenoid Displacement (ICD) Reaction," *J. Chem. Soc., Perkin Trans. 1* **1988**, 193–99.

51   Kametani, T.; Yukawa, H.; and Honda, T., "Synthesis of Pyrrolizidine Alkaloids, (±)-Trachelanthamidine, (±)-Isoretronecanol, and (±)-Supinidine, by Means of an

Intramolecular Carbenoid Displacement (ICD) Reaction," *J. Chem. Soc., Chem. Commun.* **1986**, 651–52.

52 Kametani, T.; Yukawa, H.; and Honda, T., "A Novel Synthesis of Pyrrolizidine Alkaloids by Means of an Intramolecular Carbenoid Displacement (ICD) Reaction," *J. Chem. Soc., Perkin Trans. 1* **1988**, 833–37.

53 Kametani, T.; Nakayama, A.; Itoh, A.; and Honda, T., "Studies on the Synthesis of Carbapenem Antibiotics—Stereoselective Synthesis of a Potential Intermediate for 6-Amidocarbapenem Antibiotics," *Heterocycles* **1983**, *20*, 2355–58.

54 Ernest, I., "Penicillin-Derived Diazoketones; Copper(II) Catalyzed Decomposition in Aprotic Media," *Tetrahedron* **1977**, *33*, 547–52.

55 Mak, C. P.; Baumann, K.; Mayerl, F.; Mayerl, C.; and Fliri, H., "Metal-Catalyzed Decomposition of Diazoketones and Diazoamides of β-Lactams 1. Penicillin Route to Carbapenems?," *Heterocycles* **1982**, *19*, 1647–54.

56 Kim, G. C.; Kang, S. W.; and Kim, S. N., "A Novel Route to a New Lactone Intermediate for C-Nucleosides via an Intramolecular Sulfonium Ylide Rearrangement. A Formal Synthesis of (+)-Showdomycin," *Tetrahedron Lett.* **1993**, *34*, 7627–28.

57 Kametani, T.; Kanaya, N.; Nakayama, A.; Mochizuki, T.; Yokohama, S.; and Honda, T., "Novel Construction of Penem Ring System from Penicillin Derivatives, Synthesis of 2-Carboxylpenem Derivative," *J. Org. Chem.* **1986**, *51*, 624–29.

58 Kametani, T.; Kanaya, N.; Mochizuki, T.; and Honda, T., "Ring Expansion Reactions of Penicillin Derivatives by Rhodium-Catalyzed Decomposition of Diazoketones—Stereoselective Syntheses of Eight-Membered Oxa-β-lactams," *Tetrahedron Lett.* **1983**, *24*, 221–24.

59 Kametani, T.; Kawamura, K.; Akagi, T.; Fujita, C.; and Honda, T., "Intermolecular Heteroaromatic Nucleophilic Substitution Reaction via the Sulphur Ylide Rearrangement: C-Attack vs. O-Attack," *Heterocycles* **1988**, *27*, 2531–34.

60 Chappie, T. A.; Weekly, R. M.; and McMills, M. C., "Application of Diazodecomposition Reaction in Tandem with [2,3]-Sigmatropic Rearrangement to Prepare Substituted Azabicyclic Ring Systems," *Tetrahedron Lett.* **1996**, *37*, 6523–26.

61 Padwa, A.; Hornbuckle, S. F.; Fryxell, G. E.; and Stull, P. D., "Reactivity Patterns in the Rhodium Carbenoid Induced Tandem Cyclization–Cycloaddition Reaction," *J. Org. Chem.* **1989**, *54*, 817–24.

62 Doyle, M. P.,"Catalytic Methods for Metal Carbene Transformations," *Chem. Rev.* **1986**, *86*, 919–39.

63 Doyle, M. P.,"Electrophilic Metal Carbenes as Reaction Intermediates in Catalytic Reactions," *Acc. Chem. Res.* **1986**, *19*, 348–56.

64 Maas, G.,"Transition-Metal Catalyzed Decomposition of Aliphatic Diazo Compounds—New Results and Applications in Organic Synthesis," *Top. Curr. Chem.* **1987**, *137*, 75–260.

65 Doyle, M. P.; Leusen, D. V.; and Tamblyn, W. H., "Efficient Alternative Catalysts and Methods for the Synthesis of Cyclopropanes from Olefins and Diazo Compounds," *Synthesis* **1981**, 787–89.

66 Nozaki, H.; Takaya, H.; and Noyori, R., "Reaction of Carbethoxycarbene with 2-Phenyloxirane and 2-Phenyloxetane," *Tetrahedron* **1966**, *22*, 3393–401.

67 Nozaki, H.; Takaya, H.; Moriuti, S.; and Noyori, R., "Homogeneous Catalysis in the Decomposition of Diazo Compounds by Copper Chelates," *Tetrahedron* **1968**, *24*, 3655-69.

68 Doyle, M. P.; Bagheri, V.; and Harn, N. K., "Facile Catalytic Methods for Intermolecular Generation of Allylic Oxonium Ylides and Their Stereoselective [2,3]-Sigmatropic Rearrangement," *Tetrahedron Lett.* **1988**, *29*, 5119–22.

69 Pirrung, M. C.; and Werner, J. A., "Intramolecular Generation and [2,3]-Sigmatropic Rearrangement of Oxonium Ylides," *J. Am. Chem. Soc.* **1986**, *108*, 6060–62.

70 Roskamp, E. J.; and Johnson, C. R., "Generation and Rearrangements of Oxonium Ylides," *J. Am. Chem. Soc.* **1986**, *108*, 6062–63.

71 West, F. G.; Eberlein, T. H.; and Tester, R. W., "O-Bridged Medium-Sized Rings *via* Bicyclic Oxonium Ylides," *J. Chem. Soc., Perkin Trans. 1* **1993**, 2857–59.

72 Clark, J. S., "Diastereoselective Synthesis of 2,5-Dialkyl Tetrahydrofuran-3-ones by a Copper-Catalyzed Tandem Carbenoid Insertion and Ylide Rearrangement Reaction," *Tetrahedron Lett.* **1992**, *33*, 6193–96.

73 Pirrung, M. C.; Brown, W. L.; Rege, S.; and Laughton, P., "Total Synthesis of (+)-Griseofulvin," *J. Am. Chem. Soc.* **1991**, *113*, 8561–62.

74 Clark, J. S.; and Whitlock, G. A., "A Short Synthesis of (±)-Decarestrictine L," *Tetrahedron Lett.* **1994**, *35*, 6381–82.

75 Clark, J. S.; Dossetter, A. G.; and Whittingham, W. G., "Stereoselective Synthesis of the Bicyclic Core Structure of the Highly Oxidised Sesquiterpene Neoliacinic Acid," *Tetrahedron Lett.* **1996**, *37*, 5605–08.

76 Brogan, J. B.; Bauer, C. B.; Rogers, R. D.; and Zercher, C. K., "Selectivity in the Rearrangements of Oxonium Ylides," *Tetrahedron Lett.* **1996**, *37*, 5053–56.

77 Eberlein, T. H.; West, F. G.; and Tester, R. W., "The Stevens [1,2]-Shift of Oxonium Ylides: A Route to Substituted Tetrahydrofuranones," *J. Org. Chem.* **1992**, *57*, 3479–82.

78 Padwa, A.; Krumpe, K. E.; and Weingarten, M. D., "An Unusual Example of a 6-*endo-Dig* Addition to an Unactivated Carbon–Carbon Triple Bond," *J. Org. Chem.* **1995**, *60*, 5595-603.

79 Crackett, P. H.; Sayer, P.; Stoodley, R. J.; and Greengrass, C. W., "Total Synthesis of Analogues of the β-Lactam Antibiotics. Part 6. (6R*)-4-(t-Butoxycarbonyl)-2-Methoxycarbonyl-3-Oxacepham 1,1-Dioxides," *J. Chem. Soc., Perkin Trans. 1* **1991**, 1235–43.

80 McKervey, M. A.; and Ye, T., "Asymmetric Synthesis of Substituted Chromanones via C–H Insertion Reactions of α-Diazoketones Catalysed by Homochiral Rhodium(II) Carboxylates," *J. Chem. Soc. Chem. Commun.* **1992**, 823–24.

81 Clark, J. S.; Krowiak, S. A.; and Street, L. J., "Synthesis of Cyclic Ethers from Copper Carbenoids by Formation and Rearrangement of Oxonium Ylides," *Tetrahedron Lett.* **1993**, *34*, 4385–88.

82 West, F. G.; Naidu, B. N.; and Tester, R. W., "Profound Catalyst Effects in the Generation and Reactivity of Carbenoid-Derived Cyclic Ylides," *J. Org. Chem.* **1994**, *59*, 6892–94.

83 Ye, T.; and McKervey, M. A., "Organic Synthesis with α-Diazocarbonyl Compounds," *Chem. Rev.* **1994**, *94*, 1091–160.

84 Doyle, M. P.; Bagheri, V.; and Claxton, E. E., "Synthesis of Allenes by [2,3]-Sigmatropic Rearrangement of Prop-2-yn-1-yl Oxonium Ylides Formed in Rhodium(II) Carboxylate Catalysed Reactions of Diazo Compounds," *J. Chem. Soc., Chem. Commun.* **1990**, 46–48.

85 McCarthy, N.; McKervey, M. A.; Ye, T.; McCann, M.; Murphy, E.; and Doyle, M. P., "Rhodium(II) Phos., A New Catalyst for Diazocarbonyl Reactions Including Asymmetric Synthesis," *Tetrahedron Lett.* **1992**, *33*, 5983–86.

86 Ito, K.; and Katsuki, T., "Asymmetric Carbene C–O Insertion Reaction Using Optically Active Bipyridine–Copper Complex as a Catalyst. Ring Expansion of Oxetanes to Tetrahydrofurans," *Chemistry Lett.* **1994**, 1857–60.

87 Ito, K.; Yoshitake, M.; and Katsuki, T., "Enantiospecific Ring Expansion of Oxetanes: Stereoselective Synthesis of Tetrahydrofurans," *Heterocycles* **1996**, *42*, 305–17.

88 Ito, K.; Yoshitake, M.; and Katsuki, T., "Enantioselective Synthesis of *trans*-Whiskey Lactone by Using Newly Developed Asymmetric Ring Expansion Reaction of Oxetane as a Key Step," *Chemistry Lett.* **1995**, 1027–28; "Chiral Bipyridine and Biquinoline Ligands: Their Asymmetric Synthesis and Application to the Synthesis of *trans*-Whiskey Lactone," *Tetrahedron* **1996**, *52*, 3905–20.

89 Osipov, S. N.; Sewald, N.; Kolomiets, A. F.; Fokin, A. V.; and Burger, K., "Synthesis of α-Trifluoromethyl Substituted α-Amino Acid Derivatives from Methyl 3,3,3-Trifluoro-2-Diazopropionate," *Tetrahedron Lett.* **1996**, *37*, 615–18.

90 Hata, Y.; and Watanabe, M., "Fragmentation Reaction of Aziridinium Ylids. 2," *Tetrahedron Lett.* **1972**, 4659–60.

91 West, F. G.; Glaeske, K. W.; and Naidu, B. N., "One-Step Synthesis of Tertiary α-Amino Ketones and α-Amino Esters from Amines and Diazocarbonyl Compounds," *Synthesis* **1993**, 977–80.

92 Mara, A. M.; Singh, O.; and Thomas, E. J.,"Tricyclic Products from the Reaction between Penicillin Derived Thiazoloazetidinones and Ethyl Diazoacetate. X-Ray Structure of Methyl 2-[(1*R*,3*S*,4*R*,5*S*,6*S*,8*R*)-3-Benzyl-4,5-bismethoxycarbonyl-6-ethoxy-carbonyl-9-oxo-7,10-diaza-2-thiatricyclo[6.2.0.0^{3,7}]decan-10-yl]-3-methylbut-2-enoate," *J. Chem. Soc., Perkin Trans. 1* **1982**, 2169–73.

93 Padwa, A.; Dean, D. C.; Osterhout, M. H.; Precedo, L.; and Semones, M. A., "Synthesis of Functionalized Azomethine Ylides *via* the Rh(II)-Catalyzed Cyclization of α-Diazo Carbonyls onto Imino π-Bonds," *J. Org. Chem.* **1994**, *59*, 5347–57.

94 West, F. G.; and Naidu, B. N., "New Route to Substituted Piperidines via the Stevens [1,2]-Shift of Ammonium Ylides," *J. Am. Chem. Soc.* **1993**, *115*, 1177–78.

95 West, F. G.; and Naidu, B. N., "Applications of Stevens [1,2]-Shifts of Cyclic Ammonium Ylides. A Route to Morpholin-2-Ones," *J. Org. Chem.* **1994**, *59*, 6051–56.

96 West, F. G.; and Naidu, B. N., "Piperidines via Ammonium Ylide [1,2]-Shifts: A Concise, Enantioselective Route to (−)-Epilupinine from Proline Ester," *J. Am. Chem. Soc.* **1994**, *116*, 8420–21.

97 Clark, J. S.; and Hodgson, P. B., "Intramolecular Generation and Rearrangement of Ammonium Ylides from Copper Carbenoids: A General Method for the Synthesis of Cyclic Amines," *J. Chem. Soc., Chem. Commun.* **1994**, 2701–2.

98   Clark, J. S.; and Hodgson, P. B., "An Enantioselective Synthesis of the CE Ring System of the Alkaloids Manzamine A, E and F, and Ircinal A," *Tetrahedron Lett.* **1995**, *36*, 2519–22.

99   Wright, D. L.; Weekly, R. M.; Groff, R.; and McMills, M. C., "A Metallocarbenoid Approach to the Formation of Spirocyclic Ammonium Ylides Leading to the Preparation of Medium-Sized Azacane Rings," *Tetrahedron Lett.* **1996**, *37*, 2165–68.

100   Williams, M. A.; and Miller, M. J., "Synthesis of the Carbapenam Ring System via Carbene Mediated Rearrangement of an N-Benzyloxy-β-Lactam," *Tetrahedron Lett.* **1990**, *31*, 1807-10.

101   Williams, M. A.; Hsiao, C.-N.; and Miller, M. J., "Direct Conversion of N-Alkoxy β-Lactams to Carbapenams: Application to the Synthesis of the Bicyclic PS-5 Keto Ester," *J. Org. Chem.* **1991**, *56*, 2688–94.

102   Taylor, E. C.; and Davies, H. M. L., "Approaches to the Synthesis of Aza Analogues of the β-Lactam Antibiotics: Some Anomalous Rhodium(II)-Catalyzed Carbene Insertion Reactions," *J. Org. Chem.* **1984**, *49*, 113–16.

103   Feller, D.; Davidson, E. R.; and Borden, W. T., "Allylic Resonance—When Is It Unimportant?" *J. Am. Chem. Soc.* **1984**, *106*, 2513–19.

104   Lavilla, J. A.; and Goodman, J. L., "The Determination of the Heats of Formation of Carbonyl and Nitrile Ylides by Photoacoustic Calorimetry," *Tetrahedron Lett.* **1988**, *29*, 2623–26.

105   Bekhazi, M.; and Warkentin, J., "Thermolysis of 2-Methoxy-2,5,5-trimethyl-Δ^3-1,3,4-oxadiazoline. Carbenes from Thermal Fragmentation of a Carbonyl Ylide Intermediate," *J. Am. Chem. Soc.* **1981**, *103*, 2473–74.

106   Janulis, Jr. E. P.; and Arduengo III, A. J., "Structure of an Electronically Stabilized Carbonyl Ylide," *J. Am. Chem. Soc.* **1983**, *105*, 5929–30.

107   Kharasch, M. S.; Rudy, T.; Nudenberg, W.; and Büchi, G., "Reactions of Diazoacetates and Diazoketones. I. Reaction of Ethyl Diazoacetate with Cyclohexanone and with Acetone," *J. Org. Chem.* **1953**, *18*, 1030–44.

108   Bien, S.; and Gillon, A., "Intramolecular Cyclisation of α-Diazo Ketones through Carbonyl Ylide Intermediates. A Novel Formation of the Furan-3(2H)-one System," *Tetrahedron Lett.* **1974**, 3073–74.

109   De March, P.; and Huisgen, R., "Carbonyl Ylides from Aldehydes and Carbenes," *J. Am. Chem. Soc.* **1982**, *104*, 4952.

110   Alonso, M. E.; and Jano, P., "The Syntheses of Ethoxycarbonyl-1,3-dioxoles and Oxazoles from the Copper Catalyzed Thermolysis of Ethyl Diazopyruvate in the Presence of Ketones, Aldehydes and Nitriles," *J. Heterocyclic Chem.* **1980**, *17*, 721–25.

111   Alonso, M. E.; and Chitty, A. W., "A New Convenient Synthesis of Trisubstituted 1,3-Dioxole-4-Carboxylates from Methyl 2-Diazo-3-Oxobutyrate and Aldehydes," *Tetrahedron Lett.* **1981**, *22*, 4181–84.

112   Alonso, M. E.; García, M. D.; and Chitty, A. W., "Synthesis of Polysubstituted Dioxoles from the Cycloaddition of Diazo Dicarbonyl-Compounds to Aldehydes and Ketones under Copper(II) Catalysis," *J. Org. Chem.* **1985**, *50*, 3445–49.

113   Huisgen, R.; and de March, P., "3-Component Reactions of Diazomalonic Ester, Benzaldehyde, and Electrophilic Olefins," *J. Am. Chem. Soc.* **1982**, *104*, 4953–54.

114 Alt, M.; and Maas, G., "Transition-Metal-Catalyzed Decomposition of Diazo(trialkylsilyl)-acetates: Intermolecular Formation and Trapping of Carbonyl Ylides," *Tetrahedron* **1994**, *50*, 7435–44.

115 Alt, M.; and Maas, G., "Übergangsmetall-katalysierte Reaktionen von Ungesättigten α-Diazo-α-(trimethylsilyl)essigestern mit Carbonylverbindungen," *Chem. Ber.* **1994**, *127*, 1537–42.

116 Spencer, T. A.; Villarica, R. M.; Storm, D. L.; Weaver, T. D.; Friary, R. J.; Posler, J.; and Shafer, P. R., "Total Synthesis of Racemic Methyl Vinhaticoate," *J. Am. Chem. Soc.* **1967**, *89*, 5497–99.

117 Ibata, T.; and Toyoda, J., "Formation and Reaction of Carbonyl Ylides. Production of 2:1-Cycloadducts of 2-Benzopyrylium-4-olates with Carbonyl Compounds," *Bull. Chem. Soc. Jpn.* **1986**, *59*, 2489–93.

118 Ibata, T.; Toyoda, J.; Sawada, M.; and Tanaka, T., "Formation and Reaction of Carbonyl Ylides. Structure of 2:1-Cycloadducts of 1-Methoxy-2-Benzopyrylium-4-olate with Isocyanates," *J. Chem. Soc., Chem. Commun.* **1986**, 1266–67.

119 Padwa, A., "Generation and Utilization of Carbonyl Ylides via the Tandem Cyclization–Cycloaddition Method," *Acc. Chem. Res.* **1991**, *24*, 22–28.

120 Padwa, A.; Chinn, R. L.; Hornbuckle, S. F.; and Zhang, Z. J., "Tandem Cyclization–Cycloaddition Reaction of Rhodium Carbenoids. Studies Dealing with the Geometric Requirements of Dipole Formation," *J. Org. Chem.* **1991**, *56*, 3271–78.

121 Padwa, A.; Chinn, R. L.; and Lin, Z., "Synthesis of *exo*-Brevicomin and *endo*-Brevicomin *via* the Rhodium Acetate Catalyzed Cycloaddition Reaction of 1-Diazo-2,5-Hexanedione," *Tetrahedron Lett.* **1989**, *30*, 1491–94.

122 Padwa, A.; Fryxell, G. E.; and Zhi, L., "Tandem Cyclization Cycloaddition Reaction of Rhodium Carbenoids — Scope and Mechanistic Details of the Process," *J. Am. Chem. Soc.* **1990**, *112*, 3100–9.

123 Kinder, Jr., F. R.; and Bair, K. W., "Total Synthesis of (±)-Illudin M," *J. Org. Chem.* **1994**, *59*, 6965–67.

124 Padwa, A.; Sandanayaka, V. P.; and Curtis, E. A., "Synthetic Studies toward Illudins and Ptaquilosin. A Highly Convergent Approach via the Dipolar Cycloaddition of Carbonyl Ylides," *J. Am. Chem. Soc.* **1994**, *116*, 2667–68.

125 Curtis, E. A.; Sandanayaka, V. P.; and Padwa, A., "An Efficient Dipolar-Cycloaddition Route to the Pterosin Family of Sesquiterpenes," *Tetrahedron Lett.* **1995**, *36*, 1989-92.

126 Padwa, A.; Curtis, E. A.; and Sandanayaka, V. P., "Generation and Cycloaddition Behavior of Spirocyclic Carbonyl Ylides. Application to the Synthesis of the Pterosin Family of Sesquiterpenes," *J. Org. Chem.* **1996**, *61*, 73–81.

127 Koyama, H.; Ball, R. G.; and Berger, G. D., "A Novel Synthetic Approach toward the Zaragozic Acids Core Structure," *Tetrahedron Lett.* **1994**, *35*, 9185–88.

128 Hodgson, D. M.; Bailey, J. M.; and Harrison, T., "A Cycloaddition–Rearrangement Approach to the Squalestatins," *Tetrahedron Lett.* **1996**, *37*, 4623–26.

129 Dauben, W. G.; Dinges, J.; and Smith, T. C., "A Convergent Approach toward the Tigliane Ring System," *J. Org. Chem.* **1993**, *58*, 7635–37.

130 McMills, M. C.; Zhuang, L.; Wright, D. L.; and Watt, W., "A Carbonyl-Ylide Approach to the Tigliane Diterpenes," *Tetrahedron Lett.* **1994**, *35*, 8311–14.

131 Ueda, K.; Ibata, T.; and Takebayashi, M., "Reaction of Diazoketones in the Presence of Metal Chelate. IV. Formation and Reaction of Carbonyl Ylides," *Bull. Chem. Soc. Jpn.* **1972**, *45*, 2779–82.

132 Plüg, C.; and Friedrichsen, W., "Pyryliumolates II—Generation of and Cycloaddition Reactions with Isoxazole Annulated Pyryliumolates," *J. Chem. Soc., Perkin Trans. 1* **1996**, 1035–40.

133 Padwa, A.; Carter, S. P.; Nimmesgern, H.; and Stull, P. D., "Rhodium(II) Acetate Induced Intramolecular Dipolar Cycloadditions of *o*-Carboalkoxy-diazoacetophenone Derivatives," *J. Am. Chem. Soc.* **1988**, *110*, 2894–900.

134 Hildebrandt, K.; Debaerdemaeker, T.; and Friedrichsen, W., "Intramolekulare Cycloadditionen mit Isobenzofuranen. 4. Synthese 6-Funktionalisierter 11-oxasteroide," *Tetrahedron Lett.* **1988**, *29*, 2045–46.

135 Padwa, A.; and Zhi, L., "Novel Rhodium(II)-Catalyzed Cycloaddition Reaction of α-Diazo Keto Amides," *J. Am. Chem. Soc.* **1990**, *112*, 2037–38.

136 Galt, R. H. B.; Hitchcock, P. B.; McCarthy, S. J.; and Young, D. W., "Formation of a Medium Ring Imino Ether by a Diazo Insertion Reaction," *Tetrahedron Lett.* **1996**, *37*, 8035–36.

137 Padwa, A.; Dean, D. C.; and Zhi, L., "1,3-Dipole Cascade—A New Method for Azomethine Ylide Formation," *J. Am. Chem. Soc.* **1989**, *111*, 6451–52.

138 Padwa, A.; Dean, D. C.; and Zhi, L., "Transmutation of 1,3-Dipoles—The Conversion of α-Diazo Ketones into Azomethine Ylides *via* Carbonyl Ylides," *J. Am. Chem. Soc.* **1992**, *114*, 593–601.

139 Rodgers, J. D.; Caldwell, G. W.; and Gauthier, A. D., "A Novel Carbonyl Ylide Rearrangement," *Tetrahedron Lett.* **1992**, *33*, 3273–76.

140 Hamaguchi, M.; and Ibata, T., "Synthesis of Stable Carbonyl Ylides by Intramolecular Carbenic Reaction," *Tetrahedron Lett.* **1974** 4475–76.

141 Hamaguchi, M.; and Ibata, T., "New Type of Mesoionic System. 1,3-Dipolar Cycloaddition of Isomünchnones with Ethylenic Compounds," *Chem. Lett.* **1975**, 499–502.

142 Maier, M. E.; and Evertz, K., "Intramolecular [3+2] Cyclo-Additions of Mesoionic Carbonyl Ylides," *Tetrahedron Lett.* **1988**, *29*, 1677–80.

143 Maier, M.; and Schöffling, B., "Intramolecular Cycloadditions of Mesoionic carbonyl Ylides with Alkynes. Synthesis of 5,6-Dihydro-4H-cyclopenta- and 4,5,6,7-Tetrahydrobenzo[*b*]furan Derivatives," *Chem. Ber.* **1989**, *122*, 1081–87.

144 Padwa, A.; and Hertzog, D. L., "Bimolecular Cycloaddition Reactions of Isomünchnones Derived from the Rhodium(II) Catalyzed Cyclization of Diazo Pyrrolidinones," *Tetrahedron* **1993**, *49*, 2589–600.

145 Doyle, M. P.; Pieters, R. J.; Taunton, J.; Pho, H. Q.; Padwa, A.; Hertzog, D. L.; and Precedo, L., "Synthesis of Nitrogen-Containing Polycycles via Rhodium(II)-Induced Cyclization–Cycloaddition and insertion Reactions of *N*-(Diazoacetoacetyl)amides. Conformational Control of Reaction Selectivity," *J. Org. Chem.* **1991**, *56*, 820–29.

146 González, R.; Knight, B. W.; Wudl, F.; Semones, M. A.; and Padwa, A., "The Reversible Cycloaddition of Isomünchnones to C$_{60}$," *J. Org. Chem.* **1994**, *59*, 7949–51.

147   Hertzog, D. L.; Austin, D. J.; Nadler, W. R.; and Padwa, A., "Intramolecular Cy-
      cloaddition of Isomünchnones Derived from the Rhodium(II) Catalyzed Cycliza-
      tion of Diazoimides," *Tetrahedron Lett.* **1992**, *33*, 4731–34.

148   Padwa, A.; Hertzog, D. L.; Nadler, W. R.; Osterhout, M. H.; and Price, A. T.,
      "Studies on the Intramolecular Cycloaddition Reaction of Mesoionics Derived
      from the Rhodium(II)-Catalyzed Cyclization of Diazoimides," *J. Org. Chem.*
      1994, *59*, 1418–27.

149   Padwa, A.; Austin, D. J.; and Price, A. T., "Synthesis of Polyheterocyclic Ring
      Compounds by the Intramolecular Cycloaddition of N-Alkenyl Substituted Dia-
      zoimides," *Tetrahedron Lett.* **1994**, *35*, 7159–62.

150   Padwa, A.; Brodney, M. A.; Marino, J. P., Jr.; Osterhout, M. H.; and Price, A. T.,
      "Tandem Dipolar Cycloaddition-Mannich Cyclization as an Approach to Tricyclic
      Nitrogen Heterocycles," *J. Org. Chem.,* **1997**, *62*, 67–77.

151   Padwa, A.; Brodney, M. A.; Marino, J. P., Jr.; and Sheehan, S. M., "Utilization of
      the Intramolecular Cycloaddition-Cationic π-Cyclization of an Isomünchnone
      Derivative for the Synthesis of (±)-Lycopodine," *J. Org. Chem.* **1997**, *62*, 78–87.

152   Marino, J. P., Jr.; Osterhout, M. H.; and Padwa, A., "An Approach to Lysergic
      Acid Utilizing an Intramolecular Isomünchnone Cycloaddition Pathway," *J. Org.
      Chem.* **1995**, *60*, 2704–13.

153   Padwa, A.; Hertzog, D. L.; and Nadler, W. R., "Intramolecular Cycloaddition of
      Isomünchnone Dipoles to Heteroaromatic π-Systems," *J. Org. Chem.* **1994**, *59*,
      7072-84.

154   Padwa, A.; and Price, A. T., "Tandem Cyclization–Cycloaddition Reaction of
      Rhodium Carbenoids as an Approach to the Aspidosperma Alkaloids," *J. Org.
      Chem.* **1995**, *60*, 6258–59.

155   Stork, G.; Kretchmer, R. A.; and Schlessinger, R. H., "The Stereospecific Total
      Synthesis of *dl*-Lycopodine," *J. Am. Chem. Soc.* **1968**, *90*, 1647–48.

156   Gillon, A.; Ovadia, D.; and Bien, S., "Intramolecular Cycloaddition of Carbonyl
      Ylides Generated from α-Diazo Ketones," *Tetrahedron* **1982**, *38*, 1477–84.

157   Padwa, A.; Austin, D. J.; Hornbuckle, S. F.; Semones, M. A.; Doyle, M. P.; and
      Protopopova, M. N., "Control of Chemoselectivity in Catalytic Carbenoid Reac-
      tions. Dirhodium(II) Ligand Effects on Relative Reactivities," *J. Am. Chem. Soc.*
      1992, *114*, 1874–76.

158   Padwa, A.; Austin, D. J.; and Hornbuckle, S. F., "Ligand-Induced Selectivity in
      the Rhodium(II)-Catalyzed Reactions of α-Diazo Carbonyl Compounds," *J. Org.
      Chem.* **1996**, *61*, 63–72.

159   Cox, G. G.; Moody, C. J.; Austin, D. J.; and Padwa, A., "Chemoselectivity of
      Rhodium Carbenoids. A Comparison of the Selectivity for O–H Insertion Reac-
      tions or Carbonyl Ylide Formation versus Aliphatic and Aromatic C–H Insertion
      and Cyclopropanation," *Tetrahedron* **1993**, *49*, 5109–26.

160   Prein, M.; and Padwa, A., "Ligand-Dependent Site Selectivity in the Rh(II)-
      Catalyzed Decomposition of a Glycine-Derived Diazo Acetoacetamide," *Tetrahe-
      dron Lett.* **1996**, *37*, 6981–84.

161   Takano, S.; Tomita, S.; Takahashi, T.; and Ogasawara, K., "Condensation of a
      Chiral Tetrahydro-2-furanthione with Diazocarbonyl Compounds," *Synthesis*
      1987, 1116–17.

162   Fang, F. G.; Prato, M.; Kim, G.; and Danishefsky, S. J., "The Aza-Robonson Annulation: An Application to the Synthesis of Iso-A58365A," *Tetrahedron Lett.* **1989**, *30*, 3625–28.

163   Padwa, A.; Kinder, F. R.; and Zhi, L., "Generation of Thiocarbonyl Ylides from the Rhodium(II)-Catalyzed Cyclization of Diazothiocarbonyl Compounds," *Synlett* **1991**, 287–88.

164   Padwa, A.; Kinder, F. R.; Nadler, W. R.; and Zhi, L., "Rhodium(II) Catalysed Cyclization of Diazo Thiocarbonyl Compounds for Heterocyclic Synthesis," *Heterocycles* **1993**, *35*, 367–83.

165   Honda, T.; Ishige, H.; Araki, J.; Akimoto, S.; Hirayama, K.; and Tsubuki, M., "A Synthesis of (+)-Nonactic Acid by Means of the Sulfur-Ylide Rearrangement," *Tetrahedron* **1992**, *48*, 79-88.

166   Tokitoh, N.; Suzuki, T.; Itami, A.; Goto, M.; and Ando, W., "Di-*t*-Butylthioketene S-Bis(alkoxycarbonyl)methylide: Formation, Reactions, and Its Equilibrium with 2-Alkylidene-1,3-oxathiole Isomer," *Tetrahedron Lett.* **1989**, *30*, 1249–52.

167   Nakano, H.; and Ibata, T., "The Rhodium(II) Acetate–Catalyzed Reaction of Alkenyl and Alkynyl α-Diazoacetates with Thioketenes," *Bull. Chem. Soc. Jpn.* **1995**, *68*, 1393–400.

168   Mlostón, G.; and Heimgartner, H., "Carbenoid Reactions of Dimethyl Diazomalonate with Aromatic Thioketones and 1,3-Thiazole-5(4*H*)-thiones," *Helv. Chim. Acta* **1996**, *79*, 1785–92.

169   Fang, F. G.; Maier, M. E.; Danishefsky, S. J.; and Schulte, G., "New Routes to Functionalized Benzazepine Substructures: A Novel Transformation of an α-Diketone Thioamide Induced by Trimethyl Phosphite," *J. Org. Chem.* **1990**, *55*, 831–38.

170   Kim, G. C.; Kang, S. W.; and Keum, G. C., "A Synthesis of (±)-Supinidine via an Intramolecular Carbenoid–Thioimide Coupling Reaction," *Tetrahedron Lett.* **1994**, *35*, 3747–48.

171   Kim, G. C.; Chu-Moyer, M. Y.; and Danishefsky, S. J., "Total Synthesis of *dl*-Indolizomycin," *J. Am. Chem. Soc.* **1990**, *112*, 2003–5.

172   Kim, G. C.; Chu-Moyer, M. Y.; Danishefsky, S. J.; and Schulte, G. K., "The Total Synthesis of Indolizomycin," *J. Am. Chem. Soc.* **1993**, *115*, 30–39.

173   Maggini, M.; Prato, M.; Ranelli, M.; and Scorrano, G., "Synthesis of (-)-8-Deoxy-7-hydroxy-swainsonine and (±)-6,8-Dideoxy-castanospermine," *Tetrahedron Lett.* **1992**, *33*, 6537-40.

CHAPTER 8

X–H Insertion Reactions of Diazocarbonyl Compounds (X = N, O, S, Se, P, Halogen)

8.1 INTRODUCTION

Quite apart from the many catalyzed C–H and Si–H insertion reactions discussed in Chapter 3, diazocarbonyl compounds undergo an array of X–H insertions, where X is a nitrogen, oxygen, sulfur, selenium, phosphorus, or halogen atom. These processes, which are summarized in eq. 1, are the subject of this

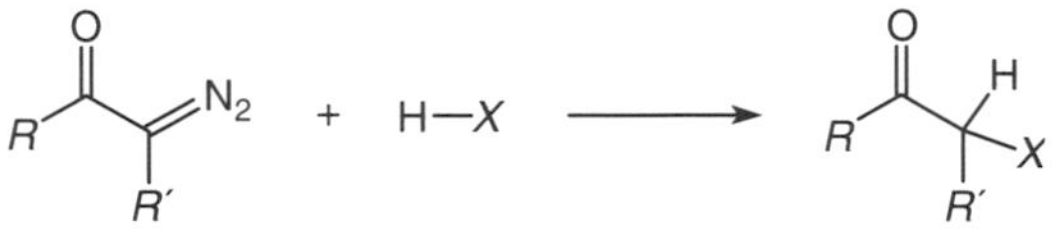

$$
\underset{\substack{R \\ R'}}{\overset{O}{\|}}\!\!\!C{=}N_2 \quad + \quad H{-}X \quad \longrightarrow \quad \underset{\substack{R \\ R'}}{\overset{O}{\|}}\!\!\!C\!\!\begin{smallmatrix}H\\X\end{smallmatrix} \tag{1}
$$

chapter. There is a less extensive group of X–Y insertion reactions, where neither X nor Y is a hydrogen atom. These processes are discussed in Chapter 12.

Scheme 1 gives some idea of the scope for variation in the reagent H–X. The process overall represents a quite general approach to the regiospecific alpha-functionalization of a ketone, in many cases under essentially neutral conditions. Furthermore, since many acyclic terminal diazoketones can be obtained from acyl chlorides and diazomethane, X–H insertion represents a method of regioselective functionalization which does not depend on the availability of the parent ketone. Although the adduct (eq. 1) is formally the product of insertion of a carbene into the X–H bond, mechanistically there probably exists a broad spectrum of processes ranging from uncatalyzed electrophilic attack on the diazocarbonyl group to ylide formation in situations where thermolysis, photolysis, or metal catalysis is employed.[1]

"Insertion" into polar X–H bonds is almost certainly a stepwise process. In the case of hydrogen halides, where catalysts are not required, the process is probably one of protonation of the diazocarbonyl compound (Scheme 2) to form a diazonium ion from which dinitrogen is expelled by attack by halide ion. This mechanism should also apply to other relatively strong acids such as sulfonic acids. In fact, sulfonic acids react very rapidly with diazocarbonyl compounds

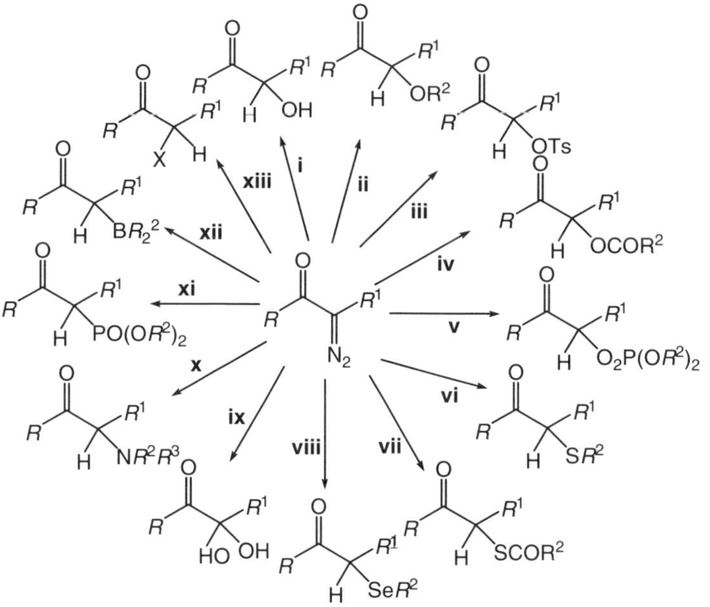

Scheme 1   **i.** H$_2$O; **ii.** *R*^{2}OH; **iii.** TsOH; **iv.** *R*^{2}CO$_2$H; **v.** (*R*^{2}O)$_2$P(O)OH; **vi.** *R*^{2}SH; **vii.** *R*^{2}COSH; **viii.** *R*^{2}SeH; **ix.** H$_2$O$_2$; **x.** *R*^{2}*R*^{3}NH; **xi.** (*R*^{2}O)$_2$P(O)H; **xii.** *R*$^2_3$B; **xiii.** H–*X* (*X* = halogen).

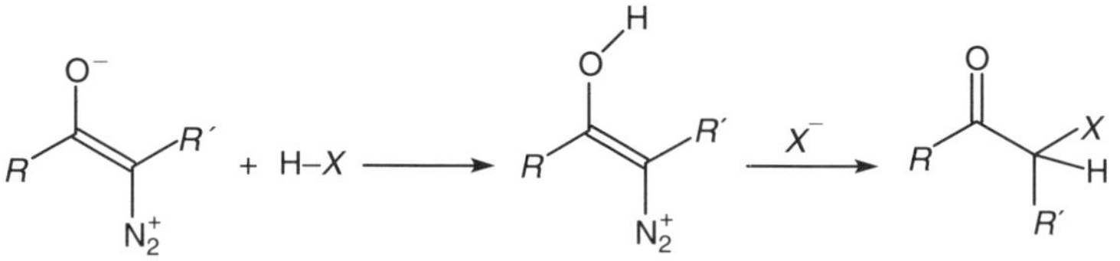

Scheme 2

without catalysis to form α-ketosulfonates. Carboxylic acids react less readily, and although there are examples where catalysts are not required, copper salts are often employed.

Alternative mechanisms for O–H insertion need to be considered for weaker acids such as water, alcohols, and phenols. Such reactions require assistance either through photolysis or catalysis. These alternative mechanisms are summarized in Scheme 3. When the reaction is performed photochemically, the precursor to O–H insertion is considered to be the free carbene (Scheme 3A). Addition of *R*^{2}OH to the carbene forms an oxonium ylide, which undergoes a 1,2-rearrangement to form the O–H insertion product. The second mechanism (Scheme 3B) illustrates a metal-catalyzed process in which nucleophilic attack by *R*^{2}OH occurs on the electrophilic metal carbene to form an oxonium ylide, which undergoes a 1,2-rearrangement to transfer a proton from oxygen to carbon with regeneration of the catalyst. The mechanism in Scheme 3C illustrates a Lewis acid, such as boron

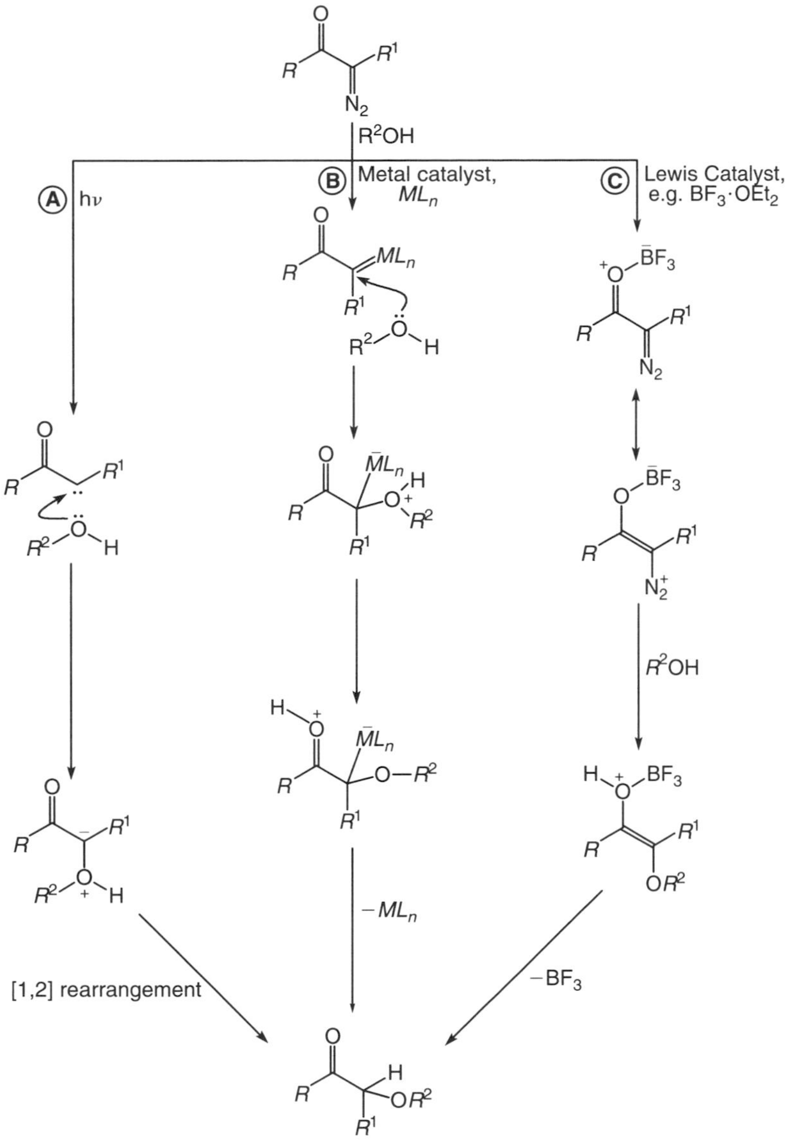

R^2OH = water, alcohol, phenol

Scheme 3

trifluoride etherate, catalyzed O–H insertion in which coordination of the carbonyl group to the Lewis acid activates the diazocarbonyl substrate for nucleophilic attack by R^2OH. Of the three alternative modes of reaction, the photochemical route is generally the least efficient because of the propensity for acylcarbenes to undergo Wolff rearrangement. Comprehensive analyses of the photochemical route to O–H insertion, in particular, the factors that influence carbene reactivity, have been published by Jorgensen[1] and by Kirmse.[2] The extent of involvement of the metal in the catalyzed route is not yet well understood. Results from studies of other ylide formation/sigmatropic rearrangements of diazocarbonyl compounds with chiral rhodium(II) catalysts suggest that the metal may play a role beyond that of generating the initial carbene species (see Chapter 7). The mechanisms of N–H, S–H, Se–H, and P–H insertion are probably quite similar in outline to those summarized in Scheme 3 for catalyzed O–H insertion.

8.2 N–H INSERTION

8.2.1 Intermolecular N–H Reactions

Insertion into N–H bonds by diazocarbonyl compounds attracted little attention as a synthetic route to α-amino ketones or esters until 1978, when a bicyclic β-lactam synthesis was published from the Merck Laboratories.[3-5] Since then there have been numerous demonstrations of the power of this reaction, especially the intramolecular version leading to nitrogen heterocycles. The history of catalyzed N–H insertion parallels closely that of other *X*–H insertions (*vide infra*), with early evidence that α-diazoacetophenone and aniline under copper bronze (*i.e.*, very finely divided copper) catalysis yield the α-anilino ketone (eq. 2).[6] Cuprous cyanide and chloride were later used as catalysts for insertion of

$$\text{Ph}\overset{\text{O}}{\underset{\text{H}}{\|}}\!\!\!\overset{}{=}\!\!N_2 \;+\; \text{PhNH}_2 \;\xrightarrow[\;33\%\;]{\text{Cu}}\; \text{Ph}\overset{\text{O}}{\|}\!\!\!\diagup\!\!\text{NHPh} \qquad (2)$$

EDA into the N–H bond of piperidine, morpholine, and *n*-butylamine.[7] Nicoud and Kagan[8] used cuprous cyanide to bring about N–H insertion of diazopropionates with benzylamines in an α-amino acid synthesis (eq. 3). Yields of up to

$$\underset{\text{CH}_3}{\overset{N_2}{\|}}\!\!\!\diagup\!\!\overset{}{\text{CO}_2R} \;+\; R^1NH_2 \;\xrightarrow{\text{CuCN}}\; \underset{\text{CH}_3}{\overset{H\;\;NHR^1}{\diagup\!\!\!\diagdown}}\!\!\overset{}{\text{CO}_2R} \;\longrightarrow\; \longrightarrow \; \underset{\text{CH}_3}{\overset{H\;\;NH_2}{\diagup\!\!\!\diagdown}}\!\!\overset{}{\text{CO}_2H}$$

$$R = C_6H_{11}, \text{ L-menthyl}$$
$$R^1 = (S)\text{-1-phenylethylamine}, (R)\text{-1-phenylethylamine},$$
$$(S)\text{-1-(1-naphthyl)ethylamine}, (R)\text{-1-(1-naphthyl)ethylamine}$$

$$(3)$$

50% were obtained for the insertion step. Introduction of rhodium(II) catalysts for N–H insertion led to major improvements with reaction between EDA and aniline in the presence of rhodium(II) acetate furnishing in 70% yield the insertion product (eq. 4).[9] Reactions are carried out either neat or in solution (benzene

$$N_2CHCO_2Et \ + \ PhNH_2 \ \xrightarrow[70\%]{Rh_2(OAc)_4} \ PhNHCH_2CO_2Et \qquad (4)$$

or ethylene glycol dimethyl ether). Among other examples of amines catalysed by rhodium(II) acetate are those in eqs. 5[10] and 6,[11] where the objective was to apply

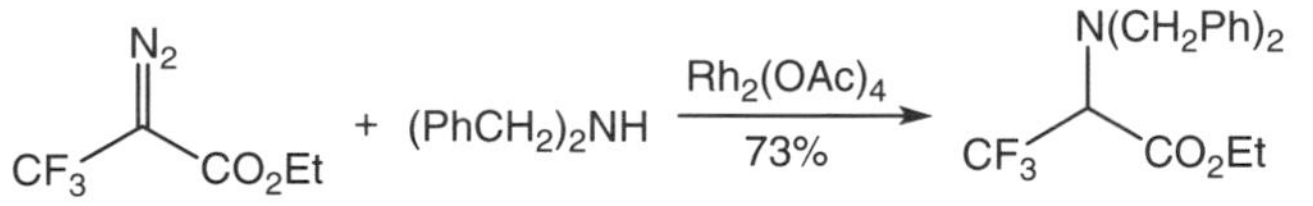

$$(5)$$

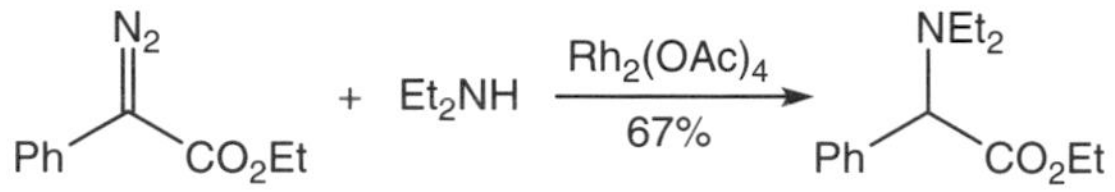

$$(6)$$

the process to the formation of novel glycine and phenylglycine derivatives. In a similar fashion Landais and Planchenault[12] used rhodium(II) acetate in benzene to bring about N–H insertion of aniline in an allylic amino ester synthesis (eq. 7).

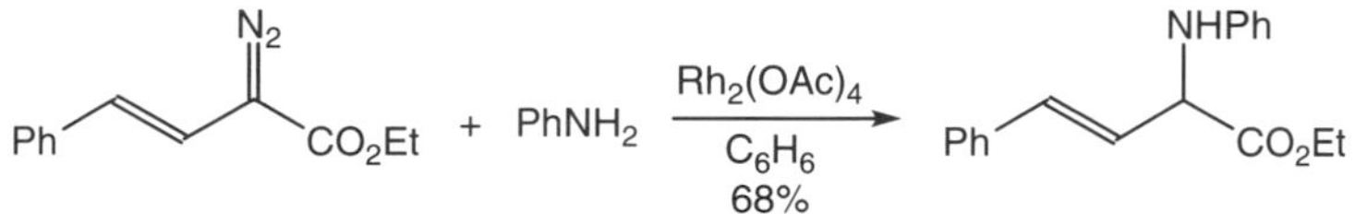

$$(7)$$

N–H insertion with diazocarbonyl compounds is not limited to amino functions. Examples of reactions involving amides, β-lactams, and carbamates are also known, and some representative examples are shown in eqs. 8,[13] 9,[10] 10[13] 11,[10]

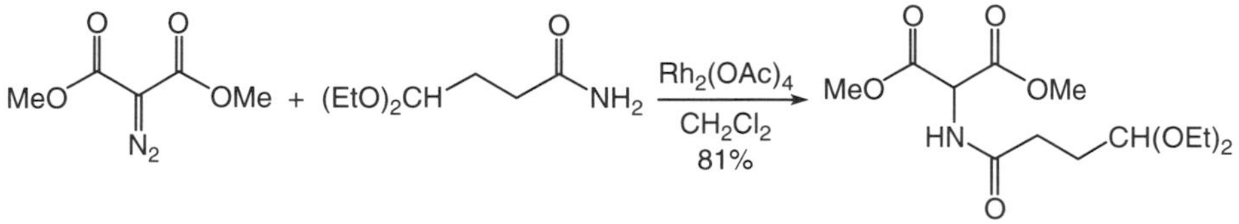

$$(8)$$

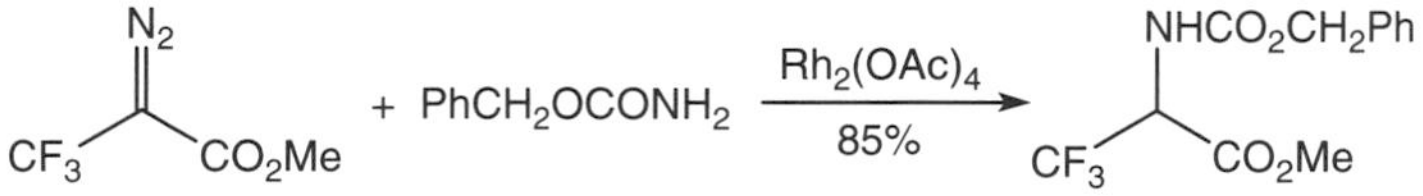

$$(9)$$

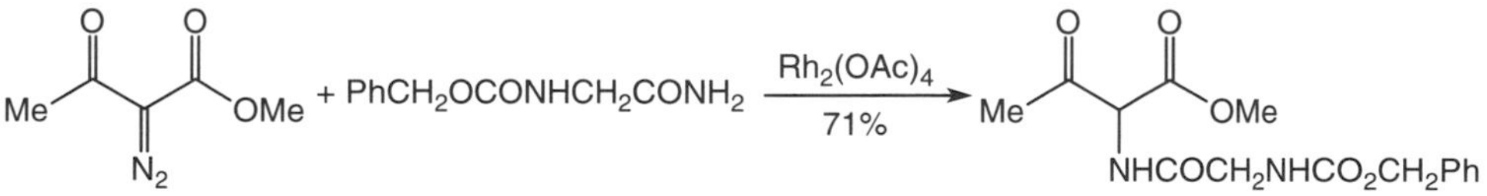

$$(10)$$

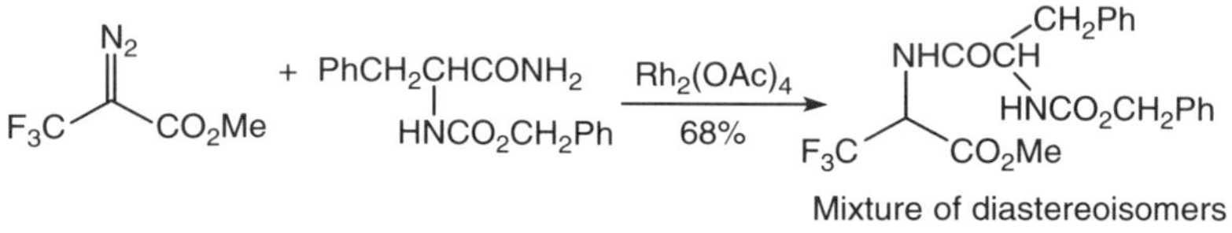

(11)

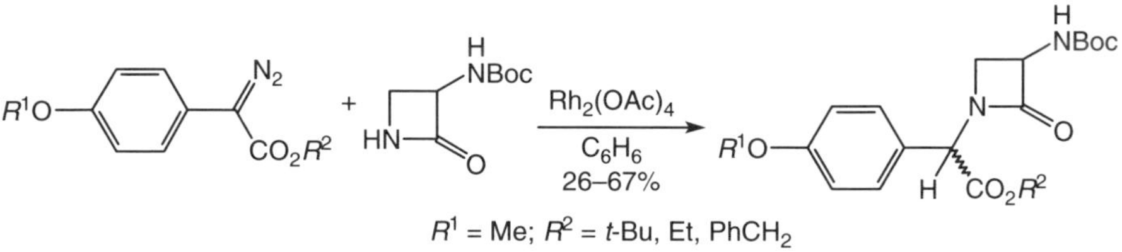

Mixture of diastereoisomers

R^1 = Me; R^2 = *t*-Bu, Et, PhCH$_2$

(12)

R^1 = Me, 4-nitrobenzyl; R^2 = 4-nitrobenzyl;
R^3 = H, Me, Et; R^4 = H, Me

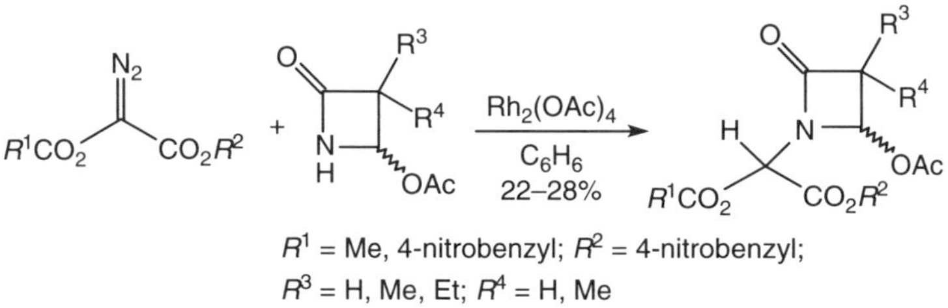

(13)

12,[14] and 13.[15] For the reactions in eqs. 10 and 11 two different N–H sites are available, one in the form of an amide and the other a carbamate. The former is preferred in both reactions. In eq. 12, where there are also two different possible insertion sites, the N–H bond of the β-lactam is favored over that of the carbamate. Similarly, *N*-malonyl β-lactams are accessible from diazomalonates and β-lactams (eq. 13). All these reactions were brought about with rhodium(II) catalysis. The products of the reactions in eqs. 11-13 were obtained as mixtures of diastereoisomers.

8.2.2   Intramolecular N–H Insertion

By far the most successful metal-catalyzed N–H insertion reactions have been intramolecular, leading to nitrogen heterocycles, though the potential of the reaction in synthesis was not appreciated until Lama and Christensen[3] reported the conversion of a penicillin analogue into the carbapenem nucleus via insertion of a metal carbene into the N–H bond of a β-lactam, a process that was later refined to become the key step in the Merck synthesis of the antibiotic (+)-thienamycin (Scheme 4).[4,5] The N–H insertion step leading to a five-membered ring fused to a preformed β-lactam was accomplished in quantitative yield by rhodium(II) acetate with a substrate:catalyst ratio of *ca.* 1000:1. This and many subsequent similar applications, a selection of which is summarized in Table 8.1, suggest that intramolecular N–H insertion is among the most efficient methods yet devised

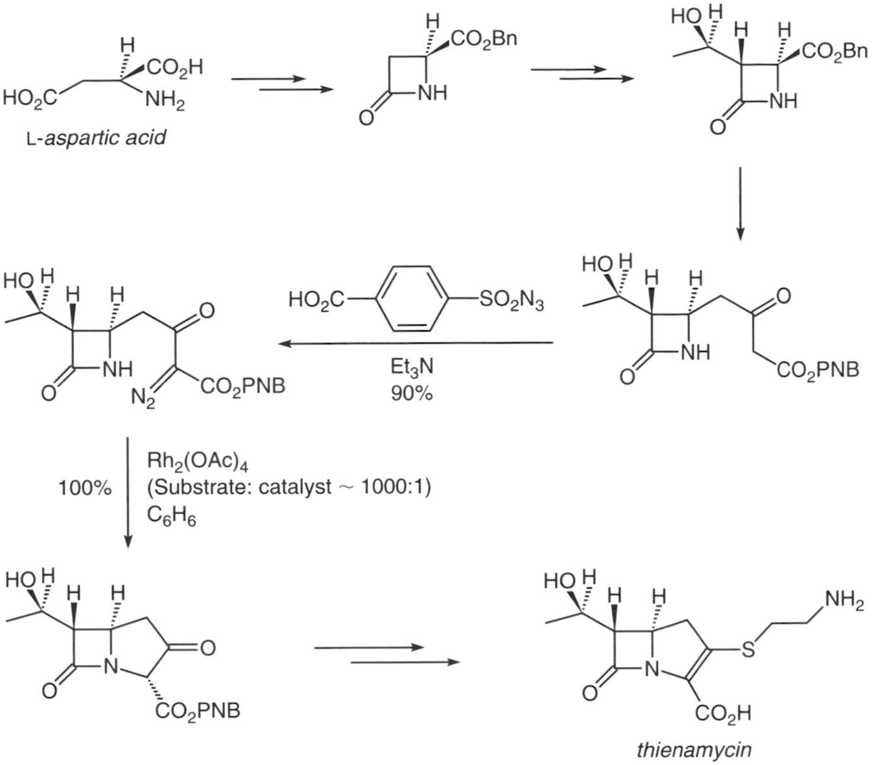

Scheme 4

TABLE 8.1 Bicyclic β-Lactam Synthesis via Rh(II)-Catalyzed N–H Insertion

Diazocarbonyl	Product	Catalyst/ Solvent	Yield (%)	Ref.
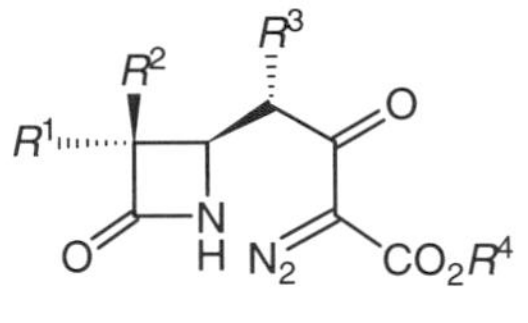	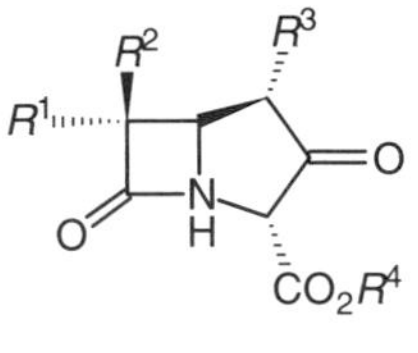	Rh$_2$(OAc)$_4$/ Toluene, Δ	76–85	4, 5, 18–42
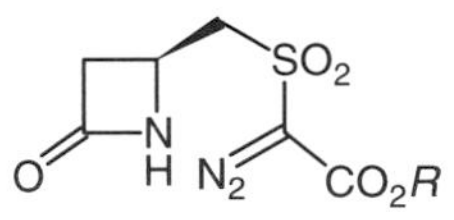	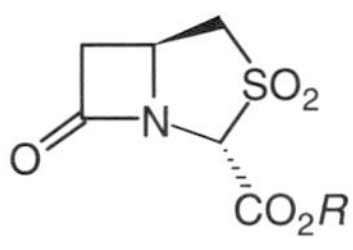	Rh$_2$(OAc)$_4$/ PhH, Δ	95	43
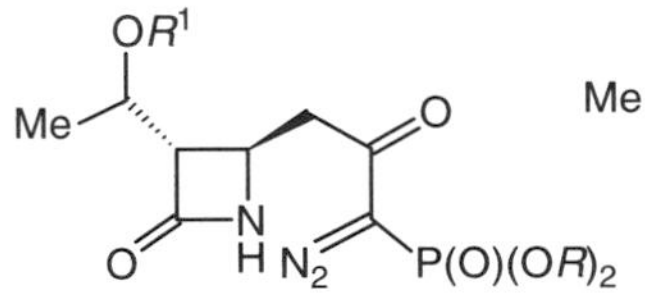	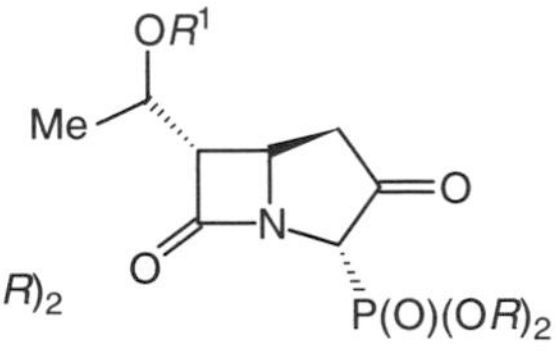	Rh$_2$(OAc)$_4$/ PhH, Δ	60–90	44

TABLE 8.1 Bicyclic β-Lactam Synthesis via Rh(II)-Catalyzed N–H Insertion (*continued*)

Diazocarbonyl	Product	Catalyst/ Solvent	Yield (%)	Ref.
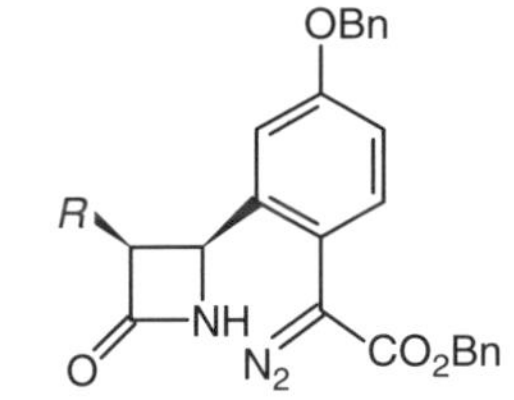	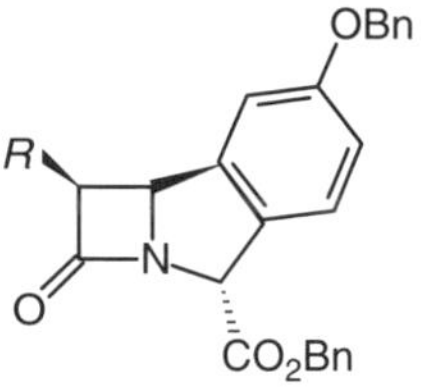	Rh$_2$(OAc)$_4$/ Toluene, Δ	5–10	45, 46
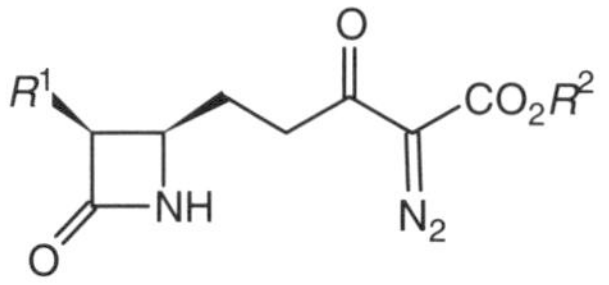	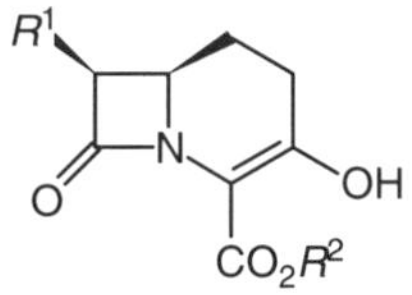	Rh$_2$(OAc)$_4$/ CHCl$_3$, Δ	70–75	47–49
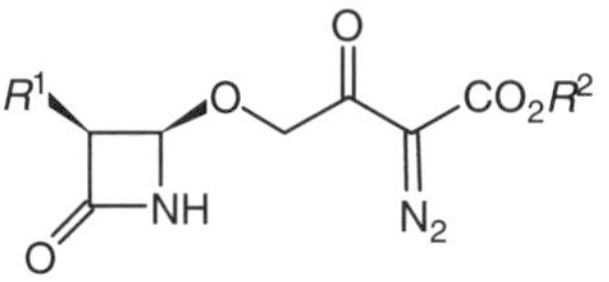	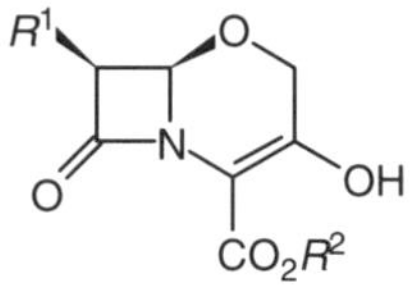	Rh$_2$(OAc)$_4$/ PhH, Δ	82–85	50–52
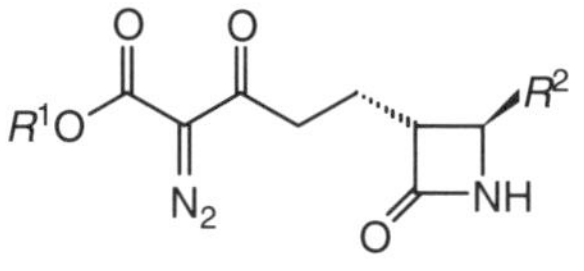	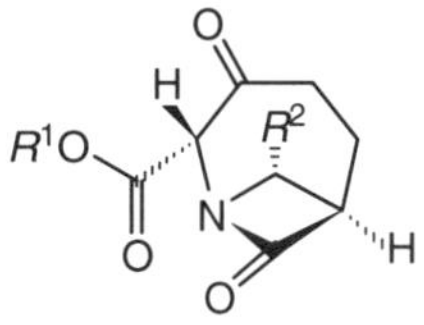	Rh$_2$(OAc)$_4$/ PhH, Δ	50	53
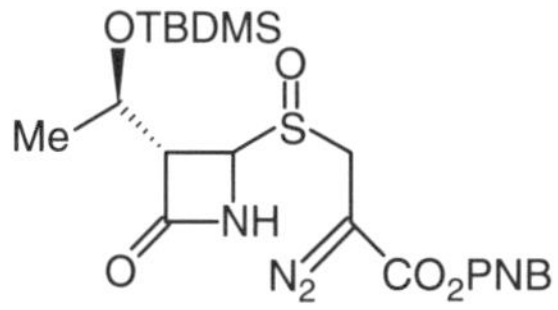	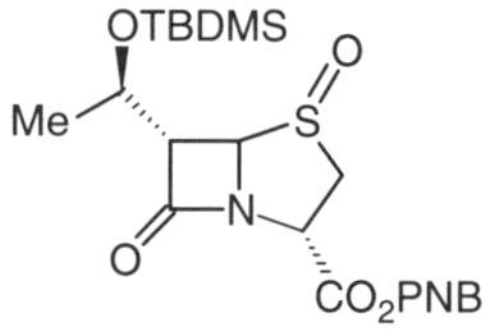	—	—	54
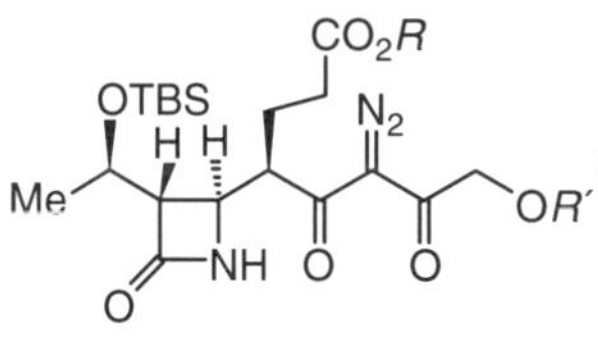	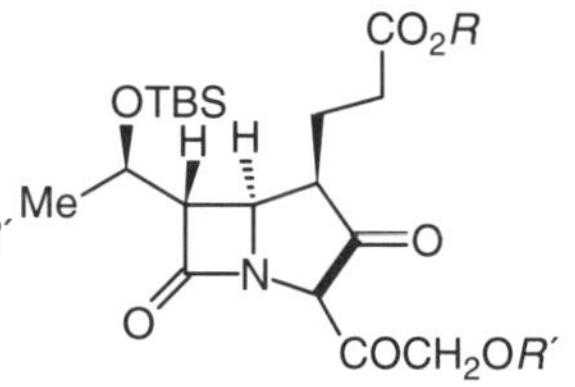	Rh(II) Octanoate/ EtOAc/Hex., Δ	46	55

for synthesizing bicyclic β-lactams from 2-azetidinones. Although the majority of examples in Table 8.1 illustrate five-membered ring formation, there are also examples of six- and seven-membered rings. Most recent studies, mainly by Rapoport and co-workers,[16] have established that while five-membered formation is preferred kinetically, as is also the case with C–H and O–H insertion, it is possible to construct four- to six-membered aza rings from simple, conformationally mobile acyclic diazoester precursors. In fact there is also one report of an N–H insertion that could be reasonably interpreted as a three-membered ring formation (Scheme 5), though the product isolated was not an azacyclopropane.[17] Rapoport and co-workers[16] constructed a series of substrates of the general type in Scheme 6 in which the nitrogen atom to undergo insertion is tethered to an

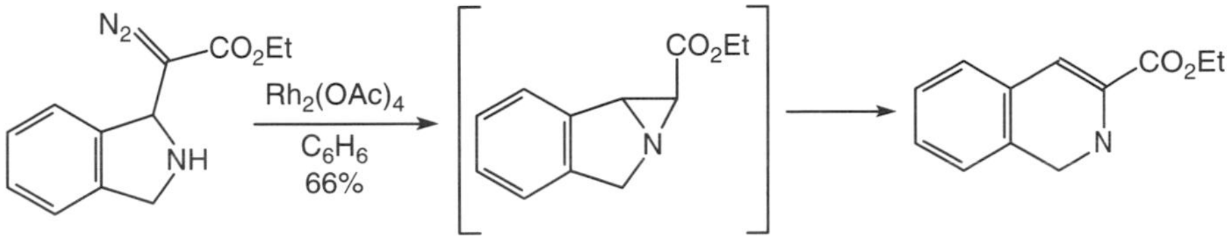

Scheme 5

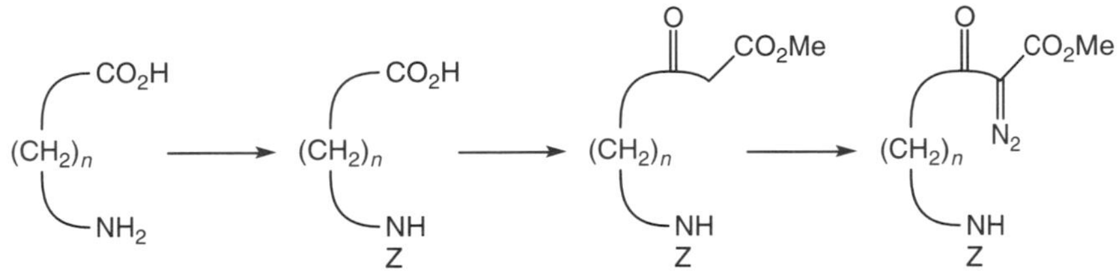

$n = 1–4$; Z = Cbz (benzyloxycarbonyl)
Scheme 6

α-diazo β-keto ester with carbon chains of various lengths to allow investigation of the effect of ring size. In each case the amino function was present as its benzyl carbamate derivative. Rhodium(II) acetate was used as catalyst throughout, and variations in solvent, temperature, and catalyst concentration were found to play a role in determining product distributions. A summary of the results is shown in Scheme 7. Cyclization to four- and five-membered rings containing N occurred efficiently and selectively (entries 1 and 2). With the diazo compound in entry 3, the opportunity exists for both intramolecular C–H and N–H insertion. While seemingly minor changes in the reaction conditions did significantly change the product distribution, the optimum conditions of benzene as solvent at 80°C and a rhodium(II) acetate concentration of 5 mol % gave the desired 3-oxopiperidine derivative in 67% with only trace amounts of other products, one of which was that of C–H insertion. Finally, the diazocompound in entry 4 presented the option of five- and six-membered ring formation from C–H insertion and seven-membered heterocycle formation from N–H insertion. In that event, this reaction yielded a cyclopentanone as the only cyclized product in 39% yield. Thus the kinetic preference for five-membered ring formation completely overwhelms any tendency for N–H insertion. However, four-membered rings are also

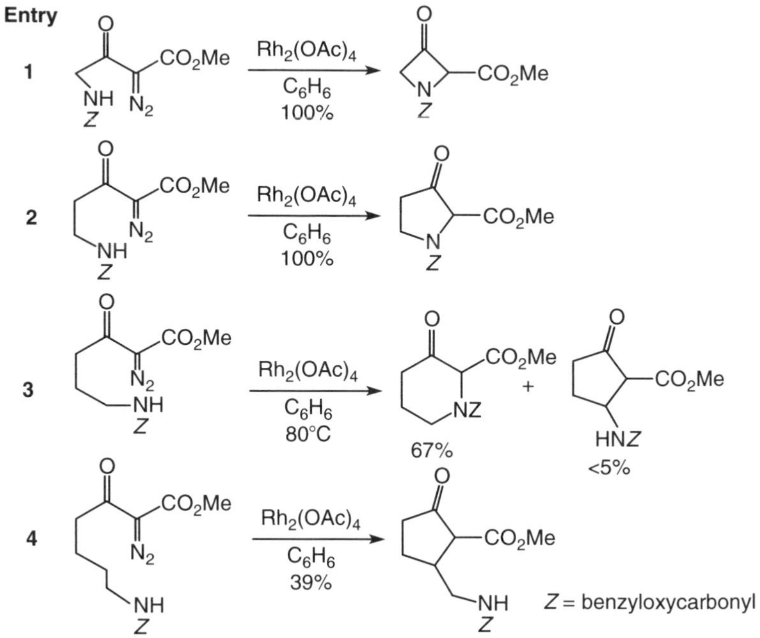

Scheme 7

directly accessible with rhodium(II) catalysts. Moody and co-workers[56] found, for example, that acyclic hydrazine derivatives (eq. 14) cyclize very efficiently

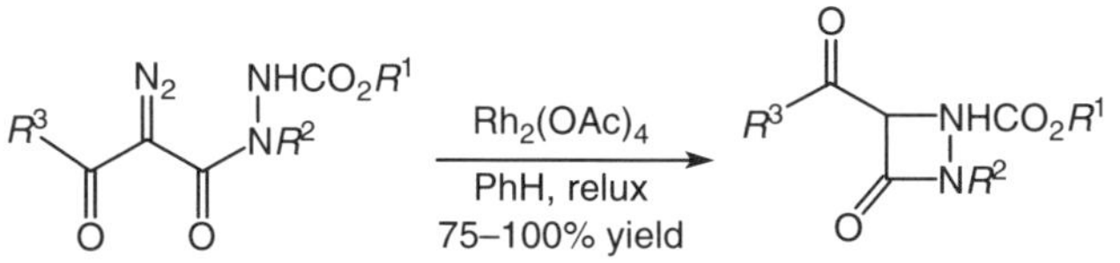

$$(14)$$

with rhodium(II) acetate to furnish diazetidinones. There are several examples of syntheses of enantiopure 2-substituted azetidin-3-ones *via* N–H insertion, illustrating further the versatility of diazocarbonyl intermediates derived from natural amino acids (Chapter 9 provides a more detailed discussion of the chemistry of diazocarbonyl compounds derived from amino acids and peptides). Seebach[57] and McKervey[58] and their collaborators have found that Boc- and Cbz-protected α-aminodiazoketones derived from L-alanine, L-valine, and L-phenylalanine cyclize very efficiently to the azetidin-3-one derivatives (eq. 15).

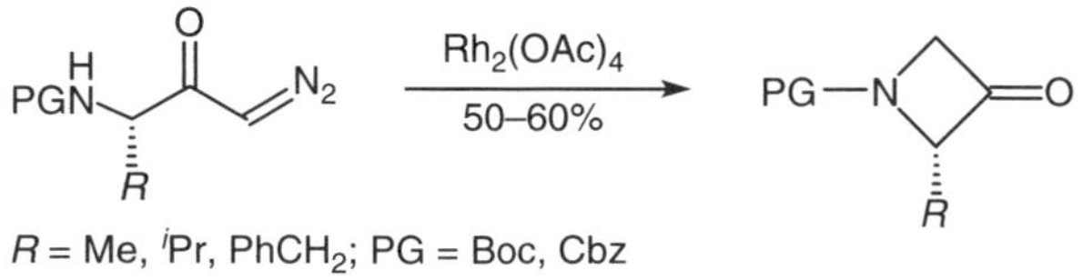

$$(15)$$

R = Me, ^{i}Pr, PhCH$_2$; PG = Boc, Cbz

Hanessian *et al.*[59] have exploited this enantiopure azetidinone synthesis by using the N- and O-protected D-serine–derived diazoketone in Scheme 8 to

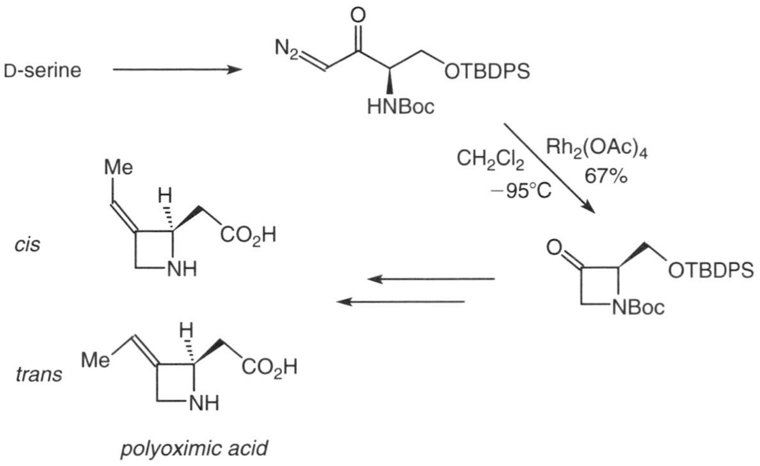

Scheme 8

produce an intermediate suitable for elaboration into optically active *cis*- and *trans*-polyoximic acid. In a related area, Emmer[60] prepared the Boc-protected azetidinone carboxylate (eq. 16) and used it to synthesize racemic polyoximic

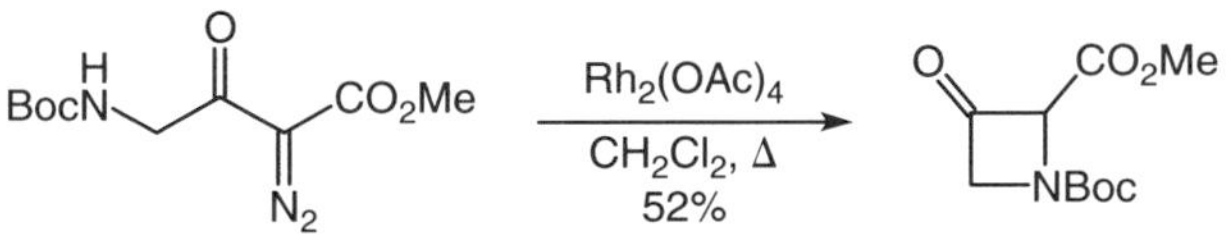

$$(16)$$

acids. Among other examples of intramolecular N–H insertion leading to five- and six-membered rings are the reactions in eqs. 17[61] and 18[62], the catalyst in each

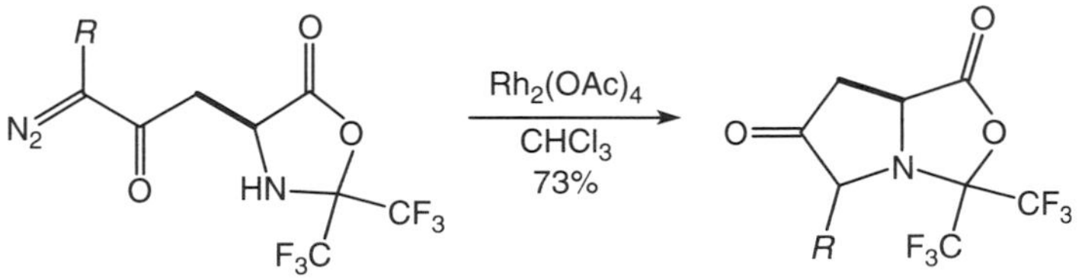

$$(17)$$

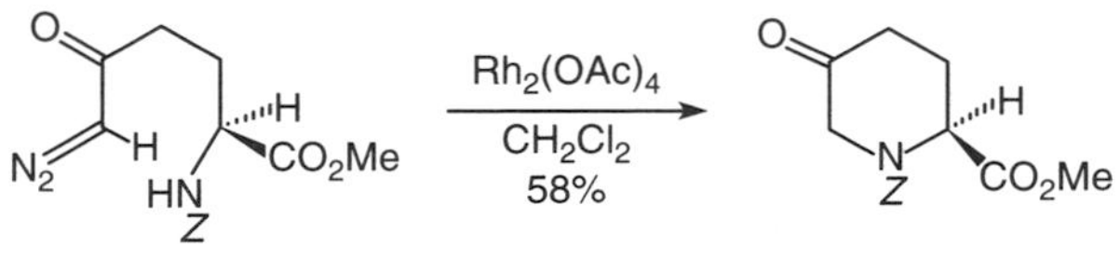

$$(18)$$

case being rhodium(II) acetate. Both processes give enantiopure products, the diazo precursors having been prepared from L-aspartic acid and L-glutamic acid, respectively. N–H insertion has also been used to produce the six-membered azalactone (eq. 19), though here the yield was poor (18%).[63]

$$\text{(19)}$$

8.2.3   Asymmetric N–H insertion

Asymmetric synthesis *via* chiral catalysts or with chiral auxiliaries involving enantioselective N–H insertion of diazocarbonyl compounds, although much less well developed than asymmetric C–H insertion (Chapter 3), is promising. An early approach by Nicoud and Kagan[8] to exploit the chiral auxiliary approach to N–H insertion through the use of chiral benzylamines, including the enantiomers of 1-phenylethylamine and 1-(1-naphthyl)ethylamine is illustrated in eq. 3. Here the objective was to achieve diastereocontrol in N–H insertion of diazopropionates and thereby effect an asymmetric synthesis of alanine after hydrogenolytic removal of the auxiliary followed by hydrolysis. Although the insertion reaction was successful, the enantioselection accompanying the creation of the new stereogenic center was poor (15–26% ee). In a more recent approach using substrates in which the chiral auxillary was present in the alkyl residue of the diazo ester, rather than in the amine, Moody and co-workers[11] found equally low levels of asymmetric induction. Garcia *et al.*[64] examined the use of chiral catalysts for asymmetric N–H insertion as a route to natural and unnatural cyclic amino acids. Suitable substrates were constructed from Cbz- or Boc-protected amino acid esters *via* diazo transfer using the Danheiser procedure (Chapter 1). These structures were chosen so as to reveal the influence of catalyst on two potentially significant competing side reactions, namely, C–H insertion and β-elimination. In all the other examples of N–H insertion discussed here, the substrates were such that β-elimination (Chapter 12) was precluded. It was therefore important to determine whether alkene formation *via* β-elimination would compete with N–H insertion. Although catalyzed decomposition of diazocarbonyls possessing β-hydrogen atoms can lead rapidly to alkene formation, Taber[65] has demonstrated that the extent of this reaction can be influenced by the electron withdrawing ability of the ligands on rhodium, with the acetate less prone to catalyze elimination than the corresponding trifluoroacetate.

Several rhodium(II) compounds were employed as catalysts for decomposition of the Cbz-protected diazoester in eq. 20. Up to three products could be isolated and identified as those of N–H insertion **1**, C–H insertion **2**, and β-elimination **3** in proportions that revealed a marked catalyst dependency. Both **2** and **3** were produced with the *cis* geometry. Rhodium(II) acetate favored N–H insertion exclusively, furnishing (±)-*N*-Cbz-pipecolic acid **1** in excellent yield. Doyle's

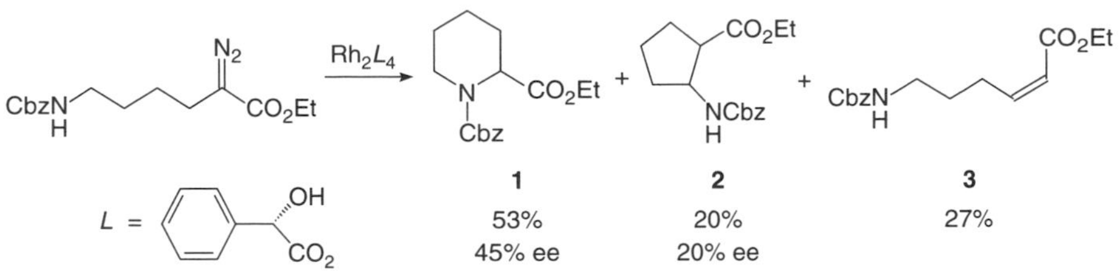

$$(20)$$

dirhodium $[4(R)\text{-MPPIM}]_4$ catalyst (Chapter 2) favored β-elimination, even though this catalyst is among the most effective for asymmetric intramolecular C–H insertion with other substrates. Of four chiral catalysts used for the reaction in eq. 20, $Rh_2[(S)\text{-mandelate}]_4$ was the most promising, yielding N-Cbz-pipecolic acid ethyl ester **1** with an ee of 45%. The product of intramolecular C–H insertion **2** had an ee of 20% with the mandelate catalyst.

8.3 O–H INSERTION WITH WATER, ALCOHOLS, AND PHENOLS

8.3.1 Intermolecular O–H Insertion

O–H insertion with diazocarbonyl compounds has been observed with water, alcohols, phenols, carboxylic acids, and sulfonic acids (eq. 21). Although examples

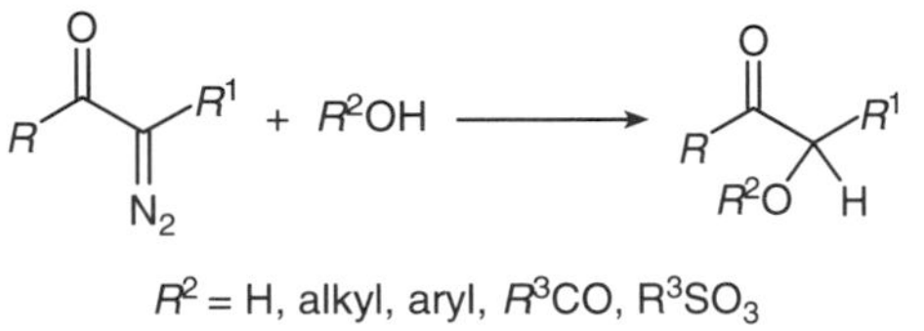

$$(21)$$

$R^2 = H$, alkyl, aryl, R^3CO, R^3SO_3

of purely thermal and photochemical processes involving Brønsted/Lewis and transition metal catalysis are known, the former are of only limited usefulness in synthesis due to competition from Wolff rearrangement (Chapter 9), which is frequently the dominant pathway to products. Other competing reactions are also possible, resulting, in some cases, in complex product mixtures, as, for example, in the photochemical reaction of EDA with 2-propanol (Scheme 9).[66] In addition to

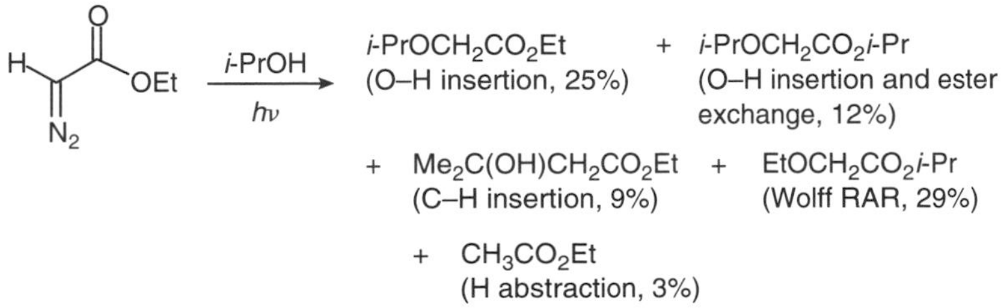

Scheme 9

two O–H insertion products (total 37%), there are products of Wolff rearrangement (29%), C–H insertion (9%), and hydrogen abstraction (3%); the product of ester exchange with O–H insertion is also the result of a photochemical process. Notwithstanding the complexity of these and similar reactions with diazoketones and diazoamides, they have provided detailed information on the electronic state and reactivity of the oxocarbenes involved as transient intermediates.[67]

Reactions with alcohols continues to attract interest, especially in the area of asymmetric synthesis of chiral α-alkoxy ketones and esters (*vide infra*). Reactions with water and phenols are less well represented, but all three can be grouped conveniently together, as they have much in common mechanistically. The topic was reviewed in 1995 by Miller and Moody.[68]

For the reasons cited above, O–H insertion is best conducted catalytically using Brønsted/Lewis acids or transition metal salts. Of the former, dilute sulfuric acid and boron trifluoride etherate are popular catalysts for intermolecular reactions, several examples of which are illustrated in eq. 22,[69] 23,[70,71] 24,[72,73] 25,[74] 26,[75] 27,[76] and 28.[77] In these and similar processes with other Brønsted/Lewis

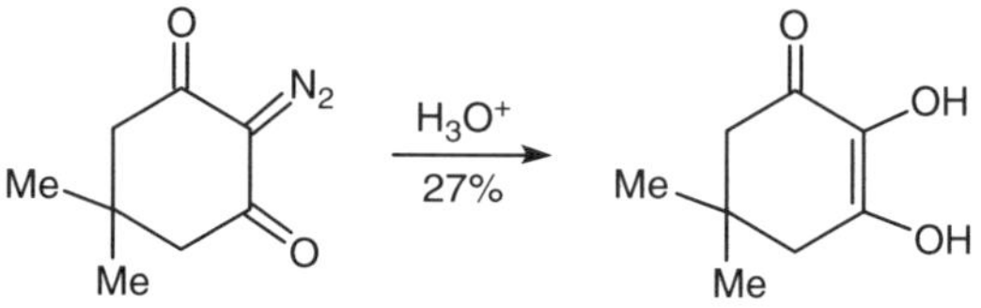

$$\text{(22)}$$

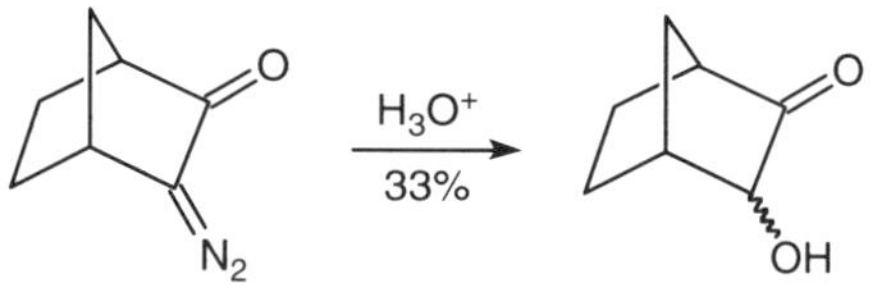

$$\text{(23)}$$

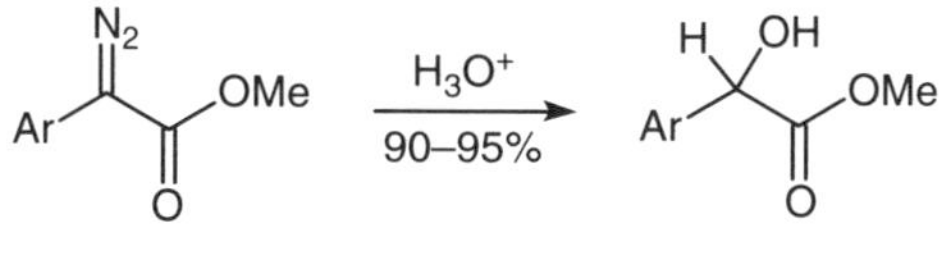

$$\text{(24)}$$

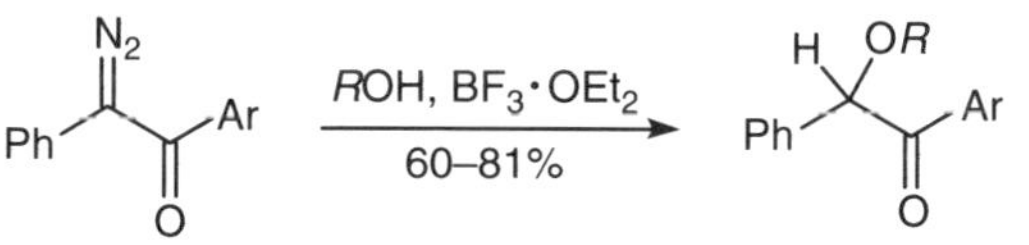

$$\text{(25)}$$

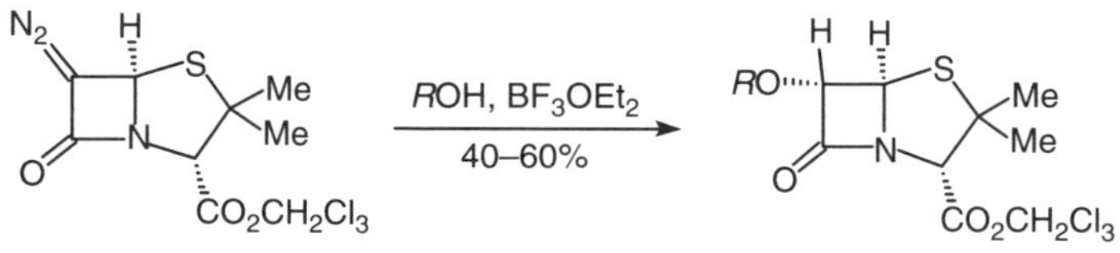

$$\text{(26)}$$

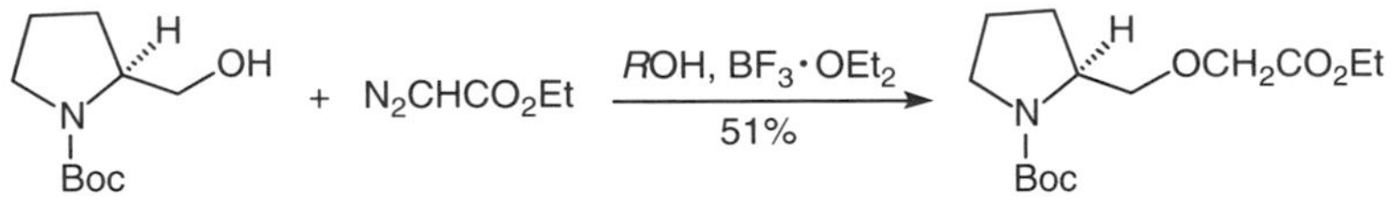

(27)

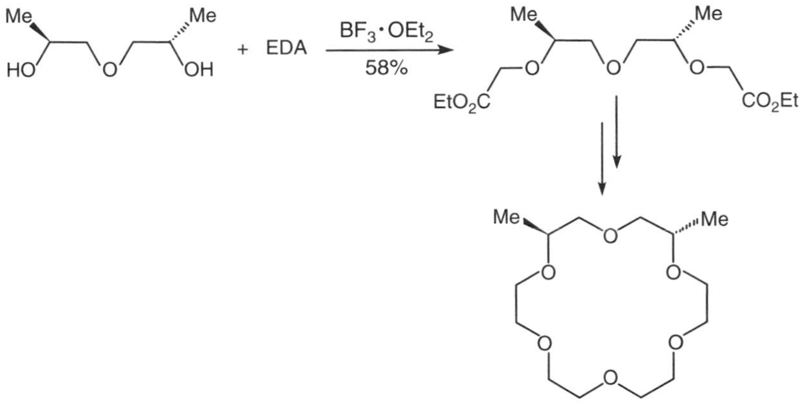

(28)

acids, protonation/coordination is believed to precede attack by the ROH species with displacement of dinitrogen (*cf.* Scheme 3). Equations 22[69] and 23[70,71] illustrate early examples of the acid-catalyzed addition of water to diazoketones, the latter also giving byproducts of rearrangements associated with carbocation intermediates.

Acid-catalyzed hydration of diazoarylacetates yields the corresponding mandelate derivatives in excellent yield (eq. 24).[72,73] Intermolecular additions of alcohols, all catalyzed by boron trifluoride etherate, are exemplified by eqs. 25,[74] 26,[75] 27,[76] and 28.[77] In the example in eq. 28, the reaction was applied to an enantiopure diol to form an adduct with EDA suitable for elaboration into a chiral dimethyl 18-crown-6 ether.[77]

Prior to the introduction of rhodium(II) catalysts for diazocarbonyl transformations, there were several examples of the use of copper in various forms for O–H insertion. Studies suggest that in general transition metal catalysts offer greater advantages over Brønsted/Lewis catalysts in terms of chemoselectivity and efficiency. Following an observation by Casanova and Reichstein[78] that copper oxide in methanol catalyzes the transformation of a steroidal side chain diazoketone (eq. 29) into an α-methoxyketone, Yates[6] in 1952 undertook a detailed

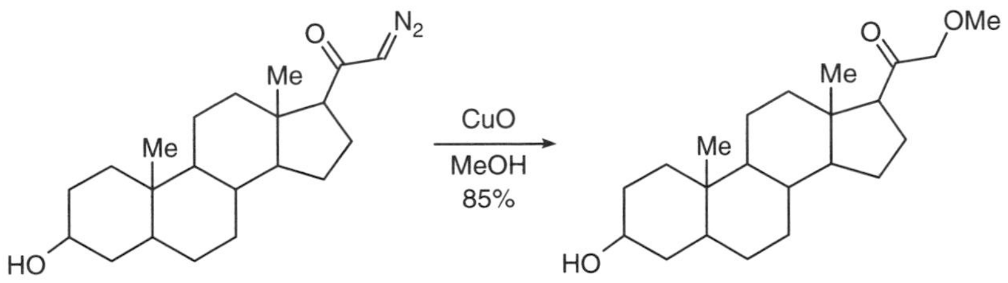

(29)

study of the reactions of two representative diazoketones, 1-diazo-2-nonade-canone (eq. 30) and α-diazoacetophenone (eq. 31), with alcohols in the presence

$$CH_3(CH_2)_{16}\overset{O}{\overset{\|}{C}}CHN_2 \xrightarrow[\substack{EtOH \\ 68\%}]{Cu\ bronze} CH_3(CH_2)_{16}\overset{O}{\overset{\|}{C}}CH_2OEt \qquad (30)$$

(31)

of copper bronze. With both substrates, α-alkoxy ketones were the preponderant reaction products with methanol, ethanol, *t*-butyl alcohol, and hexanol, though yields were generally poor. Yates[6] also studied the performance of phenol in O–H insertion and found that while α-phenoxyacetophenone was formed from α-diazoacetophenone in 63% yield (eq. 32), it was accompanied by 2-phenylben-

(32)

zofuran (23%). The latter product was reasonably formulated as resulting from *ortho*-alkylation of phenol by the diazoketone to form *o*-hydroxydes-oxybenzoin, which suffered cyclodehydration *in situ* or during isolation.

   Among other copper bronze–catalyzed reactions is the hydration of aroyl dia-zomethanes in aqueous solution to furnish α-hydroxyacetophenones.[79] Cyclic di-azoketones also respond to copper bronze in alcohols, although in the norbornyl system the course of the reaction can be very substituent dependent. Whereas the 2-diazo-3-norbornanone system (eq. 23) furnishes the *endo*-2-methoxy deriva-tive in good yield, diazocamphor under similar conditions gives cyclocam-phanone, the product of C–H insertion, in 84% yield, with only minor amounts of the O–H insertion product.[80] Copper(II) chloride has been used for O–H in-sertion of EDA with 1-butanol, benzyl alcohol, and allyl alcohol[81]; product yields were poor.

   One of the first studies published by the Belgian group in 1973 in their explora-tion of rhodium(II) acetate catalysis of diazocarbonyl chemistry was O–H inser-tion with water and alcohols.[82] The performance of rhodium(II) acetate showed that it is highly effective for decomposing EDA in water and alcohols to furnish α-hydroxy and α-alkoxy acetates (eq. 33). A short time later the Belgian group also found that rhodium(II) acetate catalyzes O–H insertion of EDA with phenol to afford ethyl phenoxyacetate in 90% yield.[83]

$$N_2CHCO_2Et + ROH \xrightarrow[82-88\%]{Rh_2(OAc)_4} ROCH_2CO_2Et \tag{33}$$

$$R = H, Et, i\text{-Pr}, t\text{-Bu}$$

With the more accessible volatile alcohols the reactions are often conducted in the neat liquid without a co-solvent. There are, however, cases where it is inappropriate to use the alcohol as solvent, either because of availability or unsuitable physical properties, and stoichiometric quantities can be used very successfully in solvents such as dichloromethane, benzene, or diethyl ether. Two examples of intermolecular O–H insertion are highlighted in eqs. 34 and Scheme 10. Reduc-

$$N_2CHCO_2Et + H_2{}^{18}O \xrightarrow[80\%]{Rh_2(OAc)_4} H^{18}OCH_2CO_2Et \xrightarrow{LiAlD_4} H^{18}OCH_2CD_2OH \tag{34}$$

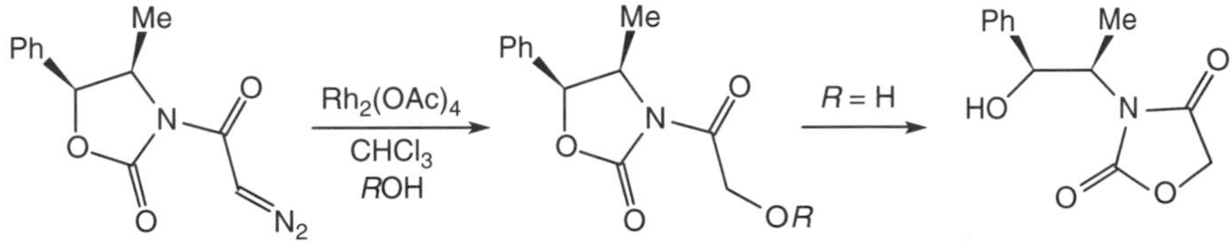

$$R = Et \quad 97\%$$

Scheme 10

tion of the resulting ethyl hydroxyacetate (eq. 34) with lithium aluminum deuteride furnished the doubly labeled ethylene glycol, this route having been chosen so as to maximize the incorporation of the ^{18}O label.[84] Similarly, rhodium(II)-catalyzed decomposition of the diazoketone of Scheme 10 in water–chloroform furnished the oxazolidinedione shown in high yield, presumably via initial O–H insertion to give the hydroxy intermediate, which undergoes internal *trans*-lactamization.[85] In ethanol with otherwise identical conditions this reaction affords the ethyl ether in 97% yield. 1-Diazo-2-cyclohexanone (eq. 35), and its 3-, 4-, and 5-methyl derivatives in eqs. 36–38, respectively, react with methanol, ethanol, 2-propanol, and *t*-butyl alcohol by rhodium(II) acetate catalysis in dichloromethane to produce α-alkoxycyclohexanones in generally excellent (for

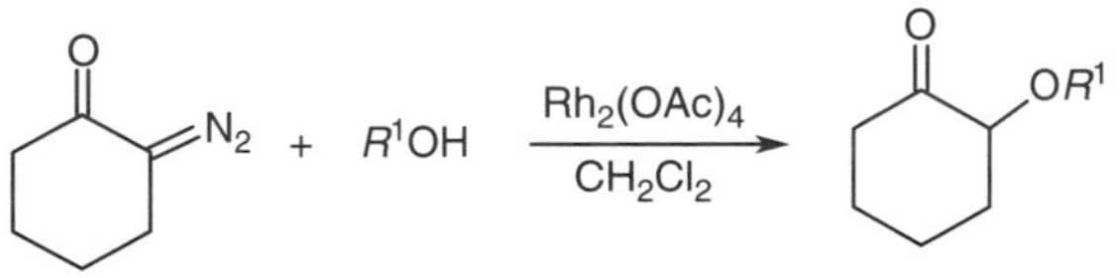

$$R^1 = Me\ (90\%);\ R^1 = Et\ (86\%);\ R^1 = i\text{-Pr}\ (75\%);\ R^1 = t\text{-Bu}\ (60\%)$$

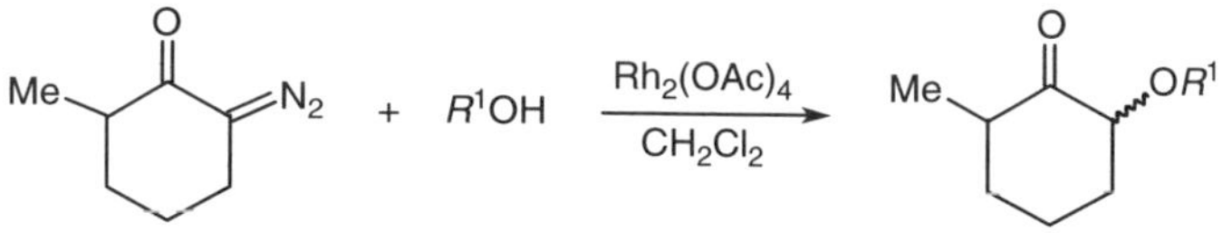

$$R^1 = Me\ (88\%);\ R^1 = Et\ (83\%);\ R^1 = i\text{-}Pr\ (69\%);\ R^1 = t\text{-}Bu\ (58\%)$$

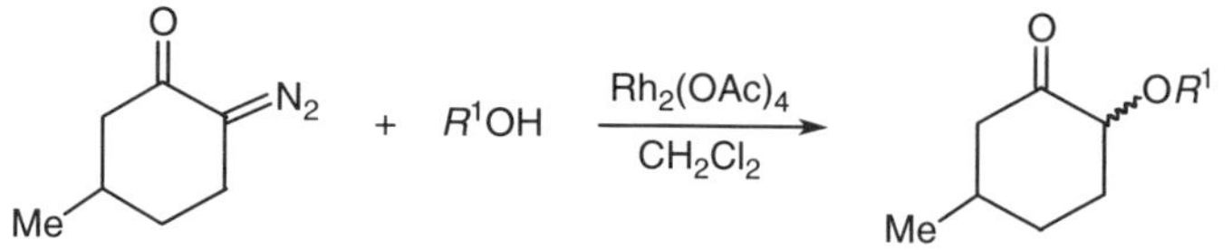

(36)

$$R^1 = Me\ (92\%);\ R^1 = Et\ (77\%);\ R^1 = i\text{-}Pr\ (55\%);\ R^1 = t\text{-}Bu\ (28\%)$$

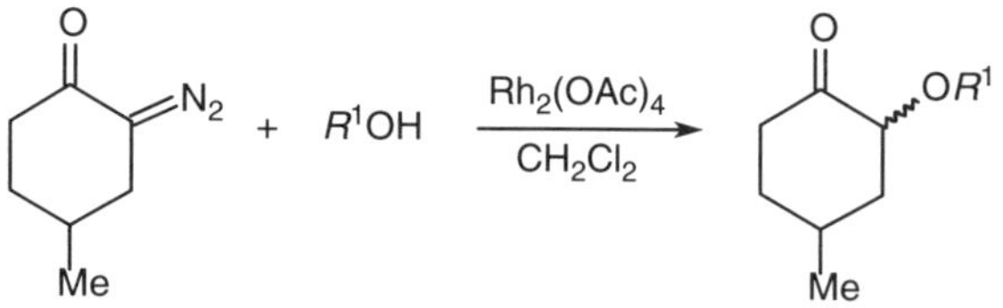

(37)

$$R^1 = Me\ (68\%);\ R^1 = Et\ (87\%);\ R^1 = i\text{-}Pr\ (74\%);\ R^1 = t\text{-}Bu\ (54\%)$$

(38)

methanol and ethanol) to good (for 2-propanol and *t*-butyl alcohol) yields.[86] In the methyl-substituted series the α-alkoxycyclohexanones were all produced as mixtures of stereoisomers. In the reaction of eq. 39[87] involving a diazoketone and

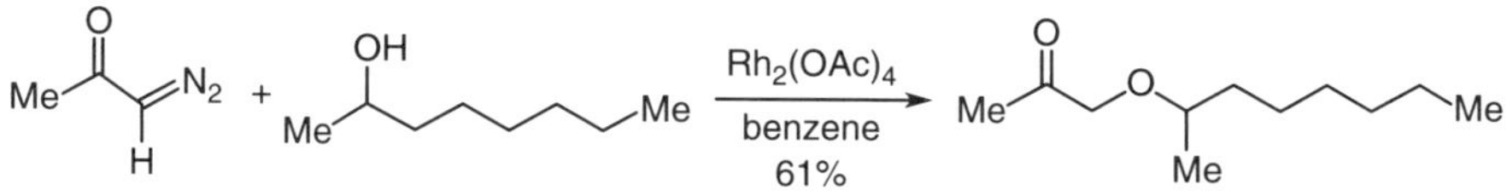

(39)

2-octanol, the use of benzene as the solvent shows that aromatic cycloaddition did not compete significantly with O–H insertion, nor was there competition from cyclopropanation in the example shown in eq. 40, where rhodium(II) acetate,

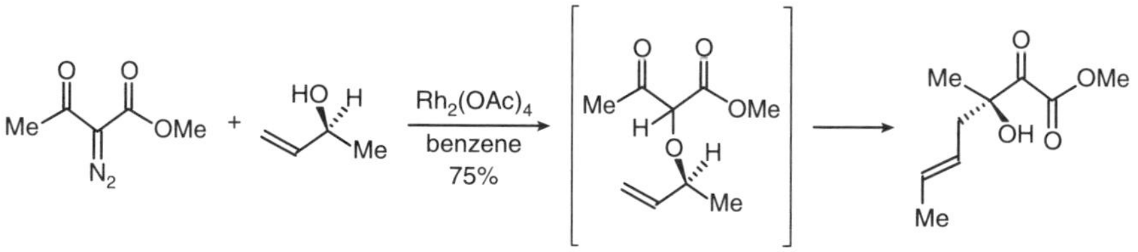

(40)

again with benzene as solvent, promoted O–H insertion into *S*-(+)-3-buten-2-ol to form an allylic ether, which *in situ* underwent a [3,3]-sigmatropic rearrange-

2-ISOPROPOXYCYCLOHEXANONE[86]

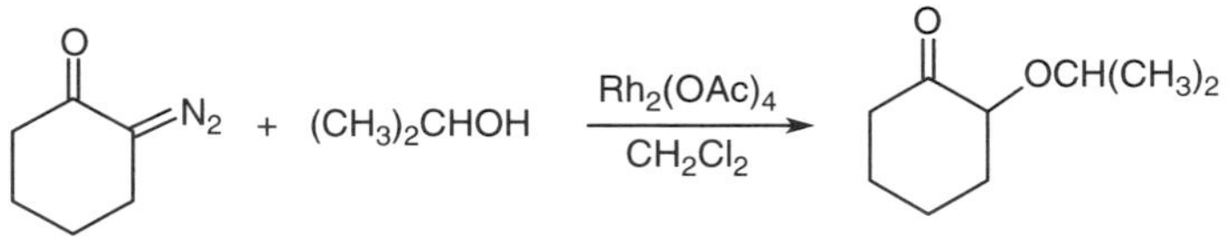

To a solution of isopropyl alcohol (0.45 g, 8 mmol) in dichloromethane (10 mL) containing rhodium(II) acetate (0.023 g, 0.08 mmol) at room temperature under nitrogen was added dropwise a solution of 2-diazocyclohexanone (0.99 g, 8 mmol) over 10 min. The reaction mixture was allowed to stir overnight, and the solution was then washed with water and dried over magnesium sulfate. After removal of the solvent, distillation of the crude product gave 2-isopropoxycyclohexanone (0.93 g, 75%) as a pale yellow oil. b.p. 55°C at 4 mm Hg. ^{1}H-NMR (CDCl$_3$): 0.95–1.24 (*m*, 6 H); 1.30–1.88 (broad *s*, 6 H); 2.11–2.59 (broad *s*, 2 H); 3.25–3.70 (*m*, 1 H); 3.74–4.05 ppm (broad *s*, 1 H). IR (film): 1730 cm^{-1}.

ment to a tertiary alcohol with an ee of 92%.[88] This product was subsequently used to supply the furanose component in a total synthesis of the naturally occurring indolocarbazole (+)-K252a. In general, competition between a double bond and the O–H function of an unsaturated alcohol for a metal carbene favors the O–H insertion product. The product distribution in the reaction of allyl alcohol with EDA (eq. 41) in Table 8.2 shows that changing the ligand in the rhodium(II) catalyst does not have a major influence on the outcome.[83] In the synthesis of (±)-azaascorbic acid (Scheme 11) *t*-butyl alcohol was both reactant and solvent with rhodium(II) acetate catalysis in the formation of a tetramic acid derivative.[89]

The fact that O–H insertion into rhodium carbenes occurs so readily under essentially neutral conditions in a nonpolar solvent, often at or below room tem-

TABLE 8.2 Product Yields for a Reaction of Allyl Alcohol with EDA

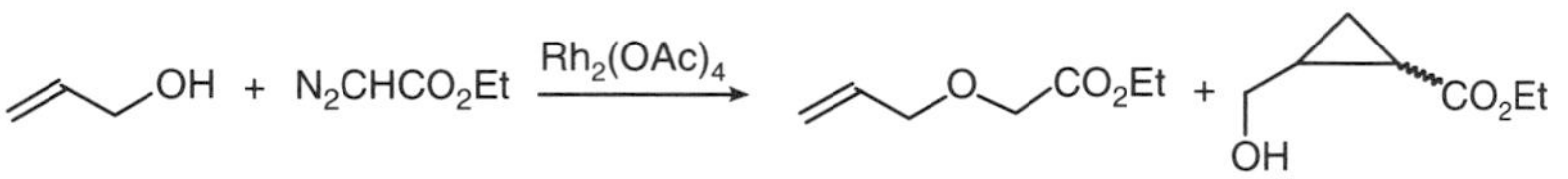

R in Rh$_2$(OCOR)$_4$	Overall yield (%)	Yield of **4** (%)	Yield of **5** (%)
CH$_3$	81	81	19
CH$_2$OCH$_3$	57	81	19
CF$_3$	70	91	9
Cu(II) triflate	73	92	8

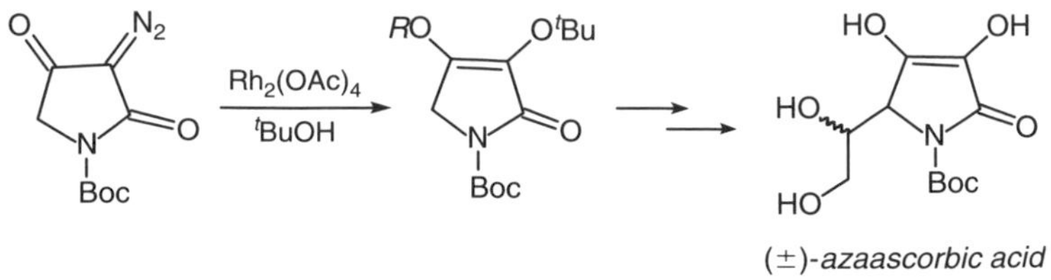

Scheme 11

perature, as compared with the alternative of Lewis or Brønsted acid catalysis, enhances further the value of the reaction, especially where stability of reactant or product may be a limiting factor. For example, Ganem and co-workers,[90,91] faced with the difficult task of attaching an enol pyruvate side chain to an already unstable cyclohexadienol (Scheme 12) in their synthesis of chorismic acid,

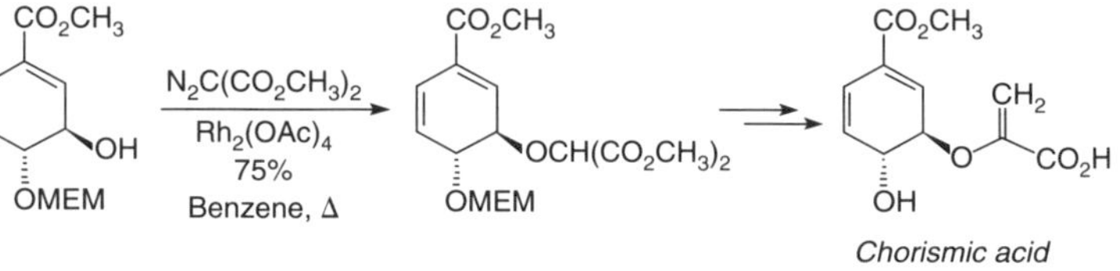

Scheme 12

found that treatment with diazomalonate in the presence of rhodium(II) acetate furnished the O–H insertion product, which was suitable for further elaboration into the natural product. The phosphonate analogues of chorismic acid have been synthesised *via* an approach involving the rhodium(II)-catalyzed insertion of ethyl diazophosphonoacetate into 2-cyclohexenol, and this is also the route by which Berchtold[92] and Bartlett[93] and their respective collaborators introduced the enol pyruvate side chain of shikimate-derived metabolites (eqs. 42 and 43). An-

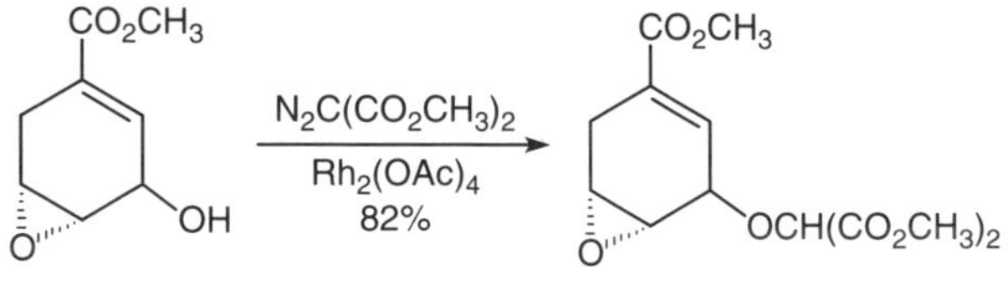

(42)

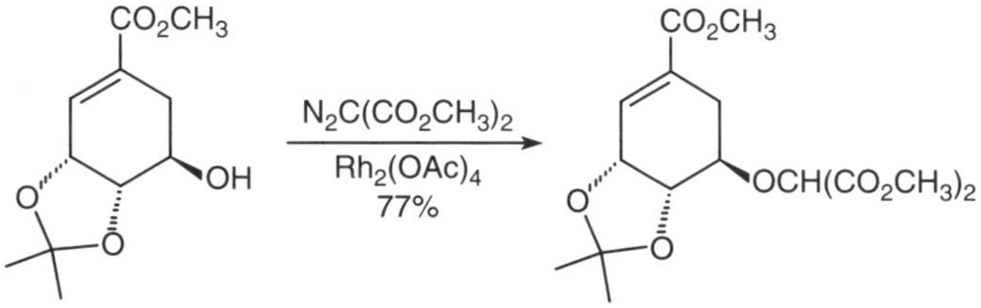

(43)

other aspect of catalyzed O–H insertion is its success in systems where steric inhibition might be expected to be a limiting factor, a case in point being that in eq. 44, where both reacting partners appear sterically crowded in the immediate

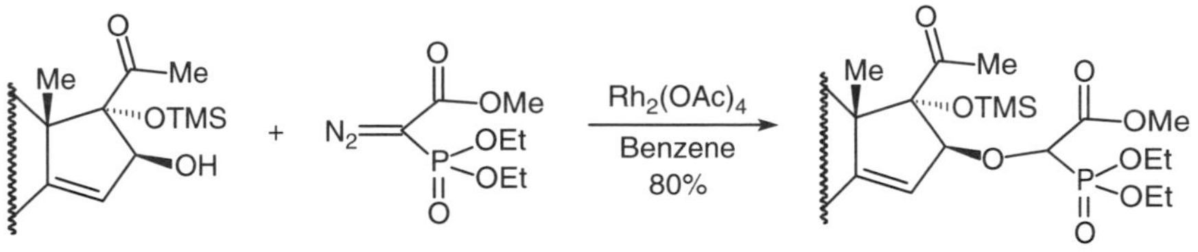

(44)

vicinity of the reaction sites[94] (eq. 44). In fact, with rhodium acetate in benzene at 80°C, this reaction proceeds to completion within hours. Among other examples of O–H insertion are those shown in Scheme 13[95] and in eqs. 45,[96] 46,[97] and

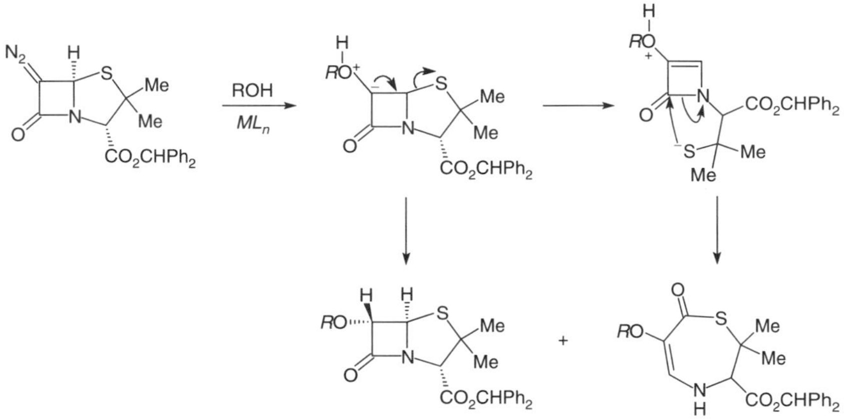

ROH	6	ML_n = Rh$_2$(OAc)$_4$	7
MeOH	55%		23%
EtOH	12%		75%
t-BuOH	6%		72%
PhCH$_2$OH	<5%		67%
CH$_2$ = CHCH$_2$OH	<5%		70%
		ML_n = Cu(acac)$_2$	
MeOH	56%		23%
EtOH	20%		29%
CH$_2$=CHCH$_2$OH	9%		56%

Scheme 13

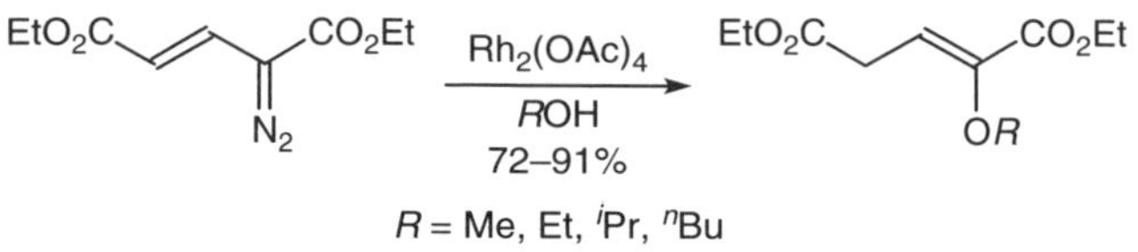

$$\text{(45)}$$

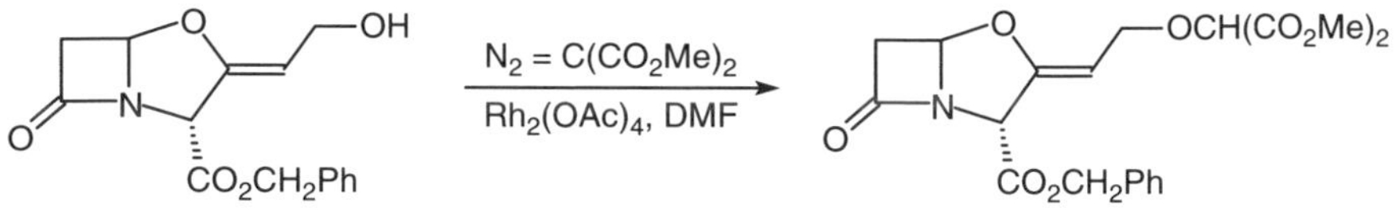

$$R = \text{Me, Et, }^i\text{Pr, }^n\text{Bu}$$

$$\text{(46)}$$

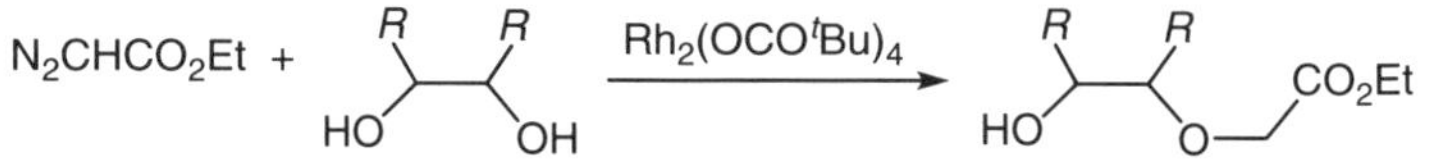

$$\text{(47)}$$

47.[98] The benzhydryl ester of 6-diazopenicillinate (Scheme 13) undergoes O–H in-sertion with a range of alcohols by rhodium(II) or copper(II) catalysis, though here there is competition with a cleavage–recyclization process to form a tetrahydro-1,4-thiazepine indicated by the arrows in Scheme 13, which is driven by release of strain on opening the β-lactam ring. The extent of O–H insertion with rhodium(II) catalysis is largest with methanol (55%), but this is exceptional, since for all other alcohol studies the tetrahydro-1,4-thiazepine is by far the preponderant product.[95] With copper(II) acetylacetonate catalysis O–H insertion is also fa-vored with methanol, but not with allyl alcohol. O–H insertion has been used to elaborate the side chains of sugar derivatives (eqs. 48 and 49).[99,100] Both reactions

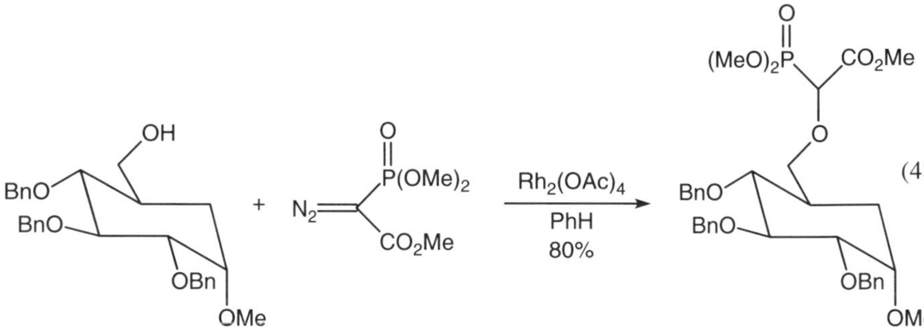

$$\text{(48)}$$

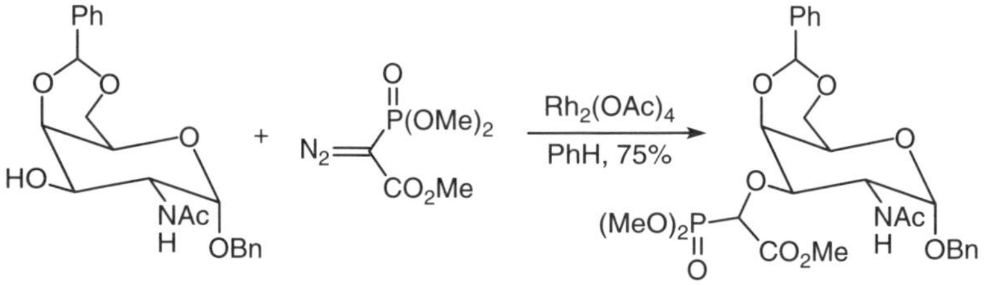

$$\text{(49)}$$

proceed in very good yield under rhodium(II) acetate catalysis in benzene; in the latter there was no competition from N–H insertion into the amide function.

Diazocarbonyl compounds with hydrogen substituents adjacent to the diazo group (β to the carbonyl group) are prone to undergo metal-catalyzed β-elimination (Chapter 12). One might expect therefore that with such substrates β-elimination would compete with other carbenoid processes, such as O–H insertion. We saw an example of this earlier with N–H insertion, but emphasized that the ratio of insertion to β-elimination is dramatically influenced by the carboxylate ligand on rhodium. A somewhat similar situation pertains with competing O–H insertion and β-elimination. Moody and co-workers[101] used methyl 2-diazo-3-(phenyl)propionate and methyl 2-diazo-3-(4-hydroxyphenyl)propionate, the latter prepared by diazotization of tyrosine methyl ester, with which to probe the incidence of β-elimination in O–H insertion reactions with a range of alcohols. The major byproduct in all cases was the cinnamate from β-elimination. To probe the elimination process more closely, Moody's group took the parent phenyldiazopropanoate and water (eq. 50) as the hydroxylic component with a range of rhodium(II) carboxylates (Table 8.3). The reactions were carried out

TABLE 8.3 Rhodium-Mediated O–H Insertion *vs.* Elimination Reactions of Diazoester

L	Ratio (O–H Insertion : H$\sim$)
2-PhC$_6$H$_4$CO$_2$–	94:6
2,4,6-Me$_3$C$_6$H$_2$CO$_2$–	91:9
1-C$_{10}$H$_7$CO$_2$–	89:11
9-C$_{14}$H$_9$CO$_2$–	88:12
MeCO$_2$–	82:18
2-HOC$_6$H$_4$CO$_2$–	64:36
CF$_3$CONH–	63:37
CF$_3$CO$_2$–	32:68
C$_3$F$_7$CO$_2$–	16:84

by dissolving the diazoester in ether saturated with water (6%), adding the catalyst (1–2 mol %), and stirring the mixture at room temperature. All catalysts promote both O–H insertion to mandelate and β-elimination to cinnamate. The cinnamate was formed largely as its Z isomer, and catalysts that gave the largest amount of cinnamate also produced the highest $Z:E$ ratios. The results, which range from 16.5:1 for OH insertion with the *o*-biphenyl based catalyst to 1.0:2.1 in favor of β-elimination with the perfluorobutyrate catalyst, demonstrate yet again the influence of catalyst on the chemoselectivity of carbenoid transformations. These results are consistent with Taber's hypothesis that intervention of a rhodium carbenoid bearing electron withdrawing ligands favors the less entropically demanding β-elimination. Bulkier, more electron-releasing ligands, on the other hand, favor the intermolecular reaction with water.

Shi and co-workers[102] have examined the special case of O–H insertion into trifluoromethyl-substituted carbenoids, hoping to develop a general method of synthesizing trifluoromethylated alkoxy and aryloxyacetic acid derivatives for which entirely satisfactory routes were not available. These workers, having developed a synthesis of ethyl trifluoromethyldiazoacetate (eq. 51), examined its

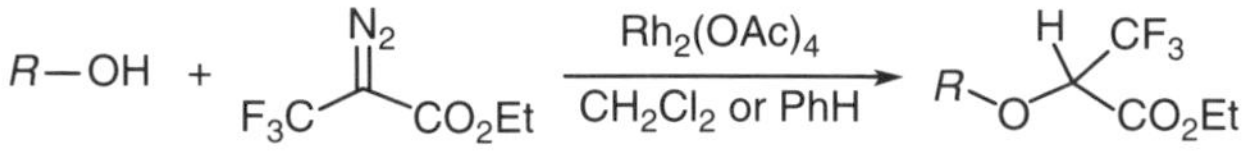

$$R-OH \ + \ \underset{F_3C}{\overset{N_2}{\underset{\|}{\longmapsto}}}\overset{}{CO_2Et} \ \xrightarrow[\text{CH}_2\text{Cl}_2 \text{ or PhH}]{\text{Rh}_2(\text{OAc})_4} \ R\overset{H}{\underset{O}{\overset{}{\bigvee}}}\overset{CF_3}{\underset{CO_2Et}{}} \tag{51}$$

reactions with a range of alcohols and phenols (2 mol equivalents) in benzene or dichloromethane in the presence of rhodium(II) acetate at room temperature (Table 8.4) and isolated the appropriate O–H insertion products in moderate to good yields. With unsaturated alcohols there was no evidence of competing cyclopropanation. With menthol there was a modest degree of asymmetric induction at the new stereogenic center to produce an 80:20 mixture of diastereoisomers. The last example in Table 8.4 refers to the use of a phenol in O–H insertion where moderate yields of α-aryloxytrifluoromethylacetates are produced. Ye and Burke[103] have used phenol O–H insertion to convert *N*-(9-fluoroenylmethoxycarbonyl)-tyrosine (eq. 52) into a malonyl derivative *via* reaction with an alkyl diazomalonate in benzene with rhodium(II) acetate catalysis.

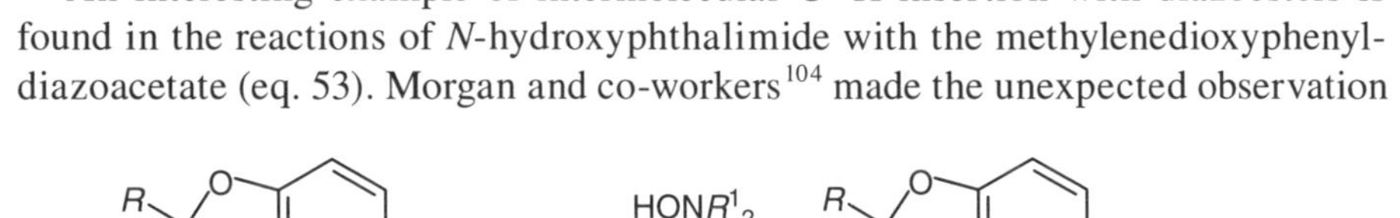

$$(52)$$

An interesting example of intermolecular O–H insertion with diazoesters is found in the reactions of *N*-hydroxyphthalimide with the methylenedioxyphenyl-diazoacetate (eq. 53). Morgan and co-workers[104] made the unexpected observation

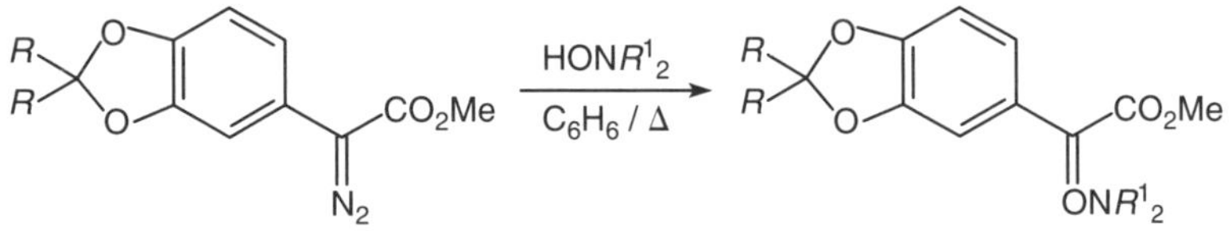

$$(53)$$

R	R^1	Product yield (%)
H	Et	49
H	Bu	61
H	Phth	57
CH_3	Phth	100

TABLE 8.4   Rhodium-Catalyzed Insertion of Metal Carbenes Derived from Trifluorodiazopropionate into O–H of Alcohols and Phenols

Entry	R–OH	Conditions	Product	Yield (%)
1	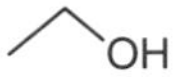	CH_2Cl_2/40°C	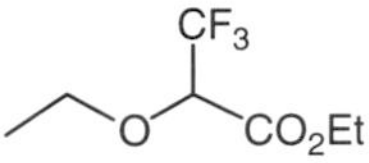	72
2	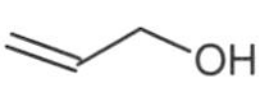	CH_2Cl_2/40°C	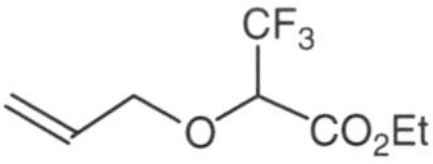	75
3	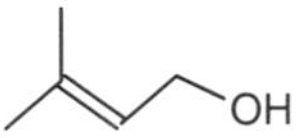	CH_2Cl_2/40°C	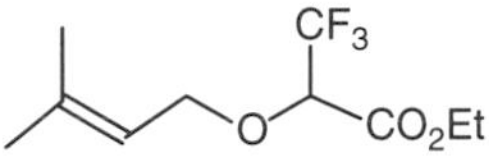	78
4	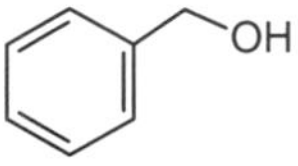	CH_2Cl_2/40°C	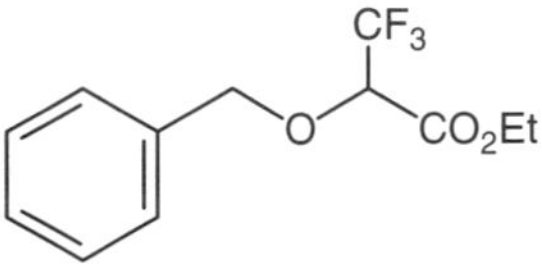	70
5	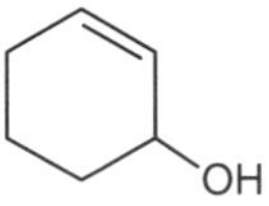	CH_2Cl_2/40°C	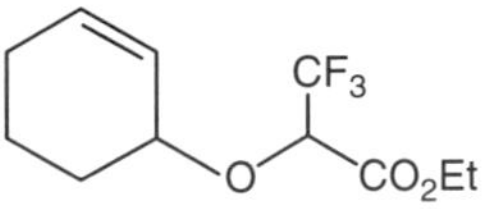	81
6	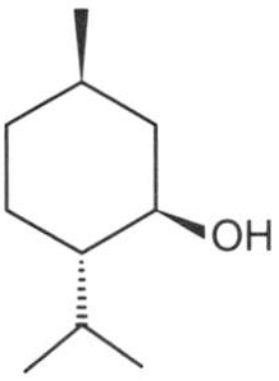	CH_2Cl_2/40°C	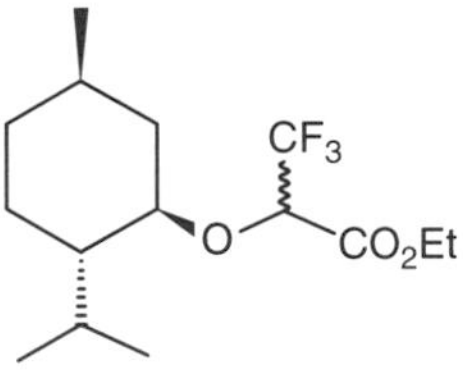	64
7	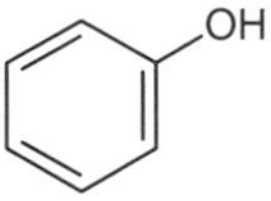	PhH/25°C	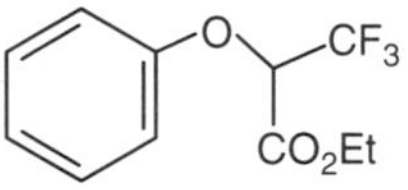	63
8	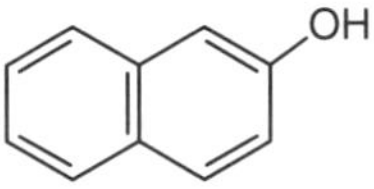	PhH/60°C	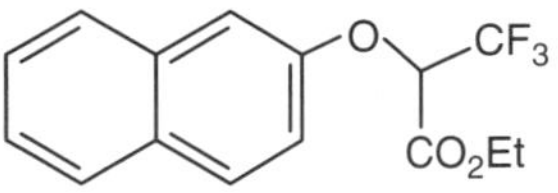	56
9	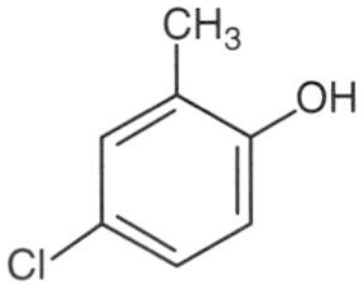	PhH/25°C	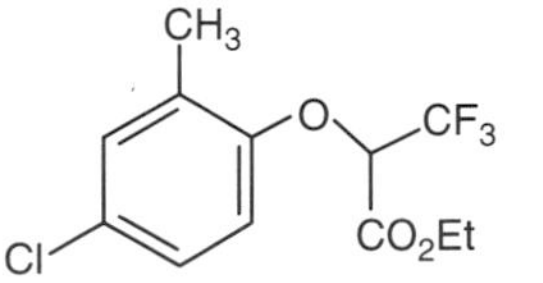	67

that this reaction was inhibited by the presence of rhodium(II) acetate, whereas a purely thermal process in benzene alone furnished the O–H insertion product in moderate to good yield. In contrast, α-diazophenylacetate was inert towards N-hydroxyphthalimide under identical conditions indicating that the oxygen substituents in the former substrate had a significant activating effect on the reaction and suggesting an ionic mechanism with a diazonium ion intermediate.

Finally, there are a few examples of 1,3-bis(diazo)ketones whose reactions with alcohols in the presence of rhodium(II) acetate have been studied. Interestingly, these reactions do not yield the expected 1,3-dialkoxyketones. For example, 1,3-bis(diazo)indan-2-one (eq. 54), on treatment with alcohols in dichlorometh-

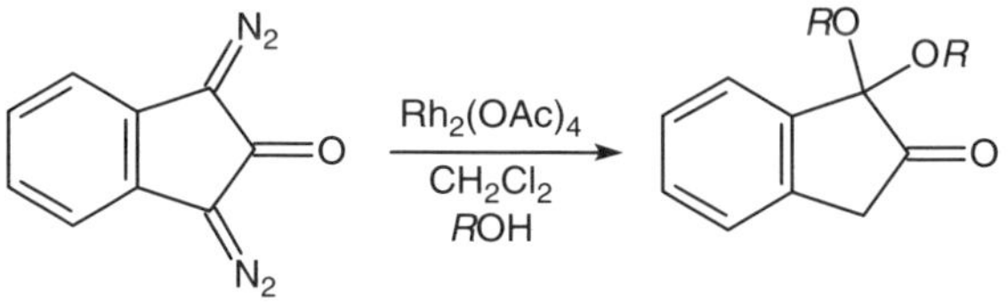

$$R = Me\ (68\%);\ R = Et\ (60\%);\ R = CH_2 = CHCH_2\ (20\%)$$

ane containing 2 mol % of rhodium(II) acetate, yield 1,1-dialkoxyindan-2-ones without formation of 1,3-dialkoxyindan-2-ones.[105]

8.3.2   Intramolecular O–H Insertion

Intramolecular versions of O–H insertion of alcohols are also known,[106] one representing a rare example of four-membered ring formation (eq. 55) in aqueous

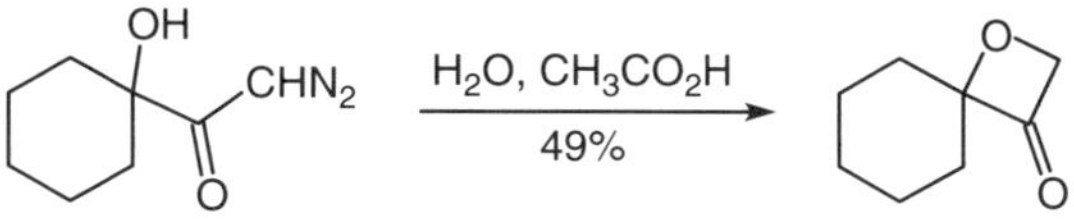

acetic acid. McClure *et al.*[107] found that L-phenylalanine could be transformed into the indenooxazinone, shown in Scheme 14, without racemization, cyclization

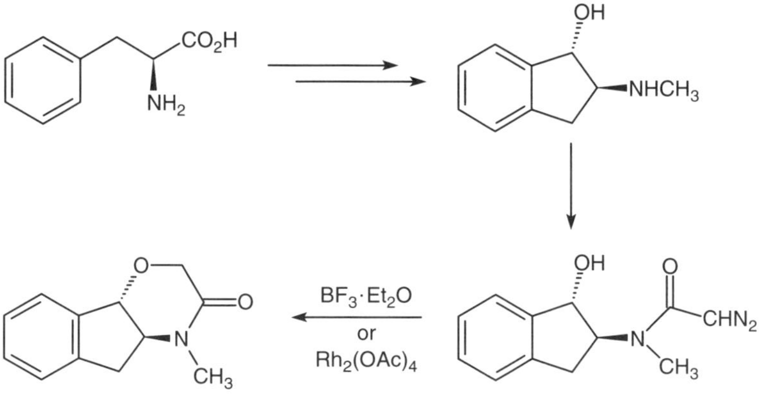

Scheme 14

of the intermediate hydroxy diazoamide having been accomplished by exposure to boron trifluoride etherate or $Rh_2(OAc)_4$.

The intramolecular O–H insertion reaction of carbenoids with alcohols is now well established as a useful route to cyclic ethers, especially some medium-ring and large-ring compounds that are otherwise difficult to prepare. In early examples of this approach to macrocycles, Kulkowit and McKervey[108] used $Cu(II)(acac)_2$ in benzene to catalyze the [1+1] union of $1,n$-oligoethylene glycols (eq. 56) and 1,12-dodecanediol (eq. 57) with α,ω-bis-diazoketones. In some of

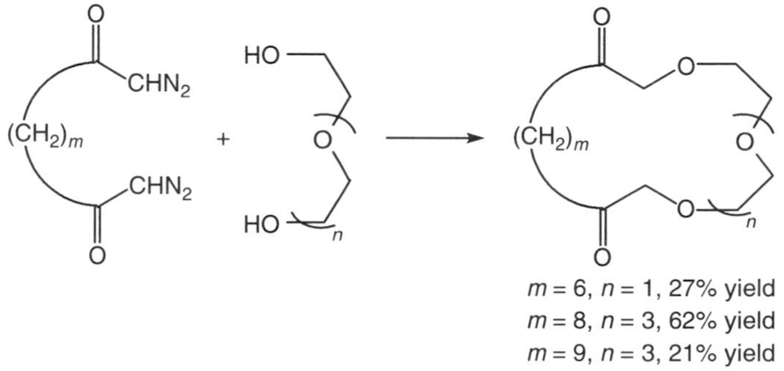

$$m = 6, n = 1, 27\% \text{ yield}$$
$$m = 8, n = 3, 62\% \text{ yield}$$
$$m = 9, n = 3, 21\% \text{ yield}$$

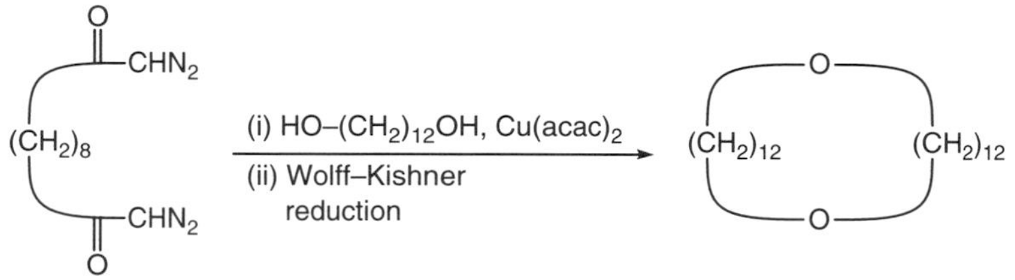

these cyclizations significant quantities of [2+2] dimeric products were also formed. The product (eq. 57) was transformed by Wolff–Kishner (WK) reduction into a 24-membered bis-ether. More recent studies show a preference for the use of rhodium(II) carboxylates, though these may not always be the most efficient, as McClure et al.[107] found in their study of the cyclization of the (diazoacetyl)aminoindanol in Scheme 14.

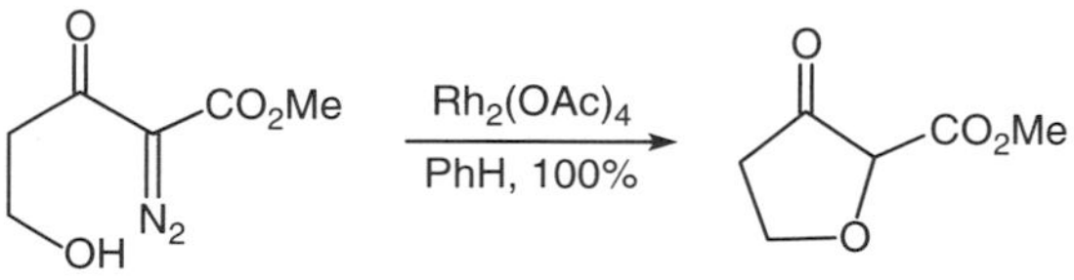

The application of the intramolecular O–H insertion to the synthesis of common- and medium-sized cyclic ethers was developed simultaneously by Moody[109] and Rapoport[16] and their respective collaborators. Rapoport demonstrated that the cyclization (eq. 58) could be brought about in benzene containing rhodium acetate to form a tetrahydrofuran derivative. Moody extended the process to include seven- and eight-membered rings[110] (with yields decreasing with increasing ring size). Several examples of the former are summarized in eq. 59;

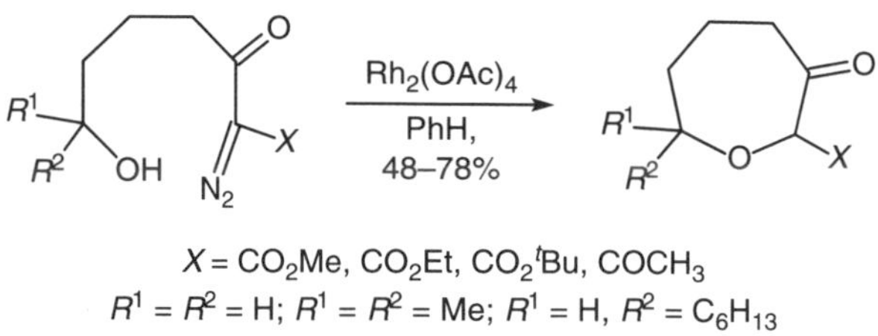

Rh2(OAc)4
PhH,
48–78%
X = CO2Me, CO2Et, CO2tBu, COCH3
R1 = R2 = H; R1 = R2 = Me; R1 = H, R2 = C6H13

(59)

in each case the acyclic precursor was synthesized by treatment of the appropriate lactone with a lithiodiazoester. Cyclizations were conducted in benzene containing rhodium(II) acetate. The reaction (eq. 60)[111] illustrates the formation of a seven-membered ring *via* phenolic O–H insertion, and eq. 61 shows eight-

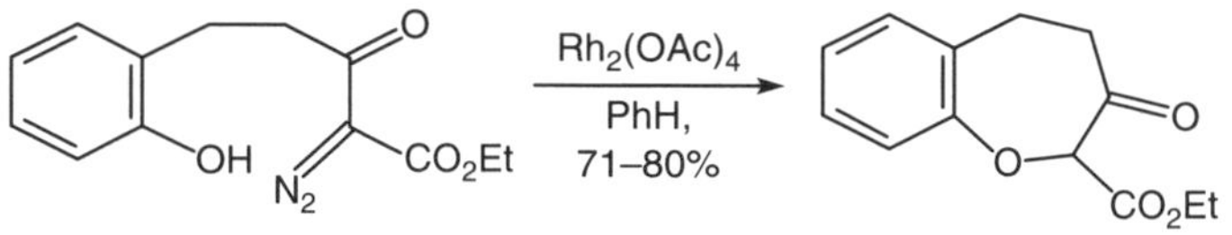

Rh2(OAc)4
PhH,
71–80%

(60)

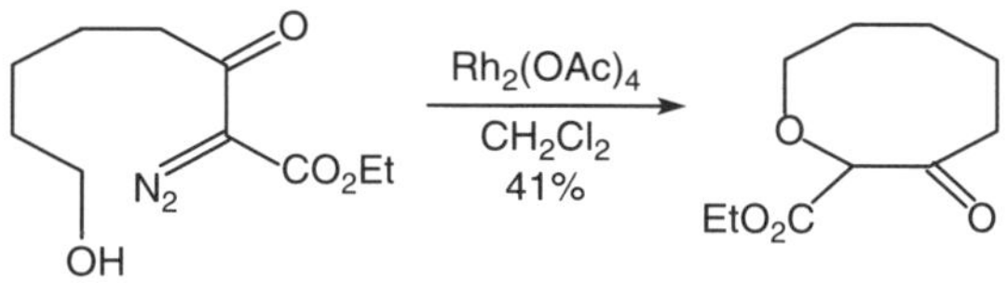

Rh2(OAc)4
CH2Cl2
41%

(61)

membered ring formation where yields vary depending on the nature of the alkyl residue in the ester group.[112] Moody's research[113] has also shown that the chemoselectivity for O–H insertion is high with rhodium(II) acetate in cases where competition from other carbenoid reactions might be expected. For example, for the three substrates in eq. 62, oxepanes were always the major product with rhodi-

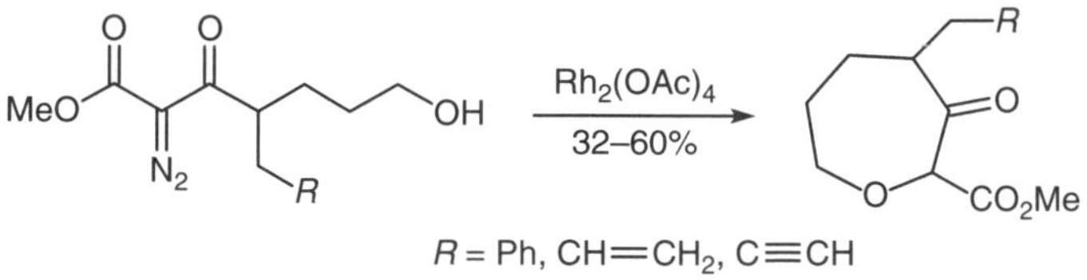

Rh2(OAc)4
32–60%
R = Ph, CH=CH2, C≡CH

(62)

um(II) acetate catalysis even when a suitably placed π-bond system was available for intramolecular cyclopropanation. However, with other rhodium(II) catalysts the situation was rather different. Thus, whereas the substrate in eq. 63 exhibited

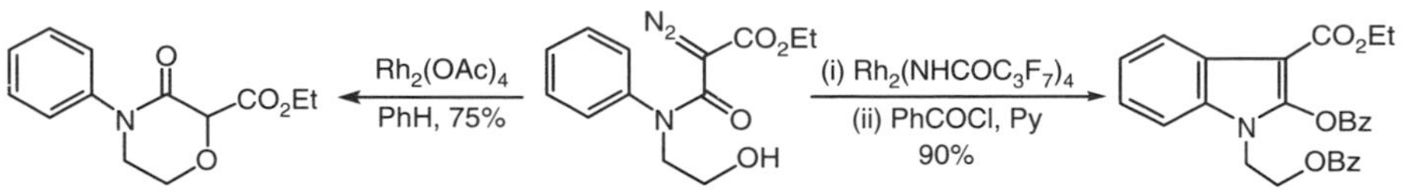

(63)

a very high preference for O–H insertion with rhodium(II) acetate, with rhodium(II) perfluorobutyramidate there was an equally high preference for aromatic substitution. This very favorable selectivity associated with the acetate catalyst enabled Moody to effect O–H insertion in some quite complex substrates. Thus various highly substituted oxepanes related to the diterpene zoapatanol (Scheme 15)[114] could be prepared, as could the cis-2,7-disubstituted oxepane skeleton of the marine natural product isolaurepinnacin (Scheme 16)[115] and several

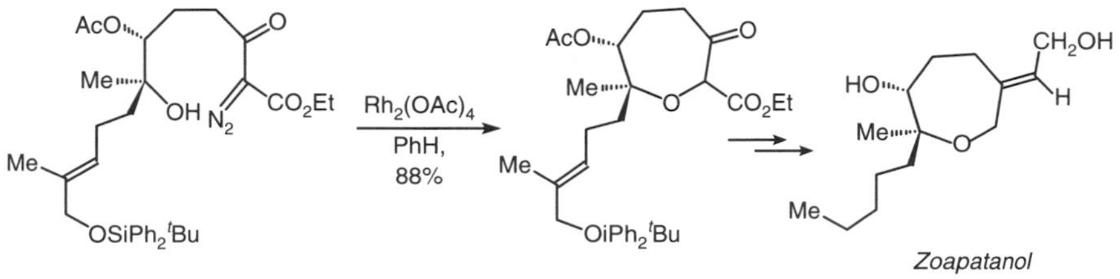

Scheme 15

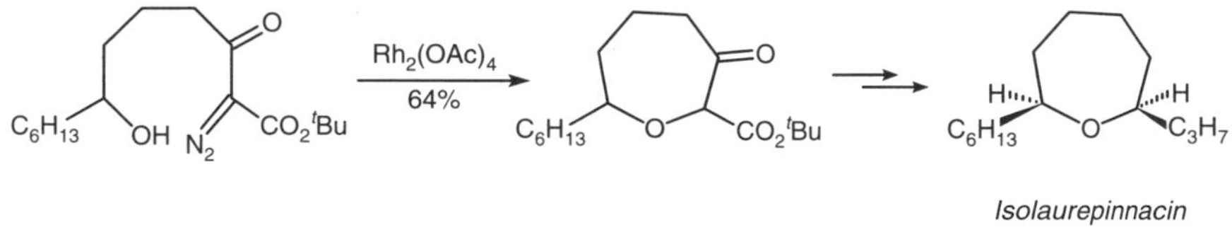

Scheme 16

pyrano-oxepane and oxepano-oxepane fused subunits of marine polyether toxins. Two final examples of O–H insertion leading to a bicyclic β-lactam derivative (eq. 64)[116] and a sugar derivative (eq. 65)[117] illustrate the application of this methodology to six-membered ring formation.

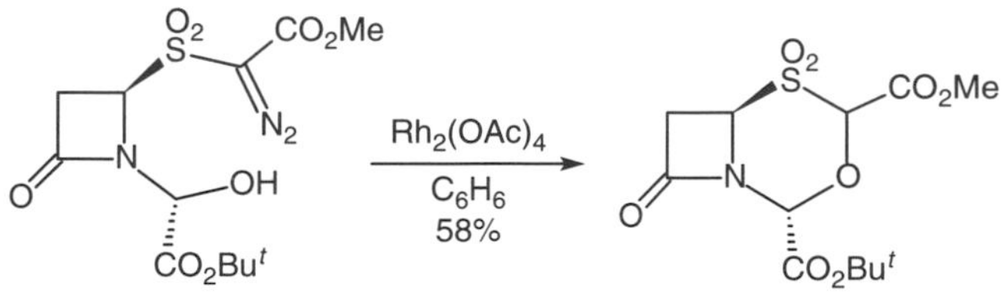

(64)

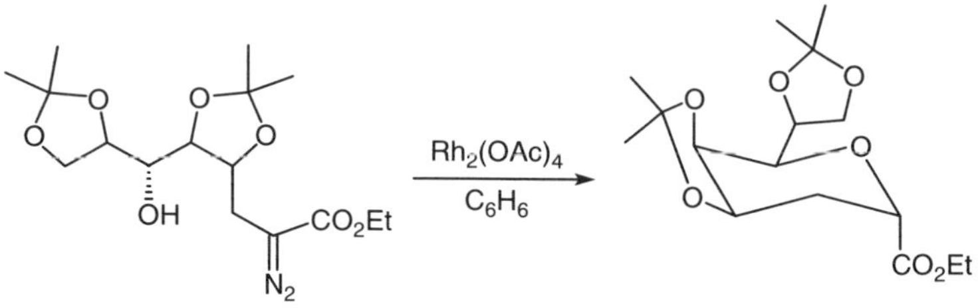

(65)

8.3.3   Asymmetric O–H Insertion Reactions

Despite the potential of asymmetric O–H insertion as an enantioselective route to chiral α-alkoxy carboxylic acid derivatives, there has been little successful research activity in the use of chiral catalysts and no reports of significant levels of enantiocontrol. The alternative chiral auxiliary approach has, however, been examined in some detail, but with limited success, by Moody and co-workers,[118] who prepared an entire series of phenyldiazoacetates (entries 1–5 in Table 8.5) derived from enantiopure (−)-borneol, (+)- and (−)-menthol, (+)-8-phenylmenthol, (+)- and (−)-2-phenylcyclohexanol, and (−)-10-(dicyclohexylsulfamoyl)-D-isoborneol, and studied their reactions with water, methanol, 2-propanol, and *t*-butyl alcohol in the presence of rhodium(II) acetate (eq. 66).

Reactions were conducted in dichloromethane with 2 molar equivalents of alcohol with respect to the diazoester, except those involving water which were carried out in diethyl ether that had been presaturated with water. In all cases, the O–H insertion products were isolated in good yields as mixtures of diastereoisomers, which could be separated by chromatography. The results show that the 8-phenylmenthyl, 2-phenylcyclohexyl, and 10-(dicyclohexylsulfamoyl)-isobornyl auxiliaries gave the best levels of diastereoselection, significantly better than the bornyl or menthyl systems, with the highest individual value achieved for the 8-phenylmenthyl ester (entry 3) and *t*-butyl alcohol (53% de). It seems reasonable to assume that the diastereoselectivity arises from attack of the nucleophilic alcohol or water on the electrophilic rhodium carbene from the face opposite the large group of the auxiliary, as shown in Scheme 17, leading to the observed configura-

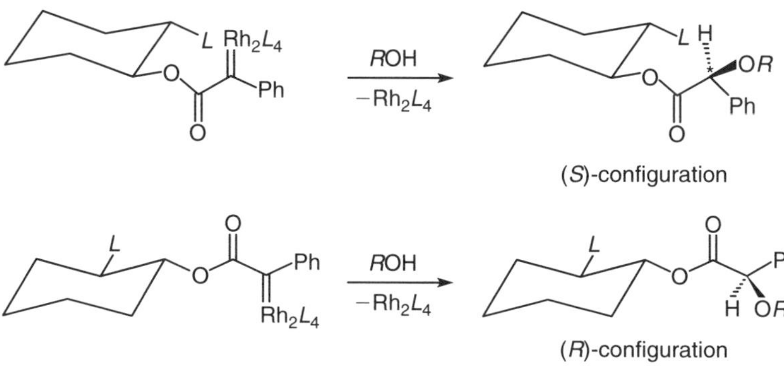

Scheme 17

TABLE 8.5   Diastereoselection in O–H Insertion of Alcohols and Chiral Phenyldiazoacetates

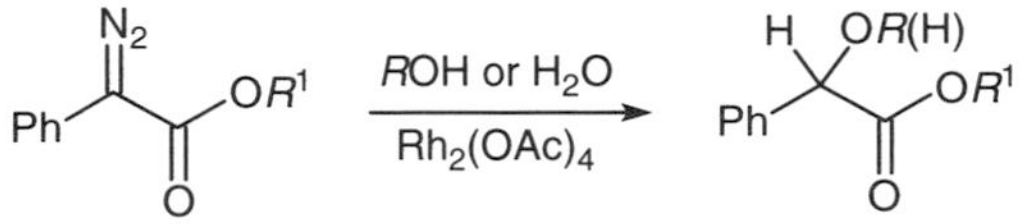

$$\text{(66)}$$

Diazoester entries	R^1	ROH	Yield (%)	dr	Major isomer
1	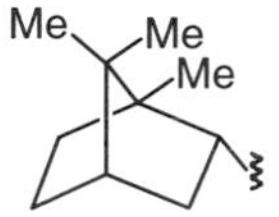 Me Me Me (structure)	MeOH	95	52:48	n.d.[b]
2	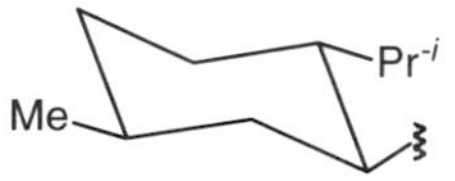 Me, Pr-i (structure)	H_2O	84	50:50	n.d.[b]
		MeOH	75	54:46	S
		i-PrOH	82	62:38	S
3	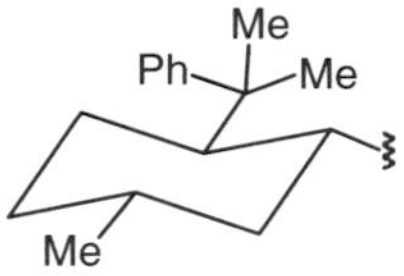 Me Ph Me Me (structure)	H_2O	79	66:34	R
		MeOH	63	72:28	R
		i-PrOH	85	68:32	R
		t-BuOH	40	76:24	(R)[a]
4	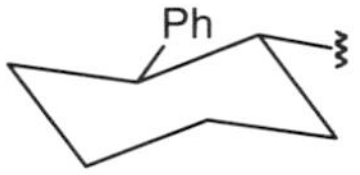 Ph (structure)	H_2O	85	75:25	S
		i-PrOH	71	71:29	R
5	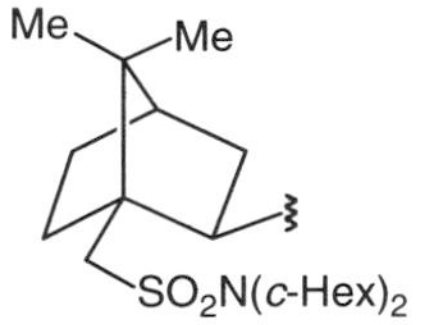 Me Me SO₂N(c-Hex)₂ (structure)	MeOH	98	66:34	R
		i-PrOH	82	74:26	R
		t-BuOH	37	75:25	(R)[a]

[a]Where the stereochemistry at the α-carbon of the major diasteromer is given in parentheses, this is inferred by analogy with that resulting from other nucleophiles used.
[b]Not determined.

tion at the α-carbon atom of the major diastereoisomer. The metal carbene was assumed to have retained the preferred *anti* arrangement about the ester O–CO bond and the (S)-*anti* conformation of its diazocarbonyl precursor, the arrangement adopted in the solid state as determined by X-ray analysis. On this basis, use of the 8-phenylmenthyl auxiliary should lead to better diastereoselection than that produced by menthol, which is indeed the case. The model (Scheme 17) predicts that the (+)-menthol-based diazoester (entry **2**) should have the (S) configuration at the new stereogenic center, whereas the diazoesters in entries **3** and **4** should have the (R) configuration at the new stereogenic center. An alternative interpretation is that the observed stereochemistry is established at a later metal-free stage in the process under the control of the chiral auxiliary. However, Moody and co-workers[118] conducted an additional study of the effect of catalyst on the reaction of diazoester (entry **2**) with isopropyl alcohol using two rhodium(II) carboxylates, three rhodium(II) carboxamidates, and five copper(II) catalysts, including Cu(acac)₂, several of which are known to be active in O–H insertion, and found that none proved superior to rhodium(II) acetate. Rhodium(II) 9-anthracenecarboxylate, for example, returned a dr of only 51:49 (71% chemical yield) for the reaction of diazoester (entry **2**) with 2-propanol. Thus the fact that the nature of the rhodium ligand does influence the diastereoselectivity may indicate the involvement of the metal does persist up to the stage at which the new sp^3 center is established.

8.4 O–H INSERTION REACTIONS WITH CARBOXYLIC ACIDS, CARBOXYLIC ESTERS, AND SULFONIC ACIDS

Carboxylic acids react with α-diazoketones on warming to form α-keto esters. Although catalysts are usually considered unnecessary, copper(II) salts, rhodium(II) carboxylates, or boron fluoride etherate are often added to accelerate the reaction, competition from the Wolff rearrangement being virtually eliminated under these conditions. Bradley and Robinson[119] observed the formation of phenacyl acetate from diazoacetophenone in glacial acetic acid (eq. 67). In a series of pa-

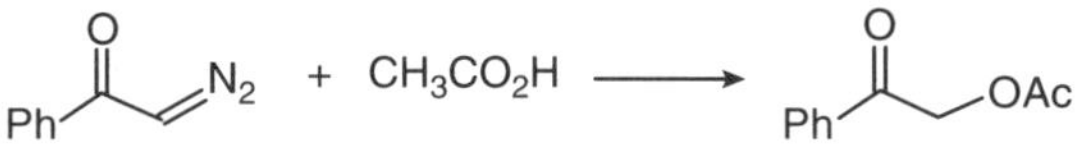

$$\text{(67)}$$

pers Wolfrom and co-workers[120] described the reactions of diazoketones derived from acylated sugars with glacial acetic acid. The process overall was described as a one-carbon (from diazomethane) homologation of sugars, for example, the conversion of the hexose carboxylic acid in Scheme 18 into the diazoketone and

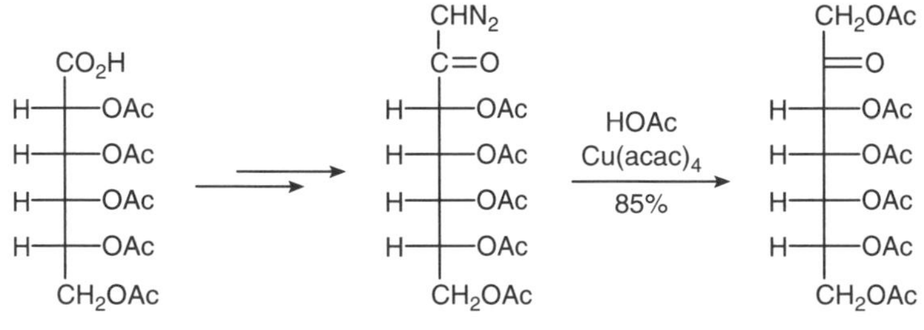

Scheme 18

hence by treatment with glacial acetic acid into a ketoheptose. In the latter step, copper bronze or copper(II) acetylacetonate was employed as catalyst.

Similar intramolecular reactions are encountered in suitably constituted diazo esters where carboxylic acid participation leads formally to O–H insertion products. Alternatively, these processes can be interpreted as taking place in a stepwise manner with hydrolysis of the ester preceding cyclization. A recent example of carboxylate participation is illustrated in Scheme 19[121] where (S)-aspartic acid,

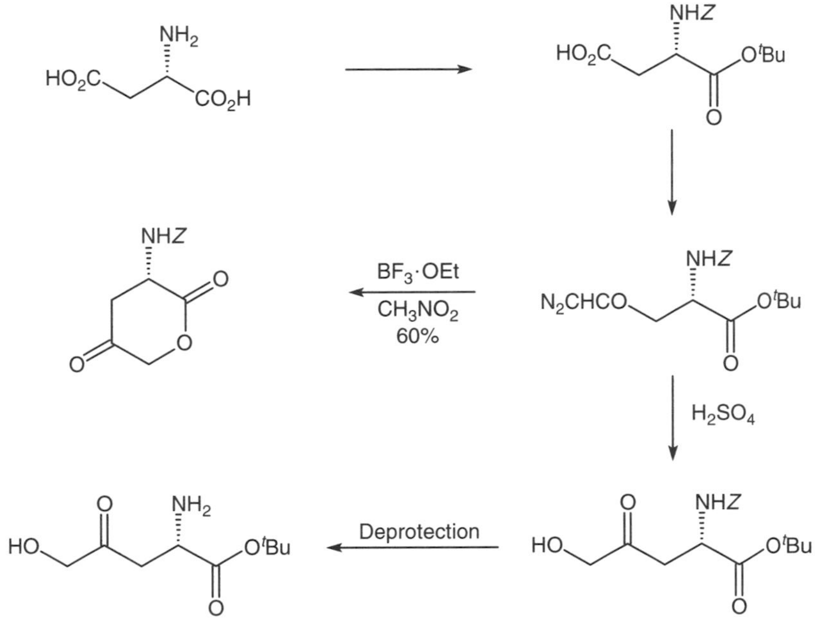

Scheme 19

doubly protected, was transformed into a diazoketone. Exposure of the latter to boron trifluoride etherate in nitromethane furnished the Z-protected aminoketolactone in 60% yield. In dilute sulfuric acid hydration of the diazoketone occurred without cyclization to form a hydroxyketo derivative that has been used to synthesize the naturally occurring antibiotic HON (2-amino-5-hydroxy-4-oxopentanoic acid).[121]

Three further examples of carboxylic ester participation are shown in eqs. 68 and 69 and Scheme 20. The first illustrates the HCl-catalyzed cyclization of a diazoester to a butyrolactone system[122]; that in eq. 69 illustrates the use of boron trifluoride etherate to form a similar product[123]; in the third process in Scheme 20

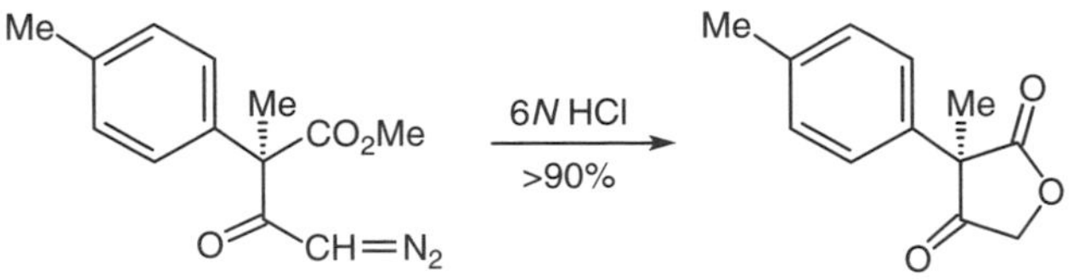

(68)

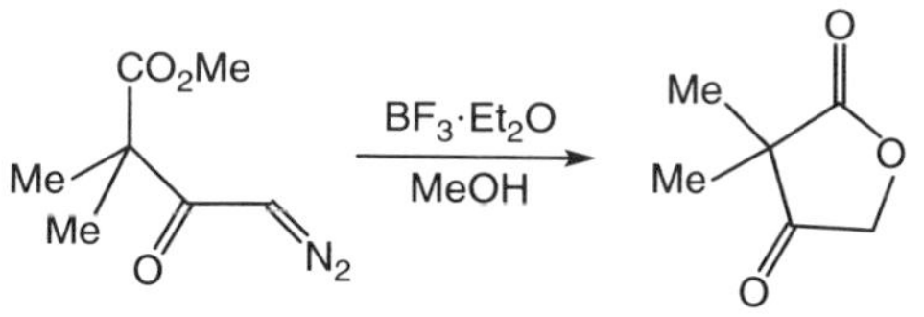

(69)

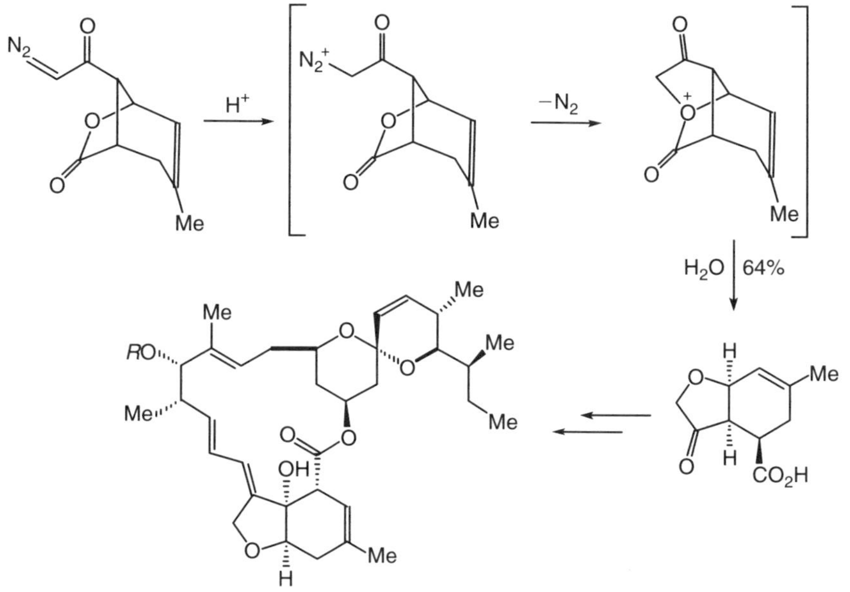

Scheme 20

the hexahydrobenzofuran moiety of the antiparisitic agent avermectin B_{1a} was constructed from a bicyclic diazoketone lactone, which on treatment with 10% sulfuric acid in dioxane rearranged to a *cis*-fused furanone.[124] The *cis* relationship between the diazocarbonyl function and the γ-lactone bridge presumably facilitates the formation of a tricyclic oxonium ion; attack by water completes the furanone formation.

The reaction of diazoketones with sulfonic acids constitutes a convenient, rapid route to α-ketosulfonates. Methane- and ethanesulfonic acid, trifluoromethanesulfonic acid, camphorsulfonic acid, and a variety of substituted benzenesulfonic acids have all been employed in reactions that generally proceed rapidly at room temperature without external catalysis.[125-131] The transformation of a cyclohexyldiazoketone into the arenesulfonate (eq. 70) on treatment with benzenesulfonic acid represents a recent application.[125] Other examples include those in eqs. 71–73.[126-132] A variation on this theme is the α,α-substitution of *p*-toluenesulfonic acid with the diazosulfone (eq. 74)[126] in a novel synthesis of sul-

(S)-3-N-(PHTHALIMIDO)-4-PHENYL-1-(METHANESULFONYLOXY)BUTAN-2-ONE[128]

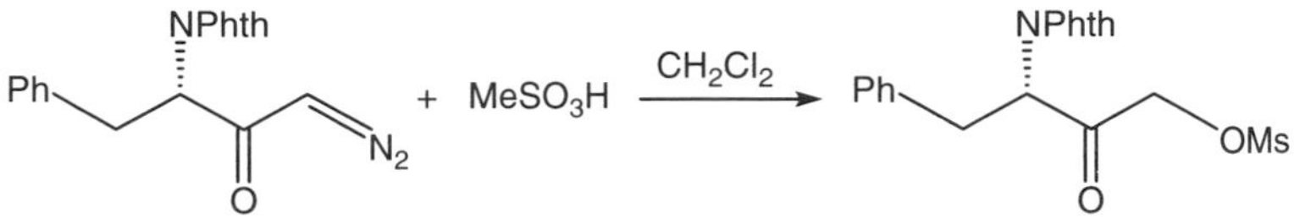
NPhth
Ph
N2
O
+ MeSO3H
CH2Cl2
NPhth
Ph
OMs
O

A flask at 0°C was charged with the *N*-protected diazoketone (0.55 g, 1.7 mmol) in dry ether (50 mL). The solution was magnetically stirred under an atmosphere maintained dry by self-indicating silica beads, and to it was added methanesulfonic acid (1.3 mL, 1.1 equiv.) in one portion. The reaction, which was monitored by TLC, was complete in less than 6 h. The ether was then washed with water (1 × 50 mL), then brine (1 × 50 mL), dried, and the solvent was removed under reduced pressure to give the crude product. Purification was effected by recrystallization from ether-dichloromethane to give the pure α-methanesulfonyloxyketone (0.62 g, 92%) as a white solid, m.p. 168–169°C. $[\alpha]_D^{20}$ −193.7 (c = 10 in CH_2Cl_2). IR (KBr): 1750 (ketone CO); 1710 (Phth. CO); 1380 cm^{-1} (-SO_2CH_3). ^{1}H-NMR ($CDCl_3$): 3.03 (3 H, *s*, SO_2CH_3); 3.40 (2 H, *d*, CH_2Ph); 4.85 (2 H, *s*, CH_2OSO_2); 5.10 (1 *H*, *m*, *CH*); 7.04 (5 H, *s*, Ar*H*); 7.70 (4 H, *s*, Phth.).

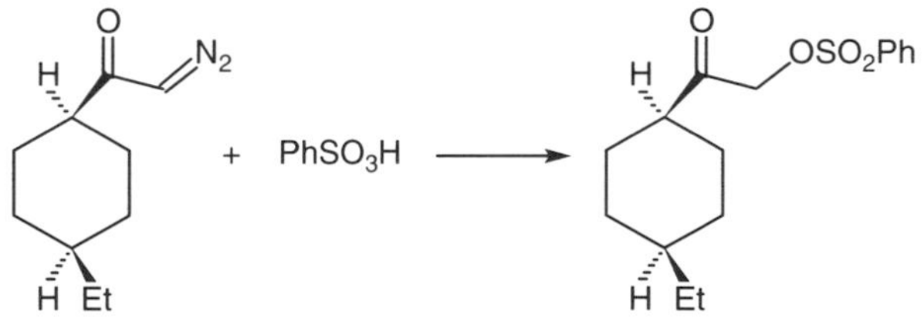
O
H
N2
H Et
+ PhSO3H
O
OSO2Ph
H
H Et

(70)

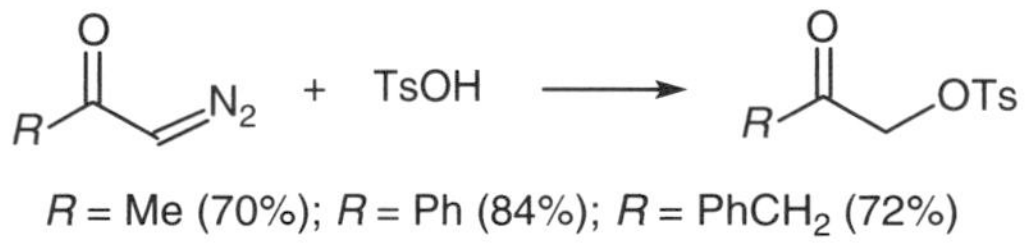
O
R
N2
+ TsOH
O
R
OTs
R = Me (70%); R = Ph (84%); R = PhCH2 (72%)

(71)

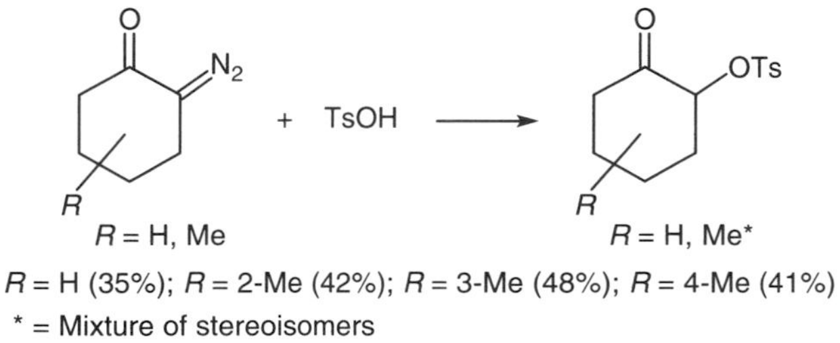
O
N2
R
R = H, Me
+ TsOH
O
OTs
R
R = H, Me*
R = H (35%); R = 2-Me (42%); R = 3-Me (48%); R = 4-Me (41%)
* = Mixture of stereoisomers

(72)

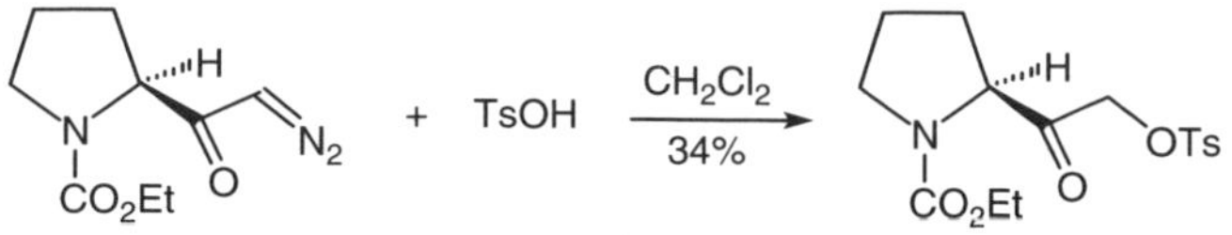

$$\text{(73)}$$

$$PhCH_2SO_2CHN_2 + TsOH \xrightarrow[\text{64\%}]{CH_2Cl_2} PhCH_2SO_2CH_2OTs \qquad (74)$$

fone α-tosylates, mixing the reactants at room temperature in dichloromethane being sufficient to bring about reaction. The diazoketone - *p*-toluenesulfonic acid reaction has been used to generate tosylates suitable for photochemical production of α-keto carbonium ions.[127] The reaction of chiral diazoketones with (+)-camphorsulfonic acid has been used to assess enantiomeric purity; the reaction of the alanine-derived diazo ketone (eq. 75) provides an illustrative example.[128]

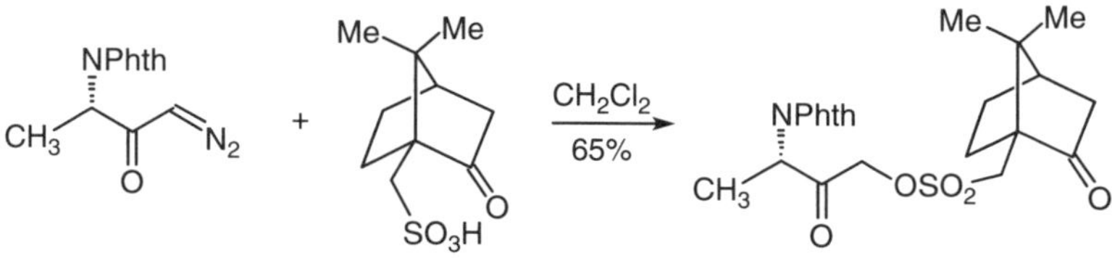

$$\text{(75)}$$

α-Hydroxyketone phosphates can be synthesized by reaction of diazoketones with dibenzyl phosphate. This transformation was the key step in Whiteside's[133] synthesis of analogues of dihydroxyacetone phosphate. Treatment of the optically active diazoketone with dibenzyl phosphate afforded the protected phosphate. Deprotection *via* hydrogenation over palladium gave the keto phosphate (eq. 76).

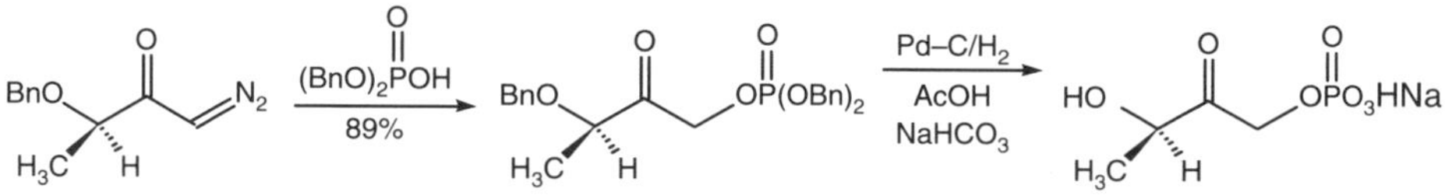

$$\text{(76)}$$

8.5 S–H INSERTION

S–H insertion of diazocarbonyls with thiols represents a useful way of placing sulfur-containing substituents adjacent to carbonyl groups in ketones and esters. Photolysis of EDA in ethanolic ethanethiol is an early example of this process, with carbene insertion into the S–H bond producing ethyl (ethylthio)acetate (eq. 77) in moderate yield.[134] The same process can also be brought about ther-

$$N_2CHCO_2Et + CH_3CH_2SH \xrightarrow[\text{ethanol}]{h\nu} CH_3CH_2SCH_2CO_2Et \qquad (77)$$

mally using AIBN as a radical initiator, under which conditions the sulfenylated product is obtained in very high yield.[134] The photochemical route to S–H insertion was used by Sheehan and co-workers[135] to produce 6β-(phenylthio)penicillanate as the major product of the reaction of thiophenol with 6-diazopenicillanate in carbon tetrachloride (eq. 78). A short time later Giddings and Thomas[73]

(78)

made the contrasting observation that boron trifluoride–catalyzed addition of benzylthiol to the same substrate produced a single adduct with the 6α configuration (eq. 78).

Metal-catalyzed insertion of diazocarbonyl compounds with thiols was first investigated by Yates,[6] who used copper to catalyze the reaction of thiophenol with diazoacetophenone (eq. 79). Although S–H insertion was observed, the need

$$\text{PhCOCHN}_2 \ + \ \text{PhSH} \ \xrightarrow[\text{67\%}]{\text{Cu}} \ \text{PhCOCH}_2\text{SPh} \qquad (79)$$

to use thiophenol as solvent or in considerable excess diminishes the practicality of this approach to ketone sulfenation. Very little subsequent work was done in this area until the Belgian group[136] in 1974 reported the high catalytic activity of rhodium(II) acetate towards reaction of thiophenol with EDA in nonpolar solvents such as dichloromethane or benzene, a discovery that coincided with a more general awareness of the versatility of α-sulfenylated carbonyl compounds in synthesis.[137] Subsequently, Ratananukul and McKervey[138] conducted a systematic study of the reaction of several acyclic and cyclic diazoketones with thiophenol in benzene and found that rhodium(II) acetate catalyzes the formation of α-phenylthioketones in good to excellent yield at room temperature. A selection of examples is illustrated in Tables 8.6 and 8.7. These include terminal mono- and bis-diazoketones (Table 8.6, entries 1–3, 6), diazoketones bearing additional sulfur-containing substituents (Table 8.6, entries 7–9), diazoketones derived from amino acids (Table 8.6, entries 10 and 11), and cycloalkyl diazoketones (Table 8.7). The nonterminal diazoketone in Table 8.6 (entry 5), prepared from the appropriate acid chloride and diazoethane, produces the adduct shown, thus demonstrating the regiochemical potential of this route to unsymmetrically sulfenylated ketones, which would be difficult to obtain by conventional enol or enolate chemistry with the parent ketone. More recent examples of the use of rhodium(II) acetate for intermolecular S–H insertion include the reactions shown in eqs. 80 and 81. The vinyl diazoester (eq. 80) undergoes S–H insertion with thiophenol to produce the allylic sulfide contaminated with an unidentified

TABLE 8.6   Products of Rhodium(II) Acetate–Catalyzed Reaction of Acyclic Diazoketones with Thiophenol in Benzene

Entry	Diazoketone	Thiophenol	Yield (%)
1	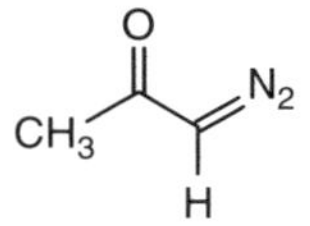	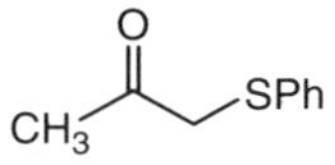	84
2	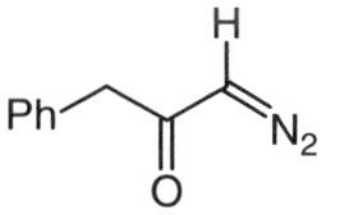	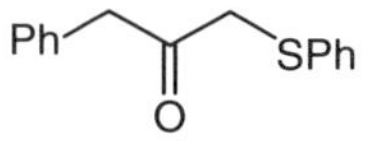	76
3	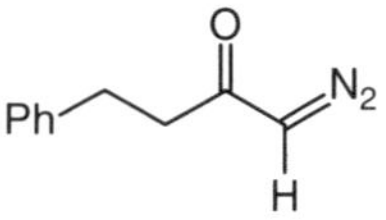	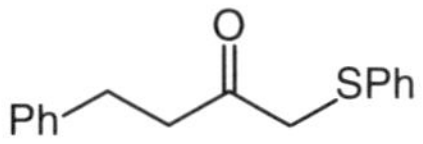	73
4	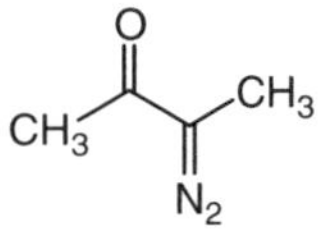	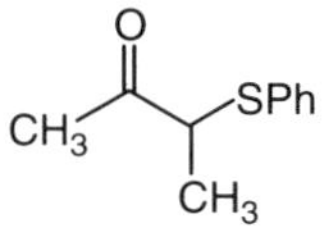	83
5	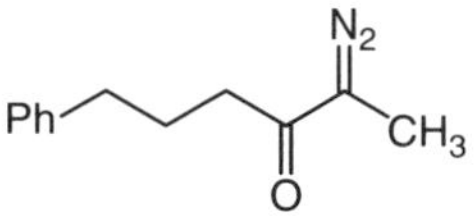	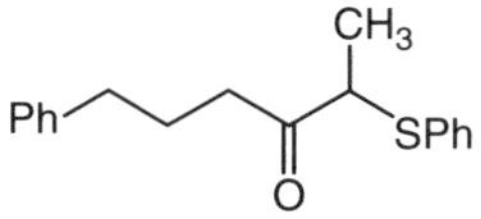	71
6	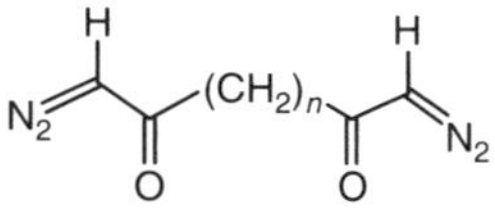 n = 4-8, 10, 12	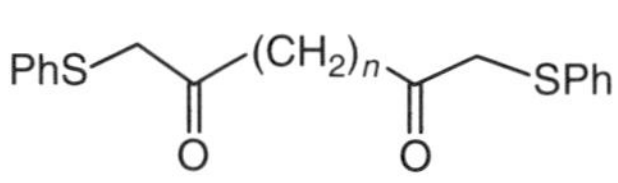	81–92
7	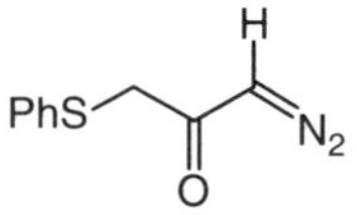	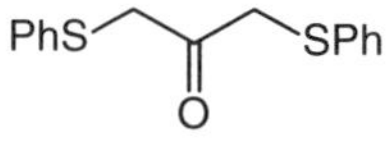	82
8	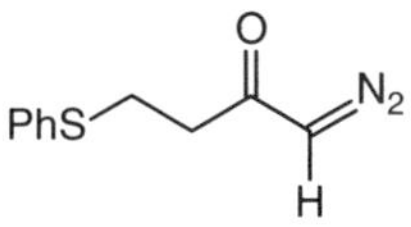	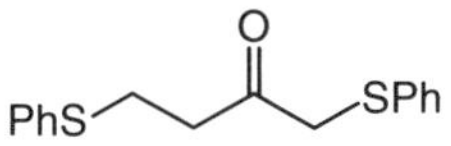	77
9	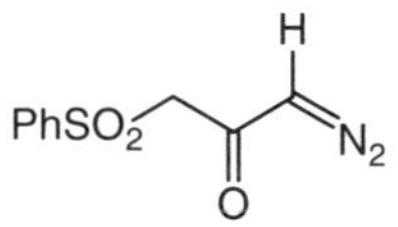	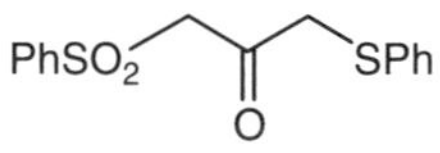	45
10	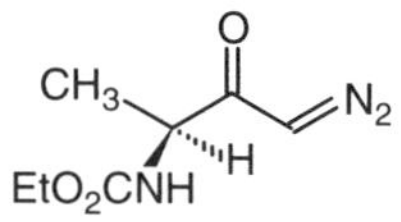	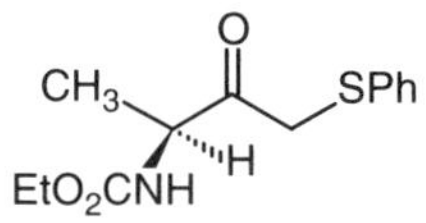	—
11	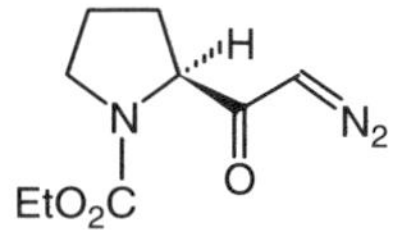	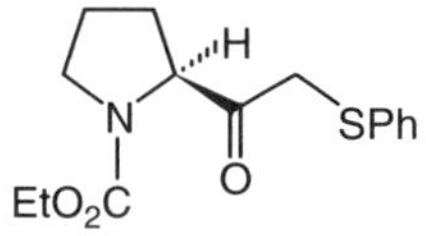	—

TABLE 8.7 Products of Rhodium(II) Acetate–Catalyzed Reaction of Cyclic Diazoketones with Thiophenol in Benzene at 20°C

Entry	Cyclic Diazoketone	Product	Yield (%)
1	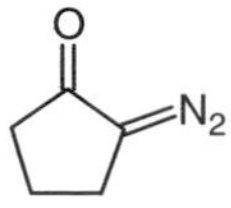	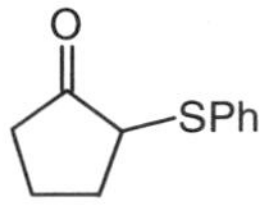	76
2	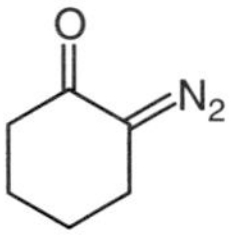	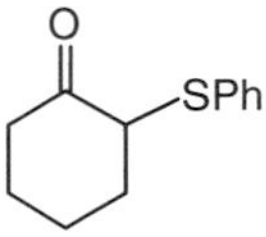	78
3	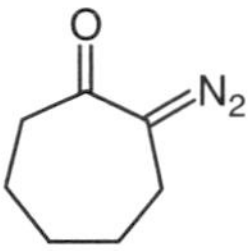	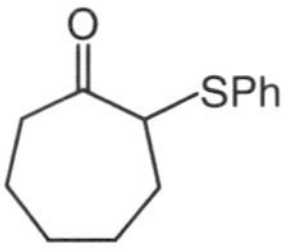	75
4	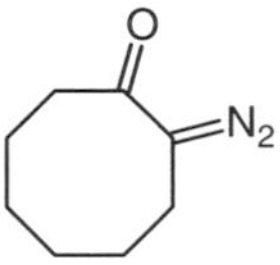	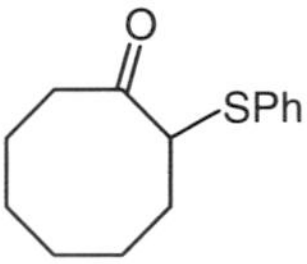	73
5	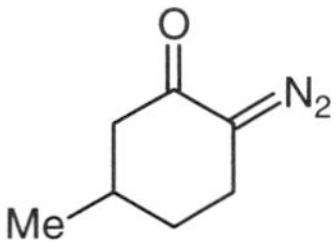	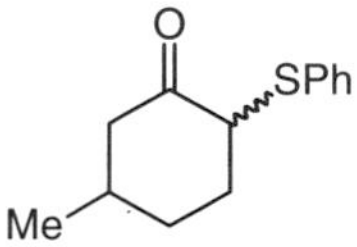	76
6	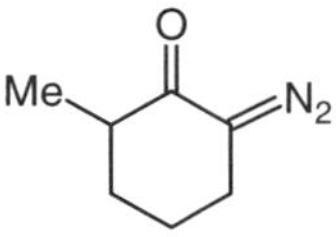	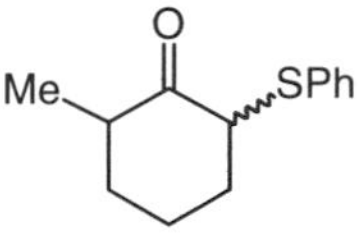	72

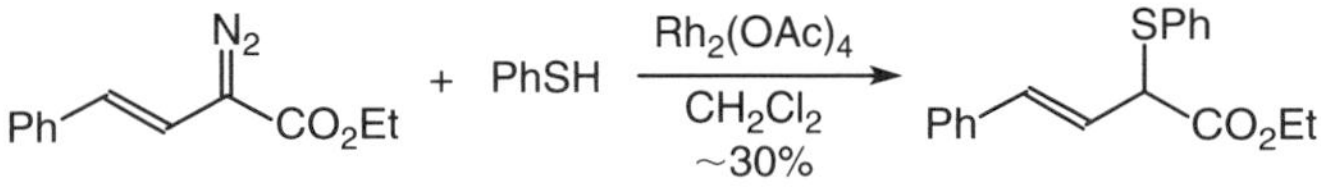

(80)

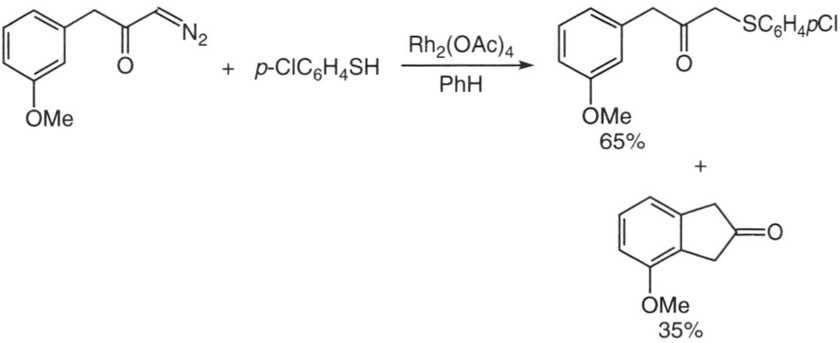

(81)

2-(PHENYLTHIO)-6-PHENYL-3-HEXANONE[138]

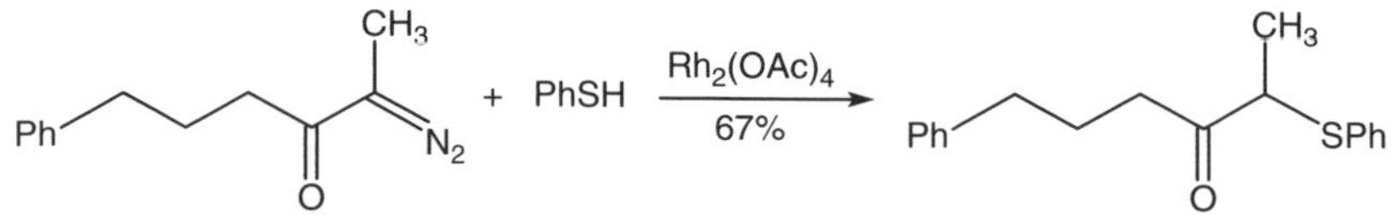

To a solution of thiophenol (0.55 g, 5 mmol) and rhodium acetate (2 mg) in dry benzene (60 mL) was added a solution of 2-diazo-6-phenyl-3-hexanone (1.01 g, 5 mmol) in dry benzene (20 mL). The reaction time was 45 min. After workup and chromatographic purification (silica gel; chloroform), the product was further purified by distillation under reduced pressure to give the pure compound (0.87 g, 61% yield). ^{1}H-NMR (CDCl$_3$): 1.35 (*d*, *J* = 7 Hz, 3 H, -CH*CH*$_3$); 1.75–1.99 (*m*, 2H, CH$_2$); 2.46–2.66 (*m*, 4 H, Ar-*CH*$_2$ and -*CH*$_2$-CO-); 3.63 (*q*, *J* = 7 Hz, 1 H, -CO-C*H*-CH$_3$); 7.21, 7.29 (2*s*, 10 H, Ar*H*). IR (film): 3070, 3035, 2940, 2870, 1715, 1610, 1590, 1500, 1455, 1440, 1375, 1245, 1020, 740, 695 cm^{-1}. m/e 284.

---

byproduct.[12] *p*-Chlorothiophenol was used for the insertion reaction (eq. 81) in which there is competition from intramolecular aromatic substitution (see Chapter 6), leading to a 35% yield of a methoxyindanone.[139]

Intramolecular versions of rhodium-catalyzed S–H insertion are also known. Rapoport[16] and Moody[111,140] and their respective collaborators have shown that the five-, six-, and seven-membered heterocycles (eq. 82) are accessible from

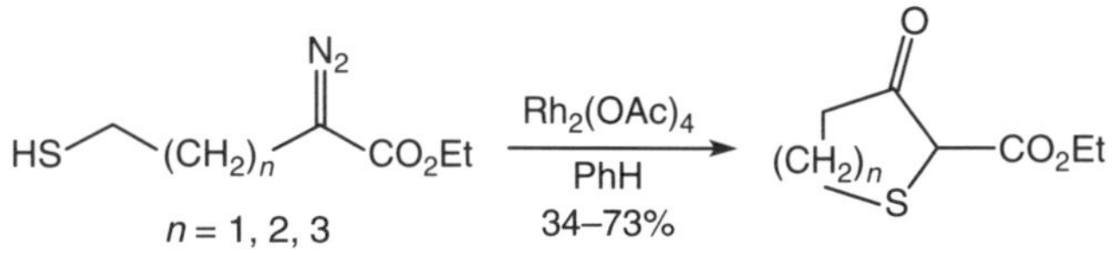

(82)

acyclic precursors. In these cases it was necessary to protect the thiol group during introduction of the diazo ester moiety. A final example of an intramolecular reaction where S–H insertion into an enzyme concerns the photochemical behavior of 1-diazo-4-undecy-2-one (DUO), which was designed as a cross-linker of a dehydrase's active-site amino acids.[141] The rationale was that as an analogue of an acetylenic thiol ester, DUO would alkylate histidine-70, as shown in Scheme 21, and once the inactivator had been tethered at the active site, photoirradiation would lead to loss of nitrogen and generation of either a 2-ketocarbene or a ketene (*via* Wolff rearrangement). Either of these species would be sufficiently reactive to modify a second active-site amino acid and in doing so cross-link the protein *via* S–H insertion. Alternative C–H, N–H, and O–H insertions were also envisaged.

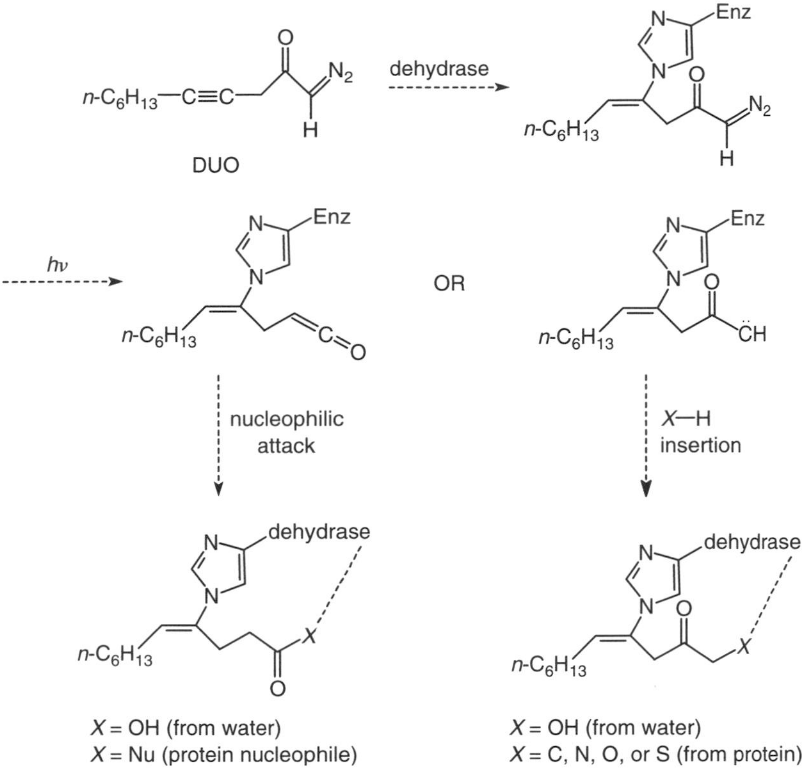

Scheme 21

8.6 Se–H INSERTION

There are few examples of Se–H insertion of diazocarbonyl compounds. Thomas and his co-workers[142] used rhodium(II) acetate to catalyze the addition of benzeneselenol to 6-diazopenicillanate (eq. 83) and this catalyst was also used by Moynihan and McKervey[143] in insertion reactions (eqs. 84 and 85).

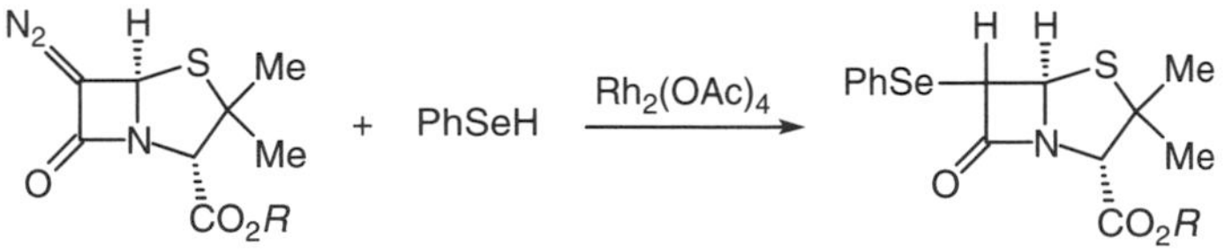

$$\text{(83)}$$

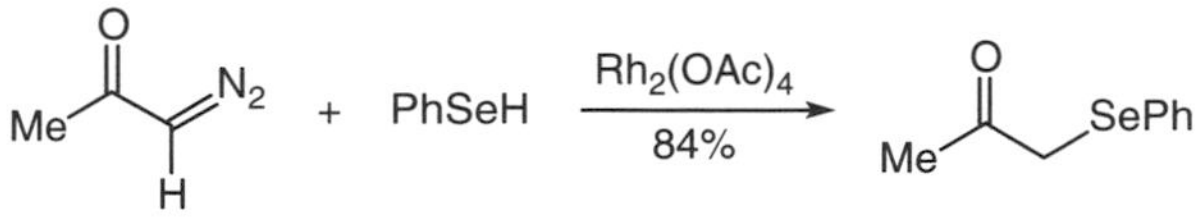

$$\text{(84)}$$

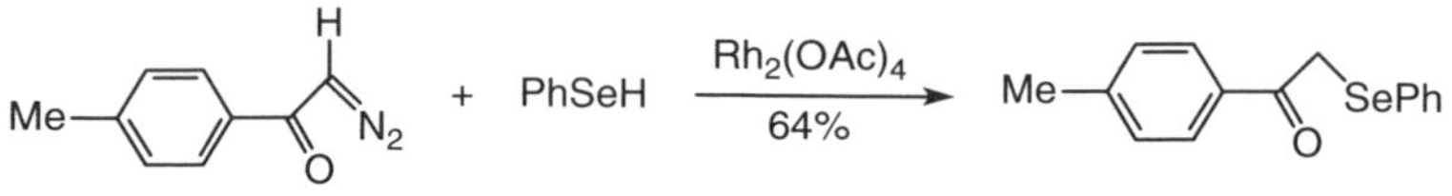

$$ (85) $$

8.7 P–H INSERTION

Polozov and co-workers[144] have reported the synthesis of several dialkyl phosphonates from diazocarbonyls and dialkyl hydrogen phosphites (eq. 86). The use

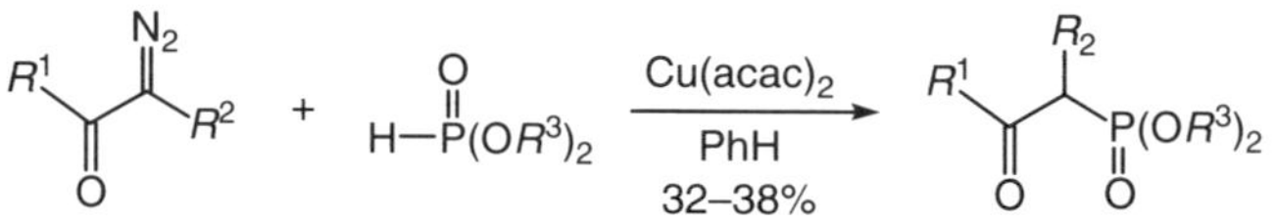

$$ (86) $$

of Cu(acac)$_2$ as the catalyst afforded a wide range of P–H insertion products in 32–38% yield. The diazo precursors included α-diazoketones, α-diazoesters, and 2-diazo-1,3-diketones. Rafferty[145] has studied the analogous reaction of diazoketones derived from N-protected α-amino acids. N-(phthaloyl)-L-alanyl diazomethane and N-(phthaloyl)-L-valyl diazomethane (eq. 87) on exposure to

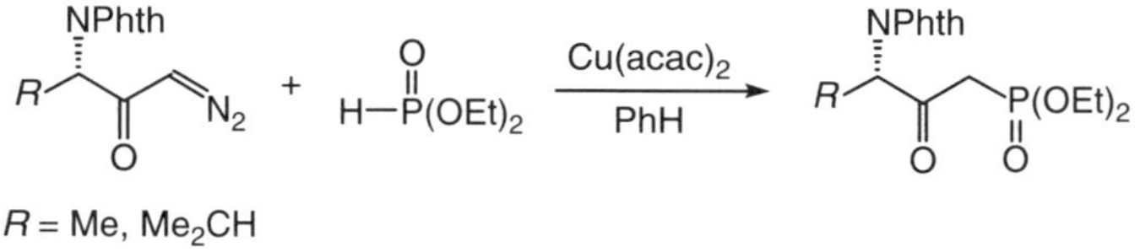

$$ (87) $$

R = Me, Me$_2$CH

diethyl hydrogen phosphite in benzene in the presence of Cu(acac)$_2$ afforded the β-ketophosphonates, eqs. 86 and 87, respectively. Efforts to extend this P–H insertion to other N-protected amino diazoketones with free N–H bonds, viz. Boc and Cbz, were unsuccessful. The products of these reactions were those of intramolecular N–H insertion (cf. Section 8.2).

8.8   X–H INSERTION (X = HALOGEN)

The hydrohalogenation–diazotization process has been used extensively with 6-aminopenicillanic acid (6-APA). Cignarella, Pifferi, and Testa[146] originally found that diazotization of 6-APA with sodium nitrite in dilute hydrochloric acid produced the 6-chloropenicillanic acid (eq. 88). The stereochemistry of the prod-

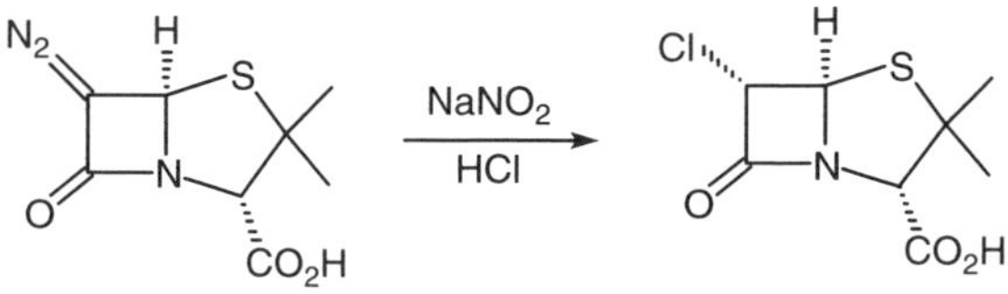

$$ (88) $$

uct was later established by McMillan and Stoodley,[147] who also confirmed that the 6-diazo derivative was an intermediate ir the reaction. The 6-bromopenicillanic acid derived from 6-APA has been extensively used as the key intermediate in the synthesis of penam antibiotics.[148–150]

Hydrohalogenation of diazoketones derived from amino acids offers an attractive route to optically active *N*-protected α-amino-α'-bromo and chloro ketones (eq. 89).[151] The diazoketone can be titrated with ethereal hydrogen chloride or hy-

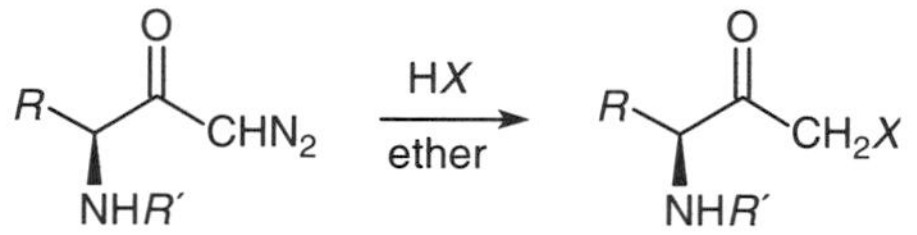

$$R' = \text{protecting group}; X = \text{Cl, Br}$$

(89)

drogen bromide without detectable amounts of racemization. This methodology has been extensively applied in the formation of chloromethyl ketones from peptides. The functionalized cyclic tetrapeptide HC-toxin analogue, *cyclo*[L-2-amino-8-oxo-9-chlorononanoyl-D-prolyl-L-alanyl-D-alanyl], was prepared *via* the above method.[152] The products of hydrohalogenation of diazoketones are useful reaction intermediates, for example, for keto azides.[151] The chloromethyl ketone prepared from the doubly protected L-aspartic acid–derived diazoketone served as a key intermediate in the synthesis of *N*-(*tert*-butoxycarbonyl)-3-(4-thiazolyl)-L-alanine (Scheme 22).[153]

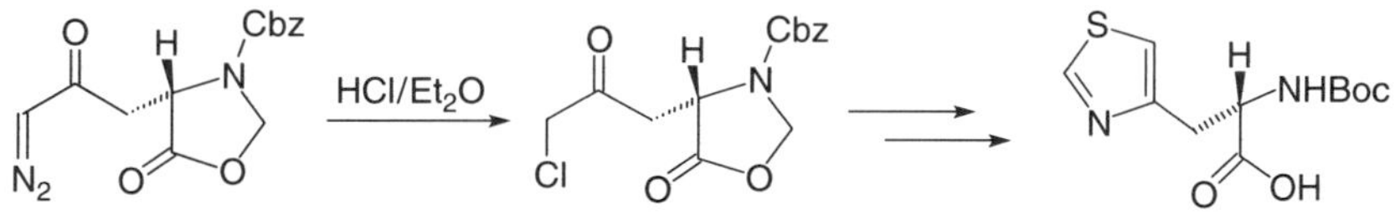

Scheme 22

α-Fluoro ketones can be prepared by reaction of α-diazoketones with 70% polyhydrogen fluoride–pyridine (eq. 88).[154] Similarly, diazoketones derived from

$$RC(O)CHN_2 \xrightarrow[\text{40–70\%}]{C_5H_5NH^+(HF)_xF^-} RC(O)CH_2F$$

(90)

N-phthaloyl amino acids can be converted into the corresponding fluoro ketones. More recent work with 6-diazopenicillinate esters indicates that this kind of transformation can also be accomplished with diethylaminosulfur trifluoride (DAST), where the α-hydroxy ketone is believed to be an intermediate (Scheme 23).[155]

Unlike the other hydrogen halides, hydrogen iodide does not normally produce α-iodo ketones from α-diazoketones. Rather, its reducing power results in the formation of methyl ketones. The process can be very efficient; it has been rec-

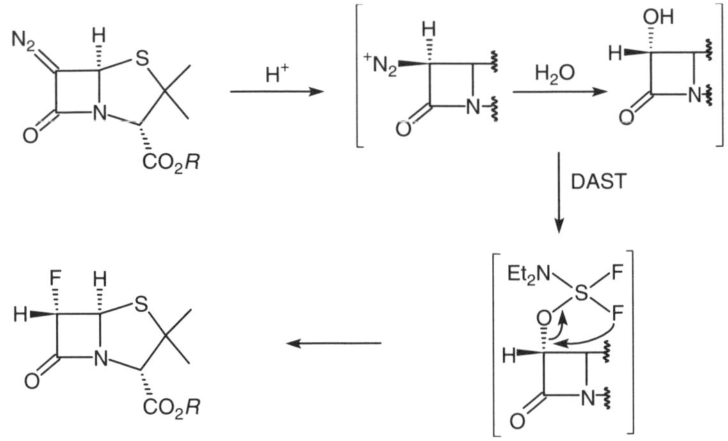

Scheme 23

ommended as a particularly mild two-step procedure for converting an acid chloride into a methyl ketone (eq. 91).[156]

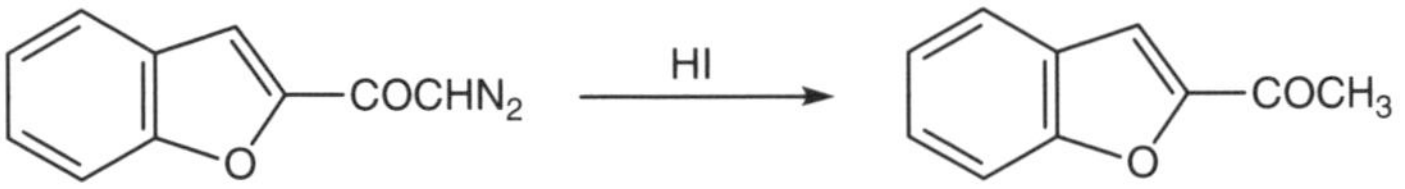

(91)

8.9 REFERENCES

1   Helson, H. E.; and Jorgensen, W. L., "Computer-Assisted Mechanistic Evaluation of Organic Reactions. 24. Carbene Chemistry," *J. Org. Chem.* **1994**, *59*, 3841–56.

2   Kirmse, W., "*Carbenes and the O–H Bond*," Advances in Carbene Chemistry, Brinker, U. Ed.; JAI Press; Vol. 1; 1–57.

3   Lama, L. D.; and Christensen, B. G., "Total Synthesis of β-Lactam Antibiotics, IX (±)-1-Oxabisnorpenicillin G," *Tetrahedron Lett.* **1978**, 4233–36.

4   Salzmann, T. N.; Ratcliffe, R. W.; Christensen, B. G.; and Bouffard, F. A., "A Stereocontrolled Synthesis of (+)-Thienamycin," *J. Am. Chem. Soc.* **1980**, *102*, 6161–63.

5   Melillo, D. G.; Shinkai, I.; Liu, T.; Ryan, K.; and Sletzinger, M., "A Practical Synthesis of (±)-Thienamycin," *Tetrahedron Lett.* **1980**, *21*, 2783–86.

6   Yates, P., "The Copper-Catalysed Decomposition of Diazoketones," *J. Am. Chem. Soc.* **1952**, *74*, 5376–81.

7   Saegusa, T.; Ito, Y.; Kobayashi, S.; Hirota, K.; and Shimizu, T., "Synthetic Reaction by a Complex-Catalyst. III. Copper Catalysed N-Alkylation of Amine with Diazoalkane," *Tetrahedron Lett.* **1966**, 6131–34.

8   Nicoud, J.-F.; and Kagan, H. B., "New Asymmetric Synthesis of Alanine by Carbene Insertion into a N–H Bond," *Tetrahedron Lett.* **1971**, 2065–68.

9 Paulissen, R.; Hayez, E.; Hubert, A. J.; and Teyssié, Ph., "Transition Metal Catalysed Reactions of Diazocompounds—Part III. A One Step Synthesis of Substituted Furans and Esters," *Tetrahedron Lett.* **1974**, 607–8.

10 Osipov, S. N.; Sewald, N.; Kolomiets, A. F.; Fokin, A. V.; and Burger, K., "Synthesis of α-Trifluoromethyl Substituted α-Amino Acid Derivatives from Methyl 3,3,3-Trifluoro-2-diazopropionate," *Tetrahedron Lett.* **1996**, *37*, 615–18.

11 Aller, E.; Buck, R. T.; Drysdale, M. J.; Ferris, L.; Haigh, D.; Moody, C. J.; Pearson, N. D.; and Sanghera, J. B., "N–H Insertion Reactions of Rhodium Carbenoids. Part 1. Preparation of α-Amino Acid and α-Aminophosphonic Acid Derivatives," *J. Chem. Soc., Perkin Trans. 1* **1996**, 2879–84.

12 Landais, Y.; and Planchenault, D., "Electronic *versus* Steric Effects in 5-*endo-trig*-Like Electrophilic Cyclizations," *Synlett* **1995**, 1191–93.

13 Bagley, M. C.; Buck, R. T.; Hind, S. L.; Moody, C. J.; and Slawin, A. M. Z., "A New Route to Functionalized Oxazoles," *Synlett.* **1996**, 825–26.

14 Gillespie, J. R.; and Porter, A. E. A., "The Reaction of Diazoalkanes with Thiophene," *J. Chem. Soc., Perkin Trans.* **1979**, 2624–28.

15 Kametani, T.; Kanaya, N.; Mochizuki, T.; and Honda, T., "Synthetic Studies on Carbapenem Antibiotics—A Carbon-Introduced Reaction at the C-4 Position of a β-Lactam by Carbene Insertion Reaction," *Heterocycles* **1982**, *19*, 1023–32.

16 Moyer, M. P.; Feldman, P. L.; and Rapoport, H., "Intramolecular N–H, O–H and S–H Insertion Reactions. Synthesis of Heterocycles from α-Diazo β-Keto Esters," *J. Org. Chem.* **1985**, *50*, 5223–30.

17 Liu, J-M.; Young, J-J.; Li, Y-J.; and Sha, C-K., "Synthesis of Substituted 1,2-Dihydroisoquinolines by the Intramolecular 1,3-Dipolar Alkyl Azide-Olefin Cycloaddition," *J. Org. Chem.* **1986**, *51*, 1120–23.

18 Ratcliffe, R. W.; Salzmann, T. N.; and Christensen, B. G., "A Novel Synthesis of Carbapen-2-em Ring System," *Tetrahedron Lett.* **1980**, *21*, 31–34.

19 Berges, D. A.; Snipes, E. R.; Chan, G. W.; Kingsbury, W. D.; and Kinzig, C. M., "A Facile Synthesis of Benzyl 3,7-Dioxo-1-azabicyclo[3.2.0]heptane-2-carboxylate—A Potential Precursor of Thienamycin and Clavulanic Acid Analogs," *Tetrahedron Lett.* **1981**, *22*, 3557–60.

20 Kametani, T.; Honda, T.; Nakayama, A.; Sasaki, J.; Terasawa, H.; and Fukumoto, K., "A Carbon–Carbon Bond Formation Reaction at the C-4 Position of a β-Lactam," *J. Chem. Soc., Perkin Trans. 1* **1981**, 1884–87; Kametani, T.; Honda, T.; Nakayama, A.; Sazakai, Y.; Mochizuki, T.; and Fukumoto, K., "A Short and Stereoselective Synthesis of the Carbapenem Antibiotic PS-5," *J. Chem. Soc., Perkin Trans. 1* **1981**, 2228–32.

21 Karady, S.; Amato, J. S.; Reamer, R. A.; and Weinstock, L. M., "Stereospecific Conversion of Penicillin to Thienamycin," *J. Am. Chem. Soc.* **1981**, *103*, 6765–67.

22 Reider, P. J.; and Grabowski, E. J. J., "Total Synthesis of Thienamycin—A New Approach from Aspartic Acid," *Tetrahedron Lett.* **1982**, *23*, 2293–96.

23 Ueda, Y.; Damas, C. E.; and Belleau, B., "Nuclear Analogs of β-Lactam Antibiotics. 18. A Short Synthesis of 2-Alkylthiocarbapen-2-em-3-carboxylate," *Can. J. Chem.* **1983**, *61*, 1996–2000.

24 Andrus, A.; Christensen, B. G.; and Heck, J. V., "Synthesis of 3-Methylphosphonyl Thienamycin and Related 3-Phosphonyl Carbapenems," *Tetrahedron Lett.* **1984**, *25*, 595–98.

25 Ona, H.; and Uyeo, S., "Total Synthesis of *dl*-Asparenomycin-A, B and C, Novel Carbapenem Antibiotics," *Tetrahedron Lett.* **1984**, *25*, 2237–40.

26 Ueda, Y.; Roberge, G.; and Vinet, V., "A Simple Method of Preparing Trimethyl-silyl-enol and *t*-Butyldimethylsilyl-enol Ethers of α-Diazoacetoacetates and Their Use in the Synthesis of a Chiral Precursor to Thienamycin Analogs," *Can. J. Chem.* **1984**, *62*, 2936–40.

27 Hirai, H.; Sawada, K.; Aratani, M.; and Hashimoto, M., "Synthetic Studies on Carbapenem Antibiotics from Penicillins. 2. Regioselective and Stereoselective Aldol Reaction of a Chiral Azetidinone—A Synthesis of Optically-Active 6-Epi-carpetimycins," *Tetrahedron Lett.* **1984**, *25*, 5075–78.

28 Knierzinger, A.; and Vasella, A., "Synthesis of 6-Epithienamycin," *J. Chem. Soc., Chem. Commun.*, **1984**, 9–11.

29 Shih, D. H.; Baker, F.; Cama, L.; and Christensen, B. G., "Synthetic Carbapenem Antibiotics. 1. 1-β-Methylcarbapenem," *Heterocycles* **1984**, *21*, 29–40.

30 Buynak, J. D.; Rao, M. N.; Pajouhesh, H.; Chandrasekaran, R. Y.; Finn, K.; De-meester, P.; and Chu, S. C., "Useful Chemistry of 3-(1-Methylethylidene)-4-acetoxy-2-azetidinone—A Formal Synthesis of (±)-Asparenomycin-C," *J. Org. Chem.* **1985**, *50*, 4245–52.

31 Aratani, M.; Hirai, H.; Sawada, K.; Yamada, A.; and Hashimoto, M., "Synthetic Studies on Carbapenem Antibiotics from Penicillins. 3. Stereoselective Radical Reduction of a Chiral 3-Isocyanoazetidinone—A Total Synthesis of Optically-Active Carpetimycins," *Tetrahedron Lett.* **1985**, *26*, 223–26.

32 Shih, D. H.; Cama, L. D.; and Christensen, B. G., "Synthetic Carbapenem Antibi-otics III. 1-Methyl Thienamycin," *Tetrahedron Lett.* **1985**, *26*, 587–90.

33 Ohtani, M.; Watanabe, F.; and Narisada, M., "Synthesis of Several New Car-bapenem Antibiotics," *J. Antibiot.*, **1985**, *38*, 610–621.

34 Chiba, T.; Nagatsuma, M.; and Nakai, T., "A Facile, Stereocontrolled Entry to Key Intermediates for Thienamycin Synthesis from Ethyl (*S*)-3-hydroxybutanoate," *Chem. Lett.* **1985**, 1343–46.

35 Ikota, N.; Shibata, H.; and Koga, K., "Stereoselective Reactions. 9. Synthetic Studies on Optically-Active β-Lactams. 1. Chiral Synthesis of Carbapenam and Carbapenem Ring-Systems Starting from (*S*)-Aspartic Acid," *Chem. Pharm. Bull. Tokyo* **1985**, *33*, 3299–306.

36 Evans, D. A.; and Sjogren, E. B., "The Asymmetric-Synthesis of β-Lactam An-tibiotics. 3. Enantioselective Synthesis of (+) PS-5," *Tetrahedron Lett.* **1986**, *27*, 3119–22.

37 Deziel, R.; and Endo, M., "An Expeditious Synthesis of a 1-β-Methylcarbapenem Key Intermediate," *Tetrahedron Lett.* **1988**, *29*, 61–64.

38 Endo, M.; and Droghini, R., "1-β-Methylcarbapenem Intermediates. 2. Stereo-selective Synthesis of (3S,4R)-3-[(1R)-1-*t*-Butyldimethylsilyloxyethyl]-4-[(1R)-3-methoxycarbonyl-1-methyl-2-oxopropyl]azetidin-2-one and Its Related Chemistry," *Can. J. Chem.* **1988**, *66*, 1400–4.

39 Georg, G. I.; and Kant, J,. "An Asymmetric-Synthesis of Carbapenem Antibiotic (+)-PS-5 from Ethyl 3-Hydroxybutanoate," *J. Org. Chem.* **1988**, *53*, 692–95.

40 Kita, Y.; Tamura, O.; Shibata, N.; and Miki, T., "Chemistry of O-Silylated Ketene Acetals. 1. A Synthesis of β-Lactam Antibiotics," *J. Chem. Soc. Perkin Trans. 1* **1989**, 1862–64.

41 Andreoli, P.; Cainelli, G.; Panunzio, M.; Bandini, E.; Martelli, G.; and Spunta, G., "β-Lactams from Ester Enolates and Silylimines—Enantioselective Synthesis of the Trans-Carbapenem Antibiotics (+)-PS-5 and (+)-PS-6," *J. Org. Chem.* **1991**, *56*, 5984–90.

42 Mastalerz, H.; Menard, M.; Ruediger, E.; and Fung-Tomc, J., "Synthesis and Antibacterial Activity of Some Novel 6-Methyl-Substituted and 6-Propenyl-Substituted Carbapenems," *J. Med. Chem.* **1992**, *35*, 953–58.

43 Brennan, J.; and Pinto, I. L., "An Efficient Sythesis of (±) 7-Oxo-3-thia-1-azabicyclo[3.2.0]heptane-2-carboxylate-3,3-dioxide," *Tetrahedron Lett.* **1983**, *24*, 4731–32.

44 Mak, C.-P.; Mayerl, C.; and Fliri, H., "Synthesis of Carbapenem-3-phosphonic Acid Derivatives," *Tetrahedron Lett.* **1983**, *24*, 347–50.

45 Heck, J. V.; and Christensen, B. G., "Nuclear Analogs of β-Lactam Antibiotics. 1. The Synthesis of 6-β-Amidocyclonocardicins," *Tetrahedron Lett.* **1981**, *22*, 5027–30.

46 Heck, J. V.; Szymonifka, M. J.; and Christensen, B. G., "Nuclear Analogs of β-Lactam Antibiotics. 2. The Synthesis of 6α-(1-Hydroxylethyl)-cyclonocardicins," *Tetrahedron Lett.* **1982**, *23*, 1519–22.

47 Salzmann, T. N.; Ratcliffe, R. W.; and Christensen, B. G., "Total Synthesis of (−)-Homothienamycin," *Tetrahedron Lett.* **1980**, *21*, 1193–96.

48 Evans, D. A.; and Sjogren, E. B., "The Asymmetric Synthesis of β-Lactam Antibiotics. 2. The First Enantioselective Synthesis of the Carbacephalosporin Nucleus," *Tetrahedron Lett.* **1985**, *26*, 3787–90.

49 Bodurow, C. C.; Boyer, B. D.; Brennan, J.; Bunnell, C. A.; Burks, J. E.; Carr, M. A.; Doecke, C. W.; Eckrich, T. M.; Fisher, J. W.; Gardner, J. P.; Graves, B. J.; Hines, P.; Hoying, R. C.; Jackson, B. G.; Kinnick, M. D.; Kochert, C. D.; Lewis, J. S.; Luke, W. D.; Moore, L. L.; Morin, J. M.; Nist, R. L.; Prather, D. E.; Sparks, D. L.; and Vladuchick, W. C., "An Enantioselective Synthesis of Loracarbef (LY163892/KT3777)," *Tetrahedron Lett.* **1989**, *30*, 2321–24.

50 Häbich, D.; and Hartwig, W., "A Convergent Scheme for the Stereoselective Synthesis of 3′-Nor-1-oxacephems," *Tetrahedron* **1984**, *40*, 3667–76.

51 Yamamoto, S.; Itani, H.; Takahashi, H.; Tsuji, T.; and Nagata, W., "An Alternative Synthesis of the 1-Oxacephem Skeleton," *Tetrahedron Lett.* **1984**, *25*, 4545–48.

52 Okonogi, T.; Shibahara, S.; Murai, Y.; Inouye, S.; Kondo, S.; and Christensen, B. G., "Novel 2-Methyl-1-oxacephalosporins. 2. Synthesis of 3-Substituted 2-Methyl-1-oxacephem Nucleus," *Heterocycles* **1990**, *31*, 797–802.

53 Williams, R. M.; Lee, B. H.; Miller, M. M.; and Anderson, O. P., "Synthesis and X-ray Crystal Structure Determination of 1,3-Bridged β-Lactams—Novel, Anti-Bredt β-Lactams," *J. Am. Chem. Soc.* **1989**, *111*, 1073–81.

54 Bissolino, P.; Alpegiani, M.; Borghi, D.; Perrone, E.; and Franceschi, G., "Synthesis of Penams from 1H-Azetidinones by Intramolecular Carbenoid Insertion," *Heterocycles* **1993**, *36*, 1529–39.

55 Bouthillier, G.; Mastalerz, H.; and Menard, M., "Synthesis of a 4β-Carboxyethyl Derivative of Thienamycin," *Tetrahedron Lett.* **1994**, *35*, 4689–92.

56 Moody, C. J.; Pearson, C. J.; and Lawton, G., "Regioselectivity in the Photochemical Ring Contraction of 4-Diazopyrazolidine-3,5-diones to Give Aza-β-Lactams," *Tetrahedron Lett.* **1985**, *26*, 3167–70.

57   Podlech, J.; and Seebach, D., "Azetidin-3-ones from (*S*)-α-Amino Acids and Their Reaction with Nucleophiles: Preparation of Some Azetidine—Containing Amino-Alcohols and Amino-Acid Derivatives," *Helv. Chim. Acta* **1995**, *78*, 1238–46.

58   McKervey, M. A.; Dilworth, B. M.; and Collins, J. C., unpublished work (see Dilworth, B. M., Ph.D. Thesis, NUI, 1987; Collins, J. C., Ph.D. Thesis, NUI, 1990).

59   Hanessian, S.; Fu, J.-M.; Chiara, J.-L.; and Di Fabio, R., "Total Synthesis of (+)-Polyoximic Acid–*cis*-3-Ethylidene-L-azetidine-2-carboxylic Acid," *Tetrahedron Lett.* **1993**, *34*, 4157–60.

60   Emmer, G., "Synthesis of (2*RS*,*E*)-3-Ethylidene-azetidine-2-carboxylic Acid (rac. Polyoximic Acid)," *Tetrahedron* **1992**, *48*, 7165–72.

61   Burger, K.; Rudolph, M.; and Fehn, S., "Synthesis of 4-Oxo-1-proline and *cis*-4-Hydroxy-L-proline Derivatives from L-Aspartic Acid," *Angew. Chem. Int. Ed. Eng.* **1993**, *32*, 285–87.

62   Ko, K.-Y.; Lee, K.-I.; and Kim, W.-J., "Synthesis of 5-Oxo-L-pipecolic Acid-Derivatives by Rhodium(II) Acetate Catalyzed Cyclization of Diazoketones," *Tetrahedron Lett.* **1992**, *33*, 6651–52.

63   Shapiro, E. A.; Lun'kova, G. V.; Dolgii, I. E.; and Nefedov, O. M., "Regioselective Interaction of Allyl Diazoacetates with Trimethylsilylisopropenylacetylene and Chemical Reactions of Esters of Trimethylsilylethynylmethylcyclopropanecarboxylic Acid," *Izv. Akad. Nauk SSR, Ser. Khim.* **1981**, 1316–19.

64   Garcia, C. F.; McKervey, M. A.; and Ye, T., "Asymmetric Catalysis of Intramolecular N–H Insertion Reactions of α-Diazocarbonyls," *J. Chem. Soc., Chem. Commun.* **1996**, 1465–66.

65   Taber, D. F.; Hennessy, M. J.; and Louey, J. P., "Rh-Mediated Cyclopentane Construction Can Compete with β-Hydride Elimination—Synthesis of (±)-Tochuinyl Acetate," *J. Org. Chem.* **1992**, *57*, 436–41.

66   Stauz, O. P.; DoMinh, T.; and Gunning, H. E., "Rearrangement and Polar Reaction of Carbethoxymethylene in 2-Propanol," *J. Am. Chem. Soc.* **1968**, *90*, 1660–61.

67   Toscano, J. P.; Platz, M. S.; Nikolaev, V.; and Popic, V., "Carboethoxycarbene—A Laser Flash-Photolysis Study," *J. Am. Chem. Soc.* **1994**, *116*, 8146–51.

68   Miller, D. J.; and Moody, C. J., "Synthetic Applications of the O–H Insertion Reaction of Carbenes and Carbenoids Derived from Diazocarbonyls and Related Diazo Compounds," *Tetrahedron* **1995**, *51*, 10811–43.

69   Eistert, B.; Elias, H.; Kosch, E.; and Wollheim, R., "Synthese und Einige Umsetzungen des 2-Hydroxydimedone," *Chem. Ber.* **1959**, *92*, 140–41.

70   Hanack, M.; and Dolde, J., "Umsetzung von Diazocampher mit Säuren," *Tetrahedron Lett.* **1966**, 321–25.

71   Kronis, J. D.; Powell, M. F.; and Yates, P., "The Acid-Induced Decomposition of 7-Substituted 3-Diazo-2-norbornanones. Kinetic Studies," *Tetrahedron Lett.* **1983**, *24*, 2423–26.

72   Bui-Nguyen, M.-H.; Dahn, H.; and McGarrity, J. F., "Substituent and Isotope Effects on Hydrolysis Rates of 2-Aryl-2-dicarboxylic Esters" *Helv. Chim. Acta* **1980**, *63*, 63–75.

73   Jones, J.; and Kresge, A. J., "Methylthio Group Migration in the Acid-Catalyzed Hydrolysis of *S*-Methyl Phenyldiazothioacetate—Kinetics and Mechanism of the Reaction," *J. Org. Chem.* **1993**, *58*, 2658–62.

74 Newman, M. S.; and Beal, P. F., "A New Synthesis of α-Alkoxy Ketones," *J. Am. Chem. Soc.* **1950**, *72*, 5161–63.

75 Giddings, P. J.; John, D. I.; and Thomas, E. J., "Some Lewis Acid Catalysed Reactions of 2,2,2-Trichloroethyl 6-Diazopenicillanate," *Tetrahedron Lett.* **1978**, 995–98.

76 Reed, P. E.; and Katzenellenbogen, J. A., "Synthesis of Proline Valine Pseudodipeptide Enol Lactones, Serine Protease Inhibitors," *J. Org. Chem.* **1991**, *56*, 2624–34.

77 Dyer, R. B.; Metcalf, D. H.; Ghirardelli, R. G.; Palmer, R. A.; and Holt, E. M., "Circular Dichroism Studies of Crown Complexed Ion Pairs: A Comparison of the Alkali and Alkaline Earth Nitrate Complexes of Chiral Crown Ethers," *J. Am. Chem. Soc.* **1986**, *108*, 3621–29.

78 Casanova, R.; and Reichstein, T., "Methoxyketone aus Diazoketonen," *Helv. Chim. Acta* **1950**, *33*, 417–22.

79 Haupter, F.; and Pucek, A., "Über die kupferkatalysierte Zersetzung von Diazoketonen in wassriger Lösung," *Chem. Ber.* **1960**, *93*, 249–52.

80 Yates, P.; Garneau, F. X.; and Hambly, G. F., "Aliphatic Diazo Compounds. Part XI. Some Observations on the Copper Induced Decomposition of α-Diazo Ketones," *Pol. J. Chem.* **1979**, *53*, 163–76.

81 Saegusa, T.; Ito, Y.; Kobayashi, S.; Hirota, K.; and Shimizu, T., "Synthetic Reactions by Copper Catalysts. VIII. Copper-Catalysed Reactions of Thiol and Alcohol with Diazoacetate," *J. Org. Chem.* **1968**, *33*, 544–47.

82 Paulissen, R.; Reimlinger, H.; Hayez, E.; Hubert, A. J.; and Teyssié, P., "Transition Metal Reactions of Diazo Compounds. II. Insertion in the Hydroxylic Bond," *Tetrahedron Lett.* **1973**, 2233–36.

83 Noels, A. F.; Demonceau, A.; Petiniot, N.; Hubert, A. J.; and Teyssié, P., "Transition-Metal-Catalyzed Reactions of Diazocompounds, Efficient Synthesis of Functionalized Ethers by Carbene Insertion into the Hydroxylic Bond of Alcohols," *Tetrahedron* **1982**, *38*, 2733–39.

84 Baumann, B. C.; and MacLeod, J. K., "Structures of the $C_2H_4O^+$ Gas-Phase Isomers—Evidence for the Formation of the $CH_2OCH_2^+$ Ion from Ethylene Carbonate," *J. Am. Chem. Soc.* **1981**, *103*, 6223–24.

85 Doyle, M. P.; Dorow, R. L.; Terpstra, J. W.; and Rodenhouse, R. A., "Synthesis and Catalytic Reactions of Chiral *N*-(Diazoacetyl)oxazolidones," *J. Org. Chem.* **1985**, *50*, 1663–66.

86 Moynihan, H.; and McKervey, M. A., unpublished results.

87 Ohira, S.; Noda, I.; Mizobata, T.; and Yamato, M., "Synthesis of Tertiary Alcohol from Secondary Alcohol via Intramolecular C–H Insertion of Alkylidenecarbene," *Tetrahedron Lett.* **1995**, *36*, 3375–76.

88 Wood, J. L.; Stoltz, B. M.; and Dietrich, H. -J., "Total Synthesis of (+)- and (−)-K252a," *J. Am. Chem. Soc.* **1995**, *117*, 10413–14.

89 Stachel, H. D.; Zeitler, K.; and Dick, S., "Synthesis of (±)-Azaascorbic Acid," *Liebigs Ann.* **1996**, 103–7.

90 Ganem, B.; Ikota, N.; Muralidharan, V. B.; Wade, W. S.; Young, S. D.; and Yukimoto, Y., "Shikimate-Derived Material. 11. Total Synthesis of (±)-Chorismic Acid," *J. Am. Chem. Soc.* **1982**, *104*, 6787–88.

91   Teng, C.-Y. P.; and Ganem, B., "Shikimate-Derived Metabolites. 13. A Key Intermediate in the Biosynthesis of Anthranilate from Chorismate," *J. Am. Chem. Soc.* **1984**, *106*, 2463–64.

92   Hoare, J. H.; Policastro, P. P.; and Berchtold, G. A., "Improved Synthesis of Racemic Chorismic Acid—Claisen Rearrangement of 4-*epi*-Chorismic Acid and Dimethyl 4-*epi*-chorismate," *J. Am. Chem. Soc.* **1983**, *105*, 6264–67.

93   Chouinard, P. M.; and Bartlett, P. A., "Conversion of Shikimic Acid to 5-Enolpyruvylshikimate 3-Phosphate," *J. Org. Chem.* **1986**, *51*, 75–78.

94   Kim, S. K.; and Fuchs, P. L., "Application of the Reich Iodoso *syn*-Elimination for the Preparation of an Intermediate Appropriate for the Synthesis of Both Hexacyclic Steroidal Units of Cephalostatin 7," *Tetrahedron Lett.* **1994**, *35*, 7163–66.

95   Matlin, S. A.; and Chan, L., "Metal-Catalyzed Reactions of Benzhydryl 6-Diazopenic-illanate with Alcohols," *J. Chem. Soc., Chem. Commun.* **1980**, 768–69.

96   Davies, H. M. L.; Smith, H. D.; and Korkov, O., "Tandem Cyclopropanation/Cope Rearrangement Sequence. Stereospecific [3+4] Cycloaddition Reaction of Vinylcarbenoids with Cyclopentadiene," *Tetrahedron Lett.* **1987**, *28*, 1853–56.

97   Beecham Group Ltd., "Chavalunic Acid Derivatives," Brit. Patent No. 1572259, 1981; *Chem Abs.* **1981**, *94*, 103343.

98   Hosten, N. G. C.; Tavernier, D.; and Anteunis, M. J. O., "A Configuration-Retaining α,β-diol to 1,4-Dioxane Conversion," *Bull. Soc. Chim. Belg.* **1985**, *94*, 183–86.

99   Paquet, F.; and Sinaÿ, P., "Intramolecular Oxymercuration–Demercuration Reaction—A New Stereocontrolled Approach to Sialic-Acid Containing Disaccharides," *Tetrahedron Lett.* **1984**, *25*, 3071–74.

100 Paquet, F.; and Sinaÿ, P., "New Stereocontrolled Approach to 3-Deoxy-D-*manno*-2-Octulosonic Acid Containing Disaccharides," *J. Am. Chem. Soc.* **1984**, *106*, 8313–15.

101 Cox, G. G.; Haigh, D.; Hindley, R. M.; Miller, D. J.; and Moody, C. J., "Competing O–H Insertion and β-Elimination in Rhodium Carbenoid Reactions," *Tetrahedron Lett.* **1994**, *35*, 3139–42.

102 Shi, G. Q.; Cao, Z. Y.; and Cai, W. L., "Rhodium-Mediated Insertion of CF₃-Substituted Carbenoid into O–H: An Efficient Method for the Synthesis of α-Trifluoromethylated Alkoxy- and Aryloxyacetic Acid Derivatives," *Tetrahedron* **1995**, *51*, 5011–18.

103 Ye, B.; and Burke, T. R., "L-O-(2-Malonyl)tyrosine (L-OMT) a New Phosphotyrosyl Mimic Suitably Protected for Solid-Phase Synthesis of Signal Transduction Inhibitory Peptides," *Tetrahedron Lett.* **1995**, *36*, 4733–36.

104 Ace, K. W.; Hussain, N.; Lathbury, D. C.; and Morgan, D. O., "Synthesis of an α-(Aminooxy)arylacetic Ester by the Reaction of an α-Diazo Ester with N-Hydroxyphthalimide," *Tetrahedron Lett.* **1995**, *36*, 8141–44.

105 Murata, S.; Kongou, C.; and Tomioka, H., "Rhodium(II)-Catalysed Reaction of 1,3-Bis(diazo)indan-2-one with Alcohols: Formation of Unexpected 1,1-Dialkoxy Ketones," *Tetrahedron Lett.* **1995**, *36*, 1499–502.

106 Marshall, J. R.; and Walker, J., "Experiments on the Synthesis of Simple C-Substituted Derivatives of Dihydroxyacetone," *J. Chem. Soc.* **1952**, 467–75.

107 McClure, D. E.; Lumma, P. K.; Arison, B. H.; Jones, J. H.; and Baldwin, J. J., "1,4-Oxazines via Intramolecular Ring-Closure of β-Hydroxydiazoacetamides — Phenylalanine to Tetrahydroindeno[1,2-b]-1,4-Oxazin-3(2*H*)-ones," *J. Org. Chem.* **1983**, *48*, 2675–79.

108 Kulkowit, S.; and McKervey, M. A., "Synthesis of Macrocyclic Oxo-Crown Ethers from α-ω-bis-Diazo-Ketones and Polyethylene Glycols," *J. Chem. Soc., Chem. Commun.* **1981**, 616–17.

109 Heslin, J. C.; Moody, C. J.; Slawin, A. M. Z.; and Williams, D. J., "Synthesis of Cyclic Ethers by Rhodium Carbenoid Cyclization," *Tetrahedron Lett.* **1986**, *27*, 1403–6.

110 Heslin, J. C.; and Moody, C. J., "Rhodium Carbenoid Mediated Cyclizations. 2. Synthesis of Cyclic Ethers," *J. Chem. Soc., Perkin Trans. 1* **1988**, 1417–23.

111 Moody, C. J.; and Taylor, R. J., "Rhodium Carbenoid Mediated Cyclizations — Use of Ethyl Lithiodiazoacetate in the Preparation of ω-Hydroxy-, -Mercapto-, and -BOC-Amino-α-diazo-β-keto Esters," *Tetrahedron Lett.* **1987**, *28*, 5351–52.

112 (a) Moody, C. J.; and Taylor, R. J., "Rhodium Carbenoid Mediated Cyclizations. Part 3. Synthesis of Cyclic Ethers from Lactones," *J. Chem. Soc., Perkin Trans. 1* **1989**, 721–31; (b) Brown, D. S.; Elliott, M. C.; Moody, C. J.; Mowlem, T. J.; Marino, J. P.; and Padwa, A., "Ligand Effects in the Rhodium(II)-Catalyzed Reactions of α-Diazoamides. Oxindole Formation is Promoted by the use of Rhodium(II) Perfluorocarboxamide Catalysts," *J. Org. Chem.* **1994**, *59*, 2447–55.

113 Cox, G. G.; Moody, C. J.; Austin, D. J.; and Padwa, A., "Chemoselectivity of Rhodium Carbenoids. A Comparison of the Selectivity for O–H Insertion Reactions or Carbonyl Ylide Formation versus Aliphatic and Aromatic C–H Insertion and Cyclopropanation," *Tetrahedron* **1993**, *49*, 5109–26.

114 Davies, M. J.; Heslin, J. C.; and Moody, C. J., "Rhodium Carbenoid Mediated Cyclizations. 4. Synthetic Approaches to Oxepanes Related to Zoapatanol," *J. Chem. Soc., Perkin Trans. 1* **1989**, 2473–84.

115 Moody, C. J.; Sie, E.-R. H. B.; and Kulagowski, J. J., "Use of Diazophosphonates in the Synthesis of Cyclic Ethers. Part 2. Synthesis of the Pyranooxepane and Oxepanooxepane Subunits of Marine Polyether Toxins," *J. Chem. Soc., Perkin Trans. 1* **1994**, 501–6.

116 Crackett, P. H.; Sayer, P.; Stoodley, R. J.; and Greengrass, C. W., "Total Synthesis of Analogs of the β-Lactam Antibiotics. 6. (6*R**)-4-(*t*-Butoxycarbonyl)-2-methoxycarbonyl-3-oxacepham 1,1-Dioxides," *J. Chem. Soc., Perkin Trans. 1* **1991**, 1235–43.

117 Sarabia-García, F.; López-Herrera, F. J.; and Pino-González, M. S., "A New Synthesis for 2-Deoxy-KDO, a Potent Inhibitor of CMP-KDO Synthetase," *Tetrahedron Lett.* **1994**, *35*, 6709–12.

118 Aller, E.; Brown, D. S.; Cox, G. G.; Miller, D. J.; and Moody, C. J., "Diastereoselectivity in the O–H Insertion Reactions of Rhodium Carbenoids Derived from Phenyldiazoacetates of Chiral Alcohols. Preparation of α-Hydroxy and α-Alkoxy Esters," *J. Org. Chem.* **1995**, *60*, 4449–60.

119 Bradley, W.; and Robinson, R., "Interaction of Benzoyl Chloride and Diazomethane Together with a Discussion of the Reaction of Diazenes," *J. Chem. Soc.* **1928**, 1310–18.

120 (a) Wolfrom, M. L.; and Brown, R. L., "The Action of Diazomethane upon Acyclic Sugar Derivatives. V. Halogen Derivatives," *J. Am. Chem. Soc.* **1943**, *65*, 1516–21; (b) Wolfrom, M. L.; Olin, S. M.; and Evans, E. F., "The Action of Diazomethane upon Acyclic Sugar Derivatives. VI. D-Sorbose," *J. Am. Chem. Soc.* **1944**, *66*, 204–6; (c) Wolfrom, M. L.; Thompson, A.; aı l Evans, E. F., "The Action of Diazomethane upon Acyclic Sugar Derivatives. VıI. D-Psicose," *J. Am. Chem. Soc.* **1945**, *67*, 1793–97; (d) Wolfrom, M. L.; and Гhompson, A., "L-Fructose," *J. Am. Chem. Soc.* **1946**, *68*, 791–93; (e) Wolfrom, M. L.; Berkebile, J. M.; and Thompson, A., "Synthesis of Persuelose (L-Galaheptose)," *J. Am. Chem. Soc.* **1949**, *71*, 2360–62; "Synthesis of Sedoheptulose (D-Altroheptulose)," *J. Am. Chem. Soc.* **1952**, *74*, 2197–98; (f) Wolfrom, M. L.; and Cooper, P. W., "Two Ketoöctoses from the D-Galaheptonic Acids," *J. Am. Chem. Soc.* **1949**, *71*, 2668–71; "D-Manno-L-*fructo*-octose," *J. Am. Chem. Soc.* **1950**, *72*, 1345–47; (g) Wolfrom, M. L.; and Wood, H. B., "L-Mannoheptulose (L-Manno-L-*tagato*-heptose," *J. Am. Chem. Soc.* **1951**, *73*, 730–33; "Two 3-Epimeric Ketononoses," *J. Am. Chem. Soc.* **1955**, *77*, 3096–98; (h) Wolfrom, M. L.; Waisbrot, S. W.; and Brown, R. L., "The Action of Diazomethane upon Acyclic Sugar Derivatives. ٦II. A New Synthesis of Ketoses and of Their Open Chain (*keto*) Acetates," *J. Am. Chem. Soc.* **1942**, *64*, 2329–31; (i) Wolfrom, M. L.; Brown, R. L.; and Evans, E. F., "The Action of Diazomethane upon Acyclic Sugar Derivatives. III. Ketose Synthesis," *J. Am. Chem. Soc.* **1943**, *65*, 1021–27.

121 Ye, Tao.; and McKervey, M. A., unpublished results.

122 Canet, J. L.; Fadel, A.; and Salaun, J., "Asymmetric Construction of Quaternary Carbons from Chiral Malonates—Selective and Versatile Total Syntheses of the Enantiomers of α-Cuparenone and β-Cuparenone from a Common Optically-Active Precursor," *J. Org. Chem.* **1992**, *57*, 3463–73.

123 Miller, R. D.; and Theis, W., "The Acid Catalysed Cyclisation of Diazoketones: Preparation of 2,4(3*H*,5*H*)Furandiones," *Tetrahedron Lett.* **1987**, *28*, 1039–42.

124 White, J. D.; Bolton, G. L.; Dantanarayana, A. P.; Fox, C. M. J.; Hiner, R. N.; Jackson, R. W.; Sakuma, K.; and Warrier, U. S., "Total Synthesis of the Antiparasitic Agent Avermectin B$_{1a}$," *J. Am. Chem. Soc.* **1995**, *117*, 1908–39.

125 Ogawa, K.; Terada, T.; Muranaka, Y.; Hamakawa, T.; Ohta, S.; Okamoto, M.; and Fujii, S., "Studies on Hypolipidemic Agents. 4. Syntheses and Biological-Activities of *trans*-2-(4-alkylcyclohexyl)-2-oxoethyl and *cis*-2-(4-alkylcyclohexyl)-2-oxoethyl Arenesulfonates," *Chem. Pharm. Bull. Tokyo* **1987**, *35*, 3276–83.

126 Hua, D. H.; Peacock, N. J.; and Meyers, C. Y., "Synthesis of a Sulfone α-Tosylate. Benzyl(tosyloxy)methyl Sulfone," *J. Org. Chem.* **1980**, *45*, 1717–19.

127 Charlton, J. L.; Lai, H. K.; and Lypka, G. N., "Photoreaction of α-Sulfonyloxyketones," *Can. J. Chem.* **1980**, *58*, 458–62.

128 Collins, J. C.; Dilworth, B. M.; Garvey, N. T.; Kennedy, M.; McKervey, M. A.; and O'Sullivan, M. B., "Homochiral Group Transfer in Organic-Synthesis via α-Diazocarbonyl Intermediates," *J. Chem. Soc., Chem. Commun.* **1990**, 362–64.

129 Sumner, T.; Ball, L. E.; and Platner, J., "Use of Substituted α-Diazoacetophenones for the Preparation of Derivatives of Sulfonic Acids, *N*-Benzoylated Aminocarboxylic and N-Benzoylated Aminophenols and a *N*-Benzoylated Amino Thiol," *J. Org. Chem.* **1959**, *24*, 2017–18.

130 Vedejs, E.; Engler, D. A.; and Mullins, M. J., "Reactive Triflate Alkylating Agents," *J. Org. Chem.* **1977**, *42*, 3109–13.

131 Flowers, W. T.; Holt, G.; and McCleery, P. P., "Competition between Oxazolium and Sulphonium Salt Formation in the Acid-Induced Interaction of 2-Diazoacetophenones with Diaryl Sulphides in Acetonitrile," *J. Chem. Soc., Perkin Trans. 1* **1979**, 1485–89.

132 Moynihan, H.; and McKervey, M.A., unpublished results.

133 Bischofberger, N.; Waldmann, H.; Saito, T.; Simon, E. S.; Lees, W.; Bednarski, M. D.; and Whitesides, G. M., "Synthesis of Analogs of 1,3-Dihydroxyacetone Phosphate and Glyceraldehyde-3-phosphate for Use in Studies of Fructose-1,6-diphosphate Aldolase," *J. Org. Chem.* **1988**, *53*, 3457–65.

134 Ando, W., in *The Chemistry of Diazonium and Diazo Groups*, Patai, S., Ed.; John Wiley & Sons: New York, 1978; Chapter 9, p. 445.

135 Sheehan, J. C.; Commons, T. J.; and Lo, Y. S., "Photochemical Conversion of β,β,β-Trichloroethyl 6-Diazopenicillanate into 6β-Thiolpennicillin Derivatives," *J. Org. Chem.* **1977**, *42*, 2224–29.

136 Paulissen, R.; Hayez, E.; Hubert, A. J.; and Teyssié, Ph., "Transition Metal Catalysed Reactions of Diazocompounds—Part III. A One-Step Synthesis of Substituted Furanes and Esters," *Tetrahedron, Lett.* **1974**, 607–8.

137 (a) Trost, B. M., "α-Sulfenylated Carbonyl Compounds in Organic Chemistry," *Chem. Rev.* **1978**, *78*, 363–82; (b) Seebach, D.; and Teschner, M., "Herstellung α-Thiolierter Carbonylverbindungen," *Chem. Ber.* **1976**, *109*, 1601–16.

138 McKervey, M. A.; and Ratananukul, P., "Regiospecific Synthesis of α-(Phenylthio)ketones via Rhodium(II) Acetate Catalyzed Addition of Thiophenol to α-Diazoketones," *Tetrahedron Lett.* **1982**, *23*, 2509–12.

139 Kota amani, H. K.; Gourdoupis, C. G.; and Stammos, I. K., "An Approach to 1-Alkyl-3-phenylpiperidine Derivatives Containing 2,5-Functionalized Groups: 1-M thyl-2-(4-chlorophenylthiomethyl)-5-(methoxycarbonyl)-piperidine," *Tetrahedron* **1994**, *50*, 10477–82.

140 Moody, C. J.; and Taylor, R. J., "Rhodium Carbenoid Mediated Cyclizations. Part 5. Synthesis and Rearrangement of Cyclic Sulfonium Ylides: Preparation of 6-Membered and 7-Membered Sulfur Heterocycles," *Tetrahedron* **1990**, *46*, 6501–24.

141 Henderson, B. S.; Larsen, B. S, and Schwab, J. M., "Chemistry and Photochemistry Attending the Inactivation of *Escherichia coli* β-hydroxydecanoyl Thiol Ester Dehydrase by an Acetylenic Diazoketone," *J. Am. Chem. Soc.* **1994**, *116*, 5025–34.

142 Giddings, P. J.; John, D. I.; and Thomas, E. J., "Reactions of 6-Diazopenicillanates with Allylic Sulphides, Selenides and Bromides," *Tetrahedron Lett.* **1980**, *21*, 395–98; "Preparation and Reduction of 6-Phenylselenylpenicillanates. A Stereoselective Synthesis of 6β-Substituted Penicillanates," *Tetrahedron Lett.* **1980**, *21*, 399–402.

143 Moynihan, M.; and McKervey, M. A., unpublished results.

144 Polozov, A. M.; Polezhaeva, N. A.; Mustaphin, A. H.; Khotinen, A. V.; and Arbuzov, B. A., "A New One-Pot Synthesis of Dialkyl Phosphonates from Diazo-Compounds and Dialkyl Hydrogen Phosphites," *Synthesis* **1990**, 515–17.

145   Rafferty, D.; McKervey, M. A.; and Walker, B. J., unpublished results.

146   Cignarella, G.; Pifferi, G.; and Testa, E., "6-Chloro and 6-Bromopenicillanic Acids," *J. Org. Chem.* **1962**, *27*, 2668–69.

147   McMillan, I.; and Stoodley, R. J., "Studies Related to Penicillins. Part I. 6-α-Chloropenicillanic Acid and its Reaction with Nucleophiles," *J. Chem. Soc. (C)* **1968**, 2533–37.

148   Ernest, I.; Gosteli, J.; and Woodward, R. B., "The Penems, a New Class of β-Lactam Antibiotics. 3. Synthesis of Optically Active 2-Methyl-(5R)-penem-3-carboxylic Acid ," *J. Am. Chem. Soc.* **1979**, *101*, 6301–05.

149   Keith, D. D.; Tengi, J.; Rossmann, P.; Todaro, L.; and Weigele, M., "A Comparison of the Anti-Bacterial and β-Lactamase Inhibiting Properties of Penam and (2,3)-β-Methylenepenam Derivatives—The Discovery of a New β-Lactamase Inhibitor—Conformational Requirements for Penicillin Anti-Bacterial Activity," *Tetrahedron* **1983**, *39*, 2445–58.

150   Brennan, J.; and Hussain, F. H. S., "The Synthesis of Penicillanate Esters from Ultrasonically Formed Organozinc Intermediates," *Synthesis* **1985**, 749–51.

151   McKervey, M. A.; O'Sullivan, M. B.; Myers, P. L.; and Green, R. H., "Reductive Acylation of α-Keto Azides Derived from L-Amino-Acids using N-Protected L-Aminothiocarboxylic S-Acids, *J. Chem. Soc., Chem Commun.* **1993**, 94–96.

152   Shute, R. E.; Dunlap, B.; and Rich, D. H., "Analogs of the Cytostatic and Antimitogenic Agents Chlamydocin and HC-Toxin—Synthesis and Biological-Activity of Chloromethyl Ketone and Diazomethyl Ketone Functionalized Cyclic Tetrapeptides," *J. Med. Chem.* **1987**, *30*, 71–78.

153   Hsiao, C.-N.; Leanna, M. R.; Bhagavatula, L.; Delara, E.; Zydowsky, T. M.; Horrom, B. W.; and Morton, H. E., "Synthesis of N-(t-Butoxycarbonyl)-3-(4-thiazolyl)-L-Alanine," *Synth. Commun.* **1990**, *20*, 3507–17.

154   Olah, G. A.; Welch, J. T.; Vankar, Y. D.; Nojima, M.; Kerekes, I.; and Olah, J. A., "Synthetic Methods and Reactions. 63. Pyridinium Poly(hydrogen fluoride) (30% Pyridine 70% Hydrogen Fluoride): A convenient Reagent for Organo Fluorination Reactions," *J. Org. Chem.* **1979**, *44*, 3872–81.

155   Rauber, P.; Angliker, H.; Walker, B.; and Shaw, E., "The Synthesis of Peptidylfluoromethanes and their Properties as Inhibitors of Serine Proteinases and Cysteine Proteinases," *Biochem., J.* **1986**, *239*, 633–40.

156   Wagner, R.; and Tome, J. M., "Derivatives of Benzofuran," *J. Am. Chem. Soc.* **1950**, *72*, 3477–78.

CHAPTER 9

The Wolff Rearrangement and Related Reactions

The Wolff rearrangement of a diazocarbonyl compound is a specific 1,2 rearrangement, accompanying or following loss of nitrogen, to a ketene that may undergo further reactions such as nucleophilic attack by water, alcohols, or amines, or cycloaddition with unsaturated systems (Scheme 1). Wolff discovered the rear-

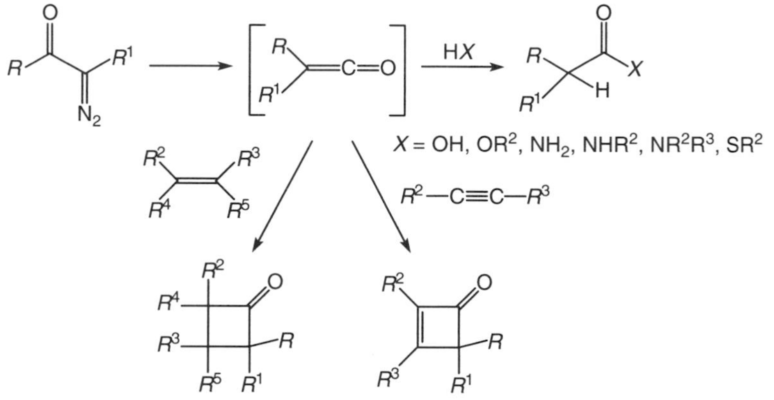

$X = OH, OR^2, NH_2, NHR^2, NR^2R^3, SR^2$

Scheme 1

rangement in 1912,[1] many years before reliable synthetic routes to diazoketones were available. When the methods of diazomethane acylation and diazo transfer were introduced (Chapter 1), the potential of the Wolff rearrangement in organic synthesis was quickly appreciated. The literature since 1940 is replete with examples of its use, not just as the key step in the Arndt–Eistert homologation reaction of carboxylic acids, but as a powerful means of imposing angle strain on cyclic systems through ring contraction and, more recently, as a means of generating highly unstable species isolated in matrices, where they can be observed spectroscopically.

The key feature of the Wolff rearrangement is the formation of a reactive ketene. The subsequent behavior of the ketene is determined by its structure, the reaction conditions, and the availability of potential reaction partners. In this chapter a selection of examples from the recent literature is grouped into four subsections to illustrate the versatility of the Wolff rearrangement in modern organic synthesis.

9.1   THE ARNDT–EISTERT HOMOLOGATION

The Arndt–Eistert synthesis is a one-carbon homologation of a carboxylic acid *via* Wolff rearrangement of a diazoketone for which many hundreds of examples are available.[2-6] Silver ion catalysis is the most effective option for effecting rearrangement to a ketene, which is then intercepted by an appropriate nucleophile such as water, alcohol, or amine. A typical application is the formation of ethyl 1-naphthylacetate from 1-(diazoacetyl)naphthalene (eq. 1).[7] Because heteroge-

$$\text{(1)}$$

neous conditions using a suspension of silver oxide in methanol can often prove erratic, Newman and Beale[7,8] developed a procedure for carrying out the reaction in homogeneous solution under mild conditions. The catalytic system involved the use of silver benzoate solubilized by triethylamine.

A recently claimed further improvement involves the use of ultrasound, which enables the reaction to occur rapidly at room temperature.[9] Alternatively, the Wolff rearrangement can be brought about photochemically, and this is preferred when the objective is to generate very strained products or ketenes for spectroscopic characterization *in situ*.

The Arndt–Eistert homologation has been outstandingly successful in amino acid and peptide modification, in particular, the transformation of α-amino acids into β-amino acids and analogues. β-Amino acids are components of some natural products,[10] including peptides,[11] which show significant biological activity. The antitumor agent taxol contains a β-amino-α-hydroxy acid.[12] Furthermore, β-lactams and β-lactam antibiotics are readily accessible from β-amino acids[13,14].

There are two important considerations in applying Arndt–Eistert homologation to α-amino acids. First, there is the nature of the N-protecting group and, second, with enantiopure amino acids, the extent of racemization in the reaction. The early studies of Balenovic *et al.*[15] established that phthaloyl-protected amino acids were suitable, and more recent extensive studies have confirmed that the synthetically more versatile carbamate-protected derivatives [Boc, Cbz (Z), ethoxycarbonyl] are equally suitable as substrates. The extent of racemization was not addressed in much of the earlier work of Balenovic *et al.*,[15] but recent studies,

ETHYL 1-NAPHTHYLACETATE[7]
(1-NAPHTHALENEACETIC ACID, ETHYL ESTER)

Caution! Diazomethane is hazardous. Follow the directions for its safe handling given in Chapter 1. The intermediate, 1-(diazoacetyl)naphthalene, is a very strong skin irritant.

A. 1-(Diazoacetyl)naphthalene.

A solution of 30.5 g (0.160 mole) of 1-naphthoyl chloride in 50 mL of dry ether is added during 30 min to a magnetically stirred, ice-cooled solution of 6.72 g (0.160 mole) of diazomethane and 16.1 g (0.160 mole) of dry triethylamine in 900 mL of dry ether. The mixture is stirred for 3 h in the cold, and the triethylamine hydrochloride is removed by filtration and washed twice with 30–50 mL portions of dry ether. The ether is removed from the combined filtrate and washings on a rotary evaporator under reduced pressure. The yellow solid residue is dissolved in 75 mL of dry ether, and the solution is cooled by a dry ice acetone mixture. The solid deposited is collected by filtration on a glass fritted-disk funnel, and the adhering ether is removed under reduced pressure as the temperature is allowed to reach room temperature. There is obtained 26.6–28.8 g (85–92%) of yellow 1-(diazoacetyl)naphthalene, m.p. 52–53°C.

B. Ethyl 1-naphthylacetate.

A solution of 15.7 g (0.080 mole) of 1-(diazoacetyl)naphthalene in 50 mL of absolute ethanol is placed in a 100-mL two-necked flask equipped with a Teflon-coated magnetic stirring bar, a serum stopper cap, and a reflux condenser connected at the top to a gas-collecting device. The solution is heated to reflux, and 1 mL of a freshly prepared catalyst solution made by dissolving 1 g of silver benzoate in 10 mL of triethylamine is added by injection through the serum cap. Evolution of nitrogen occurs, and the mixture turns black. Addition of a second milliliter of catalyst solution is made when the evolution of nitrogen almost stops. This procedure is continued until further additions cause no further evolution of nitrogen. The reaction mixture is refluxed for 1 h, cooled, and filtered. The solvents are removed from the filtrate on a rotary evaporator under reduced pressure. The residue is taken up in 75 mL of ether, and the solution is washed twice in turn with aqueous 10% sodium carbonate, water, and saturated brine. Each aqueous extract is extracted with ether, and the combined ethereal extracts and solution are dried by filtration through anhydrous magnesium sulfate. After removal of the ether, distillation affords 14.4–15.8 g (84–92%) of colorless ethyl 1-naphthylacetate, b.p. 100–105°C (0.1–0.2 mm).

the most thorough and comprehensive being those of Podlech and Seebach,[16] in which the homologation of a series of enantiopure *N*-protected α-amino acids was reinvestigated applying modern analytical techniques, have established that retention of configuration occurred in all cases except for carbamate-protected phenylglycine. Podlech and Seebach used diazoketones derived from Z-Ala-OH, Z-Phe-OH, Boc-Phe-OH, Boc-*tert*-Leu-OH, Boc-(Orn(Boc)-OH, Z-Phg-OH, and Boc-Phg-OH. NMR analysis of the Mosher derivatives of the homologated products showed that the enantiomeric purity of the β-amino acids was greater than 98%. Exceptionally, Boc-phenylglycine was transformed to the corresponding β-amino acid with an enantiomer ratio of *ca.* 9:1. The partial racemization here was traced back to the activation step for diazoketone formation and not the Wolff rearrangement step. A selection of β-amino acids, methyl esters, and dipeptides produced in this way by Podlech and Seebach[16] and by McKervey and co-workers[17] is given in Table 9.1, and a representative procedure in methanol is reproduced.

TABLE 9.1   A Selection of β-Amino Acid/Peptide Methyl Esters Prepared by Arndt–Eistert Homologation (yields, %)

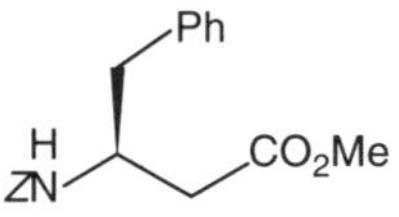

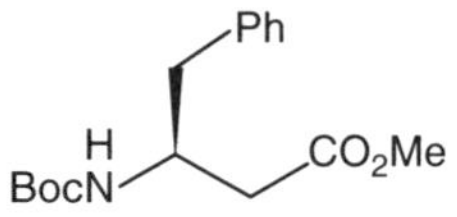

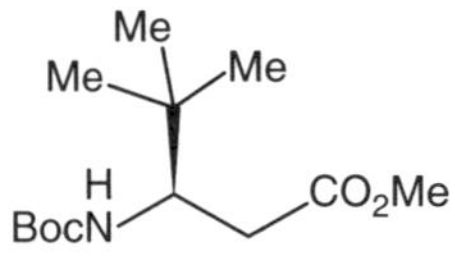

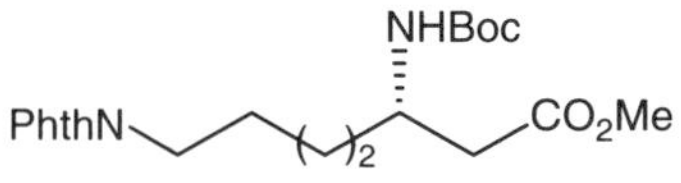

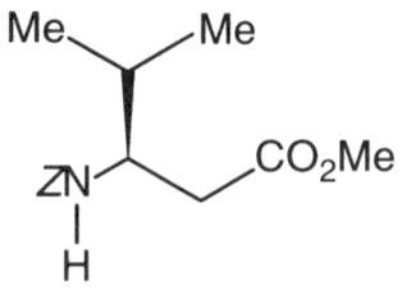

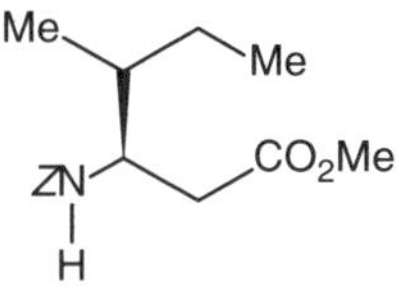

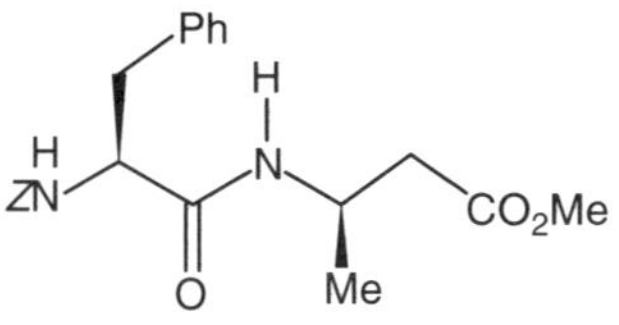

Although methanol and ethanol are the most common solvents for Arndt–Eistert homologation, it is possible to use other nucleophilic solvents that are more sterically demanding or to use an inert solvent such as THF containing the nucleophile. It is known, for example, that ketenes generated by the Wolff rearrangement can be trapped to give *tert*-butyl esters by conducting the reaction in

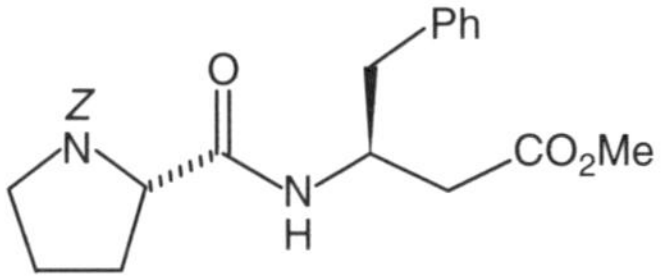

3(S)-METHYL-3-(N-TERT-BUTOXYCARBONYLAMINO)-6-(N-BIS-BENZYLOXYCARBONYL GUANIDYL) HEXANOATE

To a solution of N^α-butoxycarbonyl-N^γ-bis (benzyloxycarbonyl)-L-arginyl diazomethane (2.76 g, 4.88 mmol in MeOH:THF (1:1, 50 mL)) was added 2.5 mL of a solution of silver benzoate (1.0 g) in triethylamine (20 mL) with stirring. After 20 min an additional 1 mL of the silver benzoate solution was added and stirring was continued for 1 h. Celite and decolorizing carbon were added and, after filtration, the solution was concentrated *in vacuo* to yield the crude product. Purification by flash chromatography on silica gel using 30% ethyl acetate in hexane as eluant yielded the pure product (79%) as a white crystalline solid, m.p. 83–85°C; $[\alpha]_D^{20} = 2.3$ (c, 1.4 in CH_2Cl_2); ν_{max} (KBr)/cm^{-1} 3385, 3345 (NH), 1729, 1686 (CO); ^{1}H (300 MHz, CDCl$_3$) 1.41 (9 H, s, C(CH$_3$)$_3$), 1.64 (4 H, m, (CH$_2$)$_2$CHNH), 2.45 (2 H, d, J = 5.4 Hz, CH$_2$CO$_2$CH$_3$), 3.60 (3 H, s, OCH$_3$), 3.90 (1 H, br s, NH), 3.96 (2 H, m, CH$_2$NC = NH), 5.13 (2 H, s, OCH$_2$Ph), 5.22 (2 H, s, OCH$_2$Ph), 7.32 (10 H, m, ArH), 9.35, 9.45 (2 × 1 H, br s), HN = CH).

tert-butyl alcohol.[18] Podlech and Seebach[16] have exploited the steric demand a stage further by using ketenes generated from N-protected α-amino diazoketones to acylate sterically hindered hydroxyl groups of compounds that cannot possibly be employed as solvents (i.e., in large excess). Diazoketones derived from Z-Ala-OH **1** and Z-Ala-Ala-OH **2** were chosen as ketene precursors in THF for reaction with carbohydrates (*ca.* 4 equiv.) containing one free hydroxyl group: the bis-acetonides of α-D-galactose **3**, α-D-mannofuranose **4** and α-D-glucofuranose **5**, in which the OH groups are increasingly hindered (eqs. 2–5). The reactions were much slower than those leading to methyl esters, requiring a 48 h reaction time at *ca.* −20°C compared with a 3 h reaction time in methanol at the same temperature.

Whereas the homo-alanine esters (**6** and **7**) were formed with the least and the most hindered hydroxy groups, homodipeptide esters (**8** and **9**) were only isolated from the reactions with the two sugars, **3** and **4**, respectively, containing the less hindered hydroxy groups. With the most bulky hydroxy compound tested, the bis-acetonide of α-D-glucofuranose **5** and the Z-protected dipeptide **2**, no product could be detected even after prolonged reaction times. The general conclusion reached by Seebach and Podlech is that attack by a nucleophile on ketenes formed

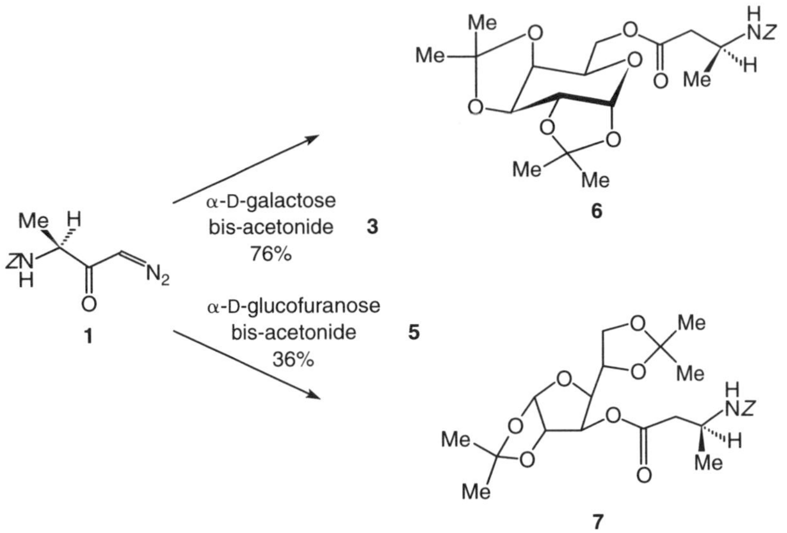

(2)

(3)

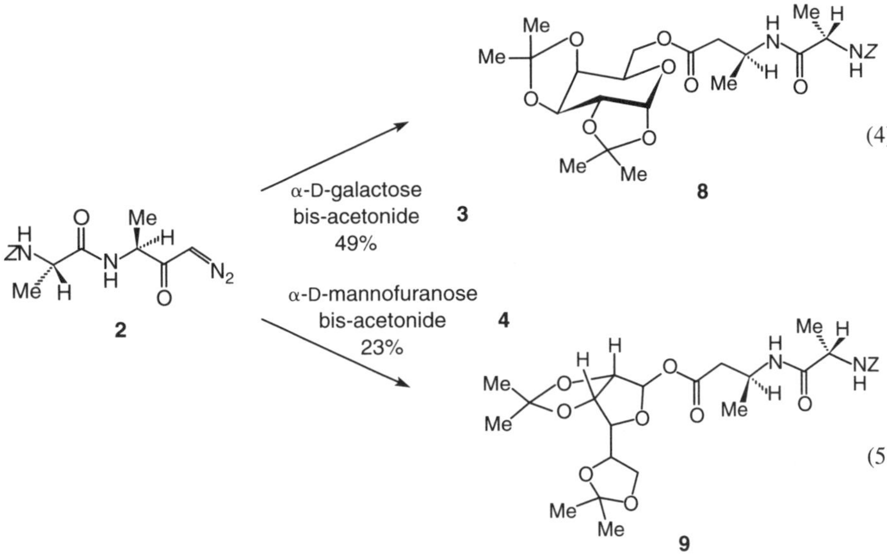

(4)

(5)

by the Wolff rearrangement of amino acid or peptide intermediates is quite sensitive to steric hindrance in either reactant, opening up the possibility of using multifunctional nucleophiles with different degrees of steric hindrance of the functional groups and different inherent nucleophilicity.[19] These nucleophiles should react selectively with the ketene (or its dihydrooxazinone equivalent, *vide infra*) to form the corresponding homologated derivatives. A simple demonstration of this type of selective reaction is provided by the rearrangement of diazoketone **10** derived from Boc-alanine in the presence of 3-methyl-1,3-butanediol

(4 equiv) in THF.[19] Although the reaction rate was very slow, there was a clear preference (19:1) for the primary ester **11** over the tertiary ester **12** (eq. 6). A

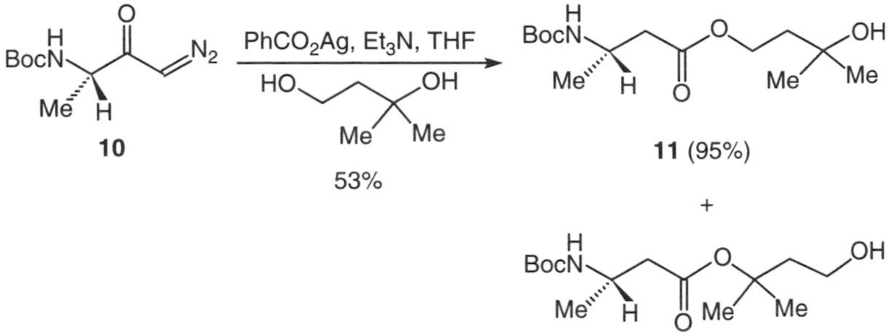

(6)

similar selectivity in favor of the primary hydroxyl function was observed with 2-valine diazoketone, an amino acid derivative with a bulkier side chain than alanine. The selectivity is even more pronounced with diazoketones derived from dipeptides with Z-Ala-AlaCHN$_2$ **2** forming the primary ester of 3-methyl-1,3-butanediol with a selectivity of >50:1. As a further example of the selective acylation of a dihydroxy compound, the diazoketone **13** of Z-valine was rearranged in the presence of isopropylidine-D-xylofuranose.[19] The ester **14** of the less hindered alcohol was formed with a selectivity of 19:1 in 55% yield (eq. 7). To test the

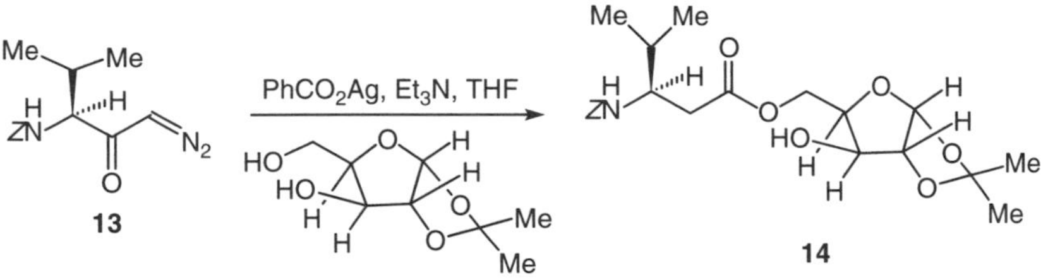

(7)

selectivity between two functional groups of inherently different nucleophilicity, 2-aminoethanol was employed as a bifunctional probe with a minimum of steric hindrance on either functional group. Rearrangement of the diazoketones of Z-Ala **1**, Boc-Ala **10**, Z-Val **13**, and Boc-Val **17** in the presence of 2-aminoethanol furnished amides **15**, **16**, **18**, **19**, respectively, in yields ranging from 74 to 87% (eqs. 8 and 9). The success of these selectivity studies with bifunctional nucle-

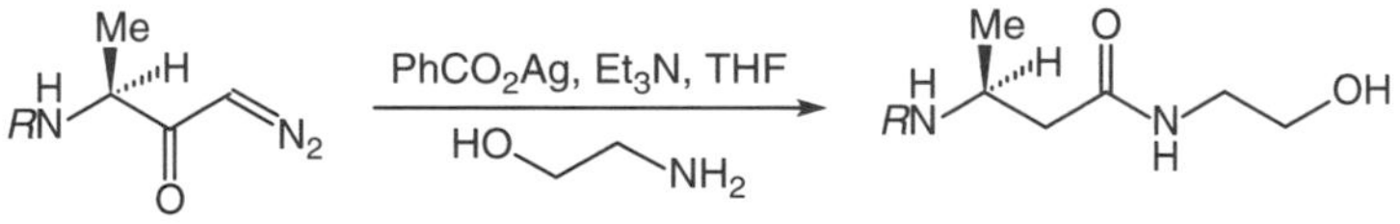

(8)

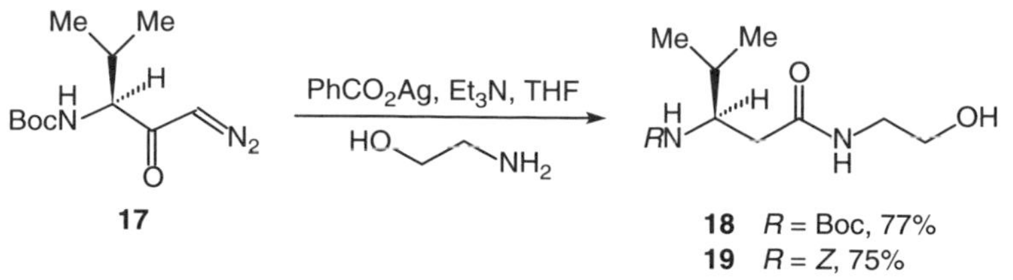

(9)

ophiles prompted the Swiss group to attempt the much more challenging problem of selective acylation of polyfunctional molecules such as totally unprotected carbohydrates,[19] for example, sucrose, which has eight different hydroxy groups. Product analysis of the reaction of the Boc-alanine diazoketone **10** with sucrose (1 equiv.) in DMF containing lithium perchlorate (to solubilize the sucrose), silver benzoate, and triethylamine revealed the presence of a large number of components, indicating poor regioselectivity. Nevertheless, one component could be isolated in 13% yield and identified as the 2-acyl derivative **20** (eq. 10). A final

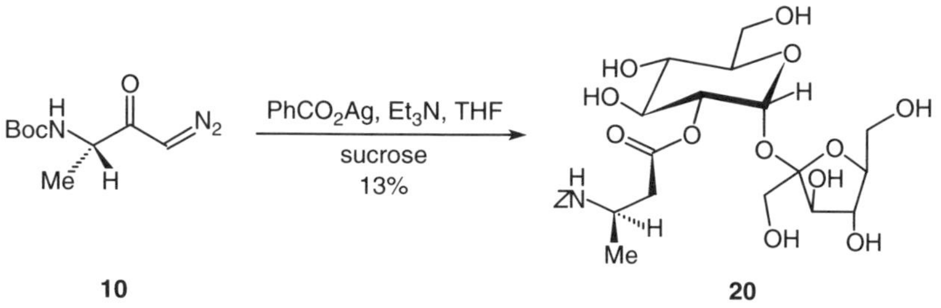

(10)

example of selective acylation *via* the Arndt–Eistert procedure is illustrated in eq. 11 by the reaction of Z-AlaCHN$_2$ **1** with thymidine, which produced in 58% yield mainly the 5′-acylated derivative **21**.[19]

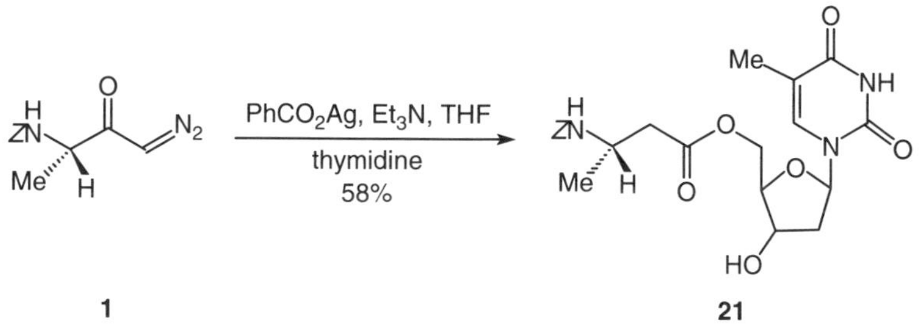

(11)

In the course of these extensive investigations of the Arndt–Eistert reaction in amino acid and peptide modification, Podlech and Seebach[20] made an additional useful observation concerning the nature of the intermediates involved, particularly the acylating agent itself, presumed to be a ketene. The question was raised as to whether a peptidic ketene **22** could be generated in a non-nucleophilic sol-

R' = amino acid/peptide residue

22

vent and trapped intermolecularly with the amino function of a second amino acid or peptide. Thus the generation of a β-amino acid might be combined with a peptide coupling step for the production of a homopeptide. In practice, a solution of Z-alanine-derived diazoketone **1** and 2–4 equiv. of valine benzyl ester in THF was treated with Et$_3$N and a catalytic amount of silver benzoate at −25°C to yield the homopeptide **23** in 81% yield (eq. 12).[20,21] In a similar manner, the pro-

PhCO$_2$Ag, Et$_3$N, THF

Valine benzyl ester

81%

(12)

1   **23**

tected di- and tripeptides Z-Ala-Ala-OH and Boc-Leu-Sar-Leu-OH were coupled *via* diazoketones **24** and **25**, respectively, with the tripeptides H-Ala-Sar-MeLeu-OBn and H-Leu-Sar-Leu-OMe (see Table 9.2). Although ketene **22** is undoubt-

TABLE 9.2 Homopeptide Derivatives Synthesized According to Equation 12

Diazoketone	Amino ester	Homopeptide (%)
Z-Ala-Ala-CHN$_2$ **24**	H-Ala-Sar-MeLeu-OBn	95
Boc-Leu-Sar-Leu-CHN$_2$ **25**	H-Leu-Sar-Leu-OMe	60

edly an intermediate in these reactions, Podlech and Seebach[20] were able to demonstrate that it is not the immediate precursor of the homopeptidic products. This became clear when diazoketone **2** was decomposed catalytically in THF in the absence of a nucleophile. Addition of methanol after 6 h produced the methyl ester **26** in virtually the same yield as that obtained when the reaction was conducted in methanol as solvent. Closer inspection of the reaction mixture by NMR spectroscopy revealed that it was not the ketene that was present prior to the addition of methanol in the former reaction, but a dihydrooxoxazinone **27** (see Scheme 2). Thus the ketene produced by Wolff rearrangement is rapidly transformed through involvement of the neighboring amide bond into the dihydrooxazinone **27**, a reactive cyclic iminoanhydride, which is readily attached intermolecularly by an amino group to form a homopeptide. Independently, Bastiaans et al.[22] developed a similar route for converting arginine into β-homoarginine dipeptides. In this conversion, an example of which is the

PROTECTED HOMOPEPTIDE

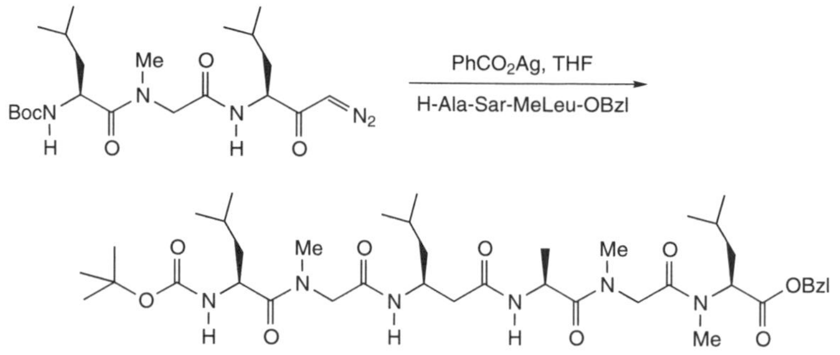

A solution of siver benzoate (16.0 mg, 70.0 μmol) in Et$_3$N (215 μl, 1.54 mmol) was added to a solution of diazoketone (238 mg, 541 μmol) and H-Ala-Sar-MeLeu-OBzl (528 mg, 1.40 mmol) in THF (10 mL) at $-25°$C under argon with exclusion of light. The reaction was warmed to room temperature over 3 h, and some Et$_2$O was added. After workup by extraction with HCl (0.2N, 2×) and saturated NaCl, NaHCO$_3$, and NaCl solutions, the organic solution was dried (MgSO$_4$) and concentrated to dryness. The residue was purified by chromatography on silica gel (ethyl acetate) to give the protected homopeptide (257 mg, 326 μmol, 60%).

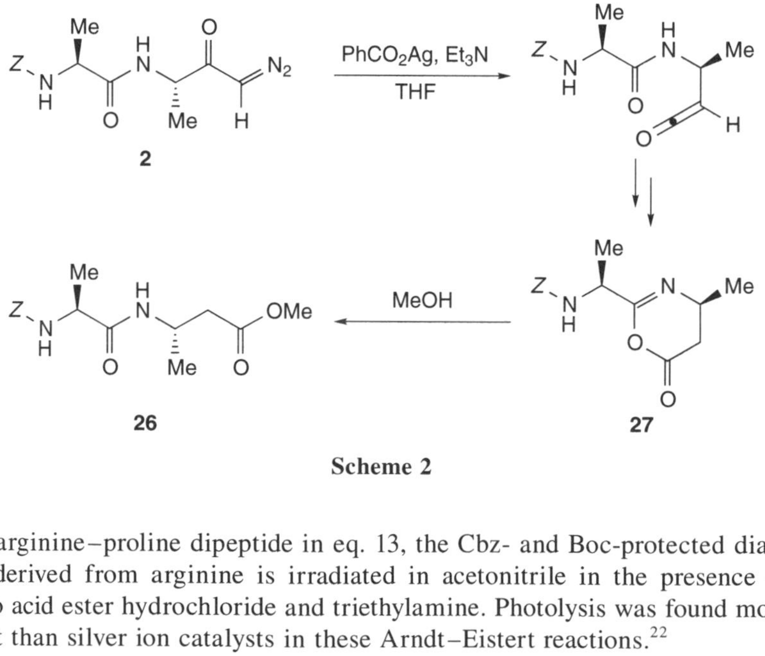

Scheme 2

homoarginine–proline dipeptide in eq. 13, the Cbz- and Boc-protected diazoketone derived from arginine is irradiated in acetonitrile in the presence of an amino acid ester hydrochloride and triethylamine. Photolysis was found more efficient than silver ion catalysts in these Arndt–Eistert reactions.[22]

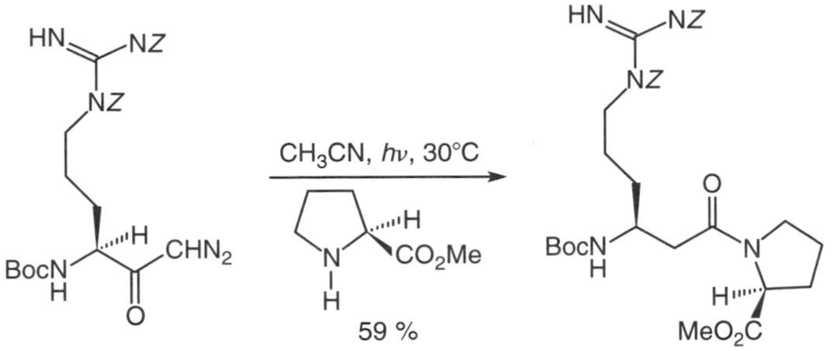

(13)

59 %

A similar approach to amide bond formation, but outside the peptide field, has been adopted by Pendrak and Chambers[23] in their synthesis of RG-14893, a high-affinity leukotriene B_4 receptor antagonist. Here an aminoethyl side chain was added by photochemical Wolff rearrangement of the 1-naphthyl-derived diazoketone **28** in the presence of methylphenylethylamine, whereupon amide **29** was obtained in 87% yield (eq. 14). Amide **29** was converted into RG-14893 via hydrolysis in aqueous sodium hydroxide.

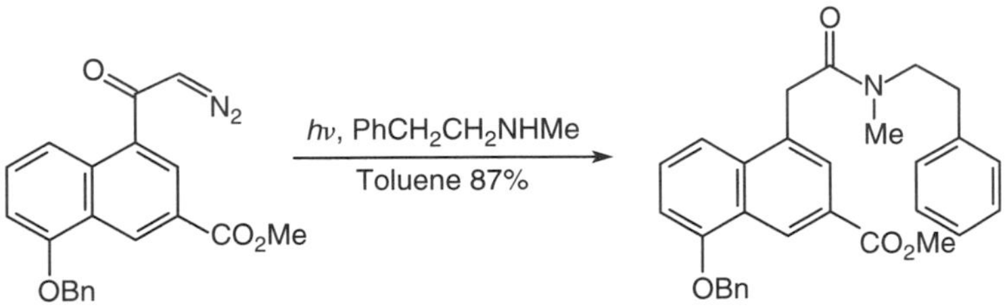

(14)

In yet another useful exploitation of ketenes generated from amino acids via Wolff rearrangement, Podlech[24] has developed an efficient route to aminoalkyl-substituted β-lactams. When generated in the presence of imines, these ketenes undergo diastereoselective [2+2] cycloaddition. Because silver ion is poisoned by imines, it was not possible to employ silver benzoate to catalyze the rearrangement. Photochemical decomposition provided a satisfactory alternative. Irradiation of the diazoketone **10** derived from Boc-alanine (eq. 15) in the presence of

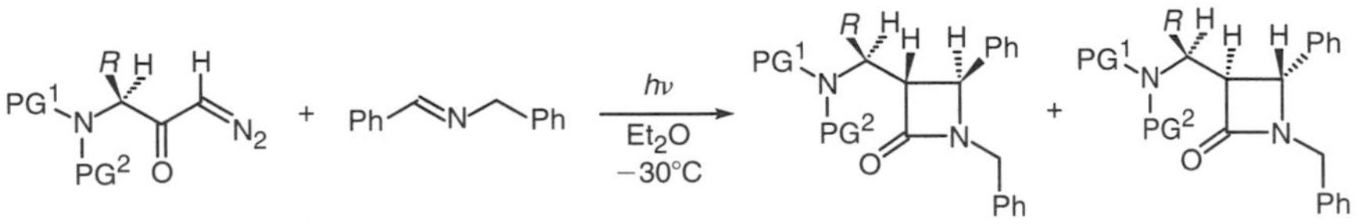

10: R = Me; PG^1 = Boc; PG^2 = H   **30**   **31a–36a**   **31b–36b**

(15)

N-benzylbenzaldimine **30** at −30°C in diethyl ether furnished two of the four possible diastereoisomeric β-lactams **31a,b** (70% yield) in a 71:29 ratio, both products having the *trans*-substitution shown in Table 3. Since the Wolff rear-

TABLE 9.3   Ketene-Imine Cycloaddition

| | | | Diazoketone | | | |
PG^1	PG^2	R	Amino Acid	Products	d.r.	Yield (%)
Z	H	Me	Ala	**32a,b**	67:33	71
	Pht	Me	Ala	**33a,b**	83:17	54
Z	H	*i*-Bu	Leu	**34a,b**	70:30	89
Z	H	*t*-Bu	Tle	**35a,b**	93:7	85
Z	H	PhCH$_2$	Phe	**36a,b**	59:41	58

rangement is known to proceed with retention of configuration, the stereogenic center present on the aminoalkyl side chain is not affected by the cycloaddition. In a similar manner, the diazoketones listed in Table 9.3 were treated with imine **30** to furnish the corresponding *trans*-substituted β-lactams **32–36** in 54–89% yield and in diastereoisomeric ratios ranging from 59:41 to 93:7. The product ratio is significantly influenced by the steric hindrance of both the side chain (*R*) and by the protecting groups (PG^1, PG^2). Thus the products derived from Z-alanine were formed in a relatively poor diastereoisomer ratio (67:33), whereas the two *tert*-leucine diazoketones provided the β-lactams in a 93:7 ratio. The bulky phthaloyl (Pht) protecting group increased the diastereoselectivity in comparison with the Z and Boc groups, but affected the product yield adversely. The exclusive formation of *trans*-substituted products was attributed by Podlech to the fact that the diazoketones were decomposed photochemically, thus generating the ketenes in an excited state. The ring closure should therefore proceed in a disrotatory fashion, leading to the observed stereochemistry.

The conversion of α-amino acids into β-amino-α-keto esters by Darkins *et al.*[17] presents another demonstration of the versatility of diazocarbonyl chemistry in stereoselective synthesis. Amino acid–derived α-keto acids and esters are effective inhibitors of the serine protease chymotrypsin,[25] and their incorporation into appropriate peptide recognition sequences has produced potent, selective inhibitors of cysteine proteases such as calpain and cathepsin B.[26] The synthesis developed by Darkins et al.[17] completely circumvents the problems of racemization encountered with earlier synthesis and is equally applicable to N-protected amino acids and peptides. The process, illustrated in Scheme 3 for elaboration of Z-Phe-Ala-OH, commences with formation of the diazoketone **37**, which is then subjected to Arndt–Eistert conditions in methanol in the usual way to form the β-amino methyl ester **38**. Next, a diazo group was reintroduced alpha to the ester function in **38** using the Danheiser procedure (see Chapter 1) to form **39**, and in the final step the diazo group was replaced by a carbonyl group as in **40** *via* oxidation with dimethyldioxirane in acetone. The oxidation of diazocarbonyl compounds by dimethyldioxirane is discussed in more detail in Chapter 12.

There are many other examples in stereoselective synthesis where the ability of the Arndt–Eistert synthesis to proceed with strict retention of configuration has been used advantageously. For example, diazoketone **41**, derived from

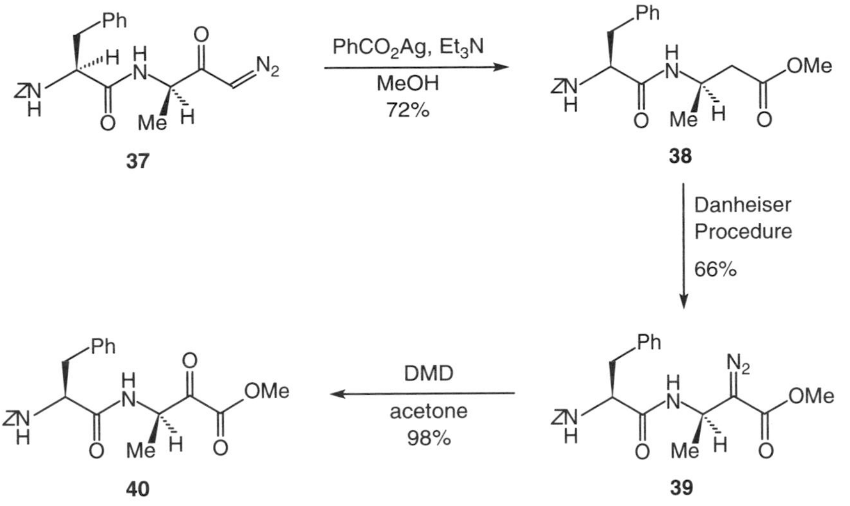

Scheme 3

Z-L-aspartic acid, is readily transformed into β-amino ester **42**, which served as an enantiopure building block in the construction of azetidinone **43** (Scheme 4).[27]

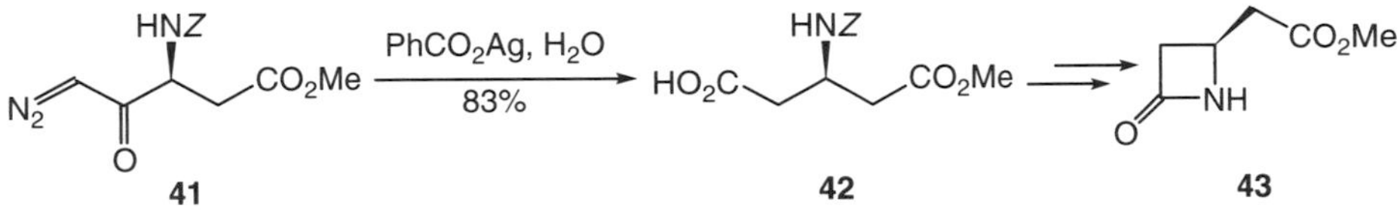

Scheme 4

Pellicciari et al.[28] have devised a similar route from L-glutamic acid to L-α-aminoadipic acid. The former, as the Z-formaldehyde adduct, was transformed into diazoketone **44**, which on Wolff rearrangement in aqueous THF furnished Z-L-aminoadipic acid **45** in quantitative yield. Deprotection of the amino function in 6N hydrochloric acid completed the synthesis in 49% overall yield from L-glutamic acid (Scheme 5). Among other examples of the Arndt–Eistert

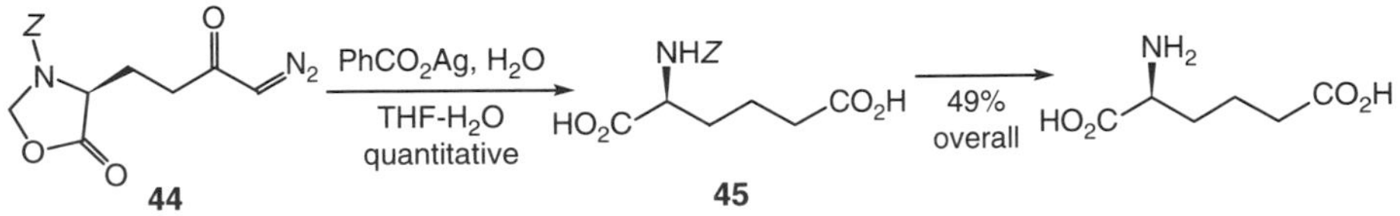

Scheme 5

homologation in the total synthesis of enantiopure natural products are those shown in Schemes 6 and 7, the former en route to the alkaloid (−)-indolizidine

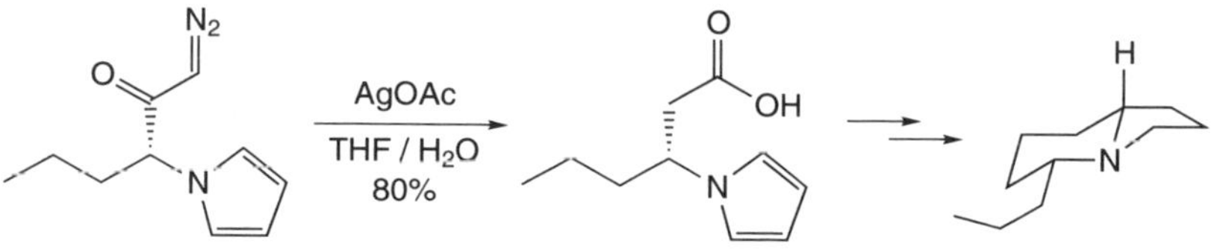

(−)-*indolizidine 167B*
46

Scheme 6

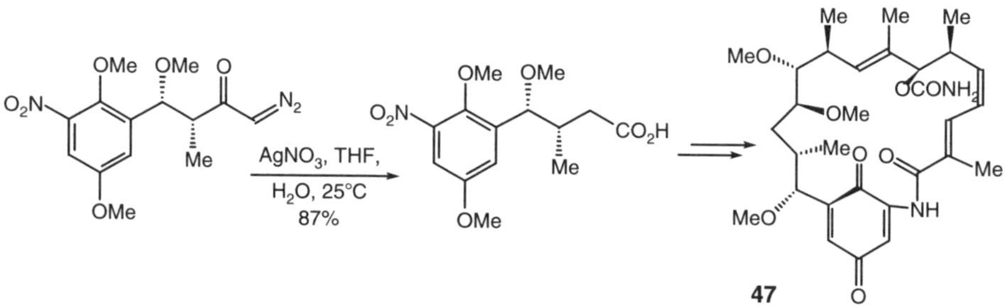

47

(+)-macbecin I

Scheme 7

46[29] and the latter, with silver nitrate as catalyst in aqueous THF, to the ansamycin antitumor antibiotic, (+)-macbecin I **47**.[30]

Leumann and coworkers[31] used the Arndt–Eistert route to synthesize a building block for a nucleic acid analog with a chiral flexible peptide backbone. Exposure of diazoketone **48** to silver benzoate in benzyl alchohol furnished the benzyl ester **49** which was subsequently transformed into bromide **50** en route to the target thiamine derivative **51** (Scheme 8). (2*S*,3*R*)-3-methylglutaric acid **52** has

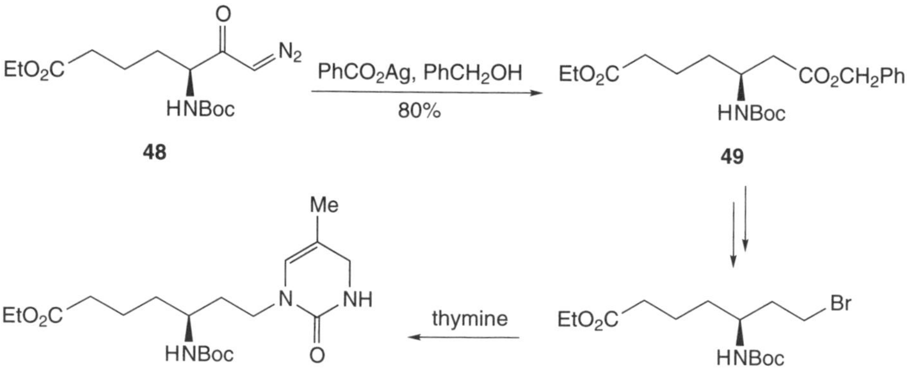

48 **49** **51** **50**

Scheme 8

been prepared *via* Arndt–Eistert homologation of (2*S*, 3*R*)-methylaspartic acid **53** (eq. 16).[32]

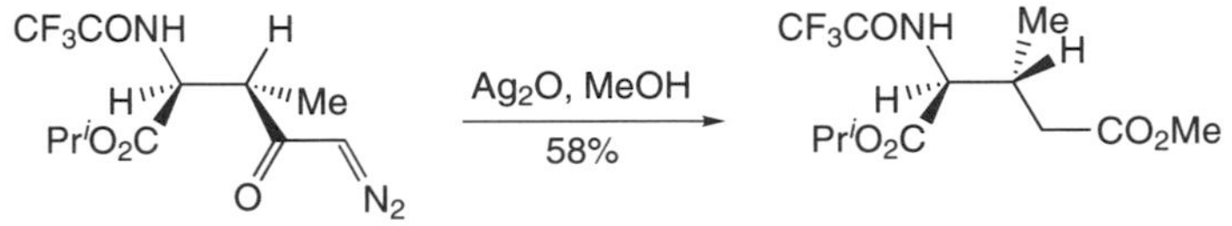

(16)

Photochemically induced Wolff rearrangement of diazoketone **54** in methanol proceeded stereospecifically to form methyl ester **55** (eq. 17); the stereochemical outcome was ascribed to the intervention of an intermediate cyclic complex involving the silicon atom.[33] Two final examples (eqs. 18 and 19) illustrate the ap-

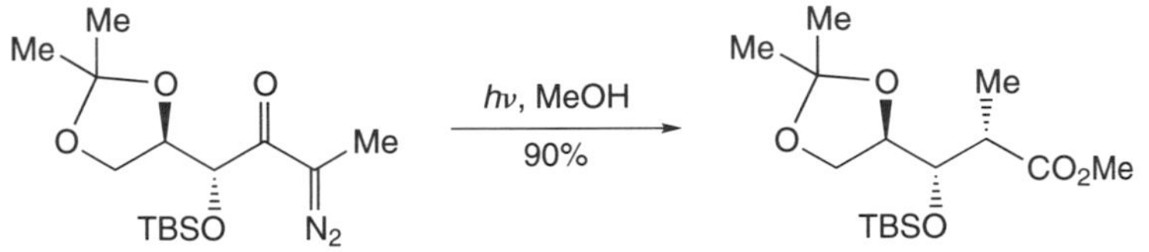

(17)

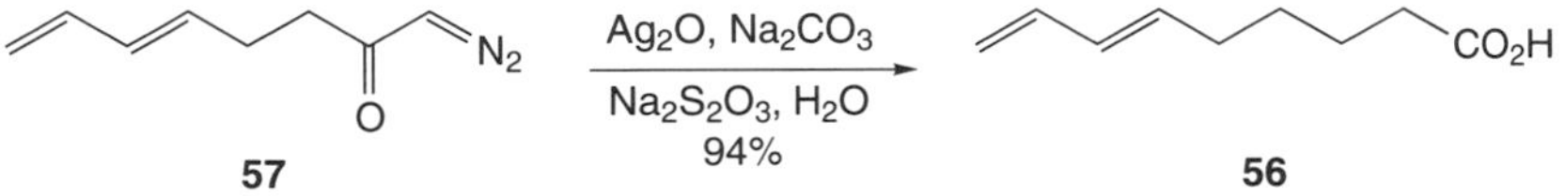

(18)

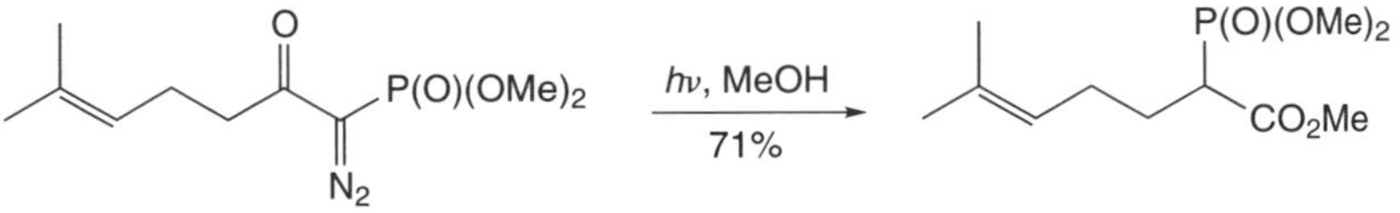

(19)

plication of the Arndt–Eistert procedure to unsaturated diazoketones where potential side reactions resulting from intramolecular cyclization are avoided. In the former silver oxide in water containing sodium carbonate and sodium thiosulfate was used to catalyze the formation of several dienic carboxylic acids **56** from terminal diazoketones **57**.[34] The latter is a photoinduced Wolff rearrangement leading from a α-diazophosphonate **58** to a synthetically useful phosphonoacetate.[35]

Epoxy diazomethyl ketones of the type in eq. 20 undergo an interesting photochemical Wolff rearrangement leading to ring opening and formation of a

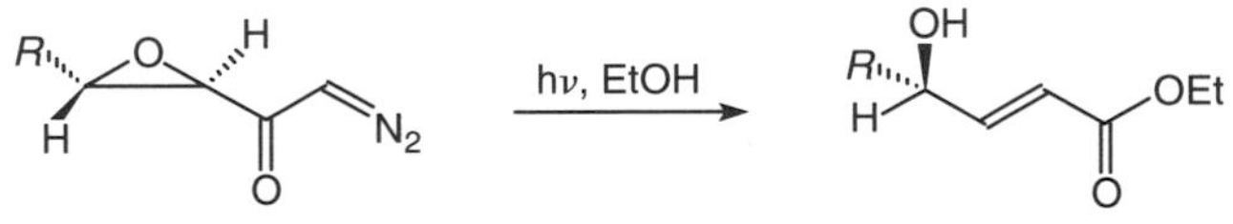

(20)

$R = C_{10}H_{21}$ (50–58%), C_4H_9 (50%), C_5H_{11} (63%), $CH_3CH(OAc)CH_2CH_2$ (61%) for products after acylation

γ-hydroxy-α,β-unsaturated ester.[36,37] The process has also been applied successfully to the stereospecific formation of more highly hydroxylated α,β-unsaturated esters, such as the product shown in eq. 21.[38] In both these examples the

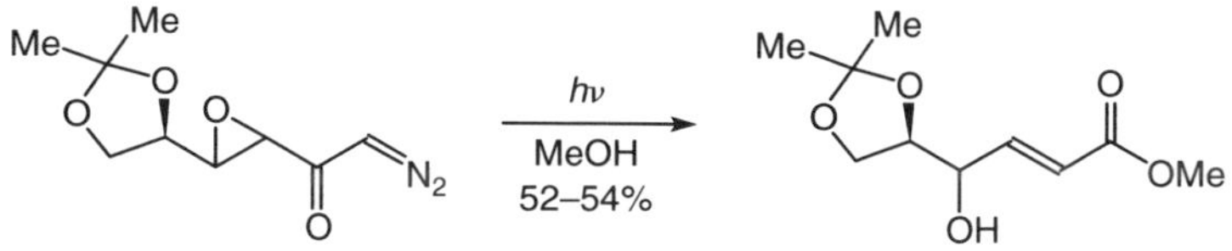

$$(21)$$

epoxy diazoketone precursors were obtained *via* Sharpless epoxidation of the appropriate allylic alcohol, which was then oxidized to the carboxylic acid in preparation for introduction of the diazomethyl group.

Extension of the Arndt–Eistert homologation to acyclic diazoesters **60** or diazoamides **61**, in general, affords low yields of products derived from Wolff rear-

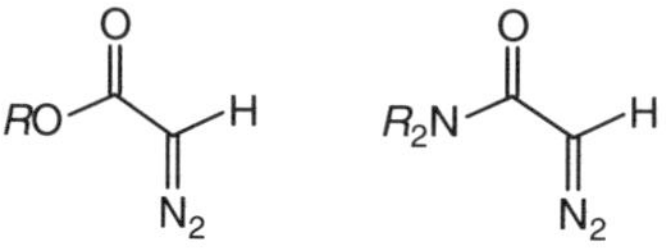

rangement, probably as a result of the relatively poor migratory power of alkoxy and amino groups, which results in competing carbenoid reactions.[39] Unsymmetrical 2-diazo-1,3-dicarbonyl compounds of general type **62** (Scheme 9) can give

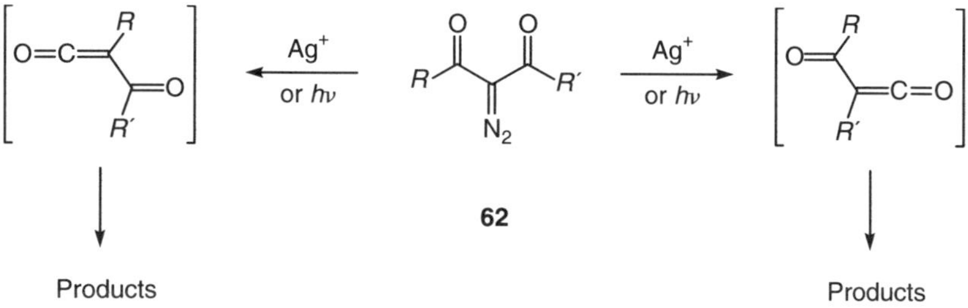

R = OMe, R' = Me; R = OMe, R' = OMe; R = OMe, R' = SPh

Scheme 9

rise to two ketenic products on the Wolff rearrangement depending on which group migrates[40-42]; examples involving ring contraction are shown in Table 9.4 (entries 16 and 21).

9.2 RING CONTRACTION *VIA* THE WOLFF REARRANGEMENT

Application of the Wolff rearrangement to ring contraction has proved to be especially effective in the synthesis of molecules with severe angle strain.

It is known that ring contraction via Wolff rearrangement is generally successful in cases where the ring strain increases by not more than 35 kcal mol^{-1}.[43,44] In an interesting test of this generalization Wiberg and co-workers[43] examined the photochemical Wolff contraction of the spiro-tricycloheptane diazoketone in eq. 23. The strain energies of the hydrocarbons corresponding to precursor and

$$(23)$$

ring-contracted product of this transformation had been calculated by molecular mechanics to differ by 37 kcal mol^{-1}, a value just beyond the upper end of the accepted range of applicability of the Wolff contraction. The reaction in methanol yielded only the product of O–H insertion with the solvent.

The photochemical version of the Wolff rearrangement has been employed repeatedly at low temperatures to generate compounds for spectroscopic examination whose extreme instability precludes isolation (*vide infra*).

Although there are no limitations in the formation of various ring sizes *via* the Wolff rearrangement protocol, very few examples of formation of three-membered rings or of large ring systems have been reported. A selection of examples from the recent literature is summarized in Table 9.4. In many cases ring contraction is the key step en route to the target molecule indicated.

Four-membered carbocyclic and heterocyclic systems are readily accessible from cyclopentyl precursors (entries 1–13). Entries 1 and 2 show how ring contraction has been employed to produce fused, polycyclic systems in which angle strain forces the central quaternary carbon atom more and more towards an energetically unfavorable planar geometry. Entry 4 is a particularly good example from the work of Eaton and co-workers, the very unfavorable dihedral angle between the 1 and 7 positions of norbornane making the resulting cyclobutane derivative exceedingly distorted. The remaining entries illustrate some of the many ways in which other bridged bi- and tricyclic carbocyclic systems can be manipulated through ring contraction. The product in entry 9 was required for an evaluation of cyclobutylcarbinyl derivatives in diterpene biosynthesis. Entry 21 illustrates a photochemical Wolff ring contraction conducted in the presence of an amine leading directly to β-ketoamides. Entry 10, ring D contraction of dehydroisoandrostone, is described with full experimental details in *Organic Synthesis*.[54] Entry 5 provides a new synthetic route to β-lactams that involves the photolytic ring contraction of 4-diazopyrrolidine-2,3-diones to 3-carboxy-2-azetidinones. Entry 15 illustrates the Wolff rearrangement in a bicyclic system con-

TABLE 9.4 Ring Contraction Strategy in Synthesis

Entry	Diazocarbonyl	Product	Target Molecule	Ref.
1	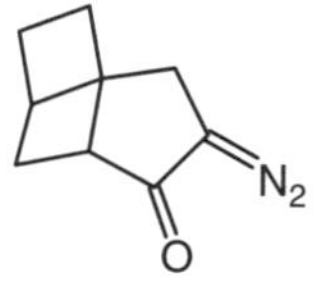	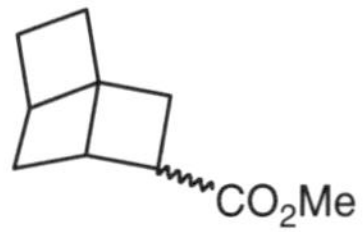 CO₂Me 42 % 3:1 ratio of isomers	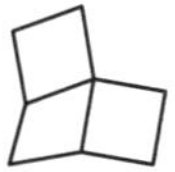	44
2	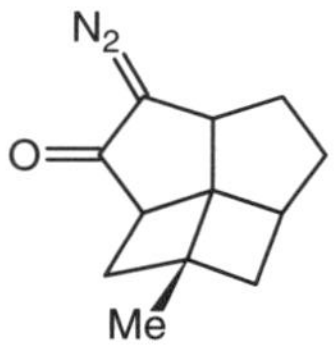	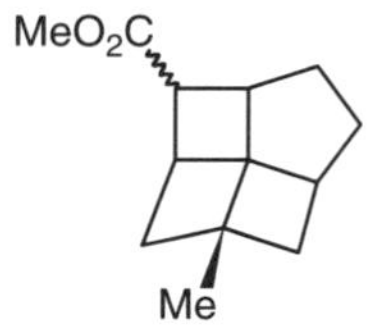 MeO₂C Me 20 % 3:1 ratio of isomers	—	45 46
3	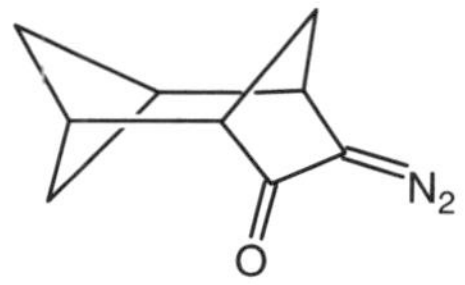	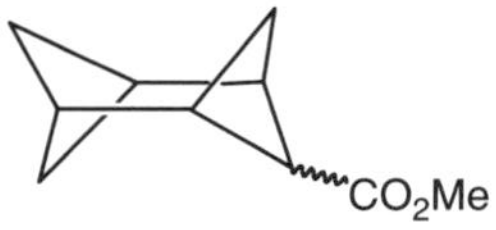 CO₂Me 51 %	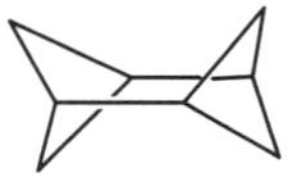	47
4	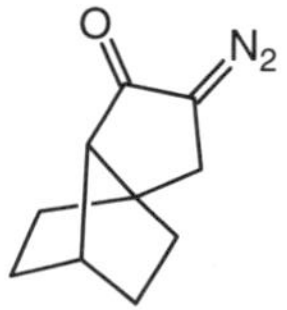	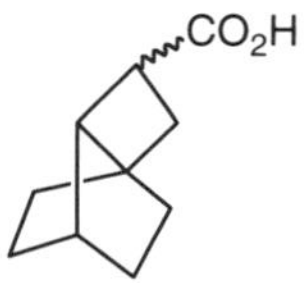 CO₂H 84 %	—	48

504

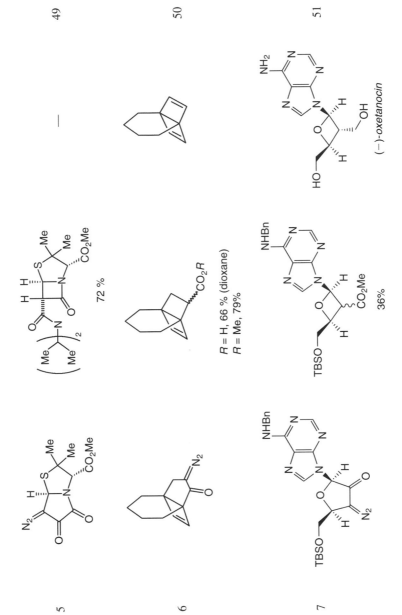

49

50

51

(−)-oxetanocin

72 %

R = H, 66 % (dioxane)
R = Me, 79%

36%

5

6

7

TABLE 9.4 *(Continued)*

Entry	Diazocarbonyl	Product	Target Molecule	Ref.
8	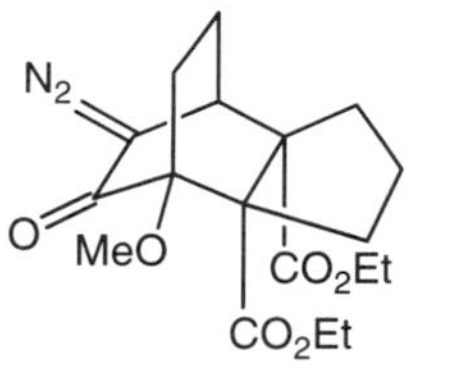	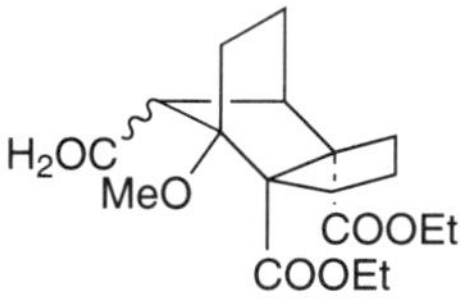	—	52
9	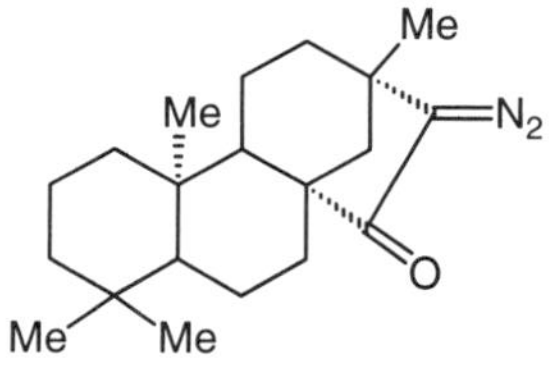	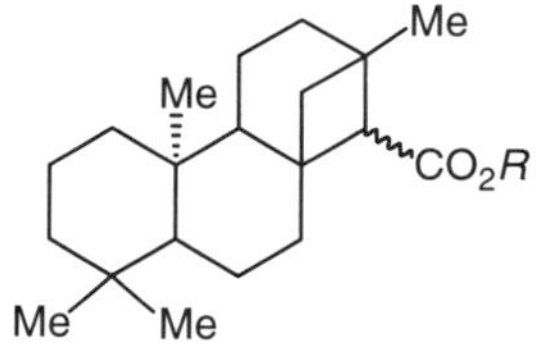 R = H; 77% yield; 63:37 ratio of isomers; dioxane R = CH₃; 64% yield; 16:84 ratio of isomers; methanol	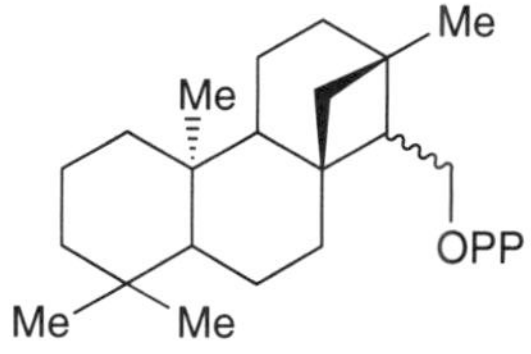	53
10	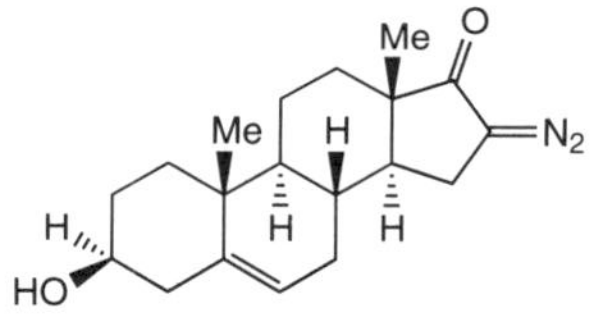	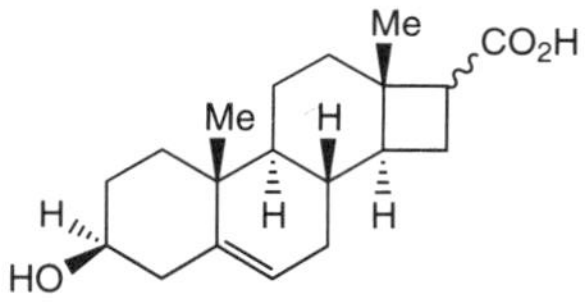	See procedure	54

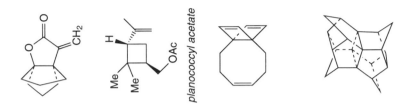

55

56

planococcyl acetate

57

58
59

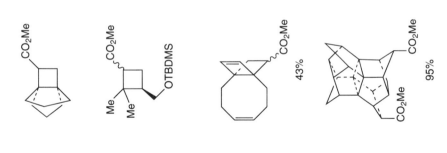

11

12

13

14

43%

95%

TABLE 9.4 *(Continued)*

Entry	Diazocarbonyl	Product	Target Molecule	Ref.
15	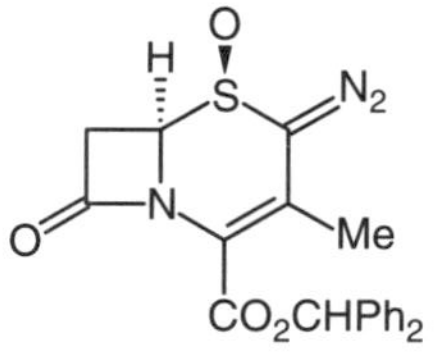	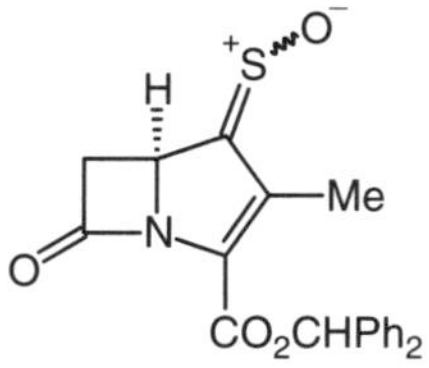 not isolated	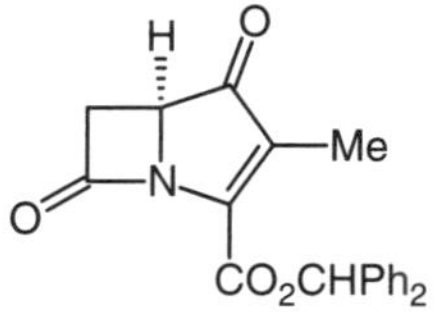	60
16	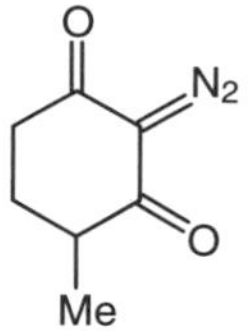	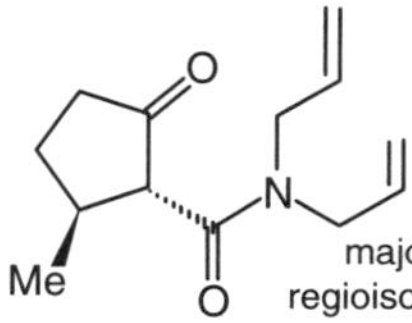 56% major regioisomer	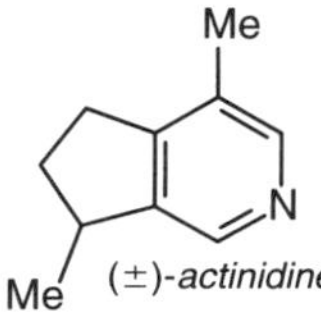 (±)-actinidine	61 62
17	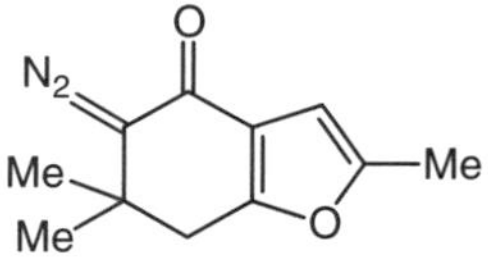	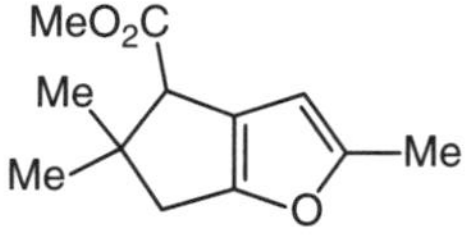 67%	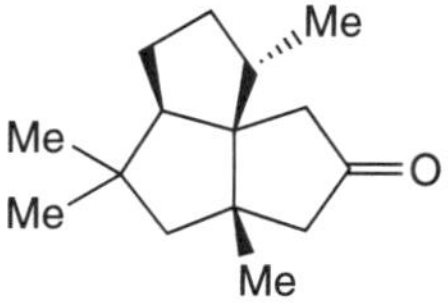	63
18	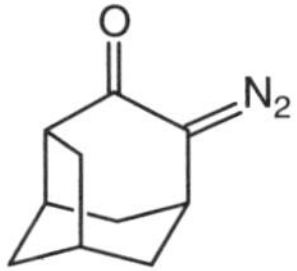	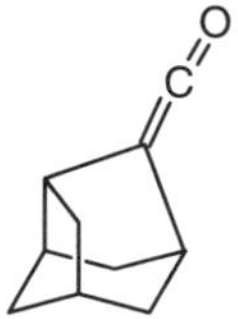	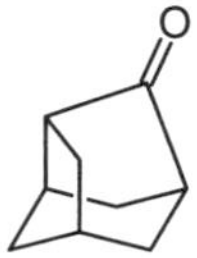	64

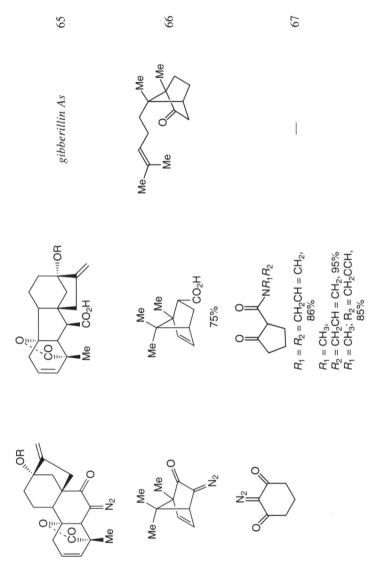

gibberillin As 65

66

67

—

19

20

21

$R_1 = R_2 = CH_2CH = CH_2,$ 86%

$R_1 = CH_3,$
$R_2 = CH_2CH = CH_2;$ 95%
$R_1 = CH_3; R_2 = CH_2CCH,$ 85%

75%

TABLE 9.4 *(Continued)*

Entry	Diazocarbonyl	Product	Target Molecule	Ref.
22	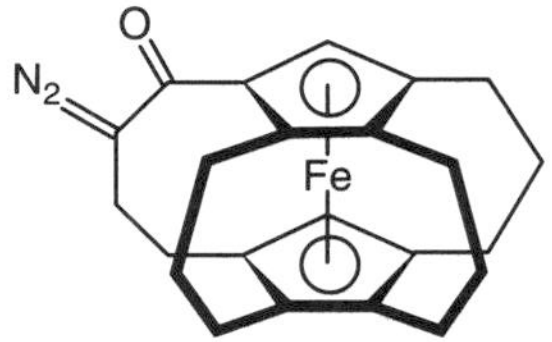	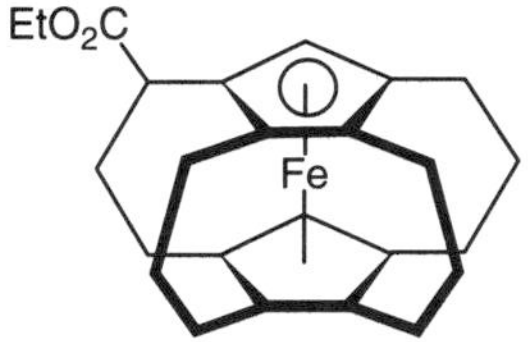 EtO$_2$C	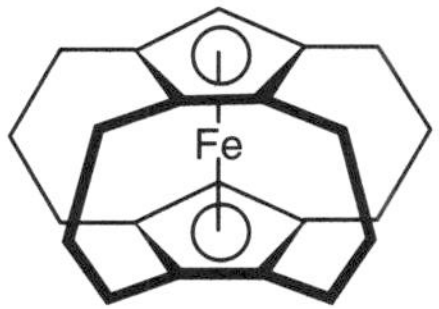	68
23	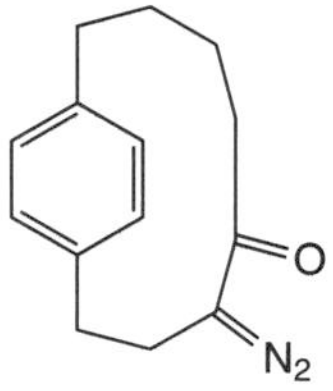	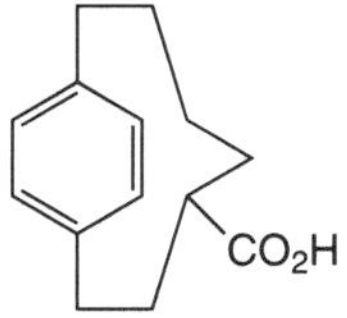 CO$_2$H 50%	—	69
24	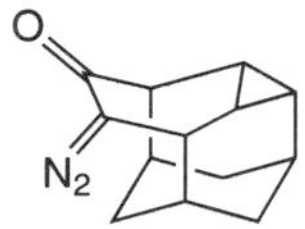	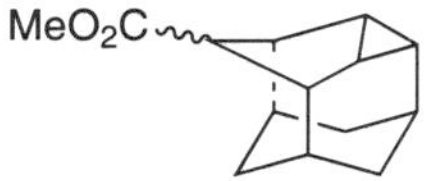 MeO$_2$C 52%	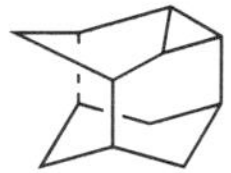 cyclohomoiceane	70
25	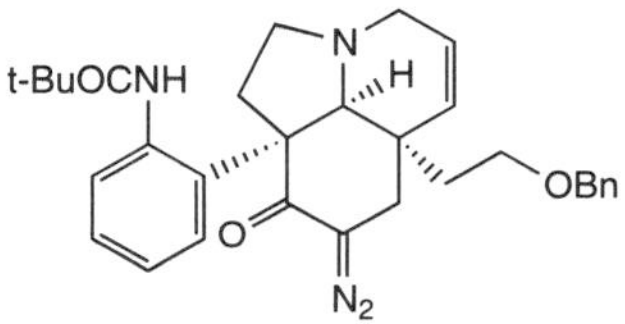 t-BuOCNH	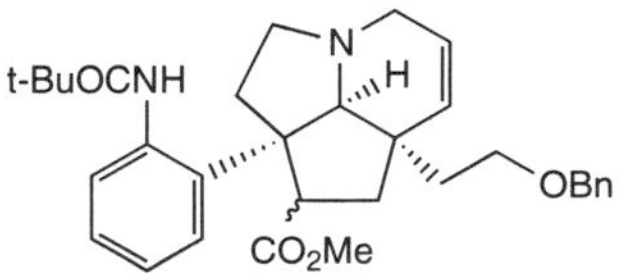 t-BuOCNH CO$_2$Me 95%	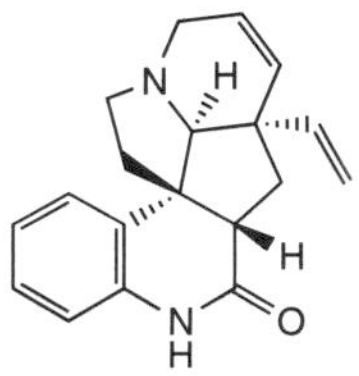 (±)-meloscine	71 72

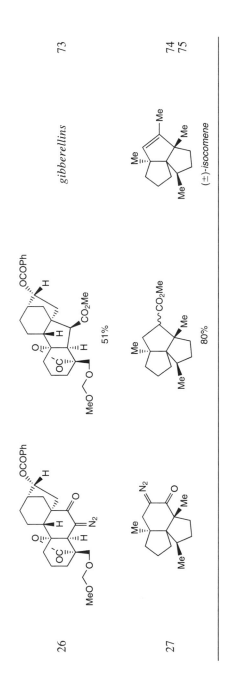

73

gibberellins

74
75

(±)-*isocomene*

51%

80%

26

27

511

FORMATION AND PHOTOCHEMICAL WOLFF REARRANGE-MENT OF CYCLIC α-DIAZOKETONES: D-NORANDROST-5-EN-3β-OL-16-CARBOXYLIC ACIDS (D-NORANDROST-5-ENE-16-CARBOXYLIC ACIDS, 3β-HYDROXY-)[54]

A. 16-oximinoandrost-5-en-3β-ol-17-one.

A 2-L three-necked, round-bottomed flask is fitted with a reflux condenser, a mechanical stirrer, and a pressure-equalized dropping funnel. To the reaction flask is added 750 mL of anhydrous *t*-butyl alcohol. As the *t*-butyl alcohol is slowly stirred, a stream of dry nitrogen is passed through the flask and 12.2 g (0.312 mole) of potassium metal is added cautiously. The flask is surrounded by a water bath maintained at 70°C to assist in dissolving the potassium metal. After 1.5 h the stirred mixture is homogeneous. The water bath is removed, and the reaction mixture is allowed to cool to room temperature. To the potassium *t*-butoxide solution is slowly added 45.0 g (0.156 mole) of dehydroisoandrosterone, and stirring is continued for 1 h until the gold-colored mixture is again homogeneous. To the reaction mixture is now added, dropwise, 42.0 mL (36.5 g, 0.312 mole) of isoamyl nitrite, and stirring is continued overnight at room temperature.

The deep orange reaction mixture is diluted with an equal volume of water, poured into a 2-L separatory funnel, and acidified with aqueous 3M hydrochloric acid. The addition of 400 mL of ether assists in effecting the separation of the clear yellow, aqueous, lower layer from the fluffy-white ethereal suspension that forms the upper layer. This suspension is filtered through a 250 mL coarse sintered glass funnel, and the precipitate of oximino ketone is washed with ether several times. After drying overnight in a vacuum desiccator at −5°C, 48.0–48.5 g of a white product, m.p. 245–247°C, is obtained whose NMR spectrum (pyridine solution) shows it to be a 1:1 solvate of the oximino ketone with *t*-butyl alcohol; the yield is 79%. This product is used without further purification in the synthesis of the α-diazoketone.

B. 16-Diazoandrost-5-en-3β-ol-17-one.

A 1-L three-necked, round-bottomed flask is fitted with a mechanical stirrer, a 50 mL pressure-equalized dropping funnel, and a thermometer. As stirring is initiated, 375 mL of methanol and 72 mL of aqueous 5*M* sodium hydroxide (0.36 mole) is added to the flask, followed by 18 g (0.046 mole) of the 1:1 solvate of 16-oximino ketone which readily dissolves to give a yellow solution. To the reaction mixture is added 28.3 mL of concentrated aqueous ammonia (0.425 mole), and the flask is surrounded by an ice bath to maintain the reaction temperature at 20°C. Through the dropping funnel

133 mL of aqueous 3.0*M* sodium hypochlorite (0.40 mole) is added drop-wise. The sodium hypochlorite solution should be kept near 0°C, so 25 mL portions could be added to the addition funnel, and the remaining solution is kept in the ice bath. It is important that the rate of addition of the sodium hypochlorite and the position of the ice bath be adjusted so as to maintain the temperature of the reaction mixture at 20 ± 1°C. As soon as all the sodium hypochlorite has been added, the ice bath is removed, and the reaction mixture is allowed to warm to room temperature and stirred for 6 h.

The reaction mixture is diluted with an equal volume of water and extracted with a 400 and a 200 mL portion of methylene chloride. The combined methylene chloride extracts are washed with three 250 mL portions of aqueous 20% sodium chloride, dried over anhydrous magnesium sulfate, and concentrated to leave a yellow solid. Recrystallization from acetone gives 8.0–9.3 g (55–64%) of crystalline α-diazoketone, m.p. 200–202°C.

C. D-Norandrost-5-en-3β-ol-16α- and 16β-caboxylic acids.

In a solution of 500 mL of 1,4-dioxane, 1250 mL of ether, and 250 mL of water contained in a 3-L three-necked, round-bottomed flask is dissolved 7.5 g (0.0239 mole) of 16-diazoandrost-5-en-3β-ol-17-one. The flask is fitted with a reflux condenser, a quartz immersion well, and a nitrogen inlet. After the reaction vessel has been flushed with nitrogen, the diazoketone solution is irradiated for 48 h with a 450-watt Hanovia lamp with a Corex filter. The photolysis mixture is decanted in portions into a 2-L separatory funnel, washed three times with 500 mL portions of water to remove the dioxane, and dried over magnesium sulfate. The ether is evaporated to leave a pale yellow residue. The residue is digested with 125 mL of boiling methylene chloride under reflux for 30 min. The methylene chloride solution is allowed to cool to room temperature and filtered to separate about 1.4 g of the crude α-isomer as a white powder. This solid is recrystallized by dissolving it in a large volume of methanol (125 mL) and concentrating the solution to a small volume (25 mL) to yield 1.2 g (17%) of D-norandrost-5-en-3β-ol-16α-carboxylic acid as a white solid, m.p. 271–274°C. The β-isomer is most readily obtained by concentrating the methylene chloride mother liquor and dissolving the residue in a mixture of 75 mL of methanol and 25 mL of ether. This solution is treated with an excess of diazomethane in ether at room temperature. After 1 h at room temperature, the excess diazomethane is removed with a stream of nitrogen, the solvent is evaporated, and the solid residue is chromatographed on 175 g of Woelm neutral alumina Activity Grade II. Elution with a benzene–ether mixture (3:1 *v/v*) gives 3.9 g of a white solid. This solid is recrystallized from ether–heptane to give 3.0–3.1g (39–41%) of white, crystalline methyl D-norandrost-5-en-3β-ol-16β-carboxylate, m.p. 161–163°C.

taining a β-lactam in which the carbapenem skeleton can be prepared through a photolytic rearrangement of readily available 2-diazoceph-3-em-1-oxide. Entry 16 shows that a Wolff rearrangement applied to the diazodiketone in the presence of diallylamine gave a 1.6:1.0 mixture of two regioisomeric amides. The major regioisomeric amide was isolated in 56% yield and could be further transformed to the monoterpene alkaloid, ($\pm$)-actinidine. Finally, through entries 3 and 14 the syntheses of diasterane and pagodane demonstrates that the ring-contraction strategy is an effective method in the construction of highly strained and chemically versatile carbon frameworks.

The photochemical version of the Wolff rearrangement has been used to great effect at low temperatures to generate molecules for spectroscopic examination whose instability precludes isolation. A notable example is Maier's[76] synthesis of cyclobutadiene by photolysis of cyclopropenyl diazomethyl ketone **63** in an argon matrix at 10K (Scheme 10). The initially formed ketene undergoes a further photo-

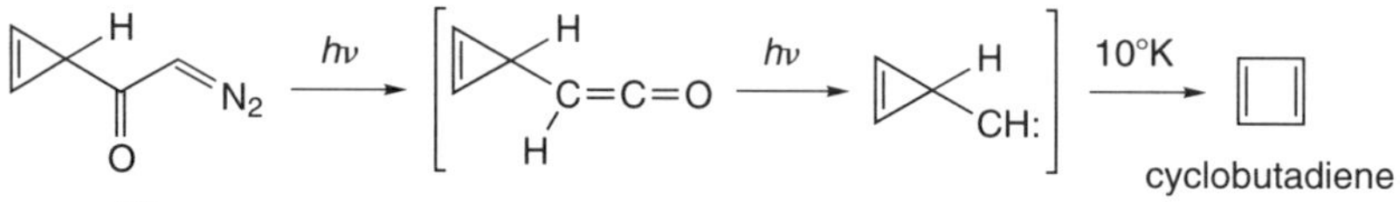

Scheme 10

chemically initiated reaction with loss of carbon monoxide to form a carbene that rearranges to cyclobutadiene. Other examples of highly unstable systems that have been generated at low temperatures photochemically from diazocarbonyl precursors *via* the Wolff rearrangement are shown in Table 9.5.

9.3 THE WOLFF REARRANGEMENT WITH CYCLOADDITION REACTIONS

Ketenes derived from the Wolff rearrangement of diazocarbonyl compounds can undergo cycloaddition reactions. [2+2] Thermal cycloadditions of ketenes to olefins are well understood and have been used for the synthesis of four-membered rings.[83] Both intermolecular[83] and intramolecular[84-89] additions of ketenes to olefins have been studied. Some representative examples are shown in eqs. 24–26.

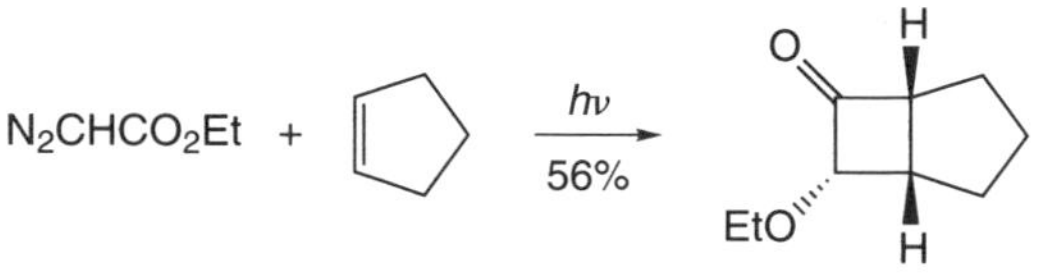

$$N_2CHCO_2Et \; + \quad \text{(cyclopentene)} \quad \xrightarrow[56\%]{h\nu} \quad \text{(product)} \qquad (24)$$

TABLE 9.5 Some Low Temperature Wolff Rearrangement Reactions

			Ref.
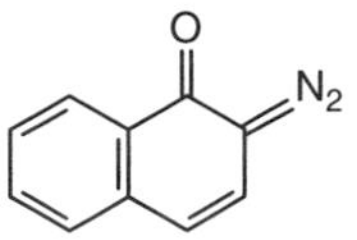	hv $\xrightarrow{}$ 77°K	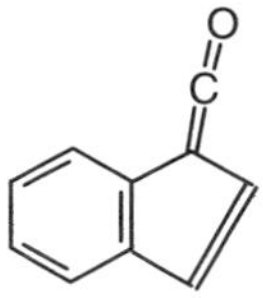	77–79
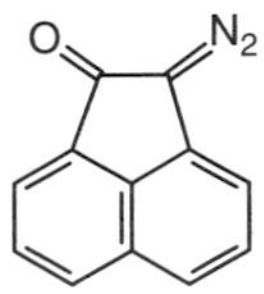	hv $\xrightarrow{}$ 10–15°K	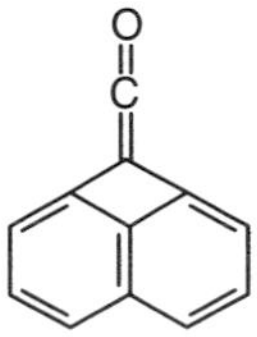	80, 81
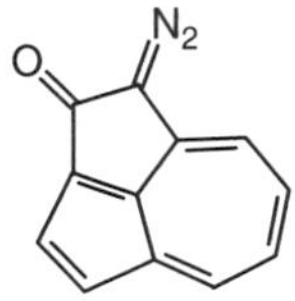	hv $\xrightarrow{}$ 10–15°K	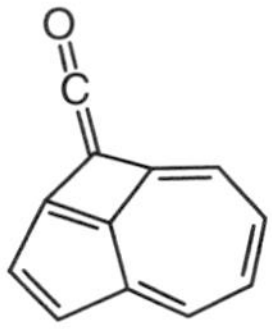	80, 81
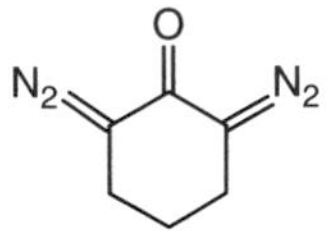	hv $\xrightarrow{}$ 8°K		81, 82
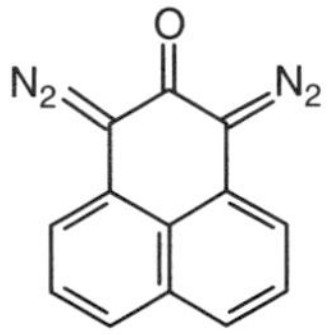	hv $\xrightarrow{}$ 15°K	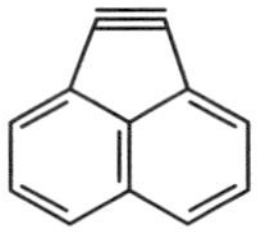	80, 81

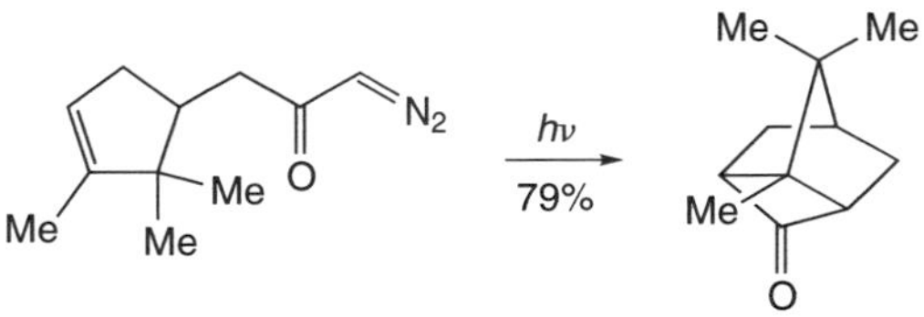

(25)

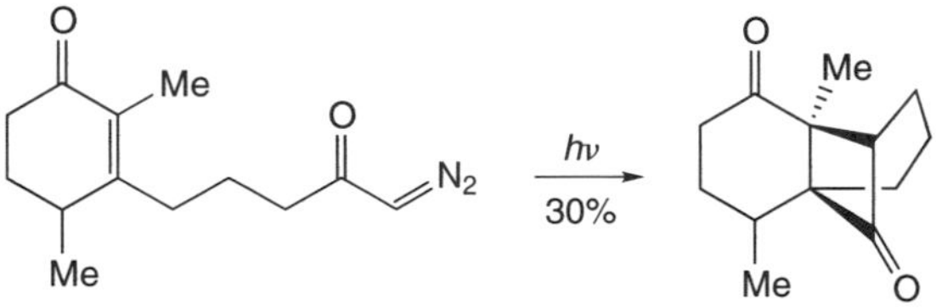

(26)

The combination of ring contraction and [2+2] cycloaddition reaction strategy has been used in Ireland's total synthesis of (±)-aphidicolin **64** (Scheme 11).[90]

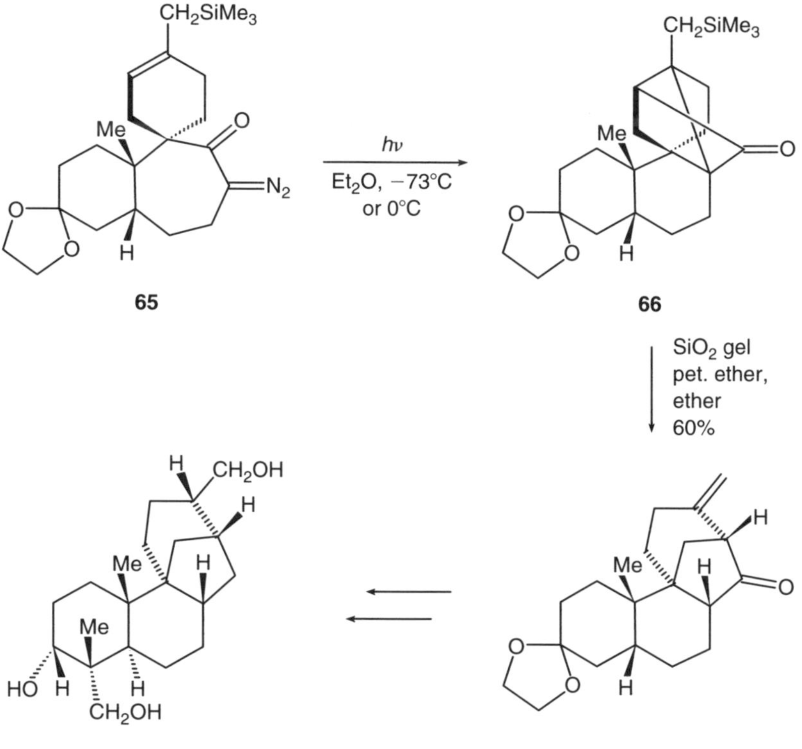

Scheme 11

The photochemical decomposition of diazoketone **65** afforded the corresponding silylcyclobutanone derivative **66**, which subsequently rearranged to ketone **67**. This ketone contains the desired bicyclo[3.2.1]octane ring system of aphidicolin **64**.

A final example of a combined Wolff rearrangement–intramolecular cycloaddition is illustrated by the behavior of some 1,2-bis(diazoketones) under thermolysis. The *trans*-cyclohexyl system **68** on heating in toluene for 6 h produced *trans*-hydroindenone **69** in 92% yield (eq. 27).[91] Danheiser and co-workers[92] have

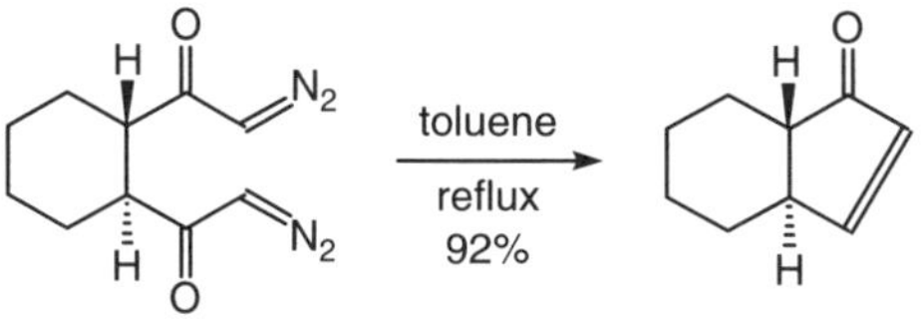

(27)

invented a new aromatic annulation process that is based on the use of a photo-chemical Wolff rearrangement of unsaturated diazoketones to generate vinyl or arylketenes (Scheme 12).[76] There follows a cascade of three pericyclic reactions:

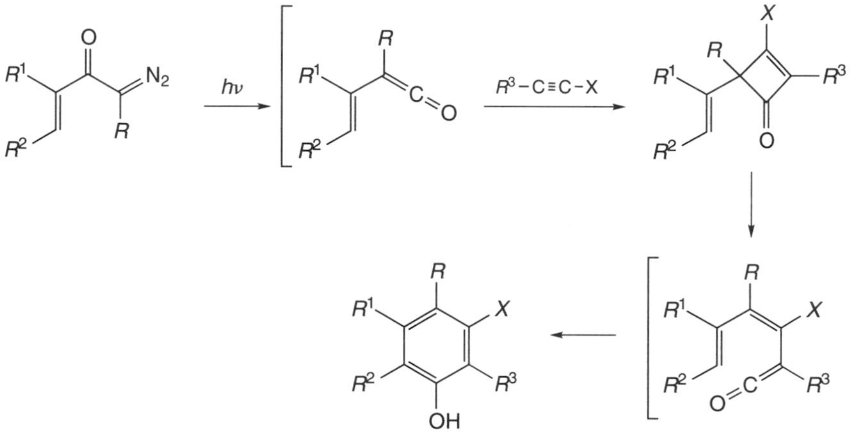

Scheme 12

[2+2] intermolecular cycloaddition to an alkyne to form a cyclobutenone; a four-electron electrocyclic ring cleavage to form a new ketene; and a six-electron elec-trocyclic ring closure (with tautomerism). In its original form the sequence was applicable to the synthesis of highly substituted monocyclic systems only (e.g., maesanin **70**, the host defence stimulant).[93] In a "second-generation" version the

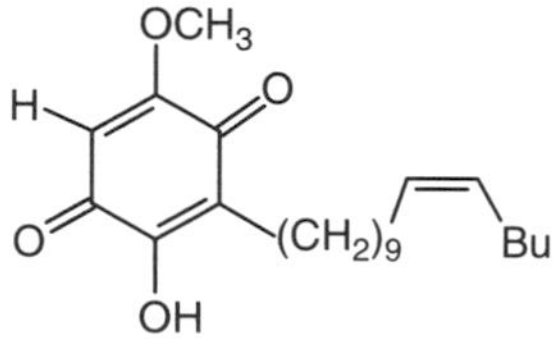

70 *maesanin*

annulation was expanded to encompass highly substituted polycyclic systems. In Scheme 13 the application of this annulation to the synthesis of the phenalenone, salvilenone **71**, is illustrated.[94] Other natural products to which the annulation has been applied successfully by Danheiser and co-workers include the Dan Shen diterpenoid quinones **72** and **73**,[95] the aegyptinones A and B, **74** and **75**,[96] (+)-neocryptotanshinone **76**,[97] and hyellazole **77**.[92] Annulation of 2-furan dia-zomethyl ketone with a suitably functionalized alkyne produced an efficient route to the linear furocoumarin bergapten **78**[98] (Table 9.6). Other types of double or triple bonds are suitable for trapping the ketene ensuing from the Wolff rearrangement. The rather labile aldoketenes react with intact α-diazoketones to

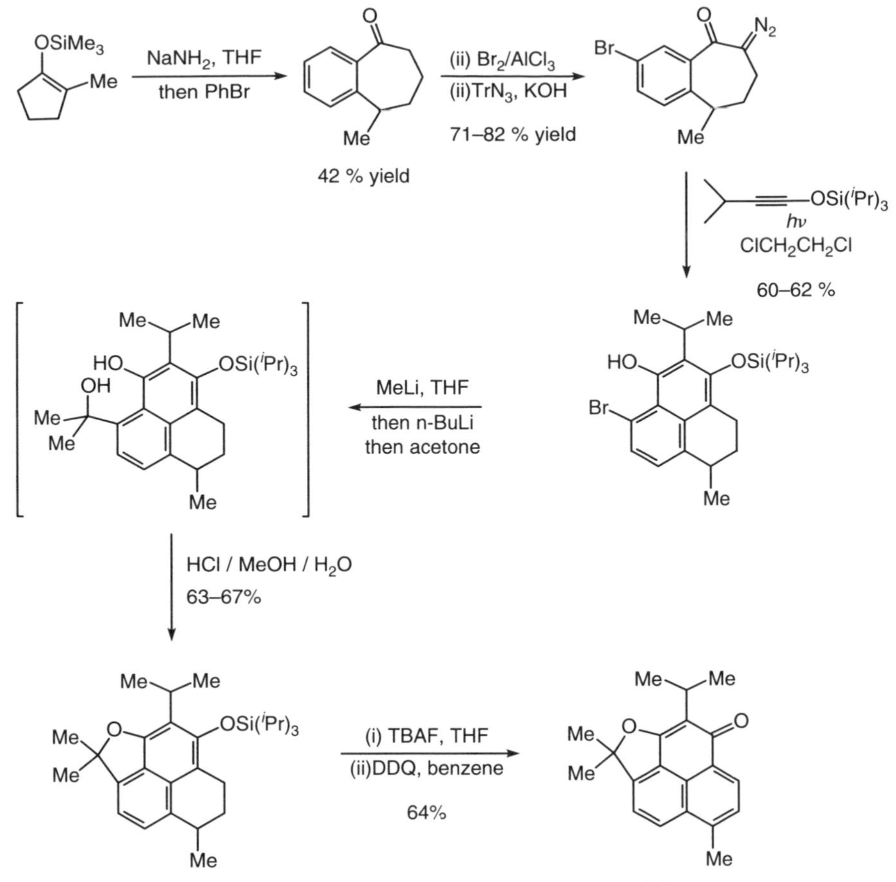

71 *salvilenone*

Scheme 13

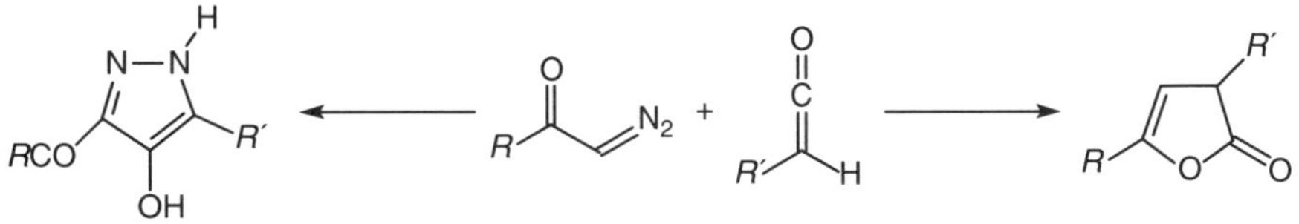

80 **79**

Scheme 14

give butenolides **79** or pyrazoles **80** (Scheme 14).[99] Staudinger[100] reported the first example of ketene addition to a carbon–nitrogen double bond to give a β-lactam. This cycloaddition reaction is a two-step process involving a zwitterionic species **81** as a reaction intermediate,[101,102] as shown in Scheme 15. The ketene **82** derived from the α-diazo thioester **83** adds to an imine acceptor, giving β-lactam **84** stereoselectively.[41,42,103] Another example involving ketenes derived

TABLE 9.6   Natural Products Accessible by the Danheiser Annulation Reaction

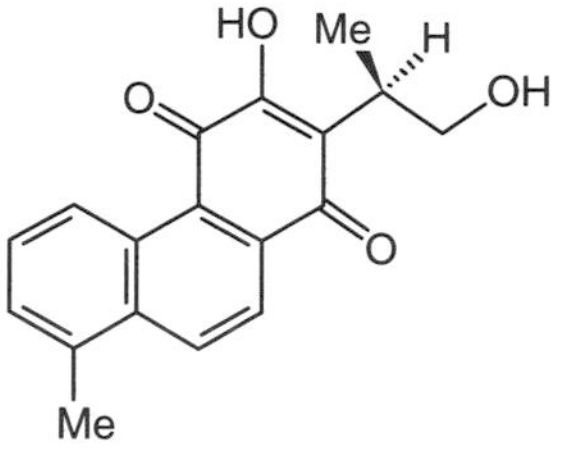

72 *danshexinkun A*

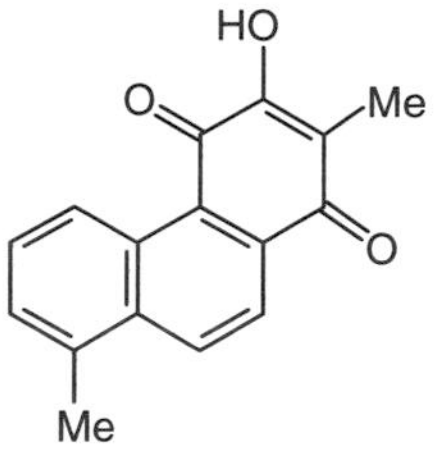

73 *danshexinkun C*

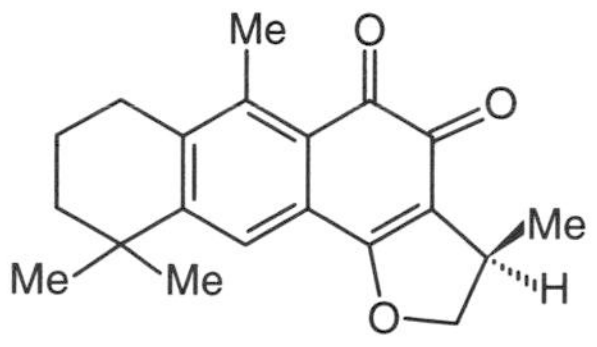

74 *aegyptinone A*

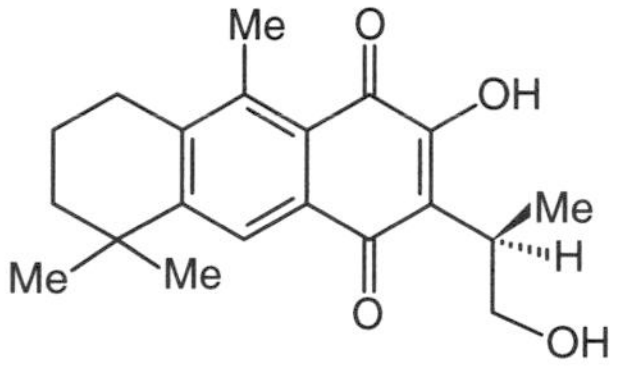

75 *aegyptinone B*

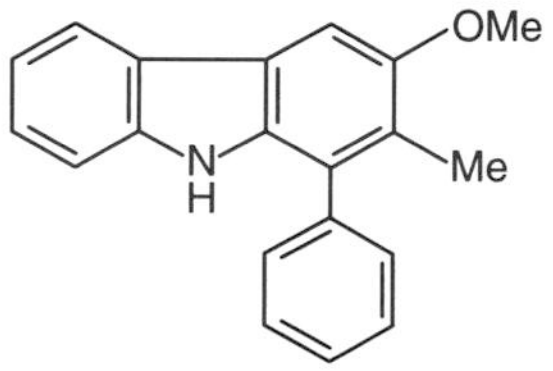

77 *hyellazole A*

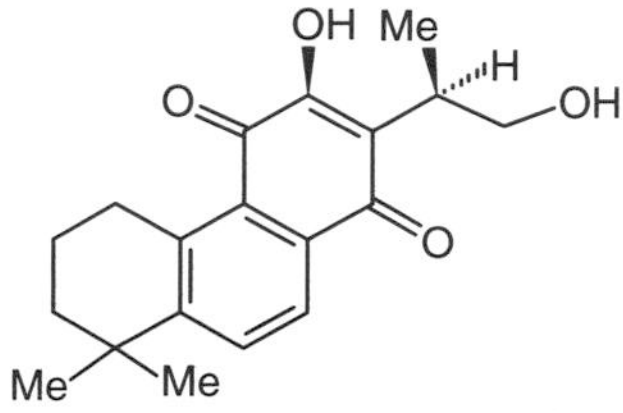

76 *neocryptotanshinone*

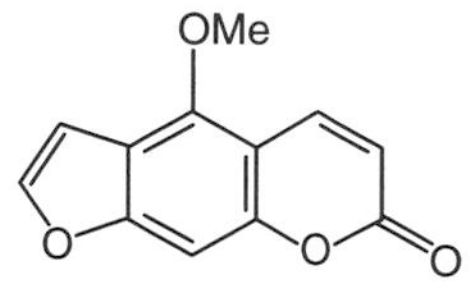

78 *bergapten*

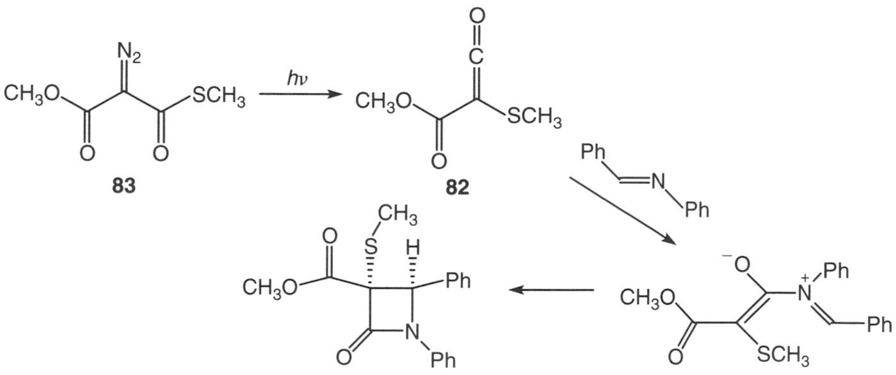

Scheme 15

from amino acids is that due to Podlech, which was discussed in Section 9.2. Otherwise, this cycloaddition to π-bonds containing heteroatoms has not been extensively exploited in synthesis.

9.4 THE VINYLOGOUS WOLFF REARRANGEMENT

From the late 1950s to the early 1970s, a few research groups discovered that β,γ-unsaturated diazoketones not only undergo the normal Wolff rearrangement leading to homologated acids but also yield, *via* a novel skeletal rearrangement, abnormal isomeric products.[104–110] A systematic study of the decomposition of β,γ-unsaturated diazoketones revealed that copper catalysts favor the abnormal product formation. This transformation, which has been termed the *vinylogous Wolff rearrangement*, represents a synthetic alternative to the Claisen-type rearrangement.[110–113]

The initial reaction of the vinylogous Wolff rearrangement is probably a cyclopropanation reaction to give a bicyclic product. Fragmentation of this intermediate leads to a β,γ-unsaturated ketene, which in turn can be captured by an available nucleophile to afford the γ,δ-unsaturated carboxylic acid derivative (Scheme 16). The copper(II) salt–catalyzed decomposition of monocyclic and

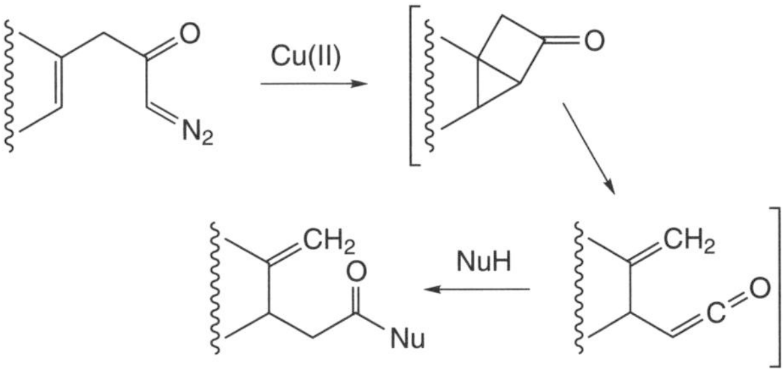

Scheme 16

acyclic β,γ-unsaturated diazoketones, in the presence of a nucleophile leading to the respective γ,δ-carboxylic acid derivatives, has been extensively studied.[114,115] Some representative examples are shown in eqs. 28–30. The vinylogous Wolff re-

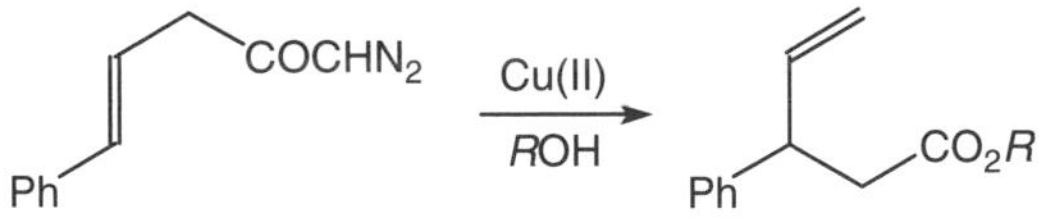

(28)

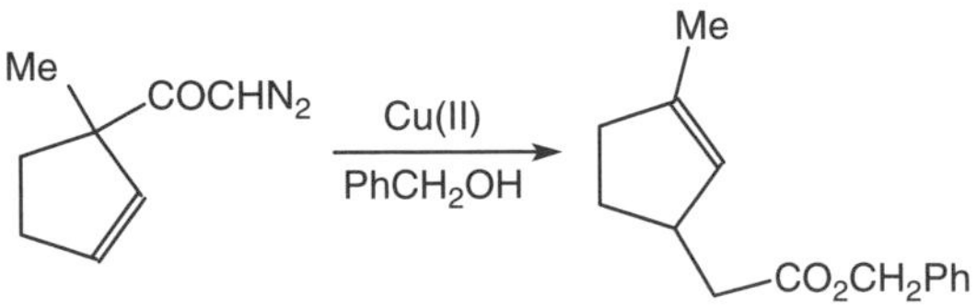

(29)

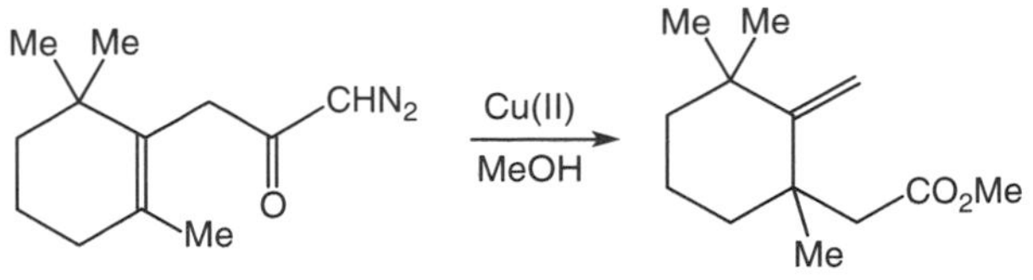

$$(30)$$

arrangement is also applicable[116-119] to rigid β,γ-unsaturated diazoketones. Equation 31 shows the flexibility of the process in the introduction of an angularly

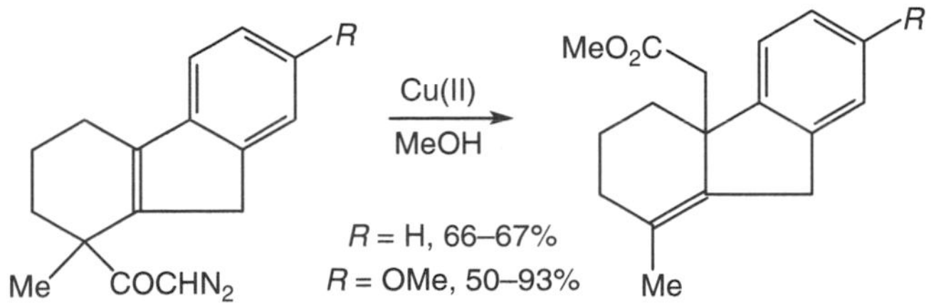

R = H, 66–67%
R = OMe, 50–93%

$$(31)$$

functionalized two-carbon residue from appropriately substituted polycyclic β,γ-unsaturated diazoketones.

β,γ-Unsaturated α-diazo-β-ketoesters also respond to metal ion catalysis in a similar manner. For example, the α-diazo-β-ketoesters of eqs. 32 and 33 in the

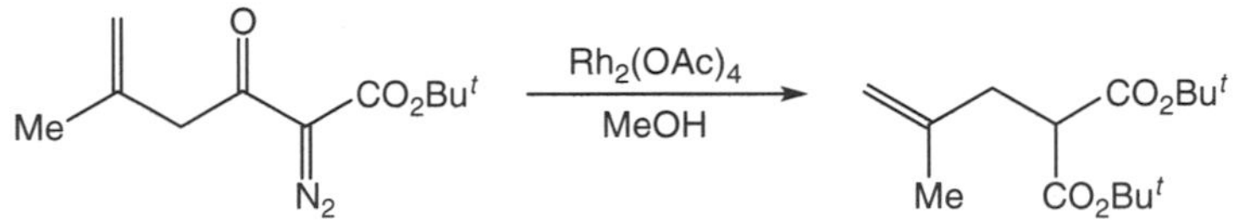

$$(32)$$

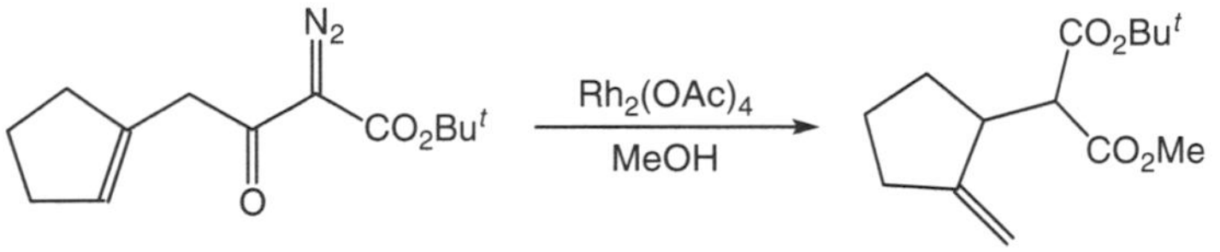

$$(33)$$

presence of rhodium(II) acetate furnish substituted malonates. In the course of a detailed study of rhodium(II)-catalyzed decomposition of β,γ-unsaturated diazoketones, Motallebi and Müller[120] observed a modest degree of enantiocontrol when chiral catalysts were used. Up to 31% ee could be realized for the vinylogous Wolff rearrangement (eq. 33) when $Rh_2[(4S)\text{-BNOX}]_4$ was the catalyst.

9.5 MISCELLANEOUS APPLICATIONS OF THE WOLFF REARRANGEMENT

Among other applications of diazocarbonyl reactions that are presumed to involve ketenes generated by the Wolff rearrangement are the formation of spiro-

cyclopropyl-$\Delta^{\alpha\beta}$-butenolides from 2-(diazoacetyl)cyclobutanones and of ene-yne-
ketenes for DNA cleavage. Miller and co-workers[121] found that thermolysis or
photolysis of 2-(diazoacetyl)butanones (Schemes 17 and 18) produced keto ketene

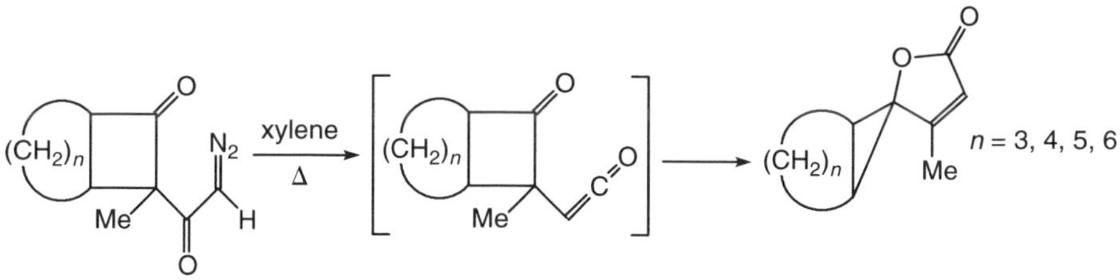

Scheme 17

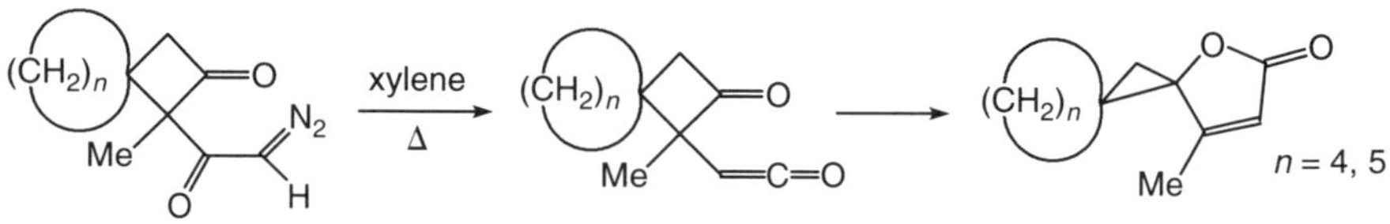

Scheme 18

intermediates, which thermally rearranged to the corresponding spirocyclopropyl
butenolides. These reactions proceeded in excellent yield and were stereospecific.
An example of photoinduced DNA cleavage by a diazoketone containing an ene-
yne moiety has been reported by Nakatani *et al.*[122,123] Diazoketones **85, 86,** and
87 were designed as mimics for the radical-generating system of neocarzinostatin

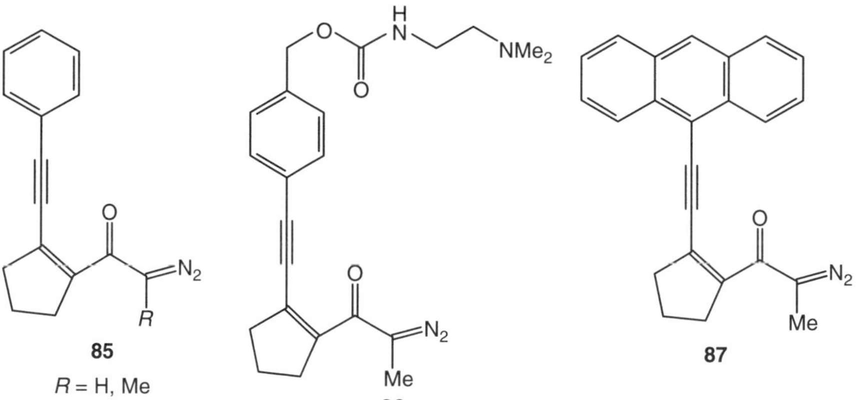

85
R = H, Me

86
Me

87

chromophore **88**. These diazoketones have the ability to generate diradicals under
thermal or photoirradiation conditions *via* cyclization (Scheme 19) of the ene-
yne-ketene intermediate resulting from the Wolff rearrangement. *Ab initio* MO

calculations revealed that the efficiency of radical generation is highly dependent on the conformation of the diazoketone, which is controlled by the substituent on the carbon directly attached to the diazo group. Diazoketones **86** and **87**, having a DNA binding moiety, greatly improved the DNA cleavage activity compared with that for diazoketone **85** under photolysis at 360 nm. DNA cleavage is probably due to hydrogen abstraction from the DNA sugar backbone by the diradical **89**

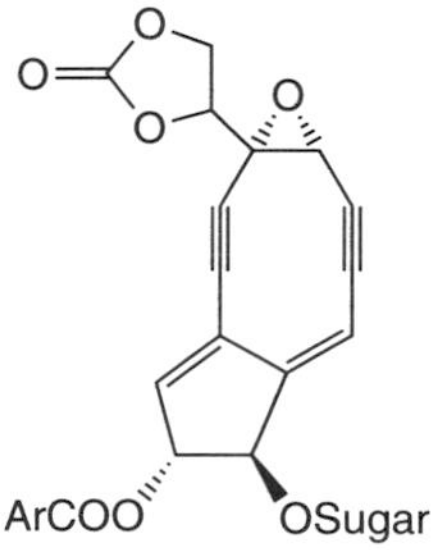

88

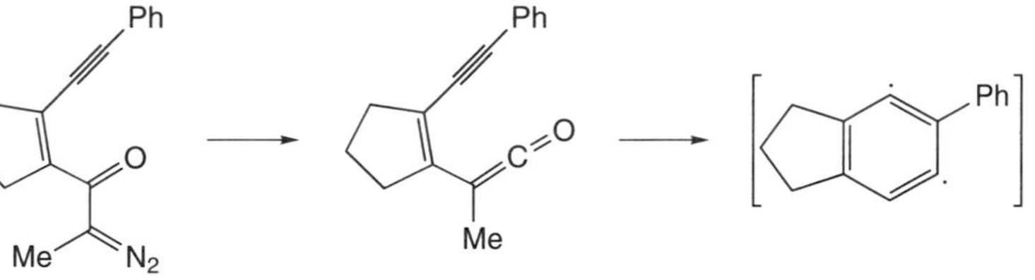

85   **90**   **89**

Scheme 19

spontaneously formed from the ene-yne-ketene **90** (Scheme 90). In a related study Nakatani et al.[124] observed that dibenzoyldiazomethanes, particularly derivatives such as **92** with a cationic side chain to improve solubility in water and enhance DNA binding ability, show DNA-cleaving activity on irradiation, again *via* formation of a highly electrophilic ketene **91**. These workers also

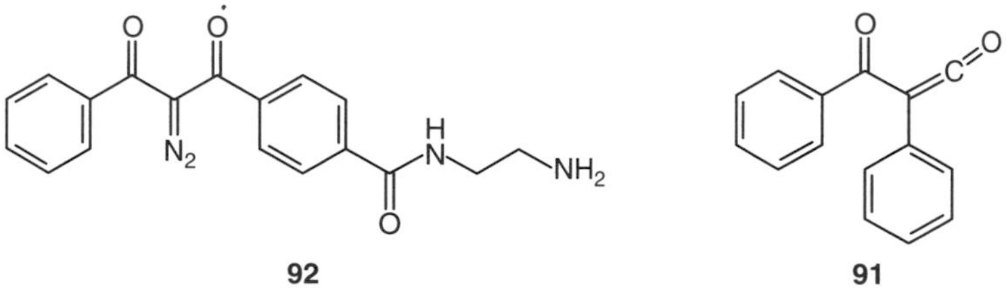

92   **91**

showed that the ketene from **92** could be generated in aqueous solution photochemically and be trapped efficiently by amino acids and amines in competition with water. Mosier and Lawton[125] have developed a family of thiol-specific photoactivatable cross-linking agents based on diazoketones. Compound **93** is a typi-

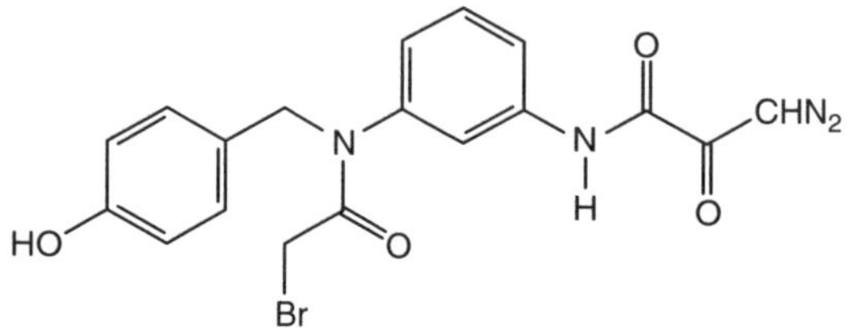

93

cal example of this type of trifunctional bioprobe. The trifunctionality is characterized by the presence of (1) a bromoacetyl function that modifies the thiol
group through alkylation, (2) a diazopyruvoyl group, which on photolysis undergoes the Wolff rearrangement generating a ketene capable of acylating nucleophiles, and (3) a phenolic function that can act as a characterization handle
through radioiodination labeling. Sodeoka et al.[126] have synthesized a novel photoaffinity ligand phorbol 12-(1-pyrene butyrate)-13-diazoacetate **94**, which when

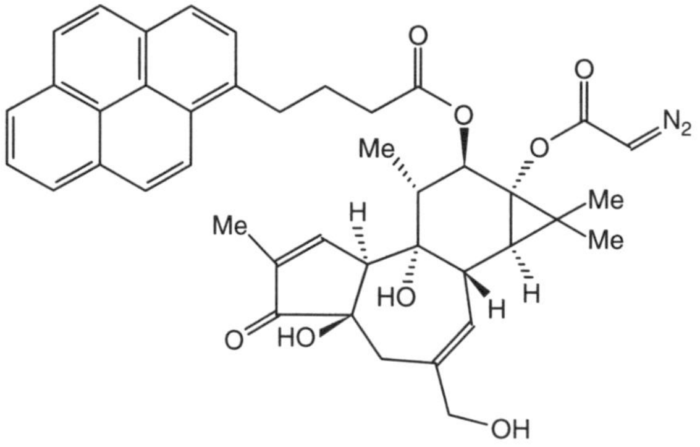

94

labeled with tritium could be used for photocross-linking experiments with protein kinase C from rat brain.

9.6 CONCLUSIONS

It is appropriate to conclude this account of applications of the Wolff rearrangement with reference to its role in the most enduring of all uses of diazocarbonyl
chemistry in modern technology. It is estimated that in 1996 some 20 billion in-
tegrated circuits will have been produced worldwide. Almost all will have been
made using a photosensitive varnish, the so-called "photoresist," that allows the
transfer of a fine line pattern from an original transparency onto a silicon wafer.
This photoresist has two components: one is a low-molecular-weight phenol-
formaldehyde condensation polymer called novolak; the other is a diazonaphthoquinone derivative, which is the photoactive part of the system. The use of these

novolak–diazoquinone resists follows from an observation made in the 1940s by Süs while studying the ease of dissolution of novolak in aqueous alkali.[127,128] Süs found that a coating of novolak containing diazoquinone that had been exposed to sunlight dissolved in alkali much faster than did either pure novolak or a novolak coating containing diazoquinone that had not been exposed to sunlight. Realizing the potential of his observation, Süs designed a lithographic printing plate that functioned like a film positive: Irradiated and unexposed areas could be distinguished by a large differential in dissolution rate in aqueous alkali. On the same principle, Süs formulated a light-sensitive varnish now known as a lithographic resist, which is now used routinely in the building of semiconductor devices. The image discrimination in these resists is based on the Wolff rearrangement of diazoquinones. On irradiation of diazonaphthoquinone within the polymer matrix, nitrogen is released, leading to a ketene that in the presence of small quantities of water is transformed into indenecarboxylic acid (eq. 34). This

$$\underset{H_2O}{\overset{h\nu}{\longrightarrow}} \tag{34}$$

product is completely soluble in aqueous alkali, unlike its precursor diazoquinone, and its presence in the polymer facilitates its dissolution. Thus the areas that have been exposed to irradiation can be stripped from the coating by aqueous alkali, leaving behind the unexposed areas of coating, which can then be used to transfer an image to another surface. Novolak–diazoquinone resists of this type have played a major role in the development of the semiconductor device industry.[129]

9.7 REFERENCES

1 Wolff, L., "Über Diazoanhyride (1,2,3-Oxydiazole oder Diazoxyde) und Diazoketone," *Liebigs Ann. Chem.* **1912**, *394*, 23–36.

2 Arndt, F.; Eistert, W.; and Partale, W., "Diazomethan und *o*-Nitroverbindungen, II[1]: *N*-Oxyisatin aus *o*-Nitrobenzoylchlorid," *Ber. Dtsch. Chem. Ges.* **1927**, *60*, 1364–70.

3 Bachmann, W. E.; and Struve, W. S., "The Arndt–Eistert Synthesis," *Org. React.*; John Wiley & Sons: New York, 1942; Vol. 1, p. 38.

4 Eistert, B., "Newer Methods of Preparative Organic Chemistry"; Interscience: New York, 1948; Vol. 1, p. 513.

5 Weygand, F.; and Bestmann, H. J., "Newer Methods of Preparative Organic Chemistry"; Academic Press: New York, 1964; Vol. 3, p. 451.

6 Gill, G. B., in *Comprehensive Organic Synthesis; Selectivity, Strategy and Efficiency in Modern Chemistry*; Trost, B. M., and Fleming, I., Eds.; Pergamon Press: Oxford, 1991; Vol. 3, p. 887.

7   Lee, V.; and Newman, M. S., "Ethyl 1-Naphthylacetate," *Org. Synth.* **1988**, *50*, 613–15.

8   Newman, M. S.; Beal III, P. F., "An Improved Wolff Rearrangement in Homogeneous Medium," *J. Am. Chem. Soc.* **1950**, *72*, 5163–65.

9   Winum, J.-Y.; Kamal, M.; Leydet, A.; Roque, J.-P.; and Montero. J.-L., "Homologation of Carboxylic Acids by Arndt–Eistert Reaction under Ultrasonic Waves," *Tetrahedron Lett.* **1996**, *37*, 1781–82.

10   Rane, D. F.; Girijavallabhan, V. K.; Ganguly, A. K.; Pike, R. E.; Saksena, A. K.; and McPhail, A. T., "Total Synthesis and Absolute Stereochemistry of the Antifungal Dipeptide SCH 37137 and Its 2*S*,3*S*-Isomer," *Tetrahedron Lett.* **1993**, *34*, 3201–4; Nomoto, S.; Teshima, T.; Wakamiya, T.; and Shiba, T., "Total Synthesis of Capreomycin," *Tetrahedron* **1978**, *34*, 921–27; Onuki, H.; Tachibana, K.; Fusetani, N., "Structure of Lipogrammistin-A, a Lipophilic Ichthyotoxin Secreted by the Soapfish *Diploprion bisfasciatum,* " *Tetrahedron Lett.* **1993**, *34*, 5609–12; Yang, L. H.; Weber, A. E.; Greenlee, W. J.; and Patchett, A. A., "Macrocyclic Renin Inhibitors: Synthesis of a Subnanomolar, Orally-Active Cysteine Derived Inhibitor," *Tetrahedron Lett.* **1993**, *34*, 7035–38.

11   Drey, C. N. C. in *Chemistry and Biochemistry of Amino Acids,* Barrett, G. C., Ed., Chapman and Hall: London, 1985, 25–54; Sone. H.; Nemoto, T.; Ishiwata, H.; Ojika, M.; and Yamada, K., "Isolation, Structure, and Synthesis of Dolastatin D, a Cytotoxic Cyclic Depsipeptide from the Sea Hare *Dolabella auricularia*," *Tetrahedron Lett.* **1993**, *34*, 8449–52.

12   Nicolaou, K. C.; Dai, W.-M.; and Guy, R. K., "Chemistry and Biology of Taxol," *Angew. Chem., Int. Ed. Engl.* **1994**, *33*, 15–44; Holton, R. A.; Kim, H.-B.; Somoza, C.; Liang, F.; Biediger, R. J.; Boatman, P. D.; Shindo, M.; Smith, C. C.; Kim, S.; Nadizadeh, H.; Suzuki, Y.; Tao, C.; Vu, P.; Tang, S.; Zhang, P.; Murthi, K. K.; Gentile, L. N.; and Liu, J. H., "First Total Synthesis of Taxol. 2. Completion of the C and D Rings," *J. Am. Chem. Soc.* **1994**, *116*, 1599–600; Nicolaou, K. C.; Ueno, H.; Liu, J.-J.; Nantermet, P. G.; Yang, Z.; Renaud, J.; Paulvannan, K.; and Chadha, R., "Total Synthesis of Taxol. 4. The Final Stages and Completion of the Synthesis," *J. Am. Chem. Soc.* **1995**, *117*, 653–59.

13   *The Organic Chemistry of β-Lactams*; Georg, G. I., Ed.; Verlag Chemie: New York. 1993.

14   Juaristic, E.; Quintana, D.; and Escalante, J., *Aldrichimica Acta* **1994**, *27*, 3–11; Cole, D. C., "Recent Stereoselective Synthetic Approaches to β-Amino Acids," *Tetrahedron* **1994**, *50*, 9517–82; Enders, D.; Bettray, W.; Raabe, G.; and Runsink, J., "Diastereoselective and Enantioselective Synthesis," *Synthesis* **1994**, 1322–26.

15   Balenovic, K., "Uber α-Diazoketone aus Phthalimido-carbonsaurechloriden," *Experientia* **1947**, *3*, 369; Balenovic, K.; Bregant, N.; Cerar, D.; and Tkalcic, M., "Reaction of N-Disubstituted Glycyl Chlorides with Diazomethane. A New Synthesis of Some β-Alanine Derivatives. Amino Acid. V," *J. Org. Chem.* **1951**, *16*, 1308–10; Balenovic, K.; and Brovet-Keglevic, D., *Arhiv za Kemiju* **1951**, *23*, 1–5; Balenovic, K.; Cerar, D.; and Fuks, "A Synthesis of (+)-β-Aminobutyric Acid from L-Alanine," *J. Chem. Soc.* **1952**, 3316–17; Balenovic, K.; and Dvornik, D., "A Synthesis of (−)-β-Amino-γ-methylvaleric Acid [(−)-β-Leucine]," *J. Chem. Soc.* **1954**, 2976; Balenovic, K.; Thaller, V.; and Filipovic, L., "Uber die dem L-Tyrosin und L-3,5-Dijod-tyrosin homologen β-Aminosauren: L-β-Amino-γ-

(p-oxyphenyl)buttersaure und L-β-Amino-γ-(3,5-dijod-4-oxyphenyl)buttersaure," *Helv. Chim. Acta* **1951**, *34*, 744–47.

16 Podlech, J.; and Seebach, D., "On the Preparation of β-Amino Acids from α-Amino Acids Using the Arndt–Eistert Reaction: Scope, Limitations and Stereoselectivity. Application to Carbohydrate Peptidation. Stereoselective α-Alkylations of Some β-Amino Acids," *Liebigs Ann.* **1995**, *7*, 1217–28.

17 Darkins, P.; McCarthy, N.; McKervey, M. A.; O'Donnell, K.; Ye, T.; and Walker, B., "First Synthesis of Enantiomerically Pure N-Protected β-Amino-α-keto Esters from α-Amino Acids and Dipeptides," *Tetrahedron: Asymmetry* **1994**, *5*, 195–98.

18 Grieco, P. A.; Hon, Y. S.; and Perez-Medrano, A., "A Convergent, Enantiospecific Total Synthesis of the Novel Cyclodepsipeptide (+)-Jasplakinolide (Jaspamide)," *J. Am. Chem. Soc.* **1988**, *110*, 1630–31.

19 Guibourdenche, C.; Podlech, J.; and Seebach, D., "Selective Acylations of Multifunctional Nucleophiles, Including Carbohydrates and Nucleosides, with Intermediates Generated by Wolff Rearrangement of Amino Acid Derived Diazo Ketones: Preparation of β-Amino Acid Derivatives," *Liebigs Ann.* **1996**, 1121–29.

20 Podlech, J.; and Seebach, D. "The Arndt–Eistert Reaction in Peptide Chemistry: A Facile Access to Homopetides," *Angew. Chem., Int. Ed. Engl.* **1995**, *34*, 471–72.

21 Seebach, D.; Overhand, M.; Kühnle, F. N. M.; Martioni, B.; Oberer, L.; Hommel, U.; and Widmer, H., "β-Peptides: Synthesis by Arndt-Eistert Homologation with Concomitant Peptide Coupling. Structure Determination by NMR and CD Spectroscopy and by X-Ray Crystallography. Helical Secondary Structure of a β-Hexapeptide in Solution and Its Stability Towards Pepsin," *Helv. Chim. Acta* **1996**, *79*, 913–41.

22 Bastiaans, H. M. M.; Alewijnse, A. E.; van der Baan, J. L.; and Ottenheijm, H. C. J., "A Facile Conversion of Arginine into β-Homoarginine Dipeptides," *Tetrahedron Lett.* **1994**, *35*, 7659–60.

23 Pendrak, I.; and Chambers, P. A., "Improved Synthesis of RG-14893, a High Affinity Leukotriene B$_4$ Receptor Antagonist, via a Photochemical Wolff Rearrangement," *J. Org. Chem.* **1995**, *60*, 3249–51.

24 Podlech, J., "Stereoselective Synthesis of Aminoalkyl-Substituted β-Lactams *via* Cycloaddition of Ketenes Generated from α-Amino Acids," *Synlett.* **1996**, 582–84.

25 Angelastro, M. R.; Mehdi, S.; Burkhart, J. P.; Peet, N. P.; and Bey, P. J., "α-Diketone and α-Keto Ester Derivatives of N-Protected Amino Acids and Peptides as Novel Inhibitors of Cysteine and Serine Proteinases," *J. Med. Chem.* **1990**, *33*, 11–13.

26 Peet, N. P.; Burkhart, J. P.; Angelastro, M. R.; Giroux, E. L.; Mehdi, S.; Bey, P.; Kolb, M.; Neises, B.; and Schirlin, D., "Synthesis of Peptidyl Fluoromethyl Ketones and Peptidyl α-Keto Esters as Inhibitors of Porcine Pancreatic Elastase, Human Neutrophil Elastase, and Rat and Human Neutrophil Cathepsin G," *J. Med. Chem.* **1990**, *33*, 394–407.

27 Pellicciari, R.; Natalini, B.; and Ursini, A., "Enantiospecific Synthesis of a Chiral Carbapenem Precursor from (R)-Aspartic Acid," *Gazz. Chim. Ital.* **1986**, *116*, 607–8.

28   Pellicciari, R.; Natalini, B.; and Marinozzi, M., "L-α-Aminoadipic Acid from L-Glutamic Acid," *Synth.Commun.* **1988**, *18*, 1707–13.

29   Jefford, C. W.; Tang, Q.; and Zaslona, A., "Short, Enantiogenic Syntheses of (−)-Indolizidine 167B and (+)-Monomorine," *J. Am. Chem. Soc.* **1991**, *113*, 3513–18.

30   Evans, D. A.; Miller, S. J.; and Ennis, M. D., "Asymmetric Synthesis of the Benzoquinoid Ansamycin Antitumor Antibiotics: Total Synthesis of (+)-Macbecin," *J. Org. Chem.* **1993**, *58*, 471–85.

31   Savithri, D.; Leumann, C.; and Scheffold, R., "Synthesis of a Building Block for a Nucleic-Acid Analog with a Chiral, Flexible Peptide Backbone," *Helv. Chim Acta* **1996**, *79*, 288–91.

32   Hartzoulakis, B.; and Gani, D., "Syntheses of (2S,3R)- and (2S,3S)-3-Methylglutamic Acid," *J. Chem. Soc., Perkin Trans. 1*, **1994**, 2525–31.

33   Lopéz-Herrera, F. J.; and Sarabia-García, F., "β-Oxy-α-Diazo Carbonyl Compounds. II. Conversion to Chiral α-Oxy-α'-Diazo Ketones and Photochemical Reaction." *Tetrahedron Lett.* **1994**, 2929–32.

34   Hudlicky, T.; and Sheth, J. P., "Synthesis of Dienic Acids. Application of Arndt–Eistert Reaction to Unsaturated Diazoketones," *Tetrahedron Lett.* **1979**, 2667–70.

35   Callant, P.; D'Haenens, L.; Van der Eycken, E.; and Vandewalle, M., "Photoinduced Wolff Rearrangement of α-Diazo-β-ketophosphonates: A Novel Entry into Substituted Phosphonoacetates," *Synth. Commun.* **1984**, *14*, 163–67.

36   van Haard, P. M. M.; Thijs, L.; and Zwanenburg, B., "Photoinduced Rearrangement of α,β-Epoxy Diazomethyl Ketones," *Tetrahedron Lett.* **1975**, 803–6.

37   Thijs, L.; Dommerholt, F. J.; Leemhuis, F. M. C.; and Zwanenburg, B., "A General Stereospecific Synthesis of γ-Hydroxy-α,β-Unsaturated Esters," *Tetrahedron Lett.* **1990**, *31*, 6589–92.

38   van Aar, M. P. M.; Thijs, L.; and Zwanenburg, B., "Stereoselective Synthesis of Enantiopure 4,5-Dihydroxy-2-alkene Esters from Simple Allylic Alcohols," *Tetrahedron* **1995**, *51*, 9699–712.

39   Davies, H. M. L.; Saikali, E.; Clark, T. J.; and Chee, E. H., "Anomalous Reactivity of Mono Substituted Rhodium Stabilized Vinylcarbenoids," *Tetrahedron Lett.* **1990**, *31*, 6299–302.

40   Meier, H.; and Zeller, K.-P., "The Wolff Rearrangement of α-Diazo Carbonyl Compounds," *Angew. Chem., Int. Ed. Engl.* **1975**, *14*, 32–43.

41   Georgian, V.; Boyer, S. K.; and Edwards, B., "A New Synthesis of β-Lactams. Rearrangements of α-Diazo Thioesters," *Heterocycles* **1977**, *7*, 1003–8.

42   Georgian, V.; Boyer, S. K.; and Edwards, B., "Photochemistry of α-Diazo Thioesters: Migratory Aptitude of Sulfur vs. Oxygen in the Photochemical Wolff Rearrangement," *J. Org. Chem.* **1980**, *45*, 1686–88.

43   Wiberg, K. B.; Snoonian, J. R.; and Lahti, P. M., "Ring Contraction of a Two-Carbon Bridged Spiropentane," *Tetrahedron Lett.* **1996**, *37*, 8285–88.

44   Wiberg, K. B.; Olli, L. K.; Golembeski, N.; and Adams, R. D., "Tricyclo[4.2.0.0^{1,4}]octane," *J. Am. Chem. Soc.* **1980**, *102*, 7467–75.

45   Rao, V. B.; Wolff, S.; and Agosta, W. C., "Synthesis of Methyl 1-Methyltetracyclo[4.3.1.0^{3,10}.0^{8,10}]decane-7-carboxylate, a Derivative of [4.4.4.5]Fenestrane," *J. Chem. Soc., Chem. Commun.* **1984**, 293–94.

46 Rao, V. B.; George, C. F.; Wolff, S.; and Agosta, W. C. "Synthetic and Structural Studies in the [4.4.4.5]Fenestrane Series," *J. Am. Chem. Soc.* **1985**, *107*, 5732–39.

47 Otterbach, A.; and Musso, H., "Diasterane (Tricyclo[3.1.1.1^{2,4}]octane)," *Angew. Chem., Int. Ed. Engl.* **1987**, *26*, 554–55.

48 Eaton, P. E.; Jobe, P. G.; and Reingold, I. D., "The 1,7-Cyclobutanonorbornane System," *J. Am. Chem. Soc.* **1984**, *106*, 6437–39.

49 Moore, H. W.; and Arnold, M.J., "Photolysis of 4-Diazopyrrolidine-2,3-diones. A New Synthetic Route to Mono- and Bicyclic β-Lactams," *J. Org. Chem.* **1983**, *48*, 3365–67.

50 Tsuji, T.; and Nishida, S., "Photochemical Generation of [4]Paracyclophanes from 1,4-Tetramethylene Dewar Benzenes: Their Electronic Absorption Spectra and Reactions with Alcohols," *J. Am. Chem. Soc.* **1988**, *110*, 2157–64.

51 Norbeck, D. W.; and Kramer, J. B., "Synthesis of (-) Oxetanocin," *J. Am. Chem. Soc.* **1988**, *110*, 7217–18.

52 Saha, G.; and Ghosh, S., "A New Route to the Synthesis of 7-Functionalised Bicyclo[2.2.1]heptane Derivatives," *Synth. Commun.* **1991**, *21*, 2129–36.

53 Coates, R. M.; and Kang, H-Y., "Synthesis and Evaluation of Cyclobutylcarbinyl Derivatives as Potential Intermediates in Diterpene Biosynthesis," *J. Org. Chem.* **1987**, *52*, 2065–74.

54 Wheeler, T. N.; and Meinwald, J., "Formation and Photochemical Wolff Rearrangement of Cyclic α-Diazoketones: D-Norandrost-5-en-3β-ol-16-carboxylic Acids," *Org. Synth.* **1988**, *52*, 53–58.

55 Kakiuchi, K.; Ue, M.; Takeda, M.; Tadaki, T.; Kato, Y.; Nagashima, T.; Tobe, Y.; Koike, H.; Ida, N.; and Odaira, Y., "Antiproliferating Polyquinanes. V. Di- and Triquinanes Involving α-Methylene or α-Alkylidene Cyclopentanone, Cyclopentenone and γ-Lactone Systems," *Chem. Pharm. Bull.* **1987**, *35*, 617–31.

56 Ghosh, A.; Banerjee, U. K.; and Venkateswaran, R. V., "Photolysis of α-Diazocyclopentanones. Ring Contraction to Functionalised Cyclobutanes and Synthesis of Junionone, Grandisol and Planococcyl Acetate," *Tetrahedron* **1990**, *46*, 3077–88.

57 Tobe, Y.; Ueda, K.; Kaneda, T.; Kakiuchi, K.; Odaira, Y.; Kai, Y.; and Kasai, N., "Synthesis and Molecular Structure of (Z)-[6]Paracycloph-3-enes," *J. Am. Chem., Soc.* **1987**, *109*, 1136–44.

58 Fessner, W. D.; Prinzbach, H.; and Richs, G., "Pagodane—An Undecacyclic C$_{20}$H$_{20}$-Polyquinane," *Tetrahedron Lett.* **1983**, *24*, 5857–60.

59 Fessner, W. D.; Sedelmeier, G.; Spurr, P. R.; Rihs, G.; and Prinzbach, H., "Pagodane: The Efficient Synthesis of a Novel, Versatile, Molecular Framework," *J. Am. Chem. Soc.* **1987**, *109*, 4626–42.

60 Rosati, R. L.; Kapili, L. V.; Morrissey, P.; Bordner, J.; and Subramanian, E., "Photochemical Transformation of Cephalosporins into Carbapenems," *J. Am. Chem. Soc.* **1982**, *104*, 4262–64.

61 Cossy, J.; and Belotti, D., "Photoreductive Cyclization: Application to the Total Synthesis of ($\pm$)-Actinidine," *Tetrahedron Lett.* **1988**, *29*, 6113–14.

62 Cossy, J.; Belotti, D.; and Leblanc, C., "Total Synthesis of ($\pm$)-Actinidine and of ($\pm$)-Isooxyskytanthine," *J. Org. Chem.* **1993**, *58*, 2351–54.

63 Rao, Y. K.; and Nagarajan, M., "Formal Total Synthesis of ($\pm$)-Silphinene *via* Radical Cyclization," *J. Org. Chem.* **1989**, *54*, 5678–83.

64 Majerski, Z.; and Vinkovic, V., "A New, One-Pot Preparation of Alicyclic Ketones via Wolff Rearrangement," *Synthesis* **1989**, *7*, 559–60.

65 Nuyttens, F.; Appendino, G.; and De Clercq, P. J., "The Intramolecular Diels–Alder Reaction with Furan-Diene: A Novel Route to (±)-Gibberellin A$_5$," *Synlett.* **1991**, *7*, 526–28.

66 Uyehara, T.; Takehara, N.; Ueno, M.; and Sato, T., "Rearrangement Approaches to Cyclic Skeletons. IX. Stereoselective Total Synthesis of (±)-Camphorenone Based on a Ring-Contraction of Bicyclo[3.2.1]oct-6-en-2-one. Reliable One-Step Diazo Transfer Followed by a Wolff Rearrangement," *Bull. Chem. Soc. Jpn.* **1995**, *68*, 2687–94.

67 Cossy, J.; Belotti, D.; Bouzide, A.; and Thellend, A., "Short and Efficient Access to β-Ketoamides," *Bull. Soc. Chim. Fr.* **1994**, *131*, 723–29.

68 Hisatome, M.; Watanabe, J.; Yamashita, R-i.; Yoshida, S.; and Yamakawa, K., "Bridge Contraction of [4]Ferrocenophanes by Wolff Rearrangement and Synthesis of [3$_4$](1,2,3,4)Ferrocenophane *via* the Contraction Reaction," *Bull. Chem. Soc. Jpn.* **1994**, *67*, 490–94.

69 Allinger, N. L.; Walter, T. J.; and Newton, M. G., "Synthesis, Structure and Properties of the [7]Paracyclophane Ring System," *J. Am. Chem. Soc.* **1974**, *96*, 4588–95.

70 Yamaguchi, R.; Honda, K.; and Kawanisi, M., "3,12-Cycloiceane (Pentacyclo[6.3.1.0^{2,4}.0^{5,10}.0^{7,8}]dodecane)," *J. Chem. Soc., Chem. Commun.*, **1987**, 83–4.

71 Overman, L. E.; Robertson, G. M.; and Robichaud, A. J., "Total Synthesis of (±)-Meloscine and (±)-Epimeloscine," *J. Org. Chem.* **1989**, *54*, 1236–38.

72 Overman, L. E.; Robertson, G. M.; and Robichaud, A. J., "Use of Aza-Cope Rearrangement-Mannich Cyclization Reactions To Achieve a General Entry to *Melodinus* and *Aspidosperma* Alkaliods. Stereocontrolled Total Syntheses of (±)-Deoxoapodine, (±)-Meloscine, and (±)-Epimeloscine and a Formal Synthesis of (±)-1-Acetylaspidoalbidine," *J. Am. Chem. Soc.* **1991**, *113*, 2598–2610.

73 Mander, L. N.; and Pyne, S. G., "Studies on Gibberellin Synthesis. Assembly of an Ethanophenanthrenoid Lactone and Conversion into a Gibbane Derivative," *Aust. J. Chem.*, **1981**, *34*, 1899–911.

74 Oppolzer, W.; Bättig, K.; and Hudlicky, T., "A Total Synthesis of (±)-Isocomene and (±) β-Isocomene by an Intramolecular Ene Reaction," *Tetrahedron* **1981**, 4359–64.

75 Oppolzer, W.; Bättig, K. and Hudlicky, T., "The Total Synthesis of (±)-Isocomene by an Intramolecular Ene Reaction," *Helve. Chim. Acta* **1979**, *62*, 1493–96.

76 Maier, G.; Hoppe, M.; Lanz, K.; and Reisenauer, H. P., "Neue Wege zum Cyclobutadien und Methylencyclopropen [1]," *Tetrahedron Lett.* **1984**, *25*, 5645–48.

77 Pacansky, J.; and Johnson, D. J., "Photochemical Studies on a Substituted Naphthalene-2,1,Diazooxide," *Electrochem. Soc.* **1977**, *124*, 862–65.

78 Pacansky, J.; and Coufal, H. J., "Electron-Beam-Induced Wolff Rearrangement," *J. Am. Chem. Soc.* **1980**, *102*, 410–12.

79 Pacansky, J.; and Johnson, D. J., "Photochemical Studies on a Substituted Naphthalene-2,1,Diazooxide," *Electrochem. Soc.* **1977**, *124*, 862–65.

80 McMahon, R. J.; Chapman, O. L.; Hayes, R. A.; Hess, T. C.; and Krimmer, H.-P., "Mechanistic Studies on the Wolff Rearrangement: The Chemistry and Spectroscopy of Some α-Ketocarbenes," *J. Am. Chem. Soc.* **1985**, *107*, 7597–606.

81 Chapman, O. L.; Gano, J.; West, P. R.; Regirtz, M.; and Maas, G., "Acenaph-thyne," *J. Am. Chem. Soc.* **1981**, *103*, 7033–36.

82 Trost, B. M.; and Whitman, P. J., "On the Chemistry of α,α'-Bis(diazo) Ketones," *J. Am. Chem. Soc.* **1974**, *96*, 7421–29.

83 DoMinh, T.; and Strausz, O. P., "Cycloaddition of Ethoxyketene to Olefins," *J. Am. Chem. Soc.* **1970**, *92*, 1766–68.

84 Yates, P.; and Fallis, A. G., "Decomposition of 1-Diazo-3-(2,2,3-Trimethylcy-clopent-3-enyl)propan-2-one. A Novel Synthesis and Rearrangement of a Tricy-clo[3.2.1.0^{3,6}]octan-4-one," *Tetrahedron Lett.* **1968**, 2493–96.

85 Becker, D.; Nagler, M.; and Birnbaum, D., "Intramolecular Photoaddition of Kete-nes to Cyclohexenones," *J. Am. Chem. Soc.* **1972**, *94*, 4771–73.

86 Geisel, M.; Grob, C. A.; Santi, W.; and Tschudi, W., "Die Cyclopropylcarbinyl-cyclobutyl-homoallyl-Umlagerung. I. Teil. Synthese von Tricyclo[3.2.1.0^{2,7}]octan-3-ol, *endo*- und *exo*-Bicyclo[3.2.1]-oct-2-en-7-ol," *Helv. Chim. Acta.* **1973**, *56*, 1046–47.

87 Becker, D.; Harel, Z.; and Birnbaum, D., "Intramolecular Photocycloadditions of Ketens and Allenes to Cyclohexenones," *J. Chem. Soc., Chem. Commun.* **1975**, 377–78.

88 Becker, D.; and Birnbaum, D., "Intramolecular Photoaddition of Ketenes to Con-jugated Cycloalkenones," *J. Org. Chem.* **1980**, *45*, 570–78.

89 Stevens, R. V.; Bisacchi, G. S.; Goldsmith, L.; and Strousse, C. E., "Synthesis of Spiro-Activated Cyclopropanes from Alkenes *via* the Irradiation of Isopropylidene Diazomalonate. A Reinvestigation," *J. Org. Chem.* **1980**, *45*, 2708–9.

90 Ireland, R. E.; Dow, W. C.; Godfrey, J. D.; and Thaisrivongs, S., "Total Synthesis of (±)-Aphidicolin and (±)-β-Chamigrene," *J. Org. Chem.* **1984**, *49*, 1001–13.

91 Nakatani, K.; Takada, K.; and Isoe, S., "Intramolecular Cooperative Reactions of 1,2-Bis(diazoketone)s. The First Syntheses of *trans*-Hydro-1*H*-2-inden-1-ones," *J. Org. Chem.* **1995**, *60*, 2466–73.

92 Danheiser, R. L.; Brisbois, R. G.; Kowalczyk, J. J.; and Miller, R. F., "An Annula-tion Method for the Synthesis of Highly Substituted Polycyclic Aromatic and Het-eroaromatic Compounds," *J. Am. Chem. Soc.* **1990**, *112*, 3093–3100.

93 Danheiser, R. L.; and Cha, D. D., "Total Synthesis of the Host Defense Stimulant Maesanin," *Tetrahedron Lett.* **1990**, *31*, 1527–30.

94 Danheiser, R. L.; and Helgason, A. L., "Total Synthesis of the Phenalenone Diter-pene Salvilenone," *J. Am. Chem. Soc.* **1994**, *116*, 9471–77.

95 Danheiser, R. L; Casebier, D. S.; and Loebach, J. L., "Total Synthesis of Dan Shen Diterpenoid Quinones," *Tetrahedron Lett.* **1992**, *33*, 1149–52.

96 Danheiser, R. L.; Casebier, D. S.; and Huboux, A. H., "Total Synthesis of Ae-gyptinenones," *J. Org. Chem.* **1994**, *59*, 4844–48.

97 Danheiser, R. L.; Casebier, D. S.; and Firooznia, F., "Aromatic Annulation Strat-egy for the Synthesis of Angularly-Fused Diterpenoid Quinones. Total Synthesis of (+)-Neocryptotanshinone, (−)-Cryptotanshinone, Tanshin-one IIA, and (±)-Royleanone," *J. Org. Chem.* **1995**, *60*, 8341–50.

98 Danheiser, R. L.; and Trova, M. P., "Synthesis of Linear Furocoumarins *via* a Photochemical Aromatic Annulation Strategy. An Efficient Total Synthesis of Bergapten," *Synlett.* **1995**, 573–74.

99   Meier, H.; and Zeller, K.-P., "The Wolff Rearrangement of α-Diazo Carbonyl Compounds," *Angew. Chem., Int. Ed. Engl.* **1975**, *14*, 32–43

100 Staudinger, H., "Diphenylketen," *Liebigs Ann. Chem.* **1907**, *356*, 51–123.

101 Haddadin, M.; and Hassner, A., "Cycloaddition of Diphenylketene to Some C=N Heterocycles. Structural Assignment and Reaction of Adducts," *J. Org. Chem.* **1973**, *38*, 2650–52.

102 Pacansky, J.; Chang, J. S.; Brown, D. W.; and Schwarz, W., "Observation of Zwitterions in the Thermal Reaction of Ketenes with Carbon–Nitrogen Double Bonds," *J. Org. Chem.* **1982**, *47*, 2233–34.

103 Blades, C. E.; and Wilds, A. L., "The Preparation of Indoles from Diazo Ketones," *J. Org. Chem.* **1956**, *21*, 1013–21.

104 Wildes, A. L.; Berghe, J. V. D.; Winestock, C. H.; von Treba, R. L.; and Woolsey, N. F., "Abnormal Acids from the Arndt–Eistert Synthesis," *J. Am. Chem. Soc.* **1962**, *84*, 1503–4.

105 Wilds, A. L.; Woolsey, N. F.; Berghe, J. V. D.; and Winestock, C. H., "Concerning the Mechanism of Wolff Rearrangement; A Dissimilarity between the Thermal Wolff and Curtius Rearrangement," *Tetrahedron Lett.* **1965**, 4841–46.

106 Wilds, A. L.; von Trebra, R. L.; and Woolsey, N. F., "Preparation and Reactions of Diazo Ketones. V. Normal and Abnormal Products from Thermal Wolff Rearrangement of 9-Phenylfluorene-9-carbonyldiazomethane," *J. Org. Chem.* **1969**, *34*, 2401–6.

107 Lokensgard, J. P.; O'Dea, J.; and Hill, E. A., "Decomposition of a β,γ-Unsaturated Diazo Ketone. Evidence for the Intermediacy of a Bicyclopentanone," *J. Org. Chem.* **1974**, *39*, 3355–57.

108 Zimmerman H. E.; and Little, R. D., "Photochemical Rearrangement of 4-Aryl-Substituted-cyclopentenones. Low-Temperature Photochemistry and Direct Observation of Reaction Intermediates," *J. Am. Chem. Soc.* **1974**, *96*, 4623–30.

109 Smith, A. B., III., "A Vinylogous Wolff Rearrangement; Copper Sulphate-Catalysed Decomposition of Unsaturated Diazomethyl Ketones," *J. Chem. Soc., Chem. Commun.* **1974**, 695–96.

110 Smith, A. B., III.; Toder, B. H.; and Branca, S. J., "Stereochemical Consequences of the Vinylogous Wolff Rearrangement," *J. Am. Chem. Soc.* **1976**, *98*, 7456–58.

111 Branca, S. J.; Lock, R.; Smith, A. B., III., "Exploitation of the Vinylogous Wolff Rearrangement. An Efficient Total Synthesis of (±)-Mayurone, (±)-Thujopsene, and (±)-Thujopsadiene," *J. Org. Chem.* **1977**, *42*, 3165–68.

112 Smith, A. B., III.; Toder, B. H.; and Branca, S. J., "Vinylogous Wolff Rearrangement. 4. General Reaction of β,γ-Unsaturated α'-Diazo Ketones," *J. Am. Chem. Soc.* **1984**, *106*, 3995–4001.

113 Thijs, L.; Stokkingreef, E. H. M.; Lemmens, J. M.; and Zwanenburg, B., "Enantiocontrolled Synthesis of the Macrocyclic C_{14}-C_{23} Subunit of Cytochalasin B," *Tetrahedron* **1985**, *41*, 2949–56.

114 Waanders, P. P.; Thijs, L.; and Zwanenburg, B., "Stereocontrolled Total Synthesis of the Macrocyclic Lactone (−)-Aspicilin," *Tetrahedron Lett.* **1987**, *28*, 2409–12.

115 Weygand, F.; and Bestmann, H. J., "Synthesen unter Verwendung von Diazoketonen," *Angew. Chem.* **1960**, *70*, 535–54.

116 Ceccherelli, P.; Curini, M.; Porter, B.; and Wenkert, E., "Construction of the Hibaene Skeleton by way of an Abnormal Wolff Reaction," *J. Org. Chem.* **1984**, *49*, 2052–54.

117 Saha, B.; Bhattacharjee, G.; and Ghatak, U. R., "A Novel Synthetic Method for Angularly Functionalized Polycyclic Systems by Vinylogous Wolff Rearrangement of β,γ-Unsaturated Diazoketones," *Tetrahedron Lett.* **1986**, *27*, 3913–14.

118 Saha, B.; Bhattacharjee, G.; and Ghatak, U. R., "Vinylogous Wolff Rearrangement of Cyclic β,γ-Unsaturated Diazomethyl Ketones: A New Synthetic Method for Angularly Functionalized Polycyclic Systems," *J. Chem. Soc., Perkin Trans. 1* **1988**, 939–44.

119 Ray, C.; Saha, B.; and Ghatak, U. R., "Synthesis of Some Angularly Cyclopentanone fused Hydrophenanthrene and Hydrofluorene Derivatives by Acid-Catalyzed Intramolecular C-Alkylation of γ,δ-Unsaturated α'-Diazomethyl Ketones," *Synth. Commun.* **1991**, *21*, 1223–42.

120 Motallebi, S.; and Müller, P., "Rhodium(II)-Catalysed Decomposition of β,γ-Unsaturated Diazo Compounds," *Helv. Chim. Acta* **1993**, *76*, 2803–13.

121 Miller, R. D.; Theis, W.; Heilig, G.; and Kirchmeyer, S., "The Generation and Rearrangement of 2-(Diazoacetyl)cyclobutanones: The Formation of 5-Spirocyclopropyl-2-(5H)-furanones," *J. Org. Chem.* **1991**, *56*, 1453–63.

122 Nakatani, K.; Isoe, S.; Maekawa, S.; and Saito, I., "Photoinduced DNA Cleavage by Designed Molecules with Conjugated Ene-Yne-Ketene Functionalities," *Tetrahedron Lett.* **1994**, *35*, 605–8.

123 Nakatani, K.; Maekawa, S.; Tanabe, K.; and Saito, I., "α-Diazoketones as Photochemical DNA Cleavers: A Mimic for the Radical Generating Systems of Neocarzinostatin Chromophore," *J. Am. Chem. Soc.* **1995**, *117*, 10635–44.

124 Nakatani, K.; Shirai, J.; Tamaki, R.; and Saito, I., "Photogeneration of Highly Electrophilic Benzoylketene from Dibenzoyldiazomethane in Aqueous Solvents: Reaction with Amino Acids and DNA Cleavage," *Tetrahedron Lett.* **1995**, *30*, 5363–66.

125 Mosier, G. G. J.; and Lawton, R. G., "Development of a New Family of Thiol Specific Photoactivatable Cross-Linking Agents," *J. Org. Chem.* **1995**, *60*, 6953–58.

126 Sodeoka, M.; Uotsu, K.; and Shibasaki, M., "Photoaffinity Labeling of PKC with a Phorbal Derivative: Importance of the 13-Acyl Group in Phorbal Esters–Pkc Interaction," *Tetrahedron Lett.* **1995**, *48*, 8795–98.

127 Süs, O., "Über die Natur der Belichtungs Produkte von Diazoverbindungen. Übergange von aromatischen 6-Ringen in 5-Ringe," *Justus Liebigs Ann. Chem.* **1944**, *556*, 65–90.

128 Süs, O.; Munder, J; and Steppan, H., "Neue Entwicklungen auf dem Gebiet der vorsensibilisierten Druckfolien," *Angew. Chem.* **1962**, *74*, 985–88.

129 Reiser, A.; Shih, H.-Y.; Yeh, T.-F.; and Huang, J.-P., "Novolak–Diazoquinone Resists: The Imaging Systems of the Computer Chip," *Angew. Chem., Int. Ed. Engl.* **1996**, *35*, 2428–40.

CHAPTER 10

Reactions of α-Diazocarbonyl Compounds with Aldehydes and Ketones

10.1 INTRODUCTION

In the reactions of aldehydes and ketones with diazocarbonyl compounds described in Chapter 7, the carbonyl group acts as a nucleophile in bonding to the electron-deficient carbene center to form an ylide of the type shown in Scheme 1.

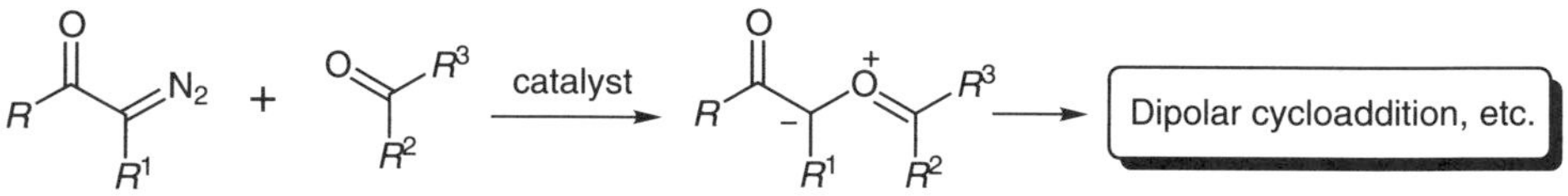

Scheme 1

Subsequent reactions of the ylide are determined by the nature and location of other functionalities present. In yet another group of reactions to be discussed in this chapter, the carbonyl component acts in an electrophilic capacity while the diazocarbonyl group acts as a nucleophile. There are two broad categories for such reactions. One is an aldol-type addition promoted by base with retention of the diazo function to form diazoketol (**1**) (eq. 1); the second is a related process of β-dicarbonyl (**2**) formation with loss of nitrogen (eq. 2). In many cases the prod-

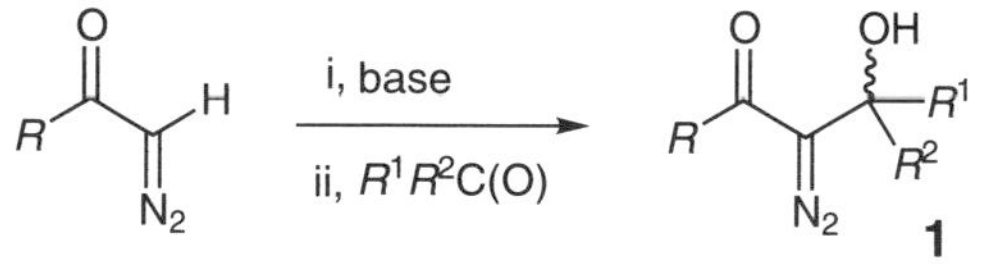

$$(1)$$

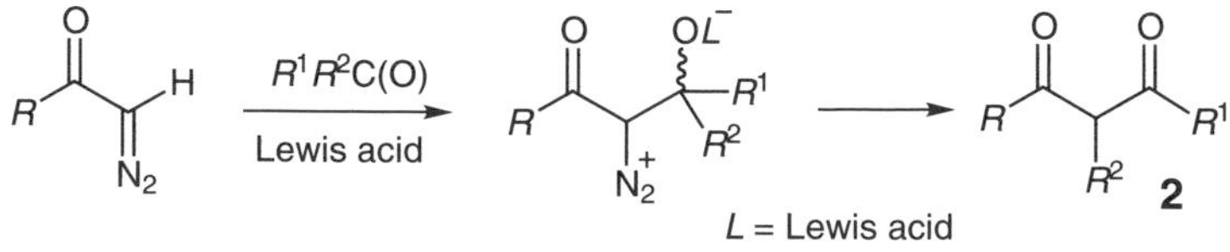

$$(2)$$

ucts of the former can be converted into those of the latter on treatment with Brønsted acid, Lewis acid, or transition metal catalysts (eq. 3). The formation of

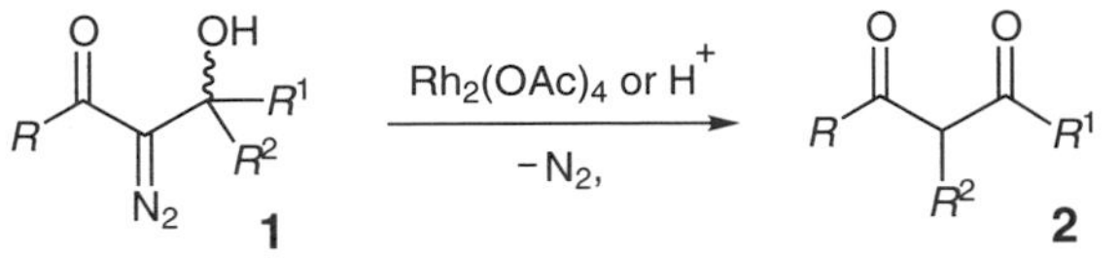

$$(3)$$

β-dicarbonyl products by either route requires a 1,2-migration of a hydrogen or other substituent from the hydroxyl-bearing carbon atom. Both types are known.

Purely thermal reactions of diazocarbonyl compounds with aldehydes and ketones are known, but they tend to be unselective and have found little applications in synthesis. A recent notable exception is the report by López-Herrera and Sarabia-García[1] that ethyl diazoacetate reacts with the arabino-derived aldehyde in Scheme 2, *without either solvent or catalyst*, to yield a diazohydroxyester

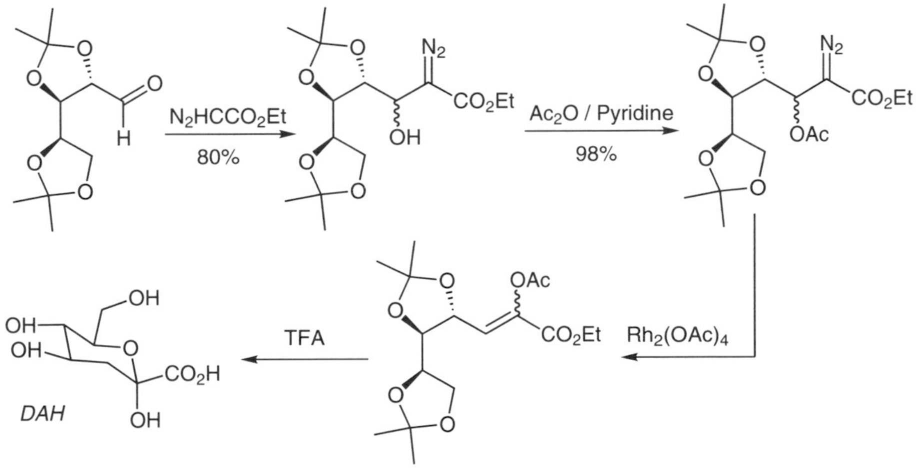

Scheme 2

adduct. Acetylation of the hydroxy function followed by rhodium(II) acetate–catalyzed rearrangement furnished an enol acetate suitable for conversion into natural 3-deoxy-D-arabino-2-heptulosinic acid (DAH) on hydrolysis.

10.2 BASE PROMOTED REACTIONS OF α-DIAZOCARBONYL COMPOUNDS WITH ALDEHYDES AND KETONES

For base-mediated aldol-type additions the diazocarbonyl precursor must be capable of deprotonation to form the α-diazocarbonyl anion or enolate. There follows addition of the anion to the aldehyde or ketone to furnish a diazoketol **1** (eq. 1). Early work by Schölkopf and co-workers[2] described the formation of ethyl lithiodiazoacetate from EDA and n-butyllithium at low temperature and its addition to several carbonyl compounds.[3] Shortly thereafter Wenkert and McPherson[4] examined the preformance of a range of bases for the deprotonation step, including n-butyllithium and lithium diisopropylamide (LDA), and recommended that a dilute solution of potassium hydroxide in methanol or ethanol was the combination of choice. However, in more recent studies, especially those following the work of Pellicciari and co-workers,[5-13] LDA has been by far the most widely employed base for diazocarbonyl deprotonation. The Italian group has used lithiated diazoesters and diazoketones to convert aldehydes and ketones into α-diazo-β-hydroxycarbonyl adducts suitable for subsequent transformations into nitrogen-free products, a particularly attractive version of the latter being the formation of β-dicarbonyl compounds by the catalytic action of rhodium(II) acetate. Examples employing this methodology for the preparation of β-ketoesters are summarized in Table 10.1. LDA was used as base to produce lithiated diazocarbonyl compound for all these examples.

Table 10.1   Synthesis of β-Ketoesters and β-Diketones via the Two-Step Sequence in Eqs. 1 and 3.

Entry	Reactants	Product (yield)	Ref.
1	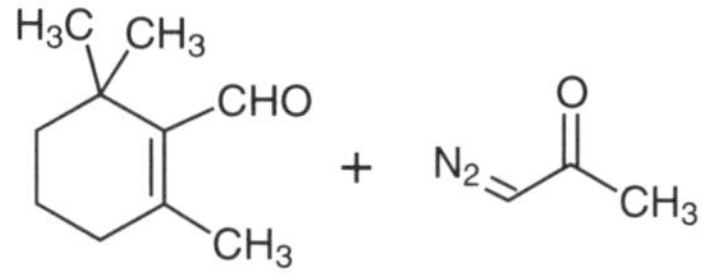	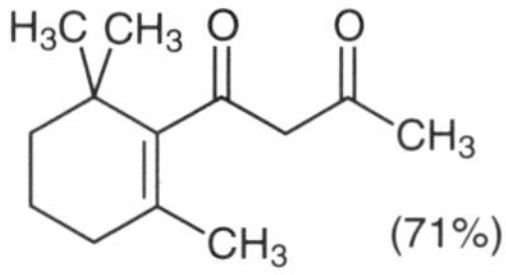 (71%)	10 11
2	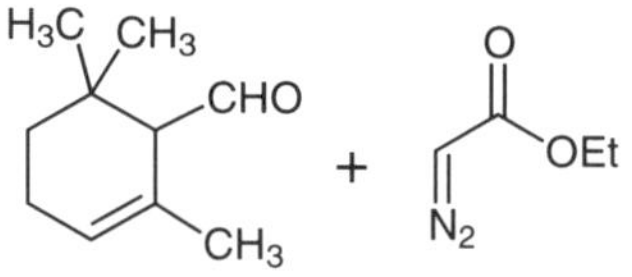	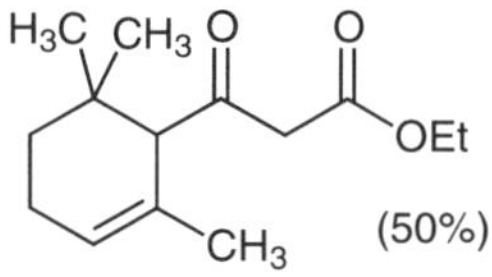 (50%)	6
3	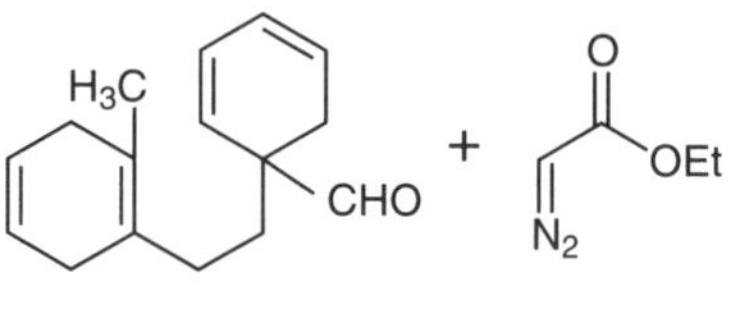	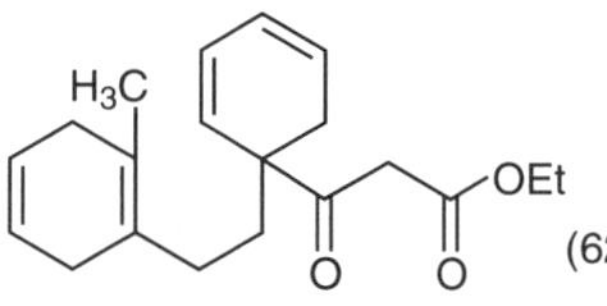 (62%)	14
4	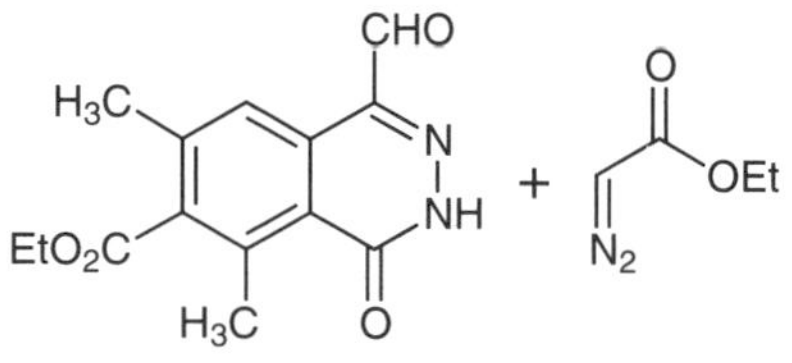	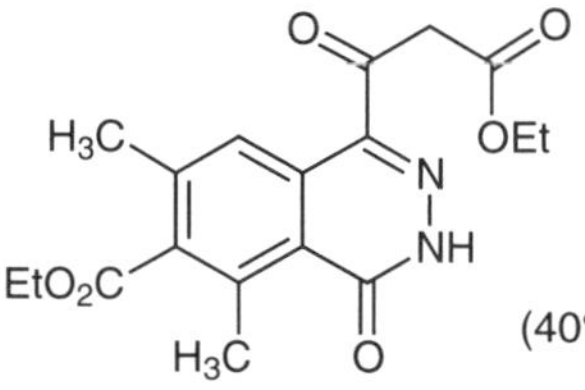 (40%)	15
5	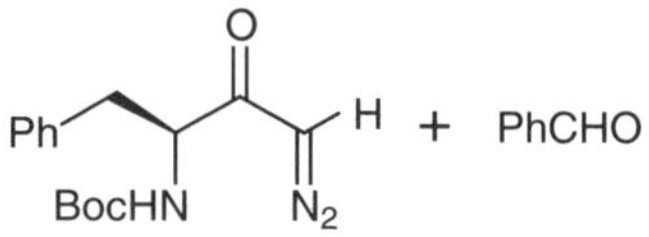	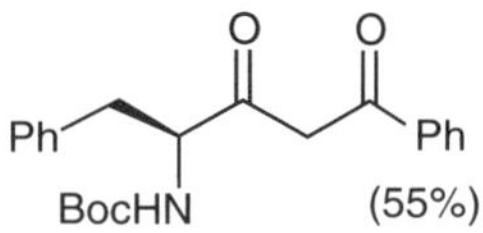 (55%)	16

Entry 1 illustrates the use of diazo lithioacetone to transform β-cyclocitral into a β-diketone suitable for elaboration into β-damascone.[10,11] α-Cyclocitral combines with ethyl diazo(lithio)acetate to afford a product that on exposure to rhodium(II) acetate produces the β-ketoester in entry 2.[6] Corey's synthesis of ($\pm$)-atractyligenin employed this two-step sequence to add a β-ketoester side chain to the bicyclic aldehyde in entry 3.[14] The β-ketoester moiety was then used to set the scene for an intramolecular copper(II)-catalyzed cyclopropanation by reintroducing a diazo group between the two carbonyl groups. A β-ketoester prepared from the corresponding aldehyde in entry 4 *via* this protocol served as a

BASE PROMOTED REACTIONS OF α-DIAZOCARBONYL COMPOUNDS WITH ALDEHYDES: SYNTHESIS OF (N-BUTOXYCARBONYL-L-PHENYLALANYL)-METHYL PHENYL KETONE[16]

A cold ($-10°C$) solution of lithium diisopropylamide (LDA) (0.69 mL of 1.5 M solution in cyclohexane,1.04 mmol) was added over 15 min to a stirred solution of diazoketone (150 mg, 0.52 mmol) in THF (10 mL) at $-84°C$. A solution of benzaldehyde (111 mg, 1.04 mmol) in THF (5 mL) was then added dropwise over 10 min, and the mixture was stirred at $-84°C$ for 1 h, after which 5 mL saturated ammonium chloride aqueous solution was added and the reaction mixture was allowed to warm to room temperature. The reaction mixture was concentrated under vacuum and then extracted with three 50 mL portions of ether. The combined organic extracts were then washed with saturated aqueous sodium bicarbonate and brine, dried, and evaporated under reduced pressure to give a brown oil. Purification by preparative thin-layer chromatography on silica gel using ethyl acetate–hexane (20–30%) as eluant furnished a pure separable mixture of the two isomeric diazoketols (186 mg, 91%) as a pale yellow oil (isomer ratio 7:3).

The resulting α-diazoketol (180 mg, 0.46 mmol) in dry dichloromethane (5 mL) was treated with rhodium(II) acetate (1.0 mg, 0.5 mol %) under nitrogen at room temperature for 1 h. The solvent was removed under vacuo, and the remaining oil was purified by preparative thin-layer chromatography eluting with ethyl acetate–hexane (30%) to yield the β-diketone (96 mg, 57%) (predominantly in enol form) as an oil that formed a crystalline solid (from ether–hexane), m.p. 142.5–143.5°C; $[\alpha]_D^{20}$ -43.5 (c, 1.19 in CH_2Cl_2).

key intermediate in the synthesis of an antiatherosclerotic agent 7-ethoxycarbonyl-4-formyl-6,8-dimethyl-1(2H)-phthalazinone.[15] Entry 5 demonstrates the application of this methodology to the elaboration of optically active amino acids using lithiated N-protected diazo aminoketones; both steps have been shown to be free of racemization.[16]

The process is also applicable to ring expansion of cyclic ketones, five examples of which are summarized in Table 10.2. Entry 1 illustrates the use of EDA to transform 3-acetoxyestrone into a D-homoestrone in high yield. Entries 2 and 3 demonstrate the application of this methodology to the ring-expansion of cyclohexanones to cycloheptanones. The excellent regioselectivity of these two ring expansion processes is noteworthy (*vide infra*). Entry 5 illustrates the elaboration of a cyclopentanone into a cyclohexanone derivative with an L-proline residue at the α-position.

Table 10.2   Ring Expansion of Cyclic Ketones via Base-Promoted Reactions

Entry	Reactants	Product (yield)	Ref.
1	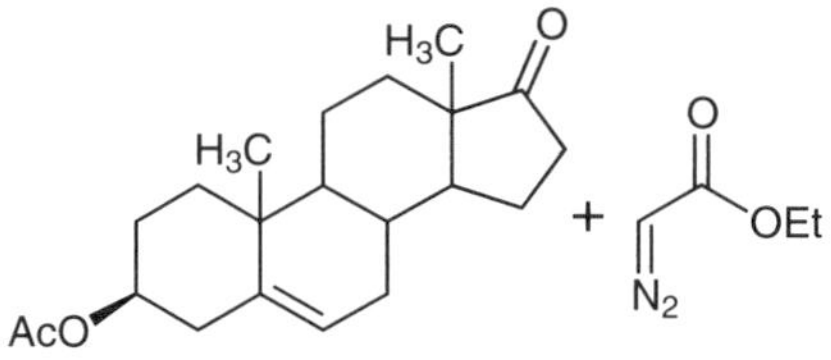	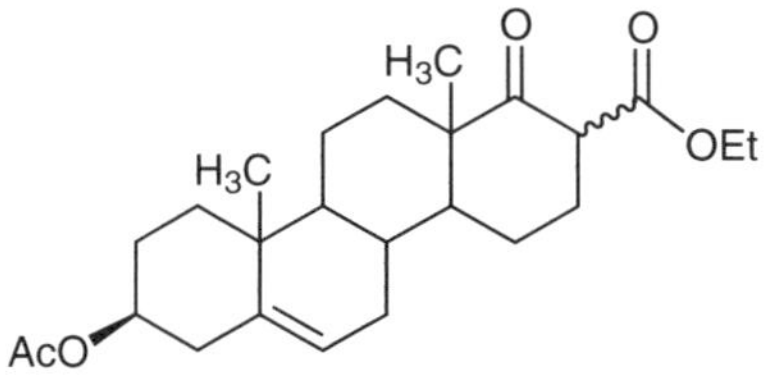	13
2	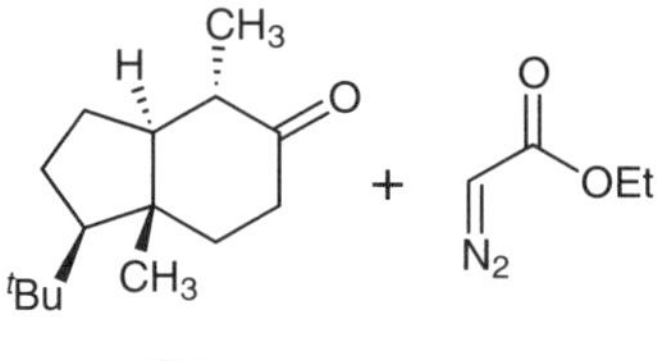	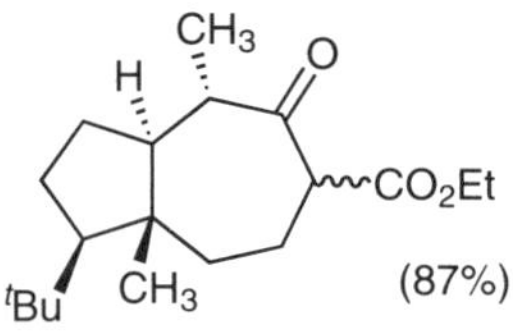 (87%)	17
3	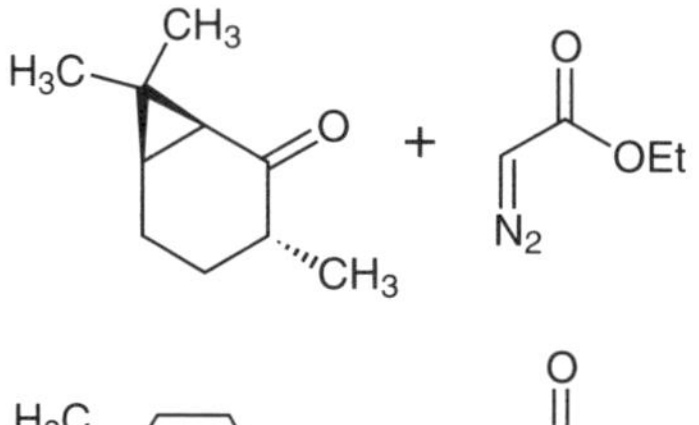	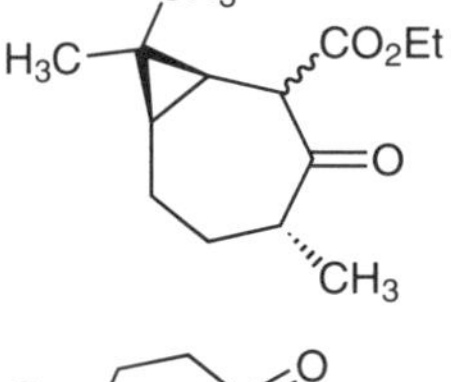 (60%)	18
4	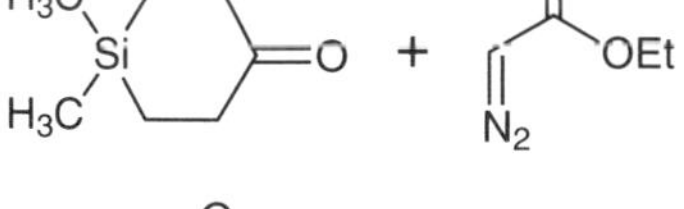	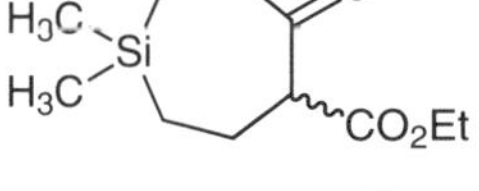 (> 83%)	19
5	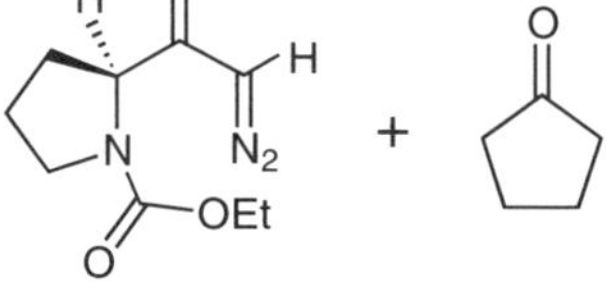	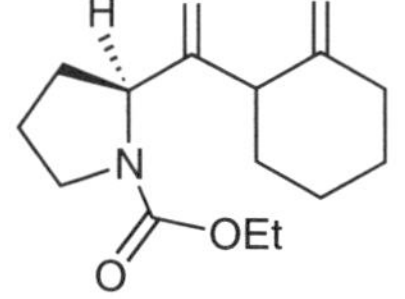 (53%)	16

Kim and co-workers[20] in a detailed study of the second, catalyzed stage of the above processes have found that catalysis is not confined to rhodium(II) acetate. Other active catalysts include palladium and cobalt chloride and tris(triphenyl-phosphine)rhodium(I) chloride. Furthermore, the migratory aptitudes of R^1 and R^2 in diazo adduct **1** (eq. 3) derived from unsymmetrical ketones were found to be catalyst dependent. Some examples of the rhodium(II) acetate–catalyzed conversion of the α-diazo-β-hydroxy esters used in Kim's study into β-ketoesters are shown in Table 10.3.

Table 10.3   Rhodium(II)-Catalyzed Decomposion of α-Diazo-β-Hydroxyesters[20]

Entry	α-Diazo-β-hydroxy esters	β-Ketoesters (yield)	

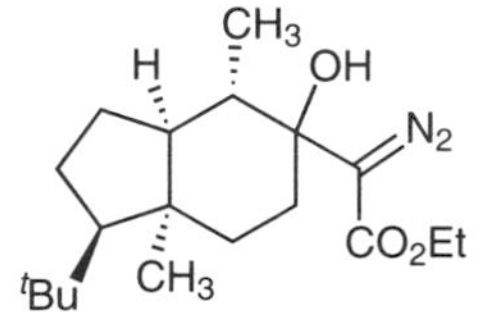

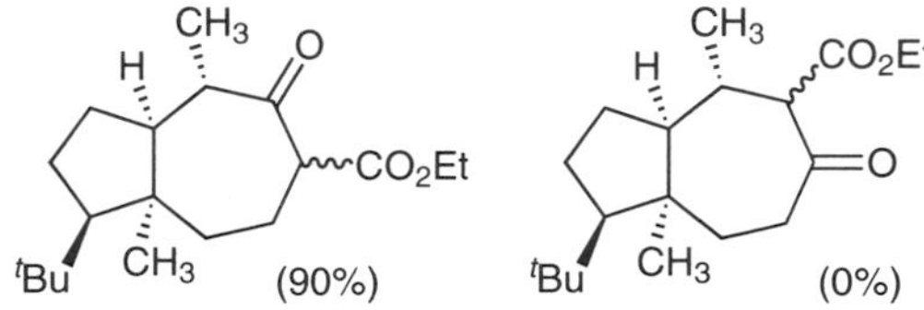

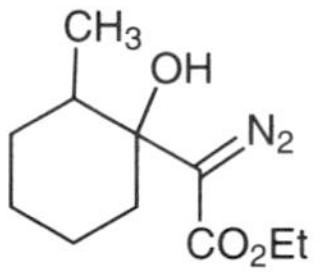

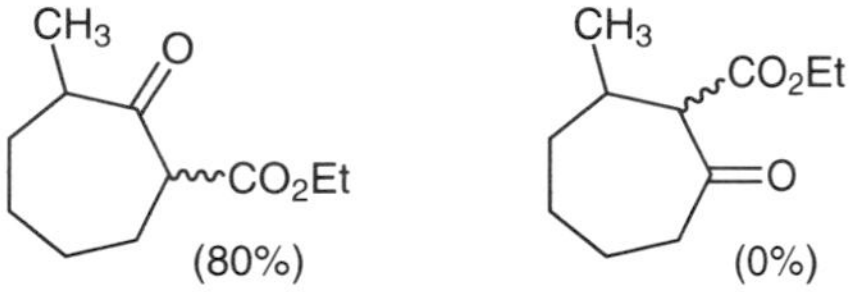

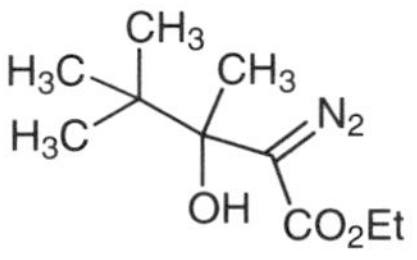

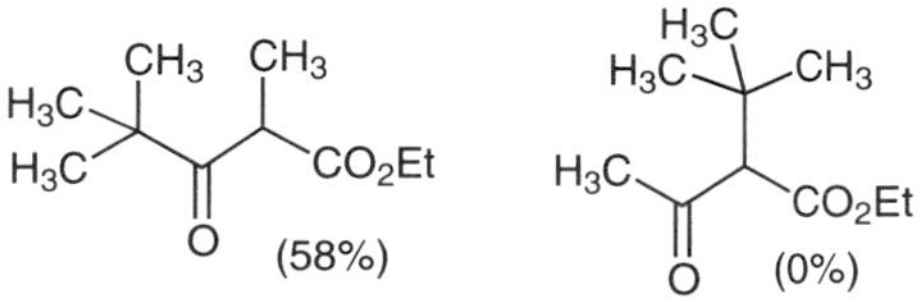

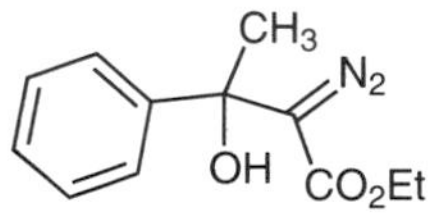

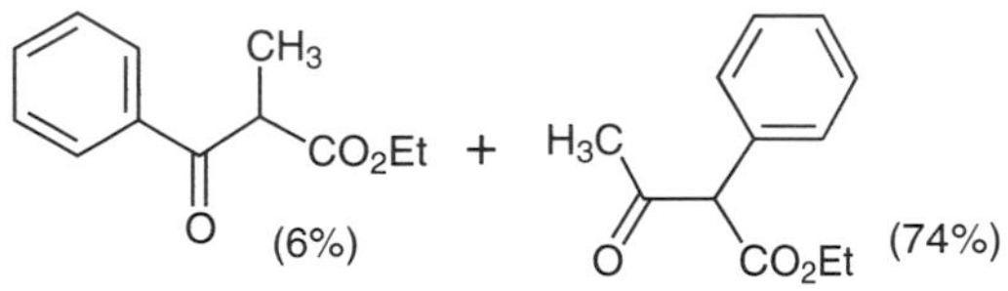

In the examples in entries 1–3 the regioselectivity for migration of the less substituted group is complete, with none of the alternative product being produced. In entry 4 the competition is between migration of a methyl group and an *n*-butyl group, and the resulting regioselectivity is very low. Entry 5 shows that a phenyl group has a significantly higher migratory aptitude than a methyl group.

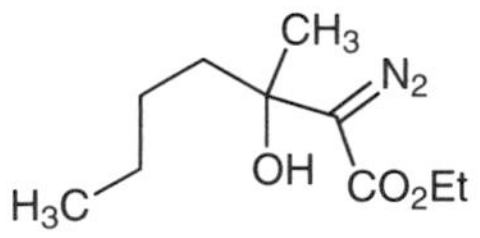

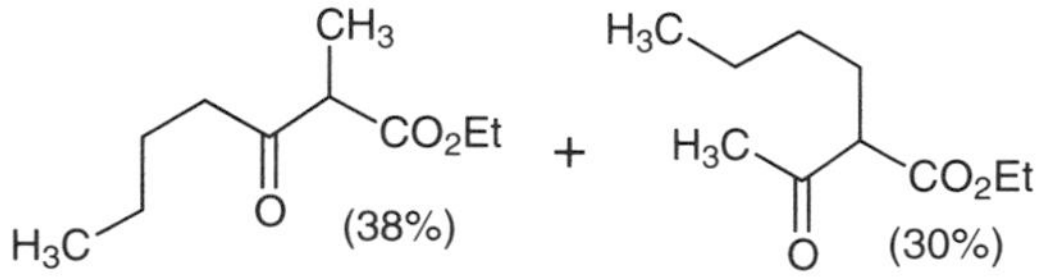

The ability of Brønsted and Lewis acids to catalyze the decomposition of α-diazo-β-hydroxyketones and esters was already known prior to the introduction of rhodium(II) catalysts. For example, Schöllkopf and co-workers[3] had shown that treatment of ethyl 2-diazo-3-hydroxy-3-methylbutanoate with hydrochloric acid furnished ethyl 2-methyl-3-oxobutanoate, the product of a 1,2-migration (eq. 4). With Lewis acid catalysts, particularly boron fluoride etherate, the decomposition pathway of α-diazo-β-hydroxy esters can be very complex, affording mixtures of products that are very substrate and solvent dependent.[21,22] Thus acyclic α-diazo-β-hydroxycarbonyl compounds **3** decompose to acetylacetylenes **4** as the major products (eq. 5) together with minor amounts of migration prod-

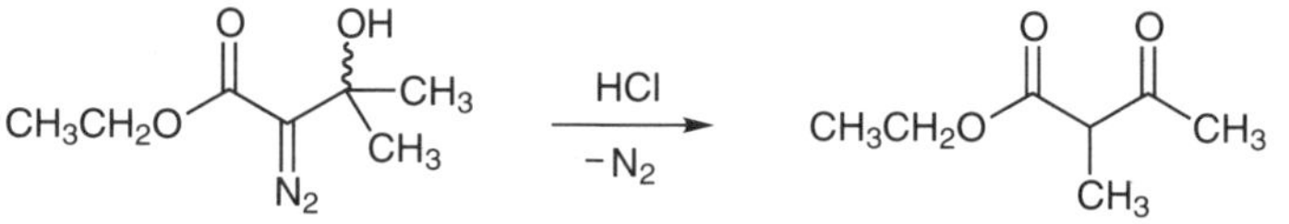

$$
\begin{array}{cc}
\text{CH}_3\text{CH}_2\text{O} & \xrightarrow[-\text{N}_2]{\text{HCl}} & \text{CH}_3\text{CH}_2\text{O} \quad \text{CH}_3 & (4)
\end{array}
$$

$$
\begin{array}{cc}
R \quad \text{3} & \xrightarrow[\text{CH}_3\text{CN}]{\text{BF}_3\cdot\text{Et}_2\text{O}} & R \quad R^1 \quad \text{4} & (5)
\end{array}
$$

ucts.[8,23,24] The formation of acetylacetylenes has been rationalized as proceeding *via* coordination of the substrate through the hydroxyl function followed by generation of an alkenyldiazonium salt, which undergoes concomitant deprotonation and loss of nitrogen. With α-diazo-β-hydroxy carbonyl compounds derived from ketones, the α-hydrogen atom, which is so crucial to the deprotonation mechanism leading to alkynes, is no longer present, and this outlet is not available. Pellicciari et al.[22] have conducted a comprehensive mechanistic study of the boron trifluoride–catalyzed decomposition of α-diazo-β-hydroxy esters derived from cycloalkanones (Scheme 3) in several solvents and have identified products from

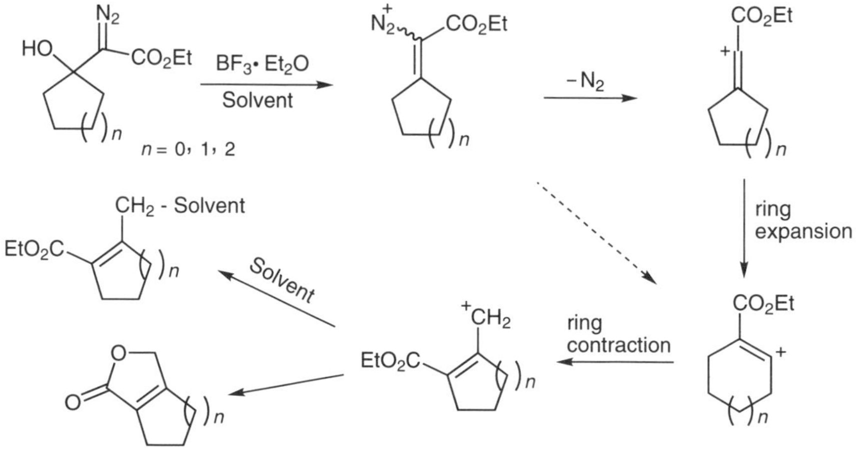

Scheme 3

mechanistic cascades involving ring expansion and ring contraction of carbocations and solvent participation.

10.3 ACID-CATALYZED REACTIONS OF α-DIAZOCARBONYL COMPOUNDS WITH ALDEHYDES

Of much greater interest from the point of view of use in synthesis is the direct, one-pot reaction of aldehydes or ketones with diazocarbonyl compounds catalyzed by Lewis acids. All these processes lead directly to nitrogen-free β-dicarbonyl products (eq. 2). An example of the use of boron fluoride etherate to catalyze the reaction of methyl diazoacetate with 2,3-O-isopropylidene-D-glyceraldehyde **5** to form β-ketoester **6** in 71% yield is shown in eq. 6.[25] The work of

Holmquist and Roskamp and of Padwa and co-workers has demonstrated that aldehydes react efficiently in the presence of tin(II) chloride with diazoesters and diazoketones to form β-keto esters or β-diketones respectively.[26,27] Several representative examples are shown in Table 10.4. Entries 1–4 illustrate the use of ethyl diazoacetate and entry 5 that of a diazopyruvate. The remaining three entries are examples of the use of diazoketones; yields are moderate to excellent. In general, the process occurs readily in dichloromethane at or below room temperature and shows a sensitivity towards aldehyde structure sufficient to allow differention between aliphatic (the more reactive) and aromatic aldehydes. For example, a reaction performed on a 1:1:1 mixture of benzaldehyde, 3-phenylpropionaldehyde, and ethyl diazoacetate gave a 1:30 ratio of β-keto esters favoring the aliphatic aldehyde at −15°C.[26] Yields of diketones were found to be significantly lower from aromatic aldehydes as compared with aliphatic aldehydes, and byproducts were more significant. For example, treatment of 1-diazo-4-phenylbutan-2-one with benzaldehyde in the presence of $SnCl_2$ afforded the H–Cl insertion product rather than the β-diketone.

In later work, Holmquist and Roskamp[29] demonstrated that this Lewis acid–catalyzed β-keto ester synthesis could be extended to include the use of alkenes as starting materials using ozonolysis to first generate the aldehyde, which in turn condenses with the ethyl diazoacetate. In this variation, tin(II) chloride plays a dual role, initially acting as a reducing agent in converting the ozonide to the aldehyde and subsequently as a Lewis acid to assist the addition of diazoester. Four examples of this two-step transformation of alkene into β-keto ester are

Table 10.4 Synthesis of β-Dicarbonyl Compounds *via* the Tin(II) Chloride–Catalyzed Reaction of Diazoesters/Diazoketones with Aldehydes

Entry	Reactants	Product (yield)	Ref.
1	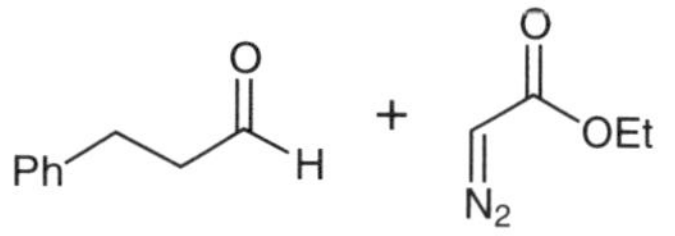	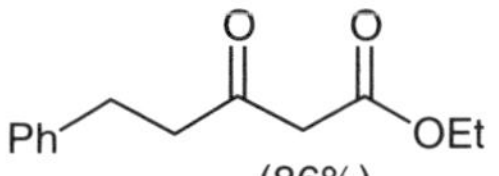 (86%)	26
2	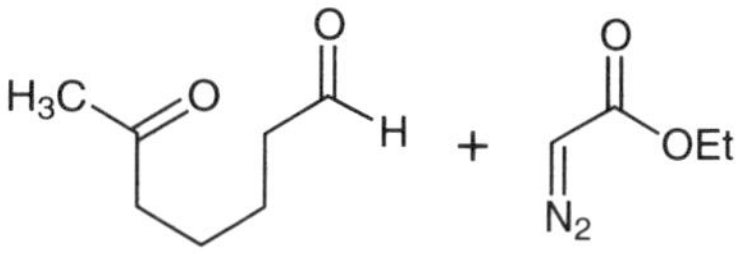	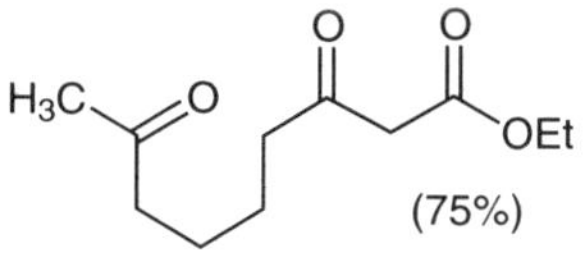 (75%)	26
3	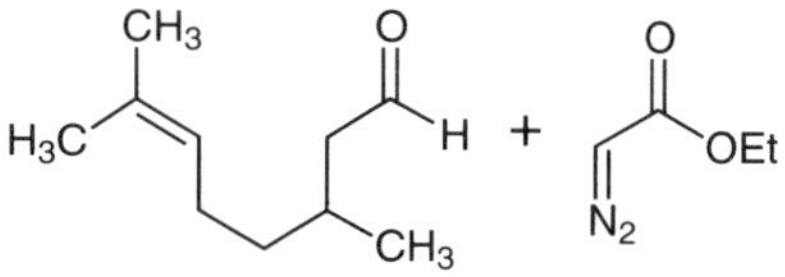	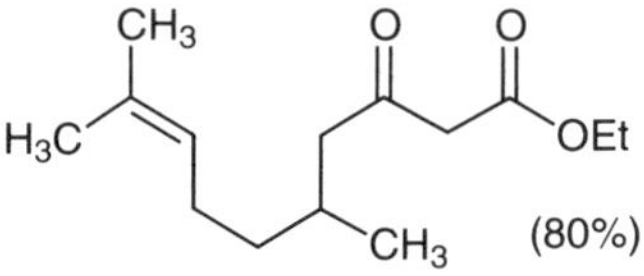 (80%)	26
4	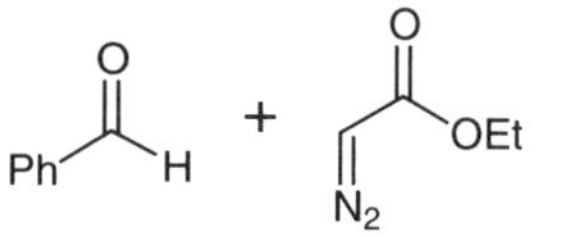	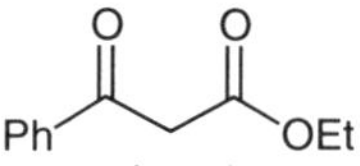 (50%)	26
5	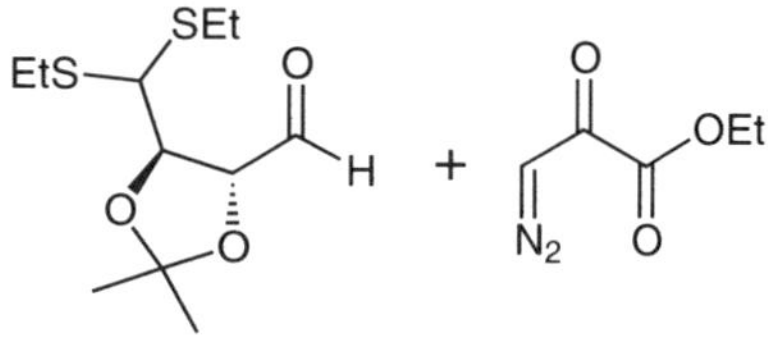	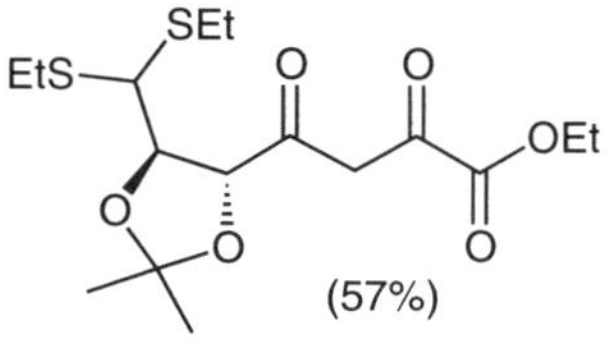 (57%)	28
6	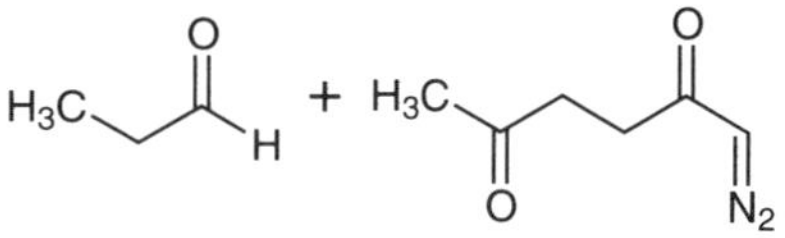	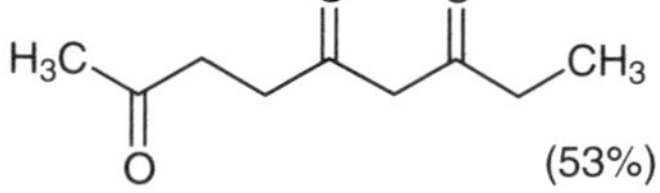 (53%)	27
7	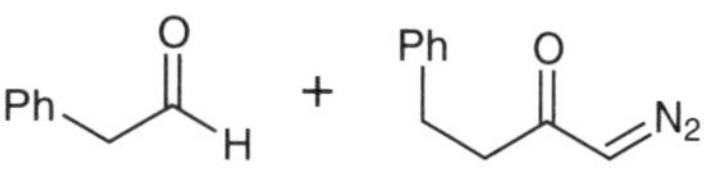	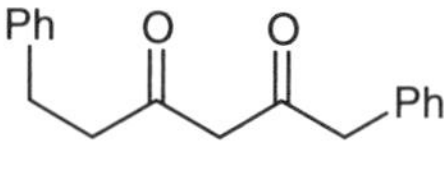 (90%)	27
8	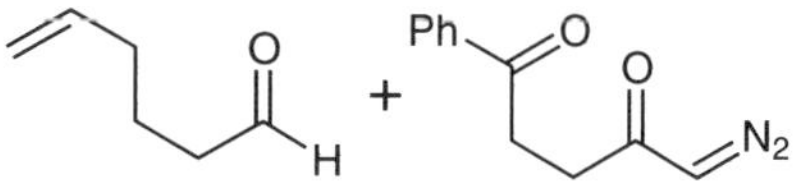	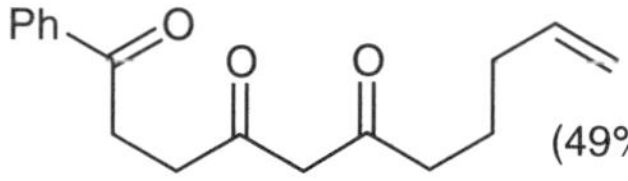 (49%)	27

shown in Table 10.5. The product yields suggest that trisubstituted alkenes (entries 1 and 2) are better precursors than 1,2-disubstituted or monosubstituted alkenes (entries 3 and 4), presumably because the former on ozonolysis produces an aldehyde and an unreactive ketone.

Table 10.5   Conversion of Olefins via Ozonolysis into β-Keto Esters Using EDA

Entry	Reactants	Product (yield)
1	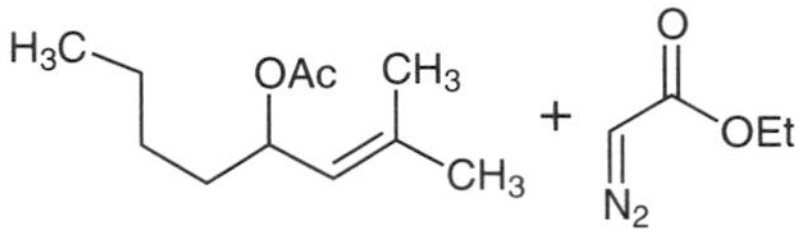	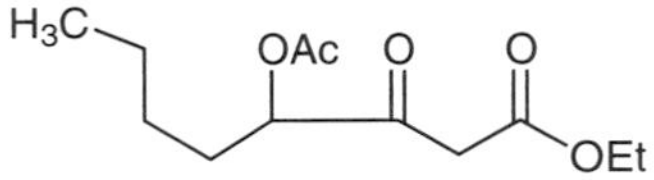 (80%)
2	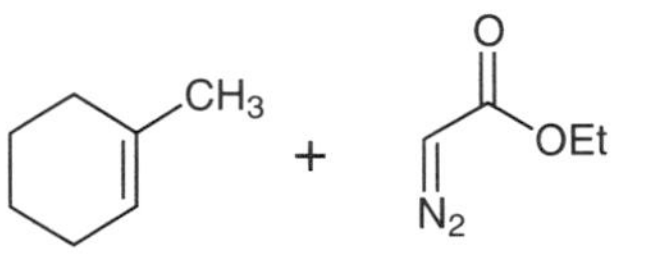	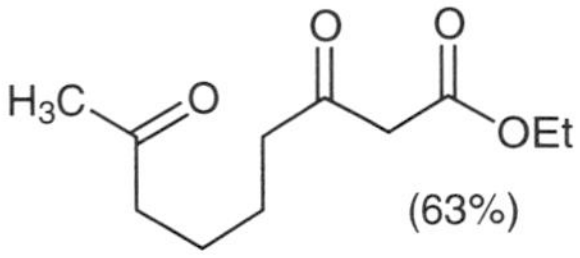 (63%)
3	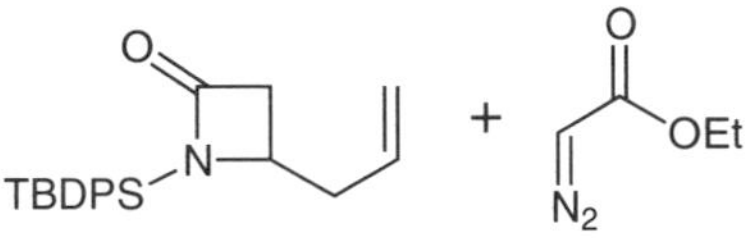	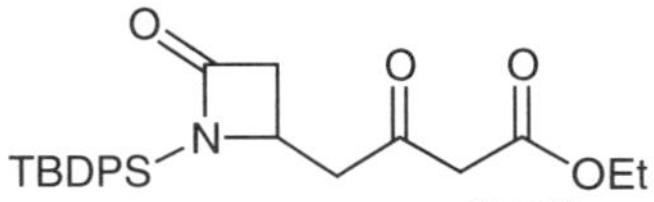 (51%)
4	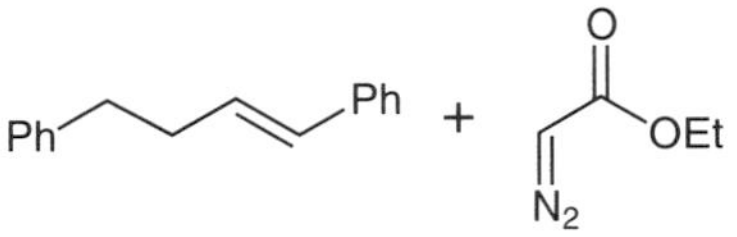	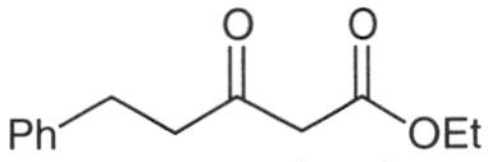 (45%)

SYNTHESIS OF A β-DICARBONYL COMPOUND VIA TIN(II) CHLORIDE–CATALYZED REACTION: CONVERSION OF HYDROCINNAMALDEHYDE INTO ETHYL 3-OXO-5-PHENYLVALERATE[26]

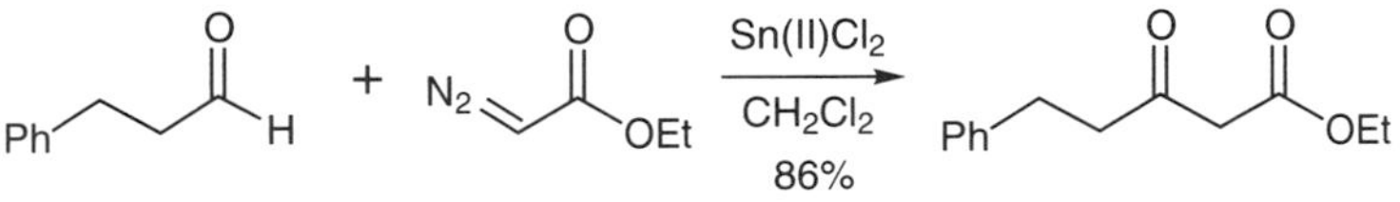

Methylene chloride (8 mL) followed by ethyl diazoacetate (0.45 g, 3.9 mmol) was added with stirring at 25°C to anhydrous tin(II) chloride (0.07 g, 0.37 mmol). To this solution were added a few drops of hydrocinnamaldehyde (0.50 g, 3.7 mmol) in methylene chloride (2 mL). When nitrogen evolution began, the remaining solution of hydrocinnamaldehyde was added dropwise over 10 min. After nitrogen evolution had stopped (~1 h), the reaction was transferred to a separatory funnel with saturated brine and extracted with diethyl ether (2 × 80 mL). The organic extracts were combined and dried and the volatiles removed under vacuum. The remaining oil was purified by silica gel chromatography eluting with ethyl acetate–hexane to yield the β-keto ester (0.71 g, 86%).

Dhavale and co-workers[30] have used a novel solid-phase reaction of aliphatic and aromatic aldehydes with ethyl diazoacetate over alumina in the absence of solvent to prepare β-keto esters in high yields. However, the need to use a large excess (tenfold) of specially activated alumina, as the reagent makes this method less attractive. More recently, Sonawane and co-workers[31] claimed that the zeolite-promoted condensation of ethyl diazoacetate with aldehydes provides a clean and practical method for the synthesis of β-keto esters in good yields (52–90%). Alkyl and aryl ketones were found to be practically unreactive when the zeolite H-beta was employed as a catalyst.

Applications of the tin(II) catalyzed β-ketoester formation from diazoesters have increased dramatically in the past few years, and there are now numerous

Table 10.6   Synthetic Applications Involving the Tin(II) Chloride–Catalyzed β-Ketoester Formation

Entry	β-Ketoester (yield)	Target Product	Ref.
1	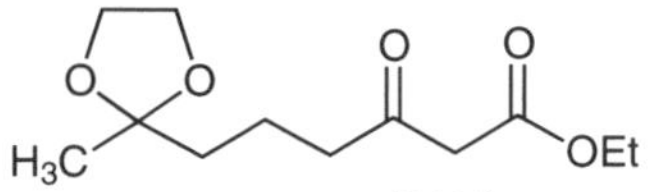 (59%)	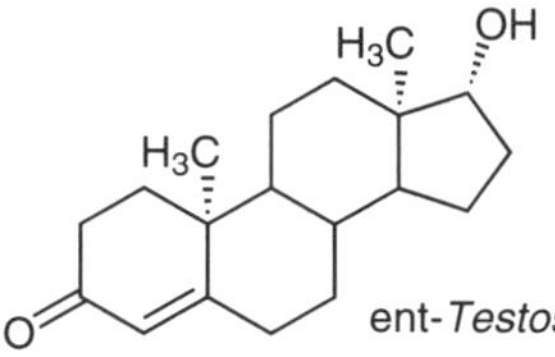 ent-*Testosterone*	32
2	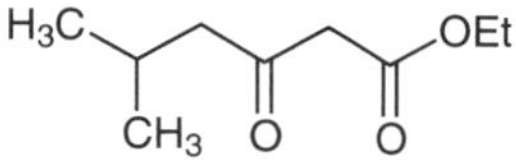 (100%)	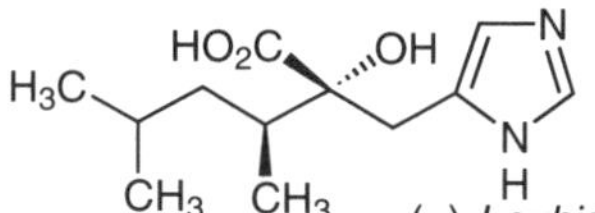 (±)-*Leuhistin*	33
3	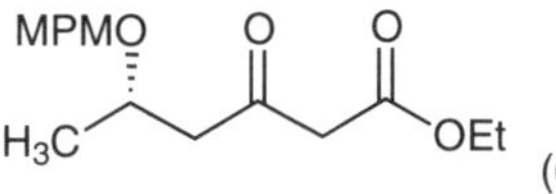 (64%)	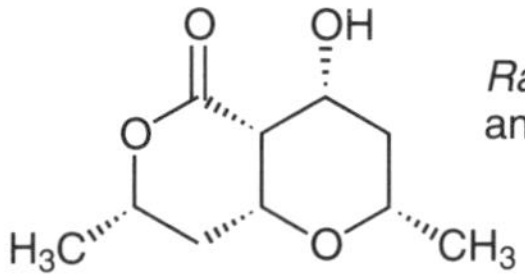 Radicinin analogue	34
4	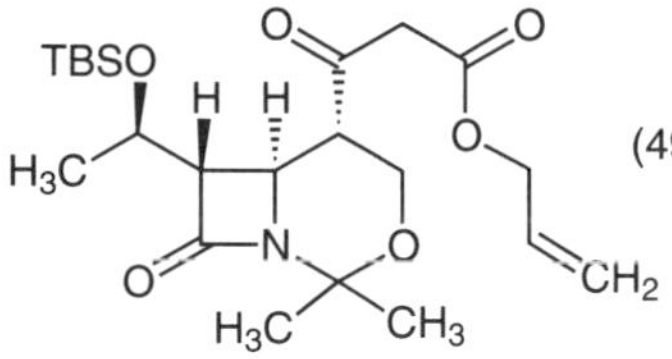 (49%)	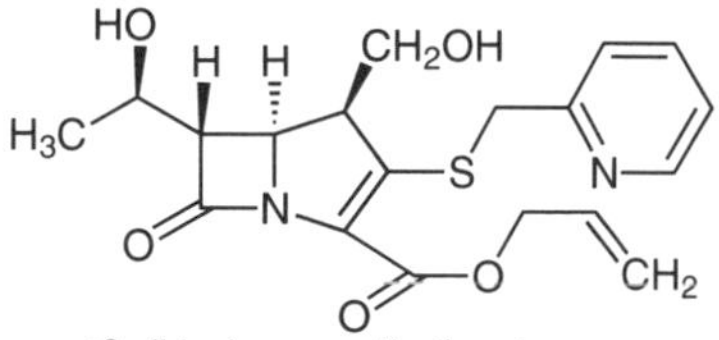 4β-(Hydroxymethyl)carbapenem	35
5	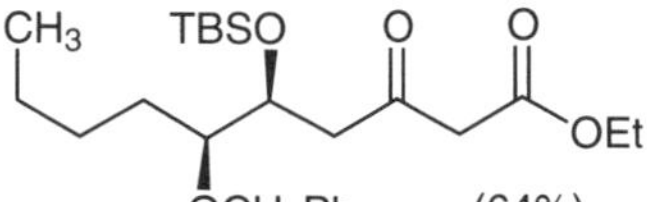 (64%)	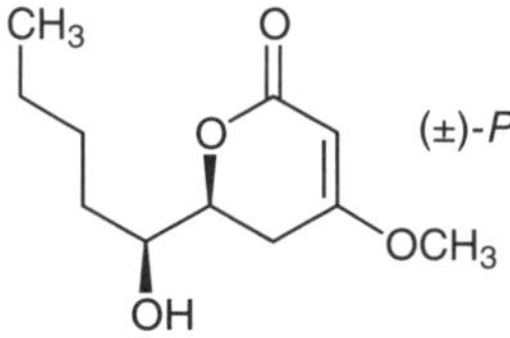 (±)-*Pestalotin*	36

Table 10.6   (Continued)

Entry	β-Ketoester (yield)	Target Product	Ref.
6	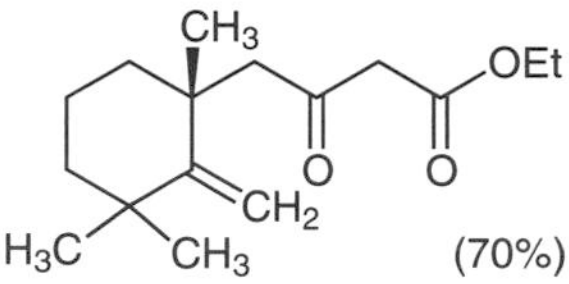 (70%)	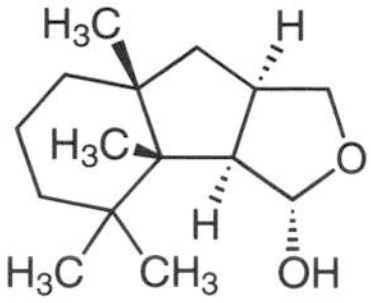 *Thapsane*	37
7	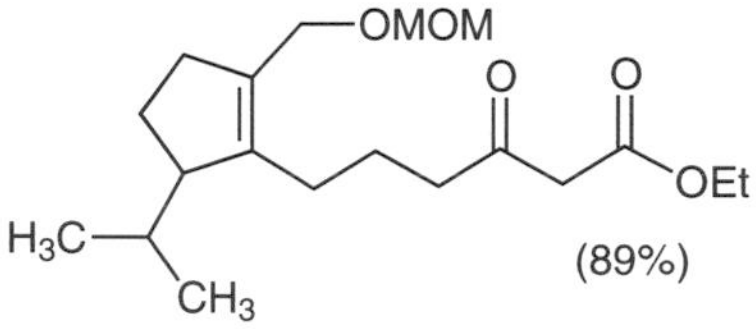 (89%)	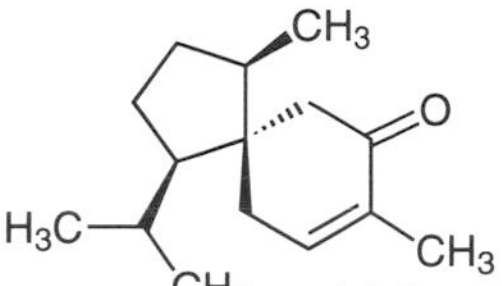 *(+)-Acorenone B*	38
8	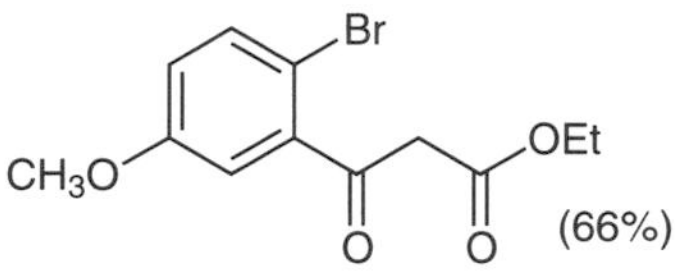 (66%)	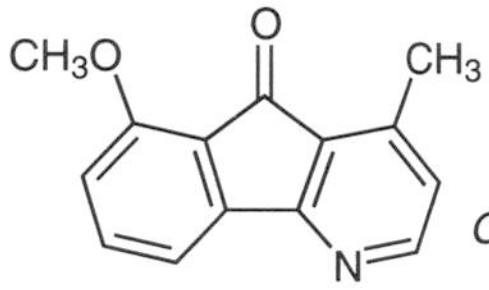 *Onychine*	39
9	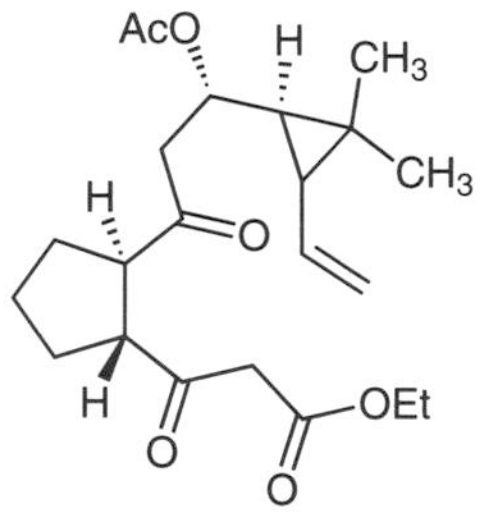 (> 61%)	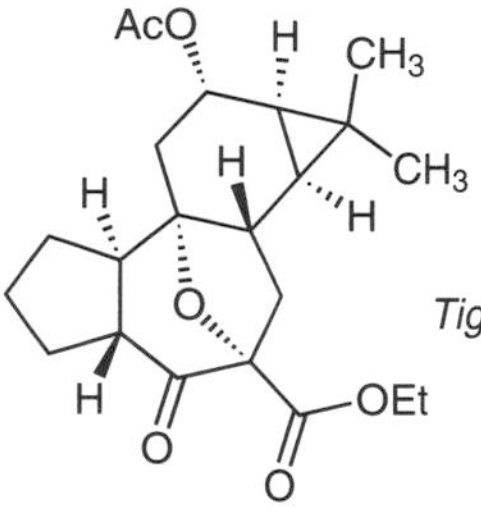 *Tigliane* Skeleton	40

examples of its use in total synthesis of natural products. The examples in Table 10.6 illustrate the diversity of structures and functional group types that have been accessed by this versatile reaction.

Yoshii and co-workers[41,42] have applied the Lewis acid–promoted β-ketoester formation protocol to the synthesis of γ-unsubstituted α-acyl-β-tetronic acids (**10**, R = OH) and α-acyl-β-tetronic amides (**10**, R = NCH$_2$Ph) (Table 10.7). These workers claimed that whereas the Sn(II)-catalyzed reactions of aldehydes **7** with diazoester **8** resulted in poor yields of dicarbonyl products **9**, ZrCl$_4$ and TiCl$_4$ catalysts were quite effective in the reaction of sterically hindered aldehydes (Table 10.7). This γ-unsubstituted α-acyl-β-tetronic acid formation protocol has been applied to the total synthesis of tetronasin (**11**), as shown in Scheme 4.[43]

Table 10.7 Synthesis of γ-Unsubstituted α-Acyl-β-tetronic Acids

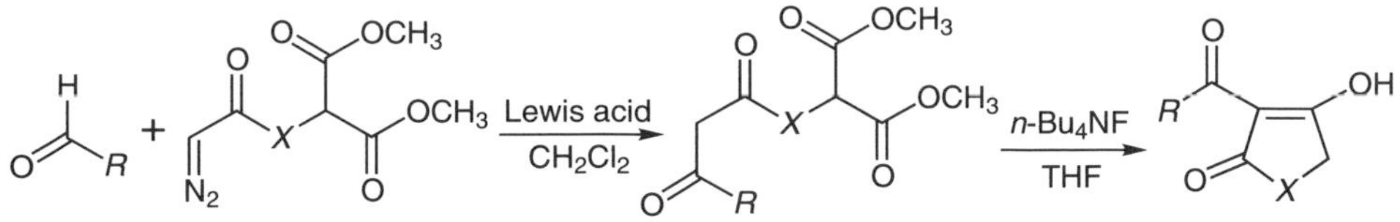

R	X	Lewis acid	9 (yield)	10 (yield)
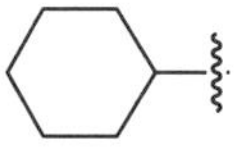 cyclohexyl	O	SnCl$_2$	78%	72%
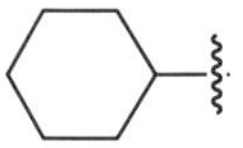 cyclohexyl	NCH$_2$Ph	ZrCl$_4$	78%	96%
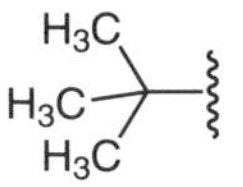 (H$_3$C)$_3$C–	O	SnCl$_2$ / ZrCl$_4$ / TiCl$_4$	29% / 73% / 75%	61%
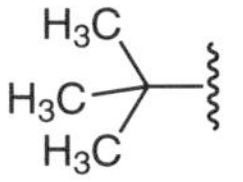 (H$_3$C)$_3$C–	NCH$_2$Ph	ZrCl$_4$	62%	98%
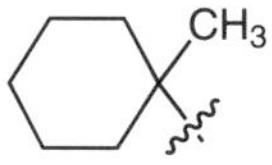 1-methylcyclohexyl	O	SnCl$_2$ / ZrCl$_4$ / TiCl$_4$	38% / 84% / 79%	61%
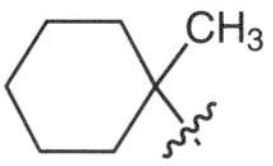 1-methylcyclohexyl	NCH$_2$Ph	ZrCl$_4$	64%	98%

10.4   SYNTHESIS OF TETRAHYDROFURANS

As shown in Tables 10.4 and 10.6, the homologation of aldehydes to β-ketoesters upon treatment with ethyl diazoacetate and stannous chloride generally proceeds in high yields. However, when there is other functionality in the aldehyde, the reaction can take a different course. For example, when the precursor is a protected β-hydroxy aldehyde, the major product is a substituted tetrahydrofuran.[44] A typical example is shown in eq. 7. Thus treatment of aldehyde **12** and ethyl diazoacetate (EDA) with SnCl$_4$ in dichloromethane at $-78°$C afforded the substituted tetrahydrofuran **14** in 76% yield and β-ketoester **15** in 12% yield. Under optimum conditions, reaction of a similar aldehyde (**13**) bearing a p-methoxybenzyl ether functional group, using 0.5 molar equivalents of SnCl$_4$, afforded the substituted tetrahydrofuran **14** in 84% yield.

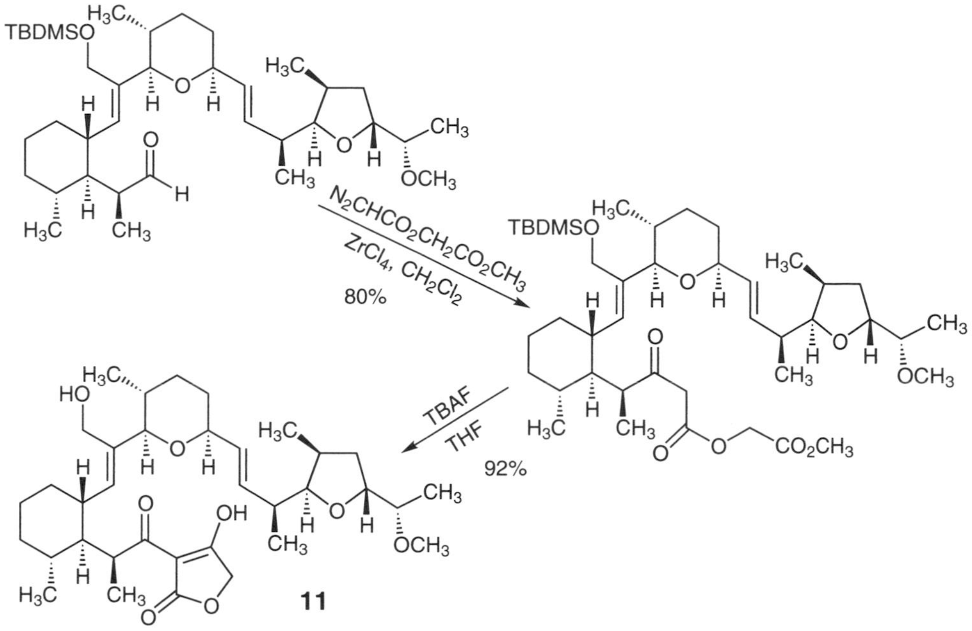

Scheme 4

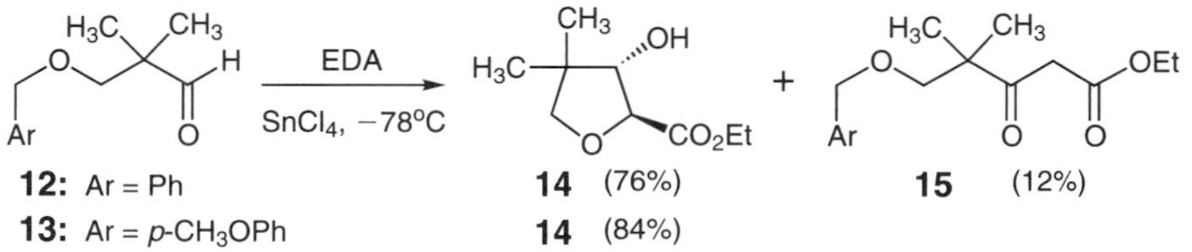

12: Ar = Ph   **14** (76%)   **15** (12%)
13: Ar = p-CH₃OPh   **14** (84%)

$$(7)$$

Additional examples of this substituted tetrahydrofuran synthesis are illustrated in Table 10.8. The presence of stereogenoic centers at either the α or β position in the starting aldehydes results in the formation of substituted tetrahydrofurans with high diastereoselectivity (entries 3 and 4). In addition, this methodology can also be extended to ketones (entry 5). A possible mechanism for tetrahydrofuran synthesis, involving an S_N2-type intramolecular displacement of N_2 and formation of an oxonium ion intermediate, is outlined in Scheme 5.

10.5 OLEFINATION OF ALDEHYDES AND KETONES WITH DIAZOCARBONYL COMPOUNDS

Amongst other novel addition reactions of diazoesters with aldehydes or ketones are Wittig-like olefinations mediated by a combination of tributylstibine and

TETRAHYDROFURAN ANNULATION WITH β-p-METHOXY-BENZYL PIVALALDEHYDE AND ETHYL DIAZOACETATE: SYNTHESIS OF SUBSTITUTED TETRAHYDROFURAN 14[44]

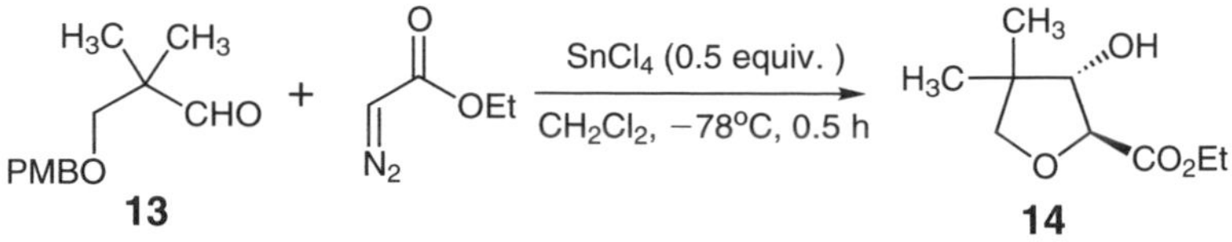

Ethyl diazoacetate (228 mg, 2 mmol) and anhydrous tin(II) chloride (95 mg, 0.5 mmol) were sequentially added to a stirred $-78°C$ solution of aldehyde **13** (222 mg, 1 mmol) in CH_2Cl_2 (0.125 M). The reaction was followed by TLC until starting material was no longer detected (30 min), and then the reaction mixture was poured into a stirred solution of $NaHCO_3$. After being stirred for 5 min, the reaction mixture was diluted with CH_2Cl_2. The aqueous layer was extracted with CH_2Cl_2 (2×), and the combined organic extracts were washed with brine, dried ($MgSO_4$), and concentrated to afford crude product. Flash chromatography on silica gel (230–400 mesh; ethyl acetate-hexane mixtures) afforded product **14** (158 mg, 84%) as a clear oil. ^1H-NMR (300 MHz, $CDCl_3$) δ 4.25 (*m*, 3 H, OCH_2 and H-2), 3.97 (*d*, $J = 5.8$ Hz, 1 H, H-3), 3.73 (*s*, 2 H, H-5), 1.98 (br *s*, 1 H, OH), 1.31 (*t*, $J = 7.1$ Hz, 3 H, OCH_2CH_3), 1.08 (*s*, 3 H, CH_3), 1.05 (*s*, 3 H, CH_3); ^{13}C-NMR (75 MHz, $CDCl_3$) δ 172.3, 83.0, 82.0, 79.4, 61.2, 42.1, 23.4, 18.7, 14.1; IR (neat) 3465, 2966, 2936, 2875, 1738, 1470, 1373, 1209, 1087, 1032, 947 cm^{-1}.

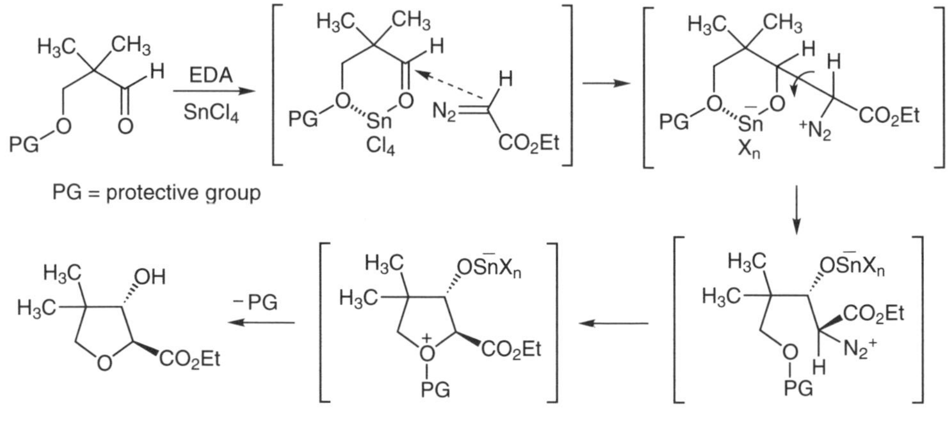

Scheme 5

Table 10.8 Tetrahydrofuran Synthesis from Aldehydes and Ketones and EDA

Entry	Carbonyl	Conditions	Tetrahydrofuran	Yield
1	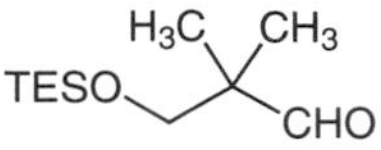 TESO, H₃C, CH₃, CHO	0.5 equiv. SnCl₄; −78°C, 0.5 h; Pyridine/HF	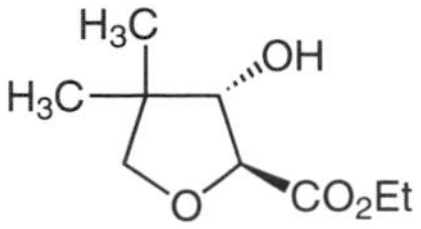 H₃C, H₃C, OH, O, CO₂Et	81%
2	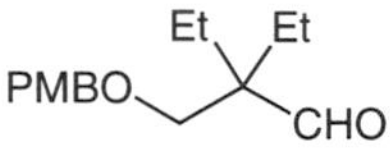 PMBO, Et, Et, CHO	0.5 equiv. SnCl₄; −78°C, 1 h; −30°C, 0.5 h	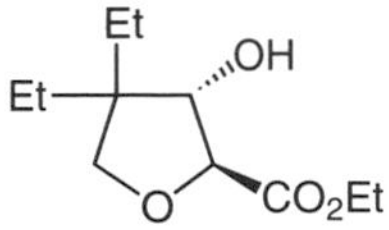 Et, Et, OH, O, CO₂Et	77%
3	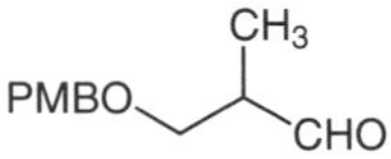 PMBO, CH₃, CHO	0.5 equiv. SnCl₄; −78°C, 1 h	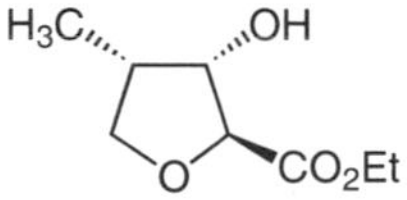 H₃C, OH, O, CO₂Et d.r. = 10:1	75%
4	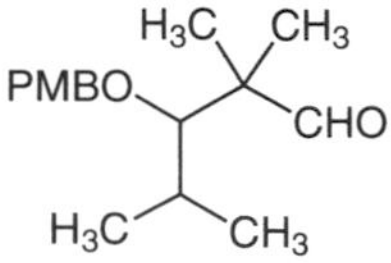 PMBO, H₃C, CH₃, CHO, H₃C, CH₃	0.5 equiv. SnCl₄; −78°C, 8 h; 0°C to rt, 12 h	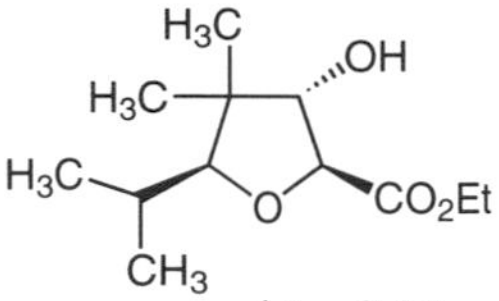 H₃C, H₃C, OH, H₃C, O, CO₂Et, CH₃ d.r. = 8.1:1	78%
5	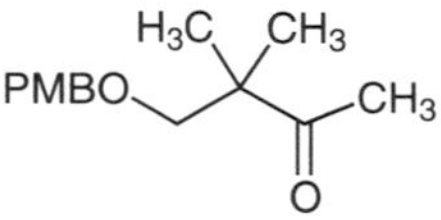 PMBO, H₃C, CH₃, CH₃, O	0.5 equiv. ZrCl₄; −78°C to rt, 12 h	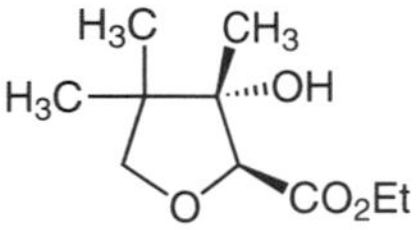 H₃C, CH₃, H₃C, OH, O, CO₂Et	80%

copper(I) iodide (Table 10.9)[45] or by a combination of a tertiary phosphine (triphenyl- or tributylphosphine) and methyltrioxorhenium (MTO) (Table 10.10).[46] The former are applicable to both aldehydes and ketones; yields are generally very good to excellent (Table 10.9). When aldehydes were used as substrates, the olefins were produced exclusively with the E geometry. The copper(I)-catalyzed olefination process is believed to proceed through the corresponding stibonium ylide. The methyltrioxorhenium-catalyzed olefination process appears to have been observed with aldehydes only. Saturated and α,β-unsaturated aldehydes react, as well as aromatic aldehydes, and yields are comparable with those in the copper(I)-catalyzed olefination process, but with slightly decreased stereoselectivity.

In the absence of a tertiary phosphine, the methyltrioxorhenium-catalyzed reaction of EDA with aldehydes or ketones afforded epoxides in good yield.[47] Aldehydes gave only the E epoxide, whereas ketones gave both geometrical isomers, with the E isomer predominating (eq. 8).

Table 10.9 Copper(I) Iodide–Catalyzed Olefination of Diazocarbonyls with Carbonyl Compounds[45]

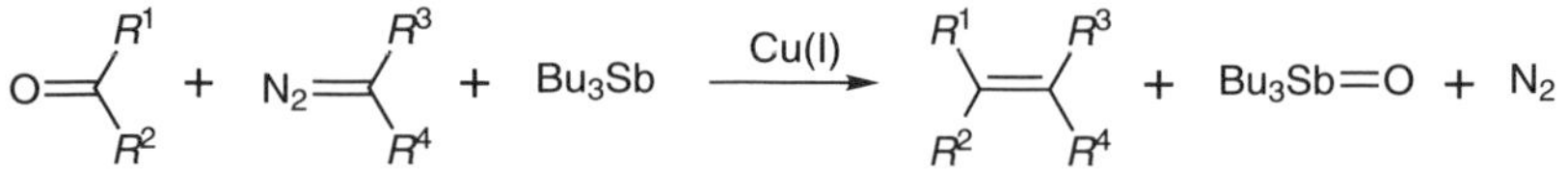

Entry	R^1	R^2	R^3	R^4	Product (yield)
1	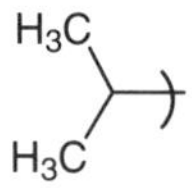 (CH₃)₂CH– structure	H	CO_2CH_3	CO_2CH_3	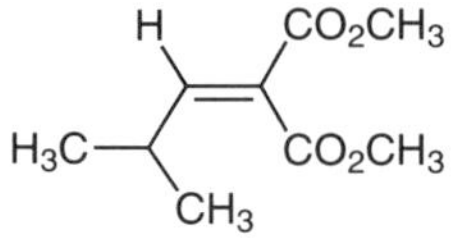 (86%)
2	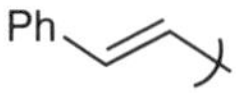 Ph–CH=CH–	H	CO_2CH_3	CO_2CH_3	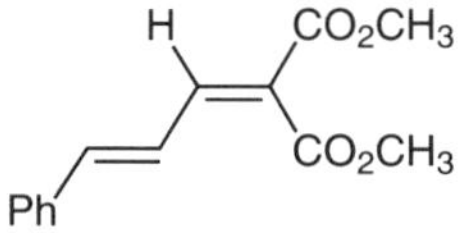 (81%)
3	Ph	H	CO_2Et	H	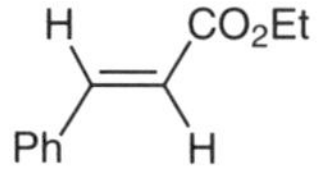 (97%)
4	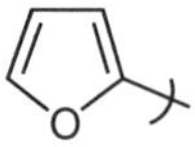 furyl	H	CO_2Et	H	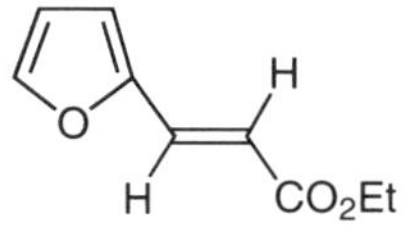 (95%)
5	Ph	H	$COCH_3$	$COCH_3$	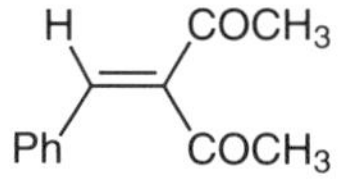 (92%)
6	Et	Et	CO_2Et	H	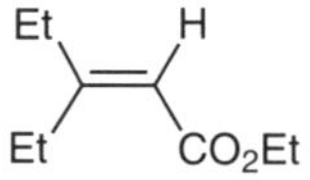 (82%)
7	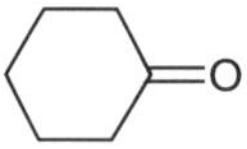 cyclohexanone	O	CO_2Et	H	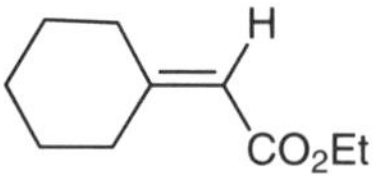 (86%)

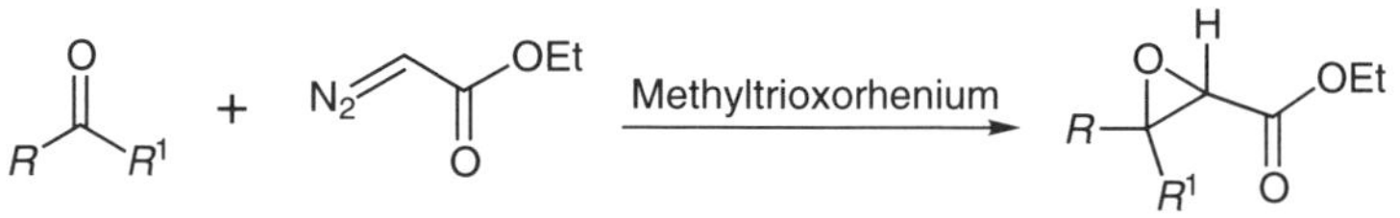

$$R = H; R^1 = Ph, \text{ yield: } 79\% \qquad R = H; R^1 = CH_3(CH_2)_2, \text{ yield: } 75\%$$
$$R = CH_3; R^1 = Ph, \text{ yield: } 57\% \qquad R = CH_3; R^1 = (CH_3)_2CH, \text{ yield: } 49\%$$

(8)

Table 10.10   Methyltrioxorhenium Catalyzed Olefination of Diazocarbonyls with Carbonyl Compounds[46]

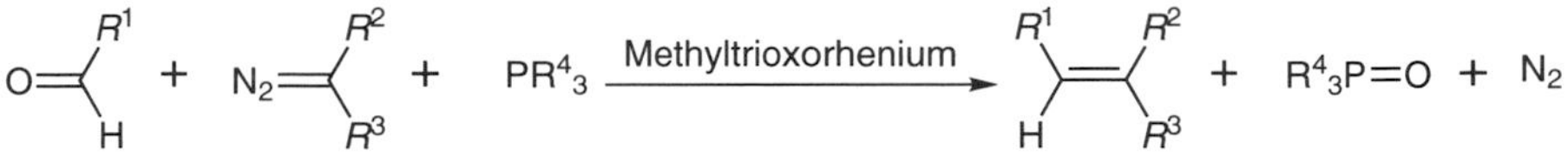

$$O=\overset{R^1}{\underset{H}{\diagup}} \ + \ N_2=\overset{R^2}{\underset{R^3}{\diagup}} \ + \ PR^4_3 \ \xrightarrow{\text{Methyltrioxorhenium}} \ \overset{R^1}{\underset{H}{\diagup}}=\overset{R^2}{\underset{R^3}{\diagup}} \ + \ R^4_3P=O \ + \ N_2$$

Entry	R^1	R^2	R^3	R^4	Product (yield)
1	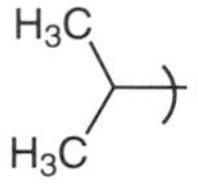 (CH₃)₂CHCH₂— (H_3C, H_3C isopropyl)	H	CO_2Et	Ph	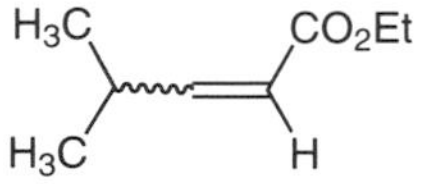 $E:Z = 85:15$ (82%)
2	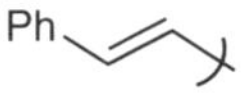 Ph—CH₂CH= (cinnamyl type)	H	CO_2Et	Ph	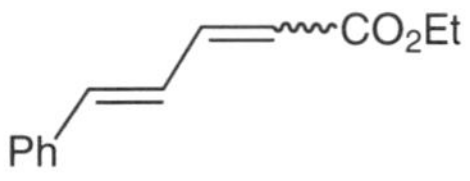 $E:Z = 60:40$ (97%)
3	Ph	H	CO_2Et	Ph	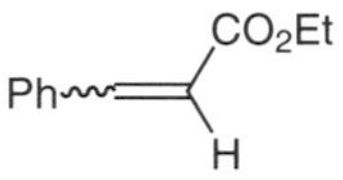 $E:Z = 89:11$ (80%)
4	$p\text{-}NO_2C_6H_4$	CO_2Me	CO_2Me	Ph	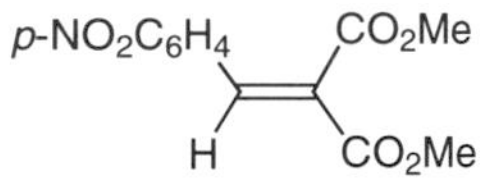 (87%)
5	$p\text{-}NO_2C_6H_4$	H	CO_2Et	$n\text{-}Bu$	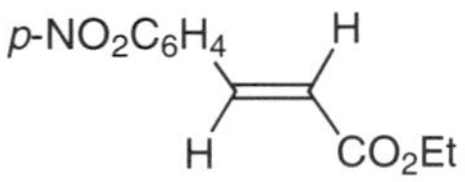 $E:Z = 98:2$ (98%)

10.6   ACID-CATALYZED REACTIONS OF α-DIAZOCARBONYL COMPOUNDS WITH KETONES

Addition of diazocarbonyl compounds to ketones promoted by Lewis acids has been used extensively as a means of homologation or ring expansion. The most widely employed Lewis acid appears to be boron trifluoride etherate, though Mock and Hartman[48] have advocated the use of triethyloxonium fluoroborate. With unsymmetrical ketones, whether acyclic or cyclic, questions of regioselectivity arise in homologation due to different migratory aptitudes of the groups flanking the carbonyl function. Liu and co-workers[49,50] studied the $BF_3 \cdot Et_2O$–catalyzed addition of ethyl diazoacetate to a series of cycloalkanones with different substitution patterns at the α and α' positions and concluded that the migratory aptitudes were uniformly such that the less substituted carbon atom migrates preferentially. Studies by Mock and Hartman[48] with triethyloxonium fluoroborate as catalyst revealed a similar trend in migratory aptitude in reac-

Table 10.11 Homologation of Acyclic Ketone and Monocyclic Ketones with Ethyl Diazoacetate

Entry	Reactants	Products	Ref.
1	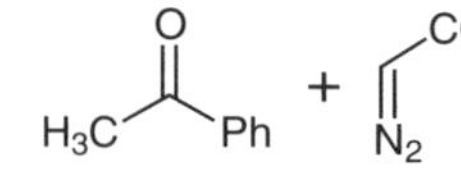	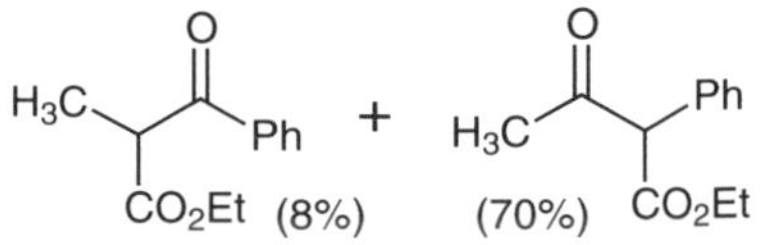	48
2	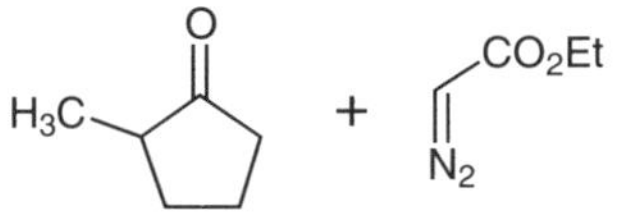	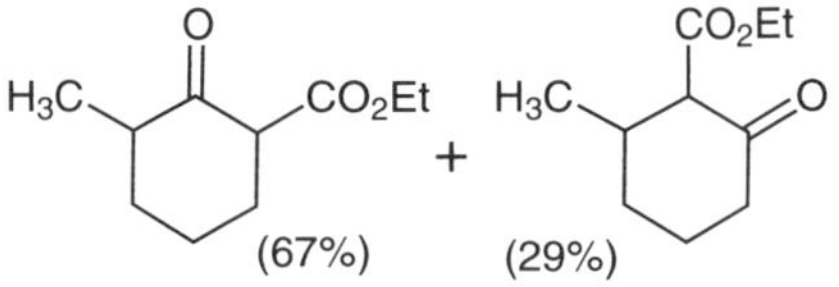	49
3	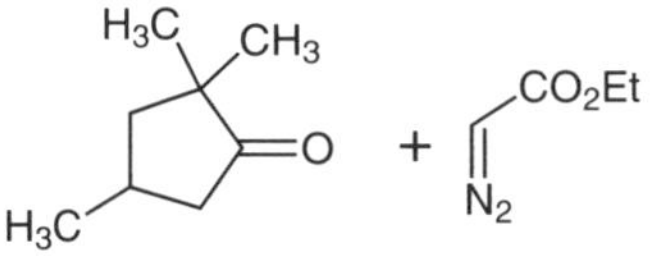	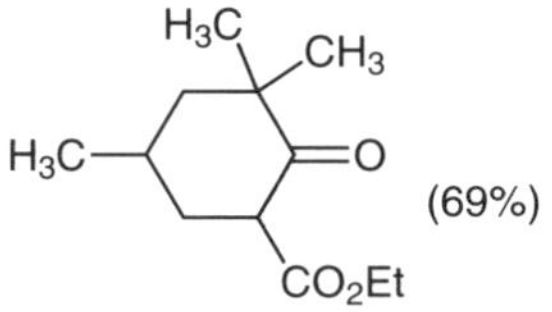	49
4	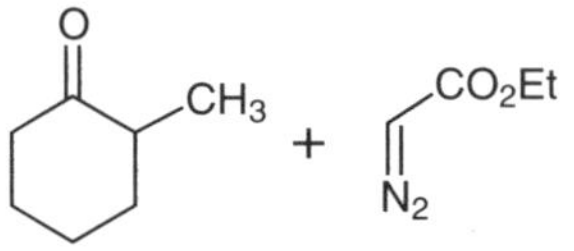	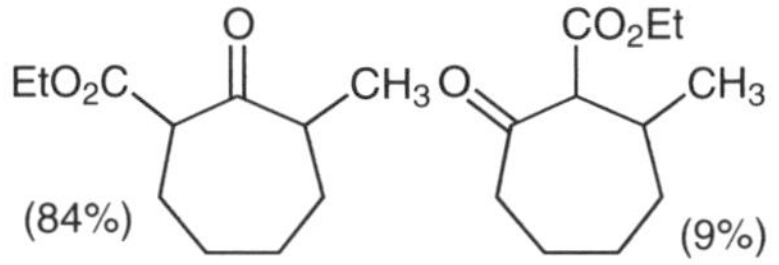	50
5	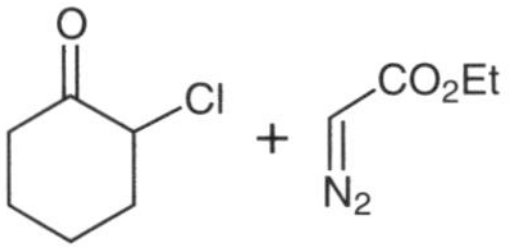	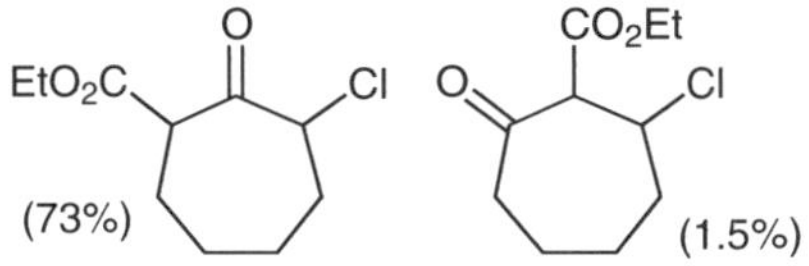	48

tions that gave good to excellent yields of homologated ketones. Some examples taken from both studies are shown in Table 10.11. However, there are cases where the regiochemical outcome is catalyst dependent.

The Lewis acid–promoted ring enlargement of a cyclopentanone to a cyclohexanone system served as the key step in a synthesis of (+)-thiathromboxane A$_2$.[51] Thus treatment of compound **16** with ethyl diazoacetate in a dichloromethane solution of boron trifluoride etherate at 0°C provided cyclohexanone derivatives **17** and **18** in 52% and 14% yields, respectively (eq. 9).

In general, ethyl diazoacetate is the reagent of choice for the Lewis acid–catalyzed homologation of ketones with diazoesters. In some cases, however, hydrolysis and decarboxylation of the resulting β-ketoester proved to be unduly difficult. Baldwin and Landmesser[52] claimed that benzyl and allyl diazoacetate are suitable alternative reagents for the homologation of ketones; the resulting β-ketoester can be cleaved by either metal/ammonia reduction or catalytic hydrogenolysis in the case of the benzyl ester or by metal/ammonia reduction alone in the case of the allyl ester.

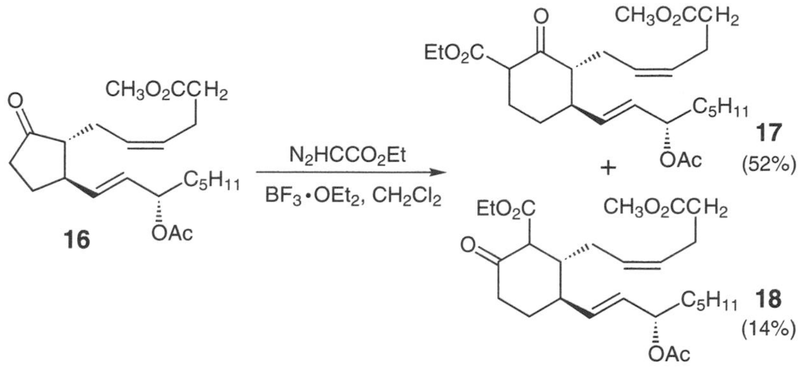

$$(9)$$

Furthermore, homologation of large-ring cycloalkanones is also possible. Cyclooctanone was transformed into 2-carboethoxycyclononanone by treatment with triethyloxonium fluoroborate (eq. 10). 2-Carboethoxycyclononanone has

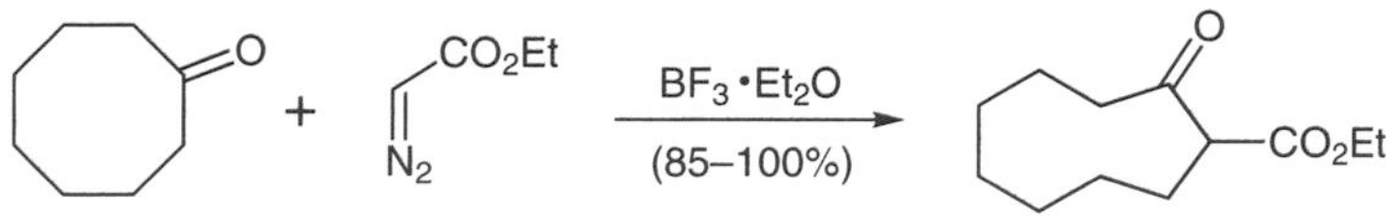

$$(10)$$

been used as the starting material in MacPherson's synthesis of an orally active macrocyclic neutral endopeptidase-24,11 inhibitor[53] and in Gribble and Silva's synthesis of the Mexican bean beetle azamacrolide allomone.[54] Zard and co-workers[55] have described an efficient synthesis of large-ring acetylenes involving the ring enlargement of cycloalkanone with ethyl diazoacetate in the presence of a Lewis acid. Two examples are illustrated in Schemes 6 and 7. The two regioiso-

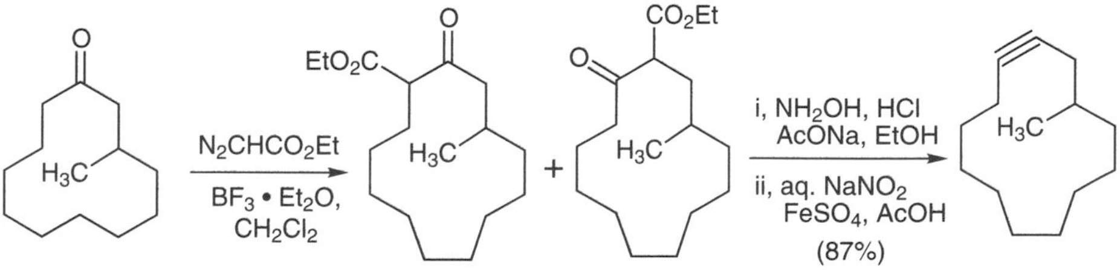

Scheme 6

meric β-ketoesters obtained by the ring-expansion process can be converted into the same cycloalkyne.

From the synthetic point of view, the Lewis acid–promoted homologation of fused or bridged polycyclic ketones is very useful. Liu and Ogino[56] have carried out a detailed investigation on the regioselectivity of boron trifluoride–catalyzed ring expansion of the bicyclo[4.2.0]octanone system with ethyl diazoacetate. The

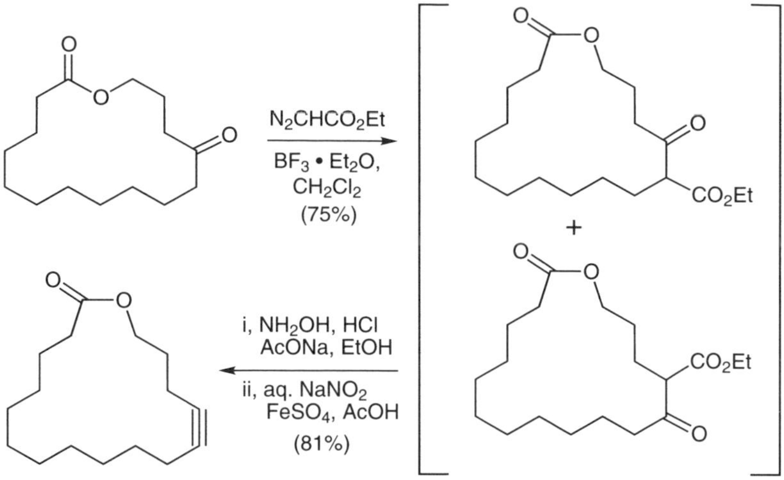

Scheme 7

results of their studies are summarized in Table 10.12. The expansion of cyclobutanone rings is highly regioselective and predictable. Where applicable, the products result either exclusively or predominantly from migration of the less substituted bond. Liu and Chan[57] have applied this approach to their synthesis of the zizaane sesquiterpenes (−)-khusimone **19** and (+)-zizanoic acid **20** (Scheme 8). Greene and co-workers[58] found, for example, that whereas addition of ethyl diazoacetate to the cyclobutanone in Scheme 9 with either boron trifluoride etherate or triethyloxonium fluoroborate as catalyst proceeded with poor selectivity, use of antimony pentachloride in dichloromethane resulted in a highly

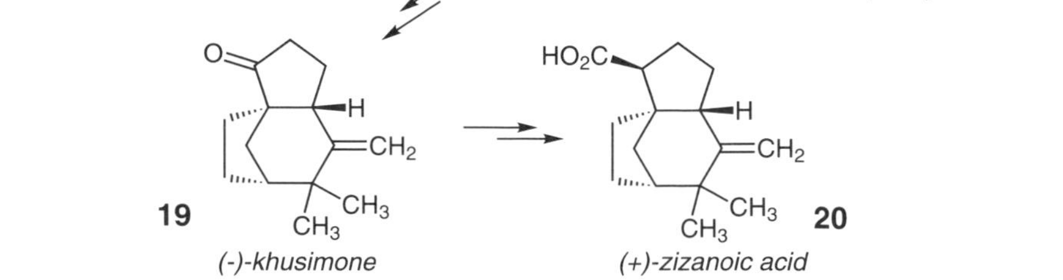

Scheme 8

Table 10.12   Homologation of Cyclobutanones with Ethyl Diazoacetate

Entry	Reactant	Product (yield)
1	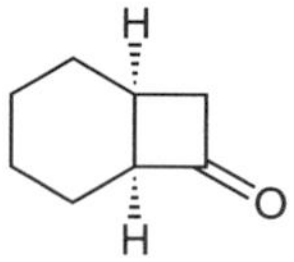	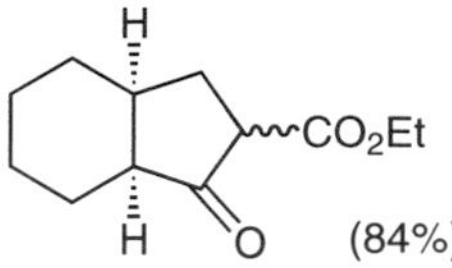 (84%)
2	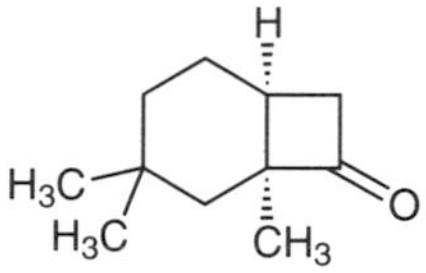	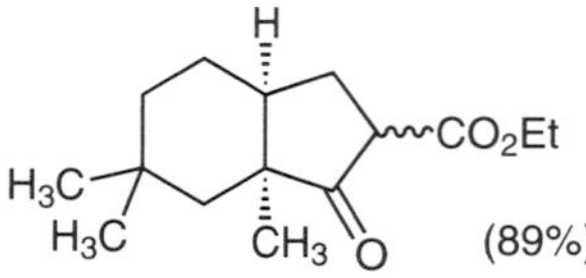 (89%)
3	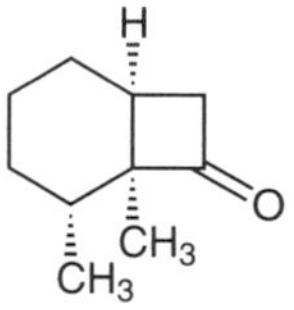	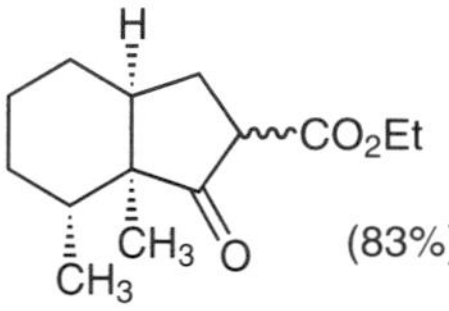 (83%)
4	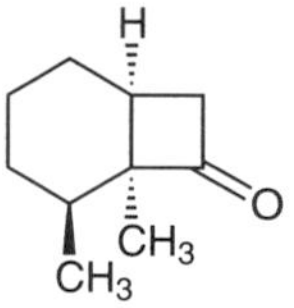	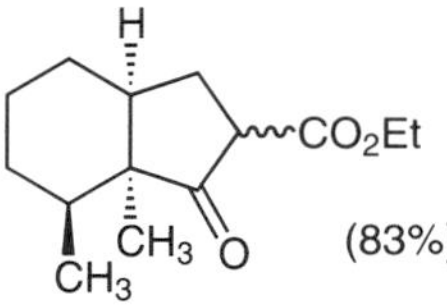 (83%)
5	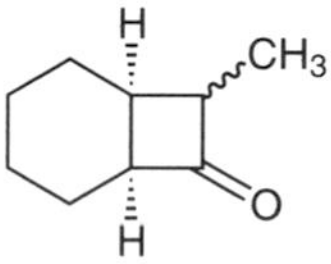	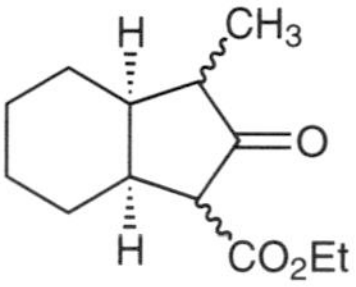 (100%)
6	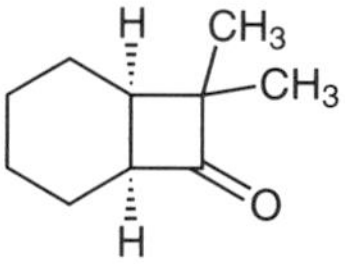	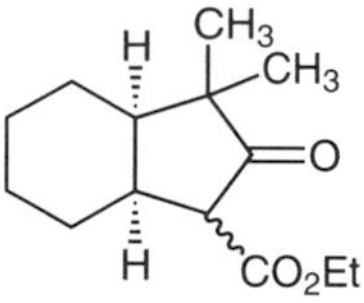 (45%) + 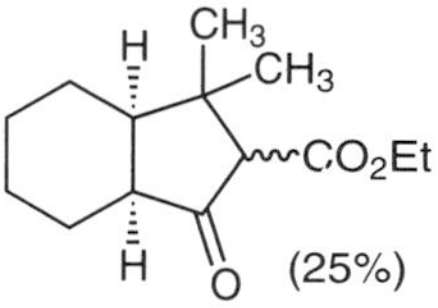 (25%)

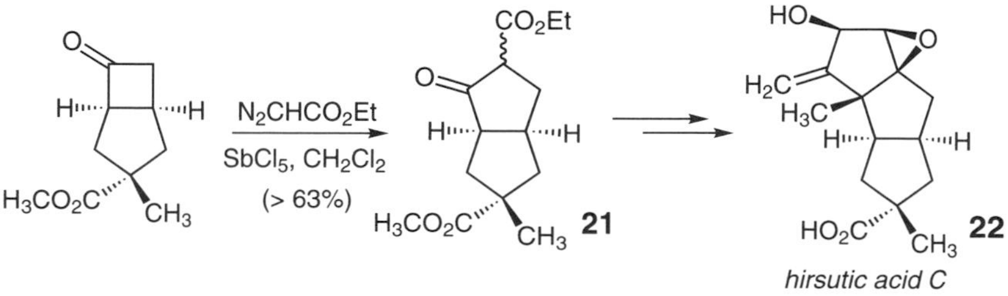

Scheme 9

regioselective ring expansion to give the product **21**. This ring-expansion adduct (**21**) was then transformed into hirsutic acid C **22**. A similar application of cyclobutanone ring expansion has been used in a total synthesis of (±)-aplysin (**23**).[59,60] In this case, the ring expansion was catalysed by $BF_3 \cdot Et_2O$ regioselectively (Scheme 10). A further example is found in three different approaches lead-

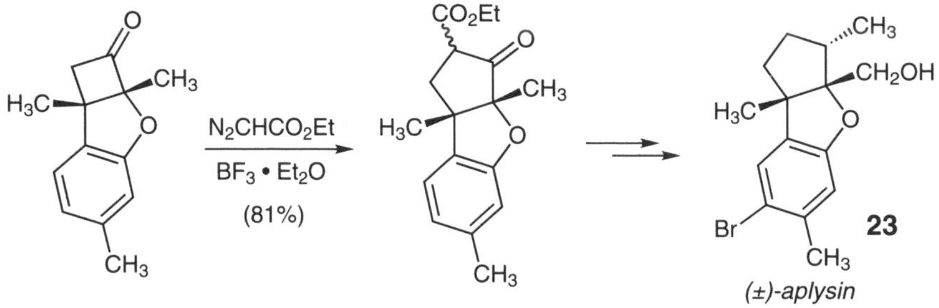

Scheme 10

ing to the synthesis of (±)-$\Delta^{(9,12)}$-capnellene (**24**),[61-64] where the fused cyclopentanone rings were again obtained from the corresponding cyclobutanones by use of the $BF_3 \cdot Et_2O$–catalyzed ring-expansion protocol (Scheme 11). In recent synthetic work, Hamelin and co-workers have used the same approach to prepare a bicyclic precursor for a synthesis of 9-acetoxyfukinanolide **25** (Scheme 12).[65]

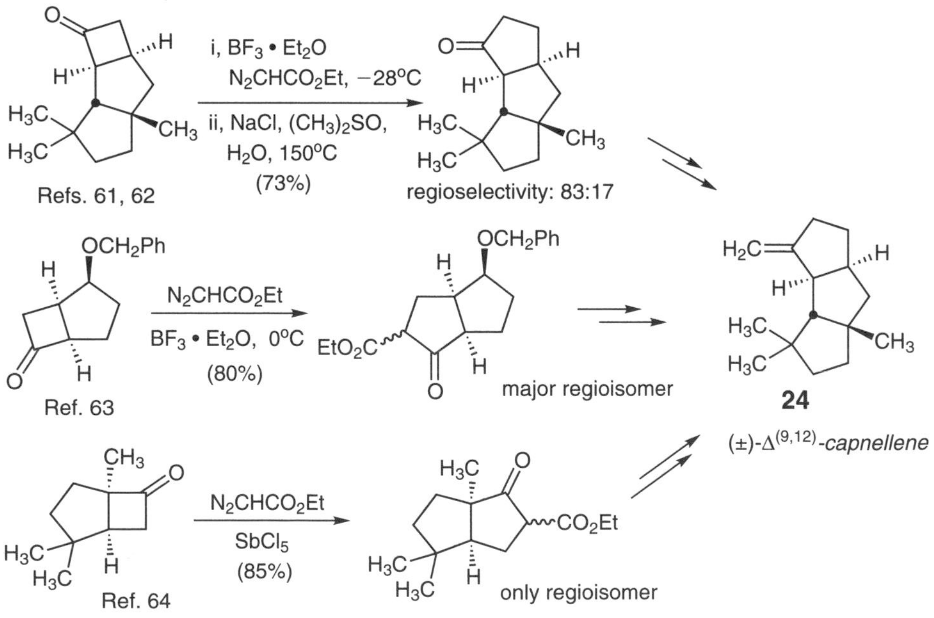

Scheme 11

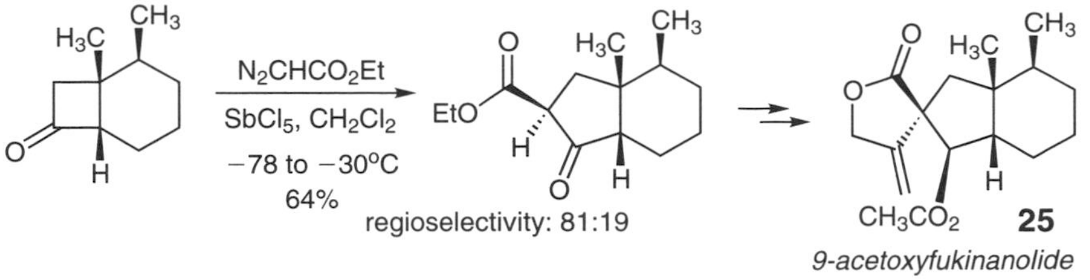

Scheme 12

HOMOLOGATION OF A CYCLOBUTANONE RING WITH ETHYL DIAZOACETATE: SYNTHESIS OF METHYL (2S, 3aR, 6aS)-1,2,3,3a,4,5,6,6a-OCTAHYDRO-2-METHYL-4-OXOPENTALENE-2-CARBOXYLATE[58]

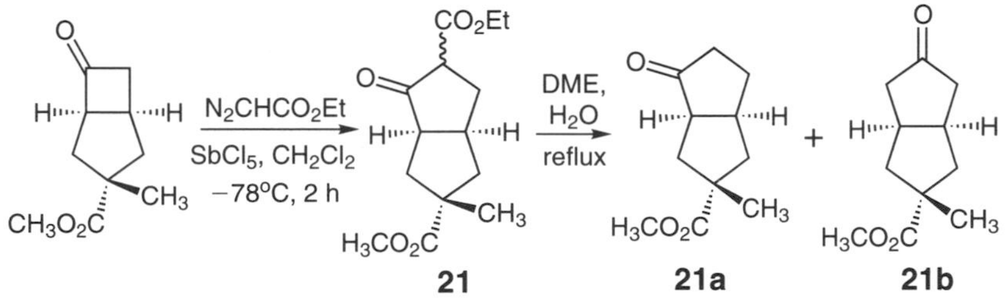

Methyl (1R,3S,5R)-3-methyl-6-oxobicyclo[3.2.0]heptane-3-carboxylate (810 mg, 4.44 mmol) was dissolved in 52 mL of methylene chloride and at −78°C treated with 290 mL (677 mg, 2.27 mmol) of antimony pentachloride and 10 min later with 970 mL (1.06 g, 9.29 mmol) of ethyl diazoacetate. After being stirred for 2 h at −78°C, the reaction mixture was allowed to warm to 0°C and was then poured into a stirred mixture of methylene chloride and aqueous sodium bicarbonate. After 45 min, the reaction mixture was filtered, and the crude product in methylene chloride was isolated in the usual fashion and then filtered with ether through a small pad of silica gel to afford impure **21**. A solution of this material in 20 mL of dimethoxyethane containing 300 μL of water was refluxed for 48 h and then concentrated under reduced pressure. GC analysis of the residue showed the presence of **21a** and **21b** in a ratio of ca. 98:2. Dry silica gel chromatography of the residue with ether in pentane (0–10%) gave 554 mg (63%) of pure **21a**: $[\alpha]_D^{20}$ −155 (c 3.3, chloroform); IR 1730, 1460, 1440, 1420, 1290, 1240, 1200, 1160, 1120, 1090 cm^{-1}; ^{1}H-NMR (CDCl$_3$) δ 1.27 (s, 3 H), 1.5–2.75 (m, 10 H), 3.68 (s, 3 H); ^{13}C-NMR (CDCl$_3$) δ 24.43 (2×), 35.81, 39.42, 40.71, 44.01, 51.60, 51.87 (2×), 177.35, 221.43; mass spectrum, m/e 196 (M$^+$).

The ready availability of bi- and tricyclobutanones from the intramolecular [2+2] cycloaddition of ketenes to an olefin makes these cyclobutanone systems attractive precursors of the corresponding cyclopentanone adducts. Both linear and angular annulated triquinanes can be obtained via a sequence involving a [2+2] cycloaddition and a Lewis acid-catalyzed ring expansion. Two examples are illustrated in Schemes 13[66] and 14.[67]

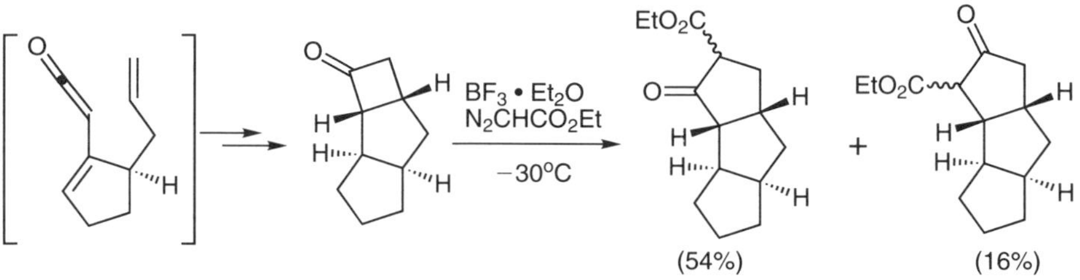

Scheme 13

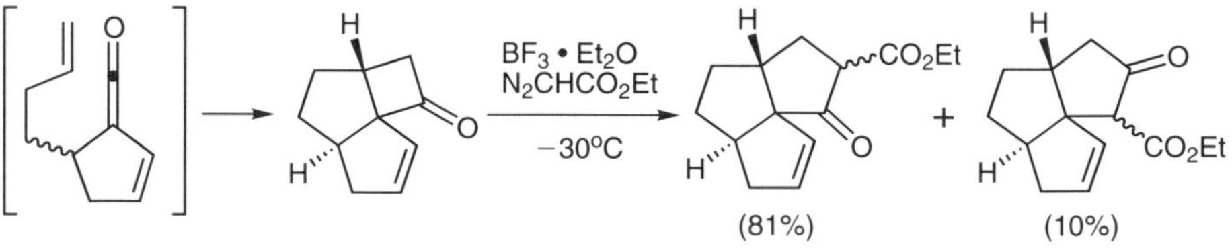

Scheme 14

Lewis acid–promoted homologation of polycyclic ketones bearing ring systems other than cyclobutanone are also effective processes. Representative examples of ring-enlargement with ethyl diazoacetate under Lewis acid–catalyzed conditions are provided in Table 10.13. Entry 1 shows the synthesis of homothujone from thujone. The homothujone subsequently served as a chiral starting material for the synthesis of the antifeedant (−)-polygodial and the ambergris fragrance (−)-ambrox.[68] The BF$_3$·Et$_2$O–catalyzed reaction of a bicyclo[4.3.0]-nonanone with EDA produced the corresponding ring-enlargement adduct as a 4:1 mixture of regioisomers (entry 3).[17] In this case, the sequential ethyl lithio-diazoacetate addition–rhodium(II)-catalyzed rearrangement proved to be advantageous and furnished the corresponding adduct as a single regioisomer (see Table 10.3, entry 1). The ring-expansion products shown in entry 4 were used in the syntheses of α-himachalene,[70] β-himachalene,[70] and isohimachalone.[71] Entries 5–8 demonstrate the application of this methodology to several bridged cyclic ketones. The ring-enlargement adducts in entry 7 were the key intermediates in a formal synthesis of (±)-longifolene.[72]

Dave and Warnoff[74] have developed an approach to regiospecific homologation of unsymmetrical, unhindered ketones based on the imposition of a bias against migration of one of the two α-substituents. The procedure consists of preparation

Table 10.13   Homologation of Polycyclic Ketones with Ethyl Diazoacetate

Entry	Ketones	Products (yields)	Ref.
1	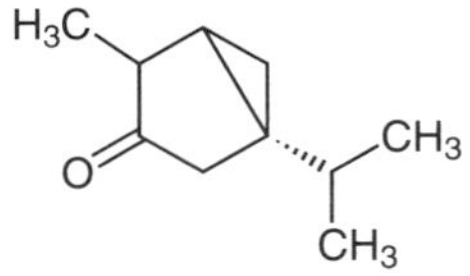	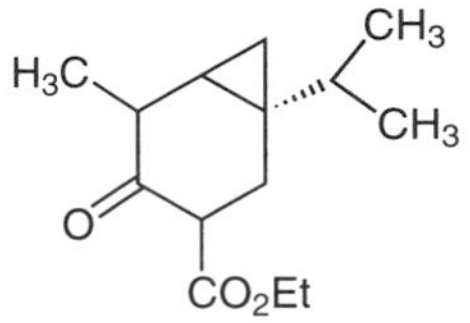 (70%)	68
2	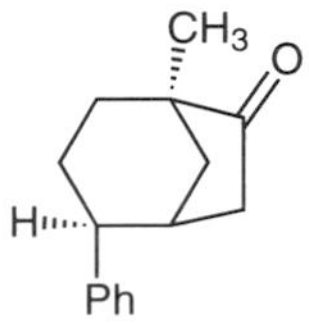	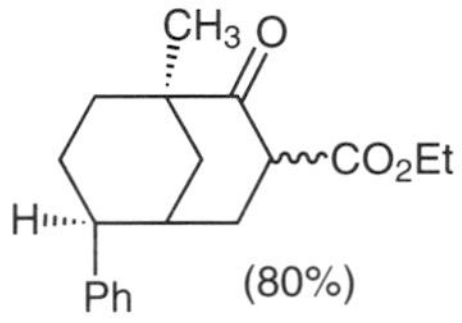 (80%) + 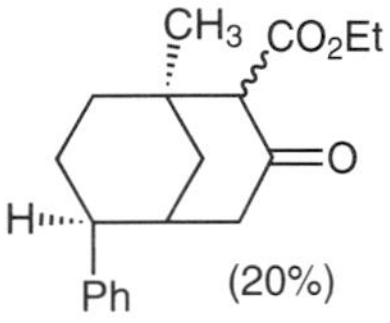 (20%)	69
3	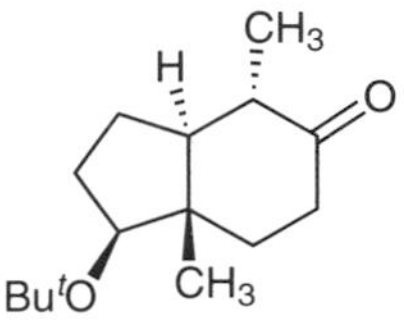	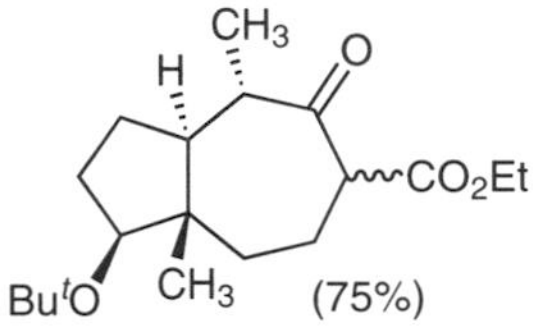 (75%) + 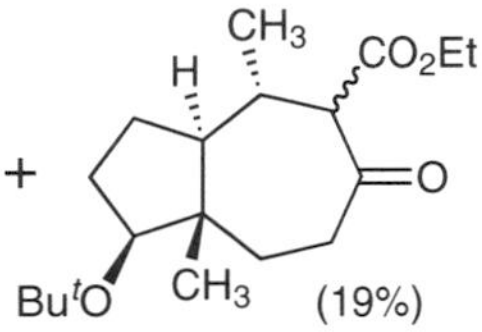 (19%)	17
4	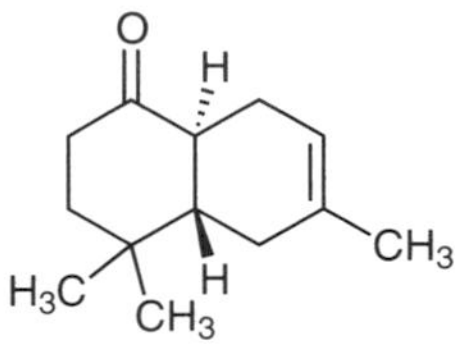	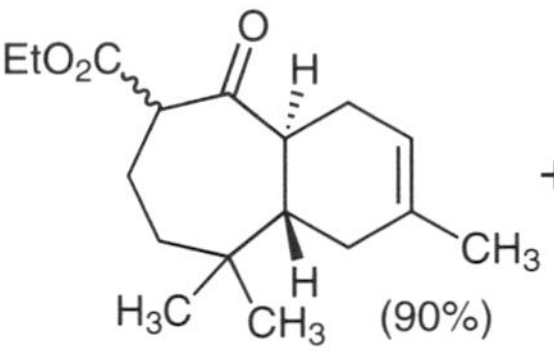 (90%) 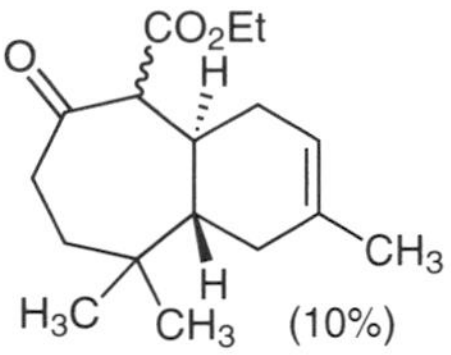 + (10%)	70 71
5	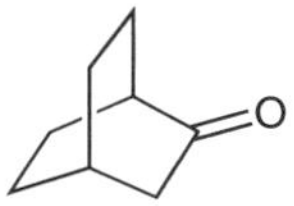	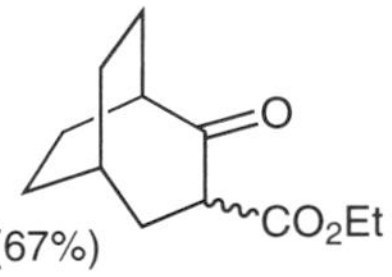 (67%) + 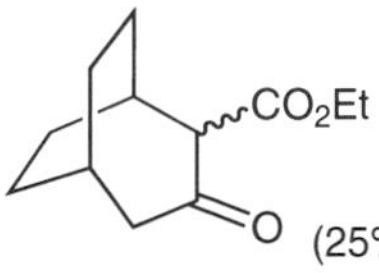 (25%)	50
6	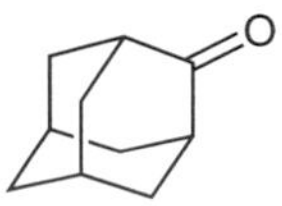	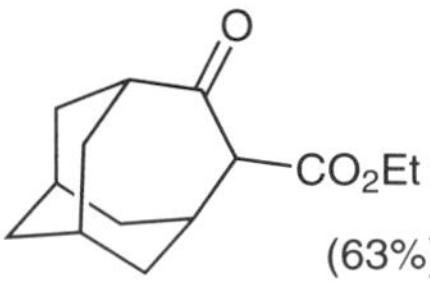 (63%)	48
7	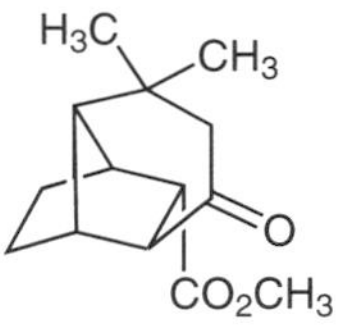	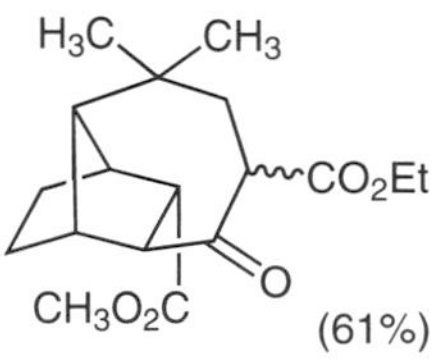 (61%) + 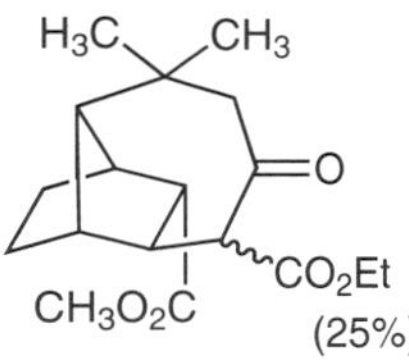 (25%)	72
8		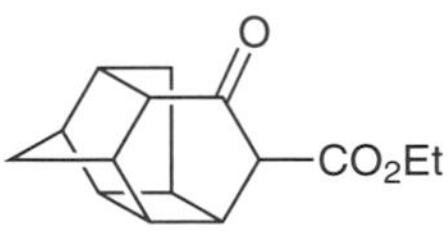 (90%)	73

of a pure α-bromo- or α-chloroketone, reaction of this derivative with ethyl diazoacetate and boron trifluoride being followed by removal of the halogen by zinc reduction and removal of the ethoxycarbonyl group by heating with water in a sealed tube at 230°C (Scheme 15). The regiospecifity of the homologation rests

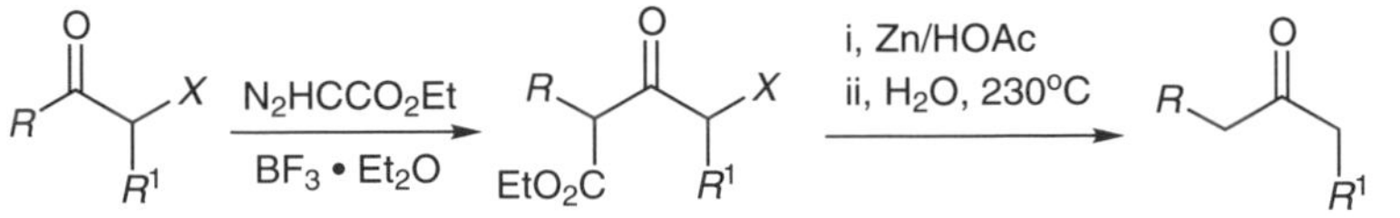

Scheme 15

on the electron-withdrawing power of the α-halogen atom, which prevents migration of the carbon atom to which it is attached. Examples of this approach are shown in Table 10.14.

Table 10.14   Regiospecific Homologation of Unsymmetrical Ketones with EDA

Reactant	Product (yield)
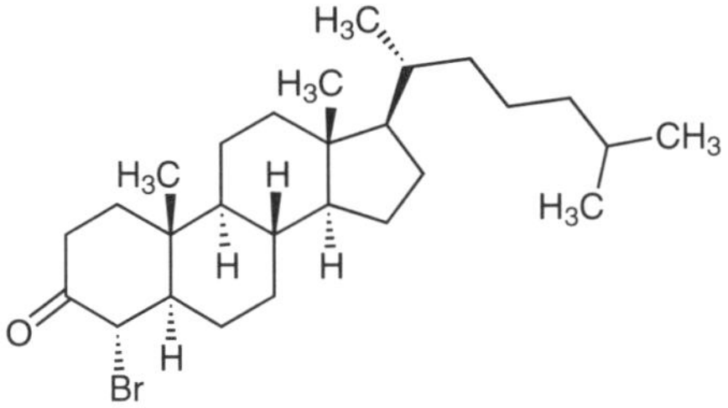	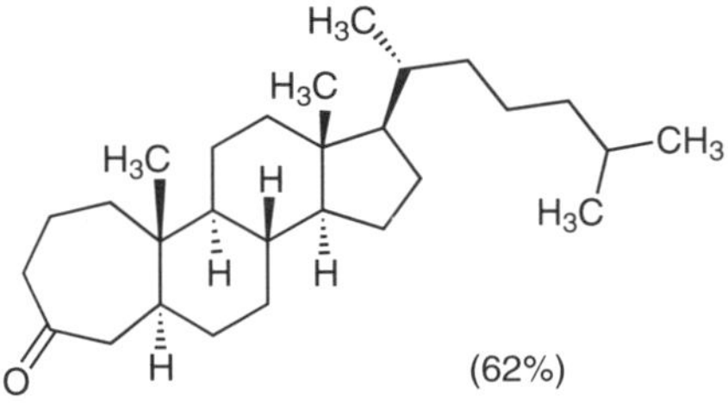 (62%)
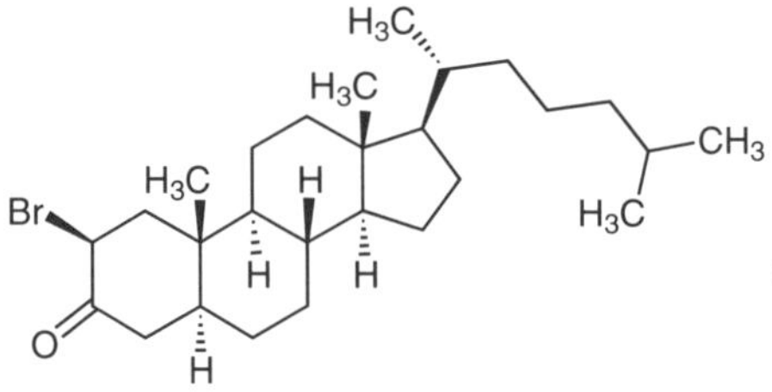	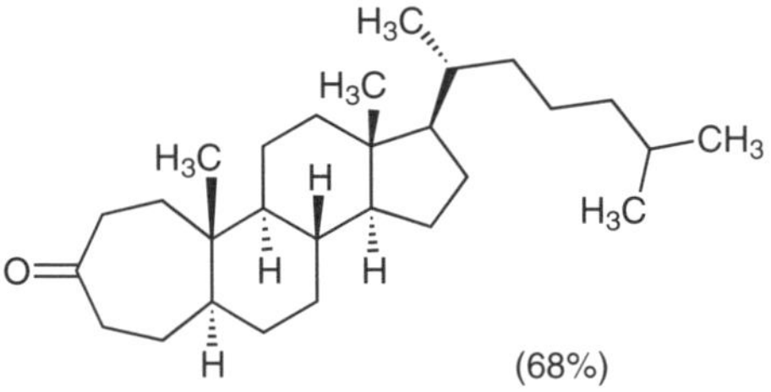 (68%)
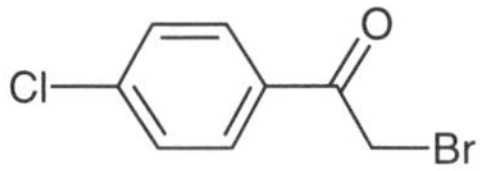	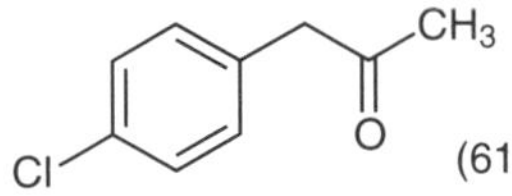 (61%)
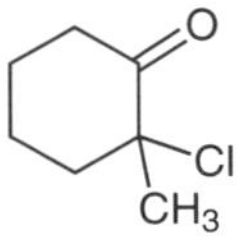	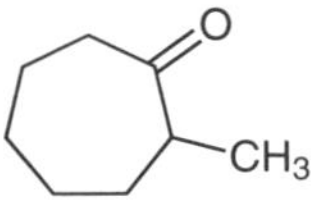 (67%)

The regiospecific homologation of unsymmetrical ketones is now finding applications in target-oriented syntheses. A good example occurs in Paquette's synthesis of [4,5]dihomotropone (**27**), where the regiospecific homologation protocol was used to establish the required tetrahydro precursor **26** (Scheme 16).[75-77] A

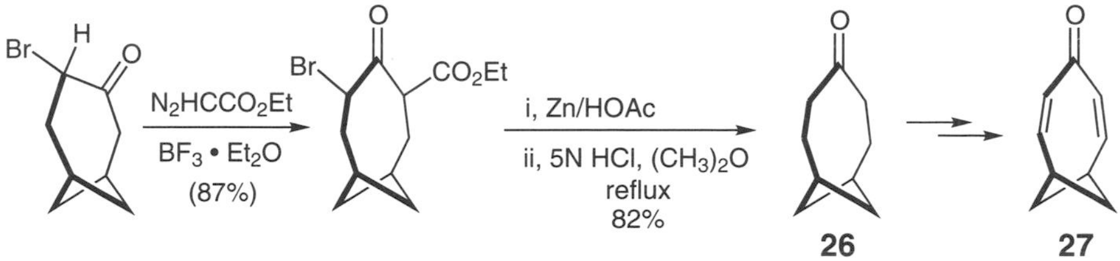

Scheme 16

further example[78] is found in the total synthesis of balanol (**29**), where the required bromoketone (**28**) was again obtained *via* the regiospecific homologation process (Scheme 17).

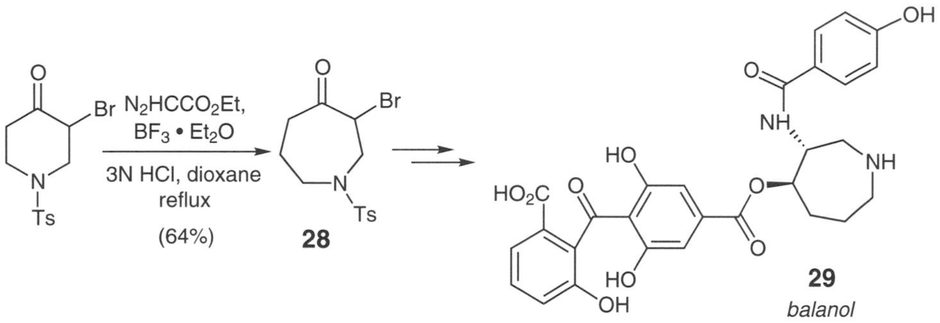

Scheme 17

α,β-Unsaturated ketones and aldehydes are not amenable to homologation with diazocarbonyl compounds.[48] Doyle's group, however, has found that the acetals of α,β-unsaturated carbonyl compounds do undergo homologation with ethyl diazoacetate in the presence of boron trifluoride etherate to produce, in good yields, β,γ-unsaturated acetals. Two examples of this transformation are illustrated in eqs. 11 and 12.[79]

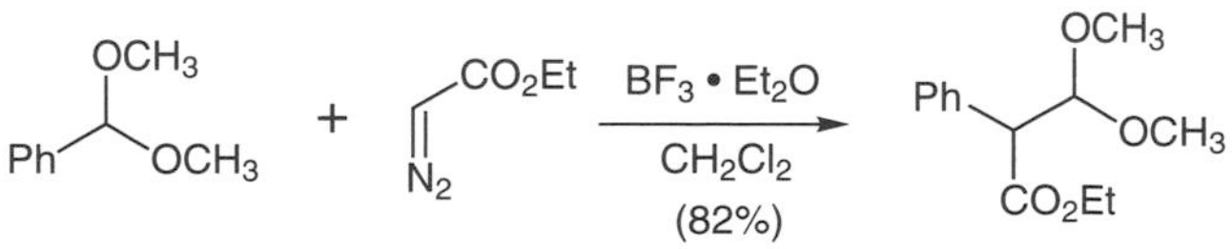

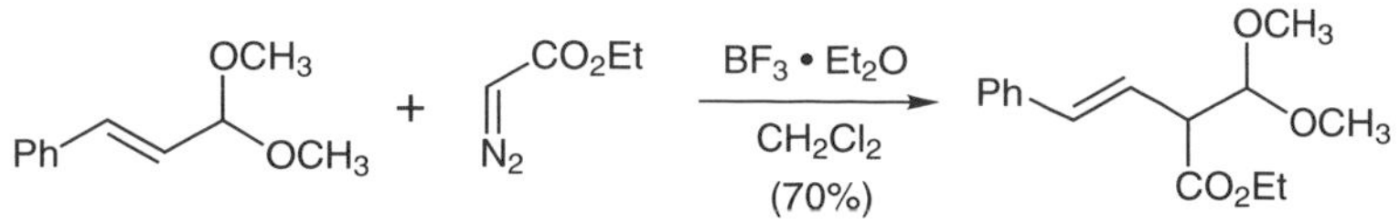

(12)

Intramolecular ring expansion of ketones is known, although the process has not yet received the same degree of exploitation in synthesis as the intermolecular version. Treatment of diazoketone **30** with triethyloxonium tetrafluoroborate[48] or tin(II) chloride[27] afforded the rearranged diketone **31** (eq. 13). Mander and

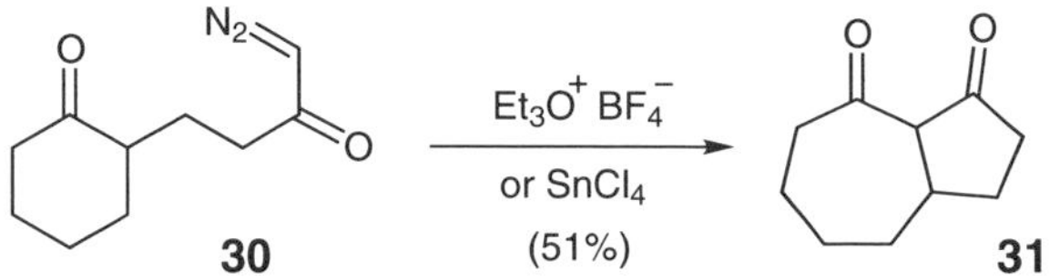

(13)

Wilshire[80] have investigated an intramolecular diazocarbonyl ring expansion reaction in the conversion of diazoketone **32** to diketone **33** (Scheme 18). Diketone

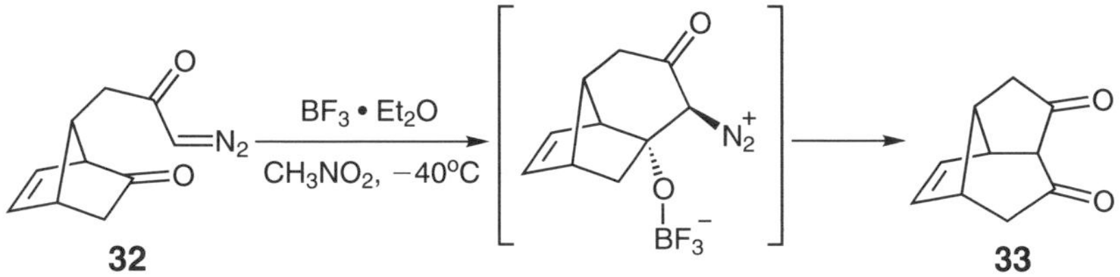

Scheme 18

33 is derived from bridgehead migration in **32**; the alternative intermediate leading to the methylene migration would require the diazonium group to adopt a less favored axial orientation.

10.7 REFERENCES

1 López-Herrera, F. J.; and Sarabia-García, F., "β-Oxo-α-diazo Carbonyl Compounds. III. Rh₂(AcO)₄ Mediated Decomposition of β-Acetoxy-α-diazo Esters. Application to the Synthesis of Natural 3-Deoxy-2-keto Aldonic Acids (KDO and DAH)," *Tetrahedron Lett.* **1994**, *35*, 6705–8.

2 Schöllkopf, U.; Frasnelli, H.; and Hoppe, D., "Ethyl 2-Oxazoline-5-carboxylate from Ethyl Isocyanoacetate and Carbonyl Compounds," *Angew. Chem., Int. Ed. Engl.* **1970**, *9*, 300–2.

3 Schöllkopf, U.; Bánhidai, B.; Frasnelli, H.; Meyer, R.; and Beckhaus, H., "α-Diazo-β-hydroxycarbonsäureester und -Ketone aus Carbonyl und Diazolithioverbindun

gen sowie ihre Umlagerung zu β-Ketocarbonsäureestern und β-Diketonen," *Liebigs Ann. Chem.* **1974**, 1767–83

4   Wenkert, E.; and McPherson, C. A., "Condensations of Acyldiazomethanes with Aldehydes, Ketones, and Their Derivatives," *J. Am. Chem. Soc.* **1972**, *94*, 8084–90.

5   Pellicciari, R.; and Natalini, B., "Ring Expansion of Thiochroman-4-one and Isothiochroman-4-one with Ethyl Diazo(lithio)acetate to Tetrahydrobenzothiepin β-Oxoesters," *J. Chem. Soc., Perkin Trans. 1* **1977**, 1822–24.

6   Pellicciari, R.; Castagnino, E.; and Corsano, S., "Reactions of α-Cyclocitral with Ethyl Diazo(lithio)acetate and with 1-Diazo-1-lithio Acetate," *J. Chem. Res.* **1979**, *(S)*, 76–77.

7   Pellicciari, R.; Natalini, B.; Taddei, M.; Ricci, A.; Bistocchi, G. A.; and De Meo, G., "Preparation of Ethyl-4-Oxotetrahydrobenzoxepin-5-Carboxylates, by Ring Expansion of Chroman-4-one and Isochroman-4-one with Ethyl Diazo(lithio)-acetate, and Their Decarboxylation to Dihydrobenzoxepinones," *J. Chem. Res.* **1979**, *(S)*, 142–43.

8   Pellicciari, R.; Castagnino, E.; Fringuelli, R.; and Corsano, S., "The Preparation of Acylacetylenic Derivatives of α-Cyclocitral on Route to Physiologically Active Terpenes," *Tetrahedron Lett.* **1979**, 481–84.

9   Pellicciari, R.; Fringuelli, R.; Ceccherelli, P.; and Sisani, E., "β-Keto Esters from the Rhodium(II) Acetate Catalysed Conversion of α-Diazo-β-hydroxy Esters," *J. Chem. Soc., Chem. Commun.* **1979**, 959–60.

10 Pellicciari, R.; Fringuelli, R.; and Sisani, E., "A New Synthesis of β-Damascone," *Tetrahedron Lett.* **1980**, 4039–42.

11 Pellicciari, R.; Fringuelli, R.; Sisani, E.; and Curini, M., "An Improved 2-Step Route for the Preparation of β-Diketones from Aldehydes and Its Application to the Synthesis of β-Damascone," *J. Chem. Soc., Perkin Trans. 1* **1981**, 2566–69.

12 Pellicciari, R.; Natalini, B.; Cecchetti, S.; and Fringuelli, R., "Reduction of α-Diazo-β-hydroxy Esters to β-Hydroxy Esters—Application in One of Two Convergent Syntheses of a (22S)-22-Hydroxy Bile Acid from Fish Bile and Its (22R)-Epimer," *J. Chem. Soc., Perkin Trans. 1* **1985**, 493–97.

13 Pellicciari, R.; Natalini, B.; and Fringuelli, R., "An Efficient Procedure for the Regiospecific Preparation of D-Homo-Steroid Derivatives," *Steroids* **1987**, *49*, 433–41.

14 Singh, A. K.; Bakshi, R. K.; and Corey, E. J., "Total Synthesis of (±)-Atractyligenin," *J. Am. Chem. Soc.* **1987**, *109*, 6187–89.

15 Eguchi, Y.; Sasaki, F.; Takashima, Y.; Nakajima, M.; and Ishikawa, M., "Studies on Antiatherosclerotic Agents. Synthesis of 7-Ethoxycarbonyl-4-formyl-6,8-dimethyl-1(2H)-phthalazinone Derivatives and Related Compounds," *Chem. Pharm. Bull.* **1991**, *39*, 795-97.

16 Ye, T.; and McKervey, M. A., "Synthesis of Chiral N-Protected α-Amino-β-diketones from α-Diazoketones Derived from Natural Amino Acids," *Tetrahedron* **1992**, *48*, 8007–22.

17 Nagao, K.; Yoshimura, I.; Chiba, M.; and Kim, S.-W., "Synthetic Studies on Pseudoguaianolides. 1. Preparation of a Key Intermediate, 1-β-*tert*-Butoxy-2,3,3aα,4,5,8a-hexahydro-4 α, 8aβ-Dimethyl-6(1H)-azulenone, for Helenanolides," *Chem. Pharm. Bull.* **1983**, *31*, 114–21.

18 Baudouy, R.; Gore, J.; and Ruest, L., "Synthese Diastereoselective de Trimethyl-4,8,8 Bicyclo[5,1,0]octanones," *Tetrahedron* **1987**, *43*, 1099–108.

19 Damour, D.; Renaudon, A.; and Mignani, S., "An Efficient Approach to the Synthesis of 1,1-Diphenyl-1-silacycloheptan-4-one," *Synlett* **1995**, 111–12.

20 Nagao, K.; Chiba, M.; and Kim, S.-W., "A New Efficient Homologation Reaction of Ketones *via* Their Lithiodiazoacetate Adducts," *Synthesis* **1983**, 197–99.

21 Pellicciari, R.; Natalini, B.; Sadeghpour, B. M.; Rosato, G. C.; and Ursini, A., "The Reaction of α-Diazo-β-hydroxy Esters with Boron Trifluoride," *J. Chem. Soc., Chem. Commun.* **1993**, 1798–800.

22 Pellicciari, R.; Natalini, B.; Sadeghpour, B. M.; Marinozzi, M.; Snyder, J. P.; Williamson, B. L.; Kuethe, J. T.; and Padwa, A., "The Reaction of α-Diazo-β-hydroxy Esters with Boron–Trifluoride Etherate Generation and Rearrangement of Destabilized Vinyl Cations—A Detailed Experimental and Theoretical Study," *J. Am. Chem. Soc.* **1996**, *118*, 1–12.

23 Wenkert, E.; and McPherson, C. A., "Synthesis of Acylacetylenes from α-Diazo-β-hydroxylcarbonyl Compounds," *Synth. Commun.* **1972**, *2*, 331–34.

24 Miyauchi, K.; Hori, K.; Hirai, T.; Takebayashi, M.; and Ibata, T., "The Acid-Catalyzed Decomposition of α-Diazo β-hydroxy Ketones," *Bull. Chem. Soc. Jpn.* **1981**, *54*, 2142-47.

25 López-Herrera, F. J.; Valpuesta-Fernández, M.; and García-Claros, S., "Reaction of Methyl Diazoacetate with 2,3 *O*-Isopropylidene-D-glyceraldehyde. Stereoselectivity in the Synthesis of 2-Deoxy-γ-D-Aldonolactones," *Tetrahedron* **1990**, *46*, 7165–74.

26 Holmquist, C. R.; and Roskamp, E. J., "A Selective Method for the Direct Conversion of Aldehydes into β-Keto-Esters with Ethyl Diazoacetate Catalyzed by Tin(II) Chloride," *J. Org. Chem.* **1989**, *54*, 3258–60.

27 Padwa, A.; Hornbuckle, S. F.; Zhang, Z.; and Zhi, L., "Synthesis of 1,3-Diketones Using α-Diazo Ketones and Aldehydes in the Presence of Tin(II) Chloride," *J. Org. Chem.* **1990**, *55*, 5297–99.

28 Herczegh, P.; Kovacs, I.; Kovacs, A.; Szilagyi, L.; and Sztaricskai, F., "Synthesis of Sugar 2,4-Diketoesters," *Synlett* **1991**, 705–6.

29 Holmquist, C. R.; and Roskamp, E. J., "The Conversion of Olefins to β-Keto-Esters—Ozonolysis of Olefins Followed by *in situ* Reduction with Tin(II) Chloride in the Presence of Ethyl Diazoacetate," *Tetrahedron Lett.* **1990**, *31*, 4991–94.

30 Dhavale, D. D.; Patil, P. N.; and Mali, R. S., "Activated Alumina-Promoted Reaction of Aldehydes with Ethyl Diazoacetate—A Simple Route to β-Oxo Esters," *J. Chem. Res. (S)* **1994**, 152–53.

31 Sudrik, S. G.; Balaji, B. S.; Singh, A. P.; Mitra, R. B.; and Sonawane, H. R., "Zeolite-Mediated Synthesis of β-Keto-Esters—Condensation of Ethyl Diazoacetate with Aldehydes," *Synlett* **1996**, 369–70.

32 Rychnovsky, S. D.; and Mickus, D. E., "Synthesis of *ent*-Cholesterol, the Unnatural Enantiomer," *J. Org. Chem.* **1992**, *57*, 2732–36.

33 Hecker, S. J.; and Werner, K. M., "Total Synthesis of (±)-Leuhistin," *J. Org. Chem.* **1993**, *58*, 1762–65.

34 Eh, M.; Schomburg, D.; Schicht, K.; and Kalesse, M., "An Efficient Synthesis of Radicinin Analogs," *Tetrahedron* **1995**, *51*, 8983–92.

35 Mastalerz, H.; and Menard, M., "Synthesis of a 4-β-(Hydroxymethyl)-carbapenem," *J. Org. Chem.* **1994**, *59*, 3223–26.

36 Honda, T.; Okuyama, A.; Hayakawa, T.; Kondoh, H.; and Tsubuki, M., "A Stereoselective Synthesis of (±)-Pestalotin," *Chem. Pharm. Bull.* **1991**, *39*, 1866–68.

37 Srikrishna, A.; and Krishnan, K., "Stereospecific Construction of Multiple Contiguous Quaternary Carbons—Total Synthesis of (±)-*cis, anti, cis*-1,8,12,12-Tetramethyl-4-oxatricyclo[6.4.0.0^{2,6}]dodecan-3-ol, a Thapsane Isolated from *Thapsia-villosa var minor*," *J. Org. Chem.* **1993**, *58*, 7751–55.

38 Kido, F.; Abiko, T.; and Kato, M., "Spiroannulation by the [2,3]Sigmatropic Rearrangement *via* the Cyclic Allylsulfonium Ylide. A Stereoselective Synthesis of (+)-Acorenone B," *J. Chem. Soc., Perkin Trans. 1* **1992**, 229–33.

39 Bracher, F., "Polycyclic Aromatic Alkaloids. 6. A Regioselective Synthesis of Azafluorenone Alkaloids," *Synlett* **1991**, 95–96.

40 Dauben, W. G.; Dinges, J.; and Smith, T. C., "A Convergent Approach toward the Tigliane Ring-System," *J. Org. Chem.* **1993**, *58*, 7635–37.

41 Nomura, K.; Iida, T.; Hori, K.; and Yoshii, E., "Synthesis of γ-Unsubstituted α-Acyl-β-tetronic Acids from Aldehydes," *J. Org. Chem.* **1994**, *59*, 488–90.

42 Iida, T.; Hori, K.; Nomura, K.; and Yoshii, E., "A New Entry to 5-Unsubstituted 3-Acyltetramic Acids from Aldehydes," *Hetreocycles* **1994**, *38*, 1839–44.

43 Hori, K.; Kazuno, H.; Nomura, K.; and Yoshii, E., "The First Total Synthesis of Tetronasin (M139603)," *Tetrahedron Lett.* **1993**, *34*, 2183–86.

44 Angle, S. R.; Wei, G. P.; Ko, Y. K.; and Kubo, K., "Stereoselective Synthesis of Tetrahydrofurans *via* the Lewis-Acid Promoted Reaction of β-Benzyloxy Aldehydes and Ethyl Diazoacetate," *J. Am. Chem. Soc.* **1995**, *117*, 8041–42.

45 Liao, Y.; and Huang, Y.-Z., "A Novel Olefination of Diazo-Compounds with Carbonyl-Compounds Mediated by Tributylstibine and Catalytic Amount of Cu(I)I," *Tetrahedron Lett.* **1990**, *31*, 5897–900.

46 Herrmann, W. A.; and Wang, M., "Methyltrioxorhenium as Catalyst of a Novel Aldehyde Olefination," *Angew. Chem., Int. Ed. Engl.* **1991**, *30*, 1641–43.

47 Zhu, Z.; and Espenson, J. H., "Reactions of Ethyl Diazoacetate Catalyzed by Methylrhenium Trioxide," *J. Org. Chem.* **1995**, *60*, 7090–91.

48 Mock, W. L.; and Hartman, M. E., "Synthetic Scope of the Triethyloxonium Ion Catalyzed Homologation of Ketones with Diazoacetic Esters," *J. Org. Chem.* **1977**, *42*, 459–65.

49 Liu, H. J.; and Majumdar, S. P., "On the Regioselectivity of Boron Trifluororide Catalyzed Ring Expansion of Cycloalkanones with Ethyl Diazoacetate," *Synth. Commun.* **1975**, *5*, 125–30.

50 Liu, H. J.; Yeh, W. L.; and Browne, E. N. C., "Activated Cycloheptenone Dienophiles—A Versatile Approach to 6,7-Fused Ring Targets," *Can. J. Chem.* **1995**, *73*, 1135–47.

51 Ohuchida, S.; Hamanaka, N.; and Hayashi, M., "Synthesis of Thromboxane A$_2$ Analogs-3 (+)-Thiathromboxane A$_2$," *Tetrahedron* **1983**, *39*, 4269–72.

52 Baldwin, S. W.; and Landmesser, N. G., "Benzyl and Allyl Diazoacetate in One Carbon Ring Expansions of Ketones," *Synth. Commun.* **1978**, *8*, 413–19.

53 Macpherson, L. J.; Bayburt, E. K.; Capparelli, M. P.; Bohacek, R. S.; Clarke, F. H.; Ghai, R. D.; Sakane, Y.; Berry, C. J.; Peppard, J. V.; and Trapani, A. J., "Design and Synthesis of an Orally-Active Macrocyclic Neutral Endopeptidase-24.11 Inhibitor," *J. Med. Chem.* **1993**, *36*, 3821-28.

54 Gribble, G. W.; and Silva, R. A., "Synthesis of a Mexican Bean Beetle Azamacrolide Allomone *via* a Novel Lactam to Lactone Ring Expansion," *Tetrahedron Lett.* **1996**, *37*, 2145–48.

55 Boivin, J.; Huppé, S.; and Zard, S. Z., "An Efficient Synthesis of Large Ring Acetylenes," *Tetrahedron Lett.* **1995**, *36*, 5737–40.

56 Liu, H. J.; and Ogino, T., "A New Approach to the Synthesis of Hydrindanonecarboxylates," *Tetrahedron Lett.* **1973**, 4937–40.

57 Liu, H.-J.; and Chan, W. H., "Total Synthesis of Zizaane Sesquiterpenes: (−)-Khusimone, (+)-Zizanoic Acid, and (−)-Epizizanoic Acid," *Can. J. Chem.* **1982**, *60*, 1081–91.

58 Greene, A. E.; Luche, M.-J.; and Serra, A. A., "An Efficient, Enantioconvergent Total Synthesis of Natural Hirsutic Acid-C," *J. Org. Chem.* **1985**, *50*, 3957–62.

59 Ghosh, A.; Biswas, S.; and Venkateswaran, R. V., "Stereocontrolled Synthesis of (±)-Debromoaplysin and (±)-Aplysin," *J. Chem. Soc., Chem. Commun.* **1988**, 1421–21.

60 Biswas, S.; Ghosh, A.; and Venkateswaran, R. V., "Stereocontrolled Synthesis of (±)-Debromoaplysin and (±)-Aplysin, (±)-Debromoaplysinol, (±)-Aplysinol, and (±)-Isoaplysin," *J. Org. Chem.* **1990**, *55*, 3498–502.

61 Stille, J. R.; and Grubbs, R. H., "Synthesis of (±)-Δ$^{(9,12)}$-Capnellene Using Titanium Reagents," *J. Am. Chem. Soc.* **1986**, *108*, 855–56.

62 Stille, J. R.; Santarsiero, B. D.; and Grubbs, R. H., "Rearrangement of Bicyclo[2.2.1]heptane Ring Systems by Titanocene Alkylidene Complexes to Bicyclo[3.2.0]heptane Enol Ethers. Total Synthesis of (±)-Δ$^{9\ (12)}$-Capnellene," *J. Org. Chem.* **1990**, *55*, 843–62.

63 Liu, H.-J.; and Kulkarni, M. G., "Total Synthesis of (±)-Δ$^{9\ (12)}$-Capnellene," *Tetrahedron Lett.* **1985**, *26*, 4847–50.

64 Sonawane, H. R.; Naik, V. G.; Bellur, N. S.; Shah, V. G.; Purohit, P. C.; Kumar, M. U.; Kulkarni, D. G.; and Ahuja, J. R., "Photoinduced Vinylcyclopropane–Cyclopentene Rearrangement: A Methodology for Chiral Bicyclo[3.2.0]heptenes. Formal Syntheses of (±)-Grandisol and Naturally Occurring (−)-Δ$^{9\ (12)}$-Capnellene and its Antipode," *Tetrahedron* **1988**, *47*, 8259–76.

65 Hamelin, O.; Deprés, J.-P.; Greene, A. E.; Tinant, B.; and Declercq, J.-P., "Highly Stereoselective First Synthesis of an A-Ring-Functionalized Bakkane: Novel Free-Radical Approach to 9-Acetoxyfukinanolide," *J. Am. Chem. Soc.* **1996**, *118*, 9992–93.

66 De Mesmaeker, A.; Veenstra, S. J.; and Ernst, B., "Intramolecular [2+2] Cycloadditions of Vinylketenes to Olefins. Part 2. The Synthesis of a Linear Annelated Triquinane Derivative," *Tetrahedron Lett.* **1988**, *29*, 459–62.

67 Veenstra, S. J.; De Mesmaeker, A.; and Ernst, B., "Intramolecular [2+2] Cycloadditions of Ketenes and Vinylketenes to Olefins. Part 3. The Synthesis of Angular Annelated Triquinane Derivative," *Tetrahedron Lett.* **1988**, *29*, 2303–6.

68 Kutney, J. P.; Chen, Y.-H.; and Rettig, S. J., "The Chemistry of Thujone. XVIII. Homothujone and its Derivatives," *Can. J. Chem.* **1996**, *74*, 666–76.

69 Chakraborti, R.; Ranu, B. C.; and Ghatak, U. R., "Stereospecific Synthesis of endo-6-Aryl-2-Oxobicyclo[3.3.1]nonanes," *Synth. Commun.* **1987**, *17*, 1539–43.

70 Liu, H.-J.; and Browne, E. N. C., "Total Synthesis of α- and β-Himachalene by an Intramolecular Diels–Alder Approach," *Can. J. Chem.* **1981**, *59*, 601–8.

71 Liu, H.-J.; and Browne, E. N. C., "Total Synthesis of a Proposed Structure of Isohimachalene," *Can. J. Chem.* **1987**, *65*, 182–88.

72 Ho, T.-L.; Yeh, W.-L.; Yule, J.; and Liu, H.-J., "Diels–Alder Approach to (±)-Longifolene: A Formal Synthesis," *Can. J. Chem.* **1992**, *70*, 1375–84.

73 Marchand, A. P.; Reddy, S. P.; Rajapaksa, D.; Ren, C.-T.; Watson, W. H.; and Kashyap, R. P., "Lewis Acid Promoted Reactions of 11-Methylenepentacyclo[5.4.0.0^{2,6}.0^{3,10}.0^{5,9}]undecan-8-one and Pentacyclo[5.4.0.0^{2,6}.0^{3,10}.0^{5,9}]undecan-8-one with Ethyl Diazoacetate," *J. Org. Chem.* **1990**, *55*, 3493–98.

74 Dave, V.; and Warnhoff, E. W., "α-Halo Ketones. 10. Regiospecific Homologation of Unsymmetrical Ketones," *J. Org. Chem.* **1983**, *48*, 2590–98.

75 Paquette, L. A.; Bacqué, E.; Bishop, R.; and Watson, T. J., "Synthesis and Properties of [4,5]Dihomotropone (Bicyclo[5.1.1]nona-2,5-dien-4-one)," *Tetrahedron Lett.* **1992**, *33*, 6559–62.

76 Paquette, L. A.; Watson, T. J.; Friedrich, D.; Bishop, R.; and Bacqué, E., "Is Through-Bond Dihydroaromaticity Attainable? Preparation of [4,5]-Dihomotropone, Investigation of Its Ground-State Properties, and an Attempt to Generate the Dihomotropylium Cation," *J. Org. Chem.* **1994**, *59*, 5700–7.

77 Paquette, L. A.; and Watson, T. J., "Through-Bond Interaction *via* Cyclobutane Relay Orbitals—Evaluation of the Question of Extended Conjugation In Belted [4,5]Dihomotropones," *J. Org. Chem.* **1994**, *59*, 5708–16.

78 Adams, C.P.; Fairway, S. M.; Hardy, C. J.; Hibbs, D. E.; Hursthouse, M. B.; Morley, A. D.; Sharp, B. W.; Vicker, N.; and Warner, I., "Total Synthesis of Balanol— A Potent Protein-Kinase-C Inhibitor of Fungal Origin," *J. Chem. Soc., Perkin Trans. 1* **1995**, 2355–62.

79 Doyle, M. P.; Trudell, M. L.; and Terpstra, J. W., "Homologation of Acetals of α,β-Unsaturated Carbonyl-Compounds with Diazo Esters—Synthesis of Acetals of β,γ-Unsaturated Carbonyl-Compounds," *J. Org. Chem.* **1983**, *48*, 5146–48.

80 Mander, L. N.; and Wilshire, C., "Studies on Intramolecular Alkylation. XI. An Adventitious Synthesis of a 1,4-Ethanopentalene Derivative from a Norbornenone-Derived Diazomethyl Ketone," *Aust. J. Chem.* **1979**, *32*, 1975–81.

CHAPTER 11

Acid-Promoted Cyclization of Unsaturated and Aromatic Diazo Ketones

11.1 INTRODUCTION

Cyclization processes in which a new bond is formed by nucleophilic addition to an ionizing center provide important routes to carbocycles and heterocycles. Diazocarbonyl compounds with suitably positioned internal nucleophiles display this type of behavior under acid catalysis. Nucleophiles include olefinic, acetylenic, and aromatic groups as well as heteroatoms such as oxygen, nitrogen, and sulfur. Of these, π-route cyclizations of alkenes and aromatics have had the most impact on organic synthesis, and this chapter is devoted to unsaturated and aromatic diazoketones. Cyclizations involving heteroatom participation are included in Chapter 8.

π-Route cyclization, as outlined in Scheme 1, is believed to proceed *via* initial protonation by a Brønsted acid (or complexation with a Lewis acid) of the diazocarbonyl function. This is followed by, or possibly accompanied by, nucleophilic attack by the π-electrons in a remote double bond, leading to loss of nitrogen and formation of a cycloalkyl cation, which proceeds to product(s), typically *via* proton elimination. Five-membered ring formation is the most commonly encountered result of π-route cyclization, which has been applied very successfully not just to simple carbocyclic systems, but to many multi-ring and bridged-ring molecules. The literature up to 1979 has been reviewed in detail by Burke and Grieco,[1] and Smith[2] has summarized his own work and that of Mander on multiple π-route cyclizations including aromatics. A few examples will suffice here to demonstrate the power of the process including recent variations.

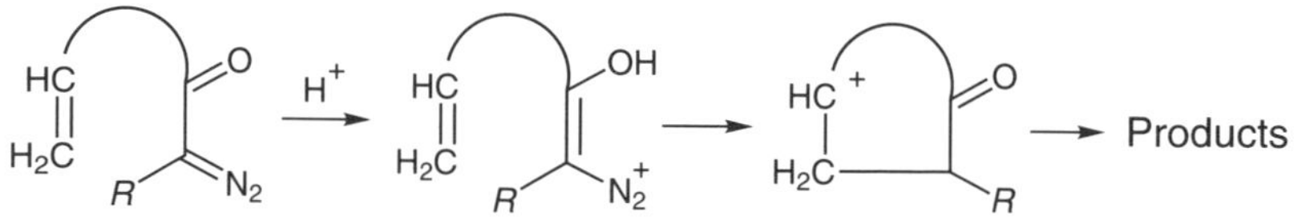

Scheme 1

11.2 CYCLIZATION OF γ,δ-UNSATURATED α-DIAZOKETONES

Acid-catalyzed intramolecular cyclization of γ,δ-unsaturated α-diazoketones provides a facile and powerful means of constructing cyclopentanones. In particular, a bicyclo[3.2.1]octanone or a bicyclo[2.2.1]heptanone moiety can be introduced into a variety of carbocyclic systems by use of this cyclopentanone annelation reaction (*vide infra*). The first example of acid-catalyzed intramolecular cyclization of γ,δ-unsaturated α-diazoketone was reported by Erman and Stone[3] in 1971. Reaction of diazoketone **1** with boron trifluoride etherate furnished bicyclic ketones **2** and **3** in 3 and 30% yields, respectively. The latter (**3**) served as the key intermediate for the total systhesis of sesquiterpenes *dl*-patchoulenone **4** and *dl*-epipatchoulenone **5**. Under the same reaction conditions, diazoketone **6** afforded bicyclic ketones **7** and **8** in higher yield (Scheme 2).

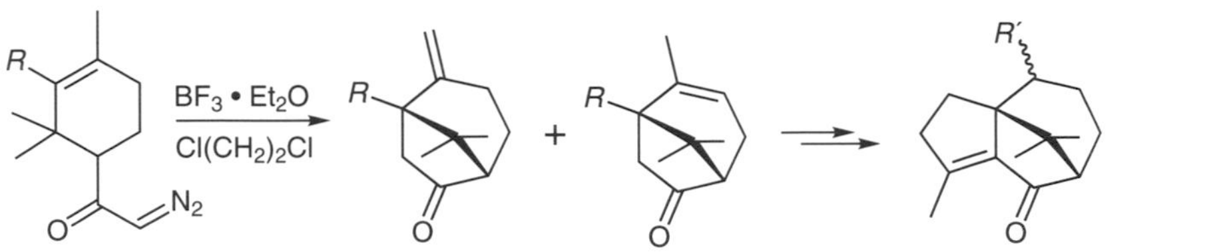

1: R = CO₂Et   **2:** R = CO₂Et (3%) **3:** R = CO₂Et (30%)   **4:** R′ = α-CH₃, *dl-patchoulenone*

6: R = H   **7:** R = H (6%)   **8:** R = H (56%)   **5:** R′ = β-CH₃, *dl-epipatchoulenone*

Scheme 2

In addition to Erman and Stone's pioneering observations, in the same year, Mander and his co-workers[4] reported their efforts in this field, including the fluoroboric acid–catalyzed decomposion of tricyclic diazoketones **9**, **10**, and **11**, as shown in eq. 1.[4]

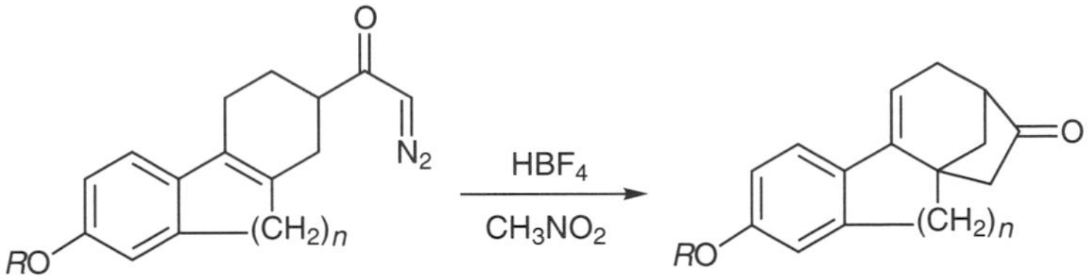

(1)

9: R = CH₃, n = 1   **12:** R = CH₃, n = 1 (97%)

10: R = CH₃, n = 2   **13:** R = CH₃, n = 2 (90%)

11: R = H, n = 2   **14:** R = H, n = 2 (78%)

As demonstrated by Erman and Stone,[3] the construction of bicyclic carbocycles based on the acid-catalyzed cyclization of monocyclic γ,δ-unsaturated α-diazoketones has assumed strategic importance in the carbon–carbon bond forming reaction. Mander and co-workers[5] have applied this type of acid-catalyzed cyclization in their total synthesis of (±)-14-norhelminthosporic acids and related compounds. Thus treatment of diazoketone **15** with boron trifluoride etherate in nitromethane furnished bicyclic ketones **16** and **17** in nearly quantitative yield as a 9:11 mixture of isomers. Hydrogenation of the ketone mixture provided the bicyclo[3.2.1]octanone derivative **18** in 91% overall yield from diazoketone **15**; standard functional group manipulation was used to convert **18** into the (±)-14-norhelminthosporic acids **19** and **20** (Scheme 3). In addition, acid-cat-

Scheme 3

alyzed decomposition of the related γ,δ-unsaturated α-diazoketone **21** to afford the corresponding bicyclic ketone **22** has been reported independently by Ghatak[6] and Mander[7] and their respective co-workers. The bicyclic ketone **22** has been converted into the helminthosporic acid analogue **23** (Scheme 4).[7] Another

Scheme 4

example is the synthesis of the hydroxylated helminthosporic acid analogue **25** from diazo ketone **24** (Scheme 5).[7]

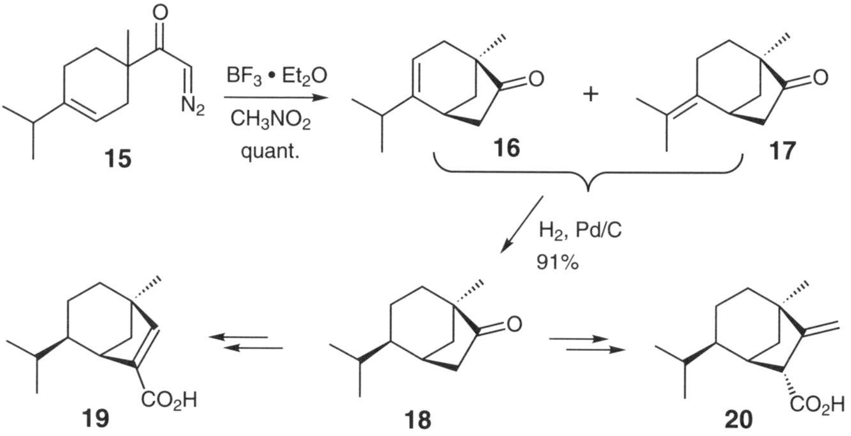

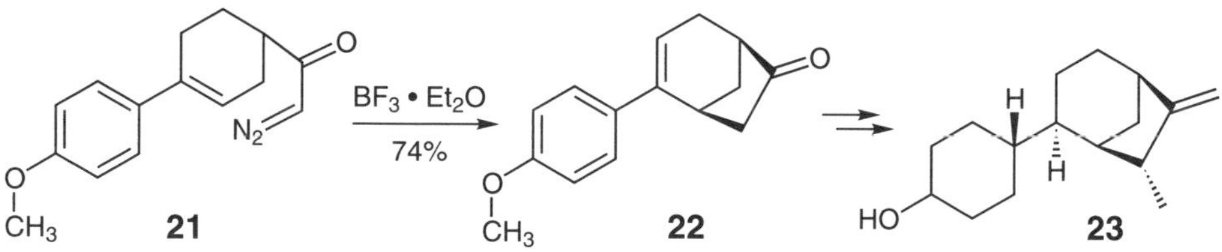

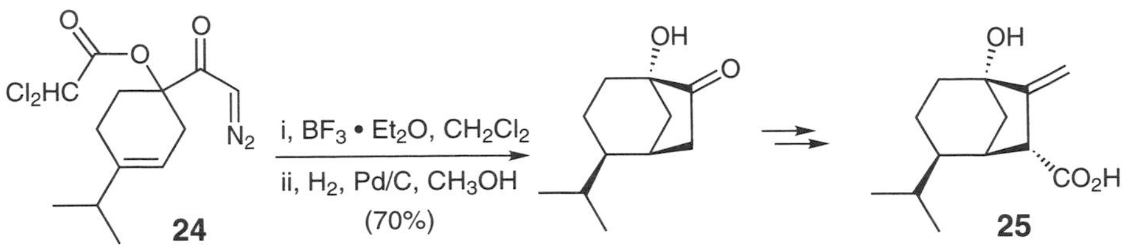

Scheme 5

Mehta and co-workers[8] have used π-route cyclization to construct the 5,6-fused bicyclic ketone **26** shown in eq. 2.[8] Similarly, brief exposure of diazoketone

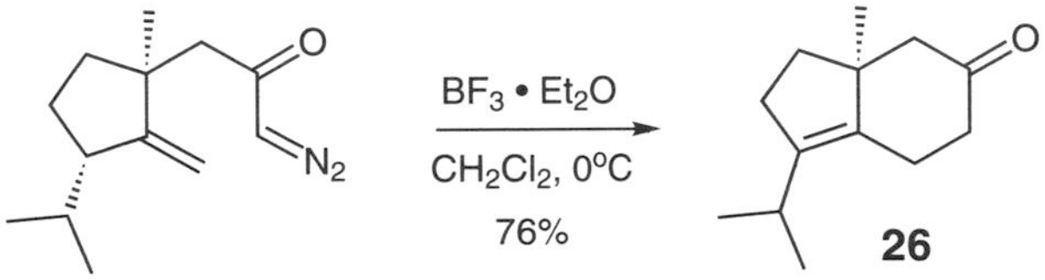

(2)

27 to BF$_3 \cdot$ Et$_2$O resulted in a facile cyclization to the bicyclic enone **28**, in which the vicinal quaternary centers were duly installed (Scheme 6). Adduct **28** was the key intermediate of the total synthesis of (−)-ceratopicanol **29** (Scheme 6).[9]

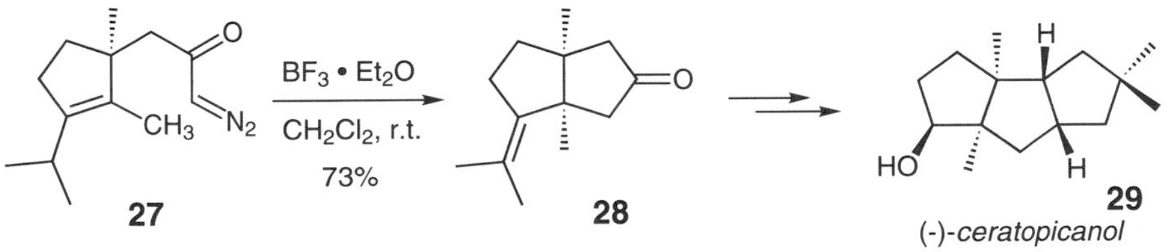

Scheme 6

A simple, general preparative route to tetracyclic gibbanes (**31**) and similar systems incorporating intermediate bridged bicyclo[3.2.1]octanone moieties, based on intramolecular acid-catalyzed cyclization of γ,δ-unsaturated diazoketones (**30**) have been developed independently by Ghatak[10,12–14] and Mander[11,15,16] and their respective co-workers. Representative examples are illustrated in eq. 3. Acid catalysts include BF$_3 \cdot$ Et$_2$O, TFA alone, or admixed with perchloric acid. The reaction yields suggest that the protonic acids are generally more efficient than the Lewis acid in promoting cyclization. The success of this route to the gibbane skeleton enabled Mander to prepare the tetracyclic ester **33**, by TFA-catalyzed cyclization of diazoketone **32**, and use it as the pivotal intermediate elegant total synthesis of gibberellic acid **34** (Scheme 7).[15,16]

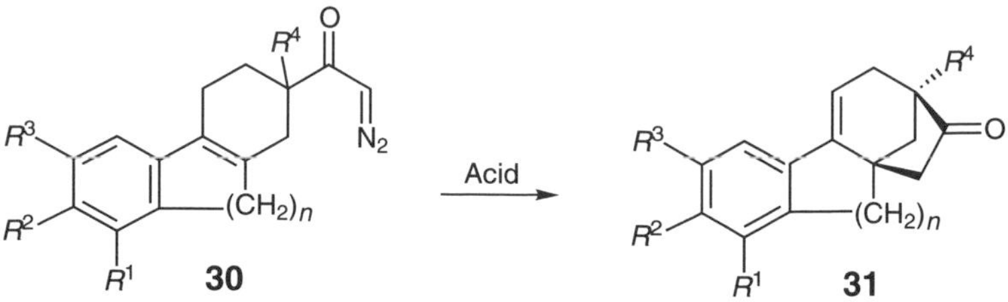

$$\text{30} \xrightarrow{\text{Acid}} \text{31} \tag{3}$$

Diazoketones	Reaction conditions	Yield	Ref.
a: $n = 1$; $R^1 = R^3 = R^4 = H$, $R^2 = CH_3O$	$BF_3 \cdot Et_2O$, CH_2Cl_2; $-10°C$	45%	10
b: $n = 2$; $R^1 = R^3 = H$, $R^2 = CH_3O$, $R^4 = CH_3$	TFA; 0°C	100%	11
c: $n = 2$; $R^1 = R^3 = H$, $R^2 = CH_3O$, $R^4 = OCOCF_3$	TFA-CH_2Cl_2 (1:10); 0°C	>60%	11
d: $n = 1$; $R^2 = R^3 = R^4 = H$, $R^1 = CH_3$	$BF_3 \cdot Et_2O$; $(ClCH_2)_2$; $-10°C$	58%	12
e: $n = 1$; $R^2 = R^3 = H$, $R^1 = R^4 = CH_3$	$BF_3 \cdot Et_2O$; $(ClCH_2)_2$; $-10°C$	80%	12
f: $n = 1$; $R^2 = R^3 = R^4 = H$, $R^1 = CH_3O$	70% $HClO_4$-TFA (1:1), $CHCl_3$; 0–5°C	94%	13
g: $n = 1$; $R^1 = R^2 = R^4 = H$, $R^3 = CH_3O$	70% $HClO_4$-TFA (1:1), $CHCl_3$; 0–5°C	94%	13
h: $n = 3$; $R^1 = R^3 = R^4 = H$, $R^2 = CH_3O$	$BF_3 \cdot Et_2O$, CH_2Cl_2; $-10°C$	50%	14

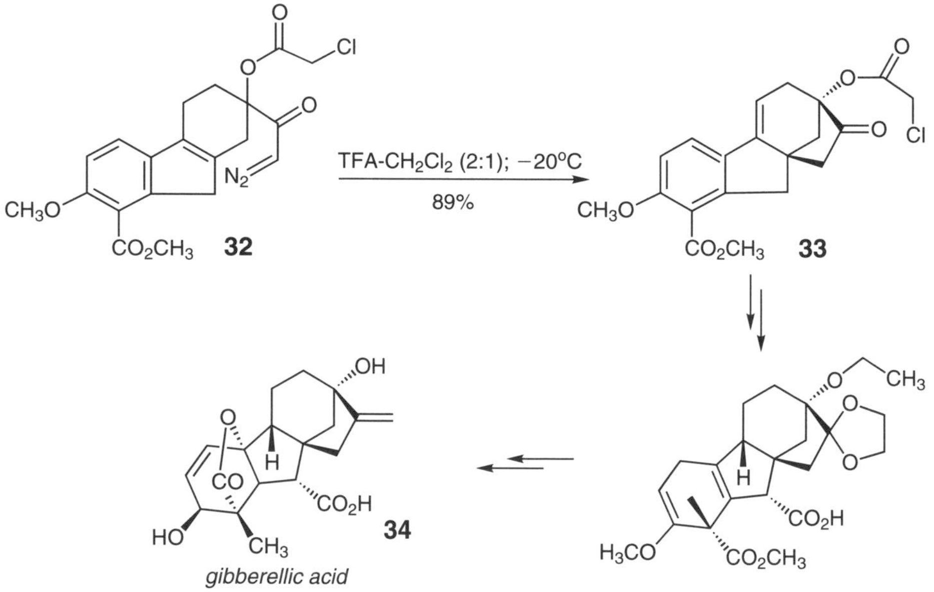

$$\text{32} \xrightarrow[\text{89\%}]{\text{TFA-CH}_2\text{Cl}_2 \text{ (2:1); } -20°C} \text{33}$$

Scheme 7

Angularly fused hydrophenanthrene and hydrofluorene derivatives have been prepared by intramolecular cyclization of the corresponding γ,δ-unsaturated α'-diazomethyl ketones.[17] Thus treatment of diazoketone **35** in dichloromethane with an aqueous $HClO_4$ (70%) and trifluoroacetic acid mixture or in dilute nitromethane solution with aqueous HBF_4 (48%) yielded the corresponding cyclopentanone **36** in good yield (eq. 4). Under the same reaction conditions decomposi-

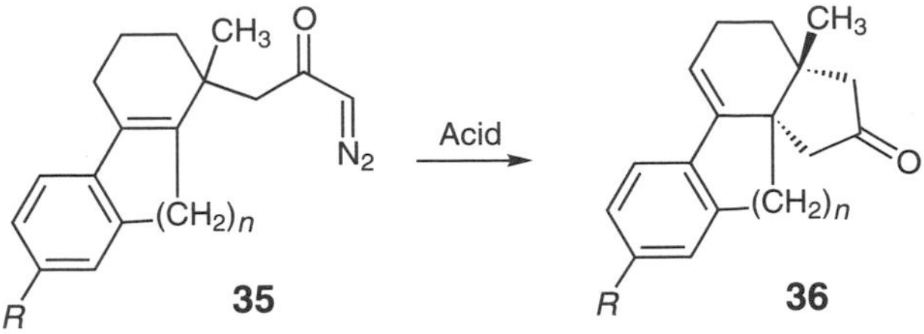

$$(4)$$

n = 2; R = H   70% HClO$_4$-TFA (1:4.5), CH$_2$Cl$_2$; 0°C   89%

n = 2; R = CH$_3$O   48% HBF$_4$, CH$_3$NO$_2$; 25–30°C   88%

n = 1; R = H   70% HClO$_4$-TFA (1:4.5), CH$_2$Cl$_2$; 0°C   95%

tion of diazoketone **37** bearing a less polarized γ,δ-double bond furnished cyclopentanone **38** in poor yield (eq. 5).[17]

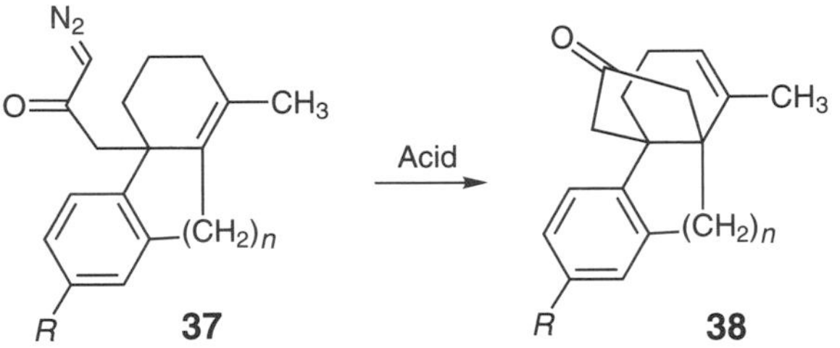

$$(5)$$

n = 1; R = H   70% HClO$_4$-TFA (1:4.5), CH$_2$Cl$_2$; 0°C   14%

n = 2; R = CH$_3$O   70% HClO$_4$-TFA (1:4.5), CH$_2$Cl$_2$; 0°C   22%

Brønsted acid–catalyzed hydrolysis of α-diazoketones is believed to proceed *via* preequilibrium protonation forming a diazonium ion, followed by rate-determining displacement of dinitrogen by a nucleophile. In general, a suitably situated olefinic bond can compete intramolecularly with a weak external nucleophile and form cyclic products. However, under certain circumstances, the substitution step involving displacement of dinitrogen is sufficiently fast that protonation of the diazoketone becomes the rate-determining step. Detailed investigation of olefin participation in the acid-catalyzed decomposition of some norbornane structures containing γ,δ-unsaturated α'-diazomethyl ketones have been carried out by Malherbe and Dahn.[18] Treatment of 5-*endo*-diazoacetyl-2-norbornene **39** with aqueous acid yielded the hydrolysis product **41** predominantly (73%) with minor amounts (27%) of the π-participation products **42**, **43**, and **44** (Scheme 8). In contrast, when the diazoacetyl group is at the 5-position of the norbornene structure, as in **40**, there is complete π-participation, leading to products **42**, **43**, and **44**. Furthermore, there is a large rate enhancement in the 7-isomer **39** as compared with the 5-isomer **40**, suggesting that the orientation of the putative oxodiazonium ions **45** and **46** is much more favorably disposed for steric reasons in the latter than the former (Scheme 8).

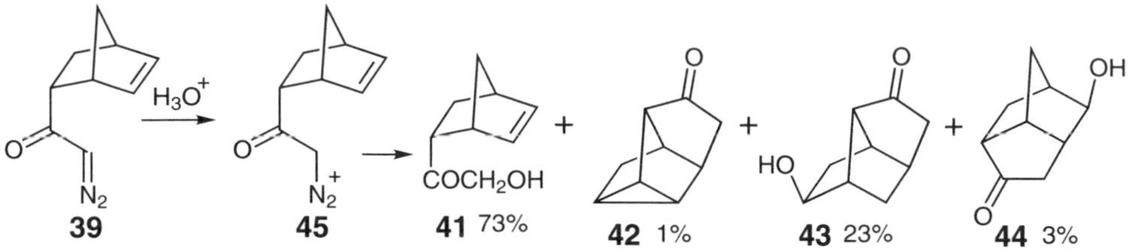

Scheme 8

Two examples that exploit the usefulness of π-route participation in the synthesis of small polycyclic hydrocarbons are shown in Schemes 9 and 10. In Scheme 9, acetic acid was used to promote the cyclization of diazoketone **47** to form a tricyclic structure (**48**) suitable for elaboration into semibullvalene **49**

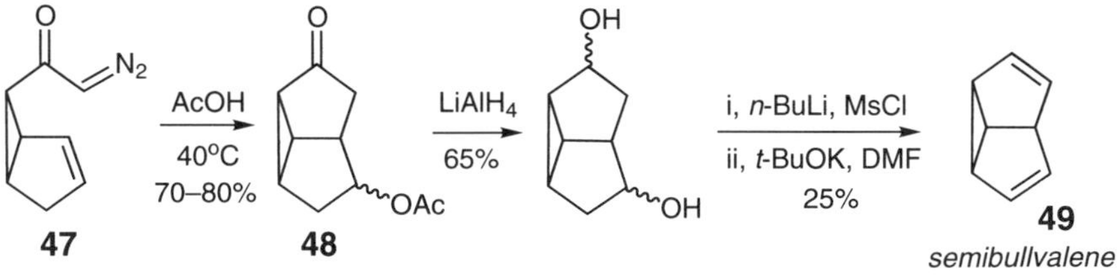

Scheme 9

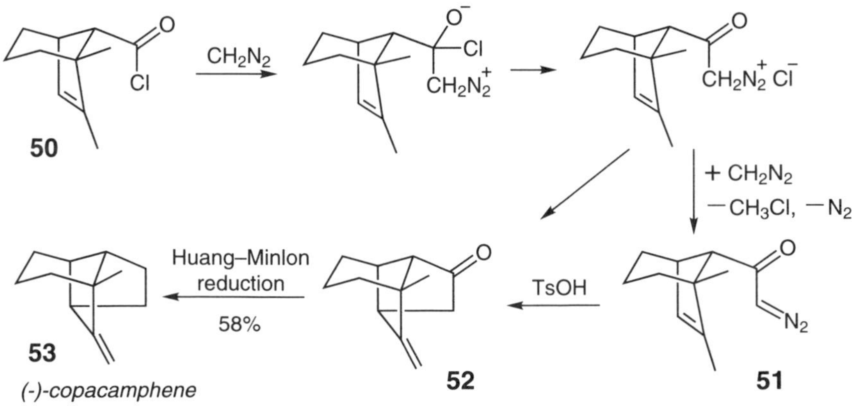

Scheme 10

(13% yield from diazoketone **47**).[19] In the second example (Scheme 10), in which acid chloride **50** was used as the precursor in a synthesis of $(-)$-copacamphene (**53**), reaction with diazomethane to generate the required diazoketone **51** led directly to the cyclization product **52** with only minor amount of **51**. Clearly the ease of π-participation in this case is such that it occurs in the diazoketone formation stage. Exposure of diazoketone **51** to toluenesulfonic acid produced the same result. To complete the synthesis, Huang–Minlon reduction of **52** afforded $(-)$-copacamphene (**53**).[20]

As illustrated in eq. 3, intramolecular acid–catalyzed cyclization of γ,δ-unsaturated diazoketones provides a general preparative route to tetracyclic gibbanes. The regioselectivity of acid-catalyzed decomposition of γ,δ-diazoketone of type **30** (eq. 3) can be attributed to the polarization in the benzylic π-bond. However, work by Ceccherelli and co-workers[21] has revealed an example where the acid-catalyzed cyclization of tricyclic α-diazoketones resulted in a mixture of the two possible regioisomeric tetracyclic ketones (eq. 6).[21] Thus treatment of pi-

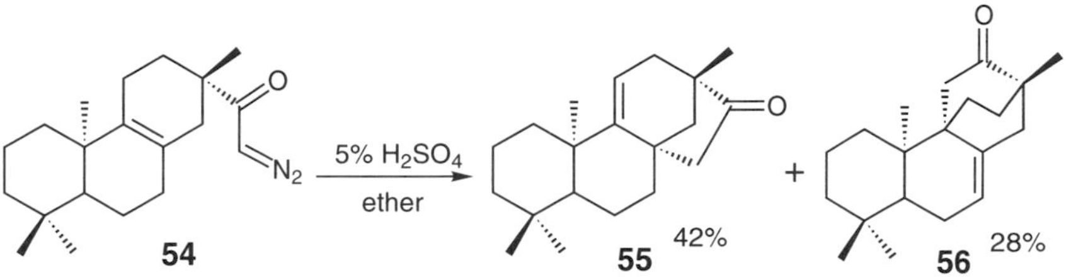

$$(6)$$

maric diazoketone **54** with 5% aqueous sulfuric acid afforded tetracyclic ketones **55** and **56** in 42% and 28% yield, respectively. The lack of regioselectivity in this cyclization reaction may be attributed to the absence of the polarization in the double bond. In contrast, decomposition of isopimaric diazoketone **57** under acid conditions afforded the six-membered ring derivative **58** exclusively (eq. 7).[21] Steric effects may also play a role in product orientation.

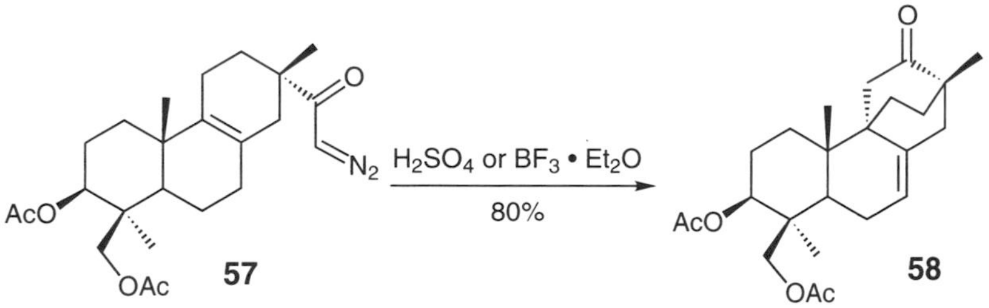

$$(7)$$

11.3 CYCLIZATION OF β,γ-UNSATURATED α-DIAZOKETONES

The acid-catalyzed cyclization of β,γ-unsaturated α-diazoketones have been investigated extensively by Smith[22–26] and Ghatak[29–34] and their respective co-workers. The π-bond participation in this acid-catalyzed cyclization process may result in either four- or five-membered ring formation (Scheme 11). Which ring size is preferred in any particular case appears to depend on both the structure of the diazoketone and on the cyclization conditions.

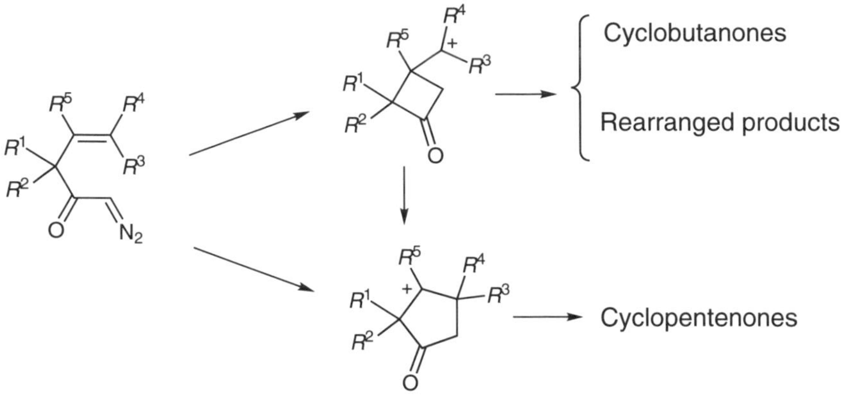

Scheme 11

Simple cyclopentenone derivatives **60** can be prepared in modest to good yield from acid-catalyzed cyclization of acyclic β,γ-unsaturated α-diazoketones of the type **59** in eq. 8.[23,24] The optimal conditions for cyclization consisted of treat-

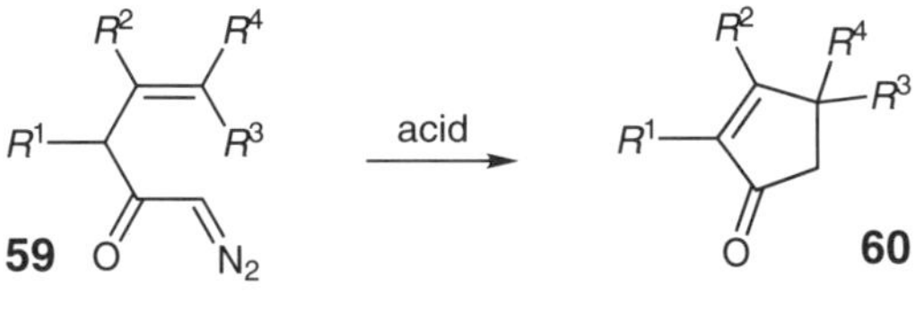

(8)

Diazoketone	Reaction condition	Yield
a: $R^1 = R^2 = R^3 = R^4 = H$	$BF_3 \cdot Et_2O$, CH_3NO_2	13%
b: $R^1 = R^3 = R^4 = H$, $R^2 = CH_3$	$BF_3 \cdot Et_2O$, CH_3NO_2	64%
b: $R^1 = R^3 = R^4 = H$, $R^2 = CH_3$	$BF_3 \cdot Et_2O$, CH_2Cl_2	73%
c: $R^1 = R^2 = CH_3$, $R^3 = R^4 = H$	$BF_3 \cdot Et_2O$, CH_3NO_2	40%
d: $R^1 = H$, $R^2 = R^3 = R^4 = CH_3$	$BF_3 \cdot Et_2O$, CH_2Cl_2	77%
e: $R^1 = C_5H_{11}$, $R^2 = CH_3$, $R^3 = R^4 = H$	$BF_3 \cdot Et_2O$, CH_3NO_2	65%

ment of the diazoketone with $BF_3 \cdot Et_2O$ in either nitromethane or dichloromethane. Reaction yields were influenced by the degree of substitution of the participating double bond. In entry **a**, where the double bond is monosubstituted, the yield is low compared with entries where there is a methyl substituent on the β-carbon atom, and yields are much higher. These trends are consistent with the formation of carbocation intermediates. This type of cyclization has been used in a synthesis of *cis*-jasmone **63** (Scheme 12). Following formation of the cyclopen-

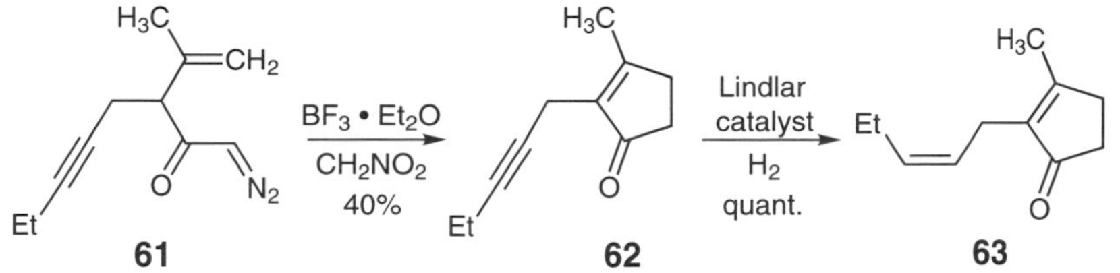

Scheme 12

tenone intermediate **62**, the *cis*-double bond was introduced by partial hydrogenation of the alkyne side chain.[23]

In all the cyclizations cited above, participation of the double bond occurred through the γ-carbon atom to afford a five-membered ring, although as we will see below, this is not always the case. Acid-catalyzed cyclization of diazoketone **64** afforded the strained cyclobutanone derivative **66** as the major product along with two cyclopentenone derivatives **68** and **70** (Scheme 13). The cyclobutanone

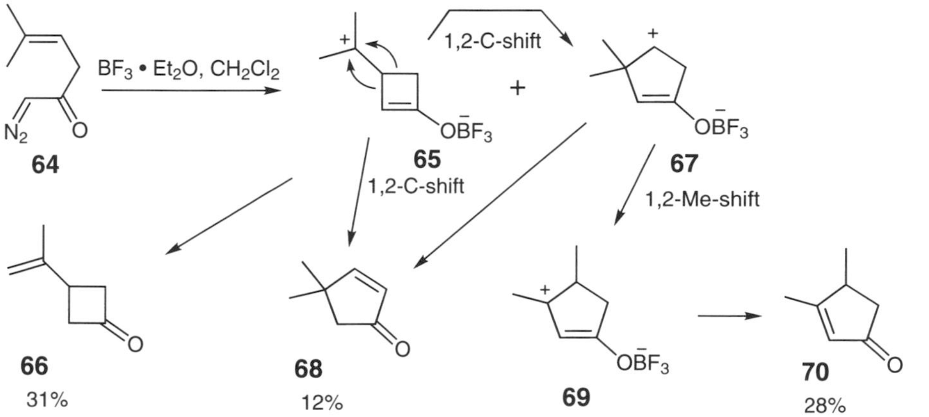

Scheme 13

derivative **66** and cyclopentenone derivatives **68** arise *via* the initial formation of a four-membered ring bearing an *exocyclic* tertiary carbocation (**65**) and a five-membered ring with an *endocyclic* secondary carbocation (**67**), respectively. A 1,2-methyl shift within the latter intermediate generates a more stable tertiary cation (**69**), from which a proton is eliminated to yield cyclopentenone derivative **70** as shown in Scheme 13.[24]

Hudlicky and Kutchan[27] have demonstrated that the bicyclic ketone **73** can be synthesized by an intramolecular acid-catalyzed alkylation of the dienic diazoketone **71** *via* initial formation of intermediate **72**. Treatment of bicyclic ketone **73** with $RhCl_3$ in aqueous ethanol yielded monoterpenoid (±)-filifolone **74** in 30–40% yield (Scheme 14).

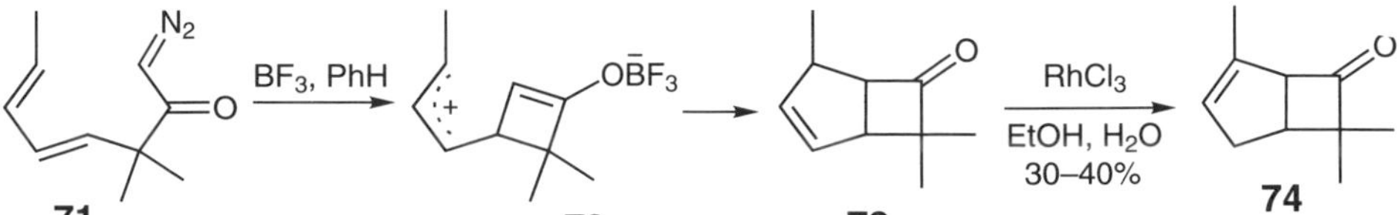

71 **72** **73** **74** (±)-*filifolone*

Scheme 14

When the β,γ-unsaturated double bond is already part of a carbocyclic system, the acid-catalyzed cyclization of monocyclic α-diazoketones affords bicyclic ketones. Some representative examples of bicyclic ketone formation from Smith's study[22,24] are shown in Schemes 15 and 16 and eq. 9. Scheme 15 demonstrates

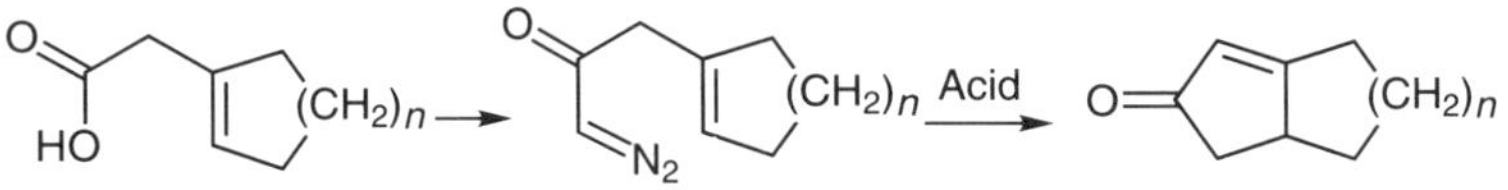

76 **77** **78**

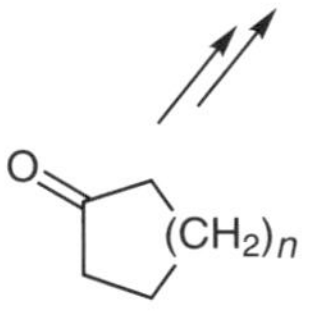

75

Diazoketones	Reaction conditions	Yield
77a: $n = 1$	$BF_3 \cdot Et_2O$, CH_2Cl_2; 0°C	50%
77b: $n = 2$	$BF_3 \cdot Et_2O$, CH_3NO_2; 0°C	50%
77c: $n = 3$	$BF_3 \cdot Et_2O$, CH_2Cl_2; 0°C	61%

Scheme 15

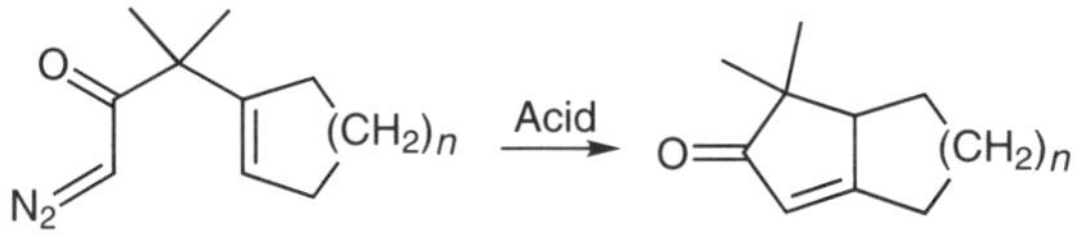

79 **80** (9)

Diazoketones	Reaction conditions	Yield
79a: $n = 1$	$BF_3 \cdot Et_2O$, CH_2Cl_2; 0°C	41%
79b: $n = 2$	$BF_3 \cdot Et_2O$, CH_3NO_2; 0°C	65%
79c: $n = 3$	$BF_3 \cdot Et_2O$, CH_2Cl_2; 0°C	68%

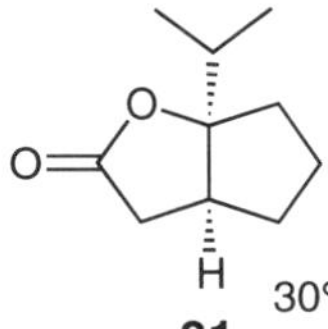

81 30%

that the acid-catalyzed decomposition of β,γ-unsaturated diazoketone (**77**), in conjunction with the numerous approaches to β,γ-unsaturated acid derivatives (i.e., **75** to **76**), provides a general cyclopentenone annulation strategy. In general, the yields of annulated cyclopentenones were quite good. Decomposition of α',α'-dimethyl β,γ-unsaturated α-diazoketone **79** also afforded the expected annulated cyclopentenones (**80**) (eq. 9). In addition to the annulated cyclopentenone, diazoketone **79a** yielded lactone **81** in 30% yield. The formation of lactone **81** arose from a ketene, which was derived from an intermediate tertiary carbocation.

The decomposition of diazoketones **82** and **83** afforded the same pair of bicyclic enones, **86** and **87** (Scheme 16). The diazo carbon of **82** attacks the olefinic

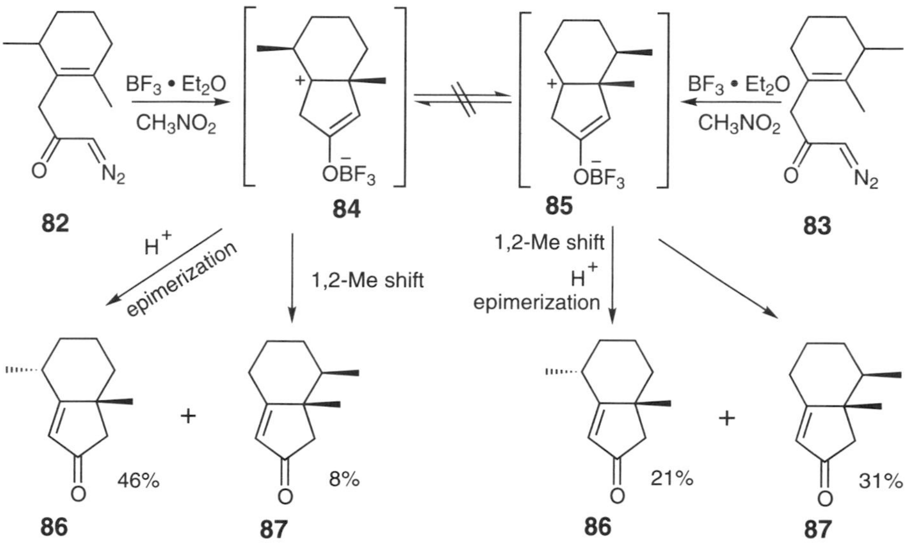

Scheme 16

linkage *anti* to the secondary methyl group and produces the intermediate **84**; whereas the nonbonded interaction between the vincinal methyl substituents of **83** accounted for formation of intermediate **85**. Compound **86**, the more stable *trans* isomer, was formed through the acid-promoted epimerization of the chiral center γ to the carbonyl functional group. Although in both cases the combined yield of **86** and **87** was identical, the ratio of **86:87** differed significantly; therefore, equilibration between intermediates **84** and **85** is apparently not involved.

Finally, it should be noted that under certain circumstances, copper(II) triflate, can act as a Lewis acid in π-route cyclization of alkenes. Doyle and Trudell[28] have shown that the diazoketone **77b** undergoes intramolecular cyclization to produce the corresponding cyclopentenone **78b** in higher yield with copper(II) triflate catalysis (59%) than with the use of boron trifluoride etherate (50%).

Ghatak and co-workers[29] found that acid-catalyzed decomposition of α-diazoketone **88** afforded mainly two isomeric ketones **93** and **94** (Scheme 17). The distribution of the various products, including the cyclobutanone (**90**) in this cy-

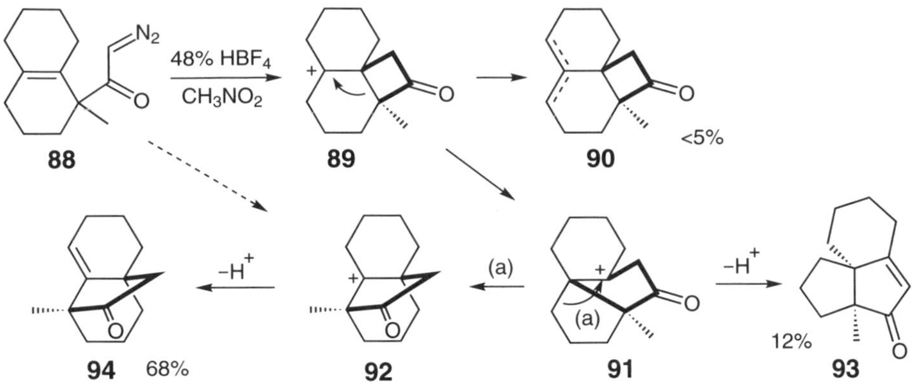

Scheme 17

clization reaction, depends upon the reactant concentration and reaction time. The authors believe that both **93** and **94** arise from the initial formation of a four-membered ring intermediate bearing a tertiary carbocation (**89**) *via* **91** and **92**, as shown in Scheme 17. However, in this particular cyclization reaction, the possibility of the formation of **94** through the direct cyclization–proton elimination sequence cannot be ruled out.

A highly efficient synthesis of angularly fused cyclobutanones based on the acid-catalyzed intramolecular cyclization of β,γ-unsaturated α-diazoketones has been introduced by Ghatak and co-workers.[29-36] Diazoketones **95** furnishes the cyclobutanones **97** on treatment with either perchloric or fluoroboric acid in benzene or chloroform (Scheme 18). In the presence of less polar solvents, the cyclization proceeds through the benzyl cation (**96**), which on deprotonation leads to the respective strained unsaturated cyclobutanones (**97**). The readily available cyclized products (**97**) can be easily transformed into the corresponding cyclopentanones (**98**) by a two-step sequence involving a remarkable stereospecific rearrangement, as shown in Scheme 18.

After an extensive investigation of the acid-catalyzed intramolecular cyclization of rigid polycyclic β,γ-unsaturated α-diazoketones, Ghatak and co-workers[34-36] concluded that it was not the structure of the substrate alone that controls the nature of the products in such a cyclization reaction but that the choice of the acid catalyst and the solvent are also critical. In cyclization of diazoketone **95** with aqueous fluoroboric acid (48%) or boron trifluoride etherate in strongly polar nitromethane, the strained cyclobutyl cation **96** undergoes bond rearrangement through the cyclopentyl cation **99**, leading ultimately to the stable tertiary benzylic cation **100**. This on subsequent attack of water results in the respective bridged hydroxyketone **101** (Scheme 19).[34-36] Alternatively, diazoketone **95** on brief treatment with an excess of p-TsOH in boiling benzene, generated the strained cyclobutanone cation **96** which underwent facile rearrangement to afford the cyclopentanone cation **99**, leading ultimately to the formation of cyclopentenone **102** (Scheme 20).[35,36]

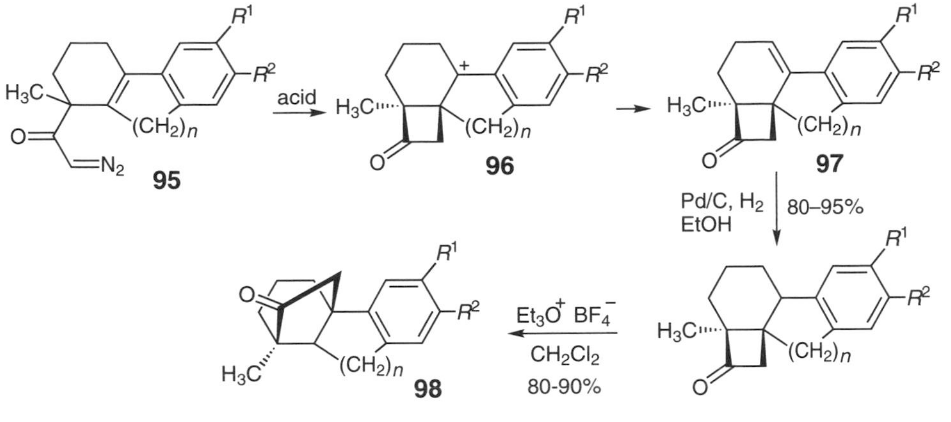

Scheme 18

Diazoketone (**95**)	Reaction conditions	Yield (**97**)	Ref.
a: $n = 2$; $R^1 = R^2 = R^3 = H$	70% HClO$_4$, CHCl$_3$; 5°C	80–99%	30
b: $n = 1$; $R^1 = R^2 = R^3 = H$	TFA-HClO$_4$, CHCl$_3$; 0°C	84%	34
c: $n = 2$; $R^1 = R^3 = H$, $R^2 = CH_3O$	48% HBF$_4$, C$_6$H$_6$; 0°C	92%	35
d: $n = 1$; $R^1 = R^3 = H$, $R^2 = CH_3O$	48% HBF$_4$, CHCl$_3$; 0°C	73%	34
e: $n = 2$; $R^1 = CH_3O$, $R^2 = R^3 = H$	70% HClO$_4$, CHCl$_3$; 0°C	88%	32
f: $n = 1$; $R^1 = CH_3O$, $R^2 = R^3 = H$	TFA, CHCl$_3$; r.t.	78%	36

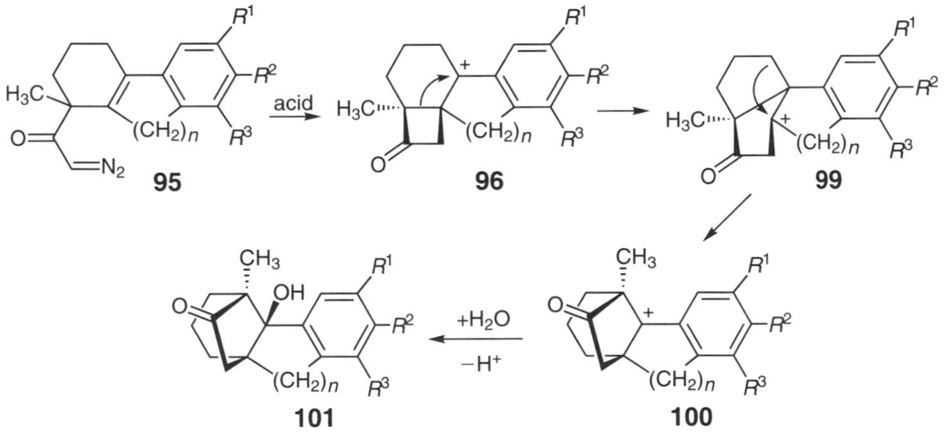

Diazoketone (**95**)	Reaction conditions	Yield (**101**)	Ref.
a: $n = 2$; $R^1 = R^2 = R^3 = H$	48% HBF$_4$, CH$_3$NO$_2$; 25°C	64%	34
c: $n = 2$; $R^1 = R^3 = H$, $R^2 = CH_3O$	H$_2$SO$_4$, CH$_3$NO$_2$; 0°C	79%	35
e: $n = 1$; $R^2 = R^3 = H$, $R^1 = CH_3O$	48% HBF$_4$, CH$_3$NO$_2$; 25°C	91%	36
e: $n = 1$; $R^1 = R^2 = H$, $R^3 = CH_3O$	48% HBF$_4$, CH$_3$NO$_2$; 25°C	91%	36

Scheme 19

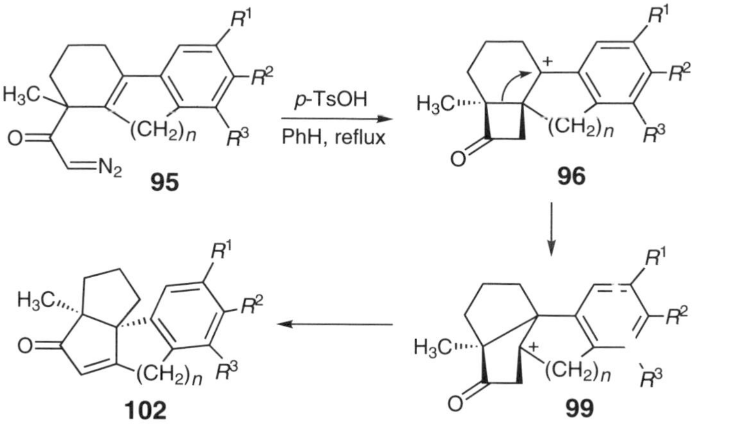

a: $n = 1$; $R^1 = R^2 = R^3 = $H (75%) **b**: $n = 2$; $R^1 = R^3 = $H, $R^2 = CH_3O$ (75%)
c: $n = 1$; $R^2 = R^3 = $H, $R^1 = CH_3O$ (81%) **d**: $n = 2$; $R^2 = R^3 = $H, $R^1 = CH_3O$ (80%)

Scheme 20

This approach to fused-ring cyclobutanone formation has been exploited by Ceccherelli et al.[37] for the preparation of the D-norsteroids **105** as illustrated in Scheme 21. In this case, the regioselective formation of cyclobutanones arises mainly through the initial formation of a four-membered ring intermediate bearing a tertiary carbocation (**104**).

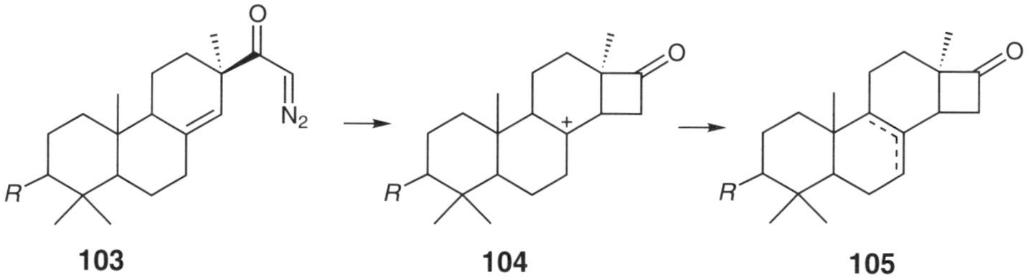

Diazoketone (**103**)	Reaction condition	Yield (**105**)
a: $R = $H	Silica gel, benzene; 25°C	>36%
b: $R = $AcO	H_2SO_4, ether; 25°C	53%

Scheme 21

Further aspects of β,γ-unsaturated α-diazocarbonyl cyclizations are worthy of note in systems where more than one olefinic group participates and in systems in which the participating π-electrons are supplied by an aromatic ring. Smith and co-workers[24-26] have provided several examples of such cyclizations of α-diazoketones. Four typical examples are shown in eqs. 10,[24] 11,[24] 12,[25] and

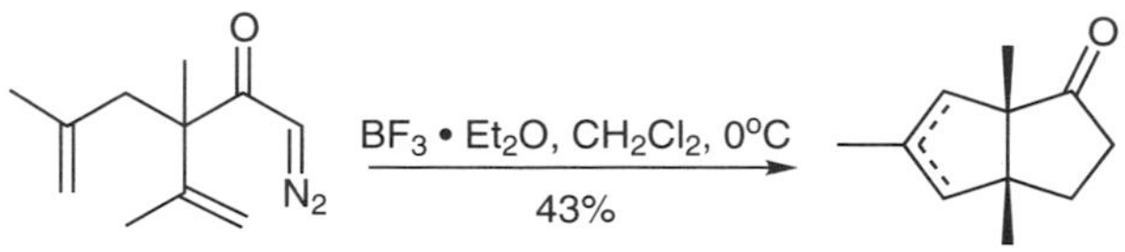
O
N2
BF3 • Et2O, CH2Cl2, 0°C
43%
O

(10)

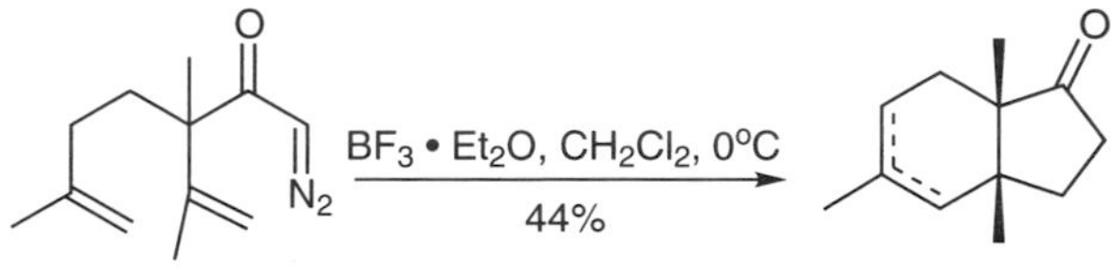
O
N2
BF3 • Et2O, CH2Cl2, 0°C
44%
O

(11)

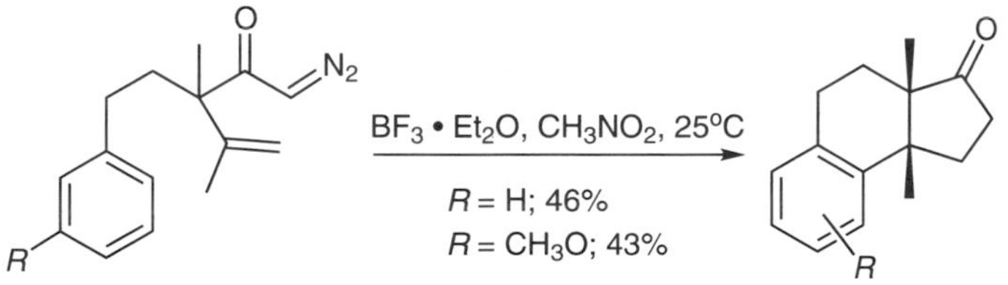
O
N2
R
BF3 • Et2O, CH3NO2, 25°C
R = H; 46%
R = CH3O; 43%
O
R

(12)

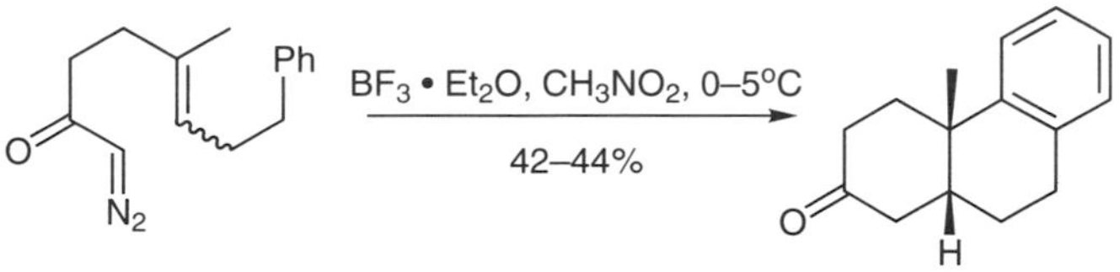
Ph
O
N2
BF3 • Et2O, CH3NO2, 0–5°C
42–44%
O
H

(13)

13.[25] The optimal combination of acid and solvent proved to be boron trifluoride etherate in nitromethane or dichloromethane.

11.4 INTRAMOLECULAR ALKYLATION OF α-DIAZOKETONES THROUGH ARYL PARTICIPATION

Aryl participation in acid-catalyzed decomposition of diazoketones with suitably positioned aromatic substituents has been known for many years. Early examples include the synthesis of 2-chrysenol **107** from the diazoketone **106** (eq. 14)[38] and

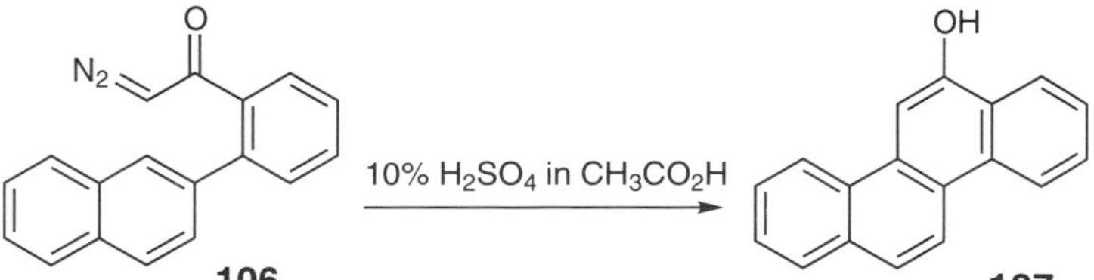
O
N2
10% H2SO4 in CH3CO2H
OH

106 **107**

(14)

of the tricyclic ketone **108** in eq. 15,[39] both reactions having been catalyzed by sulfuric acid in acetic acid.

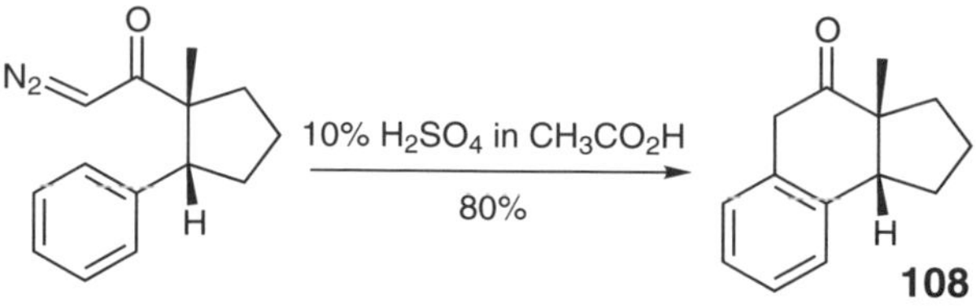

$$\text{10\% H}_2\text{SO}_4 \text{ in CH}_3\text{CO}_2\text{H} \quad 80\% \tag{15}$$

Numerous studies of π-route participation in aromatic precursors with a hydroxy or methoxy substituent confirm that this has become a very versatile route to bridged-ring and fused-ring carbocyclic systems. Cyclization products arise from either direct aryl participation or from a cyclized intermediate, which undergoes rearrangement prior to the product-forming step. In any particular substrate these processes depend both on the nature of the substituents on the aromatic ring and on the number of carbons separating the diazo function from the aromatic ring. It is convenient to adopt the notation of Heck and Winstein,[40] introduced originally to classify types of aryl participation in solvolytic reactions of tosylates and brosylates. The general notation is Ar_a-n, where Ar denotes aryl and the subscript a refers to the position in the participating aryl group involved in the creation of the new ring in the transition state. The size of the ring being made is indicated by the number n. On this basis the processes designated by Ar_1-4 and Ar_1-5 in diazocarbonyl cyclizations can be summarized by eqs. 16 and 17, respectively.

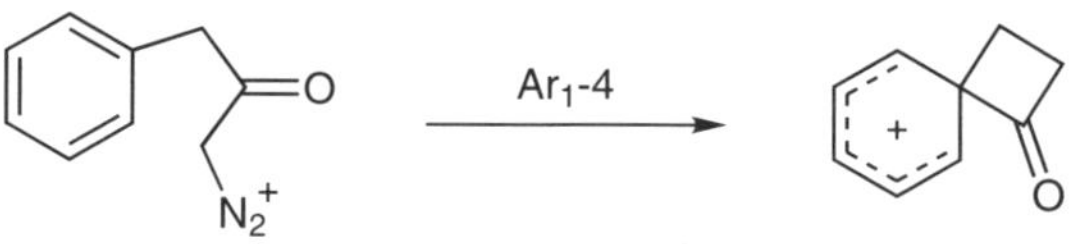

$$\text{Ar}_1\text{-4} \tag{16}$$

$$\text{Ar}_1\text{-5} \tag{17}$$

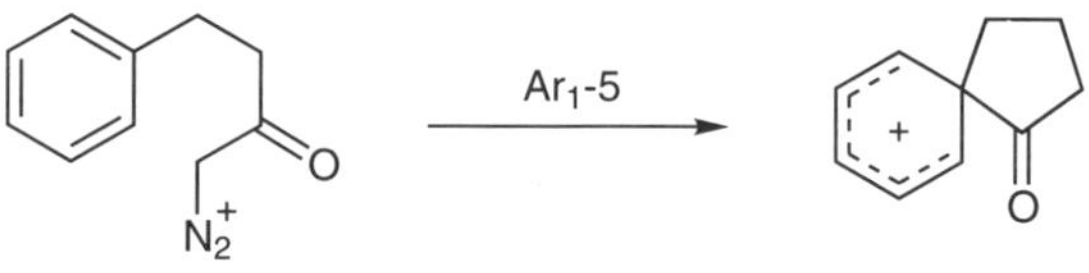

11.4.1   Ar$_1$-5 Participation in the Cyclization of Aromatic α-DiazoKetones

The acid-catalyzed cyclization of α-diazoketones bearing an aromatic function attached to a β carbon has been investigated extensively. These cyclization processes proceeded mainly *via* the Ar_1-5 pathway, depending on both the nature of the substituents, generally oxygen based, and the positions on the aromatic ring. One of the early examples of Ar_1-5 participation involving acid-catalyzed decomposition of a diazoketone is shown in eq. 18.[4] In addition, boron trifluoride–catalyzed decomposition of the simple phenolic diazoketone **109** furnished dienedione **110** in 67% yield (eq. 19).[41]

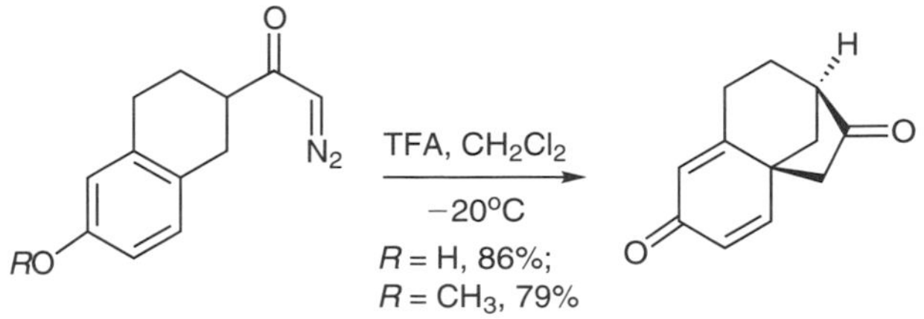

$$\text{(diagram)} \xrightarrow[\substack{-20°C \\ R = H, 86\%; \\ R = CH_3, 79\%}]{\text{TFA, CH}_2\text{Cl}_2} \text{(diagram)} \qquad (18)$$

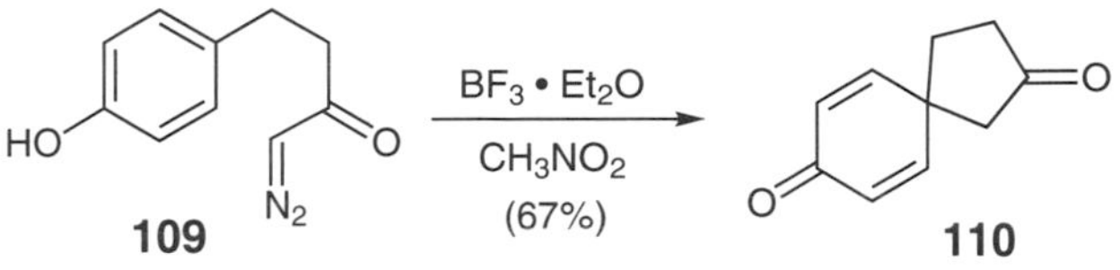

$$\underset{\textbf{109}}{\text{(diagram)}} \xrightarrow[\substack{\text{CH}_3\text{NO}_2 \\ (67\%)}]{\text{BF}_3 \cdot \text{Et}_2\text{O}} \underset{\textbf{110}}{\text{(diagram)}} \qquad (19)$$

Simple, general preparative routes to bicyclo[3.2.1]octane carbon skeleton **112** and **114** based on Ar$_1$-5 participation in acid-catalyzed cyclization of α-diazoketones **111**[42–44] and **113**[43,45,46] have been developed by Mander and co-workers. Representative examples are illustrated in eqs. 20 and 21. The optimal reaction

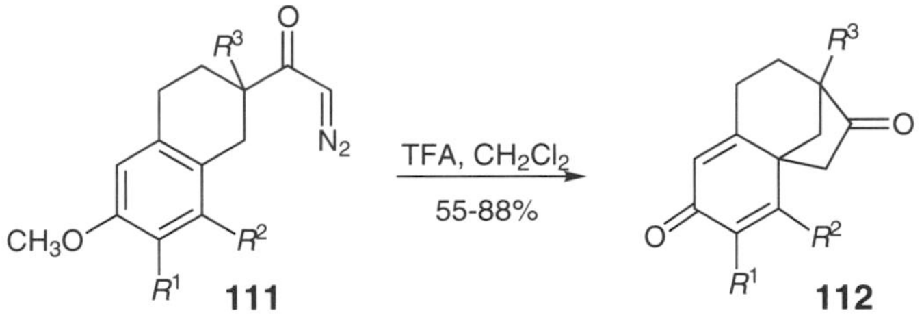

$$\underset{\textbf{111}}{\text{(diagram)}} \xrightarrow[55\text{-}88\%]{\text{TFA, CH}_2\text{Cl}_2} \underset{\textbf{112}}{\text{(diagram)}} \qquad (20)$$

$R^1 = R^2 = H, R^3 = OCOCCl_3; \; R^1 = R^2 = H, R^3 = OCOCHCl_2$
$R^1 = R^3 = H, R^2 = CH_3O; \; R^1 = R^2 = CH_3O, R^3 = H$

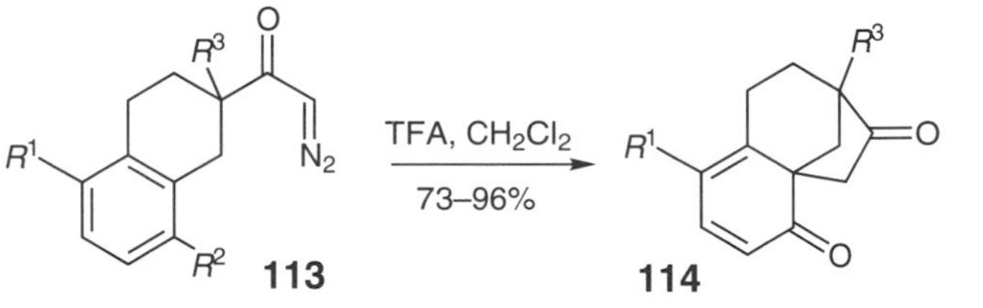

$$\underset{\textbf{113}}{\text{(diagram)}} \xrightarrow[73\text{-}96\%]{\text{TFA, CH}_2\text{Cl}_2} \underset{\textbf{114}}{\text{(diagram)}} \qquad (21)$$

$R^1 = R^3 = H, R^2 = OH; \; R^1 = (CO_2CH_3)_2CH(CH_2)_3, \; R^2 = CH_3O, \; R^3 = H$
$R^1 = H, R^2 = CH_3O, R^3 = OCOCCl_3; \; R^1 = H, R^2 = CH_3O, R^3 = OCOCHCl_2$

conditions proved to be trifluoroacetic acid in dichloromethane at low temperature. In general, the yields of these Ar$_1$-5 participations were good to excellent. The dienedione **112** prepared from **111**, as shown in eq. 20,[44] served as the starting material for the synthesis of 15-desoxyeffusin **115** (Scheme 22).[47] The cy-

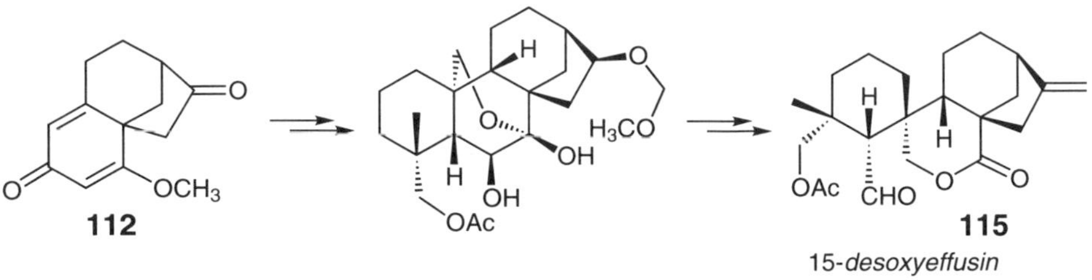

112

115

15-*desoxyeffusin*

Scheme 22

clization of diazoketone **113** to **114** proved to be the pivotal transformation in the synthetic studies towards gibbane derivatives[46,48–50]; for example, dienedione **114** was the key intermediate in Mander's elegant synthesis of (±)-gibberellin A$_1$ (**116**)[48] and gibberellic acid (**117**) (Scheme 23).[48]

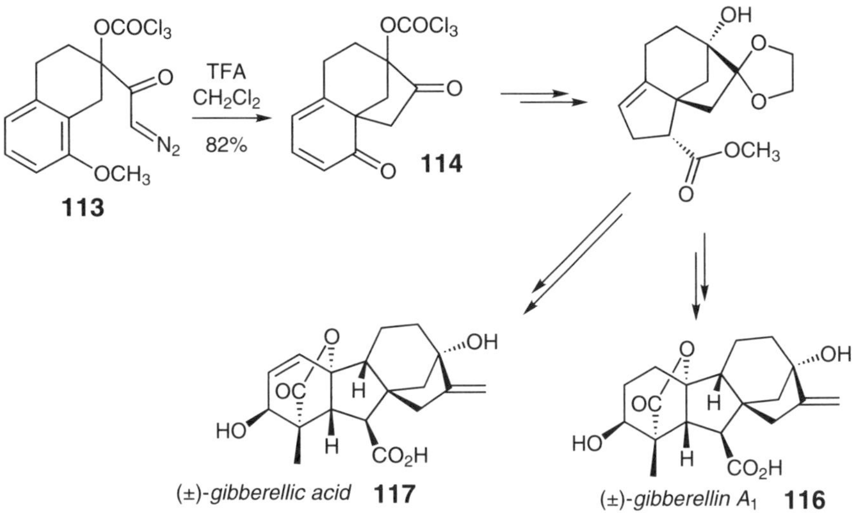

113

114

(±)-*gibberellic acid* **117**

(±)-*gibberellin A$_1$* **116**

Scheme 23

Mukherjee and co-workers[51,52] have also used the Ar$_1$-5 cyclization to synthesize tricyclic dienones **118**[51] and **119**[52] related to diterpenes. The transformations are shown in eqs. 22 and 23. Trifluoroacetic acid in dichloromethane was employed to catalyze the diazo decomposition process, and the yields were modest.

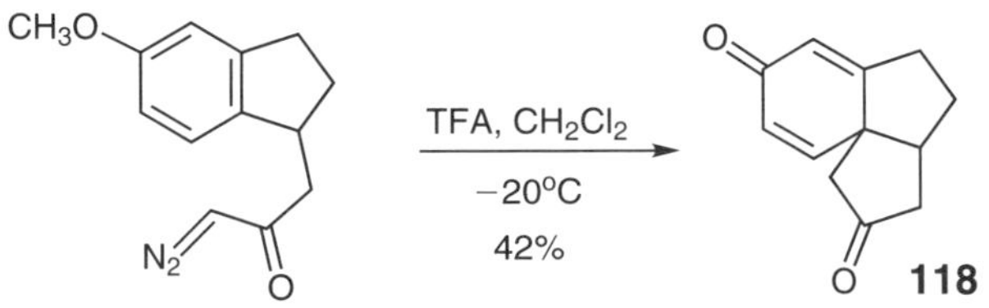

$$\text{TFA, CH}_2\text{Cl}_2$$
$$-20°\text{C}$$
$$42\%$$

118

(22)

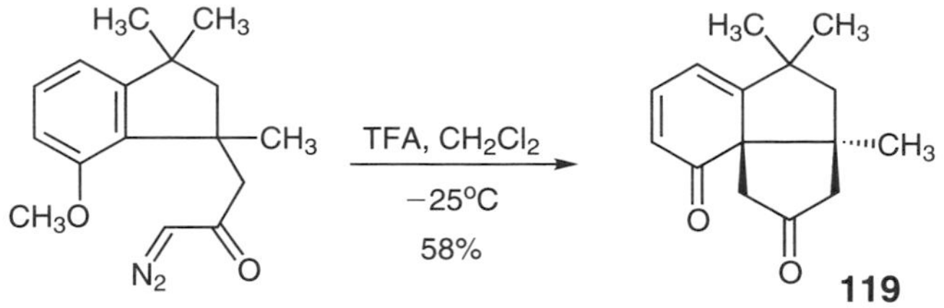

(23)

The distribution of the various products, including the enedione (**122**) and iso-meric phenanthrenones (**123, 124**) in the cyclization of diazoketone **120**, depends upon the reaction temperature. All the products arose from the initial formation of intermediate **121** *via* Ar$_1$-5 participation, followed by subsequent rearrange-ments. Reaction of **120** at 0°C yielded of a 9:1 mixture of **124:123**, whereas the reaction proceeded at −20°C with rapid quenching gave a 9:1 mixture of **122:123** (Scheme 24).[53] The temperature dependence of the trifluoroacetic acid–

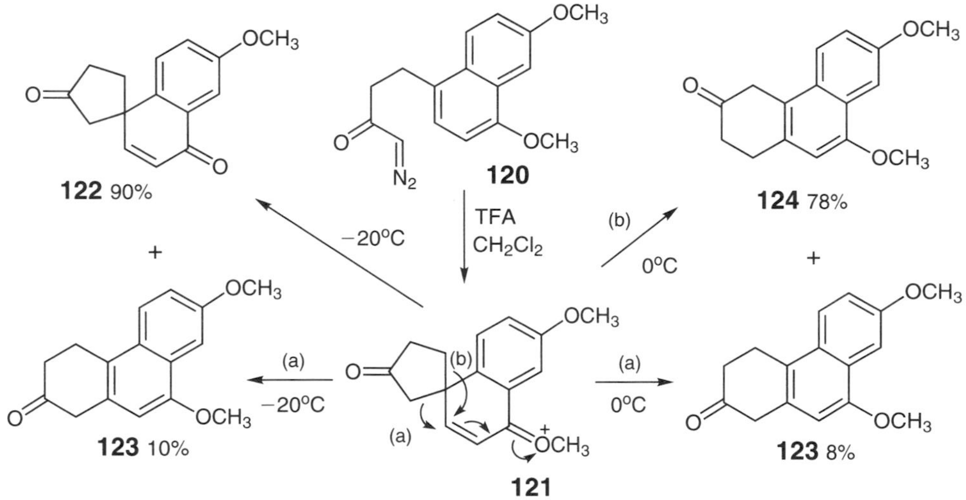

Scheme 24

catalyzed Ar$_1$-5 cyclization of α-diazoketones is not uncommon; an additional example is depicted in Scheme 25.[54]

Although the Ar$_1$-5 cyclization is favored over the Ar$_2$-6 mode, the latter cy-clization has also been used in synthesis. Acid-catalyzed decomposition of dia-zoketones **131** and **132** through Ar$_2$-6 cyclization has been used as the key step in the total synthesis of (±)-chelidonine **137**[55] and (±)-corynoline **138**.[56] Thus treat-ment of diazoketones **131** and **132** with trifluoroacetic acid afforded the ketones **133** and **134** in 19% and 42% yield, respectively (Scheme 26). The low yield of the A$_2$-6 cyclization step was due to the competing Ar$_1$-5 cyclization followed by

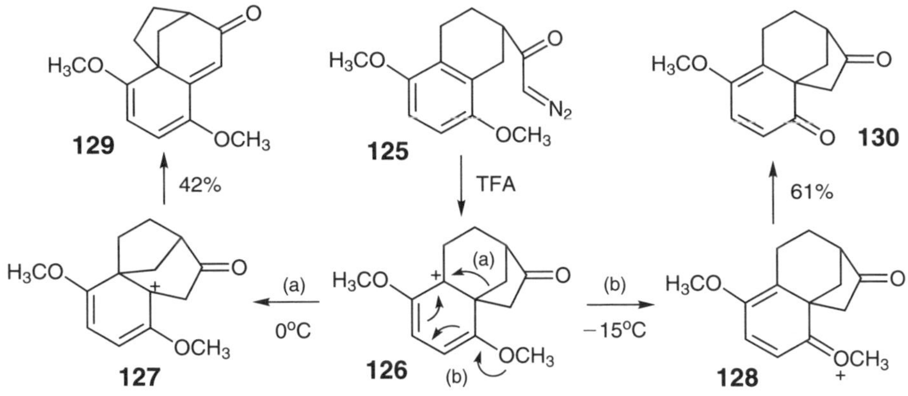

Scheme 25

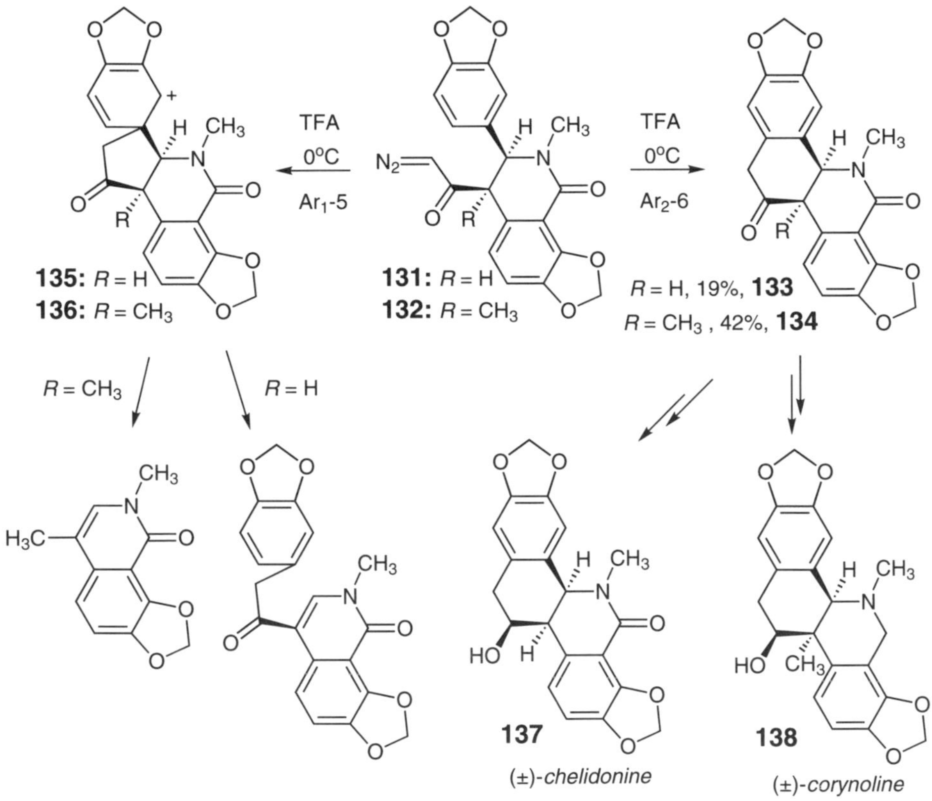

Scheme 26

fragmentation or Hayashi rearrangement of the spirocyclic cation intermediates **135** and **136**.[56]

11.4.2 Ar₁-6 Participation in the Cyclization of Aromatic α-Diazo Ketones

Clearly, an electron-donating substituent located ortho or para to the site of attack on the aryl ring will promote an Ar_1-6 cyclization. The bicyclo[2.2.2]octane carbon skeleton can be obtained through this type of cyclization, and some examples are shown in eqs. 24[42,45] and 25.[57,58] Similar examples have been reported

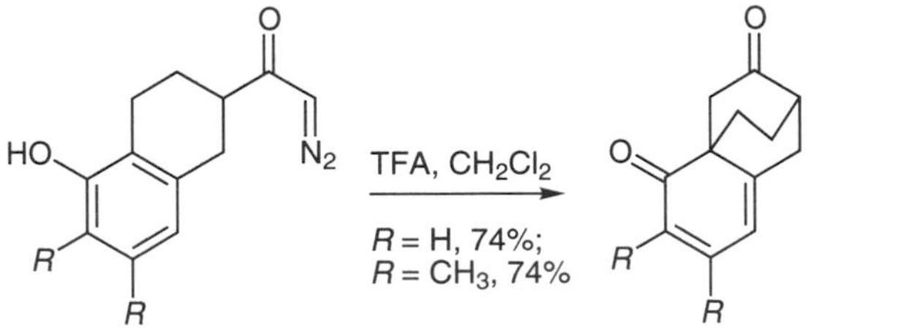

$$(24)$$

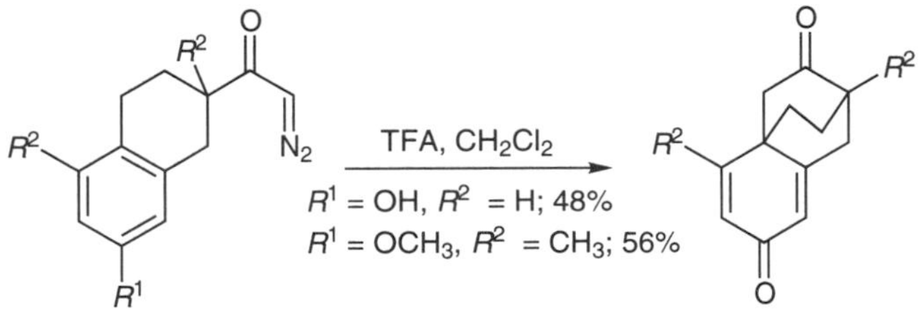

$$(25)$$

by Mukherjee and co-workers.[59-61] It should be noted that the cyclization of diazoketone **139** proceeded through both Ar_1-5 and Ar_1-6 pathways. In this case, steric and electronic factors explain the formation of dienones **140** and **141** in comparable amounts (eq. 26).[42]

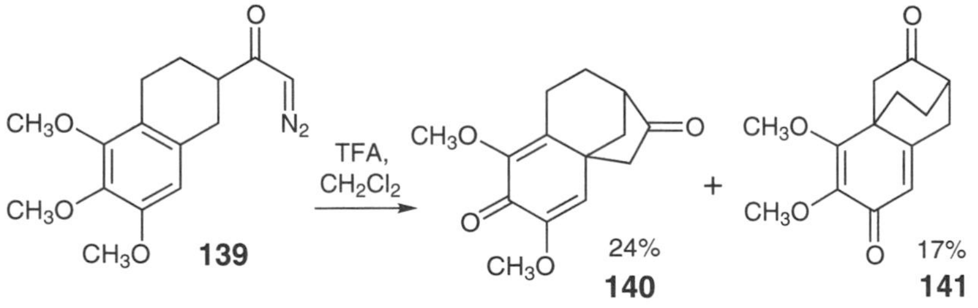

$$(26)$$

The ring system of a few diterpenes, including the bicyclo[3.2.1]octane carbon skeleton, can be synthesized through the Ar_1-6 cyclization of diazoketones.[52,60,61] Nicolaou and Zipkin[62] reported an expeditious stereocontrolled and highly efficient construction of the ring systems (**144** and **145**) of aphidicolin and related natural products involving Ar_1-6 cyclization of diazoketone **142** as the key step. Brief exposure of dienone **143** to butadiene and 1-acetoxybutadiene in CH_2Cl_2 at

0°C in the presence at SnCl$_4$ produced the tetracycles **144** and **145** in 97% and 90% yield, respectively (Scheme 27).

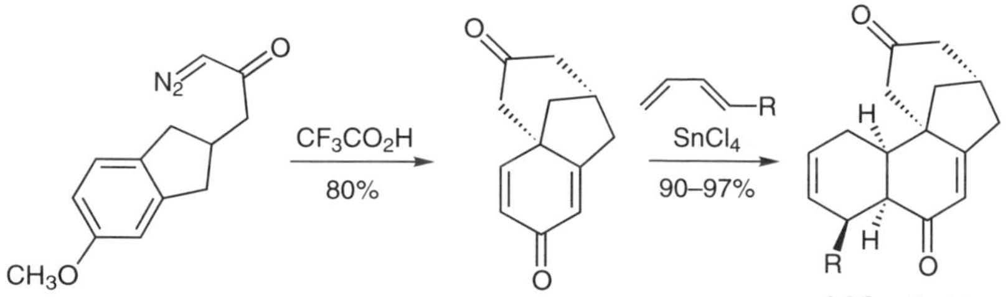

Scheme 27

11.4.3 Ar$_1$-4 Cyclization in the Cyclization of Aromatic α-Diazoketones

Two possible pathways, Ar$_1$-4 and Ar$_2$-5, could be involved in the acid-catalyzed cyclization of α-diazoketones bearing an aromatic function attached to an α carbon. Both pathways are highly dependent upon the position of the electron-donating substitutent on the aromatic ring. In general, the cyclobutane intermediate arising from the Ar$_1$-4 cyclization undergoes subsequent rearrangement to give a variety of products. Mander et al.[41] first demonstrated that an Ar$_1$-4 participation using phenolic diazoketone **146** gave the highly labile spirodienone **147** in low isolated yield. Spirodienone **147** underwent a rapid and quantitative dienone–phenol rearrangement to yield indanone **148** (Scheme 28).

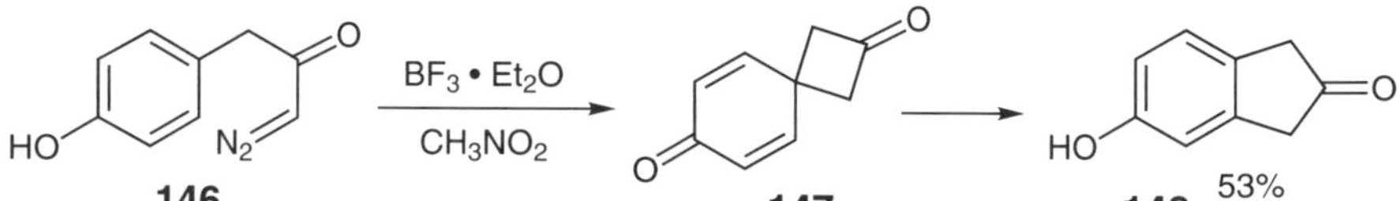

Scheme 28

Mander and co-workers[57] extended their investigation of the acid-catalyzed cyclization of the phenolic diazoketones to an examination of the reaction of a series of tetrahydronaphthyl diazoketones. Thus treatment of diazoketones **149**[57] and **150**[42] afforded tricyclic derivatives **157** and **158** in 58% and 85% yield, respectively. In both cases, the cyclization products derived from the initial formation of Ar$_1$-4 intermediates (**151** and **152**) with manifold rearrangement, as depicted in Scheme 29. A further interesting example of the Ar$_1$-4 participation/rearrangement manifold is provided by acid-catalyzed decomposition of diazoketone **159**.[63] Treatment of **159** with trifluoroacetic acid afforded a 4:1 mixture of

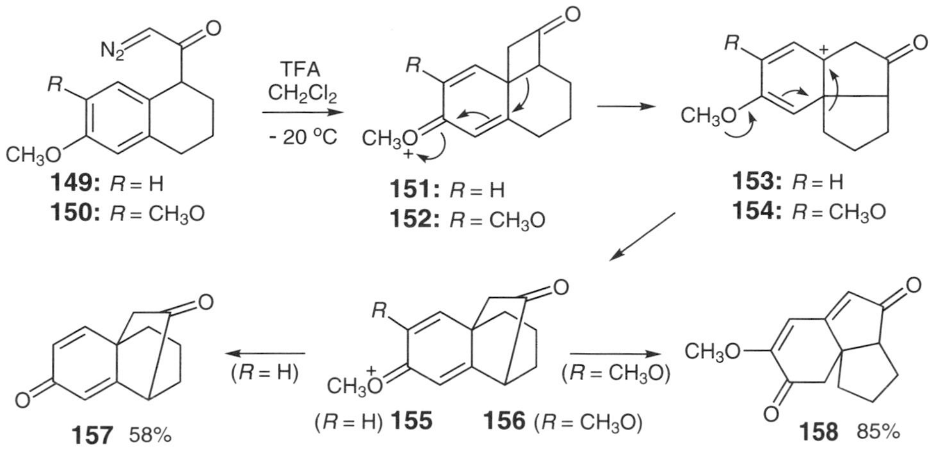

Scheme 29

162 and **163**, respectively. The preferred formation of **159** by cyclization onto the less electron-rich ring was because the Ar$_1$-4 participation reaction is favored over the Ar$_2$-5 cyclization. The intermediate **160** would be more effectively stabilized than **161** and would lead to **162** as a major product. It also becomes apparent that **163** arises *via* **161** rather than by a direct Ar$_2$-5 cyclization process (Scheme 30).

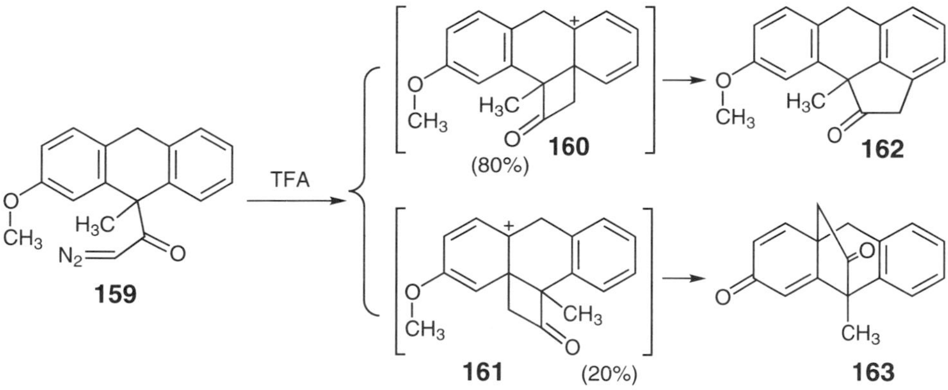

Scheme 30

11.5 SYNTHESIS OF HETEROCYCLES BY ACID-PROMOTED CYCLIZATION OF α-DIAZOCARBONYL COMPOUNDS

Several types of heterocycles are accessible by acid-catalyzed cyclization of α-diazocarbonyl precursors. These reactions provide products similar to those obtained in some rhodium(II)-catalyzed processes (see Chapter 6). For example,

Doyle and co-workers[64] found that the solid-phase sulfonic acid resin Nafion NR50® catalyzed the cyclization of diazoacetamides of the type **165** shown in eq. 27, as did rhodium(II) acetate, to form 2(3H)-indolinone derivatives. Similar cyclizations of diazocarbonyl compounds **168** catalyzed by trifluoroacetic acid resulted in two heterocycles **169** and **170** (Scheme 31). The reaction course in this

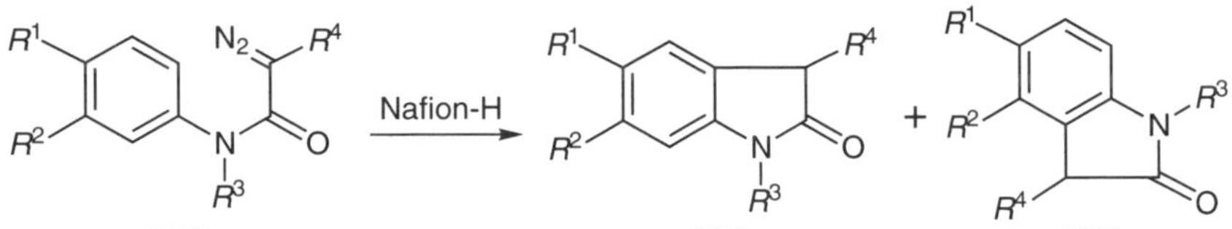

Diazoamide **165**	**166** (Yield)	**167** (Yield)
a: $R^1 = R^4 = H$, $R^2 = CH_3$, $R^3 = CH_3CH_2$	48%	48%
b: $R^1 = R^4 = H$, $R^2, R^3 = CH_2(O)_2$	82%	16%
c: $R^1 = H$, $R^2 = CH_3$, $R^3 = CH_3CH_2$, $R^4 = CH_3CO$	48%	44%
d: $R^3 = Et$, $R^1, R^2 = CH_2(O)_2$, $R^4 = CH_3CO$	73%	17%

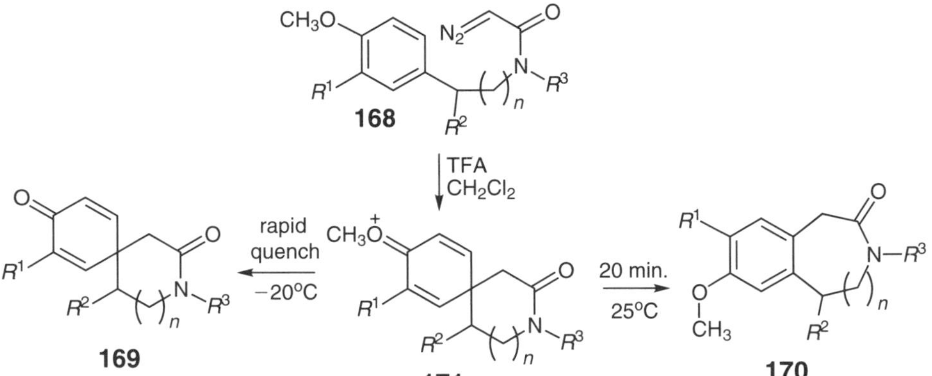

Diazoamide **168**	**169** (Yield)	**170** (Yield)
a: $n = 0$, $R^1 = TBSO$, $R^2 = H$, $R^3 = CH_3$	75%	85%
b: $n = 0$, $R^1 = CH_3O$, $R^2 = H$, $R^3 = CH_3$	80%	77%
c: $n = 0$, $R^1 = R^2 = H$, $R^3 = CH_3$	62%	—
d: $n = 1$, $R^1 = CH_3O$, $R^2 = H$, $R^3 = CH_3$	49%	70%

Scheme 31

case depends upon the reaction conditions employed. The kinetic preference for spirocyclization leads to the initial formation of a spirodienone oxonium species **171**. The isoquinolinone **170** was also derived from the intermediate **171**.[65] Furthermore, Pellicciari and co-workers[66] used boron trifluoride etherate to bring about the formation, in 92% yield, of the tricyclic indolone derivative **172** in

Scheme 32. In addition, diazoketones **173** and **174** have been converted into the tetracyclic vinylogous amides **175** and **176** in 12% and 35% yield, respectively (Scheme 33). Both **175** and **176** are synthons of functionalized aspidosperma alkaloids.[67]

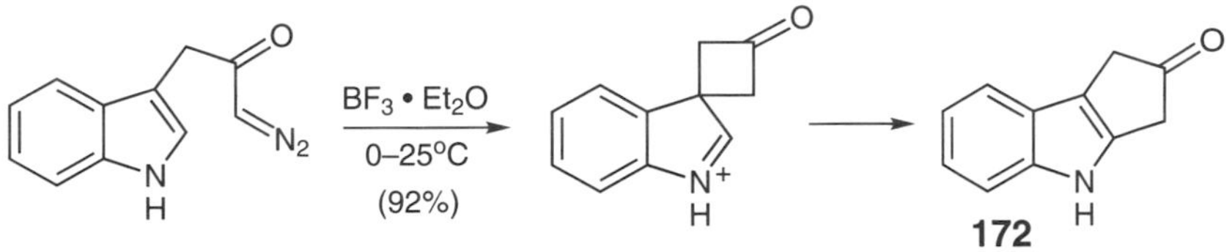

BF3 • Et2O
0–25°C
(92%)
172

Scheme 32

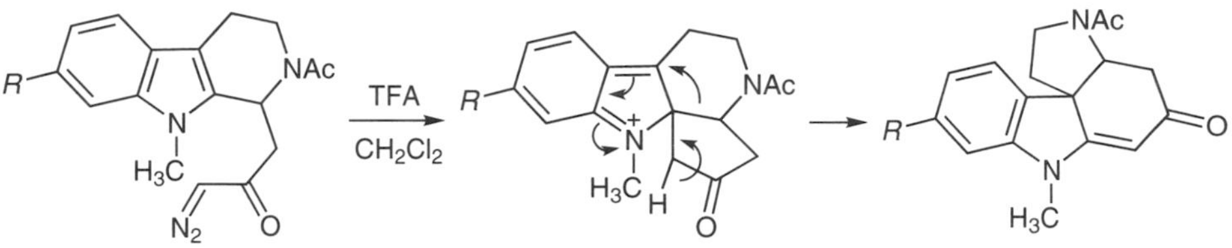

TFA
CH2Cl2
173: R = H
174: R = CH3O
175: R = H (12%)
176: R = CH3O (35%)

Scheme 33

Intermolecular acid-catalyzed reactions of α-diazocarbonyl compounds with π-electron systems are less common than intramolecular processes. Reaction with nitriles, catalyzed by boron trifluoride etherate or aluminum chloride, offers a direct route to substituted oxazoles.[68–73] Several examples are shown in eqs. 28, 29, 30,[69] and 31.[69] The process is reasonably explained by a stepwise

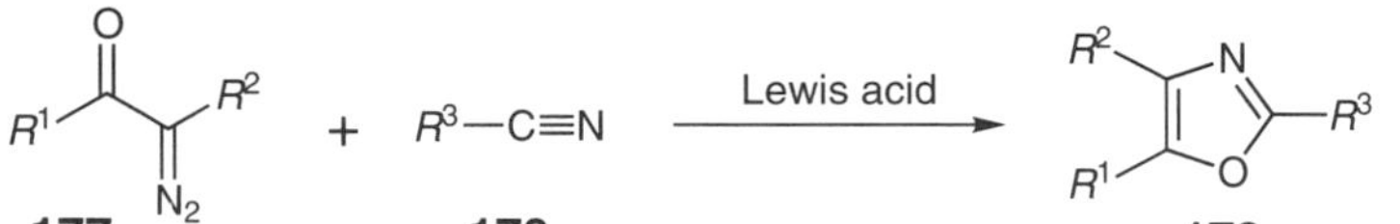

Lewis acid

177 + **178** → **179** (28)

Diazocarbonyl **177**	Nitrile **178**	Reaction condition	**179** (yield)	Ref.
a: $R^1 = p\text{-}CH_3C_6H_4$, $R^2 = H$	$R^3 = CH_3$	$AlCl_3$, 25°C	94%	68
b: $R^1 = C_6H_5$, $R^2 = H$	$R^3 = CH_3CH_2$	$BF_3 \cdot Et_2O$, 25°C	99%	69
c: $R^1 = R^2 = C_6H_5$	$R^3 = CH_3$	$BF_3 \cdot Et_2O$, 25°C	68%	69
d: $R^1 = C_6H_5$, $R^2 = C_6H_5CO$	$R^3 = CH_3$	$BF_3 \cdot Et_2O$, 25°C	79%	69
e: $R^1 = p\text{-}CH_3OC_6H_4$, $R^2 = H$	$R^3 = CH_2C(CH_3)$	$BF_3 \cdot Et_2O$, 25°C	96%	70
f: $R^1 = (CH_3)_3C$, $R^2 = H$	$R^3 = CH_3$	$BF_3 \cdot Et_2O$, 25°C	89%	70
g: $R^1 = p\text{-}NO_2C_6H_4$, $R^2 = CH_3O_2C$	$R^3 = CH_3CH_2$	$BF_3 \cdot Et_2O$, 25°C	84%	71
h: $R^1 = p\text{-}ClC_6H_4$, $R^2 = CH_3O_2C$	$R^3 = CH_3$	$BF_3 \cdot Et_2O$, 25°C	68%	71
i: $R^1 = Cl(CH_2)_3$, $R^2 = H$	$R^3 = CH_3$	$BF_3 \cdot Et_2O$, 25°C	82%	73
j: $R^1 = HCC(CH_2)_3$, $R^2 = H$	$R^3 = Cl(CH_2)_3$	$BF_3 \cdot Et_2O$, 25°C	59%	73

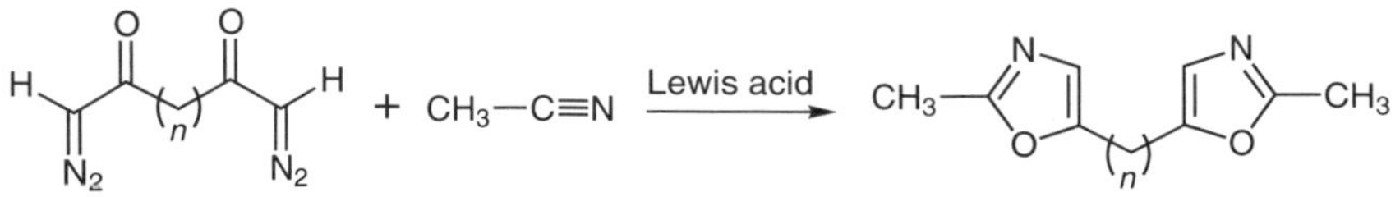

180 + $CH_3{-}C{\equiv}N$ $\xrightarrow{\text{Lewis acid}}$ **181**

(29)

Bisdiazoketone **180**	Reaction condition	**181** (yield)	Ref.
$n = 6$	$BF_3 \cdot Et_2O$, 25°C	95%	69
$n = 8$	$BF_3 \cdot Et_2O$, 25°C	94–96%	69, 70
$n = 8$	$AlCl_3$, 25°C	89%	70
$n = 10$	$BF_3 \cdot Et_2O$, 25°C	95%	69
$n = 12$	$BF_3 \cdot Et_2O$, 25°C	95%	69

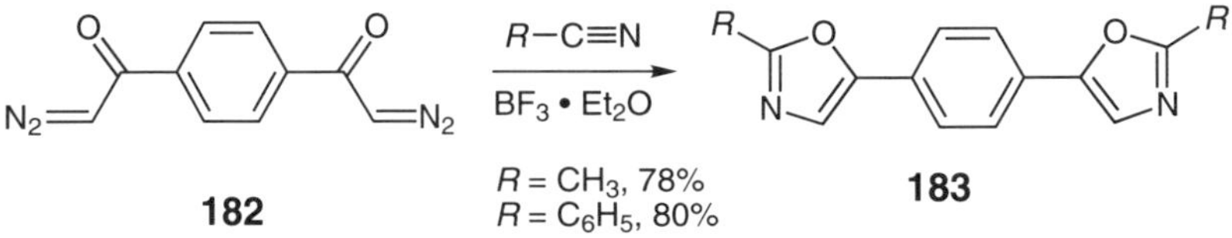

182 $\xrightarrow[\;BF_3 \bullet Et_2O\;]{R{-}C{\equiv}N}$ **183**

$R = CH_3$, 78%
$R = C_6H_5$, 80%

(30)

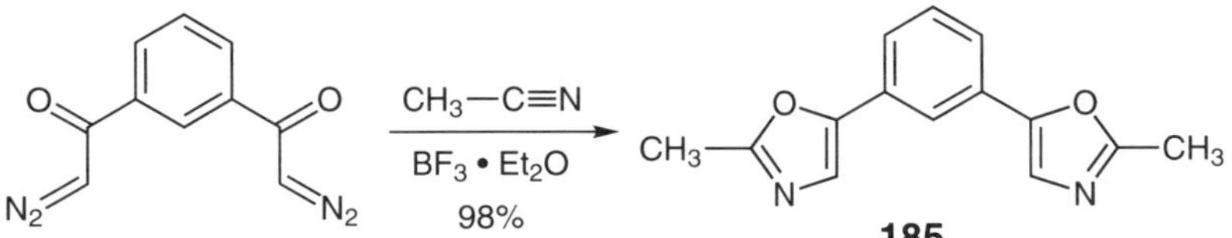

184 $\xrightarrow[\;BF_3 \bullet Et_2O\;]{CH_3{-}C{\equiv}N}$ **185**

98%

(31)

mechanism in which a Lewis acid–diazonium ion complex is attacked by the nitrogen atom of the nitrile, followed by intramolecular attack by the oxygen atom to complete the heterocyclic ring. A final example of oxazole formation is that in Scheme 34,[74] in which intermolecular reaction of acetonitrile with the indole-

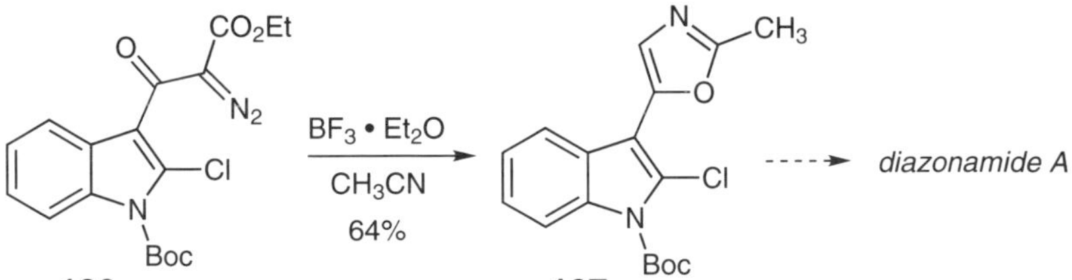

186 $\xrightarrow[\;CH_3CN\;]{BF_3 \bullet Et_2O}$ **187** $\dashrightarrow$ *diazonamide A*

64%

Scheme 34

derived diazoketo ester **186** was used to construct the oxazole side chain **187** of diazonamide A, a cytotoxic marine peptide.

11.6 REFERENCES

1 Burke, S. D.; and Grieco, P. A., "Intramolecular Reactions of Diazocarbonyl Compounds," *Org. React. (N.Y.)* **1979**, *26*, 361–475.

2 Smith, A. B., III; and Dieter, R. K., "The Acid Promoted Decomposition of α-Diazo Ketones," *Tetrahedron* **1981**, *37*, 2407–39.

3 Erman, W. F.; and Stone, L. C., "General Approach to the Synthesis of α-Patchoulane Sesquiterpenes. The Intramolecular Lewis Acid Catalyzed Addition of Diazo Ketones to Olefins," *J. Am. Chem. Soc.* **1971**, *93*, 2821–23.

4 Beames, D. J.; Klose, T. R.; and Mander, L. N., "Intramolecular C-Alkylation in Diazo-Ketones," *J. Chem. Soc., Chem. Commun.* **1971**, 773–74.

5 Mander, L. N.; Turner, J. V.; and Coombe, B. G., "Synthetic Plant Growth Regulators. 1. The Synthesis of (±)-14-Norhelminthosporic Acid and Related Compounds," *Aust. J. Chem.* **1974**, *27*, 1985–2000.

6 Ghatak, U. R.; Alam, S. K.; Chakraborti, P. C.; and Ranu, B. C., "Stereochemically Controlled Synthesis of Some *endo*-2-Aryl-6-oxobicyclo[3.2.1]octanes and Related Compounds through Intramolecular Alkylations of γ,δ-Unsaturated α'-Diazomethyl Ketones," *J. Chem. Soc., Perkin Trans. I* **1976**, 1669–72.

7 Mander, L. N.; and Palmer, L. T., "Synthetic Plant Growth Regulators. IV. The Preparation of Hydroxylated Helminthosporic Acid Analogues," *Aust. J. Chem.* **1979**, *32*, 823–32.

8 Mehta, G.; Krishnamurthy, N.; and Karra, S. R., "Terpenoids to Terpenoids—Enantioselective Construction of 5,6- Fused, 5,7-Fused, and 5,8-Fused Bicyclic Systems—Application to the Total Synthesis of Isodaucane Sesquiterpenes and Dolastane Diterpenes," *J. Am. Chem. Soc.* **1991**, *113*, 5765–75.

9 Mehta, G.; and Karra, S. R., "Polyquinanes from (*R*)-(+)-Limonene, Enantioselective Total Synthesis of the Novel Tricyclic Sesquiterpene (−)-Ceratopicanol," *J. Chem. Soc., Chem. Commun.* **1991**, 1367–68.

10 Chakrabortty, P. N.; Dasgupta, R.; Dasgupta, S. K.; Ghosh, S. R.; and Ghatak, U. R., "Synthetic Studies towards Complex Diterpenoids—VI, New Synthetic Routes to Tetracyclic Bridged-Bicyclo[3.2.1.]octane Intermediates by Intramolecular Alkylation Reactions through α-Diazomethyl Ketones of Hydroaromatic γ,δ-Unsaturated Acids," *Tetrahedron* **1972**, *28*, 4653–65.

11 Klose, T. R.; and Mander, L. N., "Studies on Intramolecular Alkylation. VII. The Preparation of Phenanthrene-Derived Tetracyclic Ketones: Intermediates for Gibberellin Synthesis," *Aust. J. Chem.* **1974**, *27*, 1287–94.

12 Ghatak, U. R.; and Chakraborti, P. C., "Synthetic Studies toward Complex Diterpenoids. 12. Stereocontrolled Total Synthesis of Some Gibbane Synthons and Degradation Products of Gibberellins," *J. Org. Chem.* **1979**, *44*, 4562–66.

13 Ranu, B. C.; Sarkar, R. M.; Chakraborti, P. C.; and Ghatak, U. R., "Synthetic Studies Directed Complex Diterpenoids. Part 15. Synthesis and Stereochemistry of the Catalytic Reduction of $\Delta^{4b(5)}$-Gibbenes and Related Compounds," *J. Chem. Soc., Perkin Trans. I* **1982**, 865–73.

14 Ghatak, U. R.; and Ray, J. K., "Condensed Cyclic and Bridged-Ring Systems. Part 6. Stereochemically Defined Synthesis of (±)-3-Methoxy-17,18,19,20-tetranor-B-homophylloclada-1(10),2,4-triene-16-one through Intramolecular Alkylation of a γ,δ-Unsaturated α'-Diazomethyl Ketone," *J. Chem. Soc., Perkin Trans. I*, **1977**, 518–20.

15 Hook, J. M.; Mander, L. N.; and Urech, R., "Total Synthesis of Gibberellic Acid. The Hydrofluorene Route," *J. Am. Chem. Soc.* **1980**, *102*, 6628–29.

16 Hook, J. M.; Mander, L. N. and Urech, R., "Studies on Gibberellin Synthesis—The Total Synthesis of Gibberellic-Acid from Hydrofluorenone Intermediates," *J. Org. Chem.* **1984**, *49*, 3250–60.

17 Ray, C.; Saha, B.; and Ghatak, U. R., "Synthesis of Some Angularly Cyclopentanone Fused Hydrophenanthrene and Hydrofluorene Derivatives by Acid-Catalyzed Intramolecular C-Alkylation of γ,δ-Unsaturated α'-Diazomethyl Ketones," *Synth. Commun.* **1991**, *21*, 1223-42.

18 Malherbe, R.; and Dahn, H., "π-Participation in Diazoketone Hydrolysis II: *Exo–endo* Cyclization Ratio in the Hydrolyses of 7-*syn*- and 5-*endo*-Diazoacetyl-2-norbornene," *Helv. Chim. Acta* **1977**, *60*, 2539–49.

19 Malherbe, R., "Acetolysis of 4-*endo*-Diazoacetylbicyclo[3.1.0]hexene: A New Synthesis of Semibullvalene," *Helv. Chim. Acta* **1973**, *56*, 2845–46.

20 Piers, E.; Geraghty, M. B.; Smillie, R. D.; and Soucy, M., "Stereoselective Total Synthesis of Copa and Ylango Sesquiterpenoids: (−)-Copacamphene, (−)-Cyclocopacamphene, (+)-Sativene, (+)-Cyclosativene," *Can. J. Chem.* **1975**, *53*, 2849–64.

21 Ceccherelli, P.; Tingoli, M.; Curini, M.; and Pellicciari, R., "Tetracyclic Diterpenoids by Intramolecular C-Alkylation of γ,δ-Unsaturated Diazo Ketones," *Tetrahedron Lett.* **1978**, 4959–62.

22 Smith, A. B., III, "Acid-Catalysed Decomposition of β,γ-Unsaturated Diazomethyl Ketones: A New Cyclopentenone Annelation," *J. Chem. Soc., Chem. Commun.* **1975**, 274–75.

23 Smith, A. B., III; Branca, S. J.; and Toder, B. H., "A New Approach to Simple Cyclopentenones: Application to the Synthesis of Dihydro and *cis*-Jasmone," *Tetrahedron Lett.* **1975**, 4225–28.

24 Smith, A. B., III, Toder, B. H.; Branca, S. J.; and Dieter, R. K., "Lewis Acid Promoted Decomposition of Unsaturated α-Diazo Ketones. 1. An Efficient Approach to Simple and Annulated Cyclopentenones," *J. Am. Chem. Soc.* **1981**, *103*, 1996–2008.

25 Smith, A. B., III; and Dieter, R. K., "Lewis Acid Promoted Decomposition of Unsaturated α-Diazo Ketones. 2. A New Initiator for Polyolefinic Cationic Cyclization," *J. Am. Chem. Soc.* **1981**, *103*, 2009–16.

26 Smith, A. B., III; and Dieter, R. K., "Lewis Acid Promoted Decomposition of Unsaturated α-Diazo Ketones. 3. Stereochemical Consequences of Polyolefinic Cyclizations Initiated by the α-Diazo Ketone Functionality," *J. Am. Chem. Soc.* **1981**, *103*, 2017–22.

27 Hudlicky, T.; and Kutchan, T., "Total Synthesis of (±)-Filifolone," *Tetrahedron Lett.* **1980**, *21*, 691–92.

28 Doyle, M. P.; and Trudell, M. L., "Catalytic Role of Copper Triflate in Lewis Acid Promoted Reactions of Diazo Compounds," *J. Org. Chem.* **1984**, *49*, 1196–99.

29 Satyanarayana, G. O. S. V.; Kanjilal, P. R.; and Ghatak, U. R., "Acid-Catalysed Intramolecular C-Alkylation Rearrangement of β,γ-Unsaturated Diazomethyl Ketones. A Novel Synthetic Entry to Pentaleo-Annelated Polycyclic Systems," *J. Chem. Soc., Chem. Commun.* **1981**, 746-47.

30 Ghatak, U. R.; and Sanyal, B., "Intramolecular C-Alkylation in β,γ-Unsaturated Diazo Ketones: A New Synthetic Route to Angularly Fused Cyclobutanones and γ-Lactones," *J. Chem. Soc., Chem. Commun.* **1974**, 876–77.

31 Ghatak, U. R.; Sanyal, B.; and Ghosh, S., "A Novel Rearrangement of Angularly Fused Cyclobutane. Stereospecific Syntheses of Intermediates to the Diterpene Alkaloids and the C₂₀ Gibberellins," *J. Am. Chem. Soc.* **1976**, *98*, 3721–22.

32 Ghatak, U. R.; Alam, S. K.; and Ray, J. K., "Synthetic Studies toward Complex Diterpenoids. 10. Stereocontrolled Total Synthesis of (±)-19α,20α-(Acetylimino)-12-hydroxy-5β,10α-podocarpa-8,11,13-triene, a Degradation Product of Atisine," *J. Org. Chem.* **1978**, *43*, 4598–604.

33 Ghatak, U. R.; Ghosh, S.; and Sanyal, B., "Synthetic Studies towards Complex Diterpenoids. Part 13. Stereospecific Synthesis of Intermediates to the Diterpene Alkaloids and the C₂₀-Gibberellins by a Novel Rearrangement of Angularly Fused Cyclobutanones," *J. Chem. Soc., Perkin Trans. I* **1980**, 2881–86.

34 Ghatak, U. R.; Sanyal, B.; Satyanarayana, G. O. S. V.; and Ghosh, S., "Acid-Catalysed Intramolecular C-Alkylation in β,γ-Unsaturated Diazomethyl Ketones. A New Synthetic Route to Angularly Fused Cyclobutanones, Bridged Cyclopentanones, and γ-Lactones," *J. Chem. Soc., Perkin Trans. I* **1981**, 1203–12.

35 Satyanarayana, G. O. S. V.; Roy, S. C.; and Ghatak, U. R., "Acid-Catalyzed Intramolecular C-Alkylation in β,γ-Unsaturated Diazomethyl Ketones. 2. A Simple New Synthetic Route to Octahydro- 4,10a-ethanophenanthren-12-ones and Octahydropentaleno[6a,1-a] naphthalen-4-ones," *J. Org. Chem.* **1982**, *47*, 5353–61.

36 Roy, S. C.; Satyanarayana, G. O. S. V.; and Ghatak, U. R., "Acid-Catalyzed Intramolecular C-Alkylation In β,γ-Unsaturated Diazomethyl Ketones. 3. A Simple Synthetic Route to Hexahydro-4,9a-ethano-1H-fluoren-11-ones, Hexahydro-6H-pentaleno[6a,1-a]indan-4-ones, and Hexahydrocyclobuta[j]fluoren-2(1H)-ones," *J. Org. Chem.* **1982**, *47*, 5361–68.

37 Ceccherelli, P.; Curini, M.; Tingoli, M.; and Pellicciari, R., "Acid-Catalysed Decomposition of β,γ-Unsaturated Diazoketones: Preparation of 4,4-Dimethyl-D-nor-Steroidal Systems from Pimaradiene and Sandaracopimaradiene Precursors," *J. Chem. Soc., Perkin Trans. I* **1980**, 1924–27.

38 Cook, J. W.; and Schoental, R.,"Polycyclic Aromatic Hydrocarbons. Part XXX. Synthesis of Chrysenols," *J. Chem. Soc.* **1945**, 288–93.

39 Newman, M. S.; Eglinton, G.; and Grotta, H. M., "Synthesis of Hydroaromatic Compounds Containing Angular Methyl Groups. III. 1,2-Cyclopentanonaphthalene Series," *J. Am. Chem. Soc.* **1953**, *75*, 349–52.

40 Heck, R.; and Winstein, S., "Neighboring Carbon and Hydrogen. XXVII. Ar₁-5 Aryl Participation and Tetralin Formation in Solvolysis," *J. Am. Chem. Soc.* **1957**, *79*, 3105–13.

41 Beames, D. J.; and Mander, L. N., "Studies on Intramolecular Alkylation. IV. The Preparation of Spirodienones from Phenolic Diazoketones," *Aust. J. Chem.* **1974**, *27*, 1257–68.

42 Johnson, D. W.; Mander, L. N.; and Masters, T. J., "Studies on Intramolecular Alkylation. XIV. The Preparation of Methoxylated Cyclohexadienone Derivatives from Intramolecular Diazoketone Cyclizations in Relationship to Isoquinoline Alkaloid Synthesis," *Aust. J. Chem.* **1981**, *34*, 1243–52.

43 Blair, I. A.; Ellis, A.; Johnson, D. W.; and Mander, L. N., "Studies on Intramolecular Alkylation. IX. The Synthesis of Tricyclic Dienones Suitable for the Synthesis of 13-Hydroxygibberellins," *Aust. J. Chem.* **1978**, *31*, 405–9.

44 Kenny, M. J.; Mander, L. N.; and Sethi, S. P., "Synthetic Studies on Rabdosia Diterpene Lactones I: The Preparation of A Key Tricyclic Intermediate," *Tetrahedron Lett.* **1986**, *27*, 3923–26.

45 Johnson, D. W.; and Mander, L. N., "Studies on Intramolecular Alkylation. VI. *ortho*-Alkylation in Phenolic Diazoketones: The Preparation of Intermediates Containing the Cyclohexa-2,4-dienone Moiety Suitable for Gibberellin Synthesis," *Aust. J. Chem.* **1974**, *27*, 1277–86.

46 Mander, L. N.; and Pyne, S. G., "A New Strategy for Gibberellin Synthesis," *J. Am. Chem. Soc.* **1979**, *101*, 3373–75.

47 Kenny, M. J.; Mander, L. N.; and Sethi, S. P., "Synthetic Studies on Rabdosia Diterpene Lactones II: The Synthesis of 15-Desoxyeffusin," *Tetrahedron Lett.* **1986**, *27*, 3927–30.

48 Lombardo, L.; Mander, L. N.; and Turner, J. V., "General Strategy for Gibberellin Synthesis: Total Syntheses of (±)-Gibberellin A_1 and Gibberellic Acid," *J. Am. Chem. Soc.* **1980**, *102*, 6626–28.

49 Lombardo, L.; and Mander, L. N., "A New Strategy for C_{20} Gibberellin Synthesis: Total Synthesis of (±)-Gibberellin A_{38} Methyl Ester," *J. Org. Chem.* **1983**, *48*, 2298–300.

50 Cossey, A. L.; Lombardo, L.; and Mander, L. N., "Total Synthesis of Gibberellin A_4," *Tetrahedron Lett.* **1980**, *21*, 4383–86.

51 Basu, B.; Maity, S. K.; and Mukherjee, D., "Studies on Intramolecular Cyclizations—Synthesis of Ring-Systems Related to Sesquiterpenoids," *Synth. Commun.* **1981**, *11*, 803–9.

52 Das, S.; Karpha, T. K.; Ghosal, M.; and Mukherjee, D., "Aryl Participated Cyclizations Involving Indane Derivatives—A Total Synthesis of (+/−)-Isolongifolene," *Tetrahedron Lett.* **1992**, *33*, 1229–32.

53 Blair, I. A.; Mander, L. N.; and Mundill, P. H. C., "Studies on Intramolecular Alkylation. XIII. The Synthesis of 3,4-Dihydrophenanthren-2-(1*H*)-ones from Pummerer Intermediates," *Aust. J. Chem.* **1981**, *34*, 1235–42.

54 Johnson, D. W.; and Mander, L. N., "Studies on Intramolecular Alkylation. X. The Acid-Catalysed Reactions of 5,8-Dialkoxy-1,2,3,4-tetrahydronaphthyl Diazomethyl Ketones," *Aust. J. Chem.* **1978**, *31*, 1561–68.

55 Cushman, M.; Choong, T.-C.; Valko, J. T.; and Koleck, M. P., "Total Synthesis of (±)-Chelidonine," *J. Org. Chem.* **1980**, *45*, 5067–73.

56 Cushman, M.; Abbaspour, A.; and Gupta, Y. P., "Total Synthesis of (±)-14-Epicorynoline, (±)-Corynoline, and (±)-6-Oxocorynoline," *J. Am. Chem. Soc.* **1983**, *105*, 2873–79.

57 Beames, D. J.; Klose, T. R.; and Mander, L. N., "Studies on Intramolecular Alkylation. V. Intramolecular Alkylation of the Aromatic Ring in Tetrahydronaphthyl Diazomethyl Ketones," *Aust. J. Chem.* **1974**, *27*, 1269–75.

58 Maity, S. K.; and Mukherjee, D., "Synthesis of Tricyclo[6.2.2.0^{1,6}]Dodecane and Tricyclo[6.3.1.0^{1,6}]Dodecane Ring-Systems Involving Intramolecular Cyclization of Diazomethyl Ketones," *Tetrahedron Lett.* **1983**, *24*, 5919–20.

59 Maity, S. K.; Bhattacharyya, S.; and Mukherjee, D., "Aryl-Participated Intramolecular Cyclisation of Diazomethyl Ketones. Stereocontrolled Synthesis of Bridged Tetracyclic Systems Related to Terpenoids," *J. Chem. Soc., Chem. Commun.* **1986**, 481–83.

60 Maity, S. K.; Basu, B.; and Mukherjee, D., "Synthesis of Bridged Tetracyclic Systems Related to Diterpenes through Acid-Induced Intramolecular Cyclization of Diazomethyl Ketones," *Tetrahedron Lett.* **1983**, *24*, 3921–22.

61 Maity, S. K.; and Mukherjee, D., "Acid-Catalyzed Intramolecular Cyclization of Diazomethyl Ketones—Synthesis of Bridged Tetracyclic Systems Related to Diterpenes," *Tetrahedron* **1984**, *40*, 757–60.

62 Nicolaou, K. C.; and Zipkin, R. E., "An Expeditious and Efficient Entry into the Aphidicolin and Related Natural-Products Ring Skeleton," *Angew. Chem. Int. Ed. Engl.* **1981**, *20*, 785–86.

63 Mander, L. N., "Exploitation of Aryl Synthons in the Synthesis of Polycyclic Natural Products," *Synlett* **1991**, 134–44.

64 Doyle, M. P.; Shanklin, M. S.; Pho, H. O.; and Mahapatro, S. N., "Rhodium(II) Acetate and Nafion-H Catalyzed Decomposition of N-Aryldiazoamides. An Efficient Synthesis of 2(3*H*)-Indolinones," *J. Org. Chem.* **1988**, *53*, 1017–22.

65 Rishton, G. M.; and Schwartz, M. A., "Acid-Catalyzed Cyclizations of Aromatic Diazoacetamides—Synthesis of Spirodienone Lactams, Isoquinolinones, and Benzazepinones," *Tetrahedron Lett.* **1988**, *29*, 2643–46.

66 Franceschetti, L.; Garzon-Aburbeh, A.; Mahmoud, M. R.; Natalini, B.; and Pellicciari, R., "Synthesis of a Novel, Conformationally Restricted Analog of Tryptophan," *Tetrahedron Lett.* **1993**, *34*, 3185–88.

67 Takano, S.; Shishido, K.; Sato, M.; Yuta, K.; and Ogasawara, K., "New Synthetic Routes to Synthons for the Synthesis of Functionalized Aspidodperma Alkaloids," *J. Chem. Soc., Chem. Commun.* **1978**, 943–45.

68 Doyle, M. P.; Oppenhuizen, M.; Elliott, R. C.; and Boelkins, M. R., "Lewis Acid–Promoted 1,3-Dipolar Addition Reactions of Diazocarbonyl Compounds. A General Synthesis of Oxazoles," *Tetrahedron Lett.* **1978**, *26*, 2247–50.

69 Ibata, T.; and Sato, R., "The Acid Catalyzed Decomposition of Diazo Compounds. I. Synthesis of Oxazoles in the BF₃ Catalyzed Reaction of Diazo Carbonyl Compounds with Nitriles," *Bull. Chem. Soc. Jpn.* **1979**, *52*, 3597–600.

70 Doyle, M. P.; Buhro, W. E.; Davidson, J. G.; Elliott, R. C.; Hoekstra, J. W.; and Oppenhuizen, M., "Lewis Acid Promoted Reactions of Diazocarbonyl Compounds. 3. Synthesis of Oxazoles from Nitriles through Intermediate β-Imidatoalkenediazonium Salts," *J. Org. Chem.* **1980**, *45*, 3657–64.

71 Ibata, T.; Yamashita, T.; Kashiuchi, M.; Nakano, S.; and Nakawa, H., "The Acid-Catalysed Decomposition of Diazo Carbonyl Compounds. II. Synthesis of 2- or 5-Heteroatom-Substituted Oxazoles," *Bull. Chem. Soc. Jpn.* **1984**, *57*, 2450–55.

72 Ibata, T.; and Isogami, Y., "Formation and Reaction of Oxazoles—Synthesis of N-Substituted 2-(Aminomethyl)oxazoles," *Bull. Chem. Soc. Jpn.* **1989**, *62*, 618–20.

73 Vedejs, E.; and Piotrowski, D. W., "Oxazole Activation for Azomethine Ylide Trapping—Singly and Doubly Tethered Substrates," *J. Org. Chem.* **1993**, *58*, 1341–48.

74 Konopelski, J. P.; Hottenroth, J. M.; Ottra, H. M.; Veliz, E. A.; and Yang, Z.-C., "Synthetic Studies on Diazonamide A; Benzofuranone-Tyrosine and Indole-Oxazole Fragment Support Studies," *Synlett* **1996**, 609–11.

CHAPTER 12

Miscellaneous Diazocarbonyl Reactions

12.1 INTRODUCTION

There remain several diazocarbonyl reactions that again demonstrate their versatility in synthesis but whose uses are less extensive than those discussed in the preceding chapters. They do, however, have applications in specific areas that justify their inclusion. These reactions are (1) oxidation, (2) β-elimination, (3) X–Y insertion in which neither X nor Y is a hydrogen atom, (4) dimerization, and (5) dipolar cycloaddition.

12.2 OXIDATIONS OF DIAZOCARBONYL COMPOUNDS

Several reagents are available for oxidation of the diazo function of α-diazocarbonyl compounds, the nature of the product depending on the choice of oxidant and the substitution pattern of the substrate. Peroxyacids react with disubstituted diazoketones to form α-diketones, which may react further with this oxidant.[1] With diazoesters and m-chloroperbenzoic acid, high-yielding transformations to keto esters have been observed, as, for example, in Scheme 1, in the course of the synthesis of the eight-carbon, naturally occurring sugar KDO.[2] *tert*-Butyl hypochlorite in ethanol oxidizes 2-diazoindanedione to a trione monoketal, which on hydrolysis furnishes ninhydrin hydrate (Scheme 2).[3] Oxidation of primary diazo ketones with ozone can lead to products resulting from C–H bond cleavage of the initially formed glyoxal.[4] However, when this complication is absent, as with 6-diazo-aminopenicillanates (eq. 1), cleavage by ozone to the α-diketone can be very efficient.[5]

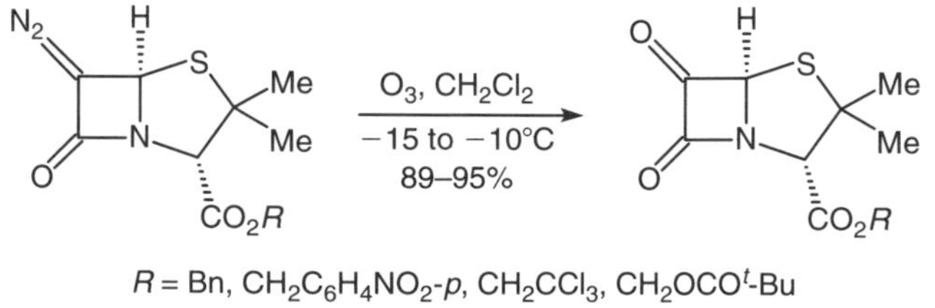

$$R = Bn, CH_2C_6H_4NO_2\text{-}p, CH_2CCl_3, CH_2OCO^t\text{-}Bu$$

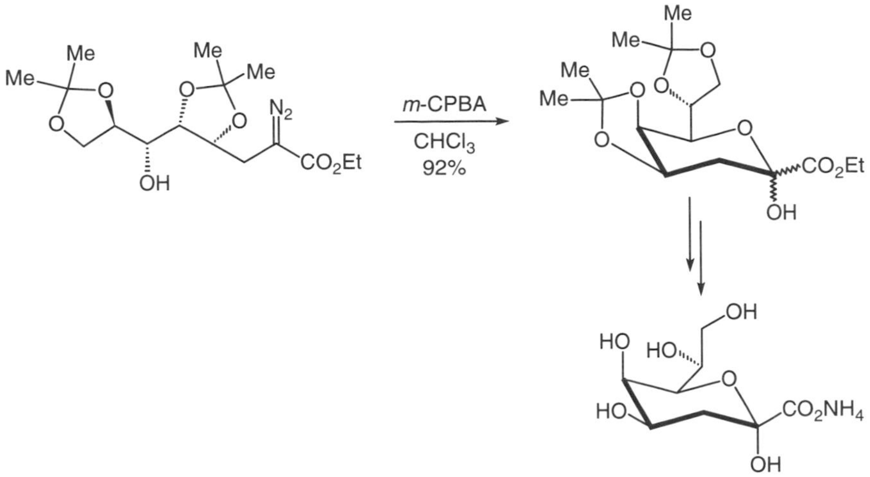

Scheme 1

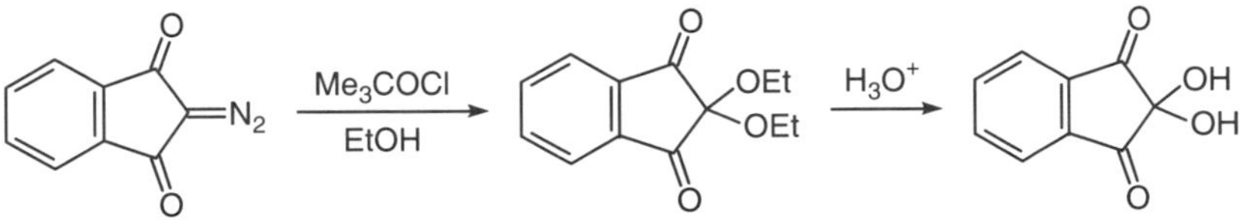

Scheme 2

Perhaps the most synthetically useful diazocarbonyl oxidation is through the use of dioxiranes, notably dimethyldioxirane (DMD).[6,7] This oxidant, which can be readily prepared and distilled as a dilute solution in acetone, converts diazoketones into diketones or glyoxals (eq. 2) in essentially quantitative yield.[8] Products

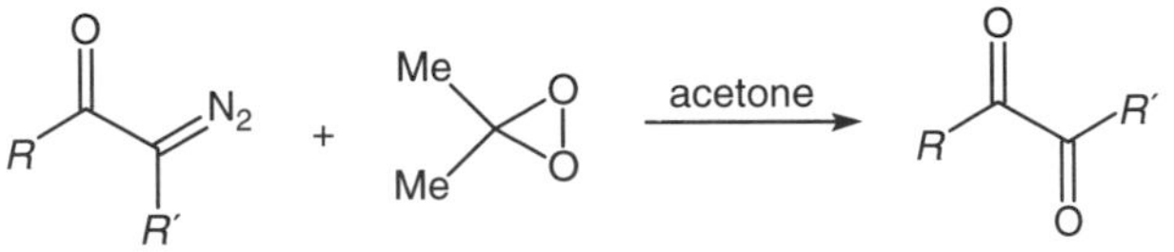

$$\tag{2}$$

are obtained in a high state of purity, since there are no byproducts other than nitrogen and acetone; glyoxals are usually isolated in hydrated form. Some examples are shown in Table 12.1.[8,9] Particularly noteworthy are transformations of diazoketones with oxidizable heterocyclic substituents, where the reaction occurs exclusively at the diazo group. This process has been used recently to produce glyoxals from homochiral N-protected amino acids and dipeptides as shown in Table 12.2.[10,11] By combining the oxidation process with a Wittig olefination routine, it is possible to bring about one-pot conversions of chiral diazoketones into optically pure enones of the type shown in Scheme 3.[12] Alternatively, the glyoxal

TABLE 12.1   Oxidation of Terminal Diazocarbonyl Compounds with DMD in Acetone at 20°C

Diazocarbonyl	Glyoxal hydrate	Yield (%)	Ref.
		100	8
		100	8
		98	8
		85	8
		100	9
		100	9

can be combined with 1,2-aminobenzene to furnish optically pure quinoxaline (Scheme 3).[10] Both these processes have been applied to the synthesis of several amino acids and dipeptides.

The oxidation of nonterminal diazocarbonyl compounds is also readily accomplished with DMD in acetone, a recent novel application being the synthesis of enantiopure N-protected β-amino-α-keto esters from natural amino acids and dipeptides.[13a] Previous syntheses of these compounds proceeded with extensive racemization. In the first step in Scheme 4 an amino acid–derived diazoketone is

TABLE 12.2 Oxidation of N-Protected Amino Diazoketones with DMD in Acetone

Diazoketone	Glyoxal hydrate	Yield
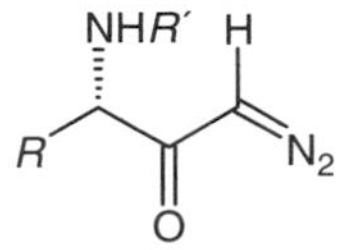 $R' = Z$, Boc, CO$_2$Et R = Me, PhCH$_2$, Me$_2$CH, MeCH$_2$CH(Me)	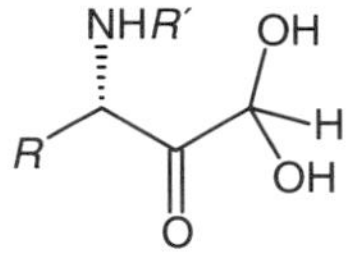	Quantitative
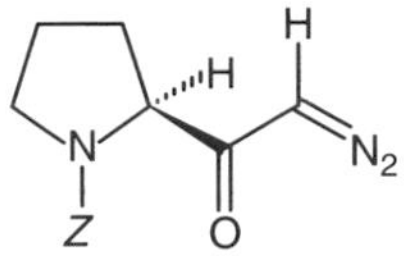	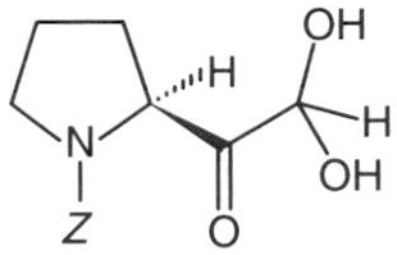	Quantitative
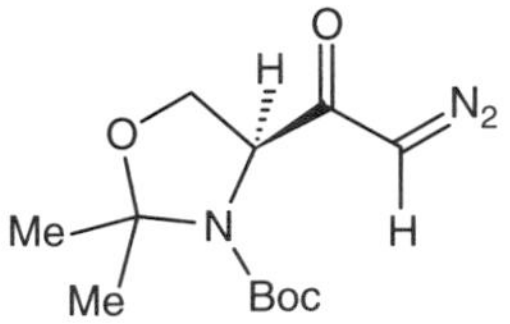	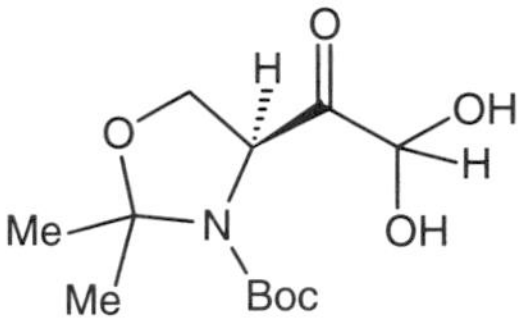	Quantitative
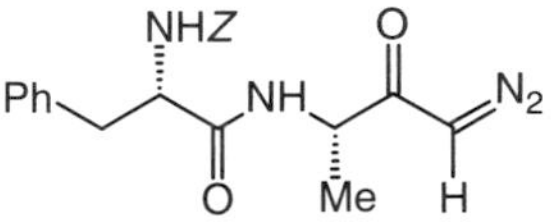	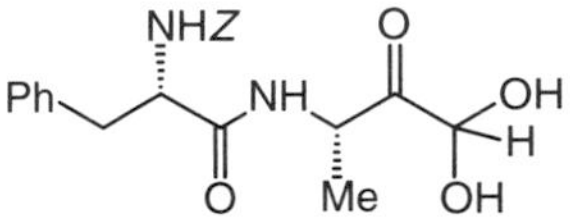	Quantitative
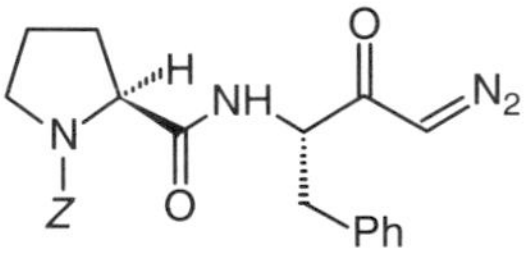	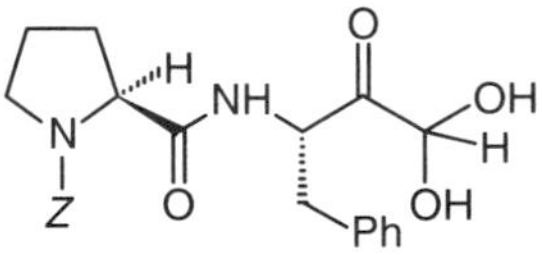	Quantitative

subjected to Wolff rearrangement using silver benzoate in methanol (Chapter 9). The resulting β-amino ester is then transformed *via* the Danheiser procedure (Chapter 1) into a diazoester. Finally, DMD oxidation converts the diazoester into the α-ketoester, none of the stages showing any measurable degree of racemization. Several examples involving dipeptides are shown in Scheme 4. All the diazocarbonyl oxidations with DMD described above were conducted at 0°C, and in most cases reactions were complete within minutes. Oxidation of α-diazo-β-dicarbonyl compounds with DMD is also effective, though reaction rates are slow by comparison with simple diazoketones or diazoesters. Nevertheless, the reaction constitutes a very efficient route to a novel group of vicinal tricarbonyl compounds, much of whose chemistry is still largely unexplored. Saba[13b] used DMD

OXIDATION OF N-BENZYLOXYCARBONYL-L-PHENYLALANYL DIAZOMETHANE WITH DIMETHYL-DIOXIRANE. PREPARATION OF AN *N*-PROTECTED AMINO GLYOXAL HYDRATE[10]

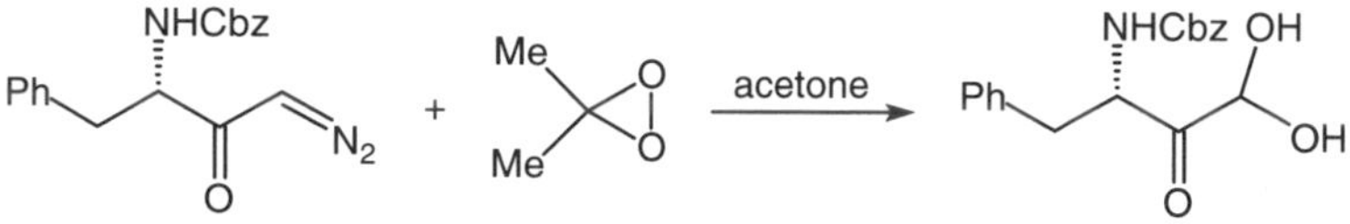

To a stirred solution of the diazoketone (311 mg, 0.96 mmol) in acetone (100 mL) at 0°C was added an equivalent amount of dimethyldioxirane in acetone (15 mL of a 0.065 M solution). After 10 min the solution was evaporated under reduced pressure to afford the glyoxal hydrate (312 mg, 100%) as a pale yellow solid, m.p. 71–73°C.

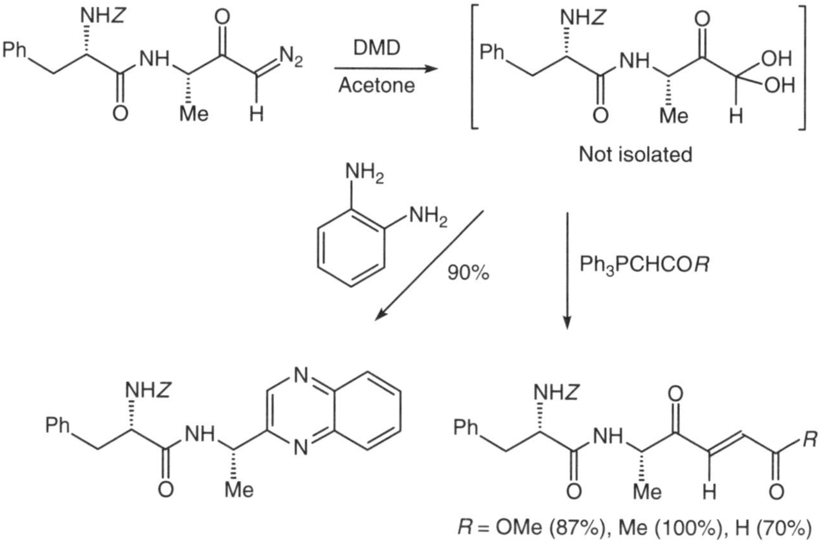

Scheme 3

to oxidize a series of α-diazo-β-ketoesters and α-diazo-β-diketones to vicinal tricarbonyl compounds at room temperature (Tables 12.3 and 12.4) in excellent yields. In some cases the product was isolated as the hydrate of the central, most electrophilic carbonyl group. Reaction times were 10–36 h.

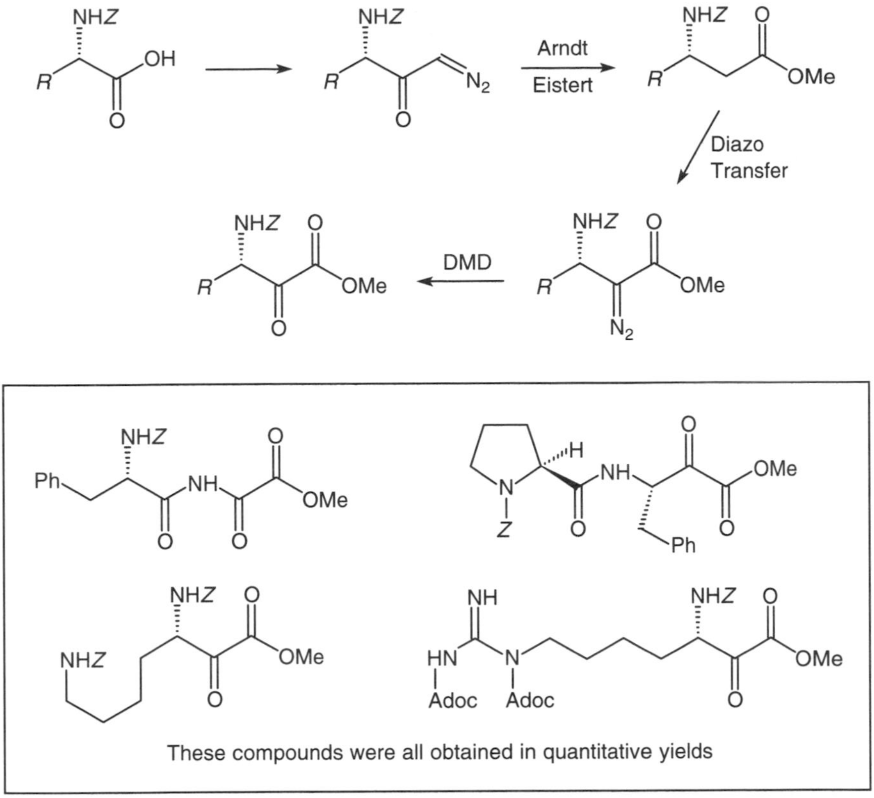

These compounds were all obtained in quantitative yields

Scheme 4

TABLE 12.3 Oxidation of α-Diazo-β-Ketoesters with DMD in Acetone

(3)

R	R'	Yield
Me	Me	95[a]
Me	Ph	98[a]
Ph	Ph	100[a]
Me	Et	100
Ph	Ph	94
OEt	Et	100
Menthyl	Menthyl	100[a]

[a]Product isolated as the hydrate.

TABLE 12.4   Oxidation of α-Diazo-β-diketones with DMD in Acetone

α-Diazo-β-carbonyl precursor	Tricarbonyl product	Yield (%)
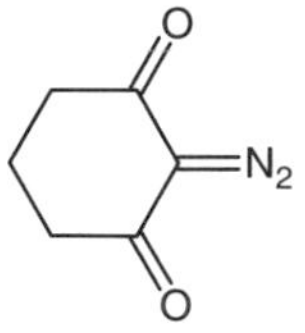	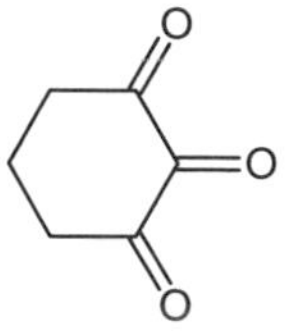	93
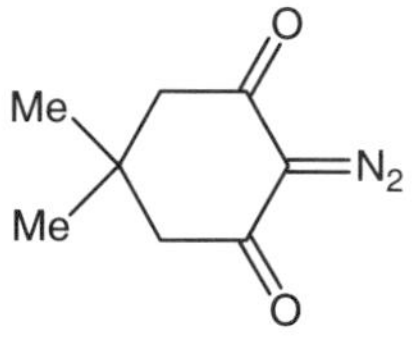	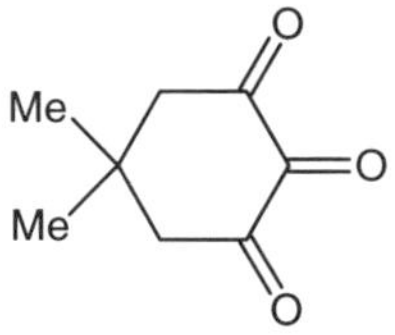	89[a]
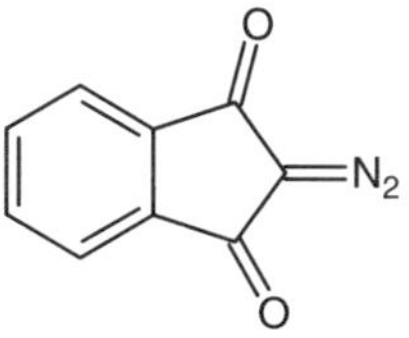	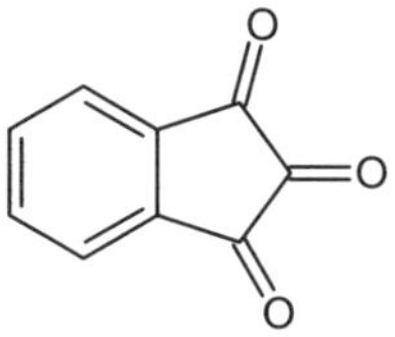	100[a]

[a]Product isolated as the hydrate.

12.3   β-HYDRIDE ELIMINATION

After the Wolff rearrangement, the most common type of rearrangement encountered in diazocarbonyl chemistry involves a 1,2-hydride or 1,2-alkyl shift. The former, which is frequently referred to as "β-hydride elimination," does have some useful applications in synthesis, though it can also represent a competing side reaction in other transformations of diazocarbonyl compounds where β-hydrogen atoms are present. The two processes are related mechanistically (Scheme 5), with the putative metal-bound carbene as a common intermediate within which a 1,2-hydride or 1,2-alkyl shift occurs. One particular type of 1,2-hydride shift that does have applications to the synthesis of β-dicarbonyl compounds occurs in the metal-catalyzed decomposition of α-diazoketols. This reaction is discussed in Chapter 10.

Early examples of β-hydride elimination in competition with Wolff rearrangement were reported by Franzen in 1957,[14] although the geometry of the alkenes formed was not discussed. A later study of several substituted diazoesters by Ganem and co-workers[15] revealed that rhodium(II) acetate in benzene catalyzed their conversions to cis-α,β-unsaturated esters (Scheme 6) in high yield. The preference for the cis geometry was ascribed to the tendency for the metal carbene shown in the Scheme to adopt a conformation in which the hydrogen atom undergoing 1,2-shift lies perpendicular to the metal-carbon bond with the R group anti to the bulky metal residue to minimize steric interactions. Hudlicky

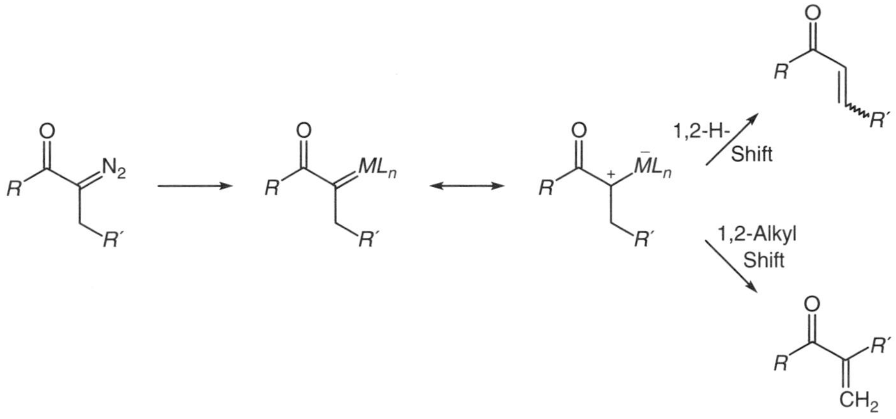

Scheme 5

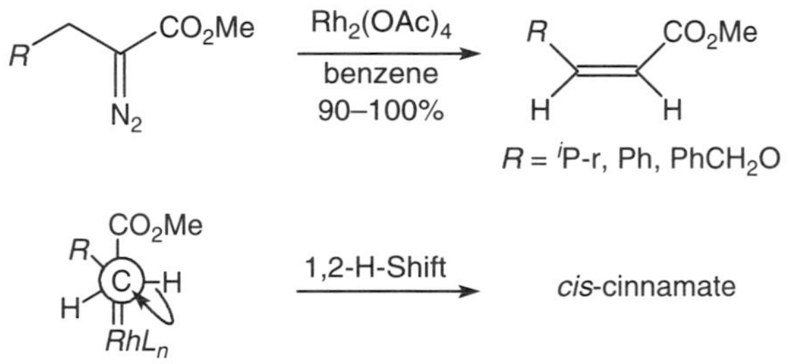

$R = {}^iP\text{-r}, Ph, PhCH_2O$

Scheme 6

and co-workers[16] have exploited β-elimination to devise a route to β-methoxy enones from acid chlorides (Scheme 7). The first step involved diazoketone formation using the previously unknown methoxydiazoethane (MeOCH₂CHN₂) fol-

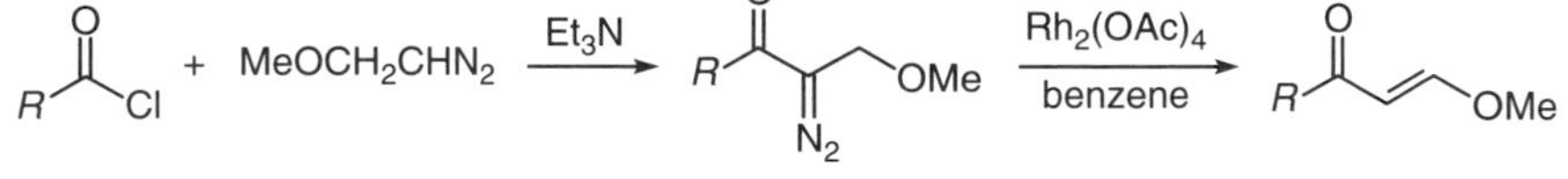

R	Diazoketone (%)	Enol ether (%)
$CH_3(CH_2)_6$	82	99
cyclohexyl	64	95
phenyl	61	75
benzyl	52	78
1-phenylethyl	52	47
3-bromophenyl	86	99

Scheme 7

lowed by exposure to rhodium(II) acetate in benzene. β-Methoxy enones were produced with the E geometry exclusively, unlike the α,β-unsaturated esters of Ganem's study, and yields were good to excellent. Although in the examples of $R = CH_3(CH_2)_6$ and $R =$ cyclohexyl competition from intramolecular C–H insertion was a possibility, this mode of reaction was not in fact observed. However, even though β-elimination is generally considered a relatively low-energy pathway for metal carbene decomposition, there are instances known where alternative modes of decomposition do become competitive. The reactions in eqs. 4[15] and 5[15] are such cases, with intramolecular cyclopropanation dominating in the first example and dimerization in the second, respectively.

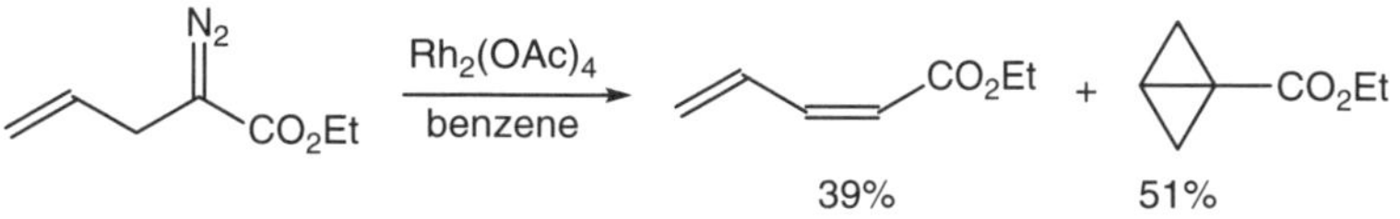

$$\text{(4)}$$

$$\text{(5)}$$

Taber and his co-workers[17,18] have made a thorough study of the factors controlling competition between β-elimination and cyclopropanation/C–H insertion. The stimulus for this study was the realization that fine tuning of chemoselectivity in the decomposition of diazocarbonyl compounds with β-methylene groups would greatly enhance their versatility in synthesis. Taber predicted that β-hydride elimination would be least likely to occur with an α-diazoethylketone, and found confirmation with the substrate in Scheme 8 and several rhodium(II) catalysts.[17] In fact, cyclopentanone formation occurred efficiently with all catalysts,

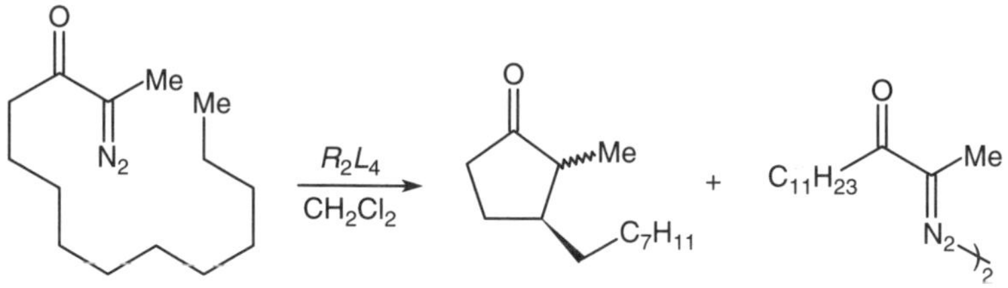

Ligand, L	Yield (%)	Insertion (%)	Azine (%)
CF_3CO_2	89	78	22
CH_3CO_2	82	77	23
$o\text{-}(PhO)C_6H_4CO_2$	78	79	21
CF_3CO_2	82	85	15

Scheme 8

and the major competing side reaction was not β-hydride elimination but dimerization to an azine. Neither the yield of cyclization nor the relative proportion of the two products was much influenced by changing the carboxylate on rhodium, but the extent of azine formation could be minimized by degassing the reaction solvent. This study was then extended to the more problematic substituted diazoketone systems. The latter, exemplified in eq. 6, with rhodium(II) catalysis gave β-hydride elimination with no trace of a cyclopentane from C–H insertion. Ganem and co-workers observed similar behavior with a diazoester (eq. 7).

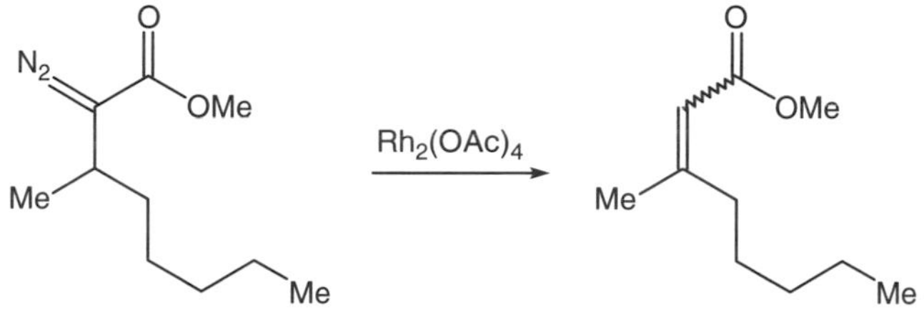

$$(6)$$

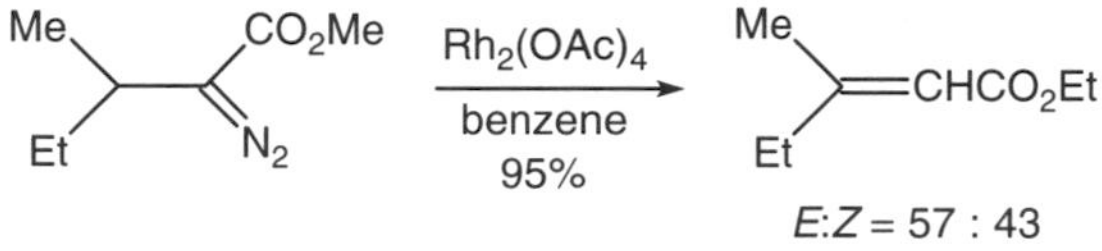

$$E:Z = 57 : 43$$

$$(7)$$

Rhodium (II) acetate promoted alkene formation in 95% yield.[15] The former was explored in both ketone and ester series, and although β-elimination was indeed now the major competing pathway, the nature of the carboxylate on the metal had a more pronounced effect on the product composition than was presented in Scheme 8. The elimination products were exclusively of the Z geometry, as was the case with Ganem's systems in Scheme 6. The results obtained in the diazoketone series are summarized in Scheme 9 and those in the diazoester series in

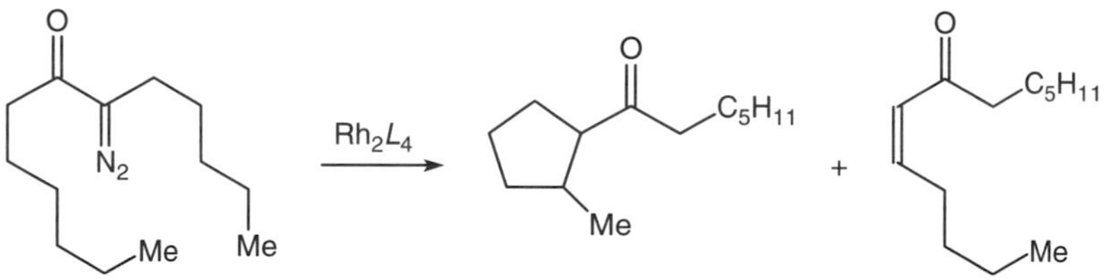

Ligand, L	Yield (%)	Insertion (%)	Elimination (%)
CH_3CO	86	70	30
$PhCO_2$	83	60	40
$(p\text{-Cl})C_6H_4CO_2$	81	40	60
CF_3CO_2	87	38	62

Scheme 9

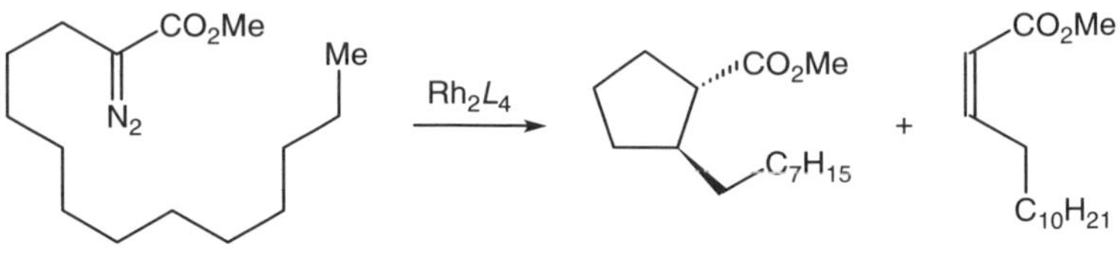

Ligand, L	Yield (%)	Insertion (%)	Elimination (%)
$(CH_3)_3CCO_2$	97	86	14
$n\text{-}C_8H_{17}CO_2$	90	78	22
$C_6H_5CO_2$	87	79	21
CH_3CO_2	92	66	34
CF_3CO_2	93	52	48

Scheme 10

Scheme 10. In both cases rhodium(II) trifluoroacetate was more effective in promoting elimination than rhodium(II) acetate, which favored C–H insertion; rhodium(II) octanoate and the corresponding pivalate were even more chemoselective with the diazoester for C–H insertion (see also Section 3.4). These trends led Taber[17] to suggest that the more electron-withdrawing ligands on rhodium favor β-elimination, while the less electron-withdrawing ligands favor C–H insertion. Interestingly, the cyclopentanone in Scheme 11 is not the only possible

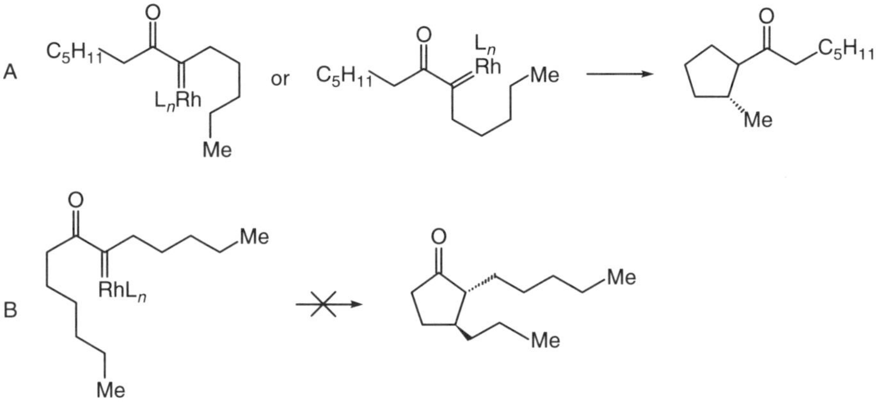

Scheme 11

product of C–H insertion since the metal carbene could also have attacked a methylene group in the chain leading away from the keto group. Taber's[19] group has fine-tuned this diazocarbonyl transformation into a useful and convenient method for the synthesis of (Z)-α,β-unsaturated carbonyl compounds. The combination of rhodium(II) trifluoroacetate as catalyst and low temperature ensured almost complete chemoselectivity for β-elimination with at most only slight traces of C–H insertion. Several representative examples are shown in Table 12.5, and an experimental procedure for the preparation of methyl (Z)-2-

TABLE 12.5 β-Hydride Elimination of α-Diazocarbonyl Compounds

α-Diazo Carbonyl	Elimination product	Yield (%)
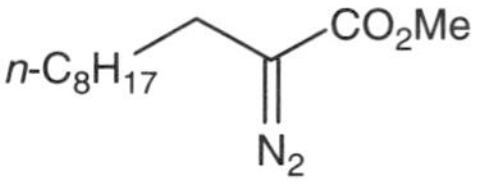	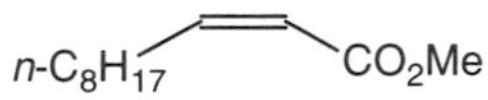	80
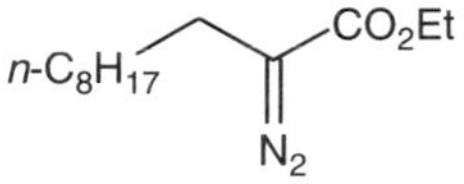	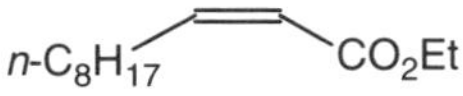	85
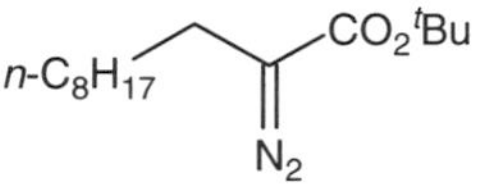	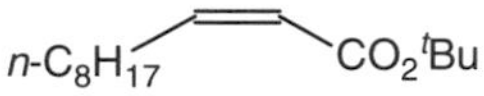	91
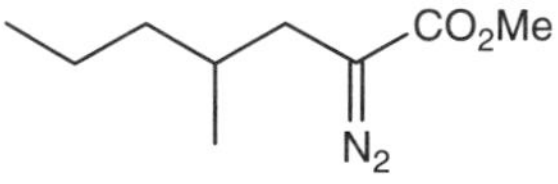	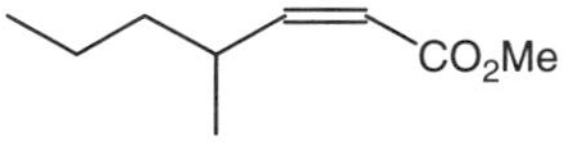	82
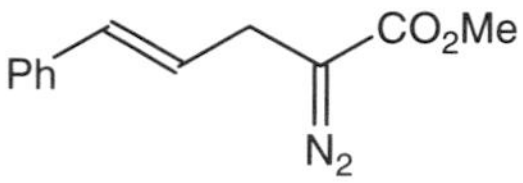	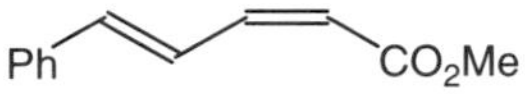	94
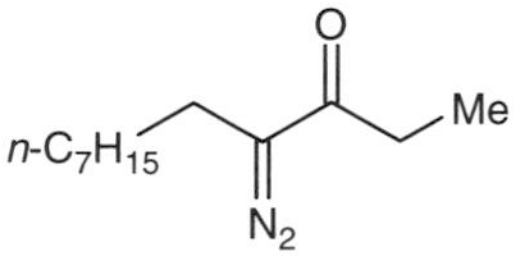	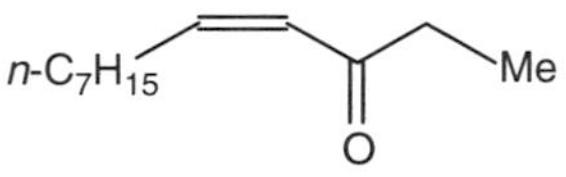	92

undecenoate is described. β-Elimination was also the sole reaction pathway in the decomposition of the diazoketone in eq. 8, even with rhodium(II) acetate.

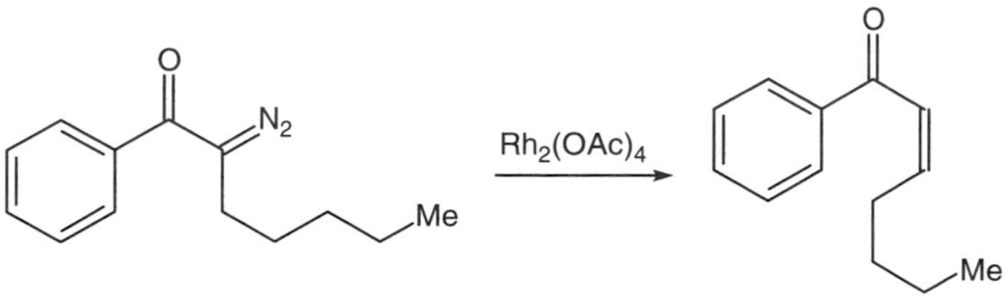

$$\text{(8)}$$

PREPARATION OF METHYL (Z)-2-UNDECENOATE FROM METHYL 2-DIAZOUNDECANOATE[19]

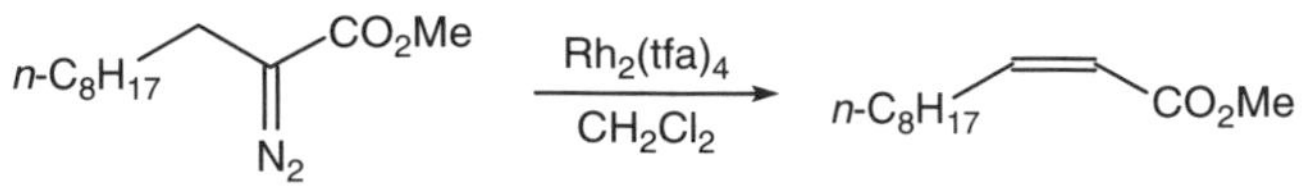

To a solution of methyl 2-diazoundecanoate (0.208 g, 0.92 mmol) in 20 mL of dry CH_2Cl_2 at $-78°C$ under nitrogen was added dropwise a solution of $Rh_2(tfa)_4$ (0.0060 g, 0.01 mmol) in 0.5 mL of dry CH_2Cl_2 (precooled to $0°C$). The mixture was stirred for 1 h, concentrated *in vacuo*, and immediately chromatographed to provide methyl (Z)-2-undecenoate as a colorless oil (0.182 g, 80%).

Taber was later able to extend this pattern of ligand-dependent chemoselectivity to situations where the competition was between β-elimination and cyclopropanation.[17] Thus, while the diazoketone in Scheme 12 did undergo elimination exclusively with rhodium trifluoroacetate, rhodium(II) acetate was chemospecific for cyclopropanation.

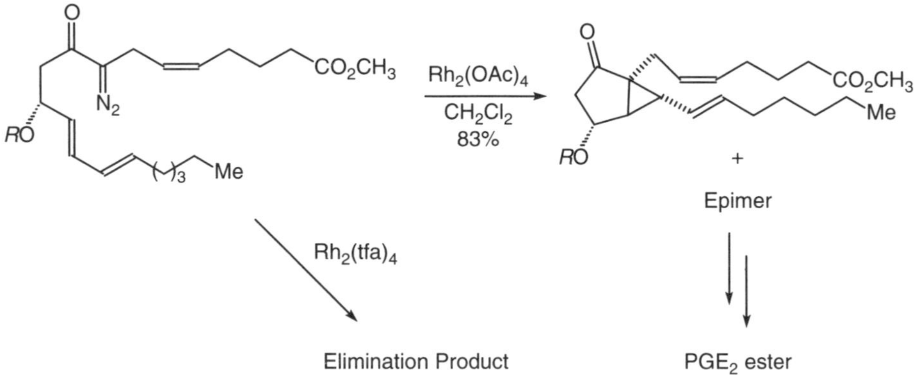

Scheme 12

There are several other examples of diazocarbonyl reactions where β-elimination competes, not just with cyclopropanation, but also with rearrangement. The three related examples in Scheme 13 illustrate just how sensitive the competition is to substituent variation in the diazocarbonyl precursor.[20] In reaction (a) where the reaction center is a diazopinacolone, the product (78%) was exclusively that of

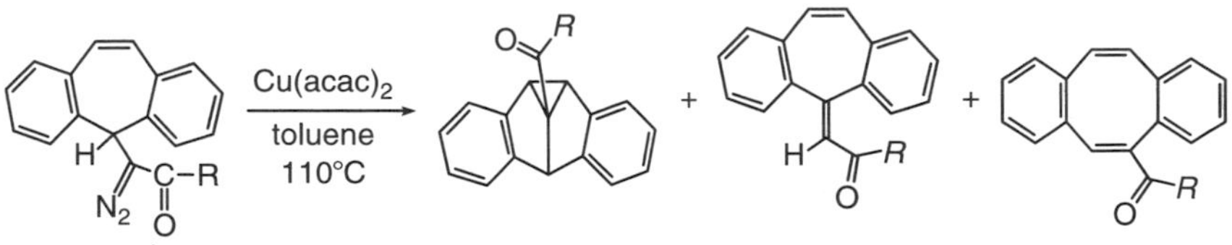

(a) $R = Bu^t$ —   78%   —
(b) $R = Ph$ 30%   —   44%
(c) $R = OMe$ 30%   —   46%

Scheme 13

β-elimination. In contrast, in reactions (b) and (c), the products were those of cy-clopropanation and rearrangement with ring expansion. A possible explanation is that the ease of cyclopropanation in the series varies inversely with steric de-mand. Rearrangement is the sole reaction pathway in palladium-catalyzed de-compositions (eq. 9), where the presence of a methyl substituent adjacent to the

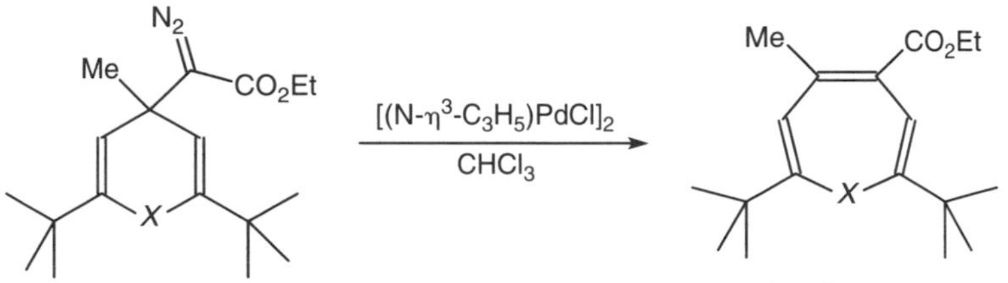

$$X = S : 99\%$$
$$X = O : 99\%$$

$$(9)$$

diazo carbon atom precludes elimination[21,22]; when this position is occupied by hydrogen, elimination is the sole reaction pathway[23] (eq. 10).

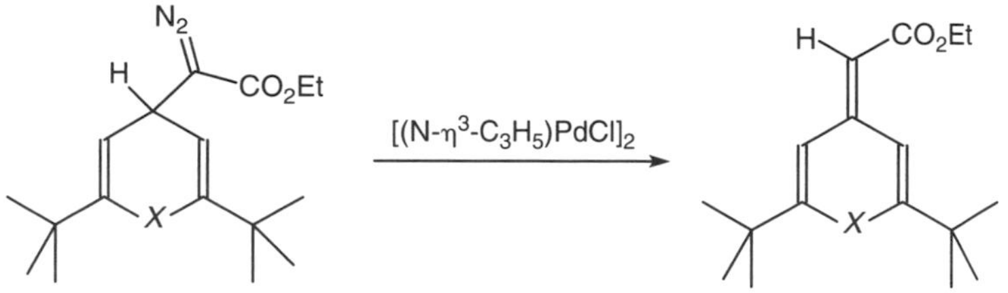

$$X = S : 98\%$$
$$X = O : 100\%$$

$$(10)$$

A final example of rearrangement with ring expansion is found in Eguchi's synthesis of homoadamentene derivatives from methyl (1-adamantyl)diazoacetate (eq. 11).[24] This reaction, catalyzed by rhodium(II) acetate, produced the unstable bridgehead alkene, which could be captured by Diels–Alder reactions with buta-diene and by oxygen, sulfur, and nitrogen nucleophiles.

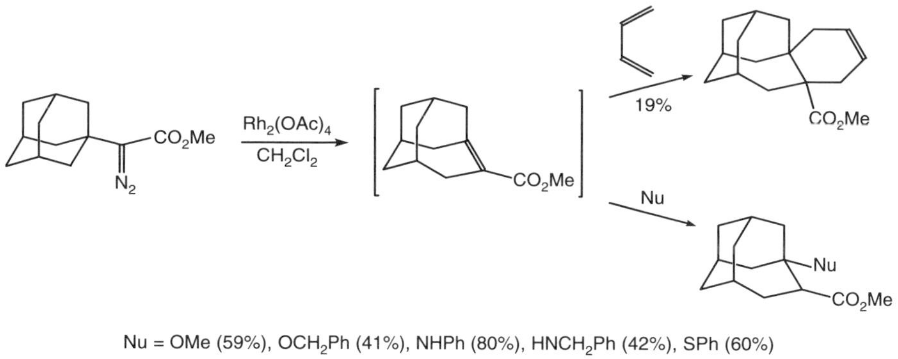

Nu = OMe (59%), OCH₂Ph (41%), NHPh (80%), HNCH₂Ph (42%), SPh (60%)

(11)

12.4 *X–Y* INSERTION REACTIONS

12.4.1 *X–Y* The Halogens

It has been known for many years that chlorine, bromine, and iodine displace nitrogen from α-diazocarbonyl compounds, furnishing α,α-dihalogenated products.[25] Although these reactions have not been employed extensively in simple systems as routes to halogenated ketones, they are central to several syntheses of novel β-lactam antibacterials and β-lactamase inhibitors such as penams and sulbactam from 6-aminopenicillanic acid (6-APA) *via* its 6-diazo derivative. In fact, in a much wider context, diazocarbonyl compounds have acquired a major role in bicyclic β-lactam chemistry, not just through extensive use of 6-diazopenicillinates for functional group interconversion in 6-APA, but as intermediates for vital ring closure *via* N–H insertion reactions in penem total synthesis (Chapter 8). Although the 6-diazo derivative of 6-APA can be isolated and is stable, it is frequently generated and used *in situ* (Scheme 14).

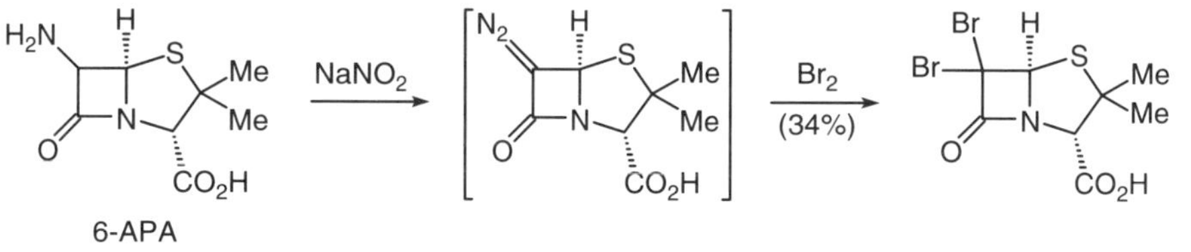

Scheme 14

Clayton[26] was among the first to observe the formation of a dibromide from 6-APA under diazotization conditions. Volkmann and his co-workers[27] obtained the dibromide in 60% yield by adding 6-APA to a cold dichloromethane–sulfuric acid mixture containing sodium nitrite and bromine. Diazotization/bromination of 6-aminopenicillanic acid *S,S*-dioxide can be significantly influenced by carrying out the reaction in the presence of an alcohol. Kapur and Fasel[28] reported that nearly 90% yield of the dibromide can be obtained by replacing sulfuric acid with hydrobromic acid in the presence of methanol. Oxidation of the dibromide to the sulfone followed by reductive removal of both halogen atoms furnishes the antibiotic sulbactam (Scheme 15). The dibromide is also a key intermediate in the

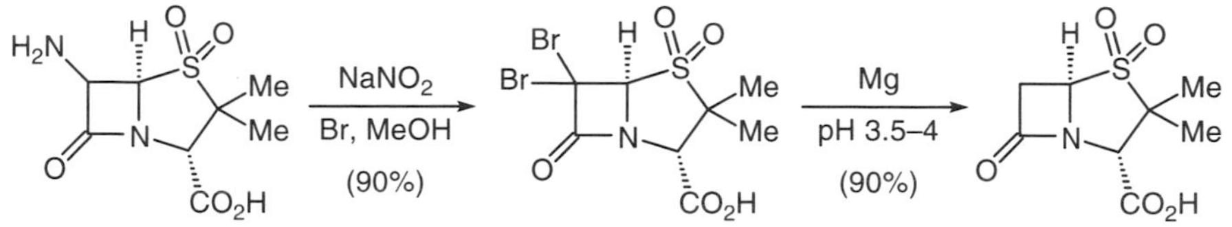

Scheme 15

synthesis of cephalosporin and carbapenam classes of β-lactam antibiotics.[29] Furthermore, with a chain reaction mediated by tributyltin radical, the ester derivative of the dibromide can be converted stereoselectively to the 6β-alkylpenicillanate.[30]

The 6,6-diiodo derivative[26] can also be produced by a diazotization procedure and mixed halides are also accessible, IBr and ICl furnishing the adducts in Scheme 16.[27,31]

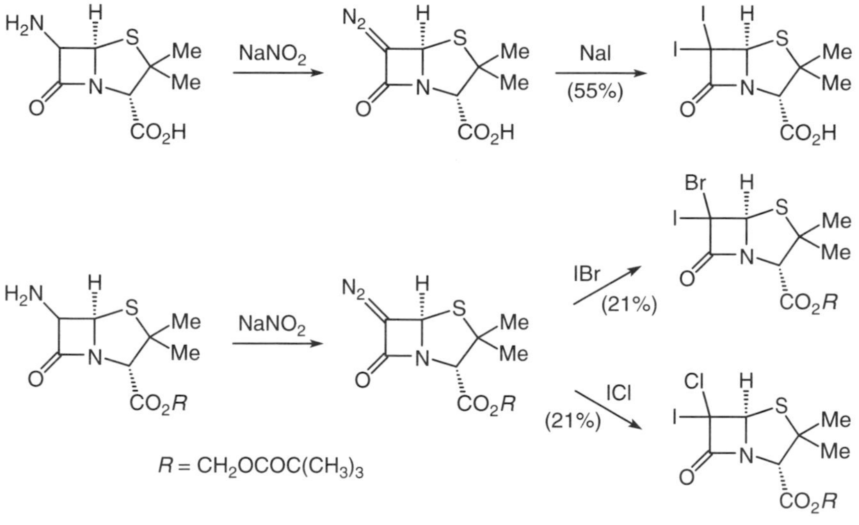

$R = CH_2OCOC(CH_3)_3$

Scheme 16

The reaction of fluorine with diazocarbonyl compounds is destructively exothermic and it was not until Leroy and Wackeslman[32] introduced the moderating device of diluting this halogen with an inert gas at low temperatures that α,α-difluoro adducts could be isolated. Patrick and co-workers[33] used Freon 11 as the inert diluent to convert 2-diazocyclohexanone into 2,2-difluorocyclohexanone in 65% yield (eq. 12). A somewhat similar set of conditions was used to prepare the difluoroanthrone in eq. 13.

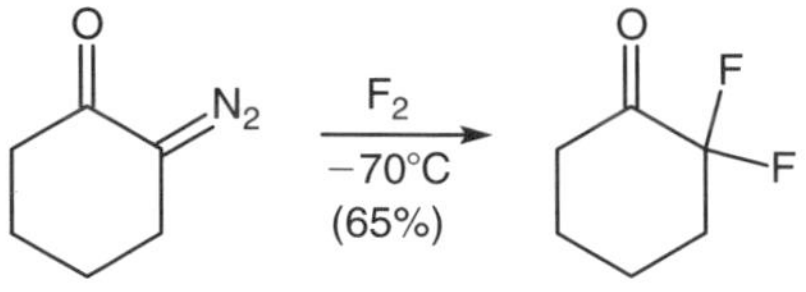

$$
\begin{array}{c}
\text{(structure)} \xrightarrow[\substack{-70°C \\ (65\%)}]{F_2} \text{(structure)}
\end{array}
\tag{12}
$$

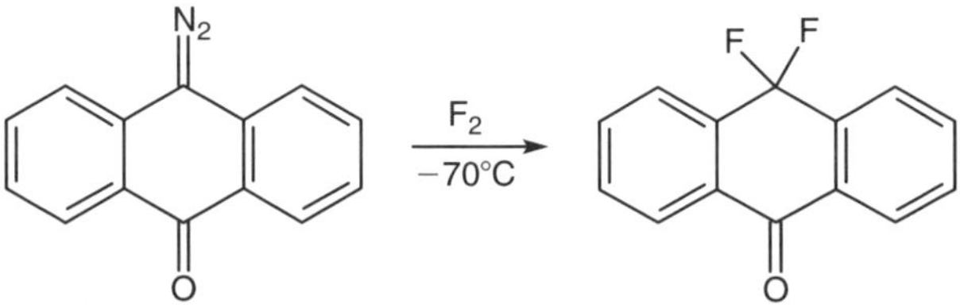

$$
\text{(structure)} \xrightarrow[-70°C]{F_2} \text{(structure)}
\tag{13}
$$

Mixed α-fluoro-α-halo ketones are formed from α-diazoketones on reaction (eq. 14) with either halide ion or N-halosuccinimide in 70% polyhydrogen fluoride

$$
RCOCHN_2 \xrightarrow[\substack{-15°C \\ (32-95\%)}]{\substack{\text{halide source} \\ 70\% \text{ HF/pyridine}}} RCOCHXF
\tag{14}
$$

$$R = Ph, C_6H_{11}, C_2H_5O \qquad\qquad X = Cl, Br, I$$

at 0°C.[34] Recently, Mascaretti[35] reported the stereoselective synthesis of 6-fluoro-6-bromopenicillanates from the 6-diazo derivative using the combined action of N-bromosuccimimide and tetrabutylammonium bifluoride in dichloromethane.

12.4.2 *X–Y* Arenesulfonyl Halides (ArS–*X*)

Of the various reactions of diazocarbonyl compounds with sulfur-containing reagents that have found applications in synthesis are those involving thiols, disulfides, and arylsulfenyl halides. Reactions with thiols have been discussed in Chapter 8. Reactions with disulfides are sparse and unselective; arylsulfenyl halides, invariably benzenesulfenyl chloride, on the other hand, do provide useful products with diazocarbonyl compounds. The original observation that benzene-sulfenyl chloride reacts spontaneously at or below room temperature with diazo-carbonyls to form α-chloro-α-phenylthio adducts, for example, eqs. 15 and 16,

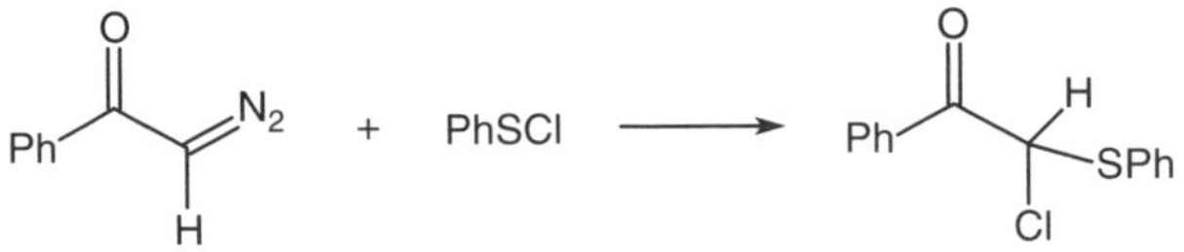

$$
\text{(structure)} + PhSCl \longrightarrow \text{(structure)}
\tag{15}
$$

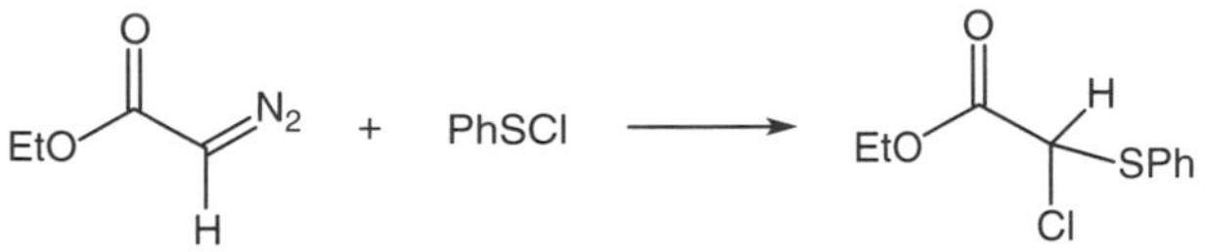

EtO—C(=O)—C(=N$_2$)—H + PhSCl ⟶ EtO—C(=O)—C(H)(Cl)(SPh)   (16)

was made in 1955 by Weygand and Bestmann.[36,37] The implications for synthesis of this double substitution, which places two reactive substituents adjacent to a carbonyl group, have been exploited in a number of ways. First, addition of PhSCl to diazoketones with β-hydrogen substituents followed by hydrogen chloride elimination offers a route to 2-sulfenylated α,β-enones. In practice, mixing the reactants at 0°C followed by the addition of triethylamine is all that is necessary to bring about such transformations (Scheme 17), examples of which are shown in Scheme 18.[38] Second, in the presence of Lewis acids, the α-chloro-α-

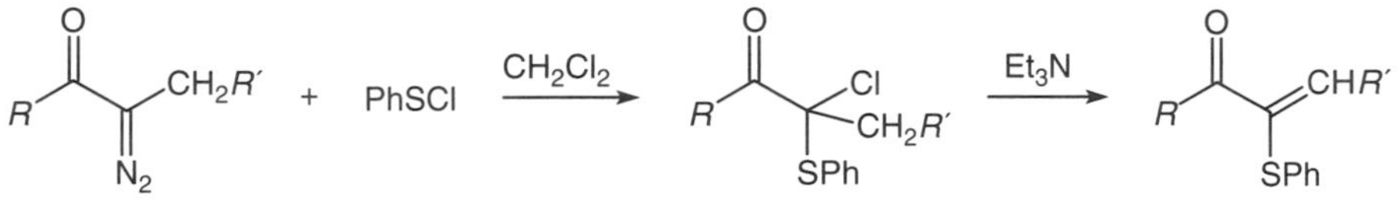

R—C(=O)—C(=N$_2$)—CH$_2$R′ + PhSCl $\xrightarrow{\text{CH}_2\text{Cl}_2}$ R—C(=O)—C(Cl)(SPh)—CH$_2$R′ $\xrightarrow{\text{Et}_3\text{N}}$ R—C(=O)—C(SPh)=CHR′

Scheme 17

$n = 1$ (75%); $n = 2$ (80%); $n = 3$ (79%); $n = 4$ (73%)

(71%)

(80%)

(81%)

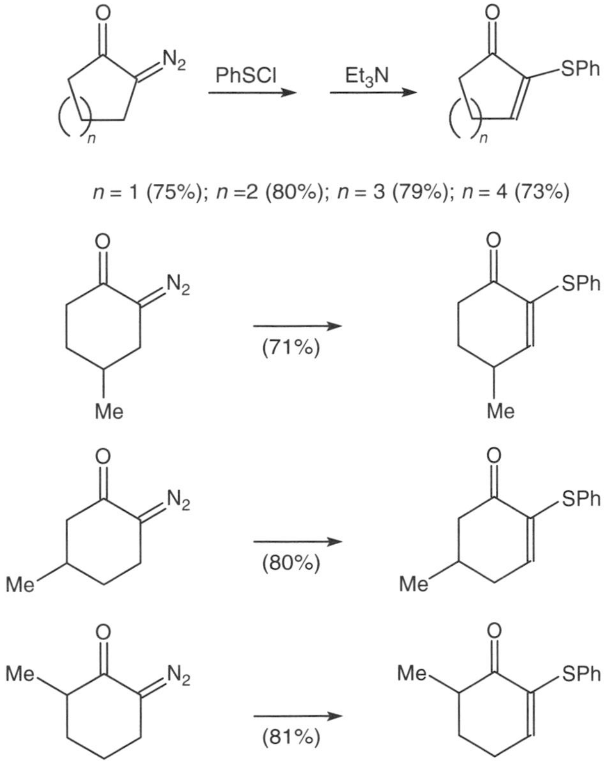

Scheme 18

(phenylthio) adducts are powerful electrophiles for intermolecular and intramolecular aromatic alkylation. For example, sequential treatment of diazoacetone (Scheme 19) with benzenesulfenyl chloride and benzene with stannic chloride ca-

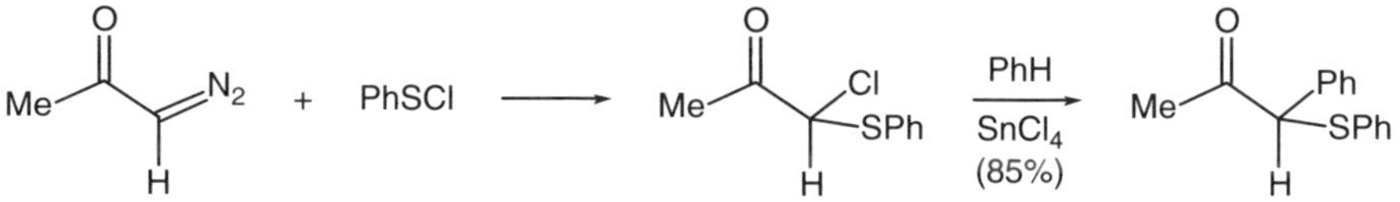

Scheme 19

talysis affords the sulfenyl benzyl ketone in 85% yield.[38] The intramolecular version of this process, leading to a sulfenylated β-tetralone is illustrated in eq. 17. Yet

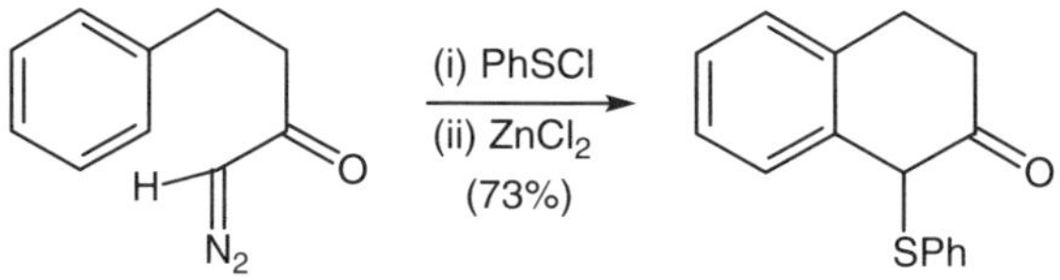

$$(17)$$

a third outlet for these α,α-adducts of diazocarbonyl compounds and PhSCl is their ready conversion to thioacetals and thioketals on exposure to thiophenol with zinc chloride catalysis.[39] Three representative examples are shown in eqs. 18–20.

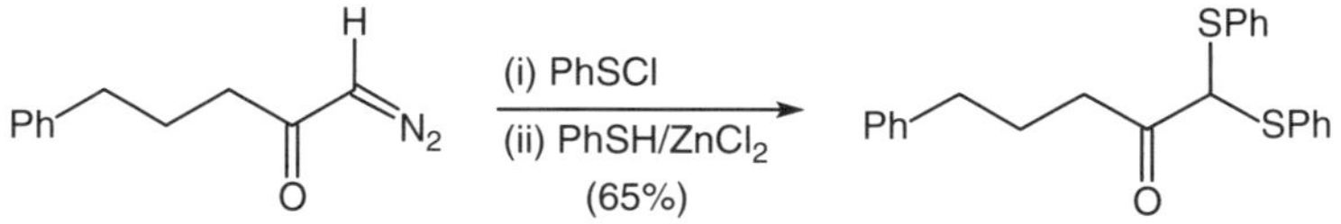

$$(18)$$

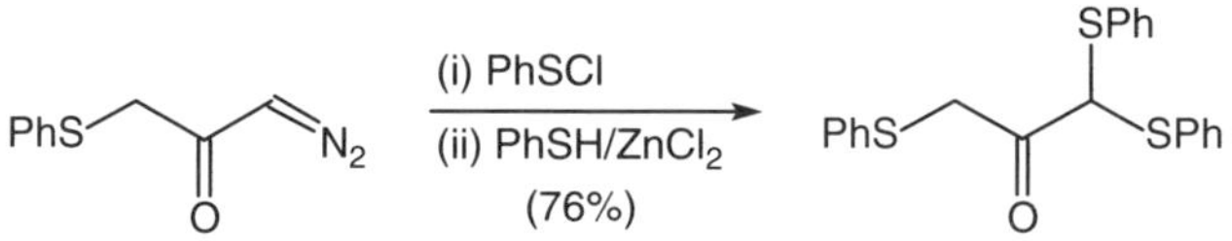

$$(19)$$

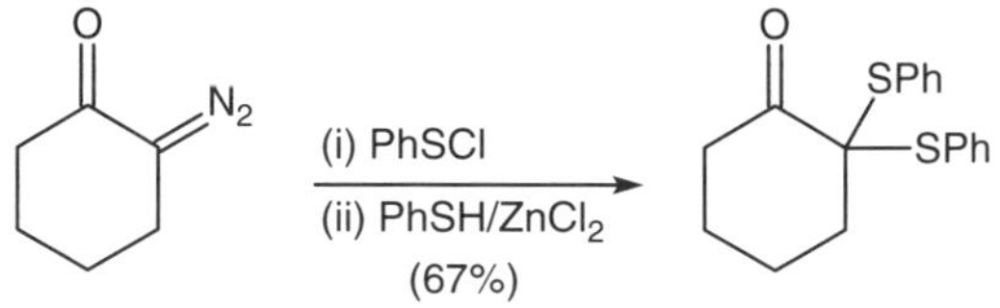

$$(20)$$

12.4.3 *X–Y* Areneselenyl Halides and Diphenyl Diselenide (ArSe–*X*)

There are close similarities between the reactions of diazocarbonyl compounds with arylsulfenyl halides and those of their organoselenium counterparts, although the latter are less well developed. Pellicciari and his co-workers[40] found that ethyl diazoacetate and diphenyl diselenide under copper catalysis produced the α,α-diphenylselenyl adduct shown in eq. 21; boron trifluoride etherate was

$$N_2CHCO_2Et \xrightarrow[\substack{Cu \\ BF_3 \cdot OEt_2}]{PhSeSePh} (PhSe)_2CHCO_2Et \quad \begin{array}{l} 46\% \\ 43\% \end{array} \tag{21}$$

active catalytically, though the yields from either process were poor. Thomas and his co-workers[41] used several selenium reagents, including diphenyl diselenide and phenylselenyl chloride, the latter not requiring a catalyst, to functionalize the 6-position of 6-diazopenicillinate esters (eq. 22). Buckley, Kulkowit, and

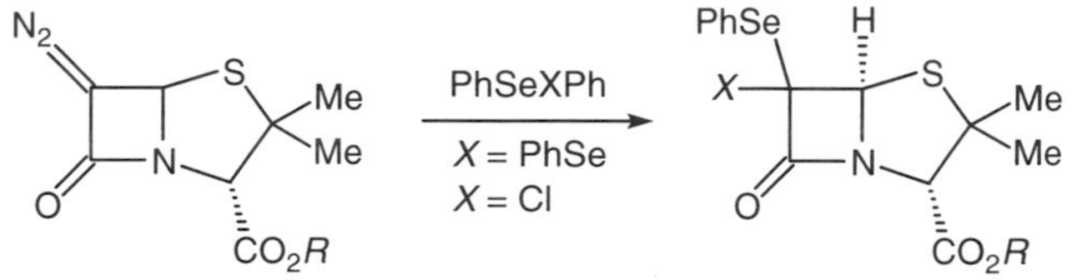

$$\xrightarrow[\substack{X = PhSe \\ X = Cl}]{PhSeXPh} \tag{22}$$

McKervey[42,43] found that several benzeneselenyl derivatives PhSe–X (X = Cl, Br, I, OAc, and SCN) react readily with α-diazoketones, furnishing α,α-adducts of the type shown in eq. 23.[44] The reactions with PhSeSCN proceeded more

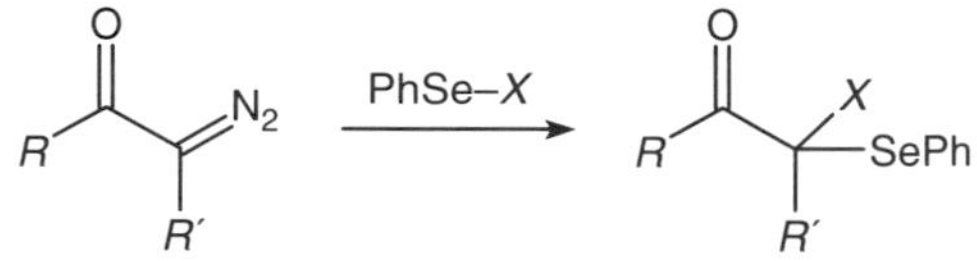

$$\xrightarrow{PhSe-X} \tag{23}$$

R' = H, alkyl, or aryl
X = Cl, Br, OAc, SCN

slowly than with PhSeCl or PhSeBr, though addition of a catalytic amount of rhodium(II) acetate did increase the reaction rate considerably. PhSeI and PhSeOAc were generated *in situ* from PhSeSePh and iodine, and PhSeCl and silver acetate, respectively, before addition of the diazoketone. Some transformations of these α,α-adducts are shown in Schemes 20–22. PhSeI was generated *in situ* from diphenyl diselenide and iodine in acetonitrile, to which solution was added the diazoketone. Upon further treatment with methanol and sodium bicarbonate the iodo adducts were transformed into α-methoxy-α-phenylselenyl ketones (Scheme 22).

12.4.4 *X–Y* Trialkylboranes (R–BR_2)

Trialkylboranes, generated from alkenes *via* hydroboration, react with α-diazocarbonyl compounds with expulsion of nitrogen to form intermediates that on hy-

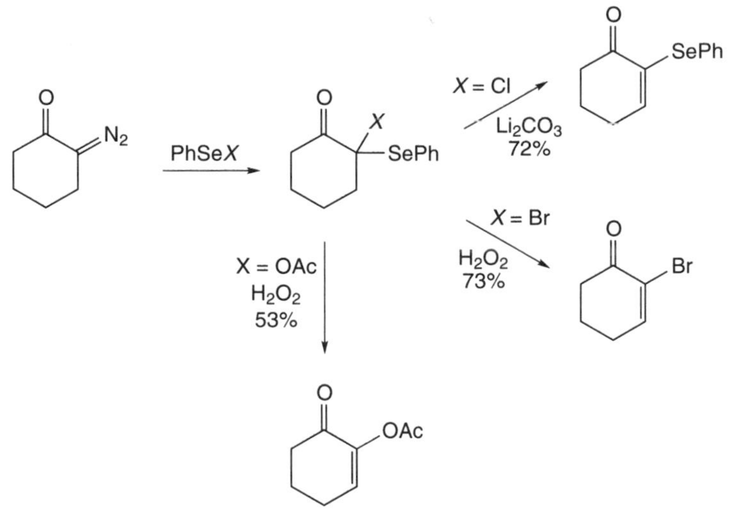

Scheme 20

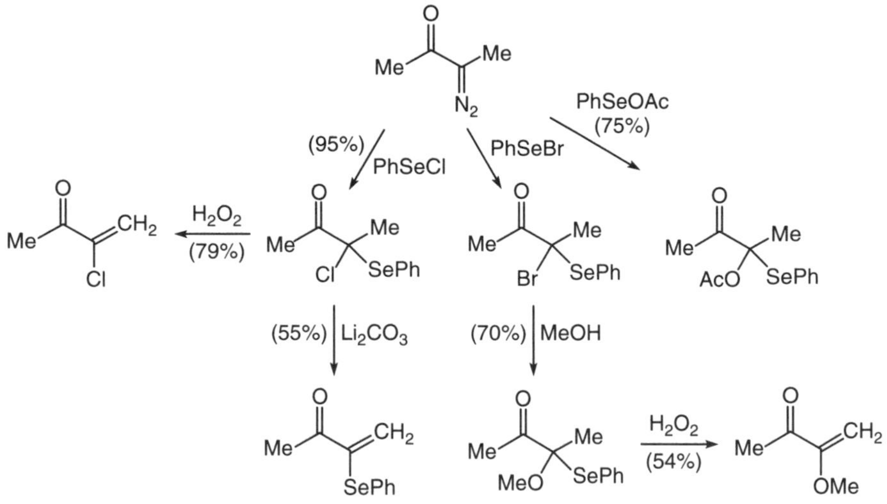

Scheme 21

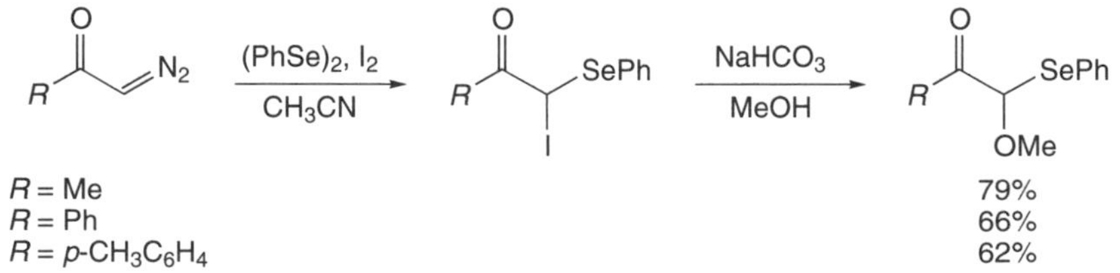

Scheme 22

R = Me
R = Ph
R = p-CH₃C₆H₄

PREPARATION OF 2-PHENYLSELENYL-2-CYCLOHEXEN-1-ONE FROM 2- DIAZOCYCLOHEXANONE[43]

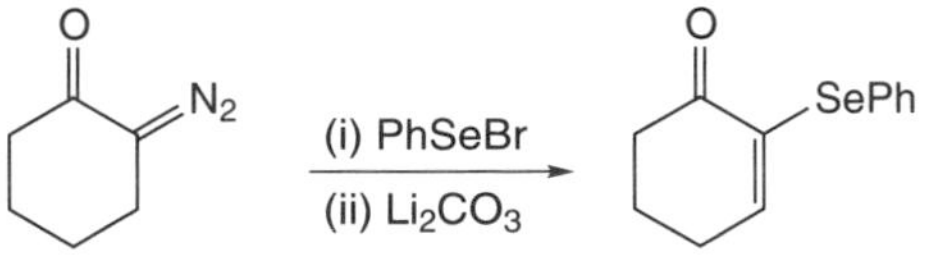

To a 5% solution of phenylselenyl chloride (1 equiv.) in CH_2Cl_2 was added dropwise with stirring at room temperature a 5% solution of 2-diazocyclohexanone in the same solvent. Evolution of nitrogen was observed over the addition period. The solvent was removed under reduced pressure, and the residue was dissolved in dry DMF to make a 5% solution. Dehydrochlorination was then accomplished by the dropwise addition of the DMF solution over approximately 15 min to a stirred suspension of anhydrous lithium carbonate (2.5 equiv.) in dry DMF at 110–120°C under nitrogen. Stirring was continued at this temperature for 1.5–2 h. The cooled mixture was filtered and the filtrate concentrated at reduced pressure to leave a solid residue that was taken up in chloroform and washed with water. After drying the chloroform solution was concentrated and the residue purified by chromatography on silica gel to yield 2-phenylselenyl-2-cyclohexen-1-one, m.p. 49.5–51.0°C, in 74% yield.

drolysis furnish α-alkylated carbonyl compounds. Hooz and his co-workers[45–48] have reported such alkylations with ethyl diazoacetate, diazoacetone, diazoacetophenone, diazoacetaldehyde, and diazoacetonitrile. The process, overall, summarized in Scheme 23, represents the homologation of an alkene to a carbonyl compound or a nitrile.

$$\text{Alkene} \longrightarrow R_3B \xrightarrow{N_2CHX} \xrightarrow{H_2O} RCH_2X$$

$X = CO_2Et$ (40–83%), $COCH_3$ (36–89%), CHO (33–88%), CN (54–100%)

Scheme 23

Alkylation of bis-diazoketones is also possible, using triethylborane to furnish symmetrical diketones of the type shown in eq. 24 in high yield.[49] Alkylation

$$(CH_3CH_2)_3B \; + \; N_2CHCO(CH_2)_nCOCHN_2 \longrightarrow CH_3CH_2CH_2CO(CH_2)_nCOCH_2CH_2CH_3$$

$$n = 2, 3$$

(24)

with boranes derived from terminal alkenes and cyclopentene proceed rapidly and in very good yields at low temperatures, while sterically hindered boranes react sluggishly to give lower yields of products.[46]

Alternatively, it is possible to use an alkyl or an aryldichloroborane derived from an alkene *via* hydroboration with dichloroborane to homologate ethyl diazoacetate. In the most recent extension of this methodology, Brown and Salunkhe[50] have shown that *trans*-1-alkenyldichloroboranes, prepared by hydroboration of 1-alkynes, react smoothly with ethyl diazoacetate at −65°C with the liberation of nitrogen. Hydrolysis of the intermediates with methanol−water furnished *trans*-β,γ-unsaturated esters stereoselectively in good yields. Similar treatment of *cis*-1-alkenyldichloroboranes yielded the corresponding *cis*-β,γ-unsaturated esters. Representative examples in the *trans* and *cis* configurations are shown in eqs. 25 and 26. Isomeric purity was ≥99% in each case.

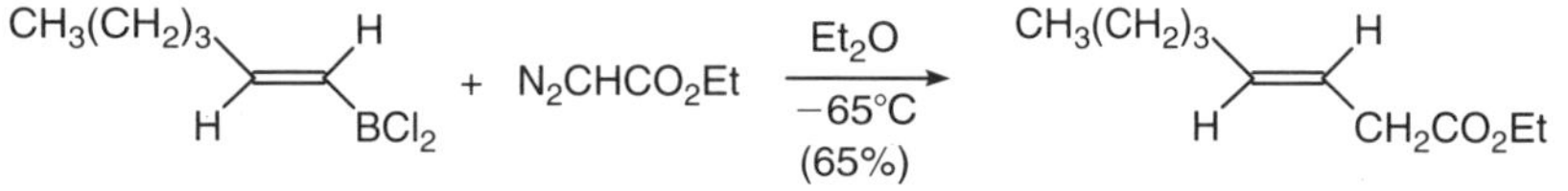

$$CH_3(CH_2)_3 \diagup\!\!=\!\!\diagdown H + N_2CHCO_2Et \xrightarrow[\substack{-65°C \\ (65\%)}]{Et_2O} CH_3(CH_2)_3 \diagup\!\!=\!\!\diagdown CH_2CO_2Et \quad (25)$$

$$CH_3(CH_2)_3 \diagup\!\!=\!\!\diagdown BCl_2 + N_2CHCO_2Et \xrightarrow[\substack{-65°C \\ (60\%)}]{Et_2O} CH_3(CH_2)_3 \diagup\!\!=\!\!\diagdown CH_2CO_2Et \quad (26)$$

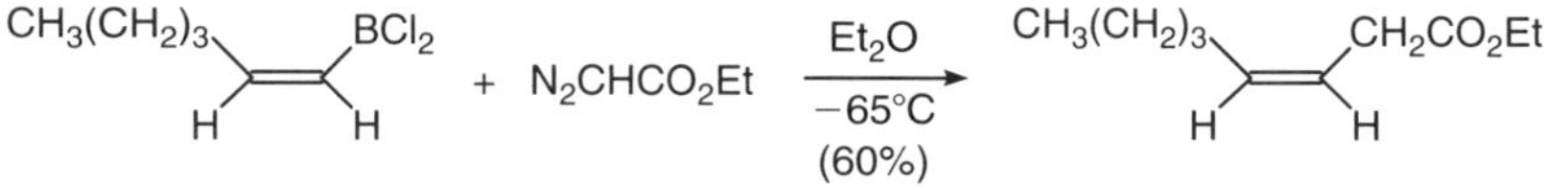

When D_2O is used instead of H_2O in the hydrolytic stage of the process, high yields of site-specific monodeuterated esters and ketones can be produced. Alkylation of ethyl diazoacetate with cyclopentene (Scheme 24) provides an illustrative example.

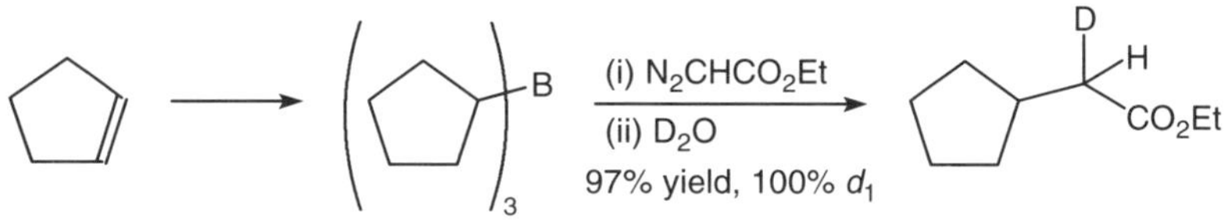

Scheme 24

Newman[51,52] has applied this methodology to the synthesis of D-*threo*-sphinganine (Scheme 25). Thus treatment of the diazoketone derived from the doubly protected L-serine with a trialkylborane gave a ketone. A stereoselective reduction of the ketone with tri-*tert*-butoxyaluminum hydride produced the desired *threo* configuration in the alcohol. Acidic methanolysis of the acetoxy group, followed by hydrazinolysis of the phthaloyl moiety, gave D-*threo*-sphinganine (Scheme 25).

The intermediates responsible for this type of behavior are believed to be enol borinates formed in a sequence commencing with coordination of the diazocarbonyl compound to the acidic trialkylborane and followed by a rapid 1,2-alkyl

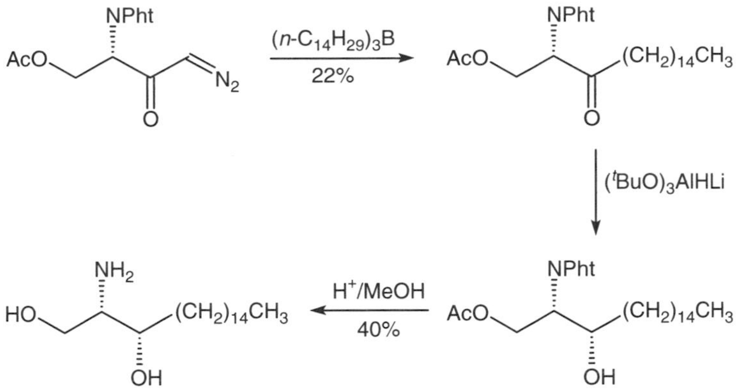

Scheme 25

shift from boron to carbon with loss of nitrogen. Migration of boron to oxygen forms the enol borinate, which is rapidly hydrolyzed by water (Scheme 26).

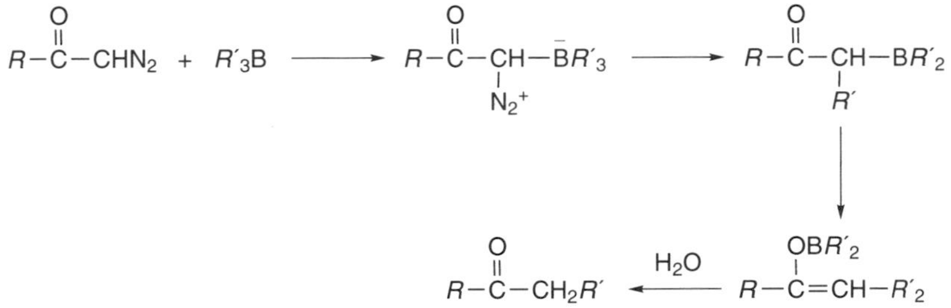

Scheme 26

Hooz and Pasto and their co-workers recognized that under anhydrous conditions these enol borinates could be intercepted in a variety of synthetically useful processes. Various ways were devised of producing both internal and terminal enol borinates in a regiospecific fashion from diazocarbonyl precursors. Significant applications of these intermediates include regiospecific synthesis of α,α-dialkylated ketones (eq. 27),[53,54] regiospecific synthesis of Mannich bases by reaction with dimethyl(methylene) ammonium iodide (eq. 28),[55] aldol addition to

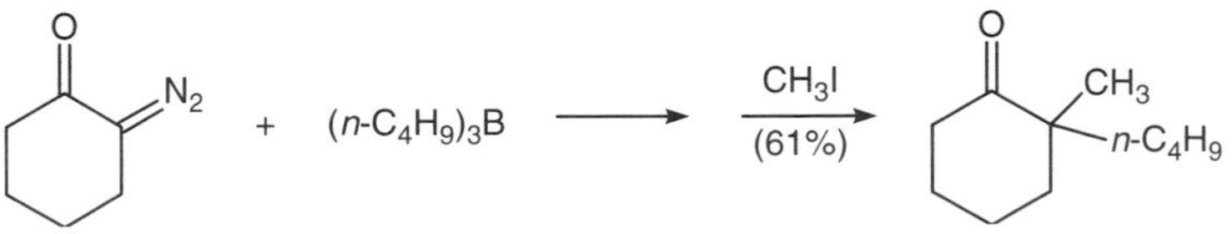

$$(27)$$

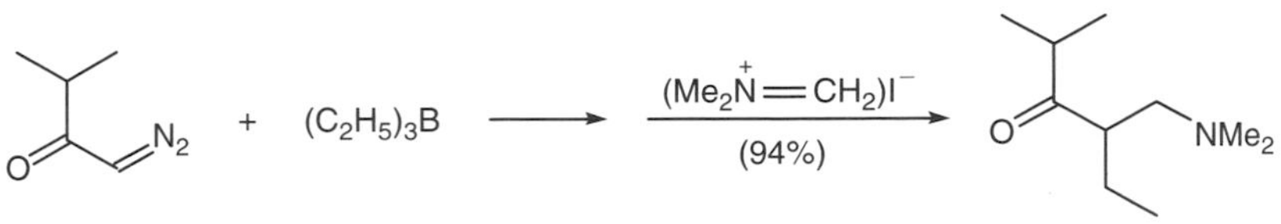

$$(28)$$

aldehydes and ketones (eq. 29),[56] regiospecific synthesis of unsymmetrical acyclic enones by addition of phenylselenyl chloride followed by oxidative elimination (eq. 30),[57] regiospecific synthesis of 1,3-diketones *via* boroxazine formation with nitriles (eq. 31),[58] and regio- and stereocontrolled formation of enol silyl ethers from *N*-(trimethylsilyl)imidazole (eq. 32).[59] In all of these processes a diazocarbonyl precursor was employed to generate the enol borinate.

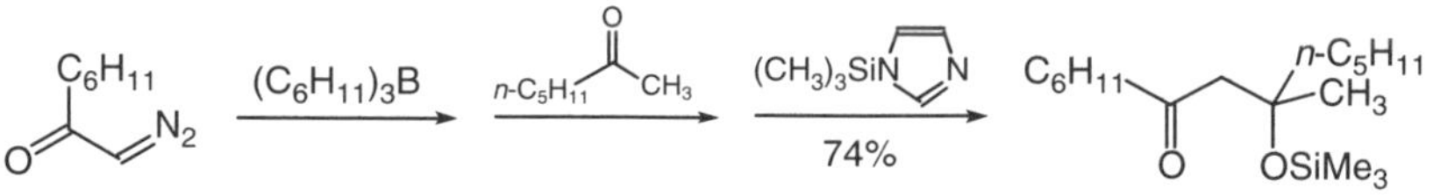

(29)

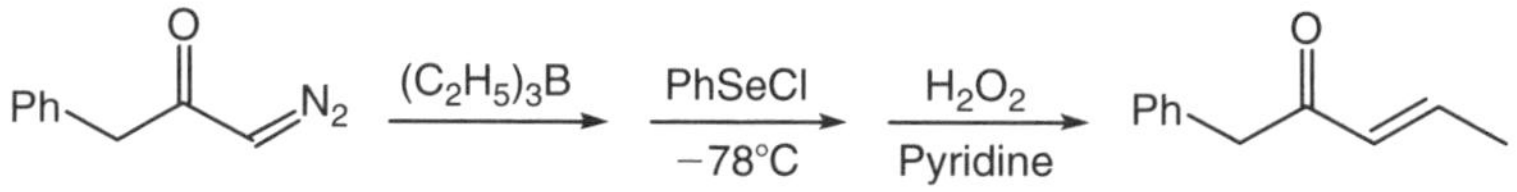

(30)

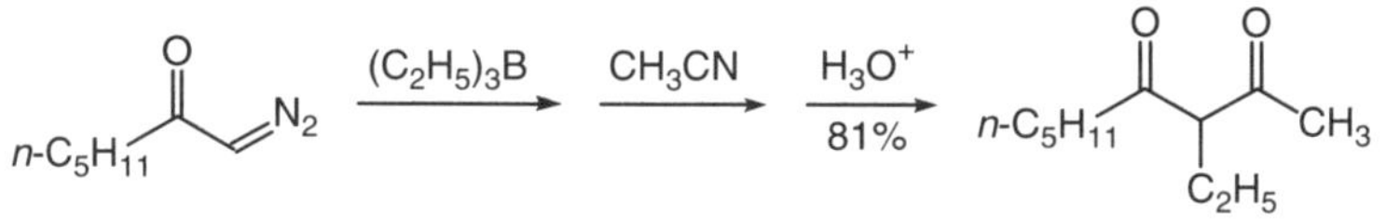

(31)

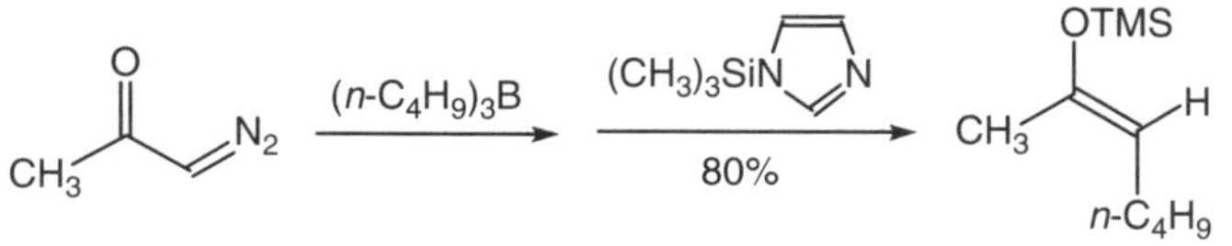

(32)

12.5 DIMERIZATION REACTIONS

The formation of alkenes, formally carbenoid dimers, is occasionally encountered as an unwanted side reaction of diazocarbonyl decomposition, particularly when the intended reaction is sluggish or is otherwise inhibited. There are, however, cases where the dimerization reaction can be quite efficient and synthetically useful. Grunmann[60] first drew attention to this reaction when he reported in 1938 that diazoketones in an inert solvent in the presence of copper oxide furnished symmetrical unsaturated 1,4-dicarbonyl products (eq. 33). Ernest and co-

$$2RCOCHN_2 \xrightarrow{\text{CuO}} RCOCH{=}CHCOR$$

(33)

10–65% yield

workers[61] applied the reaction to the synthesis of both symmetrical and unsymmetrical diketones, including products with terminal ester functions (eqs. 34 and 35), the simultaneous decomposition of two different diazoketones affording un-

$$2RO_2C(CH_2)_nCOCHN_2 \xrightarrow{\text{CuO}} RO_2C(CH_2)_nCOCH=CHCO(CH_2)_nCO_2R \quad (34)$$

$$\begin{array}{c} RO_2C(CH_2)_nCOCHN_2 \\ + \\ RC(CH_2)_nCOCHN_2 \end{array} \xrightarrow{\text{CuO}} RO_2C(CH_2)_nCOCH=CHCO(CH_2)_nR \quad (35)$$

symmetrical dimers, though in poor yield. Rhodium(II) acetate and Cu(acac)$_2$ are also active catalysts for dimer formation, a particularly efficient example being the quantitative formation of tetramethoxycarbonylethene from dimethyl diazomalonate (eq. 36).[62,63] Raney nickel has been found to decompose methyl diazo-

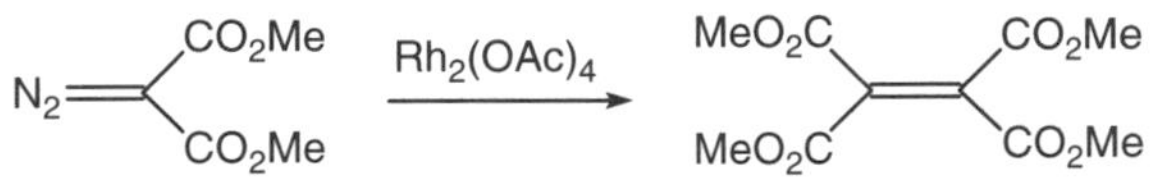

$$(36)$$

acetate quantitatively to a 7:93 mixture of dimethyl fumarate and dimethyl maleate.[64] Ruthenium tetramesitylporphyrin is also an active catalyst for EDA dimerization, affording predominantly diethyl maleate in 91% yield.[65] It seems unlikely that these and other alkene dimers result directly from two metal carbene species coming together. A mechanistic alternative, shown in Scheme 27,

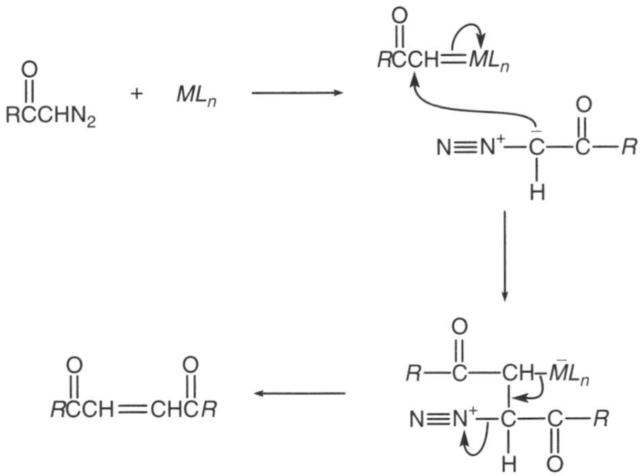

Scheme 27

involves the nucleophilic attack of an uncomplexed diazocarbonyl compound on a metal carbene to form an intermediate diazonium ion. Loss of nitrogen and regeneration of the catalyst completes double bond formation. An example of intermolecular dimerization of a diazoketone derived from the monomethyl ester of azelaic acid and its application to the synthesis of a disubstituted cyclopentenone

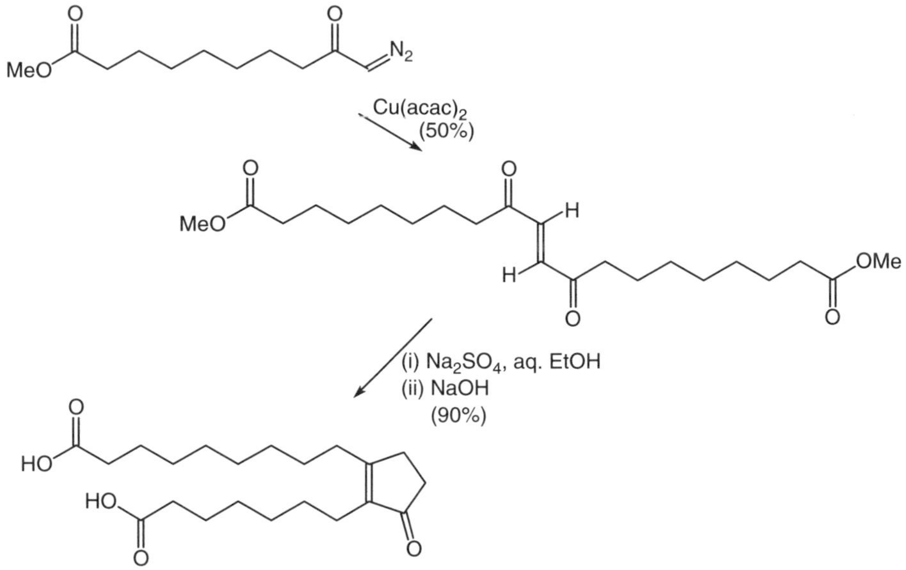

Scheme 28

are shown in Scheme 28.[66] Cu(acac)$_2$ was used as catalyst to effect dimerization in 50% yield to form a *trans*-enedione, which on successive reduction of the double bond with sodium dithionite in aqueous ethanol and aldol condensation with sodium hydroxide furnished the cyclopentenone diacid in 90% yield.

A final example of intermolecular dimerization is shown in eq. 37, in which copper bronze was used to produce the methyl ester of the unusual bis(indole) alkaloid trichotomine.[67]

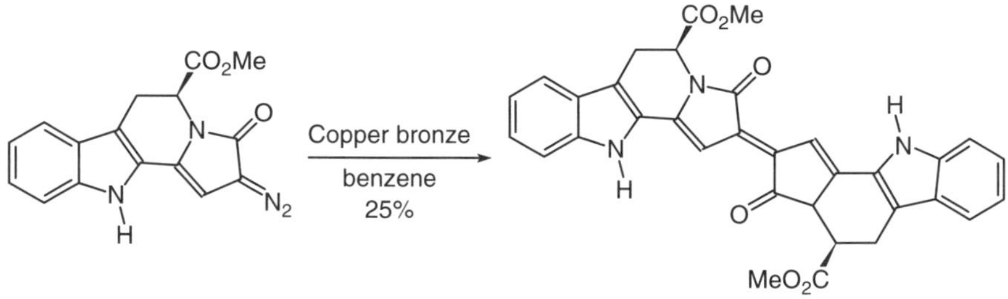

(37)

There are now several examples of intramolecular versions of carbenoid dimerizations. Serratosa and co-workers[68] used Cu(acac)$_2$ to catalyze seven-membered ring formation from the acyclic bisdiazoketone shown in Scheme 29.

Although the yield was poor (15%), the product was suitable for further elaboration into 4-hydroxytroponone on treatment with triethylamine. Kulkowit and McKervey[66] employed Cu(acac)$_2$ in benzene as the catalyst in a systematic study of medium- and large-ring formation from α,ω-bisdiazoketones, with the results shown in Scheme 30.

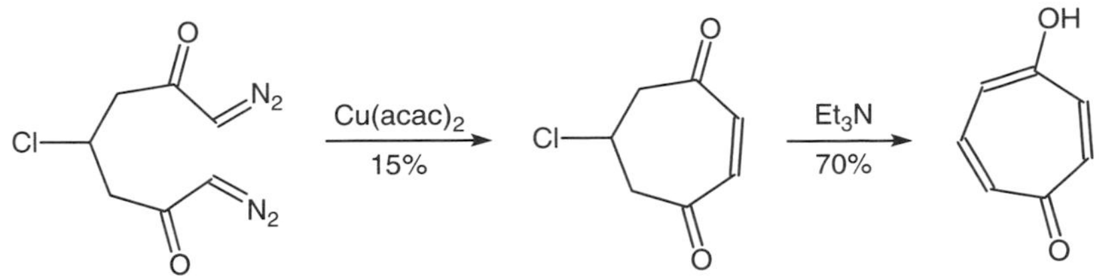

Scheme 29

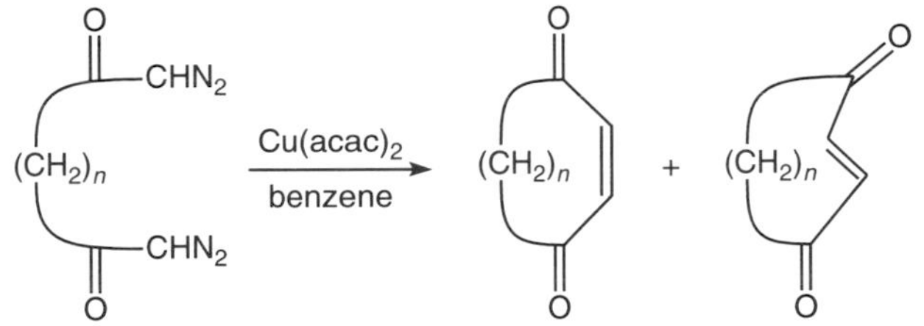

n	Ring size	Yield (%)	$E:Z$ ratio
4	8	30	$<1:20$
5	9	<10	—
6	10	—	—
7	11	30	$3:1$
8	12	71	$10:1$
9	13	25	$10:1$
10	14	80	$9:1$
12	16	44	$3:1$
16	20	29	$7:1$

Scheme 30

This intramolecular coupling worked satisfactorily for 8-, 11-, 12-, 13-, 14-, 16-, and 20-membered cycloalkenediones, but poorly for the 9-membered ring. In the 10-membered ring case coupling did apparently occur, though the product isolated in low yield was that of a subsequent Michael addition of acetylacetone to the enedione. With the exception of cyclooct-2-ene-1,4-dione ($n = 4$), which had the *cis* geometry about the double bond, the enediones were mixtures of *cis* and *trans* isomers, with the latter predominating.

12.6 DIAZOCARBONYL COMPOUNDS AS 1,3-DIPOLES IN [3+2] CYCLOADDITIONS

α-Diazocarbonyl compounds are capable of numerous intermolecular and intramolecular cycloaddition reactions, as exemplified in earlier chapters on cyclopropanation of alkenes, alkynes, and aromatics. In these processes, bond formation occurs exclusively at the carbenic carbon atom. There are two additional modes of cycloaddition in which diazocarbonyl compounds can act as 1,3-dipoles with either retention or loss of the nitrogen moiety. The chemistry of these dipoles is characterized by [3+2] cycloaddition to numerous unsaturated systems.

In the absence of catalysts in purely thermal reactions, a wide variety of α-diazocarbonyl compounds undergo 1,3-dipolar cycloaddition reactions with double bonds conjugated with carbonyl,[69-73] amine,[74] nitrile,[69] and nitro[75] groups, or double bonds as a part of a strained ring system,[76-80] without loss of nitrogen to furnish Δ^1-pyrazolines. For example, reaction of dimethyl itaconate and 1-diazo-2-propanone furnishes the Δ^1-pyrazoline, which undergoes rapid tautomerization to afford the Δ^2-pyrazoline in good yield (Scheme 31).[73] A similar type of

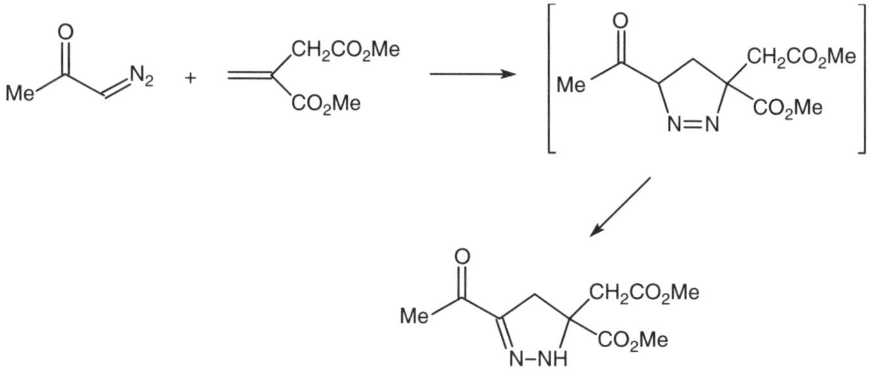

Scheme 31

reaction has been observed in the reaction of 1-diazo-2-propanone with the enolate double bond generated from β-dicarbonyl compounds in the presence of base.[81] Addition of dicyanocyclobutene to EDA gives the pyrazoline cycloadduct in 53% yield, as shown in eq. 38.[79]

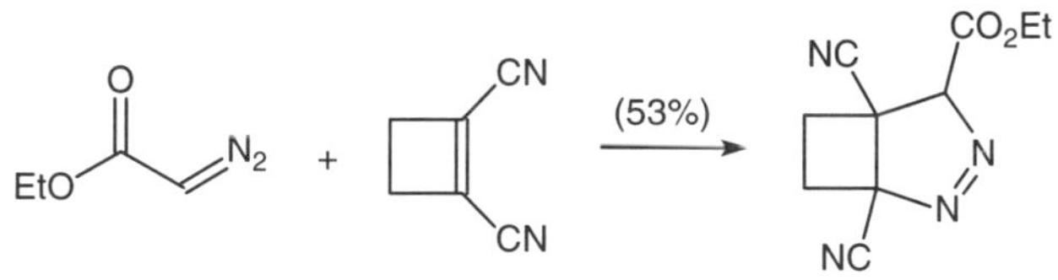

(38)

Analogous, 1,3-dipolar cycloaddition reactions occur with carbon–carbon triple bonds conjugated with carbonyl groups.[82-87] 2-Diazo-3-butanone reacts with dimethyl acetylenedicarboxylate to form the unstable pyrazolenine, which undergoes a transposition leading to $N(2)$-substituted pyrazole by a specific [1,5]-migration of the acyl group to N(2) (Scheme 32).[84] Recently, Mass and co-

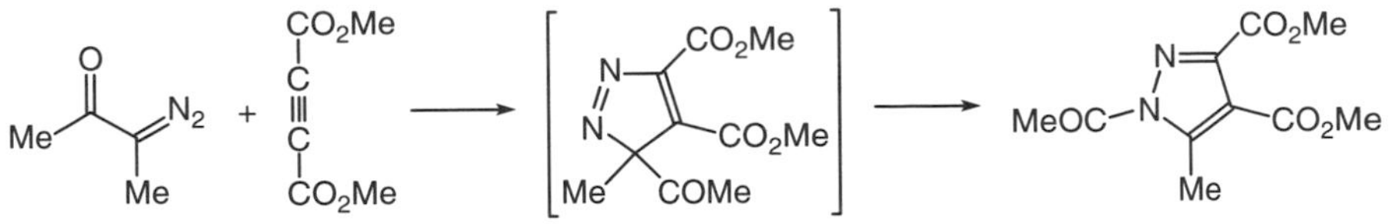

Scheme 32

workers[88] have shown that methyl diazoacetate undergoes [3+2]-cycloaddition to an (alkynyl)(phenyl)iodonium salt to give a (4-pyrazolyl)phenyl iodonium salt (Scheme 33).

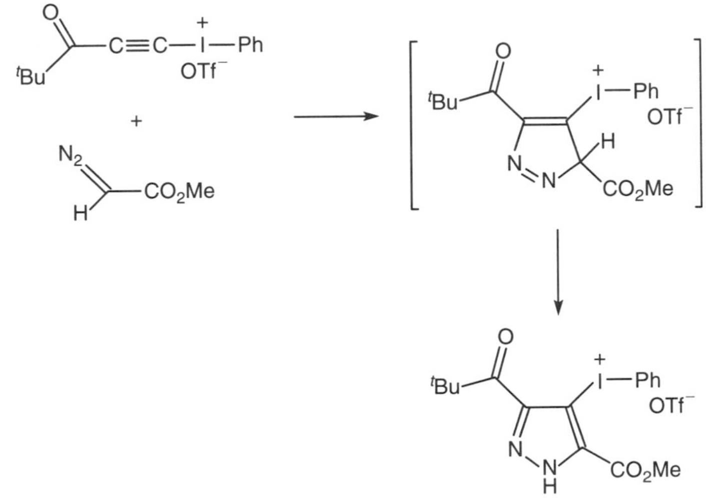

Scheme 33

Cycloaddition of α-diazocarbonyl compounds with hetero double or triple bonds is a minor process. In many cases nucleophilic attack on the electrophilic C–X bond by the diazocarbonyl compound occurs instead of 1,3-dipolar cycloaddition.[89]

There is a second catergory of cycloaddition in which, following loss of nitrogen from the diazocarbonyl compound, the carbene or its metal-bonded equivalent acts as a 1,3-dipole through participation of both the carbonyl oxygen and the carbenic carbon atoms and adds to multiple bonds to form five-membered heterocycles (Scheme 34). This process can thus also be classified as {3+2} cycloaddition. An alternative pathway that leads to the same outcome is cycloaddition of the metal carbene onto the multiple bond to form a three-membered ring, which can subsequently expand with participation of the carbonyl group. Both routes are outlined in Scheme 34. Whatever the details of the mechanism, the process has been used very extensively to synthesize oxazoles, furans, and dihydrofurans from acetylenes and enol ethers/acetates. Their reactions and their applications in total synthesis are discussed in detail in Chapter 4.

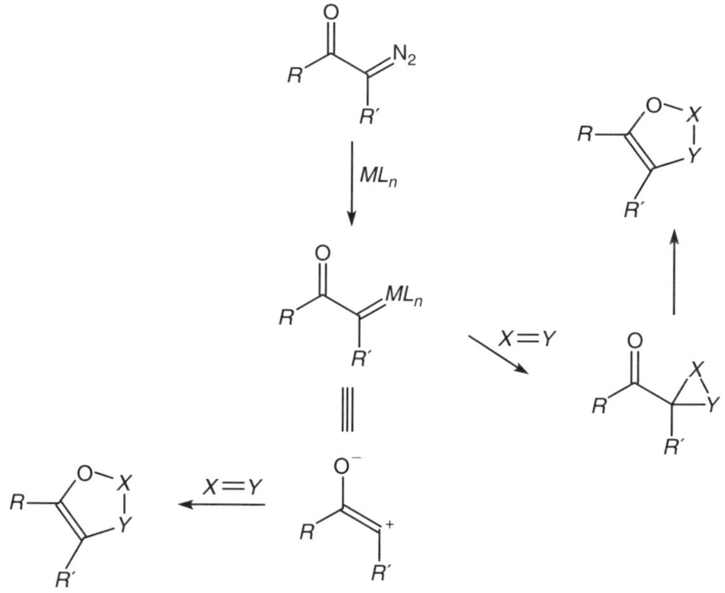

Scheme 34

A somewhat related process is the cycloaddition of diazocarbonyl compounds to nitriles to form oxazoles. Although early examples of this process involve thermal or photochemical conditions, most recent applications favor Lewis acid or transition metal catalysts.[90–102] The Lewis acid version has already been discussed in Chapter 11. The discussion here will be confined to the use of transition metal catalysts. Although the cycloaddition of a diazocarbonyl compound to a nitrile may be viewed mechanistically as involving the formation af a 1,3-dipole, which undergoes [3+2] cycloaddition with the carbon–nitrogen triple bond (cf. Scheme 34), an alternative mechanism may be the formation of a nitrile ylide, as shown in Scheme 35, with subsequent 1,5-cyclization to an oxazole. The

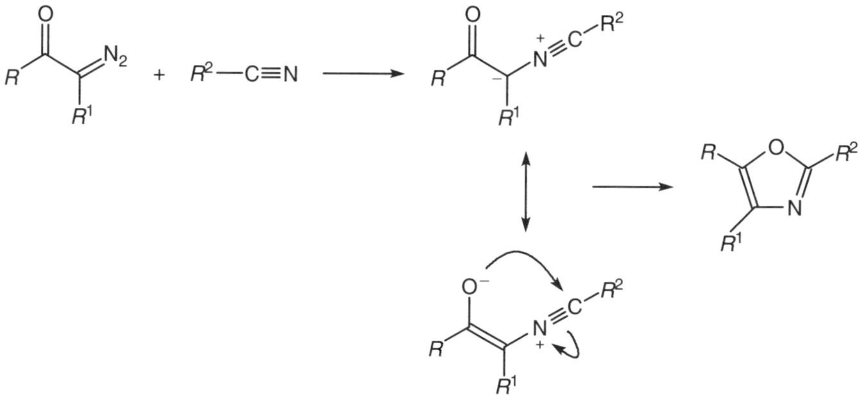

Scheme 35

latter alternative is supported by laser flash photolysis studies of diazo compounds in the presence of nitriles, which point to the transient formation of nitrile ylides.[97] In some cases stable nitrile ylides have been isolated.[98] These studies, however, involve photochemically generated free carbenes under conditions very much different from those pertaining to the formation and reactions of metal carbenes. Helquist and co-workers[91] have used rhodium(II) acetate as a catalyst for oxazole formation from diazoesters and a variety of nitriles. A selection of reactions from these and similar studies are summarised in eqs. 39,[91] 40,[92] 41,[93] 42,[94] 43,[95] 44,[95] and 45.[96] Doyle and Moody[96] studied several combinations

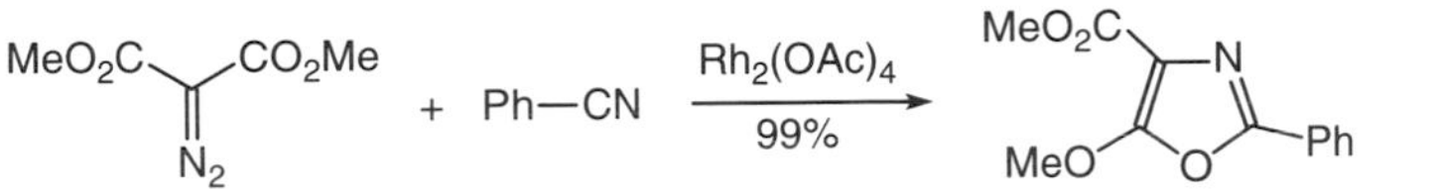

$$\text{MeO}_2\text{C} \diagdown \diagup \text{CO}_2\text{Me} \quad + \quad \text{Ph—CN} \quad \xrightarrow[99\%]{\text{Rh}_2(\text{OAc})_4} \quad (39)$$

$$\text{EtO}_2\text{C} \diagdown \diagup \text{CF}_3 \quad + \quad p\text{-ClC}_6\text{H}_4\text{CN} \quad \xrightarrow[89\%]{\text{Rh}_2(\text{OAc})_4} \quad (40)$$

$$\text{MeO}_2\text{C} \diagdown \diagup \text{CO}_2\text{Me} \quad + \quad \text{EtOCH=CH—CN} \quad \xrightarrow[97\%]{\text{Rh}_2(\text{OAc})_4} \quad (41)$$

$$\text{EtO}_2\text{C} \diagdown \diagup \text{CO}_2\text{Et} \quad + \quad \text{BrCH}_2\text{CN} \quad \xrightarrow[65\%]{\text{Rh}_2(\text{OAc})_4} \quad (42)$$

$$\text{MeO}_2\text{C} \diagdown \diagup \text{CO}_2\text{Me} \quad + \quad \text{Me}_3\text{C—C(OTBS)—CN} \quad \xrightarrow[97\%]{\text{Rh}_2(\text{OAc})_4} \quad (43)$$

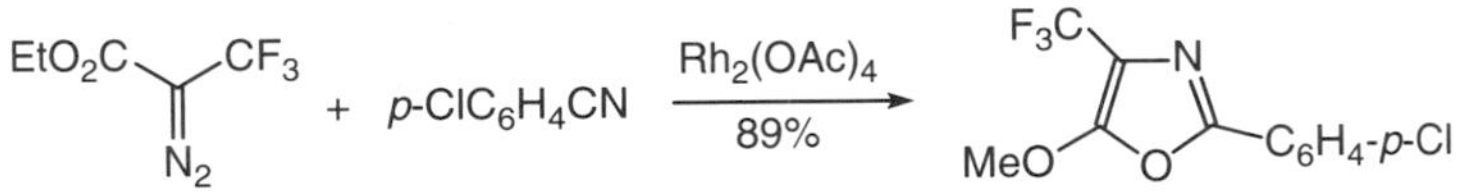

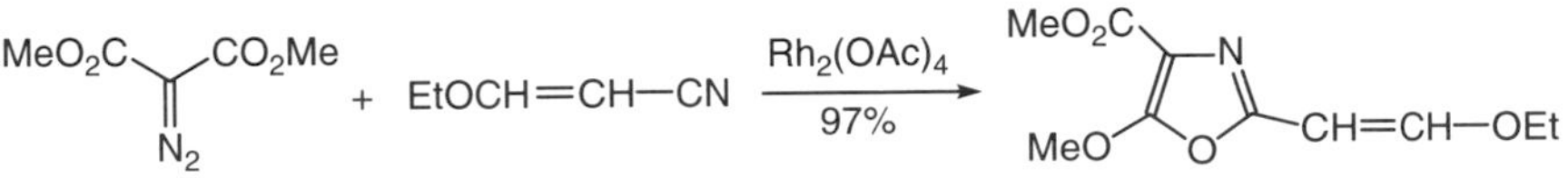

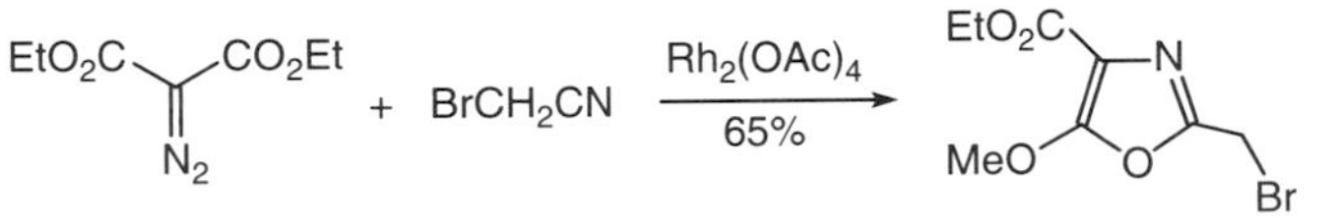

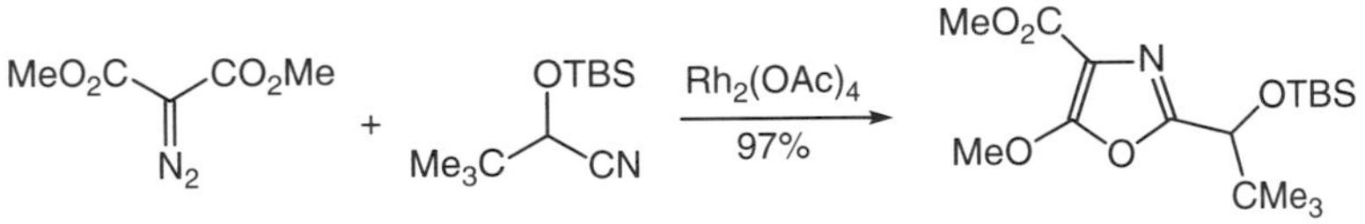

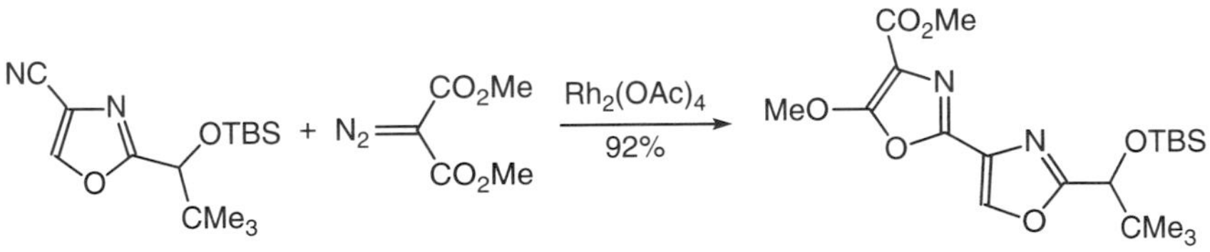

(44)

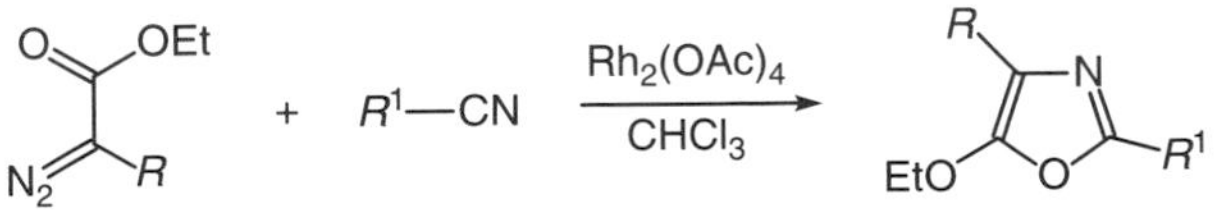

(45)

$$R = \text{Et}, R^1 = \text{SO}_2\text{Ph} \ (52\%); \quad R = \text{Ph}, R^1 = \text{SO}_2\text{Ph} \ (71\%);$$
$$R = \text{Ph}, R^1 = \text{CN} \ (25\%)$$

of nitriles and diazosulfonyl esters in the presence of rhodium(II) acetate in chloroform to form a range of oxazoles in 22–71% yield (eq. 45). These studies also revealed that oxazole yields were not greatly influenced by the ligand attached to the rhodium metal. Additionally, through a diazocyanoacetate as the precursor, Doyle and Moody were able to apply iterative cyclizations of the type shown in Scheme 36 to the formation of bis-oxazoles. Thus cyclization to benzonitrile cat-

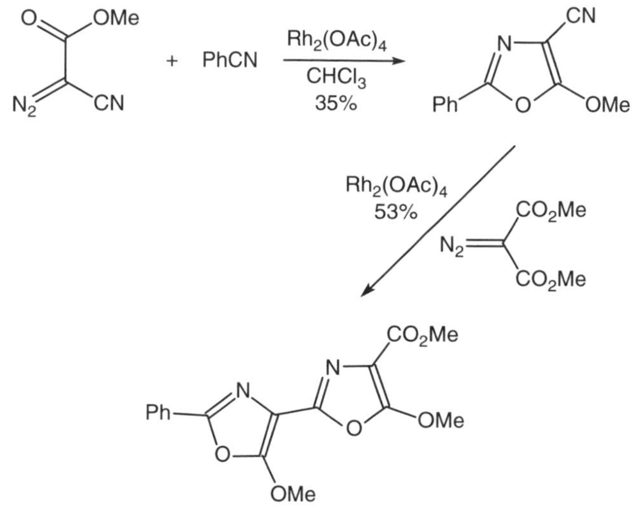

Scheme 36

alyzed by rhodium(II) acetate produced the first oxazole (35% yield) whose cyano group was used to partner diazomalonate in a rhodium(II) trifluoroacetamide–catalyzed second cycloaddition, which furnished the bis-oxazole in 53% yield. The chemoselectivity of rhodium-catalyzed oxazole formation is such that it can be used in systems where alternative aromatic cycloaddition and/or electrophilic substitution might be expected to compete. Such systems included the reactions shown in Scheme 37, which Doyle and Moody[103] used to synthesize

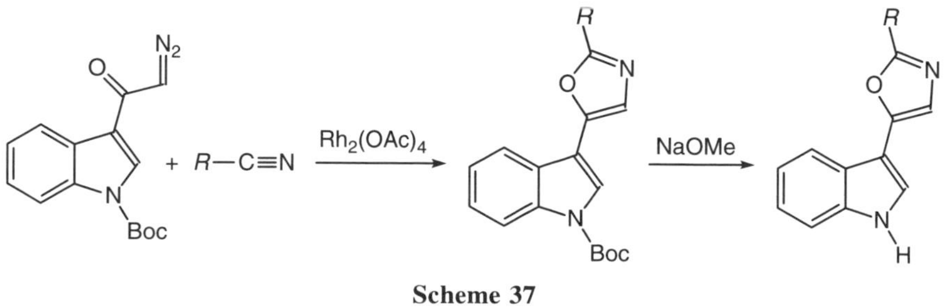

Scheme 37

oxazoylindolyl alkaloids such as pimprinine, pimprinethine, and WS-30581A. T. Ibata's[104] group in Japan has made useful variations on the diazocarbonyl route to oxazoles. By replacing nitriles by cyanamides, they were able to synthesize 2-alkylaminooxazoles of the type shown in eq. 46. The reaction was catalyzed by rhodium(II) acetate, and yields were very dependent on the degree of substitution of the amino group of the cyanamide, with tertiary derivatives more effective than their primary or secondary counterparts (Scheme 38). These work-

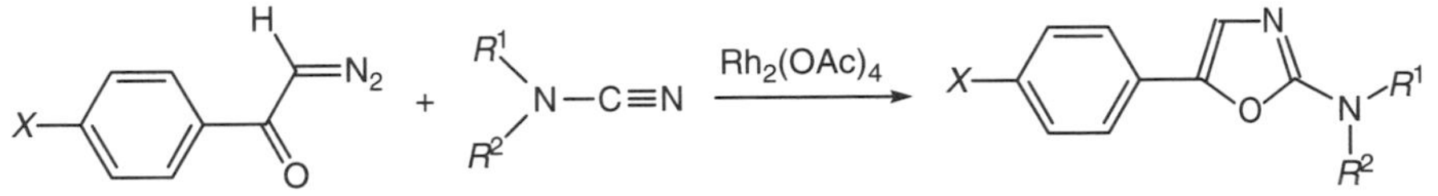

(46)

X	R^1	R^2	Yield (%)
NO$_2$	i-Pr	i-Pr	95
CN	i-Pr	i-Pr	79
H	i-Pr	i-Pr	76
NO$_2$	H	H	7
NO$_2$	H	Me	13
OMe	i-Pr	i-Pr	83

Scheme 38

ers[105] also showed that the reaction pathway from diazoketone and nitrile to oxazole could be diverted, though only to a limited extent, if dimethyl acetylenedicarboxylate was present in the reaction mixture. An example involving benzonitrile and diazoacetophenone is shown in Scheme 39. The minor

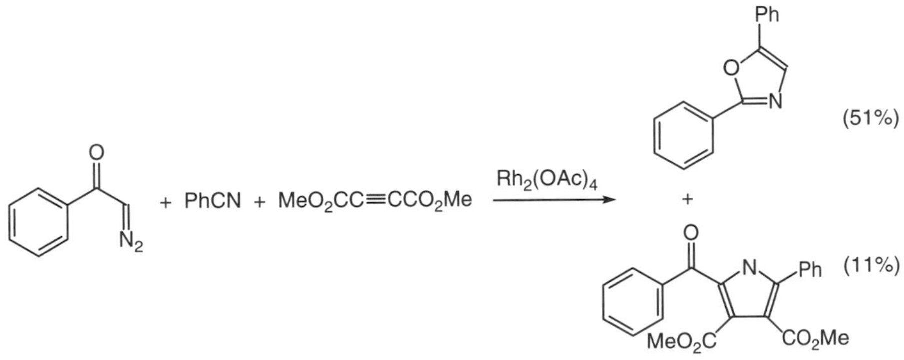

Scheme 39

pyrrole-containing product results from addition of the acetylene dicarboxylate to the dipolar adduct of the diazoketone and nitrile prior to cyclization.

12.7 REFERENCES

1   Curci, R.; Di Furia, F.; Ciabattoni, J.; and Concannon, P. W. "On Reaction of α-Diazo Ketoncs with m-Chloroperoxybenzoic Acid," *J. Org Chem.* **1974**, *39*, 3295–97.

2   López-Herrera, F. J.; Sarabia-Garcia, F. "β-Oxy-α-Diazo Carbonyl Compounds. III. Rh₂(OAc)₄ Mediated Decomposition of β-Acetoxy-α-Diazo Esters. Application to the Synthesis of Natural 3-Deoxy-2-keto Aldonic Acids (KDO and DAH)," *Tetrahedron Lett.* **1994**, *35*, 6705–8.

3   Baganz, H.; and May, H. J., "Neue Syntheses von Hydroxymalondialdehyd (Triosredukton)," *Chem. Ber.* **1966**, *99*, 3771.

4   Bailey, P. S.; Reader, A. M.; Kolsaker, P.; White, H.M.; and Barborak, J. C., "Ozonation of Monosubstituted Diazomethanes," *J. Org Chem.* **1965**, 30, 3042–44.

5   Ursini, A.; Pellicciari, R.; Tamburini, B.; Carlesso, R.; and Gaviraghi, G., "A New Synthesis of 6-Oxopenicillanates by Ozonolysis of 6-Diazopenicillanates," *Synthesis* **1992**, 363–64.

6   Murray, R. W.; and Jeyaraman, R., "Dioxiranes—Synthesis and Reactions of Methyldioxiranes," *J. Org. Chem.* **1985**, *50*, 2847–53.

7   Adam, W.; Bialas, J.; and Hadjiarapoglou, L., "A Convenient Preparation of Acetone Solutions of Dimethyldioxirane," *Chem. Ber.* **1991**, *124*, 2377.

8   Ihmels, H.; Maggini, M.; Prato, M.; and Scorrano, G., "Oxidation of Diazo-Compounds by Dimethyl Dioxirane—An Extremely Mild and Efficient Method for the Preparation of Labile α-Oxo-aldehydes," *Tetrahedron Lett.* **1991**, *32*, 6215–18.

9   McCarthy, N.; Ye, T.; and McKervey, M. A., unpublished results.

10 Darkins, P.; McCarthy, N.; McKervey, M. A.; and Ye, T., "Oxidation of α-Diazoketones Derived from L-Amino-Acids and Dipeptides Using Dimethyldioxirane. Synthesis and Reactions of Homochiral N-Protected α-Amino Glyoxals," *J. Chem. Soc., Chem. Commun.* **1993**, 1222–23.

11 Walker, B.; McCarthy, N.; Healy, A.; Ye, T.; and McKervey, M. A., "Peptide Glyoxals: A Novel Class of Inhibitors for Serine and Cysteine Protineases," *Biochem. J.* **1993**, *293*, 321–23.

12 Darkins, P.; Hawthorne S. J.; Healy, A.; Moncrieff, H.; McCarthy, N.; McKervey, M. A.; and Walker, B., "Peptidyl Vinyl Diones: A Novel Class of Selective Inactivators of Cysteine Protineases," unpublished results.

13 (a) Darkins, P.; McCarthy, N. McKervey. M. A.; O'Donnell, K.; Ye, T.; and Walker, B., "First Synthesis of Enantiomerically Pure N-Protected β-Amino-α-Keto Esters from α-Amino-Acids and Dipeptides," *Tetrahedron: Asymmetry* **1994**, *5*, 195–98; (b) Saba, A., "Synthesis of Vicinal Trioxo Compounds by Dimethyl Dioxirane Oxidation of 2-Diazo-1,3-Dioxo Derivatives," *Syn. Comm.* **1994**, *24*, 695–99.

14 Franzen, V., "A New Method for the Preparation of α, β-Unsaturated Ketones; The Decomposition of Diazoketones," *Liebigs Ann. Chem.* **1957**, 602, 199–208.

15 Ikota, N.; Takamura, N.; Young, S. D.; and Ganem, B., "Catalyzed Insertion Reactions of Substituted α-Diazoesters. A New Synthesis of cis-Enoates," *Tetrahedron Lett.* **1981**, *22*, 4163–66.

16 Hudlicky, T.; Olivo, H. F.; Natchus, M. G.; Umpierrez, E. F.; Pandolfi, E.; and Volonterio, C., "Synthesis of β-Methoxy Enones via a New 2-Carbon Extension of Carboxylic-Acids," *J. Org. Chem.* **1990**, *55*, 4767–70.

17 Taber, D. F.; and Hoerrner, R. S., "Enantioselective Rh-Mediated Synthesis of (−)-PGE$_2$ Methyl-ester," *J. Org. Chem.* **1992**, *57*, 441–47.

18 Taber, D. F.; Hennessy, M. J.; and Louey, J. P., "Rh-Mediated Cyclopentane Construction can Compete with β-Hydride Elimination: Synthesis of (±)-Tochuinyl Acetate," *J. Org. Chem.* **1992**, *57*, 436–41.

19 Taber, D. F.; Herr, R. J.; Pack, S. K.; and Geremia, J. M., "A Convenient Method for the Preparation of (Z)-α-β-Unsaturated Carbonyl Compounds," *J. Org. Chem.* **1996**, *61*, 2908–10.

20 Böhshar, M., PhD Thesis, Univ. of Kaiserslautern, **1985**.

21 Regitz, M.; and Khbeis, S. G., "Investigations on Diazo-Compounds and Azides. 44. 4-Diazomethyl-4*H*-Pyrans by Electrophilic Diazoalkane Substitution," *Chem. Ber.* **1984**, *117*, 2233–46.

22 (a) Hoffmann, K. L.; and Regitz, M., "Studies in Diazo-Compounds and Azides. 47. Metal-Catalyzed Decomposition of 4-Diazomethyl-4*H*-Pyranes—A New Arrival to the Oxepine System," *Tetrahedron Lett.* **1983**, *24*, 5355–58; (b) Hoffmann, K. L.; Maas, G.; and Regitz, M., "Carbenes. 30. Metal-Catalyzed Decomposition of 4-Diazomethyl-4*H*-pyrans—A New Access to the Oxepin System," *Chem. Ber.* **1985**, *118*, 3700–13.

23 Ando, W.; Kondo, S.; Nakayama, K.; Ichibori, K.; Kohoola, H.; Yamato, H.; Imai, I.; Nakaido, S.; and Migata, T., "Decomposition of Dimethyl Diazomalonate in Allyl Compounds Containing Heteroatoms," *J. Am. Chem. Soc.* **1972**, *94*, 3870–76.

24 Ohno, M.; Itoh, M.; Umeda, M.; Furuta, R.; Kondo, K.; and Eguchi, S., "Conjugatively Stabilised Bridgehead Olefins: Formation and Reactions of Remarkably Stable Homoadamant-3-enes Substituted with Phenyl and Methoxycarbonyl Groups," *J. Am. Chem. Soc.* **1996**, *118*, 7075–82.

25 Taylor, T. W. J.; and Forscey, L. A., "Bromine Chloride: The Action of Mixtures of Chlorine and Bromine on Aliphatic Diazo Compounds," *J. Chem. Soc.* **1930**, 2272–77.

26 Clayton, J. P., "The Chemistry of Penicillanic Acids. Part 1. 6,6-Dibromo- and 6,6-Diiodo-Derivatives," *J. Chem Soc. (C)* **1969**, 2123–27.

27 Volkmann, R. A.; Carroll, R. D.; Drolet, R. B.; Elliott, M. L.; and Moore, B. S., "Efficient Preparation of 6,6-Dihalopenicillanic Acids. Synthesis of Penicillanic Acid S,S-Dioxide (Sulbactam)," *J. Org. Chem.* **1982**, *47*, 3344–45.

28 Kapur, J. C.; and Fasel, H. P., "An Efficient Synthesis of Penicillanic Acid S,S-Dioxide in High-Yield," *Tetrahedron Lett.* **1985**, *26*, 3875–78.

29 Rosati, R. E.; Kapili, L. V.; Morrissey, P.; and Retsema, J. A., "Cephalosporins to Carbapenems: 1-Oxygenated Carbapenems and Carbapenams," *J. Med. Chem.* **1990**, *33*, 291–97.

30 Sacripante, G.; and Just, G., "Stereoselective Synthesis of α-Alkylazetidinones by a Free Radical Chain Reaction," *J. Org. Chem.* **1987**, *52*, 3659–61.

31 Belinzoni, D. U.; Mascaretti, O. A.; Alzari, P. M.; Punte, G.; Faerman, C.; and Podjarny, A., "Stereoselective Synthesis and X-Ray Crystallographic Analysis of Mixed 6,6-Dihalo Penicillanates," *Can. J. Chem.* **1985**, *63*, 3177–81.

32 Leroy, J.; and Wakselman, C., "Electrophilic Fluorination of Diazoketones," *J. Chem. Soc., Perkin Trans. 1* **1978**, 1224–27.

33 Patrick, T. B.; Scheibel, J. J.; and Cantrell, G. L., "Geminal Fluorination of Diazo-Compounds," *J. Org. Chem.* **1981**, *46*, 3917–18.

34 Olah, G. A.; and Welch, J., "Synthetic Methods and Reactions XV. Convenient Dediazoniative Hydrofluorination and Halofluorination of Diazoalkanes and Diazoketones in Pyridinium Polyhydrogen Fluoride Solution," *Synthesis* **1974**, 896–98.

35 Mata, E. G.; Setti, E. L.; and Mascaretti, O. A., "Stereoselective Synthesis of 6-Fluoropenicillanate Analogs of β-Lactamase Inhibitors," *J. Org. Chem.* **1990**, *55*, 3674–77.

36 Weygand, F.; and Bestmann, H. J., "Transformation of Diazo Ketones with Aryl or Alkylsulphur Chlorides to α-Halo-α-(aryl or alkylthio)ketones," *Z. Naturforsch.* **1955**, *10b*, 296–98.

37 Weygand, F.; Bestmann, H. J.; and Fritzsche, H., "Umsetzung Aliphatischer Diazoverbindungen mit Sulfenylchloriden," *Chem. Ber.* **1960**, *93*, 2340–44.

38 McKervey, M. A.; and Ratananukul, P., "Regiospecific Synthesis of α-(Phenylthio)cycloalkenones and of α-Phenyl-α-(phenylthio) Ketones *via* $\alpha-\alpha$ Addition of Phenylsulfenyl Chloride to α-Diazoketones," *Tetrahedron Lett.* **1983**, *24*, 117–20.

39 Cronin, J. P.; Dilworth, B. M.; and McKervey, M. A., "Organic-Synthesis with α-Chlorosulfides. Convenient Routes to Phenylthioacetals from α-Diazoketones and Alkyl Phenyl Sulphides via α-Chlorosulfides," *Tetrahedron Lett.* **1986**, *27*, 757–60.

40 Pelliciari, R.; Curini, M.; Ceccherelli, P.; and Fringuelli, R., "Catalysed Insertion Reactions of Diphenyl Diselenide with Ethyl Diazoacetate and Dimethyl Diazomalonate," *J. Chem. Soc., Chem. Commun.* **1979**, 440–41.

41 (a) Giddings, P. J.; John, D. I.; and Thomas, E. J., "Reactions of 6-Diazopenicillinates with Allylic Sulphides, Selenides and Bromides," *Tetrahedron Lett.* **1980**, *21*, 395–98; (b) Giddings, P. J.; John, D. I.; and Thomas, E. J., "Preparation and Reduction of 6-Phenylselenylpenicillanates. A Stereoselective Synthesis of 6β-Substituted Penicillinates," *Tetrahedron Lett.* **1980**, *21*, 399–402.

42 Buckley, D. J.; Kulkowit, S.; and McKervey, M. A., "Reactions of α-Diazoketones with Phenylselenyl Chloride. A New Synthesis of α-Chloro- and α-Phenylselenyl-α,β-Unsaturated Ketones," *J. Chem. Soc., Chem. Commun.* **1980**, 506–7.

43 Buckley, D. J.; and McKervey, M. A., "Reactions of α-Diazo Ketones with Selenium-based Reagents. A General-Synthesis of α-Chloro-, α-Bromo, α-Phenylseleno, α-Acetoxy-α-β-unsaturated, and α-Methoxy-$\alpha\beta$-unsaturated Ketones," *J. Chem. Soc., Perkin Trans. 1* **1985**, 2193–200.

44 Back, T. G.; and Kerr, R. G., "Insertion Reactions of Diazo-Compounds with Some Selenium(II) Electrophiles," *J. Organomet. Chem.* **1985**, *286*, 171–82.

45 Hooz, J.; and Linke, S., "The Reaction of Trialkylboranes with Diazoacetone. A New Ketone Synthesis," *J. Am. Chem. Soc.* **1968**, *90*, 5936–37.

46 Hooz, J.; and Linke, S., "The Alkylation of Diazoacetonitrile and Ethyl Diazoacetate by means of Organoboranes. A New Synthesis of Nitriles and Esters," *J. Am. Chem. Soc.* **1968**, *90*, 6891–92.

47 Hooz, J.; and Gunn, D. M., "The Reaction of β-Vinylic- and β-Alkyl-9-borabicyclo[3.3.1]nonane Derivatives with Ethyl Diazoacetate and Diazoacetone," *Tetrahedron Lett.* **1969**, 3455–56.

48 Hooz, J.; and Morrison, G. F., "Reaction of Trialkylboranes with Diazoacetaldehyde. A New Synthesis of Aldehydes," *Can. J. Chem.,* **1970**, *48*, 868–70.

49 Hooz, J.; and Gunn, D. M., "A New Diketone Synthesis *via* Alkylation of Bisdiazoketones with Trialkylboranes," *J. Chem. Soc., Chem. Commun.* **1969**, 139.

50 Brown, H. C.; and Salunkhe, A. M., "Stereoselective Synthesis of *cis*- and *trans*-β,γ-Unsaturated Carboxylic Esters *via* Reaction of 1-Alkenyldichloroboranes with Ethyl Diazoacetate," *Synlett* **1991**, 684–86.

51 Newman, H., "Short, Simple Synthetic Route to 3-Dehydrosphinganine and Related Compounds," *Chem. Phys. Lipids* **1974**, *12*, 48–52.

52 Newman, H., "Stereospecific Synthesis of D-*threo*-Sphinganine," *J. Org. Chem.* **1974**, *39*, 100-3.

53 Pasto, D. J.; and Wojtkowski, P. W., "Transfer Reactions Involving Boron. XXI. Intermediates Formed in the Alkylation of Diazo Compounds and Dimethyl Sulphoniumphenacylide via Organoboranes," *Tetrahedron Lett.* **1970**, 215–18.

54 Pasto, D. J.; Wojtkowski, P. W., "Transfer Reactions Involving Boron. XXII. The Specific Preparation of Dialkylated Ketones from Diazo Ketones and Methyl Vinyl Ketones *via* Vinyloxyboranes," *J. Org. Chem.* **1971**, *36*, 1790–92.

55 Hooz, J.; and Bridson, J. N., "A Method for the Regiospecific Synthesis of Mannich Bases. Reaction of Enol Borinates with Dimethyl(methylene)ammonium Iodide," *J. Am. Chem. Soc.* **1973**, *95*, 602–3.

56 Hooz, J.; Oudenes, J.; Roberts, J. L.; and Benderly, A., "A New Regiospecific Synthesis of Enol Boranes of Methyl Ketones," *J. Org. Chem.* **1987**, *57*, 1347–49.

57 Hooz, J.; and Oudenes, J., "Regiospecific Synthesis of α-Alkylated Ketones," *Synth. Commun.* **1980**, *10*, 139–45.

58 Hooz, J.; and Oudenes, J., "A Regiospecific Synthesis of 1,3-Diketones via Boroxazines," *J. Synth. Commun.* **1982**, *12*, 189–94.

59 Hooz, J.; and Oudenes, J., "A New Regio- and Stereocontrolled Synthesis of Enol Silyl Ethers," *Tetrahedron Lett.* **1983**, *24*, 5695–98.

60 Grundmann, C., "Decomposition of Diazoketones," *Annalen* **1938**, *536*, 2936.

61 Ernest, I.; and Stanek, J., "Zersetzung von Diazoketonenmit Kupfer(II)-Oxyd. V. EineNeue Reaktion von Aliphatischen Ungesättigten γ-Diketonen," *Coll. Czech. Chem. Commun.* **1959**, *24*, 530–35.

62 Oshima, T.; and Hagai, T., "Stereochemistry of the Copper Perchlorate or Copper Bromide–Catalysed Decomposition of Aryldiazomethanes to Stilbenes," *Tetrahedron Lett.* **1980**, *21*, 1251-54.

63 Wulfman, D. S.; Peace, B. W.; and McDaniel Jr., R. S., "Metal Salt Catalysed Carbenoids—XIV. The Mechanism of Carbene Dimer Formation from Diazoacetic Ester and Dimethyl Diazomalonate in the Presence of Some Soluble Copper Catalysts," *Tetrahedron* **1976**, *32*, 1251–55.

64 Bock, H.; Wolf, H. P; "Gas-Phase Reactions. 50. Gas Phase Reactions of Organic Compounds on Raney Nickel," *Angew. Chem. Int. Ed. Engl.,* **1985**, *23*, 418–19.

65 Collman, J. P.; Rose, E.; and Venburg G. D., "Reactivity of Ruthenium 5,10,15,20-Tetramesitylporphyrin towards Diazoesters: Formation of Olefins" *J. Chem. Soc., Chem. Commun.* **1993**, 934–35.

66 Kulkowit, S.; and McKervey, M. A., "Synthesis of Some Medium- and Large Ring Cycloalk-2-ene-1,4-diones by Intramolecular Coupling of α,ω-Bis-Diazoketones," *J. Chem. Soc., Chem. Commun.* **1983**, 1069-70.

67 Palmisano, G.; Danieli, B.; Lesma, G.; and Riva, R., "Bis(indole) Alkaloids. A Nonbiomimetic Approach to the Blue Pigment Trichotomine Dimethyl Ester," *J. Org. Chem.* **1985**, *50*, 3322–25.

68 Font, J.; Serratosa, F.; and Valls, J., "Intramolecular Cyclizations of Bis-α-Diazoketones. A New Synthesis of 4-Hydroxytropone," *J. Chem. Soc., Chem. Commun.* **1970**, 721–22.

69 Sheehan, J. C.; Chaeko, E.; Lo. Y. S.; Ponzi, D. R.; and Sato, E., "Some New Spiro Penicillins," *J. Org. Chem.* **1978**, *43*, 4856–59.

70 Noels, A. F.; Braham, J. N.; Hubert, A. J.; and Teyssié, Ph., (a) "Highly Stereospecific Dimerization of 5-Formyl-5-Methyl-1-Pyrazolines. Preparation and Characterization of Stable Carbinolamines (Amino Hemiacetals)," *J. Org. Chem.* **1977**, 42, 1527–29; (b) "Cycloadditions of Diazoesters to α,β-unsaturated Aldehydes," *Tetrahedron* **1978**, *34*, 3495–97.

71 Eberhart, P.; and Huisgen, R., "Steric Course and Regioselectivity in the Cycloadditions of Diazoacetic Ester to *trans*- and *cis*-Cinnamic Ester," *Tetrahedron Lett.* **1971**, 4337–42.

72 Doyle, M. P.; Dorow, R. L.; and Tamblyn, W. H., "Cyclopropanation of α,β-Unsaturated Carbonyl Compounds and Nitriles with Diazo Compounds. The Nature of the Involvement of Transition-Metal Promoters," *J. Org. Chem.,* **1982**, *47,* 4059–68.

73 Ghandour, N. E.; and Soulier, J. C. R., "Cycloaddition of Diazoketone to α-Methylenediesters," *Acad. Sci. Paris* **1971**, *272*, 243–45.

74 Huisgen, R., and Reissig, H.-U., "Cycloadditions of α-Diazocarbonyl Compounds to Enamines," *Angew . Chem., Int. Ed. Engl.* **1979**, *18,* 330–31.

75 Parham, W. E.; and Bleasdale, J. L., "Reactions of Diazocompounds with Nitroölefins. I. The Preparation of Pyrazoles," *J. Am. Chem. Soc.* **1950**, *72,* 3843–46.

76 Aue, D. H.; and Helwig, G. S., "Rearrangements in 1,3-Dipolar Additions to 3,3-Dimethylcyclopropene. The Effect of Ring Strain on the Rate of 1,3-Dipolar Addition," *Tetrahedron Lett.* **1974**, 721–24.

77 Aue, D. H.; Lorens, R. B.; and Helwig, G. S., "1,3-Dipolar Additions to Cyclopropenes and Methylenecyclopropene," *J. Org. Chem.* **1979**, *44,* 1202–7.

78 Franck-Neumann, M.; and Buchecker, C., "1,3-Dipolar Addition of Cyclic Diazoketones to Alkynes and Cyclopropenes. Synthesis of Heterocycles by [1,5]-Acyl Migration," *Angew. Chem., Int. Ed. Engl.* **1973**, *12,* 240–41.

79 Cobb, R. L.; and Mahan, J. E., "Chemistry of Cyclobutene-1,2-Dicarbonitrile. 2. Cycloadducts," *J. Org. Chem.* **1977**, *42,* 2597–601.

80 Wulfman, T. S.; and McDaniel, Jr., R. S., "Decomposition d'une Pyrazoline-1 par le Fluoborate Cuivrique," *Tetrahedron Lett.* **1975**, 4523–24.

81 Hampel, W., "Reactions with Diazoketones. IV. Indigo Synthesis," *J. Prackt. Chem.* **1969**, *311,* 78–81.

82   Ceccherelli, P.; Curini, M.; Porter, B.; and Wenkert, E., "Construction of the Hibaene Skeleton by Way of an Abnormal Wolff Reaction," *J. Org. Chem.* **1984**, *49*, 2052–54.

83   Huttel, R., "Über einige Aldehyde der Pyrazol- und der 1,2,3-Triazol-Reihe," *Ber. Dtsch. Chem. Ges.* **1941**, *74*, 1680–87.

84   Reid, W.; and Omran, J., "Reaktionen mit Diazocarbonylverbindungen. VI. 3-Acyl-Pyrazole aus Diazoketonen und mono-substituierten Alkinen," *Liebigs Ann. Chem.* **1963**, *666*, 144–47.

85   Franck-Neumann, M.; and Buchecker, C., "Migrations Sigmatropiques [1,5] Spontanees de Groupes Cetone, Ester et Nitrile en Série Pyraloe. Synthèse d'un Tricarbalcoxy-1,2,3-Cyclopropene," *Tetrahedron Lett.* **1972**, 937–40.

86   Korobizina, I. K.; Bulusheva, V. V.; and Rodina, L. L., "Synthesis of Pyrazoles and Indazoles Based on Aliphatic Diazocompounds," *Khim. Heterocycl. Comp.* **1978**, *5*, 579–97.

87   Huisgen, R.; Reissig, H.-U.; and Huber, H., "Diazocarbonyl Compounds and 1-Diethylaminopropyne," *J. Am. Chem. Soc.* **1979**, *101*, 3647–48.

88   Maas, G.; Regitz, M.; Moll, U.; Rahm, R.; Krebs, F.; Hector, R.; Stang, P. J.; Crittell, C. M.; and Williamson, B. L., "1,3-Dipolar Cycloaddition Reactions of α-Diazocarbonyl Compounds, Organoazides, and Ethynyl(phenyl)iodonium Triflate Salts," *Tetrahedron* **1992**, *48*, 3527–40.

89   For a comprehensive review on the cycloaddition of diazoalkanes, see: Regitz, M.; and Heydt, H., in *1,3-Dipolar Cycloaddition Chemistry I*; Padwa, A., Ed.; John Wiley & Sons, Inc.; New York, **1984**; *Vol. 1*, p. 393.

90   Yoo, S. K., "Synthesis of Poly-Oxazole Systems Found in Marine Metabolites," *Tetrahedron Lett.* **1992**, *33*, 2159–62.

91   Connell, R.; Scavo, F.; Helquist, P.; and Åkermark, B., "Functionalized Oxazoles from Rhodium-Catalyzed Reaction of Dimethyl Diazomalonate with Nitriles," *Tetrahedron Lett.* **1986**, *27*, 5559–62.

92   Shi, G. Q.; and Xu, Y., "Ethyl-3-Trifluoro-2-diazo-propionate as a Potentially Useful CF_3-Containing Building Block—Preparation and $[Rh(OAc)_2]_2$-Catalysed Reaction with Nitriles," *J. Chem. Soc., Chem. Commun.* **1989**, 607–8.

93   Connell, R. D.; Tebbe, M.; Helquist, P.; and Åkermark, B., "Direct Preparation of 4-Carboethoxy-1,3-oxazoles," *Tetrahedron Lett.* **1991**, *32*, 17–20.

94   Gangloff, A. R.; Åkermark, B.; and Helquist, P., "Generation and Use of a Zinc Derivative of a Functionalized 1,3-Oxazole—Solution of the Virginiamycin/Madumycin Oxazole Problem," *J. Org. Chem.* **1992**, *57*, 4797–99.

95   Connell, R. D.; Tebbe, M.; Gangloff, A. R.; Helquist, P.; and Åkermark, B., "Rhodium Catalyzed Heterocycloaddition Route to 1,3-Oxazoles as Building Blocks in Natural-Products Synthesis," *Tetrahedron* **1993**, *49*, 5445–59.

96   Doyle, K. J.; and Moody, C. J., "The Rhodium Carbenoid Route to Oxazoles- Synthesis of 4-Functionalized Oxazoles; Three Step Preparation of a Bis-Oxazole," *Tetrahedron* **1994**, *50*, 3761–72.

97   Abdel Wahab, A. A.; Doss, S. H.; Durr, H.; Turro, N. J. and Gould, I. R., "Mechanistic Investigations of the Cycloaddition Reactions of Thioxanthenylidene S,S-Dioxide," *J. Org. Chem.* **1987**, *52*, 429–34.

98   Janulis, E. P.; Wilson, S. R. and Arduengo III, A. J., "The Synthesis and Structure of a Stabilised Nitrilium Ylide," *Tetrahedron Lett.*, **1984**, *25*, 405–08.

99   Paulissen, R.; Moniotte, P.; Hubert, A. J.; and Teyssié, P., "Transition Metal–Catalysed 1,3-Dipolar Cycloaddition of Carbalkoxycarbene to Acrylonitriles," *Tetrahedron Lett.* **1974**, 3311–14.

100 Moniotte, P.; Hubert, A. J.; and Teyssié, P., "The Role of Copper(I) Complexes in the Selective Formation of Oxazoles from Unsaturated Nitriles and Diazoesters" *J. Organomet. Chem.* **1975**, *88*, 115–20.

101 Alonso, M. E.; and Jano, P., "The Synthesis of Ethoxycarbonyl-1,3-dioxazoles and Oxazoles from the Copper Catalyzed Thermolysis of Ethyl Diazopyruvate in the Presence of Ketones, Aldehydes and Nitriles," *J. Heterocycl. Chem.* **1980**, *17*, 721–25.

102 Ibata, T.; and Fukushima, K., "Formation and Reaction of Acyl Substituted Nitrile Ylide through the $Rh_2(OAc)_4$-Catalyzed Reaction of α-Diazocarbonyl Compounds with Benzonitrile," *Chem. Lett.* **1992**, 2197–200.

103 Doyle, K. J.; and Moody, C. J., "Synthesis of Oxazolylindolyl Alkaloids via Rhodium Carbenoids," *Synthesis* **1994**, *50*, 1021–22.

104 Fukushima, K.; and Ibata, T., "Formation of 2-Alkylaminooxazoles by the $Rh_2(OAc)_4$-Catalysed Reaction of α-Diazocarbonyl Compounds in the Presence of Cyanamides," *Heterocycles* **1995**, *40*, 149–54.

105 Fukushima, K.; and Ibata, T., "Formation of Acyl-Substituted Nitrile Ylides by $Rh_2(OAc)_4$-Catalyzed Decomposition of α-Diazocarbonyl Compounds in Nitriles," *Bull. Chem. Soc. Jpn.* **1995**, *68*, 3469–81.